FORMULAS FOR STRESS, STRAIN, AND STRUCTURAL MATRICES

FORMULAS FOR STRESS, STRAIN, AND STRUCTURAL MATRICES

Walter D. Pilkey

University of Virginia
School of Engineering and Applied Science
Department of Mechanical and Aerospace Engineering
Charlottesville, Virginia

A Wiley-Interscience Publication
JOHN WILEY & SONS, INC.
New York / Chichester / Brisbane / Toronto / Singapore

Copyright © 1994 by John Wiley & Sons, Inc.

All rights reserved. Published simultaneously in Canada.

Library of Congress Cataloging in Publication Data:

Pilkey, Walter D.
    Formulas for stress, strain, and structural matrices / Walter D.
Pilkey.
        p. cm.
    Includes index.
    ISBN 0-471-52746-7 (cloth)
    1. Strains and stresses—Tables. 2. Structural analysis
(Engineering)—Tables. 3. Structural analysis (Engineering—
Computer programs. I. Title.
TA407.2.P55 1994
624.1′76′0212—dc20   94-1008

Printed in the United States of America

10 9 8 7 6 5 4 3 2

To David Bevan

We have no theories [here], no uncertainties remain on the mind; all is demonstration and satisfaction.

—*Thomas Jefferson*

Contents

**4 MECHANICAL PROPERTIES AND TESTING
    OF ENGINEERING MATERIALS**   **141**

5 EXPERIMENTAL STRESS ANALYSIS   **211**

18 PLATES   955

19 THICK SHELLS AND DISKS   1097

20 THIN SHELLS   1151

APPENDIX

I FUNDAMENTAL MATHEMATICS   1283

Preface

This book is intended to be a succinct source of strength of material formulas. This is a reference book for the design engineer. These formulas should ease the task of the analysis and design of structural members and mechanical elements. The formulas cover stresses, displacements, buckling loads, and natural frequencies.

The intention here is to provide formulas for the designer to solve problems for which it is not necessary to employ general-purpose structural analysis computer programs. This information permits efficient static, stability, and dynamic analyses of beams, bars, plates, and shells with very general mechanical or thermal loading. Use of these formulas should be helpful to engineers who model their problems as structural members.

The universal availability of computers provides the opportunity for accurate and efficient analyses for stress and strain. The combination of the formulas of this book and the companion computer programs available from the author is a powerful tool for the designer. Not only can structural member analyses be automated by using the software, but the geometric properties, such as torsional constants, and normal and shear stresses are computed for beam cross sections with arbitrary shapes.

In addition to the formulas for stress and strain, this book contains tables of structural matrices, such as stiffness and mass matrices. Using these matrices, the reader can develop special computer programs for special problems.

The computer programs that supplement this book are user-oriented and simple to use, yet comprehensive in capability. For example, typical programs are:

BEAM The BEAM program calculates the deflection, slope, bending moment, and shear force of beams for static, steady state, and transient dynamic conditions. The critical axial load and mode shape are found for stability. The natural frequencies and mode shapes are computed for free transverse vibrations. The beam can be formed of uniform segments with any loading, in-span supports, foundations, and boundary conditions. The user can include any or all of bending, shear deformation, and rotary inertia effects. For example, the applied mechanical loading can consist of concentrated, uniformly distributed, or ramp distributed forces and moments. Acceptable in-span supports include flexible (extension or torsion) springs or branches, rigid supports, and moment or shear releases. The beam model can include classical Winkler or higher order continuous foundations. The end conditions can be fixed, simply-supported, free and guided.

SECTIONAL PROPERTIES Cross sectional properties of a bar of any cross-sectional shape are calculated. Properties include area, centroid, moments of inertia about any axes, radii of gyration, shear center, shear deformation coefficients, torsional constant, warping constant, shift in neutral axis due to curvature, and cross-sectional stability constants. Modulus weighted properties are calculated for composite sections.

STRESS ANALYSIS The normal and shear stresses are found on the cross section of a bar of any cross-sectional shape. The material can be composite. The stresses include bending stresses, shear stresses due to torsion and warping, shear stresses due to transverse loads and normal stresses due to warping. This program, along with the BEAM or THIN-WALLED BEAM programs which provide the forces and moments along a beam, can be used to perform very complete stress analyses.

There are other programs for extension and torsional systems, torsion of thin-walled beams, rotating shafts, framework, disks, plates, shells, and stress concentration.

The power of these programs can be understood through a comparison of a particular program (BEAM) with general purpose structural analysis (finite element) software. BEAM is a computer program for the static, stability, steady state, transient dynamics, and free dynamic analysis of straight beams. The deflection, slope, bending moment, and shear force are calculated for very general loading and geometric conditions. For the appropriate class of problems (beams), BEAM is significantly more comprehensive in capability than a general purpose program. More important than the superior capabilities is the engineer-oriented input of BEAM; a problem is described in simple strength-of-material terminology rather than in terms of nodes, elements, constraints, and degrees-of-freedom. Thus, an engineer can use BEAM without learning finite element terminology. The differences between the BEAM capability and those of a general purpose program include:

BEAM accepts an exact representation of mass rather than depending totally on an approximate consistent mass matrix. This means that more precise natural frequencies and mode shapes can be computed.

BEAM accepts distributed elastic foundations rather than just concentrated flexible supports.

BEAM accepts ramp (linearly distributed) applied forces and moments over any span rather than a uniform load between two nodes.

Most of the formulas of this book are derived from the work of others, which has been duly acknowledged. The general structural response formulas and matrices are the responsibility of the author and his students and colleagues.

It would be unrealistically sanguine to presume that the formulas are error-free. "For here we are not afraid to follow truth wherever it may lead, nor tolerate any error so long as reason is left free to combat it" (Thomas Jefferson). The author welcomes suggestions and corrections from the readers.

A major force in the completion of this book has been the encouragement of my wife Barbara.

WALTER D. PILKEY

Acknowledgments

Initially, this work was prepared with the support of the Association for American Railroads. In a real sense, this book was prepared in association with

Emerson Woomer
Orrin H. Pilkey, Sr.
Shizhong Han
Yongquan Liu
Heiyan Wu
Bo Suk Yang
Weize Kang

These gentlemen contributed heavily to the contents and were responsible for crafting several chapters. Frank Cerra was an important and frequent source of inspiration for this work. Professors Richard Gangloff and John Wert critically reviewed some chapters. The figures and polynomial models of tabulated data were prepared by Allen Hawkins, Jeff Chow, Debbie Pilkey, Weiwei Ding, and Xiaoquan Wu.

W. D. P.

Introduction

In this era of computers, general-purpose structural analysis computer programs are available to the engineer. However, many structures are configured so that they are analyzed more accurately as structural members than as three-dimensional systems using a general-purpose computer program. For example, because of the geometry of a train freight car, with its relatively long length and its cross section, which is symmetric about a longitudinal axis, the modeling problem can be reduced from that of a three-dimensional structure to a one-dimensional longitudinal and a two-dimensional cross-sectional analysis. These two uncoupled analyses can be treated as structural member problems that can be solved with stress–strain formulas or simple member analyses. It is the purpose of this book to provide in compact form the formulas or the analysis procedure to treat such member problems.

This book should help meet the need for engineers to have simple, accurate, and comprehensive formulas for stress analysis. The tables permit a problem to be modeled realistically and to be solved accurately.

1.1 NOTATION

The notation used in the formulas is defined in each chapter. Certain symbols are common to several chapters. Occasionally, singularity functions are employed to

1

assist in the concise expression of formulas,

$$\langle x - a \rangle^n = \begin{cases} 0 & \text{if } x < a \\ (x - a)^n & \text{if } x \geq a \end{cases} \tag{1.1}$$

$$\langle x - a \rangle^0 = \begin{cases} 0 & \text{if } x < a \\ 1 & \text{if } x \geq a \end{cases} \tag{1.2}$$

$$f\langle x - a \rangle = \begin{cases} 0 & \text{if } x < a \\ f(x - a) & \text{if } x \geq a \end{cases} \tag{1.3}$$

where $f(x - a)$ is a function of $x - a$.

1.2 CONVERSION FACTORS

Some useful conversion factors are provided in Table 1-1.

1.3 SIGN CONVENTIONS AND CONSISTENT UNITS

The sign conventions for the formulas are always evident in that the given formula corresponds to the loading direction shown. An applied load in the opposite direction requires that the load be given a negative sign in the formula.

No units are assigned to any variables in the formulas. Any consistent units can be employed. Some examples of consistent units are listed in Table 1-2.

1.4 THE SI UNITS

The International System (SI) of units is described in Table 1-3, where useful prefixes are provided. Some factors for conversion to SI units are shown in Table 1-4. Metric conversions for some commonly occurring variables are given in Table 1-5 along with some rounded-off figures that may be easy to remember. These are referred to as "recognition figures" and can be useful in quick calculations.

1.5 TYPICAL DESIGN LOADS AND STRESSES

Table 1-6 provides several typical design loads as well as values of material constants and allowable stresses.

1

Tables

TABLE 1-1 VARIOUS CONVERSION FACTORS

Multiply	By	To Obtain
acre	0.4047	ha(hectares)
acre	4047	m^2
atm	29.92	inch of mercury (32°F)
atm	101,300	N/m^2 (Pa)
atm	14.70	$lb/in.^2$ (psi)
Btu/h	12.96	ft-lb/min
cm/s	1.969	ft/min
$cm/s\,(cm/s^2)$	0.010	$m/s\,(m/s^2)$
cm/s	0.6	m/min
$cm/s\,(cm/s^2)$	0.0328	$ft/s\,(ft/s^2)$
$cm/s\,(cm/s^2)$	0.3937	$in./s\,(in./s^2)$
cm/s^2	0.00102	g
circular mil	0.7854	mil^2
cm^3/s	0.002119	ft^3/min
cup	0.24	l (liter)
ft^3/min	471.9	cm^3/s
ft^3/min	0.1247	gal/s
ft^3/min	0.4719	l/s
ft^3/s	448.8	gal/min
degree (degree/s)	0.01745	rad (rad/s)
degree	0.00273	rev
dyne	10^{-5}	N
dyne	0.000002248	lb
fathom	1.829	m
ft	0.3048	m
foot of water (60°F)	0.8843	inch of mercury (60°F)
foot of water (60°F)	2986	N/m^2
foot of water (60°F)	0.4331	$lb/in.^2$
ft/min	0.508	cm/s
ft/min	0.01136	mi/h
$ft/s\,(ft/s^2)$	12	$in./s\,(in./s^2)$
$ft/s\,(ft/s^2)$	30.48	$cm/s\,(cm/s^2)$
$ft/s\,(ft/s^2)$	0.3048	$m/s\,(m/s^2)$
ft/s^2	0.0311	g
ft/s	1.097	km/h
ft/s	0.5925	knot
ft-lb/s	0.07716	Btu/min
ft-lb	1.356	Nm
fluid ounce	29.57	ml
g (acceleration of gravity)	32.16	ft/s^2
g	386	$in./s^2$
g	980	cm/s^2
g	9.80	m/s^2
gal	3.8	l
gallon of water (60°F)	8.345	pound of water (60°F)

TABLE 1-1 Various Conversion Factors   4

TABLE 1-1 (continued)   VARIOUS CONVERSION FACTORS

Multiply	By	To Obtain
gal/s	8.021	ft^3/min
gal/s	227.1	l/min
g	980.7	dyne
g/cm^3	9807	N/m^3
g/cm^2	98.07	N/m^2
ha	2.471	acre
ha	10^4	m^2
hp	1.014	hp (metric)
hp (metric)	0.9863	hp (horsepower)
Hz	1	cycle/s, rev/s
Hz	6.283	rad/s
Hz	360	degree/s
in.	0.0254	m
inch of mercury (32°F)	0.03342	atm
inch of mercury (60°F)	1.131	foot of water (60°F)
inch of mercury (60°F)	3376	N/m^2
inch of mercury (60°F)	0.4898	$lb/in.^2$
inch of water (60°F)	0.03609	$lb/in.^2$
in./s (in./s^2)	0.0833	ft/s (ft/s^2)
in./s (in./s^2)	2.540	cm/s (cm/s^2)
in./s (in./s^2)	0.0254	m/s (m/s^2)
in./s^2	0.00259	g (acceleration of gravity)
kg	9.807	N
kg	0.6852177	slug
km/h	0.9113	ft/s
knot	1.688	ft/s
knot	1.151	mi/h
l	2.1134	pint
l	1.0567	quart
l	0.2642	gal
l/min	0.004403	gal/s
ml	0.0338	fluid ounce
m	0.1988	rod
m/min	1.667	cm/s
m/s (m/s^2)	3.28	ft/s (ft/s^2)
m/s (m/s^2)	39.37	in./s (in./s^2)
m/s (m/s^2)	100	cm/s (cm/s^2)
m/s^2	0.102	g (acceleration of gravity)
mil	0.001	in.
mil^2	1.273	circular mil
mi/h	88.0	ft/min
mi/h	0.8690	knot
N/m^2	9.872×10^{-6}	atm
N/m^2	3.349×10^{-4}	foot of water (60°F)
oz (avoirdupois)	0.9115	oz (troy)
oz (troy)	1.097	oz (avoirdupois)
oz (troy)	0.06857	lb (avoirdupois)

Various Conversion Factors   TABLE 1-1

TABLE 1-1 (continued)   VARIOUS CONVERSION FACTORS

Multiply	By	To Obtain
pint	0.4732	l
lb	4.448	N
lb (mass)	0.4535	kg
lb (avoirdupois)	14.58	oz (troy)
lb (avoirdupois)	0.031081	slug
pound of water (60°F)	0.01603	ft^3
pound of water (60°F)	0.1199	gal
$lb/in.^2$	0.06805	atm
$lb/in.^2$	2.309	foot of water (60°F)
$lb/in.^2$	6895	N/m^2
$lb/in.^2$	2.042	inch of mercury (60°F)
$lb/in.^2$	27.71	inch of water (60°F)
quart	0.9463	l
rad (rad/s)	57.30	degree (degree/s)
rad/s	0.1592	rev/s or Hz
rad/s	9.549	rpm
rev (revolution)	6.283	rad
rev/s or Hz (rev/s^2)	6.283	rad/s (rad/s^2)
rev/s or Hz	360	degree/s
rpm	0.1047	rad/s
rpm	6	degree/s
rod	5.029	m
slug	14.5939	kg
slug	32.1740	lb (avoirdupois)

TABLE 1-1 **Various Conversion Factors**   6

TABLE 1-2 CONSISTENT UNITS

Quantity	U.S. Customary (foot)	Old Metric (meter)	International Metric (SI) (meter)
Length	ft	cm	m
Force and weight, W	lb	kg	N
Time	s	s	s
Angle	rad	rad	rad
Moment of inertia	ft^4	cm^4	m^4
Mass, $= W/g$	$lb\text{-}s^2/ft$ (slug)	$kg\text{-}s^2/cm$	kg
Area	ft^2	cm^2	m^2
Mass moment of inertia	$lb\text{-}s^2\text{-}ft$	$kg\text{-}s^2\text{-}cm$	$kg \cdot m^2$
Moment	lb-ft	kg-cm	$N \cdot m$
Volume	ft^3	cm^3	m^3
Mass density	$lb\text{-}s^2/ft^4$	$kg\text{-}s^2/cm^4$	kg/m^3
Stiffness of linear spring	lb/ft	kg/cm	N/m
Stiffness of rotary spring	lb-ft/rad	kg-cm/rad	$N \cdot m/rad$
Torque	lb-ft	kg-cm	$N \cdot m$
Stiffness of torsional spring	lb-ft/rad	kg-cm/rad	$N \cdot m/rad$
Stress or pressure	lb/ft^2	kg/cm^2	N/m^2 (Pa)

TABLE 1-3 THE INTERNATIONAL SYSTEM (SI) OF UNITS

Quantity	Name of Unit	SI Symbol	Unit Formula
Units Pertinent to Structural Mechanics			
BASE			
Length	Meter	m	
Mass	Kilogram	kg	
Time	Second	s	
Temperature	Kelvin	K	
DERIVED			
Area	Square meter	m^2	
Volume	Cubic meter	m^3	
Force	Newton	N	$kg \cdot m/s^2$
Stress, pressure	Pascal	Pa	N/m^2
Work, energy	Joule	J	$N \cdot m$
Power	Watt	W	$N \cdot m/s$
SUPPLEMENTARY			
Plane angle	Radian	rad	

Prefix	Symbol	*Preferred Prefixes* Multiplication Factor	Exponential Form
Tera	T	1 000 000 000 000	10^{12}
Giga	G	1 000 000 000	10^9
Mega	M	1 000 000	10^6
Kilo	k	1 000	10^3
Milli	m	0.001	10^{-3}
Micro	μ	0.000 001	10^{-6}
Nano	n	0.000 000 001	10^{-9}
Not Recommended for Common Use[a]			
Hecto	h	100	10^2
Deka	da	10	10^1
Deci	d	0.1	10^{-1}
Centi	c	0.01	10^{-2}

[a]Except when expressing area and volume. The prefixes c and d can also be used with properties of certain standard structural sections.

TABLE 1-3 The International System (SI) of Units   8

TABLE 1-4 CONVERSION TO SI UNITS

Example: to convert from psi to pascal, multiply by $6.894\ 757 \times 10^3$. Then 1000 psi is 6.894 757 MPa.

To Convert from	To	Multiply by
Acceleration		
g	m/s^2	9.80
g	cm/s^2	980
ft/s^2	m/s^2	0.3048
ft/s^2	cm/s^2	30.48
$in./s^2$	m/s^2	0.0254
$in./s^2$	cm/s^2	2.540
Area		
ft^2	m^2	9.290304×10^{-2}
$in.^2$	m^2	6.451600×10^{-4}
Energy and Work		
Btu	J	1.055056×10^3
cal	J	4.186800
erg	J	1.000000×10^{-7}
ft-lb	J	1.355818
W-s	J	1.000000
Energy / Area (Toughness)		
erg/cm^2	J/m^2	1.000000×10^{-3}
$ft\text{-}lb/in.^2$	J/m^2	2.101522×10^3
$in.\text{-}lb/in.^2$	J/m^2	1.751268×10^2
Force		
dyne	N	1.000000×10^{-5}
kg	N	9.806650
lb	N	4.448222
poundal	N	1.382550×10^{-1}
Length		
Å	m	1.000000×10^{-10}
ft	m	3.048000×10^{-1}
in.	m	2.540000×10^{-2}
mi (U.S. nautical)	m	1.852000×10^3
mi (U.S. statute)	m	1.609344×10^3
yd	m	0.9144

TABLE 1-4 (continued)   CONVERSION TO SI UNITS

To Convert from	To	Multiply by
Mass		
grain	kg	6.479891×10^{-5}
lb (mass)	kg	4.535924×10^{-1}
slug	kg	14.59390
Mass per Volume (Density)		
g/cm^3	kg/m^3	1.000000×10^3
lb (mass)/in.^{3}	kg/m^3	2.767990×10^4
slug/ft^3	kg/m^3	5.153788×10^2
Power		
Btu/h	W	2.930711×10^{-1}
ft-lb/s	W	1.355818
hp	W	7.456999×10^2
Pressure or Stress		
atm (760 torr)	Pa	1.013250×10^5
bar	Pa	1.000000×10^5
centimeter of mercury (0°C)	Pa	1.33322×10^3
centimeter of water (4°C)	Pa	98.0638
dyne/cm^2	Pa	1.000000×10^{-1}
kg/cm^2	Pa	9.806650×10^4
kg/mm^2	Pa	9.806650×10^6
N/m^2	Pa	1.000000
lb/in.^{2} (psi)	Pa	6.894757×10^3
torr (mm Hg, 0°C)	Pa	1.33322×10^2
Temperature		
degree Celsius (°C)	K	$T\ (\mathrm{K}) = T\ (\mathrm{°C}) + 273.15$
degree Fahrenheit (°F)	K	$T\ (\mathrm{K}) = \dfrac{T\ (\mathrm{°F}) + 459.67}{1.8}$
Time		
day	s	8.640000×10^4
hour	s	3.600000×10^3
year	s	3.153600×10^7

TABLE 1-4   Conversion to SI Units   **10**

TABLE 1-4 (continued)   CONVERSION TO SI UNITS

To Convert from	To	Multiply by
Velocity		
ft/min	m/s	5.080000×10^{-3}
ft/s	m/s	0.3048
ft/s	cm/s	30.48
in./s	m/s	0.0254
in./s	cm/s	2.540
km/h	m/s	2.777778×10^{-1}
mi/h	m/s	4.470400×10^{-1}
Viscosity		
cP (centipoise)	Pa-s	1.000000×10^{-3}
P (poise)	Pa-s	1.000000×10^{-1}
lb-s/ft^2	Pa-s	47.88026
Volume		
barrel (oil, 42 U.S. gal)	m^3	1.589873×10^{-1}
fluid ounce	m^3	2.957353×10^{-5}
ft^3	m^3	2.831685×10^{-2}
gal (Imperial liquid)	m^3	4.546122×10^{-3}
gal (U.S. liquid)	m^3	3.785412×10^{-3}
in.^{3}	m^3	1.638706×10^{-5}
l	m^3	1.000000×10^{-3}
Miscellaneous Special Conversions		
ksi $\sqrt{\text{in.}}$	MPa$\sqrt{\text{m}}$	1.098843

TABLE 1-5   COMMON CONVERSION FACTORS AND SI RECOGNITION FIGURES

	Units	
U.S. Customary System (USCS)	International System (SI)	Suggested SI Recognition Figure
Length		
1 in.	25.4 mm	25 mm
10 in.	254 mm	250 mm
1 ft	0.3048 m	0.3 m
10 ft	3.048 m	3 m
1 mi	1609 m	1.6 km
1 yd	0.9144 m	0.9 m
Area		
1 ft^2	0.09290 m^2	0.1 m^2
1 ha	10^4 m^2	10^4 m^2
1 mi^2	2.59 km^2	2.6 km^2
1 yd^2	0.836 m^2	0.85 m^2
Temperature		
32°F	273 K	0°C (270 K), 1 K = 1°C; use of °C is permissible in SI
Velocity		
1 ft/min	0.00508 m/s	5 mm/s
1 mi/h	1.609 km/h = 0.447 m/s	1.6 km/h = 0.45 m/s
Power		
1 hp (550 ft-lb/s)	745.7 W	0.75 kW
ft-lb/s	1.3558 W	1.4 W
Volume		
1 ft^3	0.0283 m^3	0.03 m^3
1 yd^3	0.765 m^3	0.8 m^3
1 gal	0.003785 m^3	0.004 m^3

TABLE 1-5   Common Conversion Factors   12

Units		
U.S. Customary System (USCS)	International System (SI)	Suggested SI Recognition Figure
Weight or Force		
1 lb (force)	4.448 N	4.5 N
1 kip (1000 lb)	4.448 kN	4.5 kN
Beam Loads		
1000 lb/in.	175.13 kN/m	175 kN/m
1000 lb/ft	14.59 kN/m	15 kN/m
Mass		
1 lb (mass)	0.4536 kg	0.5 kg
1 slug	14.5939 kg	15 kg
Stress or Pressure		
1 psi (lb/in.^{2})	6.895 kN/m^2 (kPa)	7 kN/m^2
1000 psi (1 ksi)	6.895 MN/m^2 (MPa)	7 MN/m^2
1 psf	47.88 N/m^2 (Pa)	48 N/m^2
1 atm (760 torr)	1.01325×10^5 Pa	10^5 Pa

TABLE 1-6 TYPICAL VALUES OF DESIGN LOADS, MATERIAL PROPERTIES, AND ALLOWABLE STRESSES

Quantity	U.S. Customary System	International System (SI)
Design Loads		
Wind Pressure	30 lb/ft^2	1.4 kN/m^2 (kPa)
Snow		
Moderate Climate		
Flat	20 lb/ft^2	960 N/m^2 (Pa)
45° Slope	10 lb/ft^2	480 N/m^2 (Pa)
Cold Climate		
Flat	40 lb/ft^2	2 kN/m^2 (kPa)
45° Slope	10 lb/ft^2	480 N/m^2 (Pa)
Allowable Loads		
Soil		
Ordinary clay and sand mixture	2–3 tons/ft^2	200–300 kPa
Hard clay and firm coarse sand	4–6 tons/ft^2	400–600 kPa
Bed rock	> 15 tons/ft^2	> 1400 kPa
Wood, yellow pine	1600 psi	11 MPa
Concrete	1000 psi	7 MPa
Steel	20,000 psi	140 MPa
Moduli of Elasticity		
Wood, yellow pine	1.6×10^6 psi	11 GN/m^2 (GPa)
Aluminum	10.1×10^6 psi	70 GPa
Concrete	2×10^6 psi	14 GPa
Steel	30×10^6 psi	207 GPa
Weights		
Steel	490 lb/ft^3	76.98 kN/m^3
Wood	40 lb/ft^3	6.3 kN/m^3
Concrete	150 lb/ft^3	24 kN/m^3
Water	62.4 lb/ft^3	9.804 kN/m^3
Aluminum	169.3 lb/ft^3	26.60 kN/m^3
Snow		
Freshly fallen	5 lb/ft^3	800 N/m^3
Packed	12 lb/ft^3	1.9 kN/m^3
Wet	50 lb/ft^3	7.9 kN/m^3
Sand		
Dry	100 lb/ft^3	15.7 kN/m^3
Wet	115 lb/ft^3	18.1 kN/m^3

TABLE 1-6 Typical Values of Design Loads   **14**

Quantity	U.S. Customary System	International System (SI)
	Density (Mass)	
Water	0.9356×10^{-4} lb-s^2/in.^{4}	1000 kg/m^3
Steel	7.3326×10^{-4} lb-s^2/in.^{4}	7835.9 kg/m^3
	Acceleration of Gravity (g)	
	32.174 ft/s^2 (386.087 in./s^2)	9.8066 m/s^2
	Coefficients of friction	
Iron on stone	0.5	
Timber on stone	0.4	
Timber on timber	0.3	
Brick on brick	0.7	

2

Geometric Properties of Plane Areas

The geometric properties of a cross-sectional area are essential in the study of beams and bars. A brief discussion of these properties along with tables of formulas are provided in this chapter. Computer programs (Chapter 15) are available to compute these properties for cross sections of arbitrary shape.

2.1 NOTATION

A Cross-sectional area (L^2)

A_0 Area defined in Fig. 2-9

$f = Z_p/Z_e$ Shape factor

17

I, I_y, I_z	Moments of inertia of a cross section (L^4)
I_{xy}	Product of inertia (L^4)
J	Torsional constant (L^4)
J_x	Polar moment of inertia (L^4), $= I_x$
$I_{\omega y}$	Sectorial linear moment about y axis (L^5), $= \int_A \omega z \, dA$
$I_{\omega z}$	Sectorial linear moment about z axis (L^5), $= \int_A \omega y \, dA$
α_y, α_z	Shear correction factors in z, y directions
Q_y, Q_z	First moment of area with respect to y and z axes, respectively (L^3)
$\bar{q}_i$	Normalized shear flow (L^2)
Q_ω	Sectorial static moment (L^4), $= \int_{A_0} \omega \, dA$
r_y, r_z	Radii of gyration (L)
s	Coordinate along centerline of wall thickness (L)
S	Designation of shear center; elastic section modulus
t	Wall thickness (L)
T	Torque, twisting moment (FL)
x, y, z	Right-handed coordinate system
y_c, z_c	Centroids of a cross section in yz plane (L)
y_S, z_S	Shear center coordinates (L)
Z_e	Elastic section modulus (L^3), $= S$
Z_p	Plastic section modulus (L^3), $= Z$
Γ	Warping constant (sectorial moment of inertia), of a cross section (L^6), $= \int_A \omega^2 \, dA$
ω	Sectorial area, principal sectorial coordinate
$r_{S(c)}$	Perpendicular distance from the shear center (centroid) to the tangent of the centerline of the wall profile (L)
$\omega_{S(c)}$	Sectorial area or sectorial coordinate with respect to shear center (centroid) (L^2), $= \int_0^s r_{S(c)} \, ds$
ϕ	Angle of twist (rad)
ν	Poisson's ratio

2.2 CENTROIDS

Coordinates and notation* are given in Fig. 2-1, which displays "continuous" and composite shapes. The composite shape is formed of two or more standard shapes, such as rectangles, triangles, and circles, for which the geometric properties are readily available.

The *centroid* of a plane area is that point in the plane about which the area is equally distributed. It is often called the *center of gravity* of the area. For the

---

*The common properties of this chapter are discussed in detail in elementary texts such as reference 2.1 and handbooks such as reference 2.2

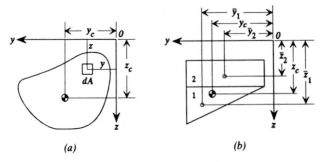

FIGURE 2-1 Coordinates and notation: (a) continuous shape; (b) composite shape.

area of Fig. 2-1a the centroid is defined as

$$
y_c = \frac{\int_A y\, dA}{\int_A dA} = \frac{\int_A y\, dA}{A} \qquad z_c = \frac{\int_A z\, dA}{\int_A dA} = \frac{\int_A z\, dA}{A} \tag{2.1}
$$

where $A = \int_A dA$.

For a composite area formed of two standard shapes, such as the one in Fig. 2-1b, the centroid is obtained using

$$
y_c = \frac{A_1 \bar{y}_1 + A_2 \bar{y}_2}{A_1 + A_2} \qquad z_c = \frac{A_1 \bar{z}_1 + A_2 \bar{z}_2}{A_1 + A_2} \tag{2.2}
$$

In general, for n standard shapes, the equations become

$$
y_c = \frac{\sum\limits_{i=1}^{n} A_i \bar{y}_i}{\sum\limits_{i=1}^{n} A_i} \qquad z_c = \frac{\sum\limits_{i=1}^{n} A_i \bar{z}_i}{\sum\limits_{i=1}^{n} A_i} \tag{2.3}
$$

where A_i $(i = 1, 2, \ldots, n)$ are the areas of identifiable simple areas and $\bar{y}_i, \bar{z}_i$ are the coordinates of the centroid of area A_i.

2.3 MOMENTS OF INERTIA

The *moment of inertia* of an area (second moment of an area) with respect to an axis is the sum of the products obtained by multiplying each element of the area dA by the square of its distance from the axis. For a section in the y, z plane

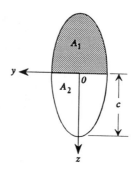

FIGURE 2-2 Notation for section moduli.

(Fig. 2-1a), the moment of inertia is defined to be

$$I_y = \int_A z^2 \, dA \quad \text{about the } y \text{ axis} \tag{2.4a}$$

$$I_z = \int_A y^2 \, dA \quad \text{about the } z \text{ axis} \tag{2.4b}$$

Section Moduli

For bending about the y axis, the *elastic section modulus* $S = Z_e$ is defined by

$$S = Z_e = \frac{I_y}{c} = \frac{\int_A z^2 \, dA}{c} \tag{2.5}$$

where c is the z distance from the centroidal (neutral) axis y to the outermost fiber.

The *plastic section modulus* $Z = Z_p$ is defined as the sum of statical moments of the areas above and below the centroidal (neutral) axis y (Fig. 2-2),

$$Z_p = -\int_{A_1} z \, dA + \int_{A_2} z \, dA \tag{2.6}$$

where A_1 and A_2 are the areas above and below, respectively, the neutral axis.

For a given shape, the ratio of the plastic section modulus to the elastic section modulus is called the *shape factor f*.

The *product of inertia* is defined as

$$I_{zy} = \int_A zy \, dA \tag{2.7}$$

In contrast to the moment of inertia, the product of inertia is not always positive. With respect to rectangular axes, it is zero if either of the axes is an axis of symmetry.

For moments and products of inertia of composite shapes the *parallel-axis formulas* are useful. These formulas relate the inertia properties of the areas about their own centroidal axes to parallel axes. For the three moments of

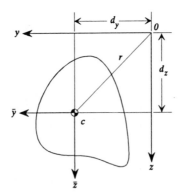

FIGURE 2-3 Geometry for transformation of axes; c is centroid of cross section.

inertia, the formulas are

$$I_y = I_{\bar{y}} + Ad_z^2 \qquad I_z = I_{\bar{z}} + Ad_y^2 \qquad I_{yz} = I_{\bar{y}\bar{z}} + Ad_y d_z \qquad (2.8)$$

It is important that the signs of the terms d_y and d_z be correct. Positive values are shown in Fig. 2-3, where d_y and d_z are the coordinates of the centroid of the cross section in the y, z coordinates.

A complicated area can often be subdivided into component areas whose moments of inertia are known. The moments of inertia of the original area are obtained by adding the individual moments of inertia, each taken about the same reference axis.

If an area is completely irregular, the moment and product of inertia can be obtained by evaluating the integrals numerically or by using a graphical technique. However, most computer programs rely on the technique of subdividing a section into standard shapes (e.g., rectangles), the more irregular the section, the finer the subdivision network required.

The *radius of gyration* is the distance from a reference axis to a point at which the entire area of a section may be considered to be concentrated and still have the same moment of inertia as the original distributed area. Thus for the y and z axes, the radii of gyration are given by

$$r_y = \sqrt{I_y/A} \qquad r_z = \sqrt{I_z/A} \qquad (2.9)$$

It is customary to express the instability criterion for a beam with an axial load in terms of one of the radii of gyration of the cross-sectional area.

2.4 POLAR MOMENT OF INERTIA

By definition, a polar axis is normal to the plane of reference, e.g., the x axis in Fig. 2-4.

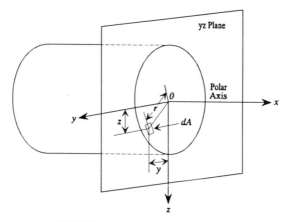

FIGURE 2-4 Polar moment of inertia.

The moment of inertia of an area (Fig. 2-4) about a point 0 in its plane is termed the *polar moment of inertia* of the area with respect to the point. It is designated by the symbol J_x and defined by the integral

$$J_x = \int_A r^2 \, dA = I_x$$

where r is the distance of the area element dA from the point 0. Since $r^2 = z^2 + y^2$, it follows from Eqs. (2.4) that

$$J_x = \int_A (z^2 + y^2) \, dA = I_y + I_z \tag{2.10}$$

With respect to a parallel axis, the polar moment of inertia is (Fig. 2-3)

$$J_{x0} = J_{xc} + Ar^2 \tag{2.11}$$

where J_{x0} and J_{xc} are the polar moments of inertial with respect to point 0 and the centroid c, respectively.

2.5 PRINCIPAL MOMENTS OF INERTIA

Moments of inertia are tensor quantities that possess properties that vary with the orientation θ (Fig. 2-5) of the reference axes. The angle θ is the angle of rotation of the centroidal reference axes and expresses the change of reference axes from the y, z system to the y', z' system.

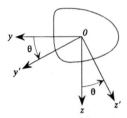

FIGURE 2-5 Geometry for rotation of axes.

If the axes are rotated an angle θ, it can be shown that the moments of inertia about the new axes y', z' vary according to

$$I_{y'} = \tfrac{1}{2}(I_y + I_z) + \tfrac{1}{2}(I_y - I_z)\cos 2\theta - I_{yz}\sin 2\theta$$
$$I_{z'} = \tfrac{1}{2}(I_y + I_z) - \tfrac{1}{2}(I_y - I_z)\cos 2\theta + I_{yz}\sin 2\theta \qquad (2.12)$$
$$I_{y'z'} = \tfrac{1}{2}(I_y - I_z)\sin 2\theta + I_{yz}\cos 2\theta$$

As the angle θ varies, several observations can be made:

1. At the value of $\theta = \theta_p$ for which $I_{y'z'} = 0$, the principal axes of inertia are defined. At this orientation, $I_{y'}$ and $I_{z'}$ will assume maximum and minimum values.

2. The value of the critical angle θ_p in terms of the known initial y, z axis properties is given by

$$\tan 2\theta_p = \frac{2I_{yz}}{I_z - I_y} \qquad (2.13)$$

3. The principal moments of inertia have the values

$$I_{max} = \tfrac{1}{2}(I_z + I_y) + \sqrt{\left[\tfrac{1}{2}(I_y - I_z)\right]^2 + I_{yz}^2} = I_1 \qquad (2.14a)$$

$$I_{min} = \tfrac{1}{2}(I_z + I_y) - \sqrt{\left[\tfrac{1}{2}(I_y - I_z)\right]^2 + I_{yz}^2} = I_2 \qquad (2.14b)$$

4. An axis of symmetry will be a principal axis and an axis of a zero product of inertia.

2.6 MOHR'S CIRCLE FOR MOMENTS OF INERTIA

The effect of a rotation of axes on the moments and product of inertia can be represented graphically using a Mohr's circle (Fig. 2-6) constructed in a manner similar to that for Mohr's circle of stress (Chapter 3).

The coordinates of a point on Mohr's circle (Fig. 2-6) are to be interpreted as representing the moment and the product of inertia of a plane area with respect

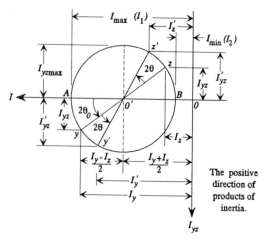

The positive direction of products of inertia.

FIGURE 2-6 Mohr's circle for moment of inertia of an area. This provides the moments of inertia with respect to the y, z system of Fig. 2-5 for an orientation of θ.

to the y axis (Fig. 2-5). The y axis is along the circle radius passing through the plotted point I_y, I_{yz}. The angle θ is measured counterclockwise from the y axis. However, the magnitudes of the angles on Mohr's circle are double those in the physical plane.

Example 2.1 Centroid Determine the centroid of the area shown in Fig. 2-7. This area is bounded by the y axis, the line $y = b$, and the parabola $z^2 = (h^2/b)y$.

To use Eqs. (2.1), choose the element of area $dA = z\,dy$ as shown in Fig. 2-7. Along the parabola, y and z are related by $z = h\sqrt{y/b}$. Then

$$y_c = \frac{\int_A y(z\,dy)}{\int_A dA} = \frac{\int_0^b y^{3/2}\,dy}{\int_0^b y^{1/2}\,dy} = \tfrac{3}{5}b \tag{1}$$

The formula $z_c = \int_A z\,dA / \int_A dA$ cannot be applied here because it is based on

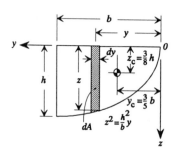

FIGURE 2-7 Example 2.1.

an element dA whose centroid is at a distance z from the y axis (Fig. 2-1a). For the dA employed here, the centroidal distance of dA from the y axis is $z/2$. Thus

$$z_c = \frac{\int_A \frac{1}{2} z \, dA}{\int_A dA} = \frac{\int_0^b \frac{1}{2} z(z \, dy)}{\int_0^b z \, dy} = \frac{3}{8} h \tag{2}$$

Other choices can be made for dA. For example, suppose $dA = dz \, dy$; then

$$z_c = \frac{\int_A z \, dA}{\int_A dA} = \frac{\int_A z \, dz \, dy}{\frac{2}{3} bh} = \frac{3}{2bh} \int_0^b \int_0^z z \, dz \, dy = \frac{3}{2bh} \int_0^b \frac{z^2}{2} \, dy = \frac{3}{8} h \tag{3}$$

Example 2.2 Moments of Inertia Compute the moments of inertia about the centroid, the angle of inclination of the principal axis, for the angle of Fig. 2-8a.

The centroid for the angle was computed and is shown in Fig. 2-8b. To compute the moments of inertia, use the parallel-axis theorem to transfer the individual shape inertias to the common reference axis of the angle's centroidal axes (Fig. 2-8b).

Begin with the product of inertia. For shape D.

$$d_z^D = -\left(\tfrac{3}{4} + \tfrac{1}{2}\right) = -\tfrac{5}{4} \text{ in.} \qquad d_y^D = \tfrac{3}{4} \text{ in.} \tag{1}$$

The negative sign occurs, since, with respect to the reference axes y, z, the

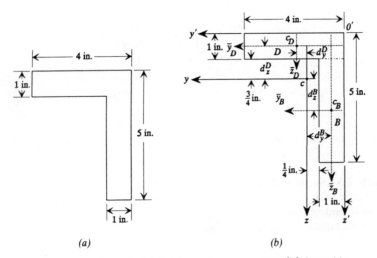

(a)   (b)

FIGURE 2-8 Example 2.2. (a) An angle cross section. (b) Centroids.

z-directed coordinate of c_D is on the negative side of the z axis:

$$I_{yz}^D = I_{\bar{y}_D \bar{z}_D} + d_z^D d_y^D A_D = 0 + \left(-\tfrac{5}{4}\right)\left(\tfrac{3}{4}\right)(4) = -3.75 \text{ in.}^4$$

For shape B,

$$d_z^B = 2 - \tfrac{3}{4} = +\tfrac{5}{4} \text{ in.} \qquad d_y^B = -\tfrac{3}{4} \text{ in.}$$

$$I_{yz}^B = I_{\bar{y}_B \bar{z}_B} + d_z^B d_y^B A_B = 0 + \left(\tfrac{5}{4}\right)\left(-\tfrac{3}{4}\right)(4) = -3.75 \text{ in.}^4 \tag{2}$$

The product of inertia for the complete angle is then

$$I_{yz} = I_{yz}^D + I_{yz}^B = -3.75 - 3.75 = -7.50 \text{ in.}^4 \tag{3}$$

The moments of inertia are computed in a similar fashion:

$$I_y^D = I_{\bar{y}_D} + \left(d_z^D\right)^2 A_D = \left(\tfrac{1}{12}\right)(4)(1^3) + \left(-\tfrac{5}{4}\right)^2(4) = 6.583 \text{ in.}^4 \tag{4}$$

$$I_z^D = I_{\bar{z}_D} + \left(d_y^D\right)^2 A_D = 7.583 \text{ in.}^4 \tag{5}$$

$$I_y^B = I_{\bar{y}_B} + \left(d_z^B\right)^2 A_B = \left(\tfrac{1}{12}\right)(1)(4^3) + \left(\tfrac{5}{4}\right)^2(4) = 11.583 \text{ in.}^4 \tag{6}$$

$$I_z^B = I_{\bar{z}_B} + \left(d_y^B\right)^2 A_B = 2.583 \text{ in.}^4 \tag{7}$$

For the entire angle

$$I_y = I_y^D + I_y^B = 18.167 \text{ in.}^4 \tag{8}$$

$$I_z = I_z^D + I_z^B = 10.167 \text{ in.}^4 \tag{9}$$

The angle of inclination with respect to the centroidal axis is given by

$$\tan 2\theta_p = \frac{2I_{yz}}{I_z - I_y} = \frac{2(-7.50)}{10.167 - 18.167} = 1.875$$

so that

$$2\theta_p = 61° 56' \quad \text{or} \quad \theta_p = 30° 58'$$

Some properties of plane sections for commonly occurring shapes are listed in Table 2-1.

Example 2.3   **Section Moduli**   Find the elastic and plastic section moduli and shape factor of a rectangular shape of width b and height h with respect to its horizontal centroidal axis.

For a rectangle $c = \tfrac{1}{2}h$ and (Table 2-1)

$$I = \tfrac{1}{12}bh^3$$

so that the elastic section modulus becomes

$$S = Z_e = \frac{I}{c} = \tfrac{1}{6}bh^2 \tag{1}$$

From Eqs. (2.6), the plastic section modulus is given by

$$Z_p = -\int_{A_1} z\, dA + \int_{A_2} z\, dA$$

$$= -\int_{-h/2}^{0} zb\, dz + \int_{0}^{h/2} zb\, dz = \tfrac{1}{4}bh^2 \tag{2}$$

The shape factor becomes

$$f = \frac{Z_p}{Z_e} = \frac{\tfrac{1}{4}bh^2}{\tfrac{1}{6}bh^2} = 1.5 \tag{3}$$

The section moduli for some selected cross sections are listed in Table 2-2.

2.7 FIRST MOMENT OF AREAS ASSOCIATED WITH SHEAR STRESSES IN BEAMS

In calculating the shear stress in a beam caused by transverse loading, a first moment Q with respect to the centroidal (neutral) axis of the beam is used. This

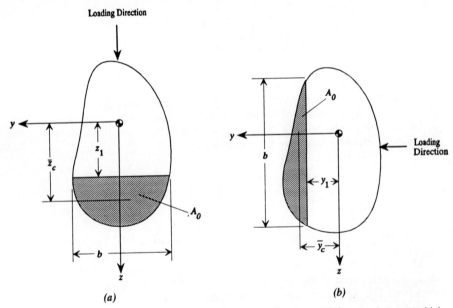

FIGURE 2-9 First moment of area: (a) y axis is centroidal axis and (b) z axis is a centroidal axis.

first moment is defined by

$$Q = Q_y = \int_{A_0} z \, dA = A_0 \bar{z}_c \qquad (2.15a)$$

where A_0 (Fig. 2-9a) is the area of that part of the section between the position z_1 at which the shear stress is to be calculated and the outer fiber and $\bar{z}_c$ is the distance from the y centroidal axis of the section to the centroid of A_0.

The formulas for Q for some sections are provided in Table 2-3.

Similarly, for loading in the y direction, the corresponding first moment of area is given by (Fig. 2-9b)

$$Q_z = \int_{A_0} y \, dA = A_0 \bar{y}_c \qquad (2.15b)$$

where $\bar{y}_c$ is the distance from the z centroidal axis of the cross section to the centroid of A_0 (Fig. 2-9b).

2.8 SHEAR CORRECTION FACTORS

Shear effects on deflection are often significant in the bending of short beams. These effects can be described in terms of shear correction factors defined by

$$\alpha_y = \frac{A}{I_y^2} \int_A \left(\frac{Q_y}{b} \right)^2 dA = \alpha_s \qquad (2.16a)$$

for loading in the z direction (Fig. 2-9a), where A is the cross-sectional area. The quantities I_y and Q_y are defined in Eqs. (2.4a) and (2.15a), respectively. Also, b is as shown in Fig. 2-9a. In general, b may vary over the cross section.

In the same fashion, for a cross section loaded in the y direction (Fig. 2-9b) the shear correction factor α_z is defined as

$$\alpha_z = \frac{A}{I_z^2} \int_A \left(\frac{Q_z}{b} \right)^2 dA \qquad (2.16b)$$

The quantities I_z and Q_z are given by Eqs. (2.4b) and (2.15b), respectively, and b is shown in Fig. 2-9b.

The shear correction factor may be viewed as the ratio of the actual beam cross-sectional area to the effective area resisting shear deformation. It can be seen from Eqs. (2.16) that the shear correction factors are always greater than or equal to zero.

The relations [Eqs. (2.16)] for α_y and α_z are approximate. More accurate determinations of shear correction factors can be made using the theory of

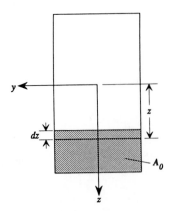

FIGURE 2-10 Example 2.4.

elasticity. For solid rectangular and circular cross sections this leads to

$$\alpha_{\text{rect}} = \frac{12 + 11\nu}{10(1 + \nu)} \qquad \alpha_{\text{circ}} = \frac{7 + 6\nu}{6(1 + \nu)}$$

In case of $\nu = 0.3$, these equations give $\alpha_{\text{rect}} = 1.18$ and $\alpha_{\text{circ}} = 1.13$. If Eqs. (2.16) are used, the corresponding shear correction factors are $\alpha_{\text{rect}} = 1.2$ and $\alpha_{\text{circ}} = 1.11$, which differ little from the more precise values.

Formulas for computing the numerical value of α_s for various beam cross sections are listed in Table 2-4. Computer programs are available with this book for calculating shear correction factors for shapes of arbitrary geometry. The inverse of the shear correction factor, called the *shear deflection constant*, is often required as an input in general-purpose finite-element analysis software. Reference 2.13 discusses problems encountered in calculating and using shear correction factors.

Example 2.4 Shear Correction Factors Determine the shear correction factors of the rectangular cross section of Fig. 2-10.

From Table 2-3, $Q_y = \frac{1}{2}b(\frac{1}{4}h^2 - z^2)$, where z_1 is replaced by z, as shown in Fig. 2-10. From Table 2-1, $I_y = \frac{1}{12}bh^3$. Substitution of these values into Eq. (2.16a) leads to

$$\alpha_y = \alpha_s = \frac{bh}{\left(\frac{1}{12}bh^3\right)^2} \int_{-h/2}^{h/2} \left[\frac{1}{2}\left(\frac{1}{4}h^2 - z^2\right)\right]^2 b\, dz = \frac{6}{5} \tag{1}$$

By the same reasoning, it can be shown that the shear correction factor for loading in the y direction is

$$\alpha_z = \frac{6}{5} \tag{2}$$

2.9 TORSIONAL CONSTANT

For a bar with circular cross section the torsional constant is the polar moment of inertia of the section. For cross sections of arbitrary shapes, the torsional constant J can be defined by the torsion formula

$$J = T/G\varphi'$$

(2.17)

where $\varphi' = d\varphi/dx$ with φ the angle of twist, T the torque, and G the shear modulus.

Thin-walled Sections

Thin-walled sections may be either open or closed. Such common structural shapes are channels, angles, I-beams, and wide-flange sections are open thin-walled sections, since the centerline of the wall does not form a closed curve. Closed sections have at least one closed curve.

Although there is no clearly defined line of demarcation between thin-walled and thick-walled sections, it is suggested that thin-walled theory may be applied with reasonable accuracy to sections if

$$t_{max}/b \leq 0.1$$

(2.18)

where t_{max} is the maximum thickness of the section and b is a typical cross-sectional dimension.

The torsional constant for a thin-walled open section (Fig. 2-11) is J, approximated by

$$J = \frac{1}{3} \int_{section} t^3 \, ds$$

(2.19a)

or for a section formed of M straight or curved segments of thickness t_i and

FIGURE 2-11 Thin-walled open section.

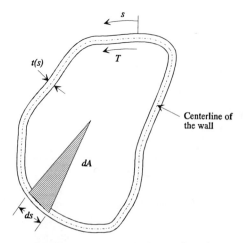

FIGURE 2-12   Thin-walled closed section.

length b_i:

$$J = \frac{\alpha}{3} \sum_{i=1}^{M} b_i t_i^3 \tag{2.19b}$$

where α is a shape factor. Use $\alpha = 1$ if no information on α is available.

For closed, thin-walled sections (one cell only), as shown in Fig. 2-12, the torsional constant of the cross section is given by

$$J = \frac{4\bar{A}^2}{\int \dfrac{ds}{t}} \tag{2.20}$$

where $\int (1/t)\, ds$ is the contour integral along the centerline s of a wall of thickness $t = t(s)$ and $\bar{A}$ is the area enclosed by the centerline of the wall.

If the hollow cross section is composed of M parts, each with the constant wall thickness t_i and the length b_i of the centerline, then the integral leads to

$$\int \frac{ds}{t} = \sum_{i=1}^{M} \frac{b_i}{t_i} \tag{2.21a}$$

For the case of a constant wall thickness t of a section with a circumference of length L, the integral becomes

$$\int \frac{ds}{t} = \frac{L}{t} \tag{2.21b}$$

Figure 2-13 shows a thin-walled cross section with four cells. In general, these cells may be interconnected in any manner, and a cross section may consist of M

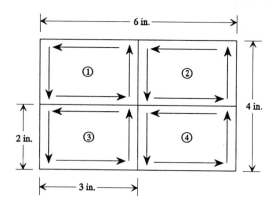

FIGURE 2-13 Example 2.5. Uniform wall thickness $t = 0.2$ in.

cells. It can be shown that the torsional constant of this type of cross section is obtained by

$$J = 4 \sum_{i=1}^{M} A_i \bar{q}_i \tag{2.22a}$$

where A_i is the centerline-enclosed area of cell i and $\bar{q}_i$ is a *normalized shear flow* (with units of area) that can be determined from the following set of equations:

$$\bar{q}_i \oint_i \frac{ds}{t} - \sum_k \bar{q}_k \int_{ik} \frac{ds}{t(s)} = A_i \tag{2.22b}$$

where $i = 1, 2, 3, \ldots, M$ and k refers to cells adjacent to the ith cell.
Table 2-5 provides torsional constants for some cross sections.

Example 2.5 Torsional Constant of Thin-Walled Section with Four Cells A thin-walled section with four cells is shown in Fig. 2-13. Determine the torsional constant using Eqs. (2.22a) and (2.22b).
From Fig. 2-13, the areas of the cells are

$$A_1 = A_2 = A_3 = A_4 = 2(3) = 6 \text{ in.}^2 \tag{1}$$

For cell 1, $i = 1$.

$$\bar{q}_i \oint_i \frac{ds}{t(s)} = \bar{q}_1 \left(\frac{2(3)}{0.2} + \frac{2(2)}{0.2} \right) = \frac{10}{0.2} \bar{q}_1 \tag{2}$$

$$\sum_k \bar{q}_k \int_{ik} \frac{ds}{t(s)} = \bar{q}_2 \frac{2}{0.2} + \bar{q}_3 \frac{3}{0.2} \tag{3}$$

Therefore, for $i = 1$, Eq. (2.22b) leads to

$$\bar{q}_1 \frac{10}{0.2} - \left(\bar{q}_2 \frac{2}{0.2} + \bar{q}_3 \frac{3}{0.2} \right) = 6 \tag{4}$$

Similarly, for cell 2, 3, and 4, $i = 2, 3, 4$, Eq. (2.22b) leads to

$$\bar{q}_2 \frac{10}{0.2} - \left(\bar{q}_1 \frac{2}{0.2} + \bar{q}_4 \frac{3}{0.2} \right) = 6 \tag{5}$$

$$\bar{q}_3 \frac{10}{0.2} - \left(\bar{q}_1 \frac{3}{0.2} + \bar{q}_4 \frac{2}{0.2} \right) = 6 \tag{6}$$

$$\bar{q}_4 \frac{10}{0.2} - \left(\bar{q}_2 \frac{3}{0.2} + \bar{q}_3 \frac{2}{0.2} \right) = 6 \tag{7}$$

respectively. Rearranging (4), (5), (6), and (7) into matrix format,

$$\begin{bmatrix} 10 & -2 & -3 & 0 \\ -2 & 10 & 0 & -3 \\ -3 & 0 & 10 & -2 \\ 0 & -3 & -2 & 10 \end{bmatrix} \begin{bmatrix} \bar{q}_1 \\ \bar{q}_2 \\ \bar{q}_3 \\ \bar{q}_4 \end{bmatrix} = \begin{bmatrix} 1.2 \\ 1.2 \\ 1.2 \\ 1.2 \end{bmatrix} \tag{8}$$

The solution of (8) is

$$\bar{q}_1 = \bar{q}_2 = \bar{q}_3 = \bar{q}_4 = 0.24 \text{ in.}^2 \tag{9}$$

Thus, the torsional constant of the cross section is, by Eq. (2.22a),

$$J = 4 \sum_{i=1}^{4} A_i \bar{q}_i = 4 \left(A_1 \bar{q}_1 + A_2 \bar{q}_2 + A_3 \bar{q}_3 + A_4 \bar{q}_4 \right)$$

$$= 4(6)(0.24 + 0.24 + 0.24 + 0.24) = 23.04 \text{ in.}^4 \tag{10}$$

2.10 SECTORIAL PROPERTIES

Sectorial properties of a cross section are useful in the study of restrained warping torsion, although they tend to be difficult to compute (Chapter 15). Some of the sectorial formulas for thin-walled cross sections are summarized below.

Sectorial Area

The sectorial area is given by

$$\omega_P = \int_0^s r_P \, ds \tag{2.23}$$

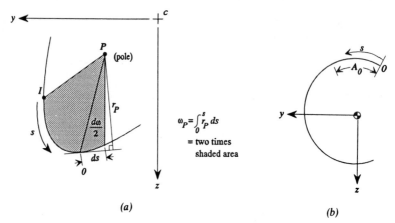

FIGURE 2-14 Sectorial properties: (a) sectorial area (sectorial coordinate); (b) first sectorial moment.

which is 2 times the shaded region in Fig. 2-14a. Point I is chosen as the origin of variable s, which lies along the centerline of the cross section, point P is the pole with respect to which ω_P is defined, and r_P is the distance between segment ds and P. Point P can be chosen arbitrarily.

Sectorial area ω_P is twice the area swept by P–0 as point 0 moves a distance s along the centerline of the cross section from initial point I (Fig. 2-14a). Consequently, the integration of Eq. (2.23) is reduced to the problem of finding the double shaded area, which is referred to as *direct integration*. Define increment $d\omega$ as positive when P–0 rotates in the counterclockwise direction. If the pole is at the centroid or the shear center, the corresponding sectorial area is, respectively,

$$\omega_c = \int_0^s r_c \, ds \tag{2.24a}$$

$$\omega_S = \int_0^s r_S \, ds \tag{2.24b}$$

First Sectorial Moment

The first sectorial moment is defined by

$$Q_\omega = \int_{A_0} \omega \, dA = \int_{A_0} \omega t \, ds \tag{2.25a}$$

where A_0 is shown in Fig. 2-14b and

$$\omega = \omega_S - \omega_0 \tag{2.25b}$$

$$\omega_0 = \frac{1}{A} \int_A \omega_S \, dA \tag{2.25c}$$

The quantity ω_S is defined with the shear center as the pole (Eq. 2.24b) and with arbitrary initial point I. Note that ω_0 is a constant that depends on ω_S. This definition of ω makes

$$\int_A \omega \, dA = 0 \tag{2.25d}$$

This follows since

$$\int_A \omega \, dA = \int_A (\omega_S - \omega_0) \, dA = \int_A \omega_S \, dA - \omega_0 A = 0$$

Sectorial area ω as defined by Eqs. (2.25b) is called the *principal sectorial coordinate* or the *principal sectorial area*.

Sectorial Linear Moments

Define the sectorial linear moments

$$I_{\omega y} = \int_A \omega_P z \, dA \qquad I_{\omega z} = \int_A \omega_P y \, dA \tag{2.26}$$

where the integration is taken over the entire cross section A.

Warping Constant (Sectorial Moment of Inertia)

The warping constant is defined as

$$\Gamma = \int_A \omega^2 \, dA \tag{2.27}$$

with the integration taken over the entire cross section A.
    If the coordinates are set at the centroid, then

$$I_{\omega y} = \int_A \omega_c z \, dA \qquad I_{\omega z} = \int_A \omega_c y \, dA \tag{2.28}$$

The choice of initial point for the sectorial coordinates is arbitrary. A different choice changes ω by a constant but leaves $I_{\omega z}$ and $I_{\omega y}$ unchanged.
    For thin-walled sections consisting of straight elements the integration in the above formulas can be performed in a piecewise manner, leading to the summation formulas of Tables 2-6 and 2-7 and some parts of Table 2-5. These formulas give the values of ω_c, ω_S, and Q_ω at each junction point (node), and the values along an element between any two junction points can be found by linear interpolation. This method is called *piecewise integration*. It reduces the task of performing the direct integration to one of finding the length of the elements and perpendicular distance r_P from the pole to that element.

2.11 SHEAR CENTER FOR THIN-WALLED CROSS SECTIONS

To bend a beam without twisting, the plane of the applied forces must pass through the shear center of every cross section of the beam. For a cross section possessing two or more axes of symmetry (e.g., an *I* section) or antisymmetry (e.g., a *Z* section), the shear center and the centroid coincide. However, for cross sections with only one axis of symmetry, the shear center and the centroid do not coincide. The location of the shear center is of greater importance for a thin-walled open cross section because of its low torsional stiffness than that for a closed section.

Open Cross Sections

If the material of the cross section remains linearly elastic and the flexure formula (Chapter 3) is valid, the shear center for thin-walled cross sections can be obtained using [2.6]

$$y_S = \frac{1}{D}\left(I_z I_{\omega y} - I_{yz} I_{\omega z}\right) \tag{2.29a}$$

$$z_S = \frac{1}{D}\left(-I_y I_{\omega z} + I_{yz} I_{\omega y}\right) \tag{2.29b}$$

with

$$D = I_y I_z - I_{yz}^2 \tag{2.30}$$

The origin of the *yz* coordinate system is at the centroid of the cross section, y_S and z_S are distances from point *P* (pole) (Fig. 2-15). Point *P* can be located arbitrarily for convenience of calculation. Normally, *P* is located at the centroid of the cross section.

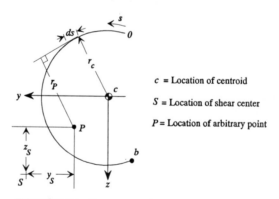

c = Location of centroid

S = Location of shear center

P = Location of arbitrary point

FIGURE 2-15 Shear center for open thin-walled section.

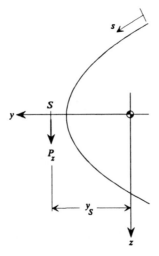

FIGURE 2-16 Thin-walled open section with one axis (y) of symmetry.

When the y, z axes are principal axes ($I_{yz} = 0$), Eq. (2.29) can be simplified as

$$y_S = \frac{1}{I_y} I_{\omega y} \qquad (2.31a)$$

$$z_S = -\frac{1}{I_z} I_{\omega z} \qquad (2.31b)$$

If the cross section has one axis of symmetry (Fig. 2-16) and if the load (P_z) is parallel to the z axis and passes through the shear center S, Eqs. (2.31) can be further simplified as

$$y_S = \frac{1}{I_y} I_{\omega y} \qquad (2.32a)$$

$$z_S = 0 \qquad (2.32b)$$

The shear centers for some common sections are given in Table 2-6.
For the shear center of closed sections, refer to [2.6].

Example 2.6 Shear Center Calculation Determine the shear center of the cross section shown in Fig. 2-17a.

In order to simplify the calculation, choose pole P at one corner and initial point I at another, as shown in Fig. 2-17b. This configuration makes $r_p = 0$ for legs I–P and P–k, so that $\omega_p = 0$. For leg I–J, $r_p = 2a$. Then

$$\omega_P = \begin{cases} -2as & (\text{legs } I\text{–}J) \\ 0 & (\text{legs } I\text{–}P \text{ and } P\text{–}k) \end{cases} \qquad (1)$$

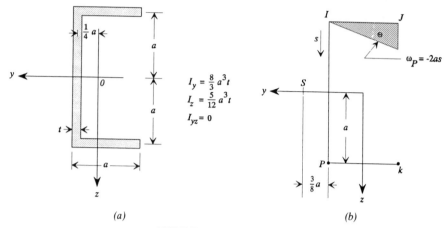

$$I_y = \tfrac{8}{3}a^3t$$
$$I_z = \tfrac{5}{12}a^3t$$
$$I_{yz} = 0$$

FIGURE 2-17 Example 2.6.

From Eqs. (2.26)

$$I_{\omega y} = \int_A \omega_p z\, dA = \int_{A_{IJ}} \omega_p z\, dA + \int_{A_{IP}} \omega_p z\, dA + \int_{A_{Pk}} \omega_p z\, dA$$

$$= \int_0^a (2as)at\, ds + 0 + 0 = a^4 t \tag{2}$$

$$I_{\omega z} = \int_A \omega_p y\, dA = \int_0^a 2as\left(s - \tfrac{1}{4}a\right)t\, ds + 0 + 0 = +\tfrac{5}{12}a^4 t \tag{3}$$

From Eqs. (2.29a) and (2.29b) with $I_{yz} = 0$

$$y_S = \frac{I_{\omega y}}{I_y} = \frac{+a^4 t}{\tfrac{8}{3}a^3 t} = +\tfrac{3}{8}a \tag{4}$$

$$z_S = -\frac{I_{\omega z}}{I_z} = -\frac{\tfrac{5}{12}a^4 t}{\tfrac{5}{12}a^3 t} = -a \tag{5}$$

This shear center S is indicated in Fig. 2-17b.

2.12 MODULUS-WEIGHTED PROPERTIES FOR COMPOSITE SECTIONS

For members of nonhomogeneous material it is useful to introduce the concept of modulus-weighted section properties. Define an increment of area as

$$dA^* = \frac{E}{E_r}\, dA \tag{2.33}$$

where E_r is an arbitrary reference modulus that can be chosen to control the magnitude of the numbers involved in the computation of modulus-weighted

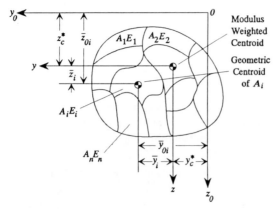

FIGURE 2-18   Cross section of a composite beam.

properties and E assumes the value of the modulus of elasticity for the point of interest on the cross section. For a homogeneous member set $E_r = E$ so that this modulus-weighted property will reduce to the ordinary geometric property of the section. The definition of Eq. (2.33) leads to the other modulus-weighted definitions:

Area:
$$A^* = \int_A dA^*$$

Centroid:
$$y_c^* = \frac{1}{A^*} \int_A y \, dA^* \qquad z_c^* = \frac{1}{A^*} \int_A z \, dA^* \quad (2.34)$$

Moments of inertia:
$$I_y^* = \int_A z^2 \, dA^* \qquad I_z^* = \int_A y^2 \, dA^*$$

$$I_{yz}^* = \int_A yz \, dA^*$$

In the case of a composite section E is piecewise constant, as shown in Fig. 2-18. Then for this section formed of n materials:

Area:
$$A_i^* = \frac{E_i}{E_r} A_i \qquad A^* = \sum_{i=1}^n A_i^* = \sum_{i=1}^n \frac{E_i}{E_r} A_i \quad (2.35a)$$

Centroid:
$$y_c^* = \frac{\displaystyle\sum_{i=1}^n \frac{E_i}{E_r} \int_{A_i} y_{0i} \, dA_i}{\displaystyle\sum_{i=1}^n \frac{E_i}{E_r} \int_{A_i} dA_i} = \frac{1}{A^*} \sum_{i=1}^n \bar{y}_{0i} A_i^*$$

$$(2.35b)$$

$$z_c^* = \frac{\displaystyle\sum_{i=1}^n \frac{E_i}{E_r} \int_{A_i} z_{0i} \, dA_i}{\displaystyle\sum_{i=1}^n \frac{E_i}{E_r} \int_{A_i} dA_i} = \frac{1}{A^*} \sum_{i=1}^n \bar{z}_{0i} A_i^*$$

where $\bar{y}_{0i}, \bar{z}_{0i}$ are the coordinates in the y_0, z_0 system of the geometric centroid of the area A_i of the ith element. The moments of inertia are calculated as follows:

$$I_y^* = I_{y_0}^* - z_c^{*2}A^* \qquad I_z^* = I_{z_0}^* - y_c^{*2}A^* \qquad I_{yz}^* = I_{y_0z_0}^* - y_c^*z_c^*A^* \quad (2.36)$$

where I_y^*, I_z^*, I_{yz}^* are the modulus-weighted moments of inertia about the modulus weighted centroidal axes y, z and $I_{y_0}^*, I_{z_0}^*, I_{y_0z_0}^*$ are the modulus-weighted moments of inertia about the reference y_0, z_0 axes:

$$I_{y_0}^* = \sum_{i=1}^{n} \frac{E_i}{E_r}\left(\bar{I}_{y_i} + \bar{z}_{0i}^2 A_i\right) \qquad I_{z_0}^* = \sum_{i=1}^{n} \frac{E_i}{E_r}\left(\bar{I}_{z_i} + \bar{y}_{0i}^2 A_i\right)$$

$$I_{y_0z_0}^* = \sum_{i=1}^{n} \frac{E_i}{E_r}\left(\bar{I}_{y_iz_i} + \bar{y}_{0i}\bar{z}_{0i} A_i\right) \tag{2.37}$$

where $\bar{I}_{y_i}, \bar{I}_{z_i}, \bar{I}_{y_iz_i}$ are the moments of inertia of area A_i about its own centroidal axis. Also

$$I_y^* = \sum_{i=1}^{n} \frac{E_i}{E_r}\left(\bar{I}_{y_i} + \bar{z}_i^2 A_i\right) \qquad I_z^* = \sum_{i=1}^{n} \frac{E_i}{E_r}\left(\bar{I}_{z_i} + \bar{y}_i^2 A_i\right)$$

$$I_{yz}^* = \sum_{i=1}^{n} \frac{E_i}{E_r}\left(\bar{I}_{y_iz_i} + \bar{y}_i\bar{z}_i A_i\right) \tag{2.38}$$

where $\bar{y}_i, \bar{z}_i$ are the coordinates in the centroidal y, z system of the geometric centroid of the area A_i of the ith element (Fig. 2-18).

The first moment is calculated as

$$Q_y^* = \int_{A_0} z \, dA^* = \int_{A_0} \frac{E}{E_r} z \, dA = \sum_{i=1}^{n_0} \bar{z}_i A_i^*$$

$$Q_z^* = \int_{A_0} y \, dA^* = \int_{A_0} \frac{E}{E_r} y \, dA = \sum_{i=1}^{n_0} \bar{y}_i A_i^* \tag{2.39}$$

where the summations using index n_0 extend over the area A_0 (Fig. 2-14b). The warping constant is derived from

$$\Gamma^* = \int_A \omega^2 \, dA^* = \int_A \frac{E}{E_r} \omega^2 \, dA = \sum_{i=1}^{n} \frac{E_i}{E_r} \int_{A_i} \omega^2 \, dA \tag{2.40}$$

and the first sectorial moment from

$$Q_\omega^* = \int_{A_0} \omega \, dA^* = \int_{A_0} \frac{E}{E_r} \omega \, dA = \sum_{i=1}^{n_0} \frac{E_i}{E_r} \int_{A_i} \omega \, dA \tag{2.41}$$

where A_0 is defined in Fig. 2-14b. The shear center is calculated as

$$y_S = y_P + \frac{I_z^* I_{\omega y}^* - I_{yz}^* I_{\omega z}^*}{I_y^* I_z^* - I_{yz}^{*2}} \qquad z_S = z_P - \frac{I_y^* I_{\omega z}^* - I_{yz}^* I_{\omega y}^*}{I_y^* I_z^* - I_{yz}^{*2}} \qquad (2.42)$$

where y_S, z_S are indicated in Fig. 2-15 and $I_{\omega y}^* = \int_A \omega_P z \, dA^*$, $I_{\omega z}^* = \int_A \omega_P y \, dA^*$.

REFERENCES

2.1. Pilkey, W. D., and Pilkey, O. H., *Mechanics of Solids*, Krieger, Malabar, Florida, 1986.

2.2. Blake, A., *Handbook of Mechanics, Materials, and Structures*, Wiley, New York, 1985.

2.3. Cowper, G. R., "The Shear Coefficient in Timoshenko's Beam Theory," *J. Appl. Mech.*, Vol. 33, 1966, pp. 335–340.

2.4. Kollbrunner, C. F., and Basler, K., *Torsion in Structures* (translated by E. C. Glauser), Springer-Verlag, New York, 1969.

2.5. Isakower, R. I., "The Shaft Book," U.S. Army ARRADCOM MISD User's Manual 80-5, March 1980.

2.6. Cook, R. D., and Young, W. C., *Advanced Mechanics of Materials*, MacMillan, New York, 1985.

2.7. Column Research Committee of Japan, *Handbook of Structural Stability*, Corona Publishing Company, Tokyo, 1971.

2.8. Reissner, E., and Tsai, W. T., "On the Determination of the Centers of Twist of Shear Cylindrical Shell Beams," ASME Paper No. 72-APM-XX, 1972.

2.9. Timoshenko, S. P., and Gere, J. M., *Mechanics of Materials*, Van Nostrand, New York, 1972.

2.10. Boresi, A. P., and Sidebottom, O. M., *Advanced Mechanics of Materials*, Wiley, New York, 1985.

2.11. Oden, J. T., *Mechanics of Elastic Structures*, McGraw-Hill, New York, 1970.

2.12. Galambos, T. V., *Structural Members and Frames*, Prentice-Hall, Englewood Cliffs, NJ, 1968.

2.13. Schramm, U., Kitis, L., Kang, W., and Pilkey, W. D., "On the Shear Deformation Coefficient in Beam Theory," *J. Fin. Elem. in Anal. & Design*, Vol. 16, 1994.

2

Tables

Shape of Section	Area, Location of Centroid (y_c, z_c)	Moments of Inertia and the Polar Moment of Inertia $(J_x = I_{\bar{y}} + I_{\bar{z}})$ with Respect to Centroidal Axial Axis	Transverse Radii of Gyration and the Polar Radius of Gyration $r_p = r_x$
1. Rectangle	$A = bh$ $y_c = \frac{1}{2}b$ $z_c = \frac{1}{2}h$	$I = I_{\bar{y}} = \frac{1}{12}bh^3$ $I_{\bar{z}} = \frac{1}{12}hb^3$ $I_{\bar{y}\bar{z}} = 0$ $J_x = \frac{1}{12}bh(b^2 + h^2)$	$r_{\bar{y}} = h/\sqrt{12}$ $r_{\bar{z}} = b/\sqrt{12}$ $r_p = \sqrt{\frac{1}{12}(b^2 + h^2)}$
2. Triangle	$A = \frac{1}{2}bh$ $y_c = \frac{1}{3}(a + b)$ $z_c = \frac{1}{3}h$	$I = I_{\bar{y}} = \frac{1}{36}bh^3$ $I_{\bar{z}} = \frac{1}{36}bh(b^2 - ab + a^2)$ $I_{\bar{y}\bar{z}} = \frac{1}{72}bh^2(2a - b)$ $J_x = \frac{1}{36}bh(h^2 + b^2 - ab + a^2)$	$r_{\bar{y}} = h/\sqrt{18}$ $r_{\bar{z}} = [\frac{1}{18}(b^2 - ab + a^2)]^{1/2}$ $r_p = \sqrt{\frac{1}{18}(b^2 + h^2 - ab + a^2)}$

TABLE 2-1 Areas, Centroids, Moments of Inertia, Radii   44

3. Trapezoid

$$A = \tfrac{1}{2}h(a + b)$$

$$y_c = \tfrac{1}{2}a$$

$$z_c = \frac{h}{3}\left(\frac{a + 2b}{a + b}\right)$$

$$I = I_{\bar{y}} = \frac{h^3}{36}\left(\frac{a^2 + 4ab + b^2}{a + b}\right)$$

$$I_{\bar{z}} = \tfrac{1}{48}h(a + b)(a^2 + b^2)$$

$$I_{\bar{y}\bar{z}} = 0$$

$$J_x = I_{\bar{y}} + I_{\bar{z}}$$

$$r_{\bar{y}} = \frac{h(a^2 + 4ab + b^2)^{1/2}}{\sqrt{18}\,(a + b)}$$

$$r_{\bar{z}} = \left[\tfrac{1}{24}(a^2 + b^2)\right]^{1/2}$$

$$r_p = \sqrt{J_x/A}$$

4. Circle

$$A = \tfrac{1}{4}\pi d^2$$

$$y_c = \tfrac{1}{2}d$$

$$z_c = \tfrac{1}{2}d$$

$$I = I_{\bar{y}} = I_{\bar{z}} = \tfrac{1}{64}\pi d^4$$

$$I_{\bar{y}\bar{z}} = 0$$

$$J_x = \tfrac{1}{32}\pi d^4$$

$$r_{\bar{y}} = r_{\bar{z}} = \tfrac{1}{4}d$$

$$r_p = d/\sqrt{8}$$

5. Annulus

$$A = \tfrac{1}{4}\pi(d_0^2 - d_i^2)$$

$$y_c = \tfrac{1}{2}d_0$$

$$z_c = \tfrac{1}{2}d_0$$

$$I = I_{\bar{y}} = I_{\bar{z}} = \tfrac{1}{4}\pi(r_0^4 - r_i^4)$$

$$I_{\bar{y}\bar{z}} = 0$$

$$J_x = \tfrac{1}{32}\pi(d_0^4 - d_i^4)$$

$$r_{\bar{y}} = r_{\bar{z}} = \tfrac{1}{4}(d_0^2 + d_i^2)^{1/2}$$

$$r_p = \sqrt{\tfrac{1}{8}(d_0^2 + d_i^2)}$$

Areas, Centroids, Moments of Inertia, Radii TABLE 2-1

Shape of Section	Area, Location of Centroid (y_c, z_c)	Moments of Inertia and the Polar Moment of Inertia $(J_x = I_{\bar{y}} + I_{\bar{z}})$ with Respect to Centroidal Axial Axis	Transverse Radii of Gyration and the Polar Radius of Gyration $r_p = r_x$
6. **Ellipse** 	$A = \pi ab$ $y_c = a$ $z_c = b$	$I = I_{\bar{y}} = \frac{1}{4}\pi ab^3$ $I_{\bar{z}} = \frac{1}{4}\pi ba^3$ $I_{\bar{y}\bar{z}} = 0$ $J_x = \frac{1}{4}\pi ab(b^2 + a^2)$	$r_{\bar{y}} = \frac{1}{2}b$ $r_{\bar{z}} = \frac{1}{2}a$ $r_p = \sqrt{\frac{1}{4}(a^2 + b^2)}$
7. **Semicircle** 	$A = \frac{1}{2}\pi r^2$ $y_c = r$ $z_c = 4r/3\pi$	$I = I_{\bar{y}} = 0.11r^4$ $I_{\bar{z}} = 0.393r^4$ $I_{\bar{y}\bar{z}} = 0$ $J_x = 0.503r^4$	$r_{\bar{y}} = 0.264r$ $r_{\bar{z}} = \frac{1}{2}r$ $r_p = 0.565r$
8. **Parallelogram** 	$A = bd$ $y_c = \frac{1}{2}(b + a)$ $z_c = \frac{1}{2}d$	$I_{\bar{y}} = \frac{1}{12}bd^3$ $I_{\bar{z}} = \frac{1}{12}bd(b^2 + a^2)$ $I_{\bar{y}\bar{z}} = -\frac{1}{12}abd^2$	$r_{\bar{y}} = 0.2887d$ $r_{\bar{z}} = 0.2887\sqrt{b^2 + a^2}$

TABLE 2-1   Areas, Centroids, Moments of Inertia, Radii   46

9. Diamond

$$A = \tfrac{1}{2}bd$$
$$y_c = \tfrac{1}{2}b$$
$$z_c = \tfrac{1}{2}d$$

$$I_{\bar y} = \tfrac{1}{48}bd^3$$
$$I_{\bar z} = \tfrac{1}{48}db^3$$

$$r_{\bar y} = 0.2041d$$
$$r_{\bar z} = 0.2041b$$

10. Sector of solid circle

$$A = \alpha R^2$$

$$y_c = R\sin\alpha$$

$$z_c = R\left(1 - \frac{2\sin\alpha}{3\alpha}\right)$$

$$I_{\bar y} = \frac{R^4}{4}\left(\alpha + \sin\alpha\cos\alpha - \frac{16\sin^2\alpha}{9\alpha}\right)$$

$$I_{\bar z} = \frac{R^4}{4}(\alpha - \sin\alpha\cos\alpha)$$

$$r_{\bar y} = \frac{R}{2}\sqrt{1 + \frac{\sin\alpha\cos\alpha}{\alpha} - \frac{16\sin^2\alpha}{9\alpha^2}}$$

$$r_{\bar z} = \frac{R}{2}\sqrt{1 - \frac{\sin\alpha\cos\alpha}{\alpha}}$$

11. Angle

$$B_1 = b_1 + \tfrac{1}{2}t$$
$$B_2 = b_2 + \tfrac{1}{2}t$$
$$c_1 = b_1 - \tfrac{1}{2}t$$
$$c_2 = b_2 - \tfrac{1}{2}t$$
$$A = t(b_1 + b_2)$$
$$y_c = \frac{B_1^2 + c_2 t}{2(b_1 + b_2)}$$
$$z_c = \frac{B_2^2 + c_1 t}{2(b_1 + b_2)}$$

$$I_{\bar y} = \tfrac{1}{3}[t(B_2 - \bar z)^3 + B_1 z^3 - c_1(\bar z - t)^3]$$

$$I_{\bar z} = \tfrac{1}{3}[t(B_1 - \bar y)^3 + B_2\bar y^3 - c_2(\bar y - t)^3]$$

$$I_{\bar y\bar z} = -\tfrac{1}{2}t[b_1\bar z(b_1 - 2\bar y) + b_2\bar y(b_2 - 2\bar z)]$$

$$r_{\bar y} = \sqrt{I_{\bar y}/A}$$
$$r_{\bar z} = \sqrt{I_{\bar z}/A}$$
$$r_p = \sqrt{J_x/A}$$

Shape of Section	Area, Location of Centroid (y_c, z_c)	Moments of Inertia and the Polar Moment of Inertia $(J_x = I_{\bar{y}} + I_{\bar{z}})$ with Respect to Centroidal Axial Axis	Transverse Radii of Gyration and the Polar Radius of Gyration $r_p = r_x$
12. I Section	$H_1 = h + t_f$ $H_2 = h - t_f$ $A = 2bt_f + H_2 t_w$ $y_c = \frac{1}{2}b$ $z_c = \frac{1}{2}H_1$	$I_{\bar{y}} = \frac{1}{12}(bH_1^3 - (b - t_w)H_2^3)$ $I_{\bar{z}} = \frac{1}{12}(H_2 t_w^3 + 2t_f b^3)$ $I_{\bar{y}\bar{z}} = 0, \ J_x = I_{\bar{y}} + I_{\bar{z}}$	$r_{\bar{y}} = \sqrt{I_y/A}$ $r_{\bar{z}} = \sqrt{I_{\bar{z}}/A}$ $r_p = \sqrt{J_x/A}$
13. Z Section	$H = h + t$ $B = b + \frac{1}{2}t$ $C = b - \frac{1}{2}t$ $A = t(h + 2b)$ $y_c = b, \ z_c = \frac{1}{2}(h + t)$	$I_{\bar{y}} = \frac{1}{12}[BH^3 - C(H - 2t)^3]$ $I_{\bar{z}} = \frac{1}{12}[H(B + C)^3 - 2hC^3 \ - 6B^2hC]$ $I_{\bar{y}\bar{z}} = -\frac{1}{2}htb^2$ $J_x = I_{\bar{y}} + I_{\bar{z}}$	$r_{\bar{y}} = \sqrt{I_{\bar{y}}/A}$ $r_{\bar{z}} = \sqrt{I_{\bar{z}}/A}$ $r_p = \sqrt{J_x/A}$

TABLE 2-1 Areas, Centroids, Moments of Inertia, Radii   48

14. Cross

$$A = ht_1 + (b - t_1)t_2$$
$$y_c = \tfrac{1}{2}b$$
$$z_c = \tfrac{1}{2}h$$

$$I_{\bar{y}} = \tfrac{1}{12}[t_1 h^3 + (b - t_1)t_2^3]$$
$$I_z = \tfrac{1}{12}[t_2 b^3 + (h - t_2)t_1^3]$$
$$I_{\bar{y}z} = 0, \quad J_x = I_{\bar{y}} + I_z$$

$$r_{\bar{y}} = \sqrt{I_{\bar{y}}/A}$$
$$r_z = \sqrt{I_z/A}$$
$$r_p = \sqrt{J_x/A}$$

15. Channel

$$B = b + \tfrac{1}{2}t_w$$
$$C = b - \tfrac{1}{2}t_w$$
$$H = h + t_f$$
$$D = h - t_f$$
$$A = ht_w + 2bt_f$$
$$y_c = \frac{2B^2 t_f + Dt_w^2}{2BH - 2DC}$$
$$z_c = \tfrac{1}{2}(h + t_f)$$

$$I_{\bar{y}} = \tfrac{1}{12}(BH^3 - CD^3)$$
$$I_z = \tfrac{1}{3}(2t_f B^3 + Dt_w^3) - A(B - y_c)^2$$
$$I_{\bar{y}z} = 0, \quad J_x = I_{\bar{y}} + I_z$$

$$r_{\bar{y}} = \sqrt{I_{\bar{y}}/A}$$
$$r_z = \sqrt{I_z/A}$$
$$r_p = \sqrt{J_x/A}$$

16. T Section

$$H = h + \tfrac{1}{2}t_f$$
$$C = b - t_w$$
$$A = bt_f + t_w D, \quad D = h - \tfrac{1}{2}t_f$$
$$y_c = \tfrac{1}{2}b, \quad D = h - \tfrac{1}{2}t_f$$
$$z_c = \frac{H^2 t_w + Ct_f^2}{2(bt_f + Dt_w)}$$

$$I_{\bar{y}} = \tfrac{1}{3}[t_w(H - z_c)^3 + bz_c^3 - C(z_c - t_f)^3]$$
$$I_z = \tfrac{1}{12}(b^3 t_f + Dt_w^3)$$
$$I_{\bar{y}z} = 0, \quad J_x = I_{\bar{y}} + I_z$$

$$r_{\bar{y}} = \sqrt{I_{\bar{y}}/A}$$
$$r_z = \sqrt{I_z/A}$$
$$r_p = \sqrt{J_x/A}$$

Areas, Centroids, Moments of Inertia, Radii — TABLE 2-1

TABLE 2-2 **Section Moduli about Centroidal Axes** 50

TABLE 2-2 SECTION MODULI ABOUT CENTROIDAL AXES

Case	Elastic Section Modulus $S = Z_e$	Plastic Section Modulus $Z = Z_p$	Shape Factor $f = Z_p/Z_e$
1. Rectangle	$Z_{ey} = \frac{1}{6}bh^2$ $Z_{ez} = \frac{1}{6}hb^2$	$Z_{py} = \frac{1}{4}bh^2$ $Z_{pz} = \frac{1}{4}hb^2$	$f_y = f_z = 1.5$
2. Hollow rectangle	$Z_{ey} = \frac{1}{6}\dfrac{bh^3 - b_i h_i^3}{h}$ $Z_{ez} = \frac{1}{6}\dfrac{hb^3 - h_i b_i^3}{b}$	$Z_{py} = \frac{1}{4}(bh^2 - b_i h_i^2)$ $Z_{pz} = \frac{1}{4}(hb^2 - h_i b_i^2)$	$f_y = 1.5\,\dfrac{h(bh^2 - b_i h_i^2)}{bh^3 - b_i h_i^3}$ $f_z = 1.5\,\dfrac{b(hb^2 - h_i b_i^2)}{hb^3 - h_i b_i^3}$

3. Triangle	$Z_{ey} = \frac{1}{24}bh^2$ $Z_{ez} = \frac{1}{24}hb^2$	$Z_{py} = 0.097bh^2$ $Z_{pz} = \frac{1}{12}hb^2$	$f_y = 2.33$ $f_z = 2.0$
4. Diamond	$Z_{ey} = \frac{1}{24}bh^2$ $Z_{ez} = \frac{1}{24}hb^2$	$Z_{py} = \frac{1}{12}bh^2$ $Z_{pz} = \frac{1}{12}hb^2$	$f_y = f_z = 2.0$
5. Circle	$Z_{ey} = Z_{ez} = \frac{1}{4}\pi r^3$	$Z_{py} = Z_{pz} = \frac{4}{3}r^3$	$f_y = f_z = 1.698$

TABLE 2-2 (continued) SECTION MODULI ABOUT CENTROIDAL AXES

Case	Elastic Section Modulus $S = Z_e$	Plastic Section Modulus $Z = Z_p$	Shape Factor $f = Z_p/Z_e$
6. Hollow circle	$Z_{ey} = Z_{ez} = \dfrac{\pi}{4}\dfrac{r_0^4 - r_i^4}{r_0}$	$Z_{py} = Z_{pz} = \tfrac{4}{3}(r_0^3 - r_i^3)$	$f_y = f_z = 1.698\,\dfrac{r_0(r_0^3 - r_i^3)}{r_0^4 - r_i^4}$
7. Ellipse	$Z_{ey} = \tfrac{1}{4}\pi ba^2$ $Z_{ez} = \tfrac{1}{4}\pi ab^2$	$Z_{py} = \tfrac{4}{3}ba^2$ $Z_{pz} = \tfrac{4}{3}ab^2$	$f_y = f_z = 1.698$

TABLE 2-2 **Section Moduli about Centroidal Axes**   52

8. **Semicircle** $\bar{x} = 0.4244\,r$	$Z_{ey} = 0.1908 r^3$ $Z_{ez} = \frac{1}{8}\pi r^3$	$Z_{py} = 0.3540 r^3$ $Z_{pz} = \frac{2}{3} r^3$	$f_y = 1.856$ $f_z = 1.698$

9.
T Section
 	
	$Z_{ey} = \dfrac{I_y}{(h + t_f - z_c)}$
	
	$Z_{ez} = \dfrac{I_z}{\frac{1}{2}b}$
	
	where I_y, I_z = moments of inertia of T section about y, z axes (Table 2-1, case 16)
		If $t_w h \le b t_f$, then
		
		$Z_{py} = \dfrac{h^2 t_w}{4} - \dfrac{b^2 t_f^2}{4 t_w} - \dfrac{b t_f(h + t_f)}{2}$
		
		Neutral axis y is distance
		$\frac{1}{2}(b t_f / t_w + h)$ from bottom
		
		If $t_w h > b t_f$, then
		$Z_{py} = \frac{1}{4} t_f^2 b + \frac{1}{2} t_w h(t_f + h - t_w h/2b)$
		Neutral axis y is
		$\frac{1}{2}(t_w h/b + t_f)$ from the top
		$Z_{pz} = \frac{1}{4}(b^2 t_f + t_w^2 h)$
			$f_y = \dfrac{Z_{py}(h + t_f - z_c)}{I_y}$
			
			$f_z = \dfrac{Z_{pz}\, b}{2 I_z}$

TABLE 2-2 (continued)   SECTION MODULI ABOUT CENTROIDAL AXES

Case	Elastic Section Modulus $S = Z_e$	Plastic Section Modulus $Z = Z_p$	Shape Factor $f = Z_p/Z_e$

10. Channel	$Z_{ey} = \dfrac{2I_y}{h}$
	
	$Z_{ez} = \dfrac{I_z}{b + t_w - y_c}$
	
	where I_y, I_z, and y_c are given in case 15 of Table 2-1
		$Z_{py} = \tfrac{1}{4}h^2 t_w + t_f b(h - t_f)$
		
		If $2t_f b \le h t_w$, then
		
		$Z_{pz} = \dfrac{b^2 t_f}{2} - \dfrac{h^2 t_w^2}{8 t_f} + \dfrac{h t_w(b + t_w)}{2}$.
		
		Neutral axis z is $\tfrac{1}{2}\left(h t_w/2t_f + b\right)$ from left side
		If $2t_f b > h t_w$, then
		$Z_{pz} = \tfrac{1}{4}t_w^2 h + t_f b(t_w + b - t_f b/h)$
		Neutral axis z is $t_f b/h + \tfrac{1}{2}t_w$ from right side
			$f_y = \dfrac{Z_{py} h}{2I_y}$
			
			$f_z = \dfrac{Z_{pz}(b + t_w - y_c)}{I_z}$

11. I Section	$Z_{ey} = \dfrac{I_y}{t_f + h/2}$ $Z_z = \dfrac{2I_z}{b}$ where I_y, I_z are given in case 12 of Table 2-1	$Z_{py} = \tfrac{1}{4}t_w h^2 + bt_f(h + t_f)$ $Z_{pz} = \tfrac{1}{2}b^2 t_f + \tfrac{1}{4}t_w^2 h$	$f_y = \dfrac{Z_{py}\left(t_f + \tfrac{1}{2}h\right)}{I_y}$ $f_z = \dfrac{Z_{pz} b}{2I_z}$

TABLE 2-2   Section Moduli about Centroidal Axes   54

TABLE 2-3 FIRST MOMENT Q ASSOCIATED WITH SHEAR STRESS IN BEAMS

Case	$Q = \int_{A_0} z \, dA = A_0 \bar{z}$	Q_{max}
1. Rectangular section	$\frac{1}{2}b\left(\frac{1}{4}h^2 - z_1^2\right)$	$\frac{1}{8}bh^2$ when $z_1 = 0$
2. Circular section	$\frac{2}{3}\left(r^2 - z_1^2\right)^{3/2}$ $\left(b = 2\sqrt{r^2 - z_1^2}\right)$	$\frac{2}{3}r^3$ when $z_1 = 0$
3. I-beam section	$\frac{1}{2}b\left(\frac{1}{4}h^2 - \frac{1}{4}h_1^2\right) + \frac{1}{2}t_w\left(\frac{1}{4}h_1^2 - z_1^2\right)$ $\left(z_1 \leq \frac{1}{2}h_1\right)$	$\frac{1}{8}bh^2 - \frac{1}{8}h_1^2(b - t_w)$ when $z_1 = 0$

TABLE 2-4 SHEAR CORRECTION FACTORS [a]

Case	Correction Factor α_s
1.	$\dfrac{7 + 6\nu}{6(1 + \nu)}$
2.	$\dfrac{(7 + 6\nu)(1 + m^2)^2 + (20 + 12\nu)m^2}{6(1 + \nu)(1 + m^2)^2}$ where $m = \dfrac{r_i}{r_0}$
3.	$\dfrac{12 + 11\nu}{10(1 + \nu)}$
4.	$\dfrac{(40 + 37\nu)b^4 + (16 + 10\nu)a^2 b^2 + \nu a^4}{12(1 + \nu)b^2(3b^2 + a^2)}$

TABLE 2-4 Shear Correction Factors   56

5.	$\dfrac{1.305 + 1.273\nu}{1 + \nu}$
6.	$\dfrac{4 + 3\nu}{2(1 + \nu)}$
7.	$\dfrac{48 + 39\nu}{20(1 + \nu)}$
8.	$\dfrac{(12 + 72m + 150m^2 + 90m^3) + \nu(11 + 66m + 135m^2 + 90m^3) + 30n^2(m + m^2) + 5\nu n^2(8m + 9m^2)}{10(1 + \nu)(1 + 3m)^2}$
	where $m = 2bt_f/ht_w, \qquad n = b/h$

TABLE 2-4 (continued)   SHEAR CORRECTION FACTORS[a]

Case	Correction Factor α_s
9.	$$\frac{(12 + 72m + 150m^2 + 90m^3) + v(11 + 66m + 135m^2 + 90m^3) + 10n^2[(3 + v)m + 3m^2]}{10(1 + v)(1 + 3m)^2}$$ where $m = bt_1/ht$, $n = b/h$
10.	$$\frac{(12 + 72m + 150m^2 + 90m^3) + v(11 + 66m + 135m^2 + 90m^3)}{10(1 + v)(1 + 3m)^2}$$ where $m = 2A_s/ht$, A_s = area of one spar
11.	$$\frac{(12 + 96m + 276m^2 + 192m^3) + v(11 + 88m + 248m^2 + 216m^3) + 30n^2(m + m^2) + 10vn^2(4m + 5m^2 + m^3)}{10(1 + v)(1 + 4m)^2}$$ where $m = bt_f/ht_w$, $n = b/h$

[a]From Cowper [2.3], with permission

TABLE 2-4   Shear Correction Factors

TABLE 2-5 TORSIONAL CONSTANT J

Shape	Torsional Constant J
	Thick Noncircular Sections
1. Ellipse 	$$J = \frac{\pi a^3 b^3}{16(a^2 + b^2)}$$
2. Hollow ellipse $k = a_i/a = b_i/b$	$$J = \frac{\pi a^3 b^3}{16(a^2 + b^2)}(1 - k^4)$$
3. Equilateral triangle 	$$J = \frac{a^4 \sqrt{3}}{80}$$

TABLE 2-5 Torsional Constant **J**

TABLE 2-5 (continued) TORSIONAL CONSTANT J

Shape	Torsional Constant J
4. Square	$J = 0.1406a^4$
5. Rectangle $a>b$	$J = \dfrac{ab^3}{3}\left(1 - 0.630\dfrac{b}{a} + 0.052\dfrac{b^5}{a^5}\right)$
	Hollow Thin-walled Sections
6.	$J = \dfrac{4\overline{A}^2}{\int_0^L (1/t(s))\,ds}$ where $\overline{A}$ = area enclosed by middle line of wall L = entire length of middle line of wall For constant t $J = \dfrac{4\overline{A}^2 t}{L}$

Circular Cross Sections

7.
Solid

$$J = \frac{1}{2}\pi r_0^4 = \frac{1}{32}\pi d_0^4$$

8.
Hollow

$$J = \frac{1}{2}\pi(r_0^4 - r_i^4) = \frac{1}{32}\pi(d_0^4 - d_i^4)$$

9.
Very Thin

$$J = 2\pi r^3 t$$

TABLE 2-5 | **Torsional Constant *J***

TABLE 2-5 (continued) TORSIONAL CONSTANT *J*

Shape	Torsional Constant *J*
	Thin-walled Open Sections
10. Any open section	$$J = \frac{\alpha}{3} \sum_{i=1}^{M} b_i t_i^3$$ where *M* is number of straight or curved segments of thickness t_i and width or height b_i composing the section; set $\alpha = 1$ except as designated otherwise

$\alpha = 1.12$ $\alpha = 1.17$ $\alpha = 1.29$ $\alpha = 1.12$

$\alpha = 1.31$ $\alpha = 1.12$ $\alpha = 1$ $\alpha = 1.12$

| **11.** | $J = \frac{1}{3} b t^3$ |

12.

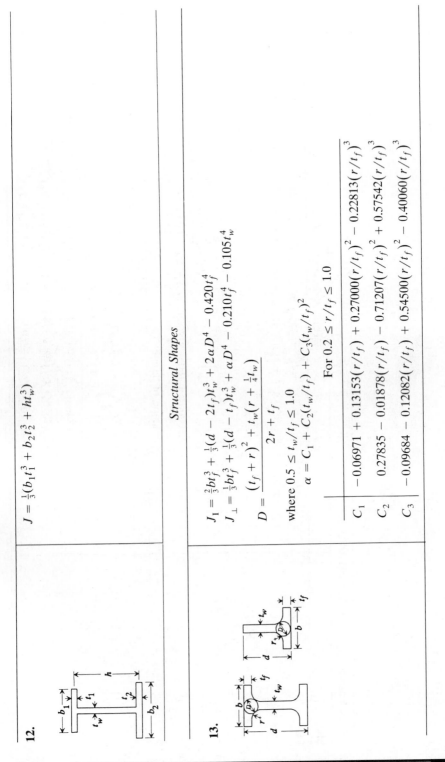

$$J = \tfrac{1}{3}(b_1 t_1^3 + b_2 t_2^3 + h t_w^3)$$

Structural Shapes

13.

$$J_\parallel = \tfrac{2}{3} b t_f^3 + \tfrac{1}{3}(d - 2t_f)t_w^3 + 2\alpha D^4 - 0.420 t_f^4$$
$$J_\perp = \tfrac{1}{3} b t_f^3 + \tfrac{1}{3}(d - t_f)t_w^3 + \alpha D^4 - 0.210 t_f^4 - 0.105 t_w^4$$

$$D = \frac{(t_f + r)^2 + t_w\left(r + \tfrac{1}{4}t_w\right)}{2r + t_f}$$

where $0.5 \le t_w/t_f \le 1.0$

$$\alpha = C_1 + C_2(t_w/t_f) + C_3(t_w/t_f)^2$$

For $0.2 \le r/t_f \le 1.0$

C_1	$-0.06971 + 0.13153(r/t_f) + 0.27000(r/t_f)^2 - 0.22813(r/t_f)^3$
C_2	$0.27835 - 0.01878(r/t_f) - 0.71207(r/t_f)^2 + 0.57542(r/t_f)^3$
C_3	$-0.09684 - 0.12082(r/t_f) + 0.54500(r/t_f)^2 - 0.40060(r/t_f)^3$

TABLE 2-5 (continued) TORSIONAL CONSTANT J

Shape	Torsional Constant J
14.	$J_1 = \frac{1}{6}(b - t_w)(t_1 + t_2)(t_1^2 + t_2^2) + \frac{2}{3}t_w t_2^3 + \frac{1}{3}(d - 2t_2)t_w^3 + 2\alpha D^4 - 4Vt_1^4$ $J_\perp = \frac{1}{12}(b - t_w)(t_1 + t_2)(t_1^2 + t_2^2) + \frac{1}{3}t_w t_2^3 + \frac{1}{3}(d - t_2)t_w^3 + \alpha D^4 - 2Vt_1^4 - 0.105t_w^4$ $D = \dfrac{(F + t_2)^2 + t_w(r + \frac{1}{4}t_w)}{F + r + t_2}$ $F = rs\left(\sqrt{\dfrac{1}{s^2} + 1} - 1 - \dfrac{t_w}{2r}\right)$ $V = 0.10504 + 0.10000s + 0.08480s^2 + 0.06746s^3 + 0.05153s^4$ $s = \dfrac{2(t_2 - t_1)}{(b - t_w)} = \text{slope of flange}$ where $0.5 \le t_w/t_2 \le 1.0$ $\alpha = C_1 + C_2(t_w/t_2) + C_3(t_w/t_2)^2$ $\text{For } 0.2 \le r/t_2 \le 1.0$

C_1	$-0.1033 + 0.3466(r/t_2) - 0.3727(r/t_2)^2 + 0.1694(r/t_2)^3$
C_2	$0.3062 - 0.7656(r/t_2) + 1.3348(r/t_2)^2 - 0.6897(r/t_2)^3$
C_3	$-0.1074 + 0.4167(r/t_2) - 0.9049(r/t_2)^2 + 0.5002(r/t_2)^3$

TABLE 2-5 Torsional Constant J 64

15.

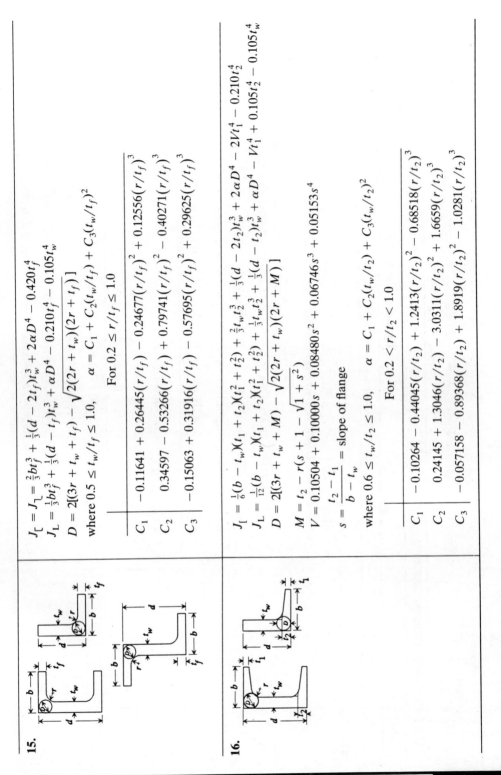

$J_C = J_L = \frac{2}{3}bt_f^3 + \frac{1}{3}(d - 2t_f)t_w^3 + 2\alpha D^4 - 0.420t_f^4$

$J_L = \frac{1}{3}bt_f^3 + \frac{1}{3}(d - t_f)t_w^3 + \alpha D^4 - 0.210t_f^4 - 0.105t_w^4$

$D = 2[(3r + t_w + t_f) - \sqrt{2(2r + t_w)(2r + t_f)}]$

where $0.5 \le t_w/t_f \le 1.0$, $\quad \alpha = C_1 + C_2(t_w/t_f) + C_3(t_w/t_f)^2$

For $0.2 \le r/t_f \le 1.0$

C_1	$-0.11641 + 0.26445(r/t_f) - 0.24677(r/t_f)^2 + 0.12556(r/t_f)^3$
C_2	$0.34597 - 0.53266(r/t_f) + 0.79741(r/t_f)^2 - 0.40271(r/t_f)^3$
C_3	$-0.15063 + 0.31916(r/t_f) - 0.57695(r/t_f)^2 + 0.29625(r/t_f)^3$

16.

$J_L = \frac{1}{6}(b - t_w)(t_1 + t_2)(t_1^2 + t_2^2) + \frac{2}{3}t_w t_2^3 + \frac{1}{3}(d - 2t_2)t_w^3 + 2\alpha D^4 - 2Vt_1^4 - 0.210t_2^4$

$J_L = \frac{1}{12}(b - t_w)(t_1 + t_2)(t_1^2 + t_2^2) + \frac{1}{3}t_w t_2^3 + \frac{1}{3}(d - t_2)t_w^3 + \alpha D^4 - Vt_1^4 + 0.105t_2^4 - 0.105t_w^4$

$D = 2[(3r + t_w + M) - \sqrt{2(2r + t_w)(2r + M)}]$

$M = t_2 - r(s + 1 - \sqrt{1 + s^2})$

$V = 0.10504 + 0.10000s + 0.08480s^2 + 0.06746s^3 + 0.05153s^4$

$s = \dfrac{t_2 - t_1}{b - t_w}$ = slope of flange

where $0.6 \le t_w/t_2 \le 1.0$, $\quad \alpha = C_1 + C_2(t_w/t_2) + C_3(t_w/t_2)^2$

For $0.2 < r/t_2 < 1.0$

C_1	$-0.10264 - 0.44045(r/t_2) + 1.2413(r/t_2)^2 - 0.68518(r/t_2)^3$
C_2	$0.24145 + 1.3046(r/t_2) - 3.0311(r/t_2)^2 + 1.6659(r/t_2)^3$
C_3	$-0.057158 - 0.89368(r/t_2) + 1.8919(r/t_2)^2 - 1.0281(r/t_2)^3$

TABLE 2-5 (continued)   TORSIONAL CONSTANT J

Shape	Torsional Constant J
	Other Cross-sectional Shapes^a

17.

$J = 2Kr_0^4$

where, for $0.1 \le r_i/r_0 \le 0.6$,

$K = C_1 + C_2(r_i/r_0) + C_3(r_i/r_0)^2 + C_4(r_i/r_0)^3$

For $0.1 \le a/r_i \le 1.0$

C_1	$0.4419 - 0.006649(a/r_i) - 0.01237(a/r_i)^2 + 0.008159(a/r_i)^3$
C_2	$-0.7992 - 0.3159(a/r_i) + 0.1041(a/r_i)^2 - 0.1180(a/r_i)^3$
C_3	$-0.05440 + 0.8918(a/r_i) - 0.2037(a/r_i)^2 + 0.2306(a/r_i)^3$
C_4	$0.4989 - 0.6441(a/r_i) + 0.1295(a/r_i)^2 - 0.1274(a/r_i)^3$

18.

$J = 2Kr^4$

where, for $0.1 \le b/r \le 0.5$,

$K = C_1 + C_2(b/r) + C_3(b/r)^2 + C_4(b/r)^3$

For $0.2 \le a/b \le 2.0$

C_1	$0.8022 - 0.05327(a/b) + 0.05212(a/b)^2 - 0.01372(a/b)^3$
C_2	$-0.2248 + 0.6236(a/b) - 0.6370(a/b)^2 + 0.1612(a/b)^3$
C_3	$-0.08032 - 3.609(a/b) + 2.481(a/b)^2 - 0.5339(a/b)^3$
C_4	$-0.2603 + 2.876(a/b) - 1.630(a/b)^2 + 0.3355(a/b)^3$

TABLE 2-5   **Torsional Constant J**

66

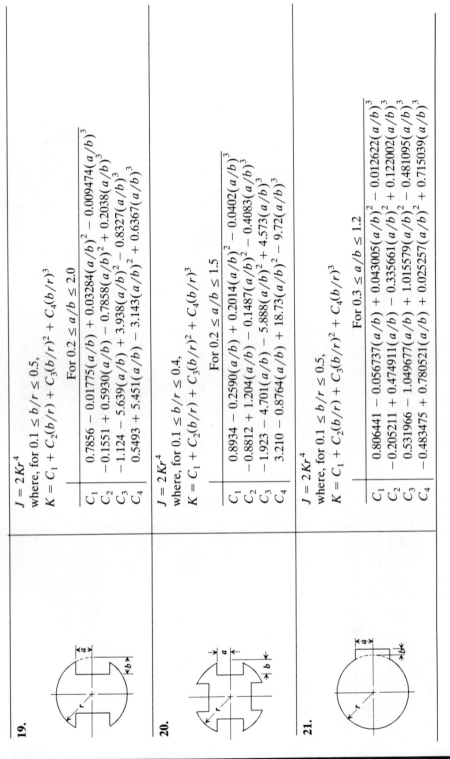

19.

$$J = 2Kr^4$$

where, for $0.1 \le b/r \le 0.5$,

$$K = C_1 + C_2(b/r) + C_3(b/r)^2 + C_4(b/r)^3$$

For $0.2 \le a/b \le 2.0$

C_1	$0.7856 - 0.01775(a/b) + 0.03284(a/b)^2 - 0.009474(a/b)^3$
C_2	$-0.1551 + 0.5930(a/b) - 0.7858(a/b)^2 + 0.2038(a/b)^3$
C_3	$-1.124 - 5.639(a/b) + 3.938(a/b)^2 - 0.8327(a/b)^3$
C_4	$0.5493 + 5.451(a/b) - 3.143(a/b)^2 + 0.6367(a/b)^3$

20.

$$J = 2Kr^4$$

where, for $0.1 \le b/r \le 0.4$,

$$K = C_1 + C_2(b/r) + C_3(b/r)^2 + C_4(b/r)^3$$

For $0.2 \le a/b \le 1.5$

C_1	$0.8934 - 0.2590(a/b) + 0.2014(a/b)^2 - 0.0402(a/b)^3$
C_2	$-0.8812 + 1.204(a/b) - 0.1487(a/b)^2 - 0.4083(a/b)^3$
C_3	$-1.923 - 4.701(a/b) - 5.888(a/b)^2 + 4.573(a/b)^3$
C_4	$3.210 - 0.8764(a/b) + 18.73(a/b)^2 - 9.72(a/b)^3$

21.

$$J = 2Kr^4$$

where, for $0.1 \le b/r \le 0.5$,

$$K = C_1 + C_2(b/r) + C_3(b/r)^2 + C_4(b/r)^3$$

For $0.3 \le a/b \le 1.2$

C_1	$0.806441 - 0.056737(a/b) + 0.043005(a/b)^2 - 0.012622(a/b)^3$
C_2	$-0.205211 + 0.474911(a/b) - 0.335661(a/b)^2 + 0.122002(a/b)^3$
C_3	$0.531966 - 1.049677(a/b) + 1.015579(a/b)^2 - 0.481095(a/b)^3$
C_4	$-0.483475 + 0.780521(a/b) + 0.025257(a/b)^2 + 0.715039(a/b)^3$

Torsional Constant J TABLE 2-5

TABLE 2-5 (continued) TORSIONAL CONSTANT J

Shape	Torsional Constant J
22. 	$J = 2Kr^4$ where, for $0.1 \le b/r \le 0.5$, $K = C_1 + C_2(b/r) + C_3(b/r)^2 + C_4(b/r)^3$ For $0.2 \le a/b \le 2.0$ C_1 | $0.820637 - 0.087139(a/b) + 0.061660(a/b)^2 - 0.019378(a/b)^3$ C_2 | $-0.321396 + 0.650230(a/b) - 0.391181(a/b)^2 + 0.182355(a/b)^3$ C_3 | $0.760392 - 1.061566(a/b) + 1.085027(a/b)^2 - 0.799552(a/b)^3$ C_4 | $-0.661936 + 0.541752(a/b) + 0.872841(a/b)^2 + 1.417918(a/b)^3$
23. 	$J = 2Kr^4$ where, for $0.1 \le b/r \le 0.5$, $K = C_1 + C_2(b/r) + C_3(b/r)^2 + C_4(b/r)^3$ For $0.2 \le a/b \le 2.0$ C_1 | $0.893973 - 0.313887(a/b) + 0.269713(a/b)^2 - 0.084344(a/b)^3$ C_2 | $-0.943550 + 2.289479(a/b) - 1.565538(a/b)^2 + 0.525646(a/b)^3$ C_3 | $2.432108 - 4.773638(a/b) + 3.659461(a/b)^2 - 1.768291(a/b)^3$ C_4 | $-1.838383 + 1.733176(a/b) + 2.880622(a/b)^2 + 2.464264(a/b)^3$

TABLE 2-5 Torsional Constant J 68

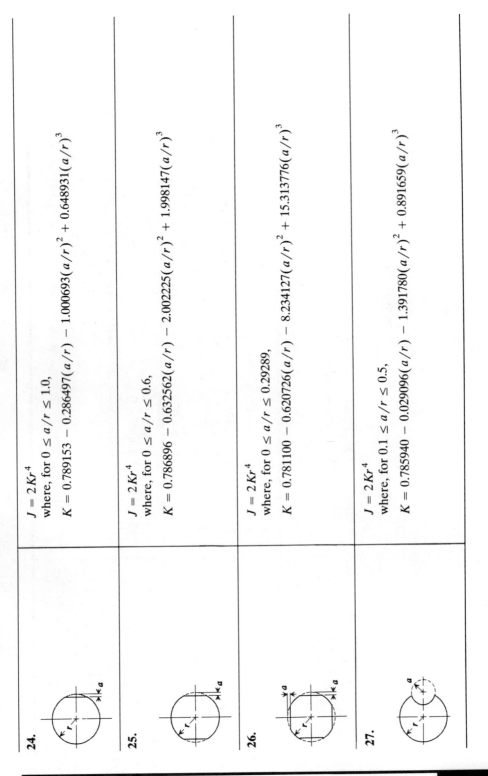

24.

$J = 2Kr^4$

where, for $0 \le a/r \le 1.0$,

$K = 0.789153 - 0.286497(a/r) - 1.000693(a/r)^2 + 0.648931(a/r)^3$

25.

$J = 2Kr^4$

where, for $0 \le a/r \le 0.6$,

$K = 0.786896 - 0.632562(a/r) - 2.002225(a/r)^2 + 1.998147(a/r)^3$

26.

$J = 2Kr^4$

where, for $0 \le a/r \le 0.29289$,

$K = 0.781100 - 0.620726(a/r) - 8.234127(a/r)^2 + 15.313776(a/r)^3$

27.

$J = 2Kr^4$

where, for $0.1 \le a/r \le 0.5$,

$K = 0.785940 - 0.029096(a/r) - 1.391780(a/r)^2 + 0.891659(a/r)^3$

Torsional Constant J   TABLE 2-5

TABLE 2-5 | Torsional Constant *J* | 70

TABLE 2-5 (continued) TORSIONAL CONSTANT J

Shape	Torsional Constant J
28.	$J = 2Kr^4$ where, for $0.1 \leq a/r \leq 0.5$, $K = 0.785300 - 0.013430(a/r) - 3.069283(a/r)^2 + 2.549998(a/r)^3$
29.	$J = 2Kr^4$ where, for $0.1 \leq a/r \leq 0.5$, $K = 0.798080 - 0.197025(a/r) - 5.683568(a/r)^2 + 6.666664(a/r)^3$
30.	$J = 2Kr^4$ where, for $0.1 \leq a/b \leq 0.8$, $K = 0.049447 + 0.043597(a/b) + 4.698128(a/b)^2 - 3.899720(a/b)^3$

[a]See Isakower [2.5].

TABLE 2-6 SHEAR CENTERS AND WARPING CONSTANTS

Shape	Location of Shear Center (S), e	Warping Constant, Γ
1. Sector of thin circle 	$$\dfrac{2r}{(\pi-\theta)+\sin\theta\cos\theta}[(\pi-\theta)\cos\theta+\sin\theta]$$ For split tube $(\theta=0)$ use $e=2r$ Ref. 2.10	$$\dfrac{tr^5}{3}\left\{2(\pi-\theta)^3-\dfrac{12[(\pi-\theta)\cos\theta+\sin\theta]^2}{(\pi-\theta)+\sin\theta\cos\theta}\right\}$$
2. Circle with different properties 	$$\dfrac{2r}{\pi}\dfrac{\left(1-\dfrac{C_G^*}{C_G}\right)\sin\theta}{\left(1-\dfrac{\theta}{\pi}\right)+\dfrac{\theta}{\pi}\dfrac{C_G^*}{C_G}\dfrac{\theta}{\pi}}$$ $$\times\dfrac{\left(1-\dfrac{\theta}{\pi}\right)\left(1-\dfrac{\theta}{\tan\theta}\right)\left(1-\dfrac{C_E^*}{C_E}\right)+\dfrac{C_E^*}{C_E}}{\left(1-\dfrac{\sin 2\theta}{2\theta}\right)\left(1-\dfrac{C_E^*}{C_E}\right)+\dfrac{C_E^*}{C_E}}$$ $C_E=Et, C_G=Gt$ Ref. 2.8	Use computer program
3. Semicircular section 	$$\dfrac{8}{15\pi}\dfrac{3+4\nu}{1+\nu}r$$	Use computer program

TABLE 2-6 (continued) SHEAR CENTERS AND WARPING CONSTANTS

Shape	Location of Shear Center (S), e	Warping Constant, Γ
4.	$-a\left(1 + \dfrac{b^2 A}{4I_z}\right) + 2h\dfrac{I_F}{I_z}$ where A = total area $\quad I_F$ = moment of inertia of each lower flange with respect to web axis $\quad I_z, I_y$ = moments of inertia with respect to z, y axes Ref. 2.7	$\dfrac{b^2}{4}\left[I_y + a^2 A\left(1 - \dfrac{b^2 A}{4I_z}\right)\right.$ $\left. + 2h^2 I_F - 2bdh^2 A_F + b^2 ahA\dfrac{I_F}{I_z} - 4h^2\dfrac{I_F^2}{I_z}\right]$ where A_F = area of each lower flange $\quad d$ = distance of centroid of lower flange from the web axis
5. Channel with unequal flanges	$z_S = e - \dfrac{b_1^2 ht}{6\left(I_y I_z - I_{yz}^2\right)}$ $\quad\times\left[-3I_{yz}(h - e) + I_y(2b_1 - 3d)\right]$ $y_S = d + \dfrac{b_1^2 ht}{6\left(I_y I_z - I_{yz}^2\right)}$ $\quad\times\left[-I_{yz}(2b_1 - 3d) + 3I_z(h - e)\right]$ $e = \dfrac{h^2 + 2b_1 h}{2h(b_1 + b_2)}, \quad d = \dfrac{b_1^2 + b_2^2}{2h(b_1 + b_2)}$ Ref. 2.9	Use computer program Ref. 2.7

TABLE 2-6 Shear Centers and Warping Constants   72

6. Thin-walled lipped angle	$e = \dfrac{b}{\sqrt{2}} \dfrac{(3-2\alpha)\alpha^2}{1+3(\alpha-\alpha^2)+\alpha^3}$ $\alpha = \dfrac{c}{b}$ Ref. 2.10	$\dfrac{tb^4c^3(4b+3c)}{6(b^3+3cb^2-3c^2b+c^3)}$
7. Thin-walled lipped channel	$t = \text{const}$ $e = \dfrac{b(3bh^2+6ah^2-8a^3)}{h^3+6bh^2+6ah^2+8a^3-12a^2h}$ Ref. 2.9	Use computer program
8. Channel	$e = \dfrac{3t_fb^2}{6bt_f+ht_w}$	$\dfrac{b^3h^2t_f}{12}\dfrac{2ht_w+3bt_f}{ht_w+6bt_f}$

TABLE 2-6 (continued) SHEAR CENTERS AND WARPING CONSTANTS

Shape	Location of Shear Center (S), e	Warping Constant, Γ
9. Thin-walled hat section	$$\dfrac{b(3bh^2 + 6ah^2 - 8a^3)}{h^3 + 6bh^2 + 6ah^2 + 8a^3 + 12a^2h}$$ Ref. 2.9	Use computer program
10. I-beam	$$\dfrac{t_1 b_1^3 h}{t_1 b_1^3 + t_2 b_2^3}$$ If $b_1 = b_2$ and $t_1 = t_2$, then $e = \frac{1}{2}h$	$t_1 = t_2 = t$ $$\dfrac{h^2 t}{12} \dfrac{b_1^3 b_2^3}{b_1^3 + b_2^3}$$ If $b_1 = b_2$, then $\Gamma = \frac{1}{24} b^3 h^2 t$
11. I-beam with unsymmetric flanges	$$\dfrac{3(b^2 - b_1^2)}{t_w h/t_f + 6(b + b_1)}, \quad b_1 < b$$ Ref. 2.10	Use computer program

TABLE 2-6 Shear Centers and Warping Constants   74

12. Thin-walled U section Ref. 2.9	$$\dfrac{4r^2 + 2b^2 + 2\pi br}{4b + \pi r}$$	Use computer program
13. Tee Ref. 2.9	S lies at intersection of centerlines of flange and web	$$\Gamma_2 = \dfrac{t_f^3 b^3}{144} + \dfrac{t_w^3 h^3}{36}$$
14. Thin-walled fork section 	$$\dfrac{3b^2(h_1^2 + h_2^2)}{h_2^3 + 6b(h_1^2 + h_2^2)}$$	Use computer program

TABLE 2-6 (continued) SHEAR CENTERS AND WARPING CONSTANTS

Shape	Location of Shear Center (S), e	Warping Constant, Γ
15. Thin-walled bowl section 	$$r \frac{D + 12\dfrac{bb_1}{r^2} + 3\pi\left(\dfrac{b_1}{r}\right)^2 - 4\left(\dfrac{b_1}{r}\right)^3 \dfrac{b}{r}}{3\pi + 12\dfrac{b + b_1}{r} + 4\left(\dfrac{b_1}{r}\right)^2\left(3 + \dfrac{b_1}{r}\right)}$$ where $D = 12 + 6\pi\dfrac{b + b_1}{r} + 6\left(\dfrac{b}{r}\right)^2$ Ref. 2.10	Use computer program
16. Unequal leg angle 	S lies at intersection of centerlines	$\Gamma_2 = \frac{1}{36}t^3(b_1^3 + b_2^3)$ If $b_1 = b_2$, then $\Gamma_2 = \frac{1}{144}A^3$ $A = b_1 t + b_2 t$
17. Zee section 	$\frac{1}{2}h$	$\dfrac{b^3 h^2 t_f}{12}\left[\dfrac{bt_f + 2ht_w}{2bt_f + ht_w}\right]$

TABLE 2-6 **Shear Centers and Warping Constants** 76

	Shear center	Warping constant
18.	$\frac{1}{2}h$	$\frac{h^2 I_z}{4}$
19. **Thin-walled lipped section**	$\frac{1}{2}h$	Ref. 2.11 $$\frac{tb^2}{6}\left[\frac{D + (6h^2 + 5bh)C + bh^2 + 0.5b^2h}{h + 2b + 2c}\right]$$ where $D = C^2 + 2(h + 2b)C^2$
20.		$\frac{t_1^3 h^3}{36} + \frac{t_2^3 b^3}{36} + \frac{\pi r^4 h^2}{4}$

TABLE 2-6 (continued)   SHEER CENTERS AND WARPING CONSTANTS

Shape	Location of Shear Center (S), e	Warping Constant, Γ
21. Cross section formed of M straight thin elements r_{ci} is perpendicular distance from centroid to tangent of wall profile for the ith element. ω_p, ω_q and ω_{cp}, ω_{cq} are the principal sectorial coordinates (warping functions) and the sectorial coordinates with respect to centroid c of the p & q ends of element i. (Ref. 2.12)	$y_S = \dfrac{I_z I_{\omega y} - I_{yz} I_{\omega z}}{I_y I_z - I_{yz}^2}$ $z_S = -\dfrac{I_y I_{\omega z} - I_{yz} I_{\omega y}}{I_y I_z - I_{yz}^2}$ $\omega_c = \displaystyle\int_0^s r_c\,ds$ $I_{\omega y} = \displaystyle\int_A \omega_c z\,dA$ $\quad = \dfrac{1}{3}\displaystyle\sum_{i=1}^{M}(\omega_{cp}z_p + \omega_{cq}z_q)t_i b_i$ $\qquad + \dfrac{1}{6}\displaystyle\sum_{i=1}^{M}(\omega_{cp}z_q + \omega_{cq}z_p)t_i b_i$ $I_{\omega z} = \displaystyle\int_A \omega_c y\,dA$ $\quad = \dfrac{1}{3}\displaystyle\sum_{i=1}^{M}(\omega_{cp}y_p + \omega_{cq}y_q)t_i b_i$ $\qquad + \dfrac{1}{6}\displaystyle\sum_{i=1}^{M}(\omega_{cp}y_q + \omega_{cq}y_p)t_i b_i$	$\displaystyle\int_A \omega^2\,dA = \dfrac{1}{3}\displaystyle\sum_{i=1}^{M}(\omega_p^2 + \omega_p\omega_q + \omega_q^2)t_i b_i$ where ω_j $(j = p$ or $q)$ is taken from case 1 of Table 2-7

TABLE 2-6   Shear Centers and Warping Constants   78

TABLE 2-7 SOME WARPING PROPERTIES[a]

Shape	Principal Sectorial Coordinate ω	
1. Cross section formed of M straight thin elements	ω (at point j) $= \omega_j = \omega_{Sj} - \omega_0$ $\omega_0 = \dfrac{1}{A}\left[\dfrac{1}{2} \displaystyle\sum_{i=1}^{M} (\omega_{Sp} + \omega_{Sq}) t_i b_i \right]$ $A = \displaystyle\sum_{i=1}^{M} t_i b_i$ $\omega_{Sj} = \displaystyle\sum_i r_{Si} b_i$ $\omega_{cj} = \displaystyle\sum_i r_{ci} b_i$ Linear distribution of ω	$\displaystyle\sum_i$ means the summation along a line of elements. Begin at an outer element and sum until reaching the point j, where the values of ω_{Sj} or ω_{cj} are desired. ω_{Sp} and ω_{Sq} are the sectorial coordinates of the p and q ends of element i. ω_0 is the average of sectorial coordinate ω_S for the whole section. r_{Si} and r_{ci} are the perpendicular distances from the shear center and centroid to element i, respectively. See Chapter 15 for examples in using these formulas.

TABLE 2-7 (continued) SOME WARPING PROPERTIES [a]

Shape		Principal Sectorial Coordinate ω
2.	$$\omega_1 = \frac{b_1 h}{2}\; \frac{1}{1 + \left(\dfrac{b_1}{b_2}\right)^3\left(\dfrac{t_1}{t_2}\right)}$$ $$\omega_2 = \frac{b_2 h}{2}\; \frac{1}{1 + \left(\dfrac{b_2}{b_1}\right)^3\left(\dfrac{t_2}{t_1}\right)}$$	$$Q_{\omega_1} = \frac{b_1^2 h t_1}{8}\; \frac{1}{1 + \left(\dfrac{b_1}{b_2}\right)^3\left(\dfrac{t_1}{t_2}\right)}$$ $$Q_{\omega_2} = \frac{b_2^2 h t_2}{8}\; \frac{1}{1 + \left(\dfrac{b_2}{b_1}\right)^3\left(\dfrac{t_2}{t_1}\right)}$$
3.	$$\omega_1 = \frac{bh}{2}\; \frac{3 + \left(\dfrac{h}{b}\right)\left(\dfrac{t_w}{t_f}\right)}{6 + \left(\dfrac{h}{b}\right)\left(\dfrac{t_w}{t_f}\right)}$$ $$\omega_2 = \frac{bh}{2}\; \frac{1}{2 + \dfrac{1}{3}\left(\dfrac{h}{b}\right)\left(\dfrac{t_w}{t_f}\right)}$$	$$Q_{\omega_1} = \frac{b^2 h t}{4}\left[\frac{3 + \left(\dfrac{h}{b}\right)\left(\dfrac{t_w}{t_f}\right)}{6 + \left(\dfrac{h}{b}\right)\left(\dfrac{t_w}{t_f}\right)}\right]^2$$ $$Q_{\omega_2} = \frac{b^2 h t}{4}\; \frac{1}{1 + \left(\dfrac{b}{h}\right)\left(\dfrac{t_f}{t_w}\right)}$$ $$Q_{\omega_3} = \frac{-b^2 h t}{8}\; \frac{1}{1 + 6\left(\dfrac{b}{h}\right)\left(\dfrac{t_f}{t_w}\right)}$$ $$a = \frac{b}{2 + \dfrac{1}{3}\left(\dfrac{h}{b}\right)\left(\dfrac{t_w}{t_f}\right)}$$

TABLE 2-7 **Warping Properties** 80

4.

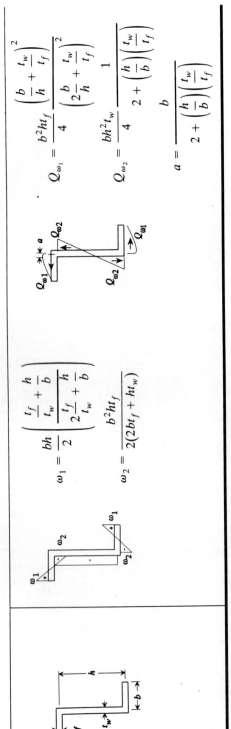

$$\omega_1 = \frac{bh}{2}\left(\frac{\dfrac{t_f}{t_w}+\dfrac{h}{b}}{2\dfrac{t_f}{t_w}+\dfrac{h}{b}}\right)$$

$$\omega_2 = \frac{b^2 h t_f}{2(2bt_f+ht_w)}$$

$$Q_{\omega_1} = \frac{b^2 h t_f}{4}\left(\frac{\dfrac{b}{h}+\dfrac{t_w}{t_f}}{2\dfrac{b}{h}+\dfrac{t_w}{t_f}}\right)^2$$

$$Q_{\omega_2} = \frac{b h^2 t_w}{4}\left(\frac{1}{2+\left(\dfrac{h}{b}\right)\left(\dfrac{t_w}{t_f}\right)}\right)$$

$$a = \frac{b}{2+\left(\dfrac{h}{b}\right)\left(\dfrac{t_w}{t_f}\right)}$$

Warping Properties TABLE 2-7

CHAPTER **3**

Stress and Strain

The concepts of stress and strain are essential to design as they characterize the mechanical properties of deformable solids.

A brief introduction to the concepts along with a discussion of theories of failure are provided in this chapter. Stress–strain formulas are given for bars subjected to extention, torsion, and bending. The equations describing the state of stress or strain in a body are applicable to any solid continuum, whether elastic or plastic, while the equations relating stress and strain depend on the material behavior.

3.1 NOTATION

A	Cross-sectional area (L^2)
A_0	Original area; shear-related area defined in Fig. 3-29 (L^2)
$\bar{A}$	Area enclosed by middle line of wall of closed thin-walled cross section (L^2)
b	Width (L)
c	Distance from centroidal (neutral) axis of beam to outermost fiber (L)
E	Modulus of elasticity, Young's modulus (F/L^2)
F	Internal force (F)
G	Shear modulus (F/L^2)
I	Moment of inertia of a member about its centroidal (neutral) axis (L^4)
J	Torsional constant; polar moment of inertia for circular cross section (L^4)
L	Length of element, original length (L)
L_s	Total length of middle line of wall of tube cross section
M	Bending moment (FL)
P	Load or axial force (F)
p	Pressure (F/L^2)
p_z	Distributed loading (F/L)
q	Shear flow (F/L)

Q	First moment of area beyond level where shear stress is to be determined (L^3)
R, r	Radius (L)
$S = Z_e$	Section modulus of beam, $S = I/c$ (L^3)
t	Wall thickness (L)
T	Torque or twisting moment (FL)
V	Shear force (F)
u, v, w	Displacements in x, y, z directions (L)
σ	Normal stress (F/L^2)
σ_m	Mean stress (F/L^2)
τ	Shear stress (F/L^2)
θ	Angle (degree or radian)
γ	Shear strain
ϵ	Normal strain
ϵ_t	Natural strain or true strain
ν	Poisson's ratio
σ_{ys}	Yield stress (F/L^2)
Δ	Increment of length (L)
φ	Angular displacement (degree or radian)

3.2 DEFINITIONS OF STRESS AND TYPES OF STRESS

Normally, forces are considered to occur in two forms: surface forces and body forces. Surface forces are forces distributed over the surface of the body, such as hydrostatic pressure or the force exerted by one body on another. Body forces are forces distributed throughout the volume of the body, such as gravitational forces, magnetic forces, or inertial forces for a body in motion. Suppose a solid is subject to external forces P_1, P_2, and P_3 (Fig. 3-1). If the body were cut, a force F would be required to maintain equilibrium. The intensity of this force, i.e., the force per unit area, is defined to be the stress.

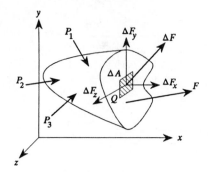

FIGURE 3-1 Stress.

The force F will not necessarily be uniformly distributed over the cut. To define the stress at some point Q in a cut perpendicular to the x axis (Fig. 3-1), suppose the resultant contribution of the internal force F on the area element ΔA at point Q is ΔF, and let the components of ΔF along the x, y, z axes be $\Delta F_x, \Delta F_y, \Delta F_z$. Stress components are defined as

$$\sigma_x = \lim_{\Delta A \to 0} \frac{\Delta F_x}{\Delta A}, \qquad \tau_{xy} = \lim_{\Delta A \to 0} \frac{\Delta F_y}{\Delta A}, \qquad \tau_{xz} = \lim_{\Delta A \to 0} \frac{\Delta F_z}{\Delta A} \qquad (3.1)$$

where σ_x is the normal stress and τ_{xy}, τ_{xz} are the shear stresses. Normal stress is the intensity of a force perpendicular to a cut while the shear stresses are parallel to the plane of the element. Tensile stresses are those normal stresses pulling away from the cut, while compressive stresses are those pushing against the cut.

3.3 STRESS COMPONENT ANALYSIS

Sign Convention

An element of infinitesimal dimensions isolated from a solid would expose the stresses shown in Fig. 3-2. The face of an element whose outward normal is directed along the positive direction of a coordinate axis is defined to be a positive face. A negative face has its normal in the opposite direction. Stress components are positive if, when acting on a positive face, their corresponding force components are in the positive coordinate direction. Also, stress components are said to be positive when their force components act on a negative face in the negative coordinate direction. Stress components not satisfying these conditions are considered as being negative.

These definitions mean that normal stress directed outward from the plane on which it acts (i.e., tension) is positive, while a normal stress directed toward the plane on which it acts (i.e., compression) is taken as being negative. Also, a shear stress is positive if the outward normal of the plane on which it acts and the

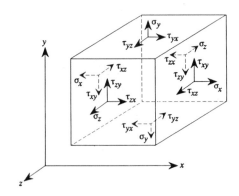

FIGURE 3-2 Stress on an element.

direction of the stress are in coordinate directions of the same sign; otherwise it is negative. The stresses of Fig. 3-2 are positive.

Stress Tensor

In Fig. 3-2, there are three normal stresses $\sigma_x, \sigma_y, \sigma_z$, where the single subscript is the axis along which the normal to the cut lies. There are also six shear stresses $\tau_{xy}, \tau_{yx}, \tau_{yz}, \tau_{zy}, \tau_{zx}, \tau_{xz}$, where the first subscript denotes the axis perpendicular to the plane on which the stress acts and the second provides the direction of the stress. For example, the shear stress τ_{xy} acts on a plane normal to the x axis and in a direction parallel to the y axis.

The conditions of equilibrium dictate that shear stresses with the same subscripts are equal:

$$\tau_{xy} = \tau_{yx}, \qquad \tau_{xz} = \tau_{zx}, \qquad \tau_{yz} = \tau_{zy} \tag{3.2}$$

In matrix form, the stress components appear as

$$\begin{bmatrix} \sigma_x & \tau_{xy} & \tau_{xz} \\ \tau_{yx} & \sigma_y & \tau_{yz} \\ \tau_{zx} & \tau_{zy} & \sigma_z \end{bmatrix} \tag{3.3}$$

This state of stress at a point is called a *stress tensor*. The stress tensor is a second-order tensor quantity.

Plane Stress

In the case of plane stress, all stress components associated with a given direction are zero. Then the state of stress can be determined by three stress components. For example, the stress in thin sheets is usually treated as being in the state of plane stress.

Variation of Normal and Shear Stress in Tension

The bar in Fig. 3-3 is in simple tension. The stresses on planes normal to an axis of the bar are considered to be uniformly distributed and are equal to P/A_0 on cross sections along the length, except near the applied load, where there may be stress concentration (Chapter 6). Here A_0 is the original cross-sectional area of

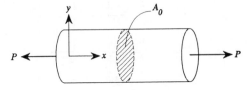

FIGURE 3-3 Axially loaded bar.

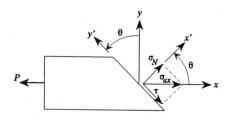

FIGURE 3-4 Stress on a cross section.

the bar. Consider the stress on an inclined face exposed by passing a plane through the bar at an angle θ, as shown in Fig. 3-4.

The stress acting in the x direction on the inclined face is $\sigma_{ax} = P/(A_0/\cos \theta)$, where $A_0/\cos \theta$ is the inclined cross-sectional area. This stress can be resolved in terms of the components σ_N and τ as though a force were being resolved since these stresses all act on the same unit of area. These relationships are as follows:

$$\text{Normal stress} \quad = \sigma_N = \frac{P \cos \theta}{A_0/\cos \theta} = \frac{P}{A_0} \cos^2 \theta \tag{3.4}$$

$$\text{Shear stress} \quad = \tau = \frac{P \sin \theta}{A_0/\cos \theta} = \frac{P}{A_0} \sin \theta \cos \theta \tag{3.5}$$

From Eqs. (3.4) and (3.5)

$$\frac{\sigma_N}{P/A_0} = \frac{\sigma_N}{\sigma_a} = \cos^2 \theta \tag{3.6}$$

$$\frac{\tau}{P/A_0} = \frac{\tau}{\sigma_a} = \sin \theta \cos \theta \tag{3.7}$$

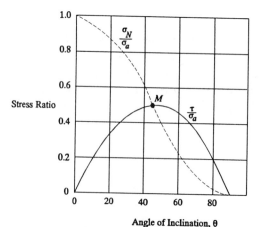

FIGURE 3-5 Variation of stress with angle of plane.

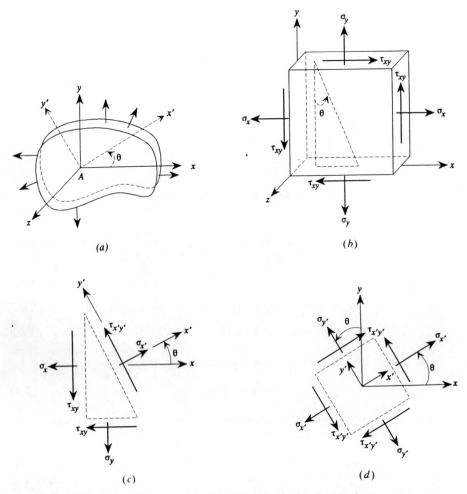

FIGURE 3-6 (a) Object under load. (b) Element at point A. (c) Element with diagonal at point A, taken from (b). (d) Element at point A. This can replace the element of (c).

where σ_a is the normal axial tensile stress on the section normal to the x axis. Equations (3.6) and (3.7) are plotted in Fig. 3-5. Note that the shear stress is a maximum at 45°, as shown at point M, and that it equals half the maximum tensile stress.

Stress at an Arbitrary Orientation for Two-Dimensional Case

Consider an element removed from a body subjected to an arbitrary loading in the xy plane (Fig. 3-6a). The stresses $\sigma_x, \sigma_y, \tau_{xy}$ will occur for the orientation of Fig. 3-6b. Once the state of stress is determined for an element with a particular orientation (such as $\sigma_x, \sigma_y, \tau_{xy}$ of Fig. 3-6b), the state of stress $\sigma_{x'}, \sigma_{y'},$ and $\tau_{x'y'}$

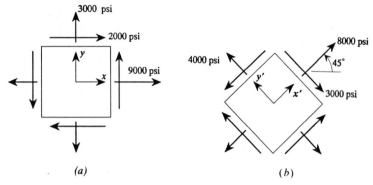

FIGURE 3-7 Two-dimensional state of stress.

at that location for an element in any orientation (Figs. 3-6c, d) can be obtained using the following transformation equations for plane stresses:

$$\sigma_{x'} = \tfrac{1}{2}(\sigma_x + \sigma_y) + \tfrac{1}{2}(\sigma_x - \sigma_y)\cos 2\theta + \tau_{xy}\sin 2\theta \qquad (3.8a)$$

$$\sigma_{y'} = \tfrac{1}{2}(\sigma_x + \sigma_y) - \tfrac{1}{2}(\sigma_x - \sigma_y)\cos 2\theta - \tau_{xy}\sin 2\theta \qquad (3.8b)$$

$$\tau_{x'y'} = -\tfrac{1}{2}(\sigma_x - \sigma_y)\sin 2\theta + \tau_{xy}\cos 2\theta \qquad (3.8c)$$

Note that it can be found from the above equations that

$$\sigma_{x'} + \sigma_{y'} = \sigma_x + \sigma_y \qquad (3.9)$$

This shows that the sum of the normal stresses is an invariant quantity, independent of the orientation of the element at the point in question.

Example 3.1 State of Stress The state of stress of an element loaded in the x, y plane is $\sigma_x = 9000$ psi, $\sigma_y = 3000$ psi, and $\tau_{xy} = 2000$ psi, as shown in Fig. 3-7a. Determine the stresses on the element rotated through an angle of 45°.

The state of stress desired can be found by substituting the given values of stresses $\sigma_x, \sigma_y, \tau_{xy}$ into Eqs. (3.8) with $\theta = 45°$. The results are $\sigma_{x'} = 8000$ psi, $\sigma_{y'} = 4000$ psi, and $\tau_{x'y'} = -3,000$ psi. This state of stress is shown in Fig. 3-7b.

═══════════════

Principal Stresses and Maximum Shear Stress for Two-Dimensional Case

The maximum value of $\sigma_{x'}$ is found by differentiating Eq. (3.8a) with respect to θ:

$$\frac{d\sigma_{x'}}{d\theta} = 0 = \frac{\sigma_x - \sigma_y}{2}(-2\sin 2\theta) + 2\tau_{xy}\cos 2\theta \qquad (3.10)$$

from which

$$\tan 2\theta_1 = 2\tau_{xy}/(\sigma_x - \sigma_y) \qquad (3.11)$$

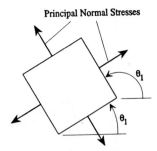

Principal Normal Stresses

FIGURE 3-8 Orientation of principal planes.

Extreme values of normal stresses occur on the orientations $\theta = \theta_1$ defined by Eq. (3.11). The two values of θ_1 are 90° apart and locate two sets of parallel planes of an element (Fig. 3-8). The maximum normal stress occurs on one of the sets of planes while the minimum normal stress occurs on the other.

Principal stresses are defined as the algebraically maximum and minimum values of the normal stresses, and the planes on which they act are called *principal planes* (Fig. 3-8). From Eq. (3.11), it follows that

$$\sin 2\theta_1 = \frac{\pm \tau_{xy}}{\sqrt{\left[\frac{1}{2}(\sigma_x - \sigma_y)\right]^2 + \tau_{xy}^2}} \tag{3.12a}$$

$$\cos 2\theta_1 = \frac{\pm \frac{1}{2}(\sigma_x - \sigma_y)}{\sqrt{\left[\frac{1}{2}(\sigma_x - \sigma_y)\right]^2 + \tau_{xy}^2}} \tag{3.12b}$$

Substitution of Eqs. (3.12a) and (3.12b) into Eqs. (3.8a) and (3.8b) gives

Algebraic maximum normal stress
$$\sigma_1 = \frac{1}{2}(\sigma_x + \sigma_y)$$
$$+ \sqrt{\frac{1}{4}\left[(\sigma_x - \sigma_y)^2\right] + \tau_{xy}^2} \tag{3.13a}$$

Algebraic minimum normal stress
$$\sigma_2 = \frac{1}{2}(\sigma_x + \sigma_y)$$
$$- \sqrt{\frac{1}{4}\left[(\sigma_x - \sigma_y)^2\right] + \tau_{xy}^2} \tag{3.13b}$$

Substitution of Eq. (3.11) into Eq. (3.8c) leads to $\tau_{x'y'} = 0$. That is, the shear stress is always zero on the principal planes.

The original stressed element can be used to determine which value of θ_1 for the orientation of principal planes corresponds to σ_1 and which to σ_2. Define the diagonal of a stressed element that passes between the heads of the arrows for the shear stresses as the *shear diagonal*. For example, if τ_{xy} is negative, it should be drawn on the element shown in Fig. 3-9, forming the indicated shear diagonal. Then the direction of σ_1 lies in the 45° arc between the algebraically larger normal stress and the shear diagonal.

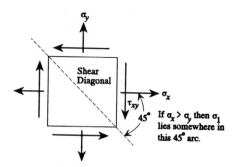

FIGURE 3-9 Shear·diagonal.

To find the maximum shear stress, set $d\tau_{x'y'}/d\theta = 0$ and find

$$\tau_{max} = \sqrt{\left[\tfrac{1}{2}(\sigma_x - \sigma_y)\right]^2 + \tau_{xy}^2} = \tfrac{1}{2}(\sigma_1 - \sigma_2) \qquad (3.14)$$

The corresponding values of θ are defined by

$$\tan 2\theta_2 = -\tfrac{1}{2}(\sigma_x - \sigma_y)/\tau_{xy} \qquad (3.15)$$

Comparison of Eqs. (3.15) and (3.11) shows that the planes of maximum shear stresses lie 45° away from the planes of the principal stresses.

The fact that the shear diagonal of the element on which the maximum shear stress occurs lies in the direction of the σ_1 stress (Fig. 3-10) assists in determining the proper directions of the maximum shear stresses.

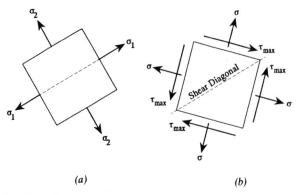

(a)   (b)

FIGURE 3-10 Direction of maximum shear stress. (a) Principal stresses. (b) Direction of maximum shear stresses for the case of (a).

On the planes of maximum shear stress, the normal stress is found by substituting θ_2 of Eq. (3.15) into Eqs. (3.8a) and (3.8b). The normal stress on each plane is

$$\sigma = \tfrac{1}{2}(\sigma_x + \sigma_y) \tag{3.16}$$

Caution must be exercised in using Eq. (3.14) to calculate the maximum shear stress. There is always a third principal stress, σ_3, although it may be equal to zero. When the three principal stresses are considered, as shown later, there are three corresponding shear stresses induced, one of which is the maximum stress.

Example 3.2 Principal Stresses For the element in Fig. 3-7a, find the principal stresses and planes and the maximum shear stress.

The principal planes are located by using Eq. (3.11):

$$\tan 2\theta_1 = \frac{2\tau_{xy}}{\sigma_x - \sigma_y} = \frac{(2)(2000)}{9000 - 3000} = 0.667$$

or $2\theta_1 = 33.7°$ and $180° + 33.7° = 213.7°$. Hence θ_1 is $16.8°$ and $106.9°$. Use of Eqs. (3.13) gives $\sigma_1 = 9605.6$ psi and $\sigma_2 = 2394.4$ psi (Fig. 3-11a). The stress σ_1 is located according to the rule for using the shear diagonal.

The maximum shear stresses are located on planes identified by θ_2 of Eq. (3.15). Thus $\tan 2\theta_2 = -(9000 - 3000)/(2 \times 2000) = -1.5$, or θ_2 is $-28.2°$ and $61.8°$ (Fig. 3-11b). Note that θ_2 can be directly located by the fact that the planes of maximum shear stress are always $45°$ from the principal planes.

From Eq. (3.14), we obtain $\tau_{max} = 3605.6$ psi. The corresponding normal stress is, by Eq. (3.16), $\sigma = 6000$ psi (Fig. 3-11b).

If $\sigma_3 = 0$, the actual maximum shear stress of the element is

$$\tau_{max} = (\sigma_1 - \sigma_3)/2 = 9605.6/2 = 4802.8 \text{ psi.}$$

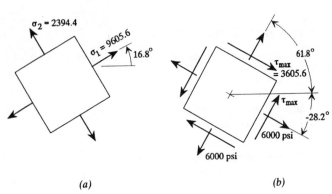

(a)   (b)

FIGURE 3-11   (a) Principal stress, and (b) maximum shear stress.

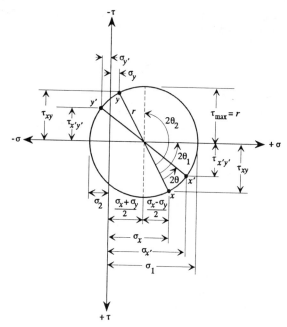

FIGURE 3-12 Mohr's circle for the two-dimensional stress of Fig. 3-6*b*. This provides the stresses of Fig. 3-6*d* for an orientation of θ.

Mohr's Circle for a Two-Dimensional State of Stress

A graphical method for representing combined stresses is popularly known as Mohr's circle method. As illustrated in Fig. 3-12, the cartesian coordinate axes represent the normal and shear stresses so that the coordinates σ, τ of each point on the circumference of a circle correspond to the state of stress at an orientation of a stressed element at a point in a body.

Construction of Mohr's Circle

For a known two-dimensional state of stress σ_x, σ_y, and τ_{xy}, Mohr's circle is drawn as follows:

1. On a horizontal axis lay off normal stresses with positive stresses to the right, and on a vertical axis place the shear stresses with positive stresses downward.

2. Find the location of the center of the circle along the σ (horizontal) axis by calculating $\frac{1}{2}(\sigma_x + \sigma_y)$. Tensile stresses are positive; compressive stresses are negative. Plot this point.

3. Plot the point $\sigma = \sigma_x$, $\tau = \tau_{xy}$. Since the positive τ axis is downward, plot a positive τ_{xy} below the σ axis.
4. Connect, with a straight line, the center of the circle from step 2 with the point plotted in step 3. This distance is the radius of Mohr's circle. Using $\frac{1}{2}(\sigma_x + \sigma_y)$ on the horizontal axis as the center, draw a circle with the radius just calculated. This is Mohr's circle.

Use of Mohr's Circle

Interpret the coordinates of a point on Mohr's circle as representing the stress components $\sigma_{x'}$ and $\tau_{x'y'}$ that act on a plane perpendicular to the x' axis (Fig. 3-6d). The x axis is along the circle radius passing through the plotted point σ_x, τ_{xy}. The angle θ is measured counterclockwise from the x axis. The magnitudes of the angles on Mohr's circle are double those in the physical plane. For example, the stresses $\sigma_{y'}$, $\tau_{x'y'}$, and the y' axis are found on the circle 180° away from $\sigma_{x'}$, $\tau_{x'y'}$ and the x' axis. It should be noted that a special sign convention of shear stress is required to interpret the $\tau_{x'y'}$ associated with $\sigma_{y'}$. That is, positive shear stress is below the σ axis for σ_x while positive shear stress corresponding to σ_y is above the σ axis. From Mohr's circle the following holds:

1. The intersections of the circle with the σ axis are the principal stresses σ_1 and σ_2. These values and their angle of orientation θ relative to the x axis can be scaled from the diagram or computed from the geometry of the figure. The shear stresses at these two points are zero.
2. The shear stress τ_{max} occurs at the point of greatest ordinate on Mohr's circle. This point has coordinates $\frac{1}{2}(\sigma_x + \sigma_y), \tau_{max}$.
3. The normal and shear stresses on an arbitrary plane for which the normal makes a counterclockwise angle θ with the x axis (Fig. 3-6d) are found by measuring a counterclockwise angle 2θ on Mohr's circle from the x axis and then determining the coordinates $\sigma_{x'}, \tau_{x'y'}$ of the circle at this angle.

Stress Acting on Arbitrary Plane in Three-Dimensional Systems

The stress components on planes that are perpendicular to the x, y, z axes are shown in Fig. 3-13, where σ_{Nx}, σ_{Ny}, and σ_{Nz} are stress components on an arbitrary oblique plane P through point 0 of a member. (In the figure the plane P is shown removed from point 0.) The direction cosines of normal N with respect to x, y, and z are l, m, and n, respectively.

If the six stress components $\sigma_x, \sigma_y, \sigma_z, \tau_{xy} = \tau_{yx}, \tau_{yz} = \tau_{zy}, \tau_{xz} = \tau_{zx}$ at point 0 are known, the stress components on any oblique plane defined by unit normal $N(l, m, n)$ can be computed using

$$\sigma_{Nx} = l\sigma_x + m\tau_{yx} + n\tau_{zx}$$

$$\sigma_{Ny} = l\tau_{xy} + m\sigma_y + n\tau_{zy} \qquad (3.17)$$

$$\sigma_{Nz} = l\tau_{xz} + m\tau_{yz} + n\sigma_z$$

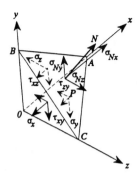

FIGURE 3-13 Stress components $\sigma_{Nx}, \sigma_{Ny}, \sigma_{Nz}$ on arbitrary plane having normal N.

Normal and Shear Stress on an Oblique Plane

The normal stress σ_N on the plane P is the sum of the projection of the stress components σ_{Nx}, σ_{Ny}, and σ_{Nz} in the direction of normal N. Therefore,

$$\sigma_N = l^2\sigma_x + m^2\sigma_y + n^2\sigma_z + 2mn\tau_{yz} + 2ln\tau_{xz} + 2lm\tau_{xy} \qquad (3.18)$$

For a particular plane through point 0, σ_N reaches a maximum value called the maximum principal stress. This maximum value along with other principal stresses are the solutions of

$$\sigma^3 - I_1\sigma^2 + I_2\sigma - I_3 = 0 \qquad (3.19a)$$

where

$$I_1 = \sigma_x + \sigma_y + \sigma_z$$

$$I_2 = \begin{vmatrix} \sigma_x & \tau_{xy} \\ \tau_{xy} & \sigma_y \end{vmatrix} + \begin{vmatrix} \sigma_x & \tau_{xz} \\ \tau_{xz} & \sigma_z \end{vmatrix} + \begin{vmatrix} \sigma_y & \tau_{yz} \\ \tau_{yz} & \sigma_z \end{vmatrix}$$

$$= \sigma_x\sigma_y + \sigma_x\sigma_z + \sigma_y\sigma_z - \tau_{xy}^2 - \tau_{xz}^2 - \tau_{yz}^2 \qquad (3.19b)$$

$$I_3 = \begin{vmatrix} \sigma_x & \tau_{xy} & \tau_{xz} \\ \tau_{xy} & \sigma_y & \tau_{yz} \\ \tau_{xz} & \tau_{yz} & \sigma_z \end{vmatrix}$$

The quantities I_1, I_2, and I_3 defined in Eq. (3.19b) are invariants of stress and must have the same values for all choices of coordinate axes (x, y, z).

The three roots $(\sigma_1, \sigma_2, \sigma_3)$ of Eq. (3.19a) are the three principal stresses at point 0. The directions of the planes corresponding to the principal stresses, called the principal planes, can be obtained from the following linear homogeneous equations in l, m, and n by setting σ in turn equal to σ_1, σ_2, and σ_3 and using the direction cosine relationship $l^2 + m^2 + n^2 = 1$:

$$l(\sigma_x - \sigma) + m\tau_{xy} + n\tau_{xz} = 0, \qquad l\tau_{xz} + m\tau_{yz} + n(\sigma_z - \sigma) = 0 \qquad (3.20)$$

The magnitude of the shear stress τ_N on plane P is given by

$$\tau_N = \sqrt{\sigma_{Nx}^2 + \sigma_{Ny}^2 + \sigma_{Nz}^2 - \sigma_N^2} \qquad (3.21)$$

The maximum value of τ_N at a point in the body plays an important role in certain theories of failure. This shear stress is zero on a principal plane.

Generally speaking, in any stressed body, there are always at least three planes on which the shear stresses are zero; these planes are always mutually perpendicular, and it is on these planes that the principal stresses act.

Maximum Shear Stress in Three-Dimensional Systems

Equations (3.14) and (3.15) deal with two-dimensional systems of stresses. In fact, there are always three principal stresses $\sigma_1, \sigma_2, \sigma_3$, where σ_3 is the principal stress in the third orthogonal direction. In this three-dimensional situation, three relative maximum shear stresses exist:

$$\tau_1 = \tfrac{1}{2}(\sigma_1 - \sigma_2), \qquad \tau_2 = \tfrac{1}{2}(\sigma_1 - \sigma_3), \qquad \tau_3 = \tfrac{1}{2}(\sigma_2 - \sigma_3) \qquad (3.22a)$$

from which the true maximum shear stress can be chosen. This maximum shear stress would be

$$\tau_{max} = \tfrac{1}{2}(\sigma_{max} - \sigma_{min}) \qquad (3.22b)$$

This, of course, is the maximum value of τ_N of Eq. (3.21). The three relative maximum shear stresses lie on planes whose normals form $45°$ angles with the principal stresses involved.

Usually σ_3 is small or zero in an assumed two-dimensional system of stresses. Then if σ_1 and σ_2 are both positive (in tension), comparison of the magnitudes of the shear stresses in Eqs. (3.22a) indicates that

$$\tau_{max} = \tfrac{1}{2}(\sigma_1 - \sigma_3) \approx \tfrac{1}{2}\sigma_1 \qquad (3.23)$$

would be the true maximum shear stress.

Mohr's Circle for Three Dimensions

Like Mohr's circle for the two-dimensional state of stress, the three mutually perpendicular principal stresses can be represented graphically. Figure 3-14 shows Mohr's circle representation of the triaxial state of stress defined by the three principal stresses in Fig. 3-15. For any section in the σ_1, σ_2 plane (i.e., planes perpendicular to plane 3) there corresponds a circle BA. In the σ_2, σ_3 plane (i.e., planes perpendicular to plane 1) there is a circle CB, and for the σ_3, σ_1 plane there exists a circle CA. From the figure, $\sigma_1 = 0A$, $\sigma_2 = 0B$, $\sigma_3 = 0C$, and $\tau_{max} = $ radius $CA = \tfrac{1}{2}(\sigma_1 - \sigma_3)$.

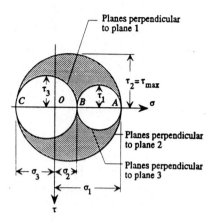

FIGURE 3-14   Mohr's circle for a three-dimensional state of stress.

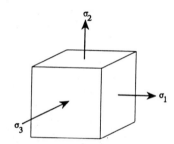

FIGURE 3-15   Triaxial state of stress.

It can be shown [3.1] that all possible stress conditions for the body fall within the shaded area between the circles in Fig. 3-14.

Mohr's circles for some common states of stress are given in Table 3-1.

Example 3.3   Mohr's Circle   For the state of stress shown in Fig. 3-16a, using Mohr's circle, determine graphically (a) the stress components on the element rotated through an angle of 45°, (b) the principal stresses and planes, and (c) the maximum shear stresses.

First, find the center $0'$ of Mohr's circle on the σ axis by using $\sigma = \frac{1}{2}(\sigma_x + \sigma_y)$ = 6000 psi, and plot the point Q with coordinates $(\sigma, \tau) = (\sigma_x, \tau_{xy}) =$ (9000, 2000). Then draw a circle with radius equal to the distance between these two points, $0'Q$. This is measured (or calculated) to be 3605.6 psi.

(a) The stress components on the element rotated through an angle of 45° are represented on Mohr's circle by rotating $0'Q$ counterclockwise $2\theta = 2 \times 45 = 90°$. This identifies the x' axis. The intersection M of the x' axis (i.e., $0'M$) with the circle gives $\sigma_{x'} = 8000$ psi and $\tau_{x'y'} = -3000$ psi. The $\sigma_{y'}$ stress, which is found 180° away from the x' axis ($0'M$), is 4000 psi. Refer to Fig. 3-16b.

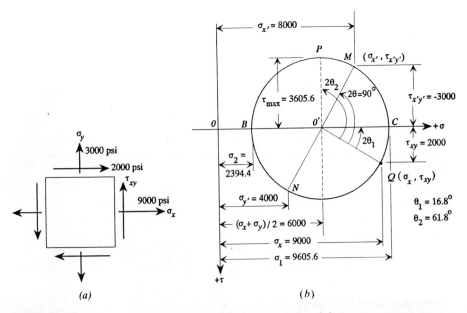

FIGURE 3-16 Example of Mohr's circle. (*a*) State of stress. (*b*) Stress components on Mohr's circle.

(b) $\sigma_1 = 0C = 00' + 0'C = 6000 + 3605.6 = 9605.6$ psi, $2\theta_1 = 33.6°$ or $\theta_1 = 16.8°$

$\sigma_2 = 0B = 00' - 0'B = 6000 - 3605.6 = 2394.4$ psi
$\theta_1 = 90° + 16.8° = 106.8°$
$\sigma_3 = 0$

(c) For a section in the σ_1, σ_2 plane, the maximum shear stresses occur on the vertical through the center of the circle (i.e., $0'P$). We measure $0'P = \tau_{max} = 3605.6$ psi and $2\theta_2 = 123.6°$ or $\theta_2 = 61.8°$. But since $\sigma_3 = 0$, the actual maximum shear stress of the element is $\tau_{max} = \frac{1}{2}(\sigma_1 - \sigma_3) = 4802.3$ psi.

Octahedral Stress

Suppose coordinate axes x, y, z are principal axes that are perpendicular to each of the principal planes, respectively. In three dimensions there are eight planes (the *octahedral* planes) that make equal angles with respect to the x, y, z directions; i.e., the absolute values of the direction cosines of the eight planes are equal, $|l| = |m| = |n| = \frac{1}{3}\sqrt{3}$. The normal and shear stress components associated with each of these planes are called the *octahedral normal stress* σ_{oct} and the *octahedral shear stress* τ_{oct}.

For this case Eqs. (3.17) and (3.18) become

$$\sigma_{Nx} = \frac{1}{3}\sqrt{3}\,\sigma_1, \qquad \sigma_{Ny} = \frac{1}{3}\sqrt{3}\,\sigma_2, \qquad \sigma_{Nz} = \frac{1}{3}\sqrt{3}\,\sigma_3$$

and

$$\sigma_{\text{oct}} = \sigma_N = \tfrac{1}{3}\sigma_1 + \tfrac{1}{3}\sigma_2 + \tfrac{1}{3}\sigma_3 = \tfrac{1}{3}I_1 \tag{3.24a}$$

Substituting σ_{Nx}, σ_{Ny}, σ_{Nz}, and σ_N into Eq. (3.21) yields

$$\tau_{\text{oct}} = \tau_N = \tfrac{1}{3}\left[(\sigma_1 - \sigma_2)^2 + (\sigma_2 - \sigma_3)^2 + (\sigma_1 - \sigma_3)^2\right]^{1/2} = \tfrac{1}{3}\left(2I_1^2 - 6I_2\right)^{1/2} \tag{3.24b}$$

In general, $\sigma_x, \sigma_y, \sigma_z$ are not principal stresses and $\tau_{xy}, \tau_{yz}, \tau_{zx}$ are not zero. However, the quantities I_1, I_2, and I_3 are invariant. The quantities σ_{oct} and τ_{oct} become

$$\sigma_{\text{oct}} = \tfrac{1}{3}I_1 = \tfrac{1}{3}(\sigma_x + \sigma_y + \sigma_z) \tag{3.25a}$$

$$
\begin{aligned}
\tau_{\text{oct}} &= \tfrac{1}{3}\left(2I_1^2 - 6I_2\right)^{1/2} \\
&= \tfrac{1}{3}\left[(\sigma_x - \sigma_y)^2 + (\sigma_x - \sigma_z)^2 + (\sigma_y - \sigma_z)^2 + 6\left(\tau_{xy}^2 + \tau_{xz}^2 + \tau_{yz}^2\right)\right]^{1/2}
\end{aligned}
\tag{3.25b}
$$

Mean and Deviator Stress

The mean stress σ_m is defined by

$$\sigma_m = \tfrac{1}{3}(\sigma_x + \sigma_y + \sigma_z) = \tfrac{1}{3}(\sigma_1 + \sigma_2 + \sigma_3) = \tfrac{1}{3}I_1 \tag{3.26}$$

It is often contended that yielding and plastic deformation of some metals are basically independent of the applied normal mean stress σ_m. As a consequence, it is useful to separate σ_m from the other stresses so that the stress tensor [Eq. (3.3)] is expressed in terms of the mean and deviator stress

$$T = T_m + T_d \tag{3.27a}$$

where

$$
T = \begin{bmatrix} \sigma_x & \tau_{xy} & \tau_{xz} \\ \tau_{yx} & \sigma_y & \tau_{yz} \\ \tau_{zx} & \tau_{zy} & \sigma_z \end{bmatrix}, \qquad
T_m = \begin{bmatrix} \sigma_m & 0 & 0 \\ 0 & \sigma_m & 0 \\ 0 & 0 & \sigma_m \end{bmatrix}
$$

and

$$
\begin{aligned}
T_d &= \begin{bmatrix} \tfrac{1}{3}(2\sigma_x - \sigma_y - \sigma_z) & \tau_{xy} & \tau_{xz} \\ \tau_{xy} & \tfrac{1}{3}(2\sigma_y - \sigma_x - \sigma_z) & \tau_{yz} \\ \tau_{xz} & \tau_{yz} & \tfrac{1}{3}(2\sigma_z - \sigma_y - \sigma_x) \end{bmatrix} \\
&= \begin{bmatrix} S_x & S_{xy} & S_{xz} \\ S_{yx} & S_y & S_{yz} \\ S_{zx} & S_{zy} & S_z \end{bmatrix}
\end{aligned}
\tag{3.27b}
$$

The matrix T_m is referred to as the *mean stress tensor* and the matrix T_d the *deviator stress tensor*. The components S_{ij} of T_d are called the *deviator stresses*. For stress tensor T, the invariants of stress, I_1, I_2, and I_3, are defined in Eq. (3.19b). Similarly, for tensors T_m, T_d, the quantities I_{1m}, I_{1d}, I_{2m}, I_{2d}, and I_{3m}, I_{3d} can also be defined. The stress invariants for principal axes x, y, z are as follows:

$$I_{1m} = I_1 = 3\sigma_m, \qquad I_{2m} = \tfrac{1}{3}I_1^2 = 3\sigma_m^2, \qquad I_{3m} = \tfrac{1}{27}I_1^3 = \sigma_m^3 \quad (\text{for } T_m) \quad (3.28a)$$

$$\left.\begin{aligned}
I_{1d} &= 0 \\
I_{2d} &= I_2 - \tfrac{1}{3}I_1^2 = \left(-\tfrac{1}{6}\right)\left[(\sigma_1 - \sigma_2)^2 + (\sigma_2 - \sigma_3)^2 + (\sigma_3 - \sigma_1)^2\right] \\
I_{3d} &= I_3 - \tfrac{1}{3}I_1 I_2 + \tfrac{2}{27}I_1^3 \\
&= \tfrac{1}{27}(2\sigma_1 - \sigma_2 - \sigma_3)(2\sigma_2 - \sigma_3 - \sigma_1)(2\sigma_3 - \sigma_1 - \sigma_2)
\end{aligned}\right\} \quad (\text{for } T_d)$$

$$(3.28b)$$

The principal values of the deviator stresses are

$$\begin{aligned}
S_1 &= \sigma_1 - \sigma_m = \tfrac{1}{3}\left[(\sigma_1 - \sigma_3) + (\sigma_1 - \sigma_2)\right] \\
S_2 &= \sigma_2 - \sigma_m = \tfrac{1}{3}\left[(\sigma_2 - \sigma_3) + (\sigma_2 - \sigma_1)\right] \\
S_3 &= \sigma_3 - \sigma_m = \tfrac{1}{3}\left[(\sigma_3 - \sigma_1) + (\sigma_3 - \sigma_2)\right]
\end{aligned} \qquad (3.29a)$$

It is apparent that $S_1 + S_2 + S_3 = 0$ $\qquad\qquad\qquad\qquad\qquad$ (3.29b)

The deviator stresses are sometimes used in theories of failure and in the theory of plasticity.

3.4 RELATIONSHIP BETWEEN STRESS AND INTERNAL FORCES

Both stress components and internal-force components are used to describe the state of the internal action of a solid. They are related in the sense that the internal forces are the resultant or total stresses. These are often referred to as *stress resultants*. Comparison of Figs. 3-17a, b for a bar cut perpendicular to the x

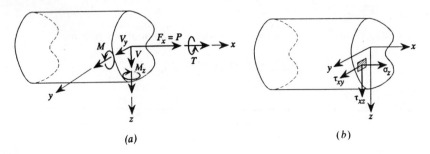

(a)   (b)

FIGURE 3-17 Internal forces and stress.

axis leads to the following relationships:

$$F_x = P = \int_A \sigma_x \, dA \tag{3.30a}$$

$$V_y = \int_A \tau_{xy} \, dA \tag{3.30b}$$

$$V = V_z = \int_A \tau_{xz} \, dA \tag{3.30c}$$

$$M_x = T = \int_A \tau_{xz} y \, dA - \int_A \tau_{xy} z \, dA \tag{3.30d}$$

$$M = M_y = \int_A \sigma_x z \, dA \tag{3.30e}$$

$$M_z = -\int_A \sigma_x y \, dA \tag{3.30f}$$

Average Shear Stress

The force acting on a plane cut in a body is called a shear force. Often an approximation for the stress acting on the plane is obtained by dividing the shear force by the area over which it acts. Thus

$$\tau = \frac{\text{force}}{\text{area}} = \frac{V}{A} \tag{3.31}$$

where τ is the shear stress, V the total force acting across and parallel to a cut plane, and A the cross-sectional area for the cut. This approximation, which is based on the assumption of a uniform distribution of stress, is called the *average shear stress*.

3.5 DIFFERENTIAL EQUATIONS OF EQUILIBRIUM

For equilibrium to exist throughout a solid for two-dimensional problems, the following differential equations must be satisfied:

$$\frac{\partial \sigma_x}{\partial x} + \frac{\partial \tau_{xy}}{\partial y} + p_x = 0, \qquad \frac{\partial \tau_{yx}}{\partial x} + \frac{\partial \sigma_y}{\partial y} + p_y = 0 \tag{3.32a}$$

In the case of three-dimensional stress, the above equations become

$$\frac{\partial \sigma_x}{\partial x} + \frac{\partial \tau_{xy}}{\partial y} + \frac{\partial \tau_{xz}}{\partial z} + p_x = 0 \tag{3.32b}$$

$$\frac{\partial \tau_{yx}}{\partial x} + \frac{\partial \sigma_y}{\partial y} + \frac{\partial \tau_{yz}}{\partial z} + p_y = 0 \tag{3.32c}$$

$$\frac{\partial \tau_{zx}}{\partial x} + \frac{\partial \tau_{zy}}{\partial y} + \frac{\partial \sigma_z}{\partial z} + p_z = 0 \tag{3.32d}$$

where p_x, p_y, p_z represent body forces per unit volume, such as those generated by weight or magnetic effects.

3.6 ALLOWABLE STRESS

Either in analyzing an existing structure or in designing a new structure, it is very important to know what constitutes a "safe" stress level. The ability of a member to resist failure is limited to a certain level. A prescribed stress level that is not to be exceeded when a member is subjected to the expected load is the *allowable* or *working* stress. The allowable stress is sometimes based on the stress level at the transition between elastic and nonelastic material behavior, i.e., yield stress. It may also be based on the occurrence of fracture (rupture) or the highest or ultimate stress that can occur in a member. In most cases the allowable stress is calculated to be lower than the yield or ultimate stress, the reduction being determined by a factor of safety. Values of allowable stress are established by local and federal agencies and by technical organizations such as the American Society of Mechanical Engineers (ASME).

3.7 RESIDUAL STRESS

Residual stress (or *lock-up stress, initial stress*) [3.3–3.7] is defined as that stress that is internal or locked into a part or assembly even though the part or assembly is free from external forces or thermal gradients. Such residual stress, whether in an individual part or in an assembly of parts, can result from a mismatch or misfit between adjacent regions of the same part or assembly.

It is often important to consider residual stresses in failure analysis and design, although residual stresses tend to be difficult to visualize, measure, and calculate [3.8]. Residual stresses are three-dimensional, self-balanced systems that need not be harmful. In fact, it may be desirable to have high compressive residual stress at the surface of parts subject to fatigue or stress corrosion.

3.8 DEFINITION OF STRAIN

As with stress, strain can be defined in terms of normal and shear strain. Normal strain is defined as the change in length per unit length of a line segment in the direction under consideration. Normal strain is a dimensionless quantity denoted by ε_i, where the subscript i indicates the given direction. Normal strain is taken as positive when the line segment elongates and negative when the line segment contracts. For the member in Fig. 3-18 with uniaxial stress,

$$\varepsilon_x = \frac{2\Delta}{2L} = \frac{\Delta}{L} = \frac{L_f - L}{L}, \qquad \varepsilon_y = -\frac{2\,\Delta h}{2h} = -\frac{\Delta h}{h} = \frac{h_f - h}{h} \quad (3.33)$$

where $2L$ and $2h$ are the original dimensions and $2L_f$ and $2h_f$ are the postdeformation dimensions.

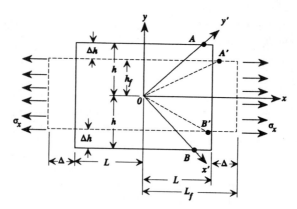

FIGURE 3-18 Elongation of an element.

Shear strain is defined as the tangent of the change in angle of a right angle in a member undergoing deformation. It is a dimensionless quantity. The symbol for the strain is γ_{ij}, where the subscripts have meanings similar to the subscripts for shear stress. For the small shear strains encountered in most engineering practice (usually less than 0.001), the tangent of the change in angle is very nearly equal to the angle change in radians. Positive shear strains are associated with positive shear stresses (Fig. 3-19a); negative shear strains correspond to negative shear stresses (Fig. 3-19b). Refer to the x', y' axes of Fig. 3-18. If this member is lengthened and thinned, A and B will move to new positions A' and B'. Angle $A'0B'$ is now less than 90°. The tangent of the total change in angle is the *shear strain*.

Another useful definition of strain is the change in length divided by the instantaneous value of the length (rather than the original length):

$$\varepsilon_t = \int_L^{L_f} \frac{d\ell}{\ell} = \ln \frac{L_f}{L} \tag{3.34}$$

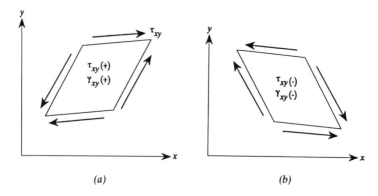

(a)   *(b)*

FIGURE 3-19 Shear strain sign.

where ε_t is referred to as the *natural* (or *true*) *strain*. The concept of true strain is very useful in handling problems in plasticity and metal forming. For the very small strains for which the equations of elasticity are valid, the two types of strains (strain and true strain) give almost the same values.

3.9 RELATIONSHIP BETWEEN STRAIN AND DISPLACEMENT

In general, the state of strain at a point in a body is determined by six strains, ε_x, ε_y, ε_z, γ_{yx}, γ_{xz}, and γ_{yz}, arranged in the same fashion as stresses. These components can be assembled into a strain tensor similar to the stress tensor.

If u, v, and w are three displacement components at a point in a body for the x, y, and z directions of coordinate axes, small strains are related to the displacements through the geometric relationships

$$\varepsilon_x = \frac{\partial u}{\partial x}, \qquad \gamma_{xy} = \frac{\partial u}{\partial y} + \frac{\partial v}{\partial x} = \gamma_{yx}$$

$$\varepsilon_y = \frac{\partial v}{\partial y}, \qquad \gamma_{xz} = \frac{\partial u}{\partial z} + \frac{\partial w}{\partial x} = \gamma_{zx} \qquad (3.35)$$

$$\varepsilon_z = \frac{\partial w}{\partial z}, \qquad \gamma_{yz} = \frac{\partial v}{\partial z} + \frac{\partial w}{\partial y} = \gamma_{zy}$$

In the case of *plane strain* (zero strains in the z direction, i.e., $\varepsilon_z = \gamma_{xz} = \gamma_{yz} = 0$), the foregoing equations become

$$\varepsilon_x = \frac{\partial u}{\partial x}, \qquad \varepsilon_y = \frac{\partial v}{\partial y}, \qquad \gamma_{xy} = \frac{\partial u}{\partial y} + \frac{\partial v}{\partial x} = \gamma_{yx} \qquad (3.36)$$

It can be shown that in order to assure unique continuous displacements, the strains cannot be independent. For example, the so-called *compatibility condition*

$$\frac{\partial^2 \varepsilon_x}{\partial y^2} + \frac{\partial^2 \varepsilon_y}{\partial x^2} = \frac{\partial^2 \gamma_{xy}}{\partial x \, \partial y} \qquad (3.37)$$

must hold. That is, the three strains of Eq. (3.36) must satisfy Eq. (3.37) to assure that the two displacements u, v are single valued and continuous.

3.10 ANALYSIS OF STRAIN

The strain components possess the same sort of tensor characteristics as the stress components. Hence, strains follow the same rules as stresses when axes are rotated. There are principal axes for strain, and a Mohr's circle for strain can be used to evaluate strain components at various orientations. The only difference is that the vertical axis is $\frac{1}{2}\gamma$ rather than τ, which is used with Mohr's circle of stress. Therefore, the normal strain ε_N at a point in the direction of N that makes a counterclockwise angle θ_N with the x axis is

$$\varepsilon_N = \varepsilon_x \cos^2 \theta_N + \varepsilon_y \sin^2 \theta_N + \gamma_{xy} \sin \theta_N \cos \theta_N \qquad (3.38)$$

In strain measurement, the majority of problems are two dimensional. The extensions (or normal strain) in one or more directions are the quantities most often measured.

3.11 ELASTIC STRESS – STRAIN RELATIONS

Poisson's Ratio

For a bar of elastic material having the same mechanical properties in all directions and under a condition of uniaxial loading, measurements indicate that the lateral compressive strain is a fixed fraction of the longitudinal extensional strain. This fraction is known as Poisson's ratio ν. In the case of the member of Fig. 3-18

$$\varepsilon_y = -\nu \varepsilon_x \qquad (3.39)$$

Like the modulus of elasticity E of the following paragraph, Poisson's ratio is a material constant that can be determined experimentally. For metals it is usually between 0.25 and 0.35. It can be as low as 0.1 for certain concretes and as high as 0.5 for rubber.

Hooke's Law

The stresses and strains are related to each other by the properties of the material. Equations of this nature are known as material laws or, in the case of elastic solids, as *Hooke's law*. For a three-dimensional state of stress and strain, Hooke's law for isotropic material appears as

$$\varepsilon_x = (1/E)\left[\sigma_x - \nu(\sigma_y + \sigma_z)\right]$$
$$\varepsilon_y = (1/E)\left[\sigma_y - \nu(\sigma_x + \sigma_z)\right]$$
$$\varepsilon_z = (1/E)\left[\sigma_z - \nu(\sigma_x + \sigma_y)\right]$$
$$\tau_{ij} = G\gamma_{ij} \qquad (i, j = x, y, z)$$

$$(3.40)$$

where G is the shear modulus, E is the modulus of elasticity, and ν is Poisson's ratio. The dimensions of G and E are force per unit area, e.g., lb/in.^{2} or N/m^2 (Pa). Typical values of E and ν for some materials are listed in Table 4-1. The bulk modulus K (also called volumetric modulus of elasticity, modulus of dilation, modulus of volume expansion, or modulus of compressibility) is a material constant defined as the ratio of the hydrostatic stress $\sigma_1 = \sigma_2 = \sigma_3$ (shear stresses are zero) to the volumetric strain (change in volume divided by the original volume). Of the many different material constants (e.g., E, ν, G, and K), only two are independent if the material is isotropic. Table 3-2 lists the relationships between commonly used material constants.

3.12   STRESS AND STRAIN IN SIMPLE CONFIGURATIONS

Direct Axial Loading (Extension and Compression)

A typical tension member is shown in Fig. 3-20. It is assumed that the force acts uniformly over the cross section so that the stress at any point is

$$\sigma_x = P/A \tag{3.41}$$

As a result of the force P, the bar elongates an amount Δ. In terms of strain ε_x along the bar,

$$\varepsilon_x = \Delta/L \tag{3.42}$$

The quantities σ_x and ε_x are called engineering stress and strain since they are based on the original dimensions of the bar.

Using Hooke's law for the axial fibers, $\sigma_x = E\varepsilon_x$, Eq. (3.41) becomes

$$\varepsilon_x = P/EA \tag{3.43}$$

or

$$\Delta = PL/AE \tag{3.44}$$

Frequently it is convenient to relate the extension of a bar to the extension of a spring. If the force in the spring of Fig. 3-21a is linearly proportional to its

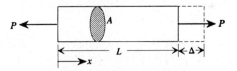

FIGURE 3-20   Extension.

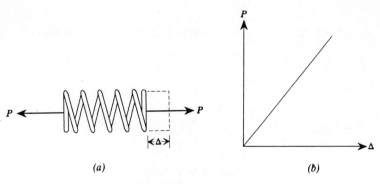

(a)

(b)

FIGURE 3-21   A spring.

displacement (Fig. 3-21b), the constant of proportionality is the *spring constant*

$$k = P/\Delta \tag{3.45}$$

The constant k is also referred to as the *stiffness coefficient*. The reciprocal of the stiffness coefficient, $1/k$, is the *flexibility coefficient*.

Example 3.4 Elongation of a Bar A steel bar with a uniform cross section of 1000 mm^2 is subject to the uniaxial forces shown in Fig. 3-22a. Calculate the total elongation of the bar ($E = 200$ GN/m^2).

The entire bar is in equilibrium since the sum of the axial forces is zero. The total elongation is determined by separating the bar into three sections, finding the elongation of each, and adding these elongations. The conditions of equilibrium give the internal force in each section. Thus for Fig. 3-22b, $\Sigma F_H = 0$: $-F_{bc} + 10 + 60 = 0$ or the internal force $F_{bc} = 70$ kN in tension. Similar manipulations give $F_{ab} = 100$ kN, $F_{cd} = 60$ kN, both in tension. Then from Eq. (3.44),

$$\Delta = \Delta_{ab} + \Delta_{bc} + \Delta_{cd} = \frac{1}{AE}\left[(FL)_{ab} + (FL)_{bc} + (FL)_{cd}\right]$$

$$= \frac{(100 \text{ kN})(3 \text{ m}) + (70 \text{ kN})(4 \text{ m}) + (60 \text{ kN})(5 \text{ m})}{(1000 \text{ mm}^2)(200 \text{ GN/m}^2)}$$

$$= 4.4 \times 10^{-3} \text{ m} = 4.4 \text{ mm}$$

There are some differences between compression and tension. First, in compression, instability failure by buckling may occur depending on the geometry, especially the length. Second, for ductile materials, there is no apparent ultimate strength in compression.

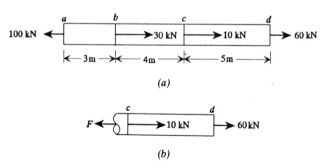

(a)

(b)

FIGURE 3-22 A bar.

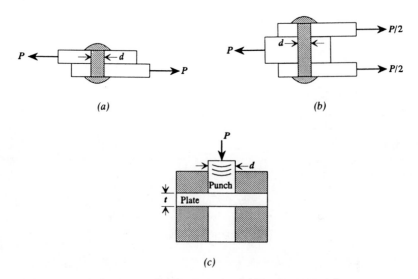

FIGURE 3-23   Examples of shear: (a) Single shear. (b) Double shear. (c) Punch on a plate.

Direct Shear in Connections

Shear may be considered to be a process whereby parallel planes move relative to one another. In direct shear, the shear stress can be calculated as an average stress. Some examples of direct shear are shown in Fig. 3-23. For the configurations in Figs. 3-23a, b, the shear stresses in the bolts are

$$\text{Single stress} \qquad \tau = \frac{P}{A} = \frac{P}{\pi d^2/4} = \frac{4P}{\pi d^2} \qquad (3.46a)$$

$$\text{Double stress} \qquad \tau = \frac{P}{2A} = \frac{P}{2\pi d^2/4} = \frac{2P}{\pi d^2} \qquad (3.46b)$$

The direct shear in Fig. 3-23c represents a punch and plate. If the punch diameter is d and the plate thickness is t, the shear stress τ in the plate is

$$\tau = P/A = P/(\pi \, d t) \qquad (3.46c)$$

Torsion

For a bar subject to an applied torque (Fig. 3-24), the torsional or shear stresses τ on a cross section of circular shape, either solid or hollow, are linearly proportional in magnitude to the distance r to the centroidal axis of the bar. This stress, which acts normal to the radius, is given by

$$\tau = Tr/J \qquad (3.47)$$

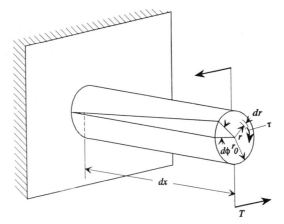

FIGURE 3-24 Torsion.

where τ = shear stress [force per unit area, psi or N/m^2 (Pa)]

T = torque or twisting moment (length × force, in.-lb or Nm)

r = radial distance from longitudinal axis (length, in. or m)

J = torsional constant (length to the fourth power, in.⁴ or mm⁴) of cross section; if cross-sectional shape is circular, $J = I_x$, polar moment of inertia about longitudinal axis

It can be seen from Eq. (3.47) that the highest stresses occur in the outer edge fibers:

$$\tau_{max} = Tr_0/J \tag{3.48}$$

The shear strain γ for any section of the bar is given by

$$\gamma = \tau/G = Tr/GJ \tag{3.49}$$

In addition, since at any distance dx from the fixed end of the bar, $\gamma = r\,d\varphi/dx$, Eq. (3.49) shows that

$$d\varphi = dx\,T/GJ \tag{3.50a}$$

which, upon integration, gives

$$\varphi = TL/GJ \tag{3.50b}$$

Torsion of Thin-walled Shafts and Tubes of Circular Cross Sections For the thin-walled circular section of Fig. 3-25, if the shear stress is assumed to be uniformly distributed across the thickness, the equilibrium conditions give

$$T = 2\pi r^2 t\tau, \qquad \tau = T/(2\pi r^2 t) \tag{3.51a}$$

where r is the radius to the midwall.

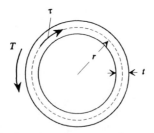

FIGURE 3-25 Thin-walled torsion.

Since the torsional constant for a thin circular section is approximately $J = 2\pi r^3 t$, the shear stress can be written as

$$\tau = Tr/J \tag{3.51b}$$

Equation (3.51a) also follows directly from Eq. (3.47). The angle of twist of this thin-walled section is still given by Eq. (3.50b).

Torsion of Thin-walled Noncircular Tubes

For thin-walled noncircular sections it is assumed that the wall thickness is small compared to the overall dimensions of the cross section and that the stress is uniform through the wall thickness. Experiments and comparisons with more exact analyses have shown this latter assumption to be reasonable for most thin-walled sections in the elastic range.

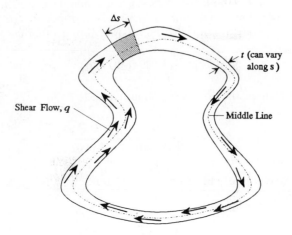

FIGURE 3-26 Thin-walled tube.

The formulas for thin-walled tubes (Fig. 3-26) are

$$q = T/2\bar{A} \tag{3.52a}$$

$$\varphi = \frac{TL}{GJ} \quad \text{or} \quad \frac{d\varphi}{dx} = \frac{T}{GJ} \tag{3.52b}$$

$$q = \tau t \tag{3.52c}$$

where q is the shear flow, J is the torsional constant, and $\bar{A}$ is the area enclosed by the middle line of the wall.

For constant t, Eq. (3.52b) becomes

$$\frac{d\varphi}{dx} = \frac{\tau L_s}{2\bar{A}G} = \frac{TL_s}{4\bar{A}^2 Gt} \tag{3.53}$$

where L_s is the total length of the middle line of the wall.

Note that although the shear flow q from Eq. (3.52a) is constant around the wall, the shear stress $\tau = q/t$ of Eq. (3.52c) can vary with t. The largest shear stress occurs where the wall is thinnest and vice versa. Also note that no distinction is made by Eq. (3.52a) between different cross-sectional shapes. According to this formula, all cross-sectional geometries with the same enclosed area $\bar{A}$ will experience the same shear flow for the same torque T.

If the walls of the hollow shaft are very thin, the possibility of buckling should be considered. Thus, a shaft safe from the standpoint of yield stress level may well be unstable.

A Useful Relation between Power, Speed of Rotation, and Torque

Power is the measure of work developed per unit time. The work done by a torque T during one revolution of a shaft is $2\pi T$. For a shaft rotating at n revolutions per minute (rpm), the work done per minute is $2\pi Tn$. In the U.S. Customary System, the usual unit of power is foot-pounds per second. In engineering work, a larger unit called horsepower (hp) is often used:

$$1 \text{ hp} = 33,000 \text{ ft-lb/min} \tag{3.54a}$$

If T is in inch-pounds, the horsepower transmitted is

$$\text{hp} = \frac{2\pi Tn}{12(33,000)} = \frac{Tn}{63,000} \tag{3.54b}$$

For the SI system, the unit of power is the watt, $W = N \cdot m/\text{sec}$. If T is in newton-meters,

$$W = \frac{2\pi Tn}{60} \tag{3.54c}$$

Normal and Shear Stress of Beams

When a simple beam bends under vertical downward load, the top fibers shorten the most and the bottom fibers lengthen the most (Fig. 3-27). Between the top and the bottom fibers there exists a layer or surface that remains neutral; neither tension nor compression is generated in it though it is curved like the rest of the layers. Hence, this layer is called the *neutral surface*. It is assumed that the fiber deformations are directly proportional to the distance from the neutral surface. This fundamental assumption about the geometry of deformation of a beam is stated as follows: *Plane sections normal to the axis of a beam remain plane as the beam is bent*.

The intersection of a cross-sectional plane with the neutral surface is called the neutral axis (NA). For example, the y axis shown in Fig. 3-27 is the neutral axis of the cross section. It can be shown that the neutral axis passes through the centroid of the cross section.

Note the sign convention here. The bending moment M is positive when tensile stress is on the bottom fiber or the center of curvature is above the beam. Positive z is taken to be downward.

On a cross section of a linearly elastic beam having the z axis as a vertical axis of symmetry, the normal stress $\sigma_x = \sigma$ acting on a longitudinal fiber at a distance z from the neutral axis is given by the *flexure formula*

$$\sigma = Mz/I \tag{3.55}$$

Here M is the net internal bending moment at the section and I is the moment of inertia of the cross section about the neutral axis (y).

The stresses, like the deformations (and strains), vary linearly with the distance from the neutral axis (Fig. 3-28). The stresses are tensile on one side of the neutral axis and compressive (negative) on the other side. The maximum stress for a cross section occurs at the outermost fibers of the beam and is given by

$$\sigma_{max} = Mc/I \tag{3.56a}$$

or

$$\sigma_{max} = M/S \tag{3.56b}$$

where c is the distance from the neutral axis to the outermost fiber. The quantity $S = I/c$ is called the *section modulus*, which is a geometric property of the cross

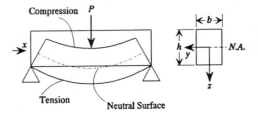

FIGURE 3-27 Beam under loading.

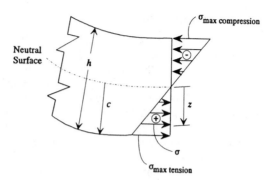

FIGURE 3-28 Stress distribution on a cross section of a beam.

section (Chapter 2) and is a measure of the resistance to the development of bending stress.

If a vertical plane is passed through a transversely loaded beam perpendicular to the axis, the vertical stresses acting along this plane are called shear stresses. Equilibrium requires that the vertical shear stress τ at any point on the cross section is numerically equal to the horizontal shear stress at the same point. These shear stresses as well as the normal stresses are assumed to be uniform across the width of the beam. However, the shear stress varies according to the shape of the cross section, as shown in Figs. 3-29a, b.

The shear stress $\tau_{xz} = \tau_{zx} = \tau$ at any point of a prescribed cross section is given by

$$\tau = VQ/Ib \qquad (3.57a)$$

where V is the shear force at the section, I is the moment of inertia about the neutral axis, b is the width of the section measured at the level at which τ is being determined, and Q is a first moment (Chapter 2) with respect to the neutral axis of the area beyond the point at which the shear stress is desired.

If the shear stress is to be determined at level z_1 of a rectangular cross section, then Q must be calculated for the shaded area A_0 of Fig. 3-29a.

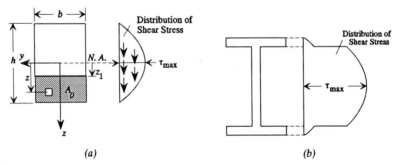

(a)   *(b)*

FIGURE 3-29 Stress distribution on different cross-sectional shapes. A_0 is the shaded area.

Equation (2.15a) gives

$$Q = A_0 \bar{z}_c = b\left(\tfrac{1}{2}h - z_1\right)\left[z_1 + \tfrac{1}{2}\left(\tfrac{1}{2}h - z_1\right)\right] = \tfrac{1}{2}b\left(\tfrac{1}{4}h^2 - z_1^2\right)$$

From Eq. (3.57a), the desired stress is

$$\tau = \frac{V}{2I}\left(\frac{h^2}{4} - z_1^2\right) \tag{3.57b}$$

and

$$\tau_{max} = \frac{Vh^2}{8I} = \frac{3}{2}\frac{V}{bh} = \frac{3V}{2A} \tag{3.57c}$$

at the neutral axis ($z_1 = 0$). This equation has been shown to be reasonably accurate for widths equal to or less than the depth ($b \leq h$), but for $b > h$, Eq. (3.57c) should be used with caution. Accurate computational solutions have been developed (Chapter 15).

For a wide-flange I-shaped structural steel, the maximum shear stress given by Eq. (3.57a) is only slightly greater than the average stress obtained by dividing the shear force by the area of the web.

A useful formula in the study of a beam formed of more than one layer, e.g., two boards nailed together, is for the shear flow q. From Eq. (3.57a),

$$q = \tau b = VQ/I \tag{3.57d}$$

This gives the horizontal shear force per unit length of beam that is transmitted between layers of the beam.

Deflection of Simple Beams The sign convention for forces and displacements of a beam is shown in Fig. 3-30. Applied forces and moments are positive if their vectors are in the direction of a positive coordinate axis. Also, internal shear forces and bending moments acting on a positive face are positive if their vectors are in positive coordinate directions. The internal forces M, V and applied loads M_1 (concentrated moment, force times length), p_z (loading intensity, force per length), and W (concentrated load, force) shown in Fig. 3-30 are positive.

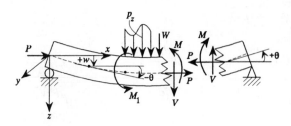

FIGURE 3-30 Positive applied loadings and internal forces.

Positive deflection w is downward, i.e., in the positive coordinate z. As shown in Fig. 3-30, θ (radians) is the angle between the axis and the tangent to the curve at a point. Positive and negative θ, which like moments adhere to the right-hand rule, are illustrated.

The basic differential equation relating the deflection w to the internal bending moment M in a beam is

$$\frac{d^2w}{dx^2} = \frac{-M}{EI} \tag{3.58a}$$

where x is the axial coordinate and EI is the flexural rigidity or bending modulus. This relationship applies to a beam that is linearly elastic and where the cross section is symmetric about the x, z plane.

For small angles,

$$\theta \approx \tan \theta = -\frac{dw}{dx}$$

i.e.,

$$\frac{dw}{dx} = -\theta \tag{3.58b}$$

and Eq. (3.58a) appears as $d\theta/dx = M/EI$. The equilibrium equations relate the internal forces M and V and the applied loading density p_z in the form $dV/dx = -p_z$, $dM/dx = V$. If these relations are gathered together,

$$\frac{dw}{dx} = -\theta, \qquad \frac{d\theta}{dx} = \frac{M}{EI}, \qquad \frac{dM}{dx} = V, \qquad \frac{dV}{dx} = -p_z \tag{3.59}$$

These equations are called *governing equations of motion* for the bending of a beam. This first-order form is convenient to handle numerically using a computer. Analytically it is frequently easier to deal with the higher order forms:

For variable EI	For constant EI
$\theta = -\dfrac{dw}{dx}$	$\theta = -\dfrac{dw}{dx}$
$M = -EI\dfrac{d^2w}{dx^2}$	$M = -EI\dfrac{d^2w}{dx^2}$
$V = \dfrac{dM}{dx} = -\dfrac{d}{dx}\left(EI\dfrac{d^2w}{dx^2}\right)$	$V = -EI\dfrac{d^3w}{dx^3}$
$p_z = -\dfrac{dV}{dx} = \dfrac{d^2}{dx^2}\left(EI\dfrac{d^2w}{dx^2}\right)$	$p_z = EI\dfrac{d^4w}{dx^4}$

$$(3.60)$$

These relations are found by successive substitution of Eqs. (3.58) into each other.

Stress in Pressure Vessels

This section discusses thin-walled containers or shells loaded with gas or liquid pressure and having the form of a surface of revolution, such as cylinders and spheres.

Cylinder Stress   On the wall of a thin-walled cylinder subjected to internal pressure, two stresses in the plane of the wall are of prime interest (Fig. 3-31). These stresses, a longitudinal stress σ_x parallel to the axis of revolution and a hoop or circumferential or cylindrical stress σ_θ perpendicular to σ_x, are called *membrane stresses*. If there are no abrupt changes in wall thickness and the wall is thin (thickness less than about one-tenth the radius r), it can be assumed that the stresses are uniformly distributed through the thickness of the wall and that no other significant stresses occur. Application of the conditions of equilibrium suffices to determine these membrane stresses (Chapter 20). For a cylinder with internal pressure p,

$$\sigma_\theta = pr/t \tag{3.61a}$$

If the ends of the cylinder are closed,

$$\sigma_x = pr/2t \tag{3.61b}$$

The results for the circumferential stress are about 5% in error on the danger side when the thickness is one-tenth the radius of the cylinder ($t = 0.1r$). Shells of greater relative thickness should be analyzed according to bending shell theories (Chapter 20).

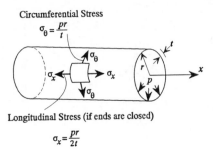

Circumferential Stress

$$\sigma_\theta = \frac{pr}{t}$$

Longitudinal Stress (if ends are closed)

$$\sigma_x = \frac{pr}{2t}$$

FIGURE 3-31   Cylinder.

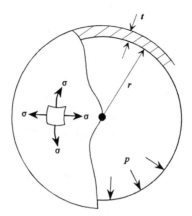

FIGURE 3-32 Sphere.

Sphere Stress The stresses σ acting in the plane of the sphere wall are the same in all directions under uniform internal pressure (Fig. 3-32):

$$\sigma = pr/2t \qquad (3.62)$$

It can be seen that they are one-half the magnitude of the circumferential stresses of the cylinder. When the thickness equals one-fifth the radius of the sphere ($t = 0.2r$), the thin-sphere formula gives values in error by about 2.5% on the danger side. If the thickness exceeds one-fifth the radius, more accurate formulas should be used (Chapter 20).

Stress for Shells of Revolution A shell of revolution is formed by rotating a plane curve, called the meridian, about an axis lying in the plane of the curve

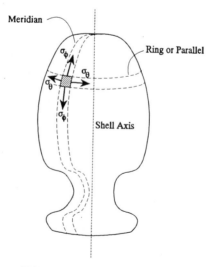

FIGURE 3-33 A shell of revolution.

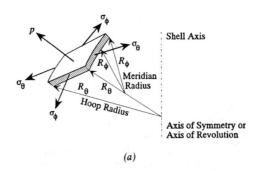

(a)

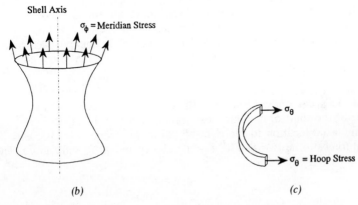

(b)   *(c)*

FIGURE 3-34 Stresses in a shell of revolution. (*a*) Shell element. (*b*) Meridional stress. (*c*) Circumferential stress.

(Fig. 3-33). The stresses on an element of a general membrane shell of revolution (Fig. 3-34*a*) are related to the pressure p by

$$\sigma_\phi/R_\phi + \sigma_\theta/R_\theta = p/t \qquad (3.63)$$

where σ_ϕ = meridional stress (psi, N/m^2 or Pa) (Fig. 3-34*b*)

σ_θ = hoop, ring, or circumferential stress (psi, N/m^2) (Fig. 3-34*c*)

R_ϕ = radius of curvature of meridian

R_θ = radius of curvature of section normal to meridian curve; i.e., R_θ is length of normal between surface and axis of revolution and originates at shell axis and in general is not perpendicular to shell axis, whereas center of curvature for R_ϕ in general will not lie on shell axis (see Fig. 3-34*a*)

3.13 COMBINED STRESSES

In the most general case, a body may be subjected to a variety of types of loadings such as a combination of tension, compression, twisting, and bending loads. In such a case, it will be assumed that each load produces the stress that it would if it were the only load acting on the body. As discussed below, the final stress is then found by careful superposition of the several states of stress.

Frequently there is little difficulty in identifying the individual states of stress composing a combined stress problem. The appropriate stress formula developed in the previous sections should be associated with each individual load. For example, in a bar subjected simultaneously to tension and torsion loads, the axial normal stress component is $\sigma_x = P/A$, where P is the tensile load and A is the cross-sectional area of the bar. Also present is a shear stress due to the torque, $\tau = Tr/J$, where T is the torque, r the radius of the section, and J the polar moment of inertia. The above case leads to one normal and one shear stress. Normal stresses, e.g., extension and bending stresses, are directly additive, as are shear stresses if they act in the same direction. If not, the methods in Section 3.3 are employed, usually to calculate the principal stresses at a point.

Note that superposition is valid if the material is linearly elastic and if the effect of one type of loading does not influence the internal force corresponding to other loadings of interest.

Example 3.5 Bar under Combined Stresses Find the maximum shear stress on the face of the shaft of circular cross section shown in Fig. 3-35.

At any axial location to the right of 120 in.-lb torque (Fig. 3-35), we find the internal forces to be $V = 800$ lb, $T = 120$ in.-lb, and $M = 800x$ in.-lb. The shear

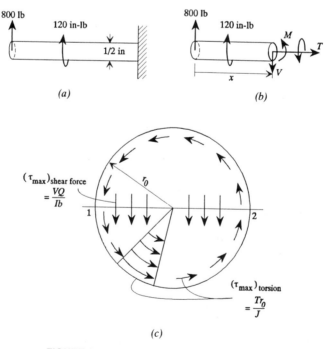

FIGURE 3-35 A bar under combined stresses.

stresses are given by Eqs. (3.47) and (3.57a). From these formulas the peak torsional stress occurs at the outer fibers and the shear stress due to V is a maximum at the diameter 1–2 in Fig. 3-35c, where Q is a maximum. The maximum combined shear stress occurs at point 1, where the two peak shear stresses act in the same direction:

$$\tau_{max} = \frac{VQ}{Ib} + \frac{Tr_0}{J} = \frac{4V}{3A} + \frac{Tr_0}{J} = \frac{4(800)}{3(0.196)} + \frac{120(0.25)}{0.00614} = 10{,}328 \text{ psi}$$

where $A = \frac{1}{4}\pi d^2 = 0.196$ in.^{2} and $J = \frac{1}{32}\pi d^4 = 0.00614$ in.^{4} Note the bar also has an axial normal stress due to bending.

Example 3.6 Eccentric Loads A cantilever beam is loaded by a force of 40 kN applied 80 mm from the centroid (Fig. 3-36). Find the maximum normal stress for a vertical cross section. Neglect the weight of the beam.

The eccentric load $P = 40$ kN is statically equivalent to the load P through the centroid and the moment $Pe = 40 \times 80$ mm $\cdot$ kN about a centroidal axis. The combined normal stress is

$$\sigma = -\frac{P}{A} - \frac{Pez}{I} = \frac{-40 \text{ kN}}{(200 \text{ mm})(50 \text{ mm})} - \frac{(40 \text{ kN})(80 \text{ mm})z}{\frac{1}{12}(50 \text{ mm})(200 \text{ mm})^3}$$

The peak bending stresses occur at the outer fibers where $z = \pm 100$ mm. Thus, at the bottom fibers,

$$\sigma = -4.0 - 9.6 = -13.6 \text{ N/mm}^2 = -13.6 \text{ MN/m}^2 \quad (\text{compression})$$

At the top fibers

$$\sigma = -4.0 + 9.6 = 5.6 \text{ N/mm}^2 = 5.6 \text{ MN/m}^2 \quad (\text{tension})$$

Example 3.7 Combined Bending and Torsion of Shafts Show that when a solid circular shaft of diameter d is subjected to a bending moment M and a torque T, (a) the maximum principal stress is equal to $16(M + \sqrt{M^2 + T^2})/\pi d^3$ and (b) the maximum shear stress is equal to $16\sqrt{M^2 + T^2}/\pi d^3$.

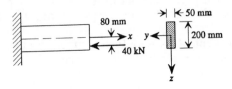

FIGURE 3-36 Eccentric load.

The maximum stresses, which occur at the outer fibers, are given by Eqs. (3.56a) and (3.47) with $J = 2I$ and $r = z = c$:

$$\sigma = Mc/I, \qquad \tau = Tc/J = Tc/2I$$

The maximum principal stress is derived using Eq. (3.13a):

$$\sigma_1 = \frac{\sigma}{2} + \sqrt{\left(\frac{\sigma}{2}\right)^2 + \tau^2} = \frac{Mc}{2I} + \sqrt{\left(\frac{Mc}{2I}\right)^2 + \left(\frac{Tc}{2I}\right)^2}$$

$$= \frac{16}{\pi d^3}\left(M + \sqrt{M^2 + T^2}\right)$$

where we have set $c = \frac{1}{2}d$. The peak shear stress is found from Eq. (3.14):

$$\tau_{max} = \sqrt{\left(\frac{\sigma}{2}\right)^2 + \tau^2} = \frac{16}{\pi d^3}\sqrt{M^2 + T^2}$$

For convenient reference, the basic stress formulas considered in this chapter for simple configurations are given in Table 3-3. The basic deformation formulas are given in Table 3-4.

3.14 UNSYMMETRIC BENDING

Normal Stress

The formula for normal stress in straight beams, $\sigma = Mz/I$, is applicable only if the bending moment acts around one of the principal axes of inertia of the cross section. That is, the bending stress theory developed thus far is appropriate for a symmetric cross section bent in its plane of symmetry.

Consider the more general case of an unsymmetric cross section with positive (tensile) axial P and bending moment components $M_y = M$ and M_z. The formula

$$\sigma = \frac{P}{A} + \frac{M_y I_z + M_z I_{yz}}{I_z I_y - I_{yz}^2}z - \frac{M_z I_y + M_y I_{yz}}{I_z I_y - I_{yz}^2}y \qquad (3.64)$$

applies. The coordinates y, z are measured from axes passing through the centroid of the cross section. The moments of inertia $I_y = I, I_z, I_{yz}$ are taken about these axes.

If the bending moment M_z is zero, Eq. (3.64) reduces to a formula applicable to an unsymmetric section loaded in a single plane.

$$\sigma = P/A + M_y(I_z z - I_{yz} y)/\left(I_z I_y - I_{yz}^2\right) \qquad (3.65)$$

Principal Axes Suppose y, z correspond to principal axes of inertia through the centroid. Then $I_{yz} = 0$ and Eq. (3.64) becomes

$$\sigma = P/A + M_y z/I_y - M_z y/I_z \tag{3.66}$$

where the bending moments have been resolved into components along the principal axes.

Bending about a Single Axis

Equation (3.64) reduces to the usual bending stress formula of Eq. (3.55) if the bending moment acts around a single principal axis of inertia through the centroid. We use Eq. (3.66) with

$$M_y = M, \qquad P = 0, \qquad M_z = 0, \qquad I_y = I$$

Then

$$\sigma = Mz/I$$

Example 3.8   Unsymmetric Bending   Consider the beam section in Fig. 3-37a. From the formulas of Chapter 2

$$I_z = \tfrac{1}{12}th^3, \qquad I_y = \tfrac{1}{3}th^3, \qquad I_{yz} = \tfrac{1}{8}th^3 \tag{1}$$

To compute the bending stresses, use Eq. (3.65) with $P = 0$, $M_y = M$,

$$\sigma = \frac{M_y(I_z z - I_{yz} y)}{I_z I_y - I_{yz}^2} = \frac{M}{th^3}\left(\tfrac{48}{7}z - \tfrac{72}{7}y\right)$$

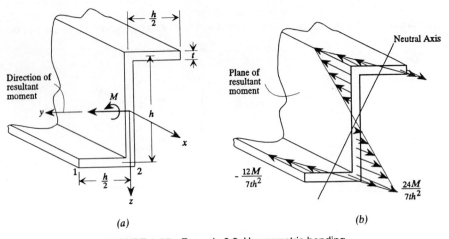

(a)   (b)

FIGURE 3-37   Example 3.8. Unsymmetric bending.

which is plotted in Fig. 3-37b. The peak stresses occur at extreme fibers. At point 1, with $z = y = \frac{1}{2}h$, we find $\sigma_1 = -12M/7th^2$. At 2, with $z = \frac{1}{2}h$ and $y = 0$, $\sigma_2 = 24M/7th^2$. See Fig. 3-37b.

If Eq. 4 in Table 3-3, which does not take the lack of symmetry into account, had been used, then

$$\sigma_1 = \sigma_2 = \left(\frac{Mz}{I_y} \right)_{z=h/2} = \frac{\frac{1}{2}Mh}{\frac{1}{3}th^3} = \frac{3M}{2th^2}$$

Comparison of this with the correct values of σ_1 and σ_2 shows that errors of 188 and 56%, respectively, would occur.

Shear Stress

The familiar formula for shear stress in straight beams, $\tau = VQ/Ib$, applies to symmetric sections in which the shear force V is along one of the principal axes of inertia of the cross section. For an unsymmetric cross section with positive shear forces V_z and V_y, the average shear stress is given by

$$\tau = \frac{I_z Q_y - I_{yz}Q_z}{b\left(I_z I_y - I_{yz}^2\right)}V_z + \frac{I_y Q_z - I_{yz}Q_y}{b\left(I_z I_y - I_{yz}^2\right)}V_y \tag{3.67}$$

where Q_y and Q_z are first moments of inertia of the area beyond the point at which τ is calculated (Fig. 3-38). These first moments are defined by Eq. (2.15).

The coordinates y, z in Eq. (3.67) are referred to axes passing through the centroid of the cross section. If the width b is chosen parallel to the y axis, Eq. (3.67) gives the stress τ_{zx}. If b is parallel to the z axis, Eq. (3.67) corresponds to τ_{xy}. Moreover, b can be chosen such that Eq. (3.67) gives the average shear stress

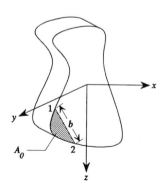

FIGURE 3-38 Shear stress.

in any direction. This is accomplished by selecting b to be the section width at the point where the stress is sought. This width is taken in a direction perpendicular to the desired stress. If the shear stress along the line 1–2 of the section in Fig. 3-38 is to be computed, then b should be selected as indicated. This fixes area A_0 and also establishes Q_y and Q_z. Note that according to this formula, the average shear stress is constant along 1–2. Hence, only when the actual shear stress is constant along 1–2 is the average shear stress of Eq. (3.67) equal to the actual shear stress on b. Equation (3.67) is normally considered to be reasonably accurate for thin-walled sections and somewhat less accurate for thick sections. More accurate stresses are provided by the computer program discussed in Chapter 15.

Equation (3.67) is usually employed to calculate the shear stress or shear flow in thin-walled open sections. This relationship reduces to $\tau = VQ/Ib$ if the loading is in the x, z plane ($V_y = 0$) and z and y are the principal axes of inertia ($I_{yz} = 0$).

Example 3.9 Shear Stress in Unsymmetric Bending Find the shear flow in the beam section in Fig. 3-39a due to the shear force V_z.

The shear flow is calculated from Eq. (3.67) using $q = \tau b$. The moments of inertia are given by Eq. (1) of Example 3.8. Equation (3.67), with $b = t$, reduces to

$$\tau = \frac{48Q_y - 72Q_z}{7t^2h^3}V_z \qquad (1)$$

The first moments Q_y and Q_z of Eq. (2.15) are taken about y, z coordinates passing through the centroid. For a point in the flange between 1 and 2 (Fig. 3-39b),

$$Q_z = \int_{A_0} y\,dA = \tfrac{1}{2}\left(\tfrac{1}{2}h + y\right)A_0$$

$$= \tfrac{1}{2}\left(\tfrac{1}{2}h + y\right)t\left(\tfrac{1}{2}h - y\right) = \tfrac{1}{2}t\left[\left(\tfrac{1}{2}h\right)^2 - y^2\right]$$

$$Q_y = \int_{A_0} z\,dA = \tfrac{1}{2}hA_0 = \tfrac{1}{2}ht\left(\tfrac{1}{2}h - y\right)$$

From (1), the stress between 1 and 2 is given by

$$\tau = \frac{12}{7th^3}\left(\frac{h^2}{4} - 2hy + 3y^2\right)V_z$$

which is a parabola in y.

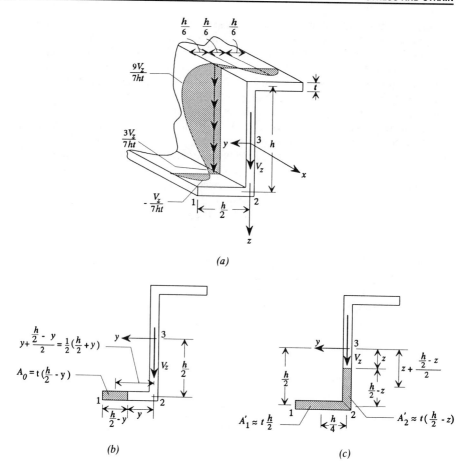

FIGURE 3-39   Example 3.9.

For a point in the web between 3 and 2 in Fig. 3-39c,

$$Q_z = \int_{A_0} y\, dA = \frac{h}{4}A_1 + (0)A_2 = \frac{h}{4}\left(t\frac{h}{2}\right) = t\frac{h^2}{8}$$

$$Q_y = \int_{A_0} z\, dA = \frac{h}{2}A_1 + \left(z + \frac{h/2 - z}{2}\right)A_2 = \frac{t}{2}\left(\frac{3h^2}{4} - z^2\right) \qquad (2)$$

$$\tau = \frac{24}{7th^3}\left(\frac{3}{8}h^2 - z^2\right)V_z$$

where A_1 is the area of the lower flange and A_2 is the area of that portion of the web beyond the point at which τ is calculated.

The distribution of shear stress is shown in Fig. 3-39a. The peak value of $9V_z/7ht$ occurs at 3, the centroid. If Eq. 5 in Table 3-3 were used to calculate the stress, the maximum value would occur at 3. Using (2), above

$$\tau = \frac{VQ}{Ib} = \frac{V_z Q_y}{I_y t} = \frac{\frac{1}{2} V_z t \left(\frac{3}{4} h^2 - z^2 \right)}{\left(\frac{1}{3} t h^3 \right) t} = \frac{3}{2 t h^3} \left(\frac{3}{4} h^2 - z^2 \right) V_z \qquad (3)$$

and at $z = 0$, $\tau_{max} = 9V_z / 8ht$. This is 12.5% in error relative to the more exact value found using Eq. (3.67).

3.15 THEORIES OF FAILURE

Concept of Failure

Structural members and machine parts may fail to perform their intended functions if excessive elastic deformation, yielding (plastic deformation), or fracture (break) occurs. For a failure safe design, the engineer must determine possible modes of failure of the structural or machine system and then establish suitable failure criteria that accurately predict the various modes of failure. The determination of modes of failure [3.8] requires extensive knowledge of the response of material or a structural system to loads. In particular, it may require a comprehensive stress analysis of the system. The mode of failure depends on the type of material used and the manner of loading, e.g., static, dynamic, and fatigue.

Two types of excessive elastic deformation result in structural failure:

1. Deformation satisfying the usual conditions of equilibrium, such as deflection of a beam or angle of twist of a shaft under gradually applied (static) loads. The ability to resist such deformation is referred to as the stiffness of a member. Furthermore, there can be excessive deformations associated with the amplitudes of the vibration of a machine member.

2. Buckling or an inordinately large displacement under conditions of unstable equilibrium that may occur in a slender column when the axial load exceeds the Euler critical load, in a thin plate when the in-plane forces exceed the critical load, or when the external pressure on a thin-walled shell exceeds a critical value. This is a form of instability.

To ascertain if it will serve its purpose, a load-carrying solid must be investigated from the standpoint of strength in addition to the possibility of the stiffness and stability failures considered above. A discussion of strength-related failure follows.

Yielding failure is due to plastic deformation of a significant part of a member, sometimes called extensive yielding in order to distinguish it from (localized) yielding of a small part of a member. Yielding under room and elevated temperatures is discussed in Chapter 4. Yielding occurs when the elastic limit (or yielding point) of the material has been exceeded. As indicated in Chapter 4, in a ductile metal under conditions of static loading at room temperature, yielding rarely results in fracture because of the strain-hardening effect. For simple tensile loading failure by excessive plastic deformation is controlled by the yield strength of the metal. However, for more complex loading conditions, the yield strength must be used with a suitable criterion, a "theory of failure," which is discussed later in this section. At temperatures significantly greater than room temperature, metals no longer exhibit significant hardening. Instead, metals can continuously deform at constant stress levels in a time-dependent yielding known as *creep*.

Members can fracture before failure defined by excessive elastic deformation or yielding can occur. The mechanisms of this fracture include the following.

1. Rapid fracture of brittle materials.
2. Fatigue or progressive fracture.
3. Fracture of flawed members.
4. Creep at elevated temperatures.

Fatigue deserves special attention because the magnitude of the repetitive load need not be high enough to cause static fracture; i.e., the stress may be relatively low. But under lengthy vibratory loading, fatigue cracks can form. Fatigue fracture is often ranked as the most serious type of fracture in machine design simply because it can occur under normal operating conditions. Fracture and fatigue are discussed in Chapter 7. Creep is discussed in Chapter 4. Failure theories for yield are treated in the next section. By replacing the yield stress by another critical stress level, e.g., the ultimate stress, these theories are often considered to be applicable to failures other than yield.

Tensile tests provide the most commonly available information about the failure level of a material. The problem arises when an attempt is made to relate this tensile data to a combined stress situation. In some combined stress cases tests can be performed to determine the yield stress. Usually it is not convenient, or even possible, to conduct a suitable model test; consequently, it is necessary to develop a relationship between stress under complicated stress conditions and the behavior of a material in simple tension or compression.

For the theories considered here, it is assumed that the tension or compression critical stresses σ_{ys} (yield stress) or σ_u (ultimate stress) are available. In developing the various failure criteria, it is convenient to use the fact that any state of stress can be reduced through a rotation of coordinates to a state of stress involving only the principal stresses σ_1, σ_2 and σ_3. Often these principal stresses are output by general-purpose structural analysis programs.

Maximum-Stress Theory In the maximum-stress, or Rankine, theory the maximum principal stress is taken as the criterion of failure. Failure, for the moment, is to be defined in terms of yielding, although the same theory applies if the yield stress is replaced by another stress level such as the ultimate stress. For the maximum-stress theory yield occurs when one of the principal stresses at a point in the structure subjected to combined stresses reaches the yield strength in simple tension (σ_{ys}) or compression for the material. According to this theory, yielding is not affected by the level of the other smaller principal stresses. Thus, for material whose tension and compression properties are the same, the failure criterion is defined as

$$\sigma_1 = \sigma_{ys} \quad \text{or} \quad |\sigma_3| = \sigma'_{ys} \tag{3.68}$$

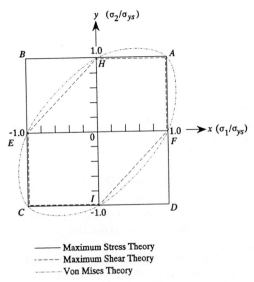

FIGURE 3-40 Graphical representation of theories of failure in two-dimensional state of stress.

where σ_{ys} and σ'_{ys} are the yield stresses in simple tension and compression, respectively. The principal stresses are so arranged that their algebraic values satisfy the relation $\sigma_1 > \sigma_2 > \sigma_3$.

Maximum-stress theory can be readily illustrated. For example, a graphical representation in a two-dimensional state of stress is shown in Fig. 3-40. The locus of failure points is the square $ABCD$.

Maximum-Strain Theory The maximum-strain theory, considered to be due to Saint Venant, postulates that a ductile material begins to yield when the maximum extensional strain reaches the yield strain in simple tension, or when the minimum strain (shortening) equals the yield point strain in simple compression. By means of Hooke's laws, for $\sigma_1 > \sigma_2 > \sigma_3$, this failure criterion is embodied in the equations

$$\sigma_1 - \nu(\sigma_2 + \sigma_3) = \sigma_{ys}, \qquad |\sigma_3 - \nu(\sigma_1 + \sigma_2)| = \sigma'_{ys} \qquad (3.69)$$

The maximum-strain theory is not considered to be reliable in many instances.

Maximum-Shear Theory The maximum-shear theory, or Tresca or Guest's theory, assumes that failure occurs in a body subjected to combined stresses when the maximum shear stress at a point, e.g., $\frac{1}{2}(\sigma_1 - \sigma_2)$, reaches the value of shear failure stress of the material in a simple tension test, e.g., $\frac{1}{2}\sigma_{ys}$. Therefore, failure under combined stresses is decided by the condition

$$\sigma_{max} - \sigma_{min} = \sigma_{ys} \qquad (3.70a)$$

where σ_{max} and σ_{min} are the maximum and minimum principal stresses, respectively.

The term $\sigma_{max} - \sigma_{min}$ can also be expressed as

$$\max(|\sigma_1 - \sigma_2|, |\sigma_2 - \sigma_3|, |\sigma_3 - \sigma_1|) \tag{3.70b}$$

The largest of these absolute values is sometimes referred to as the *stress intensity*. This quantity is often computed by general-purpose analysis software.

It is important to note that for the case $\sigma_1 > \sigma_2 > \sigma_3$, the failure criterion would be

$$\sigma_1 - \sigma_3 = \sigma_{ys} \tag{3.70c}$$

A plot of this theory for a two-dimensional state of stress is given in Fig. 3-40. The locus of failure points is the polygon *AHECIFA*.

Von Mises Theory The von Mises theory, also called the Maxwell–Huber–Hencky–von Mises theory, octahedral shear stress theory, and maximum distortion energy theory, states that failure occurs when the energy of distortion reaches the same energy for failure in tension. That is, failure takes place when the principal stresses are such that

$$(\sigma_1 - \sigma_2)^2 + (\sigma_2 - \sigma_3)^2 + (\sigma_1 - \sigma_3)^2 = 2\sigma_{ys}^2 \tag{3.71a}$$

This relation holds regardless of the relative magnitude of σ_1, σ_2, and σ_3.

The quantity

$$\left(\tfrac{1}{2}\left[(\sigma_1 - \sigma_2)^2 + (\sigma_2 - \sigma_3)^2 + (\sigma_1 - \sigma_3)^2\right]\right)^{1/2} = \sigma_e \tag{3.71b}$$

is often referred to as the *equivalent stress*. This is sometimes available as output of general-purpose structural analysis software.

In a two-dimensional state of stress ($\sigma_3 = 0$). Eq. (3.71a) becomes

$$\sigma_{ys}^2 = \sigma_1^2 + \sigma_2^2 - \sigma_1\sigma_2 \tag{3.72}$$

This relationship is plotted in Fig. 3-40.

Mohr's Theory Mohr's Theory, also called the Coulomb–Mohr theory or the internal-friction theory, is based on the results of the standard tension and compression tests, which give the tensile and compressive strengths σ_{ys} and σ_{ys}'. Two Mohr circles for these experiments can be plotted on the same diagram. A pair of lines *AB* and *CD* (Fig. 3-41) are drawn tangent to the two Mohr circles. Mohr's theory states that failure of an isotropic material, either by fracture or by the onset of yielding, will occur at a point where the largest Mohr circle for this point (having diameter $\sigma_1 - \sigma_3$, as in Fig. 3-41) touches a failure envelope. Any "interior" circle, such as the dotted one in Fig. 3-41, represents a state of stress

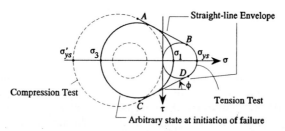

FIGURE 3-41 Mohr's theory of failure.

that is safe, while the solid circle represents a state of stress that is in failure. It can be shown that failure occurs when

$$\sigma_1/\sigma_{ys} + \sigma_3/\sigma'_{ys} \geq 1 \tag{3.73}$$

where $\sigma_{ys} > 0$ and $\sigma'_{ys} < 0$ and the maximum and minimum principal stresses σ_1 and σ_3 carry their algebraic signs. In plane stress problems, if all normal stresses are tensile, Eq. (3.73) coincides with the maximum-stress theory ($\sigma_1 \geq \sigma_{ys}$). For ductile materials, usually it is assumed that $\sigma_{ys} = -\sigma'_{ys}$, so that Eq. (3.73) becomes $\sigma_1 - \sigma_3 \geq \sigma_{ys}$.

Validity of Theories

The appropriate failure theory to be used in a given design situation depends on the mode of failure. A theory that works for ductile failure may not be appropriate for brittle failure. A single theory may not always apply to a given material because the material may behave in a ductile fashion under some conditions and in a brittle fashion under others (see Chapter 4). For the above theories, the material is assumed to be isotropic. These theories of failure pertain to material failure rather than to structural failure by such modes as buckling or excessive elastic deformation.

A comparison has been made [3.2] of experimental yield stresses for several metals under biaxial stress conditions with some of the above failure theories. The results, which are for room temperature and slow loading, seem to indicate somewhat better agreement with von Mises theory than with the maximum-shear theory.

The maximum-stress theory and the maximum-strain theory are often applicable to brittle failure of materials. The maximum-strain theory has been shown not to be reliable in many instances. The maximum-shear theory is applied frequently in machine design for ductile materials where $\sigma_{ys} = \sigma'_{ys}$. Maximum-shear theory has the advantage over the von Mises theory that the stresses are linear.

Mohr's theory is usually used for brittle materials, which are much stronger in compression than in tension, e.g., for cast iron.

3.16 APPLICATION OF FAILURE THEORIES

The following examples will illustrate the use of the failure theories discussed above.

Example 3.10 Internal Pressure of a Cylindrical Vessel A cylindrical pressure vessel 80 in. in diameter and 1 in. thick is made of steel with a yield stress in tension of 35,000 psi. Determine the internal pressure that will produce yielding by using the von Mises theory of failure as the yield criterion.

From the stress formulas for thin-walled pressure vessels presented previously, the principal stresses in a cylinder (Fig. 3-31) will be the circumferential stress σ_θ, the longitudinal stress σ_x, and the radial stress σ_r. Let $\sigma_1 = \sigma_\theta$, $\sigma_2 = \sigma_x$, and $\sigma_3 = \sigma_r$. Equations (3.61a, b) give

$$\sigma_1 = pr/t, \qquad \sigma_2 = pr/2t \tag{1}$$

The stress σ_r is small $(0 \le \sigma_r \le p)$ relative to σ_θ and σ_x and is neglected, i.e.,

$$\sigma_3 = 0 \tag{2}$$

Substituting (1) and (2) in Eq. (3.71a) gives

$$(\sigma_1 - \sigma_2)^2 + (\sigma_2 - \sigma_3)^2 + (\sigma_1 - \sigma_3)^2 = 2(\sigma_1^2 - \sigma_1\sigma_2 + \sigma_2^2)$$

$$= 2\left[\left(\frac{pr}{t}\right)^2 - \frac{pr}{t}\frac{pr}{2t} + \left(\frac{pr}{2t}\right)^2\right] = 2\sigma_{ys}^2$$

Therefore,

$$p = \sqrt{\frac{4}{3}\frac{t^2}{r^2}(\sigma_{ys}^2)} = \frac{2}{\sqrt{3}}\frac{t\sigma_{ys}}{r} = \frac{2}{\sqrt{3}}\frac{(1)(35,000)}{80/2} \approx 1010 \text{ psi} \tag{3}$$

According to von Mises theory, this p gives the pressure value that would initiate yielding of the cylinder.

If the maximum-stress theory of failure [Eq. (3.68)] and the maximum-shear theory [Eq. (3.70c)] are used, the internal pressure that will product yielding in both cases is

$$p = \frac{t\sigma_{ys}}{r} = \frac{1(35,000)}{80/2} = 875 \text{ psi} \tag{4}$$

Remember that $\sigma_3 = 0$ in establishing this relationship.

The results in (3) and (4) indicate that, in this case, using the maximum stress theory and maximum shear theory is more conservative than using the von Mises theory.

Example 3.11 Application of Tresca and von Mises Theories If the yield strength of a material in a tensile test is $\sigma_{ys} = 140$ MN/m^2, determine the largest safe shear stress τ in a cylinder of the same material in torsion.

In the simple tension test, the state of stress is $\sigma_1 = \sigma_{ys} = 140 \text{ MN/m}^2$, $\sigma_2 = \sigma_3 = 0$. From Eq. (3.70a),

$$\sigma_{max} - \sigma_{min} = \sigma_1 - 0 = \sigma_{ys} \tag{1}$$

For pure torsion of the cylinder $\tau = Tr/J$ (Table 3-3). The principal stresses are, by Eqs. (3.13),

$$\sigma_{max} = \sigma_1 = \tau, \qquad \sigma_{min} = \sigma_2 = -\tau, \qquad \sigma_3 = 0 \tag{2}$$

Therefore,

$$\sigma_{max} - \sigma_{min} = 2\tau \tag{3}$$

Use of the Tresca theory yields [see Eq.(3.70a)], from (1) and (3),

$$2\tau = \sigma_{ys} \quad \text{or} \quad \tau = \tfrac{1}{2}\sigma_{ys} = \tfrac{140}{2} = 70 \text{ MN/m}^2 \tag{4}$$

If the von Mises theory is to be used, the equivalent stress σ_e in Eq. (3.71b) is evaluated for the two states of stress. For simple tension

$$\sigma_e = \frac{1}{\sqrt{2}}\left(\sigma_{ys}^2 + \sigma_{ys}^2\right)^{1/2} = \sigma_{ys} \tag{5}$$

For torsion of the cylinder, from (2),

$$\sigma_e = \frac{1}{\sqrt{2}}\left(4\tau^2 + \tau^2 + \tau^2\right)^{1/2} = \sqrt{3}\,\tau \tag{6}$$

By the von Mises theory, equating (5) and (6)

$$\sqrt{3}\,\tau = \sigma_{ys} \quad \text{or} \quad \tau = 0.577\sigma_{ys} = 80.83 \text{ MN/m}^2 \tag{7}$$

Of course, the same result is obtained by applying Eq. (3.71a) directly.

The results in (4) and (7) indicate that for torsion of a cylinder, the Tresca (maximum-shear) theory is more conservative than the von Mises theory.

REFERENCES

3.1 Bosrei, A. P., and Sidebottom, O. M., *Advanced Mechanics of Materials*, 4th ed., Wiley, New York, 1985.

3.2 Blake, A., *Handbook of Mechanics, Materials, and Structures*, Wiley, New York, 1985.

3.3 Campus, F., "Effects of Residual Stresses on the Behavior of Structures," in *Residual Stresses in Metals and Metal Construction*, ed. by Osgood, W. R., Reinhold Publishing, 1954, pp. 1–21.

3.4 Almen, J. O., and Black, P. H., *Residual Stresses and Fatigue in Metals*, McGraw-Hill, New York, 1964.

3.5 Baldwin, W. M., Jr., "Edgar Marburg Lecture, Residual Stresses in Metals," *Proceedings*, American Society for Testing and Materials, Vol. 49, 1949, pp. 539–583.

3.6 Vande Walle, L. J., *Residual Stress for Designers and Metallurgists*, American Society for Metals, Metals Park, OH, 1981.

3.7 Rosenthal, D., in *Influence of Residual Stress on Fatigue*, ed. by G. Sines and J. L. Waisman, McGraw-Hill, New York, 1959, pp. 170–196.

3.8 Wulpi, D. J., *Understanding How Components Fail*, American Society for Metals, Metals Park, OH, 1985.

3

Tables

TABLE 3-1 MOHR'S CIRCLES FOR SOME COMMON STATES OF STRESS[a]

1.
Uniaxial compression
$(\sigma_x = \sigma_3)$,
$(\sigma_1 = \sigma_2 = 0)$

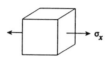

$\tau_{max} = \sigma_3/2$

2.
Uniaxial tension
$(\sigma_x = \sigma_1)$,
$(\sigma_2 = \sigma_3 = 0)$

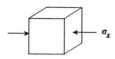

$\tau_{max} = \sigma_1/2$

3.
Pure shear
$(\sigma_x = -\sigma_y)$,
$(\sigma_x = -\sigma_y = \sigma_1 = -\sigma_3)$,
$(\sigma_2 = 0)$

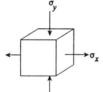

$\tau_{max} = \sigma_1$

4.
Pure shear
$(\tau_{xy} = \tau_{yx})$,
$(\sigma_2 = 0)$

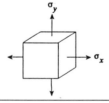

$\tau_{max} = \sigma_1$

5.
Equal biaxial tension
$(\sigma_x = \sigma_y = \sigma_1 = \sigma_2)$

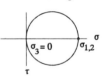

$\tau_{max} = \sigma_1/2$

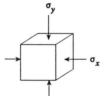

6.
Equal biaxial compression
$(\sigma_x = \sigma_y = \sigma_2 = \sigma_3)$

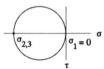

$\tau_{max} = \sigma_2/2$

TABLE 3-1 Mohr's Circles for Some Common States of Stress   136

TABLE 3-1 (continued)   MOHR'S CIRCLES FOR SOME COMMON STATES OF STRESS

7.
Equal tension (2 planes)
with lateral compression
$(\sigma_x = \sigma_y = -\sigma_z = \sigma_1 = \sigma_2 = -\sigma_3)$

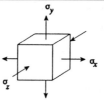

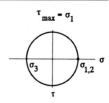

8.
Equal compression (2 planes)
with lateral tension
$(\sigma_x = -\sigma_y = -\sigma_z = \sigma_1 = -\sigma_2 = -\sigma_3)$

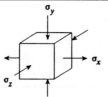

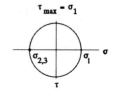

9.
Equal triaxial compression
$(\sigma_x = \sigma_y = \sigma_z = \sigma_1 = \sigma_2 = \sigma_3)$

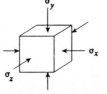

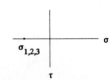

10.
Equal triaxial tension
$(\sigma_x = \sigma_y = \sigma_z = \sigma_1 = \sigma_2 = \sigma_3)$

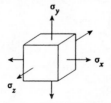

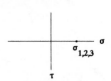

*From Blake [3.2], with permission.

TABLE 3-2 RELATIONSHIPS BETWEEN COMMONLY USED MATERIAL CONSTANTS

1. Shear modulus G (F/L^2)

$$G = \frac{E}{2(1 + \nu)}$$

2. Lamé coefficient λ (F/L^2)

$$\lambda = \frac{E\nu}{(1 + \nu)(1 - 2\nu)}$$

3. Bulk modulus K (F/L^2)

$$K = \frac{E}{3(1 - 2\nu)}$$

where E = modulus of elasticity (F/L^2)
ν = Poisson's ratio

TABLE 3-2 | Relationships Between Material Constants | 138

TABLE 3-3 BASIC STRESS FORMULAS

Bars of Linearly Elastic Material

1. Extension: $\sigma = P/A$

2. Torsion: $\tau = Tr/J$ (circular section)

3. Torsion: $\tau = T/(2\bar{A}t)$ (closed, thin-walled section)

4. Bending: $\sigma = Mz/I$

5. Shear: $\tau = VQ/(Ib)$

where σ = normal axial stress = σ_x
τ = shear stress
P = axial force
T = axial torque
V = vertical shear force = V_z
M = bending moment
   = in vertical plane = M_y
A in Eq. 1 = cross-sectional area
$\bar{A}$ in Eq. 3 = enclosed area

z = vertical coordinate
I = moment of inertia
   about neutral axis
J = torsional constant
b = width of bar
r = radius
Q = first moment with respect to
   neutral axis of area
   beyond point at which τ
   is calculated
t = wall thickness

Shells

6. Cylinder: $\sigma_\theta = pr/t$, $\sigma_x = pr/(2t)$

7. Sphere: $\sigma = pr/(2t)$

where σ_θ = hoop stress in cylinder wall
σ_x = longitudinal stress in cylinder wall
σ = membrane stress in sphere wall

p = internal pressure
t = wall thickness
r = radius

TABLE 3-4   BASIC DEFORMATION FORMULAS: BARS OF LINEARLY ELASTIC MATERIAL

1. Extension: $\Delta = PL/AE$

2. Torsion: $\phi = TL/GJ$

3. Bending: $\dfrac{d^4w}{dx^4} = \dfrac{p_z}{EI}$ $\theta = -\dfrac{dw}{dx}$ $M = -EI\dfrac{d^2w}{dx^2}$ $V = -EI\dfrac{d^3w}{dx^3}$

where A = original cross-sectional area (L^2)
Δ = elongation (L)
E = modulus of elasticity (F/L^2)
ϕ = angle of twist
G = shear modulus (F/L^2)
J = torsional constant (L^4)
L = original length (L)
I = moment of inertia about neutral axis (L^4)
P = axial force (F)
p_z = applied loading density (F/L)
T = torque (LF)
w = deflection (L)

TABLE 3-4   **Basic Deformation Formulas**   **140**

Mechanical Properties and Testing of Engineering Materials

Mechanical properties of materials are force–deformation (stress–strain) characteristics of materials. The American Society for Testing and Materials (ASTM) publishes annually many volumes of standards on characteristics and performance of materials, products, and systems. Section 3, Vol. 03.01 [4.1], of the ASTM standards applies to the physical, mechanical, and corrosion testing of metals; other volumes treat wood, plastic, rubber, cement, and other materials. Some of the specifications establish uniform standards for defining and measuring such material properties as tensile strength, offset yield strength, nil-ductility temperature, and fatigue strength. Definitions of terms relating to mechanical testing are provided in ASTM E6.

Examples of sources of values of material properties are _The Metals Handbook_, the materials reference issue of _Machine Design_ magazine, _The Materials Selector_ compiled by the publishers of _Materials Engineering_ magazine, and the product literature available from the companies that supply engineering materials. These and similar publications often include readable discussions of nomenclature, manufacturing processes, and the microstructural bases for the macroscopic behavior of materials. Texts on materials science [e.g., 4.2–4.4] offer more detailed insight into the atomic and molecular characteristics that account for the aggregate behavior of materials.

This chapter covers the tensile test, hardness tests, and impact tests as well as such mechanical properties as creep. A discussion of ferrous metals, some nonferrous metals, plastics, ceramics, and composites are also included.

Finally, tables of values of mechanical properties of various materials are presented. Data on material properties such as fracture toughness and fatigue strength are presented in Chapter 7. A discussion of important nonmetallic structural materials such as concrete, wood, and asphalt is available in several sources [e.g., 4.5].

4.1 NOTATION

A_0	Initial undeformed cross-sectional area of tensile specimen (L^2)
C	Constant in creep extension equation
C_L	Constant in equation for Larson–Miller parameter
d	Extension due to creep (L)
d_T	Total extension due to creep (L)
d_i	Extension under ith condition of stress and temperature (L)
E	Modulus of elasticity, Young's modulus (F/L^2)
F	Applied force in tension test
G	Shear modulus (F/L^2)
HB	Brinell hardness number
HK	Knoop hardness number
HR	Rockwell hardness number
HV	Vickers hardness number
ℓ	Specimen length in tensile test
ℓ_0	Initial underformed length of tensile specimen

n Constant exponent in creep extension equation

P_{MH} Manson–Haferd parameter

P_{LM} Larsen–Miller parameter

R Modulus of resilience (F/L^2)

T Temperature

t Time

T_a, t_a Constants in equation for Manson–Haferd parameter

t_i Time under ith condition of stress and temperature

t_{ki} Initial time for the kth condition of stress and temperature

Δ_i Time to failure under constant condition of ith stress and temperature

ε_e Engineering strain

ε_t True or natural strain

σ_a Amplitude of cyclic applied stress

σ_c Static creep strength

σ_e Engineering stress (F/A_0)

σ_f Fatigue strength

σ_m Mean stress level of cyclic applied stress

σ_t True stress (F/A)

σ_{ys} Yield stress (F/L^2)

4.2 THE TENSILE TEST

The tensile test serves as the basis for determining several important mechanical properties of materials. In this test, as described in ASTM E8 the yield strength, tensile strength, elongation, and reduction of area of a material specimen are determined. In addition, the modulus of elasticity, modulus of resilience, and modulus of toughness of a material are found from the stress–strain curve measured during the tensile test. (A different ASTM standard, E111 applies to the measurement of Young's modulus.) In the tensile test the specimen is loaded in uniaxial tension until the specimen fractures. Standards for testing machines, specimen types, testing speed, and determination of values of material properties are given in ASTM E8. A standard test specimen and a symbollic stress–strain curve are shown in Figs. 4-1a, b. Precise specifications of the speciman are provided in Ref. 4.1 Typical stress–strain curves obtained from tensile tests are shown in Fig. 4-2 for various metals and alloys. In these curves, stress (engineering) is defined as the applied force per unit original undeformed cross-sectional area of the specimen,

$$\sigma_e = F/A_0 \tag{4.1}$$

and strain is engineering strain,

$$\varepsilon_e = (\ell - \ell_0)/\ell_0 \tag{4.2}$$

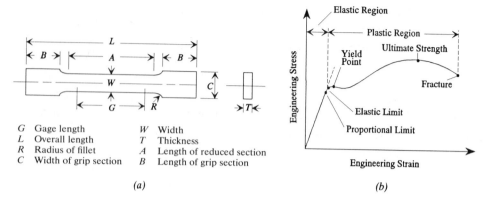

G Gage length   W Width
L Overall length   T Thickness
R Radius of fillet   A Length of reduced section
C Width of grip section   B Length of grip section

(a)   (b)

FIGURE 4-1 (a) Tensile test specimen [4.1]. Copyright ASTM. Printed with permission. (b) Conventional stress–strain diagram for a metal with a yield point.

The stress–strain curve of a metal will depend on many factors, such as chemical composition, heat treatment, prior plastic deformation, strain rate, and temperature.

Table 4-1 contains material properties of selected engineering materials.

As Fig. 4-2 shows, metals differ from each other in the shape of the stress–strain curve they produce under tensile testing; consequently, the definition of yield strength depends on the shape of the stress–strain curve. When the point at which plastic deformation begins is not clearly evident, the *offset yield strength* is employed. The construction used to find the offset yield strength of a material is shown in Fig. 4-3, where σ_{ys} is the offset yield strength and the offset is $0A$, usually taken as 0.1–0.2%. The offset yield strength is also known as the *proof strength* in Great Britain, with offset values of 0.1 or 0.5%. In reporting a value of the offset yield strength, the specified offset strain value is normally provided. Like yield strength, the offset yield strength is utilized for design and specification purposes. Because of the difficulty in determining the *elastic limit*, it is commonly replaced by the *proportional limit*, which is the stress at which the stress–strain curve is out of linearity. The *modulus of elasticity*, or *Young's modulus*, E, a measure of the stiffness of the material, is the slope of the curve below the proportional limit. The increase in load that occurs in some materials after the yield strength is reached is known as *strain hardening* or *work hardening*.

Poisson's ratio ν is the absolute value of the ratio of the transverse strain to the axial strain of a specimen under uniformly distributed axial stress below the elastic limit. The specimen for a Poisson's ratio tensile test is of rectangular cross section.

The *tensile strength* of the material is calculated by dividing the maximum applied load by the initial undeformed cross-sectional area of the specimen. For medium-carbon steels to high-carbon steels the tensile strength ranges from 45,000 psi to 140,000 psi. The tensile strength for duraluminum is 18,000 psi, for copper is 34,000 psi, and for acrylic polymer is 2000 psi. *Elongation* is the

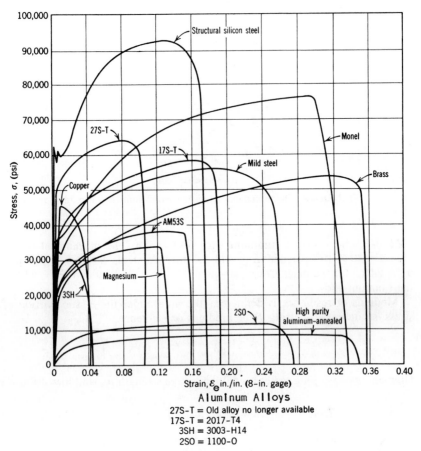

FIGURE 4-2 Engineering stress – strain curves for various metals [4.6]. Copyright © AIAA 1940. Used with permission.

percentage increase in specimen length over its initial length (the initial length is often marked on the specimen by two lines 2 in. apart). The *reduction of area* of a specimen is the maximum change in cross-sectional area at fracture expressed as a percentage of the original cross-sectional area. The *modulus of resilience* is the strain energy per unit volume absorbed up to the elastic limit for a tensile test and equals the area under the elastic part of the stress–strain curve. This quantity indicates how much energy a material can absorb without deforming plastically. For medium-carbon steels to high-carbon steels the modulus of resilience varies from 33.7 to 320. The modulus of resilience for duraluminum is 17, for copper is 5.3, and for acrylic polymer is 4.0. The *modulus of toughness* equals the total area under the stress–strain curve and measures the capacity of the material to absorb energy without fracturing.

Example 4.1   Modulus of Resilience   Compare the moduli of resilience of the American Iron and Steel Institute (AISI) 1020 carbon steel, extruded magnesium

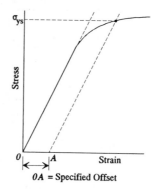

FIGURE 4-3   Determination of tensile yield strength by offset method.

AZ31B-F, and wrought[†] titanium alloy 5A1–2.5Sn. The mechanical properties are $E = 30 \times 10^6$ psi and $\sigma_{ys} = 48$ ksi for the steel, $E = 6.5 \times 10^6$ psi and $\sigma_{ys} = 24$ ksi for the magnesium alloy, and $E = 16 \times 10^6$ psi and $\sigma_{ys} = 117$ ksi for the titanium alloy.

Modulus of resilience is defined as the area under the elastic portion of the stress–strain curve. Letting R be modulus of resilience,

$$R = \tfrac{1}{2}\varepsilon\sigma_{ys} = \tfrac{1}{2}\left(\sigma_{ys}^2/E\right)$$

High-energy-absorbing materials will have high strength σ_{ys} and low stiffness (E),

Steel: $R = \tfrac{1}{2}(48 \times 10^3)^2/(30 \times 10^6) = 38.4$

Magnesium: $R = \tfrac{1}{2}(24 \times 10^3)^2/(6.5 \times 10^6) = 44.3$

Titanium: $R = \tfrac{1}{2}(117 \times 10^3)^2/(16 \times 10^6) = 427.8$

Note that this titanium alloy can absorb an order of magnitude more energy than the selected steel and magnesium alloys before plastic deformation is anticipated.

⸻

According to their ability to undergo plastic deformation under loading, materials are identified as being ductile or brittle. In a brittle material, fracture can occur suddenly because the yield strength and tensile strength are practically the same. The elongation and reduction of area give an indication of the ductility of a material specimen, and the modulus of toughness shows the energy-dissipating capacities of the material, but both ductility and capacity for energy absorption are influenced by such factors as stress concentration, specimen size, temperature, and strain rate. A normally ductile material such as mild steel will behave in a brittle manner under conditions of low temperature, high strain rate,

[†] Refer to the Section 4.6 to the discussion of carbon steels for a description of a wrought material.

and severe notching. On the other hand, normally brittle materials will behave ductily under high hydrostatic pressures and temperatures. Therefore, the assessment of the ductility and energy-absorbing capacity of a material must be made by taking into consideration the service conditions of the final product.

The curves shown in Fig. 4-2 are plots of engineering stress and strain; that is, stresses and strains are based on the undeformed dimensions of the specimen. Plots of true stress against true strain give a more realistic depiction of material behavior than do plots of engineering stress and strain. As mentioned in Chapter 3, true stress σ_t is defined as the applied load F divided by the instantaneous cross-sectional area A of the specimen at the time the load F is applied. True strain is defined by

$$\varepsilon_t = \int_{\ell_0}^{\ell} \frac{d\ell}{\ell} = \ln\left(\frac{\ell}{\ell_0}\right) \tag{4.3}$$

If the material is incompressible and the distribution of strain along the gage length is homogeneous, the true or natural stresses and strains are expressed in terms of the engineering stresses and strains as follows:

$$\varepsilon_t = \ln(1 + \varepsilon_e) \tag{4.4}$$

$$\sigma_t = \sigma_e(1 + \varepsilon_e) \tag{4.5}$$

These equations apply only until the onset of necking. A comparison of the two types of curves is shown in Fig. 4-4 for a low-carbon steel. In the true stress–true strain curve (also known as a flow curve) the curve increases continuously up to fracture.

4.3   IMPACT TESTS

In the previous section, the modulus of toughness of a material was defined as the area under the stress–strain curve obtained during the tensile test of a

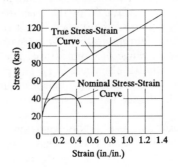

FIGURE 4-4 Comparison of engineering stress – strain curve with true stress – strain curve for mild steel.

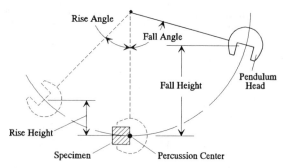

FIGURE 4-5 Charpy impact test.

specimen of the material. This modulus indicates how much energy the material can absorb in a uniaxial tensile test without fracturing. However, ferritic steels and other metals with body-centered-cubic crystallographic structures exhibit fracture behavior that cannot be deduced from the simple tensile test. Under conditions of stress concentration, low temperature, and high strain rates, these metals exhibit much less ductility than is indicated by the tension test. Several tests have been developed to measure the relative ability of materials to absorb energy under severe service conditions. Chief among these tests are the *notched-bar impact tests* (Charpy and Izod), the *drop-weight test for nil-ductility tempera-ture*, and the *dynamic tear energy test*. Another appropriate test is the *crack arrest test*.

Notched-Bar Tests

In Charpy (simply supported beam) and Izod (cantilever beam) tests, a notched material specimen is struck by a pendulum falling from a fixed height (Fig. 4-5). The energy lost by the pendulum in fracturing the specimen is the energy absorbed by the specimen before it fractures. This fracture energy obtained by the Charpy test is only a relative measure of energy and is difficult to use directly in design criteria. The test may be performed over a range of temperatures to determine the temperature at which the fracture changes from ductile to brittle. The transition point may be identified as occurring at a specified energy absorp-tion, by a change in the appearance of the fracture surface, or by a specified amount of lateral contraction of the broken bar at the root of the notch, which the specimen undergoes during testing. The Charpy test apparatus is more suitable for testing over a range of temperatures than is the Izod. The standard Charpy specimens are detailed elsewhere [4.1]. The pendulum shape, the pendu-lum head velocity, the system friction, the height of drop, and other details about the test are specified in the ASTM E23 standards. The Izod specimen is rarely used today.

The energy measured by these tests depends on specimen size, notch shape, and testing conditions. The energy values are most useful in comparing the impact properties of different materials or of the same material under different conditions. Plane strain fracture toughness (discussed in Chapter 7) may be measured for impact loads, and this quantity can be used to establish a design

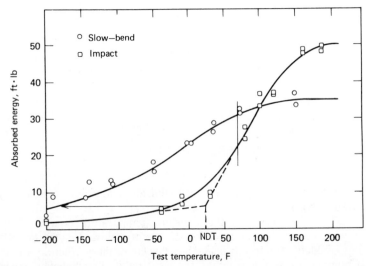

FIGURE 4-6 Typical Charpy-*V* notch curves for structural steel A36 [4.25]. (Slow-bend refers to a slow loading rate as compared with standard impact loading rates for CVN specimens.) Reprinted by permission of Prentice-Hall, Englewood Cliffs, New Jersey.

stress in a finished product. Empirical formulas have been proposed for computing values of plane strain fracture toughness from Charpy energy values, but
these formulas must be used with care [4.1]. The results of Charpy tests are
primarily useful in acceptance testing of materials when a correlation has been
established between energy values and satisfactory performance of a metal in the
finished product. Typical Charpy-V notch (CVN) curves for structural steels A36
are shown in Fig. 4-6. Note that the energy absorbed decreases with decreasing
temperatures, but for most cases the decrease does not occur sharply at a certain
temperature. Normally, the material with the lowest transition temperature is
preferred.

Drop-Weight Test for Nil-Ductility Temperature

The *nil-ductility transition temperature* (NDT) is defined as the maximum temperature at which brittle fracture occurs at a nominal stress equal to the yield point
when a "small" flaw exists in the specimen before loading.

   The drop-weight test (DWT) has been developed especially for determining
the NDT. The procedure for the NDT test is specified in ASTM E208. In the
test, a weight is dropped onto the compression side of a simple beam specimen
that has been prepared with a crack in a weld bead on the tension surface. The
brittle weld bead is fractured at near yield-stress levels as a result of the falling
weight. Because the specimen is a three-point bending beam and the anvil under
it (Fig. 4-7) restricts the deflection of the specimen ($D_C \leq D_A$), the stress on the
tension face of the specimen is limited to a value that does not exceed the yield
strength of the specimen material. Tests are conducted at various temperatures

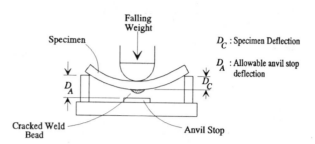

FIGURE 4-7   The NDT test method.

until the break and no-break points are found for yield point loading. This point establishes the NDT.

The NDT is also known as the transition temperature at which fracture is initiated with essentially no prior plastic deformation. Below the NDT the probability of ductile fracture is negligible. For larger flaws, fracture occurs at the NDT at nominal stresses lower than the yield point. This behavior is shown in Fig. 4-8, which shows a fracture analysis diagram for temperature, stress, and flaw size. Figure 4-8 also shows the *crack arrest temperature* (CAT) curve; at stress–temperature points to the right of the CAT curve, cracks will not propagate. The point labeled FTE in Fig. 4-8 is the *fracture transition elastic* point, the highest temperature at which a crack propagates in the elastic load range. Similarly, the *fracture transition plastic* (FTP) point is the temperature at which the fracture stress equals the material ultimate tensile stress. Above this temperature the material behaves as if it were flaw free. Thus, a crack, no matter how large, cannot propagate as an unstable fracture. The figure also shows a *stress limitation* curve; static loads below this level will not cause crack propagation in the absence of a corrosive environment.

The fracture analysis diagram has been established as a reliable means of predicting the fracture behavior of finished products under service conditions.

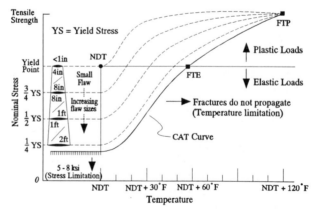

FIGURE 4-8   Fracture analysis diagram for temperature – stress-flaw size. Shown are initiation curves, indicating fracture stresses for spectrum of flaw sizes [4.1]. Copyright ASTM. Printed with permission.

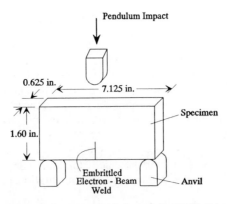

FIGURE 4-9 Dynamic tear energy test [4.1]. Copyright ASTM. Printed with permission.

Dynamic Tear Energy Test

The dynamic tear test is similar to the Charpy test, but the dynamic tear test uses a higher striking energy, a different size of specimen, and a sharper notch than the Charpy test (see ASTM E604). It is in effect a giant Charpy test (see Fig. 4-9). The dynamic tear test is used to measure energy absorption for crack progression that travels rapidly. A longer crack propagation distance is used in the dynamic test than in the Charpy. Use of the results of the dynamic test is similar to that of the Charpy.

4.4 HARDNESS TESTS

Hardness measures the resistance of a material to scratching, wear or abrasion, and indentation. For a given material a correlation exists between hardness and tensile strength; for example, the tensile strength of steel in psi is approximately 500 times the Brinell hardness number of the material. This approximation can provide a quick estimate of the tensile strength if the Brinell hardness number is available. The hardness test also serves to grade similar materials. A number of hardness tests have been developed of which the best known are the Brinell, Rockwell, Vickers, and Knoop.

Different methods are used to measure the area of indentation, and some tests measure the increase in area associated with a load increment rather than beginning at zero load (Table 4-2). Despite these differences, the basic principle for all tests is the application of a load to an indentor and the measurement of the size of the indentation.

Brinell Hardness

The test method for determining the Brinell hardness of metals is described in ASTM E10. This standard provides the details for two general classes of tests: referee tests (or verification, laboratory tests), in which a high degree of accuracy is required, and routine tests, for which a lower but adequate degree of accuracy is acceptable. The Brinell hardness test consists of indenting the metal surface with a 10 mm-diameter steel ball at a load of 3000 kgf (29 400 N). For a soft surface, to avoid a deep impression, the load is reduced to 1500 or 500 kgf (14 700 or 4900 N), and for very hard surfaces a carbide ball is employed to minimize distortion of the indentor. Different Brinell hardness numbers (BHN, or HB) may be obtained for a given material with different loads on the ball. The time interval over which the load is applied can influence the resulting hardness. The load is usually applied for 10 to 15 s in the standard test and for 30 s for soft metals.

Example 4.2   Hardness Test   An annealed aluminum alloy 7075-0 has a Brinell hardness of 60. What is the diameter of the indentation produced in the alloy during hardness testing?

For materials that are relatively soft such as aluminum, the test load is $P = 500$ kgf (4900 N). From the formula for Brinell hardness in Table 4-2 $HB = 2P/\{\pi D[D - (D^2 - d^2)^{1/2}]\}$ with $D = 10$ mm, we find $d = 3.21$ mm.

Vickers Hardness

The Vickers hardness number (HV) is defined as the load divided by the surface area of the indentation (ASTM E92). In practice, this area is determined by microscopic measurements of the length of the diagonals of the impression (Table 4-2). The test can be conducted on very thin materials. The indentor is a squared-based diamond pyramid with an angle of 136° (see Table 4-2). The load is varied over the range of 1–120 kgf (9.8–1176 N) according to the behavior of the thickness of the material. The HV test finds wide acceptance by researchers because it provides both accurate measurement and a continuous scale of hardness. The HV varies from 5 for a soft metal to 1300 (approximately 850 HB) for extremely hard metals.

Rockwell Hardness

The Rockwell hardness test is the most widely used hardness test in the United States, primarily due to the convenience of a test involving a small size of indentation. The hardness number, which is related inversely to the depth of the indentation under prescribed loading, may be read directly from a dial on the test apparatus. The standard test methods for the Rockwell hardness of metallic materials are specified in ASTM E18. This test has 15 scales, covering a rather complete spectrum of hardness (Table 4-3). Each scale has a specific indentor and major load. Indentors are either a steel ball of specified size or a sphero-conical diamond point. An initial load (called the minor load) of 10 kgf (98 N) is first applied that sets the indentor on the test specimen and holds it in position. The dial is set to zero on the black-figure scale, and the major load is applied. This major load is the total applied force. The depth measurement of the indentor depends only on the increase in depth due to the load increase from the minor to the major load. After the major load is applied and removed, according to standard procedure, the reading of the pointer is taken on the proper dial figures while the minor load is still in position. Rockwell hardness values are determined according to one of the standard scales (Table 4-3), not by a number alone. For example, 64 HRC means a hardness number of 64 on a Rockwell C scale.

Microhardness Test

The microhardness test is used to determine the hardness over very small areas or for ascertaining the hardness of a delicate machine part. The test is accomplished by forcing a diamond indentor of specific geometry under a test load of 1–1000 gf (0.0098–9.8 N) into the surface of the test material and to measure the diagonal or diagonals of indentation optically (ASTM E384-84). Usually the Knoop hardness number and the Vickers hardness number are used to represent microhardness.

Knoop hardness number (HK) is defined as the applied load P divided by the unrecovered projected area A_p of the indentation:

$$HK = P/A_p = P/(d^2 c) = 14.229 P/d^2$$

where d is the length of the long diagonal (mm), c is a manufacturer-supplied constant for each indentor, and P is the load (kgf).

Since the units normally used are grams-force and micrometers rather than kilogram-force and millimeters, the equation for the Knoop hardness number can be expressed conveniently as

$$HK = 14229 P_1/d_1^2$$

where P_1 = load, gf
d_1 = length of long diagonal, μm

The Vickers hardness number for microhardness is established using the same indentor defined previously, with loads varying from 1 to 1000 gf. Similarly it can be expressed as

$$HV = 1854.4 P_1/d_1^2$$

where P_1 = load, gf
d_1 = mean diagonal of indentation, μm

Tables for converting one hardness number to that found by a different method are available in ASTM E140.

4.5 CREEP

Creep is the occurrence of time-dependent strain in a loaded structural member, normally at elevated temperatures. In the case of metals, creep is thought to take place as the result of the competing processes of annealing due to high temperature and of work hardening caused by the load. Creep is variously attributed to grain boundary sliding and separation, vacancy migration, and dislocation cross-

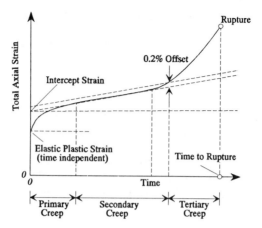

FIGURE 4-10   Typical constant temperature – stress creep curve.

slip and climb. Creep deformation continues until the part fails because of either excessive deformation or *creep rupture*. The temperature must usually be at least 40% of the melting point in degrees Kelvin for creep to occur in a metal. Little correlation exists between room temperature mechanical properties and creep properties. Creep tests are conducted by measuring deformation as a function of time when the load and temperature are held constant. It is frequently not practical to conduct full-life creep tests; however, creep tests should last a minimum of 10% of the expected life of the part under test.

The standard practice for conducting creep, creep rupture, and stress rupture tests of metallic materials is specified in ASTM E139. In the simplest creep test, a specimen is subjected to a constant uniaxial tension at constant temperature, and the strain is measured as a function of time. The test may proceed for a fixed time, to a specified strain, or to creep rupture. The results of a typical creep test are depicted in Fig. 4-10. The typical creep curve is divided into three stages, primary, secondary, and tertiary. In the first stage, the creep strain rate diminishes. In the secondary stage the strain rate is approximately constant; this constancy of strain rate is attributed to a balance between the hardening and softening processes. In the final or tertiary stage the strain rate increases until *creep rupture* occurs. Under severe conditions of loading or temperature the material may strain to the rupture point without exhibiting the secondary stage of creep behavior; this phenomenon is known as *stress rupture*. *Creep strength* is the minimum constant nominal stress that will produce a given strain rate of secondary creep under specified temperature conditions. The creep strengths of some metallic alloys are listed in Table 4-4.

A number of accelerated creep test procedures have been developed to shorten the time necessary to conduct creep tests. In one such accelerated test, specimens are tested at constant temperature, and strain is measured at various levels of constant stress for a fixed time. An acceptable design stress is found by extrapolating the curves to the desired life, as shown in Fig. 4-11. However, reliable extrapolation of this kind can be made only when no microstructural changes are anticipated that would produce a change in the slope of the curve.

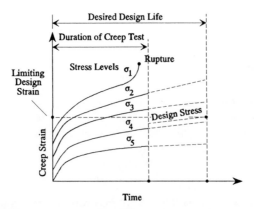

FIGURE 4-11 Constant-temperature abbreviated creep test.

Other accelerated test methods may vary strain level or temperature and measure stress as a function of time.

Two procedures for determining time–temperature test conditions that are equivalent for a given material and stress level are based on the Larsen–Miller parameter and the Manson–Haferd parameter. The Larsen–Miller parameter is

$$P_{LM} = (T + 460)(C_L + \log t) \tag{4.6}$$

where C_L is a constant, usually 20, T is the temperature in degrees Fahrenheit, and t is the test time in hours to a failure condition (either rupture or a specific level of strain).

Example 4.3 Accelerated Creep Test of a Steel Bar A bar made of 2.25Cr–1Mo (2.25% chromium, 1% molybdenum) steel must withstand 10,000 h at 1000°F with a tensile load of 15 ksi. For the same material and load find the temperature of an equivalent 24-h test.

From Eq. (4.6)

$$P_{LM} = (1000°F + 460)(20 + \log 10^4)$$
$$= 3.504 \times 10^4 \tag{1}$$

The equivalent temperature is found from

$$3.504 \times 10^4 = (T + 460)(20 + \log 24)$$
$$T = \frac{3.504 \times 10^4}{20 + \log 24} - 460 = 1179° \, F$$

The Manson–Haferd parameter is defined as

$$P_{MH} = (T - T_a)/(\log t - \log t_a) \tag{4.7}$$

where T_a and t_a are material constants and time is in hours. Values of these constants are listed in Table 4-5 for several materials.

Cumulative Creep

Several methods are available for analyzing creep behavior when the levels of stress and temperature vary with time. A linear law similar to the Palmgren Miner fatigue law (Chapter 7), has been suggested [4.12]. According to this approach, the creep failure point will be reached when the equation

$$\Sigma \frac{t_i}{\Delta_i} = 1 \tag{4.8}$$

is satisfied. The quantity t_i is the time of exposure to the ith level of stress and temperature; Δ_i is the time to failure if the point were subjected only to the ith level of stress and temperature.

In the *life-fraction* approach to cumulative creep the total creep strain is written as the sum of contributions from each level of stress and temperature:

$$\varepsilon_T = \varepsilon_1 + \varepsilon_2 + \cdots + \varepsilon_n \tag{4.9}$$

If a stress and temperature condition were applied from time zero to t_{1f}, strain ε_1 could be read from a creep–time plot for stress σ_1 and temperature T_1. When stress σ_2 and temperature T_2 act, the strain ε_2 is found from a creep–time plot for σ_2, T_2 beginning at time $t_{2i} = t_{1f}\Delta_2/\Delta_1$ and ending at time t_{2f} when a new condition of stress and temperature is applied. The creep for condition 3 is read from a creep–time curve for σ_3, T_3 beginning at $t_{3i} = t_{2f}\Delta_3/\Delta_2$ and ending at time t_{3f}. This procedure is repeated for all levels of stress and temperature. The method assumes that creep for each new load condition begins at the same fraction of the total life at the new load condition as was expended of the total lives under the previous load conditions. Thus, the effect of σ_2 and T_2 is assumed to begin at time t_{2i}, which is taken from $t_{2i}/\Delta_2 = t_{1f}/\Delta_i$.

Example 4.4   Cumulative Creep of a Tensile Specimen   A metallic alloy has the tensile creep properties shown in Fig. 4-12 at 1200°F. If a tensile specimen is loaded at 1200°F as

4000 psi for 5 h
3000 psi for 10 h
2000 psi for 20 h

and if the maximum tolerable creep strain is 0.004 in./in., determine if the specimen will complete the load cycle without reaching the maximum strain.

First we apply the cumulative creep relation of Eq. (4.8). From Fig. 4-12 the time to 0.004 in./in. strain at each load level is found to be

$$\Delta_1 = 12 \text{ h at } 4000 \text{ psi}$$
$$\Delta_2 = 26 \text{ h at } 3000 \text{ psi}$$
$$\Delta_3 = 53 \text{ h at } 2000 \text{ psi}$$

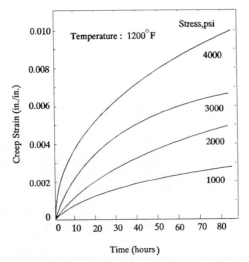

FIGURE 4-12 Creep curves for metal of Example 4.4.

Summing the fractions of creep life at each load gives

$$\frac{5}{12} + \frac{10}{26} + \frac{20}{53} = 1.18 > 1 \tag{1}$$

We conclude that failure occurs before the end of the test. By the cumulative creep law the part would reach the failure point at $t = (1 - \frac{5}{12} - \frac{10}{26})53 = 10.5$ h of the 2000-load period or 25.5 h after the beginning of the test.

Consider the same problem using the life-fraction rule. At the end of the first 5 h at 4000 psi the strain is taken from Fig. 4-12 to be $\varepsilon_1 = 0.0024$. The starting time for the second load is $t_{2i} = t_{1f}\Delta_2/\Delta_1 = 5 \times 26/12 = 10.8$ h. The strain at 3000 psi between 10.8 h and 20.8 h is read from Fig. 4-12 as $\varepsilon_2 = 0.0033 - 0.0024 = 0.0009$. The total strain after the second load period is $\varepsilon_1 + \varepsilon_2 = 0.0024 + 0.0009 = 0.0033$. The beginning time for the 2000 psi reading is $t_{3i} = t_{2f}\Delta_3/\Delta_2 = 20.8 \times 53/26 = 42.4$ h. The strain at 2000 psi between 42.4 and 62.4 h is from Fig. 4-12, $\varepsilon_3 = 0.0043 - 0.0035 = 0.0008$. The total strain for all three loads is $\varepsilon_T = 0.0024 + 0.0009 + 0.0008 = 0.0041$. Therefore, both methods predict failure, but the life-fraction method indicates the case is not as bad as the cumulative creep law shows. The life-fraction method is regarded as being superior to other schemes for estimating cumulative creep.

Simultaneous Creep and Fatigue

When a structural member is subjected to fluctuating loads at high temperatures (which may also fluctuate), the processes of creep and fatigue can occur simultaneously in the material. In an approach similar to the Goodman rule for fatigue (Chapter 7), failure for cyclic stress under creep temperature conditions occurs when the relation

$$\sigma_a/\sigma_f + \sigma_m/\sigma_c \geq 1 \tag{4.10}$$

is satisfied, where σ_a is the amplitude of the fluctuating component of the load, σ_m is the mean applied load, σ_f is the fatigue strength of the material with $\sigma_m = 0$, and σ_c is the static stress ($\sigma_a = 0$) that causes creep failure. A nonlinear version of Eq. (4.10) is sometimes applied for higher temperatures:

$$\left(\sigma_a/\sigma_f\right)^2 + \left(\sigma_m/\sigma_c\right)^2 \geq 1 \tag{4.11}$$

A number of methods have been proposed for dealing with combined creep and low cycle fatigue. A summary of these approaches, particularly the strain range partition method, is available elsewhere [4.13].

4.6 FERROUS METALS

The mechanical properties of metals and alloys depend in part on the characteristics of the pure elements from which they are made. See Table 4-6 for some properties of pure metals.

The ferrous metals consist of a group of iron alloys in which the principal alloying element is carbon. If the carbon content is at least 0.02% but not more than 2% by weight, the alloy is called *steel*. *Carbon steels* are steels in which the levels of manganese, silicon, and copper do not exceed 1.65, 0.60, and 0.60%, respectively, and there is no minimum specification for other alloying elements except carbon. *Wrought iron* is a very low carbon steel with slag inclusions. *Slag* results from the union of limestone with impurities in iron ore during the manufacture of pig iron. *Alloy steels* are steels in which the level of manganese, silicon, or copper exceeds the limits for carbon steels or in which other alloying elements are present in significant specified amounts. *Stainless steel* is an alloy steel that contains more than 10% chromium, with or without other elements. In the United States it has been customary to include with stainless steels those alloys that contain as little as 4% chromium. *Cast irons* contain more than 2% carbon and from 1 to 3% silicon. *High-strength, low-alloy steel* is a low-carbon steel that has been strengthened by the addition of manganese, cobalt, copper, vanadium, or titanium. *Tool steels* are high-alloy steels designed to have uniform properties of high strength and wear resistance. In addition, a number of specialty steels have been developed to give very high strength, e.g., maraging steels.

For a discussion of ferrous metals, several terms that describe the microstructure of steel must be defined: ferrite, cementite or iron carbide, pearlite, bainite, austenite, and martensite.

Ferrite is body-centered-cubic iron (α-iron) with a small amount of dissolved carbon. Ferrite is soft and ductile and gives steel good cold-working properties. *Body-centered-cubic* iron has a crystallographic structure composed of an iron atom at each of the eight corners of a cube plus one atom at the center of the cube. *Cold working*, which induces strain hardening, refers to mechanical deformation of a metal at ambient temperatures or at temperatures no more than one-half the recrystallization temperature. *Recrystallization* is a reversal of the effects of work hardening.

Cementite is iron carbide, Fe_3C; it is very hard and brittle at room temperature.

Pearlite is a lamellar structure of ferrite and cementite.

Bainite is a structure of ferrite and cementite in which the cementite is present in a needlelike form.

Austenite is a face-centered-cubic iron (γ-iron) with a maximum of 2% carbon in solution. The temperature at which austenite begins to form is called the lower critical temperature [1333.4°F (723°C) over most of the range of carbon content]. Face-centered-cubic iron has a crystallographic structure composed of iron atoms at the corners of a cube and one atom at the center of each of the six cube faces. The temperature at which the transformation to austenite is complete is called the upper critical temperature. The upper critical temperature varies with carbon content.

Martensite is a supersaturated solution of carbon in iron. Generally it is produced by rapidly cooling steel from above the upper critical temperature. Martensite has a body-centered-tetragonal crystal and is hard, strong, and brittle.

The microstructure of steel depends on the alloying content, the temperature, and the thermal and mechanical processing. The equilibrium phase diagram for the iron–carbon system is shown in Fig. 4-13. The regions of the diagram are labeled with the forms of material that exist under the specified conditions of temperature and carbon content. Below *the lower critical temperature* (723°C), the microstructure is either ferrite plus pearlite, or pearlite and cementite. Above the lower critical temperature, austenite begins to form. Figure 4-13 represents the equilibrium situation; the actual properties of steel depend on alloy content and the time–temperature treatments. Here equilibrium refers to thermodynamic equilibrium; that is, processes are done quasi-statically so that large gradients of temperature and concentration do not occur. In the following, terms that pertain to the heat treatment of steels (and in some cases, other materials as well) are defined.

Quenching involves heating the steel to the austenitic range and then cooling it rapidly, usually in water or oil or polymer solution, to form the martensite structure. Steel must contain at least 0.25% carbon to justify the quenching treatment.

Tempering involves heating quenched martensite steel to various temperatures below the lower critical temperature and cooling at a suitable rate to render the steel tougher, more ductile, and softer. The term *drawing* is sometimes used as a synonym for tempering.

Martempering is a process by which the steel is quenched to just above the temperature at which martensite forms, retaining it in the quenching medium until its temperature is uniform and then slowly cooling it through the martensite range. This process reduces the amount of distortion below that which occurs during normal quenching.

Austempering is similar to martempering except the steel is held above the martensite-forming temperature to form bainite; then the steel is slowly cooled.

Annealing is a process carried out on wrought or cast metals to soften them and improve ductility. Annealing of steel is usually done by heating the steel to near the lower critical temperature [1100–1400°F (593–760°C)] and then slowly cooling it in the furnace.

Normalizing is similar to annealing, but the steel is cooled at a higher rate in air rather than in a furnace, and the maximum temperature is usually about

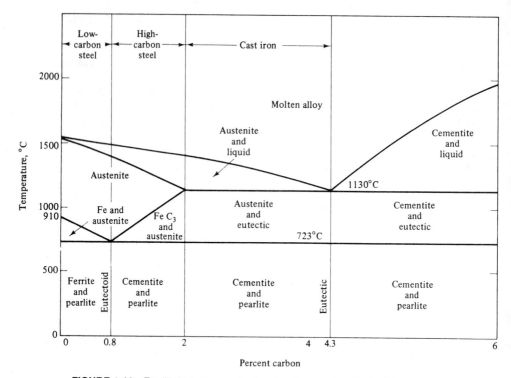

FIGURE 4-13 Equilibrium diagram for iron – carbon system [4.5], with permission.

100°F above the upper critical temperature. Normalizing is used to refine the grain structure as well as soften the material.

Spheroidize annealing produces the greatest softness by changing the shape of carbides to spheroidal. In the process the steel is held just below the lower critical temperature for an extended period of time.

Hardenability is the ease with which a uniform hardness can be attained throughout a material. The maximum hardness a heat-treated steel can reach is largely governed by its carbon content, and hardenability depends on other alloying elements in addition to carbon.

Hardening refers to increasing the hardness by suitable treatment, usually involving heating and cooling. In practice, more specific terms should be used such as surface hardening, age hardening, or quench hardening. Surface or case hardening treatments include carburizing, nitriding, flame hardening, and induction hardening. Surface hardening produces a hard wear-resistant case and leaves the core tough and ductile. The most well known of a number of tests for hardenability is the Jominy end quench test. In this test, a bar 1 in. in diameter and 4 in. long is heated above the critical temperature; one end is then quenched in water and the other end in air. The resulting hardness is measured along the bar length, and the hardness gradient is a measure of the hardenability of the steel. Some alloys of steel can be strengthened by *precipitation* or *age hardening*. In this process, the metal is held at a temperature in which a second phase precipitates in a supersaturated matrix phase. The precipitated second phase creates pinning locations that interfere with dislocation motion and so harden the material. Strain hardening techniques such as shot peening are also used, especially to increase surface resistance of materials to fatigue. In *shot peening*, a stream of high-velocity metal pellets impinge on the surface of the material and induce residual compressive stresses in the surface layer of the peened metal.

Steel Classification and Specifications

Steel can be classified on the basis of (a) chemical composition; (b) finishing methods, such as hot rolled or cold rolled; or (c) product form, such as bar, plate, sheet, strip, tubing, or structural shape. Classification by product form is very common in the steel industry.

Terms such as *grade*, *type*, and *class* are used to classify steel products in the steel industry. *Grade* is utilized to indicate chemical compositions, *type* indicates the deoxidation practice, and *class* describes some other attribute, such as strength level or surface smoothness. But in the ASTM specifications, these terms are used somewhat interchangeably. A *specification* is a written statement of attributes that a steel must possess in order to meet a particular application. A *standard specification* is a published document that describes a product acceptable for a wide range of applications and that can be produced by many manufacturers. The most comprehensive and widely used standard specifications are those of ASTM.

Designation is the specific identification of each grade, type, or class of steel by a number, letter, symbol, name, or suitable combination unique to a particular steel. Chemical composition is the most widely used basis for designation followed by mechanical property specifications. The most commonly used system of designation in the United States is that of the SAE (Society of Automotive Engineers) and the AISI (American Iron and Steel Institute). The AISI–SAE (or AISI) designations for the compositions of carbon and alloy steels are normally incorporated into the ASTM specifications for bars, wires, and billets for forging. Table 4-7 lists some of the ASTM specifications that incorporate AISI–SAE designations for compositions of the different grades of steel.

An ASTM specification consists of the letter A (for ferrous materials) and an arbitrarily assigned number. Many of the ASTM specifications have been adopted by the ASME (American Society of Mechanical Engineers) with little or no modification. ASME uses the prefix S and the ASTM specifications, e.g., ASME SA-213 and ASTM A213 are identical.

Several ASTM specifications, such as A29, contain the general requirements common to each member of a broad family of steel products. Such specifications are referred to as *generic specifications*. Table 4-8 lists several of these generic specifications, which usually must be supplemented by another specification describing a specific mill for an intermediate fabricated product.

Carbon Steels

In general, low-carbon steels have a range of carbon from 0.06 to 0.25%, medium-carbon steel from 0.25 to 0.55% carbon, and high-carbon steel above

0.55% carbon. The AISI and SAE designate many classes of carbon steels:

10xx is plain carbon with 1% maximum manganese
11xx is resulfurized steel
12xx is resulfurized and rephosphorized
15xx is plain carbon with 1–1.65% manganese

Note that the numerical designation for these carbon steels begins with 1. The second digit in some instances suggests a modification in the alloy. The xx refers to the percentage of carbon present; for example, a structural steel AISI 1020 contains 0.20% carbon. In addition to the four digits, there are various letter prefixes and suffixes that provide additional information on a particular steel, e.g., prefix E means steel made in an electric furnace. Plain carbon steels are cheaper than other types of steel, and they are the most widely used. The relatively poor hardenability of carbon steels prevents its use in many applications; the severe quenching required in hardening causes residual stresses, distortion, and sometimes quenching cracks. Manganese steel (13xx) that contains 1.75% manganese is also classed with the plain carbon steels. Manganese improves the hot-working properties of steel, and the hardness of manganese steels increases greatly with cold work. Hot working is done at a temperature at which recrystallization begins during or immediately after mechanical forming. Metals formed by mechanical deformation are referred to as *wrought*; examples of forming processes are rolling, extruding, and drawing. Metals formed by solidification of liquid metal in a mold are called *cast*. Steel is resulfurized or rephosphorized to improve machinability; these steels have poor weldability. Carbon steels with lead or boron added will contain L or B, respectively, between the second and third letters of the identifier. Boron increases the hardenability of steel without reducing its ductility, and lead improves machinability. Carbon steels may variously be described as cast, hot rolled, cold drawn, annealed, normalized, or quenched and tempered. The mechanical properties of some carbon steels are listed in Table 4-9. It should be noted that the material properties described in this book are associated with both chemical composition and heat treatment or processing conditions. In each table, the related heat treatment or processing conditions are given.

Alloy Steel

Alloys are added to steels to improve strength, hardenability, or some other properties such as machinability or toughness. The AISI designation for alloys by broad category is partly as follows:

Manganese steels
13xx
Nickel steels
23xx
25xx

Nickel–chromium steels
  31xx
  32xx
  33xx
  34xx
Molybdenum steels
  40xx
  44xx
Chromium–molybdenum steels
  41xx
Nickel–molybdenum steels
  46xx
  48xx
Chromium steels
  50xx
  51xx
  52xx
Chromium–vanadium steels
  61xx

For example, AISI 4140 has a nominal alloy content of 0.40% carbon, 0.80% chromium, and 0.25% molybdenum. The last two digits of the designator give the nominal carbon content of the alloy (percentage × 100). Additional categories of alloy steels exist, and the full classification is described elsewhere [4.15]. Table 4-10 lists the mechanical properties of some alloy steels.

Stainless Steels

The dividing line between chromium as an alloying element in steel and chromium as a corrosion inhibitor is about 10%. The stainless steels fall into five classes: austenitic, ferritic, martensitic, precipitation hardening, and duplex stainless steels. The AISI and UNS (Unified Numbering System) designations for stainless steels are shown in Fig. 4-14. Table 4-11 lists the mechanical properties of some stainless steels.

AISI uses the three-digit system to identify wrought stainless steels. The first digit indicates the classification by composition type. The 300 series is chromium–nickel alloys; the 400 series is straight chromium alloys.

Austenitic stainless steels such as 304, which is 19% chromium and 10% nickel, contain austenite, which the nickel and manganese present make stable at all temperatures. The austenitic stainless steels have good corrosion resistance and toughness. Type 304 is used widely in the chemical processing industry. Type 301 is strengthened more by cold work than is 304; type 347 is recommended for welding applications.

Ferritic stainless steels are not as tough and corrosion resistant as the austenitic, and they are difficult to harden by heat treatment or cold work. Typical applica-

AISI	UNS	AISI	UNS
		Austenitic	
201	S20100	310	S31000
202	S20200	310S	S31008
301	S30100	314	S31400
302	S30200	316	S31600
302B	S30215	316L	S31603
303	S30300	316F	S31620
303Se	S30323	316N	S31651
304	S30400	317	S31700
304L	S30403	317L	S31703
	S30430	321	S32100
304N	S30451	330	N08330
305	S30500	347	S34700
309	S30900	348	S34800
309S	S30908	384	S38400
		Ferritic	
405	S40500	430FSe	S43023
429	S42900	434	S43400
430	S43000	436	S43600
430F	S43020	442	S44200
		446	S44600
		Martensitic	
403	S40300	420F	S42020
410	S41000	431	S43100
416	S41600	440A	S44002
416Se	S41623	440B	S44003
420	S42000	440C	S44004

UNS is the unified numbering system;
AISI is the American Iron and Steel Institute

FIGURE 4-14   The AISI and UNS designations for some stainless steels [4.16].

tions of ferritic stainless steels are kitchen utensils, automotive trim, and high-temperature service. The chromium content ranges from 11.5 to about 28%.

Martensitic stainless steels have a high degree of hardenability but have lower corrosion resistance than ferritic or austenitic grades. Martensitic stainless steel is used when moderate corrosion resistance is needed along with high strength and hardness. In this group, the chromium range is from about 11.5 to 18.0%.

Cast stainless steels have a separate designation system from the wrought alloys. The most commonly used identification system is that of the Alloy Casting Institute. The ASTM provides information on the properties of cast stainless steel. Heat-resistant types have an H in the identifier and corrosion-resistant types are denoted by a C.

Cast Irons

Cast irons are alloys of iron that contain more than 2% carbon and from 1 to 3% silicon. They are divided into four basic types: gray iron, white iron, ductile iron, and malleable iron. Properties of some cast irons are listed in Table 4-12.

Gray iron contains flake graphite dispersed in the steel matrix. It is specified by two class numbers (ASTM A48) that are related to the tensile strength in ksi. Class 40, for example, has a minimum tensile strength of 40,000 psi. These specifications are based on test bars. In practice, the letters A, B, C, and S indicate the size of the tensile specimen used in measuring the tensile strength. Gray iron is cheap, is easy to cast and machine, and is wear resistant and has good vibration damping qualities. The compressive strength of gray cast iron is much larger than its tensile strength. The ductility of gray iron is very low, and it is difficult to weld and has fair corrosion resistance.

White iron contains massive iron carbides and is hard and brittle. White iron can be produced by rapidly cooling a casting containing gray or ductile iron. The white iron is wear and abrasion resistant. By controlling the cooling rate, the cast part may be produced with a white iron surface region and a core of gray or ductile iron that is tough and machinable. The most common designation system used for white iron is that of A27 of ASTM.

Ductile, or *nodular*, *iron* is alloyed with magnesium to produce spheroidal graphite dispersed in the steel structure. This shape of graphite increases the tensile strength and greatly increases the ductility over that of gray iron. Ductile iron is specified by three hyphenated numbers that give the minimum tensile strength, yield strength, and elongation. Type 80-55-06, for example, has a minimum tensile strength, yield strength, and elongation in 2 in. of 80,000 psi, 55,000 psi, and 6%, respectively. The vibration damping capacity and thermal conductivity of ductile iron is lower than that of gray iron. ASTM specification A536 covers some ductile iron grades. There are additional ASTM specifications on special-purpose ductile irons (A476, A716, A395, and A667) and on austenitic ductile irons (A439 and A571).

Malleable iron is produced by heat treating white iron to change the carbide to clumplike graphite sometimes called temper carbon. Malleable iron is stronger and more ductile than gray iron. Malleable iron has good impact and fatigue resistance, wear resistance, and machinability. Malleable irons may be ferritic, pearlitic, or martensitic in microstructure. The designation system in ASTM specifications (A47) is a five-digit number corresponding to certain mechanical properties.

High-Strength, Low-Alloy Steels

High-strength, low-alloy (HSLA) steels are low-carbon steels with small amounts of alloying elements added. These steels were developed primarily to replace plain low-carbon steels by providing equivalent loading–carrying ability with a lower weight of material. The HSLA steels have improved formability and weldability over conventional low-alloy steels. The ASTM designation number followed by the strength grade desired is used to specify these HSLA steels. The ASTM specifications A242, A440, A441, and A588 cover these steels as structural shapes. A typical designation is, e.g., steel, ASTM A242, grade 70.

Tool Steels

Tool steels have high hardness and wear resistance, often even at elevated temperatures. The AISI has established a classification system primarily based upon use. The seven broad categories of tool steels are as follows: (1) type W, water-hardening steel; (2) type S, shock-resistant steels; (3) types O, A, and D, cold-working die and tooling steels (type O is oil hardening and types A and D are air hardening); (4) type H, hot-working steels; (5) types T and M, high-speed steels; (6) types L and F, low-alloy and carbon tungsten specialty steels, respectively; and (7) types P, L, and F, special-purpose steels, e.g., for molds and dies. Each type of tool steel has its particular advantages and disadvantages, and selection must be made on that basis [4.10].

4.7 NONFERROUS METALS

Aluminum

Wrought aluminum alloys are specified by a four-digit number followed by a temper designation. The first digit indicates the alloying element, as is shown in Table 4-13. The letter H following the four-digit number indicates a strain hardening process with or without subsequent heat treatment. The letter T indicates heat treatment to cause age hardening. The partial temper code is listed in Table 4-14. Cast aluminum alloys have the same temper designation as wrought, but the designation numbering system is as shown in Table 4-15. Cast alloys are not strain hardened. Table 4-16 lists the mechanical properties of some wrought aluminum alloys.

Magnesium

Magnesium alloys are specified by two letters that designate the principal alloying elements followed by numbers that specify the amounts of the two principal alloying elements. The code AZ61 refers to a 6% aluminum, 1% zinc alloy of magnesium. The code letters used to designate these alloys are as follows:

A—Aluminum	Q—Silver
E—Rare earths	S—Silicon
H—Thorium	T—Tin
K—Zirconium	Z—Zinc
M—Manganese	

The temper designations for magnesium are identical to those for aluminum. Magnesium is difficult to cold work, and these operations should be avoided if possible. Table 4-17 lists the mechanical properties of some magnesium alloys.

Other Nonferrous Metals

Table 4-18 lists ranges of values of mechanical properties for several other nonferrous metals.

4.8 PLASTICS

Most plastics are composed of macromolecules that are polymers. These polymers are large molecules formed by joining together many smaller molecules. The mechanical behavior of plastics is quite different from that of metals. For example, plastics can continue to deform even after an imposed stress is removed. Such time-dependent behavior is termed *viscoelasticity*. Continued deformation with time can limit stresses to values significantly lower than the short-term loading allowable stresses. The designer should take the stress–time history phenomena of a plastics part into account in addition to the factors considered for metals. A further challenge for a designer arises due to the temperature-dependent mechanical properties of most plastics. In the range between -50 and $150°C$, many plastics show significant changes in material properties. A *glass transition temperature* (T_g) can be identified when plastics change during cooling from a rubbery material to a brittle state. Normally, the highest strength plastics are brittle at $20°C$.

In addition to relative ease of molding and fabrication, many plastics offer a range of important advantages in terms of high strength–weight ratio, toughness, corrosion and abrasion resistance, low friction, and excellent electrical resistance. Thus many plastics are now accepted as regular engineering materials and given the loose designation *engineering plastics*. ASTM [4.19] has established some standards for plastics. Refer to the literature [4.17] for the significance and theoretical background of the standard test for mechanical properties of plastics.

Since so many factors influence the behavior of plastics, mechanical properties (such as moduli) quoted as a single value will be applicable only for the conditions at which they are measured.

Table 4-19 lists the commonly used abbreviations for designating engineering plastics. Some typical mechanical properties of plastics are given in Table 4-20. Poisson's ratio for many brittle plastics (such as polystyrene, the acrylics, and the thermoset materials) is about 0.3. For the more flexible plasticized materials (e.g., cellulose acetate), Poisson's ratio is about 0.45. For rubber the value is 0.5. Poisson's ratio varies not only with the material itself but also with the magnitude of the strain for a given material. The Poisson ratios mentioned here are for zero strain [4.19].

Glass transition temperatures for some plastics are listed in Table 4-21.

4.9 CERAMICS

Generally speaking, there is no clear-cut boundary that separates ceramics and metals. Rather, there are intermediate compounds that behave in some aspects like ceramics and in others like metals [4.20]. In fact, *ceramics* are compounds of metals and nonmetals, or simply, ceramics are inorganic and nonmetallic solids.

Ceramic materials have become increasingly important in modern industrial and consumer technology. The traditional ceramic materials include clay products (china, brick, tile, refractories, abrasives), cement, enamels, and glasses. Traditional ceramics are brittle materials since they exhibit quite low ductility with an associated low tensile strength. They are used for furnaces and linings of furnaces, where the ceramics are often only lightly stressed, unless loading is consistently compressive. New types of ceramics are being developed for uses like gas turbines, jet engines, sandblast nozzles, nuclear plants, and high-temperature heat exchangers for which a service temperature of 1000°C and higher may be required. The carbides, borides, and nitrides of the transition elements [like silicon (Si) and magnesium (Mg)] are some examples of the new ceramics. Even at ordinary temperatures the new ceramics are widely used for their hardness and wear resistance.

A few ceramics consist of crystalline phases surrounded by glassy binders. If the ceramic content in such mixtures is reduced, hybrid materials with metallic alloys strengthened by refractory particles, called *cermets*, are obtained.

Tables 4-1, 4-22, and 4-23 list some mechanical properties of ceramics.

4.10 COMPOSITES

It has long been recognized that two or more materials judiciously combined can perform differently and sometimes more efficiently than the materials by themselves. Such combinations can occur on three different levels: a basic or elemental level at which single molecules and crystal cells are formed; a microstructural level for which crystals, phases, and compounds are formed; and a macrostructural level at which matrices, particles, and fibers are considered.

In general, a composite material is composed of reinforcements (e.g., fibers, particles, laminae or layers, flakes, fillers) embedded in a matrix (e.g., metals, polymers, ceramics). The constituents to hold the reinforcements together to form some useful shape are referred to as a *matrix*. Based on the form of the reinforcements, composite materials can be classified as follows:

1. Fiber composites, composed of continuous or chopped fibers.
2. Particulate composites, composed of particles.
3. Flake composites, composed of flat flakes.
4. Laminar composites, composed of layers or lamina constitutents.
5. Filled or skeletal composites, composed of continuous skeletal matrix filled by a second material.

These five types of composites are illustrated in Fig. 4-15.

Table 4-24 lists the mechanical properties of some composite materials. For more data and design theories for anisotropic materials, see references 4.22 and 4.23.

The mechanical properties of a composite material are often direction dependent (or anisotropic). ASTM provides numerous publications concerning the measurements of composites [4.24].

4.11 MATERIAL LAWS: STRESS – STRAIN RELATIONS

The stress–strain relations for isotropic materials are discussed in Chapter 3. An isotropic material has identical mechanical, physical, thermal, and electrical

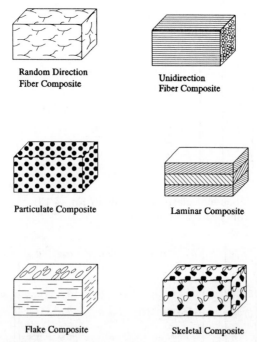

Random Direction
Fiber Composite

Unidirection
Fiber Composite

Particulate Composite

Laminar Composite

Flake Composite

Skeletal Composite

FIGURE 4-15 Various types of composite materials.

properties in every direction. However, an anisotropic material exhibits direction-dependent properties. An isotropic material may become anisotropic due to cold working and forging. Composite materials usually are anisotropic.

For an *isotropic material*, the constitutive relation [or Hooke's law, Eq. (3.40)] in matrix notation can be expressed as

$$
\begin{bmatrix} \sigma_x \\ \sigma_y \\ \sigma_z \\ \tau_{xy} \\ \tau_{xz} \\ \tau_{yz} \end{bmatrix} = \frac{E}{(1+\nu)(1-2\nu)} \begin{bmatrix} 1-\nu & \nu & \nu & & 0 & \\ \nu & 1-\nu & \nu & & & \\ \nu & \nu & 1-\nu & & & \\ & & & \dfrac{1-2\nu}{2} & 0 & 0 \\ & 0 & & 0 & \dfrac{1-2\nu}{2} & 0 \\ & & & 0 & 0 & \dfrac{1-2\nu}{2} \end{bmatrix} \begin{bmatrix} \varepsilon_x \\ \varepsilon_y \\ \varepsilon_z \\ \gamma_{xy} \\ \gamma_{xz} \\ \gamma_{yz} \end{bmatrix}
$$

$$ \boldsymbol{\sigma} \quad = \quad\quad\quad\quad\quad\quad\quad\quad\quad \mathbf{E} \quad\quad\quad\quad\quad\quad\quad\quad\quad \boldsymbol{\varepsilon} $$

$$ (4.12) $$

where the stresses are shown in Fig. 3-2 of Chapter 3.

In the most general case, for an *anisotropic material* all components in the matrix of the constitutive law are nonzero, but the symmetry still holds.

$$
\begin{bmatrix} \sigma_x \\ \sigma_y \\ \sigma_z \\ \tau_{yz} \\ \tau_{xz} \\ \tau_{xy} \end{bmatrix}
=
\begin{bmatrix}
c_{11} & c_{12} & c_{13} & \bar{c}_{14} & \bar{c}_{15} & \bar{c}_{16} \\
 & c_{22} & c_{23} & \bar{c}_{24} & \bar{c}_{25} & \bar{c}_{26} \\
 & & c_{33} & \bar{c}_{34} & \bar{c}_{35} & \bar{c}_{36} \\
 & \text{symmetric} & & c_{44} & \bar{c}_{45} & \bar{c}_{46} \\
 & & & & c_{55} & \bar{c}_{56} \\
 & & & & & c_{66}
\end{bmatrix}
\begin{bmatrix} \varepsilon_x \\ \varepsilon_y \\ \varepsilon_z \\ \gamma_{yz} \\ \gamma_{xz} \\ \gamma_{xy} \end{bmatrix}
\tag{4.13}
$$

There are 21 independent elastic constants in Eq. (4.13).
The strains in terms of stresses appear as

$$
\begin{bmatrix} \varepsilon_x \\ \varepsilon_y \\ \varepsilon_z \\ \gamma_{yz} \\ \gamma_{xz} \\ \gamma_{xy} \end{bmatrix}
=
\begin{bmatrix}
a_{11} & a_{12} & a_{13} & \bar{a}_{14} & \bar{a}_{15} & \bar{a}_{16} \\
 & a_{22} & a_{23} & \bar{a}_{24} & \bar{a}_{25} & \bar{a}_{26} \\
 & & a_{33} & \bar{a}_{34} & \bar{a}_{35} & \bar{a}_{36} \\
 & \text{symmetric} & & a_{44} & \bar{a}_{45} & \bar{a}_{46} \\
 & & & & a_{55} & \bar{a}_{56} \\
 & & & & & a_{66}
\end{bmatrix}
\begin{bmatrix} \sigma_x \\ \sigma_y \\ \sigma_z \\ \tau_{yz} \\ \tau_{xz} \\ \tau_{xy} \end{bmatrix}
\tag{4.14}
$$

A material whose properties vary in three orthogonal directions is called *orthotropic*. In this case, the barred quantities in Eqs. (4.13) and (4.14) are zero so that there are nine independent elasticity constants. The number of elastic coefficients present in both two- and three-dimensional elastic bodies and the corresponding elastic constant matrices are listed in Table 4-25.

The material constants in Eqs. (4.13) and (4.14) for orthotropic materials can be identified by performing simple tensile and shear tests. A tensile test in the x direction provides

$$
a_{11} = \frac{1}{E_x}, \qquad a_{21} = \frac{-\nu_{xy}}{E_x}, \qquad a_{31} = \frac{-\nu_{xz}}{E_x} \tag{4.15a}
$$

where E_x is the elasticity modulus in the x direction. The constants ν_{xy} and ν_{xz} are Poisson's ratios in the x, y and x, z planes, respectively. Similarly, tensile tests in the y and z directions yield

$$
a_{12} = -\frac{\nu_{yx}}{E_y}, \qquad a_{22} = \frac{1}{E_y}, \qquad a_{32} = -\frac{\nu_{yz}}{E_y} \tag{4.15b}
$$

$$
a_{13} = -\frac{\nu_{zx}}{E_z}, \qquad a_{23} = -\frac{\nu_{zy}}{E_z}, \qquad a_{33} = \frac{1}{E_z} \tag{4.15c}
$$

Since $a_{ij} = a_{ji}$, it is seen that

$$\frac{v_{ij}}{E_i} = \frac{v_{ji}}{E_j} \qquad (i, j = x, y, z) \qquad (4.15d)$$

A shear test [4.23] can provide the remaining constants a_{44}, a_{55}, and a_{66}:

$$a_{44} = \frac{1}{G_{yz}}, \qquad a_{55} = \frac{1}{G_{xz}}, \qquad a_{66} = \frac{1}{G_{yx}} = \frac{1}{G_{xy}} \qquad (4.15e)$$

Materials having the same properties in one plane (e.g., y, z) and different properties in another direction perpendicular to the plane (e.g., x direction) are called *transversely isotropic*. In this case, there are five independent elastic properties. In Eqs. (4.15), for transversely isotropic materials,

$$E_y = E_z, \qquad G_{xz} = G_{yx}, \qquad v_{yx} = v_{zx}, \qquad \frac{1}{G_{yz}} = 2\left(\frac{1}{E_y} + \frac{v_{yz}}{E_y}\right)$$

Table 4-1 gives elastic moduli values for Poisson's ratios for some important engineering materials.

REFERENCES

4.1 *Annual Book of ASTM Standards*, Vol. 03.01, ASTM, Philadelphia, PA, 1993.

4.2 Eisenstadt, M. M., *Introduction to the Mechanical Properties of Materials*, MacMillan, New York, 1971.

4.3 Flinn, R. A., and Trojan, P. K., *Engineering Material and Their Applications*, 2nd ed., Houghton Mifflin, Boston, 1981.

4.4 Jastrzebski, Z. D., *The Nature and Properties of Engineering Materials*, 2nd ed., Wiley, New York, 1976.

4.5 Cordon, W. A., *Properties Evaluation, and Control of Engineering Materials*, McGraw-Hill, New York, 1979.

4.6 Templin, R. L., and Sturm, R. G., *J. Aeronaut. Sci.*, Vol. 7, No. 7, 1940, pp. 189–198.

4.7 Blake, A., *Handbook of Mechanics, Materials and Structures*, Wiley, New York, 1985.

4.8 Pellini, W. S., and Puzak, P. P., "Fracture Analysis Diagram Procedures for the Fracture-Safe Engineering Design of Steel Structures," NRL Report 5920, NRLRA, March 15, 1963.

4.9 Hayden, H. W., Moffat, W. G., and Wulff, J., *The Structure and Properties of Materials*, Vol. 3, *Mechanical Behavior*, Wiley, New York, 1965.

4.10 "Materials Selector," *Mater. Eng.*, Vol. 74, No. 4, 1971.

4.11 Manson, S. S., and Haferd, A. M., "A Linear Time–Temperature Relation for Extrapolation of Creep and Stress Rupture Data," NACA Technical Note 2890, March 1953.

4.12 Robinson, E. S., "Effect of Temperature Variation on the Long Time Rupture Strength of Steels," *ASME Trans.*, Vol. 74, 1952, pp. 777–781.

4.13 Collins, J. A., *Failure of Materials in Mechanical Design*, 2nd ed., Wiley, New York, 1993.

4.14   Felbeck, D. K., *Strength and Fracture of Engineering Solids*, Prentice-Hall, Englewood Cliffs, NJ, 1984.

4.15a   *Metals Handbook*, 10th ed., Vol. 1, *Properties and Selection: Irons, Steels, and High Performance Alloys*, American Society for Metals, Materials Park, OH, 1990.

4.15b   Boyer, H. E., and Gall, T. L. (Eds.), *Metals Handbook*, desk ed., American Society for Metals. Materials Park, OH 1986.

4.16   *Machine Design*, Vol. 52, No. 8, 1980, p. 30.

4.17   Shah, V., *Handbook of Plastics Testing Technology*, Wiley, New York, 1984.

4.18   Kingery, W. D., *Introduction to Ceramics*, 2nd ed., Wiley, New York, 1976.

4.19   *Annual Book of ASTM Standards*, Vols. 08.01–08.03, ASTM, Philadelphia, PA, 1993.

4.20   Schwartz, M. M., *Engineering Applications of Ceramic Materials*, American Society for Metals, Materials Park, OH, 1985.

4.21   Schwartz, M. M., *Composite Materials Handbook*, 2nd ed., McGraw-Hill, New York, 1992.

4.22   Vinson, J. R., and Sierakowski, R. L., *The Behavior of Structures Composed of Composite Materials*. Martinus Nijhoff, Dordrecht, 1986.

4.23   *Annual Book of ASTM Standards*, Vol. 15.03, ASTM, Philadelphia, PA, 1992.

4.24   Hahn, T. H., "Simplified Formulas for Elastic Moduli of Unidirectional Continuous Fiber Composites," *Composite Technology Review*, Fall 1980.

4.25   Rolfe, S. T., and Barsom, J. M., *Fracture and Fatigue Control in Structures*, Prentice-Hall, New Jersey, 1977.

4.26   Richerson, D. W., *Modern Ceramic Engineering*, Marcel Dekker, New York, 1982.

4.27   Chanda, M., and Roy, S. K., *Plastics Technology Handbook*, 2nd ed., Marcel Dekker, New York, 1993.

4

Tables

TABLE 4-1 MODULI OF ELASTICITY, POISSON'S RATIOS, AND THERMAL COEFFICIENTS OF EXPANSION

The material properties provided in the table should be treated as being nominal values that may be rough estimates. In the case of thermal expansion which is generally not linear with temperature, the α values should be considered as average values over particular temperature ranges.

Material	Modulus of Elasticity, E		Poisson's Ratio, ν	Thermal Coefficient of Expansion, α	
	$\times 10^6$ psi[a]	GPa		$\times 10^{-6}/°C$[b]	$\times 10^{-6}/°F$
ABS Plastic—unfilled	0.2–0.4	1.4–2.8	—	33–72	60–130
Acrylic—cast	0.35–0.45	2.4–3.1	—	27–50	50–90
Al_2O_3	55	380	0.26	4.4	8
Alumina—fired	40	275	—	3.0	5.4
Aluminum—Alloy 2024-T4	10.6	73	0.32	12.9	23.2
Aluminum—Alloy 7075-T6	10.4	72	0.32	12.9	23.2
BeO	45	311	—	5.6	10
Beryllium	40–44	275–305	0.024–0.05	6.4	11.5
Beryllium-Copper "25"	19	131	0.28–0.30	9.3	16.7
Boron-Epoxy Composite— orthotropic (representative)	$E_1 = 40^c$ $E_2 = 4^d$	275 27.5	$0.25(\nu_{1,2}^{c,d})$	5	9
Brass—30–70	16	110	0.33	11.1	20
Bronze—Phosphor (10%)	16	110	0.31	10.2	18.4
Concrete	3–6	20–40	0.1–0.3	5.5	10
Copper	17.8	123	0.33	9.2	16.5
Duraluminum	10.5	72.4	0.33	—	—
Epoxies—unfilled	0.3–0.45	2–3	—	17–33	30–60
Fiberglass-Epoxy Composite— orthotropic (Typical)	$E_1 = 8^c$ $E_2 = 2.7^d$	55 19	$0.25(\nu_{1,2}^{c,d})$	Varies with composition	
Glass—Soda-Lime	10	70	0.21	5.1	9.2
Graphite	0.3–2.4	2–17	—	3.0	5.4
Hastelloy—C-276	24.5–29.8	169–205	0.3	6.3	11.3
Inconel—wrought	31	214	—	7.0	13.0
Inconel-X	31	214	0.26	6.7–7.8	12–14
Invar—annealed condition	21	145	—	0.7–0.8	1.3–1.4
Iron—gray cast	13–14	90–96	—	6.0–6.7	10.8–12.1
Iron—malleable	25–28	172–193	0.17	5.9–7.1	10.6–12.8

TABLE 4-1   Moduli of Elasticity, Poisson's Ratios   174

| Material | Modulus of Elasticity, E | | Poisson's Ratio, ν | Thermal Coefficient of Expansion, α | |
	$\times 10^6$ psi[a]	GPa		$\times 10^{-6}/°C$[b]	$\times 10^{-6}/°F$
Kevlar	19.4	134	—	—	—
Lead	2	14	0.4–0.45	16.3	29.3
Magnesium—AZ-31B	6.5	45	0.35	14.5	26.1
MgO	30	207	0.36	5	9
Molybdenum	47–50	325–345	0.33	2.7	4.9
Mullite ($Al_6Si_2O_{13}$)	21	145	—	2.8	5
Monel—Alloy 400	26	179	0.32	7.5–7.8	13.5–14.0
Mylar	0.55–0.80	3.8–5.5	0.38	9.4	17
Nickel-A	30	207	—	6.6–7.4	12–13
Ni-Span-C	27.7	191	0.33	4.2	7.6
Nylon-6/6—unmodified	0.16–0.41	1.1–2.8	—	44	80
Phenolics (representative)	0.4–0.5	2.7–3.4	—	14–33	25–60
Polycarbonate	0.3–0.38	2.0–2.6	—	37	67
Polyethylene	1–2	7–14	—	60–70	108–126
Polypropylene— unmodified	0.16–0.23	1.1–1.6	—	32–57	58–102
Polyurethane	0.01–0.1	0.07–0.7	—	—	—
Polyvinyl Chloride—rigid	0.42–0.52	2.9–3.6	0.26–0.34	28–33	50–60
Porcelain (high-alumina ceramic)	32–56	221–386	0.2–0.21	3.3	6.0
Quartz—fused	10.5	72.5	—	0.28	0.50
Rene-41	29.9–31.9	206–220	0.31	6.63	11.9
Rubber—neoprene	Varies with composition		0.5	340	612
SiC	60	414	—	2.5	4.5
Si_3N_4	44	304	—	1.8	3.2
SiO_2 (fused)	10	69	0.25	0.5	0.9
Spinel ($MgAl_2O_4$)	36	284	—	5	9
Steel—1008/1018	30	207	0.285	6.7	12.0
Steel—4130/4340	30	207	0.28–0.29	6.3	11.3
Steel—304 (stainless)	28	193	0.25	9.6	17.3
Steel—310 (stainless)	29–30	200–207	0.32	8.0	14.4

Moduli of Elasticity, Poisson's Ratios TABLE 4-1

Material	Modulus of Elasticity, E		Poisson's Ratio, ν	Thermal Coefficient of Expansion, α	
	$\times 10^6$ psi[a]	GPa		$\times 10^{-6}/°C$[b]	$\times 10^{-6}/°F$
Teflon—TFE	0.038–0.065	0.26–0.45	—	55	99
TiC	67	462	—	4	7.2
Titanium—6 Al-4V	16.5	115	0.34	4.9	8.8
Titanium—pure	15.1	104	0.34	4.8	8.6
Titanium—silicate	9.8	68	0.17	0.0 ($\pm$0.017)	0.0 ($\pm$0.03)
Tungsten	50	345	0.28	2.4–2.6	4.3–4.7
Tungsten-Carbide Cermet	61.6–94.3	425–650	—	2.5–3.0	4.5–5.4
Vanadium	18–20	124–138	—	4.6	8.3
Vinyl Chloride—rigid	0.3–0.5	2–3.5	0.28–0.34	28–56	50–100
Wood—structural	1–2	7–14	—	1–3	2–5
Zircaloy-2	11	76	0.37–0.41	2.9	5.2
Zirconium	13.7–14.0	95–96.5	0.37–0.41	3.1	5.6

[a]For psi, multiply tabulated values by 10^6. For example, if the entry is 40, this corresponds to 40×10^6 psi. An entry of 0.2–0.4 means that the values of E range from 0.2×10^6 psi to 0.4×10^6 psi.
[b]For α, multiply tabulated value by 10^{-6}. For example, for "Aluminum-Alloy 2024-T4", the α values are $12.9 \times 10^{-6}/°F$ and $23.2 \times 10^{-6}/C°$.
[c]E_1, ν_1 properties in fiber direction.
[d]E_2, ν_2 properties in 90° to fiber direction.

TABLE 4-1 | **Moduli of Elasticity, Poisson's Ratios** | **176**

TABLE 4-2 HARDNESS TESTING[a]

| Test | Indentor | Shape of Indentation | | Load | Formula for Hardness Number P (kgf), d and D (mm) |
		Side View	Top View		
Brinell	10-mm sphere of steel or tungsten carbide			P	$\mathrm{HB} = \dfrac{2P}{\pi D \left(D - \sqrt{D^2 - d^2} \right)}$
Vickers	Diamond pyramid			P	$\mathrm{HV} = 1.8544 P / d_1^2$
Knoop micro-hardness	Diamond pyramid			P	$\mathrm{HK} = 14.23 P / d^2$

[a] From Hayden et al. [4.9], with permission.

TABLE 4-3 ROCKWELL HARDNESS SCALES[a]

Scale Symbol	Penetrator	Major Load kgf (N)	Dial Figures	Typical Applications of Scales
B	$\frac{1}{16}$-in. (1.588-mm) ball	100 (981)	Red	Copper alloys, soft steels, aluminum alloys, malleable iron, etc.
C	Diamond	150 (1471)	Black	Steel, hard cast irons, pearlitic malleable iron, titanium, deep case-hardened steel, and other materials harder than B100
A	Diamond	60 (588)	Black	Cemented carbides, thin steel, and shallow case-hardened steel
D	Diamond	100 (981)	Black	Thin steel and medium case-hardened steel pearlitic malleable iron
E	$\frac{1}{8}$-in. (3.175-mm) ball	100 (981)	Red	Cast iron, aluminum and magnesium alloys, bearing metals
F	$\frac{1}{16}$-in. (1.588-mm) ball	60 (588)	Red	Annealed copper alloys, thin soft sheet metals
G	$\frac{1}{16}$-in. (1.588-mm) ball	150 (1471)	Red	Malleable irons, copper–nickel–zinc and cupronickel alloys; upper limit G92 to avoid possible flattening of ball
H	$\frac{1}{8}$-in. (3.175-mm) ball	60 (588)	Red	Aluminum, zinc, lead
K	$\frac{1}{8}$-in. (3.175-mm) ball	150 (1471)	Red	
L	$\frac{1}{4}$-in. (6.350-mm) ball	60 (588)	Red	Bearing metals and other very
M	$\frac{1}{4}$-in. (6.350-mm) ball	100 (981)	Red	soft or thin materials;
P	$\frac{1}{4}$-in. (6.350-mm) ball	150 (1471)	Red	use smallest ball and
R	$\frac{1}{2}$-in. (12.70-mm) ball	60 (588)	Red	heaviest load that does not
S	$\frac{1}{2}$-in. (12.70-mm) ball	100 (981)	Red	give anvil effect
V	$\frac{1}{2}$-in. (12.70-mm) ball	150 (1471)	Red	

[a]From *Annual Book of ASTM Standards* [4.1]. Copyright ASTM. Printed with permission.

TABLE 4-3 **Rockwell Hardness Scales**   178

TABLE 4-4 CREEP STRENGTHS OF SELECTED METALLIC ALLOYS[a]

Material	Creep Strength (psi)	Conditions
Ductile cast iron 60-40-18	4,000	$10^{-4}\%$ strain per hour at 1000°F
Iron superalloys 16-25-6 (25% Ni, 16% Cr, 6% Mo)	19,000	$10^{-4}\%$ strain per hour at 1200°F
Type 302 stainless wrought	17,000	1% strain in 1000 h at 1000°F
Type 316 stainless wrought	25,000	1% strain in 1000 h at 1000°F
Type 430 stainless wrought	8,500	1% strain in 10,000 h at 1000°F
Type 410 stainless wrought	9,200	1% strain in 10,000 h at 1000°F
Magnesium wrought, AZ31B-H24	1,500	0.5% strain in 100 h at 300°F

[a]Data collected from [4.10].

TABLE 4-5 CONSTANTS IN MANSON–HAFERD FORMULA[a]

Material	Creep or Rupture	T_a	$\log_{10} t_a$
25-20 stainless steel	Rupture	100	14
18-8 stainless steel	Rupture	100	15
S-590 alloy	Rupture	0	21
DM steel	Rupture	100	22
Inconel X	Rupture	100	24
Nimonic 80	Rupture or 0.2 or 0.1% plastic strain	100	17

[a]From Manson and Hafert [4.11].
T_a is a material constant. t_a is the time in hours.

TABLE 4-6 PROPERTIES OF PURE METALS AT ~ 20°C[a]

Metal	Symbol	Structure	Lattice Constant ($\times 10^{-12}$ m)[c]		Atomic Radius ($\times 10^{-12}$ m)[c]	Density		Melting Point		Elastic Modulus (GPa)	Shear Modulus (GPa)
			a	c		kg/m³	kip-s²/ft⁴[b]	°C	°F		
Aluminum	Al	fcc	405	—	143	2 700	5.239×10^{-3}	660	1220	62	24
Chromium	Cr	bcc	288	—	125	7 190	13.951×10^{-3}	1875	3407	248	95
Copper	Cu	fcc	362	—	128	8 960	17.385×10^{-3}	1083	1981	110	42
Gold	Au	fcc	408	—	144	19 300	37.448×10^{-3}	1063	1945	80	31
Iron	Fe	bcc	287	—	124	7 870	15.270×10^{-3}	1538	2800	196	76
Lead	Pb	fcc	495	—	175	11 400	22.120×10^{-3}	327	621	14	5
Magnesium	Mg	hcp	321	521	160	1 740	3.376×10^{-3}	650	1202	44	17
Molybdenum	Mo	bcc	315	—	136	10 200	19.791×10^{-3}	2610	4730	324	125
Nickel	Ni	fcc	352	—	125	8 900	17.269×10^{-3}	1453	2647	207	80
Platinum	Pt	fcc	393	—	139	21 400	41.523×10^{-3}	1769	3217	73	28
Silver	Ag	fcc	409	—	144	10 500	20.373×10^{-3}	961	1761	76	29
Tin	Sn	bct	583	318	—	7 300	14.164×10^{-3}	232	449	43	17
Titanium	Ti	hcp	295	468	—	4 510	8.751×10^{-3}	1668	3035	116	45
Tungsten	W	bcc	316	—	137	19 300	37.448×10^{-3}	3410	6170	345	133
Vanadium	V	bcc	304	—	132	6 100	11.836×10^{-3}	1900	3450	131	50
Zinc	Zn	hcp	266	495	133	7 130	13.834×10^{-3}	420	787	—	50

TABLE 4-6 Properties of Pure Metals at ~ 20°C   180

Properties of Pure Metals at ~20°C — TABLE 4-6

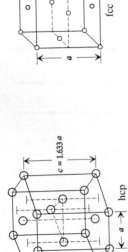

hcp — $c = 1.633a$, a

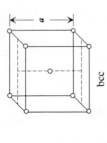

fcc — a

bcc — a

[a]From Felbeck [4.14]. Reprinted by permission of Prentice Hall, Englewood Cliffs, New Jersey. Abbreviations: fcc, face-centered-cubic; bcc, body-centered-cubic; hcp, hexagonal close-packed; bct, body-centered-tetragonal.

[b]slug/ft³.

[c]Multiply tabulated value by 10^{-12}. Equivalently, the tabulated values are given in pm.

TABLE 4-7   SELECTED ASTM SPECIFICATIONS INCORPORATING AISI–SAE DESIGNATIONS[a]

A29	Carbon and alloy steel bars, hot rolled and cold finished, generic
A108	Standard-quality cold-finished carbon steel bars
A295	High carbon–chromium ball and roller bearing steel
A304	Alloy steel bars having hardenability requirements
A322	Hot-rolled alloy steel bars
A331	Cold-finished alloy steel bars
A434	Hot-rolled or cold-finished quenched and tempered alloy steel bars
A505	Hot-rolled and cold-rolled alloy steel sheet and strip, generic
A506	Regular-quality hot-rolled and cold-rolled alloy steel sheet and strip
A507	Drawing quality hot-rolled and cold-rolled alloy steel sheet and strip
A510	Carbon steel wire rods and coarse round wire, generic
A534	Carburizing steels for antifriction bearings
A535	Special-quality ball and roller bearing steel
A544	Scrapless nut quality carbon steel wire
A545	Cold heading quality carbon steel wire for machine screws
A546	Cold heading quality medium-high-carbon steel wire for hexagon-head bolts
A547	Cold heading quality alloy steel wire for hexagon head bolts
A548	Cold heading quality carbon steel wire for tapping or sheet metal screws
A549	Cold heading quality carbon steel wire for wood screws
A575	Merchant-quality hot-rolled carbon steel bars
A576	Special-quality hot-rolled carbon steel bars
A634	Aircraft-quality hot-rolled and cold-rolled alloy steel sheet and strip
A646	Premium-quality alloy steel blooms and billets for aircraft and aerospace forgings
A659	Commercial-quality hot-rolled carbon steel sheet and strip
A680	Untempered spring-quality cold-rolled hard carbon steel strip
A682	Cold-rolled spring-quality carbon steel strip, generic
A684	Untempered spring-quality cold-rolled soft carbon steel strip
A689	Carbon and alloy steel bars for springs
A711	Carbon and alloy steel blooms, billets and slabs for forging
A713	High-carbon spring steel wire for heat-treated components

[a]From *Metals Handbook* [4.15a], with permission of ASM International

TABLE 4-8   SELECTED ASTM GENERIC SPECIFICATIONS

A6	Rolled steel structural plate, shapes, sheet piling and bars, generic
A20	Steel plate for pressure vessels, generic
A29	Carbon and alloy steel bars, hot rolled and cold finished, generic
A505	Alloy steel sheet and strip, hot rolled and cold rolled, generic
A510	Carbon steel wire rod and coarse round wire, generic
A568	Carbon and HSLA, hot-rolled and cold-rolled steel sheet and hot-rolled strip, generic
A646	Premium-quality alloy steel blooms and billets for aircraft and aerospace forgings
A711	Carbon and alloy steel blooms, billets, and slabs for forging

TABLE 4-8   **ASTM Specifications**   **182**

TABLE 4-9 MECHANICAL PROPERTIES OF SELECTED CARBON STEELS IN HOT-ROLLED, NORMALIZED AND ANNEALED CONDITION[a]

AISI Number	Treatment	Austenitizing Temperature		Tensile Strength		Yield Strength		Elongation (%)	Reduction in area (%)	Brinell Hardness	Izod Impact Strength	
		°C	°F	MPa	ksi	MPa	ksi				J	ft-lb
1015	As rolled	—	—	420.6	61.0	313.7	45.5	39.0	61.0	126	110.5	81.5
	Normalized	925	1700	424.0	61.5	324.1	47.0	37.0	69.6	121	115.5	85.2
	Annealed	870	1600	386.1	56.0	284.4	41.3	37.0	69.7	111	115.0	84.8
1020	As rolled	—	—	448.2	65.0	330.9	48.0	36.0	59.0	143	86.8	64.0
	Normalized	870	1600	441.3	64.0	346.5	50.3	35.8	67.9	131	117.7	86.8
	Annealed	870	1600	394.7	57.3	294.8	42.8	36.5	66.0	111	123.4	91.0
1030	As rolled	—	—	551.6	80.0	344.7	50.0	32.0	57.0	179	74.6	55.0
	Normalized	925	1700	520.6	75.5	344.7	50.0	32.0	60.8	149	93.6	69.0
	Annealed	845	1550	463.7	67.3	341.3	49.5	31.2	57.9	126	69.4	51.2
1040	As rolled	—	—	620.5	90.0	413.7	60.0	25.0	50.0	201	48.8	36.0
	Normalized	900	1650	589.5	85.5	374.0	54.3	28.0	54.9	170	65.1	48.0
	Annealed	790	1450	518.8	75.3	353.4	51.3	30.2	57.2	149	44.3	32.7
1050	As rolled	—	—	723.9	105.0	413.7	60.0	20.0	40.0	229	31.2	23.0
	Normalized	900	1650	748.1	108.5	427.5	62.0	20.0	39.4	217	27.1	20.0
	Annealed	790	1450	636.0	92.3	365.4	53.0	23.7	39.9	187	16.9	12.5
1060	As rolled	—	—	813.6	118.0	482.6	70.0	17.0	34.0	241	17.6	13.0
	Normalized	900	1650	775.7	112.5	420.6	61.0	18.0	37.2	229	13.2	9.7
	Annealed	790	1450	625.7	90.8	372.3	54.0	22.5	38.2	179	11.3	8.3
1080	As rolled	—	—	965.3	140.0	586.1	85.0	12.0	17.0	293	6.8	5.0
	Normalized	900	1650	1010.1	146.5	524.0	76.0	11.0	20.6	293	6.8	5.0
	Annealed	790	1450	615.4	89.3	375.8	54.5	24.7	45.0	174	6.1	4.5

TABLE 4-9 (continued) MECHANICAL PROPERTIES OF SELECTED CARBON STEELS IN HOT-ROLLED, NORMALIZED AND ANNEALED CONDITION[a]

AISI Number	Treatment	Austenitizing Temperature		Tensile Strength		Yield Strength		Elongation (%)	Reduction in area (%)	Brinell Hardness	Izod Impact Strength	
		°C	°F	MPa	ksi	MPa	ksi				J	ft-lb
1117	As rolled	—	—	486.8	70.6	305.4	44.3	33.0	63.0	143	81.3	60.0
	Normalized	900	1650	467.1	67.8	303.4	44.0	33.5	63.8	137	85.1	62.8
	Annealed	855	1575	429.5	62.3	279.2	40.5	32.8	58.0	121	93.6	69.0
1137	As rolled	—	—	627.4	91.0	379.2	55.0	28.0	61.0	192	82.7	61.0
	Normalized	900	1650	668.8	97.0	396.4	57.5	22.5	48.5	197	63.7	47.0
	Annealed	790	1450	584.7	84.8	344.7	50.0	26.8	53.9	174	49.9	36.8
1141	As rolled	—	—	675.7	98.0	358.5	52.0	22.0	38.0	192	11.1	8.2
	Normalized	900	1650	706.7	102.5	405.4	58.8	22.7	55.5	201	52.6	38.8
	Annealed	815	1500	598.5	86.8	353.0	51.2	25.5	49.3	163	34.3	25.3

[a]From *Metals Handbook* [4.15b], with permission of ASM International.

TABLE 4-9 **Properties of Selected Carbon Steels**

TABLE 4-10 MECHANICAL PROPERTIES OF SELECTED ALLOY STEELS IN HOT-ROLLED AND ANNEALED CONDITION[a]

AISI Number[b]	Treatment	Austenitizing Temperature		Tensile Strength		Yield Strength		Elongation (%)	Reduction in Area (%)	Brinell Hardness	Izod Impact Strength	
		°C	°F	MPa	ksi	MPa	ksi				J	ft-lb
1340	Normalized	870	1600	836.3	121.3	558.5	81.0	22.0	62.9	248	92.5	68.2
	Annealed	800	1475	703.3	102.0	436.4	63.3	25.5	57.3	207	70.5	52.0
3140	Normalized	870	1600	891.5	129.3	599.8	87.0	19.7	57.3	262	53.6	39.5
	Annealed	815	1500	689.5	100.0	422.6	61.3	24.5	50.8	197	46.4	34.2
4130	Normalized	870	1600	668.8	97.0	436.4	63.3	25.5	59.5	197	86.4	63.7
	Annealed	865	1585	560.5	81.3	360.6	52.3	28.2	55.6	156	61.7	45.5
4150	Normalized	870	1600	1154.9	167.5	734.3	106.5	11.7	30.8	321	11.5	8.5
	Annealed	815	1500	729.5	105.8	379.2	55.0	20.2	40.2	197	24.7	18.2
4340	Normalized	870	1600	1279.0	185.5	861.8	125.0	12.2	36.3	363	15.9	11.7
	Annealed	810	1490	744.6	108.0	472.3	68.5	22.0	49.9	217	51.1	37.7
4620	Normalized	900	1650	574.3	83.3	366.1	53.1	29.0	66.7	174	132.9	98.0
	Annealed	855	1575	512.3	74.3	372.3	54.0	31.3	60.3	149	93.6	69.0
4820	Normalized	860	1580	755.0	109.5	484.7	70.3	24.0	59.2	229	109.8	81.0
	Annealed	815	1500	681.2	98.8	464.0	67.3	22.3	58.8	197	92.9	68.5
5140	Normalized	870	1600	792.9	115.0	472.3	68.5	22.7	59.2	229	38.0	28.0
	Annealed	830	1525	572.3	83.0	293.0	42.5	28.6	57.3	167	40.7	30.0
5150	Normalized	870	1600	870.8	126.3	529.5	76.8	20.7	58.7	255	31.5	23.2
	Annealed	825	1520	675.7	98.0	357.1	51.8	22.0	43.7	197	25.1	18.5

TABLE 4-10 (continued) MECHANICAL PROPERTIES OF SELECTED ALLOY STEELS IN HOT-ROLLED AND ANNEALED CONDITION[a]

AISI Number[b]	Treatment	Austenitizing Temperature		Tensile Strength		Yield Strength		Elongation (%)	Reduction in Area (%)	Brinell Hardness	Izod Impact Strength	
		°C	°F	MPa	ksi	MPa	ksi				J	ft-lb
5160	Normalized	855	1575	957.0	138.8	530.9	77.0	17.5	44.8	269	10.8	8.0
	Annealed	815	1495	722.6	104.8	275.8	40.0	17.2	30.6	197	10.0	7.4
6150	Normalized	870	1600	939.8	136.3	615.7	89.3	21.8	61.0	269	35.5	26.2
	Annealed	815	1500	667.4	96.8	412.3	59.8	23.0	48.4	197	27.4	20.2
8620	Normalized	915	1675	632.9	91.8	357.1	51.8	26.3	59.7	183	99.7	73.5
	Annealed	870	1600	536.4	77.8	385.4	55.9	31.3	62.1	149	112.2	82.8
8650	Normalized	870	1600	1023.9	148.5	688.1	99.8	14.0	40.4	302	13.6	10.0
	Annealed	795	1465	715.7	103.8	386.1	56.0	22.5	46.4	212	29.4	21.7
8740	Normalized	870	1600	929.4	134.8	606.7	88.0	16.0	47.9	269	17.6	13.0
	Annealed	815	1500	695.0	100.8	415.8	60.3	22.2	46.4	201	40.0	29.5
9310	Normalized	890	1630	906.7	131.5	570.9	82.8	18.8	58.1	269	119.3	88.0
	Annealed	845	1550	820.5	119.0	439.9	63.8	17.3	42.1	241	78.6	58.0

[a]From *Metals Handbook* [4.15b], with permission of ASM Interantional.
[b]All grades are fine grained. Heat-treated specimens were oil quenched unless otherwise indicated.

TABLE 4-10   Properties of Selected Alloy Steels   186

TABLE 4-11   MECHANICAL PROPERTIES OF SELECTED STAINLESS STEELS[a]

AISI/UNS Grade	Condition	Tensile Strength (ksi)	Yield Strength (ksi)	Elongation (% in 2 in.)	Reduction of Area (%)	Maximum Brinell Hardness
302/S30200	Annealed	75	30	40	—	88
303/S30300	Annealed	85[b]	35[b]	50[b]	55[b]	—
304/S30400	Annealed	75	30	40	—	88
316/S31600	Annealed	75	30	40	—	95
347/S34700	Annealed	75	30	40	—	88
430/S43000	Annealed	65	30	22[c]	—	88
446/S44600	Annealed	75	40	20	—	95
442/S44200	Annealed	75	40	20	—	95
410/S41000	—	65	30	22[a]	—	95
440A/S44002	—	105[b]	60[b]	20[b]	—	95[b]

[a] From *Metals Handbook* [4.15b], with permission of ASM International.
[b] Typical values.
[c] 20% elongation for thickness of 1.3 mm (0.050 in.) or less.

TABLE 4-12 MECHANICAL PROPERTIES OF SELECTED CAST IRONS [a]

Gray Iron

ASTM Class	Tensile Strength		Torsional Shear Strength		Compressive Strength		Reversed Bending Fatigue Limit		Transverse Load on Test Bar B[j]		Brinell Hardness
	MPa	ksi	MPa	ksi	MPa	ksi	MPa	ksi	kg	lb	
20	152	22	179	26	572	83	69	10	839	1850	156
25	179	26	220	32	669	97	79	11.5	987	2175	174
30	214	31	276	40	752	109	97	14	1145	2525	210
35	252	36.5	334	48.5	855	124	110	16	1293	2850	212
40	293	42.5	393	57	965	140	128	18.5	1440	3175	235
50	362	52.5	503	73	1130	164	148	21.5	1638	3600	262
60	431	62.5	610	88.5	1293	187.5	169	24.5	1678	3700	302

Ductile Iron

Specification Number	Grade or Class	Brinell Hardness[b]	Minimum Tensile Strength[c]		Minimum Yield Strength[c]		Minimum Elongation[c] (%)
			MPa	ksi	MPa	ksi	
ASTM A395-76, ASME SA395	60-40-18	143–187	414	60	276	40	18
ASTM A476-70[d], SAE AMS5316	80-60-03	201 min	552	80	414	60	3
ASTM A536-72, MIL-I-11466B(MR)	60-40-18	—	414	60	276	40	18
	65-45-12	—	448	65	310	45	12
	80-55-06	—	552	80	379	55	6
	100-70-03	—	689	100	483	70	3
	120-90-02	—	827	120	621	90	2

TABLE 4-12 Properties of Selected Cast Irons 188

Specification Number	Grade or Class	Tensile Strength MPa	Tensile Strength ksi	Yield Strength MPa	Yield Strength ksi	Brinell Hardness	Elongation[f] (%)
SAE J434c	D4018	414	60	276	40	170 max	18
	D4512	448	65	310	45	156–217	12
	D5506	552	80	379	55	187–255	6
	D7003	689	100	483	70	241–302[d]	3[e]
	DQ & T	[e]	[e]	[e]	[e]	[e]	[e]
MIL-I-24137 (Ships)	Class A	414	60	310	45	190 max	15
	Class B	379	55	207	30	190 max	7
	Class C	345	50	172	25	175 max	20
Malleable Iron							
Ferritic ASTM A47, A338, ANSI G48.1; FED QQ-I-666c	32510	345	50	224	32	156 max	10
	35018	365	53	241	35	156 max	18
ASTM A197		276	40	207	30	156 max	5
Pearlitic and Martensitic ASTM A220; ANSI G48.2; MIL-I-11444B	40010	414	60	276	40	149–197	10
	45008	448	65	310	45	156–197	8
	45006	448	65	310	45	156–207	6
	50005	483	70	345	50	179–229	5
	60004	552	80	414	60	197–241	4
	70003	586	85	483	70	217–269	3
	80002	655	95	552	80	241–285	2
	90001	724	105	621	90	269–321	1

Properties of Selected Cast Irons TABLE 4-12

TABLE 4-12

TABLE 4-12 (continued) MECHANICAL PROPERTIES OF SELECTED CAST IRONS[a]

Malleable Iron

Specification Number	Grade or Class	Tensile Strength		Yield Strength		Brinell Hardness	Elongation[f] (%)
		MPa	ksi	MPa	ksi		
Automotive ASTM A602; SAE J158	M3210[g]	345	50	224	32	156 max	10
	M4504[h]	448	65	310	45	163–217	4
	M5003[h]	517	75	345	50	187–241	3
	M5503[i]	517	75	379	55	187–241	3
	M7002[i]	621	90	483	70	229–269	2
	M8501[i]	724	105	586	85	269–302	1

[a]From *Metals Handbook* [4.15b], with permission of ASM International.
[b]Measured at a predetermined location on casting
[c]Determined using a standard specimen taken from a separately cast test block, as set forth in the applicable specification.
[d]Range specified by mutual agreement between producer and purchaser
[e]Value must be compatible with minimum hardness specified for production castings
[f]Minimum in 50 mm or 2 in.
[g]Annealed.
[h]Air quenched and tempered.
[i]Liquid quenched and tempered.
[j]Refer to ASTM A438

TABLE 4-12 Properties of Selected Cast Irons   190

TABLE 4-13   DESIGNATION OF WROUGHT ALUMINUM ALLOYS[a]

Alloying element	Designation
Aluminum, 99.00% minimum and greater	1xxx
Aluminum alloys grouped by major alloying element(s)	
Copper	2xxx
Manganese	3xxx
Silicon	4xxx
Magnesium	5xxx
Magnesium and silicon	6xxx
Zinc	7xxx
Other element	8xxx
Unused series	9xxx

[a]From *Metals Handbook* [4.15b], with permission of ASM International.

TABLE 4-14   BASIC TEMPER DESIGNATION FOR ALUMINUM[a]

Temper Designation	Process
F	As fabricated
O	Annealed
H	Strain hardened (wrought products only)
W	Solution heat treated
T	Heat treated to produce stable tempers other than F, O, or H

[a]From *Metals Handbook* [4.15b], with permission of ASM International.

TABLE 4-15   DESIGNATION OF CAST ALUMINUM ALLOYS[a]

Major Alloying Element	Designation
Aluminum, $\geq$ 99.00%	1xx.x
Aluminum alloys grouped by major alloying element(s):	
Copper	2xx.x
Silicon, with added copper and/or magnesium	3xx.x
Silicon	4xx.x
Magnesium	5xx.x
Zinc	7xx.x
Tin	8xx.x
Other elements	9xx.x
Unused series	6xx.x

[a]From *Metals Handbook* [4.15b], with permission of ASM International.

TABLE 4-16 MECHANICAL PROPERTIES OF SELECTED NON-HEAT-TREATABLE ALUMINUM ALLOYS[a]

Alloy	Temper[b]	Tensile Strength		Yield Strength[c]		Elongation[d] (%)	Brinell Hardness[e]	Shear Strength		Fatigue Limit[f]	
		MPa	ksi	MPa	ksi			MPa	ksi	MPa	ksi
1100	O	90	13	35	5	35	23	60	9	35	5
	H14	125	18	115	17	9	32	75	11	50	7
	H18	165	24	150	22	5	44	90	13	60	9
3003	O	110	16	40	6	30	28	75	11	50	7
	H14	150	22	145	21	8	40	95	14	60	9
	H18	200	29	185	27	4	55	110	16	70	10
3004	O	180	26	70	10	20	45	110	16	95	14
	H34	240	35	200	29	9	63	125	8	105	15
	H38	285	41	250	36	5	77	145	21	110	16
	H19	295	43	285	41	2	—	—	—	—	—
3104	H19	290	42	260	38	4	—	—	—	—	—
3005	O	130	19	55	8	25	—	—	—	—	—
	H14	180	26	165	24	7	—	—	—	—	—
	H18	240	35	225	32	4	—	—	—	—	—
3105	O	115	17	55	8	24	—	85	12	—	—
	H25	180	26	160	23	8	—	105	15	—	—
	H18	215	31	195	28	3	—	115	17	—	—

TABLE 4-16 Mechanical Properties of Aluminum Alloys 192

Alloy	Temper										
5005	O	125	18	40	6	25	28	75	11	—	—
	H34	160	23	140	20	8	41	95	14	—	—
	H38	200	29	185	27	5	55	110	16	—	—
5050	O	145	21	55	8	24	36	105	15	85	12
	H34	190	28	165	24	8	53	125	18	90	13
	H38	220	32	200	29	6	63	140	20	95	14
5252	O	180	26	85	12	23	46	115	17	—	—
	H25	235	34	170	25	11	68	145	21	—	—
	H28	285	41	240	35	5	75	160	23	—	—
5154	O	240	35	115	17	27	58	150	22	115	17
	H34	290	42	230	33	13	73	165	24	130	19
	H38	330	48	270	39	10	80	195	28	145	21
	H112	240	35	115	17	25	63	—	—	115	17
5454	O	250	36	115	17	22	62	160	23	—	—
	H34	305	44	240	35	10	81	180	26	—	—
	H111	260	38	180	26	14	70	160	23	—	—
	H112	250	36	125	18	18	62	160	23	—	—
5056	O	290	42	150	22	35	65	180	26	140	20
	H18	435	63	405	59	10	105	235	34	150	22
	H38	310	60	345	50	15	100	220	32	150	22
5657	O	110	16	40	6	25	28	75	11	—	—
	H25	160	23	140	20	12	40	95	14	—	—
	H28	195	28	165	24	7	50	105	15	—	—
5082	H19	395	57	370	54	4	—	—	—	—	—
5182	O	275	40	130	19	21	—	—	—	—	—
	H19	420	61	395	57	4	—	—	—	—	—

TABLE 4-16 (continued)   MECHANICAL PROPERTIES OF SELECTED NON-HEAT-TREATABLE ALUMINUM ALLOYS[a]

Alloy	Temper[b]	Tensile Strength		Yield Strength[c]		Elongation[d] (%)	Brinell Hardness[e]	Shear Strength		Fatigue Limit[f]	
		MPa	ksi	MPa	ksi			MPa	ksi	MPa	ksi
5086	O	260	38	115	17	22	—	160	23	—	—
	H34	325	47	255	37	10	—	185	27	—	—
	H112	270	39	130	19	14	—	—	—	—	—
	H116	290	42	205	30	12	—	—	—	—	—
7072	O	70	10	—	—	15	—	—	—	—	—
	H113	75	11	—	—	15	—	—	—	—	—
8001	O	110	16	40	6	30	—	—	—	—	—
	H18	200	29	185	27	4	—	—	—	—	—
8280	O	115	17	50	7	28	—	—	—	—	—
	H18	220	32	205	30	4	—	—	—	—	—

[a]From *Metals Handbook* [4.15b], with permission of ASM International.
[b]See Table 4-14 for temper designations.
[c]At 0.2% offset.
[d]In 50 mm or 2 in.
[e]500-kg load, 10-mm ball, 30 s.
[f]Based on 500 million cycles using an R. R. Moore–type rotating-beam machine.

TABLE 4-16   **Mechanical Properties of Aluminum Alloys**   194

TABLE 4-17 NOMINAL COMPOSITIONS AND TYPICAL ROOM TEMPERATURE MECHANICAL PROPERTIES OF MAGNESIUM ALLOYS[a]

Alloy	Composition Al	Mn[b]	Th	Zn	Zr	Other	Tensile Strength MPa	ksi	Yield Strength Tensile MPa	ksi	Compressive MPa	ksi	Bearing MPa	ksi	Elongation in 50 mm or 2 in. (%)	Shear Strength MPa	ksi	Brinell Hardness[c]
Sand and Permanent Mold Castings																		
AM100A-T61	10.0	0.1	—	—	—	—	275	40	150	22	150	22	—	—	1	—	—	69
AZ63A-T6	6.0	0.15	—	3.0	—	—	275	40	130	19	130	19	360	52	5	145	21	73
AZ81A-T4	7.6	0.13	—	0.7	—	—	275	40	83	12	83	12	305	44	15	125	18	55
AZ91C-T6	8.7	0.13	—	0.7	—	—	275	40	195	21	145	21	360	52	6	145	21	66
AZ92A-T6	9.0	0.10	—	2.0	—	—	275	40	150	22	150	22	450	65	3	150	22	84
EZ33A-T5	—	—	—	2.7	0.6	3.3 RE[h]	160	23	110	16	110	16	275	40	2	145	21	50
HK31A-T6	—	—	3.3	—	0.7	—	220	32	105	15	105	15	275	40	8	145	21	55
HZ32A-T5	—	—	3.3	2.1	0.7	—	185	27	90	13	90	13	255	37	4	140	20	57
K1A-F	—	—	—	—	0.7	—	180	26	55	8	—	—	125	18	1	55	8	—
QE22A-T6	—	—	—	—	0.7	2.5 Ag, 2.1 Di[g]	260	38	195	28	195	28	—	—	3	—	—	80
QH21A-T6	—	—	60	—	0.7	2.5 Ag, 1.0 Di[g]	275	40	205	30	—	—	—	—	4	—	—	—
ZE41A-T5	—	—	—	4.2	0.7	1.2 RE[h]	205	30	140	20	140	20	350	51	3.5	160	23	62
ZE63A-T6	—	—	—	5.8	0.7	2.6 RE[h]	300	44	190	28	195	28	—	—	10	—	—	60–85
ZH62A-T5	—	—	1.8	5.7	0.7	—	240	35	170	25	170	25	340	49	4	165	24	70
ZK51A-T5	—	—	—	4.6	0.7	—	205	30	165	24	165	24	325	47	3.5	160	23	65
ZK61A-T5	—	—	—	6.0	0.7	—	310	45	185	27	185	27	—	—	—	170	25	68
ZK61A-T6	—	—	—	6.0	0.7	—	310	45	195	28	195	28	—	—	10	180	26	70
Die Castings																		
AM60A-F	6.0	0.13	—	—	—	—	205	30	115	17	115	17	—	—	6	—	—	—
AS41A-F[d]	4.3	0.35	—	—	—	1.0 Si	220	32	150	22	150	22	—	—	4	—	—	—
AZ91A and B-F[e]	9.0	0.13	—	0.7	—	—	230	33	150	22	165	24	—	—	3	140	20	63

Alloy	Composition						Tensile Strength		Yield Strength						Elongation in 50 mm or 2 in. (%)	Shear Strength		Brinell Hardness[c]
	Al	Mn[b]	Th	Zn	Zr	Other	MPa	ksi	Tensile		Compressive		Bearing			MPa	ksi	
									MPa	ksi	MPa	ksi	MPa	ksi				
Extruded Bars and Shapes																		
AZ10A-F	1.2	0.2	—	0.4	—	—	240	35	145	21	69	10	—	—	10	—	—	—
AZ21X1-F[d]	1.8	0.02	—	1.2	—	—	—	—	—	—	—	—	—	—	—	—	—	—
AZ31 B and C-F[f]	3.0	—	—	1.0	—	—	260	38	200	29	97	14	230	33	15	130	19	49
AZ61A-F	6.5	—	—	1.0	—	—	310	45	230	33	130	19	285	41	16	140	20	60
AZ80A-T5	8.5	—	—	0.5	—	—	380	55	275	40	240	35	—	—	7	165	24	82
HM31A-F	—	1.2	3.0	—	—	—	290	42	230	33	185	27	345	50	10	150	22	—
M1A-F	—	1.2	—	—	—	—	255	37	180	26	83	12	195	28	12	125	18	44
ZK21A-F	—	—	—	2.3	0.45[b]	—	260	38	195	28	135	20	—	—	4	—	—	—
ZK40A-T5	—	—	—	4.0	0.45[b]	—	276	40	255	37	140	20	—	—	4	—	—	—
ZK60A-T5	—	—	—	5.5	0.45[b]	—	365	53	305	44	250	36	405	59	11	180	26	88
Sheet and Plate																		
AZ31B-H24	3.0	—	—	1.0	—	—	290	42	220	32	180	26	325	47	15	160	23	73
HK31A-H24	—	—	3.0	—	0.6	—	255	33	200	29	160	23	285	41	9	140	20	68
HM21A-T8	—	0.6	2.0	—	—	—	235	34	170	25	130	19	270	39	11	125	18	—

[a]From *Metals Handbook* [4.15b], with permission of ASM International.
[b]Minimum.
[c]500-kg load; 10-mm ball.
[d]For battery applications.
[e]A and B are identical except that 0.30% max residual Cu is allowable in AZ91B
[f]Properties of B and C are the same except that AZ31C contains 0.15% min Mn, 0.1% max Cu, and 0.03% max Ni.
[g]Didymium.
[h]Rare earth.

TABLE 4-17   Mechanical Properties of Magnesium Alloys   196

TABLE 4-18   RANGES OF MECHANICAL PROPERTIES FOR SELECTED NONFERROUS ALLOYS[a]

Metals and Their Alloys	Tensile Strength (ksi)	Yield Strength (ksi)	Elongation in 2 in. (%)	Brinell Hardness
Titanium				
Heat treated	145–240	135–220	1–12	—
Annealed	60–170	40–150	—	—
Nickel				
Annealed and age hardened	130–190	90–120	10–25	—
Cast, annealed,			1–45 cast	—
and aged	30–145	—	1–4	300–380
Annealed	50–120	12–65		
Copper				
Hard	50–55	45	—	194
Annealed	32–35	10	35–45	40
Zinc				
Cast	25–47.6	—	1–10	—
Corrosion resistant	21–46	—	28	60–80
Heat resistant	19.5–42	—	10–65	51–61
Tin	2.8–8.7	1.3 annealed, 2–6 corrosion resistant	35 cold rolled, 55 cast	7 annealed
Lead		0.8–1.6		
Rolled	2.4–4.7		43–51	5.9–9.5
Extruded	2–3.3	—	48–75	5.1–12.4

[a]Data collected from [4.10].

TABLE 4-19 COMMONLY USED ABBREVIATIONS FOR ENGINEERING PLASTICS[a]

ABS	Acrylonitrile–butadiene–styrene terpolymer
ACPES	Acrylonitrile-chlorinated polyethylene–styrene terpolymer
BR	Butadiene rubber
CPE	Chlorinated polyethylene
CPVC	Chlorinated poly(vinyl chloride)
ECTFE	Ethylene chlorotrifluoroethylene copolymer
EPD	Ethylene–propylene–diene terpolymer
EPM	Ethylene–propylene copolymer
ETFE	Ethylene tetrafluoroethylene copolymer
EVA	Ethylene–vinyl acetate copolymer
FEP	Fluorinated ethylene propylene
HDPE	High-density polyethylene
HIPS	High-impact polystyrene
LDPE	Low-density polyethylene
LLDPE	Linear low-density polyethylene
MF	Melamine–formaldehyde resin
NBR	Acrylonitrile–butadiene rubber (nitrile rubber)
PAN	Polyacrylonitrile
PB	Polybutadiene
PBT	Poly(butylene terephthalate)
PC	Polycarbonate
PE	Polyethylene
PEG	Polyethylene glycol
PET	Poly(ethylene terephthalate)
PF	Phenol–formaldehyde resin
PIB	Polyisobutylene
PIR	Polyisocyanurate foam
PMMA	Poly(methyl methacrylate)
PP	Polypropylene
PPG	Polypropylene glycol
PPO	Poly(phenylene oxide)
PPS	Poly(phenylene sulfide)
PS	Polystyrene
PTFE	Polytetrafluoroethylene
PTMG	Polytetramethylene glycol
PTMT	Poly(tetramethylene terephthalate)
PUR	Polyurethane
PVA	Poly(vinyl acetate)
PVAL	Poly(vinyl alcohol)
PVB	Poly(vinyl butyral)
PVC	Poly(vinyl chloride)
PVF	Poly(vinyl formal)
RTV	Room temperature vulcanizing silicone rubber
SAN	Styrene–acrylonitrile copolymer
SBR	Styrene–butadiene rubber
UF	Urea–formaldehyde resin

[a]From *Annual Book of ASTM Standards* [4.19]. Copyright ASTM. Printed with permission.

TABLE 4-19 Abbreviations for Engineering Plastics   198

TABLE 4-20   TYPICAL PROPERTIES OF PLASTICS USED FOR MOLDING AND EXTRUSION[a]

	ASTM Test Method	Polyethylene		Polypropylene
		Low Density	High Density	
1. Specific gravity[b]	D792	0.91–0.925	0.94–0.965	0.900–0.910
2. Tensile modulus ($\times 10^{+5}$ psi)[c]	D638	0.14–0.38	0.6–1.8	1.6–2.25
3. Compressive modulus ($\times 10^{+5}$ psi)	D695	—	—	1.5–3.0
4. Flexural modulus ($\times 10^{+5}$ psi)	D790	0.08–0.6	1.0–2.6	1.7–2.5
5. Tensile strength ($\times 10^{+5}$ psi)	D638, D651	0.6–2.3	3.1–5.5	4.5–6.0
6. Elongation at break (%)	D638	90–800	20–130	100–600
7. Compressive strength ($\times 10^{+3}$ psi)	D695	2.7–3.6	12–18	5.5–8.0
8. Flexural yield strength ($\times 10^{+3}$ psi)	D790	—	1.0	6–8
9. Impact strength, notched Izod, (ft-lb/in.)	D256	No break	0.5–20	0.4–1.0
10. Hardness, Rockwell	D785	D40–51 (Shore)	D60–70 (Shore)	R80–102

	ASTM Test Method	Polystyrene		Poly(methyl methacrylate)
		General Purpose	Impact Resistant	
1. Specific gravity[b]	D792	1.04–1.05	1.03–1.06	1.17–1.20
2. Tensile modulus ($\times 10^{+5}$ psi)[c]	D638	3.5–4.85	2.6–4.65	3.8
3. Compressive modulus ($\times 10^{+5}$ psi)	D695	—	—	3.7–4.6
4. Flexural modulus ($\times 10^{+5}$ psi)	D790	4.3–4.7	3.3–4.0	4.2–4.6
5. Tensile strength ($\times 10^{+3}$ psi)	D638, D651	5.3–7.9	3.2–4.9	7–11
6. Elongation at break (%)	D638	1–2	13–50	2–10
7. Compressive strength ($\times 10^{+3}$ psi)	D695	11.5–16	4–9	12–18
8. Flexural yield strength ($\times 10^{+3}$ psi)	D790	8.7–14	5–12	13–19
9. Impact strength, notched Izod, (ft-lb/in.)	D256	0.25–0.40	0.5–11	0.3–0.5
10. Hardness, Rockwell	D785	M65–80	M20–80	M85–105

TABLE 4-20 Typical Properties of Plastics 200

	ASTM Test Method	Poly(vinyl chloride)		ABS Medium Impact
		Rigid	Plasticized	
1. Specific gravity[b]	D792	1.30–1.58	1.16–1.35	1.03–1.06
2. Tensile modulus ($\times 10^{+5}$ psi)[c]	D638	3.5–6	—	3–4
3. Compressive modulus ($\times 10^{+5}$ psi)	D695	—	—	2.0–4.5
4. Flexural modulus ($\times 10^{+5}$ psi)	D790	3–5	—	3.7–4.0
5. Tensile strength ($\times 10^{+3}$ psi)	D638, D651	6–7.5	1.5–3.5	6–7.5
6. Elongation at break (%)	D638	2–80	200–450	5–25
7. Compressive strength ($\times 10^{+3}$ psi)	D695	8–13	0.9–1.7	10.5–12.5
8. Flexural yield strength ($\times 10^{+3}$ psi)	D790	10–16	—	11–13
9. Impact strength, notched Izod, (ft-lb/in.)	D256	0.4–20	—	3–6
10. Hardness, Rockwell	D785	D65–85 (Shore)	A40–100 (Shore)	R107–115

TABLE 4-20 (continued) **TYPICAL PROPERTIES OF PLASTICS USED FOR MOLDING AND EXTRUSION**[a]

	ASTM Test Method	Cellulose Acetate	Cellulose Acetate-Butyrate	Fluoropolymers	
				$-CF_2-CF_2-$	$-CF_2-CFCl-$
1. Specific gravity[b]	D792	1.22–1.34	1.15–1.22	2.14–2.20	2.1–2.2
2. Tensile modulus ($\times 10^{+5}$ psi)[c]	D638	0.65–4.0	0.5–2.0	0.58	1.5–3.0
3. Compressive modulus ($\times 10^{+5}$ psi)	D695	—	—	—	—
4. Flexural modulus ($\times 10^{+5}$ psi)	D790	—	—	—	—
5. Tensile strength ($\times 10^{+3}$ psi)	D638, D651	1.9–9.0	2.6–6.9	2–5	4.5–6.0
6. Elongation at break (%)	D638	6–70	40–88	200–400	80–250
7. Compressive strength ($\times 10^{+3}$ psi)	D695	3–8	2.1–7.5	1.7	4.6–7.4
8. Flexural yield strength ($\times 10^{+3}$ psi)	D790	2–16	1.8–9.3	—	7.4–9.3
9. Impact strength, notched Izod, (ft-lb/in.)	D256	1–7.8	1–11	3.0	2.5–2.7
10. Hardness, Rockwell	D785	R34–125	R31–116	D50–55 (Shore)	R75–95

[a]From Chanda and Roy [4.27]. Reprinted from Ref. 4.27 pp. 518–525 by courtesy of Marcel Dekker.
[b]Specific gravity is defined as the ratio of the mass in air per unit volume of an impermeable portion of the material at 23°C to the mass in air of an equal volume of gas-free distilled water at the same temperature.
[c]For psi, multiply tabulated values by 10^{+5}.

TABLE 4-20 **Typical Properties of Plastics** 202

TABLE 4-21 GLASS TRANSITION TEMPERATURE FOR SELECTED PLASTICS

Thermoplastics	$T_g(°C)$	Thermosets	$T_g(°C)$
Acrylonitrile–		Alkyds, glass filled	200
butadiene–styrene, glass filled	80	Epoxies	150–250
Polyamides,	60	Epoxies, glass filled	150–250
30% glass filled		Melamines, glass filled	200
Polycarbonate, 30% glass filled	150	Phenolics, glass filled	200–300
Polyethylene		Polybutadienes, glass filled	200
Low density	−20	Polyesters, glass filled	200
High density		Polyimides, glass filled	350
Polyethylene terephthalate	67	Silicones, glass filled	300
Polymethylmethacrylate	100	Ureas	80
Poly-4-methylpentene-1	55	Urethanes, solid	100
Polyoxymethylene	−13		
Polyphenyleneoxide, glass filled	180		
Polyphenylenesulfide, 40% glass filled	150		
Polypropylene, glass filled	0		
Polystyrene	100		
Polysulfone, 30% glass filled	200		
Polytetrafluoroethylene	120		
Polyvinylchloride	80		

[a]From Felbeck [4.14]. Reprinted by permission of Prentice Hall, Englewood Cliffs, New Jersey.

TABLE 4-22 Properties of Ceramics and Glasses 204

TABLE 4-22 TYPICAL ROOM TEMPERATURE MECHANICAL PROPERTIES OF CERAMICS AND GLASSES[a]

	Melting Point (°C)	Tensile or Bending Strength σ (MPa)	Young's Modulus E (GPa)	Coefficient of Thermal Expansion $\times 10^{-6}$ (°C^{-1})[b]
Alumina (Al$_2$O$_3$)	2050	300 (700 whisker)	380	8
Magnesia (MgO)	2850	100	315	14.8
Silicon carbide (SiC)	2300 (decomposes)	200	350	4.5
Silicon nitride (Si$_3$N$_4$)	1900 (sublimes)	600 (1400 whisker)	320	2.5
Titanium carbide (TiC), Ni/Mo binder	3250	200	350	7
Tungsten carbide (WC), Co binder	2620	350	700	1
Common bulk glass	Fuses at 1600	70 (up to 4000 filaments)	70	10
Pyroceram glass ceramic	Fuses at 1600	190	126	10
Concrete, reinforced		20	35	7
Carbon graphite	3600 (sublimes)	30	6	3

[a]From Felbeck [4.14]. Reprinted by permission of Prentice Hall, Englewood Cliffs, New Jersey. *Note:* There is much more variaion in properties of these types of solid than metals, variations occurring with production methods. Properties also vary with temperature and rate of loading.
[b]Multiply tabulated values by 10^{-6}.

TABLE 4-23   TYPICAL ROOM TEMPERATURE STRENGTHS OF CERAMIC MATERIALS[a]

Material	Bending Strength (MOR)[b]		Tensile Strength	
	MPa	ksi	MPa	ksi
Sapphire (single-crystal, Al_2O_3)	620	90	—	—
Al_2O_3 (0–2% porosity)	350–580	50–80	200–310	30–45
Sintered Al_2O_3 (< 5% porosity)	200–350	30–50	—	—
Alumina porcelain (90–95% Al_2O_3)	275–350	40–50	172–240	25–35
Sintered BeO (3.5% porosity)	172–275	25–40	90–133	13–20
Sintered MgO (< 5% porosity)	100	15	—	—
Sintered stabilized ZrO_2 (< 5% porosity)	138–240	20–35	138	20
Sintered mullite (< 5% porosity)	175	25	100	15
Sintered spinel (< 5% porosity)	83–220	12–32	—	19
Hot-pressed Si_3N_4 (< 1% porosity)	620–965	90–140	350–580	50–80
Sintered Si_3N_4 (~ 5% porosity)	414–580	60–80	—	—
Reaction-bonded Si_3N_4 (15–25% porosity)	200–350	30–50	100–200	15–30
Hot-pressed SiC (< 1% porosity)	621–825	90–120	—	—
Sintered SiC (~ 2% porosity)	450–520	65–75	—	—
Reaction-sintered SiC (10–15% free Si)	240–450	35–65	—	—
Bonded SiC (~ 20% porosity)	14	2	—	—
Fused SiO_2	110	16	69	10
Vycor or pyrex glass	69	10	—	—
Glass–ceramic	245	10–35	—	—
Machinable glass–ceramic	100	15	—	—

Material	Bending Strength (MOR)[b]		Tensile Strength	
	MPa	ksi	MPa	ksi
Hot-pressed BN (< 5% porosity)	48–100	7–15	—	—
Hot-pressed B$_4$C (< 5% porosity)	310–350	45–50	—	—
Hot-pressed TiC (< 2% porosity)	275–450	40–65	240–275	35–40
Sintered WC (2% porosity)	790–825	115–120	—	—
Mullite porcelain	69	10	—	—
Steatite porcelain	138	20	—	—
Fire-clay brick	5.2	0.75	—	—
Magnesite brick	28	4	—	—
Insulating firebrick (80–85% porosity)	0.28	0.04	—	—
2600°F insulating firebrick (75% porosity)	1.4	0.2	—	—
3000°F insulating firebrick (60% porosity)	2	0.3	—	—
Graphite (grade ATJ)	28	4	12	1.8

[a]From Richerson [4.26]. Reprinted from Ref. 4.26 pp. 92–93 by courtesy of Marcel Dekker.
[b]*Bending strength* is defined as the maximum tensile stress at failure (of a 3-point loading specimen) and is often referred to as the modulus of rupture, or MOR.

TABLE 4-23   **Strengths of Ceramic Materials**   206

TABLE 4-24   MECHANICAL PROPERTIES OF FLAKE- AND FILAMENT-REINFORCED COMPOSITES[a]

Type[b]	Tensile strength		Tensile modulus		Compressive strength		Compressive modulus		Flexural strength		Flexural modulus		Interlaminar shear strength[c]	
	ksi	MPa	$\times 10^6$ psi[g]	GPa	ksi	MPa	$\times 10^6$ psi[g]	GPa	ksi	MPa	$\times 10^6$ psi[g]	GPa	ksi	MPa
Unidirectional														
Boron														
0°	110	758	43	296	164	1131	42	290	220	1517	24	165	12	83
45°	—	—	—	—	16	110	2.8	19	16	110	4	28		
90°	3	21	5	35	17	117	3	21	8	55	4	28		
E glass														
0°	160	1103	8	55	113	779	5	35	137	945	5	35	12	83
45°	25	172	—	—	19	131	—	—	11	76	1.3	9		
90°	4	28	1.4	10	17	117	1.2	8	6	41	1	7		
Pseudo-isotropic[d]														
Boron														
0°	23	159	10	69	89	614	18	124	52	357	12	83	3	21
45°	25	172	10	69	86	593	14	97	42	290	9	62		
90°	26	179	8	55	84	579	15	103	47	324	10	69		
E Glass														
0°	60	414	2	14	54	372	3	21	69	469	3	21	5	35
45°	46	317	2	14	50	345	3	21	63	434	2	14		
90°	50	345	2	14	51	350	3	21	62	427	2.5	17		

TABLE 4-24 (continued)   **MECHANICAL PROPERTIES OF FLAKE- AND FILAMENT-REINFORCED COMPOSITES[a]**

Type[b]	Tensile strength		Tensile modulus		Compressive strength		Compressive modulus		Flexural strength		Flexural modulus		Interlaminar shear strength[c]	
	ksi	MPa	×10⁶ psi[g]	GPa	ksi	MPa	×10⁶ psi[g]	GPa	ksi	MPa	×10⁶ psi[g]	GPa	ksi	MPa
AlB₂ flake composites														
Binary,[e]														
0°	35	241	33	228	23	159	19	131	56	386	20	138		
45°	35	241	33	228	—	—	19	131	56	386	20	138	5	35
90°	35	241	33	228	—	—	19	131	56	386	20	138		
Ternary,[f]														
0°	70	483	22	152	—	—	17	117	80	552	16	110		
45°	—	—	—	—	—	—	—	—	—	—	—	—	4.5	31
90°	24	165	17	117	—	—	—	—	33	228	11	76		

[a] From Schwartz [4.21], with permission.
[b] 0° = 0 rad, 45° = 0.79 rad, and 90° = 1.6 rad.
[c] Value for all orientations.
[d] Fiber axis aligned at 0, 45, and 90°.
[e] 64 vol% AlB₂ flakes in epoxy matrix.
[f] Alternate layers of flake-reinforced epoxy and glass-filament-reinforced epoxy.
[g] For psi, multiply tabulated values by 10⁶.

TABLE 4-25   MATERIAL LAW MATRICES

Material	Number of Nonzero Coefficients	Number of Independent Coefficients
Three-dimensional case		
General anisotropic	36	21
One plane of symmetry	20	13
Orthotropic	12	9
Transversely isotropic	12	5
Isotropic	12	2
Two-dimensional case		
General anisotropic	9	6
Transversely isotropic	5	4
Isotropic	5	2

General anisotropic

$$
\begin{bmatrix} \varepsilon_x \\ \varepsilon_y \\ \varepsilon_z \\ \gamma_{yz} \\ \gamma_{xz} \\ \gamma_{xy} \end{bmatrix}
=
\begin{bmatrix}
a_{11} & a_{12} & a_{13} & a_{14} & a_{15} & a_{16} \\
 & a_{22} & a_{23} & a_{24} & a_{25} & a_{26} \\
 & & a_{33} & a_{34} & a_{35} & a_{36} \\
 \text{Symmetric} & & & a_{44} & a_{45} & a_{46} \\
 & & & & a_{55} & a_{56} \\
 & & & & & a_{66}
\end{bmatrix}
\begin{bmatrix} \sigma_x \\ \sigma_y \\ \sigma_z \\ \tau_{yz} \\ \tau_{xz} \\ \tau_{xy} \end{bmatrix}
$$

One Plane $(z = 0)$ of Symmetry

$$
\begin{bmatrix} \varepsilon_x \\ \varepsilon_y \\ \varepsilon_z \\ \gamma_{yz} \\ \gamma_{xz} \\ \gamma_{xy} \end{bmatrix}
=
\begin{bmatrix}
a_{11} & a_{12} & a_{13} & 0 & 0 & a_{16} \\
 & a_{22} & a_{23} & 0 & 0 & a_{26} \\
 & & a_{33} & 0 & 0 & a_{36} \\
 \text{Symmetric} & & & a_{44} & a_{45} & 0 \\
 & & & & a_{55} & 0 \\
 & & & & & a_{66}
\end{bmatrix}
\begin{bmatrix} \sigma_x \\ \sigma_y \\ \sigma_z \\ \tau_{yz} \\ \tau_{xz} \\ \tau_{xy} \end{bmatrix}
$$

Orthotropic

$$
\begin{bmatrix} \varepsilon_x \\ \varepsilon_y \\ \varepsilon_z \\ \gamma_{yz} \\ \gamma_{xz} \\ \gamma_{xy} \end{bmatrix}
=
\begin{bmatrix}
a_{11} & a_{12} & a_{13} & 0 & 0 & 0 \\
 & a_{22} & a_{23} & 0 & 0 & 0 \\
 & & a_{33} & 0 & 0 & 0 \\
 \text{Symmetric} & & & a_{44} & 0 & 0 \\
 & & & & a_{55} & 0 \\
 & & & & & a_{66}
\end{bmatrix}
\begin{bmatrix} \sigma_x \\ \sigma_y \\ \sigma_z \\ \tau_{yz} \\ \tau_{xz} \\ \tau_{xy} \end{bmatrix}
$$

Experimental Stress Analysis

Practical problems sometimes are so complicated that there is reluctance to use simple formulas for the calculation of strains and stresses. Then experimental and numerical techniques can be helpful, with experimental methods useful both for treating complete engineering problems and for verifying the correctness of analytical or computational analyses. Since usually stress cannot be measured directly, most experimental methods serve to measure strains, making the title of this chapter somewhat of a misnomer. This chapter provides introductory information on the use of strain gages, photoelasticity, and brittle coatings.

5.1 NOTATION

b	Width of tensile specimen (L)
c	Speed of light in a vacuum (L/T)
d	Specimen thickness (L)
E	Young's modulus (F/L^2)
E_V	Output voltage (V)
E_l	Magnitude of electric field vector (F/charge)
f	Frequency (cycles/T)
F	Force (F)
f_σ	Material fringe value (F/L)
I	Current (A)
K	Stress-optic coefficient (brewsters, L^2/F)
N	Fringe order
P_g	Power dissipated by strain gage (FL/T)

R	Resistance of uniform conductor of length L, cross-sectional area A, and specific resistance ρ, $= \rho L/A$ (Ω)
S_A	Sensitivity of material of gage wire
S_c	Circuit sensitivity (V)
S_g	Gage factor
V	Applied voltage (V)
u, v, w	Displacement components (L)
α	Angle between principal axis of stress and polarizer axis
γ	Shear strain
Δ	Phase lag (rad)
ΔE	Voltage fluctuation (V)
ε	Unit extension or strain (L/L)
ε_t	Threshold strain of brittle coating (L/L)
λ	Wavelength (L)
ν	Poisson's ratio
θ	Angle to principal direction
σ	Stress (F/L^2)
σ_1, σ_2	Principal stresses (F/L^2)
σ_{ut}	Ultimate tensile strength (F/L^2)
σ_{uc}	Ultimate compressive strength (F/L^2)
ρ	Specific resistance
Ω	Resistance unit (volts/amps or ohms)
ω	Circular frequency of wave (rad/T)

5.2   INTRODUCTION

To improve upon the simple use of a micrometer to find the changes in length of a specimen after it is loaded, methods such as the Moire technique, interferometric strain gages, electric strain gages, brittle coatings, photoelasticity, X-ray diffraction, holographs, and laser speckle interferometry are employed.

A Moire pattern is defined as a visual pattern produced by the superposition of two regular motifs that geometrically interfere. These motifs are parallel lines, rectangular arrays of dots, concentric circles, or radial lines. Moire patterns are used to measure displacements, rotations, curvature, and strain.

Interferometric gages measure the change in grating pitch deposited at a desired area of specimen in terms of optical interference.

Holographics and laser speckle interferometry are relatively recent and important developments in experimental mechanics. They permit the extension of interferometry measurements of diffuse objects.

X-ray diffraction can be used to determine changes in interatomic distances. This can be very useful in analyzing stress concentration and residual stress.

Analogies are important in experimental studies. For example, they use correspondences between governing differential equations of torsion and membrane film or between differential equations of solid mechanics and electromagnetics.

The majority of current applications utilize electric strain gages. Also, photoelasticity has been demonstrated to be successful in the analysis of stress distributions within models. This chapter presents only a brief introduction of electric strain gage, photoelasticity, and brittle coating methodologies. The bases of other methods are beyond the scope of this book.

Since the 1950s, experimental stress analysis technology has developed rapidly. Developments continue in high-precision instrumentation and on-line computer processing of experimental data in real time. On-line computers can control hundreds of strain gages and process all the data automatically. This can reduce both the time and cost. Holography and laser speckle techniques are very effective experimental methods and often involve huge amounts of data processing, which can now be handled by computers. Most experimental methods can be categorized as mechanical, electrical, or optical methods. The introduction of the basic principles here of electric strain gages and photoelasticity will be helpful in understanding some of the new methods.

5.3 THE ELECTRICAL RESISTANCE STRAIN GAGE

The electrical resistance strain gage is the most frequently used device in experimental stress analysis. The gages are also used as sensors in transducers for measuring load, torque, pressure, and acceleration. The electrical resistance strain gage operates on the principle discovered by Lord Kelvin in 1856 that the electrical resistance of metal wire varies with strain. The fractional change in resistance (R) per unit extension (ε) is known as the sensitivity (S_A) of the metal or alloy of which the wire is made:

$$S_A = \frac{\Delta R/R}{\varepsilon} = 1 + 2\nu + \frac{\Delta \rho/\rho}{\varepsilon} \tag{5.1}$$

where ρ is the specific resistance.

The sensitivities of typical strain gage alloys are listed in Table 5-1. The advance or constantan alloys are widely used because the sensitivity varies little over a wide range of temperature and strains (even in the plastic region). The high sensitivity and high fatigue strength of the isoelastic alloy give it advantages in dynamic applications. The sensitivity of isoelastic gages changes with both temperature and strain, however.

The most common constructions of the modern strain gage are the bonded wire and bonded foil types. The foil gage is produced by etching a metal foil into a grid pattern. The metal foil strain gage is the most widely used gage for both general-purpose stress analysis and transducer applications. To facilitate handling, the wire or foil grid is mounted on or encapsulated in a paper or epoxy carrier or backing. The manufacturer's identifying code for a gage usually gives such information as backing type, alloy, length, and resistances. Foil gage lengths typically vary from 0.008 in. (0.20 mm) to 4 in. (102 mm) and resistances are from 120 to 350 Ω. Gages with lengths greater than 0.060 in. (1.52 mm) are also available with a resistance of 1000 Ω. The manufacturer specifies the gage factor (S_g), which is defined as

$$S_g = (\Delta R/R)/\varepsilon_a \tag{5.2}$$

where ε_a is the uniform normal strain along the axial direction of the gage. The resistance change in the definition of S_g includes effects due to transverse extensions (shear strains are negligible in measuring S_g). Manufacturer's literature usually supply values for the transverse sensitivity of gages and formulas for deriving true axial extension from the apparent extension indicated by the gage.

Backings are usually made of paper or glass-fiber-reinforced epoxy. The latter is applicable to moderate temperatures up to 750°F (400°C) or, if special precautions are taken, to even higher temperatures. Another type of gage is the weldable strain gage, which is suitable for application within the range -320 to 1200°F (-200 to 650°C) or for outdoor installation in inclement weather. See Dally and Riley [5.1] for further information.

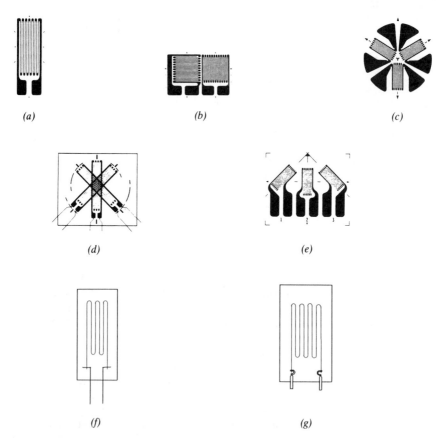

(a) (b) (c)

(d) (e)

(f) (g)

FIGURE 5-1 Examples of strain gage configurations: (a) uniaxial foil, (b) 2-element 90° "tee" rosette, (c) 60° rosette, (d) 3-element 45° stacked rosette, (e) 45° rosette, (f) uniaxial wire, (g) uniaxial wire, with ribbon leads. Courtesy of the Micro-Measurements Division of Measurement Group, Inc., Raleigh, N.C.

Several popular gage configurations are shown in Fig. 5-1. A *rosette* is the combination of two or three gages in one assembly. If nothing is known beforehand about the strain field, a three-element rosette is required for finding the elements of the small strain tensor. If the principal directions are known beforehand, a two-element 90° rosette suffices to measure the principal strains. In some cases such as uniaxial extension, bending, or torsion of rods, only one gage is necessary to find the strain.

The strain measurements are made by bonding the gage to the surface of the specimen under test and by sensing voltage changes that occur when the resistance of the strained gage changes. The application of the gage to the specimen surface is a critical step in the measurement process, and gage manufacturers provide detailed instructions for preparation of the specimen surface, bonding the gage to the surface and making electrical connections. Among the many adhesives used for applying the gage to the surface, methyl-2-cyanoacrylate, epoxy, polyimide, and several ceramics are very common. Upon completion of the installation, it is desirable to inspect the adequacy of the bonds. To test the relative completeness of the bond cure, the resistance between the gage grid and the specimen can be measured. This follows because the resistance of the adhesive layer increases as the adhesive cures. The typical resistance across the adhesive layer for strain gage installation is of the order of 10,000 MΩ [5.1].

Two basic circuits are used to measure the voltage changes across the resistance gages: the Wheatstone bridge and the potentiometer. The Wheatstone bridge is applied in both static and dynamic experiments, but the potentiometer is suitable only for dynamic signals.

The circuit of a basic Wheatstone bridge, where voltage fluctuation ΔE is to be measured in order to determine the strain, is sketched schematically in Fig. 5-2. The applied voltage V is constant. For the circuit elements in parallel with the source voltage,

$$I_1(R_1 + R_2) = V, \qquad I_2(R_3 + R_4) = V \qquad (5.3)$$

The voltage difference across BD, E_V (or V_{BD}) is

$$E_V = V_{BD} = V_{BC} - V_{DC} = I_1 R_2 - I_2 R_3 \qquad (5.4)$$

where V_{BC} and V_{DC} are the voltage differences across BC and DC, respectively.

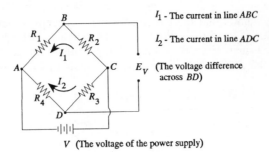

I_1 - The current in line ABC

I_2 - The current in line ADC

E_V (The voltage difference across BD)

V (The voltage of the power supply)

FIGURE 5-2 Basic Wheatstone bridge.

Using Eq. (5.3) in (5.4) gives

$$E_V = -\frac{R_1 R_3 - R_2 R_4}{(R_1 + R_2)(R_3 + R_4)} V \tag{5.5}$$

The bridge is balanced when $E_V = 0$, or $R_1 R_3 = R_2 R_4$. In the simplest cases, one resistance, say R_1, will be the strain gage. If R_1 changes by an amount ΔR_1 due to strain, the corresponding voltage fluctuation ΔE is calculated as

$$\Delta E = -\frac{(R_1 + \Delta R_1) R_3 - R_2 R_4}{(R_1 + \Delta R_1 + R_2)(R_3 + R_4)} V + \frac{R_1 R_3 - R_2 R_4}{(R_1 + R_2)(R_3 + R_4)} V$$

$$= \frac{-R_3 \Delta R_1 V}{(R_1 + R_2)(R_3 + R_4)} \tag{5.6}$$

where the products in the denominator of ΔR_1 with R_3 and R_4 have been neglected and the relation $R_1 R_3 = R_2 R_4$ has been used. (The neglected terms are small up to a strain of about 0.05). Substituting $R_3 = R_2 R_4/R_1$ in Eq. (5.6) gives

$$\Delta E = -\frac{(R_2 R_4/R_1) \Delta R_1 V}{(R_1 + R_2)(1 + R_2/R_1) R_4} = -\frac{(R_2/R_1)(\Delta R_1/R_1)}{(R_2/R_1 + 1)^2} V$$

$$= -\frac{r}{(1 + r)^2} \frac{\Delta R_1}{R_1} V \tag{5.7}$$

where $r = R_2/R_1$ and $r/(1 + r)^2$ is the circuit efficiency. The sensitivity of the circuit is the voltage change per unit extension:

$$S_c = \left|\frac{\Delta E}{\varepsilon_a}\right| = \frac{1}{\varepsilon_a} \frac{r}{(1 + r)^2} \frac{\Delta R_1}{R_1} V \tag{5.8}$$

Substituting Eq. (5.2) in (5.8) gives

$$S_c = \frac{r}{(1 + r)^2} V S_g \tag{5.9}$$

Equation (5.9) shows that the circuit sensitivity depends on the static voltage V, the gage factor S_g, and the ratio R_2/R_1. The circuit efficiency is a maximum for $R_2/R_1 = 1$. Equation (5.9) is valid if the bridge voltage V is fixed and independent of the gage current. The power dissipated by the gage is

$$P_g = I_g^2 R_g \tag{5.10}$$

Substituting Eq. (5.10) in (5.3) with $I_1 = I_g$ and $R_1 = R_g$ gives

$$V = \sqrt{P_g/R_g} (R_g + R_2) = \sqrt{P_g/R_g} R_g (1 + r) = (1 + r)\sqrt{P_g R_g} \tag{5.11}$$

Using Eq. (5.11) in (5.9) to eliminate V yields

$$S_c = rS_g\sqrt{P_gR_g}\,/(1+r)\qquad\qquad(5.12)$$

The term $S_g\sqrt{P_gR_g}$ is fixed by the gage selection. Maximum power dissipation is part of the information supplied by gage manufacturers. The term $r/(1+r)$ is determined by the design of the bridge circuit.

Figure 5-2 shows the Wheatstone circuit in its basic configuration. The above discussion is restricted to the simple case of one gage resistance in the bridge. The bridge is balanced before strains are applied to the gage in the bridge. Therefore, the voltage E is initially zero, and the strain-induced voltage ΔE can be measured directly. Since in many cases the strain gage installation is subjected to temperature change during the testing period, the effects of temperature must be eliminated. Often the Wheatstone bridge can be designed to nullify the temperature effects. Table 5-2 lists some common Wheatstone bridges in use today. The gage used to measure the strain is the *active strain gage*, whereas the *dummy gage* is mounted on a small block of material identical to that of the specimen and is exposed to the same thermal environment as the active gage. It can be shown that all but circuit 1 in Table 5-2 are temperature compensated if all the active gages in the circuit are also subject to the same thermal environment and mounted on the same material. Commercially available strain indicators have a much more complicated circuitry than shown in Table 5-2, and they give direct readout of strain.

Proper calibration of a strain gage measuring system is important. A strain-measuring system usually consists of a strain gage, a Wheatstone (or potentiometer) circuit, a power supply, circuit completion resistors, a signal amplifier, and a recording instrument. Each element contributes to overall system sensitivity. If circuit sensitivity S_c is known, the strain $|\varepsilon_a|$ can be calculated using [Eq. (5.8)] $|\varepsilon_a| = S_c|\Delta E|$. A single calibration for the complete system can be achieved by shunting a fixed resistor R_c across one arm (e.g., R_2) of the Wheatstone bridge (shown in Fig. 5-3) so that the readings from the recording instrument can be directly related to the strains that induce them. If the bridge is initially balanced, it can be shown that

$$\varepsilon_c = R_2/S_g(R_2 + R_c)\qquad\qquad(5.13)$$

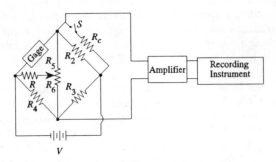

FIGURE 5-3 Typical strain-measuring system.

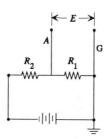

FIGURE 5-4 Potentiometer circuit.

where ε_c is the calibration strain that produces the same voltage output (ΔE) from the bridge as the calibration resistor R_c as it is placed in parallel with R_2. Thus, if the output of the recording instrument is h_c while the switch s is closed with R_c and the strain-induced output is h while the switch s is open, the strain associated with output h can be numerically calculated as

$$\varepsilon = (h/h_c)\varepsilon_c \tag{5.14}$$

This is the principle of shunt calibration. It provides an accurate and direct method for calibrating the complete system without considering the number of components in the system.

The potentiometer circuit sketched in Fig. 5-4 can be utilized to measure dynamic strains. The gage is R_1 ($R_1 = R_g$) in the figure. The circuit has the same sensitivity as the Wheatstone bridge,

$$S_c = rS_g\sqrt{P_g R_g}/(1 + r) \tag{5.15}$$

and a linear range of strain of up to 0.02–0.1 depending on the value of $r = R_2/R_1$. The circuit is only useful for dynamic strain measurement because the large static voltage E must be filtered out.

In this section, only the rudiments of strain gage technology have been discussed. In practice other complications must be considered, such as humidity, transverse sensitivity, gage heating due to electric power dissipation, stability for long-term measurement, distortion of transient strain pulses, cyclic loading, and the effect of recording instruments on the data. Many gages, which are self–temperature compensated to some extent, may also nullify temperature effects. Discussion of these refinements to strain measurement is available in Dally and Riley [5.1] and in the technical literature of manufacturers of strain gages.

Example 5.1 Delta Rosette The delta rosette utilizes three gages separated by 120°, as shown in Fig. 5-5. Gage 1 is parallel to the x direction, gage 2 is 120° counterclockwise from the x direction, and gage 3 is 240° counterclockwise. If extensions are measured of $\varepsilon_{g1} = 250 \times 10^{-6}$ in./in., $\varepsilon_{g2} = 150 \times 10^{-6}$ in./in., and $\varepsilon_{g3} = 400 \times 10^{-6}$ in./in., compute the components of the strain tensor, the principal strains, and the principal stresses. Neglect the transverse sensitivity of the gages and assume the strained specimen has $E = 30 \times 10^6$ psi, $\nu = 0.3$.

x

FIGURE 5-5 Delta rosette.

Substitution of the appropriate extensions and angles into Eq. (3.38) results in three equations for the unknowns ε_x, ε_y, and γ_{xy}. Thus,

$$250 \times 10^{-6} = \varepsilon_x$$
$$150 \times 10^{-6} = \varepsilon_x \cos^2(120°) + \varepsilon_y \sin^2(120°) + \gamma_{xy} \sin(120°)\cos(120°) \quad (1)$$
$$400 \times 10^{-6} = \varepsilon_x \cos^2(240°) + \varepsilon_y \sin^2(240°) + \gamma_{xy} \sin(240°)\cos(240°)$$

or

$$8.75 \times 10^{-5} = 0.75\varepsilon_y - 0.433\gamma_{xy}$$
$$3.375 \times 10^{-4} = 0.75\varepsilon_y + 0.433\gamma_{xy} \tag{2}$$

The solutions to these equations are

$$\gamma_{xy} = 2.8868 \times 10^{-4}, \qquad \varepsilon_y = 2.8333 \times 10^{-4}, \qquad \varepsilon_x = 2.50 \times 10^{-4} \quad (3)$$

The principal strains follow from formulas for strains similar to the principal-stress formulas of Eq. (3.13),

$$\left.\begin{matrix}\varepsilon_1 \\ \varepsilon_2\end{matrix}\right\} = \tfrac{1}{2}(\varepsilon_x + \varepsilon_y) \pm \tfrac{1}{2}\sqrt{(\varepsilon_x - \varepsilon_y)^2 + \gamma_{xy}^2}$$

$$= \tfrac{1}{2}(2.50 \times 10^{-4} + 2.8333 \times 10^{-4})$$

$$\pm \tfrac{1}{2}\left[(2.50 \times 10^{-4} - 2.8333 \times 10^{-4})^2 + (2.8868 \times 10^{-4})^2\right]^{1/2}$$

We find

$$\varepsilon_1 = 4.1196 \times 10^{-4}, \qquad \varepsilon_2 = 1.2137 \times 10^{-4}$$

From Hooke's law (Chapter 3), the principal stresses are

$$\sigma_1 = \frac{E}{1 - \nu^2}(\varepsilon_1 + \nu\varepsilon_2), \qquad \sigma_2 = \frac{E}{1 - \nu^2}(\varepsilon_2 + \nu\varepsilon_1)$$

Thus

$$\sigma_1 = \frac{30 \times 10^6}{1 - 0.3^2}\left[4.1196 \times 10^{-4} + (0.3)(1.2137 \times 10^{-4})\right] = 14{,}781.5 \text{ psi}$$

$$\sigma_2 = \frac{30 \times 10^6}{1 - 0.3^2}\left[1.2137 \times 10^{-4} + (0.3)(4.1196 \times 10^{-4})\right] = 8076 \text{ psi}$$

A summary of the equations used to determine principal strains, principal stresses, and their directions for common types of rosettes is given in Table 5-3.

5.4 BRITTLE COATING

The brittle-coating technique provides a simple and direct approach for experimental stress analysis when high precision is not necessary. In the brittle-coating method of stress analysis, a prototype of the part under study is coated with a thin layer of a material that exhibits brittle fracture. The specimen is then loaded, and when the stresses in the coating reach a certain state, a pattern of cracks is formed in the coating.

After each application of the load, the coating is examined, and the crack patterns associated with each load application are noted. The loading process is continued until the crack pattern covers the region of interest or until the part is stressed to the maximum permissible level. The brittle-coating test method is usually nondestructive, so the load must be kept below the level that would cause yield or fracture in the prototype.

Before coating, the surface of the specimen is lightly sanded and a reflective undercoat is applied to facilitate crack observation. The coating is sprayed to as nearly uniform thickness as possible. The coating may exhibit both flammability and toxicity, so suitable precautions against these dangers must be taken.

The surface coating is assumed to undergo the same strain as the specimen surface. The cracks in the coating form and propagate perpendicular to the tensile principal stresses. The cracks that form normal to principal stresses are called *isostatics*. The line enclosing a cracked area that forms during a load application is called an *isoentatic*. This line is a boundary between a cracked and uncracked region and hence is a line along which the principal stress is constant. One set of cracks will form in a field in which there is one tensile principal stress, and two will form if there are two unequal tensile principal stresses. In a uniaxial or biaxial compressive stress field the coating will not crack, but it may flake and peel off. If two equal tensile principal stresses act on the coating, the crack pattern will be random in nature. The formation of a random pattern is called

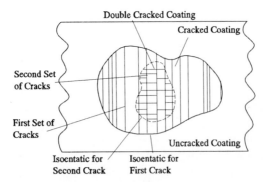

FIGURE 5-6 Crack patterns in a brittle coating.

crazing. The isostatics and isoentatics formed during two applications of a biaxial stress field are shown in Fig. 5-6.

Reference 5.2 describes brittle coating technology in more detail.

5.5 PHOTOELASTICITY

The velocity of light depends on the medium in which the light is traveling. The *index of refraction* of a material is the ratio of the velocity of light in a vacuum to that in the material. Some materials exhibit the property of *double refraction*, or *birefringence*. In these materials the index of refraction depends on the orientation of the electric vector with respect to the material specimen it is traversing. Some materials that are not normally birefringent become so when they are stressed. The phenomenon, which was discovered by Brewster in 1816, is the basis for the photoelastic measurement of stress.

According to Maxwell's theory, electromagnetic radiation consists of oscillating electric and magnetic vectors that are perpendicular to each other. For the purposes of photoelasticity it is necessary to consider only the electric field vector. Figure 5-7 shows a plot of a single wave of the electric vector as it tracks along the z axis. Mathematically the variation of the magnitude of the electric vector at a point z along the z axis is

$$E_\ell = E_0 \cos(2\pi/\lambda)(z - ct) \tag{5.16}$$

where E_0 is the amplitude of the wave, c is the velocity with which it propagates along the z axis, and λ is the wavelength, i.e., the distance between the beginning and end of one wave or the distance between any two adjacent points of the waveform that are in the same phase. Phase refers to the stage of oscillation of the vector at a fixed point. The phase of the electric vector at a point on the z axis can be represented by a rotating vector of length E_0. This representation is shown in Fig. 5-8 for a wave given by Eq. (5.16). The projection of the rotating vector on the E_ℓ, t coordinates gives the magnitude of the vibration at any instant. When a rotation of 2π radians is completed, the vector returns to its initial phase. The frequency of the wave is the number of complete cycles the

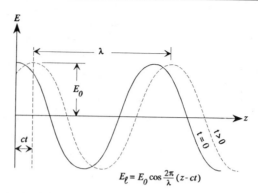

FIGURE 5-7 Magnitude of light vector as function of position along axis of propagation at two different times.

vibration executes per unit time. The wavelength, frequency f, and speed of the wave are related by

$$c = f\lambda \tag{5.17}$$

The time of one oscillation is the period T of the vibration and is defined as

$$T = 1/f \tag{5.18}$$

The circular frequency ω in radians per second is

$$\omega = 2\pi f \tag{5.19}$$

The quantity $(2\pi/\lambda)z_1$ is the initial phase of the oscillation at $z = z_1$. In photoelasticity the superposition of two waves with the same frequency that are out of phase is important. If the two waves differ by 180° in phase and have equal amplitudes and frequencies, they will cancel each other out and darkness results.

The instrument used in making photoelastic measurements is the *polariscope*. The basic components of a *plane polariscope* and a *circular polariscope* are sketched in Fig. 5-9. The plane polariscope consists of a light source (for this discussion the light is considered to be monochromatic), two polarizers with axes

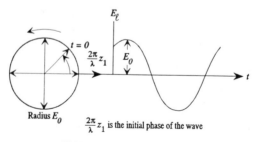

$\frac{2\pi}{\lambda}z_1$ is the initial phase of the wave

FIGURE 5-8 Electric vector.

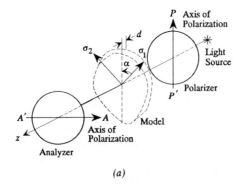

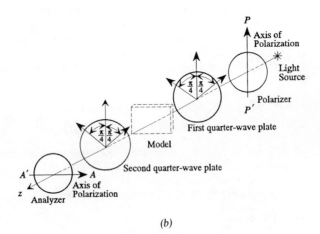

FIGURE 5-9 Arrangement of optical elements in plate polariscope and circular polariscope. (a) Plane polariscope. (b) circular polariscope.

at 90°, a transparent model of birefringent materials, and a screen. A *polarizer* is a device that blocks all light incident on it except for the component parallel to its polarizing axis. The second polarizer in the plane polariscope is called the *analyzer*. If the birefringent model were not present, the two crossed polarizers would block all light.

In Fig. 5-9a, σ_1 and σ_2 are the principal stresses at the point under consideration in the model and PP' the plane of oscillation of light as it leaves the polarizer. Line AA', the polarizing axis of the analyzer, is perpendicular to PP'. A light ray, $E_\ell = E_0 \cos \omega t$, that is vibrating parallel to PP' when it strikes the model can be resolved into components parallel to the principal stresses:

$$E_{\ell 1} = E_0 \cos \alpha \cos \omega t, \qquad E_{\ell 2} = E_0 \sin \alpha \cos \omega t \qquad (5.20)$$

The effect of the stresses is to alter the speed with which the two components of light traverse the model. Let v_1 and v_2 be the velocities of the two components of light. On leaving the model the two components will differ in phase,

$$E_{\ell 1} = E_0 \cos \alpha \cos \omega(t - t_1), \qquad E_{\ell 2} = E_0 \sin \alpha \cos \omega(t - t_2) \quad (5.21)$$

where $t_1 = d/v_1$ and $t_2 = d/v_2$ and d is the model thickness.

The difference in phase between the two components is

$$\Delta = \omega t - \omega t_1 - \omega t + \omega t_2 = \omega(t_2 - t_1)$$

According to the stress-optic law, for plane stress problems (i.e., two-dimensional photoelasticity) the phase difference is proportional to the difference in principal stresses [5.2]:

$$\Delta = (2\pi d/\lambda)K(\sigma_1 - \sigma_2) \qquad (5.22)$$

where $\sigma_1 > \sigma_2$, K is the relative stress-optic coefficient expressed in terms of brewsters (1 brewster $= 10^{-12}$ m^2/N $= 6.895 \times 10^{-9}$ in.^{2}/lb), and d is the thickness of the specimen. If the stressed model were not present, the analyzer whose polarizing direction AA' is perpendicular to PP' would permit no light to pass. With the stress-optic model present, the components of E_{1A} and E_{2A} that are parallel to AA' are

$$
\begin{aligned}
E_{1A} &= E_0 \cos \alpha \sin \alpha \cos[\omega(t - t_1)] \\
E_{2A} &= E_0 \sin \alpha \cos \alpha \cos[\omega(t - t_2)]
\end{aligned}
\qquad (5.23)
$$

Since E_{1A} and E_{2A} act in opposite directions, the resultant amplitude passing through the analyzer is

$$E_{\ell f} = E_{2A} - E_{1A} = E_0 \sin \alpha \cos \alpha \{\cos[\omega(t - t_2)] - \cos[\omega(t - t_1)]\} \quad (5.24)$$

The trigonometric identities

$$\sin 2x = 2 \sin x \cos x, \qquad \cos x - \cos y = -2 \sin \tfrac{1}{2}(x + y)\sin \tfrac{1}{2}(x - y)$$

are applied to Eq. (5.24) to give

$$E_{\ell f} = E_0 \tfrac{1}{2} \sin 2\alpha \{-2 \sin \tfrac{1}{2}[2\omega t - \omega(t_1 + t_2)]\sin \tfrac{1}{2}[\omega(t_1 - t_2)]\} \quad (5.25)$$

However,

$$\Delta = \omega(t_2 - t_1)$$

so

$$E_{\ell f} = E_0 \sin 2\alpha \sin(\Delta/2)\sin[\omega t - (\omega/2)(t_1 + t_2)] \qquad (5.26)$$

The term $\tfrac{1}{2}\omega(t_1 + t_2)$ affects only the time variation of the amplitude; hence the magnitude of the amplitude vanishes whenever

$$\sin 2\alpha \sin(\Delta/2) = 0 \qquad (5.27)$$

Whenever $\sin 2\alpha$ is zero, $\alpha = 0, \tfrac{1}{2}\pi, \pi, \ldots$, and the direction of PP' is parallel to the direction of one of the principal stresses in the model. Points at which $\sin 2\alpha = 0$ produce corresponding dark spots on the screen. The dark line that connects a series of such adjacent dark points is called an *isoclinic fringe* or an *isoclinic*.

Isoclinics are used to determine the principal-stress directions at all points of a photoelastic model. There are two procedures in practical use. One is to obtain a number of isoclinic patterns at different polariscope settings and to combine these fringe patterns to give one composite pattern showing the isoclinic parameters over the entire field of the model. The other is to find the points of interest and then to determine individually the isoclinic parameter, i.e., the polariscope setting, with respect to each of these points.

By rotating the polarizer–analyzer combination and noting the fringe pattern as it changes, the direction of the principal stresses on the model can be found. A dark point on the screen also occurs whenever

$$\sin(\Delta/2) = 0, \qquad \Delta = 2N\pi, \qquad N = 0, 1, 2, \ldots \tag{5.28}$$

A locus of adjacent dark points that corresponds to a given value of N is called an *isochromatic fringe* or simply an *isochromatic*. (For monochromatic light an isochromatic is dark. However, for white light an isochromatic will be colored due to the detection of light of a particular wavelength from the white light.) The quantity

$$N = \Delta/2\pi = (Kd/\lambda)(\sigma_1 - \sigma_2) \tag{5.29}$$

is called the order of its associated isochromatic fringe. For light of a single wavelength the stress-optic law in terms of the fringe order is

$$\sigma_1 - \sigma_2 = (N/d)f_\sigma \tag{5.30}$$

or

$$N = (d/f_\sigma)(\sigma_1 - \sigma_2) \tag{5.31}$$

where $f_\sigma = \lambda/K$, the property of the model material for a given wavelength of light known as the material fringe value. Equation (5.30) states that the principal-stress difference $\sigma_1 - \sigma_2$ can be determined in a two-dimensional model if N is measured at each point in the model. The maximum shear stress at a point in a two-dimensional stress field ($\sigma_3 = 0$) is calculated as

$$\tau_{\max} = \begin{cases} (\sigma_1 - 0)/2 & \text{for } \sigma_2 \geq 0 \\ (\sigma_1 - \sigma_2)/2 & \text{for } \sigma_2 < 0 \end{cases} \tag{5.32}$$

The stress-optic law can be written as

$$\tau_{\max} = Nf_\sigma/2d \tag{5.33}$$

Equation (5.33) gives the maximum shear stress when σ_1 and σ_2 are of opposite sign; otherwise it gives the maximum in plane shear stress. If the Tresca yield condition applies to the part being analyzed, Eq. (5.33) provides all the necessary information on the stress at a point once the fringe order is known.

Use of a plane polariscope results in the superposition of both the isoclinics and the isochromatic fringe pattern. The circular polariscope is more commonly used in making photoelastic measurements than is the plane polariscope since the circular polariscope eliminates the isoclinics from the fringe pattern.

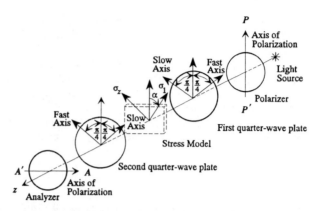

FIGURE 5-10 Stressed photoelastic model in circular polariscope (dark field, crossed polarizer and analyzer, crossed quarter-wave plates).

The principal components of a circular polariscope are sketched in Fig. 5-10. The circular polariscope adds two quarter-wave plates to the plane polariscope configuration. One *quarter-wave plate* stands between the polarizer and the stressed model and one is between the stressed model and the analyzer. A quarter-wave plate has the ability to resolve a light vector into two orthogonal components and to transmit the components with different velocities that give a relative angular retardation (or phase difference) of $\frac{1}{2}\pi$. The axis with the smaller index of refraction is called the fast axis, and the other axis is the slow axis. The light passing parallel to the slow axis trails that passing the fast axis by a phase difference of $\frac{1}{2}\pi$—hence the name quarter-wave plate. The first quarter-wave plate converts the plane-polarized light obtained by the polarizer into circularly polarized light. The second quarter-wave plate converts the circularly polarized light into plane-polarized light vibrating again in the vertical direction. As was done with the plane polariscope, the light leaving the polarizer is represented as

$$E_{PP'} = E_0 \cos \omega t \qquad (5.34)$$

The first quarter-wave plate has its axes oriented at 45° with respect to *PP'*. On leaving the first quarter-wave plate, the component of light parallel to the fast axis is

$$E_{\text{fast}} = \left(E_0/\sqrt{2}\right)\cos \omega t \qquad (5.35)$$

and that parallel to the slow axis is

$$E_{\text{slow}} = \left(E_0/\sqrt{2}\right)\cos\left(\omega t - \tfrac{1}{2}\pi\right) = \left(E_0/\sqrt{2}\right)\sin \omega t \qquad (5.36)$$

The two components E_{fast} and E_{slow} impinge on the stress-optic model and are resolved into components parallel to the principal axes of stress. These components as they leave the model are

$$E_1 = E_{\text{fast}} \cos\left(\tfrac{1}{4}\pi - \alpha\right) + E_{\text{slow}} \sin\left(\tfrac{1}{4}\pi - \alpha\right)$$
$$E_2 = E_{\text{slow}} \cos\left(\tfrac{1}{4}\pi - \alpha + \Delta\right) - E_{\text{fast}} \sin\left(\tfrac{1}{4}\pi - \alpha + \Delta\right) \qquad (5.37)$$

Substituting Eqs. (5.35) and (5.36) in (5.37) gives

$$E_1 = (E_0/\sqrt{2})[\cos \omega t \cos(\tfrac{1}{4}\pi - \alpha) + \sin \omega t \sin(\tfrac{1}{4}\pi - \alpha)]$$
$$E_2 = (E_0/\sqrt{2})[\sin \omega t \cos(\tfrac{1}{4}\pi - \alpha + \Delta) - \cos \omega t \sin(\tfrac{1}{4}\pi - \alpha + \Delta)] \tag{5.38}$$

Equations (5.38) are reduced by using the trigonometric identities

$$\cos x \cos y + \sin x \sin y = \cos(x - y)$$
$$\sin x \cos y - \cos x \sin y = \sin(x - y)$$

Thus

$$E_1 = (E_0/\sqrt{2})\cos(\omega t - \tfrac{1}{4}\pi + \alpha)$$
$$E_2 = (E_0/\sqrt{2})\sin(\omega t - \tfrac{1}{4}\pi + \alpha - \Delta) \tag{5.39}$$

When these components impinge on the second quarter wave-plate, the components parallel to the fast and slow axes are

$$E_{f2} = (E_0/\sqrt{2})[\cos(\omega t + \alpha - \tfrac{1}{4}\pi)\sin(\tfrac{1}{4}\pi - \alpha)$$
$$+ \sin(\omega t + \alpha - \tfrac{1}{4}\pi - \Delta)\cos(\tfrac{1}{4}\pi - \alpha)]$$
$$E_{s2} = (E_0/\sqrt{2})[\sin(\omega t + \alpha - \tfrac{1}{4}\pi)\cos(\tfrac{1}{4}\pi - \alpha)$$
$$+ \cos(\omega t + \alpha - \tfrac{1}{4}\pi - \Delta)\sin(\tfrac{1}{4}\pi - \alpha)] \tag{5.40}$$

where the $\tfrac{1}{2}\pi$ phase shift has been included in E_{s2}. The components E_{f2} and E_{s2} finally impinge on the analyzer. The resultant light passing through the analyzer in the AA' direction is

$$E_{\ell r} = (1/\sqrt{2})(E_{s2} - E_{f2}) \tag{5.41}$$

where Eqs. (5.40) give E_{f2} and E_{s2}. After making the substitutions and simplifying with trigonometric identities, the resultant light vector leaving the analyzer is

$$E_R = E_0 \sin(\Delta/2)\sin(\omega t + 2\alpha - \Delta/2) \tag{5.42}$$

The light will vanish whenever $\sin(\Delta/2) = 0$ or

$$\Delta = 2N\pi, \qquad N = 0, 1, 2, \dots \tag{5.43}$$

The isoclinic points have been eliminated and only the isochromatics remain. Each isochromatic fringe corresponds to a value of $N = 0, 1, 2, \dots$. Once N is known for a fringe, the principal-stress difference or the maximum shear stress along the fringe follows from Eq. (5.31). A value of N exists for each point in the field; however, N is an integer only at the dark fringes. The arrangement of the elements in the circular polariscope just described is called the dark-field configuration; i.e., the screen is dark with the model removed. If the analyzer is rotated

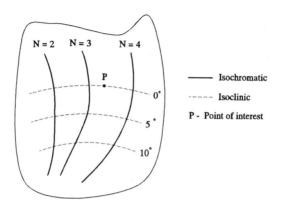

FIGURE 5-11 Fringe pattern in Tardy compensation method.

by 90° so that the axes of the polarizer and analyzer are parallel, the arrangement is called light field because the screen is light with no model present. The light-field arrangement establishes the fringe order to the nearest $\frac{1}{2}$, i.e., $N = n + \frac{1}{2}$. With interpolation this is sufficient accuracy for many applications. Fringe orders may be determined to fractions smaller than $\frac{1}{2}$ by Tardy compensation. Integer fringe orders are identified by observation of fringe formation during loading or by using white light before the monochromatic experiment. When a model is viewed with white light, the isochromatic fringe appears as a series of colored bands.

The Tardy compensation is commonly used to determine the order of the isochromatic fringe at any arbitrary point on the model. With this method, a plane polariscope is first used so that the isoclinics can be used to establish the direction of the principal stresses at the point P of interest (Fig. 5-11). The axis of the polarizer is oriented parallel to one of the principal-stress directions ($\alpha = 0, \frac{1}{2}\pi$) at the point of interest, and the other elements of the polariscope are oriented to produce a dark-field circular polariscope and the fringe order $N = 3$ near the point (P). The analyzer is then rotated through an angle γ with respect to its dark-field position until extinction occurs at the point. The fringe order of point P is

$$N = n + \gamma/\pi \tag{5.44}$$

where n is the order of the fringe that moves to the point of interest as the polariscope is in a standard dark-field circular polariscope and γ is the angle of rotation of the analyzer until extinction occurs at the point of interest. Other methods of compensation may also be applied [5.1].

To apply the photoelastic methods, the value of the material constants f_σ (or λ/K) for the model must be known accurately, and the stress difference corresponding to a fringe order of 1 must be verified. The calibration model is loaded in increments, and the fringe order and the loads are recorded. A tensile specimen may be used to determine the quantity f_σ. In the use of such a specimen $\sigma_1 = F/wd$ and $\sigma_2 = 0$, where w is the width of the neck, d its

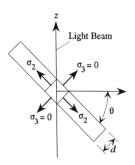

FIGURE 5-12 Oblique incidence in which σ_1 is rotation axis of model.

thickness, and F the applied force. Equation (5.31) with $N = 1$ gives

$$F/wd = f_\sigma/d \qquad (5.45)$$

or

$$f_\sigma = F/w$$

and f_σ/d is the stress difference for $N = 1$ where F is the axial load corresponding to the fringe order of 1. Other methods of calibration have been developed [5.1].

The photoelastic measurement gives only the differences in the principal stresses at a point. If the boundary can be considered free, i.e., σ_1 or $\sigma_2 = 0$, the other principal stress can be obtained from Eq. (5.31). This is also true when the boundary is not free but the applied normal load is known. Otherwise separation techniques for determining the individual stresses must be used. Of several separation techniques that have been applied, only the oblique incidence method will be briefly described here.

In the oblique incidence technique, the model is rotated so that the light strikes it not at a right angle. When the principal-stress directions are known and the model about the σ_1 axis is rotated by an angle θ (Fig. 5-12), the two principal stresses are given by

$$\sigma_1 = \frac{f_\sigma}{d} \frac{\cos\theta}{\sin^2\theta}(N_\theta - N_0 \cos\theta), \qquad \sigma_2 = \frac{f_\sigma}{d} \frac{1}{\sin^2\theta}(N_\theta \cos\theta - N_0) \quad (5.46)$$

where N_0 is the fringe order at the point in a normal incidence analysis and N_θ is the fringe order for the oblique incidence test. This method is often employed to separate stresses along a line of symmetry where one rotation of the model about the line of symmetry provides data enough to separate the stresses along the whole line. When the principal-stress directions are not known, rotation of the model is made about any axis such as $0x$ and $0y$ (Fig. 5-13). In this case three isochromatic patterns are obtained: the normal incidence pattern, the oblique incidence pattern associated with a rotation about $0x$, and the oblique incidence

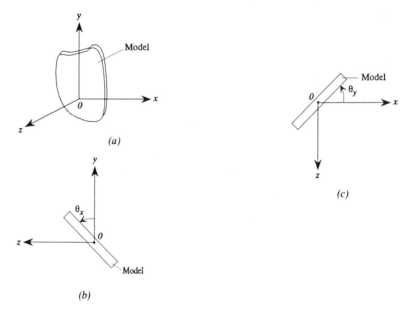

FIGURE 5-13 Oblique incidence in which $0x$ and $0y$ are rotation axes. (*a*) Point of interest 0 and the direction coordinates x, y, z. The model lies in the xy plane, (*b*) rotation of the model about x axis, (*c*) rotation of the model about y axis.

pattern associated with a rotation about $0y$. When $\theta_x = \theta_y = \theta$, it can be shown that

$$\left(\frac{\sigma_x d}{f_\sigma}\right)^2 = \frac{\cot^2 \theta}{1 - \cos^4 \theta}\left[N_{\theta x}^2 + N_{\theta y}^2 \cos^2 \theta - N_0^2(1 + \cos^2 \theta)\right]$$

$$\left(\frac{\sigma_y d}{f_\sigma}\right)^2 = \frac{\cot^2 \theta}{1 - \cos^4 \theta}\left[N_{\theta y}^2 + N_{\theta x}^2 \cos^2 \theta - N_0^2(1 + \cos^2 \theta)\right]$$

(5.47)

where $N_{\theta x}$ and $N_{\theta y}$ are the isochromatic pattern orders corresponding to the model rotation about $0x$ and $0y$, respectively.

The σ_x and σ_y from Eqs. (5.47) along with the relations $\sigma_1 + \sigma_2 = \sigma_x + \sigma_y$ presented in Chapter 3 allow separation of the physical stresses. In experimental practice, the selection of the proper material for the photoelastic model and the scaling between the model and prototype are the most important factors in photoelasticity. Refer to the literature [5.1] for detailed information.

The techniques previously described apply to transmission photoelasticity in which a transparent model of the stressed part under study is made. Photoelastic techniques may also be applied directly to the stressed part by reflection photoelasticity, in which a coating of birefringent material is applied directly to the part. A reflecting surface exists at the interface between the part and the coating. Two polariscope configurations for reflection photoelasticity are shown in Fig. 5-14. The basis of the method is analogous to that of transmission photoelasticity,

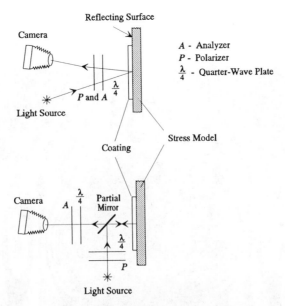

FIGURE 5-14 Reflection polariscopes.

but some modification of formulas for stress-optic relations and for separation must be made.

The methods of photoelasticity can be extended to three-dimensional stress fields by the techniques of *stress freezing* and *scattered light*.

The stress freezing method is based on the molecular behavior of the polymeric materials of which the model is constructed. As the temperature of the polymer is raised, weak secondary bonds between molecular chains break. If the model is loaded in this state, the load is carried by the elastic action of primary molecular bonds. When the material is cooled under load, the secondary bonds reform to lock the primary bonds in the deformed state. Thin slices of the stressed material may then be used for photoelastic analysis.

Example 5.2 Photoelastic Study of a Diametrically Loaded Ring The dark-field isochromatic pattern for a diametrically loaded circular ring is shown in Fig. 5-15. By observing the formation of the pattern, the fringe orders were identified. Two fringe orders are labeled on the fringe. Estimate the stresses $\sigma_x, \sigma_y, \tau_{xy}$ at points A, B, and C. The wavelength of the light used resulted in a value of $f_\sigma = 70$ lb/in., and the model thickness was 0.20 in.

At point A, the stresses σ_x and τ_{xy} must be zero because the surface at A is free of force. Because $\tau_{xy} = 0$ at A, σ_x must be a principal stress there, and σ_y is also principal at A because it is normal to σ_x. The fringe order at A is roughly $N_A = 2.3$. From Eq. (5.30),

$$\sigma_y = \frac{2.3}{0.20} \times 70 = 805 \text{ psi}, \qquad \sigma_x = \tau_{xy} = 0$$

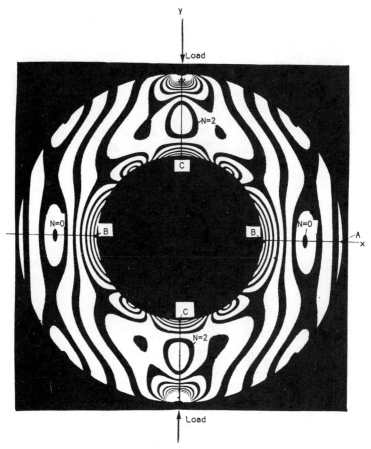

FIGURE 5-15 Dark-field isochromatic fringe patterns of ring loaded in diametrical compression. Reproduced with permission [5.3].

Applying the same argument used for point A, σ_y is a principal stress at point B. The fringe order is roughly $N_B = 7.0$:

$$\sigma_y = -\frac{(7.0)(70)}{0.20} = -2450 \text{ psi}, \qquad \sigma_x = \tau_{xy} = 0$$

By inspection, σ_y at point B is compressive.

At point C, σ_x is a principal stress and $N_c = 6.7$:

$$\sigma_x = \frac{(6.7)(70)}{0.20} = 2345 \text{ psi}, \qquad \sigma_y = \tau_{xy} = 0$$

REFERENCES

5.1   Dally, J. W., and Riley, W. F., *Experimental Stress Analysis*, 3rd ed., McGraw-Hill, New York, 1991.

5.2   Kobayashi, S. (Ed.), *Handbook on Experimental Mechanics*, Prentice-Hall, Englewood Cliffs, NJ, 1987.

5.3   Budynas, R.G., *Advanced Strength and Applied Stress Analysis*, McGraw-Hill, New York, 1977.

5

Tables

TABLE 5-1   STRAIN SENSITIVITY S_A FOR COMMON STRAIN GAGE ALLOYS[a]

Material	Composition (%)	S_A
Advance or constantan	45 Ni, 55 Cu	2.1
Nichrome V	80 Ni, 20 Cr	2.2
Isoelastic	36 Ni, 8 Cr, 0.5 Mo, 55.5 Fe	3.6
Karma	74 Ni, 20 Cr, 3 Al, 3 Fe	2.0
Armour D	70 Fe, 20 Cr, 10 Al	2.0
Alloy 479	92 Pt, 8 W	4.1

[a]From Dally and Riley [5.1], with permission.

TABLE 5-1 | Strain Sensitivity S_A 236

TABLE 5-2   CHARACTERISTICS OF SELECTED COMMON WHEATSTONE BRIDGES

Circuit	Output Voltage ΔE Due to Strain ε	Circuit Sensitivity S_c	If $R_1 = R_2 = R_3 = R_4 = R_g$ ΔE	S_c
1. Single Active in Arm R_1	$\dfrac{r}{(1+r)^2}\dfrac{\Delta R_1}{R_g}V$	$S_c = \dfrac{rS_gV}{(1+r)^2}$ $S_c = \dfrac{r}{1+r}S_g\sqrt{P_gR_g}$	$\dfrac{\Delta R_1}{4R_g}V$	$S_c = \tfrac{1}{4}S_gV$ $S_c = \tfrac{1}{2}S_g\sqrt{P_gR_g}$
2. Active Gage in Arm R_1, Dummy Gage in R_2	$\dfrac{\Delta R_1}{4R_g}V$	$S_c = \tfrac{1}{4}S_gV$ $S_c = \tfrac{1}{2}S_g\sqrt{P_gR_g}$	$\dfrac{\Delta R_1}{4R_g}V$	$S_c = \tfrac{1}{4}S_gV$ $S_c = \tfrac{1}{2}S_g\sqrt{P_gR_g}$

Circuit	Output Voltage ΔE Due to Strain ε	Circuit Sensitivity S_c	If $R_1 = R_2 = R_3 = R_4 = R_g$	
			ΔE	S_c
3. Active Gage in Arm R_1, Dummy Gage in R_4	$\dfrac{r}{(1+r)^2}\dfrac{\Delta R_1}{R_g}V$	$S_c = \dfrac{r}{(1+r)^2}VS_g$ $S_c = \dfrac{r}{1+r}S_g\sqrt{P_gR_g}$	$\dfrac{\Delta R_1}{4R_g}V$	$S_c = \tfrac{1}{4}S_gV$ $S_c = \tfrac{1}{2}S_g\sqrt{P_gR_g}$

TABLE 5-2 Characteristics of Wheatstone Bridges   238

	ΔE	S_c	ΔE	S_c
4. Four Active Gages $R_1 = R_g$, $R_2 = R_g$, $R_3 = R_g$, $R_4 = R_g$ (Active)	$\dfrac{V}{4R_g}(\Delta R_1 - \Delta R_2 + \Delta R_3 - \Delta R_4)$	$S_c = VS_g$ $S_c = 2S_g\sqrt{P_gR_g}$	$\dfrac{V}{4R_g}(\Delta R_1 - \Delta R_2 + \Delta R_3 - \Delta R_4)$ when $\Delta R_1 = \Delta R_3$ $= -\Delta R_2 = -\Delta R_4$ $\Delta E = \dfrac{\Delta R_1 V}{R_g}$	$S_c = VS_g$ $S_c = 2S_g\sqrt{P_gR_g}$
5. Active Gages in Arms R_1 and R_4 $R_1 = R_g$ (Active), R_2, $R_4 = R_g$ (Active), R_3	$\dfrac{rV(\Delta R_1 - \Delta R_4)}{(1+r)^2 R_g}$	$S_c = \dfrac{2r}{(1+r)^2}VS_g$ $S_c = \dfrac{2r}{1+r}S_g\sqrt{P_gR_g}$	$\dfrac{V}{4R_g}(\Delta R_1 - \Delta R_4)$ when $\Delta R_1 = -\Delta R_4$ $\Delta E = \dfrac{\Delta R_1 V}{2R_g}$	$S_c = \tfrac{1}{2}VS_g$ $S_c = S_g\sqrt{P_gR_g}$

Note: $r = R_2/R_1$. All the circuits except circuit 1 are temperature compensated.

Rosette	Principal Strain $(\varepsilon_1, \varepsilon_2)$ and Principal Stress (σ_1, σ_2)	Principal Angle	Condition for $0 < \theta_1 < 90°$
1. Three-element, rectangular	$$\varepsilon_{1,2} = \tfrac{1}{2}(\varepsilon_A + \varepsilon_C)$$ $$\pm \tfrac{1}{2}\sqrt{(\varepsilon_A - \varepsilon_C)^2 + (2\varepsilon_B - \varepsilon_A - \varepsilon_C)^2}$$ $$\sigma_{1,2} = \frac{E}{2}\left[\frac{\varepsilon_A + \varepsilon_C}{1-\nu}\right.$$ $$\left.\pm \frac{1}{1+\nu}\sqrt{(\varepsilon_A - \varepsilon_C)^2 + (2\varepsilon_B - \varepsilon_A - \varepsilon_C)^2}\right]$$	$$\tan 2\theta_1$$ $$= \frac{2\varepsilon_B - \varepsilon_A - \varepsilon_C}{\varepsilon_A - \varepsilon_C}$$	$$\varepsilon_B > \tfrac{1}{2}(\varepsilon_A + \varepsilon_C)$$
2. Delta	$$\varepsilon_{1,2} = \frac{\varepsilon_A + \varepsilon_B + \varepsilon_C}{3}$$ $$\pm \frac{\sqrt{2}}{3}\sqrt{(\varepsilon_A - \varepsilon_B)^2 + (\varepsilon_B - \varepsilon_C)^2 + (\varepsilon_C - \varepsilon_A)^2}$$ $$\sigma_{1,2} = \frac{E}{3}\left[\frac{\varepsilon_A + \varepsilon_B + \varepsilon_C}{1-\nu}\right.$$ $$\left.\pm \frac{\sqrt{2}}{1+\nu}\sqrt{(\varepsilon_A - \varepsilon_B)^2 + (\varepsilon_B - \varepsilon_C)^2 + (\varepsilon_C - \varepsilon_A)^2}\right]$$	$$\tan 2\theta_1$$ $$= \frac{\sqrt{3}(\varepsilon_C - \varepsilon_B)}{2\varepsilon_A - (\varepsilon_B + \varepsilon_C)}$$	$$\varepsilon_C > \varepsilon_B$$

3. Four-element, rectangular $\varepsilon_{1,2} = \tfrac{1}{4}(\varepsilon_A + \varepsilon_B + \varepsilon_C + \varepsilon_D)$ $\pm \tfrac{1}{2}\sqrt{(\varepsilon_A - \varepsilon_C)^2 + (\varepsilon_B - \varepsilon_D)^2}$ $\sigma_{1,2} = \dfrac{E}{2}\left[\dfrac{\varepsilon_A + \varepsilon_B + \varepsilon_C + \varepsilon_D}{2(1-\nu)}\right.$ $\left.\pm \dfrac{1}{1+\nu}\sqrt{(\varepsilon_A - \varepsilon_C)^2 + (\varepsilon_B - \varepsilon_D)^2}\right]$	$\tan 2\theta_1 = \dfrac{\varepsilon_B - \varepsilon_D}{\varepsilon_A - \varepsilon_C}$	$\varepsilon_B > \varepsilon_D$
4. Tee–delta $\varepsilon_{1,2} = \tfrac{1}{2}(\varepsilon_A + \varepsilon_D)$ $\pm \tfrac{1}{2}\sqrt{(\varepsilon_A - \varepsilon_D)^2 + \tfrac{4}{3}(\varepsilon_C - \varepsilon_B)^2}$ $\sigma_{1,2} = \dfrac{E}{2}\left[\dfrac{\varepsilon_A + \varepsilon_D}{1-\nu}\right.$ $\left.\pm \dfrac{1}{1+\nu}\sqrt{(\varepsilon_A - \varepsilon_D)^2 + \tfrac{4}{3}(\varepsilon_C - \varepsilon_B)^2}\right]$	$\tan 2\theta_1 = \dfrac{2(\varepsilon_C - \varepsilon_B)}{\sqrt{3}(\varepsilon_A - \varepsilon_D)}$	$\varepsilon_C > \varepsilon_B$

[a]From Dally and Riley [5.1], with permission.

Note: ε_A, ε_B, ε_C, and ε_D represent strains in directions A, B, C, and D, respectively.

Stress Concentration

Mathematical analysis and experimental measurement show that in a loaded structural member, near changes in the cross section, distributions of stress occur in which the peak stress reaches much larger magnitudes than does the average stress over the section. This increase in peak stress near holes, grooves, notches, sharp corners, and other changes in section is called *stress concentration*. The section variation that causes the stress concentration is referred to as a *stress raiser*.

6.1 NOTATION

K_ε Effective strain concentration factor

K_f Effective stress concentration factor for cyclic loading, fatigue notch factor

K_i Effective stress concentration factor for impact loads

K_σ Effective stress concentration factor

K_t Theoretical stress concentration factor in elastic range, $= \sigma_{\max}/\sigma_{\text{nom}}$

q Notch sensitivity index

q_f Notch sensitivity index for cyclic loading

q_i Notch sensitivity index for impact loading

r Notch radius (L)

ε_{nom} Nominal strain (L/L)

σ_{nom} Nominal stress (F/L^2) of notched member; e.g., for an extension member, σ_{nom} is axial load divided by cross-sectional area measured at notch, i.e., area taken remotely from notch minus area corresponding to notch

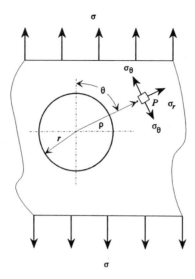

FIGURE 6-1   Infinite plate with small circular hole.

6.2 STRESS CONCENTRATION FACTORS

Figure 6-1 shows a large plate that contains a small circular hole. For an applied uniaxial tension the stress field is found from linear elasticity theory [6.1]. In polar coordinates the azimuthal component of stress at point P is given as

$$\sigma_\theta = \tfrac{1}{2}\sigma\left[1 + (r^2/\rho^2)\right] - \tfrac{1}{2}\sigma\left[1 + 3(r^4/\rho^4)\right]\cos 2\theta \qquad (6.1)$$

The maximum stress occurs at the sides of the hole where $\rho = r$ and $\theta = \tfrac{1}{2}\pi$ or $\theta = \tfrac{3}{2}\pi$. At the hole sides,

$$\sigma_\theta = 3\sigma$$

The peak stress is three times the nominal uniform stress. To account for the peak in stress near a stress raiser, the *stress concentration factor* or *theoretical stress concentration factor* is defined as the ratio of the calculated peak stress to the nominal stress that would exist in the member if the distribution of stress remained uniform, i.e.,

$$K_t = \sigma_{max}/\sigma_{nom} \qquad (6.2)$$

The nominal stress is found using basic strength-of-materials formulas, and the calculations are based on the properties of the net cross section at the stress raiser. Sometimes, because the net section is nearly the same as the overall section away from the stress raiser, the overall section is used in computing the nominal stress. This procedure was used in the case shown in Fig. 6-1, and the stress concentration factor is

$$K_t = \sigma_{max}/\sigma_{nom} = 3$$

The effect of the stress raiser is to change only the distribution of stress.

Equilibrium requirements dictate that the average stress on the section be the same in the case of stress concentration as it would be if there were a uniform stress distribution. Stress concentration results not only in unusually high stresses near the stress raiser but also in unusually low stresses in the remainder of the section.

When more than one load acts on a notched member—e.g., combined tension, torsion, and bending—the nominal stress due to each load is multiplied by the stress concentration factor corresponding to each load, and the resultant stresses are found by superposition. However, when bending and axial loads act simultaneously, superposition can be applied only when bending moments due to the interaction of axial force and bending deflections are negligible compared to bending moments due to applied loads.

The stress concentration factors for a variety of member configurations and load types are shown in Table 6-1. A general discussion of stress concentration factors and factor values for many special cases are contained in the literature [e.g., 6.2].

Example 6.1   Circular Shaft with a Groove   The circular shaft shown in Fig. 6-2 is girdled by a U-shaped groove, with $h = 10.5$ mm deep. The radius of the groove root $r = 7$ mm, and the bar diameter away from the notch $D = 70$ mm. A bending moment of 1.0 kN-m and a twisting moment of 2.5 kN-m act on the bar. The maximum shear stress at the root of the notch is to be calculated.

The stress concentration factor for bending is found from part I in Table 6-1, case 7b. Substitute

$$2h/D = \tfrac{21}{70} = 0.3, \qquad h/r = 10.5/7 = 1.5 \tag{1}$$

into the expression given for K_t:

$$K_t = C_1 + C_2(2h/D) + C_3(2h/D)^2 + C_4(2h/D)^3 \tag{2}$$

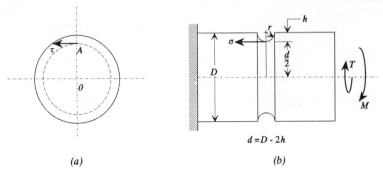

$$d = D - 2h$$

(a)   *(b)*

FIGURE 6-2   Circular shaft with U-groove.

Since $0.25 \leq h/r = 1.5 \leq 2.0$, we find, for elastic bending,

$$C_1 = 0.594 + 2.958\sqrt{h/r} - 0.520h/r$$

with C_2, C_3, and C_4 given by analogous formulas in case I7b of Table 6-1. These constants are computed as

$$C_1 = 3.44, \qquad C_2 = -8.45, \qquad C_3 = 11.38, \qquad C_4 = -5.40$$

It follows that for elastic bending

$$K_t = 3.44 - 8.45(0.3) + 11.38(0.3)^2 - 5.40(0.3)^3 = 1.78 \tag{3}$$

The tensile bending stress σ_{nom} is obtained from Eq. (3.56a) as $Md/2I$ and at the notch root the stress is

$$\sigma = K_t \frac{Md}{2I} = \frac{(1.78)(1.0 \times 10^3 \text{ N-m})(0.049 \text{ m})(64)}{2\pi(0.049)^4 \text{ m}^4} = 154.1 \text{ MPa} \tag{4}$$

The formulas from Table 6-1, part I, case 7c, for the elastic torsional load give $K_t = 1.41$. The nominal twisting stress at the base of the groove is [Eq. (3.48)]

$$\tau = \frac{K_t Td(32)}{2\pi d^4} = \frac{(1.41)(2.5 \times 10^3 \text{ N-m})16}{\pi(0.049)^3} = 152.6 \text{ MPa} \tag{5}$$

The maximum shear stress at the base of the groove is one-half the difference of the maximum and minimum principal stresses (Chapter 3). The maximum principal stress is

$$\sigma_{max} = \tfrac{1}{2}\sigma + \tfrac{1}{2}\sqrt{\sigma^2 + 4\tau^2} = \tfrac{1}{2}(154.1) + \tfrac{1}{2}\sqrt{154.1^2 + 4(152.6)^2} = 248.0 \text{ MPa}$$

and the minimum principal stress is

$$\sigma_{min} = \tfrac{1}{2}\sigma - \tfrac{1}{2}\sqrt{\sigma^2 + 4\tau^2} = \tfrac{1}{2}(154.1) - \tfrac{1}{2}\sqrt{154.1^2 + 4(152.6)^2} = -93.9 \text{ MPa}$$

Thus, the maximum shear stress is

$$\tau_{max} = \tfrac{1}{2}(\sigma_{max} - \sigma_{min}) = \tfrac{1}{2}(248.0 + 93.9) = 171.0 \text{ MPa} \tag{6}$$

6.3 EFFECTIVE STRESS CONCENTRATION FACTORS

In theory the peak stress near a stress raiser would be K_t times larger than the nominal stress at the notched cross section. However, K_t is an ideal value based on linear elastic behavior and depends only on the proportions of the dimensions of the stress raiser and the notched part. For example, in case 2a, part I, Table

6-1, if h, D, and r were all multiplied by a common factor $n > 0$, the value of K_t would remain the same. In practice, a number of phenomena may act to mitigate the effects of stress concentration. Local plastic deformation, residual stress, notch radius, part size, temperature, material characteristics (e.g., grain size, work-hardening behavior), and load type (static, cyclic, or impact) may influence the extent to which the peak notch stress approaches the theoretical value of $K_t\sigma_{nom}$.

To deal with the various phenomena that influence stress concentration, the concepts of *effective stress concentration factor* and *notch sensitivity* have been introduced.

The effective stress concentration factor of a specimen is defined to be the ratio of the stress calculated for the load at which structural damage is initiated in the specimen free of the stress raiser to the nominal stress corresponding to the load at which damage starts in the sample with the stress raiser. It is assumed that damage in the actual structure occurs when the maximum stress attains the same value in both cases. Similar to Eq. (6.2)

$$K_\sigma = \sigma_{max}/\sigma_{nom} \tag{6.3}$$

The factor K_σ is now the effective stress concentration factor as determined by the experimental study of the specimen.

For fatigue loading, the definition of experimentally determined effective stress concentration is

$$K_f = \frac{\text{fatigue strength without notch}}{\text{fatigue strength with notch}} \tag{6.4}$$

Factors determined by Eq. (6.4) should be regarded more as strength reduction factors than as quantities that correspond to an actual stress in the body. The fatigue strength is the maximum amplitude of fully reversed cyclic stress a specimen can withstand for a given number of load cycles. For static conditions the stress at rupture is computed using strength-of-materials elastic formulas even though yielding may occur before rupture. If the tests are under bending or torsion loads, the extreme fiber stress is used in the definition of K_σ and the stresses are computed using the formulas $\sigma = Mc/I$ and $\tau = Tr/J$ (Chapter 3).

No suitable experimental definition of the effective stress concentration factor in impact exists. Impact tests such as the *Charpy* or *Izod* tests (Chapter 4) measure the energy absorbed during the rupture of a notched specimen and do not yield information on stress levels.

When experimental information for a given member or load condition does not exist, the *notch sensitivity index* q provides a means of estimating the effects of stress concentration on strength. Effective stress concentration factors, which are less than the theoretical factor, are related to K_t by the equations

$$K_\sigma = 1 + q(K_t - 1) \tag{6.5}$$
$$K_f = 1 + q_f(K_t - 1) \tag{6.6}$$

A similar equation could be shown for impact loads using q_i as the notch sensitivity index. Often an explicit expression for the notch sensitivity index is

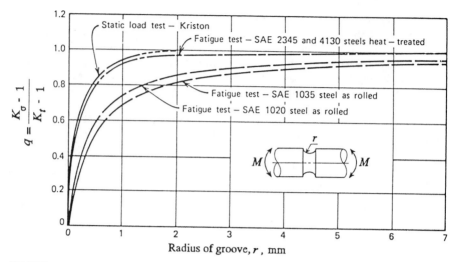

FIGURE 6-3 Influence of radii of grooves on notch sensitivity index. From Ref. 6.7, with permission.

given, e.g., $q_f = (K_f - 1)/(K_t - 1)$. The notch sensitivity index can vary from 0 for complete insensitivity to notches to 1 for the full theoretical effect. Typical values of q are shown in Fig. 6-3.

Notch sensitivity in fatigue decreases as the notch radius decreases and as the grain size increases. A larger part will generally have greater notch sensitivity than a smaller part with proportionally similar dimensions. This variation is known as the scale effect. Larger notch radii result in lower stress gradients near the notch and more material is subjected to higher stresses. Notch sensitivity in fatigue is therefore increased. Because of the low sensitivity of small notch radii, the extremely high theoretical stress concentration factors predicted for very sharp notches and scratches are not actually realized. The notch sensitivity of quenched and tempered steels is higher than that of lower strength, coarser grained alloys. As a consequence, for notched members the strength advantage of high-grade steels over other materials may be lost.

Under static loading, notch sensitivity values are recommended [6.3] as $q = 0$ for ductile metals and q between 0.15 and 0.25 for hard, brittle metals. The notch insensitivity of ductile materials is caused by local plastic deformation at the notch tip. Under conditions that inhibit plastic slip, the notch sensitivity of a ductile metal may increase. Very low temperatures and high temperatures that cause viscous creep are two service conditions that may increase the notch sensitivity of some ductile metals. The notch sensitivity of cast iron is low for static loads ($q \approx 0$) because of the presence of internal stress raisers in the form of material inhomogeneities. These internal stress raisers weaken the material to such an extent that external notches have limited additional effect.

When a notched structural member is subjected to impact loads, the notch sensitivity may increase because the short duration of the load application does not permit the mitigating process of local slip to occur. Also, the smaller sections at stress raisers decrease the capacity of a member to absorb impact energy. For

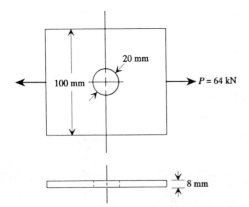

FIGURE 6-4 Tensile member with hole.

impact loads, values of notch sensitivity are recommended such as [6.3] q_i between 0.4 and 0.6 for ductile metals, $q_i = 1$ for hard, brittle materials, and $q_i = 0.5$ for cast irons. Peterson [6.2] recommends using the full theoretical factor for brittle metals (including cast irons) for both static and impact loads because of the possibility of accidental shock loads being applied to a member during handling. The utilization of fracture mechanics to predict the brittle fracture of a flawed member under static, impact, and cyclic loads is treated in Chapter 7.

Neuber's Rule

Consider the stretched plate of Fig. 6-4. For nonlinear material behavior (Fig. 6-5), where local plastic deformation can occur near the hole, the previous

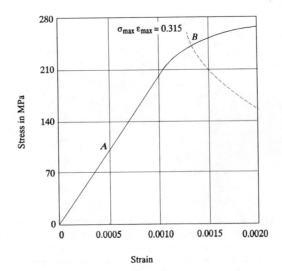

FIGURE 6-5 Stress – strain diagram for material of tensile member of Fig. 6-4.

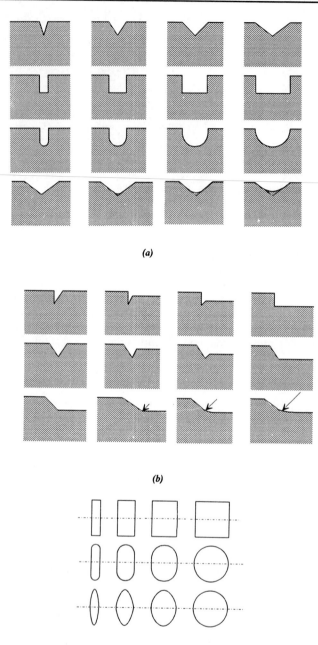

FIGURE 6-6 Effect of notches and holes on stress concentration. From Leyer [6.6], with permission. (a) Common notch shapes, arranged in order of their effect on the stress flow (decreasing from left to right and top to bottom). (b) Asymmetric notch shapes, arranged in the same way as in (a). (c) Internal notch shapes, arranged in the same way as in (a).

stress concentration formulas may not apply. Neuber [6.4] established a rule that is useful beyond the elastic limit relating the effective stress and strain concentration factors to the theoretical stress concentration factor. Neuber's rule contends that the formula

$$K_\sigma K_\varepsilon = K_t^2 \tag{6.7}$$

applies to the three factors. This relation states that K_t is the geometric mean of K_σ and K_ε, i.e., $K_t = (K_\sigma K_\varepsilon)^{1/2}$. Often, for fatigue, K_f replaces K_t. From the definition of effective stress concentration, $K_\sigma = (\sigma_{max}/\sigma_{nom})$. Also, $K_\varepsilon = \varepsilon_{max}/\varepsilon_{nom}$ defines the effective strain concentration factor, where ε_{max} is the strain obtained from the material law (perhaps nonlinear) for the stress level σ_{max}. Using these relations in Eq. (6.7) yields

$$\sigma_{max}\varepsilon_{max} = K_t^2 \sigma_{nom}\varepsilon_{nom} \tag{6.8}$$

Usually K_t and σ_{nom} are known, and ε_{nom} can be found from the stress–strain curve for the material. Equation (6.8) therefore becomes

$$\sigma_{max}\varepsilon_{max} = C \tag{6.9}$$

where C is a known constant. Solving Eq. (6.9) simultaneously with the stress–strain relation, the values of maximum stress and strain are found, and the

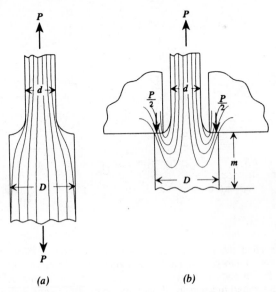

(a)   *(b)*

FIGURE 6-7 Two parts with same shape (step in cross section) but differing stress flow patterns can give totally different notch effects and widely differing stress levels at corner step. From Peterson [6.2], with permission. (a) Stress flow is smooth. (b) Sharp change in stress flow direction causes high stress.

true (effective) stress concentration factor K_σ can then be determined. In this procedure the appropriate stress–strain curve must be known.

Neuber's rule was derived specifically for sharp notches in prismatic bars subjected to two-dimensional shear, but the rule has been applied as a useful approximation in other cases, especially those in which plane stress conditions exist. The rule has been shown to give poor results for circumferential grooves in shafts under axial tension [6.5].

Example 6.2 Tensile Member with Circular Hole The member shown in Fig. 6-4 is subjected to an axial tensile load of 64 kN. The material from which the member is constructed has the stress–strain diagram of Fig. 6-5 for static tensile loading.

From Table 6-1, part II, case 2a, the theoretical stress concentration factor is computed using $d/D = \frac{20}{100}$, as

$$K_t = 3.0 - 3.140\left(\tfrac{20}{100}\right) + 3.667\left(\tfrac{20}{100}\right)^2 - 1.527\left(\tfrac{20}{100}\right)^3 = 2.51 \qquad (1)$$

The nominal stress is found using the net cross-sectional area:

$$\sigma_{\text{nom}} = \frac{P}{(D-d)t} = \frac{64}{(100-20)8} \frac{(10^3)}{(10^{-6})} = 100 \text{ MPa} \qquad (2)$$

Based on elastic behavior, the peak stress $\sigma_{\max}$ at the edge of the hole would be

$$\sigma_{\max} = K_t \sigma_{\text{nom}} = (2.51)(100) = 251 \text{ MPa} \qquad (3)$$

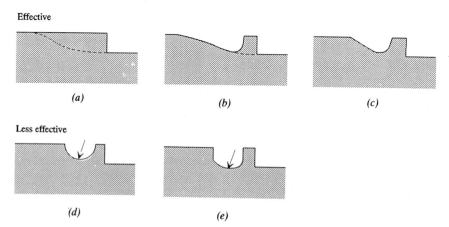

FIGURE 6-8 Guiding the lines of stress by means of notches that are not functionally essential is a useful method of reducing the detrimental effects of notches that cannot be avoided. These are termed relief notches. It is assumed here that the bearing surface of the step of case a is needed functionally. From Leyer [6.6], with permission. (a) The dotted line shows a smooth way of reducing the section, but the corner which may be functionally necessary would be eliminated. (b) Ideal way of reducing section while retaining corner. (c) Compromise because of difficulties in machining a notch such as in case (b). (d) Notch too shallow for effectiveness. (e) Notch flanks parallel which decreases effectiveness.

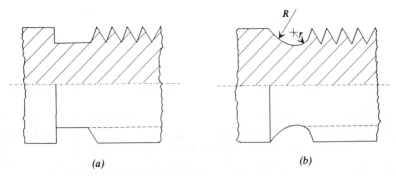

FIGURE 6-9 Relief notch where screw thread meets cylindrical body of bolt. (*a*) Considerable stress concentration can occur at the step interface. (*b*) Use of a smoother interface leads to relief of stress concentration.

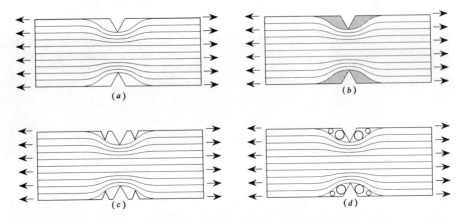

FIGURE 6-10 Alleviation of stress concentration by removal of material. From Juvinall [6.8], with permission. (*a*) It is assumed that a notch of this sort is to occur in this element. The notch alone is an unfavorable way of reducing the section. (*b*) Removal of the shaded region would be a favorable way of reducing the section, but the notch and flat top and bottom surfaces are eliminated. The machining will be much harder than that in cases (*c*) and (*d*). (*c*) Stress concentration reduced but notch retained. (*d*) Stress concentration reduced but notch and top and bottom surfaces retained.

This stress value, however, exceeds the yield point of the material. The actual peak stress and strain at the hole edge are found by using Neuber's rule. The nominal strain is read from the stress–strain curve; at $\sigma_{nom} = 100$ MPa, the strain is $\varepsilon_{nom} = 5 \times 10^{-4}$. The point $(\sigma_{nom}, \varepsilon_{nom})$ is point A in Fig. 6-5. Neuber's rule gives

$$\sigma_{max}\varepsilon_{max} = K_t^2 \, \sigma_{nom}\varepsilon_{nom} = (2.51)^2(100)(5 \times 10^{-4}) = 0.315 \text{ MPa} \qquad (4)$$

The intersection of the curve $\sigma_{max}\varepsilon_{max} = 0.315$ with the stress–strain curve (point B in Fig. 6-5) yields a peak stress of $\sigma_{max} = 243$ MPa and a peak strain of 13×10^{-4}. The effective stress concentration factor is

$$K_\sigma = \sigma_{max}/\sigma_{nom} = 243/100 = 2.43 \qquad (5)$$

The effective strain concentration factor is

$$K_\varepsilon = \frac{13 \times 10^{-4}}{5 \times 10^{-4}} = 2.6 \qquad (6)$$

In the local strain approach to fatigue analysis, fatigue life is correlated with the strain history of a point, and knowledge of the true level of strain at the point is necessary. Neuber's rule enables the estimation of local strain levels without using complicated elastic–plastic finite-element analyses.

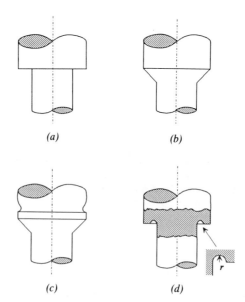

(a)   (b)

(c)   (d)

FIGURE 6-11 Reducing stress concentration in stepped shaft. From Juvinall [6.8], with permission. (a) Severe stress concentration. (b) Use large radius if possible. (c) Added groove gives further benefit. (d) Undercut shoulder helps if modifications (b) and (c) cannot be used.

6.4 DESIGNING TO MINIMIZE STRESS CONCENTRATION

A qualitative discussion of techniques for avoiding the detrimental effects of stress concentration is given by Leyer [6.6]. As a general rule, force should be transmitted from point to point as smoothly as possible. The lines connecting the force transmission path are sometimes called the "force (or stress) flow," although it is arguable if *force flow* has a scientifically based definition. Sharp transitions in the direction of the force flow should be removed by smoothing contours and rounding notch roots. When stress raisers are necessitated by functional requirements, the raisers should be placed in regions of low nominal stress if possible. Figures 6-6*a–c* (p. 250) depict forms of notches and holes in the order in which they cause stress concentration. Figure 6-7 (p. 251) shows how direction of stress flow affects the extent to which a notch causes stress concentration. The configuration in Fig. 6-7*b* has higher stress levels because of the sharp change in the direction of force flow.

When notches are necessary, removal of material near the notch can alleviate stress concentration effects. Figures 6-8 to 6-13 demonstrate instances where removal of material improves the strength of the member. Figure 6-13 shows three types of nut design in which design (*c*) is better than (b), which in turn is

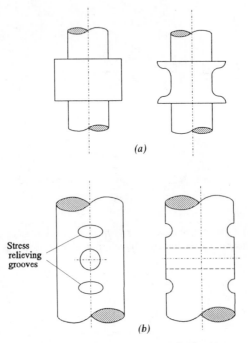

(a)

(b)

FIGURE 6-12 Bars with collars and holes. From Juvinall [6.8], with permission. (*a*) Narrow collars reduce stress concentration. The bar with the collar on the right will lead to reduced stress concentration relative to the bar on the left. (*b*) Grooves reduce stress concentration around hole.

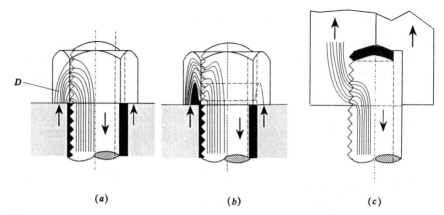

(a) (b) (c)

FIGURE 6-13 Nut designs. These are most important under fatigue loading. From Peterson [6.2], with permission. (a) Standard bolt and nut combination. The force flow near the top of the nut is sparse, but in area D the stress flow density is very high. (b) Nut with a lip. The force flow on the inner side of the lip is in the same direction as in the bolt and the force flow is more evenly distributed for the whole nut than for case (a). The peak stress is relieved. (c) "Force flow" is not reversed at all. Thus fatigue strength here is significantly higher than for the other cases.

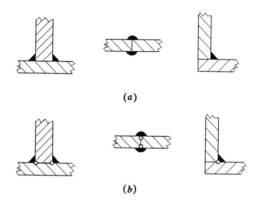

(a)

(b)

FIGURE 6-14 (a) Type of joint often used in welding practice, all of which involve interface notches. (b) Improved joints through boring out sharp edges. From Leyer [6.6], with permission.

superior to (a). In this figure, the stress flow density of area D in design (a) is the highest. Design (b) leads to a stress flow on the innerside of the lip that is in the same direction as in the bolt and, relative to the previous design, is more even in the nut, so that the peak stress is relieved. In design (c), force flow does not change direction; consequently, the stress concentration is reduced considerably.

A type of stress concentration called an *interface notch* is commonly produced when parts are joined by welding. Figure 6-14 shows examples of interface notches and one way of mitigating the effect. The surfaces where the mating plates touch without weld metal filling, form what is, in effect, a sharp crack that causes stress concentration. Stress concentration also results from poor welding techniques that create small cracks in the weld material or burn pits in the base material.

REFERENCES

6.1 Timoshenko, S. P., and Goodier, J. N., *Theory of Elasticity*, 3rd ed., McGraw-Hill, New York, 1970.

6.2 Peterson, R. E., *Stress Concentration Factors*, Wiley, New York, 1974.

6.3 Maleev, V. L., and Hartman, J. B., *Machine Design*, 3rd ed., International Textbook Co., Scranton, PA, 1954.

6.4 Neuber, H., "Theory of Stress Concentration for Shear Strained Prismatic Bodies with Nonlinear Stress–Strain Law," *J. Appl. Mech.*, Vol. 28, Series E, No. 4. 1961, pp. 544–550.

6.5 Fuchs, H. O., "Discussion: Nominal Stress or Local Strain Approaches to Cumulative Damage," *Fatigue Under Complex Loading*, Society Automotive Engineering, Warrendale, PA, 1977, pp. 203–207.

6.6 Leyer, A., *Maschinenkonstruktionslehre*, Birkhäuser Verlag, Basel, Switzerland. English language edition. *Machine Design*, Blackie & Son, London, 1974.

6.7 Boresi, A. P., Sidebottom, O. M., Seeley, F. B., and Smith, J. O., *Advanced Mechanics of Materials*, 4th ed., Wiley, New York, 1985.

6.8 Juvinall, R. C., *Stress, Strain, and Strength*, McGraw-Hill, New York, 1967.

6.9 Hooke, C. J., "Numerical Solution of Plane Elastostatic Problems by Point Matching," *J. Strain Anal.*, Vol. 3, 1968, pp. 109–115.

6.10 Liebowitz, H., Vandervelt, H., and Sanford, R. J., "Stress Concentrations Due to Sharp Notches," *Exper. Mech.*, Vol. 7, 1967.

6.11 Neuber, H., "Theory of Notch Stresses," Office of Technical Services, Department of Commerce, Washington, DC, 1961.

6.12 Atsumi, "Stress Concentrations in a Strip under Tension and Containing an Infinite Row of Semicircular Notches," *Quart. J. Mech. Appl. Math.*, Vol. 11, Part 4, 1958.

6.13 Durelli, A. J., Lake, R. L., and Phillips, E., "Stress Concentrations Produced by Multiple Semi-Circular Notches in Infinite Plates under Uniaxial State of Stress." *Proc. SESA*, Vol. 10, No. 1, 1952.

6.14 Matthews, G. J., and Hooke, C. J., "Solution of Axisymmetric Torsion Problems by Point Matching," *J. Strain Anal.*, Vol. 6, 1971, pp. 124–134.

6.15 Howland, R. C. J., "On the Stresses in the Neighborhood of a Circular Hole in a Strip under Tension," *Philos. Trans. Roy. Soc. Lond. A*, Vol. 229, 1929/1930.

6.16 Jones, N., and Hozos, D., "A Study of the Stress around Elliptical Holes in Flat Plates," *Trans. ASME, J. Eng. Ind.*, Vol. 93, series B, 1971.

6.17 Seika, M., and Ishii, M., "Photoelastic Investigation of the Maximum Stress in a Plate with a Reinforced Circular Hole under Uniaxial Tension," *Trans. ASME, J. Appl. Mech.*, Vol. 86, Series E, 1964, pp. 701–703.

6.18 Seika, M., and Amano, A., "The Maximum Stress in a Wide Plate with a Reinforced Circular Hole under Uniaxial Tension: Effects of a Boss with Fillet," *Trans. ASME, J. Appl. Mech.*, Vol. 89, Series E, 1967, pp. 232–234.

6

Tables

TABLE 6-1 STRESS CONCENTRATION FACTORS [a]

Notation

K_t	Theoretical stress concentration factor in elastic range	σ_{nom}	Nominal normal stress based on minimum cross-sectional area (F/L^2)
σ	Applied stress (F/L^2)	σ_{max}	Maximum normal stress at stress raiser (F/L^2)
P	Applied axial force (F)	τ_{nom}	Nominal shear stress based on minimum cross-sectional area (F/L^2)
M	Applied moment (FL)		
m_1, m_2, m	Applied moment per unit length (FL/L)	τ_{max}	Maximum shear stress at stress raiser (F/L^2)
T	Applied torque (FL)		

Refer to figures for the geometries of the specimens.

I. Notches and Grooves

Type of Stress Raiser	Loading Condition	Stress Concentration Factor
1. Elliptical or U-shaped notch in semi-infinite plate 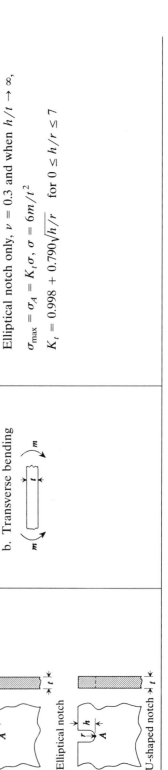	a. Uniaxial tension	$\sigma_{max} = \sigma_A = K_t \sigma$ $K_t = 0.855 + 2.21\sqrt{h/r}$ for $1 \le h/r \le 361$
	b. Transverse bending	Elliptical notch only, $\nu = 0.3$ and when $h/t \to \infty$, $\sigma_{max} = \sigma_A = K_t \sigma,\ \sigma = 6m/t^2$ $K_t = 0.998 + 0.790\sqrt{h/r}$ for $0 \le h/r \le 7$

TABLE 6-1 Stress Concentration Factors 260

2.
Opposite single U-shape notches in finite-width plate

a. Axial tension

$$\sigma_{max} = \sigma_A = K_t \sigma_{nom}, \qquad \sigma_{nom} = P/td$$

$$K_t = C_1 + C_2\left(\frac{2h}{D}\right) + C_3\left(\frac{2h}{D}\right)^2 + C_4\left(\frac{2h}{D}\right)^3$$

	$0.1 \leq h/r < 2.0$	$2.0 \leq h/r \leq 50.0$
C_1	$0.955 + 2.169\sqrt{h/r} - 0.081h/r$	$1.037 + 1.991\sqrt{h/r} + 0.002h/r$
C_2	$-1.557 - 4.046\sqrt{h/r} + 1.032h/r$	$-1.886 - 2.181\sqrt{h/r} - 0.048h/r$
C_3	$4.013 + 0.424\sqrt{h/r} - 0.748h/r$	$0.649 + 1.086\sqrt{h/r} + 0.142h/r$
C_4	$-2.461 + 1.538\sqrt{h/r} - 0.236h/r$	$1.218 - 0.922\sqrt{h/r} - 0.086h/r$

For semicircular notch ($h/r = 1.0$)

$$K_t = 3.065 - 3.472\left(\frac{2h}{D}\right) + 1.009\left(\frac{2h}{D}\right)^2 + 0.405\left(\frac{2h}{D}\right)^3$$

Stress Concentration Factors TABLE 6-1

Type of Stress Raiser	Loading Condition	Stress Concentration Factor
2. Continued Opposite single U-shape notches in finite-width plate	b. In-plane bending	$\sigma_{max} = \sigma_A = K_t \sigma_{nom}, \qquad \sigma_{nom} = 6M/d^2 t$ $$K_t = C_1 + C_2\left(\frac{2h}{D}\right) + C_3\left(\frac{2h}{D}\right)^2 + C_4\left(\frac{2h}{D}\right)^3$$

	$0.1 \le h/r < 2.0$	$2.0 \le h/r \le 50.0$
C_1	$1.024 + 2.092\sqrt{h/r} - 0.051h/r$	$1.113 + 1.957\sqrt{h/r}$
C_2	$-0.630 - 7.194\sqrt{h/r} + 1.288h/r$	$-2.579 - 4.017\sqrt{h/r} - 0.013h/r$
C_3	$2.117 + 8.574\sqrt{h/r} - 2.160h/r$	$4.100 + 3.922\sqrt{h/r} + 0.083h/r$
C_4	$-1.420 - 3.494\sqrt{h/r} + 0.932h/r$	$-1.528 - 1.893\sqrt{h/r} - 0.066h/r$

For semicircular notch ($h/r = 1.0$)

$$K_t = 3.065 - 6.637\left(\frac{2h}{D}\right) + 8.229\left(\frac{2h}{D}\right)^2 - 3.636\left(\frac{2h}{D}\right)^3$$

TABLE 6-1   Stress Concentration Factors   262

c. Transverse bending	$\sigma_{\max} = \sigma_A = K_t \sigma_{\text{nom}}, \qquad \sigma_{\text{nom}} = 6M/t^2 d$

$$K_t = C_1 + C_2\left(\frac{2h}{D}\right) + C_3\left(\frac{2h}{D}\right)^2 + C_4\left(\frac{2h}{D}\right)^3$$

$0.1 \leq h/r \leq 5.0$ and h/t is large

C_1	$1.041 + 0.839\sqrt{h/r} + 0.014\ h/r$
C_2	$-1.239 - 1.663\sqrt{h/r} + 0.118\ h/r$
C_3	$3.370 - 0.758\sqrt{h/r} + 0.434\ h/r$
C_4	$-2.162 + 1.582\sqrt{h/r} - 0.606\ h/r$

For semicircular notch ($h/r = 1.0$)

$$K_t = 1.894 - 2.784\left(\frac{2h}{D}\right) + 3.046\left(\frac{2h}{D}\right)^2 - 1.186\left(\frac{2h}{D}\right)^3$$

Stress Concentration Factors   TABLE 6-1

Type of Stress Raiser	Loading Condition	Stress Concentration Factor
3. Single U-shaped notch on one side in finite-width plate 	**a. Axial tension** 	$\sigma_{max} = \sigma_A = K_t \sigma_{nom}, \qquad \sigma_{nom} = P/td$ $K_t = C_1 + C_2\left(\dfrac{h}{D}\right) + C_3\left(\dfrac{h}{D}\right)^2 + C_4\left(\dfrac{h}{D}\right)^3$

	$0.5 \le h/r < 2.0$	$2.0 \le h/r \le 20.0$
C_1	$0.907 + 2.125\sqrt{h/r} + 0.023h/r$	$0.953 + 2.136\sqrt{h/r} - 0.005h/r$
C_2	$0.710 - 11.289\sqrt{h/r} + 1.708h/r$	$-3.255 - 6.281\sqrt{h/r} + 0.068h/r$
C_3	$-0.672 + 18.754\sqrt{h/r} - 4.046h/r$	$8.203 + 6.893\sqrt{h/r} + 0.064h/r$
C_4	$0.175 - 9.759\sqrt{h/r} + 2.365h/r$	$-4.851 - 2.793\sqrt{h/r} - 0.128h/r$

For semicircular notch $(h/r = 1.0)$.

$$K_t = 3.065 - 8.871\left(\frac{h}{D}\right) + 14.036\left(\frac{h}{D}\right)^2 - 7.219\left(\frac{h}{D}\right)^3$$

TABLE 6-1   Stress Concentration Factors   264

b. In-plane bending	$$\sigma_{\max} = \sigma_A = K_t\sigma_{\text{nom}}, \qquad \sigma_{\text{nom}} = 6M/td^2$$
	$$K_t = C_1 + C_2\left(\frac{h}{D}\right) + C_3\left(\frac{h}{D}\right)^2 + C_4\left(\frac{h}{D}\right)^3$$

	$0.5 \le h/r \le 2.0$	$2.0 \le h/r \le 20.0$
C_1	$1.795 + 1.481h/r - 0.211(h/r)^2$	$2.966 + 0.502h/r - 0.009(h/r)^2$
C_2	$-3.544 - 3.677h/r + 0.578(h/r)^2$	$-6.475 - 1.126h/r + 0.019(h/r)^2$
C_3	$5.459 + 3.691h/r - 0.565(h/r)^2$	$8.023 + 1.253h/r - 0.020(h/r)^2$
C_4	$-2.678 - 1.531h/r + 0.205(h/r)^2$	$-3.572 - 0.634h/r + 0.010(h/r)^2$

For semicircular notch ($h/r = 1.0$)

$$K_t = 3.065 - 6.643\left(\frac{h}{D}\right) + 0.205\left(\frac{h}{D}\right)^2 - 4.004\left(\frac{h}{D}\right)^3$$

4. Multiple semicircular notches in finite-width plate	Axial tension
	$$\sigma_{\max} = K_t\sigma_{\text{nom}}, \qquad \sigma_{\text{nom}} = P/td$$
	$$K_t = C_1 + C_2\left(\frac{2r}{L}\right) + C_3\left(\frac{2r}{L}\right)^2 + \left(\frac{2r}{L}\right)^3$$

$$2r/D \le 0.4, \qquad 0 \le 2r/L \le 1.0$$

C_1	$3.1055 - 3.4287\left(\dfrac{2r}{D}\right) + 0.8522\left(\dfrac{2r}{D}\right)^2$
C_2	$-1.4370 + 10.5053\left(\dfrac{2r}{D}\right) - 8.7547\left(\dfrac{2r}{D}\right)^2 - 19.6273\left(\dfrac{2r}{D}\right)^3$
C_3	$-1.6753 - 14.0851\left(\dfrac{2r}{D}\right) + 43.6575\left(\dfrac{2r}{D}\right)^2$
C_4	$1.7207 + 5.7974\left(\dfrac{2r}{D}\right) - 27.7463\left(\dfrac{2r}{D}\right)^2 + 6.0444\left(\dfrac{2r}{D}\right)^3$

Stress Concentration Factors   TABLE 6-1

TABLE 6-1 **Stress Concentration Factors** 266

TABLE 6-1 (continued) STRESS CONCENTRATION FACTORS: Notches and Grooves

Type of Stress Raiser	Loading Condition	Stress Concentration Factor
5. Opposite single V-shape notch in finite-width plate	Axial tension	$\sigma_{max} = \sigma_A = K_t \sigma_{nom}, \quad \sigma_{nom} = P/td$ For $2h/D = 0.398$ and $\alpha < 90°$ $\quad\;\; 2h/D = 0.667$ and $\alpha < 60°$ $\qquad\qquad\qquad K_t = K_{tu}$ K_{tu} is the stress concentration factor for U-shaped notch and α is notch angle in degrees. Otherwise $K_t = C_1 + C_2\sqrt{K_{tu}} + C_3 K_{tu}$ $2h/D = 0.398, \; 90° \le \alpha \le 150°, \; 1.6 \le K_{tu} \le 3.5$

C_1	$5.294 - 0.1225\alpha + 0.000523\alpha^2$
C_2	$-5.0002 + 0.1171\alpha - 0.000434\alpha^2$
C_3	$1.423 - 0.01197\alpha - 0.000004\alpha^2$

$2h/D = 0.667, \; 60° \le \alpha \le 150°, \; 1.6 \le K_{tu} \le 2.8$

C_1	$-10.01 + 0.1534\alpha - 0.000647\alpha^2$
C_2	$13.60 - 0.2140\alpha + 0.000973\alpha^2$
C_3	$-3.781 + 0.07873\alpha - 0.000392\alpha^2$

6. Single V-shaped notch on one side

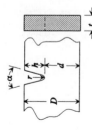

In-plane bending

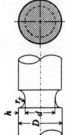

$\sigma_{max} = \sigma_A = K_t\sigma_{nom}$, $\sigma_{nom} = 6M/td^2$

For $\alpha \le 90°$

$K_t = K_{tu}$

For $90° < \alpha \le 150°$ and $0.5 \le h/r \le 4.0$

$$K_t = 1.11K_{tu} - \left[-0.0159 + 0.2243\left(\frac{\alpha}{150}\right)\right.$$
$$\left. -0.4293\left(\frac{\alpha}{150}\right)^2 + 0.3609\left(\frac{\alpha}{150}\right)^3\right]K_{tu}^2$$

K_{tu} is the stress concentration factor for U notch, case 3b, and α is notch angle in degrees

7. U-shaped circumferential groove in circular shaft

a. Axial tension

$\sigma_{max} = \sigma_A = K_t\sigma_{nom}$, $\sigma_{nom} = 4P/\pi d^2$

$$K_t = C_1 + C_2\left(\frac{2h}{D}\right) + C_3\left(\frac{2h}{D}\right)^2 + C_4\left(\frac{2h}{D}\right)^3$$

	$0.1 \le h/r < 2.0$	$2.0 \le h/r \le 50.0$
C_1	$0.89 + 2.208\sqrt{h/r} - 0.094h/r$	$1.037 + 1.967\sqrt{h/r} + 0.002h/r$
C_2	$-0.923 - 6.678\sqrt{h/r} + 1.638h/r$	$-2.679 - 2.980\sqrt{h/r} - 0.053h/r$
C_3	$2.893 + 6.448\sqrt{h/r} - 2.516h/r$	$3.090 + 2.124\sqrt{h/r} + 0.165h/r$
C_4	$-1.912 - 1.944\sqrt{h/r} + 0.963h/r$	$-0.424 - 1.153\sqrt{h/r} - 0.106h/r$

For semicircular groove$(h/r = 1.0)$

$$K_t = 3.004 - 5.963\left(\frac{2h}{D}\right) + 6.836\left(\frac{2h}{D}\right)^2 - 2.893\left(\frac{2h}{D}\right)^3$$

TABLE 6-1 (continued)   STRESS CONCENTRATION FACTORS: Notches and Grooves

Type of Stress Raiser	Loading Condition	Stress Concentration Factor

7. Continued
U-shaped circumferential groove in circular shaft

b. Bending

$$\sigma_{max} = \sigma_A = K_t\sigma_{nom}, \qquad \sigma_{nom} = 32M/\pi d^3$$

$$K_t = C_1 + C_2\left(\frac{2h}{D}\right) + C_3\left(\frac{2h}{D}\right)^2 + C_4\left(\frac{2h}{D}\right)^3$$

	$0.25 \le h/r < 2.0$	$2.0 \le h/r \le 50.0$
C_1	$0.594 + 2.958\sqrt{h/r} - 0.520h/r$	$0.965 + 1.926\sqrt{h/r}$
C_2	$0.422 - 10.545\sqrt{h/r} + 2.692h/r$	$-2.773 - 4.414\sqrt{h/r} - 0.017h/r$
C_3	$0.501 + 14.375\sqrt{h/r} - 4.486h/r$	$4.785 + 4.681\sqrt{h/r} + 0.096h/r$
C_4	$-0.613 - 6.573\sqrt{h/r} + 2.177h/r$	$-1.995 - 2.241\sqrt{h/r} - 0.074h/r$

For semicircular groove ($h/r = 1.0$)

$$K_t = 3.032 - 7.431\left(\frac{2h}{D}\right) + 10.390\left(\frac{2h}{D}\right)^2 - 5.009\left(\frac{2h}{D}\right)^3$$

c. Torsion

$$\tau_{max} = \tau_A = K_t\tau_{nom}, \qquad \tau_{nom} = 16T/\pi d^3$$

$$K_t = C_1 + C_2\left(\frac{2h}{D}\right) + C_3\left(\frac{2h}{D}\right)^2 + C_4\left(\frac{2h}{D}\right)^3$$

	$0.25 \le h/r < 2.0$	$2.0 \le h/r \le 50.0$
C_1	$0.966 + 1.056\sqrt{h/r} - 0.022h/r$	$1.089 + 0.924\sqrt{h/r} + 0.018h/r$
C_2	$-0.192 - 4.037\sqrt{h/r} + 0.674h/r$	$-1.504 - 2.141\sqrt{h/r} - 0.047h/r$
C_3	$0.808 + 5.321\sqrt{h/r} - 1.231h/r$	$2.486 + 2.289\sqrt{h/r} + 0.091h/r$
C_4	$-0.567 - 2.364\sqrt{h/r} + 0.566h/r$	$-1.056 - 1.104\sqrt{h/r} - 0.059h/r$

TABLE 6-1   Stress Concentration Factors   268

Stress Concentration Factors | TABLE 6-1

8.
Large circumferential groove in circular shaft

For semicircular groove ($h/r = 1.0$)

$$K_t = 2.000 - 3.555\left(\frac{2h}{D}\right) + 4.898\left(\frac{2h}{D}\right)^2 - 2.365\left(\frac{2h}{D}\right)^3$$

a. Axial tension

$$\sigma_{max} = \sigma_A = K_t\sigma_{nom}, \qquad \sigma_{nom} = 4P/\pi d^2$$

$$K_t = C_1 + C_2(r/d) + C_3(r/d)^2$$

	$0.3 \le r/d \le 1.0,$	$1.005 \le D/d \le 1.10$
C_1	$-81.39 + 153.10(D/d) - 70.49(D/d)^2$	
C_2	$119.64 - 221.81(D/d) + 101.93(D/d)^2$	
C_3	$-57.88 + 107.33(D/d) - 49.34(D/d)^2$	

b. Bending

$$\sigma_{max} = \sigma_A = K_t\sigma_{nom}, \qquad \sigma_{nom} = 32M/\pi d^3$$

$$K_t = C_1 + C_2(r/d) + C_3(r/d)^2$$

	$0.3 \le r/d \le 1.0,$	$1.005 \le D/d < 1.10$
C_1	$-39.58 + 73.22(D/d) - 32.46(D/d)^2$	
C_2	$-9.477 + 29.41(D/d) - 20.13(D/d)^2$	
C_3	$82.46 - 166.96(D/d) + 84.58(D/d)^2$	

Type of Stress Raiser	Loading Condition	Stress Concentration Factor
8. Continued Large circumferential groove in circular shaft	c. Torsion	$\tau_{max} = \tau_A = K_t\tau_{nom}, \qquad \tau_{nom} = 16T/\pi d^3$ $K_t = C_1 + C_2(r/d) + C_3(r/d)^2$ $0.3 \leq r/d \leq 1, \qquad 1.005 \leq D/d < 1.10$ $C_1 \quad -35.16 + 67.57(D/d) - 31.28(D/d)^2$ $C_2 \quad 79.13 - 148.37(D/d) + 69.09(D/d)^2$ $C_3 \quad -50.34 + 94.67(D/d) - 44.26(D/d)^2$
9. V-shaped groove in circular shaft	Torsion	$\tau_{max} = \tau_A = K_t\tau_{nom}, \qquad \tau_{nom} = 16T/\pi d^3$ K_{tu} = Stress concentration factor for U-shaped groove $(\alpha = 0)$, case 7c $K_t = C_1 + C_2\sqrt{K_{tu}} + C_3 K_{tu}$ $C_1 \quad 0.2026\sqrt{\alpha} - 0.06620\alpha + 0.00281\alpha\sqrt{\alpha}$ $C_2 \quad -0.2226\sqrt{\alpha} + 0.07814\alpha - 0.002477\alpha\sqrt{\alpha}$ $C_3 \quad 1 + 0.0298\sqrt{\alpha} - 0.01485\alpha - 0.000151\alpha\sqrt{\alpha}$ where α is in degrees. For $0° \leq \alpha \leq 90°$, K_t is independent of r/d $90° \leq \alpha \leq 125°$, K_t is applicable only if $r/d \leq 0.01$

TABLE 6-1 Stress Concentration Factors 270

II. Holes

1. Single circular hole in infinite plate	a. In-plane normal stress	(1) Uniaxial tension ($\sigma_2 = 0$) $\sigma_{\max} = K_t \sigma_1$ $\sigma_A = 3\sigma_1$ or $K_t = 3$ $\sigma_B = -\sigma_1$ or $K_t = -1$ (2) Biaxial tension $K_t = 3 - \sigma_2/\sigma_1$ for $-1 \leq \sigma_2/\sigma_1 \leq 1$ For $\sigma_2 = \sigma_1$, $\sigma_A = \sigma_B = 2\sigma_1$ or $K_t = 2$ For $\sigma_2 = -\sigma_1$ (pure shear stress) $\sigma_A = -\sigma_B = 4\sigma_1$ or $K_t = 4$
	b. Transverse bending	$\sigma_{\max} = K_t \sigma$, $\quad \sigma = 6m/t^2$, $\quad \nu = 0.3$ (1) Simple bending ($m_1 = m$, $m_2 = 0$) $\quad$ For $0 \leq d/t \leq 7.0$, $\sigma_{\max} = \sigma_A$ $\quad K_t = 3.000 - 0.947\sqrt{d/t} + 0.192d/t$ (2) Cylindrical bending ($m_1 = m$, $m_2 = \nu m$) $\quad$ For $0 \leq d/t \leq 7.0$, $\sigma_{\max} = \sigma_A$ $\quad K_t = 2.700 - 0.647\sqrt{d/t} + 0.129d/t$ (3) Isotropic bending ($m_1 = m_2 = m$), $\sigma_{\max} = \sigma_A$ $\quad K_t = 2$ (independent of d/t)
	c. Twisting moment (see previous figure and definitions)	$\sigma_{\max} = K_t \sigma$, $\quad \sigma = 6m/t^2$ $m_1 = m$, $\quad m_2 = -m$, $\quad \nu = 0.3$ For $0 \leq d/t \leq 7.0$ $K_t = 4.000 - 1.772\sqrt{d/t} + 0.341d/t$

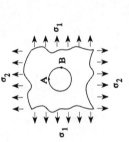

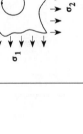

Stress Concentration Factors TABLE 6-1

TABLE 6-1 (continued) STRESS CONCENTRATION FACTORS: Holes

Type of Stress Raiser	Loading Condition	Stress Concentration Factor
2. Central single circular hole in finite-width plate	a. Axial tension	$\sigma_{max} = \sigma_A = K_t \sigma_{nom}, \qquad \sigma_{nom} = P/[t(D-d)]$ $K_t = 3.000 - 3.140(d/D) + 3.667(d/D)^2$ $\quad\quad -1.527(d/D)^3 \quad$ for $0 \le d/D \le 1$
	b. In-plane bending	(1) At edge of hole $\quad\quad \sigma_{max} = \sigma_A = K_t \sigma_{nom}, \qquad \sigma_{nom} = 6Md/(D^3 - d^3)t$ $\quad\quad K_t = 2$ (independent of d/D) (2) At edge of plate $\quad\quad \sigma_{max} = \sigma_B = K_t \sigma_{nom}, \qquad \sigma_{nom} = 6MD/(D^3 - d^3)t$ $\quad\quad K_t = 2d/D \; (\alpha = 30°)$

TABLE 6-1   Stress Concentration Factors   272

c. Transverse bending

$$\sigma_{max} = \sigma_A = K_t \sigma_{nom}, \qquad \sigma_{nom} = 6mD/(D - d)t^2$$

For $0 \le d/D \le 0.3$, $\nu = 0.3$ and $1 \le d/t \le 7$

(1) Simple bending ($m_1 = m$, $m_2 = 0$)

$$K_t = \left[1.793 + \frac{0.131}{d/t} + \frac{2.052}{(d/t)^2} - \frac{1.019}{(d/t)^3}\right]$$
$$\times \left[1 - 1.04\left(\frac{d}{D}\right) + 1.22\left(\frac{d}{D}\right)^2\right]$$

(2) Cylindrical bending ($m_1 = m$, $m_2 = \nu m$)

$$K_t = \left[1.856 + \frac{0.317}{d/t} + \frac{0.942}{(d/t)^2} - \frac{0.415}{(d/t)^3}\right]$$
$$\times \left[1 - 1.04\left(\frac{d}{D}\right) + 1.22\left(\frac{d}{D}\right)^2\right]$$

3. Eccentric circular hole in a finite width plane

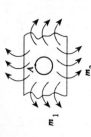

a. Axial tension

Stress on section AB is

$$\sigma_{nom} = \frac{\sigma\sqrt{1 - (d/2c)^2}}{1 - d/2c} \; \frac{1 - c/D}{1 - (c/D)\left[2 - \sqrt{1 - (d/2c)^2}\right]}$$

$$\sigma_{max} = \sigma_B = K_t \sigma_{nom}$$

$$K_t = 3.000 - 3.140\left(\frac{d}{2c}\right) + 3.667\left(\frac{d}{2c}\right)^2 - 1.527\left(\frac{d}{2c}\right)^3$$

TABLE 6-1 (continued) STRESS CONCENTRATION FACTORS: Holes

Type of Stress Raiser	Loading Condition	Stress Concentration Factor
3. Continued Eccentric circular hole in a finite width plane	b. In-plane bending	$\sigma_{max} = \max(\sigma_A, \sigma_B)$ $\sigma_B = K_{t_B}\sigma_{nom}, \qquad \sigma_{nom} = 6M/D^2t$ $K_{t_B} = C_1 + C_2\left(\dfrac{c}{e}\right) + C_3\left(\dfrac{c}{e}\right)^2$ $0 \le d/2c \le 0.5, \qquad 0 \le c/e \le 1.0$ $C_1 \quad 3.000 - 0.631(d/2c) + 4.007(d/2c)^2$ $C_2 \quad -5.083 + 4.067(d/2c) - 2.795(d/2c)^2$ $C_3 \quad 2.114 - 1.682(d/2c) - 0.273(d/2c)^2$ $\sigma_A = K_{t_A}\sigma_{nom}, \qquad \sigma_{nom} = 6M/D^2t$ $K_{t_A} = C_1' + C_2'\left(\dfrac{c}{e}\right) + C_3'\left(\dfrac{c}{e}\right)^2$ $C_1' \quad 1.0286 - 0.1638(d/2c) + 2.702(d/2c)^2$ $C_2' \quad -0.05863 - 0.1335(d/2c) - 1.8747(d/2c)^2$ $C_3' \quad 0.18883 - 0.89219(d/2c) + 1.5189(d/2c)^2$
4. Two equal circular holes in infinite plate	a. Uniaxial tension parallel to row of holes $(\sigma_1 = \sigma, \sigma_2 = 0)$	$\sigma_{max} = K_t\sigma \quad$ for $0 \le d/L \le 1$ $K_t = 3.000 - 0.712\left(\dfrac{d}{L}\right) + 0.271\left(\dfrac{d}{L}\right)^2$

TABLE 6-1   Stress Concentration Factors   **274**

b. Uniaxial tension normal to row of holes ($\sigma_2 = \sigma, \sigma_1 = 0$)

$$\sigma_{max} = \sigma_B = K_t \sigma_{nom}, \qquad \sigma_{nom} = \frac{\sigma\sqrt{1-(d/L)^2}}{1-d/L}$$

$$K_t = 3.0000 - 3.0018\left(\frac{d}{L}\right) + 1.0099\left(\frac{d}{L}\right)^2$$

for $0 \leq d/L \leq 1$

c. Biaxial tension ($\sigma_1 = \sigma_2 = \sigma$)

$$\sigma_{max} = \sigma_B = K_t \sigma_{nom}, \qquad \sigma_{nom} = \frac{\sigma\sqrt{1-(d/L)^2}}{1-d/L}$$

$$K_t = 2.000 - 2.119\left(\frac{d}{L}\right) + 2.493\left(\frac{d}{L}\right)^2 - 1.372\left(\frac{d}{L}\right)^3$$

for $0 \leq d/L \leq 1$

5. Single row of circular holes in infinite plate

a. Uniaxial tension normal to row of holes ($\sigma_1 = 0, \sigma_2 = \sigma$)

$$\sigma_{max} = \sigma_B = K_t \sigma$$

$$K_t = 3.0000 - 0.9916\left(\frac{d}{L}\right) - 2.5899\left(\frac{d}{L}\right)^2 + 2.2613\left(\frac{d}{L}\right)^3$$

for $0 \leq d/L \leq 1$

b. Uniaxial tension parallel to row of holes ($\sigma_1 = \sigma, \sigma_2 = 0$)

$$\sigma_{max} = \sigma_A = K_t \sigma_{nom}, \qquad \sigma_{nom} = \sigma/(1-d/L)$$

$$K_t = 3.000 - 3.095\left(\frac{d}{L}\right) + 0.309\left(\frac{d}{L}\right)^2 + 0.786\left(\frac{d}{L}\right)^3$$

for $0 \leq d/L \leq 1$

c. Biaxial tension ($\sigma_1 = \sigma_2 = \sigma$)

$$\sigma_{max} = \sigma_A = K_t \sigma_{nom}, \qquad \sigma_{nom} = \sigma/(1-d/L)$$

$$K_t = 2.000 - 1.597\left(\frac{d}{L}\right) + 0.934\left(\frac{d}{L}\right)^2 - 0.337\left(\frac{d}{L}\right)^3$$

for $0 \leq d/L \leq 1$

Stress Concentration Factors | **TABLE 6-1**

TABLE 6-1 (continued) STRESS CONCENTRATION FACTORS: Holes

Type of Stress Raiser	Loading Condition	Stress Concentration Factor
5. Continued Single row of circular holes in infinite plate	d. Transverse Bending ($\nu = 0.3$) 	Bending about y axis: $\sigma_{max} = K_t \sigma_{nom}$, $\quad \sigma_{nom} = 6m/t^2 \quad$ for $0 \le d/L \le 1$ (1) Simple bending ($m_1 = m$, $m_2 = 0$) $K_t = 1.787 - 0.060\left(\dfrac{d}{L}\right) - 0.785\left(\dfrac{d}{L}\right)^2 + 0.217\left(\dfrac{d}{L}\right)^3$ (2) Cylindrical bending ($m_1 = m$, $m_2 = vm$) $K_t = 1.850 - 0.030\left(\dfrac{d}{L}\right) - 0.994\left(\dfrac{d}{L}\right)^2 + 0.389\left(\dfrac{d}{L}\right)^3$ Bending about x axis $\sigma_{max} = K_t \sigma_{nom}$, $\quad \sigma_{nom} = 6m/t^2(1 - d/L)$ for $0 \le d/L \le 1$ (1) Simple bending ($m_1 = m$, $m_2 = 0$) $K_t = 1.788 - 1.729\left(\dfrac{d}{L}\right) + 1.094\left(\dfrac{d}{L}\right)^2 - 0.111\left(\dfrac{d}{L}\right)^3$ (2) Cylindrical bending ($m_1 = m$, $m_2 = vm$) $K_t = 1.849 - 1.741\left(\dfrac{d}{L}\right) + 0.875\left(\dfrac{d}{L}\right)^2 + 0.081\left(\dfrac{d}{L}\right)^3$

TABLE 6-1 Stress Concentration Factors 276

6.
Single elliptical hole in infinite plate

$$r = \frac{a^2}{b}$$

a. In-plane normal stress

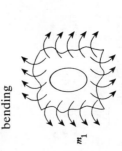

(1) Uniaxial tension ($\sigma_1 = \sigma, \sigma_2 = 0$)

$$\sigma_A = K_t \sigma$$

$$K_t = 1 + \frac{2b}{a} = 1 + 2\sqrt{\frac{b}{r}} \quad \text{for } 0 < b/a < 10$$

and $\sigma_B = -\sigma$

(2) Biaxial tension

For $-1 \leq \sigma_2/\sigma_1 \leq 1$ and $0.25 \leq b/a \leq 4$

$$\sigma_A = K_{tA}\sigma_1, \qquad K_{tA} = 1 + \frac{2b}{a} - \frac{\sigma_2}{\sigma_1}$$

$$\sigma_B = K_{tB}\sigma_1, \qquad K_{tB} = \frac{\sigma_2}{\sigma_1}\left(1 + \frac{2a}{b}\right) - 1$$

For $\sigma_1 = \sigma_2$, $K_{tA} = 2b/a$

$$K_{tB} = 2a/b$$

b. Transverse bending

$$\sigma_{max} = K_t \sigma, \qquad \sigma = 6m/t^2, \nu = 0.3$$

for $2b/t > 5$ and $0.2 \leq b/a \leq 5$

(1) Simple bending ($m_1 = m, m_2 = 0$)

$$K_t = 1 + \frac{2(1+\nu)(b/a)}{3+\nu} \quad \text{for } 2b/t > 5$$

(2) Cylindrical bending ($m_1 = m, m_2 = \nu m$)

$$K_t = \frac{(1+\nu)[2(b/a) + 3 - \nu]}{3 + \nu}$$

(3) Isotropic bending ($m_1 = m_2 = m$)

$$K_t = 2 \text{ (constant)}$$

TABLE 6-1 (continued) STRESS CONCENTRATION FACTORS: Holes

Type of Stress Raiser	Loading Condition	Stress Concentration Factor
7. Single elliptical hole in finite-width plate 	**a. Axial tension** 	$\sigma_{max} = \sigma_A = K_t \sigma_{nom}, \qquad \sigma_{nom} = \sigma/(1 - 2b/D)$ $K_t = C_1 + C_2\left(\dfrac{2b}{D}\right) + C_3\left(\dfrac{2b}{D}\right)^2 + C_4\left(\dfrac{2b}{D}\right)^3,$ $1.0 \le b/a \le 8.0$ $C_1 \quad 1.109 - 0.188\sqrt{b/a} + 2.086\,b/a$ $C_2 \quad -0.486 + 0.213\sqrt{b/a} - 2.588\,b/a$ $C_3 \quad 3.816 - 5.510\sqrt{b/a} + 4.638\,b/a$ $C_4 \quad -2.438 + 5.485\sqrt{b/a} - 4.126\,b/a$
	b. In-plane bending 	$\sigma_{max} = \sigma_A = K_t \sigma_{nom}, \qquad \sigma_{nom} = 12Mb/(D^3 - 8b^3)\,t$ $K_t = C_1 + C_2\left(\dfrac{2b}{D}\right) + C_3\left(\dfrac{2b}{D}\right)^2$ $0.4 \le 2b/D \le 1.0, \quad 1.0 \le b/a \le 2.0$ $C_1 \quad 1.509 + 0.336(b/a) + 0.155(b/a)^2$ $C_2 \quad -0.416 + 0.445(b/a) - 0.029(b/a)^2$ $C_3 \quad 0.878 - 0.736(b/a) - 0.142(b/a)^2$ for $2b/D \le 0.4$, $\sigma_{max} = \sigma_B = 6M/D^2 t$

TABLE 6-1 Stress Concentration Factors   278

8.

Eccentric elliptical hole in finite-width plate

Axial tension

Stress on section AB is

$$\sigma_{nom} = \frac{\sqrt{1 - b/c}}{1 - b/c}\; \frac{1 - c/D}{1 - (c/D)\left[2 - \sqrt{1 - (b/c)^2}\right]}$$

and

$$\sigma_{max} = K_t \sigma_{nom}$$

$$K_t = C_1 + C_2\left(\frac{b}{c}\right) + C_3\left(\frac{b}{c}\right)^2 + C_4\left(\frac{b}{c}\right)^3$$

for $1.0 \leq b/a \leq 8.0$ and $0 \leq b/c \leq 1$

Expressions for C_1, C_2, C_3, and C_4 from case 7a can be used.

9.

Infinite row of elliptical holes in infinite-width plate

Uniaxial tension

$$\sigma_{max} = K_t \sigma_{nom}, \qquad \sigma_{nom} = \sigma/(1 - 2b/L)$$

For $0 \leq 2b/L \leq 0.7$ and $1 \leq b/a \leq 10$

$$K_t = \left[1.002 - 1.016\left(\frac{2b}{L}\right) + 0.253\left(\frac{2b}{L}\right)^2\right]\left(1 + \frac{2b}{a}\right)$$

TABLE 6-1 (continued) STRESS CONCENTRATION FACTORS: Holes

Type of Stress Raiser	Loading Condition	Stress Concentration Factor
10. Circular hole with opposite semicircular lobes in finite-width plate	Axial tension	$\sigma_{\max} = K_t \sigma_{\mathrm{nom}}, \qquad \sigma_{\mathrm{nom}} = \sigma/(1 - 2b/D)$ For $0 \le 2b/D \le 1$ $K_t = K_{t0}\left[1 - \dfrac{2b}{D} + \left(\dfrac{6}{K_{t0}} - 1 \right)\left(\dfrac{2b}{D} \right)^2 + \left(1 - \dfrac{4}{K_{t0}} \right)\left(\dfrac{2b}{D} \right)^3 \right]$ where for $0.2 < r/R \le 4.0$ $K_{t0} = \dfrac{\sigma_{\max}}{\sigma} = 2.2889 + \dfrac{1.6355}{\sqrt{r/R}} - \dfrac{0.0157}{r/R}$ For infinitely wide plate, $K_t = K_{t0}$
11. Rectangular hole with rounded corners in infinite-width plate	Uniaxial tension	$\sigma_{\max} = K_t \sigma$ $K_t = C_1 + C_2\left(\dfrac{a}{b} \right) + C_3\left(\dfrac{a}{b} \right)^2 + C_4\left(\dfrac{a}{b} \right)^3$ $0.05 \le r/2a \le 0.5, \qquad 0.2 \le a/b \le 1.0$ $C_1 \quad 14.815 - 22.308\sqrt{r/2a} + 16.298(r/2a)$ $C_2 \quad -11.201 - 13.789\sqrt{r/2a} + 19.200(r/2a)$ $C_3 \quad 0.2020 + 54.620\sqrt{r/2a} - 54.748(r/2a)$ $C_4 \quad 3.232 - 32.530\sqrt{r/2a} + 30.964(r/2a)$

TABLE 6-1 Stress Concentration Factors 280

12. Slot having semicircular ends 	a. Axial tension $a_{eq} = \sqrt{rb}$ where a_{eq} is width of equivalent ellipse	If the openings such as two holes connected by a slit or an ovaloid are enveloped by an ellipse with the same $2b$ and r, K_t can be approximated by using an equivalent ellipse having the same dimensions $2b$ and r. See cases 6a and 8.
	b. In-plane bending $a_{eq} = \sqrt{rb}$	Use an equivalent ellipse. See case 6b.
13. Equilateral triangular hole with round corners in infinite-width plate 	a. Uniaxial tension $(\sigma_1 = \sigma,\ \sigma_2 = 0)$	$\sigma_{max} = K_t\sigma$ For $0.25 \leq r/R \leq 0.75$ $K_t = 6.191 - 7.215(r/R) + 5.492(r/R)^2$
	b. Biaxial tension $(\sigma_1 = \sigma,\ \sigma_2 = \sigma/2)$	$\sigma_{max} = K_t\sigma$ For $0.25 \leq r/R \leq 0.75$ $K_t = 6.364 - 8.885(r/R) + 6.494(r/R)^2$
	c. Biaxial tension $(\sigma_1 = \sigma_2 = \sigma)$	$\sigma_{max} = K_t\sigma$ For $0.25 \leq r/R \leq 0.75$ $K_t = 7.067 - 11.099(r/R) + 7.394(r/R)^2$

Stress Concentration Factors TABLE 6-1

TABLE 6-1 **Stress Concentration Factors** 282

TABLE 6-1 (continued) STRESS CONCENTRATION FACTORS: Holes

Type of Stress Raiser	Loading Condition	Stress Concentration Factor
14. Single symmetrically reinforced circular hole in finite-width plate in tension	a. Without fillet $(r = 0)$	$\sigma_{max} = \sigma_A = k_t \sigma$ where σ_{max} = maximum mean stress for thickness sliced off to plate thickness t. For $b/t = 5.0$ $K_t = C_1 + C_2\left(\dfrac{1}{h/t}\right) + C_3\left(\dfrac{1}{h/t}\right)^2$ $D/b \geq 4.0,\quad 1 \leq h/t \leq 5\quad$ and $\quad 0.3 \leq a/b \leq 0$ $C_1 \quad 1.869 + 1.196(a/b) - 0.393(a/b)^2$ $C_2 \quad -3.042 + 6.476(a/b) - 4.871(a/b)^2$ $C_3 \quad 4.036 - 7.229(a/b) + 5.180(a/b)^2$
	b. With fillet $(r \neq 0)$	For $r/t \geq 0.6$, $0.3 \leq a/b \leq 0.7$, and $h/t \geq 3.0$ $K_t = 3.000 - 2.206\sqrt{R} + 0.948R - 0.142R\sqrt{R}$ where $R = \dfrac{\text{cross-sectional area of added reinforcement}}{\text{cross-sectional area of hole}}$ $\qquad\qquad\qquad\qquad\qquad\qquad\text{(without added reinforcement)}$ $R = \left(\dfrac{b}{a} - 1\right)\left(\dfrac{h}{t} - 1\right) + (4 - \pi)\dfrac{r^2}{at}$

15. Transverse circular hole in round bar or tube

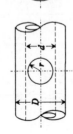

a. Axial tension

$$\sigma_{max} = \sigma_A = K_t \sigma_{nom}$$

where $\sigma_{nom} = \dfrac{4P}{\pi(D^2 - d^2)}$

$$K_t = C_1 + C_2\left(\frac{2r}{D}\right) + C_3\left(\frac{2r}{D}\right)^2$$

$$d/D \le 0.9, \quad 2r/D \le 0.45$$

C_1	3.000
C_2	$0.427 - 6.770(d/D) + 22.698(d/D)^2 - 16.670(d/D)^3$
C_3	$11.357 + 15.665(d/D) - 60.929(d/D)^2 + 16.670(d/D)^3$

b. Bending

$$\sigma_{max} = \sigma_A = K_t \sigma_{nom}$$

where $\sigma_{nom} = \dfrac{32MD}{\pi(D^4 - d^4)}$

$$K_t = C_1 + C_2\left(\frac{2r}{D}\right) + C_3\left(\frac{2r}{D}\right)^2 + C_4\left(\frac{2r}{D}\right)^3$$

$$d/D \le 0.9, \quad 2r/D \le 0.4$$

C_1	3.000
C_2	$-6.250 - 0.585(d/D) + 3.115(d/D)^2$
C_3	$41.000 - 1.071(d/D) - 6.746(d/D)^2$
C_4	$-45.000 + 1.389(d/D) + 13.889(d/D)^2$

Stress Concentration Factors TABLE 6-1

TABLE 6-1 (continued) STRESS CONCENTRATION FACTORS: Holes

Type of Stress Raiser	Loading Condition	Stress Concentration Factor
15. Continued Transverse circular hole in round bar or tube	c. Torsion	$\sigma_{max} = \sigma_A = K_t \tau_{nom}, \qquad \tau_{nom} = 16TD/\pi(D^4 - d^4)$ $K_t = C_1 + C_2\left(\dfrac{2r}{D}\right) + C_3\left(\dfrac{2r}{D}\right)^2 + C_4\left(\dfrac{2r}{D}\right)^3$ $2r/d \leq 0.4, \qquad d/D \leq 0.8$ $\begin{array}{l\|l} \hline & 2r/d \leq 0.4, \quad d/D \leq 0.8 \\ \hline C_1 & 4.000 \\ C_2 & -6.055 + 3.184(d/D) - 3.461(d/D)^2 \\ C_3 & 32.764 - 30.121(d/D) + 39.887(d/D)^2 \\ C_4 & -38.330 + 51.542\sqrt{d/D} - 27.483(d/D) \\ \hline \end{array}$ Maximum stress occurs inside hole on hole surface, near outer surface of bar. Maximum shear stress concentration factor $K_{ts} = \tau_{max}/\tau_{nom} = \tfrac{1}{2}K_t$
16. Round pin joint with closely fitting pin in finite-width plate	Tension	Nominal stress based on net section $\sigma_{max} = K_{ta}\sigma_{na}, \quad \sigma_{na} = P/(D-d)h$ Nominal stress based on bearing area $\sigma_{max} = K_{tb}\sigma_{nb}, \qquad \sigma_{nb} = P/dh$ For $0.15 \leq d/D \leq 0.75,\ L/D \geq 1.0$ $K_{ta} = 12.882 - 52.714\left(\dfrac{d}{D}\right) + 89.762\left(\dfrac{d}{D}\right)^2 - 51.667\left(\dfrac{d}{D}\right)^3$ $K_{tb} = 0.2880 + 8.820\left(\dfrac{d}{D}\right) - 23.196\left(\dfrac{d}{D}\right)^2 + 29.167\left(\dfrac{d}{D}\right)^3$

TABLE 6-1 **Stress Concentration Factors** 284

1.
Opposite shoulder fillets in stepped flat bar

III. Fillets

a. Axial tension

$$\sigma_{max} = K_t \sigma_{nom}, \qquad \sigma_{nom} = P/td$$

$$K_t = C_1 + C_2\left(\frac{2h}{D}\right) + C_3\left(\frac{2h}{D}\right)^2 + C_4\left(\frac{2h}{D}\right)^3$$

where $\dfrac{L}{D} > -1.89\left(\dfrac{r}{d} - 0.15\right) + 5.5$

	$0.1 \leq h/r \leq 2.0$	$2.0 \leq h/r \leq 20.0$
C_1	$1.006 + 1.008\sqrt{h/r} - 0.044h/r$	$1.020 + 1.009\sqrt{h/r} - 0.048h/r$
C_2	$-0.115 - 0.584\sqrt{h/r} + 0.315h/r$	$-0.065 - 0.165\sqrt{h/r} - 0.007h/r$
C_3	$0.245 - 1.006\sqrt{h/r} - 0.257h/r$	$-3.459 + 1.266\sqrt{h/r} - 0.016h/r$
C_4	$-0.135 + 0.582\sqrt{h/r} - 0.017h/r$	$3.505 - 2.109\sqrt{h/r} + 0.069h/r$

For cases where $L/D < -1.89(r/d - 0.15) + 5.5$, see ref. 6.2.

Stress Concentration Factors **TABLE 6-1**

TABLE 6-1 (continued)   STRESS CONCENTRATION FACTORS: Fillets

Type of Stress Raiser	Loading Condition	Stress Concentration Factor	
1. Continued Opposite shoulder fillets in stepped flat bar	b. In-plane bending	$\sigma_{max} = K_t \sigma_{nom}$, $\qquad \sigma_{nom} = 6M/td^2$ $$K_t = C_1 + C_2\left(\frac{2h}{D}\right) + C_3\left(\frac{2h}{D}\right)^2 + C_4\left(\frac{2h}{D}\right)^3$$ where $\dfrac{L}{D} > -2.05\left(\dfrac{r}{d} - 0.025\right) + 2.0$	
		$0.1 \leq h/r \leq 2.0$	$2.0 \leq h/r \leq 20.0$
		C_1 $1.006 + 0.967\sqrt{h/r} + 0.013h/r$	$1.058 + 1.002\sqrt{h/r} - 0.038h/r$
		C_2 $-0.270 - 2.372\sqrt{h/r} + 0.708h/r$	$-3.652 + 1.639\sqrt{h/r} - 0.436h/r$
		C_3 $0.662 + 1.157\sqrt{h/r} - 0.908h/r$	$6.170 - 5.687\sqrt{h/r} + 1.175h/r$
		C_4 $-0.405 + 0.249\sqrt{h/r} - 0.200h/r$	$-2.558 + 3.046\sqrt{h/r} - 0.701h/r$
2. Shoulder fillet in a stepped circular shaft	a. Axial tension	$\sigma_{max} = K_t \sigma_{nom}$, $\qquad \sigma_{nom} = 4P/\pi d^2$ $$K_t = C_1 + C_2\left(\frac{2h}{D}\right) + C_3\left(\frac{2h}{D}\right)^2 + C_4\left(\frac{2h}{D}\right)^3$$	
		$0.1 \leq h/r \leq 2.0$	$2.0 \leq h/r \leq 20.0$
		C_1 $0.926 + 1.157\sqrt{h/r} - 0.099h/r$	$1.200 + 0.860\sqrt{h/r} - 0.022h/r$
		C_2 $0.012 - 3.036\sqrt{h/r} + 0.961h/r$	$-1.805 - 0.346\sqrt{h/r} - 0.038h/r$
		C_3 $-0.302 + 3.977\sqrt{h/r} - 1.744h/r$	$2.198 - 0.486\sqrt{h/r} + 0.165h/r$
		C_4 $0.365 - 2.098\sqrt{h/r} + 0.878h/r$	$-0.593 - 0.028\sqrt{h/r} - 0.106h/r$

TABLE 6-1   Stress Concentration Factors   286

Stress Concentration Factors TABLE 6-1

b. Bending

$\sigma_{max} = K_t \sigma_{nom}, \qquad \sigma_{nom} = 32M/\pi d^3$

$$K_t = C_1 + C_2\left(\frac{2h}{D}\right) + C_3\left(\frac{2h}{D}\right)^2 + C_4\left(\frac{2h}{D}\right)^3$$

	$0.1 \le h/r \le 2.0$	$2.0 \le h/r \le 20.0$
C_1	$0.947 + 1.206\sqrt{h/r} - 0.131h/r$	$1.232 + 0.832\sqrt{h/r} - 0.008h/r$
C_2	$0.022 - 3.405\sqrt{h/r} + 0.915h/r$	$-3.813 + 0.968\sqrt{h/r} - 0.260h/r$
C_3	$0.869 + 1.777\sqrt{h/r} - 0.555h/r$	$7.423 - 4.868\sqrt{h/r} + 0.869h/r$
C_4	$-0.810 + 0.422\sqrt{h/r} - 0.260h/r$	$-3.839 + 3.070\sqrt{h/r} - 0.600h/r$

c. Torsion

$\tau_{max} = K_t \tau_{nom}, \qquad \tau_{nom} = 16T/\pi d^3$

$$K_t = C_1 + C_2\left(\frac{2h}{D}\right) + C_3\left(\frac{2h}{D}\right)^2 + C_4\left(\frac{2h}{D}\right)^3$$

	$0.25 \le h/r \le 4.0$
C_1	$0.905 + 0.783\sqrt{h/r} - 0.075h/r$
C_2	$-0.437 - 1.969\sqrt{h/r} + 0.553h/r$
C_3	$1.557 + 1.073\sqrt{h/r} - 0.578h/r$
C_4	$-1.061 + 0.171\sqrt{h/r} + 0.086h/r$

TABLE 6-1 (continued) STRESS CONCENTRATION FACTORS

IV. Miscellaneous Elements

Type of Stress Raiser	Loading Conditions	Stress Concentration Factor
1. Round shaft with semicircular end key seat	**a. Bending** 	$\sigma_{max} = K_t \sigma, \qquad \sigma = 32M/\pi D^3$ $b = \frac{1}{4}D, \qquad h = \frac{1}{8}D, \qquad \alpha = 10°, \qquad \beta = 15°$ (1) At location A on surface $\qquad K_{tA} = 1.6$ (2) At location B at end of keyway $\qquad K_{tB} = 1.426 + 0.1643\left(\dfrac{0.1}{r/D}\right) - 0.0019\left(\dfrac{0.1}{r/D}\right)^2$ $\qquad$ where $0.005 \le r/D \le 0.04$ $\qquad D \le 6.5 \;$ in. $\qquad h/D = 0.125$ For $D > 6.5$ in., it is suggested that the K_{tB} values for $r/D = 0.0208$ be used.
	b. Torsion 	$h = \frac{1}{8}D, \qquad b = D/r, \qquad \alpha = 15°, \qquad \beta = 50°$ $\tau = 16T/\pi D^3$ (1) At location A on surface $\qquad K_{tA} = \sigma_{max}/\tau \simeq 3.4,$ (2) At location B in fillet $\qquad K_{tB} = \sigma_{max}/\tau$ $\qquad = 1.953 + 0.1434\left(\dfrac{0.1}{r/D}\right) - 0.0021\left(\dfrac{0.1}{r/D}\right)^2$ $\qquad$ for $0.005 \le r/D \le 0.07$

TABLE 6-1 Stress Concentration Factors 288

2.
Splined shaft

a. Torsion

$$K_{tS} = \tau_{max}/\tau, \qquad \tau = 16T/\pi D^3$$

For $0.01 \le r/D \le 0.04$

$$K_{tS} = 6.083 - 14.775\left(\frac{10r}{D}\right) + 18.250\left(\frac{10r}{D}\right)^2$$

3.
Gear teeth

Bending plus some compression

A and C are points of tangency of inscribed parabola ABC with tooth profile

b = tooth width normal to plane of figure

r_f = minimum radius of tooth fillet

W = Load per unit length of tooth face

ϕ = angle between load W and normal to tooth face

Maximum stress occurs at fillet on tension side at base of tooth

$$\sigma_{max} = K_t\sigma_{nom}, \qquad \sigma_{nom} = \frac{6Wh}{bt^2} - \frac{W}{bt}\tan\phi$$

For 14.5° pressure angle

$$K_t = 0.22 + \left(\frac{t}{r_f}\right)^{0.2}\left(\frac{t}{h}\right)^{0.4}$$

For 20° pressure angle

$$K_t = 0.18 + \left(\frac{t}{r_f}\right)^{0.15}\left(\frac{t}{h}\right)^{0.45}$$

Type of Stress Raiser	Stress Concentration Factor
4. **U-Shaped member** where $\theta = 20°$	**For position A** $$K_{tA} = \frac{\sigma_{max} - P/td}{6Pe/td^2}$$ **For position B** $$K_{tB} = \frac{\sigma_{max}}{PLc_B/I_B}$$ where I_B/c_B = section modulus at section in question (section BB')

<table>
<tr><td>

$\dfrac{e}{r} = \dfrac{e}{h} = \dfrac{e}{d}$

$1.5 \leq \dfrac{e}{r} \leq 4.5$

</td><td>

(1) For square outer corners

$K_{tA} = 0.194 + 1.267\left(\dfrac{e}{r}\right) - 0.455\left(\dfrac{e}{r}\right)^2 + 0.050\left(\dfrac{e}{r}\right)^3$

$K_{tB} = 4.141 - 2.760\left(\dfrac{e}{r}\right) + 0.838\left(\dfrac{e}{r}\right)^2 - 0.082\left(\dfrac{e}{r}\right)^3$

</td></tr>
<tr><td>

$\dfrac{e}{2r} = \dfrac{e}{2h} = \dfrac{e}{d}$

$1.0 \leq \dfrac{e}{2r} \leq 2.5$

</td><td>

$K_{tA} = 0.800 + 1.147\left(\dfrac{e}{2r}\right) - 0.580\left(\dfrac{e}{2r}\right)^2 + 0.093\left(\dfrac{e}{2r}\right)^3$

$K_{tB} = 7.890 - 11.107\left(\dfrac{e}{2r}\right) + 6.020\left(\dfrac{e}{2r}\right)^2 - 1.053\left(\dfrac{e}{2r}\right)^3$

</td></tr>
</table>

TABLE 6-1 **Stress Concentration Factors**   290

Stress Concentration Factors TABLE 6-1

$\dfrac{d}{r} = \dfrac{d}{h}$

$0.75 \leq \dfrac{d}{r} \leq 2.0$

When $a = 3r$

$K_{tA} = 1.143 + 0.074\left(\dfrac{d}{r}\right) + 0.026\left(\dfrac{d}{r}\right)^3$

$K_{tB} = -1.985 + 7.841\left(\dfrac{d}{r}\right) - 6.103\left(\dfrac{d}{r}\right)^2 + 1.500\left(\dfrac{d}{r}\right)^3$

When $a = r$

$K_{tA} = 0.714 + 1.237\left(\dfrac{d}{r}\right) - 0.891\left(\dfrac{d}{r}\right)^2 + 0.239\left(\dfrac{d}{r}\right)^3$

$K_{tB} = 2.278 - 3.263\left(\dfrac{d}{r}\right) + 3.097\left(\dfrac{d}{r}\right)^2 - 0.865\left(\dfrac{d}{r}\right)^3$

$\dfrac{d}{r} = \dfrac{h}{r}$

$1.0 \leq \dfrac{d}{r} \leq 7.0$

For $a = 3r$

$K_{tA} = 0.982 + 0.303\left(\dfrac{d}{r}\right) - 0.017\left(\dfrac{d}{r}\right)^2$

$K_{tB} = 1.020 + 0.235\left(\dfrac{d}{r}\right) - 0.015\left(\dfrac{d}{r}\right)^2$

For $a = r$

$K_{tA} = 1.010 + 0.281\left(\dfrac{d}{r}\right) - 0.012\left(\dfrac{d}{r}\right)^2$

$K_{tB} = 0.200 + 1.374\left(\dfrac{d}{r}\right) - 0.412\left(\dfrac{d}{r}\right)^2 + 0.037\left(\dfrac{d}{r}\right)^3$

Type of Stress Raiser		Stress Concentration Factor
4. Continued U-Shaped Member	$\dfrac{R}{r} = \dfrac{R}{d} = \dfrac{R}{h}$ $2.0 \leq \dfrac{R}{r} \leq 2.75$	(2) For rounded outer corners For $a = 3r$ $K_{tA} = 48.959 - 60.004\left(\dfrac{R}{r}\right) + 24.933\left(\dfrac{R}{r}\right)^2 - 3.427\left(\dfrac{R}{r}\right)^3$ $K_{tB} = 79.769 - 98.346\left(\dfrac{R}{r}\right) + 40.806\left(\dfrac{R}{r}\right)^2 - 5.610\left(\dfrac{R}{r}\right)^3$ For $a = r$ $K_{tA} = 27.714 - 31.859\left(\dfrac{R}{r}\right) + 12.625\left(\dfrac{R}{r}\right)^2 - 1.648\left(\dfrac{R}{r}\right)^3$ $K_{tB} = 81.344 - 99.133\left(\dfrac{R}{r}\right) + 40.740\left(\dfrac{R}{r}\right)^2 - 5.560\left(\dfrac{R}{r}\right)^3$

TABLE 6-1 Stress Concentration Factors   292

Type of Stress Raiser	Loading Condition	Stress Concentration Factor
5. Angles and box sections 	Torsion	(1) For angle section $\tau_{\max} = \tau_A = K_t \tau$ $K_t = 6.554 - 16.077\sqrt{\dfrac{r}{t}} + 16.987\left(\dfrac{r}{t}\right) - 5.886\sqrt{\dfrac{r}{t}}\left(\dfrac{r}{t}\right)$ where $0.1 \leq r/t \leq 1.4$ (2) For box section $\tau_{\max} = \tau_B = K_t \tau$ $K_t = 3.962 - 7.359\left(\dfrac{r}{t}\right) + 6.801\left(\dfrac{r}{t}\right)^2 - 2.153\left(\dfrac{r}{t}\right)^3$ where a is 15–20 times larger than t, $0.2 \leq r/t \leq 1.4$

[a]Much of this material is based on Peterson [6.2], with permission.

CHAPTER 7

Fracture Mechanics and Fatigue

Discontinuities (sharp corners, grooves, surface nicks, and voids in welds) and material imperfections (flaws, cracks) are present in almost all engineering structures even though the structure may have been "inspected" during fabrication. However, increasing demands for optimum design and the resulting conservation of material require that structures be designed with smaller safety margins. The discipline of *fracture mechanics* helps meet the needs of accurately estimating the strength of cracked structures. In general, fracture mechanics deals with the conditions under which a load-bearing body can fail due to the enlargement of a dominant crack contained in the body.

The tendency of a structural member to fracture depends on the temperature, the microstructure of the material, the presence of corrosive agents, the thickness of the material, and the types of loading—static, impact, or cyclic—and construction practice—welded, casted, riveted and bolted, etc. Material fracture under static loading with chemically active substances present is known as *stress corrosion*; fracture under cyclic load is referred to as *fatigue*, and fracture with both cyclic loading and the presence of active substances is called *corrosion fatigue*. This chapter reviews briefly the theory of fracture mechanics and fatigue. These theories can be used in the design of structures to avoid brittle fracture.

295

More extensive and detailed treatments of fracture, fatigue, stress corrosion, and corrosion fatigue may be found in the literature [e.g., 7.1–7.3].

7.1 NOTATION

a	Flaw size, usually length or half-length of flaw (L)
a_c	Critical flaw size (L)
a_T	Flaw size at rate transition point of crack growth (L)
A	Crack growth rate under unit fluctuation of stress intensity factor (L/cycle)
E	Young's modulus (F/L^2)
E'	E for plane stress; $E/(1 - \nu^2)$ for plane strain
G	Energy release rate (F/L)
J	The J integral (FL/L)
K	Stress intensity factor ($F/L^{3/2}$)
K_c	Critical stress intensity factor for plane stress ($F/L^{3/2}$)
ΔK	Range of stress intensity factor ($F/L^{3/2}$)
K_I	Stress intensity factor for plane strain, mode I deformation ($F/L^{3/2}$)
K_{Ic}	Critical stress intensity factor for plane strain, mode I deformation, also called fracture toughness or notch toughness ($F/L^{3/2}$)
K_{II}	Mode II stress intensity factor ($F/L^{3/2}$)
K_{III}	Mode III stress intensity factor ($F/L^{3/2}$)
K_{Id}	Critical stress intensity factor for dynamic (impact) loading and plane strain conditions of maximum constraint ($F/L^{3/2}$)
K_T	Transition stress intensity factor for zero to tension loading ($F/L^{3/2}$)
ΔK_T	Fluctuation of stress intensity factor at which transition of rate of crack growth occurs ($F/L^{3/2}$)
ΔK_{th}	Threshold fluctuation of stress intensity factor below which cracks do not grow ($F/L^{3/2}$)
K_f	Fatigue strength reduction factor
K_t	Theoretical stress concentration factor
N_f	Fatigue life, number of cycles to failure (cycles)
N	Number of load cycles
q	Notch sensitivity index
r	Radius of curvature of notch (L)
R	Crack growth resistance (F/L)
r_p	True length of crack-tip plastic zone (L)
r_p^*	Apparent distance in crack-tip plastic zone (L)
t	Specimen thickness (L)
α	Material constant used in computing notch sensitivity (L)
σ	Nominal stress (F/L^2)

σ_f Fatigue strength (F/L^2)

σ_e Endurance limit (F/L^2)

σ_u Ultimate tensile strength (F/L^2)

σ_a Alternating stress level of the applied load (F/L^2)

σ_m Mean stress level of the applied load (F/L^2)

σ_{ys} Yield strength (F/L^2)

7.2 LINEAR ELASTIC FRACTURE MECHANICS AND APPLICATIONS

Linear elastic mechanics has become a practical analytical tool for studying structural fracture where the inelastic deformation surrounding a crack tip is small. Fracture mechanics deals with the conditions under which cracks form and grow. As a consequence, fracture mechanics can be used in structural design to determine acceptable stress levels, acceptable defect sizes, and material properties for certain working conditions. Linear elastic fracture mechanics is based on an analytical procedure that relates the stress field in the vicinity of the crack tip to the nominal stress of the structure; to the size, shape, and orientation of the crack; and to the material properties of the structure.

Equation (6.1) describes the distribution of stresses near a circular hole in an infinite plane under uniaxial tension. This formula was used to calculate the stress concentration factor. In a similar fashion, consider an elliptical hole of major axis $2a$ and minor axis $2b$, as shown in Fig. 7-1a. If $a \gg b$, the elliptical hole becomes a crack of length $2a$ (Fig. 7-1b). The stress formulas near the crack

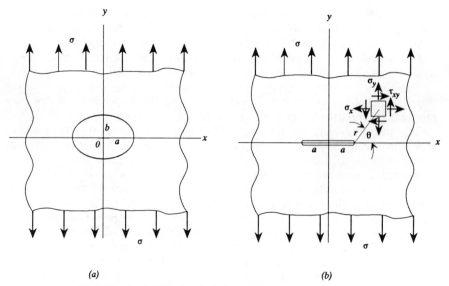

(a)   *(b)*

FIGURE 7-1   (a) Elliptical hole in an infinite plate; (b) $a \gg b$.

tip ($r \ll a$) can be shown to be [7.4]

$$\sigma_x = \sigma \frac{\sqrt{a}}{\sqrt{2r}} \cos\frac{\theta}{2}\left[1 - \sin\frac{\theta}{2}\sin\frac{3\theta}{2}\right]$$

$$\sigma_y = \sigma \frac{\sqrt{a}}{\sqrt{2r}} \cos\frac{\theta}{2}\left[1 + \sin\frac{\theta}{2}\sin\frac{3\theta}{2}\right] \qquad (7.1)$$

$$\tau_{xy} = \sigma \frac{\sqrt{a}}{\sqrt{2r}} \sin\frac{\theta}{2}\cos\frac{\theta}{2}\cos\frac{3\theta}{2}$$

where r and θ are shown in Fig. 7-1b.

The stress σ_y near the crack tip with $\theta = 0$ becomes

$$\sigma_y = \sigma\sqrt{a}/\sqrt{2r} \qquad (7.2)$$

It is clear that at the crack tip ($r = 0$) the stress is singular since $\sigma_y \rightarrow \infty$ as $r \rightarrow 0$. Because of this singularity, the usual stress concentration approach is inappropriate for this problem. Alternatively, the quantity $\sigma_y\sqrt{2r}$ is introduced, since this factor remains finite as $r \rightarrow 0$. More specifically, a factor π is introduced to this quantity, so that a new factor is defined,

$$K = \sigma_y\sqrt{2\pi r} = \left(\sigma\sqrt{a}/\sqrt{2r}\right)\sqrt{2\pi r} = \sigma\sqrt{\pi a} \qquad (7.3)$$

In a more general form, K is taken to be

$$K = C\sigma\sqrt{\pi a} \qquad (7.4)$$

The quantity K is called the *stress intensity factor* (with units MPa $\cdot \sqrt{m}$, or ksi $\cdot \sqrt{in.}$). The stress σ is the nominal stress, a is the flaw size, and C is a constant that depends on the shape and size of the flaw and specimen.

It is important to determine the stress intensity factor for the specific geometry and loading involved to assess the safety factor for a solid.

Three types of crack propagation are recognized: opening, sliding, and tearing (Fig. 7-2). These types are called modes I, II, and III, respectively. A flaw may propagate in a particular mode or in a combination of these modes.

The stress intensity factors of Eq. (7.4) corresponding to the three modes represent the most general loading environment. Formulas for stress intensity factors for various loading, and crack shapes are listed in Table 7-1.

When the combination of nominal stress and crack size attains a value such that the stress intensity factor K reaches a critical magnitude K_c that is material dependent, unstable crack propagation occurs. The critical stress intensity factor K_c for each mode of propagation is regarded as a material constant that depends, more or less, on temperature, plate thickness, and loading rate. The concept of the *critical stress intensity factor* as a material property originated with Irwin. Earlier investigators, particularly Griffith, reasoned that unstable crack propagation occurs when the elastic energy released during the formation of a unit area of crack surface exceeds the energy required to form that amount of

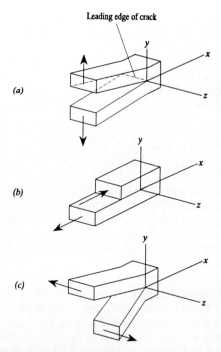

Leading edge of crack

(a)

(b)

(c)

FIGURE 7-2 Three modes of crack propagation: (a) I, opening; (b) II, sliding; (c) III, tearing.

surface. In dealing with ductile metals the energy necessary to perform plastic work at the crack tip is much more important than the surface energy [7.5]. The critical stress intensity factor of a material is also referred to as its *fracture toughness*.

The relationship between the stress intensity factor and the fracture toughness is similar to that between tensile stress and tensile strength, i.e.,

$$K \leq K_c \qquad (7.5)$$

where K_c is the critical value (fracture toughness) that depends on the degree of triaxial constraint at the crack tip. For mode I deformation under plane strain (small crack-tip plastic deformation), the critical value for fracture is designated K_{Ic}, which has units of MPa $\cdot \sqrt{m}$ (or ksi $\cdot \sqrt{in.}$). This inherent material property is measured under precisely defined procedures prescribed by the American Society for Testing and Materials (ASTM) standard E399. A useful relation between the plane strain fracture toughness K_{Ic} and the somewhat greater value of K_c is the semiempirical equation [7.6]

$$K_c = K_{Ic}\left[1 + \frac{1.4}{t^2}\left(\frac{K_{Ic}}{\sigma_{ys}}\right)^4\right]^{1/2} \qquad (7.6)$$

where t is the plate thickness and σ_{ys} is the yield stress of the material. This formula can be used to estimate the *plane stress* fracture toughness if the plane strain value is known or to obtain the plane strain fracture toughness from a K_c obtained from specimen testing. For dynamic (impact) loading and *plane strain* conditions, the critical stress intensity factor is designated K_{Id}. All of these values are also affected by temperature.

For mode I crack propagation, Eqs. (7.4) and (7.5) can be rewritten as

$$K_I = C\sigma\sqrt{\pi a} \qquad \text{(a)}$$

$$K_I \leq K_{Ic} \qquad \text{(b)}$$

where K_I is the stress intensity factor for mode I deformation under plane strain.

These equations indicate that the crack size a, stress level σ, and fracture toughness K_{Ic} are the primary factors that control the susceptibility of a structure to brittle fracture. It is also apparent that the single parameter K (or K_I) is enough to represent the stress condition in the vicinity of the crack tip. When the stress intensity factor reaches the value of the fracture toughness, failure occurs.

For steel the resistance of a material to fracture decreases as the temperature decreases and as the loading rate increases. The fracture toughness of some structural materials, such as aluminum, titanium, and some high-strength steels, changes only slightly with temperature. It is interesting that the fracture toughness decreases as the yield stress of the material increases. Depending on the service conditions, a material may fracture elastically with little plastic deformation. A particular temperature that is different for different steels is customarily used to define the transition from ductile to brittle behavior. This is known as the ductile–brittle transition temperature and as the *nil-ductility transition temperature* (NDT) (see Chapter 4). The transition temperature of a material increases markedly with loading rate and with the grain size of the material. Because of the difficulty involved in obtaining fracture toughness data directly, a number of formulas relating Charpy values (Chapter 4) and fracture toughness K_{Ic} of steels have been developed. One of these is

$$\left(\frac{K_{Ic}}{\sigma_{ys}}\right)^2 = 5\left(\frac{CVN}{\sigma_{ys}} - 0.05\right) \qquad (7.7)$$

which is based on results obtained on 11 steels having yield strengths σ_{ys} ranging from 110 to 246 ksi. Here CVN is the Charpy energy at a temperature above the transition temperature in ft-lb, σ_{ys} is in ksi, and K_{Ic} is in ksi $\cdot \sqrt{in.}$.

When two or more loads act to produce the same mode of crack propagation, the respective stress intensity factors may be added to determine if the total is under the critical value. If a crack propagates in more than one mode simultaneously, the concept of critical stress intensity factor cannot directly be applied. To deal with multimode problems, Sih [7.7] has based a theory of unstable crack propagation on a material property named the critical strain energy density factor. This theory predicts the direction of flaw propagation as well as the critical flaw size.

In most practical applications of the theory of fracture mechanics, the analysis is limited to mode I propagation only. The critical stress intensity factor for mode I deformation is shown in Table 7-2 for several materials. The specimen orientation letters refer to the relationships between the crack propagation direction in the specimen and the rolling direction of the plate. The letters L–T mean that the crack is perpendicular to the longitudinal (rolling) direction and parallel to the width (transverse) direction. The reverse of L–T conditions is designated T–L. The letters S–L mean that the crack is perpendicular to the thickness and parallel to the rolling direction. Other combinations (L–S, T–S, and S–T) are possible. An analogous code exists for round bar material with the directions being longitudinal, radial, and circumferential (L–R–C).

The above analysis is based on linear elastic fracture mechanics, assuming a stress singularity exists at the tip of the crack [Eq. (7.2)]. However, in reality, in a small region near the tip, plastic deformation probably occurs, and since the stresses are limited by yielding, a stress singularity does not occur.

Suppose the crack in the plate of finite width of Fig. 7-3a is under mode I loading. Near the crack tip where a plastic zone spreads, two zones can be identified. In the 1st zone, on the free surface, $\sigma_z = 0$ so that the plane stress state exists. However, in the 2nd zone, the strain in the z direction (parallel to the crack front) is constrained and the plane strain state exists. If the size of the

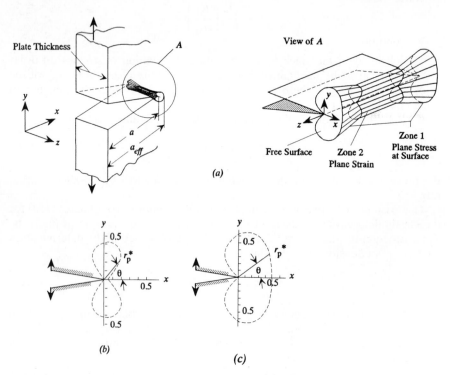

FIGURE 7-3   Plastic zone shape for mode I loading: (a) Plastic zone at the tip of the crack; (b) plane strain state; (c) plane stress state.

plastic zone (in the x or y directions) is large and on the order of the plate thickness, the crack can be modeled as being in the plane stress state. If the size of the plastic zone is much smaller than the plate thickness, the 2nd zone will dominate and the crack can be considered as being in the state of plane strain.

The size of the crack tip plastic zone in either plane stress or plane strain may be estimated by using the von Mises yield relation, along with Eqs. (7.1). The shape of the plastic zone can be expressed in terms of the boundary parameters r_p^* and θ (Fig. 7-3 b and c). For mode I deformation and a plane stress state, the relationship between r_p^* and θ is

$$r_p^* = \cos^2\frac{\theta}{2}\left(1 + 3\sin^2\frac{\theta}{2}\right)\left(\frac{1}{2\pi}\right)\left(\frac{K_I}{\sigma_{ys}}\right)^2 \tag{7.8a}$$

At $\theta = 0$

$$r_p^* = \frac{1}{2\pi}\left(\frac{K_I}{\sigma_{ys}}\right)^2 \tag{7.8b}$$

A common formula assumed for plane strain is [7.4]

$$r_p^* = \frac{1}{6\pi}\left(\frac{K_I}{\sigma_{ys}}\right)^2 \tag{7.9}$$

The plasticity at the crack tip results in a redistribution of stresses. In order for equilibrium to be maintained, the full width of the plastic zone r_p will be twice the value of r_p^*. These results are approximate because the influence of some effects is ignored.

If r_p^* is small relative to the planar dimensions, including crack size a, i.e.,

$$r_p^* \ll a, t, b - a \tag{7.10}$$

the preceding linear elastic theory based formulas tend to be reasonably accurate.

The existence of a plastic zone implies that the center of coordinates (r, θ) for the elastic field advances ahead of the real crack tip into the zone of plasticity. The correction for the "effective crack size" is often utilized in the determination of the stress intensity factor K. Define

$$a_{\text{eff}} = a + r_p^* \tag{7.11}$$

where a_{eff} is an effective crack length. Use of a_{eff} to replace a in Eqs. (7.4) to calculate the stress intensity factor is called the Irwin correction [7.8]. For example, K_I of a tension strip of infinite width with a centrally located crack (case 1 of Table 7-1, $b \gg a$) is

$$K_I = \sigma\sqrt{\pi a} \tag{7.12}$$

After introducing a_{eff}, this becomes

$$K_I = \sigma\sqrt{\pi(a + r_p^*)} \tag{7.13}$$

where r_p^* is shown in Eqs. (7.8).

Although the results tend to be quite satisfactory for r_p^* small relative to a and the other planar dimensions, for the purpose of examining trends, sometimes the "effective crack size" correction is utilized for large-scale yield even though r_p^* fails to pass the dimension constraints.

One method of comparing the resistance to fracture is to evaluate the crack toughness performance using K_{1c}/σ_{ys}. The larger this ratio, the better the resistance to fracture.

General Design by Linear Elastic Fracture Mechanics

In traditional methods, designs are based on the allowable stresses, which are usually related to, e.g., the yield strength of a tensile specimen. Such an approach applies to structures without cracklike flaws and discontinuities. In the presence of stress concentrations or discontinuities, it is assumed that the structural materials will yield locally and redistribute the load to neighboring areas. The recent development of fracture mechanics has established an analytical tool for the design of fracture-resistant structures. This fracture mechanics design refers to selection of materials and allowable stress levels based on the fact that cracklike flaws may exist or may be initiated under cyclic loads or stress corrosion and that some level of notch toughness is desirable.

In fracture mechanics design, it is assumed that the designer has the following information available:

(a) type of structure, overall and member dimensions;
(b) stress and stress fluctuating range, potential crack growth locations in the structure (e.g., welds, holes, discontinuities); and
(c) structural performance design criteria (e.g., minimum cost, maximum resistance to fracture, specified design life, working maximum loading rate).

Based on this general information, the designer can incorporate K_{1c} or K_{1d} values at the service condition (temperature and loading rate) in the fracture mechanics design.

To understand the fundamentals of fracture mechanics design, consider the case where possible crack extension in mode I has occurred by fatigue, stress corrosion, or corrosion fatigue. Combining Eqs. (7.4) and (7.5) gives the maximum flaw size a structure member can tolerate at a particular stress level,

$$a = \frac{1}{\pi}\left(\frac{K_c}{C\sigma}\right)^2 \tag{7.14}$$

After determining the values of the K_c and σ_{ys} at the service temperature and loading rate for the material of the structure and selecting the most probable type of flaw (or crack) that will exist in the member in question and the corresponding equation for K to obtain C, the designer can calculate with Eq. (7.14) the minimal unstable size of flaw at various possible stress levels. Therefore, the designer has to control three factors (σ, K_c, and a) in a fracture mechanics design. All other factors, such as temperature, loading rate, and residual stresses, only affect these three primary factors.

Once the critical flaw size has been found, quality control procedures may be established to ensure that no flaws of a critical size exist in the structure. On the other hand, if the service loads and the minimum detectable flaw size are specified, Eq. (7.14) may be used to select a material that will yield a critical flaw size greater than the minimum size of a detectable flaw. Or if the materials and minimum detectable flaw size are fixed, Eq. (7.14) enables the specification of service loads that result in a critical flaw size greater than the minimum size of a detectable flaw. The following section contains several examples of the application of Eqs. (7.4) and (7.14) to fracture computations.

Example 7.1   Imbedded Crack   A sharp penny-shaped crack of diameter 2.5 cm is completely imbedded in a solid. The applied stress is normal to the area of the imbedded crack. Catastrophic failure occurs when a stress of 500 MPa is applied.

Find the fracture toughness of the material if plane strain conditions exist at the crack perimeter.

The stress intensity factor formula of case 17 of Table 7-1 applies. For this circular crack, $c = a$:

$$K_{\mathrm{I}} = (2/\pi)\sigma\sqrt{\pi a}$$

Since failure occurred at $\sigma_c = 500$ MPa,

$$K_{\mathrm{I}c} = (2/\pi)\sigma_c\sqrt{\pi a} = (2/\pi)(500 \text{ MPa})\sqrt{\pi \times 0.0125 \text{ m}}$$

$$= 63.1 \text{ MPa} \cdot \sqrt{\text{m}} \tag{1}$$

where the crack size a is one-half of the diameter.

⸻

Example 7.2   Titanium Alloy   A titanium alloy Ti–6 Al–4 V can be heat treated to give the following mechanical properties:

$$K_{\mathrm{I}c} = 115.4 \text{ MPa} \cdot \sqrt{\text{m}}, \qquad \sigma_{ys} = 910 \text{ MPa} \tag{1}$$

If the applied stress is $0.75\sigma_{ys}$, find the dimensions of the largest stable internal elliptical flaw for $a/2c = 0.2$.

The fracture of an elliptical crack is characterized in case 17 of Table 7-1. For $a/2c = 0.2$, we find $k^2 = 1 - a^2/c^2 = 0.84$ or $k = 0.9165$.

From a mathematical handbook containing elliptic integrals

$$E(0.9165) = 1.150 \tag{2}$$

At $\theta = \pm \frac{1}{2}\pi$,

$$K_1 = \sigma\sqrt{\pi a} / E(0.9165) = \sigma\sqrt{\pi a} / 1.150 \tag{3}$$

For $K_{Ic} = 115.4$ MPa $\cdot \sqrt{m}$, $\sigma_{ys} = 910$ MPa,

$$K_1 = \sigma\sqrt{\pi a} / 1.150 \leq K_{Ic} = 115.4 \text{ MPa} \cdot \sqrt{m}$$

So that

$$a_{max} = (115.4 \times 1.15 / 0.75\sigma_{ys})^2 / \pi = 1.204 \text{ cm} \tag{4}$$

and

$$c = a/(2 \times 0.2) = 1.204/0.4 = 3.01 \text{ cm} \tag{5}$$

To account for the effect of the plastic zone for plane strain, use Eq. (7.9) and $K_I \leq K_{Ic} = 115.4$ MPa $\cdot \sqrt{m}$:

$$r_p^* = \frac{1}{6\pi}\left(\frac{115.4}{910}\right)^2 = 0.853 \text{ mm} \tag{6}$$

Similar to (4),

$$a_{max} + r_p^* = 1.204 \text{ cm} \tag{7}$$

$$a_{max} = 1.204 - 0.0853 = 1.11 \text{ cm} \tag{8}$$

and

$$c = 1.11/(2 \times 0.2) = 2.775 \text{ cm} \tag{9}$$

By comparison of (5) and (9), it follows that if the effect of the plastic zone is considered, the largest stable surface elliptical flaw will be smaller than the case in which r_p^* is not considered.

====

Example 7.3 Longitudinal Crack in a Cylindrical Tube A long circular cylindrical tube must withstand an internal pressure of $p = 10$ MPa (1450 psi). The tube has an inside radius of $r = 250$ mm and a wall thickness of $t = 12$ mm; it is constructed of AISI 4340 steel alloy that is heat treated to have a critical stress intensity factor of 59 MPa $\cdot \sqrt{m}$ (53.7 ksi $\cdot \sqrt{in.}$), a tensile yield strength of 1503 MPa (218 ksi), and an ultimate tensile strength of 1827 MPa (265 ksi).

To find the minimum size of a longitudinal crack (Fig. 7-4) that will propagate unstably, the tube wall is regarded as a wide sheet in uniform tension, and equation shown in case 1 of Table 7-1 is applied for $a/b \rightarrow 0$. The stress intensity factor is

$$K_1 = \sigma\sqrt{\pi a}\, F(0)$$

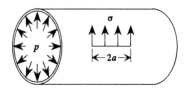

FIGURE 7-4 Longitudinal crack in pressurized cylinder.

Since $F(0) = 1$,

$$K_I = \sigma\sqrt{\pi a} \tag{1}$$

where the cylinder hoop stress is

$$\sigma = pr/t \tag{2}$$

Then

$$K_I = (pr/t)\sqrt{\pi a} \tag{3}$$

which gives the critical crack size, $a = a_c$,

$$a_c = (K_{Ic}t/pr)^2/\pi = \left\{\left[59 \times 10^6(0.012)\right]/\left[10 \times 10^6(0.25)\right]\right\}^2/\pi$$
$$= 25.5 \text{ mm} \quad (1.004 \text{ in.}) \tag{4}$$

For this material the quantity [Eq. (7.9)] $r_p^* = (1/6\pi)(K_{Ic}/\sigma_{ys})^2 = 0.0817$ mm is much less than t and a; consequently, the crack half length and material thickness both meet the condition for a plane strain distribution at the leading edge of the crack [Eq. (7.10)].

If a crack of length $2a = 8$ mm is assumed to exist in the tube, the internal pressure that would cause unstable propagation can be computed. From (3), the critical pressure $p = p_c$ would be

$$p_c = tK_{Ic}\left[\sqrt{1/(\pi a)}\,\right]/r = (0.012)(59 \times 10^6)\left[\sqrt{1/(\pi \times 0.004)}\,\right]/0.250$$
$$= 25.3 \text{ MPa} \quad (3668 \text{ psi}) \tag{5}$$

By the maximum principal stress criterion of failure, brittle fracture of the unflawed tube would occur when the circumferential stress equals the ultimate tensile strength of the material. The pressure at fracture would be

$$p_u = \sigma_u(t/r) = (1827)(12/250) = 87.7 \text{ MPa} \tag{6}$$

We see from (5) and (6) that the 8-mm crack reduces the failure pressure by 71.2%.

The Tresca maximum shear stress yielding criterion takes the form $\sigma_{max} - \sigma_{min} = p_y r/t - (-p_y) = \sigma_{ys}$. Thus, general yielding of the tube occurs when the relation

$$p_y = \sigma_{ys}/[(r/t) + 1] \tag{7}$$

is satisfied, in which the pressure at the inner wall has been included. This effect is small compared to r/t and could be ignored. For our cylinder

$$p_y = 1503/[(250/12) + 1] = 68.8 \text{ MPa} \tag{8}$$

The 8-mm crack causes fracture failure at a pressure that is 63.2% lower than this yield point pressure.

Sih [7.7] presents a solution to a cylindrical tube problem in which the crack is arbitrarily oriented with respect to the longitudinal direction. The arbitrary orientation produces simultaneous propagation of the crack in two modes, and the critical strain energy density factor is used instead of the critical stress intensity factor to predict fracture failure.

Example 7.4 Bar Subjected to Axial Force and Bending Moment A long bar with an edge crack is subjected to a concentrated force, as shown in Fig. 7-5. Stresses due to both axial force and bending moment act on the crack, but because both forces lead to mode I propagation, the two effects are additive. Let σ_T be the tensile stress due to the axial force and σ_B the stress due to bending. The bar is made of 7079-T651 aluminum alloy, which has a critical stress intensity factor of 26 MPa · $\sqrt{m}$ (23.7 ksi · $\sqrt{in.}$), a tensile yield strength of 502 MPa (72.8 ksi), and an ultimate tensile strength of 569 MPa (82.5 ksi).

Use of the stress intensity factor equations listed in cases 7 and 8 of Table 7-1 results in

$$K_I = \sigma_T\sqrt{\pi a}\,C_T + \sigma_B\sqrt{\pi a}\,C_B \tag{1}$$

in which C_T and C_B depend on $F(a/b)$. From $a = 10$ mm, $a/b = 0.1$, and case

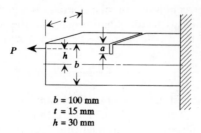

$b = 100$ mm
$t = 15$ mm
$h = 30$ mm

FIGURE 7-5 Bar in tension and bending with edge crack.

7 of Table 7-1

$$C_T = F(0.1) = \sqrt{\frac{2}{\pi \times 0.1} \tan \frac{\pi \times 0.1}{2}}$$

$$\times \frac{0.752 + 2.02 \times 0.1 + 0.37[1 - \sin(0.1\pi/2)]^3}{\cos(0.1\pi/2)}$$

$$= 1.196$$

From case 8 of Table 7-1

$$C_B = F(0.1) = \sqrt{\frac{2}{\pi \times 0.1} \tan \frac{0.1\pi}{2}} \times \frac{0.923 + 0.199[1 - \sin(0.1\pi/2)]^4}{\cos(0.1\pi/2)}$$

$$= 1.041$$

According to the standard strength of material formulas for a rectangular cross section

$$\sigma_T = P/(bt) \tag{2a}$$

At the outer fiber, $\sigma_B = M(b/2)/I$ with $I = \frac{1}{12}b^3 t$ and $M = Ph$. Thus,

$$\sigma_B = (6Ph)/(tb^2) \tag{2b}$$

Then

$$K_1 = (P/bt)\sqrt{\pi a}\,(1.196) + (6Ph/tb^2)\sqrt{\pi a}\,(1.041) \tag{3}$$

so that the critical force $P = P_c$ is calculated as

$$P_c = \left(K_{1c}bt/\sqrt{\pi a}\right)[1.196 + 6h(1.041)/b]^{-1}$$
$$= \left[(26 \times 10^6)(0.1)(0.015)/\sqrt{(0.01)\pi}\,\right]$$
$$\times [1.196 + 6(0.03) \times 1.041/0.1)]^{-1}$$
$$= 71,677 \text{ N} \tag{4}$$

At fracture failure the nominal stress at the outer edge of the bar is found from $\sigma_T + \sigma_B$ of (2) to be 134 MPa (19.4 ksi), which is much lower than the 502-MPa yield strength of the material. For this alloy the quantity $r_p^* = (1/6\pi)(K_{1c}/\sigma_{ys})^2$ = 0.142 mm and the condition for plane strain is satisfied for both a and t. Because the stress intensity factor equations contain the parametric functions C_T and C_B, the solution for a critical crack size given a known force would require an iterative procedure.

Example 7.5 Traditional and Fracture Mechanics Design of High-Strength, Thin-walled Cylinder   This example demonstrates the use of fracture mechanics concepts to select materials and to compare the results with those obtained by a traditional design where the flaw is ignored.

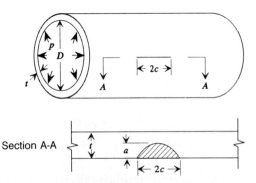

FIGURE 7-6   Example 7.5. Fracture mechanics design for thin-walled cylinder.

Suppose that a thin-walled cylinder with diameter $D = 0.75$ m is required to withstand an internal pressure $p = 34.5$ MPa and the wall thickness t must be at least 1.26 cm, as shown in Fig. 7-6. For the traditional design in which the flaw is ignored, assume a factor of safety of 2.0 against yielding. For fracture mechanics design the maximum possible flaw size is considered, and a design stress intensity of $K_1 = K_{Ic}$ is used to provide a factor of safety of 2.0 against fracture. The design stress here is thus based on resistance to fracture. If the weight of the cylinder vessel must be considered, select a steel from those available for use to meet the performance requirements and to achieve the minimal weight. Suppose the following steels are available for this design:

Steel	Yield Strength σ_{ys} (MPa)	Assumed K_{Ic} Values (MPa $\cdot$ $\sqrt{m}$)
A	1794	88
B	1516	120
C	1240	154
D	1240	240
E	965	285
F	758	185

The traditional design analysis will be discussed first. Here the flaw is ignored, and the procedure is direct. Since for the cylinder the maximum normal stress is the hoop stress $\sigma = pr/t = PD/2t$ and for design stress $\sigma = \frac{1}{2}\sigma_{ys}$, the corresponding thickness is

$$t = \frac{PD}{2\sigma} = \frac{PD}{\sigma_{ys}} \tag{1}$$

Thus the estimated weight per meter for each steel is

$$A\gamma = 7835.9\pi Dt \quad \text{kg/m} \tag{2}$$

where A is the area of the cross section of the cylinder and $\gamma = 7835.9 \text{ kg/m}^3$ is the density for steel. From (1) and (2), for steel D, where $\sigma_{ys} = 1240$ MPa,

$$t = \frac{pD}{\sigma_{ys}} = \frac{(34.5)(0.75)}{1240} = 0.0209 \text{ m} = 2.09 \text{ cm}$$

$$A = \pi D t = (3.14)(0.75)(0.0209) = 0.049 \text{ m}^2$$

$$A\gamma = (0.049)(1)(7835.9) = 386 \text{ kg/m}$$

The values for the other available steels are calculated similarly and are found to be as follows:

Steel	Yield Strength σ_{ys} (MPa)	Design Stress $\frac{1}{2}\sigma_{ys}$ (MPa)	Wall Thickness t (m)	Unit Weight (kg/m)
A	1794	897	0.0144	266
B	1516	758	0.0171	316
C	1240	620	0.0209	386
D	1240	620	0.0209	386
E	965	483	0.0268	495
F	758	379	0.0341	629

These results show that, as expected, use of a higher strength steel corresponds to a reduction in weight.

However, if fracture is considered, the following design will show that the overall safety and reliability of the vessel may be decreased by increasing the yield strength.

Consider the fracture mechanics design method. At first, the maximum possible flaw size in the cylinder wall should be estimated based on fabrication and inspection. Here we assume that a surface flaw of depth 1.26 cm and an $a/2c$ ratio of 0.25 is possible (Fig. 7-6).

For a surface flaw, the relation among K_1, σ, and a from case 18 of Table 7-1 is

$$K_1 = \sigma\sqrt{\pi a / E(k)^2} f(\theta) F\left(\frac{a}{t}, \frac{a}{c}, \frac{c}{b}\right) \tag{3}$$

Since maximum K_1 occurs at $\theta = \frac{1}{2}\pi$ and $f(\frac{1}{2}\pi) = 1$,

$$K_1 = \sigma\sqrt{\pi a / E(k)^2} F\left(\frac{a}{t}, \frac{a}{c}, \frac{c}{b}\right) \tag{4}$$

From $k^2 = 1 - a^2/c^2 = 1.0 - 0.25 = 0.75$ and a mathematical handbook,

$$E(k) = E(\sqrt{0.75}) = E(0.866) = 1.211 \tag{5}$$

Use the formulas for $F(a/t, a/c, c/b)$ from case 18 of Table 7-1 with $c/b \to 0$:

$$F = (1.13 - 0.05) + \left(\sqrt{1.211^2 \times 2} - (1.13 - 0.05)\right)\left(\frac{a}{t}\right)^{\sqrt{\pi}}$$

$$+ \sqrt{1.211^2 \times 2}\left(\sqrt{\tfrac{1}{4}\pi} - 1\right)\left(\frac{a}{t}\right)^{2\sqrt{\pi}}$$

$$= 1.08 + 0.6327\left(\frac{a}{t}\right)^{\sqrt{\pi}} - 0.1948\left(\frac{a}{t}\right)^{2\sqrt{\pi}} \tag{6}$$

Begin with $a/t = 0.5$:

$$F \approx 1.08 + 0.6327 \times 0.5^{\sqrt{\pi}} - 0.1948 \times 0.5^{2\sqrt{\pi}} = 1.249 \tag{7}$$

For crack size $a = 0.0126$ m, choose steel D with $\sigma_{ys} = 1240$ MPa and $K_{1c} = 240$ MPa $\cdot \sqrt{m}$. Consider plasticity at the crack tip and select a safety factor of 2 against fracture, i.e., $K_1 = \frac{1}{2}K_{1c}$. Equation (7.9) gives

$$r_p^* = \frac{1}{6\pi}\left(\frac{\frac{1}{2}K_{1c}}{\sigma_{ys}}\right)^2 = \frac{1}{6\pi}\left(\frac{240}{2 \times 1240}\right)^2 = 4.968 \times 10^{-4}\ \text{m} \tag{8}$$

$$K_1 = \frac{\sigma\sqrt{\pi(a + r_p^*)}}{E(k)}F = \frac{\sqrt{\pi} \times \sqrt{0.0126 + 0.00049}}{1.211} \times 1.249\sigma = 0.2092\sigma \tag{9}$$

where σ is now the design stress. From $K_1 = \frac{1}{2}K_{1c}$ and (9),

$$\sigma = \frac{K_{1c}}{2 \times 0.2092} = \frac{120}{0.2092} = 573.6\ \text{MPa} \tag{10}$$

Since $\sigma = PD/2t$,

$$t = \frac{PD}{2\sigma} = \frac{34.5 \times 0.75}{2 \times 573.6} = 2.255 \times 10^{-2}\ \text{m} \tag{11}$$

Based on the thickness of 0.02255 m, return to (6) for a second iteration:

$$a/t = 0.0126/0.02255 = 0.5588 \tag{12}$$

$$F = 1.08 + 0.6327 \times 0.5588^{\sqrt{\pi}} - 0.1948 \times 0.5588^{2\sqrt{\pi}} = 1.280 \tag{13}$$

$$K_1 = \frac{\sqrt{\pi}\ \sqrt{0.01309}}{1.211} \times 1.28\sigma = 0.2145\sigma \tag{14}$$

$$\sigma = \frac{120}{0.2145} = 559.5\ \text{MPa} \tag{15}$$

$$t = \frac{34.5 \times 0.75}{2 \times 559.5} = 2.312 \times 10^{-2}\ \text{m} = 2.31\ \text{cm} \tag{16}$$

Since the values of the thickness t in (11) and (16) are not very close, a third iteration might be in order. Let $t = 2.31$ cm. Then

$$a/t = 0.0126/0.0231 = 0.5455 \qquad (17)$$

$$F = 1.273 \qquad (18)$$

$$K_I = 0.2132\sigma \qquad (19)$$

$$\sigma = 562.88 \text{ MPa} \qquad (20)$$

$$t = 2.2298 \times 10^{-2} \text{ m} = 2.23 \text{ cm} \qquad (21)$$

It can be seen that the thickness values of (21) and (16) are quite close to each other. A fourth iteration does not lead to significant charges:

$$a/t = 0.5478$$

$$K_I = 0.2135\sigma$$

$$\sigma = 562.007 \text{ MPa}$$

$$t = 2.30 \text{ cm}$$

This numerical procedure is readily programmed for computer selection. In a manner similar to that described above, the wall thickness for the remaining steels in this example were computed and are listed below.

Results by Fracture Mechanics Based Design

Steel	Yield Strength σ_{ys} (MPa)	Design Stress σ (MPa)	Assumed K_{Ic} (MPa·$\sqrt{m}$)	Design Value K_{Ic} (MPa·$\sqrt{m}$)	Thickness t (m)	Unit Weight (kg/m)	σ/σ_{ys}
A	1794	237.7	88	44	0.0544	1004	0.13
B	1516	315.8	120	60	0.0410	757	0.21
C	1240	393.4	154	77	0.0329	607	0.32
D	1240	562.1	240	120	0.0230	425	0.45
E	965	648.6	285	142.5	0.0199	367	0.67
F	758	458.0	185	92.5	0.0282	521	0.60

If, on the premise of satisfactory performance, the least weight is the first consideration, then steel E is the choice. While for the traditional design analysis, it can be seen above that steel A, with the highest yield strength, should be the choice.

It is of interest to note from the results in the table above that on the basis of equivalent resistance to fracture in the presence of a 1.26-cm-deep surface flaw, there would be an obvious saving of weight by using a lower strength, tougher steel compared with steel A. The results show that neither the factor of safety (σ_{ys}/σ) nor the weight of a structure is necessarily related to the yield strength of the structural material. In fact, the cylinder made of the highest strength steel actually weighs the most. Further analysis of the above results indicates that for the two lowest strength steels (E and F), yielding is the most likely mode of failure and the factor of safety against fracture will be greater than 2.0.

7.3 THE ENERGY ANALYSIS OF FRACTURE

Work performed during elastic deformation is stored as strain energy U and released upon unloading. The strain energy density U_0 is defined as the strain energy per unit volume and can be expressed as

$$U_0 = U/\text{Vol} = \tfrac{1}{2}(\sigma_x \varepsilon_x + \sigma_y \varepsilon_y + \sigma_z \varepsilon_z + \tau_{yz}\gamma_{yz} + \tau_{zx}\gamma_{zx} + \tau_{xy}\gamma_{xy}) \quad (7.15)$$

During fracture, a crack needs energy to propagate, which forms the work of fracture W_f. If the crack size is a, fracture over a distance da would need a small quantity of energy dW_f that must come from either a decrease of internal strain energy U or the work due to applied loading W_e. For the crack da, the work done by the applied load is dW_e, and the change in strain energy is dU. The energy equation would be

$$\frac{d}{da}(W_e - U - W_f) = 0 \qquad (7.16)$$

or

$$\frac{d}{da}(W_e - U) = \frac{dW_f}{da} \qquad (7.17)$$

This represents a useful criterion of fracture. Fracture will propagate a crack da if enough work is released to permit dW_f to occur. Let

$$G = \frac{d}{da}(W_e - U), \qquad R = \frac{dW_f}{da}$$

so that

$$G \geq R \qquad (7.18)$$

where G is the energy release rate and R is the critical value of G, called the crack growth resistance.

Equation (7.18) is the criterion for the energy balance approach to fracture, which is known as the Griffith–Orowan–Irwin theory [7.8]. Fracture will occur according to Eq. (7.18) when the energy release rate G is sufficient to approach the critical value R at incipient crack extension. Crack propagation can be stable or unstable. If

$$\frac{dG}{da} \geq \frac{dR}{da} \qquad (7.19)$$

then unstable fracture occurs.

Equation (7.19) provides a criterion to judge unstable crack propagation from the standpoint of energy. It is possible to show a relationship between crack extension force G and the stress intensity factor K_I. For elastic bodies, including small-scale yielding, the relationship

$$G_I = K_I^2/E' \tag{7.20}$$

holds, where

$$E' = \begin{cases} E & \text{for plane stress} \\ E/(1 - \nu^2) & \text{for plane strain} \end{cases}$$

In a similar fashion G_{II} is expressed in terms of K_{II} as

$$G_{II} = K_{II}^2/E' \tag{7.21}$$

When all three modes of crack deformation are present, this becomes

$$G = \frac{1}{E'}(K_I^2 + K_{II}^2) + \frac{1 + \nu}{E'}K_{III}^2 \tag{7.22}$$

where K_I, K_{II}, and K_{III} are the stress intensity factors for modes I, II, and III, respectively.

7.4 THE J INTEGRAL

The quantities G and K describe the stress near the crack tip when the plastic zone is relatively small and the theory of elastic fracture mechanics applies. To determine the energy release rate for a specimen in which the plastic deformation must be considered, Rice introduced a contour integral J taken about the crack tip [7.10]:

$$J = \int_\Gamma \left((U_0 \, dy) - \bar{p}\frac{\partial \bar{u}}{\partial x} \, ds \right) \tag{7.23}$$

where Γ denotes the arbitrary contour enclosing the crack tip. Also, U_0 is the strain energy density [Eq. (7.15)], $\bar{p}$ is the tension vector (external surface load) on Γ normal to the contour, $\bar{u}$ is the displacement vector at the location of $\bar{p}$ in the x direction, and ds is the infinitesimal arc length shown in Fig. 7-7. Typical units of J are MN/m (Meganewtons per meter).

The J integral vanishes along any closed contour. In Fig. 7-8, the contour $\Gamma_2 + BD + (-\Gamma_1) + CA$ forms a closed path so that J along this contour equals zero, i.e., $J_{\Gamma_2} + J_{BD} + J_{(-\Gamma_1)} + J_{CA} = 0$. The quantities $\bar{p}$ and dy are equal to zero on seqments BD and CA. Hence, $J_{BD} = J_{CA} = 0$ and J along Γ_1 is the same as J along Γ_2. It is concluded that J is path independent. When the contour Γ encircles the crack tip, the J integral equals the energy release rate, i.e., $J = G$.

Denote $J_c = R$, where J_c is the critical value of the J integral. Then Eq. (7.18) appears as

$$J \geq J_c \tag{7.24a}$$

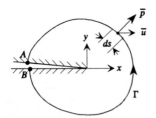

FIGURE 7-7   Parameter of J integral.

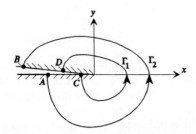

FIGURE 7-8   Two contours about crack tip.

or for mode I deformation,

$$J \geq J_{1c} \tag{7.24b}$$

where J_{1c}, the plane strain value of J at initiation of crack growth, can be determined by ASTM standard E813. Then Eq. (7.19) becomes

$$\frac{dJ}{da} \geq \frac{dJ_c}{da} \tag{7.25}$$

Equations (7.24) and (7.25) are valid for nonlinear elastic materials. For an elastic–plastic material, the J integral is uniquely defined outside the plastic region. It approximates the energy release rate for crack propagation.

When the yielding at the crack tip is limited to "small scale," Eqs. (7.20)–(7.22) can be written as

$$J_1 = G_1 = K_1^2/E' \tag{7.26a}$$

$$J_{11} = G_{11} = K_{11}^2/E' \tag{7.26b}$$

$$J = G = \frac{1}{E'}\left(K_1^2 + K_{11}^2\right) + \frac{1 + \nu}{E'}K_{111}^2 \tag{7.26c}$$

When K_1, for example, attains its critical value, J and G must also reach their critical values J_c and R. Equation (7.26a) implies that

$$J_c = R = K_{1c}^2/E' \tag{7.27}$$

In addition to the stress intensity factor and the J-integral fracture criteria (an energy method) other criteria like the R-curve method and crack-opening displacement (COD) have been developed [7.1, 7.2]. Recall that the J integral, the energy release rate G, and the intensity factor K_1 are related. Moreover, there is a relationship between these parameters and a quantity called the crack tip opening displacement δ_t. All four are equivalent fracture parameters for small-scale yielding. It can be shown that

$$J = G = K_1^2/E' = \sigma_{ys}\delta_t \tag{7.28}$$

where σ_{ys} is the yield stress. When one of the four attains its critical value, the others must also simultaneously reach their critical values.

Often the J integral is evaluated numerically using the finite-element method [7.11, 7.12]. Some general-purpose programs [e.g., 7.13, 7.14] provide J-integral evaluations. The implementation procedure and applications tend to vary from program to program.

7.5 FATIGUE FRACTURE

Although disastrous consequences may result from the fracture failure of a statically loaded structure, this type of failure is not common. A far more common mode of failure is the fracture of a structural member that has been subjected to many cycles of a fluctuating load. Failure occurs even though the load amplitude may be much less than the static yield strength of the material. This form of fracture is known as *fatigue fracture*, or simply *fatigue*. Fatigue here refers to a progressive failure of a material after many cycles of load. A form of failure known as low-cycle fatigue also occurs in which the strains are much larger than those in high-cycle fatigue.

The behavior of a material under cyclic load is influenced by many factors, but of prime importance are the amplitude of the load and the presence in the material of regions of stress concentration. The fatigue characteristics of a material are also affected by the type of loading (bending, torsion, tension, or a combination of the three); the specimen size, shape, and surface roughness; the load waveform (nonzero mean load or variations in load amplitude); and the presence of chemically active agents. Moreover, when apparently identical specimens are tested under identical conditions, significant variations often occur in the fatigue behavior of the specimens. These variations probably occur because the distributions of material microstructural properties such as the number of crack initiation sites and grain size change from specimen to specimen. Weibull [7.15] developed a statistical representation for fatigue data. McClintock has shown that if extraneous scattered data is eliminated, the variation in measured fatigue life is no greater than that in other measured mechanical properties [7.16].

Because of the uncertainty involved in predicting the fatigue behavior of a material, safety factors ranging from 1.3 to 4 are incorporated into the design of cyclically loaded members. The magnitude of the factor is chosen on the basis of the consequences of fatigue failure and on the basis of the number of imponderables involved in the problem.

Two approaches may be taken in designing to prevent fatigue failure. One method is based on graphs (*S–N* curves) that record the number of load cycles necessary to cause failure of a test specimen at various levels of reversed stress amplitude, and the second method utilizes the concepts of linear elastic fracture mechanics.

7.6 *S – N* CURVE APPROACH TO FATIGUE

In the *S–N* curve approach to fatigue, highly polished, geometrically perfect specimens of a material are tested under cyclic load, and the number of cycles to failure for each level of load is shown on a graph called an *S–N* curve as in Fig. 7-9. On a plot of log stress versus log life the curves are nearly linear. Tables 7-4 and 7-5 list several sets of *S–N* curves. A similar relationship exists for low-cycle fatigue (LCF) except that the log plastic strain amplitude is used instead of the log stress. The *fatigue strength* of a material is defined as the maximum amplitude of cyclic stress a specimen will withstand for a given number of cycles; an *S–N* curve is therefore the locus of fatigue strengths of a material over a range of cycle lives. For ferrous metals a stress amplitude exists at or below which a specimen will endure an indefinitely large number of load reversals; this level of reversed or alternating stress (σ_a) is known as the *endurance limit*. Although endurance limits are not truly characteristic of most nonferrous metals, limits for these materials are often stated for an arbitrarily chosen large number of cycles, usually 10 million to 500 million. The *endurance ratio* of a material is defined as

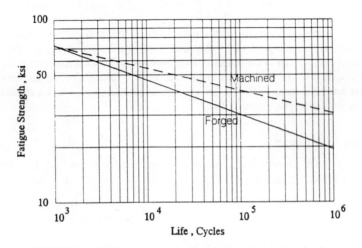

FIGURE 7-9   *S – N* curve for a typical steel under bending loads.

the ratio of the endurance limit to the ultimate tensile strength. Ranges of endurance ratios for alloys of some metals are listed in Table 7-3.

The structural member under consideration and its operating conditions will often differ in important ways from the test specimen and test conditions for which the $S-N$ curve was developed. In this case various empirical formulas and indices have been advanced as means of adjusting the test data to account for the differences. A region of stress concentration in the member, a nonzero mean value of the cyclic load, and a variation in the load amplitude are factors for which standard $S-N$ curves must frequently be adjusted.

Stress Concentration

A stress raiser in a structural member will increase the stress at the point of maximum concentration (Chapter 6). If an $S-N$ curve shows the fatigue strengths of an unnotched specimen, those strengths are divided by K_f [effective stress concentration factor defined in Eq. (6.4)] to find the values for a notched specimen. The variation of K_f with the number of load cycles can be accounted for by evaluating K_f at two points, say, 10^3 and 10^6 cycles, and joining the points with a straight line on a log–log $S-N$ curve. A more conservative approach is the application of K_f chosen for 10^6 cycles to all lower number of cycles. Here K_f is related to K_t (the theoretical stress concentration factor) by the equation

$$q_f = (K_f - 1)/(K_t - 1) \tag{7.29}$$

where q_f is the *notch sensitivity index* of the specimen [Eq. (6.6)]. For holes and for notches with nearly parallel flanks, q_f may be estimated by the so-called Neuber equation in the form [7.18]

$$q_f = 1/\left[1 + (\alpha/r)^{1/2}\right] \tag{7.30}$$

where r is the radius of curvature of the notch and $\sqrt{\alpha}$ is a material constant (Neuber constant). Values of $\sqrt{\alpha}$ for aluminum and steel are shown in Fig. 7-10.

The fatigue analysis of notched members can also be performed by a method known as the local strain approach. In this technique, Neuber's equation [Eq. (7.30)] is used to find the strain history at the notch root from the nominal stress and strain. The local notch strain history is then assumed to result in the same fatigue life as occurs when an unnotched uniaxially loaded specimen is subjected to the same history. An extensive discussion of the local strain technique is available elsewhere [7.59].

Nonzero Mean Load

The equations of Gerber and of Goodman are empirical formulas that have been proposed for use in adjusting fatigue strengths for the effects of a nonzero mean load:

$$(\sigma_a/\sigma_f) + (\sigma_m/\sigma_u) = 1 \quad \text{(Goodman)} \tag{7.31}$$

$$(\sigma_a/\sigma_f) + (\sigma_m/\sigma_u)^2 = 1 \quad \text{(Gerber)} \tag{7.32}$$

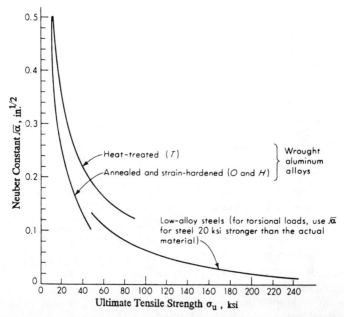

FIGURE 7-10 Neuber constants for steel and aluminum. From Juvinall [7.19], with permission.

where σ_a = alternating stress level, i.e., amplitude of applied load

σ_f = fatigue strength with zero mean load

σ_m = mean stress level, i.e., mean value of actual load

σ_u = tensile strength of material

As seen in Fig. 7-11, the applied alternating stress σ_a is superimposed on an applied static or mean stress σ_m. Because σ_f is defined for a given number of cycles or loadings, Eqs. (7.31) and (7.32) also correspond to a particular number of cycles. A typical Goodman diagram that is a plot of Eq. (7.31) is shown in Fig. 7-12. If σ_{ys} is used instead of σ_u, this is referred to as a *Soderberg diagram*. Equation (7.31) is the equation of the straight line connecting the intercept of the vertical axis, i.e., the fatigue strength σ_f corresponding to $\sigma_m = 0$, with the intercept of the horizontal axis, i.e., the static tensile strength σ_u, which corresponds to $\sigma_a = 0$. In fact, in practice the Goodman diagram is often constructed by passing a straight line through these two intercepts.

Goodman's equation usually gives good to conservative results, and it is used widely in the United States. Gerber's formula allows higher stresses than those of Goodman's and applies especially to ductile materials [7.21]. The equations apply for tensile loads because compressive mean loads usually do not affect fatigue strength. The fatigue strengths of torsion members are not usually affected by a nonzero mean load unless a region of stress concentration is present in the part. Goodman's equation and Gerber's equation both permit the peak load stress to

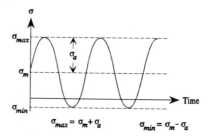

FIGURE 7-11 Fatigue stress definitions.

exceed the material yield point. Usually the peak load should be kept below the yield point. By solving the equation

$$\sigma_{\max} = \sigma_a + \sigma_m = \sigma_{ys} \tag{7.33}$$

simultaneously with Eq. (7.31) or (7.32), the amplitude at which the peak applied stress equals the yield stress can be found. This process may be carried out either algebraically or graphically. For Goodman's equation, it follows from Eq. (7.33) that the peak load stress equals the yield stress when the mean stress is at the level

$$\sigma_{my} = \sigma_u(\sigma_f - \sigma_{ys})/(\sigma_f - \sigma_u) \tag{7.34a}$$

and when the alternating stress is

$$\sigma_{ay} = \sigma_f(\sigma_{ys} - \sigma_u)/(\sigma_f - \sigma_u) \tag{7.34b}$$

Equation (7.34a) is found by setting $\sigma_m = \sigma_{my}$ in Eqs. (7.33) and (7.31) and substituting $\sigma_a = \sigma_{ys} - \sigma_{my}$ in Eq. (7.31). Equation (7.34b) follows from Eqs. (7.33) and (7.31) using $\sigma_a = \sigma_{ay}$. When the mean tensile stress equals or exceeds σ_{my}, the permissible amplitude of the cyclic load should be below

$$\sigma_a = \sigma_{ys} - \sigma_m$$

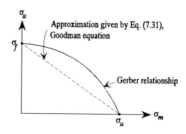

FIGURE 7-12 Typical Goodman fatigue diagram.

to avoid plastic deformation. An additional, more complex, mean stress rule has been proposed by Kececioglu [7.22].

Load with Varying Amplitude

Fackler [7.23] has compiled a summary of 20 approaches to predicting fatigue life when the applied load has a varying amplitude. These approaches are measures of cumulative damage. He presents three methods as being fundamental: the Palmgren–Langer–Miner rule [7.24, 7.25], the Corten–Dolan rule [7.26], and Shanley's method [7.27]. Here, only fully reversed loading is considered. Mean stress effects can be included by replacing cycles that have a nonzero mean load with an equivalent fully reversed load [7.28].

The Palmgren–Langer–Miner rule gives results that range from good to extremely inaccurate. This rule states that fatigue failure occurs when

$$\sum_{k=1}^{m} \frac{n_k}{N_k} = 1.0 \tag{7.35}$$

is satisfied. Here

N_k = fatigue life, i.e., number of cycles to failure at constant load applied at stress level σ_{ak}

n_k = actual number of load cycles applied at stress σ_{ak}

m = number of different stress levels σ_{ak} of applied load

Failure is anticipated if the summation is equal to or greater than 1. According to Eq. (7.35) the fatigue life is given by

$$N_f = \frac{1.0}{\sum_{k=1}^{m} (\gamma_k/N_k)} \tag{7.36}$$

where γ_k is the relative frequency of occurrence of load level σ_{ak}; i.e., γ_k is the ratio of the number of cycles at σ_{ak} to the total number of cycles ($\gamma_k = n_k/N_f$). If γ_k is given by a probability distribution, then N_f is the expected life. The Palmgren–Langer–Miner rule gives its best results if the differences between the various levels of load are small and if the different amplitudes are applied randomly instead of in a strictly increasing or decreasing sequence. The rule is based on the assumptions that fatigue damage accumulates linearly and that the amount of damage is independent of the order in which the various load levels are applied; both of these assumptions are evidently invalid. Of the two effects, the dependence of fatigue damage on the order of load application appears to be more important than nonlinear damage accumulation [7.15]. Freudenthal [7.29] has proposed to include load level interaction effects in the Palmgren–Langer–Miner rule by introducing load-dependent weighting factors W_k. These factors reduce the expected fatigue life of stress level α_{ak} by an amount that depends on the magnitude of the previously applied load:

$$N_f = \frac{1.0}{\displaystyle\sum_{k=1}^{m} (W_k \gamma_k / N_k)} \qquad (W_k \geq 1.0) \qquad (7.37)$$

An empirical approach to improving the accuracy of the Palmgren–Langer–Miner rule is the replacement of 1.0 in Eq. (7.35) by a quantity that might range from 0.5 to 2.5 and depends on the variation of the load and on the form of the part under test.

The Corten–Dolan theory results in a nonlinear law that takes into account the load history. According to this theory, the expected fatigue life N_f is calculated as

$$N_f = \frac{N_h}{\displaystyle\sum_{k=1}^{m} \gamma_k (\sigma_{ak}/\sigma_h)^d} \qquad (7.38)$$

where N_h = life if all cycles were applied at highest load level σ_h

$\quad\quad \sigma_h$ = stress amplitude of highest load level

$\quad\quad d$ = experimentally determined constant

The Corten–Dolan rule is also expressible in the form

$$\sum_{k=1}^{m} \frac{n_k}{N_h(\sigma_{ak}/\sigma_h)^d} = 1.0 \qquad (7.39)$$

A lack of knowledge of the value of d for many materials has limited the usefulness of this rule.

The Shanley 2x method is the third theory that Fackler advances as a basic approach to cumulative damage. In this theory, a constant stress level is found that will produce fatigue damage equivalent to that produced by a variable-amplitude load

$$\sigma_{eq} = \left[\frac{\displaystyle\sum_{k=1}^{m} n_k (\sigma_{ak})^{2x}}{\displaystyle\sum_{k=1}^{m} n_k} \right]^{1/2x} \qquad (7.40)$$

The constant $-1/x$ is the slope of the logarithmic S–N curve. Once σ_{eq} is found, an equivalent cycle life N_{eq} is read from the S–N curve. Then Shanley's hypothesis takes the form

$$\sum_{k=1}^{m} \frac{n_k N_{eq}}{N_k^2} = 1.0 \tag{7.41}$$

Bogdanoff and Kozin [7.30] have described a stochastic approach to cumulative damage that treats the process as a Markov chain. The model takes into account variability in manufacturing standards, duty cycles, inspection standards, and failure states. Application of the model requires knowledge of the various probabilities of occurrence of different initial states, levels of damage, etc.

The fatigue analysis of members subjected to complex load histories requires an efficient cycle counting procedure that can be implemented on a computer. Dowling [7.31] reviews a number of counting techniques. The rainflow and the range pair methods [7.30] are recommended as being superior to the other approaches.

The methods of fracture mechanics have also been applied to the problem of fatigue under variable load [7.2]. To compute the number of cycles necessary for a crack to reach critical size, a statistic of the load spectrum, such as the root-mean-square (rms) fluctuation of the stress intensity factor, is substituted in the crack propagation formula Eq. (7.47) of Section 7.7.

Effects of Load, Size, Surface, and Environment

As was previously mentioned, the load type; the size, shape, and surface finish of the test specimen; and the presence of chemically active substances can greatly influence the behavior of a cyclically loaded structural member. A brief summary of the effects of these factors is given in the subsequent paragraphs.

Load If σ_{eB} is the endurance limit of a specimen in bending, the limit for perfectly aligned axial loading will be about $0.9\sigma_{eB}$; for torsional loads the limits will be $0.58\sigma_{eB}$ for a ductile material and $0.85\sigma_{eB}$ for brittle metals. When combined bending, tensile, and torsional loads act on a member, the assumption is sometimes made that the same relation exists between endurance limits as exists between static failure loads. If σ_f is the fatigue strength of a material under uniaxial load, applying the distortion energy failure criterion results in

$$\sigma_1^2 + \sigma_2^2 + \sigma_3^2 - (\sigma_1\sigma_2 + \sigma_2\sigma_3 + \sigma_3\sigma_1) = \sigma_f^2 \tag{7.42}$$

as the fatigue failure relation for a triaxial stress state. The stresses σ_1, σ_2, and σ_3 are the amplitudes of the principal stresses at any point in the body. Similar relations exist for other static load failure theories such as the maximum shear stress criterion or the maximum normal stress rule. These approaches do not generally apply to cases in which nonzero mean stress is present. In addition, such effects as are caused by variations in size, surface, and configuration must be considered in choosing σ_f.

Hashin [7.34] proposes a quadratic polynomial in stress space to represent the locus of all stress states with the same fatigue life. This relation takes the form

$$(I_1/\sigma_u)^2 - (I_2/\tau_u)^2 = 1 \tag{7.43}$$

for completely reversed cyclic loading. The invariants of the stress tensor I_1 and I_2 [Eq. (3.19b)] are defined as

$$I_1 = \sigma_x + \sigma_y + \sigma_z \tag{7.44a}$$

$$I_2 = \sigma_x\sigma_y + \sigma_y\sigma_z + \sigma_z\sigma_x - \tau_{xy}^2 - \tau_{yz}^2 - \tau_{xz}^2 \tag{7.44b}$$

The stresses σ_u and τ_u are the fatigue strengths of the material in tension and torsion, respectively. For the most complicated cases of combined loads with different phases, different frequencies, varying amplitudes, and nonzero mean stresses, no general methods of analysis appear to be available.

Size By increasing the diameter of a specimen, the fatigue limit in bending is eventually reduced to the level for axial loads. This reduction in strength occurs because the stress gradient in the bent specimen decreases as the radius increases at constant outer fiber stress; greater volumes of material are consequently subjected to higher stresses in the larger specimens. Further reductions in endurance limits occur with increased size because the likelihood of imperfections in the surface and microstructure of the material increases with size. The shape of a member also influences fatigue life because of stress concentrations introduced by sharp corners.

Surface and Environment Because a rough surface is a form of stress raiser, surface finish can have a marked influence on fatigue life. The effect of surface finish on the endurance limit of steels is shown in Fig. 7-13. Treatments that alter the physical properties of a surface or introduce superficial residual stresses also affect the fatigue behavior of a material. The effects of shot peening and cold rolling on the endurance limits of some steels are given in [7.17]. Case-hardening treatments, such as nitriding, carburizing, and flame hardening, also improve the fatigue life of steels. Some electroplated surfaces, however, such as nickel or chromium, have an adverse influence on fatigue strength (Table 7-6). The decrease in fatigue strengths associated with these platings is thought to be related to the creation of superficial tensile stresses.

The problem of corrosion fatigue is complex, and it must be dealt with on the basis of tests of each combination of material and environment. The effect of well water and salt water on the endurance limits of various metals is shown in Tables 7-7 and 7-8. The action of the cyclic stress in destroying protective surface films and of the corrosive agent in producing surface pits and reaction products leads to reciprocal aggravation of both corrosion and fatigue of the metal. Fatigue can also be induced by the relative motion of two surfaces in contact; this form of failure is known as *fretting fatigue*.

The fatigue strengths of metals generally increase as temperatures decrease; however, the notch sensitivity and tendency to brittle fracture of some metals greatly increase at low temperatures. At higher temperatures fatigue lives generally decrease, but the analysis is complicated by the presence of the creep mechanism of failure as well as fatigue [7.36]. In the higher temperatures of the

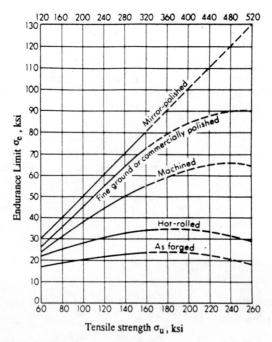

FIGURE 7-13 Effect of surface finish on endurance limit of steel. Endurance limit values for steels over 160 ksi tensile strength are realized only by proper alloy and heat treatment. From Juvinall [7.19], with permission.

creep range, creep may increase the notch sensitivity of materials that are insensitive at lower temperatures.

Stress Concentration with Nonzero Mean Load

When the effect of stress concentration is combined with that of an alternating applied load that has a nonzero mean value, the following procedure is recommended: For brittle materials, apply the fatigue strength reduction factor to both the alternating and mean values of the applied nominal stress. Goodman's formula for a brittle material becomes

$$K_f\left[(\sigma_a/\sigma_f) + (\sigma_m/\sigma_u)\right] = 1 \tag{7.45}$$

For a ductile material, apply the fatigue strength reduction factor to the alternating component of stress only. Goodman's formula for a ductile material becomes

$$(K_f\sigma_a/\sigma_f) + (\sigma_m/\sigma_u) = 1 \tag{7.46}$$

For torsion members with notches, the same formulas are used but torsional stresses and factors are substituted for the tensile values.

Example 7.6 Fatigue Analysis of a Beam The bar shown in Fig. 7-14 must withstand one million cycles of an applied bending moment that varies sinusoidally from 300 to 2000 N-m. The member is fabricated from an alloy steel that has an ultimate strength of 932 MPa (135 ksi) and a yield strength of 869 MPa (126 ksi).

The fatigue strength is taken from Table 7-4. The approximate relation $HB = \sigma_u/500$, where HB is the Brinell hardness number and σ_u is in psi, can be used to help identify the type of steel. For this case, $HB = 135(10^3)/500 = 270$ so that case Hardness of Steel: 269–285 HB of Table 7-4 can be employed, which applies for $\sigma_m = 0$. For the machined samples made from this material, the curve is $s = 10^{2.535 - 0.147 \log N}$, so that the fatigue strength is 45 ksi for one million cycles. To ensure a reliable operating life for the part, the operating stress should be safely below the fatigue strength of the material. The factor of safety is computed after modifying the S–N curve data for the effects of the notch in the bar and the nonzero mean value of the load.

For the notch shown in Fig. 7-14, Table 6-1, Part I, case 3b, gives $K_t = 3.0$. The notch sensitivity index of the material is provided by Eq. (7.30):

$$q_f = 1/\left[1 + (\alpha/r)^{1/2}\right] = 1/\left[1 + (0.04/0.34)\right] = 0.9 \tag{1}$$

where $r = 3$ mm (0.118 in.) and $\sqrt{r} = 0.34 \sqrt{\text{in.}}$ Also, from Fig. 7-10 for $\sigma_u = 135$ ksi, $\sqrt{\alpha} = 0.04\sqrt{\text{in.}}$ From Eq. (7.29)

$$K_f = 1 + (3.0 - 1.0)0.9 = 2.8 \tag{2}$$

The second deviation of the operating conditions of the member from those of a conventional fatigue test is that the mean value of the applied load differs from zero. Goodman's equation is applied as follows for a brittle material [Eq. (7.45)]:

$$K_f\left[(\sigma_a/\sigma_f) + (\sigma_m/\sigma_u)\right] = 1.0 \tag{3}$$

From Fig. 7-11, $\sigma_{\min} = \sigma_m - \sigma_a$, where $\sigma_{\min}$ is the minimum value of the applied

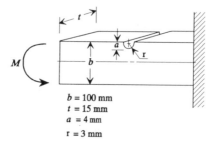

$b = 100$ mm
$t = 15$ mm
$a = 4$ mm
$r = 3$ mm

FIGURE 7-14 A U-notched bar subjected to fluctuating bending moment: depth, 4 mm; notch radius 3 mm.

nominal stress. Thus, (3) becomes

$$\frac{\sigma_a}{\sigma_f} + \frac{\sigma_{\min} + \sigma_a}{\sigma_u} = \frac{1}{K_f}$$

or

$$\sigma_a = \frac{1/K_f - \sigma_{\min}/\sigma_u}{1/\sigma_f + 1/\sigma_u} \qquad (4)$$

The minimum bending moment is 300 N-m. The resulting outer fiber stress would be

$$\sigma_{\min} = \frac{Mh}{2I} = \frac{6M}{th^2} = 6\left[(300)/0.015(0.096)^2\right] = 13 \text{ MPa} \qquad (5)$$

where the height of the beam at the notch is taken as $h = b - a = 100 - 4 = 96$ mm. Thus, the nominal stress amplitude has been computed at the notch. From (4)

$$\sigma_a = [(1/2.8) - (13/932)]/[(1/310) + 1/932)] = 79.8 \text{ MPa} \qquad (6)$$

The allowable mean stress level is

$$\sigma_m = \sigma_{\min} + \sigma_a = 13 + 79.8 = 92.8 \text{ MPa} \qquad (7)$$

and the peak stress is

$$\sigma_{\max} = \sigma_m + \sigma_a = 92.8 + 79.8 = 172.6 \text{ MPa}$$

The bending moment corresponding to $\sigma_{\max}$ is

$$M_{\max} = \tfrac{1}{6}th^2\sigma_{\max} = \tfrac{1}{6}(0.015)(0.096)^2(172.6 \times 10^6) = 3976.6 \text{ N-m} \qquad (8)$$

The peak moment of the applied load is 2000 N-m; therefore a safety factor of $(3976.6/2000) = 1.99$ exists to ensure the reliable service life of the part.

The actual peak stress in the material at the point of maximum stress concentration is

$$(\sigma_a)_{\max} = K_f(\sigma_{\max}) = K_f(\sigma_a + \sigma_m) = 2.8(172.6) = 483.8 \text{ MPa} \qquad (9)$$

which is far below the yield point of the material. With ductile materials residual stresses at the notch reduce σ_m, and the actual peak stress may be computed as $(\sigma_a)_{\max} = K_f \sigma_a + \sigma_m$.

7.7   FRACTURE MECHANICS APPROACH TO FATIGUE

A fracture mechanics based method for the prediction of fatigue is summarized in the literature [7.1–7.3]. In this approach the number of load cycles necessary for a material flaw of subcritical size to grow to critical size is calculated by integration of an empirical differential equation of the form

$$\frac{da}{dN} = A(\Delta K)^n \tag{7.47}$$

where da/dN = fatigue crack growth rate

a = flaw size

N = number of load cycles

ΔK = range of stress intensity factor during cycle

A, n = constants that depend on the material of construction

The values of A and n for several classes of steels are listed in Table 7-9. Values of the constants for some additional materials are shown in Table 7-10 in metric units.

Depending on the magnitude of the fluctuation of the stress intensity factor, the rate of fatigue crack propagation exhibits three types of behavior called region I, II, and III behavior, respectively. In region I the fluctuation of the stress intensity factor is less than a threshold value, ΔK_{th}, and subcritical size flaws do not propagate. In region II the crack propagation rate is governed by Eq. (7.47). In region III the fluctuation of the stress intensity factor exceeds a transition value, ΔK_T, and propagation occurs at a higher rate than that predicted by Eq. (7.47). For a loading fluctuating from zero to a tensile value, the transition value of the stress intensity factor is given by

$$K_T = 0.04\sqrt{E\sigma_g} \tag{7.48}$$

where E = Young's modulus (ksi)

σ_g = mean value of tensile strength and yield strength (ksi)

K_T = transition stress intensity factor (ksi $\cdot \sqrt{\text{in.}}$)

For this case of zero to tension loading, ΔK_T reduces to K_T.

For the three classes of steel listed in Table 7-9, the threshold stress intensity fluctuation is about 5.5 ksi $\cdot \sqrt{\text{in.}}$ or less; the threshold point depends on the ratio of minimum to maximum applied stress. Provided that the fluctuation in the stress intensity factor is such that region II propagation occurs, Eq. (7.47) can be used to compute the number of load cycles necessary for a flaw to propagate to a given size. This capability is a significant improvement over computations based on $S-N$ curves because a large part of the life shown on an $S-N$ curve is related to crack initiation. When a crack of significant size already exists in the material, the actual fatigue life of the specimen will be much shorter than that predicted by the $S-N$ curve.

The fatigue crack propagation formula for ferrite–pearlite steels often employed for railroad rails is given as

$$\frac{da}{dN} = 1.68 \times 10^{-10}(\Delta K)^{3.3} \tag{7.49}$$

where ΔK is in ksi · $\sqrt{\text{in.}}$ and da/dN is in in./cycle. Barsom and Imhof [7.39] present data showing that this formula is applicable to rail carbon steels as well.

Example 7.7 Fatigue Analysis of Plate with Crack A large flat plate made of 304 stainless steel is subjected to a fluctuating tensile (in-plane) load of 0–15 ksi. The properties of the steel are

$$\sigma_u = 85 \text{ ksi}, \qquad \sigma_{ys} = 40 \text{ ksi}, \qquad K_c = 85 \text{ ksi} \cdot \sqrt{\text{in.}}, \qquad E = 28 \times 10^3 \text{ ksi} \quad (1)$$

If a 0.125-in.-long ($2a_0 = 0.125$ in.) through-thickness crack oriented normal to the load is assumed to exist in the plate, the number of load cycles necessary for crack growth through region II can be computed by using Eq. (7.47).
From Eq. (7.48) the transition stress intensity factor is

$$K_T = 0.04\left[(28 \times 10^3)(40 + 85)/(2)\right]^{1/2} = 53 \text{ ksi} \cdot \sqrt{\text{in.}} \quad (2)$$

Because $K_T < K_c$, only the number of cycles to the higher growth rate region III can be computed with Eq. (7.47). If $K_c < K_T$, the complete life to critical size can be found.
For a very large plate K_I can be taken from Table 7-1, case 1, $a/b = 0$, i.e., $K_I = \sigma\sqrt{\pi a}$. This relationship can be used to find the point-of-transition (region II to III) crack length $2a_T$ by using $K_1 = K_T$. For our case with a nominal stress $\sigma = 15$ ksi,

$$a_T = (1/\pi)(53/15)^2 = 3.97 \text{ in.} \quad (3)$$

This same relationship provides the critical crack size $2a_c$ if K_1 is set equal to K_c:

$$2a_c = (2/\pi)(85/15)^2 = 20.44 \text{ in.} \quad \text{or} \quad a_c = 10.22 \text{ in.} \quad (4)$$

Equation (7.47) may be integrated directly or approximated by a finite sum; in this problem analytical integration is possible. First, we use

$$\Delta K = (\sigma_{max} - \sigma_{min})\sqrt{\pi a} = \sigma_{max}\sqrt{\pi a} \quad \text{since } \sigma_{min} = 0 \quad (5)$$

Then, from Table 7-9 for austenitic steel, we find $A = 3.0 \times 10^{-10}$, $n = 3.25$. Thus

$$\frac{da}{dN} = 3.0 \times 10^{-10}(\Delta K)^{3.25} \quad (6)$$

Rewrite Eq. (7.47) in the form

$$dN = \frac{da}{A(\Delta K)^n} \quad (7)$$

From this relationship, we can compute N_T, the number of cycles to the beginning of the higher growth rate than that predicted by Eq. (7.47). Thus, N_T is the number of cycles to reach region III. With $a_0 = 0.125/2$ and $a_T = 3.97$,

$$N_T = \int_{a_0}^{a_T} \frac{da}{A\left(\sigma_{max}\sqrt{\pi}\right)^n a^{n/2}} = \frac{1}{A\left(\sigma_{max}\sqrt{\pi}\right)^n \left(-\frac{1}{2}n + 1\right)} \left[\frac{1}{a_T^{(n/2)-1}} - \frac{1}{a_0^{(n/2)-1}} \right]$$

$$= \frac{1}{3.0(10^{-10})\left(15\sqrt{\pi}\right)^{3.25}(-0.625)} \left[\frac{1}{(3.97)^{0.625}} - \frac{1}{(0.0625)^{0.625}} \right]$$

$$= 124{,}978(5.657 - 0.422) = 654{,}260 \text{ cycles} \tag{8}$$

If growth to failure continued through region III at the rate given by Eq. (7.47), the life would be given by the above integral with a_c substituted for a_T. We find

$$N_f = 124{,}978\left[5.657 - (10.22)^{-0.625}\right] = 677{,}764 \text{ cycles} \tag{9}$$

When the final crack size is much greater than the initial length of the crack, the cycle life is almost totally dependent on the initial size. It follows from (8) and (9) that if region II behavior continued beyond a length of $2a_T = 7.94$ in., the number of cycles necessary for the last $20.44 - 7.94 = 8.5$ in. of growth would be roughly 3.6% of the number for the initial $7.94 - 0.125 = 7.82$ in. Actually, the failure life would be less than 677,764 cycles because the higher growth rate of region III would occur after the crack length reached the transition point.

———

Example 7.8 Fatigue Analysis of a Cylinder with a Flaw A hollow thin-walled cylinder is subjected to fluctuations in internal pressure that range from 30 to 36 MPa. The cylinder is fabricated from DTD 687A aluminum alloy, which has a yield strength of 495 MPa and a fracture toughness of 22 MPa · $\sqrt{m}$. The inner diameter of the tube is 15 cm, and the wall thickness is 1.5 cm. A thumbnail semicircular flaw of initial radius 5 mm exists in the tube wall with the plane of the crack normal to the hoop stress (see Fig. 7-15). If the threshold fluctuation in stress intensity for this alloy is 1.1 MPa · $\sqrt{m}$, will the 5-mm-deep flaw propagate under the given load? What is the smallest flaw that would propagate? Assuming region II growth rate occurs up to the critical point, compute the number of load cycles that will cause unstable crack growth.

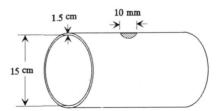

FIGURE 7-15 Cylinder with thumbnail semicircular flaw.

From case 18 of Table 7-1, for $\theta = \frac{1}{2}\pi$, $a/c = 1$ and $c/b \to 0$, the stress intensity factor for a semicircular flaw is

$$K_I = 1.1035 \times \frac{2}{\pi} \sigma \sqrt{\pi a} = \frac{2.207}{\sqrt{\pi}} \sigma \sqrt{a} \tag{1}$$

For this flaw and fluctuating loading, the range of the stress intensity factor is

$$\Delta K_I = (2.207/\sqrt{\pi})\sqrt{a}\,(\sigma_{max} - \sigma_{min}) \tag{2}$$

The hoop stress formula for a cylinder gives

$$\begin{aligned}
\sigma_{max} &= p_{max} r/t = (36 \text{ MPa})15 \text{ cm}/(2 \times 1.5 \text{ cm}) = 180 \text{ MPa} \\
\sigma_{min} &= p_{min} r/t = (30 \text{ MPa})15 \text{ cm}/(2 \times 1.5 \text{ cm}) = 150 \text{ MPa}
\end{aligned} \tag{3}$$

To find the smallest flaw that will propagate, solve (2) for a using $\Delta K_I = \Delta K_{th} = 1.1$:

$$a = \left\{\Delta K_{th}/\left[(\sigma_{max} - \sigma_{min})(2.207/\sqrt{\pi})\right]\right\}^2 = 0.00087 \text{ m} \quad \text{or} \quad a = 0.87 \text{ mm} \tag{4}$$

We conclude that the 5-mm flaw will propagate under the given load. The smallest flaw that will propagate is $a = 0.87$ mm.

The above conclusion can be reached with a different procedure. Begin by computing ΔK_I corresponding to $a = 5$ mm. From (2)

$$\Delta K_I = (2.207/\sqrt{\pi})\sqrt{0.005 \text{ m}}\,(180 - 150) \text{ MPa} = 2.64 \text{ MPa} \cdot \sqrt{m} \tag{5}$$

Since $\Delta K_I > \Delta K_{th} = 1.1$ MPa $\cdot \sqrt{m}$, the flaw will propagate.

To compute the number of load cycles to unstable crack growth, use Eq. (7.47). Table 7-10, for $R^* = 150/180 = 0.83$, lists $n = 4.8$ and $A = 1.68 \times 10^{-10}$. The initial crack size is $a_0 = 0.005$ m. To find a_c, use (1).

$$a_c = (\pi/4.87)(K_{Ic}/\sigma_{max})^2 = (\pi/4.87)(22 \text{ MPa} \cdot \sqrt{m}/180 \text{ MPa})^2 = 0.0096 \text{ m}$$

From Eq. (7.47)

$$N = \int_{a_0}^{a_c} \frac{da}{A(2.207/\sqrt{\pi})^n (a)^{n/2}(\Delta\sigma)^n} = 120{,}227 \text{ cycles}$$

where $\Delta\sigma = \sigma_{max} - \sigma_{min}$ and $n = 4.8$.

7.8 COMBINED APPROACH

The fatigue failure of a structural member is a two-step process of crack initiation followed by propagation to critical size. Fracture mechanics methods can be applied to the propagation phase of the problem. Other techniques such as the local strain approach can be used to deal with the initiation phase of fatigue failure. A method for handling crack initiation problems is presented elsewhere [7.46].

REFERENCES

7.1 Kanninen, M. F., and Popelar, C. H., *Advanced Fracture Mechanics*, Oxford University Press, New York, 1985.

7.2 Rolfe, S. T., and Barsom, J. M., *Fracture and Fatigue Control in Structures*, Prentice-Hall, Englewood Cliffs, NJ, 1977.

7.3 Hertzberg, R. W., *Deformation and Fracture Mechanics of Engineering Materials*, 3rd Ed., Wiley, New York, 1989.

7.4 Tada, H., Paris, P. C., and Irwin, G. R. *The Stress Analysis of Cracks Handbook*, Paris productions, St. Louis, MO, 1973.

7.5 Griffith, A. A., "The Phenomena of Rupture and Flow in Solids," *Trans. Roy. Soc. Lond.*, Vol. A-221, 1920.

7.6 Irwin, G. R., "Fracture Mode Transition for a Crack Traversing a Plate," *J. Basic Eng.*, Vol. 82, 1960, pp. 417–425.

7.7 Sih, G. C., *Handbook of Stress Intensity Factors for Researchers and Engineers*, Institute of Fracture and Solid Mechanics, Lehigh University, Bethlehem, PA 1973.

7.8 Gangloff, R. P., "The Fracture Mechanics of Engineering Materials," class notes, University of Virginia, Charlottesville, 1990.

7.9 Buch, A., *Fatigue Strength Calculation Methods*, Trans Tech Publications, Aedermannsdorf, Switzerland, 1988.

7.10 Rice, J. R., "A Path Independent Integral and the Approximate Analysis of Strain Concentrations by Notches and Cracks," *J. Appl. Mech.*, Vol. 35, 1968, pp. 379–386.

7.11 Henshell, R. D., and Shaw, K. G., "Crack-Tip Finite Elements are Unnecessary," *Int. J. Numer. Methods Eng.*, Vol. 9, 1975, pp. 496–507.

7.12 Barsoum, R. S., "On the Use of Isoparametric Finite Elements in Linear Elastic Fracture Mechanics," *Int. J. Numer. Methods Eng.*, Vol. 10, 1976, pp. 25–37.

7.13 *ABAQUS Theory Manual, Version 4.8*. Hibbitt, Karlsson & Sorensen, Providence, RI, 1989.

7.14 Parks, D. M., "The Virtual Crack Extension Method for Nonlinear Material Behavior," *Comput. Methods Appl. Mechan. Eng.*, Vol. 12, 1977, pp. 353–364.

7.15 Weibull, W., "A Statistical Representation of Fatigue Failures in Solids," *Roy. Inst. Tech. Trans.*, Stockholm, 1949.

7.16 McClintock, F. A., "A Criterion for Minimum Scatter in Fatigue Testing," *Trans. ASME, J. Appl. Mech.*, Vol. 22, 1955, pp. 427–431.

7.17 Lipson, C., and Juvinall, R. C., *Handbook of Stress and Strength*, Macmillan, New York, 1963.

7.18 Neuber, H., *Theory of Notch Stresses*, J. W. Edwards, Ann Arbor, MI 1946.

7.19 Juvinall, R. C., *Stress, Strain, and Strength*, McGraw-Hill Book Company, New York, 1967.

7.20 Wetzel, R. M., ed., *Fatigue under Complex Loading: Analyses and Experiments*, Society of Automotive Engineers, Warrendale, PA, 1977.

7.21 Freudenthal, A. M., "Fatigue," *Encyclopedia of Physics*, Springer-Verlag, Berlin, 1958, pp. 591–613.

7.22 Kececioglu, D. B., Chester, L. B., and Dodge, T. M., "Combined Bending—Torsion Fatigue Reliability of AISI 4340 Steel Shafting with $K = 2.34$," *Trans. ASME, J. Eng. Ind.*, Vol. 97, No. 2, 1975, pp. 748–758.

7.23 Fackler, W. C., *Equivalence Techniques for Vibration Testing*, The Shock and Vibration Information Center, U.S. Department of Defense, Washington, DC, 1972.

7.24 Palmgren, A. Z., "Die Lebensdauer von Kugellagern," *Z. Ver. Deutsch, Ing.*, Vol. 68, No. 339, 1924.

7.25 Miner, M. A., "Cumulative Damage in Fatigue," *Trans. ASME, J. Appl. Mech.*, Vol. 12, 1945, pp. A159–A164.

7.26 Dolan, T., and Corten, H., "Progressive Damage Due to Repeated Loading," *Fatigue of Aircraft Structures*, Proc. WADC-TR-59-507, Wright Air Development Center, Wright-Patterson AFB, Ohio, August 1959.

7.27 Shanley, F. R., "A Theory of Fatigue Based on Unbonding During Reversed Slip," The Rand Corporation, Report No. P-350, November 1952.

7.28 Morrow, J., "Fatigue Properties of Metals," in *Fatigue Design Handbook*, SAE, Warrendale, PA, 1968, Section 3.2.

7.29 Freudenthal, A. M., "Fatigue of Materials and Structures under Random Loading," Report No. WADC-TR-59-676, Wright Air Development Center, Wright-Patterson AFB, OH, March 1961.

7.30 Bogdanoff, J. L., and Kozin, F., *Probabilistic Models of Cumulative Damage*, Wiley, New York, 1985.

7.31 Dowling, N. E., "Fatigue Failure Predictions for Complicated Stress–Strain Histories," *J. Mater.*, Vol. 7, No. 1, 1972, pp. 71–87.

7.32 *Annual Book of ASTM Standards*, Vol. 03.01, ASTM, Philadelphia, PA 1993.

7.33 McMillan, J. C., and Pelloux, R. M. M., *Fatigue Crack Propagation under Program and Random Loads*, ASTM STP 415, American Society for Testing and Materials, Philadelphia, PA, 1967.

7.34 Hashin, Z., "Fatigue Failure Criteria for Combined Cyclic Stress," Naval Air Systems Command, Contract N00014-78-C-0544, Technical Report No. 2, Department of Materials Science and Engineering, University of Pennsylvania, March 1979.

7.35 McAdam, D. F., Jr., "Corrosion Fatigue of Non-Ferrous Metals," *Proc. ASTM*, Vol. 27, No. 2, 1927, p. 102.

7.36 Carden, A. E., McEvily, A. J., and Wells, C. H. (Eds.), *Fatigue at Elevated Temperatures*, ASTM STP 520, American Society for Testing and Materials, Philadelphia, PA, 1972.

7.37 Peterson, R. E., *Stress Concentration Design Factors*, Wiley, New York, 1954.

7.38 Pook, L. P., "Analysis and Application of Fatigue Crack Growth Data," *A General Introduction to Fracture Mechanics*, Mechanical Engineering Publications, London and New York, 1978, pp. 114–135.

7.39 Barsom, J. M., and Imhof, E. J., Jr., "Fatigue and Fracture Behavior of Carbon Steel Rails," Rail Research Vol. 5, Report No. R-301, Association of American Railroads, Chicago, IL, March 1978.

7.40 Bradt, R. C., Hasselman, D. P. H., and Lange, E. F. (Eds.), *Fracture Mechanics of Ceramics*, Vol. 2, Plenum Press, New York, 1974, p. 469.

7.41 Ingelstrom, N., and Nordberg, H., *Engineering Fracture Mechanics*, Vol. 6, Pergamon Press, New York, 1974, p. 597.

7.42 Schmidt, R. A., *Closed Loop*, Vol. 5, November 1975, p. 3.

7.43 Marshall, G. P., and Williams, J. G., "The Correlation of Fracture Data for PMMA," *J. Mater. Sci.*, Vol. 8, 1973, pp. 138–140.

7.44 Marshall, G. P., Culver, L. E., and Williams, J. G., "Fracture Phenomena in Polystyrene," *Int. J. Fract.*, Vol. 9, No. 3, 1973, pp. 295–309.

7.45 Rodon, J. C., "Influence of the Dynamic Stress Intensity Factor on Cyclic Crack Propagation in Polymers," *J. Appl. Polym. Sci.*, Vol. 17, 1973, pp. 3515–3528.

7.46 Irwin, G. R., "Analysis of Stresses and Strains Near the End of a Crack Traversing a Plate," *Trans. ASME, J. Appl. Mech.*, Vol. 24, 1957, pp. 361–364.

7.47 Brown, W. F., Jr., ed., "Review of Developments in Plane Strain Fracture Toughness Testing," ASTM STP 463, American Society for Testing and Materials, Philadelphia, PA, 1970.

7.48 Paris, C. P., and Sih, G. C., "Stress Analysis of Cracks," *Fracture Toughness Testing and Its Applications*, ASTM STP 381, American Society for Testing and Materials, Philadelphia, PA, 1965.

7.49 Brown, W. F., Jr., and Srawley, J. E., *Plane Strain Crack Toughness Testing of High Strength Metallic Materials*, ASTM STP 410, American Society for Testing and Materials, Philadelphia, PA, 1966.

7.50 Forman, R. G., and Kobayashi, A. S. "On the Axial Rigidity of a Perforated Strip and the Strain Energy Release Rate in a Centrally Notched Strip Subjected to Uniaxial Tension," *J. Basic Eng.*, Vol. 86, 1964, pp. 693–699.

7.51 Gross, B., and Srawley, J. E., "Stress Intensity Factors for Single-Edge-Notch Specimens in Bending or Combined Bending and Tension by Boundary Collocation of a Stress Function," Technical Note D-2603, NASA, January 1965.

7.52 Irwin, G. R., "Crack Extension Force for a Part-through Crack in a Plate," *Trans. ASME, J. Appl. Mech.*, Vol. 29, No. 4, 1962, pp. 651–654.

7.53 Neuman, J. C., Jr., "An Improved Method of Collocation for the Stress Analysis of Cracked Plates with Various Shaped Boundaries," Technical Note D-6376, NASA, August 1971.

7.54 McAdam, D. J., Jr., "Corrosion Fatigue of Metals as Affected by Chemical Composition, Heat Treatment, and Cold Working," *Trans. ASST*, Vol. 11, 1927, p. 355.

7.55 Novak, S. R., and Barsom, J. M., "AISI Project 168—Toughness Criteria for Structural Steels: Brittle-Fracture (K_{IIC}) Behavior of Cracks Emanating from Notches," Ninth National Symposium on Fracture Mechanics, University of Pittsburgh, August 25–27, 1975.

7.56 Matsuishi, M., and Endo, T., "Fatigue of Metals Subjected to Varying Stress," Japan Society of Mechanical Engineers, Fukuoka, Japan, March 1968.

7.57 Van Dyke, G. M., "Statistical Load Data Processing," Sixth ICAF Symposium, Miami, FL, May 1971.

7.58 Dowling, N. E., "Fatigue at Notches and the Local Strain and Fracture Mechanics Approaches," *Fracture Mechanics*, ed. by C. W. Smith, ASTM STP 677, American Society for Testing and Materials, Philadelphia, PA, 1979, pp. 247–273.

7.59 *Metals Handbook*, Vol. 10, *Failure Analysis and Prevention*, 8th ed. American Society for Metals, Metals Park, OH, 1975.

7

Tables

TABLE 7-1   STRESS INTENSITY FACTORS

Notation

K_I = mode I stress intensity factor $(F/L^{3/2})$

K_{II} = mode II stress intensity factor $(F/L^{3/2})$

K_{III} = mode III stress intensity factor $(F/L^{3/2})$

σ = tensile stress, under opening mode of loading (F/L^2)

τ = shear stress, under shearing mode of loading (F/L^2)

τ_{III} = shear stress, under tearing mode of loading, which is in out-of-plane direction (F/L^2)

Case	Intensity Factor
1. Finite width plate with center crack, tension loading 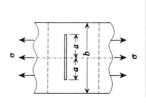	$$K_I = \sigma\sqrt{\pi a}\,F(a/b)$$ $$F(a/b) = \left[1 - 0.1(a/b)^2 + 0.96\left(\frac{a}{b}\right)^4\right]\sqrt{\sec\frac{\pi a}{b}}$$

2.
Finite width plate with center crack, mode II crack propagation (shear load along crack)

$$K_{II} = \tau\sqrt{\pi a}\, F(a/b)$$

$$F(a/b) = \left\{\left(1 - 0.1\left(\frac{a}{b}\right)^2 + 0.96\left(\frac{a}{b}\right)^4\right)\right\}\sqrt{\sec\frac{\pi a}{b}}$$

3.
Finite width plate with center crack, mode III crack propagation (out-of-plane shear loading)

$$K_{III} = \tau_{III}\sqrt{\pi a}\,\sqrt{\frac{b}{\pi a}\tan\frac{\pi a}{b}}$$

4.
Finite width plate with double-edge crack, tension loading

$$K_I = \sigma\sqrt{\pi a}\, F(a/b)$$

$$F(a/b) = \left(1 + 0.122\cos^4\frac{\pi a}{b}\right)\sqrt{\frac{b}{\pi a}\tan\frac{\pi a}{b}}$$

Stress Intensity Factors   TABLE 7-1

TABLE 7-1 | **Stress Intensity Factors** | 338

TABLE 7-1 (continued) STRESS INTENSITY FACTORS

Case	Intensity Factor
5. Finite width plate with double-edge cracks, mode II crack propagation	$K_{II} = \tau\sqrt{\pi a}\,F(a/b)$ $F(a/b)$ is the same as in case 4.
6. Finite width plate with double-edge cracks, mode III crack propagation	$K_{III} = \tau_{III}\sqrt{\pi a}\,\sqrt{\dfrac{b}{\pi a}\tan\dfrac{\pi a}{b}}$

7.
Plate with single-edge crack, tension loading

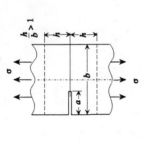

$$K_I = \sigma\sqrt{\pi a}\, F(a/b)$$

$$F(a/b) = \sqrt{\frac{2b}{\pi a}\tan\frac{\pi a}{2b}}\;\frac{0.752 + 2.02(a/b) + 0.37\left(1 - \sin\dfrac{\pi a}{2b}\right)^3}{\cos\dfrac{\pi a}{2b}}$$

8.
Plate with single-edge crack bending load $M(F \cdot L/L)$

$$\sigma = \frac{6M}{b^2}$$

$$K_I = \sigma\sqrt{\pi a}\, F(a/b)$$

$$F(a/b) = \sqrt{\frac{2b}{\pi a}\tan\frac{\pi a}{2b}}\;\frac{0.923 + 0.199\left(1 - \sin\dfrac{\pi a}{2b}\right)^4}{\cos\dfrac{\pi a}{2b}}$$

TABLE 7-1 (continued) STRESS INTENSITY FACTORS

Case	Intensity Factor
9. Beam with crack, three-point bending $P(F/L)$	$\sigma = \dfrac{6M}{b^2}$, $M = \dfrac{PL}{4}$ $K_{\mathrm{I}} = \sigma\sqrt{\pi a}\,F(a/b)$ For $L/b = 4$, $\beta = a/b$ $$F(a/b) = \frac{1}{\sqrt{\pi}}\,\frac{1.99 - \beta(1-\beta)(2.15 - 3.93\beta + 2.7\beta^2)}{(1+2\beta)(1-\beta)^{3/2}}$$ For $L/b = 8$ $$F(a/b) = 1.106 - 1.552(a/b) + 7.71(a/b)^2 - 13.53(a/b)^3 + 14.23(a/b)^4$$
10. Shaft with crack, tension loading $P(F)$	$K_{\mathrm{I}} = \sigma_{\mathrm{net}}\sqrt{\pi a}\,F_1(a/b)$ $\sigma_{\mathrm{net}} = \dfrac{P}{\pi a^2}$ $F_1(a/b) = \sqrt{1 - \dfrac{2a}{b}}\,G(a/b)$ $$G(a/b) = \frac{1}{2}\left\{1 + \frac{1}{2}\frac{2a}{b} + \frac{3}{8}\left(\frac{2a}{b}\right)^2 - 0.363\left(\frac{2a}{b}\right)^3 + 0.731\left(\frac{2a}{b}\right)^4\right\}$$

TABLE 7-1   Stress Intensity Factors   340

11.
Shaft with crack, bending load

$$K_{I_A} = \sigma_N \sqrt{\pi a} \, F_1(a/b)$$

$$\sigma_N = \frac{4M}{\pi a^3}$$

$$F_1(a/b) = \sqrt{1 - \frac{2a}{b}} \, G(a/b)$$

$$G(a/b) = \frac{3}{8}\left\{1 + \frac{1}{2}\frac{2a}{b} + \frac{3}{8}\left(\frac{2a}{b}\right)^2 + \frac{5}{16}\left(\frac{2a}{b}\right)^3 + \frac{35}{128}\left(\frac{2a}{b}\right)^4 + 0.537\left(\frac{2a}{b}\right)^5\right\}$$

12.
Shaft with crack, torsional load

$$K_{III} = \tau_N \sqrt{\pi a} \, F_1(a/b)$$

$$\tau_N = \frac{2T}{\pi a^3}$$

$$F_1(a/b) = \sqrt{1 - \frac{2a}{b}} \, G(a/b)$$

$$G(a/b) = \frac{3}{8}\left\{1 + \frac{1}{2}\frac{2a}{b} + \frac{3}{8}\left(\frac{2a}{b}\right)^2 + \frac{5}{16}\left(\frac{2a}{b}\right)^3 + \frac{35}{128}\left(\frac{2a}{b}\right)^4 + 0.208\left(\frac{2a}{b}\right)^5\right\}$$

TABLE 7-1 (continued)   STRESS INTENSITY FACTORS

Case	Intensity Factor
13. Shaft with internal circular crack, tension loading 	$$K_1 = \sigma_{\text{net}}\sqrt{\pi a}\,F_1(a/b)$$ $$\sigma_{\text{net}} = \frac{P}{\pi\left(\dfrac{b^2}{4} - a^2\right)}$$ $$F_1(a/b) = \sqrt{1 - \frac{2a}{b}}\;G(a/b)$$ $$G(a/b) = \frac{2}{\pi}\left\{1 + \frac{1}{2}\frac{2a}{b} - \frac{5}{8}\left(\frac{2a}{b}\right)^2 + 0.421\left(\frac{2a}{b}\right)^3\right\}$$
14. Shaft with internal circular crack, bending load 	$$K_{I_A} = \sigma_N\sqrt{\pi a}\,F_1(a/b)$$ $$\sigma_N = \frac{4Ma}{\pi\left(\dfrac{b^4}{16} - a^4\right)}$$ $$F_1(a/b) = \sqrt{1 - \frac{2a}{b}}\;G(a/b)$$ $$G(a/b) = \frac{4}{3\pi}\left\{1 + \frac{1}{2}\frac{2a}{b} + \frac{3}{8}\left(\frac{2a}{b}\right)^2 + \frac{5}{16}\left(\frac{2a}{b}\right)^3 - \frac{93}{128}\left(\frac{2a}{b}\right)^4 + 0.483\left(\frac{2a}{b}\right)^5\right\}$$

TABLE 7-1   Stress Intensity Factors   342

Stress Intensity Factors TABLE 7-1

15.
Shaft with internal circular crack, torsional load

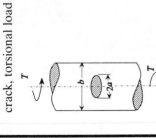

$$K_{III} = \tau_N \sqrt{\pi a}\, F_1(a/b)$$

$$\tau_N = \frac{2Ta}{\pi\left(\dfrac{b^4}{16} - a^4\right)}$$

$$F_1(a/b) = \sqrt{1 - \frac{2a}{b}}\, G(a/b)$$

$$G(a/b) = \frac{4}{3\pi}\left\{1 + \frac{1}{2}\frac{2a}{b} + \frac{3}{8}\left(\frac{2a}{b}\right)^2 + \frac{5}{16}\left(\frac{2a}{b}\right)^3 \right.$$
$$\left. - \frac{93}{128}\left(\frac{2a}{b}\right)^4 + 0.038\left(\frac{2a}{b}\right)^5\right\}$$

16.
Semi-infinite body with semi-circular crack, tension loading

$$K_{I_A} = \frac{2}{\pi}\sigma\sqrt{\pi a}\, F(\theta)$$

$$F(\theta) = 1.211 - 0.186\sqrt{\sin\theta} \quad (10° < \theta < 170°)$$

TABLE 7-1 (continued) STRESS INTENSITY FACTORS

Case	Intensity Factor
17. Infinite body with internal elliptical crack, tension loading x,y plane	$$K_{I_A} = \frac{\sigma\sqrt{\pi a}}{E(k)}\left\{\sin^2\theta + \frac{a^2}{c^2}\cos^2\theta\right\}^{1/4}$$ $$K_{1,\max} = K_1\left(\theta = \pm\tfrac{1}{2}\pi\right) = \frac{\sigma\sqrt{\pi a}}{E(k)}$$ $$K_1(c = a) = \frac{2\sigma}{\pi}\sqrt{\pi a}$$ $$K_1(c \to \infty) = \sigma\sqrt{\pi a}$$ $$E(k) = \int_0^{\pi/2}\sqrt{1 - k^2\sin^2\phi}\; d\phi$$ (Elliptic integral available in mathematical handbooks) $$k^2 = 1 - a^2/c^2$$

TABLE 7-1 Stress Intensity Factors   344

18.
Semielliptical surface crack
in finite plate, tension loading

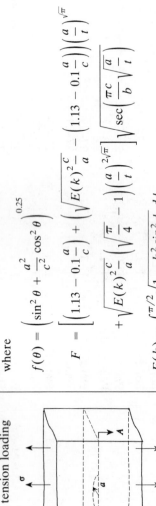

Back Face

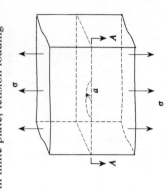

Section A–A

Front Face

$$K = \sigma \sqrt{\pi a / E(k)^2}\, f(\theta)\, F\!\left(\frac{a}{t}, \frac{a}{c}, \frac{c}{b}\right)$$

where

$$f(\theta) = \left(\sin^2\theta + \frac{a^2}{c^2}\cos^2\theta\right)^{0.25}$$

$$F = \left[\left(1.13 - 0.1\frac{a}{c}\right) + \left(\sqrt{E(k)^2\frac{c}{a}} - \left(1.13 - 0.1\frac{a}{c}\right)\right)\left(\frac{a}{t}\right)^{\sqrt{\pi}}\right.$$

$$\left. + \sqrt{E(k)^2\frac{c}{a}}\left(\sqrt{\frac{\pi}{4}} - 1\right)\left(\frac{a}{t}\right)^{2\sqrt{\pi}}\right]\sqrt{\sec\left(\frac{\pi c}{b}\sqrt{\frac{a}{t}}\right)}$$

$$E(k) = \int_0^{\pi/2}\sqrt{1 - k^2\sin^2\phi}\;d\phi$$

(Elliptic integral available in mathematical handbooks)

$$k^2 = 1 - a^2/c^2$$

TABLE 7-2 STRENGTH AND FRACTURE TOUGHNESS DATA FOR SELECTED MATERIALS[a]

Alloy	Material Supply	Specimen Orientation	Test Temperature (°C)	σ_{ys} (MPa)	K_{Ic} (MPa · $\sqrt{m}$)
Aluminum Alloys					
2014-T651	Plate	L–T	21–32	435–470	23–27
2014-T651	Plate	T–L	21–32	435–455	22–25
2014-T651	Plate	S–L	24	380	20
2014-T6	Forging	L–T	24	440	31
2014-T6	Forging	T–L	24	435	18–21
2020-T651	Plate	L–T	21–32	525–540	22–27
2020-T651	Plate	T–L	21–32	530–540	19
2024-T351	Plate	L–T	27–29	370–385	31–44
2024-T351	Plate	T–L	27–29	305–340	30–37
2024-T851	Plate	L–T	21–32	455	23–28
2024-T851	Plate	T–L	21–32	440–455	21–24
2124-T851	Plate	L–T	21–32	440–460	27–36
2124-T851	Plate	T–L	21–32	450–460	24–30
2219-T851	Plate	L–T	21–32	345–360	36–41
2219-T851	Plate	T–L	21–32	340–345	28–38

TABLE 7-2   **Strength and Fracture Toughness Data**   346

Alloy-Temper	Form	Orientation			
7049-T73	Forging	L-T	21–32	460–510	31–38
7049-T73	Forging	T-L	21–32	460–470	21–27
7050-T73651	Plate	L-T	21–32	460–510	33–41
7050-T73651	Plate	T-L	21–32	450–510	29–38
7050-T73651	Plate	S-L	21–32	430–440	25–28
7075-T651	Plate	L-T	21–32	515–560	27–31
7075-T651	Plate	T-L	21–32	510–530	25–28
7075-T651	Plate	S-L	21–32	460–485	16–21
7075-T7351	Plate	L-T	21–32	400–455	31–35
7075-T7351	Plate	T-L	21–32	395–405	26–41
7475-T651	Plate	T-L	21–32	505–515	33–37
7475-T7351	Plate	T-L	21–32	395–420	39–44
7079-T651	Plate	L-T	21–32	525–540	29–33
7079-T651	Plate	T-L	21–32	505–510	24–28
7178-T651	Plate	L-T	21–32	560	26–30
7178-T651	Plate	T-L	21–32	540–560	22–26
7178-T651	Plate	S-L	21–32	470	17

TABLE 7-2 (continued) STRENGTH AND FRACTURE TOUGHNESS DATA FOR SELECTED MATERIALS

Alloy	Material Supply	Specimen Orientation	Test Temperature (°C)	σ_{ys} (MPa)	K_{1c} (MPa · $\sqrt{m}$)
		Ferrous Alloys			
4330V(275°C temper)	Forging	L–T	21	1400	86–94
4330V(425°C temper)	Forging	L–T	21	1315	103–110
4340(205°C temper)	Forging	L–T	21	1580–1660	44–66
4340(260°C temper)	Plate	L–T	21	1495–1640	50–63
4340(425°C temper)	Forging	L–T	21	1360–1455	79–91
D6AC(540°C temper)	Plate	L–T	21	1495	102
D6AC(540°C temper)	Plate	L–T	−54	1570	62
9-4-20(550°C temper)	Plate	L–T	21	1280–1310	132–154
18 Ni(200)(480°C 6 hr)	Plate	L–T	21	1450	110
18 Ni(250)(480°C 6 hr)	Plate	L–T	21	1785	88–97
18 Ni(300)(480°C)	Plate	L–T	21	1905	50–64
18 Ni(300)(480°C 6 hr)	Forging	L–T	21	1930	83–105
AFC77 (425°C temper)	Forging	L–T	24	1530	79
		Titanium Alloys			
Ti–6 Al–4V	(Mill anneal plate)	L–T	23	875	123
Ti–6 Al–4V	(Mill anneal plate)	T–L	23	820	106
Ti–6 Al–4V	(Recrystallize anneal plate)	L–T	22	815–835	85–107
Ti–6 Al–4V	(Recrystallize anneal plate)	T–L	22	825	77–116

Ceramics[b]

Mortar	—	0.13–1.3
Concrete	—	0.23–1.43
Al_2O_3	—	3–5.3
SiC	—	3.4
Si_3N_4	—	4.2–5.2
Soda lime silicate glass	—	0.7–0.8
Electrical porcelain ceramics	—	1.03–1.25
WC(2.5–3 μm)—3 w/o Co	—	10.6
WC(2.5–3 μm)—9 w/o Co	—	12.8
WC(2.5–3.3 μm)—15 w/o Co[c]	—	16.5–18
Indiana limestone[d]	—	0.99

Polymers

PMMA[e]	—	0.8–1.75[f]
PS[g]	—	0.8–1.1[f]
Polycarbonate[h]	—	2.75–3.3[f]

[a]Unless noted otherwise, from Hertzberg [7.3], with permission. Symbols: σ_{ys}, yield strength (F/L^2); K_{Ic}, critical stress intensity factor for mode I deformation $(F/L^{3/2})$; L–T, crack is perpendicular to longitudinal direction and parallel to transverse direction; T–L, crack is perpendicular to transverse direction and parallel to longitudinal direction; S–L, crack is perpendicular to thickness and parallel to rolling direction.

[b]From Bradt et al. [7.40].

[c]From Ingelstrom and Nordberg [7.41].

[d]From Schmidt [7.42].

[e]From Marshall and Williams [7.43].

[f]K_{Ic} is a function of crack speed.

[g]From Marshall et al. [7.44].

[h]From Rodon [7.45].

TABLE 7-3 FATIGUE ENDURANCE RATIO σ_e/σ_u AND MAXIMUM FATIGUE LIMIT $(\sigma_e)_{max}$ FOR VARIOUS CLASSES OF ENGINEERING MATERIALS[a]

Notation

σ_e = fatigue endurance limit (F/L^2)
σ_u = ultimate tensile strength (F/L^2)

Material	σ_e/σ_u	$(\sigma_e)_{max}$ (MPa)
Steels	0.35–0.60	784.5
Cast irons	0.30–0.50	196.1
Al alloys	0.25–0.50	196.1
Mg alloys	0.30–0.50	147.1
Cu alloys	0.25–0.50	245.2
Ni alloys	0.30–0.50	392.3

[a] From Buch [7.9], with permission.

TABLE 7-3 | Fatigue Endurance and Maximum Fatigue Limit | 350

TABLE 7-4   S–N CURVES FOR SOME STEELS

Notation

S is completely reversed stress, or fatigue strength (ksi),

N is cycles to failure,

$S = 10^{a+b \log N}$ or $\log S = a + b \log N$ where $10^3 \le N \le 10^6$.

Load Type	Process	Coefficient a	Coefficient b
Hardness of Steel: 160–187 HB			
Bending	Polished	2.105	− 0.082
	Ground	2.159	− 0.0999
	Machined	2.231	− 0.124
	Hot rolled	2.321	− 0.154
	Forged	2.430	− 0.190
Axial	Polished	2.195	− 0.112
	Ground	2.240	− 0.127
	Machined	2.309	− 0.150
	Hot rolled	2.369	− 0.170
	Forged	2.550	− 0.230
Torsional	Polished	2.192	− 0.138
	Ground	2.258	− 0.160
	Machined	2.309	− 0.177
	Hot rolled	2.399	− 0.207
	Forged	2.555	− 0.259
Hardness of Steel: 269–285 HB			
Bending	Polished	2.336	− 0.086
	Ground	2.378	− 0.100
	Machined	2.535	− 0.147
	Hot rolled	2.648	− 0.190
	Forged	2.828	− 0.250
Axial	Polished	2.351	− 0.099
	Ground	2.387	− 0.111
	Machined	2.474	− 0.140
	Hot rolled	2.669	− 0.205
	Forged	2.804	− 0.250
Torsional	Polished	2.402	− 0.134
	Ground	2.459	− 0.153
	Machined	2.564	− 0.188
	Hot rolled	2.720	− 0.240
	Forged	2.900	− 0.300

TABLE 7-5 *S–N* CURVES FOR ALUMINUM
UNDER COMPLETELY REVERSED BENDING

Notation

S is completely reversed stress, or fatigue strength (ksi),
N is cycles to failure,
$S = 10^{a+b \log N}$ or $\log S = a + b \log N$ where $10^5 \leq N \leq 5 \times 10^8$.

Type of Aluminum			Coefficient a	Coefficient b
Wrought	1100	O	1.179	-0.05518
	1100	H12	1.391	-0.070
	1100	H14	1.394	-0.063
	1100	H16	1.471	-0.065
	1100	H18	1.627	-0.0794
	3003	O	1.306	-0.053
	3003	H12	1.545	-0.0738
	3003	H14	1.542	-0.0675
	3003	H16	1.56	-0.066
	3003	H18	1.649	-0.074
	2014	T4	2.036	-0.091
	2014	T6	2.076	-0.1036
	2017	T4	2.053	-0.085
Cast	142	T21	1.65	-0.0949
	142	T61	1.706	-0.081
	195	T4	1.964	-0.125
	220	T4	1.864	-0.117
	319	F	1.656	-0.0754
	355	T51	1.75	-0.104
	355	T6	1.708	-0.0814
	356	T51	1.876	-0.115
	356	T6	1.846	-0.0972

TABLE 7-5 *S – N* Curves for Aluminum   352

TABLE 7-6   SURFACE TREATMENTS THAT INCREASE OR DECREASE FATIGUE STRENGTH[a]

Treatment	Fatigue Limit Increase (%)	Material
Shot peening and rolling of specimen without stress concentration	10–30	Steels and Al alloys
Shot peening and rolling of specimen with stress concentration	> 50	Steels and Al alloys
Carburizing	> 30	Steels for carburizing
Nitriding of specimen without stress concentration	10–30	Steels for nitriding
Nitriding of specimen with stress concentration	> 50	Steels for nitriding
Induction and flame hardening	> 30	Steels

Treatment	Fatigue Limit Decrease (%)	Material
Decarburizing	> 30	Spring Steels
Chromium and nickel plating	> 30	Steels $S_u \geq 981$ MPa
Al clad	15–35	High-strength Al alloys
Welding	> 30	Special steels and Al alloys

[a]From Buch [7.9], with permission.

TABLE 7-7 CORROSION FATIGUE LIMITS OF SELECTED STEELS[a]

Composition (%)				Tensile Strength (ksi)	Fatigue Limit in Air (ksi)	Corrosion Fatigue Limit (ksi)	
C	Ni	Cr	Condition			Well Water	Salt Water
0.03	—	—	Annealed	42	21	15	
0.03	—	—	Quenched, drawn	44	24	20	
0.11	—	—	Annealed	46	25	16	
0.14	—	—	Quenched	70	36	23	
0.24	—	—	Annealed	59	27	16	
0.26	—	—	Quenched, drawn	85	39	23	
0.36	—	—	Annealed	79	34	25	
0.36	—	—	Quenched, drawn	104	52	19	
0.49	—	—	Annealed	83	34	23	
0.49	—	—	Quenched, drawn	111	53	20	
1.09	—	—	Annealed	103	42	23	
0.32	3.5	—	Annealed	93	49	29	
0.32	3.5	—	Quenched, drawn	125	69	25	
0.52	5.4	—	Annealed	122	55	21	
0.52	5.4	—	Quenched, drawn	133	66	19	
0.08	—	12.2	Annealed	62	39	30	
0.08	—	12.2	Quenched, drawn	90	49	31	
0.11	—	11.8	Annealed	80	41	34	14
0.33	—	11.5	Annealed	98	52	35	
0.13	—	13.4	Quenched, drawn	91	59	40	13
0.09	—	13.8	Quenched, drawn	85	50	35	18
0.09	—	15.1	—	103	—	33	
0.24	—	20.5	—	111	—	48	
0.16	8.2	17.3	—	125	50	50	25
0.38	15.9	16.0	Annealed	126	64	51	
0.39	25.3	17.7	—	115	54	45	34
0.70	25.8	17.3	—	109	57	51	32
0.24	22.9	5.4	—	98	50	32	
0.24	22.1	6.0	Normalized	85	—	—	15
0.45	28.2	8.4	Annealed	113	58	42	20
0.39	34.7	10.9	—	112	57	41	22

[a]These steels are stressed for at least 10^7 cycles at 1450 cpm. From Juvinall [7.19], with permission.

TABLE 7-8   CORROSION FATIGUE LIMITS OF SELECTED NONFERROUS METALS[a]

Composition (%)				Condition	Tensile Strength (ksi)	Fatigue Limit in Air (ksi)	Corrosion-fatigue strength, ksi	
Ni	Cu	Al	Mn				Well Water	Salt Water
99	—	—	—	Annealed, 600°F	78	33.0	23.0	21.5
99	—	—	—	Annealed, 1400°F	132	51.0	26.0	22.0
68	30	—	—	Annealed, 800°F	82	36.5	26.0	28.0
68	30	—	—	Annealed, 1400°F	127	51.5	26.0	29.0
48	48	—	—	Cold rolled	78	31.5	22.0	25.0
48	48	—	—	Annealed, 1400°F	86	36.5	22.0	26.0
21	78	—	—	Annealed, 400°F	47	19.0	18.0	18.0
21	78	—	—	Annealed, 1400°F	62	25.5	21.5	23.5
	100	—	—	Annealed, 250°F	31	9.5	10.0	10.0
	100	—	—	Annealed, 1200°F	47	16.5	17.5	17.5
		99	—	Annealed	13	5.9	—	2.1
		99	—	Half hard temper	16	7.3	—	3.0
		99	—	Hard temper	21	8.4	5.0	3.0
		98	1.2	Annealed	17	6.8	3.2	—
		98	1.2	Half hard temper	24	10.1	5.5	4.0
		98	1.2	Hard temper	30	10.7	5.5	4.0
	4	94	—	Annealed	33	13.5	7.5	6.7
	4	94	—	Tempered	69	17.0	8.0	7.0

[a]These metals are stressed for 20,000,000 cycles at 1450 cpm. From McAdam [7.35]. Copyright ASTM, reprinted with permission.

TABLE 7-9 PARAMETERS OF FATIGUE CRACK PROPAGATION EQUATION $da/dN = A(\Delta k)^n$ FOR THREE CLASSES OF STEEL

Notation

a = half-length of flaw (in.)
N = number of load cycles;
ΔK = range of stress intensity factor during a cycle (ksi $\cdot$ $\sqrt{\text{in.}}$)
da/dN = fatigue crack growth rate (in./cycle)

Class	Range of Mechanical Properties			Parameter	
	Yield Strength σ_{ys} (ksi)	Tensile Strength σ_u (ksi)	Strain-hardening Exponent[a] α'	A [in./cycle/(ksi $\cdot$ $\sqrt{\text{in}}$)^n]	n
Austenitic, stainless	$30 < \sigma_{ys} < 50$	$75 < \sigma_u < 95$	$\alpha' > 0.3$	3.0×10^{-10}	3.25
Ferrite, pearlite	$30 < \sigma_{ys} < 80$	$50 < \sigma_u < 110$	$\alpha' > 0.15, \alpha' < 0.3$	3.6×10^{-10}	3.0
Martensitic	$\sigma_{ys} > 70$	$\sigma_u > 90$	$\alpha' < 0.15$	0.66×10^{-8}	2.25

[a]Beyond the yield point, the plot of true stress σ vs. true strain ε for many materials is given as $\sigma = \beta\varepsilon^{\alpha'}$, where α' is the strain-hardening exponent and β is a constant called the strength coefficient. By definition, σ = force/(instantaneous area) and ε = ln(instantaneous length/initial length).

TABLE 7-9 | Fatigue Crack Propagation Equation   356

TABLE 7-10 FATIGUE CRACK GROWTH DATA FOR VARIOUS MATERIALS[a]

Notation

Fatigue crack propagation equation is $da/dN = A(\Delta K)^n$,
where a = half-length of flaw (m),

N = number of load cycles

ΔK = range of stress intensity factor during cycle (MPa $\cdot$ $\sqrt{m}$)

da/dN = fatigue crack growth rate (m/cycle)

R^* is ratio of minimum to maximum stress in the load cycle.

Material	Tensile Strength (MN/m²)	0.1 or 0.2% Proof Stress (MN/m²)	R^*	n	$A \left[\dfrac{\text{m/cycle}}{(\text{MPa} \cdot \sqrt{m})^n} \right]$
Mild steel	325	230	0.06–0.74	3.3	2.43×10^{-12}
Mild steel in brine[b]	435	—	0.64	3.3	2.43×10^{-12}
Cold-rolled mild steel	695	655	0.07–0.43	4.2	2.51×10^{-13}
			0.54–0.76	5.5	3.68×10^{-14}
			0.75–0.92	6.4	2.62×10^{-14}
Low-alloy steel[b]	680	—	0–0.75	3.3	4.62×10^{-12}
Maraging steel[b]	2010	—	0.67	3.0	2.33×10^{-11}
18/8 austenitic steel	665	195–255	0.33–0.43	3.1	3.33×10^{-12}
Aluminum	125–155	95–125	0.14–0.87	2.9	4.56×10^{-11}
5% Mg–aluminum alloy	310	180	0.20–0.69	2.7	2.81×10^{-10}
HS30W aluminum alloy (1% Mg, 1% Si, 0.7% Mn)	265	180	0.20–0.71	2.6	1.88×10^{-10}
HS30WP aluminum alloy (1% Mg, 1% Si, 0.7% Mn)	310	245–280	0.25–0.43	3.9	2.41×10^{-11}
			0.50–0.78	4.1	4.33×10^{-11}
L71 aluminum alloy (4.5% Cu)	480	415	0.14–0.46	3.7	3.92×10^{-11}
L73 aluminum alloy (4.5% Cu)	435	370	0.50–0.88	4.4	3.82×10^{-11}
DTD 687A aluminum alloy (5.5% Zn)	540	495	0.20–0.45	3.7	1.26×10^{-10}
			0.50–0.78	4.2	8.47×10^{-11}
			0.82–0.94	4.8	1.68×10^{-10}
ZW1 magnesium alloy (0.5% Zr)	250	165	0	3.35	1.23×10^{-9}
AM503 magnesium alloy (1.5% Mn)	200	107	0.5	3.35	3.47×10^{-9}
			0.67	3.35	4.23×10^{-9}
			0.78	3.35	6.57×10^{-9}

Fatigue Crack Growth Data | TABLE 7-10

TABLE 7-10 (continued) FATIGUE CRACK GROWTH DATA FOR VARIOUS MATERIALS

Material	Tensile Strength (MN/m^2)	0.1 or 0.2% Proof Stress (MN/m^2)	R^*	n	$A\left[\dfrac{m/cycle}{(MPa \cdot \sqrt{m})^n}\right]$
Copper	215–310	26–513	0.07–0.82	3.9	3.38×10^{-12}
Phosphor bronze[b]	370	—	0.33–0.74	3.9	3.38×10^{-12}
60/40 brass[b]	325	—	0–0.33	4.0	6.35×10^{-13}
			0.51–0.72	3.9	3.38×10^{-12}
Titanium	555	440	0.08–0.94	4.4	6.89×10^{-12}
5% Al–titanium alloy	835	735	0.17–0.86	3.8	9.56×10^{-12}
15% Mo–titanium alloy	1160	995	0.28–0.71	3.5	2.14×10^{-11}
			0.81–0.94	4.4	1.17×10^{-11}
Nickel[b]	430	—	0–0.71	4.0	1.67×10^{-13}
Monel[b]	525	—	0–0.67	4.0	6.77×10^{-13}
Inconel[b]	650	—	0–0.71	4.0	2.21×10^{-13}

[a]From Pook [7.38]. © Crown Copyright, reproduced with permission of controller HMSO.
[b]Approximate results.

TABLE 7-10 | Fatigue Crack Growth Data   358

Joints

Joints consist of separate structural elements joined with fasteners or welds. Useful formulas and tables for the analysis and design of joints are provided in this chapter. Most commonly used in engineering structures and machines are riveted, bolted, and welded connections. Figure 8-1 illustrates these three kinds of joints.

8.1 NOTATION

A Cross-sectional area (L^2)

A_b Nominal bearing area, bolt cross-sectional area (L^2)

A_g Cross-sectional gross area (L^2)

A_e Effective net area (L^2)

A_n Net sectional area (L^2)

A_w Weld area (L^2)

d Nominal or major diameter of fastener (rivet or bolt) (L)

E Elastic modulus (Young's modulus) (F/L^2)

F_i Initial tensile force (F)

f_r Nominal resultant stress in weld (F/L^2)

g Transverse spacing (gage) (L)

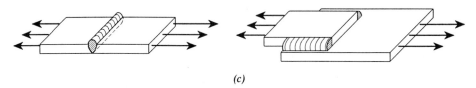

(c)

FIGURE 8-1 Common joints: (a) bolted; (b) riveted; (c) welded.

k	Stiffness constant (F/L)
ℓ	Distance (L)
M	Moment (FL)
P	Applied load (F)
P_e	External tensile load (F)
P_T	Total load-carrying capacity (F)
s	Longitudinal spacing, pitch (L)
t	Plate thickness (L)
T	Torque or twisting moment (FL)
w	Leg size of fillet weld (L)
τ	Shear stress (F/L^2)
τ_w	Allowable shear stress (F/L^2)
σ_p	Allowable bearing stress (F/L^2)
σ_{tw}	Allowable tensile stress (F/L^2)
σ_u	Ultimate tensile strength (F/L^2)
σ_{wa}	Allowable strength of particular type of weld (F/L)
σ_{ys}	Yield strength (F/L^2)

8.2 RIVETED AND BOLTED JOINTS

When joints are used for connecting members of structures such as building frames, trusses, or cranes, they are generally referred to as *connections*. Here we will not distinguish between these two terms.

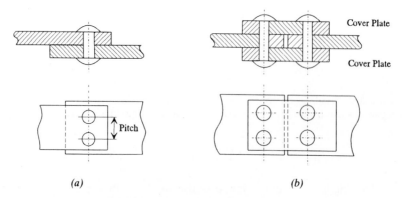

FIGURE 8-2 Rivet joints: (*a*) lap; (*b*) butt.

Rivets in connections are usually made from a soft grade of steel that does not become brittle when heated and hammered with a pneumatic riveting gun. They are manufactured with one formed head and are installed in holes that are $\frac{1}{16}$ in. larger in diameter than the nominal diameter of the rivet. When installing it, the head end of a rivet is held tightly against the pieces being joined while the opposite end is hammered until another similar head is formed. Steel rivets, as used in most connections, are usually heated to a cherry-red color (approximately 1800° F), and they can then be more easily driven. As the rivets cool, they shrink and squeeze the joined parts together. Copper and aluminum rivets used in aircraft engineering are generally driven cold.

The American Institute of Steel Construction (AISC) mandates that rivets shall conform to the provisions of "Standard Specifications for Structural Rivets," ASTM A502, Grade 1 or 2 [8.1]. The size of rivets used in general steel construction ranges in diameter from $\frac{5}{8}$ to $1\frac{1}{2}$ in. in $\frac{1}{8}$-in. increments.

Riveted joints are either lap or butt joints (Fig. 8-2). The lap joint has two plates that are lapped over each other and fastened together by one or more rows of rivets or fasteners. In the case of a butt joint, the edges of two plates are butted together and the plates are connected by cover plates.

A bolt is a threaded fastener with a head and a nut that screws on to the end without the head (Fig. 8-1*a*). The bolts most commonly used in steel construction are unfinished bolts (also called ordinary or common bolts) and high-strength bolts. Unfinished bolts are primarily used in light structures subjected to static loads. Unfinished bolts must conform to the specifications for low-carbon steel externally and internally threaded fasteners, ASTM A307 [8.1].

High-strength bolts are made from medium-carbon, heat-treated, or alloy steel and have tensile strengths several times greater than those of unfinished bolts. Specifications of the AISC state that high-strength bolts shall conform to "Specifications for Structural Joints Using ASTM A325 or A490 Bolts" [8.1].

Owing to better performance and economy in comparison to riveted joints, high-strength bolting has become the leading technique used for connecting structures in the field. In general, there are three types of connections made of

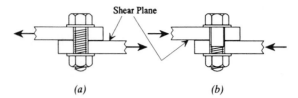

FIGURE 8-3 Bearing-type connections of high-strength bolts: (a) type N bolt, with threads included in shear plane; (b) type X bolt, with threads excluded from shear plane.

high-strength bolts:

1. The friction (F) connection, in which slip between the connected parts can not be tolerated and must be resisted by a high clamping force.
2. The bearing (N) connection, with threads included in the shear plane (Fig. 8-3a).
3. The bearing (X) connection, with threads excluded from the shear plane (Fig. 8-3b).

The allowable stresses recommended by the AISC *Manual of Steel Construction* [8.2] are given in Table 8-1. Generally, these stresses are based on the results of a large number of laboratory tests in which the ultimate strength of the rivets is determined. Division of the ultimate strength by a suitable factor of safety gives the allowable stresses.

Joint Failure Mode under Shear Loading

Four modes of failure for joints are normally considered.

(a) *Shearing* of the fastener in either single (one-sided) or double (two-sided) shear, depending on the type of joint (Fig. 8-4a): To prevent the shear failure of the fastener, the number of fasteners should be determined to limit the maximum shear stress in the critical fastener to the allowable stress listed in Table 8-1. The average shearing stress in a fastener (Fig. 8-4a) is

$$\tau = P_s/A = 4P_s/\pi d^2 \qquad (8.1)$$

where P_s is the load acting on the fastener's cross section subject to shear and A and d are the area and diameter, respectively, of the bolt or rivet cross section. It should be noticed that for double shear joints (Fig. 8-4a)

$$P_s = \tfrac{1}{2}P$$

(b) *Compression or bearing*, i.e., the crushing of either the fastener or the plate in front of it (Fig. 8-4b): To assure that no compression or bearing occurs due to the crushing force of the fasteners on the material, the minimum number of fasteners is determined.

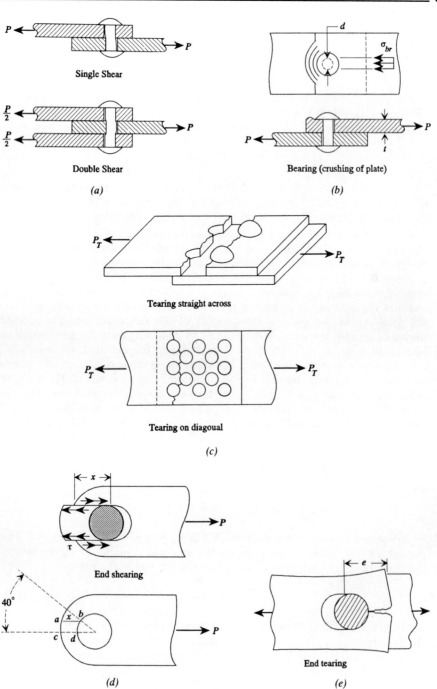

FIGURE 8-4 Failure modes of joints.

The bearing is assumed to be uniformly distributed over an area $A = td$ so that

$$\sigma_{br} = P/td \tag{8.2}$$

where P is the load, t is thickness, and d is the diameter of the shaft of the fastener, as shown in Fig. 8-4b.

The specifications of AISC recommend the allowable bearing stress σ_p to be

$$\sigma_p = \tfrac{1}{2}\sigma_u(s/d - 0.50) \le 1.50\sigma_u \tag{8.3}$$

where s is the distance between the centers of two fasteners in the direction of the stress and σ_u is the ultimate strength of the material. Equation (8.3) for σ_p has been provided in Fig. 8-5 for some ultimate strength values.

(c) *Tension* or *tearing* when a plate tears apart along some line of least resistance (Fig. 8-4c): To prevent this failure, the connected parts should be designed so that the tensile stress is less than $0.6\sigma_{ys}$ on the gross area A_g and less than $0.5\sigma_u$ on the effective net area (specifications of AISC), where the gross area A_g of a member is defined as the product of the thickness and the gross width of the member as measured normal to the tensile force σ (Fig. 8-6). The net area A_n of the plate is the product of the net width and the member thickness, and the net width is determined by deducting from the gross width the sum of all holes in the section cut. In computing the net area, the width of a fastener (rivet or bolt) hole shall be taken as $\tfrac{1}{16}$ in. larger than the normal dimension of the hole normal to the direction of applied stress. For standard holes, the hole diameter is $\tfrac{1}{16}$ in. larger than the nominal fastener size. Thus, for each hole, a value of fastener diameter $\tfrac{1}{16} + \tfrac{1}{16} = \tfrac{1}{8}$ in. should be used to calculate the net width and net area. For section 1–1 in Fig. 8-6, e.g., the net width W_n is

$$W_n = b - 2\left(d + \tfrac{1}{8}\right)$$

and

$$A_n = tW_n = t\left[b - 2\left(d + \tfrac{1}{8}\right)\right], \qquad A_g = tb$$

Sometimes a chain of holes is arranged in a zigzag pattern, as in Fig. 8-7. Then the net width is taken as the gross width minus the diameter of all the holes in the chain plus, for each "out-of-line" space in the chain, the value

$$s^2/4g \tag{8.4a}$$

where g is the gage (transverse spacing) in inches and s is the pitch (longitudinal spacing) in inches (Fig. 8-7). For example, for the chain of $ABGHI$ (or the section of $ABGHI$), where only one out-of-line spacing is in the chain, the net width is

$$W_n = b - 3\left(d + \tfrac{1}{8}\right) + s^2/4g \tag{8.4b}$$

and the net width for section $ABGCD$, where two out-of-line spacings are in the

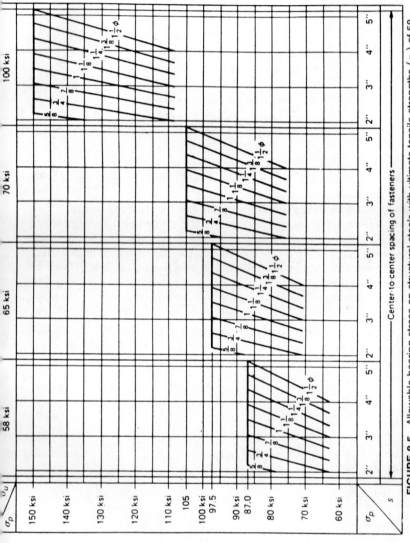

FIGURE 8-5 Allowable bearing stress on structural steels with ultimate tensile strengths (σ_u) of 58, 65, 70, and 100 ksi. The allowable bearing stress (σ_p) depends on the fastener diameter (ϕ) and center-to-center spacing (s), as well as the connected part's ultimate strength. In the figure, the numbers $\frac{5}{8}$ through $1\frac{1}{2}$ next to each of the plots, refer to the diameter of the fastener in question. From [8.7], with permission.

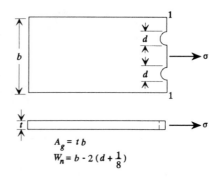

$$A_g = t\,b$$
$$W_n = b - 2\left(d + \tfrac{1}{8}\right)$$

FIGURE 8-6 Gross area A_g and net width W_n.

chain, is

$$W_n = b - 3\left(d + \tfrac{1}{8}\right) + s^2/4g + s^2/4g \qquad (8.4c)$$

The critical net section is taken at the chain that has the least net width.

If the tensile force is transmitted by fasteners (rivets or bolts) through some, but not all, of the cross-sectional elements of the member, the effective net area A_e must be computed [8.2]. This is to account for the effect of shear stress concentration in the vicinity of connections. The effective net area is defined by

$$A_e = C_t A_n \qquad (8.5)$$

where C_t is a reduction factor assumed to be 1.0 unless otherwise determined. Values of C_t recommended by AISC are listed in Table 8-2. The tensile stress can be obtained from

$$\sigma_g = P_T/A_g \qquad (8.6a)$$

$$\sigma_n = P_T/A_n \qquad (8.6b)$$

or

$$\sigma_e = P_T/A_e \qquad (8.6c)$$

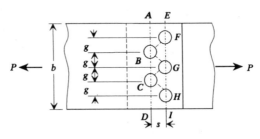

FIGURE 8-7 Zigzag pattern of holes.

where P_T is the total load on the connection member and $\sigma_g, \sigma_n, \sigma_e$ are the tensile stresses based on the gross, net, and effective areas, respectively.

(d) *End failure* including end shearing and end tearing (Figs. 8.4d, e): In the case of failure including shearing on the area xt (Fig. 8-4d), the shear stresses are

$$\tau = P/2xt \tag{8.7}$$

where P is the load acting at the hole, t is the thickness of the plate, and x is shown in Fig. 8-4d.

In actuality, the stress is probably more complicated. The AISC [8.2] recommends an experimentally determined formula. To prevent the failure mode of end tearing or shearing, the minimum edge distance e from the center of a fastener hole to the edge (Fig. 8-4e) in the direction of the force shall be greater than $2P/t\sigma_u$, i.e.,

$$e \geq 2P/t\sigma_u \tag{8.8}$$

where σ_u is the ultimate tensile strength. Table 8-3 gives some recommended e values. These edge distances are dependent on the joint type, the plate thickness, and the type of fastener.

The above analyses are based on the assumption that the stresses in fasteners or connecting members are uniform. This is not always true. When the stress is below the elastic limit, the true stresses are not necessarily equal to the average stress. Stress concentration may occur. Before the ultimate strength is reached, however, the material yields and stresses are redistributed so that they tend to approach uniform values. Because of plastic yielding of the material and because allowable stresses are obtained from tests on specimens similar to the actual structure, it is possible that this assumption is an acceptable approximation.

The four modes of failure analysis are suitable for riveted joints and N, X-type bolted joints. The only difference is in their corresponding allowable stresses for shearing and bearing (Table 8-1). For F-type connections, the design is based on the assumption that if the connection fails, the bolts will fail in shear alone, and the bearing stress of the fasteners on the connected parts need not be considered. Nevertheless, the bearing must be considered in the event the friction bolts slip and must resist bearing [8.2]. Formulas for the four modes of failure are summarized in Table 8-4.

Boiler Joints

When a riveted joint is to remain airtight under pressure, it is sometimes called a boiler joint. Special consideration is given to the analysis and design of this type of joint. The Boiler Code of the American Society of Mechanical Engineers gives the ultimate strength of boiler steel to be used for boilers and tanks and also recommends a factor of safety of 5. The efficiency of a riveted boiler joint is the ratio of the strength of the joint to the strength in tension of the unpunched plate.

Bolted Joints in Machine Design

In machine design, friction-type high-strength bolting is most commonly utilized. A bolt is tightened to develop a minimum initial tension in the bolt shank equal to about 70% of the tensile strength of the bolt. In this case, no interface slip occurs at allowable loads so that the bolts are not actually stressed in shear and are not in bearing. A discussion follows of the analysis of friction-type high-strength bolting used in machine design in which the tensile load of the bolt must be considered.

The *tightening load* is created in a bolt by exerting an initial torque on the nut or on the head of the bolt. For a *torqued-up* bolt, the tensile force in the bolt due to the torque can be approximated as [8.4]

$$T = cdF_i \tag{8.9}$$

where T is the tightening torque, c is a constant depending on the lubrication present, d is a nominal outside diameter of the shank of the bolt, and F_i is the initial tightening load in the bolt. When the threads of the bolt are well cleaned and dried, choose $c = 0.20$.

It is important to understand that when a load in the bolt shank direction is applied to a bolted connection over and above the tightening load, special consideration must be given to the behavior of the connection, which changes the allowable external load significantly. In the absence of an external load, the tensile force in the bolt is equal to the compressive force on the connected members. The external load will act to stretch the bolt beyond its initial length (Fig. 8-8). Thus, the resultant effect, which depends on the relative stiffness of the bolt and the connected members, is that only a part of the applied external load is carried by the bolt. The final tensile force F_b in the bolt and compressive force F_c in connected members can be obtained using [8.5]

$$F_b = F_i + k_b P_e / (k_b + k_c) \tag{8.10}$$

$$F_c = F_i - k_c P_e / (k_b + k_c) \tag{8.11}$$

where F_i is the initial tensile force in the bolt, P_e is the external load, and k_b and k_c are the stiffnesses of the bolt and the connected members, respectively (Fig. 8-8).

Since the external loading (P_e) is shared by the bolt and the connected members according to their relative stiffness (k_c/k_b), stiffness is usually given in the form of the ratio $k_r = k_c/k_b$, where k_r is the stiffness ratio. If the geometry for the bolt or for the connected members is simple [8.5],

$$k_b = A_b E_b / L_b \tag{8.12}$$

$$k_c = A_c E_c / L_c \tag{8.13}$$

where A_b, E_b, and L_b are the cross-sectional area, modulus of elasticity, and length of the bolt, respectively. The quantities A_c, E_c, and L_c are for the connected members.

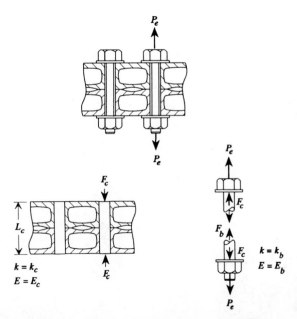

FIGURE 8-8 Inner force for bolted joint under tension.

8.3 LOAD ANALYSIS OF FASTENER GROUPS

A riveted or bolted joint may be subjected to a variety of forces. When the line of action of the resultant force that is to be resisted by the joint passes through the centroid c of the fastener group (Fig. 8-9), the joint is said to be *concentrically loaded*. Otherwise the joint is said to be *eccentrically loaded* (Fig. 8-10a). For a concentrically loaded connection, the load is assumed to be uniformly distributed among the fasteners. For an eccentrically loaded connection, the force may be

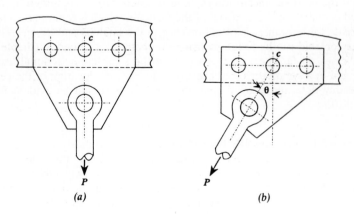

(a)   *(b)*

FIGURE 8-9 Concentrically loaded connections.

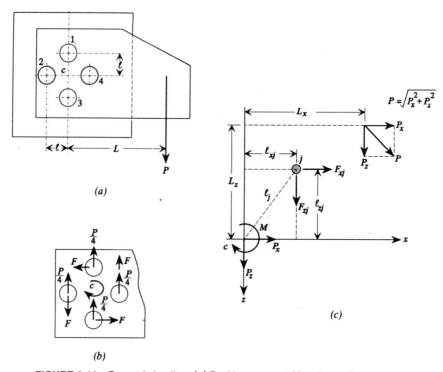

FIGURE 8-10 Eccentric loading. (c) Positive moment M and coordinate system.

replaced by an equal force at the centroid and a moment equal in magnitude to the force times its eccentricity. In this case, each fastener in the group is assumed to resist the force at the centroid uniformly and to resist the moment in proportion to its respective distance to the centroid of the fastener group.

Normally, the conditions of equilibrium, along with the above assumptions, permit the most significant forces in each fastener to be computed. In riveted and bolted connections the centroid of the fastener, which is sometimes referred to as the *center of resistance*, is of importance in the analysis and design. The location of the centroid can be found by using the method described in Chapter 2. But in most cases all fasteners of a group have the same cross-sectional area and are arranged in a symmetrical pattern; consequently, the location of the centroid can be readily determined by simple observation, as should be evident in Figs. 8-9 and 8-10a.

Example 8.1 Eccentrically Loaded Connection To illustrate the determination of fastener forces, consider the simple rivet group of the four symmetrically located rivets of Fig. 8-10a. It is assumed that each rivet takes $\frac{1}{4}P$ of the load (Fig. 8-10b). Also, a moment of magnitude PL is generated at the centroid of the rivet areas. This moment is resisted by the moment due to the rivet forces (see

Fig. 8-10b). This force in each rivet is F. Because the moments are in equilibrium,

$$PL - 4F\ell = 0 \tag{1}$$

Hence

$$F = PL/4\ell \tag{2}$$

The total force on each rivet is the vector sum of $\frac{1}{4}P$ and F. In Fig. 8-10b, it can be seen that the maximum resultant force occurs in the rivet closest to the eccentric load P, i.e.,

$$F_{max} = F + \tfrac{1}{4}P = (P/4\ell)(\ell + L) \tag{3}$$

There is another general, related method to find the moment-resisting forces F on each fastener. If the center of rotation of the eccentric moment PL is assumed to be the centroid of the fastener group, each fastener force F_j will be perpendicular to a line joining the fastener and the centroid c, and the magnitude of the force will be proportional to its distance ℓ_j from c. Therefore,

$$F_j = k\ell_j \tag{4}$$

where F_j is the force on fastener j due to eccentric loading and k is a constant for the fastener group.

From the equilibrium condition, $\Sigma M_c = 0$. For the connections shown in Fig. 8-10a,

$$PL = F_1\ell_1 + F_2\ell_2 + F_3\ell_3 + F_4\ell_4 \tag{5}$$

where F_1, F_2, F_3, F_4 are the moment-resisting forces on rivet 1, 2, 3, 4, respectively, and $\ell_1, \ell_2, \ell_3, \ell_4$ are the distances from each rivet to the centroid c.

From (4) and (5)

$$PL = k(\ell_1^2 + \ell_2^2 + \ell_3^2 + \ell_4^2) = k\sum_{j=1}^{4}\ell_j^2 \tag{6}$$

so that

$$k = \frac{PL}{\sum\limits_{j=1}^{4}\ell_j^2} \tag{7}$$

Thus, F_j in (4) can be obtained by use of k of (7). From Fig. 8-10a

$$k = PL/4\ell^2 \tag{8}$$

which can be substituted into (4) to give

$$F_1 = F_2 = F_3 = F_4 = F = PL/4\ell \qquad (9)$$

This is, of course, the same as (2).

In more general terms Eq. (7) of Example 8.1 would be written

$$k = \frac{PL}{\displaystyle\sum_{j=1}^{n} \ell_j^2} \qquad (8.14)$$

where L is the eccentric distance of load P, n is the number of fasteners in a connection, and k is the proportional constant of the fastener group under load P. It is often convenient to use the components F_x and F_z of force F in vectorial summation with the direct force P/n. The formulas are

$$k = \frac{M}{\displaystyle\sum_{j=1}^{n} \left(\ell_{xj}^2 + \ell_{zj}^2 \right)} \qquad (8.15)$$

where $M = PL$, $L = \sqrt{L_x^2 + L_z^2}$, and $\ell_j = \sqrt{\ell_{xj}^2 + \ell_{zj}^2}$ and for coordinate system xz, as shown in Fig. 8-10c,

$$F_{xj} = \frac{-M\ell_{zj}}{\displaystyle\sum_{j=1}^{n} \left(\ell_{xj}^2 + \ell_{zj}^2 \right)} \qquad (8.16a)$$

$$F_{zj} = \frac{M\ell_{xj}}{\displaystyle\sum_{j=1}^{n} \left(\ell_{xj}^2 + \ell_{zj}^2 \right)} \qquad (8.16b)$$

Therefore, the components F_{Rxj} and F_{Rzj} of the resultant force F_{Rj} on the fastener j are

$$F_{Rxj} = \frac{P_x}{n} + \frac{-M\ell_{zj}}{\displaystyle\sum_{j=1}^{n} \ell_{xj}^2 + \displaystyle\sum_{j=1}^{n} \ell_{zj}^2} \qquad (8.17a)$$

$$F_{Rzj} = \frac{P_z}{n} + \frac{M\ell_{xj}}{\displaystyle\sum_{j=1}^{n} \ell_{xj}^2 + \displaystyle\sum_{j=1}^{n} \ell_{zj}^2} \qquad (8.17b)$$

Thus

$$F_{Rj} = \sqrt{F_{Rxj}^2 + F_{Rzj}^2} \qquad (8.17c)$$

8.4 DESIGN OF RIVETED AND BOLTED CONNECTIONS

Based on the considerations and analysis described above, design of a riveted or bolted joint under a given load involves

1. determining the type of the joint and number of fasteners and
2. the use of predetermined allowable stresses in order to find the required area in shear, bearing, and tension for the fasteners and plates (connected parts) to be used.

Example 8.2 Load Capacity for a Member in Tension A bolted connection consists of a $4 \times 4 \times \frac{1}{2}$ angle with three bolts of diameter $\frac{3}{4}$ in. and a strong plate B, as shown in Fig. 8-11. Determine the tension resistance capacity of the angle. Assume the angle is made of AISI 1015 steel.
The gross area of the $4 \times 4 \times \frac{1}{2}$ angle is

$$A_g = 3.75 \text{ in.}^2 \qquad (1)$$

From failure mode 3 of Table 8-4

$$A_n = t\left[b - \left(d + \tfrac{1}{8}\right)\right] = A_g - \left(d + \tfrac{1}{8}\right)t = 3.75 - \left(\tfrac{3}{4} + \tfrac{1}{8}\right)\left(\tfrac{1}{2}\right) = 3.31 \text{ in.}^2 \quad (2)$$

For 1015 steel, from Table 4-9

$$\sigma_{ys} = 45.5 \text{ ksi}, \qquad \sigma_u = 61.0 \text{ ksi} \qquad (3)$$

and $A_e = C_t A_n = 0.85 \cdot 3.31 = 2.81 \text{ in.}^2$ ($C_t = 0.85$ from Table 8-2)
As mentioned in Section 8.2, to prevent failure of the angle, the tensile stress should be less than $0.6\sigma_{ys}$ on the gross area A_g and less than $0.5\sigma_u$ on the

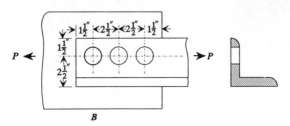

FIGURE 8-11 Example 8.2.

effective net area A_e, i.e.,

$$P/A_g \le 0.6\sigma_{ys} \quad \text{and} \quad P/A_e \le 0.5\sigma_u \qquad (4)$$

From (4)

$$P \le 0.6\sigma_{ys}A_g = (0.6)(45.5)(3.75) = 102.38 \text{ kips} \qquad (5)$$

and

$$P \le 0.5\sigma_u A_e = (0.5)(61.0)(2.81) = 85.71 \text{ kips} \qquad (6)$$

The tension resistance capacity P is the smaller value of these two: $P = 85.71$ kips.

Example 8.3   Load Capacity of a Riveted Connection   A riveted lap connection consists of two $\frac{3}{4} \times$ 12-in. plates of A36 steel and $\frac{7}{8}$-in.-diameter A502 Grade 1 rivets, as shown in Fig. 8-12. Determine the maximum tensile load P that can be resisted by the connection.

The rivets may fail in shear or bearing or the plate may fail in tension or bearing. Any failure will mean the failure of the connection. Thus each condition must be considered to determine the critical condition and the load capacity of the connection.

(a) Shear failure of the rivet: The total force P_T is calculated as

$$P_T = \tau_w AN \qquad (1)$$

where τ_w is the allowable shear stress of the rivet, A the cross-sectional area of each rivet, and N the number of the rivets in the connection.

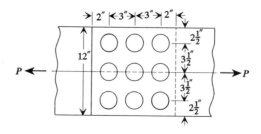

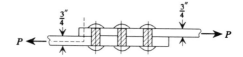

FIGURE 8-12   Example 8.3.

For A502 grade 1 rivets, Table 8-1 provides $\tau_w = 17.5$ ksi. Since

$$A = \tfrac{1}{4}\pi d^2 = \tfrac{1}{4}(3.14)(\tfrac{7}{8} \text{ in.})^2 = 0.601 \text{ in.}^2 \quad \text{and} \quad N = 9$$

it follows from (1) that

$$P_T = (17.5)(0.601)(9) = 94.66 \text{ kips} \tag{2}$$

(b) Bearing and end failure:

$$P_T = \sigma_p A_b N \tag{3}$$

where σ_p is the allowable bearing stress of the plate and A_b is the bearing area of each rivet: $A_b = dt$ (d is rivet diameter, t is thickness of the plate).
From Eq. (8.3) the allowable bearing stress σ_p is calculated as

$$\sigma_p = \tfrac{1}{2}\sigma_u(s/d - 0.5) \le 1.5\sigma_u \quad \text{between two fasteners} \tag{4}$$

For A36 steel, $\sigma_u = 58.0$ ksi,

$$\sigma_p = \left(\frac{58}{2}\right)\left(\frac{3}{7/8} - 0.5\right) = 84.9 \text{ ksi} \le (1.5)(58) = 87.0 \text{ ksi}$$

From (3)

$$P_T = 84.9(\tfrac{7}{8})(\tfrac{3}{4})\,9 = 501.44 \text{ kips} \tag{5}$$

For the fasteners near the end, from Eq. (8.8)

$$P = \tfrac{1}{2}\sigma_u te = (\tfrac{58}{2})(\tfrac{3}{4})2 = 43.5 \text{ kips}$$
$$P_T = PN = 391.5 \text{ kips} \tag{6}$$

Select the smaller value from (5) and (6), $P_T = 391.5$ kips.
(c) Tensile capacity: On the gross cross-sectional area, the allowable tensile stress σ_{tw} is [from failure mode (c) in Section 8.2 with $\sigma_{ys} = 36.0$ ksi for A36 steel)

$$\sigma_{tw} = 0.6\sigma_{ys} = (0.6)(36.0) = 21.6 \text{ ksi}$$

Since $A_g = (\tfrac{3}{4} \text{ in.})(12 \text{ in.}) = 9 \text{ in.}^2$, the tensile capacity of the gross area is

$$P_T = \sigma_{tw} A_g = (21.6)(9) = 194.4 \text{ kips} \tag{7}$$

On the effective net area

$$C_t = 1.0$$
$$\sigma_{tw} = 0.5\sigma_u = (0.5)(58) = 29.0 \text{ ksi} \quad \text{[failure mode (c) in Section 8.2]}$$
$$A_n = \left[12 - 3(\tfrac{7}{8} + \tfrac{1}{8})\right](\tfrac{3}{4}) = 6.75 \text{ in.}^2$$
$$A_e = A_n C_t = (6.75)(1.0) = 6.75 \text{ in.}^2 \quad \text{[Eq. (8.5)]}$$

Therefore, the tensile capacity of the plate for the effective net area is

$$P_T = \sigma_{tw} A_e = (29.0)(6.75) = 195.75 \text{ kips} \qquad (8)$$

Maximum load capacity of the connection in Fig. 8-12 is the least value of P_T given in (2) and (5)–(7), i.e.,

$$P_{\max} = (P_T)_{shear} = 94.66 \text{ kips}$$

Example 8.4 Bolted Connection with Bolts in Double Shear and in Zigzag Patterns   Determine the maximum value of P that the bolted connection in Fig. 8-13 can carry using 1-in.-diameter A307 bolts. Use $\tau_w = 10.0$ ksi for bolts and $\sigma_u = 58.0$ ksi and $\sigma_{ys} = 36.67$ ksi for plates. Use a procedure similar to those in Example 8.3 to investigate the critical condition in each failure mode.

(a) P_T by shear resistance of the bolts: Since the bolts are in double shear (two cross sections are subjected to shear),the total load by shear is

$$P_T = 2\tau_w AN = (2)(10.0)\left[\left(\tfrac{1}{4}\pi\right)(1^2)\right](6) = 94.2 \text{ kips} \qquad (1)$$

(b) P_T by bearing resistance of the plate: It can be seen that bearing on the $\frac{3}{4}$-in. plate is more critical than on two $\frac{1}{2}$-in. plates:

$$P_T = \sigma_p \, dt \, N$$

where σ_p is the allowable bearing stress, d is the diameter of the bolt, and $t = \frac{3}{4}$ in. is the thickness of the plate. For the bolts inside the plates, from Eq. (8.3)

$$\sigma_p = \tfrac{1}{2}\sigma_u(s/d_b - 0.5) = \tfrac{1}{2}\sigma_u(4/1.0 - 0.5) = 1.75\sigma_u$$

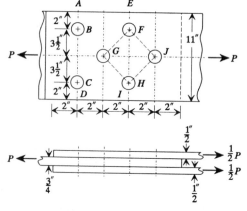

FIGURE 8-13   Example 8.4.

However, the condition $\sigma_p \leq 1.5\sigma_u$ should be met, so $\sigma_p = 1.5\sigma_u = 1.5(58) = 87.0$ ksi:

$$P_T = \sigma_p \, dt \, N = (87.0)(1.0)(\tfrac{3}{4})(6) = 391.5 \text{ kips} \tag{2}$$

For the bolts near the end, from Eq. (8.8)

$$P = \tfrac{1}{2}\sigma_u te = (\tfrac{58}{2})(\tfrac{3}{4})2 = 43.5 \text{ kips}$$
$$P_T = 43.5(6) = 261.0 \text{ kips} \tag{3}$$

(c) P_T by tension of the plate:

Cross section $ABCD$: $\quad W_n = 11 - 2(1.0 + \tfrac{1}{8}) = 8.75$ in. $\tag{4}$

Cross section $EFGHI$: $\quad W_n = 11 - 3(1.0 + \tfrac{1}{8}) + 2\left(\dfrac{2^2}{4 \times 3.5}\right) = 8.20$ in.
$$\tag{5}$$

It is seen that all other section lengths are between the values of W_n of (4) and (5). The least value is $W_n = 8.20$ in. The maximum tensile load on the gross area is

$$P_T = 0.6\sigma_{ys} A_g = (22.0)(\tfrac{3}{4})(11) = 181.5 \text{ kips} \tag{6}$$

and on the effective net area

$$P_T = 0.5\sigma_u A_e = (29.0)\left[(1.0)(\tfrac{3}{4})(8.20)\right] = 178.35 \text{ kips} \tag{7}$$

Comparison of the values of P_T in the above cases indicates that (1) governs the connection, i.e.,

$$P_{\text{max}} = 94.2 \text{ kips}$$

====

Example 8.5 Analysis of Eccentrically Loaded Riveted Joints The riveted joint shown in Fig. 8-14a is loaded with 10 kips at a distance of 8 in. from a vertical axis passing through the centroid c of the rivet group that fastens the plate to a column flange. Find the required rivet diameter for an allowable shear stress of 11,000 psi. Assume that shear failure is the critical condition and all rivets have the same diameter.

Because of the symmetrical arrangement of the rivets, the centroid c of the rivet group can be located by observation. The load P generates a moment of $PL = 80$ kip-in. about the centroid (Fig. 8-14b), so that force $P = 10$ kips ($P_x = 0$, $P_z = 10$ kips) at the centroid is being resisted equally by all six rivets, and a twisting moment is resisted by the twisting forces of the rivets.

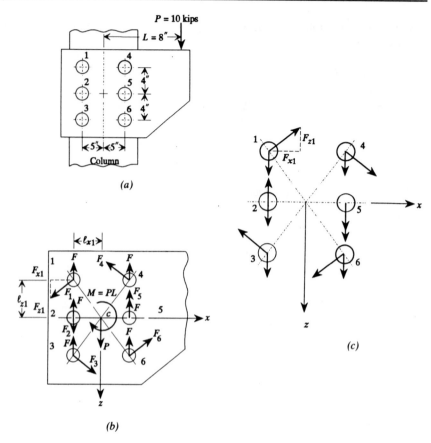

FIGURE 8-14 Eccentrically loaded riveted joint: (a) riveted joint; (b) free-body diagram of a plate; (c) rivet forces.

From the given dimensions

$$\sum_{j=1}^{6} \left(\ell_{xj}^{2} + \ell_{zj}^{2} \right) = 4(4^{2} + 5^{2}) + 2(5^{2} + 0^{2}) = 214 \text{ in.}^{2} \tag{1}$$

where ℓ_{xj} and ℓ_{zj} are the coordinates of rivet j. Therefore, from Eqs. (8.16), the components of the twisting forces on the rivets are

$$F_{x1} = \frac{-M\ell_{z1}}{\displaystyle\sum_{j=1}^{6} \left(\ell_{xj}^{2} + \ell_{zj}^{2} \right)} = \frac{-80(-4)}{214} = +1.495 \text{ kips} \tag{2}$$

$$F_{z1} = \frac{M\ell_{x1}}{\displaystyle\sum_{j=1}^{6} \left(\ell_{xj}^{2} + \ell_{zj}^{2} \right)} = \frac{80(-5)}{214} = -1.869 \text{ kips} \tag{3}$$

Similarly,

$$F_{x3} = -F_{x4} = F_{x6} = -F_{x1} = -1.495 \text{ kips} \tag{4}$$

$$F_{z3} = -F_{z4} = -F_{z6} = F_{z1} = -1.869 \text{ kips} \tag{5}$$

$$F_{x2} = F_{x5} = 0 \quad \text{and} \quad F_{z2} = -F_{z5} = 1.869 \text{ kips} \tag{6}$$

From Eqs. (8.17), the resultant force on rivet 1 is

$$F_{R1} = \left[(P_x/n + F_{x1})^2 + (P_z/n + F_{z1})^2 \right]^{1/2}$$

$$= \left[\left(\tfrac{0}{6} + 1.495 \right)^2 + \left(\tfrac{10}{6} - 1.869 \right)^2 \right]^{1/2} = 1.51 \text{ kips} \tag{7}$$

In a similar fashion, for rivets 4, 5, and 2

$$F_{R4} = 3.84 \text{ kips}, \qquad F_{R5} = 3.54 \text{ kips}, \qquad F_{R2} = 0.2 \text{ kips} \tag{8}$$

Also from Fig. 8-14c, it can be seen that

$$F_{R3} = F_{R1}, \qquad F_{R6} = F_{R4} \tag{9}$$

Thus, the maximum value of the shear force F_R on the rivet is F_{R4} or F_{R6}. The size of the rivets has to be determined for a shear force of 3.84 kips.

Assume that the rivet is in single shear with the given allowable shear stress of $\tau_w = 11$ ksi, and the shear failure is the critical condition. Since $(\tfrac{1}{4}\pi d^2)\tau_w = 3.84$, the rivet diameter d is calculated as

$$d = \sqrt{(3.84)(4)/(\pi\tau_w)} = 0.67 \text{ in.} \quad \text{or} \quad d = 17 \text{ mm} \tag{10}$$

Example 8.6 Torque Necessary to Draw Up a Nut in Machine Design A set of two bolts is to be used to provide a clamping force of 6000 lb between two bolted parts (Fig. 8-15). The joint is also subjected to an additional external load of 5000 lb. Assume (1) the forces are shared equally between the two bolts, (2) the stiffness of the bolted parts is three times that of the bolt [i.e., $k_c = 3k_b$ in Eqs. (8.10) and (8.11)], and (3) each bolt is stressed to 75% of its proof strength. Find the tensile force in the bolts and the necessary tightening torque for the nuts.

For the given conditions, the initial clamping load F_i on each bolt is $6000/2 = 3000$ lb, and the external load on each is $5000/2 = 2500$ lb. Then from Eq. (8.10), the final tensile force in one of the bolts is

$$F_b = F_i + \frac{k_b}{k_b + k_c} P_e = 3000 + \frac{k_b}{k_b + 3k_b}(2500) = 3625 \text{ lb} \tag{1}$$

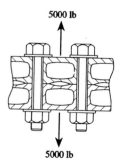

FIGURE 8-15 Bolt connection for Example 8.6.

while, from Eq. (8.11), the compressive force is

$$F_c = F_i - \frac{k_c}{k_b + k_c} P_e = 3000 - \frac{3k_b}{k_b + 3k_b}(2500) = 1125 \text{ lb} \qquad (2)$$

Thus, the compressive force in the bolted parts F_c is greater than zero, which indicates that the joint is still tight.

If a bolt made from SAE grade 4 steel (Table 8-5) is chosen, it will have a *proof strength* of 65,000 psi. Then the allowable tensile stress of the bolt is $\sigma_w = (0.75)(65,000) = 48,750$ psi and the required tensile stress area for the bolt is

$$A_t = \frac{F_b}{\sigma_w} = \frac{3625 \text{ lb}}{48,750 \text{ lb/in.}^2} = 0.0744 \text{ in.}^2 \qquad (3)$$

From Table 8-6, it can be seen that the $\frac{3}{8}$, 16 UNC thread has the required tensile stress area, since $A_t = 0.785(0.3750 - 0.9743/n)^2 = 0.07745$ in.^{2} is larger than A_t of (3). Thus, the necessary torque required would be, from Eq. (8.9),

$$T = 0.2dF_i = (0.20)(0.3750)(3000) = 225 \text{ lb-in.} \qquad (4)$$

8.5 WELDED JOINTS AND CONNECTIONS

In contrast to bolted and riveted joints, welded joints do not necessarily require an overlap in plates, thus affording more flexibility in design. Also, welded joints are usually lighter and are particularly advantageous in that they provide continuity between connected parts.

Types of Welded Joints and Typical Drawing Symbols

There are various types of welded joints, but the two main types are fillet and butt welds, shown in Fig. 8-16. The butt weld is usually loaded in tension, the strength of which is based on the net cross-sectional area of the thinner of the

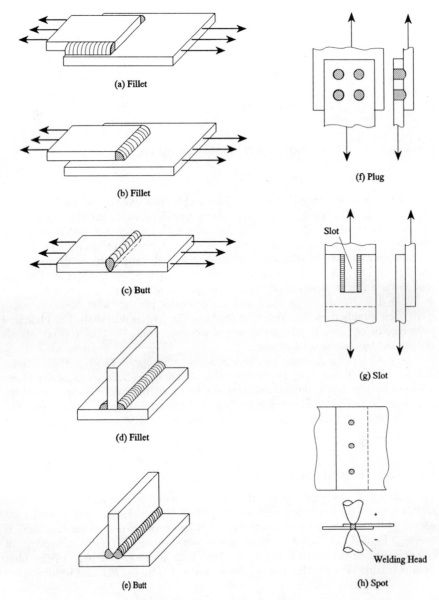

FIGURE 8-16 Types of welds.

two plates being joined. If the joint is properly made with the appropriate welding metal, the joint will be stronger than the parent metal. A fillet weld subjected to shear stress would tend to fail along the shortest dimension of the weld, which is referred to as the throat of the weld, shown in Figs. 8-17 and 8-18. In most cases the legs are of equal length, since welds with legs of different lengths are less efficient than those with equal legs.

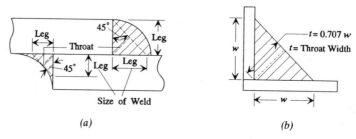

FIGURE 8-17 Notation for fillet welds.

Other types of welded joints are plug welds, slot welds, and spot welds, as shown in Fig. 8-16. Plug welds are made by punching holes into one of the two plates to be welded and then filling the holes with weld metal, which fuses into both plates. Slot welds can be used when other types of welded joints are not suitable and to provide additional strength to a fillet joint. Spot welds, which are used extensively in the fabrication of sheetmetal parts, are a quick and simple way to fasten light pieces together at intervals along a seam.

Welded joints are often used with various edge preparations, some of which have been qualified by the American Welding Society (AWS) [8.6]. The choice of joint type often directly affects the cost of welding. Thus, the choice is not always dominated by the design function.

The symbols representing the type of weld to be applied to a particular joint are shown in Table 8-7. These symbols, which have been standardized and adopted by the AWS, quickly indicate the exact welding details established for each joint to satisfy all necessary conditions of material strength and service. The symbols may be broken down into basic elements and combinations can be formed if desired.

Analysis of Welded Joints

For butt welds, as mentioned previously, the weld is stronger than the base metals and no further analysis is required [8.4]. However, the fillet welded joint needs to be analyzed to guarantee it is strong enough to sustain the applied loading. Four basic types of loading are considered here: direct tension or compression, direct vertical shear, bending, and twisting. The area of the fillet weld is calculated using leg size w, throat width t (Fig. 8-17b), and welded seam

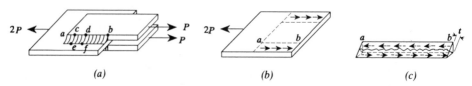

FIGURE 8-18 Stress distribution in side fillet welds. (c) Damage usually occurs along throat of weld.

length L_w. Usually, leg size w and throat width t are related, depending on the form of the welded joint. Table 8-8 gives the relationship between w and the plate thickness. Table 8-9 gives the allowable shear stresses and forces on welds.

The cross section along the welded seam, the width of which equals the throat width t, is called the effective cross section. The stress in the effective cross section should be less than the allowable stress.

The analysis of welded joints involves the following steps:

1. Draw the effective cross section of the welded connection. It is a narrow area along the weld seam width t. In the case of Fig. 8-17b, $t = 0.707w$. For example, in Fig. 8-19 the area enclosed by dashed lines represents the effective cross section of a [-shaped weld.

2. Let the centroid of the effective section be the origin and set up an orthogonal reference system x, y, z. If the normal stress is to be considered, select z, y axes as principal axes. The area and moments of inertia of the effective cross section of weld can be obtained by using Eqs. (2.1), (2.4), and (2.10), i.e.,

$$A_w = \int dA, \qquad I_y = \int z^2\, dA, \qquad I_z = \int y^2\, dA, \qquad J_x = \int (z^2 + y^2)\, dA$$

Some geometric properties of the welded connections are provided in Table 8-10.

3. Find the forces and moments that act on the welded connection. The positive directions of forces P_x, P_y, P_z and moments M_x, M_y, M_z are indicated in Fig. 8-19.

4. At any point of the connection, the stress on the weld due to a single component of load can be obtained from Eqs. (3.41), (3.46), (3.47), and (3.55). These stresses are summarized in Table 8-11.

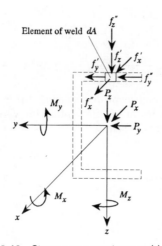

FIGURE 8-19   Stress components on weld area.

5. Determine the resultant nominal stress and the load force per unit length of weld. The nominal resultant stress f_r is the vector sum of stress components (Fig. 8-19)

$$f_r = \sqrt{f_x^2 + f_y^2 + f_z^2} = \sqrt{(f_x' + f_x'')^2 + (f_y' + f_y'')^2 + (f_z' + f_z'')^2} \quad (8.18)$$

The resultant force per unit length is $q_r = t f_r$.

It is assumed that all loads acting on a fillet weld are shear forces independent of their actual direction and the critical section is always the throat of the weld. The nominal stress f_r should be less than the allowable shear stress of the welding material (Table 8-9) to avoid failure.

The stress distribution within a fillet weld is complex due to such factors as eccentricity of the applied load, shape of the fillet, and notch effect of the root. However, the same conditions exist in the actual fillet welds tested and have been recorded as a unit force per unit length of weld as in Table 8-9.

An alternative solution to this problem is to calculate the principal stresses and the maximum shear stresses using the formulas of Chapter 3 instead of determining f_r, q_r and to check if the weld is strong enough by applying a failure theory discussed in Chapter 3. However, this method is more time consuming.

Example 8.7   Weld Joint   Determine the size of the required fillet weld for the bracket shown in Fig. 8-20, which carries a vertical load of 6000 lb.

Choose a [-shaped weld pattern. The weld will be subjected to direct vertical shear and twisting caused by the eccentric load P. Using case 5 of Table 8-10 gives the geometric properties of the weld treated as a line:

$$A_w = (2b + d)t = [(2)(5) + 8]t = 18t \text{ in.}^2$$

$$J_x = \frac{(2b + d)^3 t}{12} - \frac{b^2(b + d)^2 t}{2b + d}$$

$$= \frac{(18)^3 t}{12} - \frac{(25)(13)^2 t}{18} = 251.3t \text{ in.}^4$$

$$\bar{y} = \frac{b^2}{2b + d} = \frac{5^2}{18} = 1.39 \text{ in.}$$

Substitute these geometric properties into the proper formulas from Table 8-11 according to the types of loading to find the various forces on the weld.

The force due to vertical shear is

$$f_z' = \frac{V}{A_w} = \frac{P}{A_w} = \frac{6000}{18t} = \frac{333.3}{t} \quad \text{lb/in.}^2 \quad (1)$$

The twisting moment is

$$T = M_x = -PL = -P[6 + (5 - \bar{y})] = -(6000)(6 + 5 - 1.39)$$
$$= -57,660 \text{ lb-in.} \quad (2)$$

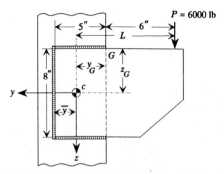

FIGURE 8-20 Weld joint for Example 8.7. Point c is centroid of weld pattern.

The moment M_x causes a force to be exerted on the weld that is perpendicular to a radial line from the centroid of the weld pattern to the point of interest. Thus, the maximum combined forces occur at point G (Fig. 8-20):

$$f_y'' = -\frac{M_x}{J_x}z_G = -\frac{(-57,660)(-4)}{251.3t} = -\frac{918}{t} \quad \text{lb/in.}^2$$

$$f_z'' = \frac{M_x}{J_x}y_G = \frac{(-57,660)(-5+1.39)}{251.3t} = \frac{828}{t} \quad \text{lb/in.}^2 \tag{3}$$

Superimpose the stress components

$$f_x = 0 \tag{4}$$

$$f_y = 0 + f_y'' = -\frac{918}{t} \quad \text{lb/in.}^2 \tag{5}$$

$$f_z = f_z' + f_z'' = \frac{333.3 + 828}{t} = \frac{1161.3}{t} \quad \text{lb/in.}^2 \tag{6}$$

so that the nominal stress becomes

$$f_r = \frac{\sqrt{(-918)^2 + 1161.3^2}}{t} = \frac{1480}{t} \quad \text{lb/in.}^2 \tag{7}$$

$$q_r = f_r t = 1480 \text{ lb/in.} \tag{8}$$

From (7) and (8) it is clear that for welded connections with uniform size, q_r can be computed by considering $t = 1$.

Suppose the base metals of the welded joints are ASTM A36 steel. From Table 8-9, if an E60 electrode is chosen for the welding, the allowable shear stress is 13,600 psi:

$$f_r = 1480/t \leq 13,600 \text{ psi}, \qquad t \geq 1480/13,600 \text{ in.}$$

Finally, the required leg size of the fillet weld connecting the bracket is

$$w = \frac{t}{0.707} = \frac{1480}{13,600 \times 0.707} = 0.154 \text{ in.}$$

Note that if the base-metal parts are thick plates, the leg size obtained above should be specified according to Table 8-8.

═══════════

REFERENCES

8.1 *Annual Book of ASTM Standards*, Vol. 15.08, American Society for Testing and Materials, Philadelphia, PA, 1986.

8.2 *Manual of Steel Construction*, 8th ed., American Institute of Steel Construction, Chicago, IL, 1980.

8.3 Jensen, A., and Chenoweth, H. H., *Applied Strength of Materials*, McGraw-Hill, New York, 1975.

8.4 Mott, R. L., *Machine Elements in Mechanical Design*, Charles E. Merrill, Columbus, OH, 1985.

8.5 Pilkey, W. D., and Pilkey, O. H., *Mechanics of Solids*, Krieger, Malabar, FL, 1986.

8.6 Blodgett, O. W., *Design of Welded Structures*, James F. Lincoln Arc Welding Foundation, Cleveland, OH, 1966.

8.7 Amon, R., Knoblock, B., and Mazumder, A., *Steel Design for Engineers and Architects*, Van Nostrand Reinhold, New York, 1982.

8.9 Faupel, J. H., and Fisher, F. E., *Engineering Design*, 2nd ed., Wiley, New York, 1981.

8

Tables

TABLE 8-1 ALLOWABLE STRESSES (SHEARING AND BEARING CAPACITIES) IN RIVETS AND BOLTS (ksi)[a]

Description of Fasteners	Allowable Tension[b] (σ_{tw})	Allowable Shears[b] (τ_w)			
		Friction-type Connections[c,d]			Bearing-type Connections[d]
		Standard Size Holes	Oversized and Short-slotted Holes	Long-slotted Holes	
A502, grade 1, hot-driven rivets	23.0[e]	—	—	—	17.5[f]
A502, grades 2, 3, hot-driven rivets	29.0[e]	—	—	—	22.0[f]
A307 bolts	20.0[e]	—	—	—	10.0[f,g]
Threaded parts meeting requirements of Sects. 1.4.1 and 1.4.4 or AISC and A449 bolts meeting the requirements of Sect. 1.4.4 of AISC when threads are not excluded from shear planes	$0.33F_u^{e,h,i}$	—	—	—	$0.17F_u^{h}$
Threaded parts meeting the requirements of Sects. 1.4.1 and 1.4.4. of AISC and A449 bolts meeting the requirements of Sect. 1.4.4 of AISC when threads are excluded from shear planes	$0.33F_u^{e,h}$	—	—	—	$0.22F_u^{h}$

TABLE 8-1 Allowable Stresses in Rivets and Bolts   388

A325 bolts, when threads are not excluded from shear planes	44.0^j	17.5	15.0	12.5	21.0^f
A325 bolts, when threads are excluded from shear planes	44.0^j	17.5	15.0	12.5	30.0^f
A490 bolts, when threads are not excluded from shear planes	54.0^j	22.0	19.0	16.0	28.0^f
A490 bolts, when threads are excluded from shear planes	54.0^j	22.0	19.0	16.0	40.0^f

[a] From AISC [8.2]. *Note:* F_u is the ultimate tensile strength, i.e., $\sigma_u = F_u$, but F_u is usually used by AISC.

[b] See Sect. 1.5.6 of AISC.

[c] When specified by the designer, the allowable shear stress, F_v, for friction-type connections having special faying surface conditions may be increased to the applicable value given in appendix E of AISC.

[d] For limitations on use of oversized and slotted holes, see Sect. 1.23.4 of AISC.

[e] Static loading only.

[f] When bearing-type connections used to splice tension members have a fastener pattern whose length, measured parallel to the line of force, exceeds 50 inches, tabulated values shall be reduced by 20 percent.

[g] Threads permitted in shear planes.

[h] See Appendix A, Table 2 of AISC, for values for specific ASTM steel specifications.

[i] The tensile capacity of the threaded portion of an upset rod, based upon the cross-sectional area at its major thread diameter, A_b, shall be larger than the nominal body area of the rod before upsetting times $0.60F_y$.

[j] For A325 and A490 bolts subject to tensile fatigue loading, see Appendix B, Sect. B3 of AISC.

Allowable Stresses in Rivets and Bolts TABLE 8-1

TABLE 8-2   REDUCTION COEFFICIENT C_t BASED ON AISC

Shape	Number of Fasteners	C_t
1. W, M, or S shapes where $b_f \geq \frac{2}{3}d$; tee sections from above shapes	3 or more	0.90
2. W, M, or S shapes not meeting above conditions and all other shapes including built-up sections	3 or more	0.85
3. All members	2	0.75

Adapted from AISC [8.2]. b_f, flange width; d, member depth.

TABLE 8-3   MINIMUM DISTANCE FROM CENTER OF STANDARD HOLE TO EDGE OF CONNECTED PART[a]

Nominal Rivet or Bolt Diameter (in.)	At Sheared Edges (in.)	At Rolled Edges of Plates, Shapes, Bars or Gas Cut Edges[b] (in.)
$\frac{1}{2}$	$\frac{7}{8}$	$\frac{3}{4}$
$\frac{5}{8}$	$1\frac{1}{8}$	$\frac{7}{8}$
$\frac{3}{4}$	$1\frac{1}{4}$	1
$\frac{7}{8}$	$1\frac{1}{2}$ [c]	$1\frac{1}{8}$
1	$1\frac{3}{4}$ [c]	$1\frac{1}{4}$
$1\frac{1}{8}$	2	$1\frac{1}{2}$
$1\frac{1}{4}$	$2\frac{1}{4}$	$1\frac{5}{8}$
Over $1\frac{1}{4}$	$1\frac{3}{4}$ × diameter	$1\frac{1}{4}$ × diameter

[a]From AISC [8.2], with permission. For oversized or slotted holes, see Sect. 1.16.5.4 of AISC.
[b]All edge distances in this column may be reduced $\frac{1}{8}$ in. when the hole is at a point where stress does not exceed 25% of the maximum allowed stress in the element.
[c]These may be $1\frac{1}{4}$-in. at the ends of beam connection angles.

TABLE 8-3   **Minimum Edge Distance**   390

TABLE 8-4   MODES OF FAILURE OF RIVETED AND BOLTED JOINTS

Notation

A = cross-sectional area of rivets or bolts
A_e = effective area
A_n = net sectional area
A_g = gross area
C_t = reduction factor (Table 8-2)
σ_e = effective stress
σ_g = stress based on gross area

P = applied force
d = diameter of rivets or bolts
t = thickness of plate
w_n = net width
σ_{br} = bearing stress
σ_u = ultimate tensile strength
τ = shear stress

Mode of Failure	Connection	Strength Formula
1. Fastener shearing	a. Single shear 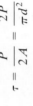	$\tau = \dfrac{P}{A} = \dfrac{4P}{\pi d^2}$
	b. Double shear 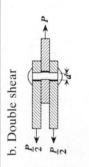	$\tau = \dfrac{P}{2A} = \dfrac{2P}{\pi d^2}$

TABLE 8-4 (continued) MODES OF FAILURE OF RIVETED AND BOLTED JOINTS

Mode of Failure	Connection	Strength Formula
2. Bearing		$$\sigma_{br} = \frac{P}{td}$$
3. Tension or tearing	a. Straight with no stagger	$\sigma_g = P/A_g$ $\sigma_n = P/A_n$ $\sigma_e = P/A_e$ where $A_e = C_t A_n$ $A_n = t w_n$ $w_n = b - \Sigma\left(d + \frac{1}{8}\right)$ $A_g = tb$

TABLE 8-4   Modes of Failure of Riveted and Bolted Joints   392

b. Stagger

$$\sigma_e = P/A_e, \quad \sigma_g = P/A_g$$

where $A_e = C_t A_n$, $A_g = tb$

$$A_n = tw_n$$

$$w_n \text{ (net width)} = b - \Sigma\left(d + \tfrac{1}{8}\right) + \Sigma \frac{s^2}{4g}$$

4. End failure

$$\tau = P/2xt$$

or

$$e \geq 2P/(t\sigma_u)$$

TABLE 8-5 SAE GRADES OF STEELS FOR BOLTS

Grade Number	Bolt Size (in.)	Tensile Strength (ksi)	Yield Strength (ksi)	Proof Strength[a] (ksi)	Head Marking
1	$\frac{1}{4}-1\frac{1}{2}$	60	36	33	None
2	$\frac{1}{4}-\frac{3}{4}$	74	57	55	None
	Over $\frac{3}{4}-1\frac{1}{2}$	60	36	33	—
4	$\frac{1}{4}-1\frac{1}{2}$	115	100	65	None
5	$\frac{1}{4}-1$	120	92	85	⬡
	Over $1-1\frac{1}{2}$	105	81	74	
	Over $\frac{1}{2}-3$	90	58	55	
5.2	$\frac{1}{4}-1$	120	92	85	⬡
7	$\frac{1}{4}-1\frac{1}{2}$	133	115	105	⬡
8	$\frac{1}{4}-1\frac{1}{2}$	150	130	120	⬡

[a]Defined as stress at which bolt will undergo permanent deformation: usually ranges between 0.90 and 0.95 times yield strength.

TABLE 8-5 SAE Grades of Steels for Bolts   394

TABLE 8-6 AMERICAN STANDARD THREAD DIMENSIONS[a]

Size	Basic Major Diameter (in.)	Coarse Threads UNC Threads per Inch	Fine Threads: UNF Threads per Inch	Extra Fine UNEF Threads per Inch
0	0.0600	—	80	
1	0.0730	64	72	
2	0.0860	56	64	
3	0.0990	48	56	
4	0.1120	40	48	
5	0.1250	40	44	
6	0.1380	32	40	
8	0.1640	32	36	
10	0.1900	24	32	
12	0.2160	24	28	32
$\frac{1}{4}$	0.2500	20	28	32
$\frac{5}{16}$	0.3125	18	24	32
$\frac{3}{8}$	0.3750	16	24	32
$\frac{7}{16}$	0.4375	14	20	28
$\frac{1}{2}$	0.5000	13	20	28
$\frac{9}{16}$	0.5625	12	18	24
$\frac{5}{8}$	0.6250	11	18	24
$\frac{3}{4}$	0.7500	10	16	20
$\frac{7}{8}$	0.8750	9	14	20
1	1.000	8	12	20
$1\frac{1}{8}$	1.125	7	12	18
$1\frac{1}{4}$	1.250	7	12	18
$1\frac{3}{8}$	1.375	6	12	18
$1\frac{1}{2}$	1.500	6	12	18
$1\frac{1}{4}$	1.750	5	—	18
2	2.000	$4\frac{1}{2}$	—	—

Abbreviations: UNC, unified coarse; UNF, unified fine; UNEF, unified extrafine. The smaller American Standard threads use a number designation from 0 to 12. The larger sizes use fractional inch designations.

[a]The tensile stress area A_t is given by

$$A_t = 0.785\left(d - \frac{0.9743}{n}\right)^2$$

where d is the basic major diameter and n is the number of threads per inch.

TABLE 8-7 TYPICAL AWS DRAFTING SYMBOLS FOR WELDED JOINTS

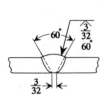

60° Angle vee groove
($\frac{3}{32}$ -in. root opening)

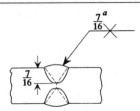

Double vee groove

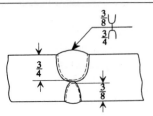

Double U-groove

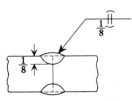

Single fillet

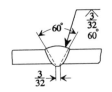

Single vee groove
(Complete penetration ;
welded both sides)

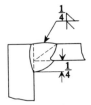

Outside single bevel
corner joint, fillet
weld

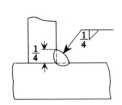

Closed square butt
joint ($\frac{1}{8}$ -in. penetration both sides)

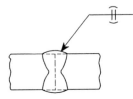

Closed square butt
joint (complete pene-
tration both sides)

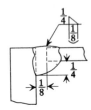

Open square-grooved
corner joint, fillet weld

Double-bevel corner
joint

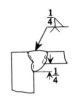

Single-vee corner joint,
fillet weld

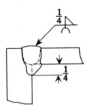

Single-U corner joint,
fillet weld

[a]Inches

TABLE 8-7   AWS Drafting Symbols for Welded Joints   396

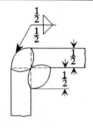

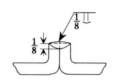

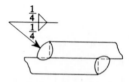

Double-fillet corner joint

Square-edge joint

Double-fillet lap joint

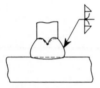

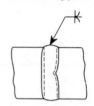

Single-bevel tee joint

Double-fillet, double-J tee joint

Double J-groove (full penetration)

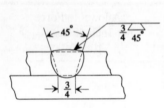

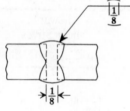

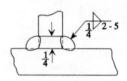

Plug weld

Open square butt joint ($\frac{1}{8}$ -in. root opening; complete penetration both sides)

Double fillet ; 2-in. welds on 5-in. centers opposite increments

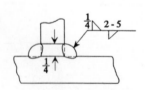

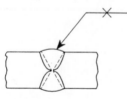

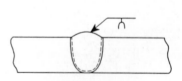

Double fillet, staggered increments

Double Vee Groove (Full penetration)

Single U-groove (full penetration)

TABLE 8-8 MINIMUM WELD SIZES FOR THICK PLATES

Plate Thickness		Minimum Leg Size (w) for Fillet Weld	
in.	mm	in.	mm
$\leq \frac{1}{2}$	≤ 12.7	$\frac{3}{16}$	4.76
$> \frac{1}{2}-\frac{3}{4}$	$> 12.7-19.1$	$\frac{1}{4}$	6.35
$> \frac{3}{4}-1\frac{1}{2}$	$> 19.1-38.1$	$\frac{5}{16}$	7.94
$> 1\frac{1}{2}-2\frac{1}{4}$	$> 38.1-57.2$	$\frac{3}{8}$	9.53
$> 2\frac{1}{4}-6$	$> 57.2-152.4$	$\frac{1}{2}$	12.70
> 6	> 152.4	$\frac{5}{8}$	15.88

TABLE 8-9 ALLOWABLE SHEAR STRESSES AND FORCES ON WELDS

Base-Metal ASTM Grade	Electrode	Allowable Shear Stress		Allowable Force per Inch of Leg (σ_{wa}) (lb/in.)
		(psi)	(MPa)	
Building-type Structures				
A36, A242, A441	E60	13 600	93.8	9 600
A36, A242, A441	E70	15 800	109	11 200
Bridge-type Structures				
A36	E60	12 400	85.5	8 800
A441, A242	E70	14 700	101	10 400

TABLE 8-9 Weld Sizes and Allowable Stresses   398

TABLE 8-10   GEOMETRIC PROPERTIES OF WELD SEAMS

Notation

M = applied moment
J_x = polar moment of inertia
T = twisting moment
Z_{ew} = elastic section modulus of the weld seam
t = width, = 1

Dimensions of Weld	Bending	Torsion
1. $A_w = d$	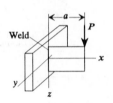 $Z_{ew} = \dfrac{1}{6}d^2$, $M = Pa$	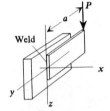 $J_x = \dfrac{1}{12}d^3$, $T = Pa$
2. $A_w = 2d$	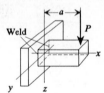 $Z_{ew} = \dfrac{1}{3}d^2$	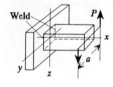 $J_x = \dfrac{(3b^2 + d^2)d}{6}$
3. 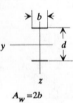 $A_w = 2b$	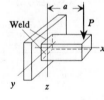 $Z_{ew} = bd$	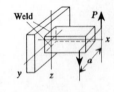 $J_x = \dfrac{1}{6}(b^3 + 3bd^2)$

TABLE 8-10 (continued)   GEOMETRIC PROPERTIES OF WELD SEAMS

Dimensions of Weld	Bending	Torsion

4.

$A_w = b + d$

$y_c = \dfrac{b^2}{2(b+d)}$

$z_c = \dfrac{d^2}{2(b+d)}$

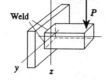
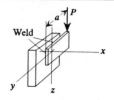

At top: $Z_{ew} = \dfrac{1}{6}(4bd + d^2)$

At bottom: $Z_{ew} = \dfrac{d^2(4b + d)}{6(2b + d)}$

$J_x = \dfrac{(b+d)^4 - 6b^2d^2}{12(b+d)}$

5.

$A_w = d + 2b$

$y_c = \dfrac{b^2}{2b+d}$

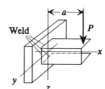
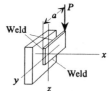

$Z_{ew} = bd + \dfrac{1}{6}d^2$

$J_x = \dfrac{1}{12}(2b+d)^3 - \dfrac{b^2(b+d)^2}{(2b+d)}$

6.

$A_w = b + 2d$

$z_c = \dfrac{d^2}{b+2d}$

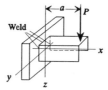
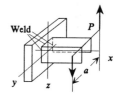

At top: $Z_{ew} = \dfrac{1}{3}(2bd + d^2)$

At bottom: $Z_{ew} = \dfrac{d^2(2b + d)}{3(b + d)}$

$J_x = \dfrac{1}{12}(b+2d)^3 - \dfrac{d^2(b+d)^2}{(b+2d)}$

7.

$z_c = \dfrac{d^2}{2(b+d)}$

$A_w = b + d$

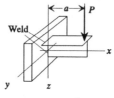
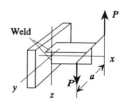

At top: $Z_{ew} = \dfrac{1}{6}(4bd + d^2)$

At bottom: $Z_{ew} = \dfrac{(4b + d)d^2}{6(2b + d)}$

$J_x = \dfrac{d^3(4b+d) + b^3(b+d)}{12(b+d)}$

TABLE 8-10   **Geometric Properties of Weld Seams**   400

TABLE 8-10 (continued)   GEOMETRIC PROPERTIES OF WELD SEAMS

Dimensions of Weld	Bending	Torsion

8.

$A_w = 2b + 2d$

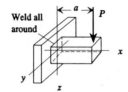

$$Z_{ew} = bd + \frac{1}{3}d^2$$

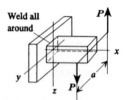

$$J_x = \frac{1}{6}(b+d)^3$$

9.

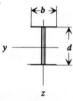

$A_w = 2d + 2b$

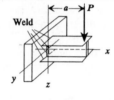

$$Z_{ew} = bd + \frac{1}{3}d^2$$

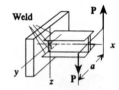

$$J_x = \frac{1}{6}(b^3 + 3bd^2 + d^3)$$

10.

$A_w = \pi d$

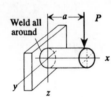

$$Z_{ew} = \frac{1}{4}\pi d^2$$

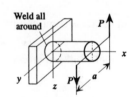

$$J_x = \frac{1}{4}\pi d^3$$

TABLE 8-11   FORMULAS FOR DETERMINING STRESSES IN WELDED JOINTS

Notation

f'_x, f'_y, f'_z = stress components of x, y, z direction due to external forces (Fig. 8-19)

f''_x, f''_y, f''_z = stress components of x, y, z direction due to external moments

f_x, f_y, f_z = algebraic sum of stress components in x, y, z direction

f_r = nominal resultant stress

q_r = resultant force per unit length

A_w = effective welded area

I_y, I_z = moments of inertia of welded area

J_x = polar moment of inertia

P_x, P_y, P_z = applied forces in x, y, z direction

$M_x = T$, M_y, M_z = applied moments in x, y, z direction

τ_w = allowable shear stress

t = effective throat dimension

Stress due to forces:

$$f'_x = P_x/A_w \qquad f'_y = P_y/A_w \qquad f'_z = P_z/A_w$$

Stress due to moments:

$$f''_x = \frac{M_y}{I_y}z - \frac{M_z}{I_z}y \qquad f''_y = -\frac{M_x}{J_x}z \qquad f''_z = \frac{M_x}{J_x}y$$

Sum of stress components:

$$f_x = f'_x + f''_x \qquad f_y = f'_y + f''_y \qquad f_z = f'_z + f''_z$$

Nominal resultant stress:

$$f_r = \sqrt{f_x^2 + f_y^2 + f_z^2}$$

Resultant force per unit of length:

$$q_r = tf_r$$

Design criterion:

$$f_r \le \tau_w$$

TABLE 8-11 | Stresses in Welded Joints   **402**

CHAPTER **9**

Contact Stresses

In most problems of stress analysis, the stress field is found within a solid body without regard for local effects caused by the application of the load; however, when two solid bodies with curved surfaces are forced together, consideration must be given to the special stress field created near the contact area. Gears and rolling-element bearings are two notable examples of machine parts in which contact stresses are of great importance in determining the operating life.

Contact problems are classified as *counterformal* if the dimensions of the area of contact are small compared to the radii of curvature of the contacting surfaces near the region of contact. If the dimensions of the contact area are not small with respect to the radii of curvature of the contacting surfaces, then the problem is classified as *conformal*. A counterformal problem is called Hertzian if the contacting surfaces can be approximated by quadratic functions in the region of contact. If the quadratic approximation is invalid, the problem is non-Hertzian. All conformal problems are non-Hertzian.

The following discussion presents an outline of the analysis of Hertzian contact stresses when two bodies with arbitrarily curved surfaces are pressed together. Charts and figures are included for use in solving various Hertzian problems. Rolling contact problems, contact stresses with friction, and the fatigue behavior of bodies subjected to repeated applications of contact loading are briefly described.

9.1 NOTATION

a semimajor axis of contact ellipse (L)

A, B Coefficients in equation for locus of contacting points initially separated by the same distance (L^{-1})

A_c Contact area (L^2)

b Semiminor axis of contact ellipse (L)

d Rigid distance of approach of contacting bodies (L); also total elastic deformation at origin

f Friction coefficient

F Force (F)

k Ratio of major to minor axis of contact ellipse

p Pressure (F/L^2)

q Line-distributed load (F/L)

R, R' Minimum and maximum radii of curvature for contacting surfaces (L)

z_s Distance below center of contact ellipse where maximum shear stress occurs (L)

σ_c Maximum compressive stress (F/L^2)

σ_{ys} Yield strength in tension (F/L^2)

τ_{max} Maximum shear stress (F/L^2)

θ Angle between planes containing principal radii of curvature for contacting bodies

Geometric Characteristics of Surfaces   Consider a surface $F(x, y, z) = 0$ (Fig. 9-1). At any point on the surface, the normal to the surface is grad(F). Let a plane pass through the length of the surface normal at point O creating a normal section. The intersection of this plane with the surface is a curve in the normal section of the surface at point O. Obviously, an infinite number of normal sections may be taken through any point on the surface. The following theorem holds [9.1]: At any point of a surface, two normal sections exist for which the radii of curvature are a minimum and a maximum; the planes that each contain one of these normal sections are perpendicular. The following terminology is adopted

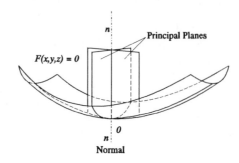

FIGURE 9-1   Geometric characteristics of surface.

here: the normal sections that have either a minimum or a maximum radius of curvature are called the *principal normal sections* of the surface at the point. The minimum and maximum radii of curvature are called the *principal radii of curvature of normal sections* at the point. The tangents to the curvatures in the principal normal sections at point O are called the *principal directions*, and the planes that create the principal normal sections are called the *principal planes* of curvature. Equations for computing the principal radii of curvature and principal directions at a point of a surface are presented in any treatise dealing with differential geometry [e.g., 9.1].

9.2 HERTZIAN CONTACT STRESSES

The first successful analysis of contact stresses is attributed to Hertz [9.2]. This analysis gave the dimensions of the contact area and the pressure distribution over that area. This information serves as a basis for computing the displacements and stresses in the neighborhood of the region of contact. Belajev [9.3] and Thomas and Hoersch [9.4] performed important calculations of the stress fields in contacting solids. Discussions of the analysis of contact stresses can be found in the literature [e.g., 9.5 and 9.6]. Tabulations of formulas applicable to special cases of contacting bodies can be found in such references as 9.7 and 9.8.

Two Bodies in Point Contact

Figure 9-2a shows the cross section of two solid bodies with curved surfaces that are in contact. Before a force is applied to press the bodies together, they touch at one point only. When a force F is applied, elastic compression occurs near the initial point of contact, and a flat area of contact is formed. This area is tangent to the undeformed surfaces of the two solids and is perpendicular to the line of action of the force F. The curvature of a surface is characterized at any point by the maximum and minimum values of the radii of curvature R' and R. The two planes are orthogonal and contain R' and R and the surface normal. A radius of curvature of the surface of a body is taken to be positive at a point if the corresponding center of curvature lies within the solid body; otherwise, the radius is negative. Quantities with the subscript 1 refer to the top body of Fig. 9-2a and those with the subscript 2 refer to the bottom solid. The two solids are assumed to be elastic, isotropic, and homogeneous; also, the contacting surfaces are smooth and free of frictional or adhesive forces. The four principal radii of curvature of the two surfaces at the point of contact are large compared to the dimensions of the contact area, and plastic deformation is ignored.

The coordinate system (x, y, z) is aligned such that the x, y plane lies tangent to the undeformed surfaces at the initial point of contact and such that the z axis coincides with the line of action of the force F. Before deformation, suppose the surfaces of the two bodies are approximately quadratic near the point of contact:

$$z_1 = A_1 x^2 + B_1 y^2 + C_1 xy \qquad (9.1)$$

$$z_2 = A_2 x^2 + B_2 y^2 + C_2 xy \qquad (9.2)$$

where z_1 and z_2 are the perpendicular distances from the tangent plane to any

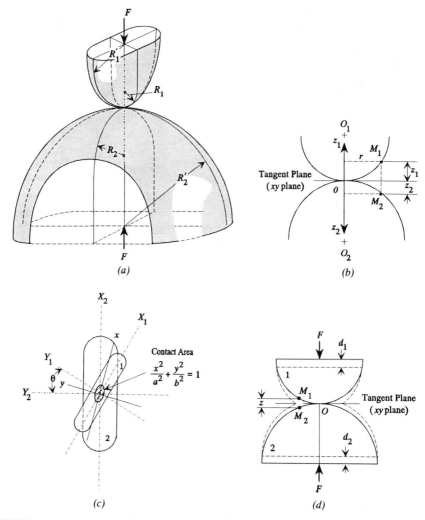

FIGURE 9-2 Two elastic solids in contact: (a) contact configuration; (b) before loading; (c) after loading, x, y axes coincide with major and minor axes of elliptical contact area (hatched area); (d) displacement of contacting points M_1 and M_2 and rigid distance of approach $d = d_1 + d_2$.

point on the surfaces of body 1 and body 2 near the point of contact, respectively, in the z direction (Fig. 9-2b). After deformation, two points that come into contact will have moved a distance

$$z_1 + z_2 = (A_1 + A_2)x^2 + (B_1 + B_2)y^2 + (C_1 + C_2)xy \qquad (9.3a)$$

Under the assumption that each pair of contacting points was initially on opposite ends of a line parallel to the z axis, all points with the same value of

$z_1 + z_2$ lie on an ellipse, and the perimeter of the contact area is elliptical. To eliminate the cross term in Eq. (9.3a), the x, y coordinates may be rotated to coincide with the major and minor axes of the elliptical contact area (Fig. 9-2c). Thus Eq. (9.3a) can be rewritten as

$$z_1 + z_2 = Ax^2 + By^2 \tag{9.3b}$$

where $A = A_1 + A_2$ and $B = B_1 + B_2$.

Far from the contact area, material points of the two bodies are unaffected by elastic compressive deformation. These two regions will approach each other by a constant distance d. This distance is a net rigid-body displacement of the two regions. Let w_1 and w_2 denote the local elastic displacements of points on the contacting surfaces. Take w_1 and w_2 as positive for compressive displacements, i.e., for displacements into the original configuration of the solid on the surface of which the point lies. The displacement of contacting points is given by

$$d - (w_1 + w_2) = z_1 + z_2 = Ax^2 + By^2 \tag{9.4}$$

where $d = d_1 + d_2$ (Fig. 9-2d). This d is referred to as the *rigid approach of two bodies*. From geometric considerations [9.9] the constants A and B are functions of the four principal radii of curvature of the two undeformed surfaces and of the orientation of the principal planes of curvature of body 1 with respect to those of body 2 (Fig. 9-2c):

$$
\begin{aligned}
A = \frac{1}{4} & \left(\frac{1}{R_1} + \frac{1}{R_2} + \frac{1}{R_1'} + \frac{1}{R_2'} \right) \\
& -\frac{1}{4} \left\{ \left[\left(\frac{1}{R_1} - \frac{1}{R_1'} \right) + \left(\frac{1}{R_2} - \frac{1}{R_2'} \right) \right]^2 \right. \\
& \left. -4 \left(\frac{1}{R_1} - \frac{1}{R_1'} \right) \left(\frac{1}{R_2} - \frac{1}{R_2'} \right) \sin^2 \theta \right\}^{1/2}
\end{aligned}
\tag{9.5}
$$

$$
\begin{aligned}
B = \frac{1}{4} & \left(\frac{1}{R_1} + \frac{1}{R_2} + \frac{1}{R_1'} + \frac{1}{R_2'} \right) \\
& +\frac{1}{4} \left\{ \left[\left(\frac{1}{R_1} - \frac{1}{R_1'} \right) + \left(\frac{1}{R_2} - \frac{1}{R_2'} \right) \right]^2 \right. \\
& \left. -4 \left(\frac{1}{R_1} - \frac{1}{R_1'} \right) \left(\frac{1}{R_2} - \frac{1}{R_2'} \right) \sin^2 \theta \right\}^{1/2}
\end{aligned}
\tag{9.6}
$$

where θ is the angle between the planes of maximum (or minimum) curvature of the two contacting bodies (Fig. 9-2c). The displacements w_1 and w_2 are found by superposition using Boussinesq's solution [9.8] for a semi-infinite body subjected to a concentrated normal force at the boundary surface (the x, y plane). This approach neglects the curvature of the surfaces outside of the contact area:

$$w_1 + w_2 = \left[\frac{1 - \nu_1^2}{\pi E_1} + \frac{1 - \nu_2^2}{\pi E_2} \right] \iint_{A_c} \frac{p\, dA_c}{r} = d - Ax^2 - By^2 \qquad (9.7)$$

In Eq. (9.7), $p\, dA_c$ is considered to be a point force acting at a point (x', y') in the contact area. The variables w_1 and w_2 are elastic compressive deformations at a point (x, y) in the contact area. The variable r is the distance between (x', y') and (x, y). Boussinesq's solution for the displacement dw_1 at (x, y) due to a point force $p\, dA_c$ at (x', y') is

$$dw_1 = \frac{1 - \nu^2}{\pi E_1} \frac{p\, dA_c}{r} = \frac{1 - \nu^2}{\pi E_1} \frac{p(x', y')\, dx'\, dy'}{\sqrt{(x - x')^2 + (y - y')^2}}$$

Of course, dw_2 is given by a similar equation.

To find the total displacement caused by the pressure p over the contact area, the elemental displacements are superposed by integrating over the contact area A_c as shown in Eq. (9.7). Hertz found that Eq. (9.7) is satisfied if $p(x, y)$ is given by

$$p = p_0 \sqrt{1 - (x^2/a^2) - (y^2/b^2)} \qquad (9.8)$$

in which a is the semimajor axis and b the semiminor axis of the contact ellipse (Fig. 9-2c). The distribution of pressure is semiellipsoid with a maximum pressure p_0 at the center of the contact area:

$$p_0 = 3F/(2\pi ab) \qquad (9.9)$$

It is apparent that the maximum pressure is 1.5 times the average pressure $(F/(\pi ab))$. In general, the determination of the axes of the contact ellipse and of the distance of approach involves the evaluation of elliptic integrals [9.9].

Reference 9.9 contains compiled graphs for computing the quantities of interest in a contact problem. Figures 9-3 and 9-4 plot coefficients used in determining these quantities for values of B/A from 1 to 10,000. The quantity c_b is used to compute b from the equation

$$b = c_b(F\Delta)^{1/3} \qquad (9.10a)$$

where

$$\Delta = \left(\frac{1 - \nu_1^2}{E_1} + \frac{1 - \nu_2^2}{E_2} \right) \frac{1}{A + B} = \gamma \frac{1}{A + B} \qquad (9.10b)$$

where

$$\gamma = \frac{1 - \nu_1^2}{E_1} + \frac{1 - \nu_2^2}{E_2}$$

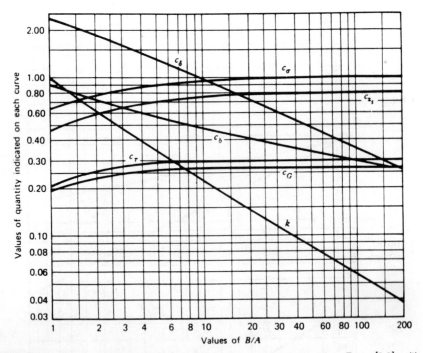

FIGURE 9-3 Stress and displacement coefficients for contacting bodies. From [9.9], with permission.

Define the quantity k to be the ratio of the minor to major axes of the contact ellipse

$$k = b/a \qquad (9.10c)$$

Once k and b are known $a = b/k$ can be obtained. The displacement d is found by using the quantity c_δ:

$$d = c_\delta(F/\pi)(A + B)/(b/\Delta) \qquad (9.11)$$

From the knowledge of the dimensions of the contact area and the pressure distribution over it, Thomas and Hoersch [9.4] derived expressions for the principal stresses along the z axis within the contacting solids. These formulas involve the evaluation of elliptic integrals. For any value of B/A, Fig. 9-3 or 9-4 can be used to compute the maximum compressive stress $(\sigma_c)_{max}$ that occurs at the origin, the maximum shear stress τ_{max} that occurs within the bodies, the maximum octahedral shear stress $(\tau_{oct})_{max}$, and the distance Z_s from the contact area at which the maximum shear stresses occur. The curves are strictly accurate when $\nu = 0.25$, but the dependence of these quantities on ν is weak:

$$\sigma_c = (\sigma_c)_{max} = -c_\sigma(b/\Delta) \qquad (9.12)$$

$$\tau_{max} = c_\tau(b/\Delta) \qquad (9.13)$$

$$\tau_{oct} = (\tau_{oct})_{max} = c_G(b/\Delta) \qquad (9.14)$$

$$Z_s = c_{zs}b \qquad (9.15)$$

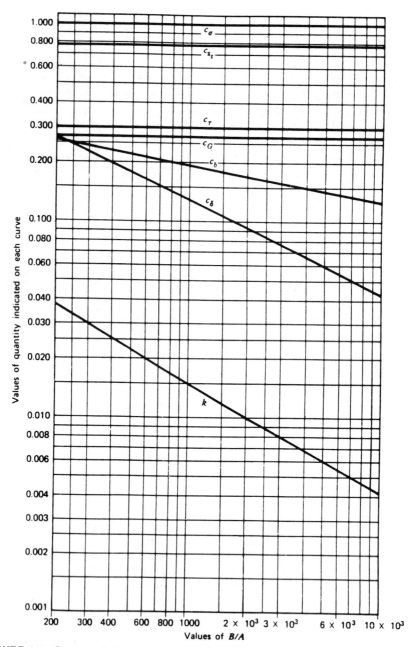

FIGURE 9-4 Stress and displacement coefficients for contacting bodies. From [9.9], with permission.

The above formulas are summarized in Table 9-1. Sometimes the values of coefficients $c_\sigma, c_\tau, c_b, \ldots$ are difficult to read from Figs. 9-3 and 9-4. This problem can be avoided by using Table 9-2 for contact stress analyses. The coefficients n_a, n_b, n_c, n_d, which appear in Table 9-2, can be taken from Table 9-3. The formulas used to calculate Table 9-3 are given in Table 9-4.

In many cases of practical interest, surface roughness, local yielding, friction, lubrication, thermal effects, and residual stresses will result in conditions that invalidate the Hertzian analysis. Consequently, the stresses computed according to Hertz's analysis must often be regarded as guidelines that are correlated with experimental failure tests to find allowable stress limits.

The following section contains several examples of the computation of Hertzian contact stresses. Formulas pertinent to a number of special cases are listed in Table 9-2. In addition, Table 9-2 also provides some solutions of problems for contact stresses when the surfaces are not curved.

Example 9.1 Wheel on Rail A steel wheel of radius 45 cm rests on a steel rail that has a radius of curvature of 35 cm (Fig. 9-5). The wheel supports a load of 40,000 N. To find the dimensions of the contact area, the maximum stresses in the contact region, and the distance below the contact surface at which the maximum shear stress and octahedral shear stress occur, the constants A and B must first be evaluated. Denoting the wheel as body 1 and the railhead as body 2, the principal radii of curvature are $R_1 = 45$ cm, $R_1' = \infty$, $R_2 = 35$ cm, and $R_2' = \infty$. The angle between the principal planes of the two bodies is $90°$. The physical constants of steel are $E = 200$ GPa, $\nu = 0.29$.
  From Eqs. (9.5) and (9.6)

$$A = \frac{1}{4}\left(\frac{1}{0.45} + \frac{1}{0.35}\right) - \frac{1}{4}\left[\left(\frac{1}{0.45} + \frac{1}{0.35}\right)^2 - 4\left(\frac{1}{0.45}\right)\left(\frac{1}{0.35}\right)\right]^{1/2}$$

$$= 1.2698 - 0.1587 = 1.111 \text{ m}^{-1} \tag{1}$$

$$B = 1.2698 + 0.1587 = 1.428 \text{ m}^{-1} \tag{2}$$

$$B/A = 1.428/1.111 = 1.285 \tag{3}$$

When both bodies have the same physical properties, Eq. (9.10b) becomes

$$\Delta = \frac{2(1 - \nu^2)}{E(A + B)} = \frac{2\left[1 - (0.29)^2\right]}{(2.0 \times 10^{11})(1.111 + 1.428)} = 3.607 \times 10^{-12} \text{ m}^3/\text{N} \tag{4}$$

From knowledge of B/A, the constants for use in determining stresses and lengths are read from Fig. 9-3, $c_b = 0.84$, $k = 0.85$, $c_\sigma = 0.69$, $c_\tau = 0.22$, $c_G = 0.21$, $c_{zs} = 0.5$, $c_\delta = 2.2$. The semiminor axis of the contact ellipse is given by [Eq. (9.10a)]

$$b = c_b[F\Delta]^{1/3} = 0.84\left[(40,000)(3.607 \times 10^{-12})\right]^{1/3}$$

$$= 0.00441 \text{ m} = 4.41 \text{ mm} \tag{5}$$

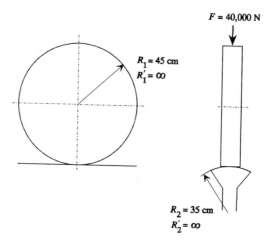

$F = 40{,}000$ N

$R_1 = 45$ cm
$R_1' = \infty$

$R_2 = 35$ cm
$R_2' = \infty$

FIGURE 9-5   Wheel on rail for Example 9.1 (crossed cylinders).

The semimajor axis of the contact ellipse is [Eq. (9.10c)]

$$a = b/k = 0.00441/0.85 = 0.00519 \text{ m} = 5.19 \text{ mm} \qquad (6)$$

The compressive stress at the center of the contact ellipse, i.e., the maximum principal stress, becomes [Eq. (9.12)]

$$\sigma_c = -c_\sigma(b/\Delta) = -0.69(0.00441/3.607 \times 10^{-12}) = -843.6 \text{ MPa} \qquad (7)$$

The maximum shear stress is [Eq. (9.13)]

$$\tau_{max} = c_\tau(b/\Delta) = 269.0 \text{ MPa} \qquad (8)$$

The maximum octahedral shear stress is given by [Eq. (9.14)]

$$\tau_{oct} = c_G(b/\Delta) = 256.8 \text{ MPa} \qquad (9)$$

The distance below the center of the contact area at which the two maximum shear stresses occur is found to be [Eq. (9.15)]

$$Z_s = c_{zs}b = 0.5(0.00441) = 0.002205 \text{ m} = 2.205 \text{ mm} \qquad (10)$$

Finally, the rigid approach of the two bodies becomes

$$d = c_\delta \frac{F}{\pi} \frac{A+B}{b/\Delta} = (2.2)(4.0 \times 10^4)\frac{1}{\pi} \frac{1.111 + 1.428}{(0.00441)/(3.607 \times 10^{-12})}$$

$$= 0.0582 \text{ mm} \qquad (11)$$

This problem also can be solved by using the formulas of Table 9-2. This is a contact stress problem of cylinders crossed at right angles. The formulas in case 2d apply. Then

$$\gamma = 2\frac{1 - \nu^2}{E} = 2\frac{1 - 0.29^2}{2 \times 10^{11}} = 9.159 \times 10^{-12} \text{ m}^2/\text{N}$$

$$K = \frac{D_1 D_2}{D_1 + D_2} = \frac{0.90 \times 0.70}{0.90 + 0.70} = 0.3938 \text{ m}$$

$$B = 1/D_2 = 1/0.70 = 1.429 \text{ m}^{-1}$$

$$A = 1/D_1 = 1/0.90 = 1.111 \text{ m}^{-1}$$

$$A/B = 0.70/0.90 = 0.7778$$

From Table 9-3

$$n_a = 1.089, \qquad n_b = 0.9212, \qquad n_c = 0.9964, \qquad n_d = 0.9964$$

The semimajor axis of the contact ellipse is

$$a = 0.909 n_a (FK\gamma)^{1/3} = 0.909(1.089)\left[40\,000(0.3938)9.159 \times 10^{-12}\right]^{1/3}$$
$$= 5.192 \times 10^{-3} \text{ m} = 5.192 \text{ mm}$$

while the semiminor axis is

$$b = 0.909 n_a (FK\gamma)^{1/3} = 0.909(0.9212)\left[40\,000(0.3938)9.159 \times 10^{-12}\right]^{1/3}$$
$$= 4.39 \times 10^{-3} \text{ m} = 4.39 \text{ mm}$$

The maximum compressive stress becomes

$$\sigma_c = 0.579 n_c \left[F/(K^2\gamma^2)\right]^{1/3}$$
$$= 0.579 \times 0.9964\left(\frac{40,000}{0.3938^2 \times 9.159^2 \times 10^{-24}}\right)^{1/3} = 838.8 \text{ MPa}$$

The rigid approach of the two bodies is given as

$$d = 0.825 n_d (F^2\gamma^2/K)^{1/3} = 0.825(0.9964)\left(\frac{40,000^2 \times 9.159^2 \times 10^{-24}}{0.3938}\right)^{1/3}$$
$$= 0.05742 \text{ mm}$$

Example 9.2 Ball Bearing For the ball bearing shown in Fig. 9-6, at the contact region find the maximum principal stress, the maximum shearing stress, the maximum octahedral shearing stress, the dimensions of the area of contact,

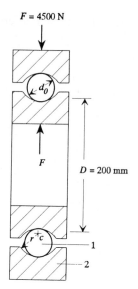

FIGURE 9-6 Single-row ball bearing: $r = 20$ mm; $d_0 = 38$ mm; c denotes center of curvature for r.

and the distance from the point of contact to the point along the force direction where the stresses occur.

The radii of concern are given in Fig. 9-6 as

$$R_1 = \tfrac{1}{2}d_0 = 19 \text{ mm}, \qquad R_1' = \tfrac{1}{2}d_0 = 19 \text{ mm}$$

$$R_2 = -r = -20 \text{ mm}, \qquad R_2' = \tfrac{1}{2}D = 100 \text{ mm}$$

From Eqs. (9.5) and (9.6)

$$A = \frac{1}{4}\left(\frac{1}{0.019} - \frac{1}{0.020} + \frac{1}{0.019} + \frac{1}{0.100} \right)$$

$$- \frac{1}{4}\left\{ \left[\left(\frac{1}{0.019} - \frac{1}{0.019} \right) + \left(-\frac{1}{0.020} - \frac{1}{0.100} \right) \right]^2 \right.$$

$$\left. - 4\left(\frac{1}{0.019} - \frac{1}{0.019} \right)\left(-\frac{1}{0.020} - \frac{1}{0.100} \right) \sin^2(0) \right\}^{1/2} = 1.316 \text{ m}^{-1} \quad (1)$$

$$B = 31.32 \text{ m}^{-1} \tag{2}$$

Then

$$B/A = 23.78 \tag{3}$$

$$\Delta = \frac{2}{A + B}\frac{1 - \nu^2}{E} = \frac{2(1 - 0.3^2)}{(31.32 + 1.316)(200 \times 10^9)} = 2.79 \times 10^{-13} \text{ m}^3/\text{N}$$

$$\tag{4}$$

From Fig. 9-3, the coefficients are found. The variables of interest are then computed using the appropriate formulas:

$$k = 0.13, \qquad c_G = 0.27, \qquad c_\tau = 0.3$$
$$c_b = 0.394, \qquad c_\sigma = 1.0, \qquad c_{zs} = 0.8$$
$$b = c_b[F\Delta]^{1/3} = 0.394[(4500)(2.79 \times 10^{-13})]^{1/3} = 4.250 \times 10^{-4} \text{ m}$$
$$= 0.425 \text{ mm}$$
$$a = 4.250 \times 10^{-4}/0.13 = 3.269 \times 10^{-3} \text{ m} = 3.269 \text{ mm}$$
$$b/\Delta = 4.250 \times 10^{-4}/(2.79 \times 10^{-13}) = 1523 \text{ MPa}$$
$$\sigma_c = -c_\sigma(b/\Delta) = (-1.0)(1523) = -1523 \text{ MPa}$$
$$\tau_{max} = c_\tau(b/\Delta) = (0.3)(1523) = 456.9 \text{ MPa}$$
$$\tau_{oct} = c_G(b/\Delta) = (0.27)(1523) = 411.2 \text{ MPa}$$
$$Z_s = c_{zs}b = (0.8)(4.250 \times 10^{-4}) = 3.40 \times 10^{-4} \text{ m} = 0.34 \text{ mm}$$

Alternatively, use the formulas of case 1e of Table 9-2:

$$\gamma = 2\frac{1 - \nu^2}{E} = 2\frac{1 - 0.3^2}{2 \times 10^{11}} = 0.91 \times 10^{-11} \text{ m}^2/\text{N}$$

$$K = \frac{1}{2/R_1 - 1/R_2 + 1/R_3} = \frac{1}{2/0.019 - 1/0.02 - 1/0.10} = 0.01532$$

$$A = \frac{1}{2}\left(\frac{1}{R_1} - \frac{1}{R_2}\right) = \frac{1}{2}\left(\frac{1}{0.019} - \frac{1}{0.020}\right) = 1.316 \text{ m}^{-1}$$

$$B = \frac{1}{2}\left(\frac{1}{R_1} + \frac{1}{R_3}\right) = \frac{1}{2}\left(\frac{1}{0.019} + \frac{1}{0.1}\right) = 31.32 \text{ m}^{-1}$$

$$A/B = 0.04202$$

From Table 9-3

$$n_a = 3.385, \qquad n_b = 0.4390, \qquad n_c = 0.6729, \qquad n_d = 0.6469$$

The semimajor axis of the contact ellipse is given by

$$a = 1.145 n_a (FK\gamma)^{1/3} = 1.145 \times 3.385 \times \left[4500(0.91 \times 10^{-11})0.01532\right]^{1/3}$$

$$= 3.31 \times 10^{-3} \text{ m} = 3.31 \text{ mm}$$

and the semiminor axis is

$$b = 1.145 n_b (FK\gamma)^{1/3} = 0.4303 \text{ mm}$$

Furthermore, the maximum compressive stress is

$$\sigma_c = 0.365 n_c \left[F/(K^2\gamma^2)\right]^{1/3} = 1508 \text{ MPa}$$

and the rigid approach of the two bodies becomes

$$d = 0.655 n_d (F^2\gamma^2/K)^{1/3} = 0.02027 \text{ mm}$$

Example 9.3 Wheel–Rail Analyses Consider again the wheel and rail shown in Fig. 9-5. In Example 9.1, the maximum octahedral shear stress was found to be 256.8 MPa and to be located 0.22 cm below the initial contact point. Suppose now that the rail steel has a tensile yield strength of 413.8 MPa.

1. Determine whether yielding occurs in the rail according to the maximum octahedral shear stress yield theory (equivalent to the von Mises–Hencky theory).

The octahedral shear stress at yield is [Chapter 3, Eq. (3.24b)]

$$(\tau_{\text{oct}})_{ys} = \tfrac{1}{3}\sqrt{2}\,\sigma_{ys} = \tfrac{1}{3}\sqrt{2}\,(413.8 \text{ MPa}) = 195.07 \text{ MPa} \tag{1}$$

Since the maximum octahedral shear stress computed by elastic theory exceeds the yield value, yielding does occur in the rail.

2. Find to what value the load must be reduced so that the computed maximum octahedral shear stress equals the yield point value. From Eqs. (9.14) and (9.10a)

$$\tau_{\text{oct}} = c_G(b/\Delta), \qquad b = c_b(F\Delta)^{1/3}$$

Hence

$$\tau_{\text{oct}} = c_G c_b (F\Delta)^{1/3}/\Delta, \qquad (\tau_{\text{oct}})^3 = (c_G c_b/\Delta)^3 \Delta F = F(c_G c_b)^3/\Delta^2$$

$$F = \left[\Delta^2/(c_G c_b)^3\right](\tau_{\text{oct}})^3 \tag{2}$$

Since c_G, c_b, and Δ do not depend on F, the yield load F_{ys} is calculated as

$$F_{ys} = \frac{(3.607 \times 10^{-12})^2 (195.07 \times 10^6)^3}{(0.2)^3 (0.84)^3} = 20{,}367 \text{ N} \qquad (3)$$

Therefore, to reduce the maximum octahedral shear stress to the yield value, the applied load must virtually be halved.

3. Suppose that the wheel–rail combination must be operated with a safety factor of 2, i.e., the maximum octahedral shear stress must be one-half the value that causes yield. Compute the maximum value the load may take under this restriction.

Since maximum octahedral shear stress varies directly as the cube root of the applied load, to halve the stress, the load must decrease by a factor of $(\frac{1}{2})^3$, or $\frac{1}{8}$. Since a load of 20,367 N corresponds to a maximum octahedral shear stress exactly at the yield point, the force

$$F_2 = \tfrac{1}{8} 20{,}367 \text{ N} = 2545.9 \text{ N} \qquad (4)$$

would result in the maximum octahedral shear stress being one-half the yield value.

4. Suppose that the operating load must be 20,367 N. Find by what common factor the radii R_1 and R_2 must be increased in order that the maximum octahedral shear stress be one-half the value that causes yielding.

Changing R_1 and R_2 by the same factor does not affect B/A, so c_G and c_b remain constant. Similarly, γ depends only on E and ν so that Δ changes only as a result of $A + B$. Let τ, A, B, R_1, R_2 be the values of variables under conditions described in question 2 and τ_{oct}^*, A^*, B^*, R_1^*, R_2^* be the conditions with R_1 and R_2 altered by a factor λ. We require that

$$\tau_{oct}^* = \tfrac{1}{2}\tau_{oct}, \qquad R_1^* = \lambda R_1, \qquad R_2^* = \lambda R_2$$

$$A + B = \frac{1}{4}\left(\frac{1}{R_1} + \frac{1}{R_2}\right) + \frac{1}{4}\left(\frac{1}{R_1} + \frac{1}{R_2}\right) = \frac{1}{2}\left(\frac{1}{R_1} + \frac{1}{R_2}\right)$$

$$A^* + B^* = \frac{1}{2}\left(\frac{1}{R_1^*} + \frac{1}{R_2^*}\right) = \frac{1}{2\lambda}\left(\frac{1}{R_1} + \frac{1}{R_2}\right)$$

$$\tau_{oct}^3 = F(c_G c_b)^3/\Delta^2, \qquad \tau_{oct}^{*3} = F(c_G c_b)^3/\Delta^{*2}$$

$$\frac{\tau_{oct}^3}{\tau_{oct}^{*3}} = \frac{\Delta^{*2}}{\Delta^2} = \frac{\left(\dfrac{\gamma}{A^* + B^*}\right)^2}{\left(\dfrac{\gamma}{A + B}\right)^2} = \frac{(A + B)^2}{(A^* + B^*)^2} = \frac{\dfrac{1}{4}\left(\dfrac{1}{R_1} + \dfrac{1}{R_2}\right)^2}{\dfrac{1}{4\lambda^2}\left(\dfrac{1}{R_1} + \dfrac{1}{R_2}\right)^2}$$

Thus

$$\tau_{oct}^3/\tau_{oct}^{*3} = \lambda^2 \quad \text{or} \quad \tau_{oct}^* = \tau_{oct}/\lambda^{2/3} \qquad (5)$$

But

$$\tau^*_{\text{oct}} = \tfrac{1}{2}\tau_{\text{oct}}$$

Therefore

$$2^3 \tau^3_{\text{oct}}/\tau^3_{\text{oct}} = \lambda^2 \quad \text{or} \quad \lambda = \sqrt{8}$$

To check, we find

$$R^*_1 = \sqrt{8}\,(45) = 127.28 \text{ cm} \qquad R^*_2 = \sqrt{8}\,(35) = 98.995 \text{ cm}$$

$$(A + B)^* = \frac{1}{2}\left(\frac{1}{1.2728} + \frac{1}{0.98995}\right) = 0.8979 \text{ m}^{-1}$$

$$\Delta^* = \frac{2\left[1 - (.29)^2\right]}{(2 \times 10^{11})(0.8979)} = 1.020 \times 10^{-11} \text{ m}^3/\text{N}$$

$$\tau^{*3} = \frac{(20{,}367)(0.2)^3(0.84)^3}{(1.02 \times 10^{-11})^2} = 9.282 \times 10^{23} \ (\text{N/m}^2)^3$$

$$\tau^* = 9.755 \times 10^7 \text{ Pa or } 97.55 \text{ MPa}$$

Since the yield value of maximum octahedral shear is 195.07 MPa and 97.55 is one-half of the yield value, increasing R_1 and R_2 by a factor of $\sqrt{8}$ decreases the maximum octahedral shear stress by one-half.

5. Suppose the operating load is fixed at 20,367 N and that the rail and wheel radii are fixed at 35 and 45 cm, respectively. Find by what factor the tensile strength of the steel must be increased to make the maximum octahedral shear stress one-half the yield point value.

Since E and ν of steel are essentially constant for steels of all strengths and $A + B$ is determined by the fixed radii of rail and wheel, the quantity Δ in Eq. (9.10b) is a fixed value. Therefore the maximum octahedral shear stress would remain at 195.07 MPa for all steels. Because tensile yield strength and octahedral shear stress at yield are directly proportional, doubling the tensile strength would result in the maximum octahedral shear stress being one-half the value that causes yield. The strength of the steel would be increased to

$$\sigma_{ys} = (3/\sqrt{2})(\tau_{\text{oct}})_{ys} = (3/\sqrt{2})(2 \times 195.07) = 827.6 \text{ MPa}$$

6. Determine in which of the three quantities (load, radii of curvature, or steel strength) would a change be most effective in producing a system with an acceptable value of maximum octahedral shear stress (one-half the value that causes yield).

Reducing the load is most ineffective in reducing maximum octahedral shear stress because the stress varies directly as the cube root of the load [Eq. (2) of this example]. When the radii of curvature are increased in constant proportion, the maximum octahedral shear stress varies inversely as the two-thirds power of the radii factor λ [see (5)]; hence changing the radii is more effective than changing the load. However, if large reductions in stress are required, it is doubtful that the necessarily large changes in radii ($\lambda = \sqrt{8} = 2.83$-fold increase for a halving of the shear stress) would be feasible. It appears that increasing the tensile strength of the material of construction is the most effective alternative when the stress is significantly higher than an acceptable level.

Two Bodies in Line Contact

Two bodies in contact along a straight line before loading are said to be in line contact. For instance, a line contact occurs when a circular cylinder rests upon a plane or when a small circular cylinder rests inside a larger hollow cylinder. In these line contact cases, Eqs. (9.5) and (9.6) become

$$A = 0, \qquad B = \tfrac{1}{2}(1/R_1 + 1/R_2)$$

and

$$B/A = \infty \tag{9.16}$$

It can be shown that in this case, the quantity k in Eq. (9.10c) approaches zero. When a distributed load q (force/length) is applied, the area of contact is a long narrow rectangle of width $2b$ in the x direction and a length $2a$ in the y direction.

The maximum principal stresses occurring at the surface of contact are

$$\sigma_x = -b/\Delta, \qquad \sigma_y = -2\nu(b/\Delta), \qquad \sigma_z = -b/\Delta \tag{9.17a}$$

Thus

$$\sigma_{\max} = -b/\Delta \tag{9.17b}$$

where

$$b = \sqrt{2q\Delta/\pi} \tag{9.18a}$$

$$\Delta = \frac{1}{1/(2R_1) + 1/(2R_2)}\left(\frac{1 - \nu_1^2}{E_1} + \frac{1 - \nu_2^2}{E_2}\right) \tag{9.18b}$$

The maximum shear stress is

$$\tau_{\max} = 0.300(b/\Delta) \tag{9.19}$$

at the depth $Z_s/b = 0.7861$.

The maximum octahedral shear stress occurs at the same point as the maximum shear. The value is

$$\tau_{\text{oct}} = 0.27(b/\Delta) \tag{9.20}$$

For the case of line contact, Eqs. (9.12)–(9.15) still apply. The coefficients

$c_\sigma, c_\tau, c_G, c_{zs}$ can also be found from Figs. 9-3 and 9-4 by selecting values of B/A greater than 50.

Contact Stress with Friction

For the case of two cylinders with longitudinal axes parallel, Smith and Liu [9.11] examined the modification of the contact stress field caused by the presence of surface friction. Mindlin [9.12] showed that the tangential stresses have the same distribution over the contact areas as have the normal stresses. For impending sliding motion, the tangential stresses are linearly related to the normal stresses by a coefficient of friction. The total stress field is the resultant of the field due to normal surface stresses plus the field due to tangential surface stresses. The degree to which tangential surface stresses change the distribution caused by normal surface stresses depends on the magnitude of the coefficient of friction. The changes in the maximum contact stresses with the coefficient of friction are provided in Table 9-5.

The presence of friction changes from a compressive stress to a stress that varies from tensile to compressive over the area. The creation of tensile stresses in the contact zone is thought to contribute to fatigue failure of bodies subject to cyclic contact stresses. Smith and Liu found in addition that if the coefficient of friction was 0.1 or greater, the point of maximum shear stress occurs on the contact surface rather than below it.

Example 9.4 Contact Stress in Cylinders with Friction Consider two steel cylinders each 80 mm in diameter and 150 mm long mounted on parallel shafts and loaded by a force $F = 80$ kN (Fig. 9-7). The two cylinders ($E = 200$ GPa and $\nu = 0.29$) are rotated at slightly different speeds so that the cylinder surfaces slide across each other. If the coefficient of sliding friction is $f = \frac{1}{3}$, determine the

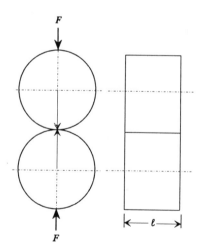

FIGURE 9-7 Example 9.4.

maximum compressive principal stress σ_c, the maximum shear stress τ_{max}, and the maximum octahedral shear stress τ_{oct}.

The value of the required quantities are obtained from Table 9-5 for $f = \frac{1}{3}$,

$$(\sigma_c)_{max} = -1.40(b/\Delta) \tag{1}$$

$$\tau_{max} = 0.435(b/\Delta) \tag{2}$$

$$\tau_{oct} = 0.368(b/\Delta) \tag{3}$$

where b and Δ are given by Eqs. (9.18a) and (9.18b):

$$\Delta = 2R\frac{1-\nu^2}{E} = \frac{2(0.040)(1-0.29^2)}{200 \times 10^9} = 3.664 \times 10^{-13} \text{ m}^3/\text{N} \tag{4}$$

$$b = \left(\frac{2F\Delta}{\ell\pi}\right)^{1/2} = \sqrt{\frac{2(80 \times 10^3)(3.664 \times 10^{-13})}{0.150\pi}} = 0.3527 \times 10^{-3} \text{ m}$$

$$= 0.3527 \text{ mm} \tag{5}$$

$$b/\Delta = 962.6 \text{ MPa} \tag{6}$$

with $q = F/\ell$

Substitution of these values into (1), (2), and (3) leads to

$$(\sigma_c)_{max} = -1.40 \times 962.6 = -1347.6 \text{ MPa}$$

$$\tau_{max} = 0.435 \times 962.6 = 418.7 \text{ MPa}, \qquad \tau_{oct} = 0.368 \times 962.6 = 354.2 \text{ MPa} \tag{7}$$

It can be seen from Table 9-5 that the friction force with $f = \frac{1}{3}$ increases the maximum compressive principal stress by 40%, the maximum shear stress by 45%, and the maximum octahedral shear stress by 35%.

9.3 CONTACT FATIGUE

A machine part subjected to contact stresses usually fails after a large number of load applications. The failure mode is that of crack initiation followed by propagation until the part fractures or until pits are formed by material flaking away. Buckingham measured the surface fatigue strengths of materials subjected to contact loads [9.13]. His results showed that hardened steel rollers did not have a fatigue limit for contact loading. Cast materials, however, did show a fatigue limit for contact loads.

9.4 ROLLING CONTACT

When two bodies roll over each other, the area of contact will in general be divided into a region of slip and a region of adhesion. In the region of slip the tangential force is related to the normal force by a coefficient of friction. Under conditions of free rolling no region of slip exists, and surface friction dissipates no energy. If gross sliding occurs, no region of adhesion exists.

When both regions are present, the motion is termed *creep, creep ratio*, or *creepage*. The creepage is resolved into three components: longitudinal, lateral, and spin. Spin creepage occurs when a relative angular velocity about an axis normal to the contact zone exists between the two contacting bodies. Longitudinal and lateral creepage occurs when a relative circumferential velocity without gross sliding exists between the contacting bodies. The forces and moments transmitted between two contacting bodies due to creepage are very important in wheel–rail contact problems. Vermeulen and Johnson [9.14] suggested a nonlinear law that does not account for spin creepage. Kalker has proposed a linear law relating creepage to the transmitted forces and moments as well as two-nonlinear creep laws [9.15].

9.5 NON-HERTZIAN CONTACT STRESS

The simplest non-Hertzian contact problem is the case in which all conditions for Hertzian contact are met except that the surfaces cannot be approximated as a second-degree polynomial near the point of contact. Singh and Paul [9.16] have described a numerical procedure for solving this type of problem. In this method a suitable contact area is first proposed; then the corresponding applied load, pressure distribution, and rigid approach are found.

REFERENCES

9.1 Smirnov, V. I., *A Course of Higher Mathematics*, Part II (translated by D. E. Brown), Addison Wesley, Reading, MA, 1964.

9.2 Hertz, H., *Miscellaneous Papers*, Macmillan, New York, 1896.

9.3 Belajev, N. M., *Memoirs on the Theory of Structures*, Petrograd, 1924.

9.4 Thomas, H. R., and Hoersch, V. A., "Stress Due to the Pressure of One Elastic Solid upon Another," University of Illinois Engineering Experimental Station Bulletin, No. 212, 1930.

9.5 Love, A. E. H., *Mathematical Theory of Elasticity*, Dover, New York, 1927.

9.6 Timoshenko, S. P., and Goodier, J. N., *Theory of Elasticity*, 3rd ed., McGraw-Hill, New York, 1970.

9.7 Lipson, C., and Juvinall, R. C., *Handbook of Stress and Strength*, Macmillan, New York, 1970.

9.8 Boussinesq, J., "Application des Potentiels a L'Etude de L'Equilibre et du Mouvement des Solides Elastiques," Gautheers-Villars, Paris, 1885.

9.9 Boresi, A. P., Sidebottom, O. M., *Advanced Mechanics of Materials*, 4th ed., Wiley, New York, 1985.

9.10 Alexander, B., *Handbook of Mechanics, Materials, and Structures*, Wiley, New York, 1985.

9.11 Smith, J. O., and Liu, C.-K., "Stresses Due to Tangential and Normal Loads on an Elastic Solid with Application to Some Contact Stress Problems," *J. Appl. Mechan.*, Vol. 20, No. 2, 1953, pp. 157–165.

9.12 Mindlin, R. D., "Compliance of Elastic Bodies in Contact," *J. Appl. Mechan.*, Vol. 71, 1949, pp. 259–268.

9.13 Buckingham, E., *Analytical Mechanics of Gears*, McGraw-Hill, New York, 1949.

9.14 Vermeulen, P. J., and Johnson, K. L., "Contact of Nonspherical Elastic Bodies Transmitting Tangential Forces," *J. Appl. Mechan.*, Vol. 31, 1964, pp. 338–340.

9.15 Kalker, J. J., "On the Rolling Contact between Two Elastic Bodies in the Presence of Dry Friction," Ph.D. Thesis, Delft University of Technology, 1967.

9.16 Singh, K. P., and Paul B., "Numerical Solution of Non-Hertzian Contact Problems," *J. Appl. Mechan.*, Vol. 41, 1974, pp. 484–490.

9.17 Lubkin, J. L., "Contact Problems," *Handbook for Engineering Mechanics*, ed. by Flugge, W., McGraw-Hill, New York, 1962.

9.18 Paul, B., and Hashemi, J., "An Improved Numerical Method for Counter formal Contact Stress Problems, Report No. FRA-ORD-78-26, Department of Transportation, Washington, DC, July 1977.

9.19 Johnson, K. L., *Contact Mechanics*, Cambridge University Press, New York, 1985.

9.20 Ruditzin, M. W., Artemov, P., and Luboshytz, M., *Handbook for Strength of Materials*, BSSR Government Publishing House, Minsk, 1958.

9

Tables

425

TABLE 9-1   SUMMARY OF GENERAL FORMULAS FOR CONTACT STRESSES

Notation

$$R_i, R'_i = \text{maximum and minimum radii of curvature}$$
$$\text{of two contacting surfaces } i = 1, 2$$
$$F = \text{applied force}$$
$$\theta = \text{angle between planes containing principal radii}$$
$$\text{of curvature}$$

$$A = \frac{1}{4}\left(\frac{1}{R_1} + \frac{1}{R_2} + \frac{1}{R'_1} + \frac{1}{R'_2}\right)$$

$$-\frac{1}{4}\left\{\left[\left(\frac{1}{R_1} - \frac{1}{R'_1}\right) + \left(\frac{1}{R_2} - \frac{1}{R'_2}\right)\right]^2 - 4\left(\frac{1}{R_1} - \frac{1}{R'_1}\right)\left(\frac{1}{R_2} - \frac{1}{R'_2}\right)\sin^2\theta\right\}^{1/2}$$

$$B = \frac{1}{4}\left(\frac{1}{R_1} + \frac{1}{R_2} + \frac{1}{R'_1} + \frac{1}{R'_2}\right)$$

$$+\frac{1}{4}\left\{\left[\left(\frac{1}{R_1} - \frac{1}{R'_1}\right) + \left(\frac{1}{R_2} - \frac{1}{R'_2}\right)\right]^2 - 4\left(\frac{1}{R_1} - \frac{1}{R'_1}\right)\left(\frac{1}{R_2} - \frac{1}{R'_2}\right)\sin^2\theta\right\}^{1/2}$$

Compute B/A and obtain coefficients $c_b, k, c_\sigma, c_\tau, c_G, c_{zs}, c_\delta$ from plots of Figs. 9-3 or 9-4. Then

$$\gamma = \frac{1 - \nu_1^2}{E_1} + \frac{1 - \nu_2^2}{E_2} \qquad \Delta = \gamma\frac{1}{A + B}$$

Formulas for Stresses and Deformation

Semiminor axis:
$$b = c_b(F\Delta)^{1/3}$$

Semimajor axis:
$$a = b/k$$

Maximum compressive stress:
$$(\sigma_c)_{\max} = -c_\sigma(b/\Delta)$$

Maximum shear stress:
$$\tau_{\max} = c_\tau(b/\Delta)$$

Maximum octahedral shear stress:
$$(\tau_{\text{oct}})_{\max} = c_G(b/\Delta)$$

Distance from contact area to location of maximum shear stress:
$$Z_s = c_{zs}b$$

Distance of approach of contacting bodies
$$d = c_\delta\frac{F}{\pi}\frac{A + B}{b/\Delta}$$

TABLE 9-1 Summary of General Formulas   426

TABLE 9-2 FORMULAS FOR CONTACT STRESSES, DIMENSIONS, AND CONTACT AREAS, AND RIGID-BODY APPROACHES[a]

Notation

$$\gamma = (1 - \nu_1^2)/E_1 + (1 - \nu_2^2)/E_2$$

a, b = semimajor axis and semiminor axis of contact ellipse, respectively

d = rigid distance of approach of contacting bodies or surface deformation

E_i = modulus of elasticity of object i, i = 1 or 2

F = force

p = pressure

q = distributed line load

σ_c = maximum compressive stress of contact area, $= (\sigma_c)_{max}$

ν_i = Poisson's ratio of object i, i = 1 or 2

τ = shear stress

Case	Formulas
	Spheres
1a Sphere on sphere 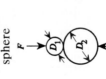 $$K = \frac{D_1 D_2}{D_1 + D_2}$$	$a = b = 0.721(FK\gamma)^{1/3}$ $\sigma_c = 0.918\left(F/(K^2\gamma^2)\right)^{1/3}$ $d = 1.040(F^2\gamma^2/K)^{1/3}$

TABLE 9-2 | Formulas for Contact Stresses 428

TABLE 9-2 (continued) FORMULAS FOR CONTACT STRESSES, DIMENSIONS AND CONTACT AREAS, AND RIGID-BODY APPROACHES

Case	Formulas
1b Sphere on flat plate $K = D_1$	$a = b = 0.721(FK\gamma)^{1/3}$ $\sigma_c = 0.918(F/(K^2\gamma^2))^{1/3}$ $d = 1.040(F^2\gamma^2/K)^{1/3}$
1c Sphere in spherical socket $K = \dfrac{D_1 D_2}{D_2 - D_1}$	$a = b = 0.721(FK\gamma)^{1/3}$ $\sigma_c = 0.918(F/(K^2\gamma^2))^{1/3}$ $d = 1.040(F^2\gamma^2/K)^{1/3}$

1d
Sphere on a cylinder

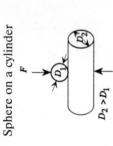

$D_2 > D_1$

$a = 0.9088 n_a (FK\gamma)^{1/3}$

$b = 0.9088 n_b (FK\gamma)^{1/3}$

$\sigma_c = 0.579 n_c (F/(K^2\gamma^2))^{1/3}$

$d = 0.825 n_d (F^2\gamma^2/K)^{1/3}$

$$A = \frac{1}{D_1} \qquad B = \frac{1}{D_1} + \frac{1}{D_2} \qquad K = \frac{D_1 D_2}{2D_2 + D_1}$$

1e
Sphere in circular race

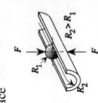

$R_3 > R_2$

$a = 1.145 n_a (FK\gamma)^{1/3}$

$b = 1.145 n_b (FK\gamma)^{1/3}$

$\sigma_c = 0.365 n_c (F/(K^2\gamma^2))^{1/3}$

$d = 0.655 n_d (F^2\gamma^2/K)^{1/3}$

$$K = \frac{1}{\dfrac{2}{R_1} - \dfrac{1}{R_2} + \dfrac{1}{R_3}} \qquad A = \frac{1}{2}\left(\frac{1}{R_1} - \frac{1}{R_2}\right) \qquad B = \frac{1}{2}\left(\frac{1}{R_1} + \frac{1}{R_3}\right)$$

1f
Sphere in cylindrical race

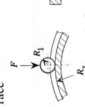

$R_2 > R_1$

$a = 1.145 n_a (FK\gamma)^{1/3}$

$b = 1.145 n_b (FK\gamma)^{1/3}$

$\sigma_c = 0.365 n_c (F/(K^2\gamma^2))^{1/3}$

$d = 0.655 n_d (F^2\gamma^2/K)^{1/3}$

$$K = \frac{R_1 R_2}{2R_2 - R_1} \qquad A = \frac{1}{2}\left(\frac{1}{R_1} - \frac{1}{R_2}\right) \qquad B = \frac{1}{2R_1}$$

Case	Formulas
	*Cylinders**

*Contact area is rectangular $(2b \times \ell)$ in cases 2a, 2b, and 2c

2a

Cylinder on cylinder
(axes parallel)

$q = \dfrac{F}{\ell}$

$b = 0.798(qK\gamma)^{1/2} \qquad \sigma_c = 0.798(q/(K\gamma))^{1/2}$

$d = \dfrac{2q}{\pi}\left[\dfrac{1-\nu_1^2}{E_1}\left(\ln\left(\dfrac{D_1}{b}\right) + 0.407\right) + \dfrac{1-\nu_2^2}{E_2}\left(\ln\left(\dfrac{D_2}{b}\right) + 0.407\right)\right]$

$K = D_1 D_2/(D_1 + D_2)$

2b

Cylinder on flat plate

$q = \dfrac{F}{\ell}$

$b = 0.798(qK\gamma)^{1/2}$

$\sigma_c = 0.798(q/(K\gamma))^{1/2}$

$K = D_1$

2c

Cylinder in cylindrical
socket

$q = \dfrac{F}{\ell}$

$b = 0.798(qK\gamma)^{1/2}$ when $E_1 = E_2$ and

$\sigma_c = 0.798(q/(K\gamma))^{1/2}$ $\qquad \nu_1 = \nu_2 = 0.3,$

$K = \dfrac{D_1 D_2}{D_2 - D_1}$ $\qquad d = 1.82\dfrac{q}{E}[1 - \ln(b)]$

TABLE 9-2 Formulas for Contact Stresses   430

2d

Cylinders crossed at right angles

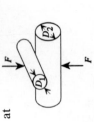

$$a = 0.909n_a(FK\gamma)^{1/3}$$
$$b = 0.909n_b(FK\gamma)^{1/3}$$
$$\sigma_c = 0.579n_c(F/(K^2\gamma^2))^{1/3}$$
$$d = 0.825n_d(F^2\gamma^2/K)^{1/3}$$

$$K = \frac{D_1 D_2}{D_1 + D_2} \qquad A = \frac{1}{D_2} \qquad B = \frac{1}{D_1}$$

Barrels

3

Barrel in a circular race

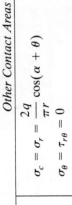

$$a = 1.145n_a(FK\gamma)^{1/3}$$
$$b = 1.145n_b(FK\gamma)^{1/3}$$
$$\sigma_c = 0.365n_c(F/(K^2\gamma^2))^{1/3}$$
$$d = 0.655n_d(F^2\gamma^2/K)^{1/3}$$

$$K = \frac{1}{\dfrac{1}{R_1} + \dfrac{1}{R_2} + \dfrac{1}{R_3} - \dfrac{1}{R_4}}$$

$$A = \frac{1}{2}\left(\frac{1}{R_2} - \frac{1}{R_4}\right) \qquad B = \frac{1}{2}\left(\frac{1}{R_1} + \frac{1}{R_3}\right)$$

Other Contact Areas

4a

Rigid knife edge on surface of semi-infinite plate line load q

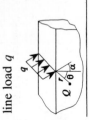

$$\sigma_c = \sigma_r = \frac{2q}{\pi r}\cos(\alpha + \theta)$$
$$\sigma_\theta = \tau_{r\theta} = 0$$

Formulas for Contact Stresses | TABLE 9-2

Case	Formulas
4b A concentrated force on surface of semi-infinite body	At an elemental area perpendicular to z axis of any point Q, the resultant stress is $$\frac{3F\cos^2\theta}{2\pi r^2}$$
4c Uniform pressure p over length ℓ on surface of semi-infinite body	At surface point O_1 outside loaded area $$d = \frac{2p}{\pi E}\left[(\ell+x_1)\ln\frac{c}{\ell+x_1} - x_1\ln\frac{d}{x_1}\right] + p\ell\frac{1-\nu}{\pi E}$$ At surface point O_2 underneath loaded area $$d = \frac{2p}{\pi E}\left[(\ell-x_2)\ln\frac{c}{\ell-x_2} + x_2\ln\frac{d}{x_2}\right] + p\ell\frac{1-\nu}{\pi E}$$ where d = displacement relative to a remote point distance c from edge of loaded area At any point Q $$\sigma_c = \frac{p}{\pi}(\alpha + \sin\alpha)$$

TABLE 9-2 **Formulas for Contact Stresses**   **432**

4d Rigid cylindrical die of radius r on surface of semi-infinite body 	$d = F(1 - \nu^2)/(2RE)$ At any point Q on surface of contact $$\sigma_c = \frac{F}{2\pi R\sqrt{R^2 - r^2}}$$ $(\sigma_c)_{\max} = \infty$ at edge $\qquad (\sigma_c)_{\min} = F/(2\pi R^2)$ at center
4e Uniform pressure p over circular area of radius R on surface of semi-infinite body 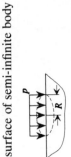	$$d_{\max} = \frac{2pR(1 - \nu^2)}{E} \quad \text{at center}, \quad d = \frac{4pR(1 - \nu^2)}{\pi E} \quad \text{on the circle}$$ $$\tau_{\max} = \frac{p}{2}\left[\frac{1 - 2\nu}{2} + \frac{2}{9}(1 + \nu)\sqrt{2(1 + \nu)}\right]$$ at point $R\sqrt{2(1 + \nu)}/(7 - 2\nu)$ below center of loaded area
4f Uniform pressure p over square area of sides $2b$ on surface of semi-infinite body 	$$d_{\max} = \frac{2.24pb(1 - \nu^2)}{E} \quad \text{at center} \qquad d = \frac{1.12pb(1 - \nu^2)}{E} \quad \text{at corners}$$ $$d_{\text{ave}} = \frac{1.90pb(1 - \nu^2)}{E}$$

[a]All diameters and radii are positive in given formulas. Values of n_a, n_b, n_c, and n_d are given in Table 9-3. Most of these formulas are adapted from Ruditzin et al. [9.20].

Formulas for Contact Stresses TABLE 9-2

TABLE 9-3 PARAMETERS FOR USE WITH FORMULAS OF TABLE 9-2

A/B	n_a	n_b	n_c	n_d
1.000000	1.000000	1.000000	1.000000	1.000000
.965467	1.013103	.987137	.999929	.999952
.928475	1.025571	.975376	.999684	.999714
.893098	1.038886	.963238	.999306	.999279
.851780	1.055557	.948729	.998565	.998574
.805934	1.075967	.931815	.997404	.997442
.767671	1.094070	.917565	.996134	.996124
.727379	1.114909	.901981	.994404	.994407
.699686	1.130480	.890831	.992984	.992988
.640487	1.166370	.866782	.989130	.989123
.594383	1.198061	.847158	.985274	.985223
.546919	1.234696	.826174	.980321	.980224
.514760	1.262333	.811413	.976303	.976163
.465921	1.309532	.788097	.968956	.968738
.416913	1.364990	.763293	.959796	.959415
.384311	1.407560	.745894	.952481	.951928
.351726	1.455592	.727797	.943952	.943170
.335659	1.481904	.718443	.939264	.938324
.319620	1.510007	.708906	.934183	.933087
.303612	1.540072	.699179	.928689	.927396
.287810	1.572253	.689231	.922811	.921331
.272153	1.606769	.679051	.916524	.914773
.264330	1.625029	.673890	.913167	.911272
.256632	1.643925	.668646	.909749	.907690
.241289	1.684020	.658001	.902457	.900066
.233648	1.705308	.652595	.898573	.895998
.226114	1.727405	.647099	.894613	.891797
.218611	1.750463	.641548	.890467	.887432
.211182	1.774490	.635922	.886181	.882914
.203782	1.799603	.630236	.881699	.878199
.196453	1.825817	.624468	.877067	.873296
.189191	1.853208	.618615	.872278	.868186
.182010	1.881858	.612675	.867328	.862885
.174887	1.911914	.606661	.862155	.857378
.167809	1.943479	.600570	.856755	.851609
.153928	2.011419	.588084	.845392	.839398
.147127	2.048051	.581680	.839412	.832932
.140401	2.086717	.575179	.833169	.826193
.133739	2.127590	.568576	.826653	.819118
.120723	2.216622	.555000	.812859	.804111
.114362	2.265238	.548013	.805555	.796099
.108093	2.316993	.540894	.797927	.787713
.095880	2.431163	.526202	.781687	.769785
.084137	2.562196	.510823	.764043	.750170
.078432	2.635404	.502862	.754578	.739598

TABLE 9-3 (continued)   PARAMETERS FOR USE WITH FORMULAS OF TABLE 9-2

A/B	n_a	n_b	n_c	n_d
.067433	2.800291	.486269	.734380	.716896
.056978	2.996047	.468688	.712143	.691708
.047131	3.232638	.449900	.687586	.663652
.043365	3.341996	.442004	.676968	.651460
.039714	3.461481	.433842	.665896	.638702
.036174	3.592781	.425404	.654286	.625283
.034451	3.663353	.421060	.648302	.618329
.031090	3.816109	.412142	.635818	.603838
.029457	3.898845	.407545	.629345	.596298
.026283	4.079175	.398062	.615852	.580534
.024745	4.177734	.393165	.608814	.572309
.023243	4.282607	.388149	.601581	.563839
.021771	4.394598	.383020	.594100	.555076
.020340	4.514196	.377744	.586438	.546068
.017583	4.780675	.366775	.570311	.527108
.016262	4.929675	.361046	.561848	.517130
.014980	5.091023	.355138	.553092	.506793
.013737	5.266621	.349046	.543983	.496048
.012536	5.458381	.342741	.534527	.484885
.011377	5.668799	.336203	.524695	.473265
.010261	5.901168	.329419	.514415	.461124
.009191	6.159118	.322351	.503678	.448458
.008168	6.447475	.314966	.492433	.435174
.007192	6.772417	.307225	.480617	.421232
.006266	7.142177	.299092	.468127	.406537
.005391	7.567233	.290498	.454903	.391004
.004570	8.062065	.281372	.440832	.374507
.003805	8.647017	.271621	.425766	.356910

TABLE 9-4   EQUATIONS FOR THE PARAMETERS OF TABLE 9-3

Definitions[a]

$$E(e) = \int_0^{\pi/2} \sqrt{1 - e^2 \sin^2 \phi}\ d\phi \qquad K(e) = \int_0^{\pi} \frac{d\phi}{\sqrt{1 - e^2 \sin^2 \phi}}$$

$$e = \sqrt{1 - (b/a)^2} \qquad k = b/a$$

Equations

$$\frac{A}{B} = \frac{K(e) - E(e)}{(1/k^2) E(e) - K(e)}$$

$$n_a = \frac{1}{k} \left(\frac{2kE(e)}{\pi} \right)^{1/3} \qquad n_b = \left(\frac{2kE(e)}{\pi} \right)^{1/3}$$

$$n_c = \frac{1}{E(e)} \left(\frac{\pi^2 kE(e)}{4} \right)^{1/3} \qquad n_d = \frac{K(e)}{[E(e)]^{1/3}} \left(\frac{2k}{\pi} \right)^{2/3}$$

[a] Elliptic integrals $E(e)$ and $K(e)$ are tabulated and readily available in mathematical handbooks. Quantities a and b are semimajor and semiminor axes of the contact ellipse, respectively.

TABLE 9-4   Equations for the Parameters of Table 9-3   436

TABLE 9-5 VALUES OF CONTACT STRESSES BETWEEN TWO LONG CYLINDRICAL BODIES SLIDING AGAINST EACH OTHER IN LINE CONTACT[a]

Notation

$2b$ = width of contact area, Eq. (9.18a)
f = friction coefficient
Δ = see Eq. (9.18b)
q = distributed load (F/L)

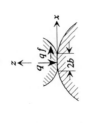

Maximum Stresses	Values of Stress in Terms of b/Δ				
	$f = 0$	$f = \frac{1}{12}$	$f = \frac{1}{9}$	$f = \frac{1}{6}$	$f = \frac{1}{3}$
Maximum tensile principal stress, occurs on surface at $x = -b$	0	$\frac{2}{12}\frac{b}{\Delta}$	$\frac{2}{9}\frac{b}{\Delta}$	$\frac{2}{6}\frac{b}{\Delta}$	$\frac{2}{3}\frac{b}{\Delta}$
Maximum compressive principal stress, occurs on surface between $x = 0$ and $x = 0.3b$	$-\frac{b}{\Delta}$	$-1.09\frac{b}{\Delta}$	$-1.13\frac{b}{\Delta}$	$-1.19\frac{b}{\Delta}$	$-1.40\frac{b}{\Delta}$
Maximum shear stress (occurs at the surface for $f \geq \frac{1}{10}$)	$0.300\frac{b}{\Delta}$	$0.308\frac{b}{\Delta}$	$0.310\frac{b}{\Delta}$	$0.339\frac{b}{\Delta}$	$0.435\frac{b}{\Delta}$
Maximum octahedral shear stress (occurs at the surface for $f \geq \frac{1}{10}$)	$0.272\frac{b}{\Delta}$	$0.265\frac{b}{\Delta}$	$0.255\frac{b}{\Delta}$	$0.277\frac{b}{\Delta}$	$0.368\frac{b}{\Delta}$

[a]From [9.9], with permission.

Dynamic Loading

Dynamic loading usually results in higher displacements and stresses than the same load would if it were applied very slowly. Most of the following chapters include formulas for natural frequencies as well as structural matrices that are used, as discussed in Appendix III, in the calculation of dynamic responses of structural members and mechanical systems. This chapter includes formulas that are useful in the dynamic design of mechanical systems subject to vibration or impact loading. Also, some fundamentals of vibration engineering are summarized, including formulas for natural frequencies and spring constants for simple systems.

10.1 NOTATION

A cross-sectional area (L^2)

B energy coefficient defined in Eq. (10.43)

E modulus of elasticity (F/L^2)

c damping coefficient (FT/L) or distance from centroid of section to its outermost fiber (L)

c_c critical damping coefficient, $2\sqrt{km} = 2m\omega_n$ (FT/L)

f frequency of vibration $(1/T)$

G	shear modulus of elasticity (F/L^2)
g	gravitational acceleration, $g = 32.16$ ft/s^2 or 386 in./s^2 or 980 cm/s^2 or 9.80 m/s^2
h	height of falling body (L)
I	moment of inertia of cross section (L^4)
I_{pi}	polar mass moment of inertia of concentrated mass at point $i(ML^2)$
k	spring constant or stiffness (F/L)
k_t	torsional stiffness (FL/rad)
L	length of member (L)
m, M_i	concentrated mass (M)
n	impact factor
P	dynamic force (F)
R	modulus of resilience of material (F/L^2)
R_u	modulus of toughness (F/L^2)
T	transmissibility or kinetic energy or torsional moment or period
U	strain energy (FL)
U_m	allowable energy absorbed in member or structure (FL)
V	potential energy (FL)
W	weight of member or structure (F)
x	displacement (L)
$\dot{x}$	velocity $= v$, (L/T)
$\ddot{x}$	acceleration $= a$, (L/T^2)
x_{st}	static displacement (L)
σ_{ys}	yield stress (F/L^2)
δ	logarithmic decrement or dynamic displacement of structure
δ_s	static deflection of beam due to its weight (L)
δ_{st}	static displacement due to weight of body and beam (L)
ω	angular frequency (rad/s)
ω_d	damped natural frequency (rad/s)
ω_n	natural frequency (rad/s)
ρ^*	mass per unit volume (M/L^3)
ρ	mass per unit length $= \rho^* A$, (M/L)
ν	Poisson's ratio
ζ	fraction of critical damping, $= c/c_c$
θ	phase angle (rad)
ψ	phase angle (rad)

10.2 CLASSIFICATION AND SOURCE OF DYNAMIC LOADINGS

Loads are often classified as static or dynamic loadings on the basis of the loading rate. In general, if the time of load application is *greater than about three times the natural period of vibration* of a structure, the loading can be specified as being *static*. If the time of load application is *less than about half the natural period of vibration*, the structure is considered to be loaded in impact or shock, i.e., the loading is *dynamic*. Another type of dynamic loading is called *inertial*-loading, which is the resisting force that must be overcome in order to cause a structure to change its velocity.

10.3 VIBRATION FUNDAMENTALS

Harmonic Motion

Vibration in general is a periodic motion. Periodic motions can be expressed as a sum of harmonic motions. A body in simple, undamped harmonic motion moves with a displacement x,

$$x = x_0 \sin \omega t = x_0 \sin(2\pi ft) \tag{10.1}$$

where f is the frequency in cycles per second, ω is the angular frequency in radians per second, and x_0 is the amplitude of the displacement. The period $T = 1/f = 2\pi/\omega$. The velocity $v = \dot{x}$ and acceleration $a = \dot{v} = \ddot{x}$ are given by

$$v = \dot{x} = x_0\omega \cos \omega t = x_0(2\pi f)\cos 2\pi ft = v_0 \cos \omega t$$
$$a = \ddot{x} = -x_0\omega^2 \sin \omega t = -x_0(2\pi f)^2 \sin 2\pi ft = a_0 \sin \omega t \tag{10.2}$$

where v_0 and a_0 are the velocity and acceleration amplitudes, respectively. Figure 10-1 exhibits the relationships between the amplitudes x_0, v_0, and a_0 as functions of frequency. The standard international units (SI) are adopted in Fig. 10-1a and the U.S. customary units in Fig. 10-1b. If two quantities of x_0, v_0, a_0, and f (frequency) are known, the other two can be found from Fig. 10-1. For example, a harmonic motion with $x_0 = 0.0001$ m and $f = 50$ Hz corresponds to point A in Fig. 10-1a. It follows from this figure that the velocity amplitude v_0 is approximately 0.032 m/s and the acceleration amplitude a_0 is about 1.02 g (9.9 m/s^2).

Single-Degree-of-Freedom System

A mass m, spring k, and damper c translational motion system is shown in Fig. 10-2. The viscous damper c, which dissipates energy, is not considered to be a very accurate representation of the actual damping in most physical systems. Analogous quantities for a rotational system are shown in Table 10-1.

General Equation of Motion   For the system of Fig. 10-2

$$m\ddot{x} + c\dot{x} + kx = P \tag{10.3}$$

Free Vibration without Damping   For a system oscillating without applied loading and damping

$$m\ddot{x} + kx = 0 \tag{10.4}$$

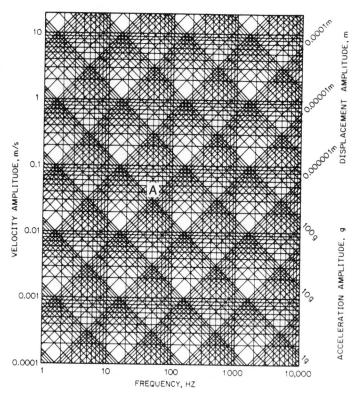

FIGURE 10-1a Relationship between frequency f and the amplitudes of displacement x_0, velocity v_0, and acceleration a_0 in Standard International Units. 1 Hz $= 1$ cycle/sec; 1 g $= 9.8$ m/sec^2.

The solution is

$$x = C_1 \sin\sqrt{k/m}\, t + C_2 \cos\sqrt{k/m}\, t = C_3 \sin(\omega_n t + \theta) \qquad (10.5)$$

where $C_3 = \sqrt{C_1^2 + C_2^2}$ and the phase angle $\theta = \tan^{-1}(C_2/C_1)$.

The natural frequency is

$$\omega_n = \sqrt{k/m} \qquad (\text{rad/s})$$

or

$$f_n = \frac{1}{T} = \frac{\omega_n}{2\pi} = \frac{1}{2\pi}\sqrt{\frac{k}{m}}$$

$$= \frac{1}{2\pi}\sqrt{\frac{kg}{W}} = \frac{1}{2\pi}\sqrt{\frac{g}{\delta_{\text{st}}}} \qquad (\text{cycles/s}) \qquad (10.6)$$

where the weight $W = mg$ and the static displacement $\delta_{\text{st}} = W/k$.

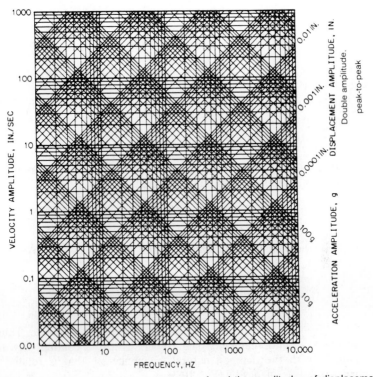

FIGURE 10-1b Relationship between frequency f and the amplitudes of displacement x_0, velocity v_0, and acceleration a_0 in U.S. Customary Units. 1 Hz = 1 cycle/sec; 1 g = 386 in/sec^2.

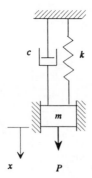

FIGURE 10-2 Translational single-degree-of-freedom system.

Free Vibration with Viscous Damping Including damping the equation of motion for the free vibration is

$$m\ddot{x} + c\dot{x} + kx = 0 \qquad (10.7)$$

The solution depends on whether c is equal to, greater than, or less than the *critical damping coefficient* c_c, where

$$c_c = 2\sqrt{km} = 2m\omega_n$$

The ratio $\zeta = c/c_c$ is known as the fraction of critical damping or the percentage of damping if it is written as a percentage.

If $c = c_c$ ($\zeta = 1$), the case of *critical damping*, there is no oscillation and the solution is

$$x = (C_1 + C_2 t)e^{-ct/2m} \tag{10.8}$$

If $c/c_c = \zeta > 1$, the system is *overdamped* so that the mass does not oscillate but returns to its equilibrium position. The solution is

$$x = e^{-ct/2m}\left(C_1 e^{\omega_n\sqrt{\zeta^2-1}\,t} + C_2 e^{-\omega_n\sqrt{\zeta^2-1}\,t}\right) \tag{10.9}$$

If $c/c_c < 1$, the system is *underdamped*. The solution is

$$x = e^{-ct/2m}(C_1 \sin \omega_d t + C_2 \cos \omega_d t)$$
$$= C_3 e^{-ct/2m} \sin(\omega_d t + \theta) \tag{10.10}$$

where $C_3 = (C_1^2 + C_2^2)^{1/2}$ and $\theta = \tan^{-1}(C_2/C_1)$ is the phase angle. Thus, after being disturbed, an underdamped system oscillates with a continuously decreasing amplitude, with the damped natural frequency ω_d that is related to the undamped natural frequency by (Fig. 10-3)

$$\omega_d = \sqrt{k/m - c^2/4m^2} = \omega_n(1 - \zeta^2)^{1/2} \tag{10.11}$$

With each cycle the amplitude of an underdamped system decreases. The *logarithmic decrement* δ is the natural logarithm of the ratio of the amplitudes of

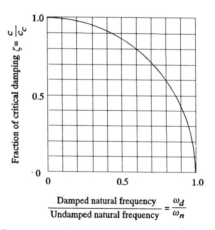

FIGURE 10-3 Damped natural frequency as it varies with critical damping.

two successive cycles.

$$\delta = \ln \frac{x_i}{x_{i+1}} \left(\text{or } \frac{x_{i+1}}{x_i} = e^{-\delta} \right) = \frac{\pi c}{m \omega_d} = \frac{2\pi \zeta}{\left(1 - \zeta^2\right)^{1/2}} \qquad (10.12)$$

For ζ less than about 0.1,

$$\delta \approx 2\pi \zeta \qquad (10.13)$$

Forced Vibration without Damping For the sinusoidal force $P = P_0 \sin \omega t$, the governing equation for an undamped system is

$$m\ddot{x} + kx = P_0 \sin \omega t \qquad (10.14)$$

with the solution

$$x = C_1 \sin \omega_n t + C_2 \cos \omega_n t + \frac{P_0/k}{1 - \omega^2/\omega_n^2} \sin \omega t \qquad (10.15)$$

where $\omega_n = \sqrt{k/m}$. The first two terms describe the oscillation at the undamped natural frequency ω_n. The coefficient C_2 is the initial displacement, and the coefficient C_1 in terms of the initial velocity is

$$C_1 = \frac{v_0}{\omega_n} - \frac{\omega P_0/(\omega_n k)}{1 - \omega^2/\omega_n^2} \qquad (10.16)$$

The third term gives the steady-state oscillation

$$x = \frac{P_0/k}{1 - \omega^2/\omega_n^2} \sin \omega t \qquad (10.17)$$

or in terms of the peak displacement (amplitude) x_0,

$$\frac{x_0}{x_{st}} = \frac{1}{1 - \omega^2/\omega_n^2} = magnification\ factor \qquad (10.18)$$

where $x_{st} = P_0/k$ is the static displacement due to force P_0. The magnification factor is plotted in Fig. 10-4. When $\omega/\omega_n < 1$, the force and motion are in phase and the magnification factor is positive. When $\omega/\omega_n > 1$, the force and motion are out of phase and the magnification is negative. The dashed line indicates the absolute value of the curve for $\omega/\omega_n > 1$. When $\omega = \omega_n$, resonance occurs and the amplitude increases steadily with time.

Force transmissibility is defined as $T = P_t/P$, where $P_t = kx$, with x of Eq. (10.17). Then

$$T = \frac{1}{1 - \omega^2/\omega_n^2} \qquad (10.19)$$

which equals x_0/x_{st}.

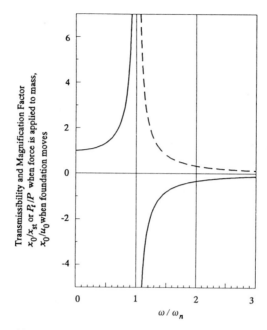

FIGURE 10-4 Magnification factor and transmissibility for an undamped system.

If the base or foundation of the system moves as $u = u_0 \sin \omega t$ and the applied force is zero, the governing equation is

$$m\ddot{x} = -k(x - u_0 \sin \omega t) \tag{10.20}$$

where x is the displacement of the mass in absolute coordinates.
The solution is the same as Eq. (10.15) with P_0/k replaced by u_0. The motion transmissibility is

$$\frac{x_0}{u_0} = \frac{1}{1 - \omega^2/\omega_n^2} = T \tag{10.21}$$

the same value as the force transmissibility.

Forced Vibration with Viscous Damping For the steady-state loading $P = P_0 \sin \omega t$, the equation of motion for a system with viscous damping is

$$m\ddot{x} + c\dot{x} + kx = P_0 \sin \omega t \tag{10.22}$$

The solution is

$$x = e^{-ct/2m}(C_1 \sin \omega_d t + C_2 \cos \omega_d t)$$
$$+ \frac{(P_0/k) \sin(\omega t - \theta)}{\sqrt{\left(1 - \omega^2/\omega_n^2\right)^2 + (2\zeta\omega/\omega_n)^2}} \tag{10.23}$$

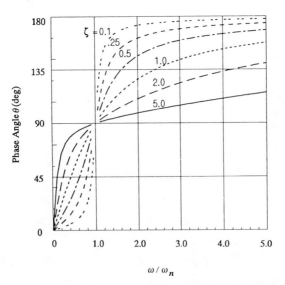

FIGURE 10-5 Phase angle θ as a function of ω/ω_n and ζ.

with

$$\theta = \tan^{-1} \frac{2\zeta\omega/\omega_n}{1 - \omega^2/\omega_n^2}$$

which is plotted in Fig. 10-5. The term of Eq. (10.23) involving C_1 and C_2 decays due to damping, leaving the steady-state motion of amplitude

$$x_0 = \frac{P_0/k}{\sqrt{\left(1 - \omega^2/\omega_n^2\right)^2 + \left(2\zeta\omega/\omega_n\right)^2}} \qquad (10.24)$$

where P_0/k is the static displacement x_{st} due to force P_0.

The magnification factor is the ratio x_0/x_{st}, which is plotted in Fig. 10-6. Force transmissibility T is obtained from

$$\frac{P_t}{P_0} = \frac{\text{force transmitted to foundation}}{\text{applied force amplitude}} = \frac{c\dot{x} + kx}{P_0} = T\sin(\omega t - \psi) \qquad (10.25a)$$

where

$$T = \sqrt{\frac{1 + \left(2\zeta\omega/\omega_n\right)^2}{\left(1 - \omega^2/\omega_n^2\right)^2 + \left(2\zeta\omega/\omega_n\right)^2}} \qquad (10.25b)$$

$$\psi = \tan^{-1} \frac{2\zeta(\omega/\omega_n)^3}{1 - \omega^2/\omega_n^2 + 4\zeta^2\omega^2/\omega_n^2} \qquad (10.25c)$$

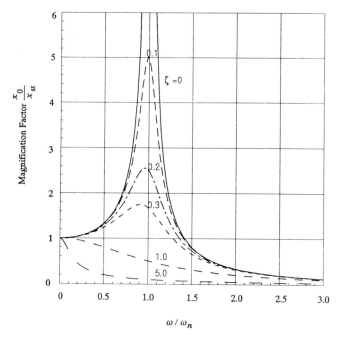

FIGURE 10-6 Magnification factor for a damped system. Maximum magnification factor occurs at $\omega / \omega_n = \sqrt{1 - 2\zeta^2}$

The transmissibility T and phase angle ψ are shown in Figs. 10-7 and 10-8.

In the case of displacement $u(t)$ applied to the base (foundation), the equation of motion is

$$m\ddot{x} + c(\dot{x} - \dot{u}) + k(x - u) = 0 \qquad (10.26)$$

For $u = u_0 \sin \omega t$, the steady-state response is

$$x = Tu_0 \sin(\omega t - \psi) \qquad (10.27)$$

where T and ψ are defined in Eqs. (10.25b) and (10.25c). The motion transmissibility T is thus given in

$$\frac{x}{u_0} = \frac{\text{displacement of mass}}{\text{applied displacement (amplitude) of base}} = T \sin(\omega t - \psi) \quad (10.28)$$

Figures 10-7 and 10-8 illustrate this transmissibility. Note that the transmissibility is less than 1 when the excitation frequency is greater than $\sqrt{2}$ times the natural frequency.

A *resonant frequency* is the frequency at which a peak response occurs. The resonant frequencies of interest, all of which differ from the damped natural

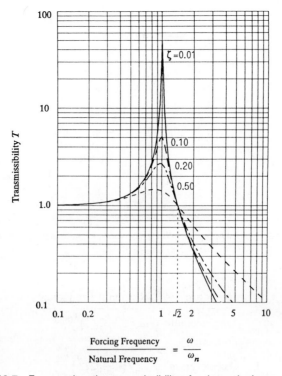

FIGURE 10-7 Force and motion transmissibility of a viscously damped system.

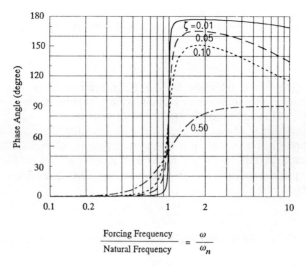

FIGURE 10-8 Phase angle ψ of force or motion transmission of a viscously damped system.

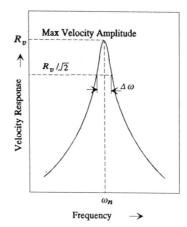

FIGURE 10-9   Velocity magnification factor showing bandwidth at half-power point.

frequency $\omega_d = \omega_n(1 - \zeta^2)^{1/2}$, are

Response	Resonant Frequency
Displacement	$\omega_n(1 - 2\zeta^2)^{1/2}$
Velocity	ω_n
Acceleration	$\omega_n/(1 - 2\zeta^2)^{1/2}$

$$(10.29)$$

For the percentage of damping in most physical systems, the differences in resonant frequencies are negligible.

Damping   The amount of damping is measured by a *quality factor Q*, the magnification factor for velocity response at the undamped natural frequency:

$$Q = \frac{1}{2\zeta} \qquad (10.30)$$

The same Q is often taken to be the magnification factor for displacement or acceleration, although these responses are slightly higher by the factor $1/(1 - \zeta^2)^{1/2}$. One refers to high-Q or low-Q systems.

The quality factor Q can be approximated from the sharpness or width of a response curve in the vicinity of a resonant frequency. Designate the width of a response curve as the frequency increment $\Delta\omega$ measured at the "half-power point," i.e., peak response/$\sqrt{2}$, as shown in Fig. 10-9. Then for $\zeta < 0.1$, Q can be approximated by

$$Q \approx \omega_n/\Delta\omega \qquad (10.31)$$

10.4 NATURAL FREQUENCIES

The natural frequencies for various structural members and mechanical systems are listed throughout this book. For more complex structures, the structural matrices provided in most of the chapters are available for computing the natural

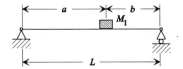

FIGURE 10-10   Simple mass beam system, Example 10.1.

frequencies. Table 10-2 contains natural frequencies for a few commonly occurring simple systems.

It was shown in the previous section that the fundamental natural frequency ω_n can be obtained using $\omega_n = \sqrt{k/m}$ if the mass m and spring constant k are known. For a rigid mass supported by massless elastic members, the spring constant is the force at the attachment point of the mass due to a unit displacement at this attachment point. Tables 10-3 and 10-4 provide spring constants of some structural members and systems. These can be used to develop models of structures and machines as single-degree-of-freedom systems and then to apply $\omega_n = \sqrt{k/m}$ to find the fundamental frequency. In the case of torsional systems $\omega_n = \sqrt{k_t/I_{pi}}$, where k_t is the torsional stiffness and I_{pi} is the polar mass moment of inertia. Mass and mass moment of inertia are provided in Table 10-5. Some damping coefficients are listed in Table 10-6.

Example 10.1   Simple Mass Beam System   Find the natural frequency for bending motion of the simple mass beam system of Fig. 10-10.

Treat the beam as being massless with a mass not having appreciable dimensions. The single degree of freedom is chosen to be the deflection of the beam at the mass. The spring constant is given by case 8 in Table 10-3 as

$$k = 3EIL/(a^2b^2) \tag{1}$$

The natural frequency for bending motion of the simple mass beam system is

$$\omega_n = \sqrt{\frac{k}{m}} = \sqrt{\frac{k}{M_1}} = \frac{1}{ab}\sqrt{\frac{3EIL}{M_1}} \tag{2}$$

If the mass is at the center of the beam, i.e., $a = b$, then the natural frequency will be

$$\omega_n = \frac{4}{L^2}\sqrt{\frac{3EIL}{M_1}} = \sqrt{\frac{48EI}{L^3M_1}} \tag{3}$$

Example 10.2   A Mass Beam System with Flexible Supports   Find the natural frequency of the system of Fig. 10-11.

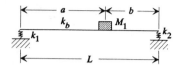

FIGURE 10-11 Mass beam system with flexible supports, Example 10.2.

The method used here is similar to that of Example 10.1. The spring constant of the beam remains

$$k_b = 3EIL/(a^2 b^2) \tag{1}$$

The spring constants of the flexible supports are k_1 and k_2. The equivalent spring constant k_s of the beam support system is from case 3 of Table 10-4:

$$k_s = \frac{(a + b)^2}{a^2/k_2 + b^2/k_1} \tag{2}$$

From case 1 of Table 10-4, the equivalent stiffness of the whole system is

$$k = k_b k_s/(k_b + k_s) \tag{3}$$

Therefore, the natural frequency of the mass beam system with flexible supports is

$$\omega_n = \sqrt{k/m} = \sqrt{k_b k_s/M_1(k_b + k_s)} \tag{4}$$

10.5 IMPACT FORMULAS

The excessively complex mathematics required for accurate analytical solutions of dynamic problems, the lack of precise knowledge of the properties of even conventional engineering materials, and the difficulties in adequately defining a complicated physical structure make it impractical in many cases to give a solution using a rigorous theory of dynamics. Therefore, approximate methods can be quite useful in engineering design. One method is to treat a dynamic load as a static load and to estimate the maximum static force and then to obtain the dynamic response by applying an "impact factor" to the static response.

The dynamic responses of simple elastic beam models subjected to impact loading are given in Table 10-7. These responses are found with an energy balance. The potential energy of the falling body changes into kinetic energy when it reaches the beam, and then the kinetic energy is absorbed by the beam when the body comes to rest. It is assumed that there is no energy loss associated with the local plastic deformation occurring at the point of impact or at the supports. Energy is thus conserved within the system. The material behaves elastically.

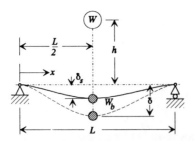

FIGURE 10-12   Beam of weight W_b and body of weight W.

The formulas of Table 10-7 apply to the beam of Fig. 10-12, whose weight (W_b) is considered to be concentrated at one location. Let δ_s be the initial deflection of the beam under its own weight and δ be the maximum total deflection of the beam after the impact of the body of weight W. The constant k represents the stiffness of the beam referenced to the point of contact of the falling body. The total maximum force P experienced by the beam is

$$P = (W + W_b) + \sqrt{W^2 + 2W(W_b + kh)} \qquad (10.32)$$

If the body W is suddenly applied to the beam from zero height, $h + \delta_s = 0$ and, from Eq. (10.32), $2W(W_b + kh) = 2Wk(\delta_s + h) = 0$. Then the maximum impact force would be

$$P = (W + W_b) + W = 2W + W_b \qquad (10.33)$$

The effect of the beam's inertia can be seen from Eq. (10.33). If the weight W_b of the beam is negligible, $W_b \approx 0$,

$$P = 2W \qquad (10.34)$$

So, this is why it is common to apply an impact factor of 2 to a static response to obtain dynamic effects.

In general, the *impact factor n* is defined as the dynamic force divided by the static load, i.e.,

$$n = P/(W + W_b)$$

where it is assumed that the weight of the beam is not negligible. From Eq. (10.33)

$$n = \frac{P}{W + W_b} = \frac{2W + W_b}{W + W_b} = 1 + \frac{1}{1 + W_b/W} \qquad (10.35)$$

The impact factor n decreases as the weight of the beam W_b increases (Fig. 10-13), because some of the energy of the falling weight is absorbed by the inertia of the beam.

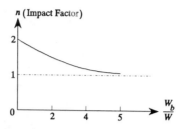

FIGURE 10-13 Effect of beam mass on impact factor.

The dynamic deflection in terms of the static load deflection is (Table 10-7)

$$\delta = \delta_{st} + \sqrt{\delta_{st}^2 + 2h(\delta_{st} - \delta_s) - \delta_s^2} \tag{10.36}$$

where

$$\cdot\ \delta = \frac{P}{k}, \qquad \delta_s = \frac{W_b}{k}, \qquad \delta_{st} = \frac{W_b + W}{k} = \delta_s + \frac{W}{k}$$

If $h + \delta_s = 0$, Eq. (10.36) becomes

$$\delta = \delta_{st} + \sqrt{\delta_{st}^2 + \delta_s(\delta_s - 2\delta_{st})} \tag{10.37}$$

Thus, when $W_b/W \approx 0$, $\delta_s \approx 0$ and, as expected, $\delta = 2\delta_{st}$, while if

$$\frac{W}{W_b} \approx 0 \tag{10.38}$$

we have

$$\delta = \delta_s \tag{10.39}$$

In reality, the mass of the supporting structure is distributed over the beam, and thus only a portion of its mass is effective in reducing the dynamic deflection δ and the impact factor n. It has been shown [10.1] that the portion of the mass of the structure to be used is

(a) for a simply supported beam with concentrated load at the midpoint

$$W_e = 0.486W_m \tag{10.40}$$

(b) for a cantilever beam with a concentrated load at the free end

$$W_e = 0.236W_m \tag{10.41}$$

where W_m is the total weight of the beam and W_e is the equivalent concentrated weight.

10.6 ENERGY-ABSORBING CHARACTERISTICS OF STRUCTURES

The strain energy for the bending of a beam is given by

$$U = \int_0^L \frac{M^2}{2EI} \, dx \qquad (10.42a)$$

Consider a beam of length L and constant I for which M is constant. Then with $\sigma = Mc/I$ (Chapter 3),

$$U = \sigma^2 IL/(2Ec^2) \qquad (10.42b)$$

This can be extended to include other cases, including nonconstant M, by expressing Eq. (10.42b) as

$$U = B\sigma^2 IL/(Ec^2) \qquad (10.43)$$

where B is a coefficient to account for particular boundary conditions and types of loading. Increasing the stress in the beam will increase its strain energy. Assume σ is limited by the material's yield stress σ_{ys}. As the stress σ reaches its maximum value σ_{ys}, the strain energy becomes

$$U_m = B\sigma_{ys}^2 IL/(Ec^2) \qquad (10.44)$$

where U_m is the maximum allowable energy, or energy that can be absorbed elastically by the member in bending. If the strain energy U in the beam exceeds the maximum allowable energy U_m, failure will probably occur. As with the relationship between stress σ and allowable stress σ_{ys} in a static stress analysis, the design condition in a dynamic analysis with this energy approach is that

$$U_{design} \le U_m \qquad (10.45)$$

In general, the potential and kinetic energy $V + T$ of the dynamic loads are converted into the internal energy of the beam and become the strain energy U. Thus

$$U = V + T \qquad (10.46)$$

Table 10-8 lists the specific formulas of Eq. (10.44) for some common member and load conditions. It is of interest to note that for the same types of member, the energy-absorbing capacity, or the maximum allowable energy U_m, with the same static strength σ_{ys} in bending depends on the section property I/c^2.

Let this section property be given as

$$N = I/c^2 \qquad (10.47)$$

For a rectangular cross section of height h and width b, $N = (\frac{1}{12}bh^3)/(\frac{1}{2}h)^2 = \frac{1}{3}bh = \frac{1}{3}A$, the section property only depends on the area of the cross section. Therefore, if the cross-sectional area is fixed, increasing the depth of a

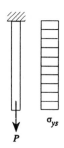

σ_{ys}

P

FIGURE 10-14 Axial impact.

cross section will increase the section's static strength (by reducing its maximum stress, e.g., $\sigma_{max} = Mc/I = 6M/Ah$ for a rectangle) but with little or no increase in impact strength, or in the energy-absorbing capacity indicated by Eq. (10.44).

Equation (10.44) also indicates that in order to obtain maximum energy-absorbing capacity for a member, we should design the member so that its maximum volume is subjected to the maximum allowable stress. For example, for a rectangular section ($I/c^2 = \frac{1}{3}bh$) for which the whole volume is stressed to σ_{ys}, Eq. (10.44) becomes

$$U_m = B\frac{bhL\sigma_{ys}^2}{3E} = \frac{B\sigma_{ys}^2}{3E}V \qquad (10.48)$$

where $V = bhL$ is the volume of the beam with constant cross section. It is evident that U_m as defined in Eq. (10.48) is the maximum value of the allowable energy absorbed in the beam. When only a portion of the beam volume V is subject to the maximum allowable stress σ_{ys}, the allowable energy absorbed from Eq. (10.43) will be less than U_m. Generally speaking, the energy-absorbing capacity will reach its highest value if this maximum allowable stress is uniform throughout the member. In the case of a member in axial tension (Fig. 10-14), a constant cross section throughout the member implies that the stress will be uniformly distributed over the entire volume. For impact loading, this corresponds to a configuration of the maximum energy-absorbing capacity; i.e.; it can withstand the maximum dynamic loading.

For a beam, a constant bending stress along its entire length can be obtained by choosing a variable depth (Fig. 10-15). In this case, the outermost fiber is stressed to the same maximum value for the entire length of the member. The energy-absorbing capacity of the beam in Fig. 10-15 (case 9 of Table 10-8) is doubled compared to that in case 1 of Table 10-8.

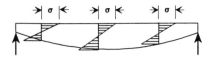

FIGURE 10-15 Constant-stress beam.

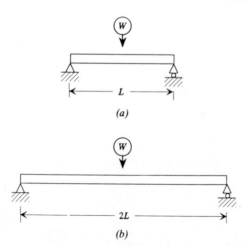

FIGURE 10-16 Dynamic characteristics of beams with identical cross sections. (*a*) case 1 with length *L*, (*b*) case 2 with length 2*L*

For a static load, increasing the length of the beam will increase the bending moment M, with a corresponding increase in the beam stress ($\sigma = Mc/I$). For an impact load, Eq. (10.44) shows that increasing the length of the beam will increase its maximum allowable energy. Thus, for the two beams in Fig. 10-16, it follows from Eq. (10.43) that the peak dynamic stresses for the two cases are related by $\sigma_1 = (1/\sqrt{2})\sigma_2$.

Since N is proportional to the area of a section, not the shape, it can be concluded that the two beams of Fig. 10-17 have the same dynamic properties. However, their static stress characteristics determined by $\sigma = Mc/I$ are distinctly different. Also, it is interesting to observe that the two tensile rods of Fig. 10-18 have the same strength under static loading. However, the rod in (*b*), with a

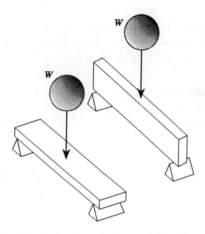

FIGURE 10-17 Two identical rectangular beams with different orientations.

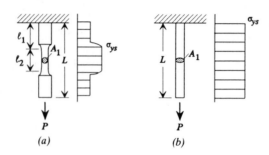

FIGURE 10-18 Stresses in rods.

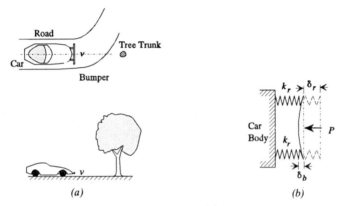

FIGURE 10-19 Example 10.3: (a) a car and a tree: (b) bumper composed of two springs and a beam.

uniform cross section, can absorb more energy and withstand a greater impact load.

The strain energy U in Eq. (10.42a) can be calculated directly from the external force by taking the beam as an elastic spring with the stiffness given by case 8 of Table 10-3. For other kinds of structures, similar reasoning can be used.

Example 10.3 Impact Force Suppose a car has a bumper formed of an elastic beam and two springs as shown in Fig. 10-19. The car hits a tree at the center of the bumper with a speed v. Analyze the impact force on the bumper, neglecting the mass of the bumper.

First consider the deformation of the bumper, i.e., the displacement of the beam and the springs, under the impact load P, as shown in Fig. 10-19b. Let k_r and k_b be the stiffness coefficients of each spring and the beam, which can be obtained from case 1 of Table 10-3 for the spring and case 8 of Table 10-3 for the beam, if its ends are simply supported. Let δ_r and δ_b be the deformations of each spring and the beam center. Assume the tree is strong enough to withstand the impact.

Let W_c be the car's weight. Then the kinetic energy of the car immediately before impact is

$$T = (W_c/2g)v^2 \tag{1}$$

The strain energy for the linear elastic deformation is given by

$$U = U_{spr} + U_{bm} = 2\left(\tfrac{1}{2}k_r\delta_r^2\right) + \tfrac{1}{2}k_b\delta_b^2 \tag{2}$$

$$= k_r(P/2k_r)^2 + \tfrac{1}{2}k_b(P/k_b)^2$$

$$= \tfrac{1}{2}P^2(1/2k_r + 1/k_b) \tag{3}$$

or by using case 4 of Table 10-4 to get the equivalent stiffness of the beam and spring system:

$$U = \tfrac{1}{2}P\delta = \frac{P}{2}\frac{P}{k} = \frac{P^2}{2}\left(\frac{1}{k_b} + \frac{1}{k_r + k_r}\right) = \frac{P^2}{2}\left(\frac{1}{2k_r} + \frac{1}{k_b}\right) \quad \text{where } \delta = \frac{P}{k}$$

Set the strain energy U of the bumper equal to the kinetic energy of the moving car when it hits the tree. Use of (1) and (3) yields

$$\frac{W_c}{2g}v^2 = \frac{P^2}{2}\left(\frac{1}{2k_r} + \frac{1}{k_b}\right)$$

Thus

$$P^2 = \frac{W_c v^2/g}{1/2k_r + 1/k_b} \tag{4}$$

or

$$P = \left(\frac{W_c v^2/g}{1/2k_r + 1/k_b}\right)^{1/2} \tag{5}$$

Let $k = (1/2k_r + 1/k_b)^{-1}$. Then

$$P = \left(kW_c v^2/g\right)^{1/2} \tag{6}$$

where k is the bumper's stiffness coefficient.

Equation (6) indicates that the higher the bumper's stiffness, the greater the impact force. Therefore, judicious selection of the stiffness of the bumper system is very important for reducing the effects of such kinds of impact loading.

10.7 DYNAMIC BEHAVIOR OF MATERIALS

In most cases, the mechanical properties of materials under impact or shock loading differ from those under static loading. For example, the static stress–strain relation for iron is quite different from its dynamic material relation. In general, increasing the loading rate will increase a material's ability to resist yielding while having little effect on its resistance to fracture. Because of the complexity of the analysis required and the rather poor knowledge of dynamic behavior of materials, energy methods similar to those of this chapter are commonly used to determine the dynamic behavior. An example is the notched-bar impact test method (Chapter 4), in which a notch is placed in a standard specimen and the maximum energy absorbed is recorded as a measure of the dynamic characteristic of the material. The results of the notched-bar impact test are considered to be of limited value and can be misleading because the standard test conditions are far from the conditions faced in practice.

As indicated in Chapter 4, the two important material properties that measure the ability to absorb energy are the modulus of resilience R and the modulus of toughness R_u. These properties can be obtained from stress–strain diagrams. Impact properties of some common engineering materials are given in Table 10-9.

The notch effect on energy absorption can also be analyzed using the formulas of Table 10-8. For the tensile members shown in Fig. 10-20a, b, assume the notch induces a stress concentration of twice the average stress (Fig. 10-20d). Then the average peak stress in the member will be reduced by $\frac{1}{2}$ and (case 4, Table 10-8)

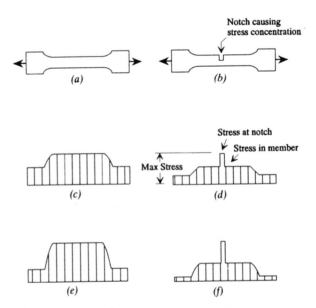

FIGURE 10-20 Notch effect on energy-absorbing capacity of tensile member [10.1]: (a) tensile member, uniform cross section; (b) tensile member with notch; (c) stress diagram for (a); (d) stress diagram for (b); (e, f) energy diagrams.

the energy absorbed (Fig. 10-20f) will be one-fourth of the energy absorbed if no notch were present (Fig. 10-20e).

10.8 INCREASING THE DYNAMIC STRENGTH OF STRUCTURES AND MINIMIZING DYNAMIC EFFECTS

It follows from the above discussions that increasing the energy-absorbing capacity of a structure amounts to improving the dynamic strength. The basic rules outlined below are useful in designing for dynamic load.

Geometric Configuration

1. For a given structure, have as much of the structure as possible stressed to the working stress level.
2. Reduce the dynamic stress in tension by increasing the volume (AL) and that in bending by increasing the value of I/c^2.
3. An increase in the length of a beam will increase the static stress but will decrease the stress due to dynamic loading. For a tensile member of uniform cross section, an increase in length will not change the static stress but will decrease the stress due to a dynamic load.
4. Avoid abrupt changes in section area and internal inhomogeneousness to minimize stress concentration.
5. Design for maximum flexibility in the vicinity of the point of impact in order to increase the energy-absorbing capacity.

Material Properties

1. Select material with a high modulus of resilience R, the energy storage capacity per unit volume, and good notch toughness. The material should be ductile enough to plastically relieve the stress in areas of high stress concentration.
2. If repeated dynamic loads are expected, choose a material with high fatigue strength.
3. For low-temperature service, the material should have a low nil ductility transition temperature (NDT). Refer to Chapter 4.
4. If possible, arrange for a metal for which the direction of hot rolling is in line with the dynamic load.

Loading

1. For dynamic forces due to inertia, decrease the mass of the structure while maintaining the proper rigidity for its particular use. In bending, a beam that is lightweight and has sufficient moment of inertia should be used.
2. For impact, minimize the impacting speed and the mass of impacting bodies, if possible.

3. To lower the acceleration of a structure and hence to reduce possible inertial forces caused by the rapid movement of a structure due to explosive energy, earthquakes, etc., employ flexible supports.

REFERENCES

10.1 Blodgett, O. W., *Design of Welded Structures*, The James F. Lincoln Arc Welding Foundation, Cleveland, OH, 1982.

10.2 Rothbart, H. A., *Mechanical Design and Systems Handbook*, 2nd ed., McGraw-Hill, New York, 1985.

10.3 Faupel, J. H., and Fisher, F. E., *Engineering Design*, 2nd ed., Wiley, New York, 1981.

10.4 Zukas, J. A., et al., *Impact Dynamics*, Wiley, New York, 1982.

10.5 Shigley, J. E., and Mischke, C. R., *Standard Handbook of Machine Design*, McGraw-Hill, New York, 1986.

10.6 Schiff, D., *Dynamic Analysis and Failure Modes of Simple Structures*, Wiley, New York, 1990.

10.7 Nashif, A. D., Jones, D. I. G., Henderson, J. P., *Vibration Damping*, Wiley, New York, 1985.

10.8 Harris, C., and Crede, C. E., *Shock and Vibration Handbook*, 3rd ed., McGraw-Hill, New York, 1988.

10

Tables

TABLE 10-1 ANALOGOUS QUANTITIES IN TRANSLATIONAL AND ROTATIONAL SYSTEMS

Translational Quantities	Rotational Quantities

Displacement	x (L)	ϕ (rad)
Spring constant	k (F/L)	k_t (FL/rad)
Damping constant	c $(F \cdot \sec/L)$	c_t $(FL \cdot \sec/\text{rad})$
Applied force	$P(t)$ (F)	$T(t)$ (FL)
Mass quantity	m (M)	I_{pi} (ML^2) (mass polar moment of inertia)
	$m = \int_V \rho^* \, dV$	$I_{pi} = \int_V \rho^* r^2 \, dV$
	$V = $ volume	$\rho^* = $ mass per unit volume
Differential		$r = $ radial distance from center of disk
equation	$m\ddot{x} + c\dot{x} + kx = P(t)$	$I_{pi}\ddot{\phi} + c_t\dot{\phi} + k_t\phi = T(t)$

TABLE 10-1 **Translational and Rotational Systems** 464

TABLE 10-2   FREQUENCIES OF COMMON SYSTEMS

Notation

ω_n = natural frequency
k = spring stiffness
I_{pi} = polar mass moment
     of inertia of lumped mass
     at point i
A = cross-sectional area
m, m_i, M_i = mass

ρ^* = mass per unit volume
ρ = mass per unit length, = $\rho^* A$
E = modulus of elasticity
I = area moment of inertia
L = length
m_b = total mass of beam
ν = Poisson's ratio
g = acceleration of gravity

See also Table 11-12

System	Natural Frequency ω_n
Extension Systems	
1.	$\sqrt{\dfrac{k}{m}}$
2. Mass suspended by spring with mass m_s = mass of spring	$\sqrt{\dfrac{k}{(m + m_s/3)}}$
3.	$\sqrt{2k/m}$
4.	$\left(\dfrac{1}{2}\left[\dfrac{k_1}{m_1} + \dfrac{k_2}{m_2}\left(1 + \dfrac{m_2}{m_1}\right) \right] \pm \sqrt{ \left[\dfrac{k_1}{m_1} + \dfrac{k_2}{m_2}\left(1 + \dfrac{m_2}{m_1}\right) \right]^2 - \dfrac{4k_1 k_2}{m_1 m_2} } \right)^{1/2}$
5.	$\sqrt{\dfrac{k(m_1 + m_2)}{m_1 m_2}}$

TABLE 10-2 (continued)   FREQUENCIES OF COMMON SYSTEMS

System	Natural Frequency ω_n
6. Rigid bar with mass 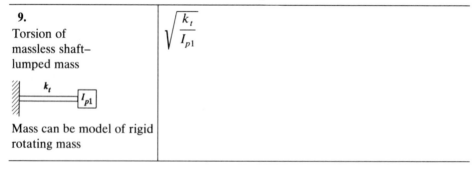	$\dfrac{a}{b}\sqrt{\dfrac{k}{m}}$
7. Longitudinal vibration of bar Free - Free Node $\longleftarrow L \longrightarrow$ Fixed - Fixed Anti - Node	$\dfrac{\pi}{L}\sqrt{\dfrac{E}{\rho^*}}$ Fundamental frequency
8. Longitudinal vibration of bar Fixed - Free $\longleftarrow L \longrightarrow$	$\dfrac{\pi}{2L}\sqrt{\dfrac{E}{\rho^*}}$ Fundamental frequency
	Torsional Systems
9. Torsion of massless shaft– lumped mass k_t I_{p1} Mass can be model of rigid rotating mass	$\sqrt{\dfrac{k_t}{I_{p1}}}$

TABLE 10-2   Frequencies of Common Systems   **466**

TABLE 10-2 (continued) FREQUENCIES OF COMMON SYSTEMS

System	Natural Frequency ω_n
10. Torsion of shaft with mass and lumped mass I_{p2} = total polar mass moment of inertia of the shaft	$\sqrt{\dfrac{k_t}{I_{p1} + \dfrac{1}{3}I_{p2}}}$
11. Torsional system	$\left(\dfrac{1}{2}\left[\dfrac{k_{t1}}{I_{p1}} + \dfrac{k_{t2}}{I_{p2}}\left(1 + \dfrac{I_{p2}}{I_{p1}}\right) \pm \sqrt{\left[\dfrac{k_{t1}}{I_{p1}} + \dfrac{k_{t2}}{I_{p2}}\left(1 + \dfrac{I_{p2}}{I_{p1}}\right)\right]^2 - \dfrac{4k_{t1}k_{t2}}{I_{p1}I_{p2}}}\,\right]\right)^{1/2}$
12. Torsional system	$\sqrt{\dfrac{k_t(I_{p1} + I_{p2})}{I_{p1}I_{p2}}}$
13. Torsional system	$\sqrt{\dfrac{1}{2}\left[C \pm \sqrt{C^2 - \dfrac{4k_{t1}k_{t2}}{I_{p1}I_{p2}I_{p3}}(I_{p1} + I_{p2} + I_{p3})}\right]}$ where $C = \dfrac{k_{t1}}{I_{p1}} + \dfrac{k_{t2}}{I_{p3}} + \dfrac{k_{t1} + k_{t2}}{I_{p2}}$
14. Torsion of geared system with massless gears	$\sqrt{\dfrac{k_{t1}k_{t2}(I_{p1} + n^2 I_{p2})}{I_{p1}I_{p2}(n^2 k_{t2} + k_{t1})}} \qquad n = \dfrac{\text{rotor 2 speed}}{\text{rotor 1 speed}}$

TABLE 10-2 (continued)   **FREQUENCIES OF COMMON SYSTEMS**

System	Natural Frequency ω_n
	Lumped Mass

15.
Beam fixed–free

Set $m_b = 0$ if beam is massless

$$\sqrt{\frac{3EI}{L^3(M_1 + 0.23m_b)}}$$

16.
Beam pinned–pinned

Set $m_b = 0$ if beam is massless

$$\sqrt{\frac{48EI}{L^3(M_1 + 0.5m_b)}}$$

17.
Beam fixed—fixed

Set $m_b = 0$ if beam is massless

$$13.86\sqrt{\frac{EI}{L^3(M_1 + 0.375m_b)}}$$

18.
Beam pinned–pinned

$$\frac{1}{ab}\sqrt{\frac{3EIL}{M_1}}$$

19.
Beam fixed–fixed

$$\frac{1}{ab}\sqrt{\frac{3EIL^3}{M_1ab}}$$

TABLE 10-2   **Frequencies of Common Systems**   **468**

TABLE 10-2 (continued) FREQUENCIES OF COMMON SYSTEMS

Beam with Uniformly Distributed Mass

$$\omega_n = \frac{\lambda^2}{L^2}\sqrt{\frac{EI}{\rho}}$$

Mode shape for the first five modes are sketched.
Nodes are located as proportion of length L measured from left.

Natural Frequency λ

System	Mode 1	Mode 2	Mode 3	Mode 4	Mode 5
20. Pinned–pinned	$\lambda^2 = 9.87$	0.500 $\lambda^2 = 39.5$	0.333 0.667 $\lambda^2 = 88.9$	0.250 0.500 0.750 $\lambda^2 = 158$	0.20 0.40 0.60 0.80 $\lambda^2 = 247$
21. Fixed–pinned	$\lambda^2 = 15.4$	0.560 $\lambda^2 = 50.0$	0.384 0.692 $\lambda^2 = 104$	0.294 0.529 0.765 $\lambda^2 = 178$	0.238 0.429 0.619 0.810 $\lambda^2 = 272$
22. Fixed–fixed	$\lambda^2 = 22.4$	0.500 $\lambda^2 = 61.7$	0.359 0.641 $\lambda^2 = 121$	0.278 0.500 0.722 $\lambda^2 = 200$	0.227 0.409 0.591 0.773 $\lambda^2 = 298$
23. Free–free	0.224 0.776 $\lambda^2 = 22.4$	0.132 0.500 0.868 $\lambda^2 = 61.7$	0.094 0.356 0.644 0.906 $\lambda^2 = 121$	0.073 0.277 0.500 0.723 0.927 $\lambda^2 = 200$	0.060 0.227 0.409 0.591 0.773 0.940 $\lambda^2 = 298$
24. Fixed–free	$\lambda^2 = 3.52$	0.774 $\lambda^2 = 22.4$	0.500 0.868 $\lambda^2 = 61.7$	0.356 0.644 0.906 $\lambda^2 = 121$	0.279 0.500 0.723 0.926 $\lambda^2 = 200$
25. Pinned–free	0.736 $\lambda^2 = 15.4$	0.446 0.853 $\lambda^2 = 50.0$	0.308 0.616 0.898 $\lambda^2 = 104$	0.235 0.471 0.707 0.922 $\lambda^2 = 178$	0.190 0.381 0.581 0.763 0.937 $\lambda^2 = 272$

TABLE 10-2 (continued)   FREQUENCIES OF COMMON SYSTEMS

Plates

a = diameter of circular plate, length of side of square plate
h = thickness of plate

$$\omega_n = \lambda \sqrt{\frac{Eh^2}{\rho^* a^4 (1 - v^2)}}$$

System	Value of λ for modes 1–6					
	1	2	3	4	5	6
26. Simply supported	5.75	16.10	29.61	34.36		
27. Clamped	11.76	24.54	40.27	45.92	58.93	70.22
28. Free	6.07	10.53	14.19	23.80	40.88	44.68
29. Clamped at center, symmetric modes	4.35	24.26	70.39	138.85	230.8	344.3
30. Simply supported	5.70	14.26	22.82	28.52	37.08	48.49

TABLE 10-2   **Frequencies of Common Systems**   470

TABLE 10-2 (continued) FREQUENCIES OF COMMON SYSTEMS

System	Value of λ for modes 1–6					
	1	2	3	4	5	6
31. One edge clamped, three edges simply supported	6.83	14.94	16.95	24.89	28.99	32.71
32. Two edges clamped, two edges simply supported	8.37	15.82	20.03	27.34	29.54	37.31
33. Clamped	10.40	21.21	31.29	38.04	38.22	47.73
34. Two edges clamped, two edges free, $\nu = 0.3$	2.01	6.96	7.74	13.89	18.25	19.00
35. One edge clamped, three edges free, $\nu = 0.3$	1.01	2.47	6.14	7.85	8.98	15.67
36. Free	4.07	5.94	6.91	10.39	17.80	18.85

TABLE 10-2 (continued)   FREQUENCIES OF COMMON SYSTEMS

System	Natural Frequency, ω_n
	Various Systems
37. Simple pendulum 	$\omega_n = \sqrt{\dfrac{g}{L}}$
38. Compound pendulum Axis of Support r = Radius of Gyration about Axis of Support	$\omega_n = \sqrt{\dfrac{ag}{r^2}}$
39. Transverse motion of massless string H = tensile force in string	$\omega_n = 2\sqrt{\dfrac{H}{M_1 L}}$
40. Pneumatic system p = pressure at each end of cylinder S = area of piston m = mass of piston V_0 = volume of each end of cylinder	$\omega_n = \sqrt{\dfrac{2pS^2}{mV_0}}$

TABLE 10-2   **Frequencies of Common Systems**

TABLE 10-2 (continued) FREQUENCIES OF COMMON SYSTEMS

System	Natural Frequency, ω_n
41. U-Tube with liquid 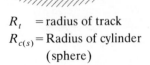	$$\omega_n = \sqrt{\frac{2g}{S}}$$
42. Plank on rotating cylinders (speed ω) μ = coefficient of friction between plank and drum	$$\omega_n = \sqrt{\frac{2\mu g}{a}}$$
43. Tanks with connecting conduit S_1 = area of tank 1 S_2 = area of tank 2 S_0 = area of conduit	$$\omega_n = \sqrt{\frac{g(1 + S_1/S_2)}{h(1 + S_1/S_2) + \ell(S_1/S_0)}}$$
44. Cylinder (c) or Sphere (s) in cylindrical track R_t = radius of track $R_{c(s)}$ = Radius of cylinder (sphere)	Cylinder: $$\omega_n = \sqrt{\frac{2g}{3(R_t - R_c)}}$$ Sphere: $$\omega_n = \sqrt{\frac{5g}{7(R_t - R_s)}}$$

TABLE 10-3   STIFFNESS OF COMMON MEMBERS

Notation

k = spring constant or stiffness, which is defined as the ratio of the applied force to the corresponding displacement

k_t = torsional stiffness

L = length

G = shear modulus of elasticity

E = elastic modulus

N = effective number of turns of spring

Member	Stiffness Formula of Member
	Elastic Members
1. Extension of helical spring $h \geq b$	For circular cross section: $$k = \frac{Gd^4}{8ND^3}.$$ For rectangular cross section: $$k = \frac{Ghb^3\eta}{ND^3}$$ $$\eta = -0.57556 + 1.231(h/b)^{1/2}$$ $$-0.5688(h/b) + 0.09336(h/b)^{3/2}$$ where d = wire diameter D = coil diameter
2. Torsion of spring T = Torque	$$k_t = \frac{Ed^4}{64ND}$$ where d = diameter of spring wire D = diameter of spring
3. Extension bar	$$k = \frac{EA}{L}$$ where A = cross-sectional area

TABLE 10-3   Stiffness of Common Members   **474**

TABLE 10-3 (continued) STIFFNESS OF COMMON MEMBERS

Member	Stiffness Formula of Member
4. Torsion bar T = Torque	$k_t = \dfrac{GJ}{L}$ J = torsional constant For hollow circular section: $J = \dfrac{\pi}{32}\left(d_0^4 - d_i^4\right)$ For rectangular cross section: $J = (bh^3/3)(1 - 0.63h/b + 0.052h^5/b^5)$
5. Bending of spring	$k_\theta = \dfrac{M}{\theta} = \dfrac{Ed^4}{64ND}\dfrac{2}{(1 + E/2G)} = \dfrac{2}{2 + \nu}\dfrac{Ed^4}{64ND}$ where d = diameter of spring wire θ = slope of deflection
6. Cantilevered beam	$k = \dfrac{3EI}{L^3}$
7. Fixed–fixed beam	$k = \dfrac{3EIL^3}{a^3b^3}$ For $a = b$: $k = \dfrac{192EI}{L^3}$
8. Simply supported beam	$k = \dfrac{3EIL}{a^2b^2}$ For $a = b = \frac{1}{2}L$: $k = \dfrac{48EI}{L^3}$

TABLE 10-3 (continued)   **STIFFNESS OF COMMON MEMBERS**

Member	Stiffness Formula of Member
9. Torsional spring 	Point A clamped: where $$k_t = \frac{EI}{L}$$ ϕ = angle of twist Point A simply supported: $k_t = T/\phi$ $$k_t = \frac{0.8EI}{L}$$ T = torque L = length of spiral I = area moment of inertia of cross section of spring
10. Transverse motion of string 	$$k = \frac{H(a+b)}{ab}$$ where H = tensile force in string
11. Beam of circular cross-section and variable depth 	$$k = \frac{\pi E d_1 d_2}{4L}$$
12. Fixed–hinged beam 	$$k = \frac{12EIL^3}{a^3 b^2 (3L+b)}$$ For $a = \frac{1}{2}L$: $k = \dfrac{768EI}{7L^3}$
13. Fixed–guided beam 	$$k = \frac{12EI}{L^3}$$

TABLE 10-3   **Stiffness of Common Members**   **476**

TABLE 10-3 (continued)   STIFFNESS OF COMMON MEMBERS

Member	Stiffness Formula of Member
14. Beam with in-span support, hinged end	$k = \dfrac{3EI}{b^2(a + b)}$
15. Fixed–free beam	$k = \dfrac{EI}{L}$
16. Beam with in-span support, fixed end	$\dfrac{1}{k} = \dfrac{b^2(a + b)\left[1 - \dfrac{a}{4(a + b)}\right]}{3EI}$
17. Beam with embedded curved end	$k = \dfrac{6EI}{k_1 L^3}$ $k_1 = 2 + 6\pi\left(\dfrac{R}{L}\right) + 24\left(\dfrac{R}{L}\right)^2 + 3\pi\left(\dfrac{R}{L}\right)^3$
18. Beam with both curved ends embedded	$k = \dfrac{24(3 + \pi R/L)EI}{k_2 L^3}$ $k_2 = 3 + 12.57\left(\dfrac{R}{L}\right) + 24\left(\dfrac{R}{L}\right)^2 + 26\left(\dfrac{R}{L}\right)^3 + 11.22\left(\dfrac{R}{L}\right)^4$

Stiffness of Common Members   TABLE 10-3

TABLE 10-3 (continued) STIFFNESS OF COMMON MEMBERS

Member	Stiffness Formula of Member
19. N-beam support 	$$k = \frac{12E}{L^3} \sum_{i=1}^{N} I_i$$ Parallel planes at ends
20. Rotation of rigid bar 	$$k_\theta = \sum_i k_i \ell_i^2$$
21. Rotation of a pinned–pinned beam 	$$k_\theta = \frac{3EIL}{L^2 - 3ab}$$ For $a = \frac{1}{2}L$: $k_\theta = \dfrac{12EI}{L}$ where $k_\theta = M/\text{slope at } x = a$
22. Rotation of a clamped–clamped beam 	$$k_\theta = \frac{EIL^3}{ab(L^2 - 3ab)}$$ For $a = \frac{1}{2}L$: $k_\theta = \dfrac{16EI}{L}$
23. Rotation of a pinned–clamped beam 	$$k_\theta = \frac{4EIL^3}{a\left[4L^3 - 3a(L+b)^2\right]}$$ For $a = \frac{1}{2}L$: $k_\theta = \dfrac{64EI}{5L}$

TABLE 10-3   **Stiffness of Common Members**   478

TABLE 10-3 (continued)   STIFFNESS OF COMMON MEMBERS

Member	Stiffness Formula of Member
24. Circular ring 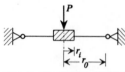	$$k = \dfrac{P}{\Delta D} = \dfrac{54.03\,EI}{D^3}$$ where ΔD is the change in diameter in the direction of P.
25. Circular membrane 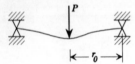	$$k = \dfrac{2\pi H}{\ln(r_0/r_i)}$$ where H = in-plane tension per unit circumference
26. Circular plate, simply supported boundary 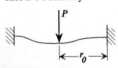	$$k = \dfrac{16\pi D}{r_0^2}\left(\dfrac{1+\nu}{3+\nu}\right) \quad \text{where}$$ h = thickness of plate ν = Poisson's ratio $D = Eh^3/\left[12(1-\nu^2)\right]$
27. Circular plate, fixed boundary 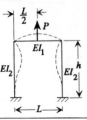	$$k = \dfrac{16\pi D}{r_0^2}$$ $D = Eh^3/\left[12(1-\nu^2)\right]$
28. Rectangular frame, fixed ends, vertical load	$$k = \dfrac{48\,EI_1}{L^3}\dfrac{2\alpha+4}{2\alpha+7}$$ $$\alpha = \dfrac{hI_1}{LI_2}$$

TABLE 10-3 (continued)   STIFFNESS OF COMMON MEMBERS

Member	Stiffness Formula of Member
29. Rectangular frame, fixed ends, horizontal load	$k = \dfrac{24EI_2}{h^3} \dfrac{6\alpha + 1}{6\alpha + 4}$ $\alpha = \dfrac{hI_1}{LI_2}$
30. Rectangular frame, pinned ends, vertical load	$k = \dfrac{48EI_1}{L^3} \dfrac{4(2\alpha + 3)}{8\alpha + 3}$ $\alpha = \dfrac{hI_1}{LI_2}$
31. Rectangular frame, pinned ends, horizontal load	$k = \dfrac{6EI_2}{h^3} \left(\dfrac{2\alpha}{2\alpha + 1} \right)^2$ $\alpha = \dfrac{hI_1}{LI_2}$
32. Rectangular frame, fixed ends, out-of-plane load	$\dfrac{1}{k} = \dfrac{L^3}{48EI_1} + \dfrac{h^3}{6EI_2} - \dfrac{L^4GJ_2}{64EI_1(2hEI_1 + LGJ_2)}$

TABLE 10-3   **Stiffness of Common Members**   **480**

TABLE 10-3 (continued)   STIFFNESS OF COMMON MEMBERS

Member	Stiffness Formula of Member

Rubber Members[a]

33.

Rectangular block
of rubber in compression

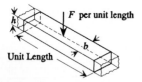

Load Applied Here

Force Free

Young's Modulus E (psi)	Shear Modulus G (psi)	Bulk Modulus K (psi)	Numerical factor c
130	43	142,000	0.93
168	53	142,000	0.89
213	64	142,000	0.85
256	76	142,000	0.80
310	90	146,000	0.73
460	115	154,000	0.64
630	150	163,000	0.57
830	195	171,000	0.54
1040	245	180,000	0.53
1340	317	189,000	0.52

$k = E_c A/(Rh)$

where

$E_c = E(1 + 2cS^2)$

E = Young's modulus

c = a numerical factor, function of
E or G

S = (loaded surface area)/(force free
area), $= Lb/[2h(L + b)]$

L = length of block

b = width of block

h = thickness of block

A = cross-sectional area Lb

$R = 1 + E_c/K$

K = bulk modulus of compression

34.

Long rubber strip compressed
normal to its length

F per unit length

Unit Length

$k = 4bE(1 + cS^2)/(3hR)$

where

b = width of strip

$S = b/2h$

c = a numerical factor, function of
E or G, see case 33

E = Young's modulus

h = thickness of strip

$R = 1 + E_c/K$

K = bulk modulus of compression; see
case 33

E_c = see case 33

TABLE 10-3 (continued)   STIFFNESS OF COMMON MEMBERS

Member	Stiffness Formula of Member
35. Rectangular section rubber ring	$k = \frac{4}{3}E\pi D(b/h)\left[1 + cb^2/(4h^2)\right]$ where $D =$ mean diameter of ring $b =$ radial width of section $h =$ thickness of section $E =$ Young's modulus $c =$ a numerical factor; see case 33
36. Circular section rubber ring	$k = 3.95ED(\delta/d)^{1/2}$ where $E =$ Young's modulus $D =$ mean diameter of ring $\delta =$ displacement due to compression $d =$ diameter of section
37. Solid rubber block in shear	$k = GA/h$ where $G =$ shear modulus $A =$ cross-sectional area of block parallel to direction of shear
38. Annulus of rubber bonded to two end plates normal to axis of annulus	$k_t = T/\phi$ $= \pi G(r_1^4 - r_0^4)/(2h) + \pi G\phi^2(r_1^6 - r_0^6)/(9h^3)$ where $T =$ torque $G =$ shear modulus $r_0 =$ inner radius of annulus $r_1 =$ outer radius of annulus $h =$ thickness of disk $\phi =$ angular rotational displacement (rad)

TABLE 10-3   **Stiffness of Common Members**   **482**

TABLE 10-3 (continued) STIFFNESS OF COMMON MEMBERS

Member	Stiffness Formula of Member
39. Rubber bush mounting under torsional load 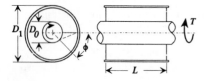	$k_t = T/\phi$ $= \pi GL/(1/D_0^2 - 1/D_1^2)$ where T = torque ϕ = angular rotational displacement (rad) G = shear modulus L = length of bush mounting D_0 = inner diameter D_1 = outer diameter
40. Rubber bush mounting under axial load 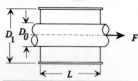	$k = \dfrac{2.73GL/\log_{10}(D_1/D_0)}{1 + \alpha(D_1/L)^2}$ where G = shear modulus L = length of bush mounting D_0 = inner diameter D_1 = outer diameter α = axial stiffness constant

Diameter Ratio D_1/D_0	Axial Stiffness Constant, α
1.05	0.0001
1.10	0.0005
1.25	0.0025
1.50	0.0068
1.75	0.0111
2	0.0148
3	0.0244
4	0.0289
5	0.0309
7	0.0321
10	0.0315
20	0.0282
100	0.0204
1000	0.0135

TABLE 10-3 (continued)   STIFFNESS OF COMMON MEMBERS

Member	Stiffness Formula of Member
41. Rubber bush mounting under radial load 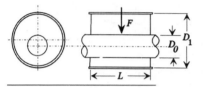	$k = \beta L G$ where $G =$ shear modulus $L =$ length of bush mounting $\beta = \beta_L$ for long bushes, β_S for short bushes

Diameter Ratio D_1/D_0	Long Bushes β_L	Short Bushes β_s
1.05	320,000	322
1.10	43,700	165
1.25	3,400	70
1.50	602	38
1.75	212	27
2	135	21
3	42	12.4
4	25	9.4
5	18.3	7.9
7	12.7	6.3
10	9.5	5.2
20	6.3	3.8
100	3.4	2.4
1000	2.1	1.5

[a]From Schiff [10.6], with permission.

TABLE 10-3   **Stiffness of Common Members**   **484**

TABLE 10-4   FORMULAS FOR EQUIVALENT STIFFNESS OF COMBINATION OF SPRINGS

Notation

k = equivalent spring constant or stiffness of the system.

Spring Combination	Stiffness Formulas for System
1.	$$\frac{1}{k} = \frac{1}{k_1} + \frac{1}{k_2}$$ $$k = \frac{k_1 k_2}{k_1 + k_2}$$
2.	$$k = k_1 + k_2$$
3.	$$k = \frac{(a+b)^2}{(a^2/k_2 + b^2/k_1)}$$
4.	$$\frac{1}{k} = \frac{1}{k_1} + \frac{1}{k_2 + k_3}$$ $$k = \frac{k_1(k_2 + k_3)}{k_1 + k_2 + k_3}$$
5.	$$k = nk_1 = \frac{n \times 192\,EI}{L^3}$$ where n = number of plate springs. Assume no friction between springs
6.	$$k_H = \sum_i k_i \cos^2 \alpha_i$$ $$k_V = \sum_i k_i \sin^2 \alpha_i$$ where k_H = horizontal stiffness k_V = vertical stiffness

TABLE 10-5 MASS AND MASS MOMENTS OF INERTIA

Notation

m, M_1 = mass (M)

m_{eq} = equivalent mass (M)

I_{pi} = polar mass moment of inertia of concentrated mass at point $i (ML^2)$

I_{eq} = equivalent polar mass moment of inertia (ML^2)

m_b = mass of beam

$\omega_n = \sqrt{k/m}$ or $\omega_n = \sqrt{k_t/I_{pi}}$

Mass	Moments of Inertia
1. Mass moment of inertia of disk about axis of rotation	$I_{pi} = \frac{1}{2}ma^2$ where m = mass of disk
2. Mass moment of inertia of bar about mass center	$I_{pi} = \frac{1}{12}m_b L^2$
3. Parallel axis theorem	$I_{pA} = I_{pG} + mL^2$ where I_{pG} = mass moment at mass center G I_{pA} = mass moment at point A

TABLE 10-5 | Mass and Mass Moments of Inertia 486

TABLE 10-5 (continued)   MASS AND MASS MOMENTS OF INERTIA

Mass	Moments of Inertia
4. Equivalent mass of geared system I_{ij} = Mass moment of the jth gear of the ith axle	$$I_{eq} = \sum_{i=1}^{n_a} \left(\frac{\Omega_i}{\Omega_1}\right)^2 \sum_{j=1}^{n_i} I_{ij} \quad \text{(about axle 1)}$$ n_i = number of gears of ith axle n_a = number of axles (for configuration shown, $n_a = 3$) $\dfrac{\Omega_i}{\Omega_1}$ = rotational speed ratio of axle i to axle 1
5. Equivalent mass for rack and gear	$I_{eq} = I_g + ma^2 \quad \text{(about gear axis)}$ $m_{eq} = m + I_g/a^2 \quad \text{(rack)}$ where I_g = mass moment of gear m = mass of rack
6. Linear spring with mass	$m_{eq} = M_1 + \frac{1}{3}m_s$ where m_s = mass of spring
7. Beam with mass and lumped mass at midspan	$m_{eq} = 0.49m_b + M_1$
8. Beam with mass and lumped mass on end	$m_{eq} = 0.24m_b + M_1$

TABLE 10-6   DAMPING COEFFICIENTS

Notation

μ = viscosity of fluid (FT/L^2), where T is time

c = damping coefficient (FT/L)

c_t = torsional damping coefficient (FLT/rad)

c_{eq} = equivalent damping coefficient (FT/L)

k = stiffness coefficient (F/L)

η = loss factor or structural damping

v = velocity (L/T)

ω = angular frequency of vibration (rad/T)

Damper	Damping Coefficients
1. Dashpot, flow past hole in piston, viscous damping	$c = \dfrac{8\mu l A^2}{\pi r^4}$ where A = area of piston r = radius of hole l = length of piston
2. Shear damper, viscous damping	$c = \mu A/h$ where A = area of plate
3. Torsional damper, viscous damping	$c_t = 2\pi\mu\left[\dfrac{r_2^3 b}{h_2} + \dfrac{1}{2}\dfrac{r_2^4 - r_1^4}{h_1}\right]$

TABLE 10-6   **Damping Coefficients**   **488**

TABLE 10-6 (continued)   DAMPING COEFFICIENTS

Damper	Damping Coefficients
4. Coulomb damper, dry friction 	$c_{eq} = 4F_f/(\pi \omega x_0)$ where $F_f = \mu_f W$ and μ_f = friction coefficient W = applied pressure force
5. Material damping, viscoelastic damping 	$c_{eq} = \eta k / \omega$
6. Serial dampers 	$\dfrac{1}{c_{eq}} = \dfrac{1}{c_1} + \dfrac{1}{c_2} + \cdots$
7. Parallel dampers 	$c_{eq} = c_1 + c_2 + \cdots$

TABLE 10-7 IMPACT EFFECT OF WEIGHT DROPPED ON ELASTIC MEMBER

Notation

W = weight of body being dropped
W_b = weight of elastic member, here a beam
h = height from which weight is dropped
k = stiffness (spring constant) of beam referenced to point of contact of falling body
v = velocity of body at instant of impact
δ_{st} = static displacement of member due to weight of body and member
δ_s = static displacement of member due only to weight of member

Case	Weight of Member Taken into Account	Body Suddenly Applied from Zero Height	Body Applied from Zero Height and Weight of Member Neglected
1. Maximum force P experienced by member	$W + W_b + \sqrt{W_b^2 + 2W(W_b + kh)}$ $= W + W_b + \sqrt{W^2 + Wkv^2/g}$	$2W + W_b$	$2W$
2. Maximum displacement δ of member after impact of body	$\delta_{st} + \sqrt{\delta_{st}^2 + 2h(\delta_{st} - \delta_s) - \delta_s^2}$ $= \delta_{st} + \sqrt{(\delta_{st} - \delta_s)^2 + v^2(\delta_{st} - \delta_s)/g}$	$\delta_{st} + \sqrt{\delta_{st}^2 + \delta_s(\delta_s - 2\delta_{st})}$	$2\delta_{st}$ δ_{st} due only to weight of body

TABLE 10-7 | Impact Effect of Weight on Elastic Member | 490

TABLE 10-8 FORMULAS FOR ENERGY ABSORBING CAPACITY OF COMMON LOAD-MEMBER CONDITIONS[a]

r = radius of gyration of cross-sectional area
I = moment of inertia
E = modulus of elasticity
G = shear modulus of elasticity
c = distance from neutral axis to outer fiber
A = cross-sectional area
σ_{ys} = yield stress
L = length of member

Conditions	Energy Absorbing Capacity
1. Simply supported beam, concentrated load, uniform section	$U_m = \dfrac{\sigma_{ys}^2 I L}{6Ec^2}$ $= \dfrac{\sigma_{ys}^2 AL}{6E}\left(\dfrac{r}{c}\right)^2$
2. Fixed–fixed Beam	$U_m = \dfrac{\sigma_{ys}^2 I L}{6Ec^2}$ $= \dfrac{\sigma_{ys}^2 AL}{6E}\left(\dfrac{r}{c}\right)^2$
3. Cantilevered beam	$U_m = \dfrac{\sigma_{ys}^2 I L}{6Ec^2}$ $= \dfrac{\sigma_{ys}^2 AL}{6E}\left(\dfrac{r}{c}\right)^2$
4. Axial tension	$U_m = \dfrac{\sigma_{ys}^2 AL}{2E}$

Conditions	Energy Absorbing Capacity
5. Torsion	$$U_m = \frac{\sigma_{ys}^2 \left(D^2 + D_1^2 \right) AL}{4GD^2}$$
6. Simply supported beam with uniform load	$$U_m = \frac{4\sigma_{ys}^2 IL}{15Ec^2}$$ $$= \frac{4\sigma_{ys}^2 AL}{15E} \left(\frac{r}{c} \right)^2$$
7. Fixed–fixed beam	$$U_m = \frac{\sigma_{ys}^2 IL}{10Ec^2}$$ $$= \frac{\sigma_{ys}^2 AL}{10E} \left(\frac{r}{c} \right)^2$$
8. Cantilevered beam	$$U_m = \frac{\sigma_{ys}^2 IL}{10Ec^2}$$ $$= \frac{\sigma_{ys}^2 AL}{10E} \left(\frac{r}{c} \right)^2$$
9. Beam of variable cross section so that $\sigma = $ const	$$U_m = \frac{\sigma_{ys}^2 IL}{3Ec^2}$$
10. Torsion	$$U_m = \frac{\sigma_{ys}^2 JL}{2Gt_{max}^2}$$ where J = torsional constant

[a]From Blodgett [10.1], with permission.

TABLE 10-8   Formulas for Energy Absorbing Capacity

TABLE 10-9

TABLE 10-9 DYNAMIC PROPERTIES OF SOME COMMON ENGINEERING MATERIALS[a]

Material	Yield Strength, σ_{ys} (lb/in.^{2})	Tensile Ultimate Strength, σ_u (lb/in.^{2})	Tensile Modulus of Elasticity, E (lb/in.^{2})	Ultimate Unit Elongation, ε_u (in./in.)	Tensile Modulus of Resilience, R (in.-lb/in.^{3})	Modulus of Toughness, R_u (in.-lb/in.^{3})
Mild steel	35,000	60,000	30×10^6	0.35	20.4	16,600
Low-alloy (under $\frac{3}{4}$ in.)	50,000	70,000	30×10^6	0.18	41.6	
($\frac{3}{4}$–$1\frac{1}{2}$ in.)	46,000	67,000	30×10^6	0.19	35.2	
(over $1\frac{1}{2}$–4 in.) steel	42,000	63,000	30×10^6	0.19	29.4	
Medium-carbon steel	45,000	85,000	30×10^6	0.25	33.7	16,300
High-carbon steel	75,000	120,000	30×10^6	0.08	94.0	5,100
T-1 steel	100,000	115,000 to 135,000	30×10^6	0.18	200.0[b]	~ 19,400
Alloy steel	200,000	230,000	30×10^6	0.12	667.0	22,000
Gray cast iron	6,000	20,000	15×10^6	0.05	1.2	70
Malleable cast iron	20,000	50,000	23×10^6	0.10	17.4	3,800

[a]From Blodgett [10.1], with permission.
[b]Based on integrator-measured area under stress-strain curve.

Beams and Columns

Formulas for the analysis and design of beams and columns are provided in this chapter. These members can be loaded statically with transverse mechanical or thermal loading. Also included are formulas for plastic design, buckling loads, natural frequencies, and the accompanying mode shapes.

Furthermore, this chapter contains tables of generalized transfer and stiffness matrices that can be utilized for the study of structural systems formed of beam members, e.g., for the analysis of frames. Computer programs, using these matrices, for the static, stability, and dynamic response of arbitrary beams have been prepared to accompany this book.

Most of the formulas are based on the technical (Euler–Bernoulli) theory of beams. For this theory it is assumed that plane cross sections remain plane, stress is proportional to strain, bending is in a principal plane, and the slope of the deformed beam is always much less than 1.

11.1 NOTATION

A	area (L^2)
a_{h_i}	location of ith plastic hinge
A_s	equivalent shear area, where α_s is shear correction factor (Table 2-4); also called shear-adjusted area, $= A/\alpha_s$
C	concentrated applied moment (FL)
c_1	magnitude of applied distributed moment, uniform in x direction (FL/L)
c_a	initial magnitude of linearly varying distributed moment (FL/L)
c_b	final magnitude of linearly varying distributed moment (FL/L)
C_c	concentrated collapse moment
E	Young's modulus of elasticity of material (F/L^2)
G	shear modulus of elasticity (F/L^2)
$I = I_y$	moment of inertia taken about neutral axis (centroidal axis) (L^4)
I, I_y, I_z, I_x, I_{yz}	moments of inertia (L^4)
k	Winkler (elastic) foundation modulus (F/L^2)
K	KL is effective column length
k^*	rotary foundation modulus (FL/L)
ℓ	length of element (L)
L	length of beam (L)
M	bending moment at any section (FL)
M_i	lumped mass (M)
m_b	mass of beam (M)
M_p	fully plastic bending moment, $= \sigma_{ys} Z_p$, (FL)
M_T	thermal moment, $= \int_A E\alpha Tz\, dA$, (FL)
P	axial force (F)
p_1	magnitude of distributed force, uniform in x direction (F/L)
p_c	uniformly distributed collapse load (F)
p_z	transverse, distributed force intensity, $= p$ (F/L)
p_1, p_a	initial magnitude of linearly varying distributed force (F/L)
p_2, p_b	final magnitude of linearly varying distributed force (F/L)
P_T	thermal axial force, $= \int_A E\alpha T\, dA$ (F)
r	radius of gyration of cross-sectional area about y axis; for buckling, r is minimum radius of gyration, $= r_y$
T	temperature change (degree), temperature rise with respect to reference temperature, $= T(x, z)$
V	shear force at any section (F)
w	transverse deflection (L)
W	concentrated force (F)
W_c	concentrated collapse load (F)

x, y, z	right-handed coordinate system
z	vertical coordinate from neutral axis (L)
Z_p	plastic modulus of cross section (L^3)
α	coefficient of thermal expansion $(L/L \cdot \text{degree})$
α_s	shear correction coefficient or shear deflection constant (Table 2-4)
λ	slenderness ratio of column, $= KL/r$
ω	natural frequency (rad/T)
ρ	mass per unit length $(M/L)\,(FT^2/L^2)$
σ	normal, bending stress (F/L^2)
τ	transverse shear stress (F/L^2)
θ	angle or slope of deflection curve (rad)

11.2 SIGN CONVENTION

Positive deflection w and positive slope θ are shown in Fig. 11-1. Positive internal bending moments M and positive internal shear forces V on the right face of a cut are illustrated in Fig. 11-2. For applied loading, the formulas provide solutions for the loading illustrated. Loadings applied in the opposite direction require the sign of the loading to be reversed in the formulas.

11.3 STRESSES

The tables in this chapter give the deflection, slope, bending moment, and shear force along a beam. The normal and shear stresses on a face of the cross section

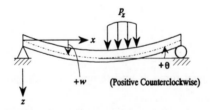

FIGURE 11-1 Positive displacement w and slope θ.

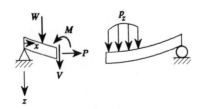

FIGURE 11-2 Positive bending moment M, shear force V and axial force P.

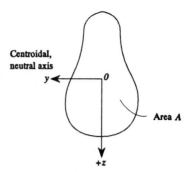

FIGURE 11-3 Cross section.

can be computed using the following formulas. Refer to Chapter 15 for more complete and more accurate stress formulas.

Normal Stress

The flexural, normal, or bending stress σ resulting from bending is

$$\sigma = Mz/I \tag{11.1}$$

where z is the vertical coordinate measured from the neutral axis. As shown in Fig. 11-3, positive z in Eq. (11.1) is taken as downward. For stresses above the neutral axis, use a negative z. Of course, the sign of stress σ also depends on the sign of the moment M taken from the tables. The moment of inertia I is given by [Eq. (2.4a)]

$$I = I_y = \int_A z^2\, dA \tag{11.2}$$

where A is the cross-sectional area. See Table 2-1 for values of I for particular cross-sectional shapes.

If a compressive axial force P and temperature change T are present, then

$$\sigma = E\alpha T - P/A + Mz/I \tag{11.3}$$

where M includes both mechanical and thermal (M_T) effects. Substitute P for $-P$ if the axial force is tensile.

Shear Stress

The average shear stress τ at any point along a width of a cross section, e.g., line 1–2 of Fig. 11-4, is

$$\tau = VQ/Ib \tag{11.4}$$

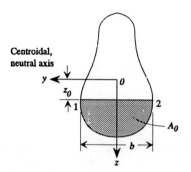

FIGURE 11-4 Definitions for shear stress.

where b is the width at the location where the stress is being computed and [Eq. (2.15a)]

$$Q = \int_{A_0} z \, dA \qquad (11.5)$$

This integral of Eq. (11.5) is taken over the area A_0 that lies between the position at which the shear stress is desired (z_0) and the outer fiber of the cross section. This area is hatched in Fig. 11-4. The quantity Q is the first moment of the area between z_0 and the outer fiber. For some common cross sections, formulas for Q are provided in Table 2-3.

11.4 SIMPLE BEAMS

The governing equations for the bending of a uniform Euler–Bernoulli beam are

$$EI \frac{d^4 w}{dx^4} = p_z = p$$

$$EI \frac{d^3 w}{dx^3} = -V$$

$$EI \frac{d^2 w}{dx^2} = -M \qquad (11.6)$$

$$\frac{dw}{dx} = -\theta$$

These relations conform to sign convention 1 presented in Appendix II. They can be solved giving the deflection, slope, bending moment, and shear force as functions of the coordinate x.

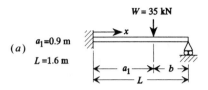

(a) $a_1 = 0.9$ m

$L = 1.6$ m

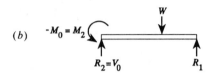

(b)

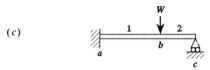

(c)

FIGURE 11-5 Simple beam model for Example 11.1: (a) the model; (b) free-body diagram; (c) notation for matrix method solutions.

Tabulated Formulas

The deflection, slope, bending moment, and shear force for uniform beams with commonly occurring end conditions and loadings are provided in Table 11-1. Included are some critical values, e.g., the peak bending moment.

Example 11.1 A Simple Beam A beam with the left end fixed and the right end simply supported, is shown in Fig. 11-5a. For this beam, $E = 200$ GN/m^2 and $I = 144$ cm^4.

The beam of Fig. 11-5 corresponds to case 11 of Table 11-1. To use the formulas of case 11, it is necessary to replace the variables x, a, b of case 11 by $L - x, b, a_1$ of Fig. 11-5a, respectively.

It follows from the formulas of case 11 that the shear and moment at the fixed end are (Fig. 11-5b)

$$R_2 = (W - R_1) = \left(W - \frac{W}{2}\frac{3a_1^2 L - a_1^3}{L^3}\right) = 21.5 \text{ kN}$$

$$M_2 = \frac{Wba_1}{2L^2}(b + L) = 9.91 \text{ kN} \cdot \text{m}$$

(1)

The reaction at the simply supported end is

$$R_1 = W - R_2 = 13.5 \text{ kN}$$

(2)

The deflections to the right side of the load of Fig. 11-5a are

$$w = -\frac{1}{6EI}\left\{R_1\left[(L-x)^3 - 3L^2(L-x)\right] + 3Wa_1^2(L-x)\right\} \qquad (3)$$

and to the left of the applied force W are

$$w = -\frac{1}{6EI}\left\{R_1\left[(L-x)^3 - 3L^2(L-x)\right] + W\left[3a_1^2(L-x) - (L-x-b)^3\right]\right\}$$
$$\qquad (4)$$

Since b (Fig. 11-5a) $> 0.414L = 0.6624$ m, the maximum deflection from case 11 is

$$w_{max} = \frac{Wba_1^2\sqrt{b/(2L+b)}}{6EI} = 4.865 \text{ mm} \qquad (5)$$

at $x = L - L\sqrt{1 - 2L/(3L - a_1)} = 0.92$ m.

Furthermore, the bending moments along the beam are

$$M = R_1(L-x) \quad (\text{for } x > a_1) \qquad (6)$$

$$M = R_1(L-x) - W(L-x-b) = R_1(L-x) - W(a_1-x) \quad (\text{for } x \le a_1) \qquad (7)$$

The maximum moment occurs at $x = 0$ and is of magnitude

$$M_{max} = M_2 = 9.91 \text{ kN-m} \qquad (8)$$

The slope at the simply supported end is

$$\theta_L = \theta_1 = \frac{W}{4EI}\left(\frac{a_1^3}{L} - a_1^2\right) = 0.01077 \text{ rad} \qquad (9)$$

The maximum shear force is equal to R_2, which occurs for $x < a_1$.

Formulas for Beams with Arbitrary Loading

If sufficient information about your uniform, single-span beams cannot be found in Table 11-1, then use Table 11-2. Table 11-2 is intended to provide the deflection, slope, bending moment, and shear force of uniform beams under arbitrary applied loading with any end conditions.

Part A in Table 11-2 lists equations for the responses. The functions F_w, F_θ, F_V, F_M are taken from part B in Table 11-2 by adding the appropriate terms for each load applied to the beam. The initial parameters w_0, θ_0, V_0, M_0, which are values of w, θ, V, and M at the left end ($x = 0$) of the beam, are evaluated using the entry in part C in Table 11-2 for the appropriate beam end conditions. In using this table, no distinction is made between statically determinate and indeterminate beams.

These general formulas are readily programmed for computer solution.

Example 11.2 A Simple Beam The simple beam of Example 11.1 can also be analyzed using the formulas of Table 11-2. This is a statically indeterminate beam. The statically indeterminate nature of the problem does not affect the methodology associated with this table.

The boundary conditions for this fixed, simply supported beam are

$$w_{x=0} = w_0 = 0, \qquad \theta_{x=0} = \theta_0 = 0, \qquad w_{x=L} = M_{x=L} = 0 \qquad (1)$$

The deflection w, slope θ, shear V, and moment M are readily obtained using part A in Table 11-2. Since $w_0 = \theta_0 = 0$,

$$w = -V_0 \frac{x^3}{3! \, EI} - M_0 \frac{x^2}{2EI} + F_w(x) \qquad (2a)$$

$$\theta = V_0 \frac{x^2}{2EI} + M_0 \frac{x}{EI} + F_\theta(x) \qquad (2b)$$

$$V = V_0 + F_V(x) \qquad (2c)$$

$$M = V_0 x + M_0 + F_M(x) \qquad (2d)$$

The loading functions F_w, F_θ, F_V, and F_M for the applied concentrated force at $x = a = 0.9m$ are (Table 11-2, part B)

$$F_w(x) = 35\langle x - 0.9\rangle^3/(3! \, EI)$$
$$F_\theta(x) = -35\langle x - 0.9\rangle^2/(2EI)$$
$$F_V(x) = -35\langle x - 0.9\rangle^0 \qquad (3)$$
$$F_M(x) = -35\langle x - 0.9\rangle$$

where

$$\langle x - a\rangle^n = \begin{cases} 0 & \text{if } x < a \\ (x-a)^n & \text{if } x \geq a \end{cases} \qquad \langle x - a\rangle^0 = \begin{cases} 0 & \text{if } x < a \\ 1 & \text{if } x \geq a \end{cases}$$

To find V_0, M_0 enter part C in Table 11-2 for a beam with a fixed left end and a pinned right end:

$$V_0 = -3EIF_{w|x=L}/L^3 - 3F_{M|x=L}/(2L), \qquad M_0 = 3EIF_{w|x=L}/L^2 + F_{M|x=L}/2 \qquad (4)$$

Substitution of F_w and F_M with $x = L$ from (3) into (4) gives

$$M_0 = -9.91 \text{ kN-m}, \qquad V_0 = 21.5 \text{ kN} \qquad (5)$$

which are the same values found in Example 11.1 since $V_0 = R_2$ and $M_0 = -M_2$. The responses along the beam of Fig. 11-5 as represented by (2) are now completely known.

Example 11.3 A Simply Supported Beam The single-span simply supported beam of Fig. 11-6 is readily analyzed with the formulas of Table 11-2. Let $EI = 1.1 \times 10^{12}$ lb-in.^{2}

Since the left end is hinged, $w_{x=0} = w_0 = 0$, $M_{x=0} = M_0 = 0$. If the tensile axial force is ignored, the deflection w, slope θ, shear force V, and moment M are given by (Table 11-2, part A)

$$w = -\theta_0 x - V_0 \frac{x^3}{3!\,EI} + F_w \tag{1a}$$

$$\theta = \theta_0 + V_0 \frac{x^2}{2\,EI} + F_\theta \tag{1b}$$

$$V = V_0 + F_V \tag{1c}$$

$$M = V_0 x + F_M \tag{1d}$$

The loading functions F_w, F_θ, F_V, and F_M for the concentrated applied moments and forces are taken from part B in Table 11-2, with the use of superposition in the case of more than one applied loading. For example,

$$F_w(x) = \frac{102{,}000(x^2)}{2\,EI} + \frac{3335\langle x - 21.5\rangle^3}{3!\,EI} + \frac{8417\langle x - 39.9\rangle^3}{3!\,EI}$$

$$+ \frac{8417\langle x - 53.1\rangle^3}{3!\,EI} + \frac{3335\langle x - 71.5\rangle^3}{3!\,EI} - \frac{102{,}000\langle x - 93\rangle^2}{2\,EI} \tag{2}$$

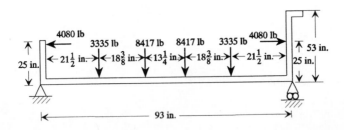

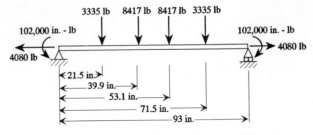

FIGURE 11-6 Simply supported beam: (a) beam; (b) model.

The boundary conditions $w_{x=L} = 0$, $M_{x=L} = 0$ are used to identify θ_0 and V_0 in (1). From part C in Table 11-2

$$\theta_0 = \frac{1}{L}\bar{F}_w + \frac{L}{6EI}\bar{F}_M = \frac{1}{L}F_{w|x=L} + \frac{L}{6EI}F_{M|x=L} = -0.6124 \times 10^{-5} \text{ rad}$$

$$V_0 = -\frac{1}{L}\bar{F}_M = -\frac{1}{L}F_{M|x=L} = 0.1175 \times 10^5 \text{ lb} \tag{3}$$

This completes the solution. The variables w, θ, V, and M are given by (1) everywhere along the beam. It will be shown in Example 11.4 that the tensile force has little influence on the results.

11.5 BEAMS WITH AXIAL FORCES ON ELASTIC FOUNDATIONS

The governing equations for a uniform beam with a compressive axial force P and resting on an elastic (Winkler) foundation of modulus k are as follows:

Without shear deformation,

$$EI\frac{d^4w}{dx^4} + P\frac{d^2w}{dx^2} + kw = p$$

$$EI\frac{d^3w}{dx^3} + P\frac{dw}{dx} = -V$$

$$EI\frac{d^2w}{dx^2} = -M \tag{11.7a}$$

$$\frac{dw}{dx} = -\theta$$

With shear deformation,

$$\frac{dw}{dx} = -\theta + \frac{V\alpha_s}{GA}$$

$$\frac{d\theta}{dx} = \frac{M}{EI}$$

$$\frac{dV}{dx} = -p_z + kw \tag{11.7b}$$

$$\frac{dM}{dx} = V - P\theta$$

where A/α_s is the equivalent shear area A_s and α_s is the shear correction factor (shear deflection constant) given in Table 2-4. For a beam with a tensile axial load, replace P by $-P$.

The formulas for the response of this type of beam are provided in Table 11-3, part A. The parameters λ, ζ, η, e_0, e_1, e_2, e_3, and e_4 are defined in Table 11-3, part B.

The F_w, F_θ, F_V, and F_M in the responses of Table 11-3, part A, are loading functions given in part C. If there are several loads on a beam, the F_w, F_θ, F_V, F_M functions are obtained by adding the terms given in part C of Table 11-3 for each load. Use the definition

$$e_i\langle x - a \rangle = \begin{cases} 0 & \text{if } x < a \\ e_i(x - a) & \text{if } x \geq a \end{cases} \tag{11.8}$$

For example, suppose $e_1 = \cosh \alpha x$ in Table 11-3, part B. Then

$$e_1\langle x - a \rangle = \begin{cases} 0 & \text{if } x < a \\ \cosh \alpha(x - a) & \text{if } x \geq a \end{cases}$$

Also, if $e_1 = 1$,

$$e_1\langle x - a \rangle = \langle x - a \rangle^0 = \begin{cases} 0 & \text{if } x < a \\ 1 & \text{if } x \geq a \end{cases} \tag{11.9}$$

The initial values w_0, θ_0, V_0, M_0 of the responses of Table 11-3, part A, are provided in Table 11-3, part D.

Example 11.4   A Simply Supported Beam   Consider again the beam of Example 11.3. This time take the tensile axial load of 4080 lb (Fig. 11-6) into account but do not include shear deformation effects.

The end conditions are still

$$w_0 = 0, \qquad M_0 = 0, \qquad w_{x=L} = M_{x=L} = 0 \tag{1}$$

According to Table 11-3, part A, the deflection, slope, shear force, and bending moment are

$$\begin{aligned} w &= -\theta_0 e_2 - V_0 e_4 / EI + F_w \\ \theta &= \theta_0 e_1 + V_0 e_3 / EI + F_\theta \\ V &= -\theta_0 \lambda EI e_3 + V_0(e_1 + \zeta e_3) + F_V \\ M &= \theta_0 EI e_0 + V_0 e_2 + F_M \end{aligned} \tag{2}$$

From Table 11-3, part B, for a beam with tensile axial load,

$$\begin{aligned} \lambda &= 0, \qquad \alpha^2 = P/EI = \zeta \\ e_0 &= \alpha \sinh \alpha x, \qquad e_1 = \cosh \alpha x \qquad e_2 = (\sinh \alpha x)/\alpha \\ e_3 &= (\cosh \alpha x - 1)/\alpha^2, \qquad e_4 = (\sinh \alpha x - \alpha x)/\alpha^3 \end{aligned} \tag{3}$$

The loading functions for the applied concentrated forces and moments are taken from Table 11-3, part C. As an example, it is seen that F_w becomes

$$F_w(x) = \frac{102{,}000\,e_3(x)}{EI} + \frac{3335\,e_4\langle x - 21.5\rangle}{EI} + \frac{8417\,e_4\langle x - 39.9\rangle}{EI}$$
$$+ \frac{8417\,e_4\langle x - 53.1\rangle}{EI} + \frac{3335\,e_4\langle x - 71.5\rangle}{EI} - \frac{102{,}000\,e_3\langle x - 93\rangle}{EI}$$

(4)

Recall how the singularity functions are handled. For example,

$$e_3(x) = e_3 = (\cosh \alpha x - 1)/\alpha^2$$

$$e_3\langle x - 93\rangle = \begin{cases} 0 & \text{if } x < 93 \\ [\cosh \alpha(x - 93) - 1]/\alpha^2 & \text{if } x \geq 93 \end{cases}$$

(5)

$$e_4\langle x - 21.5\rangle = \begin{cases} 0 & \text{if } x < 21.5 \\ [\sinh \alpha(x - 21.5) - \alpha(x - 21.5)]/\alpha^3 & \text{if } x \geq 21.5 \end{cases}$$

From Table 11-3, part D, the simply supported end conditions give

$$\theta_0 = \frac{-\bar{e}_2 \bar{F}_w - (\bar{e}_4/EI)\bar{F}_M}{\nabla} = \frac{[-e_2 F_w - (e_4/EI)F_M]_{x=L}}{\nabla}$$

$$V_0 = \frac{EI\bar{e}_0 \bar{F}_w + \bar{e}_2 \bar{F}_M}{\nabla} = \frac{(EIe_0 F_w + e_2 F_M)_{x=L}}{\nabla}$$

$$\nabla = \frac{\bar{e}_0 \bar{e}_4}{EI} - \bar{e}_2^2 = \left(\frac{e_0 e_4}{EI} - e_2^2\right)_{x=L}$$

Substitution of the appropriate values for this beam gives

$$\theta_0 = -0.6124 \times 10^{-5} \text{ rad}, \qquad V_0 = 11{,}752 \text{ lb}$$

Note that the results are virtually the same as in Example 11.3. Hence, the effect of the axial force on w, θ, V, and M is negligible.

════════════

11.6 PLASTIC DESIGN

The stresses of Section 11.3 are based on linear elastic assumptions and a design assumes that the structure is in the pure elastic state and the yield stress of the outer fiber of the beam cross sections determines the maximum load that the beam can carry. An alternative and often more realistic technique is to utilize *plastic*, or *ultimate strength*, *design*, which is based on the assumption that a beam collapses only when all fibers on a critical cross section reach a plastic state. The

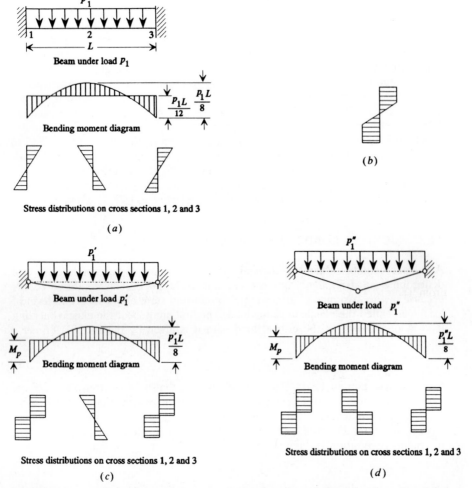

FIGURE 11-7 Collapse process of a beam with uniform load. (*a*) Beam with uniform Load p_1. (*b*) Form of stress distribution when p_1 increases. It is assumed that the material is perfectly plastic as a fiber reaches its yield stress. (*c*) Beam as cross sections 1 and 3 become completely plastic. (*d*) Beam as cross sections 1, 2, and 3 become completely plastic.

corresponding load, the *collapse load*, at this moment is considered to be the maximum load that the structure can carry.

The plastic collapse process can be illustrated by a clamped–clamped beam carrying a uniform load p_1. The moment diagram of the beam is shown in Fig. 11-7*a*. At three cross sections, the bending moment has peak values. Let p_1 be sufficiently large such that yielding begins at the upper fiber on cross sections 1 and 3. As the load p_1 increases, assume the stress distribution on these cross sections will be of the form of Fig. 11-7*b*. Continue increasing the load to p_1', the level at which cross sections 1 and 3 become completely plastic. Since this completely plastic state functions like a hinge, it is referred to as a *plastic hinge*. The beam with its two plastic hinges becomes similar to a simply supported beam (Fig. 11-7c). More load can be added until cross section 2 becomes fully plastic and the beam appears as in Fig. 11-7d. At this point the beam can collapse; consequently the load p_1'' is treated as the critical or *collapse load*. The critical load is much larger than that determined from an elastic design.

The bending moment that causes a completely plastic state on a cross section is called the *plastic moment* and is expressed as $M_p = \sigma_{ys} Z_p$, where σ_{ys} is the yielding stress of the material and Z_p is the plastic modulus of the cross section (Chapter 2).

The object of plastic design is to find the location of the plastic hinges and to determine the collapse load. Table 11-4 gives these quantities for several loads and boundary conditions. The maximum load and the location of the plastic hinge for the cases that do not appear in this table can be determined from the methods provided in many textbooks, such as reference 11.24.

11.7 BUCKLING LOADS AND COLUMNS

A slender column subject to an axial load may assume a state of large lateral deflection even if no significant lateral loads are applied. This condition occurs because of the existence of an unstable equilibrium state above a certain level of axial load called the elastic buckling load. The first analysis of the elastic buckling load is also known as the critical load, and it is given by the formula (*Euler's formula*)

$$P_{cr} = \pi^2 EI/L^2 \qquad (11.10)$$

for a pinned–pinned beam. The critical load can also be written as a critical stress by defining the *slenderness ratio* $\lambda = KL/r$ in which KL is the effective column length that varies with the end conditions and $r = \sqrt{I/A}$ is the *radius of gyration* (Chapter 2) of the beam cross section about the centroidal axis for which the radius of gyration is a minimum:

$$\sigma_{cr} = \pi^2 E/\lambda^2 \qquad (11.11)$$

The appropriate expression for the critical load is shown in Tables 11-5 to 11-9 for a variety of column configurations. These include ordinary columns (Table 11-5), columns with flexible end supports (Table 11-6), columns with in-span axial loads, columns with in-span supports (Table 11-8), and tapered columns (Table 11-9).

The buckling of complicated beams or beam structures, e.g., frames, can be determined, as explained in Appendix III, using the transfer and stiffness matrices of Sections 11.9 and 11.10. Use of the transfer matrix and corresponding generalized dynamic stiffness matrix leads to a determinant search for the buckling loads. If the geometric stiffness matrix is utilized, the buckling load can be computed as the solution to a classical eigenvalue problem.

Example 11.5 Buckling of a Cantilever Beam with a Flexibly Supported End A cantilever beam with the linear spring at the free end is shown in Fig. 11-8. For this beam, $E = 200 \text{ GN/m}^2$, and $I = 5245 \text{ cm}^4$. Find the critical axial load.

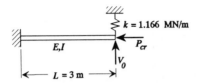

FIGURE 11-8 Cantilever beam with linear spring.

This beam corresponds to case 1 of Table 11-6, which has no rotational resistance, i.e., $k_1 = 0$. From Fig. 11-8, the elastic spring constant k has the value 1.166 MN/m.

Also, it follows from the formulas of the table that

$$c = \frac{kL}{EI/L^2} = 3.0 \tag{1}$$

and $c_1 = 0$. From the formula for case 1

$$\frac{c(m - \sin m) - m^3}{c(1 - \cos m)} = \tan m$$

it follows that $m = 2.203$. Hence, the critical load P_{cr} is calculated as

$$P_{cr} = m^2 \frac{EI}{L^2} = (2.203)^2 \frac{(200 \times 10^9)(5.245 \times 10^{-5})}{3^2} \tag{2}$$
$$= 5.65 \text{ MN}$$

Of course, methods using transfer or stiffness matrices lead to the same result.

Columns

The Euler buckling load given by Eq. (11.10) is strictly accurate for a perfectly straight slender column subject to a centered axial load. It is assumed that before the load reaches the critical value, the deflection of the column is small. When there are some imperfections or deviations such as the column not being slender or perfectly straight or the load is applied eccentrically, the justification for the use of Euler's equation is questionable. In these cases, large deformations may develop or the material may behave inelastically before the column buckles. Generally, for columns with large slenderness ratios and small load eccentricities, Euler's equation is still applicable. Aluminum and mild steel columns usually buckle elastically at slenderness ratios of about 100 or more, but this behavior depends on the imperfections in the load–column system. Some aluminum alloys tempered to high yield strength buckle elastically at slenderness ratios as low as 60. Inelastic buckling tends to occur as the slenderness ratios decrease and the imperfections in column straightness and load eccentricity increase.

In the case of inelastic buckling, the axial compressive stress P/A is greater than the proportional limit before the column buckles. The determination of P_{cr} for perfectly straight columns is closely related to the elastic modulus after the yield point. The slope of the stress–strain relationship $d\sigma/d\epsilon$ from Fig. 11-9 no

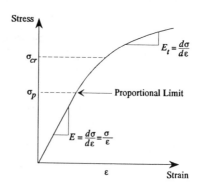

FIGURE 11-9 Stress – strain relation and moduli.

longer equals the elastic modulus E. Two theories for handling this modulus are pertinent.

The *tangent modulus theory* proposed by Engesser in 1889 describes a modulus E_t [11.3]. The tangent modulus E_t is defined as $d\sigma/d\epsilon$, the slope of the stress–strain curve above the proportional limit. Unlike the elastic modulus, the tangent modulus is not constant but, rather, is a function of applied stress. Substitute E_t for E in Euler's formula, yielding $P_{cr} = \pi^2 E_t I/L^2$ for a pinned–pinned beam.

A second theory, the *double modulus theory* or *reduced modulus theory*, assumes that the axial force remains constant during the onset of buckling [11.3]. The resulting deformation of the beam causes strain reversal on the convex side while the strain on the concave side will continue to increase. It would appear that the material laws (moduli) relating the increments of stress and strain for the two sides should be different. The convex side is chosen to be represented by the elastic modulus E and the concave to the tangent modulus E_t. This results in a value for P_{cr} that depends on both E and E_t. This yields the equation for the reduced modulus load $P_{cr} = \pi^2 E_r I/L^2$, where E_r is a function of both E_t and E. This load is the greatest load the column can bear and still remain straight. The modulus E_r is a function of both cross-sectional and material properties. See references 11.3 and 11.19 for more details of these two theories.

For the columns that are not straight and loaded eccentrically, it is difficult to derive the buckling load. However, approximate formulas have been proposed that are empirically adjusted to conform to test results. The maximum stresses P/A according to some of these formulas are given in Table 11-10.

Short Bars with Eccentric Loading

For very short bars subjected to eccentric axial loading, buckling will not occur, and the stresses can be determined from the theory of strength of materials. For the materials that do not withstand tension or compression, the load cannot be applied outside of the *kern* of the cross section; otherwise tensile stress will develop on the cross section. The kerns are shown in Table 11-11 for various shapes of cross sections. The method of obtaining the formulas in this table can be found in basic strength of material texts.

11.8   NATURAL FREQUENCIES AND MODE SHAPES

The natural frequencies ω_i (radians per second) or f_i (hertz), $i = 1, 2, \ldots,$ for the bending vibrations for uniform Euler–Bernoulli beams are presented in Table 11-12. The frequency equations for beams, including shear deformation and rotary inertia, are provided in Table 11-13. The roots ω_i, $i = 1, 2, \ldots,$ of these equations are the natural frequencies.

Table 11-14 provides frequencies for beams modeled with lumped masses. The fundamental natural frequencies for several continuous beams are tabulated in Table 11-15 and for tapered beams in Table 11-16.

Some approximate formulas for the period of vibration of buildings are given in Table 11-17. Damping information for various structures is provided in Table 11-18. Most of these results are based on observations of the response of actual structures.

The natural frequencies and mode shapes for beams more complicated than those presented in these tables can be computed using the transfer matrix and displacement methods of Appendix III. Frequencies are found using the transfer matrix and corresponding generalized dynamic stiffness matrix of Section 11.9 with a determinant frequency search, whereas the solution to a generalized eigenvalue problem will yield the frequencies if a stiffness matrix (other than a dynamic stiffness matrix) and a mass matrix are employed.

Example 11.6   Vibration of a Continuous Beam   A multispan uniform beam with rigid in-span supports and one end free with the other end pinned is shown in Fig. 11-10. For this beam, $E = 200 \text{ GN/m}^2$, $I = 144 \text{ cm}^4$, and ρ (mass per unit length) $= 19.97 \text{ kg/m}$. The fundamental natural frequency can be obtained by using Table 11-15, case 2. For three spans, $\lambda = 1.536$. Then

$$f_1 = \frac{\lambda^2 (EI/\rho)^{1/2}}{2\pi \ell^2} = 17.61 \text{ Hz} \tag{1}$$

The same result can be generated using the matrix methods discussed in the following sections.

11.9   GENERAL BEAMS

Most of the formulas provided thus far apply to single-span beams. For more general beams, e.g., those with multiple-span, variable cross-sectional properties, in-span supports, it is advisable to use the transfer matrix method or the

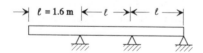

FIGURE 11-10   Multispan beam with rigid in-span supports.

displacement (stiffness) method outlined in Appendix III. These are efficient methods that can be programmed with ease.

Transfer Matrices

A few transfer matrices are tabulated in Tables 11-19 to 11-22. They are used to find the static response, buckling load, or natural frequencies as indicated in Appendix III. The transfer matrix in Table 11-22 is for a very general beam element that can be employed to take into account such effects as foundations, axial forces, and inertia.

The notation for the transfer matrix for beam element i is

$$\mathbf{U}^i = \begin{bmatrix} U_{ww} & U_{w\theta} & U_{wV} & U_{wM} & F_w \\ U_{\theta w} & U_{\theta\theta} & U_{\theta V} & U_{\theta M} & F_\theta \\ U_{Vw} & U_{V\theta} & U_{VV} & U_{VM} & F_V \\ U_{Mw} & U_{M\theta} & U_{MV} & U_{MM} & F_M \\ 0 & 0 & 0 & 0 & 1 \end{bmatrix} \tag{11.12}$$

The structural matrices here are based on the governing equations for the bending of a Timoshenko beam:

$$\frac{\partial w}{\partial x} = -\theta + \frac{V}{GA_s} \tag{11.13a}$$

$$\frac{\partial \theta}{\partial x} = \frac{M}{EI} + \frac{M_T}{EI} \tag{11.13b}$$

$$\frac{\partial V}{\partial x} = kw + \rho \frac{\partial^2 w}{\partial t^2} - p_z(x,t) \tag{11.13c}$$

$$\frac{\partial M}{\partial x} = V + (k^* - P)\theta + \rho r_y^2 \frac{\partial^2 \theta}{\partial t^2} - c(x,t) \tag{11.13d}$$

The applied distributed force and moment are indicated by $p_z(x,t)$ and $c(x,t)$, respectively. A Timoshenko beam includes the effects of shear deformation and rotary inertia as well as bending. The governing equations are reduced to those for a Rayleigh beam (bending, rotary inertia) by setting $1/GA_s$ equal to zero for a shear beam (bending, shear deformation) by making

$$\rho r_y^2 \frac{\partial^2 \theta}{\partial t^2}$$

equal to zero and for a Euler–Bernoulli beam (bending) by putting

$$1/GA_s \quad \text{and} \quad \rho r_y^2 \frac{\partial^2 \theta}{\partial t^2}$$

equal to zero. Equations (11.13) apply to beams with a compressive axial force P. Replace P by $-P$ if the axial force is tensile.

Example 11.7   A Simple Beam   Return to the simple beam of Examples 11.1 and 11.2. The same problem can be solved using transfer matrices. See the notation in Fig. 11-5c. Follow the procedure explained in Appendix III. In transfer matrix form the response at the right end is given by

$$\mathbf{z}_{x=L} = \mathbf{U}^2 \mathbf{U}_b \mathbf{U}^1 \mathbf{z}_0 \tag{1}$$

where $\mathbf{z}$ is the state vector

$$\mathbf{z} = \begin{bmatrix} w \\ \theta \\ V \\ M \\ 1 \end{bmatrix} \tag{2}$$

From Tables 11-19 and 11-21, the extended transfer matrices for the case $1/GA_s = 0$ are given by

$$\mathbf{U}^1 = \begin{bmatrix} 1 & -a_1 & -a_1^3/6EI & -a_1^2/2EI & 0 \\ 0 & 1 & a_1^2/2EI & a_1/EI & 0 \\ 0 & 0 & 1 & 0 & 0 \\ 0 & 0 & a_1 & 1 & 0 \\ 0 & 0 & 0 & 0 & 1 \end{bmatrix} \tag{3}$$

$$\mathbf{U}_b = \begin{bmatrix} 1 & 0 & 0 & 0 & 0 \\ 0 & 1 & 0 & 0 & 0 \\ 0 & 0 & 1 & 0 & -W \\ 0 & 0 & 0 & 1 & 0 \\ 0 & 0 & 0 & 0 & 1 \end{bmatrix} \tag{4}$$

$$\mathbf{U}^2 = \begin{bmatrix} 1 & -b & -b^3/6EI & -b^2/2EI & 0 \\ 0 & 1 & b^2/2EI & b/EI & 0 \\ 0 & 0 & 1 & 0 & 0 \\ 0 & 0 & b & 1 & 0 \\ 0 & 0 & 0 & 0 & 1 \end{bmatrix}, \quad b = L - a_1 \tag{5}$$

Carry out the matrix multiplications indicated in (1) using the matrices of (3), (4), and (5). This results in

$$\mathbf{z}_{x=L} = \mathbf{U}\mathbf{z}_0 \tag{6}$$

where $\mathbf{U} = \mathbf{U}^2\mathbf{U}_b\mathbf{U}^1$ is the global transfer matrix

$$\mathbf{U} = \begin{bmatrix} 1 & -L & -L^3/6EI & -L^2/2EI & b^3W/6EI \\ 0 & 1 & L^2/2EI & L/EI & -b^2W/2EI \\ 0 & 0 & 1 & 0 & -W \\ 0 & 0 & L & 1 & -bW \\ 0 & 0 & 0 & 0 & 1 \end{bmatrix} \tag{7}$$

Apply the boundary conditions to (6):

$$\begin{bmatrix} w=0 \\ \theta \\ V \\ M=0 \\ 1 \end{bmatrix}_{x=L} = \mathbf{U} \begin{bmatrix} w=0 \\ \theta=0 \\ V \\ M \\ 1 \end{bmatrix}_{x=0} \tag{8}$$

This leads to

$$w_{x=L} = 0 = -\frac{L^3 V_0}{6EI} - \frac{L^2 M_0}{2EI} + \frac{b^3 W}{6EI} \tag{9}$$

$$M_{x=L} = 0 = LV_0 + M_0 - bW$$

From (9)

$$V_0 = \tfrac{1}{2}W\left[3b/L - (b/L)^3\right] \tag{10}$$

$$M_0 = bW - LV_0$$

Use $W = 35$ kN, $L = 1.6$ m, $b = L - a_1 = 0.7$ m, and again we find that $V_0 = 21.5$ kN and $M_0 = -9.91$ kN-m. The shear force at the simple support is

$$V_{x=L} = V_0 - W = -13.5 \text{ kN} \tag{11}$$

Computer-generated results are shown in Fig. 11-11.

================

11.10 STIFFNESS AND MASS MATRICES

Stiffness Matrix

Tables 11-19 to 11-22 contain stiffness matrices for beams. Use of these matrices in static, stability, and dynamic analyses is explained in Appendix III. Textbooks covering standard structural mechanics, such as [11.23], can also be consulted.

The matrices in this section follow sign convention 2 of Appendix II. This is in contrast to most of the formulas and transfer matrices appearing earlier in this chapter, which are based on sign convention 1 of Appendix II.

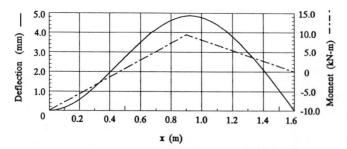

FIGURE 11-11 Computer-generated response of the beam for Example 11.7.

All of these stiffness matrices for the ith element are of the form $\mathbf{p}^i = \mathbf{k}^i\mathbf{v}^i - \overline{\mathbf{p}}^i$, where

$$
\mathbf{p}^i = \begin{bmatrix} V_a \\ M_a \\ V_b \\ M_b \end{bmatrix}^i, \qquad
\mathbf{v}^i = \begin{bmatrix} w_a \\ \theta_a \\ w_b \\ \theta_b \end{bmatrix}^i, \qquad
\overline{\mathbf{p}}^i = \begin{bmatrix} V_a^0 \\ M_a^0 \\ V_b^0 \\ M_b^0 \end{bmatrix}^i
\qquad (11.14)
$$

The format for the stiffness matrix of a plane beam element with bending deformation is

$$
\mathbf{k}^i = \begin{bmatrix}
k_{11} & k_{12} & k_{13} & k_{14} \\
k_{21} & k_{22} & k_{23} & k_{24} \\
k_{31} & k_{32} & k_{33} & k_{34} \\
k_{41} & k_{42} & k_{43} & k_{44}
\end{bmatrix}
\qquad (11.15)
$$

As explained in Appendix II, the stiffness matrices can be derived in numerous ways, including by rearranging transfer matrices.

Example 11.8 A Simple Beam Use the displacement method to find the response of the simple beam of Fig. 11-5c.

Although there is no technical reason to use more than one element as stiffness matrices for statically loaded beams are exact, we choose to discretize the beam into two elements. For each element, the stiffness matrix is given in Table 11-19.

$$
\mathbf{k}^i = \begin{bmatrix}
12 & -6\ell_i & -12 & -6\ell_i \\
-6\ell_i & 4\ell_i^2 & 6\ell_i & 2\ell_i^2 \\
-12 & 6\ell_i & 12 & 6\ell_i \\
-6\ell_i & 2\ell_i^2 & 6\ell_i & 4\ell_i^2
\end{bmatrix}
\frac{E_i I_i}{\ell_i^3} =
\begin{bmatrix}
\mathbf{k}_{aa}^i & \mathbf{k}_{ab}^i \\
\mathbf{k}_{ba}^i & \mathbf{k}_{bb}^i
\end{bmatrix}
\qquad (1)
$$

where ℓ_i, I_i, and E_i are the length, moment of inertia, and modulus of elasticity of the beam element i, respectively. For this beam $\ell_1 = 0.9$ m, $\ell_2 = 0.7$ m, $E_1 = E_2 = 200$ GN/m^2, and $I_1 = I_2 = 144$ cm^4.

Follow the procedure provided in Appendix III in which the global stiffness matrix is assembled using the element stiffness matrices.

$$
\mathbf{k}^1 = \begin{bmatrix} \mathbf{k}_{aa}^1 & \mathbf{k}_{ab}^1 \\ \mathbf{k}_{ba}^1 & \mathbf{k}_{bb}^1 \end{bmatrix} = 395\,061.73 \begin{bmatrix} 12 & -5.4 & -12 & -5.4 \\ -5.4 & 3.24 & 5.4 & 1.62 \\ -12 & 5.4 & 12 & 5.4 \\ -5.4 & 1.62 & 5.4 & 3.24 \end{bmatrix} \tag{2}
$$

$$
\mathbf{k}^2 = \begin{bmatrix} \mathbf{k}_{bb}^2 & \mathbf{k}_{bc}^2 \\ \mathbf{k}_{cb}^2 & \mathbf{k}_{cc}^2 \end{bmatrix} = 839\,650.15 \begin{bmatrix} 12 & -4.2 & -12 & -4.2 \\ -4.2 & 1.96 & 4.2 & 0.98 \\ -12 & 4.2 & 12 & 4.2 \\ -4.2 & 0.98 & 4.2 & 1.96 \end{bmatrix} \tag{3}
$$

$$
\mathbf{K} = \begin{bmatrix} \mathbf{k}_{aa}^1 & \mathbf{k}_{ab}^1 & \mathbf{0} \\ \mathbf{k}_{ba}^1 & \mathbf{k}_{bb}^1 + \mathbf{k}_{bb}^2 & \mathbf{k}_{bc}^2 \\ \mathbf{0} & \mathbf{k}_{cb}^2 & \mathbf{k}_{cc}^2 \end{bmatrix} \tag{4}
$$

$$
= 395\,061.73 \begin{bmatrix} 12 & -5.4 & -12 & -5.4 & 0 & 0 \\ -5.4 & 3.24 & 5.4 & 1.62 & 0 & 0 \\ -12 & 5.4 & 37.5044 & -3.5265 & -25.5044 & -8.9265 \\ -5.4 & 1.62 & -3.5265 & 7.4057 & 8.9265 & 2.0828 \\ 0 & 0 & -25.5044 & 8.9265 & 25.5044 & 8.9265 \\ 0 & 0 & -8.9265 & 2.0828 & 8.9265 & 4.1657 \end{bmatrix}
$$

The forces and deformations are related by

$$
\overline{\mathbf{P}} = \mathbf{K}\mathbf{V} \tag{5}
$$

where

$$
\mathbf{V} = \begin{bmatrix} w_a \\ \theta_a \\ w_b \\ \theta_b \\ w_c \\ \theta_c \end{bmatrix} = \begin{bmatrix} 0 \\ 0 \\ w_b \\ \theta_b \\ 0 \\ \theta_c \end{bmatrix} \tag{6}
$$

and

$$
\overline{\mathbf{P}} = \begin{bmatrix} V_a \\ M_a \\ V_b \\ M_b \\ V_c \\ M_c \end{bmatrix} = \begin{bmatrix} V_a \\ M_a \\ W \\ 0 \\ V_c \\ 0 \end{bmatrix} \begin{matrix} \text{(unknown} \\ \text{reactions)} \\ \\ \\ \text{(unknown} \\ \text{reaction)} \end{matrix} \tag{7}
$$

Remove the columns in (4) corresponding to the displacements that are zero [see (6)] and ignore the rows in (4) corresponding to the unknown reactions [see (7)]. Then

$$395\ 061.73 \begin{bmatrix} 37.5044 & -3.5265 & -8.9265 \\ -3.5265 & 7.4057 & 2.0828 \\ -8.9265 & 2.0828 & 4.1657 \end{bmatrix} \begin{bmatrix} w_b \\ \theta_b \\ \theta_c \end{bmatrix} = \begin{bmatrix} W \\ 0 \\ 0 \end{bmatrix} \tag{8}$$

For $W = 35$ kN, the solution of this set of equations gives

$$w_b = 4.8576 \text{ mm}$$
$$\theta_b = -7.1490 \times 10^{-4} \text{ rad} \tag{9}$$
$$\theta_c = 0.01077 \text{ rad}$$

Place these values in (5) and solve for the reactions,

$$V_a = -21.5 \text{ kN} = -V_0$$
$$M_a = 9.91 \text{ kN-m} = -M_0 \tag{10}$$
$$V_c = -13.5 \text{ kN} = V_{x=L}$$

As expected, these are the same results obtained with the other methods employed in earlier examples for the same problem.

Table 11-22 provides the generalized dynamic stiffness matrix, which includes many effects such as foundations and inertias and axial forces. In addition to incorporation in static analyses, this stiffness matrix is useful in setting up global frequency equation determinants that can be employed for the determination of exact natural frequencies and buckling loads along with the corresponding mode shapes.

Geometric Stiffness Matrix

The traditional geometric stiffness matrix for the buckling of simple beams is provided in Table 11-23.

Mass Matrix

The mass matrices contained in Tables 11-24 and 11-25 are the customary lumped-mass and consistent-mass matrices. See Appendix III for the use of mass matrices in dynamic analyses. The format for these mass matrices for a beam element is

$$\begin{bmatrix} m_{11} & m_{12} & m_{13} & m_{14} \\ m_{21} & m_{22} & m_{23} & m_{24} \\ m_{31} & m_{32} & m_{33} & m_{34} \\ m_{41} & m_{42} & m_{43} & m_{44} \end{bmatrix} \tag{11.16}$$

REFERENCES

11.1 Euler, L., "Des Curvie Elasticis," Lausanne and Geneva, 1744, translated and annotated by Oldfather, W. A., Ellis C. A., and Brown D. M. ISIS, No. 58, Vol. XX, 1, 1933.

11.2 Column Research Committee of Japan, *Handbook of Structural Stability*, Corona Publishing Company, Tokyo, 1971.

11.3 Boresi, A. P., Sidebottom, O. M., and Schmidt, R. J., *Advanced Mechanics of Materials*, 5th ed., Wiley, New York, 1993.

11.4 Blevins, R. D., *Formulas for Natural Frequency and Mode Shape*, Van Nostrand Reinhold, New York, 1979.

11.5 Culver, C. G., "The Influence of Shear Deformations and Rotary Inertia on the Natural Frequencies of Beams," Marine Engineering Laboratory, Technical Memorandum 4/66, 1966.

11.6 Shaker, F. J., "Effect of Axial Load on Mode Shapes and Frequencies of Beams," Lewis Research Center Report NASA-TN-8109, December 1975.

11.7 Ratzersdorfer, J., *Die Knickfestigkeit von Staben und Stabwerken*, Springer, Wien, 1936.

11.8 Nakagawa, Y., "Effect of the End Fixity on the Buckling of the Column," *Trans. Jpn Soc. Mech. Eng.*, Vol. 4, No. 14, 1938.

11.9 Bleich, F., *Theories und Berechnung der Eisernen Brucken*, Springer, Wien, 1936.

11.10 Moyar, G. J., Pilkey, W. D., and Pilkey, B. F. (Eds.), *Track Train Dynamics and Design: Advanced Techniques*, Pergamon, New York, 1977.

11.11 Nakagawa, K., "Vibrational Characteristics of Reinforced Concrete Buildings Existing in Japan," *Proceedings of the Second World Conference on Earthquake Engineering*, Tokyo and Kyoto, Japan, 1960.

11.12 Blume, J. A., and Binder, R. W., "Records of a Modern Multistory Office Building during Construction," *Proceedings of the Second World Conference on Earthquake Engineering*, Tokyo and Kyoto, Japan, 1960.

11.13 Housner, G. W., and Brody, G. A., "Natural Periods of Vibration of Buildings," *J. Eng. Mech. Div. ASCE*, Vol. 89, No. EM4, 1963, pp. 31–65.

11.14 Sachs, P., *Wind Forces in Engineering*, 2nd ed., Pergamon, Elmsford, NY, 1978.

11.15 Adams, O. E., Jr., and Gilbert, W. D., *Seismic Resistance Criteria*, HN-65931, July 8, 1960.

11.16 Abbett, R. W. (Ed.), "Earthquake and Earthquake-resistant Design," *Am. Civ. Eng. Pract.*, Vol. III, Wiley, New York, 1957, pp. 34-02 to 34-23.

11.17 Tsui, E. Y. W., "Aseismic Design of Structures by Rigidity Criterion," *J. Structural Div.*, *ASCE*, Vol. 85, No. ST2, 1959, pp. 81-106.

11.18 Selberg, A., "Dampening Effects on Suspension Bridges," *Int. Assoc. Bridge Struct. Eng.*, Vol. 10, 1970.

11.19 Chen, W. F., and Lui, E. M., *Structural Stability, Theory and Implementations*, Elsevier, New York, 1987.

11.20 Esling, F. K., "A Problem Relating to Railway—Bridge Piers of Masonry or Brickwork," *Proc. Inst. Civil Eng.*, Vol. 165, 1905–1906, pp. 219–230.

11.21 Ruditzin, M. W., Artemov, P., and Luboshytz, M., *Handbook for Strength of Materials*, BSSR Government Publishing House, Minsk, 1958.

11.22 Schiff, D., *Dynamic Analysis and Failure Modes of Simple Structures*, Wiley, New York, 1990.

11.23 Pilkey, W. D., and Wunderlich, W., *Structural Mechanics, Computational and Variational Methods*, CRC, Florida, 1993.

11.24 Cook, R. D., and Young, W. C., *Advanced Mechanics of Materials*, Macmillan, New York, 1985.

11

Tables

TABLE 11-1 UNIFORM BEAMS

Positive deflection w, slope θ, moment M, and shear force V

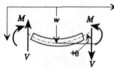

The positive directions of the reactions (R_1, R_2, M_1, M_2) are shown in the figures for each case.

Type of Beam	Reactions	Deflection at Any Point x
1. Coordinate x is measured from the left hand end for all entries in this table.	$R_1 = W$ $M_1 = Wa$	For $x < a$ $\dfrac{W}{6EI}(-x^3 + 3x^2a)$ For $x \geq a$ $\dfrac{W}{6EI}(3a^2x - a^3)$
2.	$R_1 = p_1 L$ $M_1 = \frac{1}{2}p_1 L^2$	$\dfrac{p_1 x^2}{24EI}(x^2 + 6L^2 - 4Lx)$
3.	$R_1 = \frac{1}{2}pL$ $M_1 = \frac{1}{6}pL^2$	$\dfrac{px^2}{120LEI}(10L^3 - 10L^2x + 5Lx^2 - x^3)$
4.	$R_1 = \frac{1}{2}pL$ $M_1 = \frac{1}{3}pL^2$	$\dfrac{px^2}{120LEI}(20L^3 - 10L^2x + x^3)$
5.	$R_1 = 0$ $M_1 = M^*$	For $x < a$ $\dfrac{x^2 M^*}{2EI}$ For $x \geq a$ $\dfrac{M^*a}{2EI}(2x - a)$

TABLE 11-1 Uniform Beams 520

Maximum Deflection	Moment at Any Point x	Maximum Moment	Important Slope	Maximum Shear Force
$\dfrac{Wa^2(3L-a)}{6EI}$ at $x=L$	For $x < a$ $-W(a-x)$ For $x \geq a$ 0	$Wa = M_1$ at $x=0$	$\theta_{max} = \dfrac{Wa^2}{2EI}$ at $x=L$	W at $x<a$
$\dfrac{p_1 L^4}{8EI}$ at $x=L$	$-\frac{1}{2}p_1(L^2 - 2Lx + x^2)$	$\frac{1}{2}p_1 L^2 = M_1$ at $x=0$	$\theta_{max} = \dfrac{p_1 L^3}{6EI}$	$p_1 L$ at $x=0$
$\dfrac{pL^4}{30EI}$ at $x=L$	$-\dfrac{p}{6L}(L^3 - 3L^2 x + 3Lx^2 - x^3)$	$\frac{1}{6}pL^2 = M_1$ at $x=0$	$\theta_{max} = \dfrac{pL^3}{24EI}$ at $x=L$	$\frac{1}{2}pL$ at $x=0$
$\dfrac{11pL^4}{120EI}$ at $x=L$	$-\dfrac{p}{6L}(2L^3 - 3L^2 x + x^3)$	$\frac{1}{3}pL^2 = M_1$ at $x=0$	$\theta_{max} = \dfrac{pL^3}{8EI}$ at $x=L$	$\frac{1}{2}pL$ at $x=0$
$\dfrac{M^* a}{2EI}(2L - a)$ at $x=L$	For $x < a$ $-M^*$ For $x \geq a$ 0	M^* at all $x<a$	$\theta_{max} = \dfrac{M^* a}{EI}$ at $x=L$	0

Uniform Beams TABLE 11-1

TABLE 11-1 (continued)   UNIFORM BEAMS

Type of beam	Reactions	Deflection at any point x	Maximum deflection
6.	$R_1 = \dfrac{Wb}{L}$ $R_2 = \dfrac{Wa}{L}$	For $x < a$ $\dfrac{Wb}{6LEI}\left[-x^3 + (L^2 - b^2)x\right]$ For $x \geq a$ $\dfrac{Wa}{6LEI}\left[-(L-x)^3 \right.$ $\left. + (L^2 - a^2)(L-x)\right]$	When $a > \dfrac{L}{2}$ $\dfrac{Wb(L^2 - b^2)^{3/2}}{9\sqrt{3}\,LEI}$ at $x = \sqrt{\tfrac{1}{3}(L^2 - b^2)}$
7.	$R_1 = \tfrac{1}{2}p_1 L$ $R_2 = \tfrac{1}{2}p_1 L$	$\dfrac{p_1 x}{24EI}(L^3 - 2Lx^2 + x^3)$	$\dfrac{5p_1 L^4}{384EI}$ at $x = \tfrac{1}{2}L$
8.	$R_1 = \tfrac{1}{6}pL$ $R_2 = \tfrac{1}{3}pL$	$\dfrac{px}{360LEI}$ $\times (3x^4 - 10L^2x^2 + 7L^4)$	$\dfrac{0.00652\,pL^4}{EI}$ at $x = 0.519L$
9.	$R_1 = \dfrac{M^*}{L}$ $R_2 = \dfrac{M^*}{L}$	$\dfrac{M^* Lx}{6EI}\left(1 - \dfrac{x^2}{L^2}\right)$	$\dfrac{M^* L^2}{9\sqrt{3}\,EI}$ at $x = \dfrac{L}{\sqrt{3}}$
10.	$R_1 = \dfrac{M^*}{L}$ $R_2 = \dfrac{M^*}{L}$	For $x < a$ $-\dfrac{M^*}{6EIL}(6axL - 3a^2x$ $-2L^2x - x^3)$ For $x \geq a$ $-\dfrac{M^*}{6EIL}(3a^2L + 3x^2L - x^3$ $-2L^2x - 3a^2x)$	Maximum deflection occurs at $x = x_1$ and/or $x = x_2$, $x_1 =$ $\left(2aL - a^2 - \tfrac{2}{3}L^2\right)^{1/2}$ $x_2 =$ $L - \left(\tfrac{1}{3}L^2 - a^2\right)^{1/2}$

TABLE 11-1   Uniform Beams   **522**

Moment at any point x	Maximum moment	Important slope	Maximum shear force
For $x < a$ $\dfrac{Wbx}{L}$ For $x \geq a$ $\dfrac{Wa}{L}(L - x)$	$\dfrac{Wab}{L}$ at $x = a$	$\theta_1 = \dfrac{Wab}{6LEI}(2L - a)$ $\theta_2 = \dfrac{Wab}{6LEI}(2L - b)$	If $a > b$ $\dfrac{Wa}{L}$ at $x > a$ If $a < b$ $\dfrac{Wb}{L}$ at $x < a$
$\dfrac{1}{2}p_1L\left(x - \dfrac{x^2}{L}\right)$	$\dfrac{1}{8}p_1L^2$ at $x = \dfrac{1}{2}L$	$\theta_1 = \theta_2$ $= \dfrac{p_1L^3}{24EI}$	$\dfrac{1}{2}p_1L$ at $x = 0, L$
$-\dfrac{1}{6}pL\left(\dfrac{x^3}{L^2} - x\right)$	$0.064pL^2$ at $x = \dfrac{1}{3}\sqrt{3}\,L$	$\theta_1 = \dfrac{7}{360}\dfrac{pL^3}{EI}$ $\theta_2 = \dfrac{1}{45}\dfrac{pL^3}{EI}$	$\dfrac{1}{3}pL$ at $x = L$
$\dfrac{M^*x}{L}$	M^* at $x = L$	$\theta_1 = \dfrac{M^*L}{6EI}$ $\theta_2 = \dfrac{M^*L}{3EI}$	$\dfrac{M^*}{L}$ at all x
For $x < a$ $-\dfrac{M^*x}{L}$ For $x > a$ $\dfrac{M^*(L - x)}{L}$	$-R_1a$ at $x = a^-$ $R_1(L - a)$ at $x = a^+$	$\theta_1 = -\dfrac{M^*}{6EIL}$ $\times(2L^2 - 6aL + 3a^2)$ $\theta_2 = \dfrac{M^*}{6EIL}(L^2 - 3a^2)$ $\dfrac{M^*}{3EIL}$ $\times(3aL - 3a^2 - L^2)$ at $x = a$	$\dfrac{M^*}{L}$ at all x

Uniform Beams TABLE 11-1

TABLE 11-1 (continued)   UNIFORM BEAMS

Type of beam	Reactions	Deflection at any point x	Maximum deflection
11.	$R_1 =$ $\dfrac{W}{2}\dfrac{3b^2L - b^3}{L^3}$ $R_2 = W - R_1$ $M_2 =$ $\dfrac{Wab}{2L^2}(a + L)$	For $x < a$ $-\dfrac{Wa(L - x)^2}{12EIL^3}$ $\times (2La^2 - 3L^2x + a^2x)$ $+\dfrac{W(a - x)^3}{6EI}$ For $x \geq a$ $-\dfrac{Wa(L - x)^2}{12EIL^3}$ $\times (2La^2 - 3L^2x + a^2x)$	If $a > 0.414L$ $\dfrac{Wab^2\sqrt{a/(2L + a)}}{6EI}$ at $x = L\sqrt{1 - \dfrac{2L}{3L - b}}$ If $a < 0.414L$ $\dfrac{Wa(L^2 - a^2)^3}{3EI(3L^2 - a^2)^2}$ at $x = \dfrac{L(L^2 + a^2)}{3L^2 - a^2}$ If $a = 0.414L$ $\dfrac{0.0098WL^3}{EI}$ at $x = a$ (peak possible deflection)
12.	$R_1 = \frac{3}{8}p_1L$ $R_2 = \frac{5}{8}p_1L$ $M_2 = \frac{1}{8}p_1L^2$	$-\dfrac{p_1}{48EI}(3Lx^3 - 2x^4 - L^3x)$	$\dfrac{0.0054p_1L^4}{EI}$ at $x = 0.4215L$
13.	$R_1 = \frac{1}{10}pL$ $R_2 = \frac{2}{5}pL$ $M_2 = \frac{1}{15}pL^2$	$-\dfrac{p}{120EIL}$ $\times (2L^2x^3 - L^4x - x^5)$	$\dfrac{0.00238pL^4}{EI}$ at $x = \dfrac{L}{\sqrt{5}}$
14.	$R_1 = \frac{11}{40}pL$ $R_2 = \frac{9}{40}pL$ $M_2 = \frac{7}{120}pL^2$	$-\dfrac{p}{240EIL}$ $\times (11L^2x^3 - 3L^4x$ $- 10x^4L + 2x^5)$	$\dfrac{0.00304L^4}{EI}$ at $x = 0.402L$

TABLE 11-1   **Uniform Beams**   **524**

Moment at any point x	Maximum moment	Important slope	Maximum shear force
For $x < a$ R_1x For $x \geq a$ $R_1x - W(x - a)$	$M_2 = \dfrac{Wab}{2L^2}(L + a)$ at $x = L$	$\theta_1 = -\dfrac{W}{4EI}\left(\dfrac{b^3}{L} - b^2\right)$	If $a > 0.348L$ R_2 at $x > a$ If $a < 0.348L$ R_1 at $x < a$
$-p_1L\left(\dfrac{1}{2}\dfrac{x^2}{L} - \dfrac{3}{8}x\right)$	$M = \frac{9}{128}p_1L^2$ at $x = \frac{3}{8}L$ $M = \frac{1}{8}p_1L^2 = M_2$ at $x = L$	$\theta_1 = \dfrac{p_1L^3}{48EI}$	$\frac{5}{8}p_1L$ at $x = L$
$-\frac{1}{2}pL\left(\dfrac{x^3}{3L^2} - \dfrac{x}{5}\right)$	$M = 0.03pL^2$ at $x = 0.4474L$ $+M = M_2$ at $x = L$	$\theta_1 = \dfrac{pL^3}{120EI}$	$\frac{2}{5}pL$ at $x = L$
$-\frac{1}{2}pL\left(\dfrac{x^2}{L} - \dfrac{11}{20}x - \dfrac{1}{3}\dfrac{x^3}{L^2}\right)$	$-M = -0.0423pL^2$ at $x = 0.3292L$ $M = \frac{7}{120}pL^2 = M_2$ at $x = L$	$\theta_1 = \dfrac{pL^3}{80EI}$	$\frac{11}{40}pL$ at $x = 0$

Uniform Beams TABLE 11-1

TABLE 11-1 (continued)   UNIFORM BEAMS

Type of beam	Reactions	Deflection at any point x	Maximum deflection
15.	$R_1 =$ $\dfrac{3M^*}{2L}\dfrac{L^2 - a^2}{L^2}$ $R_2 =$ $\dfrac{3M^*}{2L}\left(\dfrac{L^2 - a^2}{L^2}\right)$ $M_2 =$ $\dfrac{M^*}{2}\left(1 - \dfrac{3a^2}{L^2}\right)$	For $x < a$ $-\dfrac{M^*}{EI}\left[\dfrac{L^2 - a^2}{4L^3}(3L^2x - x^3)\right.$ $\left. -(L-a)x\right]$ For $x \geq a$ $-\dfrac{M^*}{EI}\left[\dfrac{L^2 - a^2}{4L^3}(3L^2x - x^3)\right.$ $\left. -Lx + \dfrac{x^2 + a^2}{2}\right]$	Maximum deflection occurs at $x = x_1$ and/or $x = x_2$ at $x_1 = L\sqrt{\dfrac{3a - L}{3(L + a)}}$ $\dfrac{M^*}{6EI}$ $\times \dfrac{(a - L)(3a - L)^{3/2}}{(3(L + a))^{1/2}}$ at $x_2 = L\dfrac{L^2 + 3a^2}{3(L^2 - a^2)}$ $-\dfrac{M^*}{27EI}\dfrac{(3a^2 - L^2)^3}{(a^2 - L^2)^2}$
16.	$R_1 = \dfrac{Wb^2}{L^3}(3a + b)$ $R_2 = \dfrac{Wa^2}{L^3}(3b + a)$ $M_1 = \dfrac{Wab^2}{L^2}$ $M_2 = \dfrac{Wa^2b}{L^2}$	For $x < a$ $-\dfrac{Wb^2x^2}{6L^3EI}(3ax + bx - 3aL)$ For $x \geq a$ $-\dfrac{Wa^2(L - x)^2}{6L^3EI}$ $\times[(3b + a)(L - x) - 3bL]$	If $a \geq b$ $\dfrac{2W}{3EI}\dfrac{a^3b^2}{(3a + b)^2}$ at $x = \dfrac{2aL}{3a + b}$ If $a < b$ $\dfrac{2W}{3EI}\dfrac{a^2b^3}{(3b + a)^2}$ at $x = \dfrac{L^2}{a + 3b}$

TABLE 11-1 | Uniform Beams   **526**

Moment at any point x	Maximum moment	Important slope	Maximum shear force
For $x < a$ $\quad -R_1 x$ For $x \geq a$ $\quad -R_1 x + M^*$	$M =$ $M^*\left[1 - \dfrac{3a(L^2 - a^2)}{2L^3}\right]$ at $x = a^+$ $+M = M_2$ at $x = L$ when $a < 0.257L$ $M = -R_1 a$ at $x = a^-$ when $a > 0.257L$	$\theta_1 =$ $\dfrac{M^*}{EI}\left(a - \dfrac{L}{4}\right.$ $\left. - \dfrac{3a^2}{4L}\right)$	$\dfrac{3M^*}{2L}\dfrac{L^2 - a^2}{L^2}$
For $x < a$ $\quad -\dfrac{Wab^2}{L^2} + R_1 x$ For $x \geq a$ $\quad -\dfrac{Wab^2}{L^2} + R_1 x$ $\quad -W(x - a)$	$-M = \dfrac{Wab^2}{L^2} - R_1 a$ at $x = a$ $+M = M_1$ at $x = 0$ $+M = M_2$ at $x = L$		If $a > b$ R_2 at $x > a$ If $a < b$ R_1 at $x < a$

TABLE 11-1 (continued)   UNIFORM BEAMS

Type of beam	Reactions	Deflection at any point x
17.	$R_1 = \frac{1}{2}p_1L$ $R_2 = \frac{1}{2}p_1L$ $M_1 = \frac{1}{12}p_1L^2$ $M_2 = \frac{1}{12}p_1L^2$	$-\dfrac{p_1x^2}{24EI}(2Lx - L^2 - x^2)$
18.	$R_1 = \frac{3}{20}pL$ $R_2 = \frac{7}{20}pL$ $M_1 = \frac{1}{30}pL^2$ $M_2 = \frac{1}{20}pL^2$	$-\dfrac{pL}{120EI}$ $\times\left(3x^3 - 2Lx^2 - \dfrac{x^5}{L^2}\right)$
19.	$R_1 =$ $\dfrac{6M^*}{L^3}(aL - a^2)$ $R_2 =$ $\dfrac{6M^*}{L^3}(aL - a^2)$ $M_1 = \dfrac{M^*}{L^2}$ $\times(4La - 3a^2 - L^2)$ $M_2 = \dfrac{M^*}{L^2}$ $\times(2La - 3a^2)$	For $x < a$ $-\dfrac{1}{6EI}(3M_1x^2 - R_1x^3)$ For $x \geq a$ $-\dfrac{1}{6EI}[(M^* + M_1)$ $\times(3x^2 - 6Lx + 3L^2)$ $+ R_1(3L^2x - x^3 - 2L^3)]$

TABLE 11-1 Uniform Beams   **528**

Maximum deflection	Moment at any point x	Maximum moment	Important slope	Maximum shear force
$\dfrac{p_1 L^4}{384 EI}$ at $x = \frac{1}{2}L$	$-\dfrac{p_1 L}{2}\left(\dfrac{x^2}{L} + \dfrac{L}{6} - x\right)$	$M = \frac{1}{24}p_1 L^2$ at $x = \frac{1}{2}L$ $+M = \frac{1}{12}p_1 L^2$ at $x = 0, L$		$\frac{1}{2}p_1 L$ at $x = 0, L$
$\dfrac{0.001308\,pL^4}{EI}$ at $x = 0.525L$	$-\dfrac{pL}{2}\left(\dfrac{L}{15} + \dfrac{x^3}{3L^2} - \dfrac{3}{10}x\right)$	$M = 0.0215\,pL^2$ at $x = 0.548L$ $+M = \frac{1}{20}pL^2 = M_2$ at $x = L$		$\frac{7}{20}pL$ at $x = L$
If $a > \frac{1}{3}L$ peak deflection at $x = \left(1 - \dfrac{L}{3a}\right)L$ If $a < \frac{2}{3}L$ peak deflection at $x = \dfrac{L}{3}$ $\times\left(1 + \dfrac{a}{L-a}\right)$	For $x < a$ $M_1 - R_1 x$ For $x \geq a$ $M_1 - R_1 x + M^*$	$M =$ $M^*\left(\dfrac{4a}{L} - \dfrac{9a^2}{L^2} + \dfrac{6a^3}{L^3}\right)$ at $x = a^+$ $+M =$ $M^*\left(\dfrac{4a}{L} - \dfrac{9a^2}{L^2}\right.$ $\left. + \dfrac{6a^3}{L^3} - 1\right)$ at $x = a^-$		R_1

TABLE 11-2, PART A SIMPLE BEAMS WITH ARBITRARY LOADINGS: GENERAL RESPONSE EXPRESSIONS

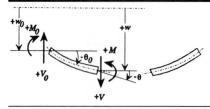

Response

1.
Deflection

$$w = w_0 - \theta_0 x - V_0 \frac{x^3}{3!EI} - M_0 \frac{x^2}{2EI} + F_w$$

2.
Slope

$$\theta = \theta_0 + V_0 \frac{x^2}{2EI} + M_0 \frac{x}{EI} + F_\theta$$

3.
Shear force

$$V = V_0 + F_V$$

4.
Bending moment

$$M = M_0 + V_0 x + F_M$$

TABLE 11-2, Part A   Simple Beams with Arbitrary Loadings   530

TABLE 11-2, PART B   SIMPLE BEAMS WITH ARBITRARY LOADINGS: LOADING FUNCTIONS[a]

	W	p_1	$p_1,\ p_2$	C	$p(x)$
$F_W(x)$	$\dfrac{W\langle x-a\rangle^3}{3!EI}$	$\dfrac{p_1}{4!EI}(\langle x-a_1\rangle^4 - \langle x-a_2\rangle^4)$	$\dfrac{p_2-p_1}{5!EI(a_2-a_1)}(\langle x-a_1\rangle^5 - \langle x-a_2\rangle^5)$ $+\dfrac{1}{4!EI}(p_1\langle x-a_1\rangle^4 - p_2\langle x-a_2\rangle^4)$	$\dfrac{C\langle x-a\rangle^2}{2EI}$	$\dfrac{1}{EI}\int_{a_1}^x dx\int^x dx\int^x dx\int^x p\,dx$
$F_\theta(x)$	$-\dfrac{W\langle x-a\rangle^2}{2EI}$	$-\dfrac{p_1}{3!EI}(\langle x-a_1\rangle^3 - \langle x-a_2\rangle^3)$	$-\dfrac{p_2-p_1}{4!EI(a_2-a_1)}(\langle x-a_1\rangle^4 - \langle x-a_2\rangle^4)$ $-\dfrac{1}{3!EI}(p_1\langle x-a_1\rangle^3 - p_2\langle x-a_2\rangle^3)$	$-\dfrac{C\langle x-a\rangle}{EI}$	$-\dfrac{1}{EI}\int_{a_1}^x dx\int^x dx\int^x p\,dx$
$F_V(x)$	$-W\langle x-a\rangle^0$	$-p_1(\langle x-a_1\rangle - \langle x-a_2\rangle)$	$-\dfrac{p_2-p_1}{2(a_2-a_1)}(\langle x-a_1\rangle^2 - \langle x-a_2\rangle^2)$ $-p_1\langle x-a_1\rangle + p_2\langle x-a_2\rangle$	0	$-\int_{a_1}^x p\,dx$
$F_M(x)$	$-W\langle x-a\rangle$	$-\tfrac{1}{2}p_1(\langle x-a_1\rangle^2 - \langle x-a_2\rangle^2)$	$-\dfrac{p_2-p_1}{3!(a_2-a_1)}(\langle x-a_1\rangle^3 - \langle x-a_2\rangle^3)$ $-\tfrac{1}{2}(p_1\langle x-a_1\rangle^2 - p_2\langle x-a_2\rangle^2)$	$-C\langle x-a\rangle^0$	$-\int_{a_1}^x dx\int^x p\,dx$

[a]By definition:

$$\langle x-a\rangle^n = \begin{cases} 0 & \text{if } x < a \\ (x-a)^n & \text{if } x \geq a \end{cases} \qquad n \geq 1$$

$$\langle x-a\rangle^0 = \begin{cases} 0 & \text{if } x < a \\ 1 & \text{if } x \geq a \end{cases}$$

TABLE 11-2, PART C SIMPLE BEAMS WITH ARBITRARY LOADING: INITIAL PARAMETERS[a]

Left End \ Right End	1. Pinned, hinged, or on rollers	2. Fixed	3. Free	4. Guided	5. Partially fixed
1. Pinned, hinged, or on rollers $w_0=0,\ M_0=0$	$\theta_0 = \dfrac{1}{L}\bar{F}_w + \dfrac{L}{6EI}\bar{F}_M$ $V_0 = -\dfrac{1}{L}\bar{F}_M$	$\theta_0 = \dfrac{3}{2L}\bar{F}_w + \dfrac{1}{2}\bar{F}_\theta$ $V_0 = -\dfrac{3EI}{L^3}\bar{F}_w - \dfrac{3EI}{L^2}\bar{F}_\theta$	Subject to rigid body motion; therefore kinematically unstable	$\theta_0 = \dfrac{L^2}{2EI}\bar{F}_V - \bar{F}_\theta$ $V_0 = -\bar{F}_V$	$\theta_0 = \dfrac{A_1\bar{F}_w - A_2 L/6EI}{A_3}$ $V_0 = \dfrac{(k_2^*/L^2)\,\bar{F}_w + A_2/L}{A_3}$
2. Fixed $w_0=0,\ \theta_0=0$	$V_0 = -\dfrac{3EI}{L^3}\bar{F}_w - \dfrac{3}{2L}\bar{F}_M$ $M_0 = \dfrac{3EI}{L^2}\bar{F}_w + \dfrac{1}{2}\bar{F}_M$	$V_0 = -\dfrac{12EI}{L^3}\bar{F}_w - \dfrac{6EI}{L^2}\bar{F}_\theta$ $M_0 = \dfrac{6EI}{L^2}\bar{F}_w + \dfrac{2EI}{L}\bar{F}_\theta$	$V_0 = -\bar{F}_V$ $M_0 = L\bar{F}_V - \bar{F}_M$	$V_0 = -\bar{F}_V$ $M_0 = -\dfrac{EI}{L}\bar{F}_\theta + \dfrac{1}{2}L\bar{F}_V$	$V_0 = \dfrac{-(3EIA_5/L^3)\,\bar{F}_w + 3A_2/2L}{A_4}$ $M_0 = \dfrac{(3EI/L^2)A_6\bar{F}_w - \tfrac{1}{2}A_2}{A_4}$
3. Free $V_0=0,\ \theta_0=0$	Subject to rigid body motion; therefore kinematically unstable	$w_0 = -\bar{F}_w - L\bar{F}_\theta$ $\theta_0 = -\bar{F}_\theta$	Subject to rigid body motion; therefore kinematically unstable	Subject to rigid body motion; therefore kinematically unstable	$w_0 = -\bar{F}_w - A_2\dfrac{L}{k_2^*}$ $\theta_0 = -A_2\dfrac{1}{k_2^*}$

TABLE 11-2, Part C | Simple Beams: Initial Parameters | 532

Case			Subject to rigid body motion; therefore kinematically unstable	Subject to rigid body motion; therefore kinematically unstable	
4. Guided	$w_0 = -\overline{F}_w - \dfrac{L^2}{2EI}\overline{F}_M$ $M_0 = -\overline{F}_M$	$w_0 = -\overline{F}_w - \dfrac{1}{2}L\overline{F}_\theta$ $M_0 = -\dfrac{EI}{L}F_\theta$			$w_0 = \dfrac{-A_6\overline{F}_w + (L^2/2EI)A_2}{A_6}$ $M_0 = A_2/A_6$
$\theta_0 = 0,\; V_0 = 0$					
5. Partially fixed	$\theta_0 = \dfrac{(1/L)\overline{F}_w + (L/6EI)\overline{F}_M}{A_7}$ $V_0 = \dfrac{(k_1^*/L^2)\overline{F}_w + A_8\overline{F}_M}{A_7}$	$\theta_0 = \dfrac{(3/2L)\overline{F}_w + \tfrac{1}{2}\overline{F}_\theta}{A_9}$ $V_0 = \dfrac{-(A_{11}\overline{F}_w + (3EI/L)A_8\overline{F}_\theta)}{A_9}$	$\theta_0 = -\dfrac{1}{k_1^*}$ $\times\left(\overline{F}_M - L\overline{F}_V\right)$ $V_0 = -\overline{F}_V$	$\theta_0 = \dfrac{-\overline{F}_\theta + (L^2/2EI)\overline{F}_V}{A_{10}}$ $V_0 = -\overline{F}_V$	$\theta_0 = \dfrac{(A_5/L)\overline{F}_w - LA_2/6EI}{A_{13}}$ $V_0 = \dfrac{A_{12}\overline{F}_w + A_2 A_8}{A_{13}}$

$$w_0 = 0,\; M_0 = k_1^*\theta_0$$

[a]Although the response due to static loading cannot be determined for a kinematically unstable beam, the natural frequencies and buckling load can be calculated.

Note: $\overline{F}_w = F_w|_{x=L},\; \overline{F}_\theta = F_\theta|_{x=L},\; \overline{F}_M = F_M|_{x=L},\; \overline{F}_V = F_V|_{x=L}$

$$A_1 = \frac{1}{L} - \frac{k_2^*}{2EI} \qquad A_4 = 1 - \frac{k_2^* L}{4EI} \qquad A_7 = 1 + \frac{k_1^* L}{3EI} \qquad A_{10} = 1 + \frac{k_1^* L}{EI} \qquad A_{12} = \frac{k_2^* - k_1^*}{L^2} + \frac{k_1^* k_2^*}{EIL}$$

$$A_2 = k_2^*\overline{F}_\theta - \overline{F}_M \qquad A_5 = 1 - \frac{k_2^* L}{2EI} \qquad A_8 = \frac{1}{L} + \frac{k_1^*}{2EI} \qquad A_{11} = \frac{3EI}{L^3} + \frac{3k_1^*}{L^2} \qquad A_{13} = 1 + \frac{1}{3}\frac{k_1^* L}{EI} - \frac{1}{3}\frac{k_2^* L}{EI} - \frac{k_1^* k_2^* L^2}{12(EI)^2}$$

$$A_3 = 1 - \frac{k_2^* L}{3EI} \qquad A_6 = 1 - \frac{k_2^* L}{EI} \qquad A_9 = 1 + \frac{k_1^* L}{4EI}$$

TABLE 11-3, PART A   BEAMS WITH AXIAL FORCES AND ELASTIC FOUNDATIONS: GENERAL RESPONSE EXPRESSIONS

Notation

E = Modulus of elasticity

P = compressive axial force

    replace P by $-P$ for tensile force

A_s = equivalent shear area, $= A/\alpha_s$

I = moment of inertia

α_s = shear correction factor

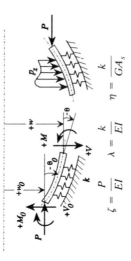

$$\zeta = \frac{P}{EI} \qquad \lambda = \frac{k}{EI} \qquad \eta = \frac{k}{GA_s}$$

Set $\dfrac{1}{GA_s} = 0$ for beams without shear deformation effects.

Response	
1. Deflection	$w = w_0(e_1 + \zeta e_3) - \theta_0 e_2 - V_0 \left(\dfrac{e_4}{EI} - \dfrac{e_2 + \zeta e_4}{GA_s} \right) - M_0 \dfrac{e_3}{EI} + F_w$
2. Slope	$\theta = w_0 \lambda e_4 + \theta_0(e_1 - \eta e_3) + V_0 \dfrac{e_3}{EI} + M_0 \dfrac{e_2 - \eta e_4}{EI} + F_\theta$
3. Shear force	$V = w_0 \lambda EI(e_2 + \zeta e_4) - \theta_0 \lambda EI e_3 + V_0(e_1 + \zeta e_3) - M_0 \lambda e_4 + F_V$
4. Bending moment	$M = w_0 \lambda EI e_3 + \theta_0 EI(e_0 - \eta e_2) + V_0 e_2 + M_0(e_1 - \eta e_3) + F_M$

TABLE 11-3, Part A   **Axial Forces and Elastic Foundations**   534

Values of e_i

Ordinary Beam with or without Shear Deformation	Beam with Compressive Axial Force P	Beam with Tensile Axial Force P	Beam On Elastic Foundation, k, with Shear Deformation Effects[a] $\lambda = k/(EI),\ \zeta = 0,\ \eta = ka_s/(GA) = k/(GA_s)$	
			$\lambda \geq \frac{1}{4}\eta^2$	$\lambda < \frac{1}{4}\eta^2$
$\zeta = \lambda = \eta = 0$	$\lambda = 0,\ \eta = 0,$ $\alpha^2 = P/(EI) = \zeta$	$\lambda = 0,\ \eta = 0,$ $\alpha^2 = P/(EI),\ \zeta = \alpha^2$	$a^2 = \frac{1}{2}\sqrt{\lambda} + \frac{1}{4}\eta$ $\quad b^2 = \frac{1}{2}\sqrt{\lambda} - \frac{1}{4}\eta$	$a^2 = \frac{1}{2}\eta + \sqrt{\frac{1}{4}\eta^2 - \lambda}$ $b^2 = \frac{1}{2}\eta - \sqrt{\frac{1}{4}\eta^2 - \lambda}$ $\quad g = a^2 - b^2$
$e_0 = 0$	$e_0 = -\alpha \sin \alpha x$	$e_0 = \alpha \sinh \alpha x$	$e_0 = -\lambda e_4 + \eta e_2$	$e_0 = -\dfrac{1}{g}(b^3 \sinh bx - a^3 \sinh ax)$
$e_1 = 1$	$e_1 = \cos \alpha x$	$e_1 = \cosh \alpha x$	$e_1 = \cosh ax \cos bx$ $+ \dfrac{\eta}{4ab}\sinh ax \sin bx$	$e_1 = -\dfrac{1}{g}(b^2 \cosh bx - a^2 \cosh ax)$
$e_2 = x$	$e_2 = \dfrac{1}{\alpha}\sin \alpha x$	$e_2 = \dfrac{1}{\alpha}\sinh \alpha x$	$e_2 = \dfrac{1}{2ab}$ $\times (a \cosh ax \sin bx$ $+ b \sinh ax \cos bx)$	$e_2 = -\dfrac{1}{g}(b \sinh bx - a \sinh ax)$
$e_3 = \frac{1}{2}x^2$	$e_3 = \dfrac{1}{\alpha^2}(1 - \cos \alpha x)$	$e_3 = \dfrac{1}{\alpha^2}(\cosh \alpha x - 1)$	$e_3 = \dfrac{1}{2ab}\sinh ax \sin bx$	$e_3 = \dfrac{1}{g}(\cosh ax - \cosh bx)$
$e_4 = \frac{1}{6}x^3$	$e_4 = \dfrac{1}{\alpha^3}(\alpha x - \sin \alpha x)$	$e_4 = \dfrac{1}{\alpha^3}(\sinh \alpha x - \alpha x)$	$e_4 = \dfrac{1}{2(a^2 + b^2)}\left(\dfrac{1}{b}\cosh ax \sin bx\right.$ $\left. - \dfrac{1}{a}\sinh ax \cos bx\right)$	$e_4 = \dfrac{1}{g}\left(\dfrac{1}{a}\sinh ax - \dfrac{1}{b}\sinh bx\right)$
$e_5 = \frac{1}{24}x^4$	$e_5 = \dfrac{1}{\alpha^4}\left(\dfrac{\alpha^2 x^2}{2}\right.$ $\left. + \cos \alpha x - 1\right)$	$e_5 = \dfrac{1}{\alpha^4}\left(-\dfrac{\alpha^2 x^2}{2}\right.$ $\left. + \cosh \alpha x - 1\right)$	$e_5 = \dfrac{1}{\lambda}(1 - e_1 + \eta e_3)$	$e_5 = -\dfrac{1}{g}\left(\dfrac{1}{b^2}\cosh bx\right.$ $\left. - \dfrac{1}{a^2}\cosh ax\right) + \dfrac{1}{\lambda}$
$e_6 = \frac{1}{120}x^5$	$e_6 = \dfrac{1}{\alpha^5}\left(\dfrac{\alpha^3 x^3}{6}\right.$ $\left. + \sin \alpha x - \alpha x\right)$	$e_6 = \dfrac{1}{\alpha^5}\left(-\dfrac{\alpha^3 x^3}{6}\right.$ $\left. + \sinh \alpha x - \alpha x\right)$	$e_6 = \dfrac{1}{\lambda}(x - e_2 + \eta e_4)$	$e_6 = -\dfrac{1}{g}\left(\dfrac{1}{b^3}\sinh bx\right.$ $\left. - \dfrac{1}{a^3}\sinh ax\right) + \dfrac{x}{\lambda}$

[a]Note: if $\lambda = \frac{1}{4}\eta^2$, set $(\sin bx)/b = x$

TABLE 11-3, PART B BEAMS WITH AXIAL FORCES AND ELASTIC FOUNDATIONS: LOADING FUNCTIONS

By definition:

$$e_i\langle x - a \rangle = \begin{cases} 0 & \text{if } x < a \\ e_i(x - a) & \text{if } x \geq a \end{cases}$$

Also, if $e_i = 1$, then

$$e_i\langle x - a \rangle = \langle x - a \rangle^0 = \begin{cases} 0 & \text{if } x < a \\ 1 & \text{if } x \geq a \end{cases}$$

	$W \downarrow$ $\vdash a \rightarrow$	$P_1 \downarrow\downarrow\downarrow\downarrow\downarrow$ $\vdash a_1 \rightarrow$ $\vdash\!\!\!-\!\!\! a_2 \longrightarrow$
$F_w(x)$	$W\left(\dfrac{e_4\langle x - a \rangle}{EI} - \dfrac{e_2\langle x - a \rangle + \zeta e_4\langle x - a \rangle}{GA_s} \right)$	$P_1\left(\dfrac{e_5\langle x - a_1 \rangle - e_5\langle x - a_2 \rangle}{EI} - \dfrac{e_3\langle x - a_1 \rangle + \zeta e_5\langle x - a_1 \rangle - e_3\langle x - a_2 \rangle - \zeta e_5\langle x - a_2 \rangle}{GA_s} \right)$
$F_\theta(x)$	$-W\dfrac{e_3\langle x - a \rangle}{EI}$	$-P_1\dfrac{e_4\langle x - a_1 \rangle - e_4\langle x - a_2 \rangle}{EI}$
$F_V(x)$	$-W(e_1\langle x - a \rangle + \zeta e_3\langle x - a \rangle)$	$-P_1[(e_2\langle x - a_1 \rangle - e_2\langle x - a_2 \rangle) - \zeta(e_4\langle x - a_1 \rangle - e_4\langle x - a_2 \rangle)]$
$F_M(x)$	$-We_2\langle x - a \rangle$	$-P_1(e_3\langle x - a_1 \rangle - e_3\langle x - a_2 \rangle)$

TABLE 11-3, Part B   Axial Forces and Elastic Foundations   536

For example:

if $e_i = \cos \alpha x$, then

$$e_i\langle x - a \rangle = \cos \alpha \langle x - a \rangle = \begin{cases} 0 & \text{if } x < a \\ \cos \alpha (x - a) & \text{if } x \geq a \end{cases}$$

$\dfrac{p_2 - p_1}{a_2 - a_1}\left[\dfrac{e_6\langle x - a_1\rangle - e_6\langle x - a_2\rangle}{EI}\right.$ $\left. -\dfrac{e_4\langle x - a_1\rangle + \zeta e_6\langle x - a_1\rangle - e_4\langle x - a_2\rangle - \zeta e_6\langle x - a_2\rangle}{GA_s}\right]$ $+\dfrac{1}{EI}(p_1 e_5\langle x - a_1\rangle - p_2 e_5\langle x - a_2\rangle)$ $-\dfrac{1}{GA_s}[p_1(e_3\langle x - a_1\rangle + \zeta e_5\langle x - a_1\rangle)$ $-p_2(e_3\langle x - a_2\rangle + \zeta e_5\langle x - a_2\rangle)]$	$C\dfrac{e_3\langle x - a\rangle}{EI}$
$-\dfrac{p_2 - p_1}{a_2 - a_1}\dfrac{e_5\langle x - a_1\rangle - e_5\langle x - a_2\rangle}{EI}$ $+\dfrac{1}{EI}(p_2 e_4\langle x - a_2\rangle - p_1 e_4\langle x - a_1\rangle)$	$-C\dfrac{e_2\langle x - a\rangle - \eta e_4\langle x - a\rangle}{EI}$
$-\dfrac{p_2 - p_1}{a_2 - a_1}(e_3\langle x - a_1\rangle + \zeta e_5\langle x - a_1\rangle - e_3\langle x - a_2\rangle$ $-\zeta e_5\langle x - a_2\rangle) + p_2(e_2\langle x - a_2\rangle + \zeta e_4\langle x - a_2\rangle)$ $-p_1(e_2\langle x - a_1\rangle - \zeta e_4\langle x - a_1\rangle + 2\zeta e_4\langle x - a_2\rangle)$	$C\lambda e_4\langle x - a\rangle$
$-\dfrac{p_2 - p_1}{a_2 - a_1}(e_4\langle x - a_1\rangle - e_4\langle x - a_2\rangle)$ $+p_2 e_3\langle x - a_2\rangle - p_1 e_3\langle x - a_1\rangle$	$-C(e_1\langle x - a\rangle - \eta e_3\langle x - a\rangle)$

TABLE 11-3, PART C BEAMS WITH AXIAL FORCES AND ELASTIC FOUNDATIONS: INITIAL PARAMETERS

$\bar{F}_w = F_{w|x=L}$

$\bar{F}_\theta = F_{\theta|x=L}$

$\bar{F}_V = F_{V|x=L}$

$\bar{F}_M = F_{M|x=L}$

$A_1 = EI(\bar{e}_0 - \eta\bar{e}_2) + k_1^*(\bar{e}_1 - \eta\bar{e}_3)$

$A_2 = \bar{e}_2 + k_1^*\bar{e}_3/EI$

$A_3 = \bar{e}_1 - \eta\bar{e}_3 + k_1^*(\bar{e}_2 - \eta\bar{e}_4)/EI$

$A_4 = EI\bar{e}_3 + k_1^*\bar{e}_4$

$A_5 = k_2^*\bar{e}_3 - EI\bar{e}_2$

$A_6 = EI(\bar{e}_0 - \eta\bar{e}_2) - k_2^*(\bar{e}_1 - \eta\bar{e}_3)$

$A_7 = k_2^*(\bar{e}_2 - \eta\bar{e}_4) - EI(\bar{e}_1 - \eta\bar{e}_3)/EI$

$A_8 = EI\bar{e}_3 - k_2^*\bar{e}_4$

$A_9 = k_2^*\bar{F}_\theta - \bar{F}_M$

$\bar{e}_i = e_{i|x=L}$

$i = 0,1,\dots,5$

Left End \ Right End	1. Pinned, hinged, or on rollers	2. Fixed or infinite to the right
		Fixed or infinite to the right
		Fixed
		Infinite to the right
		All loading, changes in cross sectional properties, etc., must be placed between $x = 0$ and $x = L$
1. Pinned, hinged, or on rollers $w_0 = 0,\ M_0 = 0$	$\theta_0 = \{-\bar{e}_2\bar{F}_w - [\bar{e}_4/EI - (\bar{e}_2 + \zeta\bar{e}_4)/GA_s]\bar{F}_M\}/\nabla$ $V_0 = (EI\bar{e}_0\bar{F}_w + \bar{e}_2\bar{F}_M)/\nabla$ $\nabla = \bar{e}_0[\bar{e}_4/EI - (\bar{e}_2 + \zeta\bar{e}_4)/GA_s] - \bar{e}_2^2$	$\theta_0 = \{-\bar{F}_w\bar{e}_3/EI - \bar{F}_\theta[\bar{e}_4/EI - (\bar{e}_2 + \zeta\bar{e}_4)/GA_s]\}/\nabla$ $V_0 = [\bar{e}_2\bar{F}_\theta + (\bar{e}_1 - \eta\bar{e}_3)\bar{F}_w]/\nabla$ $\nabla = (\bar{e}_1 - \eta\bar{e}_3)[\bar{e}_4/EI - (\bar{e}_2 + \zeta\bar{e}_4)/GA_s]$ $\qquad - \bar{e}_2\bar{e}_3/EI$
2. Fixed $w_0 = 0,\ \theta_0 = 0$	$V_0 = [\bar{F}_M\bar{e}_3/EI - \bar{F}_w(\bar{e}_1 - \eta\bar{e}_3)]/\nabla$ $M_0 = \{\bar{F}_w\bar{e}_2 + \bar{F}_M[\bar{e}_4/EI + (\bar{e}_2 + \zeta\bar{e}_4)/GA_s]\}/\nabla$ $\nabla = \bar{e}_2\bar{e}_3/EI$ $\qquad - [\bar{e}_4/EI - (\bar{e}_2 + \zeta\bar{e}_4)/GA_s](\bar{e}_1 - \eta\bar{e}_3)$	$V_0 = [\bar{e}_3\bar{F}_\theta - (\bar{e}_2 - \eta\bar{e}_4)\bar{F}_w]/\nabla$ $M_0 = \{\bar{e}_3\bar{F}_w - [\bar{e}_4/EI - (\bar{e}_2 + \zeta\bar{e}_4)/GA_s]\bar{F}_\theta EI\}/\nabla$ $\nabla = \bar{e}_3^2 - (\bar{e}_2 - \eta\bar{e}_4)[\bar{e}_4/EI - (\bar{e}_2 + \zeta\bar{e}_4)/GA_s]$

TABLE 11-3, Part C **Axial Forces and Elastic Foundations** 538

3. Free or infinite to the left

Free:

$M_0 = 0, V_0 = 0$

Infinite to the left

All loadings, changes in cross sectional properties, etc., must be placed between $x = 0$ and $x = L$

$w_0 = [\bar{F}_M \bar{e}_2/EI - (\bar{e}_0 - \eta\bar{e}_2)\bar{F}_w]/\nabla$

$\theta_0 = [\lambda\bar{e}_3\bar{F}_w - (\bar{e}_1 + \zeta\bar{e}_3)\bar{F}_M/EI]/\nabla$

$\nabla = (\bar{e}_1 + \zeta\bar{e}_3)(\bar{e}_0 - \eta\bar{e}_2) + \lambda\bar{e}_2\bar{e}_3$

$w_0 = -[(\bar{e}_1 - \eta\bar{e}_3)\bar{F}_w + \bar{e}_2\bar{F}_\theta]/\nabla$

$\theta_0 = [\lambda\bar{e}_4\bar{F}_w - (\bar{e}_1 + \zeta\bar{e}_3)\bar{F}_\theta]/\nabla$

$\nabla = \lambda\bar{e}_2\bar{e}_4 + (\bar{e}_1 + \zeta\bar{e}_3)(\bar{e}_1 - \eta\bar{e}_3)$

4. Guided

$\theta_0 = 0, V_0 = 0$

$w_0 = [-(\bar{e}_1 - \eta\bar{e}_3)\bar{F}_w - (\bar{e}_3/EI)\bar{F}_M]/\nabla$

$M_0 = [\lambda EI\bar{e}_3\bar{F}_w - (\bar{e}_1 + \zeta\bar{e}_3)\bar{F}_M]/\nabla$

$\nabla = (\bar{e}_1 - \eta\bar{e}_3)(\bar{e}_1 + \zeta\bar{e}_3) + \lambda\bar{e}_3^2$

$w_0 = [-(\bar{e}_2 - \eta\bar{e}_4)\bar{F}_w - \bar{e}_3\bar{F}_\theta]/\nabla$

$M_0 = [\lambda EI\bar{e}_4\bar{F}_w - EI(\bar{e}_1 + \zeta\bar{e}_3)\bar{F}_\theta]/\nabla$

$\nabla = (\bar{e}_2 - \eta\bar{e}_4)(\bar{e}_1 + \zeta\bar{e}_3) + \lambda\bar{e}_3\bar{e}_4$

5. Partially fixed

k_1^*

$w_0 = 0, M_0 = k_1^*\theta_0$

$\theta_0 = \{-\bar{e}_2\bar{F}_w - \bar{F}_M[\bar{e}_4 - (\bar{e}_2 + \zeta\bar{e}_4)/GA_s]/EI\}/\nabla$

$V_0 = (A_1\bar{F}_w + A_2\bar{F}_M)/\nabla$

$\nabla = A_1[\bar{e}_4 - (\bar{e}_2 + \zeta\bar{e}_4)/GA_s]/EI - A_2\bar{e}_2$

$\theta_0 = [-(\bar{e}_3/EI)\bar{F}_w$

$\quad - \bar{F}_\theta[\bar{e}_4 - (\bar{e}_2 + \zeta\bar{e}_4)/GA_s]/EI\}/\nabla$

$V_0 = (A_3\bar{F}_w + A_2 F_\theta)/\nabla$

$\nabla = A_3[\bar{e}_4 - (\bar{e}_2 + \zeta\bar{e}_4)/GA_s]/EI - A_2\bar{e}_3/EI$

Left End	Right End →	3. Free $\longleftarrow L \longrightarrow$
1. Pinned, hinged, or on rollers $\longrightarrow x$ $w_0 = 0,\ M_0 = 0$		$\theta_0 = \left\{ -\left[(\bar{e}_1 + \zeta\bar{e}_3)/EI\right]\bar{F}_M + (\bar{e}_2/EI)\bar{F}_V \right\}/\nabla$ $V_0 = \left[-\lambda\bar{e}_3\bar{F}_M - (\bar{e}_0 - \eta\bar{e}_2)\bar{F}_V \right]/\nabla$ $\nabla = (\bar{e}_0 - \eta\bar{e}_2)(\bar{e}_1 + \zeta\bar{e}_3) + \lambda\bar{e}_2\bar{e}_3$
2. Fixed $\longrightarrow x$ $w_0 = 0,\ \theta_0 = 0$		$V_0 = \left[-\lambda\bar{e}_4\bar{F}_M - (\bar{e}_1 - \eta\bar{e}_3)\bar{F}_V \right]/\nabla$ $M_0 = \left[-(\bar{e}_1 + \zeta\bar{e}_3)\bar{F}_M + \bar{e}_2\bar{F}_V \right]/\nabla$ $\nabla = (\bar{e}_1 - \eta\bar{e}_3)(\bar{e}_1 + \zeta\bar{e}_3) + \lambda\bar{e}_2\bar{e}_4$

TABLE 11-3, Part C   Axial Forces and Elastic Foundations   540

Axial Forces and Elastic Foundations TABLE 11-3, Part C

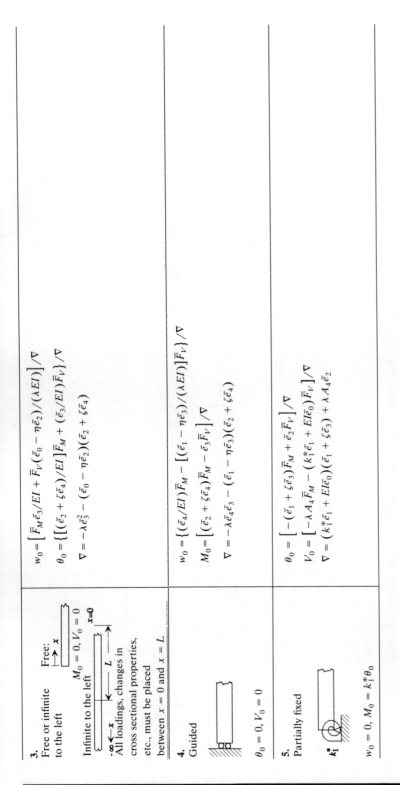

3. Free or infinite to the left

Free: diagram, x, $x=0$

$M_0 = 0,\ V_0 = 0$

Infinite to the left: diagram, $-\infty$, x, L

All loadings, changes in cross sectional properties, etc., must be placed between $x = 0$ and $x = L$

$w_0 = \left[\bar{F}_M \bar{e}_3/EI + \bar{F}_V(\bar{e}_0 - \eta\bar{e}_2)/(\lambda EI)\right]/\nabla$

$\theta_0 = \left\{[(\bar{e}_2 + \zeta\bar{e}_4)/EI]\bar{F}_M + (\bar{e}_3/EI)\bar{F}_V\right\}/\nabla$

$\nabla = -\lambda\bar{e}_3^2 - (\bar{e}_0 - \eta\bar{e}_2)(\bar{e}_2 + \zeta\bar{e}_4)$

4. Guided

$\theta_0 = 0,\ V_0 = 0$

$w_0 = \left\{(\bar{e}_4/EI)\bar{F}_M - [(\bar{e}_1 - \eta\bar{e}_3)/(\lambda EI)]\bar{F}_V\right\}/\nabla$

$M_0 = \left[(\bar{e}_2 + \zeta\bar{e}_4)\bar{F}_M - \bar{e}_3\bar{F}_V\right]/\nabla$

$\nabla = -\lambda\bar{e}_4\bar{e}_3 - (\bar{e}_1 - \eta\bar{e}_3)(\bar{e}_2 + \zeta\bar{e}_4)$

5. Partially fixed

k_1^*

$w_0 = 0,\ M_0 = k_1^*\theta_0$

$\theta_0 = \left[-(\bar{e}_1 + \zeta\bar{e}_3)\bar{F}_M + \bar{e}_2\bar{F}_V\right]/\nabla$

$V_0 = \left[-\lambda A_4\bar{F}_M - (k_1^*\bar{e}_1 + EI\bar{e}_0)\bar{F}_V\right]/\nabla$

$\nabla = (k_1^*\bar{e}_1 + EI\bar{e}_0)(\bar{e}_1 + \zeta\bar{e}_3) + \lambda A_4\bar{e}_2$

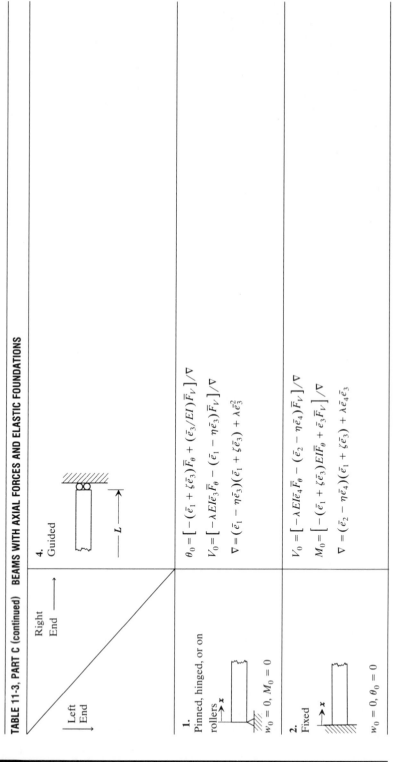

TABLE 11-3, Part C   Axial Forces and Elastic Foundations   542

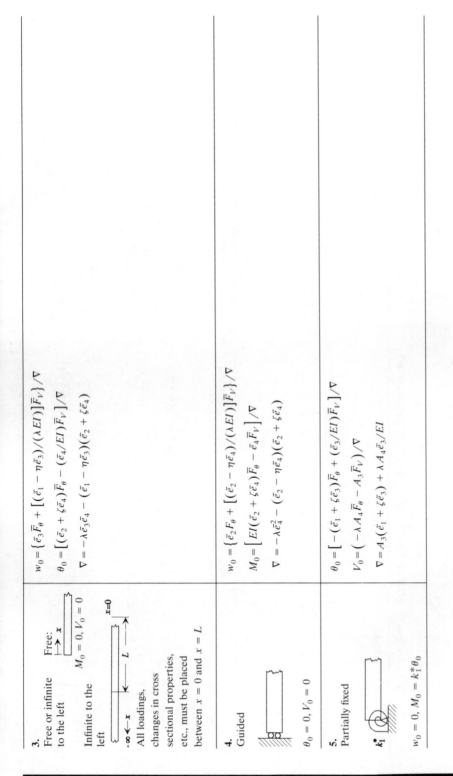

3.
Free or infinite to the left   Free:

$M_0 = 0, V_0 = 0$

Infinite to the left

All loadings, changes in cross sectional properties, etc., must be placed between $x = 0$ and $x = L$

$$w_0 = \{\bar{e}_3\bar{F}_\theta + [(\bar{e}_1 - \eta\bar{e}_3)/(\lambda EI)]\bar{F}_V\}/\nabla$$

$$\theta_0 = [(\bar{e}_2 + \zeta\bar{e}_4)\bar{F}_\theta - (\bar{e}_4/EI)\bar{F}_V]/\nabla$$

$$\nabla = -\lambda\bar{e}_3\bar{e}_4 - (\bar{e}_1 - \eta\bar{e}_3)(\bar{e}_2 + \zeta\bar{e}_4)$$

4.
Guided

$\theta_0 = 0, V_0 = 0$

$$w_0 = \{\bar{e}_2\bar{F}_\theta + [(\bar{e}_2 - \eta\bar{e}_4)/(\lambda EI)]\bar{F}_V\}/\nabla$$

$$M_0 = [EI(\bar{e}_2 + \zeta\bar{e}_4)\bar{F}_\theta - \bar{e}_4\bar{F}_V]/\nabla$$

$$\nabla = -\lambda\bar{e}_4^2 - (\bar{e}_2 - \eta\bar{e}_4)(\bar{e}_2 + \zeta\bar{e}_4)$$

5.
Partially fixed

k_1^*

$w_0 = 0, M_0 = k_1^*\theta_0$

$$\theta_0 = \left[-(\bar{e}_1 + \zeta\bar{e}_3)\bar{F}_\theta + (\bar{e}_3/EI)\bar{F}_V \right]/\nabla$$

$$V_0 = \left(-\lambda A_4\bar{F}_\theta - A_3\bar{F}_V \right)/\nabla$$

$$\nabla = A_3(\bar{e}_1 + \zeta\bar{e}_3) + \lambda A_4\bar{e}_3/EI$$

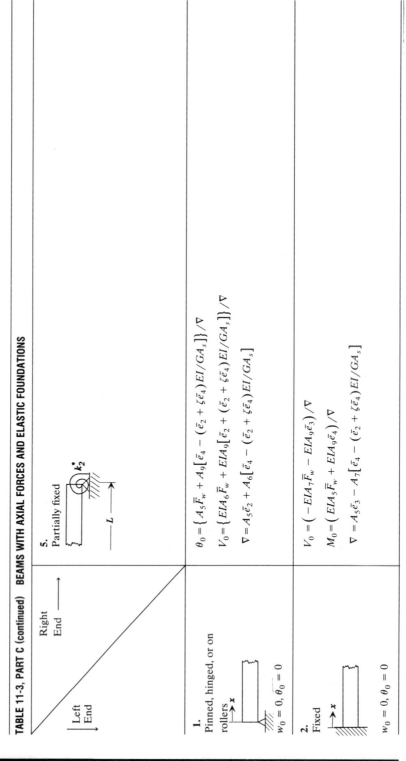

Left End	Right End
	5. Partially fixed

$$\theta_0 = \left\{ A_5 \overline{F}_w + A_9 \left[\overline{e}_4 - (\overline{e}_2 + \zeta \overline{e}_4) EI/GA_s \right] \right\} / \nabla$$

$$V_0 = \left\{ EIA_6 \overline{F}_w + EIA_9 \left[\overline{e}_2 + (\overline{e}_2 + \zeta \overline{e}_4) EI/GA_s \right] \right\} / \nabla$$

$$\nabla = A_5 \overline{e}_2 + A_6 \left[\overline{e}_4 - (\overline{e}_2 + \zeta \overline{e}_4) EI/GA_s \right]$$

1. Pinned, hinged, or on rollers

$$w_0 = 0, \quad \overline{\theta}_0 = 0$$

2. Fixed

$$w_0 = 0, \quad \theta_0 = 0$$

$$V_0 = \left(-EIA_7 \overline{F}_w - EIA_9 \overline{e}_3 \right) / \nabla$$

$$M_0 = \left(EIA_5 \overline{F}_w + EIA_9 \overline{e}_4 \right) / \nabla$$

$$\nabla = A_5 \overline{e}_3 - A_7 \left[\overline{e}_4 - (\overline{e}_2 + \zeta \overline{e}_4) EI/GA_s \right]$$

TABLE 11-3, Part C   Axial Forces and Elastic Foundations   544

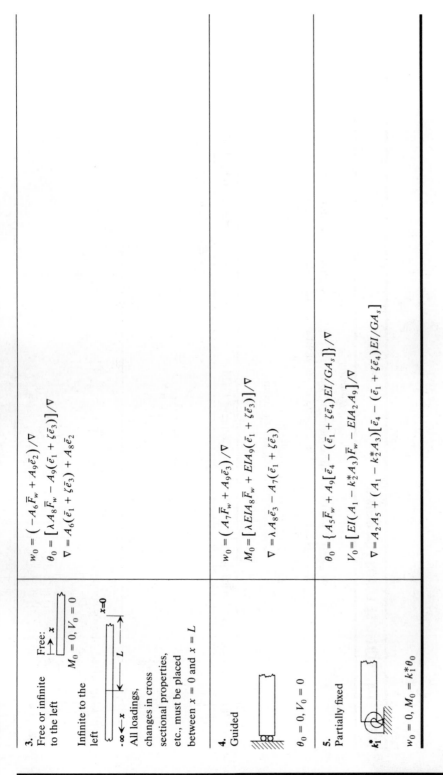

3.
Free or infinite to the left

Free:

Infinite to the left

All loadings, changes in cross sectional properties, etc., must be placed between $x = 0$ and $x = L$

$M_0 = 0, V_0 = 0$

$w_0 = \left(-A_6\bar{F}_w + A_9\bar{e}_2\right)/\nabla$

$\theta_0 = \left[\lambda A_8\bar{F}_w - A_9(\bar{e}_1 + \zeta\bar{e}_3)\right]/\nabla$

$\nabla = A_6(\bar{e}_1 + \zeta\bar{e}_3) + A_8\bar{e}_2$

4.
Guided

$\theta_0 = 0, V_0 = 0$

$w_0 = \left(A_7\bar{F}_w + A_9\bar{e}_3\right)/\nabla$

$M_0 = \left[\lambda EI A_8\bar{F}_w + EI A_9(\bar{e}_1 + \zeta\bar{e}_3)\right]/\nabla$

$\nabla = \lambda A_8\bar{e}_3 - A_7(\bar{e}_1 + \zeta\bar{e}_3)$

5.
Partially fixed

k_1^*

$w_0 = 0, M_0 = k_1^*\theta_0$

$\theta_0 = \left\{A_5\bar{F}_w + A_9\left[\bar{e}_4 - (\bar{e}_1 + \zeta\bar{e}_4)EI/GA_s\right]\right\}/\nabla$

$V_0 = \left[EI(A_1 - k_2^* A_3)\bar{F}_w - EI A_2 A_9\right]/\nabla$

$\nabla = A_2 A_5 + (A_1 - k_2^* A_3)\left[\bar{e}_4 - (\bar{e}_1 + \zeta\bar{e}_4)EI/GA_s\right]$

TABLE 11-4 COLLAPSE LOADS FOR BEAMS

Notation

M_p = full plastic bending moment, where Z_p (plastic section modulus)
is taken from Table 2-2 and σ_{ys} is yield stress of material, $= \sigma_{ys} Z_p$

W_c, C_c, p_c = collapse loads

a_{h_i} = location of ith plastic hinge

ε = short length along axis of beam

L = length of beam

Boundary Conditions and Loading	Collapse Loads	Plastic Hinge Locations
1. Fixed–fixed w $\;a\;$	$W_c = \dfrac{2M_p L}{a(L-a)}$	$a_{h_1} = 0,\; a_{h_2} = a,\; a_{h_3} = L$
2. Simply supported–simply supported w $\;a\;$	$W_c = \dfrac{M_p L}{a(L-a)}$	$a_{h_1} = a$

TABLE 11-4   Collapse Loads for Beams   546

3. Simply supported–fixed	$W_c = \dfrac{M_p(L + a)}{a(L - a)}$	$a_{h_1} = a,\ a_{h_2} = L$
4. Free–fixed	$W_c = \dfrac{M_p}{L - a}$	$a_{h_1} = L$
5. Guided–fixed	$W_c = \dfrac{2M_p}{L - a}$	$0 \leq a_{h_1} \leq a,\ a_{h_2} = L$

TABLE 11-4 (continued) COLLAPSE LOADS FOR BEAMS

Boundary Conditions and Loading	Collapse Loads	Plastic Hinge Locations
6. Guided–simply supported 	$$W_c = \dfrac{M_p}{L - a}$$	$0 \le a_{h_1} \le a$
7. Fixed–fixed 	$C_c = 2M_p$	$a_{h_1} = a - \varepsilon,\ a_{h_2} = a + \varepsilon$ For $0 < a < \frac{1}{2}L,\ a_{h_3} = L$ For $\frac{1}{2}L < a < L,\ a_{h_3} = 0$ For $a = \frac{1}{2}L,\ 0 < a_{h_1},\ a_{h_2} < \frac{1}{2}L$ $\frac{1}{2}L < a_{h_3} < L$
8. Simply supported–simply supported 	For $0 \le a \le \frac{1}{2}L,\ C_c = \dfrac{M_p L}{L - a}$ $\frac{1}{2}L \le a \le L,\ C_c = \dfrac{M_p L}{a}$ $a = \frac{1}{2}L,\ C_c = 2M_p$	For $0 \le a \le \frac{1}{2}L,\ a_{h_1} = a + \varepsilon$ $\frac{1}{2}L \le a \le L,\ a_{h_1} = a - \varepsilon$ $a = \frac{1}{2}L,\ a_{h_1} = a + \varepsilon,\ a_{h_2} = a - \varepsilon$

TABLE 11-4 Collapse Loads for Beams 548

9. Simply supported–fixed C $\leftarrow a \rightarrow$	For $\frac{1}{3}L \leq a \leq L, C_c = 2M_p$ For $0 \leq a \leq \frac{1}{3}L, C_c = \dfrac{M_p(L + a)}{L - a}$	For $\frac{1}{3}L \leq a \leq L, a_{h_1} = a - \varepsilon$, $a_{h_2} = a + \varepsilon$ For $0 \leq a \leq \frac{1}{3}L, a_{h_1} = a + \varepsilon$, $a_{h_2} = L$
10. Free–fixed C $\leftarrow a \rightarrow$	$C_c = M_p$	$a < a_{h_1} < L$
11. Guided–fixed C $\leftarrow a \rightarrow$	$C_c = 2M_p$	$0 \leq a_{h_1} \leq a, a \leq a_{h_2} \leq L$

Collapse Loads for Beams   TABLE 11-4

TABLE 11-4 (continued) COLLAPSE LOADS FOR BEAMS

Boundary Conditions and Loading	Collapse Loads	Plastic Hinge Locations
12. Guided–simply supported	$C_c = M_p$	$0 \le a_{h1} \le a$
13. Fixed–fixed	$P_c =$ $$\frac{16 M_p L^2}{a_1^4 + a_2^4 - 4a_1^2 L^2 + 4a_1^2 a_2 L - 2a_1^2 a_2^2 + 4a_2^2 L^2 - 4a_2^3 L}$$	$a_{h_1} = 0,\ a_{h_2} = \dfrac{a_1^2 + 2a_2 L - a_2^2}{2L}$ $a_{h_3} = L$
14. Simply supported–simply supported	$P_c =$ $$\frac{8 M_p L^2}{a_1^4 + a_2^4 + 4a_2^2 L^2 - 4a_1^2 L^2 + 4a_1^2 a_2 L - 2a_1^2 a_2^2 - 4a_2^3 L}$$	$a_{h_1} = \dfrac{a_1^2 + 2a_2 L - a_2^2}{2L}$

TABLE 11-4 Collapse Loads for Beams 550

15. Simply supported–fixed		$$p_c = \frac{2M_p(L + a_{h_1})}{La_{h_1}^2 - a_1^2(L - a_{h_1}) + 2La_{h_1}(a_2 - a_{h_1}) - a_2^2 a_{h_1}}$$	$a_{h1} = \left[(L^2 - a_2^2) + 2(a_2 L + a_1^2)\right]^{1/2} - L$ $a_{h_2} = L$
16. Free–fixed		$$p_c = \frac{2M_p}{a_1^2 + 2La_2 - 2La_1 - a_2^2}$$	$a_{h_1} = L$
17. Guided–fixed		$$p_c = \frac{4M_p}{\left(a_1^2 - 2a_1 L + 2a_2 L - a_2^2\right)}$$	$a_{h_1} = a_1, \; a_{h_2} = L$
18. Guided–simply supported		$$p_c = \frac{2M_p}{(a_2 - a_1)(2L - a_2 - a_1)}$$	$0 \le a_{h_1} \le a_1$

TABLE 11-5 ELASTIC BUCKLING LOADS AND MODE SHAPES FOR AXIALLY LOADED COLUMNS WITH IDEAL END CONDITIONS

Notation

E = modulus of elasticity
I = moment of inertia
L = length of column
P_{cr} = Buckling load

End Conditions	P_{cr}	Buckling Mode Shape
1. Free–free	$\dfrac{\pi^2 EI}{L^2}$	$\sin \dfrac{\pi x}{L}$
2. Free–guided	$\dfrac{\pi^2 EI}{4L^2}$	$\sin \dfrac{\pi x}{2L}$
3. Clamped–free	$\dfrac{\pi^2 EI}{4L^2}$	$1 - \cos \dfrac{\pi x}{2L}$
4. Free–pinned	$\dfrac{\pi^2 EI}{L^2}$	$\sin \dfrac{\pi x}{L}$
5. Pinned–pinned	$\dfrac{\pi^2 EI}{L^2}$	$\sin \dfrac{\pi x}{L}$
6. Clamped–pinned	$\dfrac{2.05 \pi^2 EI}{L^2}$	$\dfrac{EI}{k^3}\left[\sin kx - kL \cos kx + kL\left(1 - \dfrac{x}{L}\right)\right]$ where $k = 1.4318\dfrac{\pi}{L}$
7. Clamped–clamped	$\dfrac{4\pi^2 EI}{L^2}$	$1 - \cos \dfrac{2\pi x}{L}$
8. Clamped–guided	$\dfrac{\pi^2 EI}{L^2}$	$1 - \cos \dfrac{\pi x}{L}$
9. Guided–pinned	$\dfrac{\pi^2 EI}{4L^2}$	$\cos \dfrac{\pi x}{2L}$
10. Guided–guided	$\dfrac{\pi^2 EI}{L^2}$	$\cos \dfrac{\pi x}{L}$

TABLE 11-5 **Buckling Loads of Uniform Columns** 552

End Conditions	Critical Load
1. Upper end is elastically restrained, lower end is fixed. Spring constants are $$k = \frac{V_L}{w_L} \qquad k_1 = \frac{M_L}{\theta_L}$$	$$P_{cr} = m^2 \frac{EI}{L^2}$$ Find m by solving $$\frac{c(m - \sin m) - m^3}{c(1 - \cos m)} = \frac{m \sin m + c_1(1 - \cos m)}{m \cos m + c_1 \sin m}$$ where $c = \dfrac{kL^3}{EI} \qquad c_1 = \dfrac{k_1 L}{EI}$

End Conditions	Critical Load
2. Both ends restrained 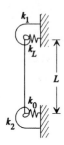 For the upper end: $k_L = V_L/w_L$ $k_1 = M_L/\theta_L$ For the lower end: $k_0 = V_0/w_0$ $k_2 = M_0/\theta_0$ 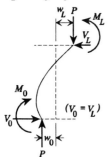	$P_{cr} = m^2 \dfrac{EI}{L^2}$ Find m by solving $$\frac{m - \sin m + m^3(1 + \alpha)/c_L}{1 - \cos m + \beta m^2/c_1 - \beta m^4(1 + \alpha)/(c_1 c_L)}$$ $$= \frac{(m/c_1)\sin m + 1 - \cos m}{(m/c_1)(\beta + \cos m) + \sin m}$$ where $$c_L = \frac{k_L L^3}{EI}$$ $$c_1 = \frac{k_1 L}{EI} \qquad \frac{k_L}{k_0} = \alpha \qquad \frac{k_1}{k_2} = \beta$$

TABLE 11-7 ELASTIC BUCKLING LOADS FOR COLUMNS WITH IN-SPAN AXIAL LOADS

Notation

$$(P + p_x L)_{cr} = \eta \frac{EI}{L^2}$$

p_x = Uniformly distributed axial load (F/L)

End Condition	η
1. Pinned–Pinned $P \downarrow$ $p_x = P_0$ $P + p_x L$	$\dfrac{16.7 p_x L}{p_x L + 1.36 P} + 9.87$
2. Pinned–fixed $P \downarrow$ $p_x = P_0$ $P + p_x L$	$\dfrac{40.6 (p_x L)^{1.35}}{(p_x L)^{1.35} + 2.6 P^{1.35}} + 19.74$
3. Fixed–pinned $P \downarrow$ $p_x = P_0$ $P + p_x L$	$\dfrac{24.5 (p_x L)^{1.35}}{(p_x L)^{1.35} + 1.68 P^{1.35}} + 19.74$

End Condition	η
4. Fixed–fixed P $p_x = P_0$ $P + p_x L$	$\dfrac{46.1(p_x L)^{1.35}}{(p_x L)^{1.35} + 2.4P^{1.35}} + 39.48$
5. Free–fixed P $p_x = P_0$ $P + p_x L$	$\dfrac{5.38(p_x L)^{1.35}}{(p_x L)^{1.35} + 2.7P^{1.35}} + 2.47$

TABLE 11-7   **Buckling of Columns with In-Span Loading**   556

TABLE 11-8 BUCKLING LOADS FOR COLUMNS WITH IN-SPAN SUPPORTS

Notation

$$P_{cr} = \eta \frac{EI}{L^2}$$

End Conditions	Restrictions	$\eta = C_1 + C_2\left(\dfrac{b}{L}\right) + C_3\left(\dfrac{b}{L}\right)^2$		
1. Free–pinned	$0 \leq \dfrac{b}{L} \leq 0.5$	$C_1 = 2.4825$	$C_2 = 2.6372$	$C_3 = 6.4821$
	$0.5 < \dfrac{b}{L} \leq 1$	$C_1 = -4.7167$	$C_2 = 25.432$	$C_3 = -10.7286$
2. Guided–pinned	$0 \leq \dfrac{b}{L} \leq 0.6$	$C_1 = 9.68072$	$C_2 = 17.11428$	$C_3 = 6.21428$
	$0.7 \leq \dfrac{b}{L} \leq 1.0$	$C_1 = 33.50997$	$C_2 = -22.84993$	$C_3 = 9.4996$
3. Pinned–pinned	$0 \leq \dfrac{b}{L} \leq 0.4$	$C_1 = 20.15514$	$C_2 = 27.79714$	$C_3 = 34.85716$
	$0.5 \leq \dfrac{b}{L} \leq 1.0$	$C_1 = 63.08144$	$C_2 = -48.77858$	$C_3 = 5.50000$
4. Clamped–pinned	$0 \leq \dfrac{b}{L} \leq 0.3$	$C_1 = 39.44701$	$C_2 = 56.31997$	$C_3 = 29.00003$
	$0.4 \leq \dfrac{b}{L} \leq 1.0$	$C_1 = 109.70282$	$C_2 = -149.23325$	$C_3 = 59.52375$

TABLE 11-8 (continued)   BUCKLING LOADS FOR COLUMNS WITH IN-SPAN SUPPORTS

End Conditions	Restrictions	$\eta = C_1 + C_2\left(\dfrac{b}{L}\right) + C_3\left(\dfrac{b}{L}\right)^2$		
5. Free– clamped *P*	$0 \le \dfrac{b}{L} \le 0.6$	$C_1 = \quad 2.55579$	$C_2 = 1.31893$	$C_3 = 12.62499$
	$0.7 \le \dfrac{b}{L} \le 1.0$	$C_1 = -17.17765$	$C_2 = 43.67535$	$C_3 = -6.25020$
6. Guided– clamped *P*	$0 \le \dfrac{b}{L} \le 0.6$	$C_1 = \quad 10.01666$	$C_2 = \quad 9.72857$	$C_3 = 38.47618$
	$0.7 \le \dfrac{b}{L} \le 1.0$	$C_1 = -31.24563$	$C_2 = 151.40530$	$C_3 = -80.75015$
7. Pinned– clamped *P*	$0 \le \dfrac{b}{L} \le 0.6$	$C_1 = 19.99333$	$C_2 = \quad 30.18574$	$C_3 = 59.52375$
	$0.7 \le \dfrac{b}{L} \le 1.0$	$C_1 = 124.76723$	$C_2 = -114.32053$	$C_3 = 29.00031$
8. Clamped– clamped *P*	$0 \le \dfrac{b}{L} \le 0.4$	$C_1 = 39.38744$	$C_2 = \quad 61.88135$	$C_3 = 70.07159$
	$0.5 \le \dfrac{b}{L} \le 1.0$	$C_1 = 128.82428$	$C_2 = -96.99966$	$C_3 = 6.80359$

TABLE 11-8   Buckling of Columns with In-Span Supports   **558**

TABLE 11-9 BUCKLING LOADS FOR TAPERED COLUMNS

Notation

$$P_{cr} = \eta \frac{EI}{L^2}$$

$I_0, I_x, I =$ moments of inertia (L^4)
$x =$ coordinate from virtual vertex (L)
$a =$ length of taper, beginning at virtual vertex
$\frac{1}{2}(L - \ell) =$ length of taper

Boundary Conditions	Taper	Restrictions	$\eta = C_1 + C_2(I_0/I) + C_3(I_0/I)^2$
1. Simply supported– simply supported	$I_x = I(x/a)$	$0.1 \leq I_0/I \leq 0.8$ $0.0 \leq a/L \leq 0.8$	$C_1 = 5.930731 + 7.494147(a/L) - 3.20702(a/L)^2$ $C_2 = 5.376184 - 7.29300(a/L) + 0.62360(a/L)^2$ $C_3 = -1.50574 - 0.39881(a/L) + 2.93134(a/L)^2$
2. Fixed—fixed	$I_x = I(x/a)$	$0.2 \leq I_0/I \leq 0.8$ $0.0 \leq a/L \leq 0.8$	$C_1 = 14.26143 + 3.89573(a/L) + 10.14281(a/L)^2$ $C_2 = 33.71884 + 4.71145(a/L) - 16.21420(a/L)^2$ $C_3 = -9.02855 - 9.71431(a/L) + 7.14275(a/L)^2$
3. Simply supported– simply supported	$I_x = I(x/a)^2$	$0.1 \leq I_0/I \leq 0.8$ $0.0 \leq a/L \leq 0.8$	$C_1 = 4.48594 + 9.27104(a/L) - 3.17469(a/L)^2$ $C_2 = 9.54820 - 12.13186(a/L) + 0.15266(a/L)^2$ $C_3 = -4.59559 + 2.96185(a/L) + 3.60045(a/L)^2$

TABLE 11-9 (continued) BUCKLING LOADS FOR TAPERED COLUMNS

Boundary Conditions	Taper	Restrictions	$\eta = C_1 + C_2(I_0/I) + C_3(I_0/I)^2$
4. Fixed–fixed	$I_x = I(x/a)^2$	$0.2 \le I_0/I \le 0.8$ $0.0 \le a/L \le 0.8$	$C_1 = 11.71678 + 10.39221(a/L) + 1.99100(a/L)^2$ $C_2 = 40.10596 - 15.17206(a/L) + 9.39316(a/L)^2$ $C_3 = -13.13751 + 5.31273(a/L) - 12.50024(a/L)^2$
5. Simply supported–simply supported	$I_x = I(x/a)^3$	$0.1 \le I_0/I \le 0.8$ $0.0 \le a/L \le 0.8$	$C_1 = 3.95433 + 9.71412(a/L) - 2.91616(a/L)^2$ $C_2 = 11.17553 - 13.09159(a/L) - 1.11514(a/L)^2$ $C_3 = -5.78722 + 3.53687(a/L) + 4.64377(a/L)^2$
6. Fixed–fixed	$I_x = I(x/a)^3$	$0.2 \le I_0/I \le 0.8$ $0.0 \le a/L \le 0.8$	$C_1 = 10.88200 + 12.03998(a/L) + 0.00004(a/L)^2$ $C_2 = 42.20943 - 19.96425(a/L) + 15.39278(a/L)^2$ $C_3 = -14.5500 + 9.39996(a/L) - 17.49989(a/L)^2$
7. Simply supported–simply supported	$I_x = I(x/a)^4$	$0.1 \le I_0/I \le 0.8$ $0.0 \le a/L \le 0.8$	$C_1 = 3.68757 + 9.69379(a/L) - 2.45495(a/L)^2$ $C_2 = 11.90928 - 12.34295(a/L) - 3.26411(a/L)^2$ $C_3 = -6.30219 + 2.60257(a/L) + 6.67416(a/L)^2$
8. Fixed–fixed	$I_x = I(x/a)^4$	$0.2 \le I_0/I \le 0.8$ $0.0 \le a/L \le 0.8$	$C_1 = 10.45043 + 12.80821(a/L) - 0.85714(a/L)^2$ $C_2 = 43.28855 - 21.81049(a/L) + 17.41934(a/L)^2$ $C_3 = -15.17141 + 9.83910(a/L) - 18.08010(a/L)^2$

TABLE 11-9   **Buckling of Tapered Columns**   560

TABLE 11-10   FORMULAS FOR MAXIMUM STRESS IN IMPERFECT COLUMNS

Notation

A = area

$C = \begin{cases} 1 & \text{for simply supported ends} \\ 4 & \text{for fixed ends} \end{cases}$

c = distance between outside fiber and neutral axis

E = modulus of elasticity

e = eccentricity

KL = effective length of column

L = length of column

P = maximum axial force

r = least radius of gyration

α = constant

λ = eccentricity ratio

σ_{ys} = yield stress

σ = ultimate stress

Formula	Explanation
1. Secant formula: $$\sigma = \frac{\sigma_{yc}}{1 + \lambda \sec\left(\dfrac{KL}{2r}\sqrt{\dfrac{P}{AE}}\right)}$$ with $\lambda = ec/r^2$	This formula allows the direct calculation of maximum stress for an eccentrically loaded column. The eccentricity factor λ is determined by conforming the equation to the experimental data.
2. Rankine formula: $$\frac{P}{A} = \frac{\sigma}{1 + \varphi\left(\dfrac{L}{r}\right)^2}$$	The factor φ can be calculated from $\varphi = \sigma/(C\pi^2 E)$. But more often σ and φ are adjusted empirically to make P/A agree with the test data in the range of most L/r values.
3. Simple polynomial formula: $$\frac{P}{A} = \sigma - \alpha\left(\frac{L}{r}\right)^n$$	This is an empirical equation. For most steels, $n = 2$, and for cast irons and many aluminum alloys, $n = 1$. The factor α is experimentally determined. When $n = 2$, σ is the yielding stress and α makes the parabola that the equation defines intersect tangent to the Euler curve for a long column. When $n = 1$, σ is the maximum stress at failure and α is experimentally determined to make the straight line that the equation defines the tangent to the Euler curve.
4. Exponential formula: $\beta \le 1.5$: $$\frac{P}{A} = 0.658^{\beta^2}\sigma_{ys}$$ $\beta > 1.5$: $$\frac{P}{A} = 0.877\beta^{-2}\sigma_{ys}$$ with $$\beta = \frac{KL}{r\pi}\left(\frac{\sigma_{ys}}{E}\right)^{1/2}$$	This formula was suggested by the American Institute of Steel Construction and uses terms from the secant formula.

TABLE 11-11 FORMULAS FOR KERNS OF SHORT PRISMS LOADED ECCENTRICALLY

Notation

y_1, z_1, r, r_{min} = dimensions of kerns, shown in the shaded areas

kern = area in which compressive axial force is applied to produce no tensile stress on cross section

Kern (Shaded) of Cross Section	Expressions for y_1, z_1, r, r_{min}
1. Solid square	$y_1 = z_1 = \frac{1}{6}h$ $y_2 = z_2 = \frac{1}{3}h$ $r_{min} = 0.0589h$
2. Solid rectangle	$y_1 = \frac{1}{6}b \qquad z_1 = \frac{1}{6}h$ $y_2 = \frac{1}{3}b \qquad z_2 = \frac{1}{3}h$ $r_{min} = \dfrac{bh}{6\sqrt{b^2 + h^2}}$
3. Solid isosceles triangle	$y_1 = \frac{1}{8}b \qquad z_1 = \frac{1}{12}h$ $y_2 = \frac{1}{6}h$
4. Hollow rectangle	$y_1 = \dfrac{1}{6}\dfrac{hb^3 - h_1 b_1^3}{b(bh - b_1 h_1)}$ $z_1 = \dfrac{1}{6}\dfrac{bh^3 - b_1 h_1^3}{h(bh - b_1 h_1)}$ when $h = b$, $h_1 = b_1$ $r_{min} = 0.0589h\left[1 + \left(\dfrac{h_1}{h}\right)^2\right]$

TABLE 11-11   Formulas for Kerns of Short Prisms   562

Kern (Shaded) of Cross Section	Expressions for y_1, z_1, r, r_{min}
5. Solid octagon	$r_{min} = 0.2256R$ If the octagon is hollow, $$r_{min} = 0.2256R\left[1 + \left(\frac{R_1}{R}\right)^2\right]$$ R_1 is the inside radius
6. Solid circle	$r = \frac{1}{8}d$
7. Hollow circle	$$r = \frac{1}{8}\lambda d_0\left[1 + \left(\frac{d}{d_0}\right)^2\right]$$
8. Thin-walled circle	$r = \frac{1}{4}d$

TABLE 11-12 NATURAL FREQUENCIES AND MODE SHAPES FOR UNIFORM BEAMS

Notation

E = Modulus of elasticity $\qquad\qquad$ ρ = Mass per unit length
I = Moment of inertia $\qquad\qquad\qquad$ L = Length of the beam

Natural frequency: ω_i (rad/s) = $\dfrac{\lambda_i^2}{L^2}\left(\dfrac{EI}{\rho}\right)^{1/2}$ $\qquad$ f_i (Hz) = $\dfrac{\lambda_i^2}{2\pi L^2}\left(\dfrac{EI}{\rho}\right)^{1/2}$

Boundary Conditions	λ_i, $i = 1, 2, 3, \ldots$	Mode Shapes	β_i, $i = 1, 2, 3, \ldots$
1. Pinned–pinned	$i\pi$	$\sin\dfrac{i\pi x}{L}$	
2. Fixed–pinned	3.92660231 7.06858275 10.21017612 13.35176878 16.49336143 $(4i + 1)\pi/4,\ i > 5$	$\cosh\dfrac{\lambda_i x}{L} - \cos\dfrac{\lambda_i x}{L}$ $-\beta_i\left(\sinh\dfrac{\lambda_i x}{L} - \sin\dfrac{\lambda_i x}{L}\right)$	$\dfrac{\cosh\lambda_i - \cos\lambda_i}{\sinh\lambda_i - \sin\lambda_i}$
3. Fixed–fixed	4.73004074 7.85320462 10.99560790 14.13716550 17.27875970 $(2i + 1)\pi/2,\ i > 5$	$\cosh\dfrac{\lambda_i x}{L} - \cos\dfrac{\lambda_i x}{L}$ $-\beta_i\left(\sinh\dfrac{\lambda_i x}{L} - \sin\dfrac{\lambda_i x}{L}\right)$	$\dfrac{\cosh\lambda_i - \cos\lambda_i}{\sinh\lambda_i - \sin\lambda_i}$
4. Free–free	4.73004074 7.85320462 10.99560780 14.13716550 17.27875970 $(2i + 1)\pi/2,\ i > 5$	$\cosh\dfrac{\lambda_i x}{L} + \cos\dfrac{\lambda_i x}{L}$ $-\beta_i\left(\sinh\dfrac{\lambda_i x}{L} + \sin\dfrac{\lambda_i x}{L}\right)$	$\dfrac{\cosh\lambda_i - \cos\lambda_i}{\sinh\lambda_i - \sin\lambda_i}$
5. Free–guided	2.36502037 5.49780392 8.63937983 11.78097245 14.92256510 $(4i - 1)\pi/4,\ i > 5$	$\cosh\dfrac{\lambda_i x}{L} + \cos\dfrac{\lambda_i x}{L}$ $-\beta_i\left(\sinh\dfrac{\lambda_i x}{L} + \sin\dfrac{\lambda_i x}{L}\right)$	$\dfrac{\sinh\lambda_i - \sin\lambda_i}{\cosh\lambda_i + \cos\lambda_i}$
6. Fixed–free	1.87510407 4.69409113 7.85475744 10.99554073 14.13716839 $(2i - 1)\pi/2,\ i > 5$	$\cosh\dfrac{\lambda_i x}{L} - \cos\dfrac{\lambda_i x}{L}$ $-\beta_i\left(\sinh\dfrac{\lambda_i x}{L} - \sin\dfrac{\lambda_i x}{L}\right)$	Same as free–guided

Boundary Conditions	$\lambda_i, i = 1, 2, 3, \ldots$	Mode Shapes	$\beta_i, i = 1, 2, 3, \ldots$
7. Free–pinned 	3.92660231 7.06858275 10.21017612 13.35176878 16.49336143 $(4i + 1)\pi/4, i > 5$	$\cosh \dfrac{\lambda_i x}{L} + \cos \dfrac{\lambda_i x}{L}$ $-\beta_i \left(\sinh \dfrac{\lambda_i x}{L} + \sin \dfrac{\lambda_i x}{L} \right)$	Same as free–free
8. Fixed–guided 	2.36502037 5.49780392 8.63937983 11.78097245 14.92256510 $(4i - 1)\pi/4, i > 5$	$\cosh \dfrac{\lambda_i x}{L} - \cos \dfrac{\lambda_i x}{L}$ $-\beta_i \left(\sinh \dfrac{\lambda_i x}{L} - \sin \dfrac{\lambda_i x}{L} \right)$	Same as free–guided
9. Guided–pinned 	$(2i - 1)\pi/2$	$\cos \dfrac{(2i - 1)\pi x}{2L}$	
10. Guided–guided 	$i\pi$	$\cos \dfrac{i\pi x}{L}$	

TABLE 11-13 FREQUENCY EQUATIONS FOR UNIFORM BEAMS

Notation

E = modulus of elasticity
G = shear modulus of elasticity
I = moment of inertia
A_s = equivalent shear area

End Conditions	Bending (Euler–Bernoulli Beam)	Bending and Shear Deformation (Shear Beam)
1. Hinged–hinged	$\omega_n^2 \dfrac{\rho}{EI} - \dfrac{n^4 \pi^4}{L^4} = 0$	$\omega_n^2 \left[\dfrac{\rho}{EI} + \dfrac{\rho}{GA_s} \dfrac{n^2 \pi^2}{L^2} \right] - \dfrac{n^4 \pi^4}{L^4} = 0$ Pure shear: $\sin\left(\omega_n L \sqrt{\dfrac{\rho}{GA_s}} \right) = 0$
2. Free–free or fixed–fixed	$\cosh \beta_1 L \cos \beta_1 L - 1 = 0$	$\cosh \beta_1 L \cos \beta_2 L$ $\quad + \frac{1}{2} c_1 \sinh \beta_1 L \sin \beta_2 L - 1 = 0$ Pure shear: same as hinged–hinged
3. Fixed–hinged	$\tanh \beta_1 L - \tan \beta_1 L = 0$	$\cosh \beta_1 L \sin \beta_2 L$ $\quad + c_5 \sinh \beta_1 L \cos \beta_2 L = 0$ Pure shear: same as hinged–hinged
4. Fixed–free	$\cosh \beta_1 L \cos \beta_1 L + 1 = 0$	$\cosh \beta_1 L \cos \beta_2 L$ $\quad + c_2 c_3 \sinh \beta_1 L \sin \beta_2 L$ $\quad - 2 c_3 = 0$ Pure shear: $\cos\left(\omega_n L \sqrt{\dfrac{\rho}{GA_s}} \right) = 0$

ρ = mass per unit length
r_y = radius of gyration about y axis
L = length of beam

Bending and Rotary Inertia (Rayleigh Beam)	Bending, Shear Deformation, and Rotary Inertia (Timoshenko Beam)
$$\omega_n^2\left[\frac{\rho}{EI} + \frac{\rho r_y^2}{EI}\frac{n^2\pi^2}{L^2}\right] - \frac{n^4\pi^4}{L^4} = 0$$	$$\omega_n^4 r_y^2 - \left[\frac{n^2\pi^2}{L^2}\frac{EI}{\rho} + \frac{n^2\pi^2}{L^2}r_y^2\frac{A_sG}{\rho} + \frac{A_sG}{\rho}\right]\omega_n^2$$ $$+ \frac{n^4\pi^4}{L^4}\frac{GA_s EI}{\rho^2} = 0$$
$\cosh\beta_1 L \cos\beta_2 L - 1$ $+\frac{1}{2}c_2 \sinh\beta_2 L \sin\beta_2 L = 0$	$\cosh\beta_1 L \cos\beta_2 L - 1$ $+\frac{1}{2}c_1 \sinh\beta_1 L \sin\beta_2 L = 0$
$\cosh\beta_1 L \sin\beta_2 L$ $-\frac{\beta_2}{\beta_1}\sinh\beta_1 L \cos\beta_2 L = 0$	$\cosh\beta_1 L \sin\beta_2 L$ $+c_5 \sinh\beta_1 L \cos\beta_2 L = 0$
$\cosh\beta_1 L \cos\beta_2 L$ $+\frac{1}{2}c_2 \sinh\beta_1 L \sin\beta_2 L$ $+\frac{1}{2}c_4 = 0$	$\cosh\beta_1 L \cos\beta_2 L$ $+c_2 c_3 \sinh\beta_1 L \sin\beta_2 L - 2c_3 = 0$

TABLE 11-13 (continued)   FREQUENCY EQUATIONS FOR UNIFORM BEAMS

$$n = 1, 2, \ldots$$

Euler–Bernoulli Beam

$$\beta_1^2 = \beta_2^2 = \sqrt{\frac{\rho \omega_n^2}{EI}}$$

Rayleigh Beam

$$\beta_1^2 = \frac{1}{2}\left[\frac{-\rho \omega_n^2 r_y^2}{EI} + \sqrt{\left(\frac{\rho \omega_n^2 r_y^2}{EI}\right)^2 + \frac{4\rho \omega_n^2}{EI}} \right]$$

$$\beta_2^2 = \frac{1}{2}\left[\frac{\rho \omega_n^2 r_y^2}{EI} + \sqrt{\left(\frac{\rho \omega_n^2 r_y^2}{EI}\right)^2 + \frac{4\rho \omega_n^2}{ET}} \right]$$

$$c_1 = \frac{\beta_2}{\beta_1}\left(\frac{\rho \omega_n^2 \mp A_s G \beta_2^2}{\rho \omega_n^2 \pm A_s G \beta_2^2} \right) - \frac{\beta_1}{\beta_2}\left(\frac{\rho \omega_n^2 \pm A_s G \beta_1^2}{\rho \omega_n^2 \mp A_s G \beta_1^2} \right)$$

(upper sign for free–free,
lower for fixed–fixed.)

$$c_2 = \frac{\beta_2}{\beta_1} - \frac{\beta_1}{\beta_2}$$

$$c_3 = \frac{(\rho \omega_n^2 - A_s G \beta_2^2)(\rho \omega_n^2 + A_s G \beta_1^2)}{(\rho \omega_n^2 + A_s G \beta_1^2)^2 + (\rho \omega_n^2 - A_s G \beta_2^2)^2}$$

$$c_4 = \left(\frac{\beta_1}{\beta_2}\right)^2 + \left(\frac{\beta_2}{\beta_1}\right)^2$$

$$c_5 = \frac{\beta_1}{\beta_2}\left(\frac{\rho \omega_n^2 - A_s G \beta_2^2}{\rho \omega_n^2 + A_s G \beta_1^2} \right)$$

Timoshenko Beam

$$\beta_1^2 = \frac{1}{2}\left\{ -\rho \omega_n^2\left(\frac{1}{GA_s} + \frac{r_y^2}{EI}\right) + \sqrt{\left[\rho \omega_n^2\left(\frac{1}{GA_s} + \frac{r_y^2}{EI}\right)\right]^2 + \frac{4\rho \omega_n^2}{EI}\left(1 - \frac{\rho \omega_n^2 r_y^2}{GA_s}\right)} \right\}$$

$$\beta_2^2 = \frac{1}{2}\left\{ \rho \omega_n^2\left(\frac{1}{GA_s} + \frac{r_y^2}{EI}\right) + \sqrt{\left[\rho \omega_n^2\left(\frac{1}{GA_s} + \frac{r_y^2}{EI}\right)\right]^2 + \frac{4\rho \omega_n^2}{EI}\left(1 - \frac{\rho \omega_n^2 r_y^2}{GA_s}\right)} \right\}$$

Shear Beam

$$\beta_1^2 = \frac{1}{2}\left[-\frac{\rho \omega_n^2}{GA_s} + \sqrt{\left(\frac{\rho \omega_n^2}{GA_s}\right)^2 + \frac{4\rho \omega_n^2}{EI}} \right]$$

$$\beta_2^2 = \frac{1}{2}\left[\frac{\rho \omega_n^2}{GA_s} + \sqrt{\left(\frac{\rho \omega_n^2}{GA_s}\right)^2 + \frac{4\rho \omega_n^2}{EI}} \right]$$

TABLE 11-13   **Frequency Equations for Uniform Beams**   568

TABLE 11-14   NATURAL FREQUENCIES FOR BEAMS WITH CONCENTRATED MASSES

Notation

m_b = Total mass of beam E = Modulus of elasticity
M_i = Concentrated mass I = Moment of inertia
L = Length of beam

See Tables 10-2 and 17-1 for additional cases.

Description	Fundamental Natural Frequency (Hz)
1. Center mass, pinned–pinned beam	$\dfrac{2}{\pi}\left[\dfrac{3EI}{L^3(M_i + 0.4857 m_b)}\right]^{1/2}$
2. End mass, cantilever beam	$\dfrac{1}{2\pi}\left[\dfrac{3EI}{L^3(M_i + 0.2357 m_b)}\right]^{1/2}$
3. End masses, free–free beam	$\dfrac{\pi}{2}\left\{\dfrac{EI}{L^3 m_b}\left[1 + \dfrac{5.45}{1 - 77.4(M_i/m_b)^2}\right]\right\}^{1/2}$

TABLE 11-14 (continued) NATURAL FREQUENCIES FOR BEAMS WITH CONCENTRATED MASSES

Description	Fundamental Natural Frequency (Hz)
4. Off-center mass, pinned–pinned beam	$\dfrac{1}{2\pi}\left\{\dfrac{3EI(a+b)}{a^2b^2[M_i+(\alpha+\beta)m_b]}\right\}^{1/2}$ $\alpha = \dfrac{a}{a+b}\left[\dfrac{(2b+a)^2}{12b^2}+\dfrac{a^2}{28b^2}-\dfrac{a(2b+a)}{10b^2}\right]$ $\beta = \dfrac{b}{a+b}\left[\dfrac{(2a+b)^2}{12a^2}+\dfrac{b^2}{28a^2}-\dfrac{b(2a+b)}{10a^2}\right]$
5. Center mass, clamped–clamped beam	$\dfrac{4}{\pi}\left[\dfrac{3EI}{L^3(M_i+0.37m_b)}\right]^{1/2}$
6. Off-center mass, clamped–clamped beam	$\dfrac{4}{\pi}\left(\dfrac{3EIL^3}{a^3b^3[M_i+(\alpha+\beta)m_b]}\right)^{1/2}$ $\alpha = \dfrac{a}{a+b}\left[\dfrac{(3a+b)^2}{28b^2}+\dfrac{9(a+b)^2}{20b^2}-\dfrac{(a+b)(3a+b)}{4b^2}\right]$ $\beta = \dfrac{b}{a+b}\left[\dfrac{(3b+a)^2}{28a^2}+\dfrac{9(a+b)^2}{20a^2}-\dfrac{(a+b)(3b+a)}{4a^2}\right]$

TABLE 11-14 Frequencies of Lumped Mass Beams   570

TABLE 11-15 FUNDAMENTAL NATURAL FREQUENCY BY NUMBER OF SPANS OF MULTISPAN BEAMS WITH RIGID IN-SPAN SUPPORTS[a]

Notation

Fundamental natural frequency: $\omega_1 \text{ (rad/s)} = \dfrac{\lambda_1^2}{\ell^2} \left(\dfrac{EI}{\rho} \right)^{1/2}$ $\qquad f_1 \text{ (Hz)} = \dfrac{\lambda_1^2}{2\pi \ell^2} \left(\dfrac{EI}{\rho} \right)^{1/2}$

E = modulus of elasticity $\qquad I$ = moment of inertia $\qquad \rho$ = mass per unit length

Boundary Conditions	Number of Spans									
	1	2	3	4	5	6	7	8	9	10
	λ_1									
1. Free–free	4.730	1.875	1.412	1.506	1.530	1.537	1.538	1.539	1.539	1.539
2. Free–pinned	3.927	1.505	1.536	1.539	1.539	1.539	1.539	1.539	1.539	1.539
3. Pinned–pinned	3.142	3.142	3.142	3.142	3.142	3.142	3.142	3.142	3.142	3.142
4. Fixed–free	1.875	1.570	1.541	1.539	1.539	1.539	1.539	1.539	1.539	1.539
5. Fixed–pinned	3.927	3.393	3.261	3.210	3.186	3.173	3.164	3.159	3.156	3.153
6. Fixed–fixed	4.730	3.927	3.557	3.393	3.310	3.260	3.230	3.210	3.196	3.186

[a]Each span is a beam segment of length ℓ. A beam with one span has no in-span supports.

TABLE 11-16 FUNDAMENTAL NATURAL FREQUENCIES OF TAPERED BEAMS

Notation

$$\omega_1 = \frac{\lambda h_a}{L^2}\sqrt{E/\rho^*} \quad \text{rad/s}$$

$$f_1 = \frac{\lambda h_a}{2\pi L^2}\sqrt{E/\rho^*} \quad \text{Hz}$$

E = modulus of elasticity

ρ^* = mass per unit volume

Elevation: direction of vibration ↕   Plan: vibration perpendicular to plane of paper

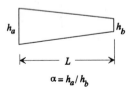

$\alpha = h_a/h_b$

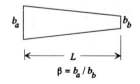

$\beta = b_a/b_b$

Boundary Conditions		λ			
		$\alpha = 2$	$\alpha = 3$	$\alpha = 4$	$\alpha = 10$
1. Fixed–fixed	$\beta = 1$	4.713	4.041	3.670	2.853
2. Fixed–simply supported	$\beta = 1$	3.551	3.182	2.972	2.502
3. Fixed–free	$\beta = 1$	1.104	1.162	1.206	1.337
4. Simply supported–fixed	$\beta = 1$	2.937	2.361	2.046	1.382
5. Simply supported–simply supported	$\beta = 1$	2.057	1.732	1.544	1.122
6. Simply supported–free	$\beta = 1$	3.608	3.310	3.183	2.993
7. Free–fixed	$\beta = 1$	0.473	0.304	0.224	0.0818
8. Free–simply supported	$\beta = 1$	2.973	2.425	2.127	1.487

9. Fixed–free
$1 \le \alpha \le 5$
$1 \le \beta \le 5$

$$\lambda = C_1 + C_2(\beta) + C_3(\beta^2)$$
$$C_1 = 0.66592 + 0.10686(\alpha) - 0.00886(\alpha)^2$$
$$C_2 = 0.28430 + 0.00161(\alpha) - 0.00004(\alpha)^2$$
$$C_3 = -0.02546 - 0.00008(\alpha) - 0.00001(\alpha)^2$$

TABLE 11-16   Natural Frequencies of Tapered Beams   572

TABLE 11-17   FORMULAS FOR APPROXIMATE PERIODS OF VIBRATION OF BUILDINGS

Notation

$$\text{Period} = \frac{1}{\text{frequency (cycles/s)}} = \frac{2\pi}{\text{frequency (rad/s)}}$$

N = number of stories

Type of Building	Fundamental Natural Period (s)	Source
1. Reinforced concrete	$0.07\text{--}0.09N$	11.11
2. Rigid frame	$0.1N$	11.12
3. Space frame	$0.5\sqrt{N} - 0.4$	11.13
4. Low- and medium-height	~ 0.1	11.14
5. Tall	$\sim 1\text{--}10$	11.14

TABLE 11-18 DAMPING OF STRUCTURES[a]

Structure	Damping (%)	Source
1. Welded steel structure without fireproofing	0.5–2	11.15, 11.16
2. Riveted steel structure without fireproofing	2–3	11.15, 11.16
3. Steel frames	3–6	11.17
4. Concrete buildings	7–14	11.17
5. Brick walls	> 14	11.17
6. Masonry structures	15–40	11.15, 11.16
7. Fluid containers, ground supported	0.5	11.15, 11.16
Elevated Water Tanks		
8. Riveted	5	11.18
9. Welded	2	11.18
Bridges		
10. With concrete decks	1.1–2.5 (V)[b] 0.78–2.86 (T)[c]	11.14, 11.19 —
11. With steel decks	0.32–0.78 for (V) & (T)	—
12. With timber decks	1.59–3.5 (V) 2.55–4.8 (T)	—

TABLE 11-18 Damping of Structures — 574

TABLE 11-18 (continued)   DAMPING OF STRUCTURES

Structure	Damping (%)	Source
Chimneys or Towers		
13. Bolted or riveted: open lattice	0.32–2.86	11.14
14. Bolted or riveted: unlined, closed, circular	0.32–1.59	11.14
15. Bolted or riveted: lined, closed, circular	0.48–1.43	11.14
16. Welded: unlined, closed, circular	0.16–1.9	—
17. Welded: lined, closed, circular	0.48–0.96	—
18. Reinforced concrete: unlined, closed, circular	0.96–1.9	—

[a]Approximate percentage of critical damping of the fundamental mode of vibration is listed.
[b](V) = Vertical damping.
[c](T) = Torsional damping.

Notation

E = modulus of elasticity
ℓ = length of element
r_y = radius of gyration about y axis
A_s = equivalent shear area
A = area

i = ith element
I = moment of inertia
G = shear modulus of elasticity
ν = Poisson's ratio

Transfer Matrix (Sign Convention 1)

$$z_b = U^i z_a$$

$$
\begin{bmatrix}
1 & -\ell & -\dfrac{\ell^3}{6EI} + \dfrac{\ell}{GA_s} & -\dfrac{\ell^2}{2EI} & F_w \\[2mm]
0 & 1 & \dfrac{\ell^2}{2EI} & \dfrac{\ell}{EI} & F_\theta \\[2mm]
0 & 0 & 1 & 0 & F_V \\[2mm]
0 & 0 & \ell & 1 & F_M \\[2mm]
0 & 0 & 0 & 0 & 1
\end{bmatrix}
\begin{bmatrix}
w_a \\ \theta_a \\ V_a \\ M_a \\ 1
\end{bmatrix}
$$

$$U^i \qquad\qquad z_a$$

Set $1/GA_s = 0$ if shear deformation is not to be considered.

Stiffness Matrices (Sign Convention 2)

$$p^i = k^i v^i - \bar{p}^i$$

1. Effects of shear deformation neglected

a.

$$
\begin{bmatrix}
V_a \\ M_a/\ell \\ V_b \\ M_b/\ell
\end{bmatrix}
= \frac{EI}{\ell^3}
\begin{bmatrix}
12 & -6 & -12 & -6 \\
-6 & 4 & 6 & 2 \\
-12 & 6 & 12 & 6 \\
-6 & 2 & 6 & 4
\end{bmatrix}
\begin{bmatrix}
w_a \\ \theta_a \ell \\ w_b \\ \theta_b \ell
\end{bmatrix}
- \bar{p}^i
$$

$$p^i \qquad\qquad\quad k^i \qquad\qquad\qquad v^i$$

b.

$$
\begin{bmatrix}
V_a \\ M_a \\ V_b \\ M_b
\end{bmatrix}
=
\begin{bmatrix}
12EI/\ell^3 & -6EI/\ell^2 & -12EI/\ell^3 & -6EI/\ell^2 \\
-6EI/\ell^2 & 4EI/\ell & 6EI/\ell^2 & 2EI/\ell \\
-12EI/\ell^3 & 6EI/\ell^2 & 12EI/\ell^3 & 6EI/\ell^2 \\
-6EI/\ell^2 & 2EI/\ell & 6EI/\ell^2 & 4EI/\ell
\end{bmatrix}
\begin{bmatrix}
w_a \\ \theta_a \\ w_b \\ \theta_b
\end{bmatrix}
- \bar{p}^i
$$

$$p^i \qquad\qquad\qquad\qquad k^i \qquad\qquad\qquad\qquad v^i$$

TABLE 11-19   Transfer and Stiffness Beam Matrices   576

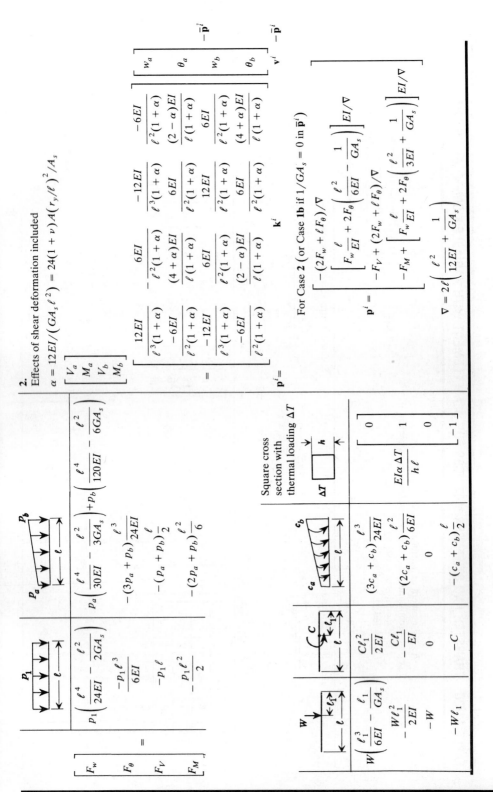

2.

Effects of shear deformation included

$$\alpha = 12EI/(GA_s\ell^2) = 24(1+\nu)A(r_y/\ell)^2/A_s$$

$$\begin{bmatrix} V_a \\ M_a \\ V_b \\ M_b \end{bmatrix}$$

$$\mathbf{p}^i = \mathbf{k}^i \qquad \mathbf{v}^i \qquad -\bar{\mathbf{p}}^i$$

$$\mathbf{k}^i = \begin{bmatrix} \dfrac{12EI}{\ell^3(1+\alpha)} & \dfrac{6EI}{\ell^2(1+\alpha)} & \dfrac{-12EI}{\ell^3(1+\alpha)} & \dfrac{-6EI}{\ell^2(1+\alpha)} \\[2mm] \dfrac{6EI}{\ell^2(1+\alpha)} & \dfrac{(4+\alpha)EI}{\ell(1+\alpha)} & \dfrac{-6EI}{\ell^2(1+\alpha)} & \dfrac{(2-\alpha)EI}{\ell(1+\alpha)} \\[2mm] \dfrac{-12EI}{\ell^3(1+\alpha)} & \dfrac{-6EI}{\ell^2(1+\alpha)} & \dfrac{12EI}{\ell^3(1+\alpha)} & \dfrac{6EI}{\ell^2(1+\alpha)} \\[2mm] \dfrac{6EI}{\ell^2(1+\alpha)} & \dfrac{(2-\alpha)EI}{\ell(1+\alpha)} & \dfrac{6EI}{\ell^2(1+\alpha)} & \dfrac{(4+\alpha)EI}{\ell(1+\alpha)} \end{bmatrix} \begin{bmatrix} w_a \\ \theta_a \\ w_b \\ \theta_b \end{bmatrix}$$

$$\mathbf{v}^i = [\,w_a \quad \theta_a \quad w_b \quad \theta_b\,]$$

For Case **2** (or Case **1b** if $1/GA_s = 0$ in $\bar{\mathbf{p}}^i$)

$$\mathbf{p}^i = \begin{bmatrix} -(2F_w + \ell F_\theta)/\nabla \\[2mm] \left[F_w\dfrac{\ell}{EI} + 2F_\theta\left(\dfrac{\ell^2}{6EI} - \dfrac{1}{GA_s}\right)\right]EI/\nabla \\[2mm] -F_V + (2F_w + \ell F_\theta)/\nabla \\[2mm] -F_M + \left[F_w\dfrac{\ell}{EI} + 2F_\theta\left(\dfrac{\ell^2}{3EI} + \dfrac{1}{GA_s}\right)\right]EI/\nabla \end{bmatrix}$$

$$\nabla = 2\ell\left(\dfrac{\ell^2}{12EI} + \dfrac{1}{GA_s}\right)$$

$$\begin{bmatrix} F_w \\ F_\theta \\ F_V \\ F_M \end{bmatrix} =$$

$p_1\left(\dfrac{\ell^4}{24EI} - \dfrac{\ell^2}{2GA_s}\right)$	$p_a\left(\dfrac{\ell^4}{30EI} - \dfrac{\ell^2}{3GA_s}\right) + p_b\left(\dfrac{\ell^4}{120EI} - \dfrac{\ell^2}{6GA_s}\right)$
$\dfrac{-p_1\ell^3}{6EI}$	$-(3p_a + p_b)\dfrac{\ell^3}{24EI}$
$-p_1\ell$	$-(p_a + p_b)\dfrac{\ell}{2}$
$-\dfrac{p_1\ell^2}{2}$	$-(2p_a + p_b)\dfrac{\ell^2}{6}$

Square cross section with thermal loading ΔT

$W\left(\dfrac{\ell_1^3}{6EI} - \dfrac{\ell_1}{GA_s}\right)$	$\dfrac{C\ell_1^2}{2EI}$	$(3c_a + c_b)\dfrac{\ell^3}{24EI}$	0
$-\dfrac{W\ell_1^2}{2EI}$	$-\dfrac{C\ell_1}{EI}$	$-(2c_a + c_b)\dfrac{\ell^2}{6EI}$	$\dfrac{EI\alpha\,\Delta T}{h\ell}$
$-W$	0	0	1
$-W\ell_1$	$-C$	$-(c_a + c_b)\dfrac{\ell}{2}$	-1

TABLE 11-20 TRANSFER AND STIFFNESS MATRICES FOR BEAM ELEMENT WITH VARIABLE MOMENT OF INERTIA

Notation

$I(x)$ = moment of inertia at x

I_i, c_1, c_2 = constants that define the variable cross section

E = modulus of elasticity

ℓ = length of element

$\alpha_1 = c_1 - c_2\ell$

$\alpha_2 = \ln c_1 - \ln \alpha_1$

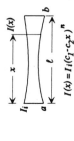

$$I(x) = I_i(c_1 - c_2 x)^n$$

Transfer Matrix (Sign Convention 1)

$$\mathbf{z}_b = \mathbf{U}^i \mathbf{z}_a$$

$$\mathbf{U}^i = \begin{bmatrix} 1 & -\ell & U_{wV} & U_{wM} & F_w \\ 0 & 1 & U_{\theta V} & U_{\theta M} & F_\theta \\ 0 & 0 & 1 & 0 & F_V \\ 0 & 0 & \ell & 1 & F_M \\ 0 & 0 & 0 & 0 & 1 \end{bmatrix} \begin{bmatrix} w_a \\ \theta_a \\ V_a \\ M_a \\ 1 \end{bmatrix}$$

Stiffness Matrix (Sign Convention 2)

$$\mathbf{p}^i = \mathbf{k}^i \mathbf{v}^i - \bar{\mathbf{p}}^i$$

$$\begin{bmatrix} V_a \\ M_a \\ V_b \\ M_b \end{bmatrix} = \begin{bmatrix} k_{11} & k_{12} & k_{13} & k_{14} \\ k_{21} & k_{22} & k_{23} & k_{24} \\ k_{31} & k_{32} & k_{33} & k_{34} \\ k_{41} & k_{42} & k_{43} & k_{44} \end{bmatrix} \begin{bmatrix} w_a \\ \theta_a \\ w_b \\ \theta_b \end{bmatrix} - \bar{\mathbf{p}}^i$$

$$\mathbf{p}^i = \mathbf{k}^i \mathbf{v}^i - \bar{\mathbf{p}}^i$$

$$\bar{\mathbf{p}}^i = \begin{bmatrix} k_{13}F_w + k_{14}F_\theta \\ k_{23}F_w + k_{24}F_\theta \\ -F_V + k_{33}F_w + k_{34}F_\theta \\ -F_M + k_{43}F_w + k_{44}F_\theta \end{bmatrix}$$

$k_{11} = U_{\theta M}/\Delta$ $\Delta = U_{wV}U_{\theta M} - U_{wM}U_{\theta V}$ $k_{13} = -k_{11}$ $k_{14} = U_{wM}/\Delta$

$k_{21} = -U_{\theta V}/\Delta$ $k_{22} = (\ell U_{\theta V} + U_{wV})/\Delta$ $k_{23} = -k_{21}$ $k_{24} = -U_{wV}/\Delta$

$k_{33} = -k_{13}$ $k_{34} = -k_{13}$ $k_{44} = (U_{wV} - \ell U_{wM})/\Delta$ $k_{ij} = k_{ji}(i, j = 1, 2, 3, 4)$

TABLE 11-20   **Element Matrices for Beams with Variable Inertia**   **578**

TABLE 11-20 (continued) TRANSFER AND STIFFNESS MATRICES FOR BEAM ELEMENT WITH VARIABLE MOMENT OF INERTIA

	$n = 1$	$n = 2$	$n = 3$	$n > 3$
U_{wM}	$\dfrac{\alpha_1\alpha_2 - c_2\ell}{EI_ic_2^2}$	$\dfrac{c_2\ell - c_1\alpha_2}{EI_ic_1c_2^2}$	$-\dfrac{\ell^2}{2EI_ic_1c_2^2\alpha_1}$	$\dfrac{\alpha_1^{n-2}[\alpha_1 + (n-1)c_2\ell] - c_1^{n-1}}{(n-1)(n-2)EI_ic_1^{n-1}c_2^2\alpha_1^{n-2}}$
U_{wV}	$\dfrac{c_2^2\ell^2 + 2c_1\alpha_1 - 2c_1c_2\ell}{2EI_ic_2^3}$	$\dfrac{2c_2\ell - (c_1 + \alpha_1)\alpha_2}{EI_ic_2^3}$	$\dfrac{c_2^2\ell^2 - 2c_1c_2\ell + 2c_1\alpha_1\alpha_2}{2EI_ic_1c_2^3\alpha_1}$	$\dfrac{c_1^{n-2}[2c_1 - (n-1)c_2\ell] - \alpha_1^{n-2}[2c_1 + (n-3)c_2\ell]}{(n-1)(n-2)(n-3)EI_ic_1^{n-2}c_2^3\alpha_1^{n-2}}$
$U_{\theta M}$	$\dfrac{\alpha_2}{EI_ic_2}$	$\dfrac{\ell}{EI_ic_2\alpha_1}$	$\dfrac{2c_1\ell - c_2\ell^2}{2EI_ic_1^2\alpha_1^2}$	$\dfrac{c_1^{n-1} - \alpha_1^{n-1}}{(n-1)EI_ic_1^{n-1}c_2\alpha_1^{n-1}}$
$U_{\theta V}$	$\dfrac{c_1\alpha_2 - c_2\ell}{EI_ic_2^2}$	$\dfrac{c_2\ell - \alpha_1\alpha_2}{EI_ic_2^2\alpha_1}$	$\dfrac{\ell^2}{2EI_ic_1\alpha_1^2}$	$\dfrac{\alpha_1^{n-1} - (n-1)c_1^{n-2}\alpha_1 + (n-2)c_1^{n-1}}{(n-1)(n-2)EI_ic_1^{n-2}c_2^2\alpha_1^{n-1}}$

Loading Functions:

$$F_M = -\tfrac{1}{6}\ell^2(2p_a + p_b), \qquad F_V = -\tfrac{1}{2}\ell(p_a + p_b)$$

	F_w
$n = 1$	$\dfrac{1}{2EI_ic_2^5}\left[p_a\left(-\dfrac{1}{4}c_2^4\ell^3 + \dfrac{8}{9}c_1c_2^3\ell^2 - \dfrac{1}{6}c_1^2c_2^2\ell - \dfrac{1}{3}c_1^3c_2 + c_1c_2\alpha_1^2\alpha_2 + \dfrac{c_1\alpha_1^3\alpha_2}{3\ell} \right) \right.$ $\left. + p_b\left(-\dfrac{1}{12}c_2^4\ell^3 + \dfrac{11}{18}c_1c_2^3\ell^2 - \dfrac{5}{6}c_1^2c_2^2\ell + \dfrac{1}{3}c_1^3c_2 - \dfrac{c_1\alpha_1^3\alpha_2}{3\ell} \right) \right]$
$n = 2$	$\dfrac{1}{EI_ic_2^5}\left\{ p_a\left[-\dfrac{7}{9}c_2^3\ell^2 + \dfrac{1}{3}c_1c_2^2\ell + \dfrac{2}{3}c_1^2c_2 - (c_1 + \tfrac{1}{2}\alpha_1)c_2\alpha_1\alpha_2 - \dfrac{1}{6\ell}(3c_1 + \alpha_1)\alpha_1^2\alpha_2 \right] \right.$ $\left. + p_b\left[-\dfrac{17}{36}c_2^3\ell^2 + \dfrac{7}{6}c_1c_2^2\ell - \dfrac{2}{3}c_1^2c_2 + \dfrac{(3c_1 + \alpha_1)\alpha_1^2\alpha_2}{6\ell} \right] \right\}$
$n = 3$	$\dfrac{1}{2EI_ic_1c_2^5}\left\{ p_a\left[\dfrac{1}{3}c_2^3\ell^2 - c_1c_2^2\ell - 2c_1^2c_2 + (c_1^2 + 2c_1\alpha_1)c_2\alpha_2 + \dfrac{c_1(c_1 + \alpha_1)\alpha_1\alpha_2}{\ell} \right] \right.$ $\left. + p_b\left[\dfrac{1}{6}c_2^3\ell^2 - 2c_1c_2^2\ell + 2c_1^2c_2 - \dfrac{c_1}{\ell}(c_1 + \alpha_1)\alpha_1\alpha_2 \right] \right\}$
$n = 4$	$\dfrac{1}{6EI_ic_1^2c_2^5}\left\{ p_a\left[\dfrac{1}{3}c_2^3\ell^2 + \left(c_1 + \dfrac{1}{\alpha_1}\right)c_2^2\ell + 4c_1^2c_2 - 3c_1^2c_2\alpha_2 - \dfrac{(c_1 + 3\alpha_1)c_1^2\alpha_2}{\ell} \right] \right.$ $\left. + p_b\left[\dfrac{1}{6}c_2^3\ell^2 + c_1c_2^2\ell - 4c_1^2c_2 + \dfrac{(c_1 + 3\alpha_1)c_1^2\alpha_2}{\ell} \right] \right\}$
$n = 5$	$\dfrac{p_a}{(n-1)(n-2)(n-3)EI_ic_1^{n-2}c_2^4}\left[\dfrac{n-3}{2}c_2^2\ell^2 + 2c_1c_2\ell + \dfrac{c_1^2(c_1^{n-3} - \alpha_1^{n-3})}{\alpha_1^{n-3}} \right.$ $\left. - \dfrac{(n-1)(c_1^{n-4} - \alpha_1^{n-4})c_1^2}{(n-4)\alpha_1^{n-4}} \right]$ $+ \dfrac{p_b - p_a}{24EI_ic_1^3c_2^5}\left[\dfrac{1}{3}c_2^3\ell^2 + c_1c_2^2\ell + \dfrac{(c_1 + 3\alpha_1)c_1^2c_2}{\alpha_1} - \dfrac{4c_1^3\alpha_2}{\ell} \right]$
$n \ge 6$	This entry consists of two terms, one multiplied by p_a and one by $p_b - p_a$. For the p_a term, use the F_w entry above in the $n = 5$ row. The $p_b - p_a$ term follows. $\dfrac{p_b - p_a}{(n-1)(n-2)(n-3)EI_ic_1^{n-2}c_2^5}\left[\dfrac{n-3}{6}c_2^3\ell^2 + c_1c_2^2\ell - \dfrac{n-1}{n-5}\dfrac{(c_1^{n-5} - \alpha_1^{n-5})c_1^3}{\ell\alpha_1^{n-5}} \right.$ $\left. + \dfrac{(n-3)c_1 + (n-1)\alpha_1}{n-4}\dfrac{(c_1^{n-4} - \alpha_1^{n-4})c_1^2}{\ell\alpha_1^{n-4}} - \dfrac{(c_1^{n-3} - \alpha_1^{n-3})c_1^2}{\ell\alpha_1^{n-3}} \right]$

TABLE 11-20 Element Matrices for Beams with Variable Inertia

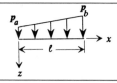

$$F_\theta$$

$$\frac{1}{2EI_ic_2^4}\left[p_a\left(\frac{2}{3}c_2^3\ell^2 - \frac{1}{2}c_1c_2^2\ell + 2c_1c_2\alpha_1\alpha_2 - c_1^2c_2 + \frac{c_1\alpha_1^2\alpha_2}{\ell}\right)\right.$$
$$\left. +p_b\left(\frac{1}{3}c_2^3\ell^2 - \frac{3}{2}c_1c_2^2\ell + c_1^2c_2 - \frac{c_1\alpha_1^2\alpha_2}{\ell}\right)\right]$$

$$\frac{1}{2EI_ic_2^4}\left\{p_a\left[\frac{3}{2}c_2^2\ell - 2(c_1 + \alpha_1)c_2\alpha_2 + 3c_1c_2 - \frac{(2c_1 + \alpha_1)\alpha_1\alpha_2}{\ell}\right]\right.$$
$$\left. +p_b\left[\frac{5}{2}c_2^2\ell - 3c_1c_2 + \frac{(2c_1 + \alpha_1)\alpha_1\alpha_2}{\ell}\right]\right\}$$

$$\frac{1}{2EI_ic_1c_2^4\alpha_1}\left\{p_a\left[\frac{1}{2}c_2^3\ell^2 + \frac{3}{2}c_1c_2^2\ell - 3c_1^2c_2 + 2c_1c_2\alpha_1\alpha_2 + \frac{c_1(c_1 + 2\alpha_1)\alpha_1\alpha_2}{\ell}\right]\right.$$
$$\left. +p_b\left[\frac{1}{2}c_2^3\ell^2 - \frac{7}{2}c_1c_2^2\ell + 3c_1^2c_2 - \frac{c_1(c_1 + 2\alpha_1)\alpha_1\alpha_2}{\ell}\right]\right\}$$

$$\frac{p_a}{(n-1)(n-2)EI_ic_1^{n-2}c_2^3}\left[-c_2\ell + \frac{(n-1)(c_1^{n-3} - \alpha_1^{n-3})c_1}{(n-3)\alpha_1^{n-3}} - \frac{(c_1^{n-2} - \alpha_1^{n-2})c_1}{\alpha_1^{n-2}}\right]$$
$$+\frac{p_b - p_a}{6EI_ic_1^2c_2^4}\left(-\frac{1}{2}c_2^2\ell - 3c_1c_2 - \frac{c_1c_2^2\ell}{\alpha_1} + \frac{3c_1^2\alpha_2}{\ell}\right)$$

This entry consists of two terms, one multiplied by p_a and one by $p_b - p_a$.
For the p_a term use the F_θ entry above in the $n = 4$ row. The $p_b - p_a$ term follows.

$$\frac{p_b - p_a}{(n-1)(n-2)EI_ic_1^{n-2}c_2^4}\left[-\frac{1}{2}c_2^2\ell + \frac{(n-1)(c_1^{n-4} - \alpha_1^{n-4})c_1^2}{\ell(n-4)\alpha_1^{n-4}}\right.$$
$$\left. -\frac{(n-2)c_1 + (n-1)\alpha_1}{n-3}\frac{(c_1^{n-3} - \alpha_1^{n-3})c_1}{\ell\alpha_1^{n-3}} + \frac{(c_1^{n-2} - \alpha_1^{n-2})c_1}{\ell\alpha_1^{n-3}}\right]$$

For the p_a term use the F_θ entry above in the $n = 4$ row.
For the $p_b - p_a$ term use the F_θ entry above in the $n = 5$ row.

Element Matrices for Beams with Variable Inertia TABLE 11-20

TABLE 11-21 POINT MATRICES FOR CONCENTRATED OCCURRENCES[a]

Case	Transfer Matrix (Sign Convention 1)	Stiffness Matrix (Sign Convention 2)
1. Transverse force and moment		Traditionally, these applied loads are implemented as nodal conditions.
2. Abrupt change in slope		
3. Abrupt change in beam axis		
4. Linear hinge		
5. Rotary hinge		

$$U_i = \left[\begin{array}{cccc:c} 1 & 0 & \dfrac{1}{k_2} & 0 & w_1 \\[2mm] 0 & 1 & 0 & \dfrac{1}{k_2^*} & -\alpha \\[2mm] \hdashline 0 & 0 & 1 & 0 & -W \\ 0 & 0 & 0 & 1 & -C \\ \hdashline 0 & 0 & 0 & 0 & 1 \end{array}\right]$$

k_2 – force / length

k_2^* – force - length /length

TABLE 11-21   Point Matrices for Concentrated Occurrences   582

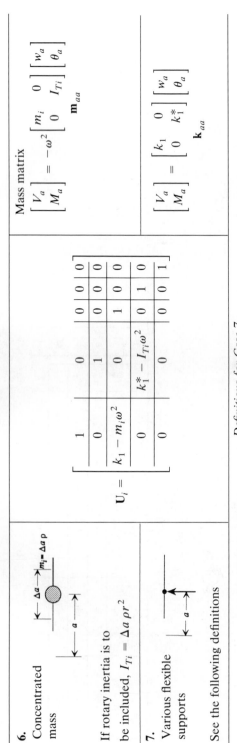

6.

Concentrated mass

If rotary inertia is to be included, $I_{Ti} = \Delta a\,\rho r^2$

$$U_i = \begin{bmatrix} 1 & 0 & 0 & 0 & 0 \\ 0 & 1 & 0 & 0 & 0 \\ k_1 - m_i\omega^2 & 0 & 1 & 0 & 0 \\ 0 & k_1^* - I_{Ti}\omega^2 & 0 & 1 & 0 \\ 0 & 0 & 0 & 0 & 1 \end{bmatrix}$$

Mass matrix

$$\begin{bmatrix} V_a \\ M_a \end{bmatrix} = -\omega^2 \underbrace{\begin{bmatrix} m_i & 0 \\ 0 & I_{Ti} \end{bmatrix}}_{\mathbf{m}_{aa}} \begin{bmatrix} w_a \\ \theta_a \end{bmatrix}$$

7.

Various flexible supports

$$\begin{bmatrix} V_a \\ M_a \end{bmatrix} = \underbrace{\begin{bmatrix} k_1 & 0 \\ 0 & k_1^* \end{bmatrix}}_{\mathbf{k}_{aa}} \begin{bmatrix} w_a \\ \theta_a \end{bmatrix}$$

See the following definitions

Definitions for Case 7

	k_1	k_1^*	k_3–k_4	E,I fixed	E,I pinned	E,I	E,I	N_B, M_B	k_3, m_i	k_3, m_i, k_4
k_1 (F/L)	k_1	0	$\dfrac{k_3 k_4}{k_3 + k_4}$	$\dfrac{EA}{\ell_b}$	$\dfrac{EA}{\ell_b}$	$\dfrac{EA}{\ell_b}$	$\dfrac{EA}{\ell_b}$	$\dfrac{u_B(a_i)}{N_B(a_i)}$	$-\dfrac{k_3 m_i\omega^2}{k_3 - m_i\omega^2}$	$\dfrac{k_3(k_4 - m_i\omega^2)}{k_3 + k_4 - m_i\omega^2}$
k_1^* (FL/rad)	0	k_1^*	0	$\dfrac{4EI}{\ell_b}$	$\dfrac{3EI}{\ell_b}$	0	0	$\dfrac{\theta_B(a_i)}{M_B(a_i)}$	0	0

[a]Units: k_1, k_2, k_3, k_4 are force/length (F/L).

TABLE 11-22 TRANSFER AND STIFFNESS MATRICES FOR A GENERAL BEAM SEGMENT

Notation

$\lambda = (k - \rho\omega^2)/EI$

$\xi = EI/GA_s$

$k^* = $ rotary foundation modulus

$\ell = $ element length

$G = $ shear modulus of elasticity

$A_s = $ equivalent shear area, A/α_s

$\eta = (k - \rho\omega^2)/GA_s$

$k = $ elastic foundation modulus

$\omega = $ natural frequency

$I = $ moment of inertia

$r_y = $ radius of gyration

$\alpha_s = $ shear correction factor (Table 2-4)

$\zeta = (P - k^* + \rho r_y^2\omega^2)/EI$

$\rho = $ mass per unit length

$E = $ elastic modulus

$P = $ axial force, compressive; replace by $-P$ for tensile axial force

To use these matrices, follow the steps:

1. Calculate the three parameters λ, ζ, η. If shear deformation is not to be considered, set $1/GA_s = 0$.
2. Compare the magnitude of these parameters and look up the appropriate e_i using the definitions for e_i given below.
3. Substitute these expressions in the matrices below.

Transfer Matrix (Sign Convention 1)

$z_b = U^i z_a$

$e_1 + \zeta e_3$	$-e_2$	$-e_4/EI$	$-e_3/EI$	F_w	w_a
λe_1	$e_1 - \eta e_3$	$+(e_2 + \zeta e_4)/GA_s$	$(e_2 - \eta e_4)/EI$	F_θ	θ_a
$\lambda EI(e_2 + \zeta e_4)$	$-\lambda EI e_3$	e_3/EI	$-\lambda e_4$	F_V	V_a
$\lambda EI e_3$	$EI(e_0 - \eta e_2)$	$e_1 + \zeta e_3$	e_2	$e_1 - \eta e_3$	F_M	M_a
0	0	0	0	1	1

U^i z_a

Stiffness Matrix (Sign Convention 2)

$p^i = k^i v^i - \bar{p}^i$

$$\begin{bmatrix} V_a \\ M_a \\ V_b \\ M_b \end{bmatrix} = \begin{bmatrix} k_{11} & k_{12} & k_{13} & k_{14} \\ k_{21} & k_{22} & k_{23} & k_{24} \\ k_{31} & k_{32} & k_{33} & k_{34} \\ k_{41} & k_{42} & k_{43} & k_{44} \end{bmatrix} \begin{bmatrix} w_a \\ \theta_a \\ w_b \\ \theta_b \end{bmatrix} - \bar{p}^i$$

p^i = k^i v^i $-\bar{p}^i$

TABLE 11-22 Matrices for General Beam   584

$$F_w = [p_a(e_5 - e_6/\ell) + p_be_6/\ell + c_a(e_4 - e_5/\ell) + c_be_5/\ell - M_{Ta}(e_3 - e_4/\ell)$$
$$- M_{Tb}e_4/\ell]/EI - [p_a(e_3 + \zeta e_5) - (e_4 + \zeta e_6)/\ell) + p_b(e_4 + \zeta e_6)/\ell]/GA_s$$

$$F_\theta = [p_a(-e_4 + e_5/\ell) - p_be_5/\ell + c_a(-e_3 + \eta e_5 + (e_4 - \eta e_6)/\ell)$$
$$- c_b(e_4 - \eta e_6)/\ell + M_{Ta}((e_2 - \eta e_4) - (e_3 - \eta e_5)/\ell) + M_{Tb}(e_3 - \eta e_5)/\ell]/EI$$

$$F_V = p_a(-(e_2 + \zeta e_4) + (e_3 + \zeta e_5)/\ell) - p_b(e_3 + \zeta e_5)/\ell$$
$$+ \lambda[c_a(e_5 - e_6/\ell) + c_be_6/\ell + M_{Ta}(-e_4 + c_5/\ell) - M_{Tb}e_5/\ell]$$

$$F_M = p_a(-e_3 + e_4/\ell) - p_be_4/\ell + c_a[(-e_2 + e_3/\ell) + \eta(e_4 - e_5/\ell)]$$
$$- c_b(e_3 - \eta e_5)/\ell + M_{Ta}[(e_1 - 1 - e_2/\ell) + \eta(-e_3 + e_4/\ell)]$$
$$+ M_{Tb}(e_2 - \ell - \eta e_4)/\ell$$

$$k_{11} = [(e_2 - \eta e_4)(e_1 + \zeta e_3) + \lambda e_3 e_4]EI/\Delta$$
$$k_{12} = [e_3(e_1 - \eta e_3) - e_2(e_2 - \eta e_4)]EI/\Delta$$
$$k_{13} = -(e_2 - \eta e_4)EI/\Delta$$
$$k_{14} = -e_3EI/\Delta$$
$$k_{21} = k_{12}$$
$$k_{22} = (-(e_1 - \eta e_3)(e_4 - \xi(e_2 + \zeta e_4)] + e_2 e_3)EI/\Delta$$
$$k_{23} = e_3EI/\Delta = -k_{14}$$
$$k_{24} = [e_4 - \xi(e_2 + \zeta e_4)]EI/\Delta$$
$$k_{31} = k_{13}, \quad k_{41} = k_{14}, \quad k_{42} = k_{24},$$
$$k_{32} = k_{23}, \quad k_{43} = k_{34}$$
$$k_{33} = [(e_1 + \zeta e_3)(e_2 - \eta e_4) + \lambda e_3 e_4]EI/\Delta = k_{11}$$
$$k_{34} = [(e_1 + \zeta e_3)e_3 + \lambda e_4[e_4 - \xi(e_2 + \zeta e_4)]]EI/\Delta$$
$$k_{44} = [e_2 e_3 - (e_1 - \eta e_3)(e_4 - \xi(e_2 + \zeta e_4))]EI/\Delta = k_{22}$$
$$\Delta = e_3^2 - (e_2 - \eta e_4)(e_4 - \xi(e_2 + \zeta e_4)]$$

$$\overline{\mathbf{p}}^i = \begin{bmatrix} V_a^0 \\ M_a^0 \\ V_b^0 \\ M_b^0 \end{bmatrix}$$

$$V_a^0 = -[(e_2 - \eta e_4)F_w + e_3F_\theta]EI/\Delta$$
$$M_a^0 = \{e_3F_w + [e_4 - \xi(e_2 + \zeta e_4)]F_\theta\}EI/\Delta$$
$$V_b^0 = -F_V + \{[(e_1 + \zeta e_3)(e_2 - \eta e_4) + \lambda e_3 e_4]F_w$$
$$+ [(e_1 + \zeta e_3)e_3$$
$$+ \lambda e_4[e_4 - \xi(e_2 + \zeta e_4)]]F_\theta\}EI/\Delta$$
$$M_b^0 = -F_M + \{[(e_1 + \zeta e_3)e_3 + \zeta e_4[e_4 - \xi(e_2 + \zeta e_4)]]F_w$$
$$+ [e_2 e_3 - (e_1 - \eta e_3)(e_4 - \xi(e_2 + \zeta e_4)]]F_\theta\}EI/\Delta$$

Matrices for General Beam TABLE 11-22

TABLE 11-22 (continued)   TRANSFER AND STIFFNESS MATRICES FOR A GENERAL BEAM SEGMENT

Definitions for e_i ($i = 0, 1, 2, \ldots 6$)

	1. $\lambda < 0$	**2.** $\lambda = 0, \lambda - \zeta\eta = 0$; $\zeta = \eta = 0$	**3.** $\eta = 0, \zeta \neq 0$	**4.** $\lambda - \zeta\eta = \frac{1}{4}(\zeta - \eta)^2$	**5.** $\lambda > 0, \lambda - \zeta\eta > 0$; $\lambda - \zeta\eta < \frac{1}{4}(\zeta - \eta)^2$, $\zeta - \eta \neq 0$	**6.** $\lambda - \zeta\eta > \frac{1}{4}(\zeta - \eta)^2$
e_0	$\frac{1}{g}(d^3C - q^3D)$	0	$-\zeta B$	$-\frac{\zeta - \eta}{4}(3C + A\ell)$	$-\frac{1}{g}(q^3D - d^3C)$	$-(\lambda - \zeta\eta)e_4 - (\zeta - \eta)e_2$
e_1	$\frac{1}{g}(d^2A + q^2B)$	1	A	$\frac{1}{2}(2A - B\ell)$	$\frac{p}{g}(q^2B - d^2A)$	$AB - \frac{q^2 - d^2}{2\,dq}CD$
e_2	$\frac{1}{g}(dC + qD)$	ℓ	B	$\frac{1}{2}(C + A\ell)$	$\frac{p}{g}(qD - dC)$	$\frac{1}{2\,dq}(dAD + qBC)$
e_3	$\frac{1}{g}(A - B)$	$\frac{\ell^2}{2}$	$\frac{1}{\zeta}(1 - A)$	$\frac{C\ell}{2}$	$\frac{1}{g}(A - B)$	$\frac{1}{2\,dq}CD$
e_4	$\frac{1}{g}\left(\frac{C}{d} - \frac{D}{q}\right)$	$\frac{\ell^3}{6}$	$\frac{1}{\zeta}(\ell - B)$	$\frac{1}{(\zeta - \eta)}(C - A\ell)$	$\frac{1}{g}\left(\frac{C}{d} - \frac{D}{q}\right)$	$\frac{1}{2(d^2 + q^2)}\left(\frac{AD}{q} - \frac{BC}{d}\right)$
e_5	$\frac{1}{g}\left(\frac{A}{d^2} + \frac{B}{q^2}\right) - \frac{1}{d^2q^2}$	$\frac{\ell^4}{24}$	$\frac{1}{\zeta}\left(\frac{\ell^2}{2} - e_3\right)$	$\frac{2}{(\zeta - \eta)^2} \times (-2A - B\ell + 2)$	$\frac{p}{g}\left(\frac{B}{q^2} - \frac{A}{d^2}\right) + \frac{1}{d^2q^2}$	$\frac{1 - e_1}{\lambda - \zeta\eta} - \frac{\zeta - \eta}{\lambda - \zeta\eta}e_3$
e_6	$\frac{1}{g}\left(\frac{C}{d^3} + \frac{D}{q^3}\right) - \frac{\ell}{d^2q^2}$	$\frac{\ell^5}{120}$	$\frac{1}{\zeta}\left(\frac{\ell^3}{6} - e_4\right)$	$\frac{2}{(\zeta - \eta)^2} \times (-3C + A\ell + 2\ell)$	$\frac{p}{g}\left(\frac{D}{q^3} - \frac{C}{d^3}\right) + \frac{\ell}{d^2q^2}$	$\frac{\ell - e_2}{\lambda - \zeta\eta} - \frac{\zeta - \eta}{\lambda - \zeta\eta}e_4$

TABLE 11-22 Matrices for General Beam   586

Definitions for A, B, C, D, g, d, q

$\lambda < 0$	$\lambda = 0, \lambda - \zeta\eta = 0$	$\lambda - \zeta\eta = \frac{1}{4}(\zeta-\eta)^2$	$\lambda > 0, \lambda - \zeta\eta > 0$	
			$\lambda - \zeta\eta < \frac{1}{4}(\zeta-\eta)^2, \zeta - \eta \neq 0$	$\lambda - \zeta\eta > \frac{1}{4}(\zeta-\eta)^2$
1. $A = \cosh d\ell$ $B = \cos q\ell$ $C = \sinh d\ell$ $D = \sin q\ell$ $g = d^2 + q^2$	**2.** $\zeta > 0: \alpha^2 = \zeta$ $A = \cos \alpha\ell$ $B = (\sin \alpha\ell)/\alpha$	**4.** $\zeta - \eta > 0: \beta^2 = \frac{1}{2}(\zeta-\eta)$ $A = \cos \beta\ell, B = \beta \sin \beta\ell$ $C = (\sin \beta\ell)/\beta$	**6.** $\zeta - \eta > 0: g = q^2 - d^2, p = 1$ $A = \cos d\ell, B = \cos q\ell$ $C = \sin d\ell, D = \sin q\ell$ $d^2 = \frac{1}{2}(\zeta-\eta) - \sqrt{\frac{1}{4}(\zeta+\eta)^2 - \lambda}$ $q^2 = \frac{1}{2}(\zeta-\eta) + \sqrt{\frac{1}{4}(\zeta+\eta)^2 - \lambda}$	**8.** $A = \cosh d\ell, B = \cos q\ell$ $C = \sinh d\ell, D = \sin q\ell$ $d^2 = \frac{1}{2}\sqrt{\lambda - \zeta\eta} - \frac{1}{4}(\zeta - \eta)$ $q^2 = \frac{1}{2}\sqrt{\lambda - \zeta\eta} + \frac{1}{4}(\zeta - \eta)$
$d^2 = \sqrt{\beta^4 + \frac{1}{4}(\zeta+\eta)^2}$ $\quad -\frac{1}{2}(\zeta-\eta)$ $q^2 = \sqrt{\beta^4 + \frac{1}{4}(\zeta+\eta)^2}$ $\quad +\frac{1}{2}(\zeta-\eta)$ $\beta^4 = -\lambda$	**3.** $\zeta < 0: \alpha^2 = -\zeta$ $A = \cosh \alpha\ell$ $B = (\sinh \alpha\ell)/\alpha$	**5.** $\zeta - \eta < 0: \beta^2 = -\frac{1}{2}(\zeta-\eta)$ $A = \cosh \beta\ell, B = -\beta \sinh \beta\ell$ $C = (\sinh \beta\ell)/\beta$	**7.** $\zeta - \eta < 0: g = d^2 - q^2, p = -1$ $A = \cosh d\ell, B = \cosh q\ell$ $C = \sinh d\ell, D = \sinh q\ell$ $d^2 = -\frac{1}{2}(\zeta-\eta) + \sqrt{\frac{1}{4}(\zeta+\eta)^2 - \lambda}$ $q^2 = -\frac{1}{2}(\zeta-\eta) - \sqrt{\frac{1}{4}(\zeta+\eta)^2 - \lambda}$	

Matrices for General Beam TABLE 11-22

TABLE 11-23   GEOMETRIC STIFFNESS MATRIX (CONSISTENT)

Notation

ℓ = length of element
$\mathbf{k}_G^i$ = geometric stiffness matrix for ith element

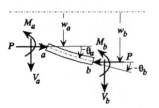

Sign Convention 2

$$
\begin{bmatrix}
6/(5\ell) & -1/10 & -6/(5\ell) & -1/10 \\
-1/10 & 2\ell/15 & 1/10 & -\ell/30 \\
-6/(5\ell) & 1/10 & 6/(5\ell) & 1/10 \\
-1/10 & -1\ell/30 & 1/10 & 2\ell/15
\end{bmatrix}
\begin{bmatrix}
w_a \\ \theta_a \\ w_b \\ \theta_b
\end{bmatrix}
$$
$$\qquad\qquad\qquad \mathbf{k}_G^i \qquad\qquad\qquad\quad \mathbf{v}^i$$

TABLE 11-23   **Geometric Stiffness Matrix (Consistent)**   588

TABLE 11-24 LUMPED MASS MATRICES

Notation

m_a, m_b = lumped masses at points a and b
I_{Ta}, I_{Tb} = lumped rotary inertia of mass at points a and b
ρ = mass per unit length
ℓ = length of element
r_y = radius of gyration about y axis

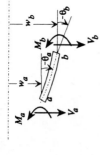

Sign Convention 2

(a) Mass Lumped at One Point

$$\begin{bmatrix} V_a \\ M_a \end{bmatrix} = -\omega^2 \begin{bmatrix} m_a & 0 \\ 0 & I_{Ta} \end{bmatrix} \begin{bmatrix} w_a \\ \theta_a \end{bmatrix}$$

$m_a = \Delta a\, \rho$

If rotary inertia is to be included, $I_{Ta} = \Delta a\, \rho\, r_y^2$

(b) Mass Lumped at Two End Points

$$\begin{bmatrix} m_a & 0 & 0 & 0 \\ 0 & I_{Ta} & 0 & 0 \\ 0 & 0 & m_b & 0 \\ 0 & 0 & 0 & I_{Tb} \end{bmatrix} \begin{bmatrix} w_a \\ \theta_a \\ w_b \\ \theta_b \end{bmatrix}$$

$\mathbf{m}^i$ $\qquad$ $\mathbf{v}^i$

Set $I_{Ta} = I_{Tb} = 0$ if the rotary inertia
of the mass is to be ignored

TABLE 11-24 (continued) LUMPED MASS MATRICES

Inertial Properties of Lumped Mass Matrix

Mass Distribution	m_a	m_b	I_{Ta}	I_{Tb}
1. Uniformly distributed ρ	$\frac{1}{2}\rho\ell$	$\frac{1}{2}\rho\ell$	$\frac{1}{24}\rho\ell^3 + \frac{1}{2}\left(\rho\ell r_y^2\right)$	$\frac{1}{24}\rho\ell^3 + \frac{1}{2}\rho\ell r_y^2$
2. Linearly distributed $\rho = x\rho_0/\ell$	$\frac{1}{6}\rho_0\ell$	$\frac{1}{3}\rho_0\ell$	$\frac{4}{81}\rho_0\ell^3 + \frac{1}{6}\rho_0\ell r_y^2$	$\frac{1}{108}\rho_0\ell^3 + \frac{1}{3}\rho_0\ell r_y^2$
3. Arbitrarily distributed $\rho = \rho(x)^a$	$\dfrac{\ell - x_0}{\ell}\int_0^\ell \rho(x)\,dx$	$\dfrac{x_0}{\ell}\int_0^\ell \rho(x)\,dx$	$I_{Ta} = \int_0^{x_0} x^2\rho(x)\,dx + r_y^2 m_a$	$I_{Tb} = \int_{x_0}^\ell (\ell - x)^2\rho(x)\,dx + r_y^2 m_b$

Note: For arbitrary mass distribution, the mass center can be calculated by

$$x_0 = \frac{\int_0^\ell x\rho\,dx}{\int_0^\ell \rho\,dx}$$

TABLE 11-24   Lumped Mass Matrices   590

TABLE 11-25 CONSISTENT MASS MATRIX AND GENERAL MASS MATRIX

Notation

ρ = mass per unit length
ℓ = length of element
r_y = radius of gyration about y axis

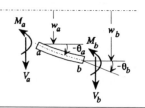

Sign Convention 2

(a) Without Rotary Inertia

$$\mathbf{m}^i = \frac{\rho\ell}{420} \begin{bmatrix} \begin{matrix} w_a & \theta_a & w_b & \theta_b \end{matrix} \\ \begin{matrix} 156 & -22\ell & 54 & 13\ell \\ -22\ell & 4\ell^2 & -13\ell & -3\ell^2 \\ 54 & -13\ell & 156 & 22\ell \\ 13\ell & -3\ell^2 & 22\ell & 4\ell^2 \end{matrix} \end{bmatrix}$$

(b) With Rotary Inertia

$$+ \frac{\rho\ell}{30} \left(\frac{r_y}{\ell}\right)^2 \begin{bmatrix} \begin{matrix} w_a & \theta_a & w_b & \theta_b \end{matrix} \\ \begin{matrix} 36 & -3\ell & -36 & -3\ell \\ -3\ell & 4\ell^2 & 3\ell & -\ell^2 \\ -36 & 3\ell & 36 & 3\ell \\ -3\ell & -\ell^2 & 3\ell & 4\ell^2 \end{matrix} \end{bmatrix}$$

(c) General Mass Matrix

$$\mathbf{m}^i = \int_0^\ell \rho \mathbf{\Lambda} \mathbf{\Lambda}^T \, dx, \qquad \mathbf{\Lambda} = \begin{bmatrix} H_1 \\ H_2 \\ H_3 \\ H_4 \end{bmatrix}$$

where H_j are the components of the displacement shape function

$$w(x) = H_1 w_a + H_2 \theta_a + H_3 w_b + H_4 \theta_b.$$

If, for constant ρ, the static responses

$$H_1 = 1 - 3(x/\ell)^2 + 2(x/\ell)^3, \qquad H_2 = \ell\left[-x/\ell + 2(x/\ell)^2 - (x/\ell)^3\right],$$

$$H_3 = 3(x/\ell)^2 - 2(x/\ell)^3, \qquad H_4 = -(x^2/\ell)(x/\ell - 1)$$

are employed, $\mathbf{m}^i$ is the consistent mass matrix of case (a) with rotary inertia ignored.

If other shape functions $\mathbf{\Lambda}$ are used, other mass matrices will be found. In particular, more exact, e.g., frequency-dependent, shape functions will lead to more exact mass matrices.

Torsion and Extension of Bars

Formulas for the analysis and design of bars subjected to torsion or extension are given in this chapter. The bars undergoing torsion, strictly speaking, must have circular cross sections although the formulas can and are frequently utilized for other shapes. For thin-walled cross sections the formulas of Chapter 14 give more accurate answers than those of this chapter. Here the bars cannot be restrained against warping.

12.1 NOTATION

Torsion

The formulas of this chapter that apply for bars subjected to torsion utilize the following notation. As noted in subsequent sections, with a change in notation the formulas also apply for the extension of bars.

A	area of cross section (L^2)
$\overline{A}$	enclosed area of thin-walled section (L^2)
G	shear modulus of elasticity (F/L^2)

I_{pi}	polar mass moment of inertia of concentrated mass at point $i(ML^2)$; to calculate, use $I_{pi} = \Delta a\, \rho r_p^2$, where Δa is the length of shaft lumped at point i
I_x	polar moment of inertia, $r_p^2 A$
I_p	polar mass moment of inertia per unit length of bar (ML); for hollow circular section $I_p = \frac{1}{2}\rho(r_{outer}^2 + r_{inner}^2)$, where r is radius measured from bar axis, $I_p = \rho^* I_x = \rho r_p^2$
J	torsional constant (L^4); for circular cross section, J is polar moment of inertia I_x of cross-sectional area with respect to axis x of bar
k	torsional spring constant (FL)
k_t	elastic foundation modulus (FL/L)
L	length of bar (L)
ℓ	length of element (L)
m_x	distributed torque (FL/L)
m_{x1}	magnitude of distributed torque that is uniform in x direction (FL/L)
m_{xa}, m_{xb}	initial and final magnitudes, respectively, of linearly varying distributed torque (FL/L)
r_p	polar radius of gyration (L), i.e., radius of gyration of cross-sectional area with respect to axis x of bar
t	thickness, thin-walled section (L)
T	twisting moment, torque (FL)
T_1, T_i	applied torque, concentrated (FL)
q	shear flow (F/L)
ω	natural frequency (rad/T)
ϕ	angle of twist, rotation (rad)
ρ	mass per unit length of bar, $\rho^* = \rho/A$ (M/L)
ρ^*	mass per unit volume (M/L^3)
τ	torsional shear stress (F/L^2)

Extension

The torsion formulas apply as well for extension of bars if the following notation adjustments are abided by:

Extension	Torsion
u	ϕ
P	T
E	G
A	J
ρ	ρr_p^2
p_x	m_x
k_x	k_t
M_i	I_{pi}

A summary of the notation for the extension of bars follows:

A	cross-sectional area (L^2)
E	Young's modulus (F/L^2)
k	spring constant (F/L)
k_x	elastic foundation modulus (F/L^2)
M_i	concentrated mass (M)
P	axial force (F)
P_i	applied concentrated axial force (F)
p_x	applied distributed axial force (F/L)
p_{x1}	magnitude of applied uniform axial force in x direction (F/L)
p_{xa}, p_{xb}	initial and final magnitudes, respectively, of linearly varying distributed axial force (F/L)
T	change in temperature (degrees), i.e., temperature rise with respect to reference temperature
u	axial displacement (L)
α	coefficient of thermal expansion $(L/L \cdot \text{degree})$
ω	natural frequency (rad/T)
ρ	mass per unit length (M/L)
ρ^*	mass per unit volume (M/L^3)
σ, σ_x	axial stress (F/L^2)

12.2 SIGN CONVENTIONS

Positive displacements and forces are illustrated in the various tables of this chapter.

12.3 STRESSES

The tables of this chapter give the torque along a bar for torsion and the axial force for extensions. These variables can be placed in the formulas of this section to find the stresses.

Torsional Stresses The shear stress τ due to torsion takes the form

$$\tau = \frac{Tr}{J} \tag{12.1}$$

where r is the radial distance from the central longitudinal axis (x). This formula applies to either solid or hollow cross sections. For circular cross sections $J = I_x$, where I_x is the polar moment of inertia of the cross section about the central axis of a shaft. In this case, J is given by [Eq. (2.10)]

$$J = J_x = I_x = \int_A r^2 \, dA = \int_A (z^2 + y^2) \, dA = I_y + I_z \tag{12.2}$$

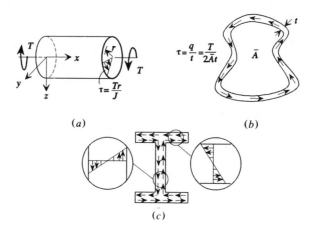

FIGURE 12-1 Stress patterns on various cross sections: (*a*) solid section; (*b*) thin-walled closed section; (*c*) thin-walled open section.

where A is the cross-sectional area. Figure 12-1*a* shows the cross-sectional distribution of this stress.

Hollow Thin-walled Cross Sections

The torsional stress in a hollow thin-walled shaft is given as in Chapter 3 by (Fig. 12-1*b*)

$$\tau = \frac{q}{t} = \frac{T}{2\overline{A}t} \tag{12.3}$$

where q is the *shear flow* and $\overline{A}$ is the area enclosed by the middle line of the wall. In deriving this formula, it is assumed that the shear stress is uniformly distributed through the thickness t.

Thin-walled Open Sections

The torsional stress in a thin-walled open section appears to vary linearly across the thickness of a cross section. It acts parallel to the edges as shown in Fig. 12-1*c*. The stresses on the two edges of the thin wall are equal in magnitude and opposite in direction. Along the edges, where the maximum stresses occur,

$$\tau = Tt/J \tag{12.4}$$

where t is the thickness at the location the stress is being calculated. For this thin-walled section, the torsional constant, which is a geometric property, is given by Eq. (2.19*b*):

$$J = \frac{\alpha}{3} \sum_{i=1}^{M} b_i t_i^3 \tag{12.5}$$

where α is a shape factor (case 10 in Table 12-1 or in Table 2-5) and M is the number of straight or curved segments of thickness t_i and width or height b_i that make up the section. Use $\alpha = 1$ if no information on α is available.

Specific Stress Formulas

Table 12-1 gives values of torsional shear stress on a variety of cross-sectional shapes. These values are provided at some important points. The stress distribution on the entire cross section can be obtained by using a computer program of the sort described in Chapter 15. The distribution of the stress can also be viewed from the *membrane analogy*, which reflects that the governing equation for the torsion problem closely resembles the equilibrium equation of a flat membrane in the shape of the cross section subjected to a lateral pressure. The membrane lies in the y, z plane. The shear stress in any direction is proportional to the slope of the membrane in the direction perpendicular to the shear stress, and the torque carried by the cross section is proportional to 2 times the volume under the membrane. From this analogy, it can be reasoned that the peak stresses are found at the boundaries of the thicker portions of the cross section. For thin-walled open cross sections, the shape of the membrane over the wall thickness is parabolic, so the maximum shear stress is on the boundary in the direction of the wall contour. On the middle line of the wall the shear stress is zero. At the reentrant corners, i.e., the interior corners where the thin-walled segments are connected, stress concentration may develop. See Table 6-1 for an indication of the stress that may occur at these corners.

The membrane analogy can also be used to compare the torsional constant of different cross sections. Since the torque that the cross section carries is proportional to 2 times the volume under the membrane, it is apparent that those cross-sectional shapes with smaller volumes under the membrane are stiffer. This follows because in terms of the effect on ϕ ($= TL/GJ$), a smaller volume, and hence a smaller T, corresponds to a larger J. Generally, for a cross section of a given area, the closer the shape is to being circular, the stiffer it is and the higher the value of J. Solid cross sections are stiffer than the thin-walled ones, and thin branches attached to the solid part have little affect on the torsional rigidity. For the same r [Eq. (12.1)], a higher J gives smaller shear stresses.

Extensional Stress The normal or axial stress in a bar subjected to extension or compression is given by Eq. (3.41):

$$\sigma = P/A = \sigma_x \tag{12.6}$$

This stress is distributed uniformly over the cross section.

12.4 SIMPLE BARS

Torsion

For a single uniform shaft element of length L, the most commonly used formula for the angle of twist is

$$\phi = \frac{TL}{GJ} \tag{12.7a}$$

where the torsional constant J can be taken from Table 2-5 for various shapes.

The governing differential equations for the torsion of a uniform bar are [12.1]

$$GJ\frac{d^2\phi}{dx^2} - k_t\phi = -m_x, \qquad GJ\frac{d\phi}{dx} = T \qquad (12.7b)$$

Extension

For a uniform bar the governing equations for a bar undergoing extension are [12.1]

$$AE\frac{d^2u}{dx^2} - k_x u = -p_x, \qquad AE\frac{du}{dx} = P + \alpha AE\,\Delta T \qquad (12.8)$$

These relations can be solved for the displacements and the forces as functions of the coordinate x.

Tabulated Formulas

The angle of twist and torque for the torsion of uniform members with various applied loadings and end conditions are provided in Table 12-2. The same table gives the axial displacement and force for the extension of a bar.

Table 12-2, part A, lists equations for the responses. The loading functions are taken from Table 12-2, part B, by adding the appropriate terms for each load applied to the member. The initial parameters are evaluated using the entry in Table 12-2, part C, for the appropriate end conditions of the member.

12.5 NATURAL FREQUENCIES

The natural frequencies ω_i, $i = 1, 2, \ldots,$ and mode shapes for torsion and extension of uniform bars are presented in Tables 12-3 and 12-4. Table 12-5 gives polar mass moments of inertia for lumped masses.

Example 12.1 Natural Frequencies of Torsional Vibration Find the first three torsional natural frequencies of a uniform shaft with a torsional spring at the left end and an unconstrained (free) right end.

From Table 12-3, the natural frequencies are given by

$$\omega_i = (\lambda_i/L)\sqrt{GJ/\rho^* I_x} \qquad \text{rad/s} \qquad (1)$$

where λ_i are the roots of the equation (case 4) $\lambda_i \tan \lambda_i = kL/GJ$. Let $E = 210$ GN/m^2, $\rho^* = 7850$ kg/m^3, $L = 1$ m, $\nu = 0.3$, and $k = 1$ GN-m/rad for a shaft of 0.1 m diameter:

$$G = \frac{E}{2(1 + \nu)} = \frac{2.1 \times 10^{11}}{2(1 + 0.3)} = 0.8077 \times 10^{11} \text{ N/m}^2 \qquad (2)$$

$$\frac{kL}{GJ} = \frac{10^9 \times 1 \times 32}{0.8077 \times 10^{11}(\pi)0.1^4} = 1.261 \times 10^3$$

$$\lambda_i \tan \lambda_i = 1.261 \times 10^3 \qquad (3)$$

The lowest values for λ_i for which (3) holds are

$$\lambda_1 = 1.56955, \qquad \lambda_2 = 4.70865, \qquad \lambda_3 = 7.84776$$

Finally, the natural frequencies of the first three modes are as follows:

Mode	Natural Frequencies	
	ω_i (rad/s)	f_i (Hz)
1	5034.60	801.28
2	15,103.81	2403.85
3	25,173.06	4006.42

12.6 GENERAL BARS

The formulas of Table 12-2 apply to uniform bars. For more general members, e.g., those with variable-section properties, it is advisable to use the displacement method or the transfer matrix procedure, which are explained technically at the end of this book (Appendices II and III).

Several transfer and stiffness matrices are tabulated in Tables 12-6 to 12-9. Mass matrices for use in a displacement method analysis are given in Tables 12-10 and 12-11.

The torsional responses are based on the governing equations [12.1]

$$\frac{\partial \phi}{\partial x} = \frac{T}{GJ} \qquad (12.9a)$$

$$\frac{\partial T}{\partial x} = k_t \phi + \rho r_p^2 \frac{\partial^2 \phi}{\partial t^2} - m_x(x, t) \qquad (12.9b)$$

For extension replace ϕ by u, T by P, GJ by EA, k_t by k_x, ρr_p^2 by ρ, and m_x by p_x.

Example 12.2 Torsional System with Branch Consider the gear-branched system of Fig. 12-2, where $m = 2$ is the speed ratio between gears 4 and 2.

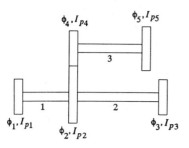

I_{pi}	kg · m²
I_{p1}	10
I_{p2}	24
I_{p3}	5
I_{p4}	10
I_{p5}	5

FIGURE 12-2 Branched torsional system.

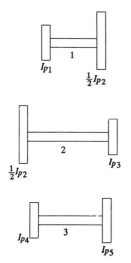

Element	$\dfrac{GJ}{\ell}$ (N · m)	$I_p \ell$ (kg · m²)
1	30000	2.4
2	50000	1.5
3	100000	1.2

FIGURE 12-3 Three elements composing system of Fig. 12-2.

Treat the gears as lumped masses. Discretize the system into three elements as shown in Fig. 12-3. Each element is composed of a bar with lumped masses attached at the ends. Hence, the mass matrices can be formed as the summation of cases 3 and 4 of Table 12-10. For gear 2, half of the mass is attached to element 1 and another half to element 2.

For element 1, the consistent mass matrix for the bar is (Table 12-10, case 4)

$$\mathbf{m}_{bar}^1 = \begin{bmatrix} \dfrac{2.4}{3} & \dfrac{2.4}{6} \\ \dfrac{2.4}{6} & \dfrac{2.4}{3} \end{bmatrix}$$

and the lumped mass matrix is (Table 12-10, case 3)

$$\mathbf{m}^1_{\text{lumped}} = \begin{bmatrix} 10 & 0 \\ 0 & 12 \end{bmatrix}$$

The mass matrix for the whole element is

$$\mathbf{m}^1 = \mathbf{m}^1_{\text{bar}} + \mathbf{m}^1_{\text{lumped}} = \begin{bmatrix} 10 + \dfrac{2.4}{3} & \dfrac{2.4}{6} \\[2ex] \dfrac{2.4}{6} & 12 + \dfrac{2.4}{3} \end{bmatrix}$$

Similarly, the other mass matrices are

$$\mathbf{m}^2 = \begin{bmatrix} 12 + \dfrac{1.5}{3} & \dfrac{1.5}{6} \\[2ex] \dfrac{1.5}{6} & 5 + \dfrac{1.5}{3} \end{bmatrix}, \qquad \mathbf{m}^3 = \begin{bmatrix} 10 + \dfrac{1.2}{3} & \dfrac{1.2}{6} \\[2ex] \dfrac{1.2}{6} & 5 + \dfrac{1.2}{3} \end{bmatrix}$$

The consistent stiffness matrices are (Table 12-6, case 1)

$$\mathbf{k}^1 = 10^4 \begin{bmatrix} 3 & -3 \\ -3 & 3 \end{bmatrix}, \qquad \mathbf{k}^2 = 10^4 \begin{bmatrix} 5 & -5 \\ -5 & 5 \end{bmatrix}, \qquad \mathbf{k}^3 = 10^4 \begin{bmatrix} 10 & -10 \\ -10 & 10 \end{bmatrix} \tag{1}$$

Note that the angles of twist ϕ_2 and ϕ_4 are related and so are the moments T_2 and T_4. For element 3,

$$\mathbf{p}^3 = [\mathbf{k}^3 - \omega^2 \mathbf{m}^3]\mathbf{v}^3 \tag{2}$$

in which

$$\mathbf{p}^3 = \begin{bmatrix} T_4 \\ T_5 \end{bmatrix} \quad \text{and} \quad \mathbf{v}^3 = \begin{bmatrix} \phi_4 \\ \phi_5 \end{bmatrix}$$

Since the speed ratio of gears 2 and 4 is m,

$$\phi_4 = -m\phi_2$$

and

$$T_4 = -\frac{1}{m}T_2$$

Then $\mathbf{p}^3$ and $\mathbf{v}^3$ become

$$\mathbf{p}^3 = \begin{bmatrix} -1/m & 0 \\ 0 & 1 \end{bmatrix}\begin{bmatrix} T_2 \\ T_5 \end{bmatrix} = (\boldsymbol{\tau}^T)^{-1}\mathbf{p}_{\text{new}}^3, \qquad \mathbf{v}^3 = \begin{bmatrix} -m & 0 \\ 0 & 1 \end{bmatrix}\begin{bmatrix} \phi_2 \\ \phi_5 \end{bmatrix} = \boldsymbol{\tau}\mathbf{v}_{\text{new}}^3 \quad (3)$$

Thus (2) becomes

$$\mathbf{p}_{\text{new}}^3 = [\boldsymbol{\tau}^T\mathbf{k}^3\boldsymbol{\tau} - \omega^2\boldsymbol{\tau}^T\mathbf{m}^3\boldsymbol{\tau}]\mathbf{v}_{\text{new}}^3 = \left[\mathbf{k}_{\text{new}}^3 - \omega^2\mathbf{m}_{\text{new}}^3\right]\mathbf{v}_{\text{new}}^3$$

with

$$\mathbf{k}_{\text{new}}^3 = 10^4\begin{bmatrix} 10m^2 & 10m \\ 10m & 10 \end{bmatrix} = 10^4\begin{bmatrix} 40 & 20 \\ 20 & 10 \end{bmatrix}$$

$$\mathbf{m}_{\text{new}}^3 = \begin{bmatrix} \left(10 + \dfrac{1.2}{3}\right)m^2 & -\dfrac{1.2}{6}m \\ -\dfrac{1.2}{6}m & 5 + \dfrac{1.2}{3} \end{bmatrix} = \begin{bmatrix} 41.6 & -0.4 \\ -0.4 & 5.4 \end{bmatrix} \quad (4)$$

Define the vector of degrees of freedom

$$\boldsymbol{\phi} = \begin{bmatrix} \phi_1 & \phi_2 & \phi_3 & \phi_5 \end{bmatrix} \quad (5)$$

and expand each element matrix into the global nodal numbering system

$$\mathbf{m}^1 = \begin{bmatrix} 10.8 & 0.4 & 0 & 0 \\ 0.4 & 12.8 & 0 & 0 \\ 0 & 0 & 0 & 0 \\ 0 & 0 & 0 & 0 \end{bmatrix}, \qquad \mathbf{m}^2 = \begin{bmatrix} 0 & 0 & 0 & 0 \\ 0 & 12.5 & 0.25 & 0 \\ 0 & 0.25 & 5.4 & 0 \\ 0 & 0 & 0 & 0 \end{bmatrix}$$

$$\mathbf{m}^3 = \begin{bmatrix} 0 & 0 & 0 & 0 \\ 0 & 41.6 & 0 & -0.4 \\ 0 & 0 & 0 & 0 \\ 0 & -0.4 & 0 & 5.4 \end{bmatrix} \tag{6}$$

$$\mathbf{k}^1 = 10^4\begin{bmatrix} 3 & -3 & 0 & 0 \\ -3 & 3 & 0 & 0 \\ 0 & 0 & 0 & 0 \\ 0 & 0 & 0 & 0 \end{bmatrix}, \qquad \mathbf{k}^2 = 10^4\begin{bmatrix} 0 & 0 & 0 & 0 \\ 0 & 5 & -5 & 0 \\ 0 & -5 & 5 & 0 \\ 0 & 0 & 0 & 0 \end{bmatrix}$$

$$\mathbf{k}^3 = 10^4\begin{bmatrix} 0 & 0 & 0 & 0 \\ 0 & 40 & 0 & 20 \\ 0 & 0 & 0 & 0 \\ 0 & 20 & 0 & 10 \end{bmatrix} \tag{7}$$

The global mass and stiffness matrices are obtained by adding (6) and (7),

$$
\mathbf{M} = \begin{bmatrix} 10.8 & 0.4 & 0 & 0 \\ 0.4 & 66.9 & 0.25 & -0.4 \\ 0 & 0.25 & 5.4 & 0 \\ 0 & -0.4 & 0 & 5.4 \end{bmatrix}, \qquad \mathbf{K} = \begin{bmatrix} 3 & -3 & 0 & 0 \\ -3 & 48 & -5 & 20 \\ 0 & -5 & 5 & 0 \\ 0 & 20 & 0 & 10 \end{bmatrix} \quad (8)
$$

The natural frequencies and mode shapes for the branched system are found by solving the eigenvalue problem

$$
\mathbf{K}\phi = \omega^2 \mathbf{M}\phi \tag{9}
$$

Some results, in radians per second, are

$$
\omega_1 \approx 0 \ \ (\text{rigid-body mode}), \qquad \omega_2 = 55.58, \qquad \omega_3 = 98.74, \qquad \omega_4 = 159.54
$$

REFERENCES

12.1 Pilkey, W. D., and Wunderlich, W., *Structural Mechanics Variational and Computational Methods*, CRC, Florida, 1993.

12.2 Pilkey, W. D., and Okada, Y., *Matrix Methods in Mechanical Vibrations*, Corona Publishing Co., Tokyo, 1989.

12.3 Darwish, I. A., and Johnston, B. G., "Torsion of Structural Shapes," *J. Struct. Div. ASCE*, Vol. 91, ST1, 1965, pp. 203–227.

12.4 Kollbrunner, C. F., and Basler, K., *Torsion in Structures*, Springer-Verlag, New York, 1969.

12.5 Cook, R. D., and Young, W. C., *Advanced Mechanics of Materials*, MacMillan, New York, 1985.

12.6 Isakower, R. I., "The Shaft book," U.S. Army ARRADCOM MISD User's Manual 80-5, March 1980.

12

Tables

TABLE 12-1 IMPORTANT TORSIONAL STRESS VALUES ON VARIOUS CROSS SECTIONAL SHAPES

Cross Sectional Shapes	Torsional Stress Values
	Thick Noncircular Sections
1. Ellipse	$\tau_{max} = \tau_A = \dfrac{16T}{\pi ab^2}$ $a > b$
2. Hollow ellipse	$\tau_{max} = \tau_A = \dfrac{16T}{\pi ab^2(1 - k^4)}$ $a > b$ where $k = a_i/a = b_i/b$
3. Equilateral triangle	$\tau_{max} = \tau_A$ $= \dfrac{20T}{a^3}$
4. Square 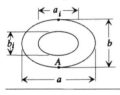	$\tau_{max} = \tau_A = \dfrac{4.81T}{a^3}$
5. Rectangle	$\tau_{max} = \tau_A = \dfrac{3T}{ab^2\left(1 - 0.630\dfrac{b}{a} + 0.250\dfrac{b^2}{a^2}\right)}$

TABLE 12-1 | Important Torsional Stress Values | **606**

Cross Sectional Shapes	Torsional Stress Values
6. Hollow thin-walled sections $\overline{A}$ = Area enclosed by the middle line wall	$$\tau_{max} = \frac{T}{2\overline{A}t}$$ at location of minimum thickness t
	Circular Cross Sections
7. Solid	$$\tau_{max} = \frac{2T}{\pi r_0^3}$$
8. Hollow $\alpha = \dfrac{d_i}{d_0}$	$$T_{max} = \frac{2T}{\pi r_0^3(1 - \alpha^4)}$$
9. Very thin	$$\tau_{max} = \frac{T}{2\pi r^2 t}$$

Important Torsional Stress Values   TABLE 12-1

Thin-Walled Open Sections

$$\tau^i_{max} = \frac{T t_i}{J}$$

where t_i = thickness of segment i, τ^i_{max} = maximum shear stress in segment i

Cross Sectional Shapes	Torsional Stress Values
10. Any open section $\alpha=1.31$ $\alpha=1.12$ $\alpha=1.12$ $\alpha=1.17$ $\alpha=1$ $\alpha=1.29$ $\alpha=1.12$ $\alpha=1.12$	$J = \dfrac{\alpha}{3}\displaystyle\sum_{i=1}^{M} b_i t_i^3$ where M is the number of the straight or curved segments of thickness t_i and width or height b_i comprising the section. Set $\alpha = 1$ except as designated otherwise.
11. t b b	$J = \dfrac{b t^3}{3}$
12. b_1, t_w, t_1, h, t_2, b_2 Set $\alpha = 1$ unless specified otherwise	$J = \frac{1}{3}\alpha(b_1 t_1^3 + b_2 t_2^3 + h t_w^3)$

TABLE 12-1 | Important Torsional Stress Values   608

Cross Sectional Shapes	Torsional Stress Values
13.	$\tau_{max} = \dfrac{TKt_f}{J} = \tau_A$ See Table 2-5, case 13 for J. For $0 \le t_w/t_f \le 1.0$ $K = C_1 + C_2(t_w/t_f) + C_3(t_w/t_f)^2$

	For $0 \le r/t_f \le 1.0$
C_1	$1.00124 + 0.05540(r/t_f) + 0.05540(r/t_f)^2$
C_2	$0.00401 + 0.12065(r/t_f) + 0.05280(r/t_f)^2$
C_3	$0.13890 + 0.11549(r/t_f) - 0.15337(r/t_f)^2$

Ref. 12.3

14. r = radius of fillet	$\tau_{max} = \tau_A = \dfrac{TKt_2}{J}$ See Table 2-5, case 14 for J. For $0 \le t_w/t_2 \le 1.0$ $K = C_1 + C_2(t_w/t_2) + C_3(t_w/t_2)^2$

	For $0 \le r/t_2 \le 1.0$
C_1	$0.92424 + 0.05486(r/t_2) + 0.04533(r/t_2)^2$
C_2	$0.06852 + 0.01174(r/t_2) + 0.13874(r/t_2)^2$
C_3	$0.16321 + 0.03151(r/t_2) - 0.10696(r/t_2)^2$

Ref. 12.3

15.	$\tau_{max} = \tau_A = \dfrac{TKt}{J}$ See Table 2-5, case 15 for J. $K = 3.73 - 9.264\dfrac{r}{t} + 10.24\left(\dfrac{r}{t}\right)^2$ $0.1 \le \dfrac{r}{t} \le 0.3$ $K = 1.7261 - 0.1800\dfrac{r}{t} + 0.09761\left(\dfrac{r}{t}\right)^2$ $0.3 \le \dfrac{r}{t} \le 2.0$ Ref. 12.5

Cross Sectional Shapes	Torsional Stress Values
	Other Shapes

16.

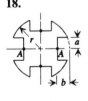

$$\tau_{max} = \tau_A = \frac{TKt}{J}$$

For $0.2 \le \dfrac{r}{t} \le 1.5$

$$K = 3.736 - 6.206\left(\frac{r}{t}\right) + 5.182\left(\frac{r}{t}\right)^2 - 1.487\left(\frac{r}{t}\right)^3$$

Ref. 12.5

17.

$$\tau_{max} = \tau_A = TK/r^3$$

For $0.1 \le b/r \le 0.5$

$$K = C_1 + C_2(b/r) + C_3(b/r)^2 + C_4(b/r)^3$$

	For $0.5 \le a/b \le 2.0$
C_1	$0.8040 + 0.5667(a/b) - 0.7522(a/b)^2 + 0.2277(a/b)^3$
C_2	$3.955 - 10.116(a/b) + 9.670(a/b)^2 - 2.715(a/b)^3$
C_3	$-6.831 + 19.717(a/b) - 22.226(a/b)^2 + 7.176(a/b)^3$
C_4	$2.126 + 0.8590(a/b) + 13.468(a/b)^2 - 6.542(a/b)^3$

Ref. 12.6

18.

$$\tau_{max} = \tau_A = TK/r^3$$

For $0.2 \le b/r \le 0.4$

$$K = C_1 + C_2(b/r) + C_3(b/r)^2 + C_4(b/r)^3$$

	For $0.3 \le a/b \le 1.2$
C_1	$1.923 - 5.435(a/b) + 7.183(a/b)^2 - 2.939(a/b)^3$
C_2	$-13.63 + 86.26(a/b) - 117.4(a/b)^2 + 47.96(a/b)^3$
C_3	$86.23 - 459.88(a/b) + 594.08(a/b)^2 - 235.43(a/b)^3$
C_4	$-126.24 + 671.3(a/b) - 811.29(a/b)^2 + 313.75(a/b)^3$

Ref. 12.6

TABLE 12-1   **Important Torsional Stress Values**   610

Cross Sectional Shapes	Torsional Stress Values

19.

$$\tau_{max} = \tau_A = \frac{TK}{r_0^3}$$

For $0.2 \le r_i/r_0 \le 0.6$

$$K = C_1 + C_2(r_i/r_0) + C_3(r_i/r_0)^2 + C_4(r_i/r_0)^3$$

For $0.1 \le a/r_i \le 1.0$

C_1	$2.005 - 0.5442(a/r_i) + 1.135(a/r_i)^2 - 1.123(a/r_i)^3$
C_2	$2.739 + 7.664(a/r_i) - 10.99(a/r_i)^2 + 11.15(a/r_i)^3$
C_3	$-14.86 - 22.52(a/r_i) + 35.12(a/r_i)^2 - 34.16(a_i/r_i)^3$
C_4	$27.67 + 27.70(a/r_i) - 42.81(a/r_i)^2 + 41.69(a/r_i)^3$

Ref. 12.6

20.

$$\tau_{max} = \tau_A = TK/r^3$$

For $0.1 \le b/r \le 0.5$

$$K = C_1 + C_2(b/r) + C_3(b/r)^2 + C_4(b/r)^3$$

For $0.3 \le a/b \le 2.0$

C_1	$1.103 - 0.1886(a/b) - 0.05643(a/b)^2 + 0.03359(a/b)^3$
C_2	$3.042 - 8.194(a/b) + 6.508(a/b)^2 - 1.556(a/b)^3$
C_3	$-7.027 + 17.65(a/b) - 11.88(a/b)^2 + 2.596(a/b)^3$
C_4	$4.459 - 6.721(a/b) + 3.416(a/b)^2 - 0.6402(a/b)^3$

Ref. 12.6

21.

$$\tau_{max} = \tau_A = TK/r^3$$

For $0.1 \le b/r \le 0.5$

$$K = C_1 + C_2(b/r) + C_3(b/r)^2 + C_4(b/r)^3$$

For $0.2 \le a/b \le 2.0$

C_1	$0.642688 - 0.023769(a/b) + 0.036964(a/b)^2 - 0.018478(a/b)^3$
C_2	$-0.111483 + 0.509322(a/b) - 0.728092(a/b)^2 + 0.341498(a/b)^3$
C_3	$0.616841 - 2.741024(a/b) + 3.582741(a/b)^2 - 1.706508(a/b)^3$
C_4	$-0.889778 + 4.072625(a/b) - 6.123450(a/b)^2 + 2.675014(a/b)^3$

Ref. 12.6

Cross Sectional Shapes	Torsional Stress Values
22. 	$\tau_{\max} = \tau_A = TK/r^3$ For $0.1 \leq b/r \leq 0.5$ $$K = C_1 + C_2(b/r) + C_3(b/r)^2 + C_4(b/r)^3$$ For $0.2 \leq a/b \leq 2.0$ C_1 $0.604248 + 0.107977(a/b) - 0.108329(a/b)^2 + 0.029090(a/b)^3$ C_2 $0.441822 - 1.474396(a/b) + 1.537202(a/b)^2 - 0.406942(a/b)^3$ C_3 $-1.556588 + 5.486226(a/b) - 6.406379(a/b)^2 + 1.555353(a/b)^3$ C_4 $1.850184 - 6.684652(a/b) + 6.467717(a/b)^2 - 1.359464(a/b)^3$ Ref. 12.6
23. 	$\tau_{\max} = \tau_A = TK/r^3$ For $0.1 \leq a/r \leq 0.5$ $$K = 0.650320 + 7.648135(a/r) - 30.943539(a/r)^2$$ $$+ 57.049957(a/r)^3$$ Ref. 12.6
24. 	$\tau_{\max} = \tau_j = TK/r^3 \qquad j = A \text{ or } B$ For $0.1 \leq a/b \leq 0.8$ At A $$K = -0.245866 + 0.622510(b/a) - 0.021154(b/a)^2$$ $$+ 0.000204(b/a)^3$$ At B $$K = 0.327176 + 0.257521(b/a) - 0.033676(b/a)^2$$ $$+ 0.001221(b/a)^3$$ Ref. 12.6
25. 	$\tau_{\max} = \tau_A = TK/r^3$ For $0.1 \leq b/r \leq 0.5$ $$K = C_1 + C_2(b/r) + C_3(b/r)^2 + C_4(b/r)^3$$ For $0.2 \leq a/b \leq 2.0$ C_1 $0.574531 + 0.241436(a/b) - 0.298341(a/b)^2 + 0.108557(a/b)^3$ C_2 $0.695576 - 2.766746(a/b) + 3.589467(a/b)^2 - 1.291860(a/b)^3$ C_3 $-2.435372 + 10.207893(a/b) - 14.264500(a/b)^2 + 4.659470(a/b)^3$ C_4 $2.651900 - 11.083944(a/b) + 12.897306(a/b)^2 - 2.957086(a/b)^3$ Ref. 12.6

TABLE 12-1   Important Torsional Stress Values

Cross Sectional Shapes	Torsional Stress Values
26.	$\tau_{\max} = \tau_A = TK/r^3$ For $0.1 \leq a/r \leq 1.0$ $K = 0.623960 + 1.617474(a/r) - 2.605508(a/r)^2$ $\quad + 3.239703(a/r)^3$ Ref. 12.6
27.	$\tau_{\max} = \tau_A = TK/r^3$ For $0.1 \leq a/r \leq 0.5$ $K = 0.6382 + 2.165(a/r) - 4.849(a/r)^2 + 12.92(a/r)^3$ For $0.5 < a/r \leq 0.8$ $K = -87.91 + 467.5(a/r) - 817.94(a/r)^2 + 486.2(a/r)^3$ Ref. 12.6
28.	$\tau_{\max} = \tau_j = TK/b^3 \qquad j = A \text{ or } B$ For $0.1 \leq a/b \leq 0.4$ At A $K = 75.74442 - 679.83777(a/b) + 2136.45068(a/b)^2$ $\quad - 2242.53418(a/b)^3$ For $0.1 \leq a/b \leq 0.8$ At B $K = 0.707437 - 0.273233(b/a) + 0.165791(b/a)^2$ $\quad - 0.005774(b/a)^3$ Ref. 12.6
29.	$\tau_{\max} = \tau_A = TK/r^3$ For $0.1 \leq a/r \leq 0.5$ $K = 1.025961 + 1.180013(a/r) - 2.788884(a/r)^2$ $\quad + 3.708282(a/r)^3$ Ref. 12.6
30.	$\tau_{\max} = \tau_A = TK/r^3$ For $0.1 \leq a/r \leq 0.5$ $K = 1.005380 + 1.543949(a/r) - 2.954627(a/r)^2$ $\quad + 7.058312(a/r)^3$ Ref. 12.6

TABLE 12-2   UNIFORM BARS WITH ARBITRARY LOADING

A. General Response Expressions

TORSION

Positive Angle of Twist ϕ and Torque T

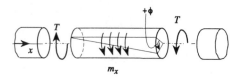

Case	Response
Angle of twist	$\phi = \phi_0 + T_0 \dfrac{x}{GJ} + F_\phi$
Torque	$T = T_0 + F_T$

EXTENSION

Positive Elongation u and Axial Force P

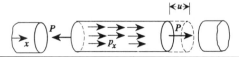

Case	Response
Axial displacement	$u = u_0 + P_0 \dfrac{x}{EA} + F_u$
Axial force	$P = P_0 + F_P$

B. Loading Functions

$$\langle x - a \rangle^n = \begin{cases} 0 & \text{if } x < a \\ (x - a)^n & \text{if } x \ge a \end{cases} \qquad \langle x - a \rangle^0 = \begin{cases} 0 & \text{if } x < a \\ 1 & \text{if } x \ge a \end{cases}$$

TORSION: $F_\phi(x)$, $F_T(x)$

	Concentrated Torque	Uniformly Distributed Torque
$F_\phi(x)$	$\dfrac{-T_1 \langle x - a \rangle}{GJ}$	$-\dfrac{m_{x1}}{2GJ}(\langle x - a_1 \rangle^2 - \langle x - a_2 \rangle^2)$
$F_T(x)$	$-T_1 \langle x - a \rangle^0$	$-m_{x1}(\langle x - a_1 \rangle - \langle x - a_2 \rangle)$

TABLE 12-2   Uniform Bars with Arbitrary Loading   614

TABLE 12-2 (continued)   UNIFORM BARS WITH ARBITRARY LOADING

EXTENSION: $F_u(x), F_P(x)$ where α = Thermal expansion coefficient

	Point Force	Uniformly Distributed Force	Temperature Change ΔT
	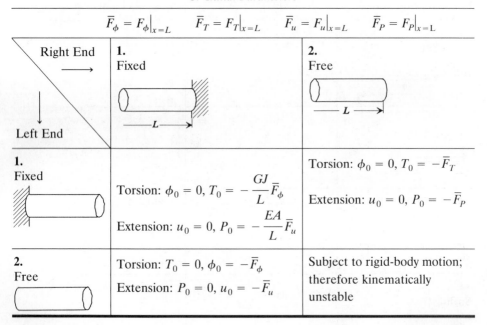		
$F_u(x)$	$-\dfrac{P_1\langle x - a\rangle}{EA}$	$-\dfrac{p_{x_1}}{2EA}(\langle x - a_1\rangle^2 - \langle x - a_2\rangle^2)$	$\alpha\,\Delta T x$
$F_P(x)$	$-P_1\langle x - a\rangle^0$	$-p_{x1}(\langle x - a_1\rangle - \langle x - a_2\rangle)$	0

C. Initial Parameters

$$\bar{F}_\phi = F_\phi\big|_{x=L} \qquad \bar{F}_T = F_T\big|_{x=L} \qquad \bar{F}_u = F_u\big|_{x=L} \qquad \bar{F}_P = F_P\big|_{x=L}$$

Right End ⟶ Left End ↓	1. Fixed	2. Free
1. Fixed	Torsion: $\phi_0 = 0$, $T_0 = -\dfrac{GJ}{L}\bar{F}_\phi$ Extension: $u_0 = 0$, $P_0 = -\dfrac{EA}{L}\bar{F}_u$	Torsion: $\phi_0 = 0$, $T_0 = -\bar{F}_T$ Extension: $u_0 = 0$, $P_0 = -\bar{F}_P$
2. Free	Torsion: $T_0 = 0$, $\phi_0 = -\bar{F}_\phi$ Extension: $P_0 = 0$, $u_0 = -\bar{F}_u$	Subject to rigid-body motion; therefore kinematically unstable

TABLE 12-3 NATURAL FREQUENCIES AND MODE SHAPES FOR TORSION OF UNIFORM BARS

Notation

J = torsional constant
I_x = polar moment of inertia
L = length of bar
G = shear modulus of elasticity

I_p = polar mass moment of inertia per unit length
I_{pi} = polar mass moment of inertia of concentrated mass

$$\text{Natural Frequencies} \quad \omega_i \ (\text{rad/s}) = \left(\frac{\lambda_i}{L}\right)\left(\frac{GJ}{\rho^* I_x}\right)^{1/2} = \frac{\lambda_i}{L}\left(\frac{GJ}{I_p}\right)^{1/2}$$

$$f_i \ (\text{Hz}) = \frac{\omega_i}{2\pi} \qquad i = 1, 2, 3, \ldots$$

Boundary Conditions	λ_i	Mode Shapes, $\xi = x/L$
1. Free–free	$\lambda_i = \pi, 2\pi, 3\pi, \ldots$	$\cos \lambda_i \xi$
2. Clamped–free	$\lambda_i = \frac{1}{2}(2i - 1)\pi$	$\sin \lambda_i \xi$
3. Clamped–clamped	$\lambda_i = \pi, 2\pi, 3\pi, \ldots$	$\sin \lambda_i \xi$
4. Spring–free	$\lambda_i \tan \lambda_i = \dfrac{kL}{GJ}$	$\cot \lambda_i \cos \lambda_i \xi + \sin \lambda_i \xi$
5. Clamped–spring	$\lambda_i = -\dfrac{kL}{GJ} \tan \lambda_i$	$\sin \lambda_i \xi$

TABLE 12-3 Natural Frequencies for Torsion of Bars   616

Boundary Conditions	λ_i	Mode Shapes, $\xi = x/L$
6. Spring–spring k_1 k_2 L	$\lambda_i^2 - K_1 \dfrac{\lambda_i(1 + \alpha)}{\tan \lambda_i} - \alpha K_1^2 = 0$ $K_1 = \dfrac{k_1 L}{GJ} \qquad K_2 = \dfrac{k_2 L}{GJ}$ $\alpha = \dfrac{k_2}{k_1}$	$\sin \lambda_i \xi + \dfrac{\lambda_i}{K_1} \cos \lambda_i \xi$
7. Free with mass I_{p1} L	$\lambda_i = -I_1 \tan \lambda_i$ $I_1 = \dfrac{\rho L J}{I_{p1}}$	$\cos \lambda_i \xi$
8. Clamped with mass I_{p1} L	$\lambda_i = I_1 \cot \lambda_i$ $I_1 = \dfrac{\rho L J}{I_{p1}}$	$\sin \lambda_i \xi$
9. Spring with mass k I_{p1} L	$\lambda_i^2 + I_1 \lambda_i (1 + \alpha) \tan \lambda_i - \alpha I_1^2 = 0$ $\alpha = \dfrac{K}{I_1} \qquad K = \dfrac{kL}{GJ} \qquad I_1 = \dfrac{\rho L J}{I_{p1}}$	$\sin \lambda_i \xi + \dfrac{\lambda_i}{K} \cos \lambda_i \xi$
10. Free with spring and mass I_{p1} L	$\lambda_i^2 + I_1 \lambda_i \tan \lambda_i - \alpha I_1^2 = 0$ $I_1 = \dfrac{\rho L J}{I_{p1}}$ $\alpha = \dfrac{K}{I_1} \qquad K = \dfrac{kL}{GJ}$	$\cos \lambda_i \xi$
11. Fixed–mass with spring I_{p1} L	$\lambda_i^2 - I_1 \dfrac{\lambda_i}{\tan \lambda_i} - \alpha I_1^2 = 0$ $I_1 = \dfrac{\rho L J}{I_{p1}}$ $\alpha = \dfrac{K}{I_1} \qquad K = \dfrac{kL}{GJ}$	$\sin \lambda_i \xi$

Boundary Conditions	λ_i	Mode Shapes, $\xi = x/L$
12. Masses at both ends 	$\lambda_i^2 - I_{p1}\lambda_i \dfrac{1 + \alpha}{\tan \lambda_i} - I_{p1}^2\alpha = 0$ $\alpha = \dfrac{I_{p2}}{I_{p1}}$	$\sin \lambda_i\xi - \dfrac{I_{p1}}{\lambda_i}\cos \lambda_i\xi$
13. 	$\dfrac{\lambda_i^4}{I_1 I_2} - \dfrac{\lambda_i^3(I_1 + I_2)\cot \lambda_i}{I_1 I_2}$ $-\lambda_i^2\left(1 + \dfrac{K_1}{I_2} + \dfrac{K_2}{I_1}\right)$ $+\lambda_i(K_1 + K_2)\cot \lambda_i$ $+K_1 K_2 = 0$ $K_1 = \dfrac{k_1 L}{GI_p} \qquad K_2 = \dfrac{k_2 L}{GI_p}$ $I_1 = \dfrac{I_p L}{I_{p1}} \qquad I_2 = \dfrac{I_p L}{I_{p2}}$	$\cos \lambda_i\xi + \delta \sin \lambda_i\xi$ $\delta = \lambda_i/\left(K_1 - \lambda_i^2/I_1\right)$

TABLE 12-3   Natural Frequencies for Torsion of Bars   618

TABLE 12-4 NATURAL FREQUENCIES AND MODE SHAPES FOR EXTENSION OF UNIFORM BARS

Notation

ρ^* = mass per unit volume $\qquad$ L = length of bar

E = modulus of elasticity $\qquad$ A = cross-sectional area

$$\text{Natural Frequencies} \qquad \omega_i \text{ (rad/s)} = \left(\frac{\lambda_i}{L}\right)\left[\frac{E}{\rho^*}\right]^{1/2}$$

$$f_i \text{ (Hz)} = \frac{\omega_i}{2\pi} \qquad i = 1, 2, 3, \dots$$

Boundary Conditions	λ_i	Mode Shapes, $\xi = x/L$
1. Free–free	$\lambda_i = i\pi$	$\cos \lambda_i \xi$
2. Clamped–free	$\lambda_i = \frac{1}{2}\pi(2i - 1)$	$\sin \lambda_i \xi$
3. Clamped–clamped	$\lambda_i = i\pi$	$\sin \lambda_i \xi$
4. Spring–free	$\lambda_i \tan \lambda_i = \dfrac{kL}{AE}$	$\cot \lambda_i \cos \lambda_i \xi + \sin \lambda_i \xi$

Boundary Conditions	λ_i	Mode Shapes, $\xi = x/L$
5. Spring–clamped	$\lambda_i = -\dfrac{kL}{EA}\tan \lambda_i$	$\sin \lambda_i \xi$
6. Clamped–mass	$\lambda_i \tan \lambda_i = \dfrac{\rho^* AL}{M_1}$	$\sin \lambda_i \xi$
7. Free–mass	$\dfrac{1}{\lambda_i}\tan \lambda_i = -\dfrac{M_1}{\rho^* AL}$	$\cos \lambda_i \xi$
8. Spring–mass	$\lambda_i^2 + I_1\lambda_i(1+\alpha)\tan \lambda_i - \alpha I_1^2 = 0$ $\alpha = \dfrac{K}{I_1} \quad K = \dfrac{kL}{EA} \quad I_1 = \dfrac{\rho LA}{M_1}$	$\sin \lambda_i \xi + \dfrac{\lambda_i}{K}\cos \lambda_i \xi$

TABLE 12-4   **Natural Frequencies for Extension of Bars**   **620**

TABLE 12-5   POLAR MASS MOMENTS OF INERTIA FOR CONCENTRATED MASSES

Notation

ρ = mass per unit length
ρ^* = mass per unit volume
M = total mass of concentrated mass

Mass Shape	Polar Mass Moment of Inertia I_{pi} (ML^2) about x Axis
1. Disk: hollow circular cylinder	$\frac{1}{2}\rho h(r_{\text{outer}}^2 + r_{\text{inner}}^2)$
2. Solid circular cylinder	$\frac{1}{2}\pi h\rho^* r^4 = \frac{1}{2}\rho h r^2$
3. Rectangular prism	Three cases: 1. x passes through center of gravity (as shown) $(M/12)(a^2 + b^2)$ 2. x coincides with A–A $(M/12)(a^2 + 4b^2)$ 3. x coincides with B–B $(M/3)(a^2 + b^2)$

Mass Shape	Polar Mass Moment of Inertia I_{pi} (ML^2) about x Axis
4. Torus	$\frac{1}{64}\pi^2\rho^*(r_{outer} + r_{inner})[4(r_{outer}^2 - r_{inner}^2)^2 + 3(r_{outer} - r_{inner})^4]$
5. Cone	$\frac{1}{10}\pi\rho^*r^4h$
6. Sphere	$\frac{8}{15}\pi\rho^*r^5$
7. Hollow sphere	$\frac{8}{15}\pi\rho^*(r_{outer}^5 - r_{inner}^5)$

TABLE 12-5   **Polar Mass Moments of Inertia**

Mass Shape	Polar Mass Moment of Inertia I_{pi} (ML^2) about x Axis
8. Ellipsoid	$\frac{4}{15}\pi\rho^*abc(a^2 + c^2)$
9. Hemisphere	$\frac{4}{15}\pi\rho^*r^5$
10. Paraboloid of revolution	$\frac{1}{6}\pi\rho^*ab^4$
11. Right rectangular pyramid	$\frac{1}{60}\rho^*abc(c^2 + a^2)$

TABLE 12-6 STRUCTURAL MATRICES FOR TORSION OF BARS

Notation
J = torsional constant
G = shear modulus of elasticity
ℓ = length of element

Case (ith element)	Transfer Matrices	Linearly Varying Distributed Applied Torque	Stiffness Matrices
1. Simple static bar	$z_b = U^i z_a$ $U^i = \begin{bmatrix} 1 & \dfrac{\ell}{GJ} & F_\phi \\ 0 & 1 & F_T \\ 0 & 0 & 1 \end{bmatrix}\begin{bmatrix} \phi_a \\ T_a \\ 1 \end{bmatrix} = z_a$	$F_\phi = -\dfrac{1}{GJ}\left(m_a\dfrac{\ell^2}{3} + m_b\dfrac{\ell^2}{6}\right)$ $F_T = -m_a\dfrac{\ell}{2} - m_b\dfrac{\ell}{2}$	$p^i = k^i v^i - \bar{p}^i$ $k^i = \begin{bmatrix} \dfrac{GJ}{\ell} & -\dfrac{GJ}{\ell} \\ -\dfrac{GJ}{\ell} & \dfrac{GJ}{\ell} \end{bmatrix}$ $\qquad v^i = \begin{bmatrix} \phi_a \\ \phi_b \end{bmatrix}$ $\bar{p}^i = \begin{bmatrix} -\dfrac{GJ}{\ell}F_\phi \\ -F_T + \dfrac{GJ}{\ell}F_\phi \end{bmatrix}$ $\qquad p^i = \begin{bmatrix} T_a \\ T_b \end{bmatrix}$

TABLE 12-6   Structural Matrices for Torsion of Bars   624

2. On elastic foundation

$$\beta^2 = k_t/GJ$$

$$\begin{bmatrix} \cosh\beta\ell & \dfrac{\sinh\beta\ell}{GJ\beta} & F_\phi \\[2mm] GJ\beta\sinh\beta\ell & \cosh\beta\ell & F_T \\[2mm] 0 & 0 & 1 \end{bmatrix}$$

$$F_\phi = -\frac{m_a}{GJ}\left[\frac{-1+\cosh\beta\ell}{\beta^2}\right] - \frac{m_b}{GJ}\left[\frac{-\beta\ell+\sinh\beta\ell}{\beta^3\ell}\right]$$

$$F_T = -m_a\left[\frac{\sinh\beta\ell}{\beta}\right] - m_b\left[\frac{-1+\cosh\beta\ell}{\beta^2\ell}\right]$$

$$\mathbf{k}^i = \begin{bmatrix} GJ\beta\,\dfrac{\cosh\beta\ell}{\sinh\beta\ell} & -\dfrac{GJ\beta}{\sinh\beta\ell} \\[3mm] -\dfrac{GJ\beta}{\sinh\beta\ell} & GJ\beta\,\dfrac{\cosh\beta\ell}{\sinh\beta\ell} \end{bmatrix}$$

$$\bar{\mathbf{p}}^i = \begin{bmatrix} -\dfrac{GJ\beta}{\sinh\beta\ell}F_\phi \\[3mm] -F_T + \dfrac{\cosh\beta\ell}{\sinh\beta\ell}GJ\beta F_\phi \end{bmatrix}$$

3. With mass and foundation

Dynamic stiffness matrix

$$\begin{bmatrix} \cos\beta\ell & \dfrac{\sin\beta\ell}{GJ\beta} & F_\phi \\[2mm] -GJ\beta\sin\beta\ell & \cos\beta\ell & F_T \\[2mm] 0 & 0 & 1 \end{bmatrix}$$

$$F_\phi = -\frac{m_a}{GJ}\left[\frac{1-\cos\beta\ell}{\beta^2}\right] - \frac{m_b}{GJ}\left[\frac{\beta\ell-\sin\beta\ell}{\beta^3\ell}\right]$$

$$F_T = -m_a\left[\frac{\sin\beta\ell}{\beta}\right] - m_b\left[\frac{1-\cos\beta\ell}{\beta^2\ell}\right]$$

$$\mathbf{k}^i = \begin{bmatrix} GJ\beta\,\dfrac{\cos\beta\ell}{\sin\beta\ell} & -\dfrac{GJ\beta}{\sin\beta\ell} \\[3mm] -\dfrac{GJ\beta}{\sin\beta\ell} & GJ\beta\,\dfrac{\cos\beta\ell}{\sin\beta\ell} \end{bmatrix}$$

$$\bar{\mathbf{p}}^i = \begin{bmatrix} -\dfrac{GJ\beta}{\sin\beta\ell}F_\phi \\[3mm] -F_T + \dfrac{\cos\beta\ell}{\sin\beta\ell}GJ\beta F_\phi \end{bmatrix}$$

$$\beta^2 = (\omega^2\rho r_p^2 - k_t)/GJ$$
$$= (\omega^2 I_p - k_t)/GJ$$

ρ = mass/length

If $\beta^2 < 0$, use the formulas of Case 2 with $\beta^2 =$ $(k_t - \omega^2 I_p)/GJ$

TABLE 12-7 SELECTED TRANSFER MATRICES FOR TORSION

Notation

I_{pi} = polar mass moment of inertia at point i

T_i = applied torque

k_b = torsional spring constant[a] of branch system (FL)

k_t = elastic foundation modulus

ω = natural frequency

r_p = polar radius of gyration

Case	Transfer Matrix[a]
1. Point matrix (Branch System)	$$\underbrace{\begin{bmatrix} 1 & 1/k & 0 \\ k_b - I_{pi}\omega^2 & 1 & -T_i \\ 0 & 0 & 1 \end{bmatrix}}_{\mathbf{U}_i} \underbrace{\begin{bmatrix} \phi \\ T \\ 1 \end{bmatrix}}_{\mathbf{z}_i}$$
2. Rigid bar	$$\mathbf{U}^i = \begin{bmatrix} 1 & 0 & 0 \\ \ell\left(k_t - \rho r_p^2 \omega^2\right) & 1 & -\tfrac{1}{2}\ell\left(m_a + m_b\right) \\ 0 & 0 & 1 \end{bmatrix}$$

[a]The spring constant k can be considered to be equivalent to GJ/ℓ of Table 12-6, case 1. In the case of a coil spring, $k = Ed^4/(32ND)$, where E = Young's modulus, d = spring wire diameter, D = mean coil diameter, and N = number of coils.

TABLE 12-7 Selected Transfer Matrices for Torsion 626

TABLE 12-8   STRUCTURAL MATRICES FOR EXTENSION OF BARS

Notation

A = cross-sectional area
E = modulus of elasticity
ℓ = length of element
α = thermal expansion coefficient

Case (ith element)	Transfer Matrices	Linearly Varying Distributed Axial Force	Stiffness Matrices
1. Simple static bar 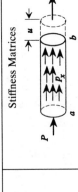	$\mathbf{z}_b = \mathbf{U}^i \mathbf{z}_a$ $$\mathbf{U}^i = \begin{bmatrix} 1 & \dfrac{\ell}{AE} & F_u \\ 0 & 1 & F_P \\ 0 & 0 & 1 \end{bmatrix}\begin{matrix} u_a \\ P_a \\ 1 \end{matrix} \;\; \mathbf{z}_a$$	$F_u = -p_a \dfrac{\ell^2}{3AE} - p_b \dfrac{\ell^2}{6AE} + \alpha\,\ell\,\Delta T$ $F_P = -p_a \dfrac{\ell}{2} - p_b \dfrac{\ell}{2}$	$\mathbf{p}^i = \mathbf{k}^i \mathbf{v}^i - \bar{\mathbf{p}}^i$ $$\mathbf{k}^i = \begin{bmatrix} \dfrac{AE}{\ell} & -\dfrac{AE}{\ell} \\ -\dfrac{AE}{\ell} & \dfrac{AE}{\ell} \end{bmatrix}$$ $$\bar{\mathbf{p}}^i = \begin{bmatrix} -\dfrac{AE}{\ell}F_u \\ -F_P + \dfrac{AE}{\ell}F_u \end{bmatrix}$$ $\mathbf{v}^i = \begin{bmatrix} u_a \\ u_b \end{bmatrix}$ $\mathbf{p}^i = \begin{bmatrix} p_a \\ p_b \end{bmatrix}$

TABLE 12-8 (continued) STRUCTURAL MATRICES FOR EXTENSION OF BARS

Case	Transfer Matrices	Linearly Varying Distributed Axial Force	Stiffness Matrices

Transfer Matrices

$$\mathbf{z}_b = \mathbf{U}^i \mathbf{z}_a$$

Stiffness Matrices

$$\mathbf{p}^i = \mathbf{k}^i \mathbf{v}^i - \bar{\mathbf{p}}^i$$

(ith element)

---

2. On elastic foundation k_x

$$\beta^2 = k_x/AE$$

Transfer:

$$\begin{bmatrix} \cosh \beta\ell & \dfrac{\sinh \beta\ell}{AE\beta} & F_u \\ AE\beta \sinh \beta\ell & \cosh \beta\ell & F_P \\ 0 & 0 & 1 \end{bmatrix}$$

Linearly Varying Distributed Axial Force:

$$F_u = -\frac{p_a}{AE}\left[\frac{-1+\cosh\beta\ell}{\beta^2} - \frac{-\beta\ell+\sinh\beta\ell}{\beta^3\ell}\right] - \frac{p_b}{AE}\left[\frac{-\beta\ell+\sinh\beta\ell}{\beta^3\ell}\right] + \alpha\,\Delta T\,\frac{\sinh\beta\ell}{\beta}$$

$$F_P = -p_a\left[\frac{\sinh\beta\ell}{\beta} - \frac{-1+\cosh\beta\ell}{\beta^2\ell}\right] - p_b\left[\frac{-1+\cosh\beta\ell}{\beta^2\ell}\right] - AE\alpha\,\Delta T(1-\cosh\beta\ell)$$

Stiffness Matrices:

$$\mathbf{k}^i = \begin{bmatrix} AE\beta\dfrac{\cosh\beta\ell}{\sinh\beta\ell} & -\dfrac{AE\beta}{\sinh\beta\ell} \\ -\dfrac{AE\beta}{\sinh\beta\ell} & AE\beta\dfrac{\cosh\beta\ell}{\sinh\beta\ell} \end{bmatrix}$$

$$\bar{\mathbf{p}}^i = \begin{bmatrix} -\dfrac{AE\beta}{\sinh\beta\ell}F_u \\ -F_p + \dfrac{\cosh\beta\ell}{\sinh\beta\ell}AE\beta F_u \end{bmatrix}$$

---

3. With mass and foundation

$$\rho = \text{mass/length}$$
$$\beta^2 = (\rho\omega^2 - k_x)/AE$$
If $\beta^2 < 0$, use case 2
with $\beta^2 = (k_x - \rho\omega^2)/AE$

Transfer:

$$\begin{bmatrix} \cos \beta\ell & \dfrac{\sin \beta\ell}{AE\beta} & F_u \\ -AE\beta \sin \beta\ell & \cos \beta\ell & F_P \\ 0 & 0 & 1 \end{bmatrix}$$

Linearly Varying Distributed Axial Force:

$$F_u = -\frac{p_a}{AE}\left[\frac{1-\cos\beta\ell}{\beta^2} - \frac{\beta\ell-\sin\beta\ell}{\beta^3\ell}\right] - \frac{p_b}{AE}\left[\frac{\beta\ell-\sin\beta\ell}{\beta^3\ell}\right] + \alpha\,\Delta T\,\frac{\sin\beta\ell}{\beta}$$

$$F_p = -p_a\left[\frac{\sin\beta\ell}{\beta} - \frac{1-\cos\beta\ell}{\beta^2\ell}\right] - p_b\left[\frac{1-\cos\beta\ell}{\beta^2\ell}\right] - AE\alpha\,\Delta T(1-\cos\beta\ell)$$

Dynamic stiffness matrix:

$$\mathbf{k}^i = \begin{bmatrix} AE\beta\dfrac{\cos\beta\ell}{\sin\beta\ell} & -\dfrac{AE\beta}{\sin\beta\ell} \\ -\dfrac{AE\beta}{\sin\beta\ell} & AE\beta\dfrac{\cos\beta\ell}{\sin\beta\ell} \end{bmatrix}$$

$$\bar{\mathbf{p}}^i = \begin{bmatrix} -\dfrac{AE\beta}{\sin\beta\ell}F_u \\ -F_P + AE\beta\dfrac{\cos\beta\ell}{\sin\beta\ell}F_u \end{bmatrix}$$

TABLE 12-8 Structural Matrices for Extension of Bars 628

TABLE 12-9   SELECTED TRANSFER MATRICES FOR EXTENSION

Notation

k_b = spring constant[a] of branch system (F/L)

k_x = elastic foundation modulus

M_i = concentrated mass

P_i = applied concentrated axial force

ω = natural frequency

ρ = mass per unit length

Case	Transfer Matrix
1. Point matrix P_i M_i k_b (Branch System)	$$\mathbf{U}_i = \begin{bmatrix} 1 & 1/k & 0 \\ k_b - M_i\omega^2 & 1 & -P_i \\ 0 & 0 & 1 \end{bmatrix}$$
2. Rigid bar P_a P_b ℓ k_x	$$\mathbf{U}^i = \begin{bmatrix} 1 & 0 & 0 \\ \ell\left(k_x - \rho\omega^2\right) & 1 & -\tfrac{1}{2}\ell\left(p_a + p_b\right) \\ 0 & 0 & 1 \end{bmatrix}$$

[a]The spring constant k can be considered to be equivalent to EA/ℓ of Table 12-8, case 1. In the case of springs in parallel $k = k_1 + k_2$. In the case of a coil spring $k = Ed^4/(8ND^3)$, where E = Young's modulus, d = spring wire diameter, D = mean coil diameter, and N = number of coils. Other equivalent spring constants are given in Table 10-3.

TABLE 12-10   MASS MATRICES FOR BARS IN TORSION

Notation

I_{pi} = polar mass moment of inertia of concentrated mass
at point i

I_p = polar mass moment of inertia per unit length

Bars in Torsion	Mass Matrix $\mathbf{m}^i$
1. Lumped-mass model for distributed mass moment of inertia I_p	$\dfrac{I_p \ell}{2} \begin{bmatrix} 1 & 0 \\ 0 & 1 \end{bmatrix}$
2. Lumped-mass single-point model for distributed mass moment of inertia I_p	1×1 mass matrix $I_p \ell$
3. Lumped-mass matrix for two concentrated masses	$\begin{bmatrix} I_{pa} & 0 \\ 0 & I_{pb} \end{bmatrix}$
4. Consistent mass matrix	$-\dfrac{I_p \ell}{6} \begin{bmatrix} 2 & 1 \\ 1 & 2 \end{bmatrix}$
5. General mass matrix	$\mathbf{m}^i = \int_0^\ell \dfrac{J}{A} \rho \boldsymbol{\Lambda} \boldsymbol{\Lambda}^T \, dx \qquad \boldsymbol{\Lambda} = \begin{bmatrix} H_1 \\ H_2 \end{bmatrix}$ with the displacement shape function $\phi(x) = H_1 \varphi_a + H_2 \varphi_b$. If for constant J the static response $H_1 = 1 - x/\ell$, $H_2 = x/\ell$ is employed, then $\mathbf{m}^i$ is the consistent mass matrix of case 4.

TABLE 12-10   Mass Matrices for Bars in Torsion   630

TABLE 12-11 MASS MATRICES FOR BARS IN EXTENSION

Notation

ρ = mass per unit length
ℓ = length of element

Bars in Extension	Mass Matrix $\mathbf{m}^i$
1. Lumped-mass two-point model for distributed mass 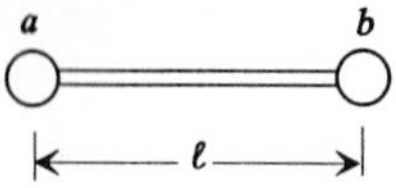	$\frac{\rho\ell}{2}\begin{bmatrix} 1 & 0 \\ 0 & 1 \end{bmatrix}$
2. Lumped-mass single-point model for distributed mass ρ 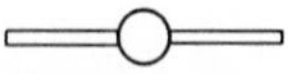	1×1 mass matrix $\rho\ell$
3. Lumped-mass matrix for two concentrated masses 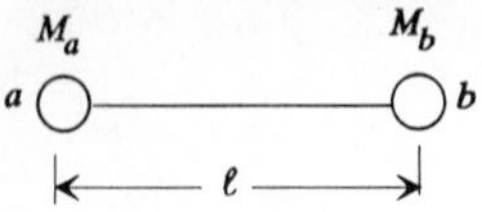	$\begin{bmatrix} M_a & 0 \\ 0 & M_b \end{bmatrix}$
4. Consistent mass matrix 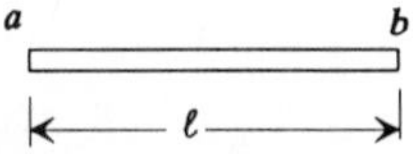	$\frac{\rho\ell}{6}\begin{bmatrix} 2 & 1 \\ 1 & 2 \end{bmatrix}$
5. General mass matrix 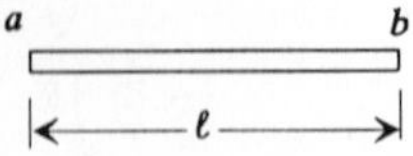	$\mathbf{m}^i = \int_0^\ell \rho \mathbf{\Lambda}\mathbf{\Lambda}^T \, dx \qquad \mathbf{\Lambda} = \begin{bmatrix} H_1 \\ H_2 \end{bmatrix}$ with the displacement shape function $u(x) = H_1 u_a + H_2 u_b$. If for constant ρ the static response $H_1 = 1 - x/\ell$, $H_2 = x/\ell$ is employed, then $\mathbf{m}^i$ is the consistent mass matrix of case 4.

Frames

Structures formed of bars that are rigidly connected are referred to as *frames*, while those of bars that are pin connected are *trusses*. Analytically, trusses are treated as being a special case of frames. For the frames of this chapter, it is assumed that there is no interaction between axial, torsional, and flexural deformations, i.e., the responses are based on uncoupled extension, torsion, and bending theory.

Formulas are provided for several simple frame configurations with simple loadings. Also, structural matrices required for more complicated frames are listed. Many commercially available general-purpose structural analysis computer programs can be used to analyze complicated frames.

Entries in most of the tables of this chapter give salient values of reactions, forces, and moments. Also, a moment diagram is shown. This moment can be used to calculate the bending stresses using the technical beam theory flexural stress formula. Formulas for buckling loads and natural frequencies are tabulated.

Special attention is given to gridworks, which are flat networks of beams with transverse loading.

Collapse loads are provided for plastic design.

13.1 NOTATION

e	$= h/L$, where h is the length of the vertical members and L is the length of the horizontal members
E	Modulus of elasticity of material (F/L^2)
H	Horizontal reaction; H_A is horizontal reaction at location A (F)
I	Moment of inertia of member about its neutral axis (L^4)
I_1	Moment of inertia of vertical members (L^4)
I_2	Moment of inertia of horizontal members (L^4)
I_x	Polar moment of inertia, $= r_p^2 A$ (L^4)
J	Torsional constant (L^4)
L	Length of member (L)
M	Bending moment (LF). A bending moment is taken as positive when it causes tension on the inner side of the frame and compression on the outer side. Opposing bending moments are taken to be negative.
p	Applied distributed loading (F/L)
R	Vertical reaction; R_A is vertical reaction at location A (F)
$\tilde{u}, \tilde{v}, \tilde{w}$	Displacements in x, y, and z directions, respectively
u_X, u_Y, u_Z	Displacements in X, Y, and Z directions, respectively
v	Displacement; v_{Ax} is displacement at location A in the x (horizontal) direction; other displacements defined similarly (L)
x, y, z	Local coordinates
X, Y, Z	Global coordinates
β	$= I_h$ (horizontal beam)$/I_v$ (vertical member)
ω	Natural frequency
$\theta = \theta_y$	Rotation angle of cross section about y axis
θ_z	Rotation angle of cross section about z axis

Notation for Gridworks

g, s	Index for girders and stiffeners, respectively
I_g, I_s	Moments of inertia of girders and stiffeners, respectively (L^4)
L_g, L_s	Length of girders and stiffeners, respectively (L)
n_g, n_s	Total number of girders and stiffeners, respectively
P_g, P_s	Axial forces in girders and stiffeners, respectively (F)
p_s	Loading intensity along sth stiffener (F/L)
w_g, θ_g, M_g, V_g	Deflection, slope, bending moment, and shear force of gth girder
W_{sg}	Concentrated force at intersection x_s, y_g (F)
ρ_g, ρ_s	Mass per unit length of girders and stiffeners, respectively $(M/L, FT^2/L^2)$

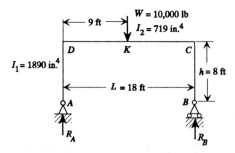

FIGURE 13-1 Statically determinate frame.

13.2 FRAMES

Formulas

Tables 13-1 to 13-3 provide formulas for the static response of simple frameworks. More complicated loading configurations can be obtained by superimposing the formulas for the cases given in the tables. This is illustrated in Example 13.5. Formulas for frames of more complicated geometries are to be found in standard references [e.g., 13.1, 13.2]. Readily available structural analysis computer programs can be used to find the forces and displacements as well as buckling loads and natural frequencies in frameworks of any complexity.

Example 13.1 Statically Determinate Frame with Concentrated Force The frame of Fig. 13-1 is hinged at the lower end of the left-hand member and is roller supported at the lower end of the right-hand member.
    From case 1 of Table 13-1

$$R_A = R_B = \tfrac{1}{2}W = 5000 \text{ lb}, \qquad M_{max} = \tfrac{1}{4}WL = 45{,}000 \text{ ft-lb}$$

Example 13.2 Statically Indeterminate Frame with Concentrated Force Suppose for the frame of Fig. 13-1 that the lower end of the right-hand member is hinged. Then the frame is statically indeterminate so that case 1 of Table 13-2 applies. Use $a = \tfrac{1}{2}L = 108$ in., $e = h/L = \tfrac{8}{18} = 0.444$, $\beta = I_2/I_1 = \tfrac{719}{1890} = 0.380$. From Table 13-2

$$H_A = H_B = \frac{3Wa}{2hL}\frac{L-a}{2\beta e + 3} = 2528 \text{ lb}$$

$$R_A = R_B = \tfrac{1}{2}W = 5000 \text{ lb}$$

$$M_C = M_D = \frac{3Wa}{2L}\frac{L-a}{2\beta e + 3} = 242{,}700 \text{ in.-lb}$$

$$M_K = \frac{Wa(L-a)}{2L}\frac{4\beta e + 3}{2\beta e + 3} = 297{,}300 \text{ in.-lb}$$

The moment diagram is sketched in case 1 of Table 13-2.

Example 13.3 Frame with Fixed Legs If the lower ends of the legs of the frame of Fig. 13-1 are fixed, then the reactions and moment distribution can be calculated using case 6 of Table 13-2. As in Example 13.2, $a = 108$ in., $e = 0.444$, and $\beta = 0.380$. The reactions are

$$R_A = R_B = \tfrac{1}{2}W = 5000 \text{ lb}$$

$$H_A = H_B = 3WL/[8h(\beta e + 2)] = 3891 \text{ lb}$$

$$M_A = M_B = WL/[8(\beta e + 2)] = H_A h/3 = 124{,}497 \text{ in.-lb}$$

$$M_C = M_D = WL/[4(\beta e + 2)] = 2M_A = 248{,}995 \text{ in.-lb}$$

$$M_K = \frac{WL}{4}\frac{\beta e + 1}{\beta e + 2} = 291{,}005 \text{ in.-lb}$$

Case 6 of Table 13-2 illustrates the moment distribution.

Example 13.4 Laterally Loaded Frame Suppose the vertical load is removed from the frame of Example 13.2 and replaced by a lateral load acting at half height as shown in Fig. 13-2.

Use the formulas of Case 3 of Table 13-2 with $W = 8000$ lb, $h = 96$ in., $a = \tfrac{1}{2}h = 48$ in., $\beta = 0.380$, and $L = 216$ in. Define a constant

$$A = a\beta(2h - a)/[h(2h\beta + 3L)] = 0.0379494$$

Then we find

$$R_A = R_B = W(h - a)/L = 1778 \text{ lb}$$

$$H_A = (W/2h)[h + a - (h - a)A] = 5924 \text{ lb}$$

$$H_B = [W(h - a)/2h](1 + A) = 2076 \text{ lb}$$

$$M_C = H_B h = 199{,}296 \text{ in.-lb}$$

$$M_D = \tfrac{1}{2}W(h - a)(1 - A) = 184{,}714 \text{ in.-lb}$$

$$M_K = (h - a)H_A = 284{,}352 \text{ in.-lb}$$

The moment diagram is given in Table 13-2, case 3.

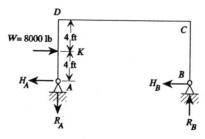

FIGURE 13-2 Statically indeterminate frame of Example 13.4. The dimensions and section properties are given in Fig. 13-1.

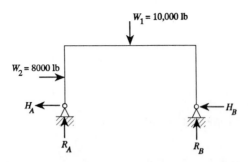

FIGURE 13-3   Frame of Fig. 13-2 with horizontal and vertical loading.

Example 13.5   Superposition of Solutions for a Frame with Several Loadings
Suppose the frame of Fig. 13-2 is simultaneously subjected to the loads of
Examples 13.2 and 13.4. This configuration is shown in Fig. 13-3.

Since these frame formulas are based on linear theory, superposition holds.
See cases 1 and 3 of Table 13-2 for the directions of the reactions. Superposition
gives, for the frame of Fig. 13-3,

$$H_A = 5924 - 2528 = 3396 \text{ lb}, \qquad H_B = 2076 + 2528 = 4604 \text{ lb}$$
$$R_A = 5000 - 1778 = 3222 \text{ lb}, \qquad R_B = 5000 + 1778 = 6778 \text{ lb}$$

The directions of these reactions are shown in Fig. 13-3.

The moment diagram can be obtained by superimposing the moment diagrams
of cases 1 and 3, Table 13-2, with due regard being given to the signs of the
moments. Alternatively, the moment diagram can be calculated using the applied
loading and the computed reactions. Thus,

$$M_C = H_B h = 441,984 \text{ in.-lb}$$
$$M_{K_1} = -H_B h + \tfrac{1}{2} R_B L = +290,040 \text{ in.-lb}$$
$$M_{K_2} = H_A \times 48 = 163,008 \text{ in.-lb}$$
$$M_D = W_2 \times 48 - H_A h = 57,984 \text{ in.-lb}$$

The combined moment diagram is illustrated in Fig. 13-4.

Buckling Loads

The buckling loads for some frames are given in Table 13-4. Reference 13.3
provides more cases. Methods for obtaining buckling loads of simple frames are
described in reference 13.4. For more complicated frames use the matrix methods given in Section 13.4.

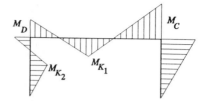

FIGURE 13-4 Moment diagram for frame of Fig. 13-3.

Natural Frequencies

Table 13-5 provides the fundamental natural frequencies for some simple framework configurations. The computational methods of Section 13.4 can be used to obtain the natural frequencies for more general frames.

Plastic Design

As in the case of beams, the concept of plastic design can be applied to frames. The primary objective of the design is to find the collapse load and the location of the plastic hinges. Normally, these plastic designs are restricted to proportional loading such that all loads acting on a frame remain in fixed proportion as their magnitudes are varied. The common factor that multiples all loads as they vary in fixed proportion is called the *load factor*. The procedure for finding the load factor is as follows [13.5]:

1. Find the locations of the plastic hinges in each component of the frame using the same method as for beams.
2. Form possible failure modes called mechanisms by different combinations of plastic hinges. The number of hinges in each mechanism is equal to the number of redundancies plus 1.
3. Calculate the collapse load factor for each individual mechanism.
4. Calculate the moments in the frame for each collapse load factor to determine the correct load factor. The true load factor should be such that the moment in the frame due to this load should not exceed the plastic moment M_p.

In addition to the collapse load factors that can be determined, a safe-load region can be established. Table 13-6 shows safe-load regions for several frameworks. In Table 13-6, a combination of forces applied on the frame define a point on the x, y plane. When this point falls inside the safe region, no collapse occurs. When the point falls on the boundary of the region, collapse occurs and the collapse mode is identified by the location on the boundary, as indicated by the figures in Table 13-6. Loadings leading to points outside of the region correspond to a collapsed framework. In fact, an attempt to increase the applied loads beyond that necessary to reach the boundary results in further movements of the plastic hinges without an increase in the collapse loads. See reference 13.5 for techniques for calculating the safe-load region.

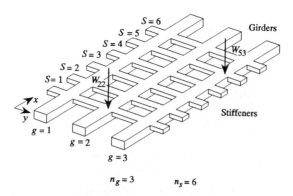

FIGURE 13-5 Typical gridwork.

13.3 GRIDWORKS [13.6]

A special case of frames is a *gridwork*, or *grillage*, which is a network of beams rigidly connected at the intersections, loaded transversely. That is, a gridwork is a network of closely spaced beams with out-of-plane loading. It may be of any shape and the network of beams may intersect at any angle. These beams need not be uniform.

The gridworks treated here are plane structures (Fig. 13-5), with the beams lying in one direction called *girders* and those in the perpendicular direction called *stiffeners*. Either set of gridwork beams can be selected to be the girders. In practice, the wider spaced and heavier set is usually designated as girders while the closer spaced and lighter beams are stiffeners. For a *uniform gridwork*, the girders are identical in size, end conditions, and spacing. Also, the stiffeners are identical in size, end conditions, and spacing. However, the set of stiffeners may differ from the set of girders.

For the formulas here, the cross section of the beams may be open or closed, although torsional rigidity is not taken into account. For closed cross sections this may lead to an error of up to 5%.

Stresses in the girders and stiffeners can be calculated using the formulas for beams of Chapter 11.

For gridworks not covered by the formulas here, use can be made of a framework computer program. The structural matrices, including transfer, stiffness, and mass matrices, for a grillage are provided in Section 13.4.

The sign convention of the transfer matrix method for displacements and forces for the beams of Chapter 11 apply to the gridwork beams here.

Static Loading

The deflection, slope, bending moment, and shear force of the gth girder of the gridwork are given in Table 13-7. The ends of both the girders and stiffeners are simply supported. Table 13-8 provides the parameters K_j for particular loadings. Sufficient accuracy is usually achieved if only M terms, where $M \ll \infty$, are

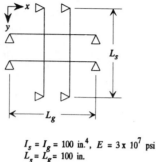

$I_s = I_g = 100 \text{ in.}^4,\ E = 3 \times 10^7 \text{ psi}$
$L_s = L_g = 100 \text{ in.}$

FIGURE 13-6 Grillage for Examples 13.6–13.8.

included in the formulas of Tables 13-7 and 13-8, i.e.,

$$\sum_{j=1}^{\infty} = \sum_{j=1}^{M}$$

Example 13.6 Deflection of a Gridwork with Uniform Force The grillage of Fig. 13-6 is loaded with a uniform force of 10 psi. Use the formulas of Tables 13-7 and 13-8 to find the deflections at the intersections of the beams. Assume the axial forces in both the girders and stiffeners are zero.

As indicated in case 3, Table 13-8, only a single term is needed in the summation of the formulas of Table 13-7. It is reasonable to assume that the loading intensity along either of the stiffeners will be $p_s = (10 \text{ psi})L_g/(n_s + 1) = 10(\frac{100}{3}) = 333.33 \text{ lb/in.}$ Use one term of case 1, Table 13-7:

$$w_g = \sin \frac{\pi g}{n_g + 1} K_1 \sin \frac{\pi x}{L_g} = K_1 \sin \frac{\pi g}{3} \sin \frac{\pi x}{100} \tag{1}$$

where, from case 3 of Table 13-8, since $P_g = P_s = 0$,

$$K_1 = \frac{\dfrac{4L_s^4}{EI_s\pi^5} \displaystyle\sum_{s=1}^{2} p_s \sin \dfrac{\pi s}{3}}{\dfrac{3}{2} + \dfrac{3}{2}} = \frac{\dfrac{4L_s^4 p_s}{EI_s\pi^5}(\sqrt{3}/2 + \sqrt{3}/2)}{\dfrac{3}{2} + \dfrac{3}{2}} = \frac{4L_s^4 p_s}{EI_s\pi^5}\frac{\sqrt{3}}{3} \tag{2}$$

Then

$$w_1|_{x=L_g/3} = w_2|_{x=L_g/3} = w_1|_{x=2L_g/3} = w_2|_{x=2L_g/3}$$

$$= \frac{4L_s^4 p_s}{EI_s\pi^5}\frac{\sqrt{3}}{3} \sin \frac{\pi}{3} \sin \frac{\pi}{3} = 0.062886 \text{ in.} \tag{3}$$

Example 13.7 Moment in a Gridwork with Uniform Force and Axial Loads Find the maximum bending moment in the grillage of Fig. 13-6. The grillage is loaded with a transverse uniform force of 10 psi. In addition, the girders are subject to compressive axial forces of 5000 lb.

The bending moments in the girders are given by case 3, Table 13-7. As noted in case 3 of Table 13-8, only one term in case 3, Table 13-7, is required. Thus

$$M_g = EI_g \sin \frac{\pi g}{n_g + 1} K_1 \frac{\pi^2}{L_g^2} \sin \frac{\pi x}{L_g} \tag{1}$$

The coefficient K_1 is taken from case 3, Table 13-8. Use the data $L_s = L_g = 100$ in., $E = 3 \times 10^7$ psi, $I_s = I_g = 100$ in.^{4}, $P_s = 0$, $P_g = 5000$ lb, $n_s = 2$, $n_g = 2$, $p_s = 333.33$ lb/in. (Example 13.6).

$$P_e = \frac{\pi^2(3 \times 10^7)100}{100^2} = 2,960,881 = P_c, \qquad \frac{P_g}{P_c} = 1.69 \times 10^{-3} \tag{2}$$

$$K_1 = \frac{\dfrac{4L_s^4 p_s}{EI_s \pi^5} \displaystyle\sum_{s=1}^{2} \sin \dfrac{\pi s}{3}}{\tfrac{3}{2}(0.99831) + \tfrac{3}{2}} = \frac{4L_s^4 p_s}{EI_s \pi^5}(0.57784) \tag{3}$$

It follows from symmetry that the maximum moment occurs at $x = \tfrac{1}{2}L_g$. Then, for $g = 1$,

$$M_{1,\,\text{max}} = M_g|_{x=L_g/2} = EI_g \sin\left(\frac{\pi}{3}\right) \frac{4L_s^4 p_s}{EI_s \pi^5}(0.57784)\frac{\pi^2}{L_g^2} = 215,190 \text{ in.-lb} \tag{4}$$

Example 13.8 Deflections Due to Concentrated Forces Consider again the grillage of Fig. 13-6. Assume there are no distributed or in-plane axial forces. Suppose concentrated forces of 10,000 lb act at each intersection.

With equal concentrated forces, sufficient accuracy is usually achieved with one term of the formulas of Table 13.7:

$$w_g = K_1 \sin \frac{\pi g}{3} \sin \frac{\pi x}{100} \tag{1}$$

with (case 1 of Table 13-8)

$$K_1 = \frac{\dfrac{2L_s^3}{EI_s \pi^4} \times 10,000 \displaystyle\sum_{s=1}^{2}\sum_{g=1}^{2} \sin \dfrac{\pi g}{3} \sin \dfrac{\pi s}{3}}{\tfrac{3}{2} + \tfrac{3}{2}} = \frac{2L_s^3}{EI_s \pi^4} \times 10,000 \tag{2}$$

Substitute (2) into (1),

$$w_1|_{x=L_g/3} = w_2|_{x=L_g/3} = w_1|_{x=2L_g/3} = w_2|_{x=2L_g/3} = 0.0514 \text{ in.} \tag{3}$$

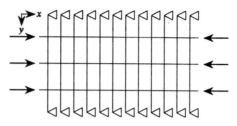

FIGURE 13-7   Example 13.9.

Buckling Loads

The buckling or critical axial loads in the girders of uniform gridworks are given in Tables 13-9 and 13-10. That is, these are formulas for $P_g = P_{cr}$. The formulas that apply for girders and stiffeners with fixed or simply supported ends are accurate for gridworks with more than five stiffeners. In some cases, the formulas will be sufficiently accurate for as few as three stiffeners.

Example 13.9   **Buckling Loads**   Compute the critical axial forces in the girders of the gridwork of Fig. 13-7 if the girders can be simply supported or fixed. The stiffeners are simply supported. Suppose $I_g = I_s$ and $L_g = L_s = L$. From Fig. 13-7, $n_g = 3$ and $n_s = 12$.

The girder buckling loads P_{cr} are given by the formulas of Table 13-9 for girders with fixed or simply supported ends. These formulas involve the constant C_1, which is taken from Table 13-10 according to the stiffener end conditions. To use Table 13-9, first calculate D_1. For simply supported stiffeners and $n_g = 3$, the constant C_1 is given as 0.041089 in Table 13-10. Thus

$$D_3 = \sqrt{C_1 L_g L_s^3 I_g / [I_s(n_s + 1)]} = \sqrt{C_1 L^4 / 13} = L^2 \sqrt{C_1/13}$$

and

$$D_1 = 0.0866 L_g^2 / D_3 = 0.0866 \sqrt{13/C_1} = 1.54$$

$$D_2 = 0.202 L_g^2 / D_3 = 3.5930$$

Since $D_1 > 1$, cases 2 and 4 in Table 13-9 are used. These give $P_{cr} = D_2 P_e = 3.5930 P_e$ for simply supported girders and $P_{cr} = 6.5930 P_e$ for fixed girders.

Natural Frequencies

Designate the natural frequencies of a gridwork as ω_{mn} where the subscript m indicates the number of mode shape half waves in the y (stiffener) direction and n indicates the number of half waves in the x (girder) direction. Figure 13-8 illustrates typical mode shapes associated with ω_{mn}.

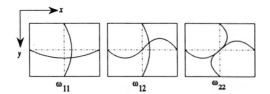

ω_{11} ω_{12} ω_{22}

FIGURE 13-8 Mode shapes corresponding to frequencies ω_{mn}.

For a uniform grillage with simply supported stiffeners, the lower natural frequencies (radians per time) are given by

$$
\omega_{mn}^2 = \frac{EI_s L_g \left(\dfrac{\pi m}{L_s}\right)^4 + EI_g \dfrac{n_g + 1}{C_n L_g^3} - P_s \left(\dfrac{m\pi}{L_s}\right)^2 L_g}{\rho_s L_s + \rho_g L_g}
\tag{13.1}
$$

where
n_g = number of girders
I_g, I_s = moments of inertia of girders and stiffeners, respectively
L_g, L_s = length of girders and stiffeners, respectively
ρ_g, ρ_s = mass per unit length of girders and stiffeners, respectively $(M/L, FT^2/L^2)$
E = modulus of elasticity

The stiffener axial force P_s is simply set equal to zero if the stiffeners are not subject to axial forces. The parameter C_n is given in Table 13-11 for girders with fixed or simply supported ends. Recall that either set of grillage beams can be selected to be the girders.

If each of the girders is subjected to an axial force P_g, Eq. (13.1) still provides the natural frequencies if C_n is replaced by

$$
C_n \frac{P_e}{P_e - P_g}
\tag{13.2}
$$

where $P_e = \pi^2 EI_g/L_g^2$.

Example 13.10 Natural Frequencies of a Simply Supported Gridwork Find the lower natural frequencies of a grillage for which all beam ends are simply supported. This is a 3×3 grillage, i.e., $n_g = n_s = 3$, with $I_g = I_s = 100$ in.[4], $\rho_g = \rho_s = 1$ lb-s^2/in.^{2}, $L_g = L_s = 100$ in., $E = 3 \times 10^7$ psi. There are no axial forces, i.e., $P_s = 0$, $P_g = 0$. From Eq. (13.1)

$$
\omega_{mn}^2 = \frac{(3 \times 10^7)100\left[m^4\pi^4 + (3 + 1)/C_n\right]/100^3}{2 \times 100} = 15\left(m^4\pi^4 + \frac{4}{C_n}\right)
\tag{1}
$$

To calculate ω_{11}, ω_{21}, ω_{12}, and ω_{22}, enter Table 13-11 for $n_s = 3$ and find $C_1 = 0.041089$ and $C_2 = 0.0026042$. Use (1):

$$\omega_{11}^2 = 15(\pi^4 + 4/C_1) = 2921.37 \quad \text{or} \quad \omega_{11} = 54 \text{ rad/s}$$

$$\omega_{21}^2 = 15(16\pi^4 + 4/C_1) = 24{,}838.347 \quad \text{or} \quad \omega_{21} = 157.6 \text{ rad/s}$$

$$\omega_{12}^2 = 15(\pi^4 + 4/C_2) = 24{,}500.831 \quad \text{or} \quad \omega_{12} = 156.5 \text{ rad/s} \tag{2}$$

$$\omega_{22}^2 = 15(16\pi^4 + 4/C_2) = 46{,}417.89 \quad \text{or} \quad \omega_{22} = 215 \text{ rad/s}$$

Other frequencies can be calculated in a similar fashion.

General Grillages

The formulas for uniform gridworks are provided in this section. Since gridworks are a special case of frameworks, use a computer program for the analysis of frames to find the response of complicated grillages. The structural matrices for grillages are listed in Section 13.4 under plane frames with out-of-plane loading.

13.4 MATRIX METHODS

Frames and trusses (both generally referred to as frames) can be considered as assemblages of beams and bars. As a consequence they can be analyzed using the matrix methods (transfer and displacement) of Appendix III. The displacement method can be employed to obtain the nodal responses, while the displacements and forces between the nodes along the members can be obtained using the transfer matrix method. Such references as 13.7–13.10 contain frame analysis formulations.

Frames are often classified as being plane (two-dimensional) and spatial (three-dimensional) in engineering practice.

Transfer Matrix Method

The transfer matrices provided in Chapters 11 and 12 can be combined properly to obtain the transfer matrices for the analysis of frames or frame members. See Appendices II and III.

Stiffness and Mass Matrices

In general, the analysis of plane frames requires the inclusion of the axial effects (extension or torsion) as well as bending in the stiffness matrix. As discussed in Appendix III, the analysis also requires a transformation of many variables from local to global coordinates. Then the global system matrix can be assembled. For dynamic problems, the mass matrices can be treated similarly to get the system mass matrix. The nodal displacements are found by introducing the boundary conditions and solving resulting equations. See the examples in Appendix III.

The stiffness matrices for plane and space trusses and frames are presented in Tables 13-12 through 13-15. Mass matrices for frames are listed in Tables 13-16 to 13-17. All of these matrices use sign convention 2 of Appendix II. Use a frame analysis to analyze a truss for dynamic responses. Stiffness matrices for more complex members can be constructed from the general stiffness matrices of Chapter 11. For example, it is possible to introduce a 4×4 beam stiffness matrix that includes the effect of an axial force on bending. Also, if thin-walled cross sections are of concern, the 4×4 structural matrices of Chapter 14 can replace the 2×2 torsional matrices of this chapter.

Stability Analysis

The stiffness matrices listed in the tables of this chapter do not include the interactions between bending and axial forces. However, in some analyses, e.g., a stability analysis, this interaction must be considered in that the bending moment caused by the axial forces must be included. To do so, introduce the stiffness matrix of Table 11-22 with $P \neq 0$. The buckling loading can be obtained using a determinant search after the global stiffness matrix is assembled and the boundary conditions applied. The details of this instability procedure follow.

1. Perform a static analysis of the frame using the stiffness matrices given in Tables 13-12 to 13-15 to determine the axial forces in each element resulting from a given load.
2. Use element stiffness matrices, such as that given in Table 11-22, that include the effects of bending and the axial force interaction.
3. Assemble the element matrices to form the global stiffness matrix, and impose the boundary conditions on the global matrix using the procedure described in Appendix III.
4. Note that all internal axial forces remain in the same fixed proportions to each other throughout the search for the critical applied load. These fixed proportions are determined in Step 1. Introduce a single *load factor* λ that holds for global structural matrices that model the entire structure. This λ is a common factor that multiplies all loads as they vary in fixed proportion.
5. Let the determinant of the global stiffness matrix be zero and determine λ, usually employing a numerical search technique. This λ is the critical load factor.

The stability analysis can also be conducted approximately, but efficiently, by employing the geometric stiffness matrix given in Table 11-23 and using the displacement method of Appendix III.

REFERENCES

13.1 Leontovich, V., *Frames and Arches*, McGraw Hill, New York, 1959.

13.2 Kleinlogel, A., *Rigid Frame Formulas*, F. Ungar, New York, 1952.

13.3 *Handbook of Structural Stability*, Column Research Committee of Japan, Corona Publishing Company, Tokyo, 1971.

13.4 Timoshenko, S., *Theory of Elastic Stability*, McGraw-Hill, New York, 1936.

13.5 Baker, J., and Heyman, J., *Plastic Design of Frames*, Cambridge University Press, London, 1969.

13.6 Pilkey, W. D., and Chang, P. Y., *Modern Formulas for Statics and Dynamics*, McGraw-Hill, New York, 1978.

13.7 Weaver, W., and Johnston, P. R., *Structural Dynamics by Finite Elements*, Prentice-Hall, Englewood Cliffs, NJ, 1987.

13.8 Paz, M., *Structural Dynamics*, Van Nostrand Reinhold, New York, 1980.

13.9 Ross, C. T. F., *Finite Element Methods in Structural Mechanics*, Ellis Harwood, England, 1985.

13.10 Pilkey, W. D., and Okada, Y., *Matrix Methods in Mechanical Vibration*, Corona Publishing, Tokyo, 1989.

13.11 Schiff, D., *Dynamic Analysis and Failure Modes of Simple Structures*, Wiley, New York, 1990.

13.12 Blevins, R. D., *Formulas for Natural Frequency and Mode Shape*, Van Nostrand, New York, 1979.

13

Tables

TABLE 13-1 STATICALLY DETERMINATE RECTANGULAR SINGLE BAY FRAMES OF CONSTANT CROSS SECTION

The direction of the reaction forces are shown in the figures of the configurations. The signs of the moments are shown in the moment diagrams. A bending moment is indicated as positive when it causes tension on the inner side of the frame and compression on the outer side. Opposing moments are negative. The formulas in the table give the magnitudes of these quantities. The horizontal and vertical coordinate axes are x and y, respectively. v_{jk} is the displacement of point j in the k direction. θ_j is the slope at j.

Configuration	Moment Diagram	Important Values
1.		$H_A = 0$ $R_A = R_B = \frac{1}{2}W$ $v_{Bx} = \dfrac{WhL^2}{8EI}$ $M_{max} = \frac{1}{4}WL$ at point K
2.		$H_A = W \qquad R_A = R_B = W\dfrac{h}{L}$ $v_{Bx} = \dfrac{Wh^2}{6EI}(3L + 2h)$ $v_{Cy} = 0 \qquad v_{Cx} = \dfrac{Wh^2}{3EI}(L + h)$ $M_{max} = Wh$ at point D
3.		$H_A = W \qquad R_A = R_B = 0$ $v_{Bx} = \dfrac{wh^2}{3EI}(3L + 2h)$ $M_{max} = Wh$
4.		$H_A = 0 \qquad R_A = R_B = \dfrac{M_0}{L}$ $v_{Bx} = \dfrac{M_0 hL}{2EI}$ $M_{max} = M_0$ at point C

TABLE 13-1 | Statically Determinate Rectangular Frames 648

Configuration	Moment Diagram	Important Values
5.		$H_A = 0 \qquad R_A = R_B = \dfrac{M_0}{L}$ $\theta_K = \dfrac{M_0 L}{12 EI}$ $M_{max} = \tfrac{1}{2} M_0 \quad$ at point K
6.		$H_A = 0 \qquad R_A = R_B = \tfrac{1}{2} p_1 L$ $v_{Bx} = \dfrac{p_1 h L^3}{12 EI}$ $M_{max} = \tfrac{1}{8} p_1 L^2 \quad$ at $x = \tfrac{1}{2} L$
7.		$H_A = p_1 h \qquad R_A = R_B = \dfrac{p_1 h^2}{2L}$ $v_{Bx} = \dfrac{p_1 h^3}{24 EI}(6L + 5h)$ $M_{max} = \tfrac{1}{2} p_1 h^2 \quad$ at point D
8.		$H_A = p_1 h \qquad R_A = R_B = \dfrac{p_1 h^2}{2L}$ $v_{Bx} = \dfrac{p_1 h^3}{24 EI}(18L + 11h)$ $M_{max} = p_1 h^2 \quad$ at point D
9.		$H_A = W \qquad R_A = 0 \qquad M_A = 0$ $v_{Dx} = \dfrac{Wh^2}{3 EI}(3L + 4h)$ $v_{Dy} = -\dfrac{WhL}{2 EI}(L + h)$ $M_{max} = Wh \quad$ at points B, C

Configuration	Moment Diagram	Important Values
10.		$H_A = 0 \qquad R_A = W \qquad M_A = WL$ $$v_{Dx} = -\frac{WhL}{2EI}(L + 2h)$$ $$v_{Dy} = \frac{WL^2}{3EI}(L + 3h)$$ $M_{\max} = WL$
11.		$H_A = W \qquad R_A = 0 \qquad M_A = Wh$ $$v_{Dx} = -\frac{Wh^3}{2EI} \qquad v_{Dy} = \frac{WLh^2}{2EI}$$ $$v_{Cx} = \frac{Wh^3}{3EI} \qquad v_{Cy} = \frac{WLh^2}{2EI}$$ $M_{\max} = Wh \quad$ at point A
12.		$H_A = 0 \qquad R_A = 0 \qquad M_A = M_0$ $$v_{Dx} = \frac{M_0 h}{EI}(L + 3h)$$ $$v_{Dy} = -\frac{M_0 L}{2EI}(L + 2h)$$ $\theta_D = \dfrac{M_0}{EI}(L + 2h) \qquad M_{\max} = M_0$
13.		$H_A = 0 \qquad R_A = p_1 L$ $M_A = \frac{1}{2}p_1 L^2$ $$v_{Dx} = -\frac{p_1 L^2 h}{6EI}(L + 3h)$$ $$v_{Dy} = \frac{p_1 L^3}{8EI}(L + 4h)$$ $M_{\max} = \frac{1}{2}p_1 L^2$

TABLE 13-1 | **Statically Determinate Rectangular Frames** | **650**

Configuration	Moment Diagram	Important Values
14.		$H_A = 0 \qquad R_A = W \qquad M_A = WL$ $$v_{Cx} = \frac{WLh^2}{2EI}$$ $$v_{Cy} = \frac{WL^2}{3EI}(L + 3h)$$ $$\theta_C = \frac{WL}{2EI}(L + 2h)$$ $M_{max} = WL$
15.		$H_A = W \qquad R_A = 0 \qquad M_A = Wh$ $$v_{Cx} = \frac{Wh^3}{3EI}$$ $$v_{Cy} = \frac{Wh^2 L}{2EI}$$ $M_{max} = Wh \quad$ at point A
16.		$H_A = 0 \qquad R_A = 0 \qquad M_A = M_0$ $$v_{Cx} = \frac{M_0 h^2}{2EI}$$ $$v_{Cy} = \frac{M_0 L}{2EI}(L + 2h)$$ $$\theta_C = \frac{M_0}{EL}(L + h)$$ $M_{max} = M_0$
17.		$H_A = 0 \qquad R_A = p_1 L$ $M_A = \frac{1}{2}p_1 L^2$ $$v_{Cx} = \frac{p_1 h^2 L^2}{4EI}$$ $$v_{Cy} = \frac{p_1 L^3}{8EI}(L + 4h)$$ $M_{max} = \frac{1}{2}p_1 L^2$

Configuration	Moment Diagram	Important Values
18.		Free-end relative displacement $$v = v_{Ax} - v_{Bx} = \frac{Wa^2}{3EI}(2a + 3L)$$ $$M_{max} = Wa$$
19.		Free-end relative displacement $$v = v_{Ax} - v_{Bx} = \frac{M_0 a}{EI}(a + L)$$ $$M_{max} = M_0$$
20.		Free-end relative displacement $$v = v_{Ax} - v_{Bx} = \frac{p_1 a^3}{4EI}(a + 2L)$$ $$M_{max} = \tfrac{1}{2}p_1 a^2$$

TABLE 13-1   **Statically Determinate Rectangular Frames**   **652**

TABLE 13-2   STATICALLY INDETERMINATE RECTANGULAR FRAMES

The directions of the reaction forces are shown in the figures of the configurations. The signs of moments are shown in the moment diagrams. A bending moment is indicated as positive when it causes tension on the inner side of the member and compression on the outer side. Opposing moments are negative. The formulas in the table give the magnitudes of the forces and moments.

Definitions

$$e = h/L$$
$$\beta = I_h \text{ (horizontal beam)}/I_v \text{ (vertical members)}.$$

Configuration	Moment Diagram	Important Values
1.		$R_A = W\dfrac{L-a}{L}$ $R_B = W\dfrac{a}{L}$ $H_A = H_B = \dfrac{3Wa}{2hL}\dfrac{L-a}{2\beta e + 3}$ $M_C = M_D = \dfrac{3Wa}{2L}\dfrac{L-a}{2\beta e + 3}$ $M_K = \dfrac{Wa(L-a)}{2L}\dfrac{4\beta e + 3}{2\beta e + 3}$
2.		$R_A = R_B = W\dfrac{h}{L}$ $H_A = H_B = \tfrac{1}{2}W$ $M_C = M_D = \tfrac{1}{2}Wh$

TABLE 13-2 (continued)   STATICALLY INDETERMINATE RECTANGULAR FRAMES

Configuration	Moment Diagram	Important Values
3.		$R_A = R_B = W\dfrac{h-a}{L}$
		$H_A = \dfrac{W}{2h}\left[h + a - (h-a)\dfrac{a\beta(2h-a)}{h(2h\beta + 3L)}\right]$
		$H_B = \dfrac{W(h-a)}{2h}\left[1 + \dfrac{a\beta(2h-a)}{h(2h\beta + 3L)}\right]$
		$M_C = \tfrac{1}{2}W(h-a)\left[1 + \dfrac{a\beta(2h-a)}{h(2h\beta + 3L)}\right]$
		$M_D = \tfrac{1}{2}W(h-a)\left[1 - \dfrac{a\beta(2h-a)}{h(2h\beta + 3L)}\right]$
		$M_k = \dfrac{W(h-a)}{2h}$
		$\times \left[h + a - (h-a)\dfrac{a\beta(2h-a)}{h(2h\beta + 3L)}\right]$
4.		$R_A = R_B = \tfrac{1}{2}p_1 L$
		$H_A = H_B = \dfrac{p_1 L}{4e(2\beta e + 3)}$
		$M_C = M_D = \dfrac{p_1 L^2}{4(2\beta e + 3)}$
		$M_K = \dfrac{p_1 L^2}{8}\dfrac{2\beta e + 1}{2\beta e + 3}$
5.		$R_A = R_B = \dfrac{p_1 h^2}{2L}$
		$H_A = \dfrac{p_1 h}{8}\dfrac{11\beta e + 18}{2\beta e + 3}$
		$H_B = \dfrac{p_1 h}{8}\dfrac{5\beta e + 6}{2\beta e + 3}$
		$M_C = \dfrac{p_1 h^2}{8}\dfrac{5\beta e + 6}{2\beta e + 3}$
		$M_D = \dfrac{3p_1 h^2}{8}\dfrac{\beta e + 2}{2\beta e + 3}$

TABLE 13-2   Statically Indeterminate Rectangular Frames   **654**

TABLE 13-2 (continued)   STATICALLY INDETERMINATE RECTANGULAR FRAMES

Configuration	Moment Diagram	Important Values
6.		$R_A = R_B = \frac{1}{2}W$ $H_A = H_B = \dfrac{3WL}{8h(\beta e + 2)}$ $M_A = M_B = \dfrac{WL}{8(\beta e + 2)}$ $M_C = M_D = \dfrac{WL}{4(\beta e + 2)}$ $M_K = \dfrac{WL}{4}\dfrac{\beta e + 1}{\beta e + 2}$
7.		$R_A = R_B = \frac{1}{2}p_1 L$ $H_A = H_B = \dfrac{p_1 L^2}{4h(\beta e + 2)}$ $M_A = M_B = \dfrac{p_1 L^2}{12(\beta e + 2)}$ $M_C = M_D = \dfrac{p_1 L^2}{6(\beta e + 2)}$ $M_K = \dfrac{p_1 L^2(3\beta e + 2)}{24(\beta e + 2)}$
8.		$R_A = R_B = p_1 h \dfrac{\beta e^2}{6\beta e + 1}$ $H_A = \dfrac{p_1 h}{4}\left[\dfrac{8\beta e + 17}{2(\beta e + 2)} - \dfrac{4\beta e + 3}{6\beta e + 1}\right]$ $H_B = \dfrac{p_1 h}{4}\left[\dfrac{4\beta e + 3}{6\beta e + 1} - \dfrac{1}{2(\beta e + 2)}\right]$ $M_A = \dfrac{p_1 h^2}{4}\left[\dfrac{4\beta e + 1}{6\beta e + 1} + \dfrac{\beta e + 3}{6(\beta e + 2)}\right]$ $M_B = \dfrac{p_1 h^2}{4}\left[\dfrac{4\beta e + 1}{6\beta e + 1} - \dfrac{\beta e + 3}{6(\beta e + 2)}\right]$ $M_C = p_1 h^2 \dfrac{\beta e}{4}\left[\dfrac{2}{6\beta e + 1} + \dfrac{1}{6(\beta e + 2)}\right]$ $M_D = p_1 h^2 \dfrac{\beta e}{4}\left[\dfrac{6}{6\beta e + 1} - \dfrac{1}{6(\beta e + 2)}\right]$

TABLE 13-2 (continued)   STATICALLY INDETERMINATE RECTANGULAR FRAMES

Configuration	Moment Diagram	Important Values
9.		$R_A = \dfrac{Wa\left[L^2(2\beta e + 3) - a^2\right]}{2L^3(\beta e + 1)}$ $R_B = W - R_A$ $H_A = H_B = \dfrac{Wa(L^2 - a^2)}{2hL^2(\beta e + 1)}$ $M_C = \dfrac{Wa(L^2 - a^2)}{2L^2(\beta e + 1)}$ $M_D = \dfrac{a}{L}\left[W(L - a) - M_C\right]$
10.		$R_A = \dfrac{p_1 L}{8}\,\dfrac{4\beta e + 5}{\beta e + 1}$ $R_B = \dfrac{p_1 L}{8}\,\dfrac{4\beta e + 3}{\beta e + 1}$ $H_A = H_B = \dfrac{p_1 L^2}{8h(\beta e + 1)}$ $M_C = \dfrac{p_1 L^2}{8(\beta e + 1)}$
11.		$R_A = \dfrac{Wa}{L}\left(1 + \dfrac{2}{L^2}\dfrac{L^2 - a^2}{3\beta e + 4}\right)$ $R_B = \dfrac{W(L - a)}{L}\left(1 - \dfrac{2a}{L^2}\dfrac{L + a}{3\beta e + 4}\right)$ $H_A = H_B = \dfrac{3Wa}{hL^2}\dfrac{L^2 - a^2}{3\beta e + 4}$ $M_A = \dfrac{Wa}{L^2}\dfrac{L^2 - a^2}{3\beta e + 4}$ $M_C = \dfrac{2Wa}{L^2}\dfrac{L^2 - a^2}{3\beta e + 4}$ $M_D = \dfrac{Wa(L - a)}{L}\left(1 - \dfrac{2a}{L^2}\dfrac{L + a}{3\beta e + 4}\right)$

TABLE 13-2   **Statically Indeterminate Rectangular Frames**   **656**

TABLE 13-2 (continued) STATICALLY INDETERMINATE RECTANGULAR FRAMES

Configuration	Moment Diagram	Important Values
12.		$R_A = R_B = \dfrac{3Wa(h-a)^2}{hL^2}\dfrac{\beta}{3\beta e + 4}$ $H_A = \dfrac{Wa}{h}\left[1 + \dfrac{h-a}{h^2}\dfrac{3a\beta e + 2(h+a)}{3\beta e + 4}\right.$ $\left. - \dfrac{3(h-a)^2}{hL}\dfrac{\beta}{3\beta e + 4}\right]$ $H_B = W - H_A$ $M_A = \dfrac{Wa(h-a)}{h^2}\dfrac{3a\beta e + 2(h+a)}{3\beta e + 4}$ $M_C = \dfrac{3Wa(h-a)^2}{hL}\dfrac{\beta}{3\beta e + 4}$ $M_D = H_A(h-a) - M_A$
13.		$R_B = \tfrac{3}{2}p_1 L\dfrac{\beta e + 1}{3\beta e + 4}$ $R_A = \tfrac{1}{2}p_1 L\dfrac{3\beta e + 5}{3\beta e + 4}$ $H_A = H_B = \dfrac{3p_1 L^2}{4h(3\beta e + 4)}$ $M_A = \dfrac{p_1 L^2}{4(3\beta e + 4)}$ $M_C = \dfrac{p_1 L^2}{2(3\beta e + 4)}$
14.		$R_A = R_B = \tfrac{1}{4}p_1 h\dfrac{\beta e^2}{3\beta e + 4}$ $H_A = \tfrac{1}{2}p_1 h\dfrac{3\beta e + 5}{3\beta e + 4}$ $H_B = \tfrac{3}{2}p_1 h\dfrac{\beta e + 1}{3\beta e + 4}$ $M_A = \tfrac{1}{4}p_1 h^2\dfrac{\beta e + 2}{3\beta e + 4}$ $M_C = \tfrac{1}{4}p_1 h^2\dfrac{\beta e}{3\beta e + 4}$

TABLE 13-2 (continued)   STATICALLY INDETERMINATE RECTANGULAR FRAMES

Configuration	Moment Diagram	Important Values
15.		$R_A = \dfrac{Wa^2}{2L^3(\beta e + 1)}$ $\times [\beta e(3L - a) + 2(3L - 2a)]$ $R_B = W - R_A$ $H_A = H_B = \dfrac{3Wa^2}{2hL^2}\dfrac{L - a}{\beta e + 1}$ $M_A = \dfrac{Wa^2}{2L^2}\dfrac{L - a}{\beta e + 1}$ $M_B = \dfrac{Wa(L - a)}{2L^2}$ $\times \left[\dfrac{\beta e(2L - a) + 2(L - a)}{\beta e + 1}\right]$ $M_C = \dfrac{Wa^2}{L^2}\dfrac{L - a}{\beta e + 1}$ $M_D = R_B a - M_B$
16.		$R_A = \tfrac{1}{8}p_1 L\,\dfrac{3\beta e + 4}{\beta e + 1}$ $R_B = \tfrac{1}{8}p_1 L\,\dfrac{5\beta e + 4}{\beta e + 1}$ $H_A = H_B = \dfrac{p_1 L^2}{8h(\beta e + 1)}$ $M_A = \dfrac{p_1 L^2}{24(\beta e + 1)}$ $M_B = \tfrac{1}{24}p_1 L^2\,\dfrac{3\beta e + 2}{\beta e + 1}$ $M_C = \dfrac{p_1 L^2}{12(\beta e + 1)}$

TABLE 13-2   **Statically Indeterminate Rectangular Frames**   **658**

TABLE 13-3 NON-RECTANGULAR SINGLE BAY FRAMES

The direction of the reaction forces are shown in the figures of the configurations. The signs of moments are shown in the moment diagrams. A bending moment is indicated as positive when it causes tension on the inner side of the member and compression on the outer side. Opposing moments are negative. The formulas in the table give the magnitudes of these quantities.

Symmetrical Gable Frames

$$k = \frac{I_1 a}{I_2 h} \qquad \phi = \frac{f}{h} \qquad \alpha = 4\left(3 + 3\phi + \phi^2 + \frac{1}{k}\right)$$

$$\gamma = \frac{3(1 - k\phi)}{2(1 + k\phi^2)} \qquad \lambda = \frac{6(1 + k)}{1 + k\phi^2} \qquad \eta = 12[2 + 2k - \gamma(1 - k\phi)]$$

Configuration	Moment Diagram	Important Values
1.		$H_A = H_B = \dfrac{WL(3 + 2\phi)}{2\alpha h}$ $R_A = R_B = \frac{1}{2}W$ $M_E = M_C = H_B h$ $M_D = \frac{1}{4}WL - H_B h(1 + \phi)$
2.		$H_B = \dfrac{W}{\alpha}\left(6 + 3\phi + \dfrac{2}{k}\right)$ $H_A = W - H_B$ $R_A = R_B = \dfrac{Wh}{L}$ $M_E = h(W - H_B)$ $M_C = H_B h$ $M_D = H_B h(1 + \phi) - \frac{1}{2}Wh$

TABLE 13-3 (continued)   NON-RECTANGULAR SINGLE BAY FRAMES

Configuration	Moment Diagram	Important Values
3.		$H_A = H_B = \dfrac{p_1 L^2}{8\alpha h}(8 + 5\phi)$ $R_A = R_B = \frac{1}{2}p_1 L$ $M_E = M_C = H_B h$ $M_D = \frac{1}{8}p_1 L^2 - H_B h(1 + \phi)$
4.		$W = p_1(f + h)$ $H_B = \dfrac{p_1 h}{4\alpha}\left(12 + \dfrac{8\phi}{k} + 30\phi \right.$ $\left. + 20\phi^2 + 5\phi^3 + \dfrac{5}{k}\right)$ $H_A = W - H_B$ $R_A = R_B = \dfrac{p_1(h + f)^2}{2L}$ $M_E = H_A h - \frac{1}{2}p_1 h^2$ $M_C = -H_B h$ $M_D = -\frac{1}{4}p_1(h + f)^2$ $\qquad + H_B h(1 + \phi)$
5.		$H_A = H_B = \dfrac{WLk}{\eta h}(3\gamma + \lambda\phi)$ $R_A = R_B = \frac{1}{2}W$ $M_E = M_C = \dfrac{WLk}{\eta}(3 + 2\gamma\phi)$ $M_A = M_B = -M_E + H_A h$ $M_D = -M_E + \frac{1}{4}WL - H_B f$

TABLE 13-3   **Non-Rectangular Single Bay Frames**   **660**

TABLE 13-3 (continued)   NON-RECTANGULAR SINGLE BAY FRAMES

Configuration	Moment Diagram	Important Values
6.		$H_B = \dfrac{2W}{\eta}(\lambda - 3\gamma)$ $H_A = W - H_B$ $R_A = R_B = \dfrac{3Wh}{2(3+k)L}$ $M_E = 4Wh\left(\dfrac{3-2\gamma}{2\eta}\right.$ $\left. + \dfrac{3}{16(3+k)}\right)$ $M_C = 4Wh\left(\dfrac{-3+2\gamma}{2\eta}\right.$ $\left. + \dfrac{3}{16(3+k)}\right)$ $M_A = h(W - H_B) - M_E$ $M_B = -M_C + H_B h$ $M_D = H_B f - \dfrac{2Wh}{\eta}(3 - 2\gamma)$
7.		$S = 2 + \frac{5}{4}\gamma\phi \qquad T = 2\gamma + \frac{5}{8}\lambda\phi$ $H_A = H_B = \dfrac{p_1 L^2 T k}{\eta h}$ $R_A = R_B = \frac{1}{2}p_1 L$ $M_A = M_B = \dfrac{p_1 L^2 k}{\eta}(T - S)$ $M_E = M_C = \dfrac{p_1 L^2 S k}{\eta}$ $M_D = -\dfrac{p_1 L^2 S k}{\eta} + \frac{1}{8}p_1 L^2 - H_B f$ For the left half of girder $M_x = \left(-M_E + \frac{1}{4}p_1 L x\right)$ $\times\left(1 - \dfrac{2x}{L}\right) + M_D \dfrac{2x}{L}$

TABLE 13-3 (continued) NON-RECTANGULAR SINGLE BAY FRAMES

Configuration	Moment Diagram	Important Values
8.		$S = 6 - 4\gamma - k\phi\left(4 + \frac{5}{2}\gamma\phi\right)$ $T = 2\lambda + k\phi\left(4\gamma + \frac{5}{4}\lambda\phi\right) - 6\gamma$ $R = \dfrac{Sf}{h + f} + \dfrac{h}{h + f}\left(2 - \frac{3}{2}\gamma\right)$ $Q = \dfrac{4h}{h + f} + \dfrac{f}{h + f}(12 - k\phi)$ $W = p_1(f + h)$ $H_B = \dfrac{W}{\eta(h + f)}\left(Tf + \frac{3}{4}\lambda h - 2\gamma h\right)$ $H_A = W - H_B$ $R_A = R_B = \dfrac{Wh}{32(3 + k)L}$ $\times\left(4Q + 16(3 + k)\dfrac{f}{h + f}\phi\right)$ $M_E = Wh\left(\dfrac{R}{\eta} + \dfrac{Q}{16(3 + k)}\right)$ $M_C = -Wh\left(\dfrac{R}{\eta} - \dfrac{Q}{16(3 + k)}\right)$ $M_A = -M_E - H_Bh + \frac{1}{2}Wh\dfrac{h + 2f}{h + f}$ $M_D = -\frac{1}{2}(M_E - M_C) + H_Bf$ $\qquad - \dfrac{Wf^2}{4(h + f)}$ $M_B = -M_C + H_Bh$

TABLE 13-3 Non-Rectangular Single Bay Frames **662**

TABLE 13-3 (continued)   NON-RECTANGULAR SINGLE BAY FRAMES

Symmetrical Arched Frames

$$k = \frac{I_1 L}{I_2 h} \qquad \phi = \frac{f}{h} \qquad \alpha = 8[1 + k(1.5 + 2\phi + 0.8\phi^2)]$$

$$\beta = \frac{1.5 - k\phi}{1 + 0.8k\phi^2} \qquad \gamma = \frac{3 + 1.5k}{1 + 0.8k\phi^2} \qquad \eta = 12(2 + k) - 4\beta(3 - 2k\phi)$$

Configuration	Moment Diagram	Important Values
9.		$H_A = H_B = \dfrac{WLk}{\alpha h} \dfrac{6 + 5\phi}{4}$ $R_A = R_B = \frac{1}{2}W$ $M_E = M_C = H_A h$ $M_D = \frac{1}{4}WL - H_A(h + f)$
10.		$e = \dfrac{4}{\alpha}(1 + 1.5k + k\phi)$ $H_B = We \qquad H_A = W - H_B$ $R_A = R_B = \dfrac{Wh}{L}$ $M_E = h(W - H_B)$ $M_C = H_B h$
11.		$H_A = H_B = \dfrac{p_1 L^2 k}{\alpha h}\left(1 + \frac{4}{5}\phi\right)$ $R_A = R_B = \frac{1}{2}p_1 L$ $M_E = M_C = H_A h$ $M_D = \frac{1}{8}p_1 L^2 - H_A(f + h)$

TABLE 13-3 (continued)   NON-RECTANGULAR SINGLE BAY FRAMES

Configuration	Moment Diagram	Important Values
12.		$e = 4(1 + 1.5k + k\phi)/\alpha$ $H_B = \dfrac{p_1 h}{2\alpha}(1 + \alpha e)$ $H_A = p_1 h - H_B$ $R_A = R_B = \dfrac{p_1 h^2}{2L}$ $M_E = \frac{1}{2}p_1 h^2 - H_B h$ $M_C = H_B h$
13.		$H_A = H_B = \dfrac{WLk}{\eta h}\,\dfrac{6\beta + 5\gamma\phi}{4}$ $R_A = R_B = \frac{1}{2}W$ $M_E = M_C = \dfrac{WLk}{\eta}\,\dfrac{6 + 5\beta\phi}{4}$ $M_A = M_B = -M_E + H_A h$ $M_D = \frac{1}{4}WL - M_E - H_A f$
14.		$H_B = \dfrac{2W}{\eta}(2\gamma - 3\beta)$ $H_A = W - H_B$ $R_A = R_B = \dfrac{3Wh}{(6 + k)L}$ $M_E = \dfrac{Wh}{\eta}(6 - 4\beta)$ $\qquad + \dfrac{3Wh}{2(6 + k)}$ $M_C = \dfrac{Wh}{\eta}(6 - 4\beta)$ $\qquad + \dfrac{3Wh}{2(6 + k)}$ $M_A = h(W - H_B) - M_E$ $M_B = -M_C + H_B h$

TABLE 13-3   **Non-Rectangular Single Bay Frames**   **664**

TABLE 13-3 (continued)   NON-RECTANGULAR SINGLE BAY FRAMES

Configuration	Moment Diagram	Important Values
15.		$H_A = H_B$ $= \dfrac{p_1 L^2 k}{5\eta h}(5\beta + 4\gamma\phi)$ $R_A = R_B = \frac{1}{2}p_1 L$ $M_E = M_C$ $= \dfrac{p_1 L^2 k}{5\eta}(5 + 4\beta\phi)$ $M_A = M_B = -M_E + H_A h$ $M_D = \frac{1}{8}p_1 L^2 - M_E - H_B f$
16.		$H_B = \dfrac{p_1 h}{2\eta}(3\gamma - 4\beta)$ $H_A = p_1 h - H_B$ $R_A = R_B = \dfrac{p_1 h^2}{(6 + k)L}$ $M_E = \dfrac{p_1 h^2}{2\eta}(4 - 3\beta)$ $\qquad + \dfrac{p_1 h^2}{2(6 + k)}$ $M_C = -\dfrac{p_1 h^2}{2\eta}(4 - 3\beta)$ $\qquad + \dfrac{p_1 h^2}{2(6 + k)}$ $M_A = -M_E - H_B h + \frac{1}{2}p_1 h^2$ $M_B = -M_C + H_B h$

TABLE 13-3 (continued)   NON-RECTANGULAR SINGLE BAY FRAMES

Symmetrical Polygonal Frames

$$k_1 = \frac{I_3 a}{I_1 e} \qquad k_2 = \frac{I_3 d}{I_2 e} \qquad B_0 = \frac{2a}{h}(k_1 + 1) + 1 \qquad C_0 = \frac{a}{h} + 2 + 3k_2$$

$$N_0 = \frac{aB_0}{h} + C_0 \qquad C_1 = \frac{b}{a}(2 + 3k_2) \qquad C_2 = 1 + \frac{h}{a}(2 + 3k_2)$$

$$C_3 = 1 + \frac{d}{L}(2 + k_2) \qquad R = \frac{b}{a}C_2 - k_1 \qquad N_1 = k_3 k_4 - R^2$$

$$\beta = 3k_1 + 2 + \frac{d}{L} \qquad N_2 = 3k_1 + \beta + \frac{d}{L}C_3 \qquad k_3 = 2(k_1 + 1) + \frac{h}{a}(1 + C_2)$$

$$k_4 = 2k_1 + \frac{b}{a}C_1$$

Configuration	Moment Diagram	Important Values
17.		$X = \dfrac{WcC_0 + \left(\frac{3}{4}Wd\right)k_2}{2N_0}$ $H_A = H_B = \dfrac{X}{h}$ $R_A = R_B = \frac{1}{2}W$ $M_E = M_D = \frac{1}{2}Wc - X$ $M_F = M_C = \dfrac{a}{h}X$ $M_k = \frac{1}{4}Wd + M_E$
18.		$X = \dfrac{p_1 dcC_0 + \frac{1}{2}p_1 d^2 k_2}{2N_0}$ $H_A = H_B = \dfrac{X}{h}$ $R_A = R_B = \frac{1}{2}p_1 d$ $M_E = M_D = \frac{1}{2}p_1 dc - X$ $M_F = M_C = \dfrac{a}{h}X$ $M_K = \frac{1}{8}p_1 d^2 + M_E$

TABLE 13-3   Non-Rectangular Single Bay Frames   666

TABLE 13-3 (continued)   NON-RECTANGULAR SINGLE BAY FRAMES

Configuration	Moment Diagram	Important Values
19. 		$$X = \frac{Wa(B_0 + C_0)}{2N_0}$$ $$H_B = \frac{X}{h} \qquad H_A = W - H_B$$ $$R_A = R_B = \frac{Wa}{L}$$ $$M_F = Wa - \frac{a}{h}X$$ $$M_C = \frac{a}{h}X$$ $$M_E = \left(1 - \frac{c}{L}\right)Wa - X$$ $$M_D = \frac{c}{L}Wa - X$$
20. 		$$X = \frac{p_1 a^2\left[2(B_0 + C_0) + \dfrac{a}{h}k_1\right]}{8N_0}$$ $$H_B = \frac{X}{h} \qquad H_A = p_1 a - H_B$$ $$R_A = R_B = \frac{p_1 a^2}{2L}$$ $$M_F = \tfrac{1}{2}p_1 a^2 - \frac{a}{h}X$$ $$M_C = \frac{a}{h}X$$ $$M_E = R_B(L - c) - X$$ $$M_D = X - R_B c$$

TABLE 13-3 (continued)   NON-RECTANGULAR SINGLE BAY FRAMES

Configuration	Moment Diagram	Important Values
21.		$$B_1 = WcC_1 + \frac{3b}{4a}Wdk_2$$ $$B_2 = WcC_2 + \frac{3h}{4a}Wdk_2$$ $$X_1 = \frac{B_1k_3 - B_2R}{2N_1}$$ $$X_2 = \frac{B_2k_4 - B_1R}{2N_1}$$ $$H_A = H_B = \frac{1}{a}(X_1 + X_2)$$ $$R_A = R_B = \tfrac{1}{2}W$$ $$M_A = M_B = X_1$$ $$M_F = M_C = X_2$$ $$M_E = M_D = \tfrac{1}{2}Wc - \frac{b}{a}X_1$$ $$- \frac{h}{a}X_2$$ $$M_K = \tfrac{1}{4}Wd + M_E$$
22.		$$B_1 = p_1dcC_1 + \frac{p_1d^2b}{2a}k_2$$ $$B_2 = p_1dcC_2 + \frac{p_1d^2h}{2a}k_2$$ $$X_1 = \frac{B_1k_3 - B_2R}{2N_1}$$ $$X_2 = \frac{B_2k_4 - B_1R}{2N_1}$$ $$H_A = H_B = \frac{1}{a}(X_1 + X_2)$$ $$R_A = R_B = \tfrac{1}{2}p_1d$$ $$M_A = M_B = X_1$$ $$M_F = M_C = X_2$$ $$M_E = M_D = \tfrac{1}{2}p_1dc - \frac{b}{a}X_1$$ $$- \frac{h}{a}X_2$$ $$M_K = \tfrac{1}{8}p_1d^2 + M_E$$

TABLE 13-3   **Non-Rectangular Single Bay Frames**   **668**

TABLE 13-3 (continued)   NON-RECTANGULAR SINGLE BAY FRAMES

Configuration	Moment Diagram	Important Values
23.		$B_1 = bC_1W \qquad B_2 = bC_2W$ $$B_3 = Wa\left(\beta + \frac{d}{L}C_3\right)$$ $$X_1 = \frac{B_1k_3 - B_2R}{2N_1}$$ $$X_2 = \frac{B_2k_4 - B_1R}{2N_1}$$ $$X_3 = \frac{B_3}{2N_2}$$ $$H_B = \frac{W}{2} - \frac{X_1 + X_2}{a}$$ $$H_A = W - H_B$$ $$R_A = R_B = \frac{2}{L}\left(\frac{Wa}{2} - X_3\right)$$ $$M_A = X_1 + X_3$$ $$M_B = -X_1 + X_3$$ $$M_F = X_2 + \frac{Wa}{2} - X_3$$ $$M_C = X_2 - \frac{Wa}{2} + X_3$$ $$M_E = -\frac{Wb}{2} + \frac{b}{a}X_1 + \frac{h}{a}X_2$$ $$+ \frac{d}{L}\left(\frac{Wa}{2} - X_3\right)$$ $$M_D = \frac{Wb}{2} - \frac{b}{a}X_1 - \frac{h}{a}X_2$$ $$+ \frac{d}{L}\left(\frac{Wa}{2} - X_3\right)$$

TABLE 13-3 (continued)   NON-RECTANGULAR SINGLE BAY FRAMES

Configuration	Moment Diagram	Important Values

24.

$$B_1 = \frac{p_1 ab}{2} C_1 + \frac{p_1 a^2}{4} k_1$$

$$B_2 = \frac{p_1 ab}{2} C_2 - \frac{p_1 a^2}{4} k_1$$

$$B_3 = \frac{p_1 a^2}{2} \left(\beta + \frac{d}{L} C_3 + k_1 \right)$$

$$X_1 = \frac{B_1 k_3 - B_2 R}{2 N_1}$$

$$X_2 = \frac{B_2 k_4 - B_1 R}{2 N_1}$$

$$X_3 = \frac{B_3}{2 N_2}$$

$$H_B = \frac{p_1 a}{4} - \frac{X_1 + X_2}{a}$$

$$H_A = p_1 a - H_B$$

$$R_A = R_B = \frac{2}{L} \left(\frac{p_1 a^2}{4} - X_3 \right)$$

$$M_A = X_1 + X_3$$

$$M_B = -X_1 + X_3$$

$$M_F = X_2 + \left(\frac{p_1 a^2}{4} - X_3 \right)$$

$$M_C = X_2 - \left(\frac{p_1 a^2}{4} - X_3 \right)$$

$$M_E = -\frac{p_1 ab}{4} + \frac{b}{a} X_1$$

$$+ \frac{h}{a} X_2 + \frac{d}{L} \left(\frac{p_1 a^2}{4} - X_3 \right)$$

$$M_D = \frac{p_1 ab}{4} - \frac{b}{a} X_1 - \frac{h}{a} X_2$$

$$+ \frac{d}{L} \left(\frac{p_1 a^2}{4} - X_3 \right)$$

TABLE 13-3   **Non-Rectangular Single Bay Frames**   **670**

TABLE 13-4   BUCKLING LOADS FOR FRAMES

Notation

E = modulus of elasticity

I = moment of inertia

I_h, I_v = moments of inertia of horizontal and vertical members

A = area of cross section

A_h = area of the cross section of horizontal member

A_{vi} = area of the cross section of ith (from left to right) vertical member;
$\quad A_{vi} = A_v$ if all vertical members are identical

L = width of frame

h = height of frame

P_{cr} = buckling load; unless specified otherwise, $P_{cr} = \pi^2 EI_v/(\alpha h)^2$

α = constant given in table

$$k = \frac{I_v L}{I_h h} \qquad n = \frac{P_1 + P}{2P} \qquad m = \begin{cases} \dfrac{4I_v}{L^2 A_v} & \text{for cases 1 and 2} \\[2mm] \dfrac{I_v}{L^2}\left(\dfrac{1}{A_{v1}} + \dfrac{1}{A_{v2}}\right) & \begin{array}{l}\text{for cases 3, 4, 5,} \\ \text{and 6}\end{array} \\[2mm] \dfrac{4EI_h}{L} & \text{for cases 7 and 8} \end{cases}$$

$$\zeta(\eta) = \frac{3}{\eta}\left(\frac{1}{\sin 2\eta} - \frac{1}{2\eta}\right) \qquad \beta(\eta) = \frac{3}{2\eta}\left(\frac{1}{2\eta} - \frac{1}{\tan 2\eta}\right) \qquad \eta = \frac{h}{2}\sqrt{\frac{P}{EI_v}}$$

For the cases where two forces P_1 and P are applied, the ratio n is predetermined. Calculate α and then find P_{cr}. Then P_{1cr} can be calculated using $P_{1cr} = (2n - 1)P_{cr}$.

Configuration	Buckling Loads
1.	$\alpha = \sqrt{n} \cdot \sqrt{1 + 0.35k + 2.1m - 0.017(k + 6m)^2}$ $m \le 0.2 \qquad n \le 1 \qquad k \le 10$
2.	$\alpha = \sqrt{n} \cdot \sqrt{4 + 1.4k + 8.4m + 0.02(k + 6m)^2}$ $n \le 1 \qquad k \le 10 \qquad m \le 0.2$

TABLE 13-4 (continued)   BUCKLING LOADS FOR FRAMES

Configuration	Buckling Loads
3.	$$\alpha = \sqrt{1 + 0.7k + 2.1m - 0.068(k + 3m)^2}$$
4.	$$\alpha = \sqrt{(0.14 + 1.72n)\left[1 + 0.7k + 2.1m - 0.068(k + 3m)^2\right]}$$ $$n \leq 1.5$$
5.	$$\alpha = \sqrt{4 + 2.8k + 8.4m + 0.08(k + 3m)^2}$$
6.	$$\alpha = \sqrt{(0.04 + 1.92n)\left[4 + 2.8k + 8.4m + 0.08(k + 3m)^2\right]}$$ $$n \geq 1.5$$
7.	P_{cr} is determined by solving $$\frac{1}{m} + \frac{L\beta(\eta)}{3EI_v} = 0$$ Ref. 13.4

TABLE 13-4 Buckling Loads for Frames   **672**

TABLE 13-4 (continued)   BUCKLING LOADS FOR FRAMES

Configuration	Buckling Loads
8.	P_{cr} is obtained by solving $$\left[\frac{3EI_v}{mL} + \beta(\eta)\right]\beta(\eta) = \tfrac{1}{4}[\zeta(\eta)]^2$$ Ref. 13.4
9.	$\alpha = 0.558\pi$
10.	$\alpha = 0.623\pi$
11.	$\alpha = 0.701\pi$
12.	$\alpha = 0.9\pi$
13.	$\alpha = 0.627\pi$

TABLE 13-5 FUNDAMENTAL NATURAL FREQUENCIES OF FRAMES

Notation

E_h, E_v = moduli of elasticity of horizontal and vertical beams
G = shear modulus of elasticity
E = modulus of elasticity
I_h, I_v = moments of inertia of horizontal and vertical beams
J_v = torsional constants of vertical beam
ρ_i = mass per unit length of vertical beams; $\rho_i = \rho_v$, all vertical beams are identical
ρ_h = mass per unit length of horizontal beam
W = total weight of frame

$$f = \frac{\lambda^2}{2\pi h^2}\left(\frac{E_v I_v}{\rho_v}\right)^{1/2} \qquad \text{Hz (cycles/s) for cases 1, 2, 3, and 4.}$$

TABLE 13-5 **Fundamental Natural Frequencies of Frames** 674

Configuration	Natural Frequency
1. First symmetric in-plane mode, pinned	$\lambda = a_1 + a_2\sqrt{c_2} + a_3(\sqrt{c_2})^2 + a_4(\sqrt{c_2})^3 + a_5(\sqrt{c_2})^4$

Natural Frequency (continued):

$0.1 \leq c_2 \leq 10.0 \qquad 1.5 < c_1 \leq 10.0$

a_1	$0.05881 + 3.7774\dfrac{1}{c_1} + 4.4214\left(\dfrac{1}{c_1}\right)^2 - 4.5495\left(\dfrac{1}{c_1}\right)^3$
a_2	$0.006772 - 0.08744\dfrac{1}{c_1} + 1.8371\left(\dfrac{1}{c_1}\right)^2 - 16.9061\left(\dfrac{1}{c_1}\right)^3 + 15.9685\left(\dfrac{1}{c_1}\right)^4$
a_3	$0.1265 - 2.1961\dfrac{1}{c_1} + 6.139\left(\dfrac{1}{c_1}\right)^2 - 3.07026\left(\dfrac{1}{c_1}\right)^3$
a_4	$-0.04549 + 0.7259\dfrac{1}{c_1} - 1.4984\left(\dfrac{1}{c_1}\right)^2 + 0.4223\left(\dfrac{1}{c_1}\right)^3$
a_5	$0.00545 - 0.08128\dfrac{1}{c_1} + 0.1277\left(\dfrac{1}{c_1}\right)^2 + 0.00154\left(\dfrac{1}{c_1}\right)^3$

$0.1 \leq c_2 \leq 10.0 \qquad 0.1 \leq c_1 \leq 1.5$

a_1	$2.9505 + 0.01426c_1 + 0.3933c_1^2 - 0.1953c_1^3$
a_2	$2.09517 - 5.5922c_1 + 5.7203c_1^2 - 2.2752c_1^3$
a_3	$-1.6907 + 6.8916c_1 - 8.2798c_1^2 + 3.09981c_1^3$
a_4	$0.5590 - 2.6368c_1 + 3.3057c_1^2 - 1.2275c_1^3$
a_5	$-0.06605 + 0.3357c_1 - 0.4296c_1^2 + 0.1592c_1^3$

Mode shape

$c_1 = \dfrac{L}{h}\left(\dfrac{E_v I_v}{E_h I_h}\dfrac{\rho_h}{\rho_v}\right)^{1/4}\left(\dfrac{E_h I_h}{E_v I_v}\right)^{3/4}$

$c_2 = \left(\dfrac{\rho_h}{\rho_v}\right)^{1/4}\left(\dfrac{E_h I_h}{E_v I_v}\right)$

TABLE 13-5 (continued)   FUNDAMENTAL NATURAL FREQUENCIES OF FRAMES

Configuration	Natural Frequency
2. First symmetric in-plane mode, clamped [figure: E_h, I_h, ρ_h; E_v, I_v, ρ_v; h; L] **Mode shape** [figure] $c_1 = \dfrac{L}{h}\left(\dfrac{E_v I_v}{E_h I_h}\dfrac{\rho_h}{\rho_v}\right)^{1/4}$ $c_2 = \left(\dfrac{\rho_h}{\rho_v}\right)^{1/4}\left(\dfrac{E_h I_h}{E_v I_v}\right)^{3/4}$	$\lambda = a_1 + a_2\sqrt{c_2} + a_3\left(\sqrt{c_2}\right)^2 + a_4\left(\sqrt{c_2}\right)^3$ $0.1 \le c_2 \le 10.0 \qquad 1.2 < c_1 \le 10.0$ a_1: $18.33 - 23.028\sqrt{c_1} + 11.843\left(\sqrt{c_1}\right)^2 - 2.8164\left(\sqrt{c_1}\right)^3 + 0.25598\left(\sqrt{c_1}\right)^4$ a_2: $-6.951 + 8.992\sqrt{c_1} - 4.364\left(\sqrt{c_1}\right)^2 + 0.9325\left(\sqrt{c_1}\right)^3 - 0.07345\left(\sqrt{c_1}\right)^4$ a_3: $3.728 - 5.64\sqrt{c_1} + 3.169\left(\sqrt{c_1}\right)^2 - 0.7878\left(\sqrt{c_1}\right)^3 + 0.07319\left(\sqrt{c_1}\right)^4$ a_4: $-0.5991 + 0.9657\sqrt{c_1} - 0.5712\left(\sqrt{c_1}\right)^2 + 0.1485\left(\sqrt{c_1}\right)^3 - 0.01437\left(\sqrt{c_1}\right)^4$ $0.1 \le c_2 \le 10.0 \qquad 1.2 \ge c_1 \ge 0.1$ a_1: $2.1037 + 13.649\sqrt{c_1} - 37.686\left(\sqrt{c_1}\right)^2 + 42.2\left(\sqrt{c_1}\right)^3 - 16.218\left(\sqrt{c_1}\right)^4$ a_2: $1.8503 - 3.4236\sqrt{c_1} + 4.852\left(\sqrt{c_1}\right)^2 - 3.5313\left(\sqrt{c_1}\right)^3 - 0.34975\left(\sqrt{c_1}\right)^4$ a_3: $0.0647 - 5.8812\sqrt{c_1} + 19.008\left(\sqrt{c_1}\right)^2 - 21.873\left(\sqrt{c_1}\right)^3 + 8.8495\left(\sqrt{c_1}\right)^4$ a_4: $-0.06883 + 1.3714\sqrt{c_1} - 4.2867\left(\sqrt{c_1}\right)^2 + 4.8985\left(\sqrt{c_1}\right)^3 - 1.931\left(\sqrt{c_1}\right)^4$

TABLE 13-5   Fundamental Natural Frequencies of Frames   676

3.
First asymmetric
in-plane mode,
pinned

$$\lambda = a_1 + a_2(\sqrt{c_1}) + a_3 c_1 + a_4(\sqrt{c_1})^3$$

$$12.0 \geq h/L \geq 0.25 \qquad 12.0 \geq c_1 \geq 0.25$$

$$c_2 = 0.25$$

a_1	$0.5270 + 0.7587\sqrt{\dfrac{h}{L}} - 0.2330\dfrac{h}{L} + 0.02650\left(\sqrt{\dfrac{h}{L}}\right)^3$
a_2	$-0.7049 + 0.9064\sqrt{\dfrac{h}{L}} - 0.3750\dfrac{h}{L} + 0.04973\left(\sqrt{\dfrac{h}{L}}\right)^3$
a_3	$0.2644 - 0.4642\sqrt{\dfrac{h}{L}} + 0.2145\dfrac{h}{L} - 0.02996\left(\sqrt{\dfrac{h}{L}}\right)^3$
a_4	$-0.03382 + 0.06720\sqrt{\dfrac{h}{L}} - 0.03350\dfrac{h}{L} + 0.004914\left(\sqrt{\dfrac{h}{L}}\right)^3$

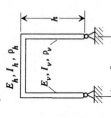

E_h, I_h, ρ_h

E_v, I_v, ρ_v

Mode shape

$$c_1 = \frac{E_v I_v}{E_h I_h}$$

$$c_2 = \frac{\rho_v}{\rho_h}$$

Fundamental Natural Frequencies of Frames   TABLE 13-5

TABLE 13-5 (continued) FUNDAMENTAL NATURAL FREQUENCIES OF FRAMES

Configuration	Natural Frequency

3. Continued

$c_2 = 0.75$

a_1
$$0.7608 + 0.7983\sqrt{\frac{h}{L}} - 0.2993\left(\sqrt{\frac{h}{L}}\right)^2 + 0.03833\left(\sqrt{\frac{h}{L}}\right)^3$$

a_2
$$-1.09597 + 1.8224\sqrt{\frac{h}{L}} - 1.1311\left(\sqrt{\frac{h}{L}}\right)^2 + 0.3092\left(\sqrt{\frac{h}{L}}\right)^3 - 0.03122\left(\sqrt{\frac{h}{L}}\right)^4$$

a_3
$$0.4930 - 1.1224\sqrt{\frac{h}{L}} + 0.8202\left(\sqrt{\frac{h}{L}}\right)^2 - 0.2500\left(\sqrt{\frac{h}{L}}\right)^3 + 0.02730\left(\sqrt{\frac{h}{L}}\right)^4$$

a_4
$$-0.06778 + 0.1709\sqrt{\frac{h}{L}} - 0.1329\left(\sqrt{\frac{h}{L}}\right)^2 + 0.04220\left(\sqrt{\frac{h}{L}}\right)^3 - 0.004738\left(\sqrt{\frac{h}{L}}\right)^4$$

$c_2 = 1.5$

a_1
$$0.8222 + 1.1944\sqrt{\frac{h}{L}} - 0.8201\left(\sqrt{\frac{h}{L}}\right)^2 + 0.2544\left(\sqrt{\frac{h}{L}}\right)^3 - 0.02879\left(\sqrt{\frac{h}{L}}\right)^4$$

a_2
$$-1.3211 + 2.3536\sqrt{\frac{h}{L}} - 1.5610\left(\sqrt{\frac{h}{L}}\right)^2 + 0.4528\left(\sqrt{\frac{h}{L}}\right)^3 - 0.0481\left(\sqrt{\frac{h}{L}}\right)^4$$

a_3
$$0.5721 - 1.3439\sqrt{\frac{h}{L}} + 1.0166\left(\sqrt{\frac{h}{L}}\right)^2 - 0.3193\left(\sqrt{\frac{h}{L}}\right)^3 + 0.03571\left(\sqrt{\frac{h}{L}}\right)^4$$

a_4
$$-0.07699 + 0.1987\sqrt{\frac{h}{L}} - 0.1587\left(\sqrt{\frac{h}{L}}\right)^2 + 0.05152\left(\sqrt{\frac{h}{L}}\right)^3 - 0.005889\left(\sqrt{\frac{h}{L}}\right)^4$$

TABLE 13-5 Fundamental Natural Frequencies of Frames 678

$c_2 = 3.0$

a_1	$1.2461 + 0.3113\sqrt{\dfrac{h}{L}} - 0.09981\left(\sqrt{\dfrac{h}{L}}\right)^2 + 0.007269\left(\sqrt{\dfrac{h}{L}}\right)^3 + 0.0009349\left(\sqrt{\dfrac{h}{L}}\right)^4$
a_2	$-2.002781 + 4.6441\sqrt{\dfrac{h}{L}} - 3.7526\left(\sqrt{\dfrac{h}{L}}\right)^2 + 1.2573\left(\sqrt{\dfrac{h}{L}}\right)^3 - 0.1480\left(\sqrt{\dfrac{h}{L}}\right)^4$
a_3	$1.1303 - 3.3717\sqrt{\dfrac{h}{L}} + 3.01029\left(\sqrt{\dfrac{h}{L}}\right)^2 - 1.06090\left(\sqrt{\dfrac{h}{L}}\right)^3 + 0.1286\left(\sqrt{\dfrac{h}{L}}\right)^4$
a_4	$-0.2818 + 0.9551\sqrt{\dfrac{h}{L}} - 0.9064\left(\sqrt{\dfrac{h}{L}}\right)^2 + 0.3304\left(\sqrt{\dfrac{h}{L}}\right)^3 - 0.04087\left(\sqrt{\dfrac{h}{L}}\right)^4$
a_5	$0.02631 - 0.09744\sqrt{\dfrac{h}{L}} + 0.09638\left(\sqrt{\dfrac{h}{L}}\right)^2 - 0.03596\left(\sqrt{\dfrac{h}{L}}\right)^3 + 0.004509\left(\sqrt{\dfrac{h}{L}}\right)^4$

TABLE 13-5 (continued) FUNDAMENTAL NATURAL FREQUENCIES OF FRAMES

Configuration	Natural Frequency
3. Continued	$c_2 = 6.0$

a_1	$1.4901 - 0.03882\sqrt{\dfrac{h}{L}} + 0.1021\left(\sqrt{\dfrac{h}{L}}\right)^2 - 0.04520\left(\sqrt{\dfrac{h}{L}}\right)^3 + 0.005995\left(\sqrt{\dfrac{h}{L}}\right)^4$
a_2	$-1.9893 + 4.5893\sqrt{\dfrac{h}{L}} - 3.6732\left(\sqrt{\dfrac{h}{L}}\right)^2 + 1.2198\left(\sqrt{\dfrac{h}{L}}\right)^3 - 0.1426\left(\sqrt{\dfrac{h}{L}}\right)^4$
a_3	$1.01408 - 3.06477\sqrt{\dfrac{h}{L}} + 2.7202\left(\sqrt{\dfrac{h}{L}}\right)^2 - 0.9512\left(\sqrt{\dfrac{h}{L}}\right)^3 + 0.1145\left(\sqrt{\dfrac{h}{L}}\right)^4$
a_4	$-0.2289 + 0.8108\sqrt{\dfrac{h}{L}} - 0.7698\left(\sqrt{\dfrac{h}{L}}\right)^2 + 0.2790\left(\sqrt{\dfrac{h}{L}}\right)^3 - 0.03429\left(\sqrt{\dfrac{h}{L}}\right)^4$
a_5	$0.01929 - 0.0778\sqrt{\dfrac{h}{L}} + 0.0776\left(\sqrt{\dfrac{h}{L}}\right)^2 - 0.02886\left(\sqrt{\dfrac{h}{L}}\right)^3 + 0.003601\left(\sqrt{\dfrac{h}{L}}\right)^4$

TABLE 13-5 Fundamental Natural Frequencies of Frames 680

Fundamental Natural Frequencies of Frames   TABLE 13-5

$c_2 = 12.0$

a_1	$1.7059 - 0.4443\sqrt{\dfrac{h}{L}} + 0.4067\left(\sqrt{\dfrac{h}{L}}\right)^2 - 0.1448\left(\sqrt{\dfrac{h}{L}}\right)^3 + 0.01784\left(\sqrt{\dfrac{h}{L}}\right)^4$
a_2	$-1.9940 + 4.6785\sqrt{\dfrac{h}{L}} - 3.8064\left(\sqrt{\dfrac{h}{L}}\right)^2 + 1.2804\left(\sqrt{\dfrac{h}{L}}\right)^3 - 0.1511\left(\sqrt{\dfrac{h}{L}}\right)^4$
a_3	$0.9691 - 3.03021\sqrt{\dfrac{h}{L}} + 2.7507\left(\sqrt{\dfrac{h}{L}}\right)^2 - 0.9771\left(\sqrt{\dfrac{h}{L}}\right)^3 + 0.1189\left(\sqrt{\dfrac{h}{L}}\right)^4$
a_4	$-0.2119 + 0.7952\sqrt{\dfrac{h}{L}} - 0.7783\left(\sqrt{\dfrac{h}{L}}\right)^2 + 0.2876\left(\sqrt{\dfrac{h}{L}}\right)^3 - 0.03583\left(\sqrt{\dfrac{h}{L}}\right)^4$
a_5	$0.0174083 - 0.07637\sqrt{\dfrac{h}{L}} + 0.07910\left(\sqrt{\dfrac{h}{L}}\right)^2 - 0.03009\left(\sqrt{\dfrac{h}{L}}\right)^3 + 0.0038131\left(\sqrt{\dfrac{h}{L}}\right)^4$

TABLE 13-5 (continued) FUNDAMENTAL NATURAL FREQUENCIES OF FRAMES

Configuration	Natural Frequency
4. First asymmetric in-plane mode, clamped E_h, I_h, ρ_h E_v, I_v, ρ_v **Mode shape** $c_1 = \dfrac{E_v I_v}{E_h I_h}$ $c_2 = \dfrac{\rho_v}{\rho_h}$	$\lambda = a_1 + a_2\left(\sqrt{c_1}\right) + a_3 c_1 + a_4\left(\sqrt{c_1}\right)^3$ $12 \geq h/L \geq 0.25 \qquad 12.0 \geq c_1 \geq 0.25$ $c_2 = 0.25$ a_1 : $0.4687 + 1.8309\sqrt{\dfrac{h}{L}} - 0.9885\left(\sqrt{\dfrac{h}{L}}\right)^2 + 0.2793\left(\sqrt{\dfrac{h}{L}}\right)^3 - 0.03053\left(\sqrt{\dfrac{h}{L}}\right)^4$ a_2 : $-0.7082 + 0.6148\sqrt{\dfrac{h}{L}} - 0.06704\left(\sqrt{\dfrac{h}{L}}\right)^2 - 0.05671\left(\sqrt{\dfrac{h}{L}}\right)^3 + 0.01210\left(\sqrt{\dfrac{h}{L}}\right)^4$ a_3 : $0.3999 - 0.6247\sqrt{\dfrac{h}{L}} + 0.2986\left(\sqrt{\dfrac{h}{L}}\right)^2 - 0.05446\left(\sqrt{\dfrac{h}{L}}\right)^3 + 0.002969\left(\sqrt{\dfrac{h}{L}}\right)^4$ a_4 : $-0.06957 + 0.1329\sqrt{\dfrac{h}{L}} - 0.08131\left(\sqrt{\dfrac{h}{L}}\right)^2 + 0.02082\left(\sqrt{\dfrac{h}{L}}\right)^3 - 0.001938\left(\sqrt{\dfrac{h}{L}}\right)^4$

TABLE 13-5 **Fundamental Natural Frequencies of Frames** 682

$c_2 = 0.75$

a_1	$0.6517 + 2.3508\sqrt{\dfrac{h}{L}} - 1.4862\left(\sqrt{\dfrac{h}{L}}\right)^2 + 0.4412\left(\sqrt{\dfrac{h}{L}}\right)^3 - 0.04870\left(\sqrt{\dfrac{h}{L}}\right)^4$
a_2	$-1.0348 + 1.2196\sqrt{\dfrac{h}{L}} - 0.4609\left(\sqrt{\dfrac{h}{L}}\right)^2 + 0.05260\left(\sqrt{\dfrac{h}{L}}\right)^3 + 0.001087\left(\sqrt{\dfrac{h}{L}}\right)^4$
a_3	$0.5720 - 1.01757\sqrt{\dfrac{h}{L}} + 0.6004\left(\sqrt{\dfrac{h}{L}}\right)^2 - 0.14999\left(\sqrt{\dfrac{h}{L}}\right)^3 + 0.01366\left(\sqrt{\dfrac{h}{L}}\right)^4$
a_4	$-0.09659 + 0.1992\sqrt{\dfrac{h}{L}} - 0.1351\left(\sqrt{\dfrac{h}{L}}\right)^2 + 0.03858\left(\sqrt{\dfrac{h}{L}}\right)^3 - 0.003989\left(\sqrt{\dfrac{h}{L}}\right)^4$

$c_2 = 1.5$

a_1	$0.8888 + 2.4200\sqrt{\dfrac{h}{L}} - 1.6779\left(\sqrt{\dfrac{h}{L}}\right)^2 + 0.5212\left(\sqrt{\dfrac{h}{L}}\right)^3 - 0.05889\left(\sqrt{\dfrac{h}{L}}\right)^4$
a_2	$-1.311 + 1.8646\sqrt{\dfrac{h}{L}} - 0.9713\left(\sqrt{\dfrac{h}{L}}\right)^2 + 0.2194\left(\sqrt{\dfrac{h}{L}}\right)^3 - 0.01811\left(\sqrt{\dfrac{h}{L}}\right)^4$
a_3	$0.6995 - 1.3471\sqrt{\dfrac{h}{L}} + 0.8811\left(\sqrt{\dfrac{h}{L}}\right)^2 - 0.2464\left(\sqrt{\dfrac{h}{L}}\right)^3 + 0.02515\left(\sqrt{\dfrac{h}{L}}\right)^4$
a_4	$-0.1146 + 0.2476\sqrt{\dfrac{h}{L}} - 0.1776\left(\sqrt{\dfrac{h}{L}}\right)^2 + 0.05346\left(\sqrt{\dfrac{h}{L}}\right)^3 - 0.005787\left(\sqrt{\dfrac{h}{L}}\right)^4$

TABLE 13-5 (continued) FUNDAMENTAL NATURAL FREQUENCIES OF FRAMES

Configuration	Natural Frequency

4. Continued

$c_2 = 3.0$

$$a_1 \quad 1.2508 + 2.09185\sqrt{\frac{h}{L}} - 1.5600\left(\sqrt{\frac{h}{L}}\right)^2 + 0.5042\left(\sqrt{\frac{h}{L}}\right)^3 - 0.05829\left(\sqrt{\frac{h}{L}}\right)^4$$

$$a_2 \quad -1.5471 + 2.4771\sqrt{\frac{h}{L}} - 1.4901\left(\sqrt{\frac{h}{L}}\right)^2 + 0.3968\left(\sqrt{\frac{h}{L}}\right)^3 - 0.03920\left(\sqrt{\frac{h}{L}}\right)^4$$

$$a_3 \quad 0.7925 - 1.6060\sqrt{\frac{h}{L}} + 1.1121\left(\sqrt{\frac{h}{L}}\right)^2 - 0.3282\left(\sqrt{\frac{h}{L}}\right)^3 + 0.03510\left(\sqrt{\frac{h}{L}}\right)^4$$

$$a_4 \quad -0.1253 + 0.2790\sqrt{\frac{h}{L}} - 0.2066\left(\sqrt{\frac{h}{L}}\right)^2 + 0.06396\left(\sqrt{\frac{h}{L}}\right)^3 - 0.007083\left(\sqrt{\frac{h}{L}}\right)^4$$

$c_2 = 6.0$

$$a_1 \quad 1.6631 + 1.4804\sqrt{\frac{h}{L}} - 1.1766\left(\sqrt{\frac{h}{L}}\right)^2 + 0.39396\left(\sqrt{\frac{h}{L}}\right)^3 - 0.04649\left(\sqrt{\frac{h}{L}}\right)^4$$

$$a_2 \quad -1.6468 + 2.7901\sqrt{\frac{h}{L}} - 1.7805\left(\sqrt{\frac{h}{L}}\right)^2 + 0.5016\left(\sqrt{\frac{h}{L}}\right)^3 - 0.05210\left(\sqrt{\frac{h}{L}}\right)^4$$

$$a_3 \quad 0.8142 - 1.6929\sqrt{\frac{h}{L}} + 1.2020\left(\sqrt{\frac{h}{L}}\right)^2 - 0.3626\left(\sqrt{\frac{h}{L}}\right)^3 + 0.03949\left(\sqrt{\frac{h}{L}}\right)^4$$

$$a_4 \quad -0.1237 + 0.2797\sqrt{\frac{h}{L}} - 0.2098\left(\sqrt{\frac{h}{L}}\right)^2 + 0.06565\left(\sqrt{\frac{h}{L}}\right)^3 - 0.0073295\left(\sqrt{\frac{h}{L}}\right)^4$$

TABLE 13-5 Fundamental Natural Frequencies of Frames 684

$c_2 = 12.0$

a_1 $2.0122 + 0.8737\sqrt{\dfrac{h}{L}} - 0.7544\left(\sqrt{\dfrac{h}{L}}\right)^2 + 0.2641\left(\sqrt{\dfrac{h}{L}}\right)^3 - 0.03196\left(\sqrt{\dfrac{h}{L}}\right)^4$

a_2 $-1.6071 + 2.7565\sqrt{\dfrac{h}{L}} - 1.7815\left(\sqrt{\dfrac{h}{L}}\right)^2 + 0.5078\left(\sqrt{\dfrac{h}{L}}\right)^3 - 0.05327\left(\sqrt{\dfrac{h}{L}}\right)^4$

a_3 $0.7787 - 1.6318\sqrt{\dfrac{h}{L}} + 1.1642\left(\sqrt{\dfrac{h}{L}}\right)^2 - 0.3526\left(\sqrt{\dfrac{h}{L}}\right)^3 + 0.03850\left(\sqrt{\dfrac{h}{L}}\right)^4$

a_4 $0.1144 + 0.2604\sqrt{\dfrac{h}{L}} - 0.1955\left(\sqrt{\dfrac{h}{L}}\right)^2 + 0.06116\left(\sqrt{\dfrac{h}{L}}\right)^3 - 0.0068249\left(\sqrt{\dfrac{h}{L}}\right)^4$

Approximate formula

$$f = \frac{\sqrt{g}}{2\pi}\left\{\frac{W}{2}\left[\frac{L^3}{24EI_h} + \frac{h^3}{3EI_v} - \frac{L^4 GJ_v}{32EI_h(2hEI_h + LGJ_v)}\right]\right\}^{-1/2} \quad \text{Hz}$$

where g is the gravitational acceleration constant

Ref. 13.11

5.
First out-of-plane
mode

Mode shape

TABLE 13-5 (continued) FUNDAMENTAL NATURAL FREQUENCIES OF FRAMES

Configuration	Natural Frequency
6. Rigid beam supported by n slender legs, in-plane mode 	Approximate formula $f = \dfrac{1}{2\pi}\left[\dfrac{12\Sigma E_i I_i}{h^3(M_h + 0.37\Sigma M_i)}\right]^{1/2}$ Hz [13.12] M_h = Mass of the top beam M_i = Mass of the ith vertical beam

TABLE 13-5   Fundamental Natural Frequencies of Frames   686

TABLE 13-6   SAFE-LOAD REGIONS

The combination of loadings describes a region on the x-y plane. If a prescribed loading defines a point inside the safe region, no collapse occurs. If a point falls on the boundary, collapse occurs according to the collapse mode indicated. Fully plastic bending moment is defined as $M_p = \sigma_{ys} Z_p$, where Z_p is the plastic section modulus taken from Table 2-2 and σ_{ys} is the yield stress of the material.

Frame and Loading	Safe Load Region
1.	Mode 1: $x = 3$ Mode 2: $y = 8$ Mode 3: $2x + y = 10$ $x = \dfrac{W_H h}{M_p}$ $y = \dfrac{W_V L}{M_p}$
2.	Mode 1: $x = 4$ Mode 2: $y = 8$ Mode 3: $2x + y = 12$ $x = \dfrac{W_H h}{M_p}$ $y = \dfrac{W_V L}{M_p}$
3.	Mode 1: $x = 8$ Mode 2: $y = 8$ Mode 3: $x + y = 10$ $x = \dfrac{W_H h}{M_p}$ $y = \dfrac{M_V L}{M_p}$

TABLE 13-6 (continued)   SAFE-LOAD REGIONS

Frame and Loading	Safe Load Region
4.	Mode 1: $x = 3$ Mode 2: $y = 16$ Mode 3: $2x + y - 4\sqrt{y} - 2 = 0$ $x = \dfrac{W_H h}{M_p}$ $y = \dfrac{p_V L^2}{M_p}$
5.	Mode 1: $x = 4$ Mode 2: $y = 16$ Mode 3: $2x + y - 4\sqrt{y} - 4 = 0$ $x = \dfrac{W_H h}{M_p}$ $y = \dfrac{p_V L^2}{M_p}$
6.	Mode 1: $y = 8$ Mode 2: $2x + y = 10$ $x = \dfrac{W_H h}{M_p}$ $y = \dfrac{W_V L}{M_p}$
7.	Mode 1–4: $x + y = 8$ $x = \dfrac{W_H h}{M_p}$ $y = \dfrac{W_V L}{M_p}$

TABLE 13-6   Safe Load Regions   688

TABLE 13-6 (continued)   SAFE-LOAD REGIONS

Frame and Loading		Safe Load Region
8.		Mode 1: $x = 4$ Mode 2: $y = 18\sqrt{3}$ Mode 3: $y - 6x + 12$ $+ (9x - 36)\sqrt{\dfrac{3y}{y - 6x + 12}} = 0$ $x = \dfrac{W_H h}{M_p}$ $y = \dfrac{p_0 L^2}{M_p}$
9.		Mode 1: $x = 3$ Mode 2: $y = 18\sqrt{3}$ Mode 3: $(y - 6x - 6)^3$ $- (27 - 9x)^2 3y = 0$ $x = \dfrac{W_H h}{M_p}$ $y = \dfrac{p_0 L^2}{M_p}$

TABLE 13-7 UNIFORM GRIDWORKS[a]

Notation

The ends of both the girders and stiffeners are simply supported.

Girders: beams that lie parallel to the x axis.

Stiffeners: beams that lie parallel to the y axis.

n_g, n_s = total number of girders and stiffeners, respectively

g, s = index for girders and stiffeners, respectively

w_g, θ_g, M_g, V_g = deflection, slope, bending moment, and shear force of gth girder

I_g, I_s = moments of inertia of girders and stiffeners, respectively

L_g, L_s = length of girders and stiffeners, respectively

M = number of terms chosen by user to be included in summation

$$\langle x - x_s \rangle^0 = \begin{cases} 0 & \text{if } x < x_s \\ 1 & \text{if } x \geq x_s \end{cases}$$

K_j = Take from Table 13-8.

	Response
1. Deflection	$w_g = \sin\dfrac{\pi g}{n_g + 1} \displaystyle\sum_{j=1}^{\infty} K_j \sin\dfrac{j\pi x}{L_g}$
2. Slope	$\theta_g = -\sin\dfrac{\pi g}{n_g + 1} \displaystyle\sum_{j=1}^{\infty} K_j \dfrac{j\pi}{L_g}\cos\dfrac{j\pi x}{L_g}$
3. Bending moment	$M_g = EI_g \sin\dfrac{\pi g}{n_g + 1} \displaystyle\sum_{j=1}^{\infty} K_j \left(\dfrac{j\pi}{L_g}\right)^2 \sin\dfrac{j\pi x}{L_g}$
4. Shear force	$V_g = EI_g \sin\dfrac{\pi g}{n_g + 1} \displaystyle\sum_{j=1}^{\infty} K_j$ $\times\left[\left(\dfrac{j\pi}{L_g}\right)^3 \cos\dfrac{j\pi x}{L_g} + \dfrac{\pi^4 I_s}{(n_g + 1)L_s^3 I_g}\displaystyle\sum_{s=1}^{M}\langle x - x_s \rangle^0 \sin\dfrac{j\pi x_s}{L_g}\right]$

[a]From Modern Formulas for Statics and Dynamics [13.6]

TABLE 13-7 | **Uniform Gridworks**   **690**

TABLE 13-8 PARAMETERS K_j OF TABLE 13-7 FOR THE STATIC RESPONSE OF GRIDWORKS[a]

Notation

P_g, P_s = axial forces in girders and stiffeners, respectively

p_s = loading intensity along the sth stiffener (F/L)

W_{sg} = concentrated force at intersection x_s, y_g

$$P_e = \frac{\pi^2 EI_s}{L_s^2} \qquad P_c = \frac{\pi^2 EI_g}{L_g^2}$$

Loading	K_j
1. For concentrated loads W_{sg} at x_s, y_g	$\dfrac{\dfrac{2L_s^3}{EI_s\pi^4}\dfrac{P_e}{P_e - P_s}\sum\limits_{s=1}^{n_s}\sum\limits_{g=1}^{n_g}W_{sg}\sin\dfrac{\pi g}{n_g+1}\sin\dfrac{j\pi s}{n_s+1}}{\dfrac{n_g+1}{2}j^4\left(\dfrac{L_s}{L_g}\right)^3\dfrac{I_g}{I_s}\left(1-\dfrac{P_g}{jP_c}\right)+\dfrac{n_s+1}{2}}$
2. For uniform force p_s along sth stiffener	$\dfrac{\dfrac{4L_s^4}{EI\pi^5}\dfrac{P_e}{P_e - P_s}\sum\limits_{s=1}^{n_s}p_s\sin\dfrac{j\pi s}{n_s+1}}{\dfrac{n_g+1}{2}j^4\left(\dfrac{L_s}{L_g}\right)^3\dfrac{I_g}{I_s}\left(1-\dfrac{P_g}{jP_c}\right)+\dfrac{n_s+1}{2}}$
3. If uniform force p_s is same for all stiffeners	Only the first term $(j = 1)$ in the equations of Table 13-7 is required: $$K_1 = \frac{\dfrac{4L_s^4}{EI_s\pi^5}\dfrac{P_e}{P_e - P_s}\sum\limits_{s=1}^{n_s}p_s\sin\dfrac{\pi s}{n_s+1}}{\dfrac{n_g+1}{2}\left(\dfrac{L_s}{L_g}\right)^3\dfrac{I_g}{I_s}\left(1-\dfrac{P_g}{P_c}\right)+\dfrac{n_s+1}{2}}$$

TABLE 13-9 CRITICAL AXIAL LOADS IN GIRDERS[a]

Notation

n_s = number of stiffeners

L_g, L_s = length of girders and stiffeners, respectively

E = modulus of elasticity

I_g, I_s = moments of inertia of girders and stiffeners, respectively

P_{cr} = unstable value of P_g, axial force in girders

$$D_1 = \frac{0.0866 L_g^2}{D_3} \qquad D_2 = \frac{0.202 L_g^2}{D_3} \qquad D_3 = \sqrt{\frac{C_1 L_g L_s^3 I_g}{I_s(n_s + 1)}}$$

$$P_e = \frac{\pi^2 E I_g}{L_g^2}$$

Take C_1 from Table 13-10.

End Conditions of Girders	Case	D_1	P_{cr}
Simply supported	1	≤ 1	$(1 + D_1)P_e$
	2	> 1	$D_2 P_e$
Fixed	3	≤ 1	$(4 + D_1)P_e$
	4	> 1	$(3 + D_2)P_e$

[a] From Modern Formulas for Statics and Dynamics [13.6]

TABLE 13-9 | Critical Axial Loads in Girders | 692

TABLE 13-10   VALUES OF C_1 OF TABLE 13-9 FOR STABILITY

Number of Girders n_g	End Conditions of Stiffeners, C_1	
	Simply supported	Fixed
1	0.020833	0.0052083
2	0.030864	0.0061728
3	0.041089	0.0080419
4	0.051342	0.010009
5	0.061603	0.011997
6	0.071866	0.013990
7	0.082131	0.015986
8	0.092396	0.017982
9	0.10266	0.019979
10	0.11293	0.021976

Note: For simply supported stiffeners the formula

$$C_1 = \frac{n_g + 1}{\pi^4}\left(1 + \sum_{j=1}^{\infty}\left\{\left[2j(n_g + 1) + 1\right]^{-4} + \left[2j(n_g + 1) - 1\right]^{-4}\right\}\right)$$

applies for any n_g.

TABLE 13-11 VALUES OF NATURAL FREQUENCY PARAMETERS C_n OF EQS. (13.1) AND (13.2)

Number of Stiffeners n_s	C_1	C_2	C_3	C_4	C_5	C_6	C_7	C_8	C_9	C_{10}
				Girders with Simply Supported Ends						
1	0.020833									
2	0.030864	0.0020576								
3	0.041089	0.0026042	0.00057767							
4	0.051342	0.0032240	0.00065790	0.0002462						
5	0.061603	0.0038580	0.00077160	0.00025720	0.00012564					
6	0.071866	0.0044962	0.00089329	0.00028895	0.00012688	0.000073890				
7	0.082131	0.0051361	0.0010177	0.00032552	0.00013769	0.000072209	0.000047321			
8	0.092396	0.0057767	0.0011431	0.00036387	0.00015157	0.000076208	0.000045226	0.000032215		
9	0.10266	0.0064178	0.0012691	0.00040301	0.00016667	0.000082237	0.000046681	0.000030328	0.000022963	
10	0.11293	0.0070590	0.0013954	0.00044252	0.00018233	0.000089133	0.000049521	0.000030753	0.000021400	0.000016967

$$C_n = \frac{n_s+1}{\pi^4}\left[\frac{1}{n^4} + \sum_{j=1}^{\infty}\left\{[2j(n_s+1)+n]^{-4} + [2j(n_s+1)-n]^{-4}\right\}\right]$$

Number of Stiffeners n_s	C_1	C_2	C_3	C_4	C_5	C_6	C_7	C_8	C_9	C_{10}
				Girders with Fixed Ends						
1	0.0052083									
2	0.0061728	0.0011431								
3	0.0080419	0.0011393	0.00042165							
4	0.010009	0.0013459	0.00039075	0.00020078						
5	0.011997	0.0015917	0.00043081	0.00018009	0.00011111					
6	0.013990	0.0018480	0.00048904	0.00018923	0.000098217	0.000067910				
7	0.015986	0.0021078	0.00055303	0.00020779	0.000099794	0.000059682	0.000044545			
8	0.017982	0.0023691	0.00061925	0.00022977	0.00010668	0.000059226	0.000039097	0.000030804		
9	0.019970	0.0026311	0.00068645	0.00025320	0.00011572	0.000061961	0.000038155	0.000027067	0.000022193	
10	0.021976	0.0028934	0.00075415	0.00027732	0.00012573	0.000066109	0.000039232	0.000026101	0.000019547	0.000016522

TABLE 13-11 Values of the Natural Frequency Parameters 694

TABLE 13-12 STIFFNESS MATRIX FOR PLANE TRUSSES

Notation

$$E = \text{modulus of elasticity}$$
$$A = \text{area of the cross section}$$
$$\ell = \text{length of element}$$
$$xX = \text{angle between } x \text{ and } X \text{ axes}$$
$$xZ = \text{angle between } x \text{ and } Z \text{ axes}$$

Right-handed global XYZ and local xyz coordinate systems are employed. The identity $\cos^2 xX + \cos^2 xZ = 1$ is useful.

The relationships of this table should be used for the static analysis of trusses. For dynamic analyses of trusses, use the frame formulas.

LOCAL COORDINATES

$$\begin{bmatrix} \tilde{N}_a \\ \tilde{N}_b \end{bmatrix} = \frac{EA}{\ell} \begin{bmatrix} 1 & -1 \\ -1 & 1 \end{bmatrix} \begin{bmatrix} \tilde{u}_a \\ \tilde{u}_b \end{bmatrix}$$

$$\tilde{\mathbf{p}}^i = \tilde{\mathbf{k}}^i \quad \tilde{\mathbf{v}}^i$$

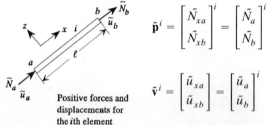

$$\tilde{\mathbf{p}}^i = \begin{bmatrix} \tilde{N}_{xa} \\ \tilde{N}_{xb} \end{bmatrix}^i = \begin{bmatrix} \tilde{N}_a \\ \tilde{N}_b \end{bmatrix}^i$$

$$\tilde{\mathbf{v}}^i = \begin{bmatrix} \tilde{u}_{xa} \\ \tilde{u}_{xb} \end{bmatrix}^i = \begin{bmatrix} \tilde{u}_a \\ \tilde{u}_b \end{bmatrix}^i$$

Positive forces and displacements for the ith element

GLOBAL COORDINATES

$$\mathbf{p}^i = \mathbf{k}^i \mathbf{v}^i$$

$$\mathbf{v}^i = \begin{bmatrix} u_{Xa} \\ u_{Za} \\ u_{Xb} \\ u_{Zb} \end{bmatrix}^i \qquad \mathbf{p}^i = \begin{bmatrix} F_{Xa} \\ F_{Za} \\ F_{Xb} \\ F_{Zb} \end{bmatrix}^i$$

$$\mathbf{k}^i = \mathbf{T}^{iT} \tilde{\mathbf{k}}^i \mathbf{T}^i = \frac{EA}{\ell} \begin{bmatrix} \mathbf{A} & -\mathbf{A} \\ -\mathbf{A} & \mathbf{A} \end{bmatrix}$$

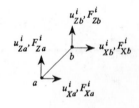

Global Coordinates

Global $\mathbf{v}^i$ and $\mathbf{p}^i$

$$\mathbf{T}^i = \begin{bmatrix} \cos xX & \cos xZ & 0 & 0 \\ 0 & 0 & \cos xX & \cos xZ \end{bmatrix} \qquad \mathbf{A} = \begin{bmatrix} \cos^2 xX & \cos xX \cos xZ \\ \cos xX \cos xZ & \cos^2 xZ \end{bmatrix}$$

TABLE 13-13   STIFFNESS MATRIX FOR SPACE TRUSSES

Notation

E = modulus of elasticity
ℓ = length of element
A = area of cross section
xX = angle between x axis and X axis, and so on.

The relationships of this table should be used for the static analysis of trusses. For dynamic analyses of trusses, use the frame formulas. See Table 13-12 for coordinate system and other definitions.

LOCAL COORDINATES

$$
\begin{bmatrix} \tilde{N}_a \\ \tilde{N}_b \end{bmatrix} = \frac{EA}{\ell} \begin{bmatrix} 1 & -1 \\ -1 & 1 \end{bmatrix} \begin{bmatrix} \tilde{u}_a \\ \tilde{u}_b \end{bmatrix}
$$

$$
\tilde{\mathbf{p}}^i \quad = \qquad \tilde{\mathbf{k}}^i \qquad \tilde{\mathbf{v}}^i
$$

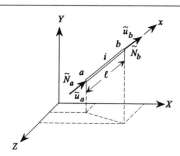

GLOBAL COORDINATES

$$\mathbf{p}^i = \mathbf{k}^i \mathbf{v}^i$$

$$
\mathbf{v}^i = \begin{bmatrix} u_{Xa} \\ u_{Ya} \\ u_{Za} \\ u_{Xb} \\ u_{Yb} \\ u_{Zb} \end{bmatrix} = \begin{bmatrix} u_a \\ v_a \\ w_a \\ u_b \\ v_b \\ w_b \end{bmatrix} \qquad \mathbf{p}^i = \begin{bmatrix} F_{Xa} \\ F_{Ya} \\ F_{Za} \\ F_{Xb} \\ F_{Yb} \\ F_{Zb} \end{bmatrix}
$$

$$\mathbf{k}^i = \mathbf{T}^{iT} \tilde{\mathbf{k}}^i \mathbf{T}^i$$

$$
\mathbf{T}^i = \begin{bmatrix} \cos xX & \cos xY & \cos xZ & 0 & 0 & 0 \\ 0 & 0 & 0 & \cos xX & \cos xY & \cos xZ \end{bmatrix}
$$

TABLE 13-13   **Stiffness Matrix for Space Trusses**   **696**

TABLE 13-14 STIFFNESS MATRICES FOR PLANE FRAMES

Notation

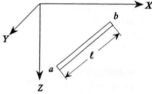

E = modulus of elasticity

I, I_z = moments of inertia about local y and z axes

$$I_z = \int_A y^2 \, dA \qquad I = \int_A z^2 \, dA$$

ℓ = length of element

G = shear modulus of elasticity

J = torsional constant

A = area of the cross section

xX = angle between x and X axes; and so on
for xZ, zX, and zZ

Frame lies in the XY plane

Right-handed global *XYZ* and local *xyz* coordinate systems are employed. The identities $\cos^2 xX + \cos^2 xZ = 1$ and $\cos^2 zX + \cos^2 zZ = 1$ are useful. Bending is modeled using Euler-Bernoulli beams.

In-Plane Loading (Bending and Extension)	Out-of-Plane Loading (Bending and Torsion)
DISPLACEMENTS AND FORCES	DISPLACEMENTS AND FORCES
$\tilde{\mathbf{v}}^i = [\tilde{u}_a \quad \tilde{w}_a \quad \tilde{\theta}_a \quad \tilde{u}_b \quad \tilde{w}_b \quad \tilde{\theta}_b]^T$	$\tilde{\mathbf{v}}^i = [\tilde{\phi}_a \quad \tilde{v}_a \quad \tilde{\theta}_{za} \quad \tilde{\phi}_b \quad \tilde{v}_b \quad \tilde{\theta}_{zb}]^T$
$\tilde{\mathbf{p}}^i = [\tilde{N}_a \quad \tilde{V}_a \quad \tilde{M}_a \quad \tilde{N}_b \quad \tilde{V}_b \quad \tilde{M}_b]^T$	$\tilde{\mathbf{p}}^i = [\tilde{T}_a \quad \tilde{V}_{ya} \quad \tilde{M}_{za} \quad \tilde{T}_b \quad \tilde{V}_{yb} \quad \tilde{M}_{zb}]^T$
POSITIVE FORCES AND DISPLACEMENTS	POSITIVE FORCES AND DISPLACEMENTS
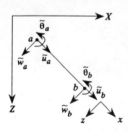	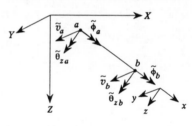
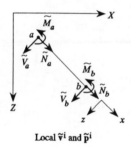	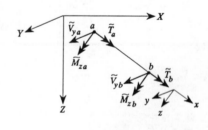
Local $\tilde{\mathbf{v}}^i$ and $\tilde{\mathbf{p}}^i$	Local $\tilde{\mathbf{v}}^i$ and $\tilde{\mathbf{p}}^i$

TABLE 13-14 (continued)   STIFFNESS MATRICES FOR PLANE FRAMES

In-Plane Loading (Bending and Extension)	Out-to-Plane Loading (Bending and Torsion)

LOCAL COORDINATES

$\bar{\mathbf{p}}^i = \bar{\mathbf{k}}^i \bar{\mathbf{v}}^i$

$\bar{\mathbf{k}}^i =$

$$\frac{EI}{\ell^3}\begin{bmatrix} A\ell^2/I & & & & \text{Symmetric} \\ 0 & 12 & & & & \\ 0 & -6\ell & 4\ell^2 & & & \\ -A\ell^2/I & 0 & 0 & A\ell^2/I & & \\ 0 & -12 & 6\ell & 0 & 12 & \\ 0 & -6\ell & 2\ell^2 & 0 & 6\ell & 4\ell^2 \end{bmatrix}$$

LOCAL COORDINATES

$\bar{\mathbf{p}}^i = \bar{\mathbf{k}}^i \bar{\mathbf{v}}^i$

$\bar{\mathbf{k}}^i =$

$$\frac{EI_z}{\ell^3}\begin{bmatrix} GJ\ell^2/EI_z & & & & \text{Symmetric} \\ 0 & 12 & & & & \\ 0 & -6\ell & 4\ell^2 & & & \\ -GJ\ell^2/EI_z & 0 & 0 & GJ\ell^2/EI_z & & \\ 0 & -12 & 6\ell & 0 & 12 & \\ 0 & -6\ell & 2\ell^2 & 0 & 6\ell & 4\ell^2 \end{bmatrix}$$

GLOBAL COORDINATES

$\mathbf{p}^i = \mathbf{k}^i \mathbf{v}^i$

$\mathbf{v}^i = \begin{bmatrix} u_{Xa} & u_{Za} & \theta_a & u_{Xb} & u_{Zb} & \theta_b \end{bmatrix}^T$

$\mathbf{p}^i = \begin{bmatrix} F_{Xa} & F_{Za} & M_a & F_{Xb} & F_{Zb} & M_b \end{bmatrix}^T$

$\mathbf{k}^i = \mathbf{T}^{iT}\bar{\mathbf{k}}^i\mathbf{T}^i$

GLOBAL COORDINATES

$\mathbf{p}^i = \mathbf{k}^i \mathbf{v}^i$

$\mathbf{v}^i = \begin{bmatrix} \theta_{Xa} & u_{Ya} & \theta_{Za} & \theta_{Xb} & u_{Yb} & \theta_{Zb} \end{bmatrix}^T$

$\mathbf{p}^i = \begin{bmatrix} M_{Xa} & F_{Ya} & M_{Za} & M_{Xb} & F_{Yb} & M_{Zb} \end{bmatrix}^T$

$\mathbf{k}^i = \mathbf{T}^{iT}\bar{\mathbf{k}}^i\mathbf{T}^i$

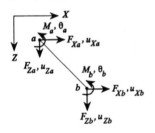

Global $\mathbf{v}^i$ and $\mathbf{p}^i$

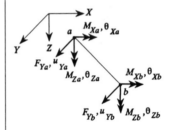

Global $\mathbf{v}^i$ and $\mathbf{p}^i$

$$\mathbf{T}^i = \begin{bmatrix} \cos xX & \cos xZ & 0 & 0 & 0 & 0 \\ \cos zX & \cos zZ & 0 & 0 & 0 & 0 \\ 0 & 0 & 1 & 0 & 0 & 0 \\ 0 & 0 & 0 & \cos xX & \cos xZ & 0 \\ 0 & 0 & 0 & \sin zX & \cos zZ & 0 \\ 0 & 0 & 0 & 0 & 0 & 1 \end{bmatrix}$$

$$\mathbf{T}^i = \begin{bmatrix} \cos xX & 0 & \cos xZ & 0 & 0 & 0 \\ 0 & 1 & 0 & 0 & 0 & 0 \\ \cos zX & 0 & \cos zZ & 0 & 0 & 0 \\ 0 & 0 & 0 & \cos xX & 0 & \cos zZ \\ 0 & 0 & 0 & 0 & 1 & 0 \\ 0 & 0 & 0 & \cos zX & 0 & \cos zZ \end{bmatrix}$$

TABLE 13-14   Stiffness Matrices for Plane Frames   **698**

TABLE 13-15   STIFFNESS MATRIX FOR BAR IN SPACE

Notation

E = modulus of elasticity
I, I_z = moments of inertia about y and z axes

$$I_z = \int_A y^2\, dA \qquad I_y = I = \int_A z^2\, dA$$

xX = angle between x and X axes; similarly for xY, xZ, yX, yY, yZ, zX, zY, and zZ.

G = shear modulus of elasticity
A = area of cross section
J = torsional constant
ℓ = length of element

The identities $\cos^2 jX + \cos^2 jY + \cos^2 jZ = 1$, $j = x, y, z$ are useful.

DISPLACEMENTS AND FORCES

LOCAL COORDINATES $\bar{\mathbf{p}}^i = \bar{\mathbf{k}}^i \bar{\mathbf{v}}^i$

$$\bar{\mathbf{v}}^i = [\bar{u}_a \;\; \bar{v}_a \;\; \bar{w}_a \;\; \phi_a \;\; \bar{\theta}_{ya} \;\; \bar{\theta}_{za} \;\; \bar{u}_b \;\; \bar{v}_b \;\; \bar{w}_b \;\; \phi_b \;\; \bar{\theta}_{yb} \;\; \bar{\theta}_{zb}]^T$$

$$\bar{\mathbf{p}}^i = [\bar{N}_a \;\; \bar{V}_{ya} \;\; \bar{V}_{za} \;\; T_a \;\; \bar{M}_{ya} \;\; \bar{M}_{za} \;\; \bar{N}_b \;\; \bar{V}_{yb} \;\; \bar{V}_{zb} \;\; T_b \;\; \bar{M}_{yb} \;\; \bar{M}_{zb}]^T$$

$$\bar{\mathbf{k}}^i = \begin{bmatrix}
EA/\ell & 0 & 0 & 0 & 0 & 0 & -EA/\ell & 0 & 0 & 0 & 0 & 0 \\
 & 12EI_z/\ell^3 & 0 & 0 & 0 & 6EI_z/\ell^2 & 0 & -12EI_z/\ell^3 & 0 & 0 & 0 & 6EI_z/\ell^2 \\
 & & 12EI_y/\ell^3 & 0 & -6EI_y/\ell^2 & 0 & 0 & 0 & -12EI_y/\ell^3 & 0 & -6EI_y/\ell^2 & 0 \\
 & & & GJ/\ell & 0 & 0 & 0 & 0 & 0 & -GJ/\ell & 0 & 0 \\
 & & & & 4EI_y/\ell & 0 & 0 & 0 & 6EI_y/\ell^2 & 0 & 2EI_y/\ell & 0 \\
 & & & & & 4EI_z/\ell & 0 & -6EI_z/\ell^2 & 0 & 0 & 0 & 2EI_z/\ell \\
 & & & & & & EA/\ell & 0 & 0 & 0 & 0 & 0 \\
 & & & & \text{Symmetric} & & & 12EI_z/\ell^3 & 0 & 0 & 0 & -6EI_z/\ell^2 \\
 & & & & & & & & 12EI_y/\ell^3 & 0 & 6EI_y/\ell^2 & 0 \\
 & & & & & & & & & GJ/\ell & 0 & 0 \\
 & & & & & & & & & & 4EI_y/\ell & 0 \\
 & & & & & & & & & & & 4EI_z/\ell
\end{bmatrix}$$

TABLE 13-15 (continued)   STIFFNESS MATRIX FOR BAR IN SPACE

GLOBAL COORDINATES $\mathbf{p}^i = \mathbf{k}^i \mathbf{v}^i$

$\mathbf{v}^i = \begin{bmatrix} u_{Xa} & u_{Ya} & u_{Za} & \theta_{Xa} & \theta_{Ya} & \theta_{Za} & u_{Xb} & u_{Yb} & u_{Zb} & \theta_{Xb} & \theta_{Yb} & \theta_{Zb} \end{bmatrix}^T$

$\mathbf{p}^i = \begin{bmatrix} F_{Xa} & F_{Ya} & F_{Za} & M_{Xa} & M_{Ya} & M_{Za} & F_{Xb} & F_{Yb} & F_{Zb} & M_{Xb} & M_{Yb} & M_{Zb} \end{bmatrix}^T$

$\mathbf{k}^i = \mathbf{T}^{iT} \tilde{\mathbf{k}}^i \mathbf{T}^i$

$\tau_0 = \begin{bmatrix} \cos xX & \cos xY & \cos xZ \\ \cos yX & \cos yY & \cos yZ \\ \cos zX & \cos zY & \cos zZ \end{bmatrix}$

$\mathbf{T}^i = \begin{bmatrix} \tau_0 & & & \\ & \tau_0 & & \\ & & \tau_0 & \\ & & & \tau_0 \end{bmatrix}$

POSITIVE FORCES AND DISPLACEMENTS

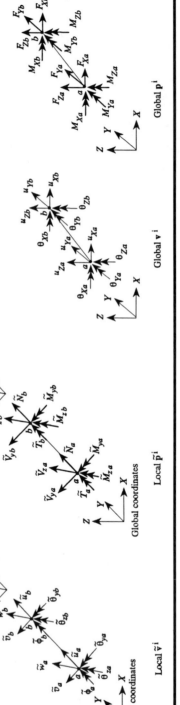

TABLE 13-15   Stiffness Matrix for Bar in Space   700

TABLE 13-16 MASS MATRICES FOR PLANE FRAMES

Notation

ρ = mass per unit length

I_x = polar moment of inertia, $I_x = J_x$

$I_{xxj}, I_{yyj}, I_{zzj}$ = rotary inertia of lumped mass at point j about the x, y, z axes, respectively

A = area of cross section

r_y, r_z = radius of gyration about y and z axes

$r_y = \sqrt{I_y/A}, r_z = \sqrt{I_z/A}$

I_y, I_z = moments of inertia about y and z axes

ℓ = length of element

See Table 13-14 for coordinate systems, displacement vectors, and force vectors

---

In-Plane Loading (Bending and Extension) | **Out-of-Plane Loading (Bending and Torsion)**

Mass Lumped at Both Ends of Element

$$a \;\underline{\qquad \ell \qquad}\; b$$

In-Plane Loading (Bending and Extension)

LOCAL COORDINATES

$$\bar{\mathbf{m}}^i = \frac{\rho\ell}{2}
\begin{bmatrix}
1 & & & & & \text{Symmetric} \\
0 & 1 & & & & \\
0 & 0 & \frac{\ell^2}{12} + r_y^2 & & & \\
0 & 0 & 0 & 1 & & \\
0 & 0 & 0 & 0 & 1 & \\
0 & 0 & 0 & 0 & 0 & \frac{\ell^2}{12} + r_y^2
\end{bmatrix}$$

$$=
\begin{bmatrix}
m_a & & & & & \\
 & m_a & & & & \\
 & & I_{yya} & & & \\
 & & & m_b & & \\
 & & & & m_b & \\
 & & & & & I_{yyb}
\end{bmatrix}$$

Set $I_{yya} = I_{yyb} = 0$ if rotary inertia is neglected.

Out-of-Plane Loading (Bending and Torsion)

LOCAL COORDINATES

$$\bar{\mathbf{m}}^i = \frac{\rho\ell}{2}
\begin{bmatrix}
I_x/A & & & & & \text{Symmetric} \\
0 & 1 & & & & \\
0 & 0 & \frac{\ell^2}{12} + r_z^2 & & & \\
0 & 0 & 0 & I_x/A & & \\
0 & 0 & 0 & 0 & 1 & \\
0 & 0 & 0 & 0 & 0 & \frac{\ell^2}{12} + r_z^2
\end{bmatrix}$$

$$=
\begin{bmatrix}
I_{xxa} & & & & & \\
 & m_a & & & & \\
 & & I_{zza} & & & \\
 & & & I_{xxb} & & \\
 & & & & m_b & \\
 & & & & & I_{zzb}
\end{bmatrix}$$

Set $I_{zza} = I_{zzb} = 0$ if rotary inertia is neglected.

TABLE 13-16 (continued) MASS MATRICES FOR PLANE FRAMES

In-Plane Loading (Bending and Extension)	Mass Lumped at Point a	Out-of-Plane Loading (Bending and Torsion)

Mass Lumped at Point a — In-Plane Loading:

$$\bar{\mathbf{m}}^i = \rho\ell \begin{bmatrix} 1 & & & & & \\ & 1 & & & & \\ & & \dfrac{\ell^2}{12}+r_y^2 & & & \\ & & & m_a & & \\ & & & & m_a & \\ & & & & & I_{yya} \end{bmatrix}$$

Mass Lumped at Point a — Out-of-Plane Loading:

$$\bar{\mathbf{m}}^i = \rho\ell \begin{bmatrix} I_x/A & & & & & \\ & \dfrac{\ell^2}{12}+r_y^2 & & & & \\ & & 1 & & & \\ & & & I_{xxa} & & \\ & & & & I_{zza} & \\ & & & & & m_a \end{bmatrix}$$

Consistent Mass Matrices for Uniform Beams

In-Plane Loading:

$$\bar{\mathbf{m}}^i = \frac{\rho\ell}{420}\begin{bmatrix} 140 & & & & & \\ 0 & 156 & & \text{Symmetric} & & \\ 0 & -22\ell & 4\ell^2 & & & \\ 70 & 0 & 0 & 140 & & \\ 0 & 54 & -13\ell & 0 & 156 & \\ 0 & 13\ell & -3\ell^2 & 0 & 22\ell & 4\ell^2 \end{bmatrix}$$

$$+\;\frac{\rho A\ell}{30}\left(\frac{r_y}{\ell}\right)^2\begin{bmatrix} 0 & & & & & \\ 0 & 36 & & \text{Symmetric} & & \\ 0 & -3\ell & 4\ell^2 & & & \\ 0 & 0 & 0 & 0 & & \\ 0 & -36 & 3\ell & 0 & 36 & \\ 0 & -3\ell & -\ell^2 & 0 & 3\ell & 4\ell^2 \end{bmatrix}$$

(rotary inertia)

Out-of-Plane Loading:

$$\bar{\mathbf{m}}^i = \frac{\rho\ell}{420}\begin{bmatrix} 140I_x/A & & & & & \\ 0 & 156 & & \text{Symmetric} & & \\ 0 & 22\ell & 4\ell^2 & & & \\ 70I_x/A & 0 & 0 & 140I_x/A & & \\ 0 & 54 & 13\ell & 0 & 156 & \\ 0 & -13\ell & -3\ell^2 & 0 & -22\ell & 4\ell^2 \end{bmatrix}$$

$$+\;\frac{\rho A\ell}{30}\left(\frac{r_z}{\ell}\right)^2\begin{bmatrix} 0 & & & & & \\ 0 & 36 & & \text{Symmetric} & & \\ 0 & 3\ell & 4\ell^2 & & & \\ 0 & 0 & 0 & 0 & & \\ 0 & -36 & -3\ell & 0 & 36 & \\ 0 & 3\ell & -\ell^2 & 0 & -3\ell & 4\ell^2 \end{bmatrix}$$

(rotary inertia)

GLOBAL COORDINATES
$\mathbf{m}^i = \mathbf{T}^{iT}\bar{\mathbf{m}}^i\mathbf{T}^i$ where $\mathbf{T}^i$ is given in Table 13-14

GLOBAL COORDINATES
$\mathbf{m}^i = \mathbf{T}^{iT}\bar{\mathbf{m}}^i\mathbf{T}^i$ where $\mathbf{T}^i$ is given in Table 13-14

TABLE 13-16   Mass Matrices for Plane Frames   702

TABLE 13-17   MASS MATRICES FOR SPACE FRAMES

Notation

ρ = mass per unit length
I_x = polar moment of inertia, $I_x = J_x$
$I_{xxj}, I_{yyj}, I_{zzj}$ = rotary inertia of lumped mass
at point j about the x, y, z
axes, respectively.
A = area of cross section

r_y, r_z = radius of gyration about y
and z axes
$r_y = \sqrt{I_y/A}$, $r_z = \sqrt{I_z/A}$
I_y, I_z = moments of inertia about y
and z axes
ℓ = length of element

See Table 13-15 for coordinate systems force vector, and displacement vector definitions. Formulas for
$m_j, I_{xxj}, I_{yyj}, I_{zzj}$ are defined in Table 13-16

Mass Lumped at Both Ends of Element

$$\bar{\mathbf{m}}^i = \frac{\rho l}{2}
\begin{bmatrix}
1 \\
0 & 1 & & & & & & & & & & \text{Symmetric} \\
0 & 0 & 1 \\
0 & 0 & 0 & I_x/A \\
0 & 0 & 0 & 0 & \dfrac{\ell^2}{12}+r_y^2 \\
0 & 0 & 0 & 0 & 0 & \dfrac{\ell^2}{12}+r_z^2 \\
0 & 0 & 0 & 0 & 0 & 0 & 1 \\
0 & 0 & 0 & 0 & 0 & 0 & 0 & 1 \\
0 & 0 & 0 & 0 & 0 & 0 & 0 & 0 & 1 \\
0 & 0 & 0 & 0 & 0 & 0 & 0 & 0 & 0 & I_x/A \\
0 & 0 & 0 & 0 & 0 & 0 & 0 & 0 & 0 & 0 & \dfrac{\ell^2}{12}+r_y^2 \\
0 & 0 & 0 & 0 & 0 & 0 & 0 & 0 & 0 & 0 & 0 & \dfrac{\ell^2}{12}+r_z^2
\end{bmatrix}$$

$$=
\begin{bmatrix}
m_a \\
& m_a \\
& & m_a \\
& & & I_{xxa} \\
& & & & I_{yya} \\
& & & & & I_{zza} \\
& & & & & & m_b \\
& & & & & & & m_b \\
& & & & & & & & m_b \\
& & & & & & & & & I_{xxb} \\
& & & & & & & & & & I_{yyb} \\
& & & & & & & & & & & I_{zzb}
\end{bmatrix}$$

TABLE 13-17 (continued)   MASS MATRICES FOR SPACE FRAMES

Mass Lumped at Point a

$$\mathbf{m}^i = \rho\ell \begin{bmatrix} 1 \\ & 1 \\ & & 1 \\ & & & I_x/A \\ & & & & \dfrac{\ell^2}{12} + r_y^2 \\ & & & & & \dfrac{\ell^2}{12} + r_z^2 \end{bmatrix} = \begin{bmatrix} m_a \\ & m_a \\ & & m_a \\ & & & I_{xxa} \\ & & & & I_{yya} \\ & & & & & I_{zza} \end{bmatrix}$$

Consistent Mass for Uniform Space Bars

$$\tilde{\mathbf{m}}^i = \frac{\rho\ell}{420} \begin{bmatrix}
140 \\
0 & 156 \\
0 & 0 & 156 & & & & & & \text{Symmetric} \\
0 & 0 & 0 & 140I_x/A \\
0 & 0 & -22\ell & 0 & 4\ell^2 \\
0 & 22\ell & 0 & 0 & 0 & 4\ell^2 \\
70 & 0 & 0 & 0 & 0 & 0 & 140 \\
0 & 54 & 0 & 0 & 0 & 13\ell & 0 & 156 \\
0 & 0 & 54 & 0 & -13\ell & 0 & 0 & 0 & 156 \\
0 & 0 & 0 & 70I_x/A & 0 & 0 & 0 & 0 & 0 & 140I_x/A \\
0 & 0 & 13\ell & 0 & -3\ell^2 & 0 & 0 & 0 & 22\ell & 0 & 4\ell^2 \\
0 & -13\ell & 0 & 0 & 0 & -3\ell^2 & 0 & -22\ell & 0 & 0 & 0 & 4\ell^2
\end{bmatrix}$$

$$+ \frac{\rho}{30\ell} \begin{bmatrix}
0 \\
0 & 36r_z^2 \\
0 & 0 & 36r_y^2 & & & & & & \text{Symmetric} \\
0 & 0 & 0 & 0 \\
0 & 0 & -3\ell r_y^2 & 0 & 4\ell^2 r_y^2 \\
0 & 3\ell r_z^2 & 0 & 0 & 0 & 4\ell^2 r_z^2 \\
0 & 0 & 0 & 0 & 0 & 0 & 0 \\
0 & -36r_z^2 & 0 & 0 & 0 & 0 & 3\ell r_z^2 & 0 & 36r_z^2 \\
0 & 0 & -36r_y^2 & 0 & 3\ell r_y^2 & 0 & 0 & 0 & 36r_y^2 \\
0 & 0 & 0 & 0 & 0 & 0 & 0 & 0 & 0 & 0 \\
0 & 0 & -3\ell r_y^2 & 0 & -\ell^2 r_y^2 & 0 & 0 & 0 & 3\ell r_y^2 & 0 & 4\ell^2 r_y^2 \\
0 & 3\ell r_z^2 & 0 & 0 & 0 & -\ell^2 r_z^2 & 0 & 3\ell r_z^2 & 0 & 0 & 0 & 4\ell^2 r_z^2
\end{bmatrix}$$

(rotary inertia)

GLOBAL COORDINATES

$\mathbf{m}^i = \mathbf{T}^{iT}\tilde{\mathbf{m}}^i\mathbf{T}^i$ where $\mathbf{T}^i$ is given in Table 13-15

TABLE 13-17   **Mass Matrices for Space Frames**   704

Torsion of Thin-Walled Beams

The torsion of beams having noncircular cross sections and in particular thin-walled beams are treated in this chapter. A thin-walled beam is made from thin plates joined along their edges. If restrained warping occurs, it is essential to employ the formulas of this chapter rather than simple torsion formulas. The term *warping* is defined as the out-of-plane distortion of the cross section of a beam in the direction of the beam's longitudinal axis. Restrained warping will be significant during the twisting of a thin-walled beam when the applied twisting moment or the boundary conditions create an internal twisting moment that varies along the beam axis. For example, this situation occurs if the in-span conditions prevent cross sections from warping freely. The shear stresses and strains in a thin-walled beam tend to be higher than those in a beam of solid cross section.

14.1   NOTATION

A Cross-sectional area (L^2)

b_x Distributed bimoment (FL)

B Bimoment, warping moment (FL^2)

C^2 $= GJ/E\Gamma \ (1/L^2)$

C_P

$$\begin{cases} = (I_y + I_z)/A \text{ if shear center and centroid coincide,} \\ = (I_y + I_z)/A + (z_S/I_y)\int_A z(y^2 + z^2)\, dA \\ \quad + (y_S/I_z)\int_A y(y^2 + z^2)\, dA - (y_S^2 + z_S^2) \text{ if shear} \\ \quad \text{center and centroid do not coincide} \\ \quad \text{and the axial force passes through shear center} \end{cases}$$

E	Modulus of elasticity (F/L^2)
$E\Gamma$	Warping rigidity (FL^4)
G	Shear modulus of elasticity (F/L^2)
GJ	Torsional rigidity (FL^2)
I_y	Moment of inertia of cross section about y axis, $= I\ (L^4)$
I_z	Moment of inertia of cross section about z axis (L^4)
I_{pi}	Polar mass moment of inertia of concentrated mass at location $i\ (ML^2)$; can be computed as $I_{pi} = \Delta a\, \rho r_p^2$, where Δa is length of shaft lumped at location i
J	Torsional constant (L^4); for circular cross sections, J is polar moment of inertia (I_x) of cross sectional area with respect to centroidal axis of bar
k_t	Elastic foundation modulus (FL/L)
L	Length of beam (L)
m_x	Distributed torque, twisting moment intensity (FL/L)
m_{x1}	Magnitude of torque that is uniformly distributed in x direction (FL/L)
m_{x0}	Initial magnitude of linearly varying distributed torque (FL/L)
$m_{x\ell}$	Final magnitude of linearly varying distributed torque (FL/L)
M, M_z	Bending moments about y and z axes (FL)
P	Compressive axial force passing through shear center; replace P by $-P$ for tensile axial forces (F)
Q_ω	First sectorial moment (L^4)
r_S	Perpendicular distance from shear center to tangent of centerline of wall profile (L)
r_p	Polar radius of gyration; i.e., r_p is radius of gyration of cross-sectional area about longitudinal (x) axis of beam (L)
S	Shear center
s	Arc length measured from outer edge of wall profile (L)
t	Wall thickness (L)
T	Twisting moment, torque (FL)
T_ω	Warping torque (FL)
T_t	Torque due to pure torsion (FL)
v, w	Displacements in y and z directions
V_y, V	Shear forces in y and z directions
y_S, z_S	Distance along y, z directions between shear center and centroid (L)
Γ	Warping constant (L^6)
ψ	Rate of change of angle of twist ϕ with respect to x axis (rad/L)

ω	Warping (L^2) of cross section with respect to plane of average warping, or principal sectorial coordinate with respect to shear center; also, natural frequency (rad/T)
ω_S	Sectorial coordinate with respect to shear center S (L^2)
ϕ	Angle of twist, rotation (rad)
ϕ_a, ϕ_b	Angles of twist at points a and b
ρ	Mass per unit length (M/L, FT^2/L^2)
σ_ω	Normal stress caused by warping, or normal warping stress (F/L^2)
τ_ω	Shear stress caused by warping, or shear warping stress (F/L^2)
θ, θ_z	Rotations about y and z axes

14.2 SIGN CONVENTION AND DEFINITIONS

For the torsion of thin-walled beams, where restrained warping may be important, the response variables are ϕ, the angle of twist; ψ, the rate of change of ϕ with respect to the x axis; B, the bimoment; T, the torsional torque, and T_ω, the warping torque. A bimoment can be considered to be two equal and opposite moments M_c acting about the same axis and separated from one another (Fig. 14-1). Its value is the product of the moment and the separation distance. The effect of a bimoment is to warp cross sections and twist the beam. Positive twisting displacements and moments are illustrated as part of the tables of this chapter.

14.3 STRESSES

The response formulas of this chapter give the internal bimoment and warping torque along a beam. The normal and shear warping stresses (σ_ω and τ_ω) on a face of the cross section (Fig. 14-2) can be computed using the formulas of this section. For thin-walled open sections, the normal and shear stresses due to restrained warping or nonuniform torsion should be taken into account as they are often higher than the nonwarping stresses. Analytical and numerical procedures for the calculation of warping sectional properties and stresses are described in Chapters 2 and 15.

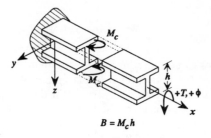

FIGURE 14-1 Bimoment B due to a twisting moment T is shown. A positive angle of twist ϕ is also illustrated.

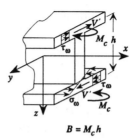

$$B = M_c h$$

FIGURE 14-2 Normal warping stress σ_ω and shear warping stress in an I-beam cross section.

Normal Warping Stress

The normal stress σ_x caused by warping is

$$\sigma_x = \sigma_\omega = \frac{B\omega}{\Gamma} \tag{14.1}$$

This stress acts perpendicular to the surface of the cross section (Fig. 14-2). It is assumed to be constant through the thickness of the thin-walled section. The quantity ω is the principal sectorial coordinate defined as $\omega = \omega_S - \omega_0$ where

$$\omega_0 = \frac{1}{A}\int_A \omega_S \, dA = \frac{1}{A}\int_A \omega_S t \, ds \tag{14.2a}$$

$$\omega_S = \begin{cases} \displaystyle\int_0^s r_S \, ds & \text{for open cross sections} \\[2ex] \displaystyle\int_0^s r_S \, ds - \frac{\oint r_S \, ds}{\oint (1/t) \, ds}\int_0^s \frac{1}{t} \, ds & \text{for closed cross sections} \end{cases} \tag{14.2b}$$

The integration $\int_0^s$ in Eq. (14.2b) is taken from the free edge to the point at which the stress σ_x is desired. The symbol $\oint$ indicates that the integration is taken completely around the closed section. See Chapters 2 and 15 for the computation of ω. The bimoment is defined by

$$B = \int_A \sigma_x \omega \, dA \tag{14.3a}$$

The torsional moment can cause warping (longitudinal displacement) of the cross section. Consequently, the longitudinal displacement of the cross section caused by axial forces on the cross section may result in the twist of the beam. Thus a bimoment may develop. In this case, the bimoment is expressed as

$$B = \sum_i P_i \omega_i \tag{14.3b}$$

where P_i is a concentrated longitudinal axial load in the cross section where B is sought and ω_i is the corresponding principal sectorial coordinate at the location where P_i is applied. Here Σ indicates summation of all the $P_i\omega_i$ for the cross section under consideration.

The warping constant Γ is defined as

$$\Gamma = \begin{cases} \displaystyle\int_A \omega^2\, dA & \text{for open cross section} \\[2em] \displaystyle\int_A \omega^2\, dA - \frac{\displaystyle\oint r_s\, ds}{\displaystyle\oint (1/t)\, ds}\oint(Q_\omega/t)\, ds & \text{for closed cross sections} \end{cases} \qquad (14.4)$$

and can be found in Table 2-6 for some common cross sections.

Shear Warping Stress

The shear stress τ due to warping is

$$\tau_\omega = \begin{cases} \displaystyle\frac{T_\omega Q_\omega}{t\Gamma} & \text{for open cross sections} \\[2em] \displaystyle\frac{T_\omega}{t\Gamma}\left[Q_\omega - \frac{\displaystyle\oint(Q_\omega/t)\, ds}{\displaystyle\oint(1/t)\, ds}\right] & \text{for closed cross sections} \end{cases} \qquad (14.5)$$

This stress acts parallel to the edges of the cross section (Fig. 14-2). Like the normal warping stress, the shear warping stress is assumed to be constant through the thickness of the thin-walled section. The quantity

$$Q_\omega = \int_{A_0} \omega\, dA = \int_0^{s_0} \omega t\, ds \qquad (14.6)$$

is the first sectorial moment (Chapter 2). For an open cross section, the area A_0 lies between the position (s_0) at which the stress is desired and the outer fiber of the cross section. For a closed section, the integration of Eq. (14.6) should be taken as though the section were open at an arbitrary point. This point then replaces the outer fiber.

14.4 SIMPLE BEAMS

The governing equations for the twisting of a thin-walled beam are

$$E\Gamma\frac{d^4\phi}{dx^4} - GJ\frac{d^2\varphi}{dx^2} = m_x$$

$$E\Gamma\frac{d^3\phi}{dx^3} - GJ\frac{d\phi}{dx} = -T$$

$$E\Gamma\frac{d^2\phi}{dx^2} = -B \qquad (14.7)$$

$$\frac{d\phi}{dx} = -\psi$$

These relations can be solved giving the angle of twist ϕ; rate of angle of twist ψ; bimoment B, twisting moment T, and warping torque T_ω as functions of the coordinate x. Then, the warping torque T_ω and pure torsion torque T_t can be calculated from $T_\omega = -E\Gamma d^3\phi/dx^3$ and $T_t = GJ d\phi/dx$.

Formulas for Beams with Arbitrary Loading

The angle of twist, bimoment, torsional moment, and warping torque of beams under arbitrary loading with any end conditions are provided in Table 14-1.

Table 14-1, part A, lists equations for the responses. The functions F_ϕ, F_ψ, F_B, and F_T are taken from Table 14-1, part B, by adding the appropriate terms for each load applied to the beam. The initial parameters ϕ_0, ψ_0, B_0, and T_0, which are values of ϕ, ψ, B, and T at the left end ($x = 0$) of the beam, are evaluated using the entry in Table 14-1, part C, for the appropriate beam end conditions.

The end conditions for the twisting of a thin-walled beam are frequently referred to as simply supported, fixed, free, and guided. An end is said to be *simply supported* or *pinned* if the cross section at the end is allowed to warp freely but is prevented from rotating. The angle of twist ϕ and the bimoment B are taken as being zero at the simply supported end. If the rotation and the warping of an end cross section are completely constrained, the end is said to be *fixed*. At a fixed end the angle of twist ϕ and the rate of angle of twist ψ are zero. If neither the rotation nor the warping is restrained, the end is considered to be *free*. For free ends the bimoment B and the twisting moment T are zero. A *guided end* twists but does not warp. For a guided end the rate of angle of twist ψ and the torque T are zero.

Example 14.1 Response and Stresses of an I-beam Calculate the displacements, moments, and stresses of the clamped–free I-beam of Fig. 14-3 under uniform twisting moment m_x.

The cross sectional properties of the beam are obtained from Tables 2-5 and 2-6 as

$$J = \frac{1.29}{3}\sum_{i=1}^{3} b_i t_i^3 = 12.9 \text{ in.}^4, \qquad \Gamma = \tfrac{1}{24}b^3h^2t = 4166.667 \text{ in.}^6 \qquad (1)$$

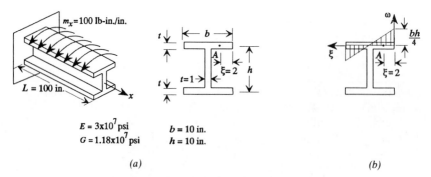

$E = 3\text{x}10^7 \text{ psi}$ $b = 10$ in.
$G = 1.18\text{x}10^7 \text{ psi}$ $h = 10$ in.

(a) (b)

FIGURE 14-3 Cantilevered I-beam: (a) Beam characteristics, (b) Warping function.

and

$$C = (GJ/E\Gamma)^{1/2} = 3.49 \times 10^{-2} \text{ in.}^{-1} \qquad (2)$$

The loading functions are taken from Table 14-1, part B:

$$F_\phi(x) = \frac{m_x}{C^2 GJ}\left(\cosh Cx - 1 - \tfrac{1}{2}C^2 x^2\right) = 5.394 \times 10^{-4}\left(\cosh Cx - 1 - \tfrac{1}{2}C^2 x^2\right)$$

$$F_\psi(x) = \frac{m_x}{CGJ}(Cx - \sinh Cx) = 1.88 \times 10^{-5}(Cx - \sinh Cx)$$

$$(3)$$

$$F_B(x) = -\frac{m_x}{C^2}(\cosh Cx - 1) = -82{,}118(\cosh Cx - 1)$$

$$F_T(x) = -m_x x = -100x$$

The four initial parameters are based on the formulas of Table 14-1, part C:

$$\phi_0 = \psi_0 = 0$$

$$B_0 = \frac{\bar{F}_T}{C}\tanh CL - \frac{\bar{F}_B}{\cosh CL} = -\frac{m_x L}{C}\tanh CL + \frac{m_x}{C^2}\frac{\cosh CL - 1}{\cosh CL}$$

$$(4)$$

$$= -208{,}902.85 \text{ lb-in.}^2$$

$$T_0 = 10{,}000 \text{ lb-in.}$$

Then the responses are (Table 14-1, part A)

$$\phi = B_0\frac{1 - \cosh Cx}{GJ} + T_0\frac{Cx - \sinh Cx}{CGJ} + F_\phi(x)$$

$$\psi = B_0\frac{\sinh Cx}{CE\Gamma} - T_0\frac{1 - \cosh Cx}{GJ} + F_\psi(x)$$

$$(5)$$

$$B = B_0\cosh Cx + T_0\frac{\sinh Cx}{C} + F_B(x)$$

$$T = T_0 + F_T(x)$$

$$T_\omega = B_0 C\sinh Cx + T_0\cosh Cx + F_T(x) + GJF_\psi(x)$$

At particular locations along the beam (20, 40, 60, 80) values for these responses are as follows:

	20	40	60	80
ϕ (rad)	2.442×10^{-4}	7.195×10^{-4}	0.0012	0.001597
ψ (rad/in.)	-2.064×10^{-5}	-2.509×10^{-5}	-2.228×10^{-5}	-1.744×10^{-5}
B (lb-in.^{2})	$-66,077$	1577	28,380	27,931
T (lb-in.)	8000	6000	4000	2000
T_ω (lb-in.)	4857.8	2180.05	608.24	-655.01

With B known, the cross-sectional stresses can be calculated. Consider the cross section at $x = 60$. From case 2 of Table 2-7, along the top flange the warping function ω is linearly distributed and is zero at the middle. At point A (Fig. 14-3b)

$$\omega = \tfrac{1}{2}h\left(\tfrac{1}{2}b - \xi\right)\big|_{\xi=2} = 15 \text{ in.}^2 \tag{6}$$

From Eq. (14.6)

$$Q_{\omega A} = \int_0^2 \tfrac{1}{2}h\left(\tfrac{1}{2}b - \xi\right) d\xi = 40 \text{ in.}^4 \tag{7}$$

The stresses at point A are [Eqs. (14.1) and (14.5)]

$$\sigma_\omega = \frac{B\omega}{\Gamma} = \frac{28,380(15)}{4166.667} = 102.2 \text{ lb/in.}^2$$

$$\tau_\omega = \frac{T_\omega Q_\omega}{t\Gamma} = \frac{608.24(40)}{4166.667} = 5.84 \text{ lb/in.}^2 \tag{8}$$

For this cross section, the maximum normal stress is found at the tips of the flanges and the maximum shear stresses are at the intersections of the flanges and the web. Note that the normal stress vanishes in the web.

14.5 BUCKLING LOADS

A thin-walled beam can buckle in a bending mode as presented in Chapter 11. At certain levels of axial force P, it can also undergo torsional instability. The critical axial force for torsional buckling of uniform bars with axial forces passing through the shear center is given by

$$P_{cr} = \frac{1}{C_p}\left(\frac{C_1\pi^2}{L^2}E\Gamma + GJ\right) \tag{14.8}$$

where

$$
C_1 = \begin{cases} \frac{1}{4} & \text{fixed–free ends} \\ 1 & \text{simply supported at both ends} \\ 2.045 & \text{fixed–simply supported ends} \\ 4 & \text{fixed at both ends} \end{cases}
$$

For bars with axial forces located at the eccentricity e_y, e_z from the centroid, critical loads are provided in Table 14-2.

The formulas in the table are based on the solution of the differential equations [14.1]

$$
EI_z v^{iv} = q_y, \qquad EI w^{iv} = q_z, \qquad EI_\omega \phi^{iv} - GJ\phi'' = m_x \qquad (14.9)
$$

where

$$
q_y = -Pv'' - \left(z_S P + M_y\right)\phi'', \qquad q_z = -Pw'' + \left(y_S P - M_z\right)\phi'' \qquad (14.10)
$$
$$
m_x = -\left(y_S P + M_y\right)v'' + \left(y_S P - M_z\right)w'' + \left(-r^2 P + 2\beta_z M_y - 2\beta_y M_z\right)\phi''
$$

These loadings are obtained from the projections of the stress of the unbuckled state in the y and z directions when the beam is undergoing buckling deformations. In Eqs. (14.10), β_y, β_z, and r^2 are called *stability parameters* and are expressed as

$$
\beta_y = \int_A y\left(y^2 + z^2\right) dA/2I_z - y_S, \qquad \beta_z = \int_A z\left(y^2 + z^2\right) dA/2I - z_S
$$
$$
r^2 = \frac{I + I_z}{A} + y_S^2 + z_S^2 \qquad (14.11)
$$

Thin-walled beams are also susceptible to local buckling.

14.6 NATURAL FREQUENCIES

Values of the fundamental natural torsional frequencies of uniform, open-section, thin-walled beams can be taken from Fig. 14-4 for various end conditions. For these plots, the shear center must coincide with the centroid.

Table 14-3 provides the frequency equations for a variety of uniform thin-walled beams. These can be solved iteratively for the natural frequencies.

14.7 GENERAL BEAMS

The formulas provided thus far apply to single-span thin-walled beams. For more general beam systems, e.g., those with multiple spans, it is advisable to use the displacement method or the transfer matrix procedure of Appendix III. The transfer, stiffness, and mass matrices can be employed to find the static response, buckling load, or natural frequencies.

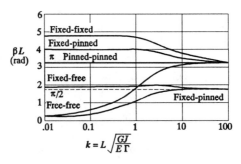

Definition:

$$\beta = (\{-GJ+[(GJ)^2+ 4E\,\Gamma\rho r_p^2\,\omega^2]^{1/2}\}/2\,E\,\Gamma)^{1/2}$$

ω = fundamental natural frequency

FIGURE 14-4 Value of frequency parameter βL for first mode vibration of thin-walled beam under torsion [14.2]. Shear center and centroid coincide. To use this figure, enter the figure at the appropriate k, read βL from the plots, and calculate the first natural frequency ω_1 from the definition of β.

For thin-walled beams where the centroid and shear center coincide, the transfer, stiffness, and mass matrices can be obtained from those of Chapter 11 with the change of notation of Table 14-4. The point matrices of Table 14-5 can be used to incorporate point occurrences in the solution. The notation for the transfer, stiffness, and mass matrices for the twisting of the thin-walled beam is

$$
\begin{bmatrix} \phi_b \\ \psi_b \\ B_b \\ T_b \\ 1 \end{bmatrix} = \mathbf{U}^i \begin{bmatrix} \phi_a \\ \psi_a \\ B_a \\ T_a \\ 1 \end{bmatrix}, \qquad
\begin{bmatrix} T_a \\ B_a \\ T_b \\ B_b \end{bmatrix} = \mathbf{k}^i \begin{bmatrix} \phi_a \\ \psi_a \\ \phi_b \\ \psi_b \end{bmatrix} + \begin{bmatrix} T_a^0 \\ B_a^0 \\ T_b^0 \\ B_b^0 \end{bmatrix}
$$

$$
\mathbf{U}^i = \begin{bmatrix}
U_{\phi\phi} & U_{\phi\psi} & U_{\phi B} & U_{\phi T} & F_\phi \\
U_{\psi\phi} & U_{\psi\psi} & U_{\psi B} & U_{\psi T} & F_\psi \\
U_{B\phi} & U_{B\psi} & U_{BB} & U_{BT} & F_B \\
U_{T\phi} & U_{T\psi} & U_{TB} & U_{TT} & F_T \\
0 & 0 & 0 & 0 & 1
\end{bmatrix}
\tag{14.12}
$$

$$
\mathbf{k}^i = \begin{bmatrix}
k_{11} & k_{12} & k_{13} & k_{14} \\
k_{21} & k_{22} & k_{23} & k_{24} \\
k_{31} & k_{32} & k_{33} & k_{34} \\
k_{41} & k_{42} & k_{43} & k_{44}
\end{bmatrix}
$$

The sign convention for the angle of twist and moment for transfer matrices is shown in Fig. 14-5a. Figure 14-5b gives the sign convention for stiffness and mass matrices.

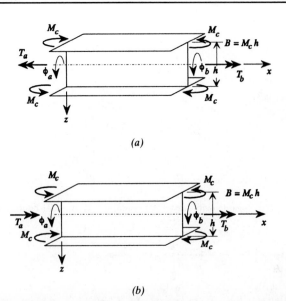

FIGURE 14-5 Sign conventions for transfer and stiffness matrices: (a) transfer matrices (sign convention 1): (b) stiffness matrices (sign convention 2).

The responses that these matrices represent are based on the governing equations for a thin-walled beam under torsion:

$$\frac{\partial \phi}{\partial x} = -\psi \tag{14.13a}$$

$$\frac{\partial \psi}{\partial x} = \frac{B}{E\Gamma} \tag{14.13b}$$

$$\frac{\partial B}{\partial x} = T + (GJ - C_p P)\psi = T_\omega \tag{14.13c}$$

$$\frac{\partial T}{\partial x} = k_t \phi + \rho r_p^2 \frac{\partial^2 \phi}{\partial t^2} - m_x(x, t) \tag{14.13d}$$

These governing equations apply to thin-walled beams of either constant or variable cross sections with the centroid and shear center coinciding with a compressive force P ($-P$ for tension) passing through the shear center of the cross section of the element.

For the thin-walled beams with arbitrary cross sections, i.e., the centroid and the shear center do not coincide (Fig. 14-6), the static deformation as well as the dynamic responses are coupled for torsion and bending. These are referred to as torsional–flexural responses. In these cases, the static, dynamic responses and buckling loads can be obtained from the stiffness, mass, and geometric stiffness matrices of Tables 14-6 and 14-7. The element variables corresponding to these

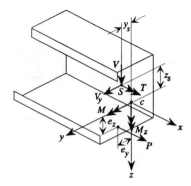

FIGURE 14-6 Thin-walled beam for which centroid c and shear center S do not coincide.

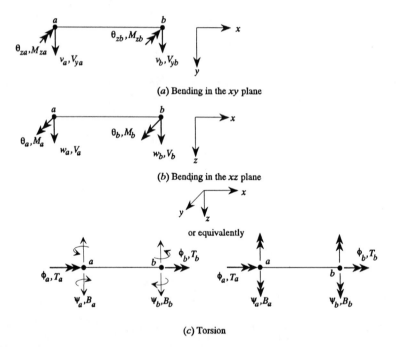

(a) Bending in the xy plane

(b) Bending in the xz plane

or equivalently

(c) Torsion

FIGURE 14-7 Element variables of thin-walled beam element under torsional–flexural deformation: (a) bending in the x, y plane; (b) bending in the x, z plane; (c) torsion and warping.

matrices are (Fig. 14-7)

$$\mathbf{v}^i = \begin{bmatrix} v_a & \theta_{za} & v_b & \theta_{zb} & w_a & \theta_a & w_b & \theta_b & \phi_a & \psi_a & \phi_b & \psi_b \end{bmatrix}^T \quad (14.14a)$$

and the forces are

$$\mathbf{p}^i = \begin{bmatrix} V_{ya} & M_{za} & V_{yb} & M_{zb} & V_a & M_a & V_b & M_b & T_a & B_a & T_b & B_b \end{bmatrix}^T \quad (14.14b)$$

The sign convention for v, θ_z, w, and θ as well as for V_y, M_z, V, and M are the same as that in Chapter 13, and the sign convention for ϕ and T is given in Fig. 14-5b.

The stiffness matrix and loading vector in Table 14-6 give exact results for static responses and approximate results for dynamic and stability analyses.

REFERENCES

14.1 Vlasov, V. Z., *Thin-walled Elastic Beams*, Israel Program for Scientific Translation, Jerusalem, Israel, 1961.

14.2 Gere, J. M., "Torsional Vibrations of Beams of Thin-walled Open Section," *Trans. ASME J. Appl. Mechan.*, Vol. 76, 1954, pp. 381–387.

14.3 Krajcinovic, D., "A Consistent Discrete Elements Technique for Thin-walled Assemblages," *Int. J. Solids Struct.*, Vol. 5, 1969, pp. 639–662.

14

Tables

TABLE 14-1, PART A SIMPLE BEAMS WITH ARBITRARY LOADINGS: GENERAL RESPONSE EXPRESSIONS

Definitions

$$C = \sqrt{GJ/E\Gamma}$$

$$\bar{F}_\Psi = F_\Psi|_{x=L}$$

$$\bar{F}_T = F_T|_{x=L}$$

$$\bar{F}_\phi = F_\phi|_{x=L}$$

$$\bar{F}_B = F_B|_{x=L}$$

$$\langle x - a \rangle^n = \begin{cases} 0 & \text{if } x < a \\ (x - a)^n & \text{if } x \geq a \end{cases}$$

$$\langle x - a \rangle^0 = \begin{cases} 0 & \text{if } x < a \\ 1 & \text{if } x \geq a \end{cases}$$

$$\cosh C\langle x - a \rangle = \begin{cases} 0 & \text{if } x < a \\ \cosh C(x - a) & \text{if } x \geq a \end{cases} \quad \text{etc.}$$

Positive angle of twist ϕ and torque T are shown in Fig. 14-1.

Response

1.
Angle of twist

$$\phi = \phi_0 - \Psi_0 \frac{\sinh Cx}{C} + T_0 \frac{Cx - \sinh Cx}{CGJ}$$
$$+ B_0 \frac{1 - \cosh Cx}{GJ} + F_\phi(x)$$

2.
Rate of angle of twist

$$\Psi = \Psi_0 \cosh Cx - T_0 \frac{1 - \cosh Cx}{GJ}$$
$$+ B_0 \frac{\sinh Cx}{CE\Gamma} + F_\Psi(x)$$

3.
Twisting moment

$$T = T_0 + F_T(x)$$

4.
Bimoment

$$B = \Psi_0 CE\Gamma \sinh Cx + T_0 \frac{\sinh Cx}{C}$$
$$+ B_0 \cosh Cx + F_B(x)$$

5.
Warping torque

$$T_\omega = \Psi_0 CJ \cosh Cx + T_0 \cosh Cx$$
$$+ B_0 C \sinh Cx + F_T(x) + GJF_\Psi(x)$$

TABLE 14-1, Part A   Simple Beams with Arbitrary Loadings   720

TABLE 14-1, PART B   SIMPLE BEAMS WITH ARBITRARY LOADING: LOADING FUNCTIONS

		Uniformly Distributed Torque
	$-a \rightarrow \|T_1$ $-a \rightarrow$ $W \downarrow c$ $T_1 = Wc$	$-a_2 \rightarrow$ $-a_1 \rightarrow$ m_{x1} $-a_2 \rightarrow$ $-a_1 \rightarrow P_1$ c $m_{x1} = cp_1$
$F_\phi(x)$	$\dfrac{T_1}{CGJ}(-C\langle x - a\rangle + \sinh C\langle x - a\rangle)$	$\dfrac{m_{x1}}{C^2 GJ}\left(\cosh C\langle x - a_1\rangle - \langle x - a_1\rangle^0 - C^2 \dfrac{\langle x - a_1\rangle^2}{2} \right.$ $\left. - \cosh C\langle x - a_2\rangle + \langle x - a_2\rangle^0 + C^2 \dfrac{\langle x - a_2\rangle^2}{2} \right)$
$F_\psi(x)$	$\dfrac{T_1}{GJ}(\langle x - a\rangle^0 - \cosh C\langle x - a\rangle)$	$\dfrac{m_{x1}}{CGJ}(C\langle x - a_1\rangle - \sinh C\langle x - a_1\rangle$ $- C\langle x - a_2\rangle + \sinh C\langle x - a_2\rangle)$
$F_T(x)$	$-T_1\langle x - a\rangle^0$	$-m_{x1}(\langle x - a_1\rangle - \langle x - a_2\rangle)$
$F_B(x)$	$-\dfrac{T_1}{C}\sinh C\langle x - a\rangle$	$-\dfrac{m_{x1}}{C^2}\left(\cosh C\langle x - a_1\rangle - \langle x - a_1\rangle^0 \right.$ $\left. - \cosh C\langle x - a_2\rangle + \langle x - a_2\rangle^0 \right)$

	Concentrated Bimoment $B_1 = M_1 c$	
$F_\phi(x)$	$\dfrac{B_1}{GJ}(-\langle x-a\rangle^0 + \cosh C\langle x-a\rangle)$	$\dfrac{m_{x0}}{C^2 GJ}\bigg(\cosh C\langle x-a_1\rangle - \langle x-a_1\rangle^0$ $\qquad - C^2\dfrac{\langle x-a_1\rangle^2}{2} - \cosh C\langle x-a_2\rangle$ $\qquad + \langle x-a_2\rangle^0 + C^2\dfrac{\langle x-a_2\rangle^2}{2}\bigg)$ $\quad + \dfrac{m_{x\ell}-m_{x0}}{E\Gamma(a_2-a_1)}\bigg(\dfrac{1}{C}\sinh C\langle x-a_1\rangle$ $\qquad - \langle x-a_1\rangle - C^2\dfrac{\langle x-a_1\rangle^2}{6}$ $\qquad - \dfrac{1}{C}\sinh C\langle x-a_2\rangle + \langle x-a_2\rangle$ $\qquad + C^2\dfrac{\langle x-a_2\rangle^2}{6}\bigg)$
$F_\psi(x)$	$-\dfrac{B_1}{CE\Gamma}\sinh C\langle x-a\rangle$	$\dfrac{m_{x0}}{CGJ}(C\langle x-a_1\rangle - \sinh C\langle x-a_1\rangle$ $\qquad - C\langle x-a_2\rangle + \sinh C\langle x-a_2\rangle)$ $\quad - \dfrac{m_{x\ell}-m_{x0}}{CJG(a_2-a_1)}\bigg(\dfrac{1}{C}\cosh C\langle x-a_1\rangle$ $\qquad - \dfrac{\langle x-a_1\rangle^0}{C} - \dfrac{C}{2}\langle x-a_1\rangle^2$ $\qquad - \dfrac{1}{C}\cosh C\langle x-a_2\rangle$ $\qquad + \dfrac{\langle x-a_2\rangle^0}{C} + \dfrac{C}{2}\langle x-a_2\rangle^2\bigg)$
$F_T(x)$	0	$-m_{x0}(\langle x-a_1\rangle - \langle x-a_2\rangle)$ $\quad - \dfrac{m_{x\ell}-m_{x0}}{2(a_2-a_1)}\big(\langle x-a_1\rangle^2 - \langle x-a_2\rangle^2\big)$
$F_B(x)$	$-B_1\cosh C\langle x-a\rangle$	$-\dfrac{m_{x0}}{C^2}\big(\cosh C\langle x-a_1\rangle - \langle x-a_1\rangle^0$ $\qquad - \cosh C\langle x-a_2\rangle + \langle x-a_2\rangle^0\big)$ $\quad + \dfrac{m_{x\ell}-m_{x0}}{a_2-a_1}\bigg(\langle x-a_1\rangle - \dfrac{1}{C}\sinh C\langle x-a_1\rangle$ $\qquad - \langle x-a_2\rangle + \dfrac{1}{C}\sinh C\langle x-a_2\rangle\bigg)$

TABLE 14-1, Part B   Arbitrary Loadings: Loading Functions   722

TABLE 14-1, PART C   SIMPLE BEAMS WITH ARBITRARY LOADING: INITIAL PARMETERS

Left End \ Right End →	1. Warps but does not twist (pinned or simply supported)	2. No warp and no twist (fixed)	3. Warps and twists (free)	4. Twists but does not warp (guided)
1. Warps but does not twist (pinned or simply supported) $\phi_0=0, B_0=0$	$\psi_0 = \dfrac{\bar{F}_B}{GJL}\left(-1 + \dfrac{CL}{\sinh CL}\right)$ $+ \dfrac{\bar{F}_\phi}{L}$ $T_0 = -\dfrac{1}{L}\left(GJ\bar{F}_\phi + \bar{F}_B\right)$	$\psi_0 = \dfrac{1}{\nabla}\left[(CL - \sinh CL)\bar{F}_\psi\right.$ $\left. - C(\cosh CL - 1)\bar{F}_\phi\right]$ $T_0 = \dfrac{GJ}{\nabla}\left(\bar{F}_\phi C \cosh CL \right.$ $\left. + \bar{F}_\psi \sinh CL\right)$ $\nabla = \sinh CL - CL \cosh CL$	$\psi_0 = \dfrac{1}{GJ}\left[\bar{F}_T - \dfrac{C\bar{F}_B}{\sinh CL}\right]$ $T_0 = -\bar{F}_T$	$\psi_0 = \dfrac{-1}{\cosh CL}$ $\times \left[\dfrac{\bar{F}_T}{GJ}(1 - \cosh CL) + \bar{F}_\psi\right]$ $T_0 = -\bar{F}_T$
2. No warp and no twist (fixed) $\phi_0=0, \psi_0=0$	$B_0 = \dfrac{1}{\nabla}\left[(CL - \sinh CL)\bar{F}_B\right.$ $\left. - GJ\bar{F}_\phi \sinh CL\right]$ $T_0 = \bar{F}_\phi ETC^3 \cosh CL$ $- C(1 - \cosh CL)\bar{F}_B$ $\nabla = \sinh CL - CL \cosh CL$	$B_0 = \dfrac{1}{\nabla}\left[ETC(CL - \sinh CL)\bar{F}_\psi\right.$ $\left. - GJ(\cosh CL - 1)\bar{F}_\phi\right]$ $T_0 = \dfrac{1}{\nabla}\left[GJC\bar{F}_\phi \sinh CL\right.$ $\left. - GJ(1 - \cosh CL)\bar{F}_\psi\right]$ $\nabla = \cosh CL$	$B_0 = \dfrac{\bar{F}_T}{C}\tanh CL$ $- \dfrac{\bar{F}_B}{\cosh CL}$ $T_0 = -\bar{F}_T$	$B_0 = \dfrac{-1}{\sinh CL}$ $\times \left[\dfrac{\bar{F}_T}{C}(1 - \cosh CL) + CET\bar{F}_\phi\right]$ $T_0 = -\bar{F}_T$

Left End ↓ / Right End →	1. Warps but does not twist (pinned or simply supported)	2. No warp and no twist (fixed)	3. Warps and twists (free)	4. Twists but does not warp (guided)
3. Warps and twists (free) $B_0 = 0,\ T_0 = 0$	$\phi_0 = -\dfrac{\bar{F}_B}{GJ} - \bar{F}_\phi$ $\psi_0 = -\dfrac{\bar{F}_B}{E\Gamma C \sinh CL}$	$\phi_0 = -\dfrac{\bar{F}_\psi}{C}\tanh CL - \bar{F}_\phi$ $\psi_0 = \dfrac{-\bar{F}_\psi}{\cosh CL}$	Kinematically unstable	Kinematically unstable
4. Twists but does not warp (guided) $\psi_0 = 0,\ T_0 = 0$	$\phi_0 = \dfrac{\bar{F}_B}{GJ}\dfrac{(1-\cosh CL)}{\cosh CL} - \bar{F}_\phi$ $B_0 = \dfrac{-\bar{F}_B}{\cosh CL}$	$\phi_0 = \dfrac{\bar{F}_\psi}{C}\dfrac{(1-\cosh CL)}{\sinh CL} - \bar{F}_\phi$ $B_0 = \dfrac{-\bar{F}_\psi CE\Gamma}{\sinh CL}$	Kinematically unstable	Kinematically unstable

TABLE 14-1, Part C   Arbitrary Loadings: Initial Parameters   724

TABLE 14-2   CRITICAL ELASTIC FLEXURAL–TORSIONAL LOADS FOR THIN-WALLED COLUMNS

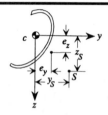

Notation

$$\lambda = n\pi/L \qquad n = 1, 2, \ldots \text{ is modal number of buckled mode}$$

e_y, e_z = eccentricities of appplied axial force from y and z axes

y_S, z_S = y and z coordinates of shear center

β_y, β_z, r = elastic flexural–torsional parameters for thin-walled beams

$$\beta_y = \frac{U_z}{2I_z} - y_S$$

$$\beta_z = \frac{U_y}{2I} - z_S$$

$$r^2 = \frac{I + I_z}{A} + y_S^2 + z_S^2$$

$$U_z = \int_A y^3 \, dA + \int_A z^2 y \, dA$$

$$U_y = \int_A z^3 \, dA + \int_A y^2 z \, dA$$

M_y, M_z = bending moments about y and z axes due to eccentricity of applied axial force P

$$M_y = Pe_z$$
$$M_z = -Pe_y$$

I, I_z = moments of inertia about y and z axes

J = torsional constant

E = modulus of elasticity

G = shear modulus of elasticity

c = centroid

S = shear center

The critical buckling loads are determined from the equation

$$\begin{vmatrix} B_{11} & 0 & B_{13} \\ 0 & B_{22} & B_{23} \\ B_{31} & B_{32} & B_{33} \end{vmatrix} = 0$$

Configuration	Parameters in Buckling Equation	
1. Pinned–pinned or guided–guided or pinned–guided	$B_{11} = EI_z\lambda^2 - P$ $B_{13} = -(M_y + z_S P)$ $B_{22} = EI\lambda^2 - P$ $B_{23} = -(M_z - y_S P)$	$B_{31} = -(M_y + z_S P)$ $B_{32} = -(M_z - y_S P)$ $B_{33} = E\Gamma\lambda^2$ $\quad -(r^2 P + 2\beta_y M_z - 2\beta_z M_y - GJ)$
2. Fixed–fixed	$B_{11} = 4EI_z\lambda^2 - P$ $B_{13} = -(M_y + z_S P)$ $B_{22} = 4EI\lambda^2 - P$ $B_{23} = -(M_z - y_S P)$	$B_{31} = -(M_y + z_S P)$ $B_{32} = -(M_z - y_S P)$ $B_{33} = 4E\Gamma\lambda^2$ $\quad -(r^2 P + 2\beta_y M_z - 2\beta_z M_y - GJ)$
3. Pinned–fixed	$B_{11} = \frac{1}{4}EI_z\lambda^2 - P$ $B_{13} = -(M_y + z_S P)$ $B_{22} = EI\lambda^2 - P$ $B_{23} = -(M_z - y_S P)$	$B_{31} = -(M_y + z_S P)$ $B_{32} = -(M_z - y_S P)$ $B_{33} = \frac{1}{4}E\Gamma\lambda^2 - (r^2 P + 2\beta_y M_z)$

TABLE 14-2 **Buckling Loads for Thin-walled Columns**   **726**

TABLE 14-3   FREQUENCY EQUATIONS FOR TORSIONAL VIBRATION OF BEAMS OF THIN-WALLED OPEN SECTION WITH SHEAR CENTER COINCIDING WITH CENTROID[a]

Notation

E = modulus of elasticity
G = shear modulus of elasticity
L = length of beam
r_p = polar radius of gyration

Γ = warping constant
J = torsional constant
ρ = mass per unit length

$$\alpha = \sqrt{\frac{GJ + \sqrt{(GJ)^2 + 4E\Gamma\rho r_p^2 \omega_n^2}}{2E\Gamma}} \qquad \beta = \sqrt{\frac{-GJ + \sqrt{(GJ)^2 + 4E\Gamma\rho r_p^2 \omega_n^2}}{2E\Gamma}}$$

End Conditions	Frequency Equation
1. Pinned–pinned	$\sin\beta l = 0 \qquad \omega_n = \dfrac{n\pi}{L^2}\sqrt{\dfrac{n^2\pi^2 E\Gamma + L^2 GJ}{\rho r_p^2}}$
2. Fixed–fixed	$2(\alpha L)(\beta L)[1 - \cosh(\alpha L)\cos(\beta L)]$ $+[(\alpha L)^2 - (\beta L)^2]\sinh(\alpha L)\sin(\beta L) = 0$
3. Fixed–pinned	$(\beta L)\tanh(\alpha L) = (\alpha L)\tan(\beta L)$
4. Fixed–free	$\dfrac{(\alpha L)^4 + (\beta L)^4}{(\alpha L)^2(\beta L)^2}\cosh(\alpha L)\cos(\beta L)$ $+ \dfrac{(\alpha L)^2 - (\beta L)^2}{(\alpha L)(\beta L)}\sinh(\alpha L)\sin(\beta L) + 2 = 0$
5. Pinned–free	$(\alpha L)^3 \tanh(\alpha L) = (\beta L)^3 \tan(\beta L)$
6. Free–free	$[(\alpha L)^6 - (\beta L)^6]\sinh(\alpha L)\sin(\beta L)$ $+ 2(\alpha L)^3(\beta L)^3[\cosh(\alpha L)\cos(\beta L) - 1] = 0$

[a]Adopted from Gere [14.2].

TABLE 14-4   CHANGE OF NOTATION NECESSARY FOR TABLE 11-22 TO BE USED FOR TWISTING OF THIN-WALLED BEAMS

Bending of Beams	Twisting of Thin-walled Beams
w	ϕ
θ	ψ
M	B
V	T
p_a	m_{x0}
p_b	$m_{x\ell}$
$\dfrac{1}{GA_s}$	0
$\lambda = (k - \rho\omega^2)/EI$	$\lambda = (k_t - \rho r^2\omega^2)/E\Gamma$
$\eta = (k - \rho\omega^2)/GA_s$	$\eta = 0$
$\zeta = (P + \rho r_y^2\omega^2 - k^*)/EI$	$\zeta = (C_P P - GJ)/E\Gamma$

TABLE 14-4   Correlation of Twisting and Bending of Beams   728

TABLE 14-5 POINT MATRICES FOR CONCENTRATED OCCURRENCES

Case	Transfer Matrices (Sign Convention 1)	Stiffness Matrices (Sign Convention 2)
1. Concentrated applied torque	$$U_i = \begin{bmatrix} 1 & & & & \vdots & 0 \\ & 1 & & & \vdots & 0 \\ & & 1 & & \vdots & -B_i \\ & & & 1 & \vdots & -T_i \\ \hdashline 0 & 0 & 0 & 0 & \vdots & 1 \end{bmatrix}$$	Traditionally, these applied loads are implemented as nodal conditions.
2. Concentrated applied moments equivalent to bimoment B_i		
3. Concentrated mass $I_{pi} = \rho r_p^2 \Delta a$	$$U_i = \begin{bmatrix} 1 & 0 & 0 & 0 & \vdots & 0 \\ 0 & 1 & 0 & 0 & \vdots & 0 \\ 0 & 0 & 1 & 0 & \vdots & 0 \\ k_{ti} - I_{pi}\omega^2 & 0 & 0 & 1 & \vdots & 0 \\ \hdashline 0 & 0 & 0 & 0 & \vdots & 1 \end{bmatrix}$$	$T_a = -\omega^2 I_{pi}\phi_a$
4. Torsional spring		$T_a = k_{ti}\phi_a$

TABLE 14-6 STIFFNESS AND MASS MATRICES FOR TORSIONAL–FLEXURAL DEFORMATION OF THIN-WALLED BEAMS

Notation

E = modulus of elasticity
J = torsional constant
ℓ = length of element
$\alpha = \sqrt{GJ/E\Gamma}\,\ell = C\ell$
r_p = polar radius of gyration

y_S, z_S = distance between shear center and centroid
$I_y = I, I_z$ = moments of inertia about y and z axes
G = shear modulus of elasticity
Γ = warping constant

ρ = mass per unit length
$D = 2(1 - \cosh \alpha) + \alpha \sinh \alpha$
m_x = distributed torque
ϕ = angle of twist, rotation about x axis
$\psi = d\phi/dx$

The centroid and shear center of the cross section do not necessarily coincide. Refer to reference 14.3 for a discussion of coupled torsion and bending of beams.

The matrices $\mathbf{k}^i_{yy}$, $\mathbf{k}^i_{zz}$, $\mathbf{m}^i_{yy}$ and $\mathbf{m}^i_{zz}$ in this table can be obtained from the matrices in Tables 11-19 and 11-25. $\mathbf{k}^i_{zz}$ and $\mathbf{m}^i_{zz}$ are the same as the stiffness and mass matrices in those two tables. $\mathbf{k}^i_{yy}$ and $\mathbf{m}^i_{yy}$ are obtained by replacing I by I_z, r_y by r_z, and changing the signs of the elements in rows 2 and 4 of the matrices in those tables and then changing the signs of the elements in columns 2 and 4 of the resulting matrices. The same matrices can be obtained from the stiffness and mass matrices for bars in Chapter 13.

The element displacement and force vectors corresponding to the stiffness and mass matrices of this table are (Eq. 14.14)

$$\mathbf{v}^i = \begin{bmatrix} v_a & \theta_{za} & v_b & \theta_{zb} & w_a & \theta_a & w_b & \theta_b & \phi_a & \psi_a & \phi_b & \psi_b \end{bmatrix}^T$$

$$\mathbf{p}^i = \begin{bmatrix} V_{ya} & M_{za} & V_{yb} & M_{zb} & V_a & M_a & V_b & M_b & T_a & B_a & T_b & B_b \end{bmatrix}^T$$

See Fig. 14-7 for the sign convention for the displacements and forces.

Stiffness Matrix

$$\mathbf{k}^i = \begin{bmatrix} \mathbf{k}^i_{yy} & 0 & 0 \\ 0 & \mathbf{k}^i_{zz} & 0 \\ 0 & 0 & \mathbf{k}^i_{\phi\phi} \end{bmatrix}$$

$\mathbf{k}^i_{yy}$ and $\mathbf{k}^i_{zz}$ are stiffness matrices for bending and can be obtained from Table 11-19:

$$\mathbf{k}^i_{\phi\phi} = \frac{E\Gamma}{D\ell^3} \begin{bmatrix} \alpha^3 \sinh \alpha & -\alpha^2(1 - \cosh \alpha)\ell & -\alpha^3 \sinh \alpha & -\alpha^2(1 - \cosh \alpha)\ell \\ & \alpha(\alpha \cosh \alpha - \sinh \alpha)\ell^2 & -\alpha^2(\cosh \alpha - 1)\ell & \alpha(\sinh \alpha - \alpha)\ell^2 \\ & & \alpha^3 \sinh \alpha & -\alpha^2(\cosh \alpha - 1)\ell \\ \text{Symmetric} & & & \alpha(\alpha \cosh \alpha - \sinh \alpha)\ell^2 \end{bmatrix}$$

The loading vectors corresponding to $\mathbf{k}^i_{yy}$ and $\mathbf{k}^i_{zz}$ are given in Table 11-19. The loading vector for $\mathbf{k}^i_{\phi\phi}$ is $\bar{p}_i = \int_0^\ell \gamma_i(x)\, m_x(x)\, dx$, $i = 1, 2, 3, 4$ with

$$\gamma_1(x) = -\frac{1}{D}[(1 - \cosh \alpha)\cosh Cx + \sinh \alpha \sinh Cx - Cx \sinh \alpha + 1 - \cosh \alpha + \alpha \sinh \alpha]$$

$$\gamma_2(x) = \frac{1}{\alpha D}[(\alpha \cosh \alpha - \sinh \alpha)\cosh Cx + (\cosh \alpha - 1 - \alpha \sinh \alpha)\sinh Cx$$
$$+ Cx(\cosh \alpha - 1) + \sinh \alpha - \alpha \cosh \alpha]$$

$$\gamma_3(x) = -\frac{1}{D}[(\cosh \alpha - 1)\cosh Cx - \sinh \alpha \sinh Cx + Cx \sinh \alpha + (1 - \cosh \alpha)]$$

$$\gamma_4(x) = \frac{1}{\alpha D}[(\sinh \alpha - \alpha)\cosh Cx + (1 - \cosh \alpha)\sinh Cx + Cx + Cx(\cosh \alpha - 1) + \alpha - \sinh \alpha]$$

TABLE 14-6 Stiffness and Mass Matrices for Thin-walled Beams 730

Mass Matrix

$$\mathbf{m}^i = \begin{bmatrix} \mathbf{m}^i_{yy} & 0 & \mathbf{m}^i_{y\phi} \\ 0 & \mathbf{m}^i_{zz} & \mathbf{m}^i_{z\phi} \\ \mathbf{m}^i_{y\phi} & \mathbf{m}^i_{z\phi} & \mathbf{m}^i_{\phi\phi} \end{bmatrix}$$

$\mathbf{m}^i_{yy}$ and $\mathbf{m}^i_{zz}$ are consistent mass matrices for bending and can be obtained from Table 11-25:

$$\mathbf{m}^i_{y\phi} = 2\rho z_S \overline{\mathbf{m}} \qquad \mathbf{m}^i_{z\phi} = -2\rho y_S \overline{\mathbf{m}} \qquad \mathbf{m}^i_{\phi\phi} = \rho r_p^2 \hat{\mathbf{m}}$$

The elements in $\overline{\mathbf{m}}$ and $\hat{\mathbf{m}}$:

$$\overline{m}_{11} = \overline{m}_{33} = -\frac{1}{D\alpha^4}[(12\alpha - \alpha^3 + 0.35\alpha^5)\sinh\alpha + (24 + 0.5\alpha^4)(1 - \cosh\alpha)]$$

$$\overline{m}_{21} = -\overline{m}_{23} = -\frac{\ell}{D\alpha^4}[(6\alpha - 0.05\alpha^5)\sinh\alpha + (12 + \alpha^2 + 0.083\alpha^4)(1 - \cosh\alpha)]$$

$$\overline{m}_{31} = \overline{m}_{13} = -\frac{1}{D\alpha^4}[(-12\alpha + \alpha^3 + 0.15\alpha^5)\sinh\alpha + (-24 + 0.5\alpha^4)(1 - \cosh\alpha)]$$

$$\overline{m}_{41} = -\overline{m}_{43} = \frac{\ell}{D\alpha^4}[(-6\alpha + 0.033\alpha^5)\sinh\alpha + (-12 - \alpha^2 + 0.083\alpha^4)(1 - \cosh\alpha)]$$

$$\overline{m}_{12} = -\overline{m}_{34} = \frac{\ell}{D\alpha^4}[(-6\alpha + 1.5\alpha^3)\sinh\alpha + (12 - \alpha^2 - 0.35\alpha^4)\cosh\alpha - 12 + \alpha^2 - 0.15\alpha^4]$$

$$\overline{m}_{22} = \overline{m}_{44} = -\frac{\ell^2}{D\alpha^4}[(5\alpha - 0.083\alpha^3)\sinh\alpha - (8 + \alpha^2 - 0.05\alpha^4)\cosh\alpha + 8 + 0.033\alpha^4]$$

$$\overline{m}_{32} = -\overline{m}_{14} = \frac{\ell}{D\alpha^4}[(6\alpha + 0.5\alpha^3)\sinh\alpha - (12 + \alpha^2 + 0.15\alpha^4)\cosh\alpha + 12 + \alpha^2 - 0.35\alpha^4]$$

$$\overline{m}_{42} = \overline{m}_{24} = -\frac{\ell^2}{D\alpha^4}[(\alpha + 0.083\alpha^3)\sinh\alpha - (4 + 0.033)\cosh\alpha + 4 + \alpha^2 - 0.05\alpha^4]$$

$$\hat{m}_{11} = \hat{m}_{33} = \frac{1}{D^2\alpha}[(-5\sinh\alpha + 2\alpha - \alpha\cosh\alpha + \alpha^2\sinh\alpha)(1 - \cosh\alpha) + (0.33\alpha^3 - 2\alpha)\sinh^2\alpha]$$

$$\hat{m}_{12} = -\frac{\ell}{D^2\alpha^2}[(3.5\alpha\sinh\alpha - 0.5\alpha^2)(1 - \cosh\alpha) + 2\alpha^2\sinh^2\alpha - 0.1667\alpha^3\sinh\alpha(1 + 2\cosh\alpha)]$$

$$\hat{m}_{13} = 0.5\ell^2 - \hat{m}_{11}$$

$$\hat{m}_{14} = \frac{-\ell}{D^2\alpha^2}[-(8 + 2.5\alpha\sinh\alpha + 0.5\alpha^2)(1 - \cosh\alpha) - 4\sinh^2\alpha + 0.1667\sinh\alpha(2 + \cosh\alpha) - \alpha^2\sinh^2\alpha]$$

$$\hat{m}_{22} = \hat{m}_{44} = \frac{\ell^2}{D^2\alpha^3}[(3\alpha + 3\sinh\alpha)(1 - \cosh\alpha) + 0.1667\alpha^3(7 + 2\cosh\alpha) - \alpha^2\sinh\alpha(2 + 2.5\cosh\alpha)(6\alpha + 0.1667\alpha^3\sinh^2\alpha)]$$

$$\hat{m}_{23} = \frac{-\ell}{D\alpha^2}[(2 - 0.5\alpha^2)(1 - \cosh\alpha) + \alpha(2 + \sinh\alpha - \alpha\cosh\alpha)] - \hat{m}_{12}$$

$$\hat{m}_{24} = \frac{\ell^2}{D^2\alpha^3}[(5\alpha + 0.667\alpha^3 - 3\sinh\alpha)(1 - \cosh\alpha) - 0.5\alpha^3\cosh\alpha + \alpha^2\sinh\alpha(3.5 + \cosh\alpha) - \alpha^3 - (2\alpha + 0.1667)\sinh^2\alpha]$$

$$\hat{m}_{34} = \frac{-\ell}{D\alpha^2}[-(2 + 0.5\alpha^2)(1 - \cosh\alpha) + \alpha(\alpha - 2\sinh\alpha)] - \hat{m}_{14}$$

All the $\overline{m}_{ij}$ and $\hat{m}_{ij}$ that are not specified here are zero.

TABLE 14-7 CONSISTENT GEOMETRIC STIFFNESS MATRIX FOR TORSIONAL–FLEXURAL DEFORMATION OF THIN-WALLED BEAMS [a]

Notation

β_y, β_z, r^2 = elastic flexural-torsional parameters for thin-walled beams. See Table 14-2.

e_y, e_z = eccentricities of the applied force, measured from y and z axes.

$\alpha = \sqrt{GJ/ET}\, \ell$

$D = 2(1 - \cosh \alpha) + \alpha \sinh \alpha$

ℓ = length of element

See Table 14-6 for definitions of other quantities in these matrices, including the displacement and force vectors.

The matrices $\mathbf{k}^i_{yyG}$ and $\mathbf{k}^i_{zzG}$ in this table can be obtained from Table 11-23. $\mathbf{k}^i_{yyG}$ is the same as the matrix given in Table 11-23. $\mathbf{k}^i_{zzG}$ is obtained by changing the signs of the elements in rows 2 and 4 of the geometric stiffness matrix in Table 11-23 and then changing the signs of the elements of columns 2 and 4 of the resulting matrix.

$$
\mathbf{k}^i_G = \begin{bmatrix} \mathbf{k}^i_{yyG} & 0 & \mathbf{k}^i_{y\phi G} \\ 0 & \mathbf{k}^i_{zzG} & \mathbf{k}^i_{z\phi G} \\ \mathbf{k}^i_{y\phi G} & \mathbf{k}^i_{z\phi G} & \mathbf{k}^i_{\phi\phi G} \end{bmatrix}
$$

TABLE 14-7 | Consistent Geometric Stiffness Matrix   732

$$\mathbf{g}_G = -2\begin{bmatrix} g_{11} & g_{12} & g_{13} & g_{14} \\ g_{21} & g_{22} & g_{23} & g_{24} \\ g_{31} & g_{32} & g_{33} & g_{34} \\ g_{41} & g_{42} & g_{43} & g_{44} \end{bmatrix}$$

$\mathbf{k}^i_{y\phi G} = 2(z_S + e_z)\mathbf{g}_G$

$\mathbf{k}^i_{z\phi G} = -2(y_S + e_y)\mathbf{g}_G$

$$g_{11} = -g_{31} = -g_{13} = g_{33} = \frac{12}{\ell\alpha^2}\left\{1 - \frac{\alpha^3}{12[\alpha - 2\tanh(0.5\alpha)]}\right\}$$

$$g_{21} = g_{41} = -g_{23} = -g_{43} = g_{12} = -g_{32} = g_{14} = -g_{34}$$
$$= \frac{6}{\alpha^2}\left\{1 + \frac{\alpha^2}{12} - \frac{\alpha^3}{12[\alpha - 2\tanh(0.5\alpha)]}\right\}$$

$$g_{22} = g_{44} = \frac{4\ell}{\alpha^2}\left[-1 + \frac{\alpha}{4D}(\alpha\cosh\alpha - \sinh\alpha)\right]$$

$$g_{42} = g_{24} = -\frac{2\ell}{\alpha^2}\left[-1 - \frac{\alpha}{2D}(\alpha - \sinh\alpha)\right]$$

$$\mathbf{g}^\phi_G = \begin{bmatrix} g^\phi_{11} & g^\phi_{12} & g^\phi_{13} & g^\phi_{14} \\ g^\phi_{21} & g^\phi_{22} & g^\phi_{23} & g^\phi_{24} \\ g^\phi_{31} & g^\phi_{32} & g^\phi_{33} & g^\phi_{34} \\ g^\phi_{41} & g^\phi_{42} & g^\phi_{43} & g^\phi_{44} \end{bmatrix}$$

$\mathbf{k}^i_{\phi\phi G} = (r^2 - 2\beta_y e_y - 2\beta_z e_z)\mathbf{g}^\phi_G$

$$g^\phi_{11} = -g^\phi_{13} = g^\phi_{33} = \frac{\alpha}{D^2\ell}[(1 - \cosh\alpha)(3\sinh\alpha - \alpha) + \alpha\sinh^2\alpha]$$

$$g^\phi_{12} = -\frac{1}{D^2}\left[\left(4 + \frac{\alpha^2}{2} + \frac{\alpha}{2}\sinh\alpha\right)(1 - \cosh\alpha) + 2\sinh^2\alpha\right]$$

$$g^\phi_{22} = g^\phi_{44} = \frac{\ell}{\alpha D^2}\left[(\cosh\alpha - 1)(\sinh\alpha + \alpha) + \alpha\sinh\alpha(\alpha - 2\sinh\alpha) - \frac{\alpha^2}{2}(\alpha - \sinh\alpha\cosh\alpha)\right]$$

$$g^\phi_{24} = \frac{\ell}{\alpha D^2}\left[(\sinh\alpha - 3\alpha)(1 - \cosh\alpha) + \frac{\alpha^2}{2}(\alpha\cosh\alpha - 3\sinh\alpha)\right]$$

$g^\phi_{34} = -g^\phi_{14} = -g^\phi_{12}$

$g^\phi_{23} = -g^\phi_{12}$

[a]Adopted from Krajcinovic [14.3]. All g_{ij} and g^ϕ_{ij} that are not specified are zero.

CHAPTER **15**

Cross-sectional Stresses: Combined Stresses

Formulas for the stress analysis of bars are presented here. The stresses can result from combinations of bending and torsional loadings. In the earlier chapters, the stresses due to these loadings were considered separately. In the case of bending, unsymmetrical bending is treated; thus, it is no longer necessary to consider only symmetrical cross sections bent in the plane of symmetry. Most of the formulas encompass composite materials.

All of the stresses can now be reliably computed for arbitrary cross-sectional shapes using computer programs based on finite-element methodology. Formulas essential for finite-element calculations are tabulated here.

15.1 NOTATION

A_i	Area of a segment composed of material i in composite bar (L^2)
E_r	Reference modulus (F/L^2)
$I_{\omega y}, I_{\omega z}$	Sectorial products of inertia (L^5)
$Q_\omega^{(i)}$	First sectorial moment for element i (L^4)
$Q_{\omega j}$	First sectorial moment at node j (L^4)
Q_y, Q_z	First moments of area (L^3)
T	Temperature change (degrees)

y_c, z_c	Coordinates of geometric centroid (L)
α_y, α_z	Shear correction coefficients
α	Coefficient of thermal expansion $(L/L \cdot \deg)$
ω	Principal sectorial coordinate, warping function (L^2)
$\omega_P^{(i)}$	Sectorial coordinate for element i with respect to a point (or pole) P; usually P is shear center S, which results in ω_S, or geometric centroid c, which leads to ω_c (L^2)
ω_{Pj}	Sectorial coordinate at node j (L^2)
$\omega^{(i)}$	Principal sectorial coordinate for element i (L^2)
ω_j	Principal sectorial coordinate at node j (L^2)
Π	Potential energy (FL)
τ_t	Torsional shear stress (F/L^2)
τ_b	Bending shear stress (F/L^2)
τ_ω	Warping shear stress (F/L^2)
*	Superscript indicating a property of a composite bar

15.2 SIGN CONVENTION

It is essential in using the formulas that a particular sign convention be employed. The right-handed Cartesian coordinates of Fig. 15-1 are used throughout. The coordinate directions shown are defined to be positive. The exposed internal face whose outward normal points along the positive direction of the x axis is defined to be positive (Fig. 15-1a), while the other opposing face is known as the

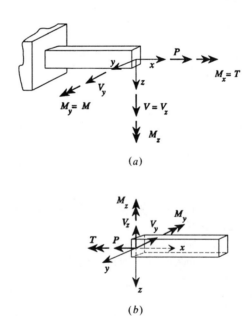

(a)

(b)

FIGURE 15-1 Sign convention for internal forces: (a) positive internal forces on left face of a cut; (b) positive internal forces on right face of a cut.

negative face (Fig. 15-1*b*). The internal force and moment components shown in Fig. 15-1*a* are defined to be positive because they are acting on a positive face with their vectors in the positive coordinate directions. Also, internal forces and moments acting on a negative face are positive if their vectors are in the negative coordinate directions. Thus, the forces and moments on the negative face of Fig. 15-1*b* are positive.

Unless defined otherwise, applied forces and moments are positive if their vectors are in the direction of a positive coordinate axis.

15.3 WARPING PROPERTIES

The geometric properties of plane areas associated with warping are discussed in Chapter 2. Some of these properties are essential for the use of the formulas for normal and shear stresses that are provided in the tables of this chapter. The determination of these important properties, which are tabulated in Chapter 2, will be illustrated in this section.

In all calculations for the sectorial coordinates, e.g., ω_S and ω_c, a special sign convention must be employed. Integration or summations are positive if their paths are counterclockwise with respect to a pole (usually the shear center or geometric centroid). Clockwise paths should be assigned a negative sign. This sign convention does not apply during the calculation of other warping characteristics.

Example 15.1 Calculation of Geometric Warping Characteristics for I-Section
Derive expressions for the sectorial coordinate ω_S for the thin-walled, wide-flange section shown in Fig. 15-2.

The direct integration method will be carried out first followed by the piecewise integration method to show how to handle the formulas of case 21 of Table 2-6 and case 1 of Table 2-7.

For the thin-walled wide-flange section, the location of the shear center and the centroid are both at the center of the web.

Direct Integration Method
Let node 1 be the origin of the coordinate s (Fig. 15-2*a*) and let $\omega_{S1} = 0$. From Eq. (2.24b) we calculate

$$\omega_S^{(1)}(s) = (-)\int_0^s \frac{h}{2}\, d\xi = -\frac{hs}{2}, \qquad \omega_{S2} = \omega_S^{(1)}\left(\frac{b}{2}\right) = -\frac{bh}{4}$$

$$\omega_S^{(2)}(s) = \omega_{S2} + (-)\int_{b/2}^s \frac{h}{2}\, d\xi = -\frac{hs}{2}, \qquad \omega_{S3} = \omega_S^{(2)}(b) = -\frac{bh}{2}$$

$$\omega_S^{(3)}(s) = \omega_{S2} + \int_{b/2}^s (0)\, d\xi = -\frac{bh}{4}, \qquad \omega_{S4} = \omega_S^{(3)}\left(\frac{b}{2} + h\right) = -\frac{bh}{4}$$

$$\omega_S^{(4)}(s) = \omega_{S4} + (-)\int_{b/2+h}^s \frac{h}{2}\, d\xi = \frac{h}{2}(h - s), \qquad \omega_{S5} = \omega_S^{(4)}(b + h) = -\frac{bh}{2}$$

$$\omega_S^{(5)}(s) = \omega_{S4} + \int_{b/2+h}^s \frac{h}{2}\, d\xi = \frac{h}{2}(s - b - h), \qquad \omega_{S6} = \omega_S^{(5)}(b + h) = 0$$

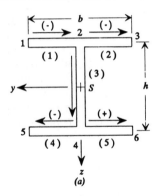

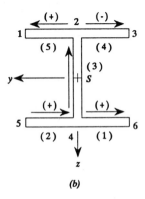

FIGURE 15-2 Example 15.1. Node numbers represented by 1, 2, 3, 4, 5, 6; element numbers by (1), (2), (3), (4), (5).

Note that the integration is performed along the wall profile from the origin of s to the desired point. Also, the sign convention for integration must be carefully applied.

If node 5 were to be chosen as the origin of the coordinate s (Fig. 15-2b), then the values of $\omega_S^{(i)}(s)$ could be calculated in a similar manner. Let $\omega_{S5} = 0$:

$$\omega_S^{(2)}(s) = \int_0^s \frac{h}{2} d\xi = \frac{hs}{2},$$

$$\omega_{S4} = \omega_S^{(2)}\left(\frac{b}{2}\right) = \frac{bh}{4}$$

$$\omega_S^{(1)}(s) = \omega_{S4} + \int_{b/2}^s \frac{h}{2} d\xi = \frac{hs}{2},$$

$$\omega_{S6} = \omega_S^{(1)}(b) = \frac{bh}{2}$$

$$\omega_S^{(3)}(s) = \omega_{S4} + \int_{b/2}^s (0) d\xi = \frac{bh}{4},$$

$$\omega_{S2} = \omega_S^{(3)}\left(\frac{b}{2} + h\right) = \frac{bh}{4}$$

$$\omega_S^{(5)}(s) = \omega_{S2} + \int_{(b/2)+h}^s \frac{h}{2} d\xi = \frac{h}{2}(s - h),$$

$$\omega_{S1} = \omega_S^{(5)}(b + h) = \frac{bh}{2}$$

$$\omega_S^{(4)}(s) = \omega_{S2} + (-)\int_{b/2+h}^s \frac{h}{2} d\xi = \frac{h}{2}(b + h - s),$$

$$\omega_{S3} = \omega_S^{(4)}(b + h) = 0$$

The values of $\omega_S^{(i)}$ and ω_{Sj} depend upon the choice of the origin of s. This is due to the fact that the sectorial coordinate $\omega_S^{(i)}$ or ω_{Sj} is the relative warping between the initial points of s ($s = 0$) and the final point of integration involved in the expression for $\omega_S^{(i)}$ or ω_{Sj}.

Piece-wise Integration Method

The formula for ω_{Sj} is given in Eq. (2.24b) and Table 2-7

$$\omega_{Sj} = \int_0^{s_j} r_S \, ds = \sum_i (\pm) r_{Si} b_i$$

The notation in this formula, which can be used to find ω_S at any node j, deserves special attention. The subscript i refers to the ith element. The signs ($\pm$) in the parentheses indicate counter-clockwise ($+$) or clockwise ($-$) directions for coordinate s with respect to the shear center S. The symbol $\sum_i$ means a summation along a line of elements. Begin at an outer element and sum until reaching the point j, where the value of ω_{Sj} is desired. The summation on i does not always involve a numerically increasing sequence of i values. Also, the initial value of i need not always be 1. The summation sequence follows the path of conventional integration along the wall profile, as with the direct integration method mentioned above. If ω_S is to be calculated at node j in Fig. 15-3, begin the summation at element 1 and continue until reaching node j, but not including elements not along the primary path, i.e., elements such as 4 and 5 are not to be included in the summation.

In Figs. 15-2a, b two alternative paths of summation (integration) are indicated by the arrows parallel to the wall profile. The node and element numbers may be arbitrarily assigned, but it is recommended that they be assigned with increasing numbers along the flow of the arrows (path of summation). The directions of the arrows are determined by choosing any free edge of the wall profile as the initial point, e.g., node 1 in case a and node 5 in case b. The main path of summation is established such as path $1 \to 2 \to 4 \to 5$ in case a and path

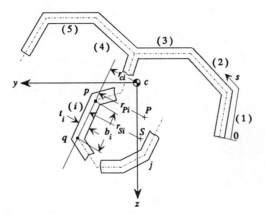

FIGURE 15-3 Cross-sectional notation system. The tangential coordinate s, which is measured along the center line of the wall profile, is positive if the direction is counter-clockwise with respect to the pole P (P can be c or S).

$5 \rightarrow 4 \rightarrow 2 \rightarrow 1$ in case b. Any intermediate branch elements, e.g., $2 \rightarrow 3$ in case a and $4 \rightarrow 6$ in case b, have outward paths of summation.

The signs in parentheses in Fig. 15-2 indicate counterclockwise $(+)$ or clockwise $(-)$ paths of summation with respect to the shear center S. If ω_{cj} is sought, the clockwise or counterclockwise paths are considered with respect to the centroid.

To establish a systematic calculation for complicated cross sections, it is recommended that a table such as the following be set up:

CASE a (Fig. 15-2a)

Node No.	Element No.	b_i	r_{Si}	$(\pm)r_{Si}b_i$	$\omega_{Sj} = \sum_i (\pm)r_{Si}b_i$
1					0
	(1)	$\frac{1}{2}b$	$\frac{1}{2}h$	$(-)\frac{1}{4}bh$	
2					$-\frac{1}{4}bh$
	(3)	h	0	0	
4					$-\frac{1}{4}bh$
	(4)	$\frac{1}{2}b$	$\frac{1}{2}h$	$(-)\frac{1}{4}bh$	
5					$-\frac{1}{2}bh$
2					$-\frac{1}{4}bh$
	(2)	$\frac{1}{2}b$	$\frac{1}{2}h$	$(-)\frac{1}{4}bh$	
3					$-\frac{1}{2}bh$
4					$-\frac{1}{4}bh$
	(5)	$\frac{1}{2}b$	$\frac{1}{2}h$	$\frac{1}{4}bh$	
6					0

CASE b (Fig. 15-2b)

Node No.	Element No.	b_i	r_{Si}	$(\pm)r_{Si}b_i$	$\omega_{Sj} = \sum_i (\pm)r_{Si}b_i$
5					0
	(2)	$\frac{1}{2}b$	$\frac{1}{2}h$	$\frac{1}{4}bh$	
4					$\frac{1}{4}bh$
	(3)	h	0	0	
2					$\frac{1}{4}bh$
	(5)	$\frac{1}{2}b$	$\frac{1}{2}h$	$\frac{1}{4}bh$	
1					$\frac{1}{2}bh$
4					$\frac{1}{4}bh$
	(1)	$\frac{1}{2}b$	$\frac{1}{2}h$	$\frac{1}{4}bh$	
6					$\frac{1}{2}bh$
2					$\frac{1}{4}bh$
	(4)	$\frac{1}{2}b$	$\frac{1}{2}h$	$(-)\frac{1}{4}bh$	
3					0

In calculating $r_{Si}b_i$, the notation for i of Fig. 15-3 is employed. The sign in parentheses is chosen according to the direction of summation with respect to the pole, e.g., the shear center for Fig. 15-2. For example, in the case of Fig. 15-3, moving along element 1 in the s direction is counterclockwise with respect to the pole P. Hence, the sign in parentheses for $r_{P_ib_i}$ $(= r_{P_1b_1})$ would be positive.

Example 15.2 Calculation of Warping-related Constants for a Z-Section Derive expressions for the principal sectorial coordinate ω, the shear center location e_y, e_z, the first sectorial moment Q_ω, and the warping constant Γ for the thin-walled Z-section shown in Fig. 15-4. The use of the direct integration method will be demonstrated first followed by the piecewise integration method.

Direct Integration Method

Begin by computing the location of the shear center, which for this simple cross section is known at the outset to coincide with the centroid. See Section 2.11.

Step 1: Calculate ω_c (with respect to the geometric centroid). The first step is to find the sectorial coordinate ω_c. Choose node 1 as the origin of the coordinate s and let $\omega_{c1} = 0$. We calculate ω_c by using Eq. (2.24a) as follows:

$$\omega_c^{(1)}(s) = (-)\int_0^s \frac{h}{2}\, d\xi = -\frac{hs}{2}, \qquad \omega_{c2} = \omega_c^{(1)}(b) = -\frac{bh}{2}$$

$$\omega_c^{(2)}(s) = \omega_{c2} + \int_b^s (0)\, d\xi = -\frac{bh}{2}, \qquad \omega_{c3} = \omega_c^{(2)}(b+h) = -\frac{bh}{2}$$

$$\omega_c^{(3)}(s) = \omega_{c3} + \int_{b+h}^s \frac{h}{2}\, d\xi = \frac{h}{2}(s - 2b - h), \qquad \omega_{c4} = \omega_c^{(3)}(2b+h) = 0$$

Step 2: Calculate the remaining constants in the shear center location relation. The formulas for the shear center coordinates y_S and z_S are given by Eq. (2.29). Using $\omega_c(s)$ computed in Step 1 and Eq. (2.28), we calculate $I_{\omega y}$ and $I_{\omega z}$ as

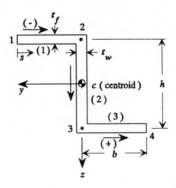

FIGURE 15-4 Arrows along wall profile show directions of summation. Signs depend on direction of integration.

follows:

$$I_{\omega y} = \int_0^b \omega_c^{(1)}(s)\left(-\frac{h}{2}\right)t_f\,ds + \int_b^{b+h}\omega_c^{(2)}(s)\left(b + \frac{h}{2} - s\right)t_w\,ds$$

$$+ \int_{b+h}^{2b+h}\omega_c^{(3)}(s)\left(\frac{h}{2}\right)t_f\,ds$$

$$= \frac{h^2}{4}\int_0^b s t_f\,ds - \frac{bh}{2}\int_b^{b+h}\left(b + \frac{h}{2} - s\right)t_w\,ds + \frac{h^2}{4}\int_{b+h}^{2b+h}(s - 2b - h)t_f\,ds$$

$$= \frac{b^2 h^2}{8}t_f - 0 - \frac{b^2 h^2}{8}t_f = 0$$

$$I_{\omega z} = \int_0^b \omega_c^{(1)}(s)(b - s)t_f\,ds + \int_b^{b+h}\omega_c^{(2)}(s)(0)t_w\,ds$$

$$+ \int_{b+h}^{2b+h}\omega_c^{(3)}(s)(b + h - s)t_f\,ds$$

$$= -\frac{h}{2}\int_0^b s(b - s)t_f\,ds + 0 + \frac{h}{2}\int_{b+h}^{2b+h}(s - 2b - h)(b + h - s)t_f\,ds$$

$$= -\frac{b^3 h}{12}t_f + \frac{b^3 h}{12}t_f = 0.$$

Substitution of $I_{\omega y} = I_{\omega z} = 0$ into the expressions for y_S and z_S [Eqs. (2.29)] yields $y_S = z_S = 0$. This shows that the geometric centroid and the shear center are at the same point.

To find the principal sectorial coordinate ω and the warping constant Γ, we proceed with the following steps.

Step 3: Calculate ω_S (with respect to the shear center). This involves repeating the calculation of step 1 using the shear center as a pole. However, for this example, the shear center and centroid are at the same location ($y_S = z_S = 0$). Hence, $\omega_S^{(i)}(s) = \omega_c^{(i)}(s)$. That is, $\omega_S^{(1)}(s) = -\frac{1}{2}hs$, $\omega_S^{(2)}(s) = -\frac{1}{2}bh$, $\omega_S^{(3)}(s) = \frac{1}{2}h(s - 2b - h)$, and $\omega_{S1} = 0$, $\omega_{S2} = -\frac{1}{2}bh$, $\omega_{S3} = -\frac{1}{2}bh$, and $\omega_{S4} = 0$.

Step 4: Calculate the principal sectorial coordinate. According to Eq. (2.25b), the principal sectorial coordinate ω is given by $\omega = \omega_S - \omega_0$. The sectorial coordinate ω_S has been calculated in step 3. For ω_0 we find (Eq. (2.25c))

$$\omega_0 = \frac{1}{A}\int_A \omega_S\,dA$$

$$= \frac{1}{A}\left[\int_0^b \omega_S^{(1)}t_f\,ds + \int_b^{b+h}\omega_S^{(2)}t_w\,ds + \int_{b+h}^{2b+h}\omega_S^{(3)}t_f\,ds\right]$$

$$= \frac{1}{2bt_f + ht_w}\left(-\frac{ht_f}{2}\int_0^b s\,ds - \frac{bht_w}{2}\int_b^{b+h}ds + \frac{ht_f}{2}\int_{b+h}^{2b+h}(s - 2b - h)\,ds\right)$$

$$= \frac{1}{2bt_f + ht_w}\left(-\frac{b^2 ht_f}{4} - \frac{bh^2 t_w}{2} - \frac{b^2 ht_f}{4}\right)$$

$$= -\frac{bh(ht_w + bt_f)}{2(2bt_f + ht_w)}$$

Hence,

$$\omega^{(1)} = \omega_S^{(1)} - \omega_0 = -\frac{hs}{2} + \frac{bh(ht_w + bt_f)}{2(2bt_f + ht_w)} \qquad (0 \le s \le b)$$

$$\omega^{(2)} = \omega_S^{(2)} - \omega_0 = -\frac{b^2 ht_f}{2(2bt_f + ht_w)} \qquad (b \le s \le b + h)$$

$$\omega^{(3)} = \omega_S^{(3)} - \omega_0 = -\frac{h(2b + h - s)}{2} + \frac{bh(ht_w + bt_f)}{2(2bt_f + ht_w)} \quad (b + h \le s \le 2b + h)$$

Step 5: Calculate the warping constant Γ. The formula for warping constant Γ is given by Eq. (2.27) using ω as calculated in step 4. It is found that

$$\Gamma = \int_0^b (\omega^{(1)})^2 t_f \, ds + \int_b^{b+h} (\omega^{(2)})^2 t_w \, ds + \int_{b+h}^{2b+h} (\omega^{(3)})^2 t_f \, ds$$

$$= \int_0^b \left(\frac{hs}{2} + \omega_0\right)^2 t_f \, ds + \int_b^{b+h} \left(\frac{bh}{2} + \omega_0\right)^2 t_w \, ds$$

$$+ \int_{b+h}^{2b+h} \left[\frac{h(2b + h - s)}{2} + \omega_0\right]^2 t_f \, ds$$

$$= \frac{b^3 h^2 t_f}{12} \frac{(bt_f + 2ht_w)}{2bt_f + ht_w}$$

Step 6: Calculate the first sectorial moment Q_ω. Use the formula for Q_ω of Eq. (2.25a). The value of the principal sectorial coordinate ω calculated in step 4 can be inserted in the formula for Q_ω. Then for the upper flange

$$Q_\omega^{(1)} = \int_0^s \omega^{(1)} t_f \, d\xi = \int_0^s \left(-\frac{h\xi}{2} - \omega_0\right) t_f \, d\xi$$

$$= \left(-\frac{hs^2}{4} - \omega_0 s\right) t_f \qquad (0 \le s \le b)$$

$$Q_{\omega 2} = Q_\omega^{(1)} \text{ at node } 2 = \left(-\frac{b^2 h}{4} - \omega_0 b\right) t_f$$

$$Q_\omega^{(2)} = Q_{\omega 2} + \int_b^s \omega^{(2)} t_w \, d\xi = -\left(\frac{b^2 h}{4} + \omega_0 b\right) t_f + \int_b^s \left(-\frac{bh}{2} - \omega_0\right) t_w \, d\xi$$

$$= -\frac{b^2 h t_f t_w}{2(2bt_f + ht_w)} \left(s - b - \frac{h}{2}\right) \qquad (b \le s \le b + h)$$

$$Q_{\omega 3} = Q_{\omega}^{(2)} \text{ at node } 3 = -\frac{b^2 h^2 t_f t_w}{4(2bt_f + ht_w)}$$

$$Q_{\omega}^{(3)} = Q_{\omega 3} + \int_{b+h}^{2b+h} \omega^{(3)} t_f \, d\xi = -\frac{b^2 h^2 t_f t_w}{4(2bt_f + ht_w)}$$

$$+ \int_{b+h}^{s} \left[-\frac{h(2b + h - \xi)}{2} - \omega_0 \right] t_f \, d\xi$$

$$= \left[\left(-\frac{h^2}{2} - bh - \omega_0 \right) s + \frac{hs^2}{4} \right] t_f - \frac{b^2 h^2 t_f t_w}{4(2bt_f + ht_w)}$$

$$+ t_f \left[\left(\frac{h^2}{2} + bh + \omega_0 \right)(b + h) - \frac{h(b + h)^2}{4} \right] \qquad (b + h \leq s \leq 2b + h)$$

$$Q_{\omega 4} = Q_{\omega}^{(3)} \text{ at node } 4 = 0$$

These expressions for Q_{ω} are ready for use in computing the shear stress of Table 15-2.

Piecewise Integration Method

This method proves to be useful in calculating the sectorial properties and the warping characteristics for thin-walled beams with complicated cross sections formed of straight elements. The integration is then reduced to a summation. It should be emphasized that the sequence of summation for sectorial coordinates follows the direct integration and must adhere to the same sign convention. All of the calculations for the sectorial properties and warping characteristics for the Z-section will be repeated in order to demonstrate the use of the formulas in this method. We begin, as before, by computing the location of the shear center.

Step 1: *Calculate* ω_{ci} (*with respect to the geometric centroid*).

Node No.	Element No.	r_{ci}	b_i	$r_{ci}b_i$	$(\pm)\omega_{ci} = \sum_i (\pm)r_{ci}b_i$
1					0
	(1)	$\frac{1}{2}h$	b	$(-)\frac{1}{2}bh$	
2					$-\frac{1}{2}bh$
	(2)	0	h	0	
3					$-\frac{1}{2}bh$
	(3)	$\frac{1}{2}h$	b	$\frac{1}{2}bh$	
4					0

Step 2: *Calculate the remaining constants in the shear center location relations.* The formulas for shear center y_S and z_S are taken from case 21 of Table 2-6 as

$$y_S = \frac{I_z I_{\omega y} - I_{yz} I_{\omega z}}{I_y I_z - I_{yz}^2}, \qquad z_S = -\frac{I_y I_{\omega z} - I_{yz} I_{\omega y}}{I_y I_z - I_{yz}^2}$$

where

$$I_{\omega y} = \int_A \omega_c z\, dA = \frac{1}{3} \sum_{i=1}^{M} (\omega_{cp}z_p + \omega_{cq}z_q)t_i b_i + \frac{1}{6} \sum_{i=1}^{M} (\omega_{cp}z_q + \omega_{cq}z_p)t_i b_i$$

$$I_{\omega z} = \int_A \omega_c y\, dA = \frac{1}{3} \sum_{i=1}^{M} (\omega_{cp}y_p + \omega_{cq}y_q)t_i b_i + \frac{1}{6} \sum_{i=1}^{M} (\omega_{cp}y_q + \omega_{cq}y_p)t_i b_i$$

M is the total number of elements. The quantities ω_{cp}, (y_p, z_p) and ω_{cq}, (y_q, z_q) are the sectorial areas and coordinates of the two ends of element i. For example, the first end (p) of element 2 of Fig. 15-4 is node 2 and the second end (q) is node 3 in the node sequence. So for element 2, $\omega_{cp} = \omega_{c2}$ and $\omega_{cq} = \omega_{c3}$.

| | Node No. | | Coordinate | | | | | | From Step 1 | |
Element No.	p	q	y_p	z_p	y_q	z_q	b_i	t_i	ω_{cp}	ω_{cq}
(1)	1	2	b	$-\frac{1}{2}h$	0	$-\frac{1}{2}h$	b	t_f	0	$-\frac{1}{2}bh$
(2)	2	3	0	$-\frac{1}{2}h$	0	$\frac{1}{2}h$	h	t_w	$-\frac{1}{2}bh$	$-\frac{1}{2}bh$
(3)	3	4	0	$\frac{1}{2}h$	$-b$	$\frac{1}{2}h$	b	t_f	$-\frac{1}{2}bh$	0

In expanded form, $I_{\omega y}$ and $I_{\omega z}$ appear as

$$I_{\omega y} = \frac{1}{3}\Bigg[\underbrace{(\omega_{c1}z_1 + \omega_{c2}z_2)t_1 b_1}_{\text{Element 1}} + \underbrace{(\omega_{c2}z_2 + \omega_{c3}z_3)t_2 b_2}_{\text{Element 2}}$$
$$+ \underbrace{(\omega_{c3}z_3 + \omega_{c4}z_4)t_3 b_3}_{\text{Element 3}}\Bigg]$$
$$+ \frac{1}{6}\Bigg[\underbrace{(\omega_{c1}z_2 + \omega_{c2}z_1)t_1 b_1}_{\text{Element 1}} + \underbrace{(\omega_{c2}z_3 + \omega_{c3}z_2)t_2 b_2}_{\text{Element 2}}$$
$$+ \underbrace{(\omega_{c3}z_4 + \omega_{c4}z_3)t_3 b_3}_{\text{Element 3}}\Bigg]$$

$$I_{\omega z} = \frac{1}{3}\Bigg[\underbrace{(\omega_{c1}y_1 + \omega_{c2}y_2)t_1 b_1}_{\text{Element 1}} + \underbrace{(\omega_{c2}y_2 + \omega_{c3}y_3)t_2 b_2}_{\text{Element 2}}$$
$$+ \underbrace{(\omega_{c3}y_3 + \omega_{c4}y_4)t_3 b_3}_{\text{Element 3}}\Bigg]$$
$$+ \frac{1}{6}\Bigg[\underbrace{(\omega_{c1}y_2 + \omega_{c2}y_1)t_1 b_1}_{\text{Element 1}} + \underbrace{(\omega_{c2}y_3 + \omega_{c3}y_2)t_2 b_2}_{\text{Element 2}}$$
$$+ \underbrace{(\omega_{c3}y_4 + \omega_{c4}y_3)t_3 b_3}_{\text{Element 3}}\Bigg]$$

Substitution of the appropriate values from the above table give $I_{\omega y} = I_{\omega z} = 0$. The formulas for y_S and z_S then provide $y_S = z_S = 0$.

To find the principal sectorial coordinates ω and warping constant Γ, we proceed with the following steps.

Step 3: Calculate ω_{Si} (with respect to the shear center). This involves repeating the tabular calculation of step 1 using r_{Si} instead of r_{ci}. However, for this example, the shear center and centroid are at the same location ($y_S = z_S = 0$). Hence, $r_{Si} = r_{ci}$, which gives $\omega_{Sj} = \omega_{cj}$. That is, $\omega_{S1} = 0$, $\omega_{S2} = -\frac{1}{2}bh$, $\omega_{S3} = -\frac{1}{2}bh$, $\omega_{S4} = 0$.

Step 4: Evaluate the remaining constants in the expressions for the principal coordinates. From Eq. (2.25c) and case 1 of Table 2-7, the expression ω_0 is given by

$$\omega_0 = \frac{1}{A}\int_A \omega_S\, dA = \frac{1}{A}\left[\frac{1}{2}\sum_{i=1}^{M}(\omega_{Sp} + \omega_{Sq})t_i b_i\right]$$

In expanded form, noting, for example, that for element 2 $\omega_{Sp} = \omega_{S2}$ and $\omega_{Sq} = \omega_{S3}$,

$$\omega_0 = \frac{1}{A}\frac{1}{2}\left[\underbrace{(\omega_{S1} + \omega_{S2})t_1 b_1}_{\text{Element 1}} + \underbrace{(\omega_{S2} + \omega_{S3})t_2 b_2}_{\text{Element 2}} + \underbrace{(\omega_{S3} + \omega_{S4})t_3 b_3}_{\text{Element 3}}\right]$$

with $A = 2bt_f + ht_w$. Substitution of the values of ω_{Sj} from step 3 into the expression for ω_0 leads to

$$\omega_0 = -\frac{bh(t_f b + t_w h)}{2(2bt_f + ht_w)}$$

Then using the notation ω_j to represent ω at node j, $\omega_j = \omega_{Sj} - \omega_0$,

$$\omega_1 = \omega_{S1} - \omega_0 = \frac{bh(ht_w + bt_f)}{2(2bt_f + ht_w)} = \omega_4$$

$$\omega_2 = \omega_{S2} - \omega_0 = -\frac{b^2 ht_f}{2(2bt_f + ht_w)} = \omega_3$$

The value of the principal sectorial coordinate ω between the two nodes of any element varies linearly and can be expressed in terms of the nodal values ω_p, ω_q. For example, for element 1 (Fig. 15-5), ω between nodes 1 and 2 is found

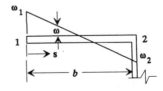

FIGURE 15-5 Variation of ω along an element.

to be

$$\omega^{(1)} = \omega \text{ in element } 1 = \frac{\left(b^2 h t_f + b h^2 t_w\right) - \left(2 b h t_f + h^2 t_w\right)s}{2\left(2 b t_f + h t_w\right)}$$

$$= -\frac{hs}{2} + \frac{bh\left(h t_w + b t_f\right)}{2\left(2 b t_f + h t_w\right)} \qquad (0 \le s \le b)$$

The expressions for ω_j or $\omega^{(i)}$ calculated above can be inserted in Table 15-1 for the computation of the normal stress.

Step 5: Calculate the warping constant Γ. The formula for the warping constant Γ is given in Eq. (2.27) and case 21, Table 2-6:

$$\Gamma = \int_A \omega^2 \, dA = \frac{1}{3} \sum_{i=1}^M \left(\omega_p^2 + \omega_p \omega_q + \omega_q^2\right) t_i b_i$$

where ω_p and ω_q are the principal sectorial coordinates at the ends of element i. This can be expanded as

$$\Gamma = \frac{1}{3}\left[\underbrace{\left(\omega_1^2 + \omega_1\omega_2 + \omega_2^2\right)t_1 b_1}_{\text{Element 1}} + \underbrace{\left(\omega_2^2 + \omega_2\omega_3 + \omega_3^2\right)t_2 b_2}_{\text{Element 2}} \right.$$

$$\left. + \underbrace{\left(\omega_3^2 + \omega_3\omega_4 + \omega_4^2\right)t_3 b_3}_{\text{Element 3}} \right]$$

Substitution of the values of ω_j calculated in step 4 and $t_1 = t_f = t_3$, $b_1 = b = b_3$, $t_2 = t_w$, and $b_2 = h$ yields

$$\Gamma = \frac{b^3 h^2 t_f}{12} \frac{b t_f + 2 h t_w}{2 b t_f + h t_w}$$

Step 6: Calculate the first sectorial moment Q_ω. The formula for Q_ω is given in Eq. (2.25a). The value of the principal sectorial coordinate ω calculated in step 4 can be inserted into the expression for Q_ω. Then for the upper flange

$$Q_\omega^{(1)} = \int_0^s \omega^{(1)} t_f \, ds = \int_0^s \left[-\frac{hs}{2} + \frac{bh\left(h t_w + b t_f\right)}{2\left(2 b t_f + h t_w\right)} \right] t_f \, ds$$

The integration leads to

$$Q_\omega^{(1)} = \left[-\frac{hs^2}{4} + \frac{bh\left(h t_w + b t_f\right)}{2\left(2 b t_f + h t_w\right)} s \right] t_f \qquad (0 \le s \le b)$$

A similar calculation yields

$$Q_\omega^{(2)} = -\frac{b^2 h t_f t_w}{2(2bt_f + ht_w)}\left(s - b - \frac{h}{2}\right) \qquad (b \le s \le b + h)$$

These expressions for Q_ω are ready for use in computing the shear stress of Table 15-2.

15.4 NORMAL STRESSES

Formulas for normal stresses on the face of a cross section are given in Table 15-1. These formulas are based on the following assumptions:

The Euler–Bernoulli assumption, regarding the axial and flexural modes of deformations, that a cross-sectional plane normal to the centroidal axis (or the modulus-weighted centroidal axis in the case of a composite beam) remains plane after deformation.

The sectorial concept (see Chapter 2), regarding the restrained warping mode of deformation, that the shear center is used as a pole.

Neutral Axis

The neutral axis is, by definition, a line through the cross section along which the normal stress is zero. To find the equation of this line for the general case of bending, set $\sigma = 0$ in case 1 of Table 15-1:

$$z = \frac{M_z I_y^* + M_y I_{yz}^*}{M_y I_z^* + M_z I_{yz}^*} y \qquad (15.1)$$

where the axial force $\bar{P}$, thermal effects, and bimoment B have been neglected. The neutral axis defined by Eq. (15.1) passes through the centroid. It is the line about which the "plane section" rotates. The neutral axis is not, in general, perpendicular to the plane of the resultant internal moment, nor does it usually coincide with either of the principal axes of inertia. An exception is the case of simple bending using case 5 of Table 15-1 in which the largest stress occurs at the point most removed from the neutral axis.

Example 15.3 Normal Stresses in Composite Beam Find the normal stresses at points A and B on the composite cross section shown in Fig. 15-6. The section is subjected to a bending moment $M = 10$ kN-m in the direction shown. Also locate the neutral axis.

The stresses are found using the formula of case 1 of Table 15-1 with B, $\bar{P}$, and T equal to zero. First the location of modulus-weighted centroid of the cross section must be found.

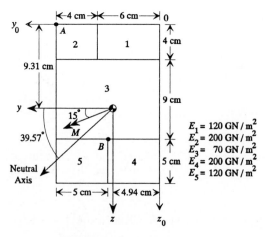

FIGURE 15-6 Example 15.3.

The desired centroid can be located in the y_0, z_0 reference frame using Eq. (2.35b). The modulus-weighted area A^* is given by Eq. (2.35a). Choose $E_r = 100 \text{ GN/m}^2$. Then we find

$$y_c^* = \frac{\sum \bar{y}_{0i} A_i^*}{A^*} = 4.94 \text{ cm}, \qquad z_c^* = \frac{\sum \bar{z}_{0i} A_i^*}{A^*} = 9.31 \text{ cm} \qquad (1)$$

The moments of inertia of the cross section about the centroidal y, z axes are [Eqs. (2.38)]

$$I_y^* = \sum_{i=1}^{5} \frac{E_i}{E_r} \left(\bar{I}_{y_i} + \bar{z}_i^2 A_i \right) = 7028.52 \text{ cm}^4$$

$$I_z^* = \sum_{i=1}^{5} \frac{E_i}{E_r} \left(\bar{I}_{z_i} + \bar{y}_i^2 A_i \right) = 1723.27 \text{ cm}^4 \qquad (2)$$

$$I_{yz}^* = \sum_{i=1}^{5} \frac{E_i}{E_r} \left(\bar{I}_{y_i z_i} + \bar{y}_i \bar{z}_i A_i \right) = -590.23 \text{ cm}^4$$

The bending moments about the y, z axes are

$$\bar{M}_y = M_y = M \cos 15° = 9.659 \text{ kN-m}, \qquad \bar{M}_z = M_z = M \sin 15° = 2.588 \text{ kN-m} \qquad (3)$$

Then, from case 1 of Table 15-1,

$$\sigma = \frac{E}{E_r}\left(\frac{\overline{M}_y I_z^* + \overline{M}_z I_{yz}^*}{I_y^* I_z^* - I_{yz}^{*2}}z - \frac{\overline{M}_z I_y^* + \overline{M}_y I_{yz}^*}{I_y^* I_z^* - I_{yz}^{*2}}y\right)$$

$$= \frac{E}{100}(0.001285z - 0.001062y) \qquad (4)$$

with E in giganewtons per meter squared and y, z in centimeters.

At point A, $y = 5.06$ cm, $z = -9.31$ cm, $E = E_2 = 200$ GN/m^2, and (4) yields

$$\sigma = 2[0.001285(-9.31) - 0.001062(5.06)] = -0.0347 \text{ GN/m}^2 \qquad (5)$$

At point B of material 5, $y = 0.06$ cm, $z = 3.69$ cm, $E = E_5 = 120$ GN/m^2, and $\sigma = 1.2[0.001285(3.69) - 0.001062(0.06)] = 0.0056$ GN/m^2.

The position of the neutral axis is obtained by setting $\sigma = 0$ in (4). This leads to

$$\tan\phi = \frac{z}{y} = \frac{0.001062}{0.001285} = 0.8265 \quad \text{or} \quad \phi = 39.57° \qquad (6)$$

Example 15.4 Thin-walled Composite Section with Stringers and Thermal Loading Find the bending stresses in the longitudinal stringers of the simplified representation of an aircraft wing shown in Fig. 15-7. The section is subjected to bending moment $M_y = 5 \times 10^5$ in.-lb. Columns 2, 3, 4, 5, and 6 of Fig. 15-8 indicate geometric and material properties of the section along with the applied thermal loading. Assume $\alpha = 1.2 \times 10^{-5}$ in./in.-°F.

Typically, flight structures of this sort are comprised of thin metal skins connecting longitudinal stringers. It is often assumed that the skin panels do not resist bending. The moments of inertia are then based on the areas of the stringers, not including the area of the skins.

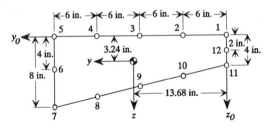

FIGURE 15-7 Example 15.4.

Most of the calculations can be placed in tabular form, as shown in Fig. 15-8. Let $E_r = 10^7$ psi. We find

$$y_c^* = \frac{\sum \bar{y}_{0i} A_i^*}{A^*} = \frac{186.12}{13.61} = 13.68 \text{ in.}, \qquad z_c^* = \frac{\sum \bar{z}_{0i} A_i^*}{A^*} = \frac{44.14}{13.61} = 3.24 \text{ in.}$$

$$I_y^* = \sum \bar{z}_i^2 A_i^* = 130.12 \text{ in.}^4, \qquad I_z^* = \sum \bar{y}_i^2 A_i^* = 1270.01 \text{ in.}^4,$$

$$I_{yz}^* = \sum \bar{y}_i \bar{z}_i A_i^* = 110.51 \text{ in.}^4$$

$$\bar{P} = P + \int E\alpha T \, dA = 0 + \sum (E\alpha T A^*)_i = 370.4 \times 10^3 \text{ lb}$$

$$\bar{M}_y = M_y + \int E\alpha T z \, dA = 5 \times 10^5 + \sum (E\alpha T z A^*)_i = 618.2 \times 10^3 \text{ in.-lb}$$

$$\bar{M}_z = M_z - \int E\alpha T y \, dA = 0 - \sum (E\alpha T y A^*)_i = -52.15 \times 10^3 \text{ in.-lb}$$

$$C_1 = \frac{\bar{P}}{A^*} = \frac{370.4 \times 10^3}{13.61} = 27.22 \times 10^3 \text{ psi}$$

$$C_2 = \frac{\bar{M}_z I_y^* + \bar{M}_y I_{yz}^*}{I_y^* I_z^* - I_{yz}^{*2}} = \frac{-52.15(10^3)130.12 + 618.2(10^3)110.51}{130.12(1270.01) - (110.51)^2}$$

$$= 0.402 \times 10^3 \text{ lb/in.}^3$$

$$C_3 = \frac{\bar{M}_y I_z^* + \bar{M}_z I_{yz}^*}{I_y^* I_z^* - I_{yz}^{*2}} = \frac{618.2(10^3)1270.01 - 52.15(10^3)110.51}{130.12(1270.01) - (110.51)^2}$$

$$= 5.092 \times 10^3 \text{ lb/in.}^3$$

From case 1 of Table 15-1, the stresses are calculated using

$$\sigma_i = \frac{E_i}{E_r}(C_1 + C_3 z_i - C_2 y_i - E_r \alpha T_i)$$

$$= \frac{E}{E_r}(27.22 \times 10^3 + 5.092 \times 10^3 z_i - 0.402 \times 10^3 y_i - E_r \alpha T_i)$$

The normal stress in each stringer is listed in the final column of Fig. 15-8.

15.5 SHEAR STRESSES

The formulas for the shear stress on the face of the cross section are listed in Table 15-2. For case 1 for a solid beam the stresses are positive if they point in the directions of the y or z coordinates. For the thin-walled section of case 2 the stresses are positive if they point in the direction of the positive s coordinate. Considerable care must be exercised in applying these formulas. Especially

1	2	3	4	5	6	7	8	9	10	11	12
i station	A_i in²	$\bar{y}_{oi}$ in	$\bar{z}_{oi}$ in	T_i °F	E_i 10^6 psi	E_i/E_r	A_i^* in² 2×7^a	$\bar{y}_{oi}A_i^*$ in³ 3×8	$\bar{z}_{oi}A_i^*$ in³ 4×8	$\bar{y}_i$ in $3-13.68$	$\bar{z}_i$ in $4-3.24$
1	1.5	0	0	180	9.2	0.92	1.38	0	0	−13.68	−3.24
2	0.8	6	0	230	10.0	1.00	0.80	4.80	0	−7.68	−3.24
3	0.8	12	0	230	10.0	1.00	0.80	9.60	0	−1.68	−3.24
4	0.8	18	0	230	10.0	1.00	0.80	14.40	0	4.32	−3.24
5	2.0	24	0	205	9.5	0.95	1.90	45.60	0	10.32	−3.24
6	1.2	24	4	180	10.3	1.03	1.24	29.76	4.96	10.32	0.76
7	2.0	24	8	205	9.5	0.95	1.90	45.60	15.20	10.32	4.76
8	1.0	18	7	330	10.1	1.01	1.01	18.18	7.07	4.32	3.76
9	1.0	12	6	330	10.1	1.01	1.01	12.12	6.06	−1.68	2.76
10	1.0	6	5	330	10.1	1.01	1.01	6.06	5.05	−7.68	1.76
11	1.2	0	4	180	9.5	0.95	1.14	0	4.56	−13.68	0.76
12	0.6	0	2	130	10.2	1.02	0.612	0	1.24	−13.68	−1.24
Σ							13.60	186.12	44.14		

station i	13 $\bar{y}_i^2 A_i^*$ in^4 $11^2 \times 8$	14 $\bar{z}_i^2 A_i^*$ in^4 $12^2 \times 8$	15 $\bar{y}_i \bar{z}_i A_i^*$ in^4 $11 \times 12 \times 8$	16 $(E\alpha TA^*)_i$ 10^3 lb $6 \times \alpha \times 5 \times 8$	17 $(E\alpha T\bar{y}A^*)_i$ 10^3 in-lb 11×16	18 $(E\alpha T\bar{z}A^*)_i$ 10^3 in-lb 12×16	19 $C_2\bar{y}_i$ 10^3 psi	20 $C_3\bar{z}_i$ 10^3 psi	21 $E_r\alpha T_i$ 10^3 psi	22 σ 10^3 psi $7 \times (C_1 - 19 + 20 - 21)$
1	258.26	14.49	61.17	27.42	-407.80	-96.58	-5.50	-16.50	21.60	-4.95
2	47.16	8.40	19.91	22.08	-169.57	-71.54	-3.09	-16.50	27.60	-13.79
3	2.26	8.40	4.35	22.08	-37.09	-71.54	-0.68	-16.50	27.60	-16.20
4	14.93	8.40	-11.20	22.08	95.39	-71.54	1.74	-16.50	27.60	-18.62
5	202.35	19.95	-63.53	44.40	482.36	-151.44	4.15	-16.50	24.60	-17.13
6	132.06	0.72	9.73	27.59	275.54	20.29	4.15	3.87	21.60	5.50
7	202.35	43.05	93.33	44.40	482.36	222.48	4.15	24.24	24.60	21.57
8	18.85	14.28	16.41	40.40	172.80	150.40	1.74	19.16	39.60	5.09
9	2.85	7.09	-4.68	40.40	-67.20	110.40	-0.68	14.05	39.60	2.37
10	59.57	3.13	-13.65	40.40	-307.20	70.40	-3.09	8.96	39.60	-0.33
11	213.34	0.66	-11.85	23.39	-336.80	18.71	-5.50	3.87	21.60	14.24
12	116.03	0.95	10.52	9.86	-130.64	-11.84	-5.50	-6.31	15.60	11.13
Σ	1270.01	130.12	110.51	370.40	52.15	118.20				

^{a}The notation 2×7 indicates that the entry in column 2 is multiplied by the entry in column 7 to obtain the entry in column 8.

FIGURE 15-8 Data for Example 15.4.

delicate is the need to add the terms in cases 1 and 2 vectorially if the particular shear stresses, (torsional, transverse, and restrained warping shear stresses) act in different directions. The y, z coordinates for case 3 are taken relative to axes passing through the centroid of the cross section. Case 3 provides the average shear stress along b, which is the width in any direction on a cross section, e.g., see b in Fig. 2-9.

Shear Center

Loading on a beam will usually produce combined bending and twisting. Some of the formulas of this chapter are based on the assumption that no twisting moment is developed. It is possible to locate a point in the cross-sectional plane through which the resultant forces must pass (sometimes in a particular direction) if there is to be no twisting. This point is called the *shear center*. Formulas for shear center locations are given in Chapter 2. For several specific cross sections, see Table 2-6.

It is possible to avoid the shear center formulas and to locate the shear center easily by balancing the internal shear forces V_y, V_z with the resultant of the shear stresses τ. This is accomplished by setting the summation of moments about any convenient point equal to zero. If the resultant of V_y and V_z as caused by external loading is not equal, opposite, and collinear to the resultant of the internal shear stresses, then bending is accompanied by twisting of the beam.

The following characteristics of a shear center can be demonstrated:

(a) The shear center for a section consisting of two intersecting rectangular elements is at the point of intersection.
(b) The shear center for a section with one axis of symmetry lies on this axis.
(c) The shear center for a section with two axes of symmetry is at the intersection of the two axes, i.e., at the centroid.

Example 15.5 Shear Center The shear center for a channel (Fig. 15-9) lies on the axis of symmetry, i.e., along the y axis. To find the shear center coordinate e

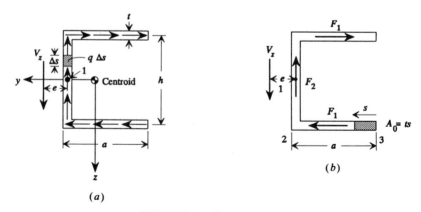

FIGURE 15-9 Example 15.5.

along the y axis, we could use the formulas of Chapter 2 or simply sum moments about point 1. The resultant forces in the section are designated F_1 and F_2. Then

$$\sum M_1 = 0: \qquad V_z e - F_1 h = 0 \tag{1}$$

so that

$$e = \frac{F_1 h}{V_z} = \frac{h}{V_z} \int_2^3 q\, ds = \frac{ht}{V_z} \int_2^3 \tau\, ds = \frac{ht}{V_z} \int_2^3 \frac{Q_y V_z}{t I_y}\, ds = \frac{h}{I_y} \int_2^3 Q_y\, ds \tag{2}$$

where τ is given by case 3 of Table 15-2 with $b = t$, $V_y = 0$, and $I_{yz} = 0$ (the section has an axis of symmetry). By definition of a first moment, $Q_y = \bar{y} A_0 = (h/2)ts$. From (2),

$$e = \frac{h}{I_y} \int_2^3 \frac{h}{2} ts\, ds = \frac{t h^2 a^2}{4 I_y} \tag{3}$$

The channel in this example has the y axis as an axis of symmetry. If there were no axes of symmetry, special care would have to be taken. In particular, each term in (1) would contain either V_z or V_y. [Note that in this example F_1 is written in terms of V_z, i.e., $F_1 = V_z \int_2^3 (Q_y/I_y)\, ds$.] The coordinates y_S, z_S of the shear center are found by equating the coefficients of V_y and V_z terms to zero. This yields two equations for the two unknowns y_S and z_S. This manipulation is equivalent to taking moments about a point in the section with V_y and V_z applied separately.

Example 15.6 Thin-walled Composite Section with Stringers and Thermal Loading Find the shear stress in the panels of the thin-walled section of Fig. 15-10. Other than the lack of a web connecting stringers 3 and 4, this section has the same physical and material properties as the section in Example 15.4. The section is subjected to a downward vertical shear force of 1500 lb at the shear center in addition to an applied longitudinal thermal gradient $T_i' = dT_i/dx = -1.2 \times 10^{-3}\, T_i$ (°F/in.) where T_i is the temperature rise in the ith stringer (column 5 of Fig. 15-8).

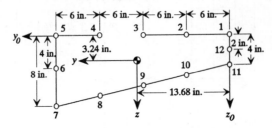

FIGURE 15-10 Example 15.6.

1	2	3	4	5	6	7	8	9	10
i	$\bar{y}_i$ in	$\bar{z}_i$ in	A_i^* in²	$y_i A_i^*$ in³ 2×4^a	Q_z^* in³ $\Sigma 5$	$\bar{z}_i A_i^*$ in³ 3×4	Q_y^* in³ $\Sigma 7$	A_0^* in² $\Sigma 4$	$E_i A_i^*$ 10^6 lb
station		Fig. 15-8							
4	4.32	−3.24	0.80	3.46	3.46	−2.59	−2.59	0.80	8.0
5	10.32	−3.24	1.90	19.61	23.07	−6.16	−8.75	2.70	18.05
6	10.32	0.76	1.24	12.80	35.87	0.94	−7.81	3.94	12.77
7	10.32	4.76	1.90	19.61	55.48	9.04	1.23	5.84	18.05
8	4.32	3.76	1.01	4.36	59.84	3.80	5.03	6.85	10.2
9	−1.68	2.76	1.01	−1.70	58.14	2.79	7.82	7.86	10.2
10	−7.68	1.76	1.01	−7.76	50.38	1.78	9.60	8.87	10.2
11	−13.68	0.76	1.14	−15.60	34.78	0.87	10.47	10.01	10.83
12	−13.68	−1.24	0.612	−8.48	26.30	−0.77	9.70	10.63	6.24
1	−13.68	−3.24	1.38	−18.88	7.42	−4.47	5.23	12.01	12.67
2	−7.68	−3.24	0.80	−6.14	1.28	−2.59	2.64	12.81	8.0
3	−1.68	−3.24	0.80	−1.34	0.06	−2.59	0.05	13.61	8.0
Σ			13.60						

station i	11 $\alpha T_i'$ 10^{-6} in^{-1}	12 $E_i A_i \alpha T_i'$ lb/in	13 $E_i A_i^* \bar{y} \alpha T_i'$ lb 2 × 12	14 $E_i A_i^* \bar{z} \alpha T_i'$ lb 3 × 12	15 $\dfrac{A_0^*}{A^*}\bar{P}$ lb/in	16 $C_2' Q_z^*$ lb/in	17 $C_3' Q_y^*$ lb/in	18 $\Sigma(EA^*\alpha T')_i$ $\Sigma 12$	19 q lb/in $-15 + 16$ $-17 + 18$
4	-3.31	-26.48	-114.39	85.80	-25.7	3.58	-29.14	-26.48	31.94
5	-2.95	-53.25	-549.54	172.53	-86.74	23.85	-98.45	-79.73	129.2
6	-2.59	-33.07	-341.28	-25.13	-126.57	37.09	-87.87	-112.8	138.7
7	-2.95	-53.25	-549.54	-253.47	-187.61	57.37	13.84	-166.05	65.1
8	-4.75	-48.45	-209.30	-182.17	-220.06	61.87	56.59	-214.5	10.8
9	-4.75	-48.45	81.4	-133.72	-252.5	60.12	87.98	-262.95	-38.3
10	-4.75	-48.45	372.1	-85.27	-284.95	52.09	108.01	-311.4	-82.4
11	-2.59	-28.05	383.7	-21.32	-321.57	35.96	117.80	-339.45	-99.7
12	-1.87	-11.67	159.65	14.47	-341.49	27.19	109.13	-351.12	-91.3
1	-2.59	-32.82	448.98	106.34	-385.82	7.67	58.94	-383.94	-49.3
2	-3.31	-26.48	203.37	85.80	-411.5	1.32	29.70	-410.42	-27.3
3	-3.31	-26.48	44.49	85.80	-437.22	0.06	0.56	-436.9	-0.18
Σ		-436.90	-70.36	-150.34					

FIGURE 15-11 Data for Example 15.6.

[a]The notation 2 × 4 indicates that the entry in column 2 is multiplied by the entry in column 4 to obtain the entry in column 5.

Since the shear force passes through the shear center, torsional effects will be neglected. Also, warping effects are to be ignored. The shear stress τ or shear flow $q = \tau b$ will be calculated using case 1 of Table 15-2. As in the case of Example 15.4, the first area moments Q_y^*, Q_z^* are based on the areas of the stringers alone. The thin webs are ignored for this computation.

Most of the calculations are indicated in Fig. 15-11. Other necessary computations are

$$\bar{P}' = \frac{d}{dx} \int E\alpha T \, dA = \sum_{i=1}^{12} E_i A_i^* \alpha T_i' = -436.9 \text{ lb/in.}$$

$$\bar{M}_z' = -V_y - \frac{d}{dx} \int E\alpha T \bar{y} \, dA = -\sum_{i=1}^{12} E_i A_i^* \bar{y}_i \alpha T_i' = 70.36 \text{ lb} \tag{1}$$

$$\bar{M}_y' = V_z + \frac{d}{dx} \int E\alpha T \bar{z} \, dA = 1500 + \sum_{i=1}^{12} E_i A_i^* \bar{z}_i \alpha T_i'$$

$$= 1500 - 150.34 = 1349.66 \text{ lb}$$

For tabulated calculations of the sort required here, it is convenient in case 2 of Table 15-2 to gather together those terms that remain constant for all points on the cross section. Thus, we write

$$q = -\frac{A_0^*}{A^*} \bar{P}' + C_2' Q_z^* - C_3' Q_y^* + \int_{A_0} E(\alpha T)' \, dA \tag{2}$$

where

$$C_2' = \frac{\bar{M}_z' I_y^* + \bar{M}_y' I_{yz}^*}{I_y^* I_z^* - I_{yz}^{*2}} \quad \text{and} \quad C_3' = \frac{\bar{M}_y' I_z^* + \bar{M}_z' I_{yz}^*}{I_y^* I_z^* - I_{yz}^{*2}}$$

do not vary over a cross section. For our case

$$C_2' = \frac{\bar{M}_z' I_y^* + \bar{M}_y' I_{yz}^*}{I_y^* I_z^* - I_{yz}^{*2}} = \frac{70.36(130.12) + 1349.66(110.51)}{130.12(1270.01) - (110.51)^2} = 1.034 \text{ lb/in.}^4$$

$$\tag{3}$$

$$C_3' = \frac{\bar{M}_y' I_z^* + \bar{M}_z' I_{yz}^*}{I_y^* I_z^* - I_{yz}^{*2}} = \frac{1349.66(1270.01) + 70.36(110.51)}{130.12(1270.01) - (110.51)^2} = 11.251 \text{ lb/in.}^4$$

The final calculations for q are listed in column 19 of Fig. 15-11. In passing over the third stringer the shear flow should drop to zero. The value 0.18 lb/in. shown is a result of round-off error accumulation.

15.6 COMBINED NORMAL AND SHEAR STRESSES

A complete cross-sectional stress analysis usually involves both normal and shear stresses. The following example illustrates such an analysis.

Example 15.7   Normal and Shear Stress Calculation including Warping Stresses
Determine the normal and shear stresses on the cross section of the cantilevered channel beam shown in Fig. 15-12a.

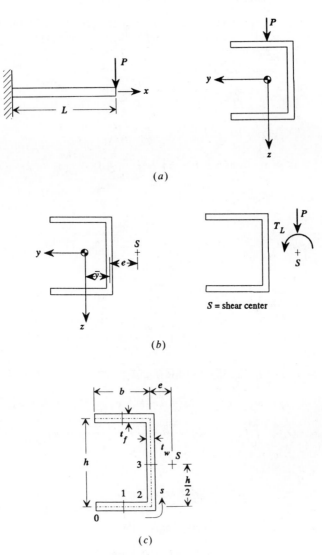

FIGURE 15-12   Example 15.7.

The eccentric force P is resolved into a force P applied through the shear center and a torque $T_L = P(\bar{y} + e)$ (Fig. 15-12b). The stresses due to P and T_L can be determined independently and then, for a point, added together to obtain the combined stress at the point.

The geometric properties required for normal stresses can be obtained from Table 2-1. Although many of the properties for warping-related stresses are listed in Table 2-6, we choose to detail most of the calculations here. Also, the properties needed for transverse shear are computed.

Cross-sectional Properties

Select the origin of the s coordinate to be at the free edge of the lower flange (Fig. 15-12c). Then the sectorial coordinate with respect to the shear center is found to be [Eq. (2.24)]

$$\omega_S = \int_0^s \frac{h}{2}\,d\xi = \frac{h}{2}s \quad \text{for } 0 \le s \le b$$

$$\omega_S = \frac{h}{2}b + (-)\int_b^s e\,d\xi = \frac{bh}{2} - e(s - b) \quad \text{for } b \le s \le b + h$$

$$\omega_S = \frac{bh}{2} - e(b + h - b) + \int_{b+h}^s \frac{h}{2}\,d\xi = \frac{h}{2}(s - 2e - h)$$

$$\text{for } b + h \le s \le 2b + h$$

From Eq. (2.25c)

$$\omega_0 = \frac{1}{A}\int_0^{2b+h} \omega_S\,dA$$

$$= \frac{1}{2bt_f + ht_w}\left\{\int_0^b \frac{h}{2}st_f\,ds + \int_b^{b+h}\left[\frac{bh}{2} - e(s - b)\right]t_w\,ds\right.$$

$$\left. + \int_{b+h}^{2b+h} \frac{h}{2}(s - 2e - h)t_f\,ds\right\} = \frac{h}{2}(b - e)$$

It follows from Eq. (2.25a) that for $0 \le s \le b$

$$\omega = \omega_S - \omega_0 = \tfrac{1}{2}hs - \tfrac{1}{2}h(b - e) = \tfrac{1}{2}h(s - b + e)$$

$$Q_\omega = \int_0^s (\omega_S - \omega_0)t_f\,d\xi = \tfrac{1}{2}ht_f s(\tfrac{1}{2}s - b + e) \tag{1}$$

for $b \le s \le b + h$

$$\omega = \omega_S - \omega_0 = e(\tfrac{1}{2}h - s + b)$$

$$Q_\omega = (Q_\omega)_{s=b} + \int_b^s (\omega_S - \omega_0)t_w\,d\xi \tag{2}$$

$$= \tfrac{1}{2}bht_f(e - \tfrac{1}{2}b) - et_w\left[\tfrac{1}{2}(s^2 - b^2) - (b + \tfrac{1}{2}h)(s - b)\right]$$

and for $b + h \le s \le 2b + h$

$$\omega = \omega_S - \omega_0 = \frac{h}{2}(s - e - h - b)$$

$$Q_\omega = (Q_\omega)_{s=b+h} + \int_{b+h}^s \frac{h}{2}(\xi - e - h - b)t_f \, d\xi \qquad (3)$$

$$= \frac{bht_f}{2}\left(e - \frac{b}{2}\right) - \frac{ht_f}{2}\left\{(b + e + h)(s - b - h) - \frac{1}{2}\left[s^2 - (b + h)^2\right]\right\}$$

At point 1, $s = \frac{1}{2}b$

$$Q_{\omega 1} = \tfrac{1}{2}ht_f\left[\tfrac{1}{2}b(\tfrac{1}{4}b - b + e)\right] = \tfrac{1}{4}bht_f\left(e - \tfrac{3}{4}b\right) \qquad (4)$$

At point 2, $s = b$,

$$Q_{\omega 2} = \tfrac{1}{2}ht_f b(\tfrac{1}{2}b - b + e) = \tfrac{1}{4}bht_f(2e - b) \qquad (5)$$

At point 3, $s = b + \tfrac{1}{2}h$,

$$Q_{\omega 3} = \tfrac{1}{4}bht_f(2e - b) - et_w\left(-\tfrac{1}{8}h^2\right) = \tfrac{1}{4}bht_f(2e - b) + \tfrac{1}{8}eh^2t_w \qquad (6)$$

The warping constant is, from Eq. (2.27),

$$\Gamma = \int_0^{2b+h} \omega^2 t \, ds = \int_0^{2b+h} (\omega_S - \omega_0)^2 t \, ds$$

$$= \int_0^b \frac{h^2}{4}(s - b + e)^2 t_f \, ds + \int_b^{b+h} e^2\left(\frac{h}{2} - s + b\right)^2 t_w \, ds$$

$$+ \int_{b+h}^{2b+h} \frac{h^2}{4}(s - e - h - b)^2 t_f \, ds$$

$$= \frac{h^2 t_f}{6}(b^3 - 3b^2 e + 3be^2) + \frac{e^2 h^3 t_w}{12}$$

$$= \frac{h^2}{12}\left[2t_f(b^3 - 3b^2 e) + e^2(6t_f b + t_w h)\right]$$

The shear center location e can be taken from case 8 of Table 2-6 as

$$e = \frac{3b^2 t_f}{6t_f b + t_w h} \qquad (7)$$

Thus, we get

$$\Gamma = \frac{h^2 b^3 t_f}{12} \frac{3t_f b + 2t_w h}{6t_f b + t_w h} \qquad (8)$$

The first moments at points 1, 2, and 3 needed for shear stresses due to transverse forces are, from (Eq. 2.15),

$$(Q_y)_1 = \tfrac{1}{4}bht_f, \qquad (Q_y)_2 = \tfrac{1}{2}bht_f, \qquad (Q_y)_3 = \tfrac{1}{2}bht_f + \tfrac{1}{8}h^2 t_w \qquad (9)$$

Internal Moments

The formulas of Chapter 14 are used to determine the internal bimoment and torque. We use the equations of Table 14-1, part A, with the loading functions for the concentrated torque at $x = L$ taken from Table 14-1, part B. The initial parameters are given in Table 14-1, part C. The initial parameters can also be computed by applying the end conditions of $\phi = \psi = 0$ at $x = 0$ and $B = T = 0$ at $x = L$ to the equations of Table 14-1, part A:

$$\phi = \frac{T_L}{GJC}[(\cosh Cx - 1)\tanh CL - \sinh Cx + Cx]$$

$$B = -\frac{T_L}{C}(\tanh CL \cosh Cx - \sinh Cx) \qquad (10)$$

$$T_\omega = -T_L(\tanh CL \sinh Cx - \cosh Cx)$$

where $C^2 = GJ/E\Gamma$.

At x = 0

(a) At $x = 0$, it can be demonstrated that T_t (pure torsional torque) $= 0$. Substitute the end condition $\psi = -d\phi/dx = 0$ into $T_t = GJ\,d\phi/dx$ (Section 14.4). Since $T_t = 0$, it follows that $\tau_t = 0$.

(b) The warping shear stress is given by $\tau_\omega = Q_\omega T_\omega/t\Gamma$, where t is the thickness of the wall. At points 1, 2, and 3 on the cross section

$$\tau_{\omega 1} = \frac{Q_{\omega 1}T_L}{t_f\Gamma}, \qquad \tau_{\omega 2} = \frac{Q_{\omega 2}T_L}{t_f\Gamma}, \qquad \tau_{\omega 3} = \frac{Q_{\omega 3}T_L}{t_w\Gamma}$$

where $Q_{\omega 1}$, $Q_{\omega 2}$, and $Q_{\omega 3}$ are given by (4), (5), (6).

(c) At $x = 0$, $V_z = P$ and using case 4, Table 15-2 (with thickness $b = t$), the bending shear stress becomes

$$\tau = \tau_b = -\frac{V_z Q_y}{tI_y} = -\frac{PQ_y}{tI_y}$$

The shear stress is positive in the direction of positive s coordinate. At points 1, 2, 3

$$\tau_{b1} = -\frac{Pbh}{4I_y}, \qquad \tau_{b2} = -\frac{Pbh}{2I_y}, \qquad \tau_{b3} = -\frac{Ph}{t_w I_y}\left(\tfrac{1}{2}bt_f + \tfrac{1}{8}ht_w\right)$$

(d) At $x = 0$, $M_y = -PL$ and, from case 5 of Table 15-1, the bending normal stress is given by

$$\sigma = \sigma_b = \frac{M_y z}{I_y} = -\frac{PLz}{I_y}$$

At points $1, 2, 3$

$$\sigma_{b1} = -\frac{PhL}{2I_y}, \qquad \sigma_{b2} = -\frac{PhL}{2I_y}, \qquad \sigma_{b3} = 0$$

(e) Warping normal stress

$$\sigma_\omega = \frac{B\omega}{\Gamma} = -\frac{T_L \tanh CL}{C\Gamma}\omega$$

where the principal sectorial coordinate ω for the lower flange and the web are given by (1) and (2). Then at points $1, 2, 3$

$$\sigma_{\omega 1} = -\frac{T_L \tanh CL}{C\Gamma}\frac{h}{4}(2e - b), \qquad \sigma_{\omega 2} = -\frac{T_L \tanh CL}{C\Gamma}\left(\frac{eh}{2}\right),$$

$$\sigma_{\omega 3} = -\frac{T_L \tanh CL}{CL}(0) = 0$$

At x = L

From (10) at $x = L$ we have the bimoment $B = 0$, the warping torque $T_\omega = T_L/\cosh CL$, and the rate of angle of twist

$$-\psi = \phi' = \frac{T_L}{GJ}\left(1 - \frac{1}{\cosh CL}\right)$$

(a) From Eq. (14.7) and Table 15-2, the pure torsional stress at the edges of the wall is calculated as

$$T_t = GJ\phi' = T_L\left(1 - \frac{1}{\cosh CL}\right)$$

$$\tau_t = \frac{T_t t}{J} = \frac{tT_L}{J}\left(1 - \frac{1}{\cosh CL}\right)$$

The pure torsional stress varies linearly across the thickness. It is a maximum at both edges and zero at the middle. The stresses on the two edges of the thin wall are equal in magnitude and opposite in direction.

(b) The warping shear stress is calculated as

$$\tau_\omega = \frac{Q_\omega T_\omega}{t\Gamma} = \frac{Q_\omega T_L}{t\Gamma \cosh CL}$$

with Q_ω for points 1, 2, and 3 taken from (4), (5), and (6) and by using the appropriate wall thickness.

(c) Since the shear force V_z is constant along the beam length, the bending shear stress at $x = L$ is the same as at $x = 0$.

(d) Since at $x = L$, $M_y = 0$, the bending normal stress is $\sigma = 0$.

(e) The warping normal stress is calculated as

$$\sigma = B\omega/\Gamma = 0$$

Combined Results

The normal and shear stresses at points 1, 2, and 3 on the cross section are obtained as follows.

At x = 0

(a) Normal stresses:

$$\sigma_1 = \sigma_{b1} + \sigma_{\omega 1} = -\frac{PhL}{2I_y} - \frac{h(2e - b)}{4}\frac{T_L \tanh CL}{C\Gamma}$$

$$\sigma_2 = \sigma_{b2} + \sigma_{\omega 2} = -\frac{PhL}{2I_y} - \frac{eh}{2}\frac{T_L \tanh CL}{C\Gamma}$$

$$\sigma_3 = \sigma_{b3} + \sigma_{\omega 3} = 0$$

(b) Shear stresses:

$$\tau_1 = \tau_t + \tau_{b1} + \tau_{\omega 1} = -\frac{Pbh}{4I_y} + \frac{Q_{\omega 1}T_L}{t_f \Gamma}$$

$$\tau_2 = \tau_t + \tau_{b2} + \tau_{\omega 2} = -\frac{Pbh}{2I_y} + \frac{Q_{\omega 2}T_L}{t_f \Gamma}$$

$$\tau_3 = \tau_t + \tau_{b3} + \tau_{\omega 3} = -\frac{Ph}{t_w I_y}\left(\frac{bt_f}{2} + \frac{ht_w}{8}\right) + \frac{Q_{\omega 3}T_L}{t_w \Gamma}$$

At x = L

(a) Normal stresses:

$$\sigma_1 = \sigma_2 = \sigma_3 = 0$$

(b) Shear stresses:

$$\tau = \tau_t + \tau_b + \tau_\omega$$

where τ_b and τ_ω are uniform through thickness t, τ_t varies linearly through thickness t and is in opposite directions on the two sides (i.e., $\tau_t = \pm t T_L/J$). The extreme shear stresses at points 1, 2, and 3 can be written as:

$$\tau_1 = \tau_t + \tau_{b1} + \tau_{\omega 1} = \pm\frac{tT_L}{J}\left(1 - \frac{1}{\cosh CL}\right) - \frac{Phb}{4I_y} + \frac{T_L Q_{\omega 1}}{t_f \Gamma \cosh CL}$$

$$\tau_2 = \pm\frac{tT_L}{J}\left(1 - \frac{1}{\cosh CL}\right) - \frac{Pbh}{2I_y} + \frac{T_L Q_{\omega 2}}{t_f \Gamma \cosh CL}$$

$$\tau_3 = \pm\frac{tT_L}{J}\left(1 - \frac{1}{\cosh CL}\right) - \frac{Ph}{t_w I_y}\left(\frac{bt_f}{2} + \frac{ht_w}{8}\right) + \frac{T_L Q_{\omega 3}}{t_w \Gamma \cosh CL}$$

15.7 FINITE-ELEMENT ANALYSIS

Characteristics such as normal stresses are readily found using the formulas presented in Table 15-1. The cross-sectional properties and stresses related to torsional moments and shear forces are often so difficult to calculate that they have to be computed with a numerical method, especially for sections of arbitrary shape. The most common technique for determining such properties and stresses is the finite-element method. Commercial software packages using finite elements are now available to calculate all important cross-sectional properties and stresses.

The finite-element formulation of the warping and shear problems can be based on the elasticity solutions of a cantilever beam with the $x = 0$ end fixed and the $x = L$ end free and loaded by transverse end (shear) force and twisting moment [15.1]–[15.5]. An appeal to Saint Venant's principle permits these solutions to be applied to more general cases of loadings. More specifically, it is assumed that the shear stresses on a particular cross section of a beam depend only upon the forces at that cross section provided that the cross section of interest is far enough away from any points of rapid variation in the shear force. The elasticity solutions for shear stresses from the torsional moments and shear forces are expressed in terms of unknown displacement functions (warping functions), as shown in Table 15-3. The finite-element displacement formulation begins with the establishment of the principle of virtual work for the problem as (Appendices II and III)

$$\delta W = \delta(W_i + W_e) = 0$$

The principles for the torsion and shear problems are shown in Table 15-3. The cross section is discretized into elements, and in each element approximate solutions (shape functions) are introduced for the warping functions in the form

$$\omega = \sum_{i=1}^{n} N_i \omega_i$$

where n is the number of nodes in the element, N_i are the shape functions, and ω_i are the values of ω at the nodes. Substitution of the shape functions into the variational principles leads to the element stiffness matrices and loading vectors

$$\mathbf{kv} = \mathbf{p} \quad \text{or} \quad \mathbf{k\omega} = \mathbf{p}$$

The expressions for the elements in the stiffness matrices and loading vectors are given in Table 15-3. Assemble the element stiffness equations to form a system of simultaneous linear equations that can be solved for the warping functions at the nodal points. The cross-sectional properties and stresses related to the warping functions can then be computed. Table 15-3 summarizes the fundamental formulas for the finite-element formulation for homogeneous, isotropic material, starting with the principle of virtual work. Examples of the use of finite elements to solve cross-sectional problems are provided in the documentation of the cross-sectional properties computer program that accompanies this book.

Reference 15.5 provides a thorough discussion of the underlying theory and computational solution methodology for the shear correction coefficients.

REFERENCES

15.1 Herrmann, L. R., "Elastic Torsional Analysis of Irregular Shapes," *J. Eng. Mechan. Div. ASCE*, Vol. 91, 1965, pp. 11–19.

15.2 Mason, W. E., and Herrmann, L. R., "Elastic Shear Analysis of General Prismatic Beams," *J. Eng. Mechan. Div. ASCE*, Vol. 94, 1968, pp. 965–983.

15.3 Chang, P. Y., Thasanatorn, C., and Pilkey, W. D., "Restrained Warping Stresses in Thin-walled Open Sections," *J. Struct. Div. ASCE*, Vol. 101, 1975, pp. 2467–2472.

15.4 Pilkey, W. D., and Wunderlich, W., *Structural Mechanics, Variational and Computational Methods*, CRC, Florida, 1993.

15.5 Schramm, U., Kitis, L., Kang, W., and Pilkey, W. D., "On the Shear Deformation Coefficient in Beam Theory," *Fin. Elem. in Anal. & Design*, Vol. 16, 1994.

15

Tables

TABLE 15-1 NORMAL STRESSES ON BEAM CROSS SECTION

Notation

σ = normal stress
E = modulus of elasticity
A = area of cross section
$I = I_y, I_z$ = moments of inertia about y and z axes
I_{yz} = product moment of inertia
Γ = warping constant
α = thermal coefficient of expansion
T = temperature change on cross section
ω = warping function
B = bimoment = $\int_A \sigma\omega\, dA$
P = axial force, positive in tension

$$\bar{P} = P + P_T, \qquad \bar{M}_y = M_y + M_{Ty} \qquad \bar{M}_z = M_z + M_{Tz}$$

with

$$P_T = \int E\alpha T\, dA \qquad M_{Ty} = \int E\alpha Tz\, dA \qquad M_{Tz} = \int E\alpha Ty\, dA$$

E_r = reference modulus of elasticity for composite section
$E_r = E$ for homogeneous material
y, z = coordinates measured from centroid
$B\omega/\Gamma$ = applies only to thin-walled beams

Superscript asterisk indicates a property of a composite bar
Ignore the $*$ for homogeneous material

Case	Normal Stresses
1. General	$\sigma = \dfrac{E}{E_r}\left[\dfrac{\bar{P}}{A^*} + \dfrac{\bar{M}_y I_z^* + \bar{M}_z I_{yz}^*}{I_y^* I_z^* - I_{yz}^{*2}} z - \dfrac{\bar{M}_z I_y^* + \bar{M}_y I_{yz}^*}{I_y^* I_z^* - I_{yz}^{*2}} y - E_r\alpha T + \dfrac{B\omega}{\Gamma} \right]$
2. Homogeneous section without warping and thermal loading	$\sigma = \dfrac{P}{A} + \dfrac{M_y I_z + M_z I_{yz}}{I_y I_z - I_{yz}^2} z - \dfrac{M_z I_y + M_y I_{yz}}{I_y I_z - I_{yz}^2} y$
3. Unsymmetrical section with $M_z = 0$.	$\sigma = \dfrac{P}{A} + \dfrac{M_y\left(I_z z - I_{yz} y\right)}{I_y I_z - I_{yz}^2}$
4. Bending about principal axes	$\sigma = \dfrac{P}{A} + \dfrac{M_y z}{I_y} - \dfrac{M_z y}{I_z}$
5. Bending about single (y) principal axis	$\sigma = \dfrac{P}{A} + \dfrac{M_y z}{I}$

TABLE 15-1 Normal Stresses on Beam Cross Sections 768

TABLE 15-2   SHEAR STRESSES ON BEAM CROSS SECTION

Notation

τ = shear stress
τ_t = shear stress due to pure torsion
A = area of cross section
$I_y = I, I_z$ = moments of inertia about y and z axes
I_{yz} = Product of inertia
$Q = Q_y, Q_z$ = first moments of inertia about y and z axes
Γ = warping constant
Q_ω = first sectorial moment
α = thermal coefficient of expansion
T = temperature change on cross section
B = bimoment
T_t = pure torsion torque
T_ω = warping torque
V_y, V_z = V − shear forces along y and z axes
b = width of cross section at point where
        shear stress is calculated
t = thickness of thin-walled cross section
r = radial distance from the centroidal longitudinal axis

$$\bar{P}' = \frac{\partial}{\partial x}(P + P_T), \qquad (\alpha T)' = \frac{\partial}{\partial x}(\alpha T)$$

$$\bar{M}'_y = \frac{\partial}{\partial x}(M_y + M_{Ty}), \qquad \bar{M}'_z = -\frac{\partial}{\partial x}(M_z + M_{Tz})$$

$$B' = \frac{\partial B}{\partial x}$$

For beams with no axial force, rotary foundation, rotary inertia, and applied distributed moment:

$$\bar{M}'_y = V_z + \frac{\partial}{\partial x}M_{Ty}, \qquad \bar{M}'_z = -V_y - \frac{\partial}{\partial x}M_{Tz}$$

$$B' = T_\omega$$

A_0 = area defined in chapter 2, Fig. 2-9.

For composite bars $A_0^* = \int_{A_0} dA^* = \int_{A_0}(E/E_r)\, dA$

Superscript asterisk indicates a property of a composite bar. See Section 2.12.

See Chaper 2 for detailed definitions. Also, see definitions of Table 15-1.

Shear Stresses on Beam Cross Section   TABLE 15-2

TABLE 15-2 (continued)   SHEAR STRESSES ON BEAM CROSS SECTION

Case	Shear Stresses
1. Solid beam cross section $\tau_t = \dfrac{T_t r}{J}$	$\tau = \tau_t + \dfrac{1}{b}\left[\dfrac{A_0^*}{A^*}\overline{P}' + \dfrac{Q_y^* I_z^* - Q_z^* I_{yz}^*}{I_y^* I_z^* - I_{yz}^{*2}}\overline{M}_y' \right.$ $\left. - \dfrac{Q_z^* I_y^* - Q_y^* I_{yz}^*}{I_y^* I_z^* - I_{yz}^{*2}}\overline{M}_z' + \displaystyle\int_{A_0} E(\alpha T)'\, dA\right]$
2. Thin-walled open cross section $\tau_t = \dfrac{T_t t}{J}$	$\tau = \tau_t - \dfrac{1}{t}\left[\dfrac{A_0^*}{A^*}\overline{P}' + \dfrac{Q_y^* I_z^* - Q_z^* I_{yz}^*}{I_y^* I_z^* - I_{yz}^{*2}}\overline{M}_y' \right.$ $- \dfrac{Q_z^* I_y^* - Q_y^* I_{yz}^*}{I_y^* I_z^* - I_{yz}^{*2}}\overline{M}_z'$ $\left. - \displaystyle\int_{A_0} E(\alpha T)'\, dA + \dfrac{Q_\omega^*}{\Gamma^*}B'\right]$
3. Homogeneous section with no torsion, no axial force, and no thermal loading	$\tau = \dfrac{1}{b}\left(\dfrac{Q_y I_z - Q_z I_{yz}}{I_y I_z - I_{yz}^2}V_z + \dfrac{Q_z I_y - Q_y I_{yz}}{I_y I_z - I_{yz}^2}V_y\right)$
4. Symmetric section, bending about single (y) axis	$\tau = \dfrac{VQ}{Ib}$

TABLE 15-2   **Shear Stresses on Beam Cross Section**   **770**

TABLE 15-3 FINITE-ELEMENT SOLUTION FOR SHEAR STRESSES

Notation

E = modulus of elasticity

G = shear modulus of elasticity

ν = Poisson's ratio

A = area of cross section

I, I_z = moments of inertia about y and z axes

I_{yz} = product of inertia

ω = warping function

ψ = warping function for direct shear

ω_i, ψ_i = values of ω, ψ at the nodes

$\boldsymbol{\omega}, \boldsymbol{\psi}$ = element vectors containing ω_i, ψ_i

$\mathbf{k}$ = element stiffness matrix (k_{ij})

$\mathbf{p}$ = element nodal force vector

$\mathbf{v}$ = element nodal displacement vector

V_y, V_z = shear forces in y and z directions

n = number of nodes in an element

J = torsional constant

$\alpha_y = \alpha_{yy}, \alpha_{yz}, \alpha_z = \alpha_{zz}$ = shear correction coefficients (see Table 2-4)

Γ = warping constant

θ = derivative of angle of twist along bar

$$K_y = \frac{V_y I - V_z I_{yz}}{E\left(II_z - I_{yz}^2\right)} \qquad K_z = \frac{V_z I_z - V_y I_{yz}}{E\left(II_z - I_{yz}^2\right)}$$

$$S_1 = \nu[\tfrac{1}{2}K_y(y^2 - z^2) + yzK_z] \qquad S_2 = \nu[\tfrac{1}{2}K_z(z^2 - y^2) + yzK_y]$$

$$K = II_z - I_{yz}^2$$

$$q_1 = \frac{\nu}{4(1+\nu)}[I(I_z - I) - 2I_{yz}^2] \qquad q_2 = \frac{\nu}{4(1+\nu)}[I_z(I - I_z) - 2I_{yz}^2]$$

N_i = shape functions

TABLE 15-3 (continued) FINITE-ELEMENT SOLUTION FOR SHEAR STRESSES

Case	Torsion	Direct Shear
1. Principle of virtual work	$\delta W = \int_A \left(\dfrac{\partial \omega}{\partial y} \delta \dfrac{\partial \omega}{\partial z} + \dfrac{\partial \omega}{\partial z} \delta \dfrac{\partial \omega}{\partial z} \right) dA$ $+ \int_A \left(-z \delta \dfrac{\partial \omega}{\partial y} + y \delta \dfrac{\partial \omega}{\partial z} \right) dA = 0$	$\delta W = \int_A \left[\left(\dfrac{\partial \psi}{\partial y} - S_1 \right) \delta \dfrac{\partial \psi}{\partial y} + \left(\dfrac{\partial \psi}{\partial z} - S_2 \right) \delta \dfrac{\partial \psi}{\partial z} \right] dA$ $+ K_y \int_A y \, \delta \psi \, dA + K_z \int_A z \, \delta \psi \, dA = 0$
2. Shape functions	$\omega = \sum_{i=1}^{n} N_i \omega_i$	$\psi = \sum_{i=1}^{n} N_i \psi_i$
3. Stiffness matrices and loading vectors, $\mathbf{k v} = \mathbf{p}$	$k_{ij} = \int_A \left(\dfrac{\partial N_i}{\partial y} \dfrac{\partial N_j}{\partial y} + \dfrac{\partial N_i}{\partial z} \dfrac{\partial N_j}{\partial z} \right) dA$ $p_i = \int_A \left(z \dfrac{\partial N_i}{\partial y} - y \dfrac{\partial N_i}{\partial z} \right) dA$ $\mathbf{k} \boldsymbol{\omega} = \mathbf{p}$	$k_{ij} = \int_A \left(\dfrac{\partial N_i}{\partial y} \dfrac{\partial N_j}{\partial y} + \dfrac{\partial N_i}{\partial z} \dfrac{\partial N_j}{\partial z} \right) dA$ $p_i = \int_A \left(S_1 \dfrac{\partial N_i}{\partial y} + S_2 \dfrac{\partial N_i}{\partial z} \right) dA$ $+ \int_A 2(1 + \nu) N_i (y K_y + z K_z) \, dA$ $\mathbf{k} \boldsymbol{\psi} = \mathbf{p}$

4. **Stresses**	$\tau_{xy} = G\theta\left(\dfrac{\partial\omega}{\partial y} - z\right) = G\theta\left[\displaystyle\sum_{i=1}^{n}\left(\dfrac{\partial N_i}{\partial y}\omega_i\right) - z\right]$ $\tau_{xz} = G\theta\left(\dfrac{\partial\omega}{\partial z} + y\right) = G\theta\left[\displaystyle\sum_{i=1}^{n}\left(\dfrac{\partial N_i}{\partial z}\omega_i\right) + y\right]$	$\tau_{xy} = G\left\{\dfrac{\partial\psi}{\partial y} - \nu\left[\dfrac{1}{2}K_y(y^2 - z^2) + K_z yz\right]\right\}$ $\quad = G\left\{\displaystyle\sum_{i=1}^{n}\dfrac{\partial N_i}{\partial y}\psi_i - \nu\left[\dfrac{1}{2}K_y(y^2 - z^2) + K_z yz\right]\right\}$ $\tau_{xz} = G\left\{\dfrac{\partial\psi}{\partial z} - \nu\left[K_y yz + \dfrac{1}{2}K_z(z^2 - y^2)\right]\right\}$ $\quad = G\left\{\displaystyle\sum_{i=1}^{n}\dfrac{\partial N_i}{\partial z}\psi_i - \nu\left[K_y yz + \dfrac{1}{2}K_z(z^2 - y^2)\right]\right\}$
5. **Typical** **cross-** **sectional** **properties**	$J = \displaystyle\int_A\left[y^2 + z^2 + z\dfrac{\partial\omega}{\partial y} - y\dfrac{\partial\omega}{\partial z}\right]dA$ $\Gamma = \displaystyle\int_A\left\{\omega - \left[\dfrac{I_\omega}{A} + \left(\dfrac{I_{\omega z}I - I_{\omega y}I_{yz}}{II_z - I_{yz}^2}\right)y\right.\right.$ $\qquad\qquad \left.\left. + \left(\dfrac{I_{\omega y}I_z - I_{\omega z}I_{yz}}{II_z - I_{yz}^2}\right)z\right]\right\}^2 dA$ where $I_{\omega y}$, $I_{\omega z}$ are given by Eq. (2.28) and $I_\omega = \displaystyle\int_A \omega_c\, dA$	$\alpha_{yy} = \dfrac{A}{V_y^2}\displaystyle\int_A(\tau_{xy}^2 + \tau_{xz}^2)_{V_z=0}\, dA$ $\alpha_{yz} = \dfrac{A}{V_y V_z}\displaystyle\int_A\left(\tau_{xy}\big|_{V_y=0}\,\tau_{xy}\big|_{V_z=0} + \tau_{xz}\big|_{V_y=0}\,\tau_{xz}\big|_{V_z=0}\right)dA$ $\alpha_{zz} = \dfrac{A}{V_z^2}\displaystyle\int_A(\tau_{xy}^2 + \tau_{xz}^2)_{V_y=0}\, dA$

CHAPTER **16**

Curved Bars

Formulas for static, stability, and dynamic responses of curved bars lying in a plane are treated in this chapter. The stresses and the equations of motion of in-plane and out-of-plane deformations are provided. The formulas of this chapter correspond to coupled extension and bending in the case of in-plane motion and coupled torsion and bending for out-of-plane motion.

The theory of deformations for circular curved bars is based on the classical theory of arches, which relies on the assumption that plane cross sections remain plane, stress is proportional to strain, rotations and translations are small, and the thickness of the bar must be small in comparison to the radius of curvature of the bar.

775

16.1 NOTATION

All Curved Bars

A Cross-sectional area (L^2)

A_s Equivalent shear area, $= A/\alpha_s$ (L^2)

α_s Shear correction coefficient

E Modulus of elasticity of material (F/L^2)

G Shear modulus of elasticity (F/L^2)

g Gravitational acceleration (L/T^2)

k Winkler (elastic) foundation modulus (F/L^2)

k^* Rotary foundation modulus (FL/L)

L Length of bar (L)

ℓ Length of segment along bar, span of structural matrix

m_i Concentrated mass (M)

R Radius of curvature of centroidal axis along bar (L)

t Time

T Change of temperature (degrees), i.e., temperature rise with respect to reference temperature

α Coefficient of thermal expansion $((L/L)/\text{degree})$

ω Natural frequency (rad/T)

ρ Mass per unit length $(M/L, FT^2/L^2)$

In-Plane Stress and Deformation

Bars with a cross section symmetric about the plane of curvature are considered. The loadings and deformation lie in the same plane as the bar. The bar, which can be formed of straight or circular segments, undergoes extension and bending. In the case of straight segments, the formulas are essentially a combination of the bending formulas of Chapter 11 and the extension formulas of Chapter 12, so that such effects as shear deformation and rotary inertia can be taken into account.

c Applied bending moment intensity; positive if vector, according to right-hand rule, is in positive y direction (FL/L); c_1 designates uniform moment

e Shift in location of neutral axis (L)

h Height (thickness) of cross section (L)

I^* Moment of inertia modified for curvature of bar, $= \int_A [z^2/(1 - z/R)\, dA$ (L^4)

I Moment of inertia about centroidal y axis (L^4)

I_{Ti} Rotary inertia, transverse or diametrical mass moment of inertia of concentrated mass at station i; can be calculated as $I_{Ti} = \Delta a\, \rho r_y^2$, where Δa is length of beam lumped at station i (ML^2)

k_x Extension elastic foundation modulus (F/L^2)

k_i, k_o Correction factors for use of straight-beam formulas to calculate stresses in curved bars

ℓ_b Length of branch

M Bending moment at any section (FL)

M_T Thermal moment, $= \int_A E\alpha Tz\, dA$ (FL)

P Axial force (F)

p Transverse loading intensity (F/L); p_1 designates uniform load

p_x Distributed axial force, loading intensity (F/L); p_{x1} designates uniform force

r_y Radius of gyration of cross-sectional area about y axis (L)

r Point on cross section measured from center of curvature (L)

u Axial displacement (L)

V Shear force at any section (F)

w Transverse displacement (L)

W Concentrated applied transverse force (F)

σ_x Circumferential stress on cross section of curved bars or normal stress, or fiber stress (F/L^2)

σ_z Radial stress (F/L^2)

τ Shear stress (F/L^2)

θ Slope of deflection curve (rad)

Out-of-Plane Stress and Deformation

In this part of the chapter, bars lying in a plane with torsional loading and deformation, along with out-of-plane transverse loading and bending deformation, are treated. The formulas are essentially a combination of the bending formulas of Chapter 11 and the torsion formulas of Chapter 12, with some adjustments for the curvature of the bar. This means that the limitations on the applicability of the torsional theory of that chapter apply here as well. For example, for both straight and circular segments the torsional effects of restraints against warping are not taken into account.

c_z Applied bending moment intensity, positive if vector according to right-hand rule is in positive z direction (FL/L): c_{z1} designates uniform moment

I_z Moment of inertia taken about neutral (z) axis (L^4)

I_p Polar moment of inertia about x axis (L^4)

I_{pi} Polar mass moment of inertia of concentrated mass at station i; can be calculated as $I_{pi} = \Delta a\, \rho r_p^2$, where Δa is length of beam lumped at station i (ML^2)

J Torsional constant; for a circular cross section J is polar moment of inertia I_x of cross-sectional area with respect to axis of bar (L^4)

k_t Torsional elastic foundation modulus (FL/L)

m_x Distributed torque, twisting moment intensity (FL/L); m_{x1} is uniform torque

M_z Bending moment at any section (FL)

M_{Tz} Thermal moment, $= -\int_A E\alpha\, T y\, dA$ (FL)

p_y Transverse loading intensity (F/L); p_{y1} designates uniform load

r_z Radius of gyration of cross-sectional area about z axis (L)

r_p Polar radius of gyration (L)

T Twisting moment, torque (FL). Also, change of temperature

v Transverse displacement (L)

V_y Shear force at any section (F)

W_y Concentrated applied transverse force (F)

ϕ Angle of twist, rotation (rad)

θ_z Slope of deflection curve (rad)

16.2 IN-PLANE STRESS AND DEFORMATION

Sign Convention

Positive displacements u, w, and slope θ are shown in Fig. 16-1. Positive internal bending moment M and forces V, P on the right face of an element are also illustrated in Fig. 16-1. For applied loadings, the formulas provide solutions for the loading illustrated. Loadings applied in the opposite direction require the sign of the loading to be reversed in the formulas.

Stresses

The tables of this chapter provide the internal axial force, bending moment, and shear force at any point along a bar. The normal and shear stresses on a face of the cross section can be calculated using the stress formulas given in this section.

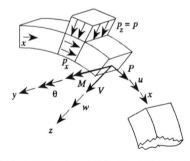

FIGURE 16-1 Positive displacements, internal forces and moments, and applied loadings for in-plane motion.

Normal Stress During bending, the cross sections of a straight beam are assumed to remain plane, and the strains of compression and extension fibers equidistant from the centroid of the cross section are equal in magnitude so that the normal stress is distributed linearly on the cross section. For a curved beam, the cross sections are also assumed to remain plane when the beam is bent, but the strains at two points on opposite sides of and equidistant from the centroid are no longer equal in magnitude. Although the magnitude of the extension and compression of the fibers at these points are the same, the "original lengths" of the fibers are different. The normal stress distribution is no longer linear. Figure 16-2 shows the stress distribution on a cross section. Also, the neutral axis of the cross section does not coincide with the centroid but is shifted. For pure bending, the distance of neutral axis shift is [16.1]

$$e = R - A/A_m = R - r_n \qquad 0.6 < R/h < 8 \qquad (16.1a)$$

with

$$R = \frac{1}{A} \int_A r \, dA \qquad A_m = \int_A \frac{dA}{r} \qquad (16.1b)$$

where A is the cross-sectional area, R is the distance from the center of curvature to the centroid of the cross section, h is the height of the cross section along the direction of R, r_n is the distance from the center of curvature to the neutral axis, and r locates a point on the section measured from the center of curvature (Fig. 16-2). Analytical expressions for these quantities for common cross sections are given in Table 16-1. Note that e is a cross-sectional property and not related to the applied forces. Equations (16.1) are applicable for the range $0.6 < R/h < 8$. When $R/h < 0.6$, the stress values can have a large error. In the cases when $R/h > 8$, i.e., for slender curved beams, round-off errors or small inconsistencies in treating a cross section of complicated shape may have a large effect on the calculated value of e. To avoid this, e can be computed using [16.1]

$$e \cong I/(RA) \qquad R/h > 8 \qquad (16.1c)$$

where I is the moment of inertia about the centroidal axis.

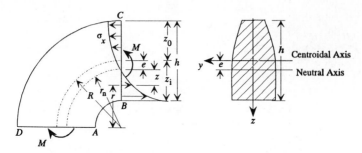

FIGURE 16-2 Normal stress distribution of curved beam.

If the tensile axial force P is applied,

$$e = R - \frac{AM}{A_m M + P(A - RA_m)} = R - r_n \qquad (16.1d)$$

where M is the bending moment about the y axis (Fig. 16-2).
The normal (circumferential) stress on the cross section is

$$\sigma_x = \frac{Mz}{Aer} = \frac{M(r_n - r)}{Aer} = \frac{M(A - rA_m)}{Ar(RA_m - A)} \qquad (16.2a)$$

where z is the distance from the neutral axis to the point of interest. When a tensile axial force P through the centroidal axis occurs on the cross section, the term P/A should be added to Eq. (16.2a):

$$\sigma_x = P/A + Mz/(Aer) \qquad (16.2b)$$

The expression P/A implies that the normal stress due to P is taken to be constant over the cross section, an assumption that is usually reasonable considering that the stress due to P is normally much smaller than the stress due to M. Also, Eq. (16.2b) is more accurate for pure bending than for shear loading (Fig. 16-3). When the first term (P) is comparable in magnitude to the second term (M) or R/h is small, the error of using Eq. (16.2b) increases significantly.

Cook [16.2] introduced two formulas for the circumferential stress σ_x to cope with these inaccuracies:

$$\sigma_x = \frac{M(r_n - r)}{Aer} + \frac{P}{A}\left[\frac{r_n}{r} + \frac{Ae}{I}(r - R)\right] \qquad (16.3a)$$

and

$$\sigma_x = Pr_n/(Ar) + Mz/(Aer) \qquad (16.3b)$$

Equation (16.3a) is more accurate than Eq. (16.3b) since the stress distribution expressed by Eq. (16.3b) does not satisfy the equilibrium condition on the cross section. Despite this, Eq. (16.3b) is preferred for its simplicity and adequate accuracy. Equations (16.2) and (16.3) are derived based on the assumptions that

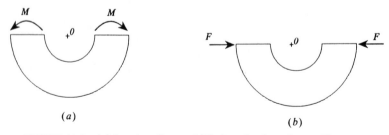

FIGURE 16-3 (a) Pure bending and (b) shear loading of curved beams.

the shear and radial stresses vanish, so they are best suited for those parts of the cross section where these stresses are not significant. Also, for I- and T-section curved beams, use of Eqs. (16.2) or (16.3) may cause nonnegligible errors due to the distortion of the profile of the cross section. This problem will be considered later.

A comparison of Eqs. (16.2) and (16.3) with the straight-beam flexure formula $(\sigma = Mz/I)$ indicates that the straight-beam solution is significantly in error for small values of R/h and the error is not conservative. Generally, for curved beams, with $R/h > 5.0$, the straight-beam flexure formula can be used [16.3]. It can be shown that when $R/h \to \infty$, Eq. (16.2a) becomes $\sigma_x = Mz/I$.

The normal stresses at the extreme fibers of the cross section can be calculated by using the formulas for the normal stress for straight beams multiplied by a factor. Thus the stress in the inside fiber (fiber AB of Fig. 16-2) is

$$\sigma_i = k_i(P/A + Mz_i/I) \tag{16.4a}$$

while the stress in the outer fiber (fiber CD) is

$$\sigma_o = k_o(P/A - Mz_o/I) \tag{16.4b}$$

where z_i, z_o are the distances from the centroid to the inner and outer fibers. Table 16-1 gives values for k_i and k_o. These formulas give the same results at the inside and outer fibers as Eq. (16.2).

When the cross section of a curved beam is composed of two or more of the regular shapes listed in Table 16-1, the values of A, A_m, and R in Eq. (16.2) for the composite section are given as

$$A = \sum_{i=1}^{n} A_i \tag{16.5a}$$

$$A_m = \sum_{i=1}^{n} A_{mi} \tag{16.5b}$$

$$R = \frac{\sum_{i=1}^{n} R_i A_i}{\sum_{i=1}^{n} A_i} \tag{16.5c}$$

where n is the number of regular shapes that form the composite section.

The stress formulas are summarized in Table 16-2.

Example 16.1   Stress in Curved Beam   The curved beam in Fig. 16-4 has a circular cross section 50 mm in diameter. The inside radius r_i of the curved beam is 40 mm. Determine the stress at B when $F = 20$ kN.

The radius R is obtained from the geometry: $R = r_i + b = 40 + 25 = 65$ mm. Values of A and A_m for the curved beam are calculated using the formulas in

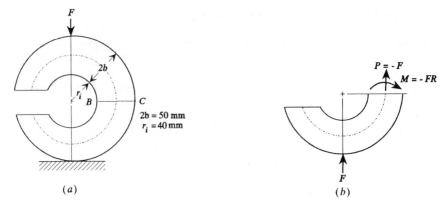

FIGURE 16-4 Example 16.1: (a) external loading; (b) free-body diagram.

case 4 of Table 16-1. For $2b = 50$ mm

$$A = \pi b^2 = (3.1416)(25)^2 = 1963.5 \text{ mm}^2$$
$$A_m = 2\pi\left(R - \sqrt{R^2 - b^2}\right)$$
$$= 2(3.1416)\left(65 - \sqrt{65^2 - 25^2}\right) \tag{1}$$
$$= 31.416 \text{ mm}$$
$$e = R - A/A_m = 65 - 1963.5/31.416 = 2.5 \text{ mm}$$
$$r_n = R - e = 65 - 2.5 = 62.5 \text{ mm}$$

On the cross section BC, the axial force $P = -F = -20\,000$ N and the moment is calculated as

$$M = -FR$$
$$= -20\,000 \times 65 = -1\,300\,000$$
$$= -1300 \text{ N} \cdot \text{m} \tag{2}$$

Therefore, the stress at B is, by Eq. (16.2) with $r = 40$ mm,

$$(\sigma_x)_B = \frac{P}{A} + \frac{M(A - rA_m)}{Ar(RA_m - A)}$$
$$= -\frac{20\,000}{1963.5} - \frac{1\,300\,000 \times [1963.5 - (40)(31.416)]}{(1963.5)(40)[(65)(31.416) - (1963.5)]}$$
$$= -159.15 \text{ MPa} \tag{3}$$

Example 16.2 Stress in a Crane Hook For a large number of manufactured crane hooks, section BC is the critically stressed section (Fig. 16-5a). The cross-sectional area can be closely modeled by a trapezoidal area A_2, with half of an ellipse A_1, and the area A_3 contained by an arc of a circle. If the dimensions of the critical section are shown in Fig. 16-5b and the hook is subjected to an axial load $F = 100$ kN, determine the circumferential stresses at the inner and outer radii.

The circumferential stresses σ_x are calculated using Eq. (16.2). The geometric property values A, R, and A_m of the cross sections are obtained using Eq. (16.1) and Table 16-1.

For the semiellipse area A_1, refer to case 10 in Table 16-1. For the geometry shown, $a = 65.0 + 30.0 = 95.0$ mm, $2b = 100.0$, $b = 50.0$ mm, and $h = 30.0$ mm. Then

$$A_1 = \frac{1}{2}\pi bh = 2356.2 \text{ mm}^2, \qquad R_1 = a - \frac{4h}{3\pi} = 82.3 \text{ mm}$$

$$A_{m1} = 2b + \frac{\pi b}{h}\left(a - \sqrt{a^2 - h^2}\right) - \frac{2b}{h}\sqrt{a^2 - h^2}\sin^{-1}\left(\frac{h}{a}\right) = 30.6 \text{ mm} \tag{1}$$

For the trapezoidal area A_2 use case 3 in Table 16-1. With

$$a = 65.0 + 30.0 = 95.0 \text{ mm}, \qquad c = 95.0 + 60.0 = 155.0 \text{ mm},$$
$$b_1 = 100.0 \text{ mm}, \qquad b_2 = 60.0 \text{ mm, we obtain}$$

$$A_2 = \frac{b_1 + b_2}{2}(c - a) = 4800.0 \text{ mm}^2$$

$$R_2 = \frac{a(2b_1 + b_2) + c(b_1 + 2b_2)}{3(b_1 + b_2)} = 122.5 \text{ mm} \tag{2}$$

$$A_{m2} = \frac{b_1 c - b_2 a}{c - a}\ln\frac{c}{a} - b_1 + b_2 = 39.95 \text{ mm}$$

Case 8 of Table 16-1 corresponds to area A_3. From Fig. 16-5b,

$$b^2 = 30^2 + (b - 10)^2 \qquad \text{or} \qquad b = 50 \text{ mm} \tag{3}$$

$$\theta = \tan^{-1}\left(\frac{30}{b - 10}\right) = 0.6435$$

and

$$a = 65.0 + 30.0 + 60.0 - b\cos 0.6435 = 115.0 \text{ mm}$$

Thus,

$$A_3 = b^2\theta - \frac{b^2}{2}\sin 2\theta = 408.75 \text{ mm}^2$$

$$R_3 = a + \frac{4b\sin^3\theta}{3(2\theta - \sin 2\theta)} = 159.04 \text{ mm} \tag{4}$$

$$A_{m3} = 2a\theta - 2b\sin\theta - \pi\sqrt{a^2 - b^2} + 2\sqrt{a^2 - b^2}\sin^{-1}\frac{b + a\cos\theta}{a + b\cos\theta}$$

$$= 2.57 \text{ mm} \quad \text{(for } a > b)$$

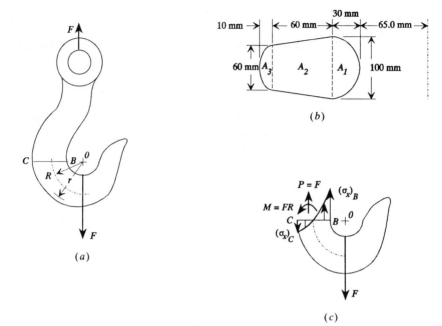

FIGURE 16-5   A crane hook, Example 16.2.

From Eqs. (16.5), we have

$$A = A_1 + A_2 + A_3 = 7564.95 \text{ mm}^2 \qquad A_m = A_{m1} + A_{m2} + A_{m3} = 73.12 \text{ mm}$$

$$R = \frac{R_1 A_1 + R_2 A_2 + R_3 A_3}{A} = 111.95 \text{ mm} \tag{5}$$

As shown in Fig. 16-5c, the internal load on section BC is $P = F$ and the moment is $M = FR$. The maximum tension and compression values of the circumferential stresses σ_x occur at points B and C, respectively.

For B and C, from the given dimensions in Fig. 16-5b

$$r_B = 65.0 \text{ mm} \qquad r_C = 65.0 + 30.0 + 60.0 + 10.0 = 165.0 \text{ mm} \tag{6}$$

Substitution of the appropriate values into Eq. (16.2) yields

$$(\sigma_x)_B = \frac{P}{A} + \frac{M(A - r_B A_m)}{A r_B (R A_m - A)} = 103.13 \text{ MPa}$$

$$(\sigma_x)_C = \frac{P}{A} + \frac{M(A - r_C A_m)}{A r_C (R A_m - A)} = -51.8 \text{ MPa} \tag{7}$$

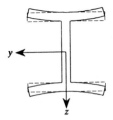

FIGURE 16-6   Distortion of an I-section curved beam.

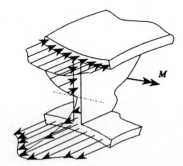

FIGURE 16-7   Stress distribution in I-section curved beam.

Circumferential Stresses for Thin-Flange Cross Sections   Generally speaking, the cross sections of curved beams with thin flanges tend to distort when the beams are subjected to bending moments. Often this is referred to as *profile distortion*. In the case shown in Fig. 16-6, the thin flanges are bent and tend to deflect radially as shown. As a consequence, the circumferential stress (normal stress) distribution is not constant along the flanges. The maximum stress occurs at the center of the inner flange (Fig. 16-7). Since the curved beam formula of Eq. (16.2) assumes that the normal stress is constant in the flange, corrections are required if the formula is to be used in the design of curved beams having thin-flange (e.g., I or T) cross sections. One approximate approach is to "correct" the curved beam physically to prevent the distortion of the cross section by welding radial stiffeners to the curved beams and then to use the curved beam formula. Another is Bleich's [16.3] method, which suggests that for the same bending moment the actual maximum circumferential stresses in the flanges with distortion for the I- or T-section curved beam (Fig. 16-8a) may be calculated by applying Eq. (16.2) to an undistorted I- or T-section curved beam with reduced flange width (Fig. 16-8b).

The reduced flange width is obtained by multiplying the original width of the flange by the factor α given in Table 16-3. To use the table, calculate the ratio $f = b_{pi}^2/\bar{r}_i t_{fi}$ using the dimensions for the ith flange of the original cross section, where b_{pi} and t_{fi} are as shown in Fig. 16-8a. Here $\bar{r}_i$ is the radius of curvature to the center of the ith flange. The reduced widths of the flange (Fig. 16-8b) are

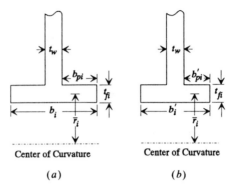

FIGURE 16-8 Bleich's method for flanges of I- or T-sections: (a) original flange and web; (b) modified flange and web.

given by

$$b'_{pi} = \alpha b_{pi} \tag{16.6a}$$

$$b'_i = 2b'_{pi} + t_w \tag{16.6b}$$

where α is obtained from Table 16-3 according to the computed value of f, b'_{pi} is the reduced width of the projecting part of each flange, b'_i is the reduced width of each flange, and t_w is the thickness of the web. Use these new dimensions in the standard stress formulas to compute the peak circumferential stresses.

Due to the bending of the flanges (Fig. 16-6), stress component σ_y is developed. Bleich provided an approximate relation for σ_y in the inner flange, the flange that is the closest to the center of curvature:

$$\sigma_y = -\beta\bar{\sigma}_x \tag{16.7}$$

where β is obtained from Table 16-3 for the computed ratio f and $\bar{\sigma}_x$ is the circumferential stress at midthickness of the flange at the junction, calculated based on the corrected cross section. The negative sign indicates that the sign of σ_y is opposite to that of $\bar{\sigma}_x$. The stress σ_y is assumed to be uniformly distributed in the flange.

The radial stress given later can be calculated using either the original or the modified cross section.

Example 16.3 Bleich Method for T-Section Curved Beam A T-section curved beam has the cross section shown in Fig. 16-9a. The center of curvature of the curved beam lies 40 mm from the flange. If the curved beam is subjected to a positive bending moment $M = 2.50$ kN · m, determine (a) the stresses at points A and B and (b) the maximum shear stress in the curved beam. Use Bleich's method.

(a) The width dimensions of the modified cross section (Fig. 16-9b) are calculated by Bleich's method. Since there is only one flange, the subscript i will

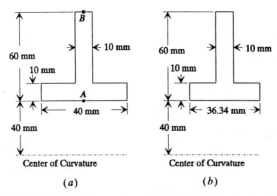

FIGURE 16-9 Example 16.3: (a) original section; (b) modified section.

be dropped. The coefficient f is calculated as

$$f = \frac{b_p^2}{\bar{r}t_f} = \frac{[(40 - 10)/2]^2}{(40 + 10/2)(10)} = 0.5$$

From Table 16-3 we obtain $\alpha = 0.878$ and $\beta = 1.238$. Thus, by Eqs. (16.6) the modified flange width is $b_p' = \alpha b_p = (0.878)(15) = 13.17$ mm and $b' = 2b_p' + t_w = 2(13.17) + 10 = 36.34$ mm (Fig. 16-9b).

The values of A, A_m, and R of the modified section are computed by Eqs. (16.5) and case 1 in Table 16-1:

$$A = (36.34)(10) + (50)(10) = 863.4 \text{ mm}^2$$

$$R = \frac{1}{863.4}[(36.34)(10)(45) + (50)(10)(75)] = 62.37 \text{ mm}$$

$$A_m = (36.34)\left(\ln\frac{50}{40}\right) + (10)\left(\ln\frac{100}{50}\right) = 15.04 \text{ mm}$$

At the inner radius of the modified section with $r = 40$ mm and $P = 0$ (pure bending), Eq. (16.2) gives the stress

$$(\sigma_x)_i = \frac{M(A - rA_m)}{Ar(RA_m - A)}$$

$$= \frac{2.50[863.4 - (40)(15.04)]}{(863.4)(40)[(62.37)(15.04) - 863.4]} = 253.9 \text{ MPa} \qquad (1)$$

Similarly, at the outer radius of the modified section, with $r = 100$ mm and $P = 0$,

$$(\sigma_x)_o = \frac{2.50[863.4 - (100)(15.04)]}{(863.4)(100)[(62.37)(15.04) - 863.4]} = -247.0 \text{ MPa} \qquad (2)$$

(b) Use Eq. (3.14) to compute the peak shear stress. The maximum principal stress occurs at the inner radius of the cross section and is given by (Eq. 3.13a) $\sigma_1 = (\sigma_x)_i = 253.9$ MPa, while the minimum principal stress (Eq. 16.13b) at this point is obtained from Eq. (16.7):

$$\sigma_3 = \sigma_y = -\beta\bar{\sigma}_x = -(1.238)\frac{2.50[863.4 - (45)(15.04)]}{(863.4)(45)[(62.37)(15.04) - 863.4]}$$

$$= -199.1 \text{ MPa} \tag{3}$$

Thus, the maximum shearing stress in the curved beam is (Eq. 3.14)

$$\tau_{max} = \tfrac{1}{2}(\sigma_1 - \sigma_3) = \tfrac{1}{2}[253.9 - (-199.1)] = 226.5 \text{ MPa} \tag{4}$$

This stress can be used to evaluate the strength of the beam. It is different from the shear stress τ of the next section in that τ is the shear stress on the plane of the cross section while τ_{max} is the maximum shear stress occurring on a plane oriented at a particular angle (Fig. 3-10).

Shear Stress The average shear stress τ across a width of a cross section, e.g., line 1–2 of Fig. 11-4, is

$$\tau = Vr_n Q/\left(bAe(R - z)^2\right) \tag{16.8}$$

where

$$Q = \int_{A'} r\,dA \tag{16.9}$$

The integration is taken over the area A' that lies between the position at which the stress is desired and the inner fiber of the cross section nearest to the center of curvature. The distance z is measured from the neutral axis.

Equation (16.8) is derived from the assumption that the shear stress is parallel to the shear force, and the shear stresses at points on an element perpendicular to the shear force are equal to the parallel shear stresses where the shear force is applied. Although there are some cases in practice that do not coincide with these assumptions, the assumptions tend to be good approximations and Eq. (16.8) can often be used without serious error. A comparison of the results of $\tau_{max}/\tau_{average}$, with the exact elasticity solution for a rectangular cross section, shows that when $1.25 \leq R/h \leq 5.5$, the error of the value $\tau_{max}/\tau_{average}$ does not exceed 1% [16.4].

Radial Stress Radial stress is usually not a major consideration for the design of the curved beams with solid cross sections because the magnitude of the radial stress is small compared to the circumferential stress. But for curved beams that have flanged cross sections with thin webs, the radial as well as circumferential stresses may be large at the junctions of the flanges and webs. As a consequence,

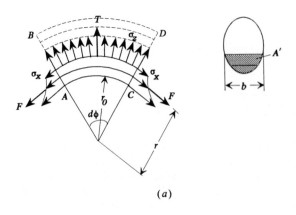

(a)

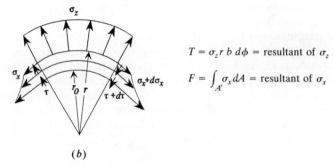

$T = \sigma_z r\, b\, d\phi$ = resultant of σ_z

$F = \int_{A'} \sigma_x dA$ = resultant of σ_x

(b)

FIGURE 16-10   Radial stresses: (a) equilibrium of segment of curved beam; (b) resultants.

the shear stress may also be large, and hence yielding may occur. A large radial stress in a thin web may also cause the web to buckle. In such cases, the radial stress cannot be neglected.

The radial stress is expressed by [16.3]

$$\sigma_z = \frac{1}{br}\left[\frac{A'}{A}P + \frac{AA'_m - A'A_m}{A(RA_m - A)}M\right] \tag{16.10a}$$

where

$$A'_m = \int_{r_0}^{r}\frac{r\,dA}{r} \quad \text{and} \quad A' = \int_{r_0}^{r} dA$$

with b, A', r_0 and r shown in Fig. 16-10. A moment M that tends to straighten the curved beam generates a tensile radial stress. This expression is obtained from the equilibrium of the beam segment in Fig. 16-10a, where the resultants F of σ_x, which takes the form of Eq. (16.2b) and is assumed to be constant along the beam segment, and T of σ_z form an equilibrium system.

If shear stresses τ [Eq. (16.8)] on the cross sections are considered and σ_x and τ are assumed to vary along the beam segment (Fig. 16-10b), the conditions of equilibrium for the beam segment result in the expression for the radial stress [16.5]:

$$\sigma_z = \frac{r_n}{Aebr}\left[(M - PR)\left(A'_m - \frac{A'}{r_n}\right) + \frac{P}{r}(RA' - Q)\right] \qquad (16.10b)$$

Equation (16.10b) is more accurate than Eq. (16.10a). Equations (16.10) are reasonably accurate approximations for the radial stress σ_z in curved beams although σ_x is derived with the assumption that the shear and radial stresses vanish. This is similar to the case of a straight beam where the normal stress is based on the assumption that a plane cross section remains plane, which implies that the shear stresses vanish. Then the straight-beam shear stresses are obtained from the equilibrium of the resultants of the normal and shear stresses. A comparison of Eq. (16.10a) [16.3] for rectangular cross-sectional beams subjected to shear loading (Fig. 16-3) with a corresponding theory of elasticity solution indicates that Eq. (16.10a) is conservative, and it remains conservative to within 6% for values of $R/h > 1.0$ even without considering the P term.

As in the case of the cross sections with thin flanges where the circumferential stresses on the cross sections should be corrected, the expressions for the radial stress on the cross sections with thick flanges and thin webs should also be corrected since the flanges tend to rotate about their own neutral axes during deformation and larger radial and shear stress are developed. See reference 16.6 for a method of correction.

Simple Curved Bars

The response of curved bars can be obtained by solving the fundamental equations of motion in first-order form for the in-plane deformation of a circular bar [16.7]:

$$\frac{\partial u}{\partial x} = \frac{P}{AE} - \frac{M}{ARE} + \frac{w}{R} + \alpha T$$

$$\frac{\partial w}{\partial x} = -\theta - \frac{u}{R} + \frac{V}{GA_s}$$

$$\frac{\partial \theta}{\partial x} = \frac{M}{EI^*} + \frac{w}{R^2} + \frac{M_T}{EI^*}$$

$$\frac{\partial M}{\partial x} = V + k^*\theta + \rho r_y^2 \frac{\partial^2 \theta}{\partial t^2} - c(x, t) \qquad (16.11)$$

$$\frac{\partial V}{\partial x} = kw - \frac{P}{R} + \rho \frac{\partial^2 w}{\partial t^2} - p(x, t)$$

$$\frac{\partial P}{\partial x} = k_x u + \frac{V}{R} + \rho \frac{\partial^2 u}{\partial t^2} - p_x(x, t)$$

These relations conform to sign convention 1 of Appendix II. They can be solved for a variety of loading and end conditions.

Tabulated Formulas   The extension, deflection, slope, bending moment, shear force, and axial force for uniform circular bars with various end conditions and loadings are provided in Table 16-4. The deflection formulas apply only to uniform beams with $R/z_0 \geq 4$, where z_0 is the distance from the centroid of the cross section to the outermost fiber. Included are some values at critical points. Table 16-5 contains formulas for uniform circular rings. The formulas of Table 16-5 are based on the assumptions that (1) the radius of curvature is very large compared to the dimensions of the cross section so that the deflection theory of straight bars is used in deriving these expressions, (2) the effect of axial and shear forces on the displacements is negligible, and (3) the deformations are small. Using superposition, these responses can be combined to cover more complicated applied loadings.

Formulas for Bars with Arbitrary Loading   Part A of Table 16-6 provides the displacements, slope, moment, and force responses for uniform bars based on the solution of Eqs. (16.11) without consideration of shear deformation. The assumptions underlying Tables 16-4 and 16-5 are no longer involved. Loading functions needed by the formulas of Table 16-6, part A, are listed in part B of the table. For the boundary conditions in part C of the table, the initial parameters can be determined by the method provided in Appendix III.

Buckling Loads

Table 16-7 gives the buckling loads for several uniform circular arches and rings. Although all the loads are applied in the planes of the bars, some buckling modes are out of plane (case 3). The formulas in this table apply to thin bars, i.e., the radius of gyration of the cross section is negligible compared to the radius of curvature of the bar.

Reference 16.10 provides more cases of buckling forces for various structures and load types.

Natural Frequencies

The fundamental natural frequencies are given for arches in Table 16-8. For rings, Table 16-9 provides natural frequencies and mode shapes. Table 16-10 gives frequency information for various structures with curved members. In all of these tables, both in-plane and out-of-plane motion are considered.

16.3   OUT-OF-PLANE STRESS AND DEFORMATION

Sign Convention

Positive rotations ϕ, θ_z, displacement v, moments T, M_z, force V_y, and applied loads p_y, m_x are shown in Fig. 16-11. The formulas of this section provide responses for the loading illustrated. Loadings applied in the opposite direction require the sign of the loading to be reversed in the formulas.

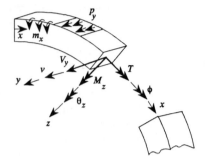

FIGURE 16-11 Positive rotations, displacements, internal moments and forces, and applied loadings for out-of-plane motion.

Stresses

The tables of this chapter provide the internal moments and force along a bar. Once these variables are known, the bending and direct shear stresses can be calculated using the stress formulas in Chapter 11. The torsional shear stress is found from the formulas in Chapter 12.

Simple Curved Bars

The fundamental equations of motion in first-order form for out-of-plane deformation of a circular bar are

$$\frac{\partial \phi}{\partial x} = \frac{T}{GJ} - \frac{\theta_z}{R}$$

$$\frac{\partial v}{\partial x} = \theta_z + \frac{V_y}{GA_s}$$

$$\frac{\partial \theta_z}{\partial x} = \frac{\phi}{R} + \frac{M_z}{EI_z} + \frac{M_{Tz}}{EI_z}$$

$$\frac{\partial M_z}{\partial x} = -V_y + k^*\theta_z - \frac{T}{R} + \rho r_z^2 \frac{\partial^2 \theta_z}{\partial t^2} - c_z(x,t) \qquad (16.12)$$

$$\frac{\partial V_y}{\partial x} = kv + \rho \frac{\partial^2 v}{\partial t^2} - p_y(x,t)$$

$$\frac{\partial T}{\partial x} = \frac{M_z}{R} + k_t\phi + \rho r_p^2 \frac{\partial^2 \phi}{\partial t^2} - m_x(x,t)$$

These relations conform to sign convention 1 of Appendix II.

Tabulated Formulas The internal forces and displacements at the tips of some simple curved bars are given in Table 16-11.

Formulas for Bars with Arbitrary Loading If Table 16-11 does not provide sufficient information, then use Table 16-12, which gives the rotation, deflection, moments, and shear force responses for uniform bars under more general applied loading with any end conditions.

Table 16-12, part A, lists equations for the responses. The functions $F_\phi, F_v, F_{\theta_z}, F_{M_z}, F_{V_y}, F_T$ are taken from Table 16-12, part B, by adding appropriate terms for each load applied to the bar. The initial parameters $\phi_0, v_0, \theta_{z_0}, M_{z_0}, V_{y_0}, T_0$, which are values of $\phi, v, \theta_z, M_z, V_y, T$ at the left end ($x = 0$) of the bar, are evaluated for the end conditions shown in Table 16-12, part C, using the procedure outlined in Appendix III.

Buckling Loads

See Table 16-7 for the buckling loads of out-of-plane modes of some curved bars.

Natural Frequencies

Tables 16-8 to 16-10 give natural frequencies and mode shapes for various configurations of curved bars with out-of-plane motion.

16.4 GENERAL BARS

Most of the formulas provided thus far apply to single-span, uniform bars. For more general bars, it is advisable to use the displacement method or the transfer matrix procedure, which are explained technically at the end of this book (Appendices II and III).

Several transfer and stiffness matrices are tabulated in Tables 16-13 to 16-16. Mass matrices for use in a displacement method analysis are given in Table 16-17.

Frameworks containing curved elements are handled by using the stiffness matrices of this chapter, as appropriate, in conjunction with the stiffness matrices of the frame chapter (Chapter 13) and following the displacement method of analysis.

Rings

Rings are structural members that connect to themselves, a characteristic that requires special techniques to be employed in handling the boundaries. For the transfer matrix method, the initial parameters at a chosen point (say $x = 0$) are equal to the state variables at the same point after moving around the loop (say $x = L$). That is, for the transfer matrix method, using extended matrices including applied loading terms,

$$\mathbf{z}_0 = \mathbf{z}_{x=L} \quad \text{or} \quad \mathbf{z}_0 = \mathbf{U}\mathbf{z}_0$$

This gives

$$(\mathbf{U} - \mathbf{I})\mathbf{z}_0 = 0 \tag{16.13}$$

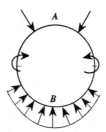

FIGURE 16-12 Ring.

where $\mathbf{I}$ is a 7×7 diagonal matrix. This expression can then be solved for $\mathbf{z}_0$. In most cases for a ring, the applied loadings are symmetric about some axis so that only half of the ring needs to be analyzed. Some of the variables at the ends of the half ring are known by inspection. For example, for the in-plane deformation of the ring of Fig. 16-12,

$$u_A = \theta_A = 0 \qquad u_B = \theta_B = 0$$

Consider $x = 0$ to be at A and $x = L$ to be at B. For this symmetric, in-plane deformation the remaining boundary conditions are $V_A = V_B = 0$. Then the responses of the ring can be obtained by applying these end conditions to the equations of Table 16-13, part A.

It may be desirable to model a whole ring as a single element. If the displacement method is used for an analysis, the stiffness equations can be expressed as (Fig. 16-13)

$$\begin{bmatrix} \mathbf{k}_{aa} & \mathbf{k}_{ab} \\ \mathbf{k}_{ba} & \mathbf{k}_{bb} \end{bmatrix} \begin{bmatrix} \mathbf{v}_a \\ \mathbf{v}_b \end{bmatrix} = \begin{bmatrix} \bar{\mathbf{p}}_a \\ \bar{\mathbf{p}}_b \end{bmatrix} + \begin{bmatrix} \mathbf{p}_a \\ \mathbf{p}_b \end{bmatrix} \tag{16.14}$$

where $\bar{\mathbf{p}}$ and $\mathbf{p}$ are the loading vectors due to the applied loads and internal loads, respectively. Since the ring is in equilibrium, $\mathbf{v}_a = \mathbf{v}_b$ and $\mathbf{p}_a = -\mathbf{p}_b$, Eq. (16.14) can be rearranged as

$$(\mathbf{k}_{aa} + \mathbf{k}_{ab} + \mathbf{k}_{ba} + \mathbf{k}_{bb})\mathbf{v}_a = \bar{\mathbf{p}}_a + \bar{\mathbf{p}}_b \tag{16.15}$$

to find the displacement $\mathbf{v}_a$. If more elements are used, the conventional displacement method should be employed.

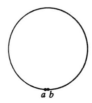

FIGURE 16-13 One-element model for analysis of ring.

In the case of a closed-loop bar the eigenvalues are found with the transfer matrix method as the roots of the determinant,

$$|\mathbf{U} - \mathbf{I}| = 0 \qquad (16.16)$$

When the displacement method is used to find the natural frequencies of a ring formed of a single element, the first three natural frequencies can be found from

$$\mathbf{K}' - \omega^2 \mathbf{M}' = 0 \qquad (16.17)$$

where, from Eq. (16.15), $\mathbf{K}' = \mathbf{k}_{aa} + \mathbf{k}_{ab} + \mathbf{k}_{ba} + \mathbf{k}_{bb}$ and similarly $\mathbf{M}' = \mathbf{m}_{aa} + \mathbf{m}_{ab} + \mathbf{m}_{ba} + \mathbf{m}_{bb}$ where $\mathbf{k}_{ij}$ and $\mathbf{m}_{ij}$ are given in Tables 16-14 and 16-17 with $b = a$. Only three natural frequencies are computed because the number of degrees of freedom of Eq. (16.17) is 3. If more natural frequencies are required, the ring should be divided into elements and a standard eigensolution solver should be used to extract the eigenfunctions from

$$(\mathbf{K} - \omega^2 \mathbf{M})\mathbf{V} = 0 \qquad (16.18)$$

where $\mathbf{K}$ and $\mathbf{M}$ are the assembled mass and stiffness matrices.

REFERENCES

16.1 Cook, R. D., and Young, W. C., *Advanced Mechanics of Materials*, Macmillan, New York, 1985.

16.2 Cook, R. D., "Circumferential Stresses in Curved Beams," *J. Appl. Mechan.*, Vol. 59, No. 1, 1992, pp. 224–225.

16.3 Boresi, A. P., and Sidebottom, O. M., *Advanced Mechanics of Materials*, 4th ed., Wiley, New York, 1985.

16.4 Young, W. C., and Cook, R. D., "Radial Stress Formula for Curved Beams," *ASME J. Vibrat. Acoust. Stress Reliabil. Design*, Vol. 111, No. 4, 1989, pp. 491–492.

16.5 Wang, T. S., "Shear Stresses in Curved Beams," *Machine Design*, Vol. 39, No. 28, 1967, pp. 175–178.

16.6 Broughton, D. C., Clark, M. E., and Corten, H. T., "Tests and Theory of Elastic Stresses in Curved Beams Having I- and T-Sections," *Proceedings of the Society of Experimental Stress Analysis*, Vol. 8, No. 1, 1950.

16.7 Oden, J. T., and Ripperger, E. A., *Mechanics of Elastic Structures*, 2nd ed., McGraw-Hill, New York, 1981.

16.8 Hopkins, R. B., *Design Analysis of Shafts and Beams*, McGraw-Hill, New York, 1970.

16.9 Griffel, W., *Handbook of Formulas for Stress and Strain*, Frederick Ungar, New York, 1966.

16.10 Column Research Committee of Japan, Ed., *Handbook of Structural Stability*, Corona, Tokyo, 1971.

16.11 Blevins, R. D., *Formulas for Natural Frequency and Mode Shape*, Van Nostrand Reinhold, New York, 1979.

16.12 Lee, L. S. S., "Vibrations of an Intermediately Supported U-Bend Tube," *ASME J. Eng. Ind.*, Vol. 97, 1975, pp. 23–32.

16.13 Davis, R., Henshell, R. D., and Warburton, G. B., "Constant Curvature Finite Element for In-Plane Vibration," *J. Sound Vibrat.*, Vol. 25, 1972, pp. 561–576.

16.14 Davis, R., Henshell, R. D., and Warburton, G. B., "Curved Beam Finite Elements for Coupled Bending and Torsional Vibration," *Earthquake Eng. Struct. Dynam.*, Vol. 1, 1972, pp. 165–175.

16.15 Timoshenko, S., *Theory of Elastic Stability*, 2nd Ed., McGraw-Hill, N.Y., 1961.

16.16 Huang, N. C. and Vehidi, G., "Dynamic Snap-Through of an Elastic Imperfect Simple Shallow Truss," *ZAMP*, Vol. 19, 1968, pp. 501–509.

16.17 Cheney, J. A., "Bending and Buckling of Thin Walled Open-Section Rings," *Proc. ASCE (EM)*, Vol. 89, EM5, 1963, pp. 17–44.

16.18 Ratzersdorfer, J., "Über die Stabilität des Kreisringes in Seiner Ebene," *Z.d. Österreichischen Ingenieur u. Architekten Vereines*, 1938, pp. 141–148.

16

Tables

TABLE 16-1   SOME GEOMETRIC PROPERTIES OF CROSS SECTIONS

Notation

A = area of cross section

R = radius of curvature of centroidal axis of cross section; values given are for circular bars, where R is constant, and can be considered to be reasonable approximations for many noncircular bars, where bar is modeled as being formed of short circular segments

$A_m = \int_A (1/r)\,dA$

$e = R - A/A_m$ = shift of neutral axis from centroidal axis

$\sigma_x = Mz/(Aer)$ = normal stress on cross section [Eq. (16.2)]

z = distance from neutral axis to point of interest on cross section

r = distance from center of curvature to point of interest

k_i, k_o = factors used to multiply straight beam stress formulas to calculate stress in extreme fibers of curved bars; $\sigma_i = k_i(P/A + Mz_i/I)$, $\sigma_o = k_o(P/A - Mz_o/I)$, where i and o indicate extreme fibers on inner and outer sides

z_i = distance from centroidal axis to extreme inner fiber

z_o = distance from centroidal axis to extreme outer fiber

P = tensile axial force applied at centroid; replace P by $-P$ if axial force is compressive

M = bending moment about y axis

I = moment of inertia about y centroidal axis

Case	A	A_m	R
1. Rectangle	$b(c - a)$	$b \ln \dfrac{c}{a}$	$\frac{1}{2}(a + c)$
2. Triangle	$\frac{1}{2}b(c - a)$	$\dfrac{bc}{c - a} \ln \dfrac{c}{a} - b$	$\frac{1}{3}(2a + c)$
3. Trapezoid	$\frac{1}{2}(b_1 + b_2)(c - a)$	$\dfrac{b_1 c - b_2 a}{c - a} \ln \dfrac{c}{a} - b_1 + b_2$	$[a(2b_1 + b_2) + c(b_1 + 2b_2)]/3(b_1 + b_2)$

TABLE 16-1   Some Geometric Properties of Cross Sections   798

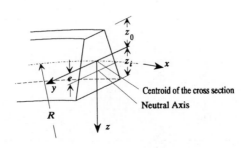

Centroid of the cross section

Neutral Axis

k_i	k_o
$\dfrac{c-a}{6e}\dfrac{c-a-2e}{2R-c+a}$	$\dfrac{c-a}{6e}\dfrac{c-a+2e}{2R+c-a}$
$\dfrac{c-a}{6e}\dfrac{c-a-3e}{3R-c+a}$	$\dfrac{c-a}{12e}\dfrac{c-a+3e}{3R+c-a}$
$\dfrac{(c-a)(b_1+2b_2)}{6e(b_1+b_2)}$ $\times\dfrac{(c-a)(b_1+2b_2)-3e(b_1+b_2)}{3R(b_1+b_2)-(c-a)(b_1+2b_2)}$ $\times\dfrac{1+4b_2/b_1+(b_2/b_1)}{(1+2b_2/b_1)^2}$	$\dfrac{(c-a)(2b_1+b_2)}{6e(b_1+b_2)}$ $\times\dfrac{(c-a)(2b_1+b_2)+3e(b_1+b_2)}{3R(b_1+b_2)+(c-a)(2b_1+b_2)}$ $\times\dfrac{1+4b_2/b_1+(b_2/b_1)^2}{(2+b_2/b_1)^2}$

TABLE 16-1 (continued)   SOME GEOMETRIC PROPERTIES OF CROSS SECTIONS

Case	A	A_m
4. Circle	πb^2	$2\pi(R - \sqrt{R^2 - b^2})$
5. Ellipse	πbh	$\dfrac{2\pi b}{h}(R - \sqrt{R^2 - h^2})$
6. Hollow circle	$\pi(b_1^2 - b_2^2)$	$2\pi\left(\sqrt{R^2 - b_2^2} - \sqrt{R^2 - b_1^2}\right)$
7. Hollow ellipse	$\pi(b_1 h_1 - b_2 h_2)$	$2\pi\left(\dfrac{b_1 R}{h_1} - \dfrac{b_2 R}{h_2} - \dfrac{b_1}{h_1}\sqrt{R^2 - h_1^2} + \dfrac{b_2}{h_2}\sqrt{R^2 - h_2^2}\right)$
8. Portion of circle	$b^2\theta - \frac{1}{2}b^2 \sin 2\theta$	For $a > b$ $2a\theta - 2b\sin\theta - \pi\sqrt{a^2 - b^2} + 2\sqrt{a^2 - b^2} \times \sin^{-1}\dfrac{b + a\cos\theta}{a + b\cos\theta}$ For $b > a$ $2a\theta - 2b\sin\theta + 2\sqrt{b^2 - a^2} \times \ln\left[\dfrac{b + a\cos\theta}{a + b\cos\theta} + \dfrac{\sqrt{b^2 - a^2}\,\sin\theta}{a + b\cos\theta}\right]$

TABLE 16-1   Some Geometric Properties of Cross Sections   800

R	k_i	k_o
As shown in figure	$\dfrac{b}{4e}\dfrac{b-e}{R-b}$	$\dfrac{b}{4e}\dfrac{e+b}{R+b}$
As shown in figure	$\dfrac{h}{4e}\dfrac{h-e}{R-h}$	$\dfrac{h}{4e}\dfrac{e+h}{R+h}$
As shown in figure	$\dfrac{b_1}{4e}\dfrac{b_1-e}{R-b_1}$ $\times\left[1+(b_2/b_1)^2\right]$	$\dfrac{b_1}{4e}\dfrac{b_1+e}{R+b_1}$ $\times\left[1+(b_2/b_1)^2\right]$
As shown in figure	$\dfrac{h_1}{4e}\dfrac{h_1-e}{R-h_1}$ $\times\dfrac{1-(b_2/b_1)(h_2/h_1)^3}{1-(b_2/b_1)(h_2/h_1)}$	$\dfrac{h_1}{4e}\dfrac{h_1+e}{R+h_1}$ $\times\dfrac{1-(b_2/b_1)(h_2/h_1)^3}{1-(b_2/b_1)(h_2/h_1)}$
$a+\dfrac{4b\sin^3\theta}{3(2\theta-\sin 2\theta)}$	$\dfrac{I}{Ade}\dfrac{d-e}{R-d}$ $d=\begin{cases} b\left(\dfrac{2\sin^3\theta}{3(\theta-\sin\theta\cos\theta)}-\cos\theta\right)\\ \qquad\qquad \theta\le\dfrac{\pi}{4}\\ 0.2R\theta^2(1-0.0619\theta^2+0.0027\theta^4)\\ \qquad\qquad \theta>\dfrac{\pi}{4}\end{cases}$	$\dfrac{I}{Ae}\dfrac{e+a+b-R}{(a+b-R)(a+b)}$

801   **Some Geometric Properties of Cross Sections**   TABLE 16-1

TABLE 16-1 (continued)   SOME GEOMETRIC PROPERTIES OF CROSS SECTIONS

Case	A	A_m	R
9. Portion of circle 	$b^2\theta - \tfrac{1}{2}b^2\sin 2\theta$	$2a\theta + 2b\sin\theta - \pi\sqrt{a^2 - b^2}$ $-2\sqrt{a^2 - b^2}$ $\times \sin^{-1}\dfrac{b - a\cos\theta}{a - b\cos\theta}$	$a - \dfrac{4b\sin^3\theta}{3(2\theta - \sin 2\theta)}$
10. Solid semicircle or semiellipse 	$\tfrac{1}{2}\pi bh$	$2b + \dfrac{\pi b}{h}\left(a - \sqrt{a^2 - h^2}\right)$ $-\dfrac{2b}{h}\sqrt{a^2 - h^2}\,\sin^{-1}\left(\dfrac{h}{a}\right)$	$a - \dfrac{4h}{3\pi}$

TABLE 16-1 **Some Geometric Properties of Cross Sections**   **802**

k_i	k_o
$\dfrac{I(R - a + b - e)}{Ae(R - a + b)(a - b)}$	$\dfrac{I(a - b\cos\theta - R - e)}{Ae(a - b\cos\theta - R)(a - b\cos\theta)}$
$\dfrac{0.2109d}{e}\quad\dfrac{d - e}{R - d}$ $d = \dfrac{h(3\pi - 4)}{3\pi}$	$\dfrac{0.2860d}{e}\quad\dfrac{e + 0.7374d}{R + 0.7374d}$ $d = \dfrac{h(3\pi - 4)}{3\pi}$

TABLE 16-2 SUMMARY OF IN-PLANE LOADED CURVED BAR STRESSES

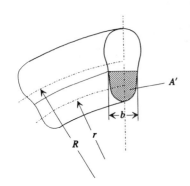

Notation

R = radius of curvature of centroidal axis
A = cross-sectional area
e = shift of neutral axis from centroidal axis
  $= R - r_n = R - A/A_m$
r_n = radius of curvature of neutral axis along the bar
M = bending moment about y axis
V = shear force
P = tensile axial force applied at centroid.
  Replace P by $-P$ if axial force is compressive

$$A_m = \int_A \frac{dA}{r} \qquad A'_m = \int_{A'} \frac{dA}{r} \qquad Q = \int_{A'} r\, dA$$

b = width of cross section where stresses
  are calculated
A' = area of part of cross section below b
I = moment of inertia about centroidal y axis
z = distance in z direction from centroid
  to point where stresses are calculated

Circumferential Stress σ_x

General $(e = r - r_n = R - A/A_m)$	Slender Beams $[e = I/(RA)]$	With Thin Flanges
1. Most accurate $$\frac{P}{A}\left[\frac{r_n}{r} + \frac{Ae}{I}(r - R)\right]$$ $$+ \frac{M(r_n - r)}{Aer}$$ Ref. 16.2	**1.** Most accurate $$\frac{P}{A}\left(\frac{r_n}{r} + \frac{r}{R} - 1\right)$$ $$+ \frac{M(r_n - r)}{Aer}$$ Ref. 16.2	Calculate modified flange width using Eq. (16.6) and then calculate stress using formulas to left
2. More accurate than traditional but less accurate and easier to apply than case 1 $$\frac{Pr_n}{Ar} + \frac{M(r_n - r)}{Aer}$$ Ref. 16.2	**2.** Traditional formula $$\frac{P}{A} + \frac{M(r_n - r)}{Aer}$$	
3. Traditional formula $$\frac{P}{A} + \frac{M(r_n - r)}{Aer}$$		

TABLE 16-2 | **Curved Bar Stresses**   **804**

TABLE 16-2 (continued) SUMMARY OF IN-PLANE LOADED CURVED BAR STRESSES

Radial Stress σ_z

1.
Most accurate

$$\frac{r_n}{Aebr}\left[(M - PR)\left(A'_m - \frac{A'}{r_n}\right)\right] + \frac{P}{r}(RA' - Q)$$

Ref. 16.4

2.
Traditional formula

$$\frac{1}{br}\left[\frac{A'}{A}P + \frac{AA'_m - A'A_m}{A(RA_m - A)}M\right]$$

Shear Stress τ

1.

$$\frac{Vr_nQ}{bAe(R - z)^2}$$

TABLE 16-3 BLEICH'S CORRECTION FACTORS α AND β FOR CALCULATING EFFECTIVE WIDTH AND LATERAL BENDING STRESS OF I- OR T-SECTIONS

Notation

b_{pi}, t_{fi}, and $\bar{r}_i$ are shown in Fig. 16-8 [16.3].

$f = b_{pi}^2 / (\bar{r}_i t_{fi})$.

$\alpha = C_0 + C_1 f + C_2 f^2 + C_3 f^3 + C_4 f^4 + C_5 f^5$

$\beta = D_0 + D_1 f + D_2 f^2 + D_3 f^3 + D_4 f^4 + D_5 f^5$

where C_j and D_j are taken from

j	C_j	D_j
0	1.0934663695	-0.10845356
1	-0.5142515924	4.0551477436
2	0.1284598483	-3.306290153
3	-0.0112111533	1.2434264877
4	0	-0.219079535
5	0	0.0146197753

TABLE 16-3   Bleich's Correction Factors   806

TABLE 16-4   IN-PLANE RESPONSE OF UNIFORM CIRCULAR BARS

Notation

u, w = displacements in x and z directions
θ = slope about y axis
u_X, u_Z = displacements in X and Z directions
E = modulus of elasticity
M = bending moment
V = shear force
I = moment of inertia about the centroidal y axis
P = axial force R = radius of bar

Case	Moments and Forces at Angle α	Displacements at Angle α
		$(u = u_X \sin \alpha + u_Z \cos \alpha,$
		$w = -u_X \cos \alpha + u_Z \sin \alpha)$
1. Radial load	$M = WR \sin \alpha$ $P = W \sin \alpha$ $V = -W \cos \alpha$	$u_Z = -\dfrac{WR^3}{4EI}(\cos 2\psi - \cos 2\alpha + 4\cos^2 \alpha - 4\cos \psi \cos \alpha)$ $u_X = \dfrac{WR^3}{4EI}\big[(2(\psi - \alpha) + \sin 2\alpha - \sin 2\psi - 4(\cos \alpha - \cos \psi)\sin \alpha\big]$ At $\alpha = 0$ $u = -\dfrac{WR^3}{4EI}(\cos 2\psi - 4\cos \psi + 3)$ $w = -\dfrac{WR^3}{4EI}(2\psi - \sin 2\psi)$ $\theta = \dfrac{WR^2}{EI}(1 - \cos \psi)$

TABLE 16-4 (continued)   IN-PLANE RESPONSE OF UNIFORM CIRCULAR BARS

Case	Moments and Forces at Angle α	Displacements at Angle α
2. Tangential load	$M = -WR(1 - \cos\alpha)$ $P = W\cos\alpha$ $V = W\sin\alpha$	$\left(u = u_X \sin\alpha + u_Z \cos\alpha,\right.$ $\left.w = -u_X \cos\alpha + u_Z \sin\alpha\right)$ $u_Z = \dfrac{WR^3}{4EI}\left[4(\psi - \alpha)\cos\alpha - 4(1 + \cos\alpha)(\sin\psi - \sin\alpha)\right.$ $\left. + 2(\psi - \alpha) + \sin 2\psi - \sin 2\alpha\right]$ $u_X = -\dfrac{WR^3}{4EI}\left[4\sin\alpha\sin\psi + 4\cos\alpha - 4\cos\psi\right.$ $\left. -4(\psi - \alpha)\sin\alpha - 4\sin 2\alpha + \cos 2\psi - \cos 2\alpha\right]$ At $\alpha = 0$ $u = \dfrac{WR^3}{4EI}(6\psi + \sin 2\psi - 8\sin\psi)$ $w = \dfrac{WR^3}{4EI}(\cos 2\psi - 4\cos\psi + 3)$ $\theta = -\dfrac{WR^2}{EI}(\psi - \sin\psi)$

TABLE 16-4 **In-Plane Response of Uniform Circular Bars**

| 3.
End moment
 | $M = M_0$ | $u_z = -\dfrac{M^*R^2}{EI}[(\psi - \alpha)\cos\alpha - \sin\psi + \sin\alpha]$

$u_x = \dfrac{M^*R^2}{EI}(\cos\alpha + \alpha\sin\alpha - \cos\psi - \psi\sin\alpha)$

At $\alpha = 0$

$u = -\dfrac{M^*R^2}{EI}(\psi - \sin\psi)$

$w = -\dfrac{M^*R^2}{EI}(1 - \cos\psi)$

$\theta = \dfrac{M^*R}{EI}\psi$ |
| 4.
Radial load
guided end
 | $M = WR\sin\alpha - M_0$

$P = W\sin\alpha + \dfrac{M_0\cos\alpha}{R(1 - \cos\alpha)}$

$V = -W\sin\alpha + \dfrac{M_0\sin\alpha}{R(1 - \cos\alpha)}$

where

$M_0 = \dfrac{WR}{4}\dfrac{\cos 2\psi - 4\cos\psi + 3}{\psi - \sin\psi}$ | At $\alpha = 0$

$u = 0$

$w = -\dfrac{WR^3}{4EI}\left[2\psi - \sin 2\psi + \dfrac{\cos 2\psi - 4\cos\psi + 3}{\psi - \sin\psi}(\cos\psi - 1)\right]$

$\theta = 0$ |

TABLE 16-4 (continued)   IN-PLANE RESPONSE OF UNIFORM CIRCULAR BARS

Case	Moments and Forces at Angle α	Displacements at Angle α $(u = u_X \sin \alpha + u_Z \cos \alpha,$ $w = -u_X \cos \alpha + u_Z \sin \alpha)$
5. Eccentric tangential load	$M = -W[R(\cos \alpha - 1) - a]$ $P = W \cos \alpha$ $V = W \sin \theta$	At $\alpha = 0$ $u = \dfrac{WR^2}{4EI}[2\psi(3R + 2a) - 4(2R + a)\sin \psi + R \sin 2\psi]$ $w = \dfrac{WR^2}{4EI}[3R + 4a - 4(R + a)\cos \psi + R \cos 2\psi]$ $\theta = -\dfrac{WR}{EI}[\psi(R + a) - R \sin \psi]$
6. Uniform vertical load	$M = pR^2(\alpha \cos \alpha - \sin \alpha)$ $P = pR\alpha \cos \alpha$ $V = PR\alpha \sin \alpha$	$u_Z = \dfrac{pR^4}{8EI}[3(\cos 2\psi - \cos 2\alpha) + 16 \cos \alpha(\cos \alpha - \cos \psi)$ $\quad + 2(\psi^2 - \alpha^2) - 8\psi \sin \psi \cos \alpha$ $\quad + 2\psi \sin 2\psi + 2\alpha \sin 2\alpha]$ $u_X = -\dfrac{pR^4}{8EI}[4(\psi - \alpha) + 2\psi \cos 2\psi - 2\alpha \cos 2\alpha - 3 \sin 2\psi$ $\quad - 5 \sin 2\alpha + 8 \sin \alpha(2 \cos \psi + \psi \sin \psi - \alpha \sin \alpha)]$ At $\alpha = 0$ $u = \dfrac{pR^4}{8EI}(2\psi \sin 2\psi - 8\psi \sin \psi + 3 \cos 2\psi - 16 \cos \psi + 2\psi^2 + 13)$ $w = \dfrac{pR^4}{8EI}(4\psi + 2\psi \cos 2\psi - 3 \sin 2\psi)$

TABLE 16-4   In-Plane Response of Uniform Circular Bars   810

**7.
Uniform
radial load**

$M = -p_1 R^2 (1 - \cos \alpha)$
$P = p_1 R(\cos \alpha - 1)$
$V = p_1 R \sin \alpha$

$$w = \frac{p_1 R^4}{EI}\left[1 + \left(\tfrac{1}{2}\psi - \sin\psi - \tfrac{1}{4}\sin 2\psi - \tfrac{1}{2}\alpha\right)\sin\alpha - \left(\cos\psi + \tfrac{1}{2}\sin^2\psi\right)\cos\alpha\right]$$

If $\psi = 180°$

$$w = \frac{p_1 R^4}{EI}\left[1 + \tfrac{1}{2}(\pi - \alpha)\sin\alpha + \cos\alpha\right]$$

If $\psi = 180°$ and $\alpha = 90°$

$$w = 1.7854 \frac{p_1 R^4}{EI}$$

**8.
Split ring**

$M = WR(\cos\gamma - \cos\alpha)$
$P = -W\cos\alpha$
$V = -W\sin\alpha$

$$w = -\frac{WR^3}{EI}\left[\tfrac{1}{2}(\pi - \alpha)\sin\alpha + (1 + \cos\alpha)\cos\gamma\right]$$

At $\alpha = 90°$

$$w = \frac{WR^3}{EI}\left(\frac{\pi}{4} + \cos\gamma\right)$$

Total change in end gap:

$$\delta = \frac{WR^3}{EI}\left[(\pi - \gamma)(1 + 2\cos^2\gamma) + \tfrac{3}{2}\sin 2\gamma\right]$$

TABLE 16-4 (continued) IN-PLANE RESPONSE OF UNIFORM CIRCULAR BARS

Case	Moments and Forces at Angle α	Displacements at Angle α $(u = u_X \sin\alpha + u_Z \cos\alpha,$ $w = -u_X \cos\alpha + u_Z \sin\alpha)$
9. Vertical load	$M = WR(\cos\alpha - \cos\beta)$ $V = W\sin\alpha$ $P = W\cos\alpha$	At $\alpha = 0$ $u = K_u\left(\dfrac{WR^3}{EI}\right) \times 10^{-4}$ $w = K_w\left(\dfrac{WR^3}{EI}\right) \times 10^{-4}$

$$
K_u = \begin{cases}
-6.4935 - 0.517\psi - 0.3577\psi^2 + 0.018\psi^3 - 7.01 \times 10^{-5}\psi^4 & \beta = 0 \\[4pt]
-4.33 + 22.9\psi - 1.473\psi^2 + 0.02815\psi^3 - 9.82 \times 10^{-5}\psi^4 & \beta = 30° \\[4pt]
6.11\psi + 0.05082\psi^2 - 0.0177\psi^3 + 3.72 \times 10^{-4}\psi^4 - 2.229 \\
\quad \times 10^{-6}\psi^5 + 4.19 \times 10^{-9}\psi^6 & \beta = 60° \\[4pt]
-3.79 - 11.35\psi + 0.933\psi^2 - 0.02346\psi^3 + 2.163 \times 10^{-4}\psi^4 \\
\quad - 6.0 \times 10^{-7}\psi^5 & \beta = 90° \\[4pt]
31.66\psi - 2.366\psi^2 + 0.062886\psi^3 - 7.587 \times 10^{-4}\psi^4 + 4.201 \\
\quad \times 10^{-6}\psi^5 - 8.573 \times 10^{-9}\psi^6 & \beta = 120° \\[4pt]
-34{,}500 + 396.67\psi - 1.111\psi^2 & \beta = 150°
\end{cases}
$$

TABLE 16-4 In-Plane Response of Uniform Circular Bars 812

$$K_w = \begin{cases}
21.1 - 21.96\psi + 1.11\psi^2 - 0.027364\psi^3 + 3.36 \times 10^{-4}\psi^4 \\
\quad - 9.43 \times 10^{-7}\psi^5 & \beta = 0° \\[6pt]
1.62 + 10.6\psi - 0.304\psi^2 - 7.47 \times 10^{-3}\psi^3 + 1.976 \\
\quad \times 10^{-4}\psi^4 - 6.0 \times 10^{-7}\psi^5 & \beta = 30° \\[6pt]
15.15 - 19.0\psi + 0.896\psi^2 - 0.0172\psi^3 + 1.636 \\
\quad \times 10^{-4}\psi^4 - 3.43 \times 10^{-7}\psi^5 & \beta = 60° \\[6pt]
-8.117 + 1.63\psi + 0.247\psi^2 - 0.01126\psi^3 + 1.18 \\
\quad \times 10^{-4}\psi^4 - 2.572 \times 10^{-7}\psi^5 & \beta = 90° \\[6pt]
34.6 - 34.2\psi + 1.414\psi^2 - 0.0174\psi^3 + 6.55 \times 10^{-5}\psi^4 & \beta = 120° \\[6pt]
-1.077 + 23.167\psi - 1.563\psi^2 + 0.03664\psi^3 - 3.84 \times 10^{-4}\psi^4 \\
\quad + 1.83 \times 10^{-6}\psi^5 - 3.2 \times 10^{-9}\psi^6 & \beta = 150°
\end{cases}$$

$$\psi \leq 180°$$

TABLE 16-4 (continued)   **IN-PLANE RESPONSE OF UNIFORM CIRCULAR BARS**

Case	Moments and Forces at Angle α	Displacements at Angle α $(u = u_X \sin \alpha + u_Z \cos \alpha,$ $w = -u_X \cos \alpha + u_Z \sin \alpha)$
10. Horizontal load	$M = -WR(\sin \alpha - \sin \beta)$ $V = W \cos \alpha$ $P = -W \sin \alpha$	At $\alpha = 0$ $u = K_u \left(\dfrac{WR^3}{EI} \right) \times 10^{-4}$ $w = K_w \left(\dfrac{WR^3}{EI} \right) \times 10^{-4}$

$$K_u = \begin{cases}
32.46 - 48.2\psi + 2.19\psi^2 - 7.23 \times 10^{-3}\psi^3 - 4.2 \times 10^{-6}\psi^4 & \beta = 0° \\
\quad -5.195 + 29.07\psi - 2.078\psi^2 + 4.7517 \times 10^{-2}\psi^3 - 3.287 & \\
\quad \times 10^{-4}\psi^4 + 7.2 \times 10^{-7}\psi^5 & \beta = 30° \\
-31.472\psi + 2.592\psi^2 - 7.72 \times 10^{-2}\psi^3 + 1.0284 & \\
\quad \times 10^{-3}\psi^4 - 6.0271 \times 10^{-6}\psi^5 + 1.2669 \times 10^{-8}\psi^6 & \beta = 60° \\
-21.33\psi + 1.593\psi^2 - 4.228 \times 10^{-2}\psi^3 + 5.092 & \\
\quad \times 10^{-4}\psi^4 - 2.812 \times 10^{-6}\psi^5 + 5.716 \times 10^{-9}\psi^6 & \beta = 120° \\
-4.329 + 26.3\psi - 1.528\psi^2 + 2.871 \times 10^{-2}\psi^3 - 2.114 & \\
\quad \times 10^{-4}\psi^4 + 5.144 \times 10^{-7}\psi^5 & \beta = 90° \\
-0.39083 - 39.0514\psi + 3.15292\psi^2 - 9.186 \times 10^{-2}\psi^3 & \\
\quad + 1.20157 \times 10^{-3}\psi^4 - 6.96541 \times 10^{-6}\psi^5 + 1.45645 & \\
\quad \times 10^{-8}\psi^6 & \beta = 150°
\end{cases}$$

TABLE 16-4   **In-Plane Response of Uniform Circular Bars**   **814**

$$K_w = \begin{cases} -23.8 - 3.89\psi + 0.1389\psi^2 + 4.63 \times 10^{-3}\psi^3 + 5.144 \\ \qquad \times 10^{-5}\psi^4 - 3.429 \times 10^{-7}\psi^5 \qquad\qquad\qquad \beta = 0° \\[4pt] -13.528 - 18.674\psi + 1.16\psi^2 - 1.95 \times 10^{-2}\psi^3 + 1.88 \\ \qquad \times 10^{-4}\psi^4 - 6.0 \times 10^{-7}\psi^5 \qquad\qquad\qquad \beta = 30° \\[4pt] 4.33 + 13.065\psi - 0.795\psi^2 + 1.3 \times 10^{-2}\psi^3 - 5.61 \times 10^{-5}\psi^4 \quad \beta = 60° \\[4pt] -6.49 + 12.31\psi - 0.612\psi^2 + 9.3 \times 10^{-3}\psi^3 - 4.44 \times 10^{-5}\psi^4 \quad \beta = 90° \\[4pt] 1.08 + 10.16\psi - 0.558\psi^2 + 8.65 \times 10^{-3}\psi^3 - 3.975 \\ \qquad \times 10^{-5}\psi^4 \qquad\qquad\qquad\qquad\qquad\qquad \beta = 120° \\[4pt] 2.923 - 26.344\psi + 1.657\psi^2 - 3.66 \times 10^{-2}\psi^3 + 3.574 \\ \qquad \times 10^{-4}\psi^4 - 1.5385 \times 10^{-6}\psi^5 + 2.278 \times 10^{-9}\psi^6 \quad \beta = 150° \end{cases}$$

$$\psi \le 180°$$

11.
Uniform
longitudinal
load

$M = -p_{x1}R^2(\alpha - \sin\alpha)$
$P = p_{x1}R\sin\alpha$
$V = p_{x1}R(1 - \cos\alpha)$

At $\alpha = 0$

$$u = \frac{p_{x1}R^4}{EI}\left[\sin\psi + \tfrac{1}{4}\sin 2\psi - \tfrac{1}{2}\psi - \psi\cos\psi\right]$$

$$w = \frac{p_{x1}R^4}{EI}\left[\tfrac{1}{2}\psi^2 - \psi\sin\psi + \tfrac{1}{2}\sin\psi\right]$$

$$\theta = -\frac{p_{x1}R^3}{EI}\left(\tfrac{1}{2}\psi^2 + \cos\psi - 1\right)$$

TABLE 16-5 IN-PLANE RESPONSE OF UNIFORM CIRCULAR RINGS

Notation

δ_x, δ_z = change in horizontal and vertical diameters; increase is positive

ΔR = change in upper half of vertical diameter of ring, decrease is negative

E = modulus of elasticity

I = moment of inertia about the centroidal y axis

M = bending moment

P = axial tensile force

V = shear force

$M_1, V_1, P_1, M_2, V_2, P_2$ = moments, shear forces, and axial forces at bottom and top, respectively

R = radius of ring

ρ^* = mass of liquid per unit volume

e = shift of neutral axis from centroid

g = gravitational acceleration

Case	$\delta_x, \delta_z, \Delta R, M, P, V$
1. 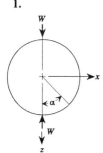	$\delta_x = 0.137WR^3/EI \qquad \delta_z = -0.149WR^3/EI$ $\max(+M) = 0.3183WR \qquad$ at bottom $\quad (\alpha = 0)$ $\max(-M) = -0.1817WR \quad$ at side $\quad \left(\alpha = \tfrac{1}{2}\pi\right)$ $P_1 = 0 \qquad V_1 = -\tfrac{1}{2}W$
2. 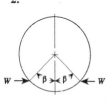	$\delta_x = (K_x WR^3/EI) \times 10^{-6} \quad \delta_z = (K_z WR^3/EI) \times 10^{-5}$ $\Delta R = (K_R WR^3/EI) \times 10^{-6} \quad M = K_M WR \times 10^{-5}$ $P = K_P W \times 10^{-4} \qquad\qquad V = K_V W \times 10^{-5}$ $K_x = -37.72 - 11.137\beta + 4.3146\beta^2 - 0.945\beta^3 + 0.007735\beta^4$ $K_z = 59.7 - 33.676\beta + 2.898\beta^2 + 0.030688\beta^3 - 0.00045\beta^4$ $K_R = -15.4174 + 9.0944\beta - 1.7172\beta^2 + 0.35\beta^3 - 0.00265\beta^4$ $K_M = 96.3 - 42.8\beta + 2.619\beta^2 + 0.1522\beta^3 - 0.0014852\beta^4$ $K_P = -9988.79 - 5.328\beta + 1.8837\beta^2 + 0.0064\beta^3 - 0.000145\beta^4$ $K_V = 155.337 - 94.085\beta + 8.89\beta^2 - 0.29125\beta^3 + 0.00149\beta^4$ $0 \le \beta < 90°$

TABLE 16-5 (continued) IN-PLANE RESPONSE OF UNIFORM CIRCULAR RINGS

Case	$\delta_x, \delta_z, \Delta R, M, P, V$
3. 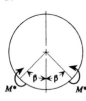	$\delta_x = (K_x M^* R^2 / EI) \times 10^{-5}$ $\delta_z = (K_z M^* R^2 / EI) \times 10^{-5}$ $\Delta R = (K_R M^* R^2 / EI) \times 10^{-6}$ $M = K_M M^* \times 10^{-5}$ $P = (K_P M^* / R) \times 10^{-5}$ $V = (K_V M^* / R) \times 10^{-6}$ $K_x = -9.65 - 628.15\beta - 0.55\beta^2 + 0.1068\beta^3 - 0.000255\beta^4$ $K_z = -10.3376 + 1115.77\beta - 15.62\beta^2 + 0.011\beta^3 + 0.000276\beta^4$ $K_R = 3.2423 + 2363.26\beta + 1.828\beta^2 - 0.3414\beta^3 + 0.000823\beta^4$ $K_M = -99979 + 1650.43\beta + 2.05254\beta^2 - 0.3165\beta^3 + 0.001760\beta^4$ $K_P = 19 + 1094\beta + 2.0747\beta^2 - 0.3168\beta^3 + 0.001763\beta^4$ $K_V = 1121.34 - 638.56\beta + 250.35\beta^2 - 1.699\beta^3 - 0.001563\beta^4$ $\qquad\qquad\qquad\qquad\qquad\qquad\qquad\qquad\qquad 0 \le \beta < 90°$
4.	$\delta_x = -0.1366 WR^3 / EI$ $\delta_z = 0.1488 WR^3 / EI$ $\Delta R = 0.0554 WR^3 / EI$ At $\alpha = 0$ to $\alpha = \frac{1}{2}\pi$ $M = K_M WR \times 10^{-5}$ $P = K_P W \times 10^{-5}$ $V = K_V W \times 10^{-5}$ $K_M = -49283.5 + 1490.52\alpha + 10.845\alpha^2 - 0.4216\alpha^3 + 0.002871\alpha^4$ $\qquad - 0.0000060366\alpha^5$ $0 \le \alpha < 180°$ $K_P = \begin{cases} 31835.1 + 1741.02\alpha - 4.4\alpha^2 - 0.103865\alpha^3 + 0.0003478\alpha^4 \\ \qquad\qquad\qquad\qquad\qquad\qquad 0 \le \alpha < 90° \\ 41025.3 - 180.548\alpha - 4.8662\alpha^2 + 0.02014\alpha^3 \\ \qquad\qquad\qquad\qquad\qquad\qquad 90° \le \alpha < 180° \end{cases}$ $K_V = \begin{cases} 99887.7 - 518.163\alpha - 17.57\alpha^2 - 0.078\alpha^3 \\ \qquad\qquad\qquad\qquad\qquad\qquad 0 \le \alpha < 90° \\ -29732.8 + 1500.6\alpha - 10.735\alpha^2 + 0.0184\alpha^3 \\ \qquad\qquad\qquad\qquad\qquad\qquad 90° \le \alpha < 180° \end{cases}$
5. 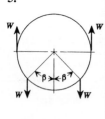	$\delta_x = (K_x WR^3 / EI) \times 10^{-5}$ $\delta_z = (K_z WR^3 / EI) \times 10^{-5}$ $\Delta R = (K_R WR^3 / EI) \times 10^{-5}$ At $\alpha = \beta$ $M = K_M WR \times 10^{-5}$ $P = K_P W \times 10^{-5}$ $V = -K_V W \times 10^{-5}$ $K_x = -13593.9 - 36.7665\beta + 8.83016\beta^2 - 0.1004\beta^3$ $\qquad + 0.0002787\beta^4$ $K_z = 14848.2 + 21.2347\beta - 11.9295\beta^2$ $\qquad + 0.17515\beta^3 - 0.000728\beta^4$ $K_R = 5525.18 + 12.155\beta - 3.169\beta^2 + 0.03308\beta^3 - 0.0000754\beta^4$ $K_M = -50172.6 + 1850.02\beta - 29.129\beta^2 + 0.22465\beta^3$ $\qquad - 0.000663\beta^4$ $K_P = 31657.9 + 103.53\beta - 24.26\beta^2 + 0.315\beta^3 - 0.00112\beta^4$ $K_V = -95.94 + 597.267\beta - 2.48\beta^2 - 0.1856\beta^3 + 0.00156\beta^4$ $\qquad\qquad\qquad\qquad\qquad\qquad\qquad\qquad\qquad 0 \le \beta < 90°$

TABLE 16-5 (continued)   IN-PLANE RESPONSE OF UNIFORM CIRCULAR RINGS

Case	$\delta_x, \delta_z, \Delta R, M, P, V$
6.	$\delta_x = (K_x WR^3/EI) \times 10^{-5}$ $\delta_z = (K_z WR^3/EI) \times 10^{-5}$ $\Delta R = (K_R WR^3/EI) \times 10^{-5}$ At $\alpha = \beta$ $M = K_M WR \times 10^{-5} \qquad P = K_P W \times 10^{-5}$ $V = K_V W \times 10^{-5}$ $K_x = -66.0127 + 36.7665\beta - 8.83\beta^2 + 0.100477\beta^3$ $\qquad -0.000279\beta^4$ $K_z = 31.75 - 21.2347\beta + 11.93\beta^2 - 0.175\beta^3 + 0.000728\beta^4$ $K_R = 43043 - 2068.6\beta + 41.97\beta^2 - 0.3328\beta^3$ $\qquad +0.0008599\beta^4$ $K_M = 180.54 - 108.476\beta + 24.7\beta^2 - 0.328\beta^3 + 0.001\beta^4$ $K_P = 179.02 + 1638.54\beta + 19.8\beta^2 - 0.41778\beta^3$ $\qquad +0.00146\beta^4$ $K_V = 99896.1 + 47.1116\beta - 18.235\beta^2 - 0.1413\beta^3$ $\qquad +0.001762\beta^4$ $\qquad\qquad\qquad\qquad\qquad\qquad\qquad 0 \le \beta < 90°$
7. 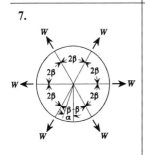	Radial displacement of each load point: $\left(K_s WR^3/2EI\right) \times 10^{-5}$ (outward) Radial displacement at $\alpha = 0, 2\beta, 4\beta, \ldots$: $\left(K_s' WR^3/4EI\right) \times 10^{-5}$ (inward) At $\alpha = 0, 2\beta, 4\beta, \ldots$: $\max(+M) = \frac{1}{2}K_M WR \times 10^{-5}$ At each load: $\max(-M) = -\frac{1}{2}K_M' WR \times 10^{-5}$ At $\alpha = 0, 2\beta, 4\beta, \ldots$: $P = \frac{1}{2}K_P W \times 10^{-4}$ At each load: $P = \frac{1}{2}K_P' W \times 10^{-5}$ $K_s = \begin{cases} -2112779.99 + 261138.9\beta - 14555.9\beta^2 + 421.84\beta^3 \\ \quad -6.5346\beta^4 + 0.0512583\beta^5 - 0.00015983\beta^6 \\ \qquad\qquad\qquad\qquad\qquad\qquad 0 \le \beta < 90° \\ -436680.99 + 13990.1\beta - 164.09\beta^2 + 0.8548\beta^3 \\ \quad -0.001548\beta^4 \qquad\qquad\qquad 90° \le \beta < 180° \end{cases}$ $K_s' = \begin{cases} 2112770 - 261425.99\beta + 14555.7\beta^2 - 421.749\beta^3 \\ \quad +6.53425\beta^4 - 0.0512493\beta^5 + 0.00015978\beta^6 \\ \qquad\qquad\qquad\qquad\qquad\qquad 0 \le \beta < 90° \\ -379655000 + 19386100\beta - 407977\beta^2 + 4529.78\beta^3 \\ \quad -27.9914\beta^4 + 0.0913129\beta^5 - 0.000122919\beta^6 \\ \qquad\qquad\qquad\qquad\qquad\qquad 90° \le \beta < 180° \end{cases}$

TABLE 16-5   **In-Plane Response of Uniform Circular Rings**   **818**

TABLE 16-5 (continued) IN-PLANE RESPONSE OF UNIFORM CIRCULAR RINGS

Case	$\delta_x, \delta_z, \Delta R, M, P, V$
7. Continued	$K_M = \begin{cases} 57.9248 + 278.769\beta + 0.732387\beta^2 - 0.00672\beta^3 \\ \quad + 0.000154986\beta^4 \qquad\qquad\qquad 0 \le \beta < 90° \\ 593102000 - 29371900\beta + 600602.9\beta^2 - 6491.33\beta^3 \\ \quad + 39.118\beta^4 - 0.124651\beta^5 + 0.000164149\beta^6 \\ \qquad\qquad\qquad\qquad\qquad\qquad 90° \le \beta < 180° \end{cases}$
	$K'_M = \begin{cases} -195.475 + 612.305\beta - 1.1217\beta^2 + 0.0246\beta^3 \\ \qquad\qquad\qquad\qquad\qquad\qquad 0 \le \beta < 90° \\ 593089000 - 29370999.9\beta + 600591\beta^2 - 6491.21\beta^3 \\ \quad + 39.1173\beta^4 - 0.124649\beta^5 + 0.000164147\beta^6 \\ \qquad\qquad\qquad\qquad\qquad\qquad 90 \le \beta < 180° \end{cases}$
	$K_P = \begin{cases} 211277 - 25909.8\beta + 1455.57\beta^2 - 42.1996\beta^3 \\ \quad + 0.65343\beta^4 - 0.00512423\beta^5 + 0.0000159778\beta^6 \\ \qquad\qquad\qquad\qquad\qquad\qquad 0 \le \beta < 90° \\ 14589499 - 735481.9\beta + 15343.6\beta^2 - 169.466\beta^3 \\ \quad + 1.04532\beta^4 - 0.00341539\beta^5 + 0.00000462055\beta^6 \\ \qquad\qquad\qquad\qquad\qquad\qquad 90° \le \beta < 170° \end{cases}$
	$K'_P = \begin{cases} 2112750 - 259966.9\beta + 14555.3\beta^2 - 421.999\beta^3 \\ \quad + 6.53388\beta^4 - 0.051238\beta^5 + 0.00015975\beta^6 \\ \qquad\qquad\qquad\qquad\qquad\qquad 0 \le \beta < 90° \\ -145215000 + 7337399.9\beta - 153198\beta^2 + 1692.69\beta^3 \\ \quad - 10.4435\beta^4 + 0.0341276\beta^5 - 0.0000461753\beta^6 \\ \qquad\qquad\qquad\qquad\qquad\qquad 90° \le \beta < 170° \end{cases}$
8. $2pR\sin\beta$	$\delta_x = (K_x 2pR^4/EI) \times 10^{-5} \qquad \delta_z = (K_z 2pR^4/EI) \times 10^{-5}$ $\Delta R = (K_R pR^4/EI) \times 10^{-5} \qquad M_1 = K_M pR^2 \times 10^{-5}$ at bottom At $\alpha = \beta$ $\quad P = -K_P pR \times 10^{-5} \qquad V = -K_V pR \times 10^{-5}$ $K_x = 31522.7 - 761.555\beta + 10.2634\beta^2 - 0.057365\beta^3 + 0.000124\beta^4$ $K_z = -26140.6 + 553.141\beta - 7.597\beta^2 + 0.04243\beta^3 - 0.00007107\beta^4$ $K_R = -25649.8 + 522.245\beta - 6.4062\beta^2 + 0.03126\beta^3$ $\qquad - 0.0000409\beta^4$ $K_M = 56051.4 - 686.78\beta + 18.98\beta^2 - 0.15\beta^3 + 0.0003169\beta^4$ $K_P = -363585.9 + 12617.9\beta - 107.08\beta^2 + 0.26787\beta^3$ $K_V = 151482 - 10368.6\beta + 174.6\beta^2 - 1.0304\beta^3 + 0.001969\beta^4$ $\qquad\qquad\qquad\qquad\qquad\qquad 90° \le \beta < 180°$

TABLE 16-5 (continued)   IN-PLANE RESPONSE OF UNIFORM CIRCULAR RINGS

Case	$\delta_x, \delta_z, \Delta R, M, P, V$

9.	$\delta_x = -(K_x pR^4/EI) \times 10^{-5}$ $\delta_z = -(K_z pR^4/EI) \times 10^{-5}$
	 $M_1 = K_M pR^2 \times 10^{-5}$ at bottom
	 At $\alpha = \beta$
	    $P = -K_P pR \times 10^{-5}$ $V = -K_V pR \times 10^{-5}$
	 $K_x = 52.42 - 501.15\beta + 1.66645\beta^2 + 0.06487\beta^3 - 0.0004992\beta^4$
	 $K_z = -82.634 + 565.013\beta - 3.8146\beta^2 - 0.0413\beta^3$
	       $+ 0.0004167\beta^4$
	 $K_M = -59.498 + 1149.25\beta - 18.9\beta^2 + 0.1281\beta^3 - 0.000281\beta^4$
	 $K_P = 178.8 - 100.5\beta + 39.367\beta^2 - 0.2678\beta^3 - 0.0002397\beta^4$
	 $K_V = 35.0613 + 1718.95\beta + 3.229\beta^2 - 0.497\beta^3 + 0.0027665\beta^4$
	                                                $0 \le \beta < 90°$

10.	$\delta_x = 0.1228 p_1 R^4/EI$ $\delta_z = -0.1220 p_1 R^4/EI$
	 At bottom ($\alpha = 0$)
	   $M_1 = 0.305 p_1 R^2$ $P_1 = -0.0265 p_1 R$ $V_1 = -0.5000 p_1 R$
	 At $\alpha = \pi/2$
	   $M = -0.165 p_1 R^2$ $P = -0.500 p_1 R$ $V = -0.0265 p_1 R$
	 At $\alpha = \pi$
	   $M = 0.1914 p_1 R^2$ $P = 0.0265 p_1 R$ $V = 0$

11.	$M_1 = K_M pR^2 \times 10^{-5}$ at bottom
	   $P = K_P pR \times 10^{-5}$ at $\alpha = \beta$ $V = K_V pR \times 10^{-5}$ at $\alpha = \beta$
	 $K_M = -162.73 + 21.46\beta + 0.97334\beta^2 - 0.09627\beta^3$
	       $+ 0.000879\beta^4 - 0.00000223957\beta^5$
	 $K_P = -96.97 + 116.72\beta - 11.45\beta^2 + 0.34\beta^3 - 0.0034138\beta^4$
	       $+ 0.00000982\beta^5$
	 $K_V = 154.534 - 118.746\beta + 12.831\beta^2 - 0.4956\beta^3$
	       $+ 0.0082537\beta^4 - 0.0000527838\beta^5 + 0.00000011193\beta^6$
	                                                $0 \le \beta < 180°$

12.	$\delta_x = (K_x 2p_1 R^4/EI) \times 10^{-5}$ $\delta_z = (K_z 2p_1 R^4/EI) \times 10^{-5}$
	 $M_1 = K_M p_1 R^2 \times 10^{-5}$ at bottom
	 At $\alpha = \beta$
	    $P = K_P p_1 R \times 10^{-5}$ $V = K_V p_1 R \times 10^{-5}$
	 $K_x = 3.943 - 2.6611\beta + 0.275\beta^2 - 0.026175\beta^3$
	       $+ 0.0001559\beta^4$
	 $K_z = 23.748 - 10.215\beta + 0.702\beta^2 + 0.013\beta^3 - 0.0001038\beta^4$
	 $K_M = 32.28 - 5.0077\beta - 15.3897\beta^2 + 0.10335\beta^3$
	 $K_P = -21.797 + 8.691\beta - 15.9\beta^2 + 0.08046\beta^3 + 0.000021\beta^4$
	 $K_V = 24.8136 - 21.14\beta + 2.6929\beta^2 - 0.39\beta^3 + 0.0025\beta^4$
	                                                $0 \le \beta < 90°$

TABLE 16-5   **In-Plane Response of Uniform Circular Rings**   **820**

TABLE 16-5 (continued) IN-PLANE RESPONSE OF UNIFORM CIRCULAR RINGS

Case	$\delta_x, \delta_z, \Delta R, M, P, V$
13. 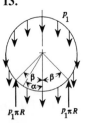 Ring symmetrically supported and loaded by its own weight p_1 (force per unit length)	$\delta_x = \left(K_x 2p_1 R^4/EI \right) \times 10^{-5} \qquad \delta_z = \left(K_z p_1 R^4/EI \right) \times 10^{-5}$ $M_1 = K_M p_1 R^2 \times 10^{-4}$ at bottom At $\alpha = \beta$ $P = K_P p_1 R \times 10^{-5} \qquad V = K_V p_1 R \times 10^{-5}$ $K_x = 22005.2 + 18.3033\beta - 15.9884\beta^2 + 0.2397\beta^3 - 0.00104\beta^4$ $K_z = -46067.4 - 292.218\beta + 52.942\beta^2 - 0.93\beta^3 + 0.006129\beta^4$ $\qquad - 0.00001366\beta^5$ $K_M = 15435.3 - 655.64\beta + 9.7253\beta^2 - 0.06153\beta^3 + 0.0001345\beta^4$ $K_P = -47188.8 - 1164.71\beta + 152.709\beta^2 - 2.1833\beta^3$ $\qquad + 0.012524\beta^4 - 0.0000267856\beta^5$ $K_V = 809.224 + 2108.65\beta + 51.143\beta^2 - 2.7738\beta^3 + 0.033555\beta^4$ $\qquad - 0.000169\beta^5 + 3.1099 \times 10^{-7}\beta^6$ $\qquad\qquad\qquad\qquad\qquad\qquad\qquad\qquad 0 \le \beta < 180°$
14. 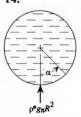 Unit length of pipe filled with liquid of mass density ρ^* and supported at base	$\delta_x = 0.2146 g\rho^* R^5/EI \qquad \delta_z = -0.2337 g\rho^* R^5/EI$ $\Delta R = -0.0938 g\rho^* R^5/EI$ $\max(+M) = M_1 = 0.750 g\rho^* R^3/EI$ at bottom $\max(-M) = 0.321 g\rho^* R^3$ at $\alpha = 75°$ $M = K_M g\rho^* R^3 \times 10^{-5} \qquad P = K_P g\rho^* R^2 \times 10^{-5}$ $V = K_V g\rho^* R^2 \times 10^{-5}$ $K_M = \begin{cases} 12318\alpha & \alpha < 5° \\ 77375.6 - 3059.47\alpha + 22.126\alpha^2 + 0.0112\alpha^3 \\ \quad - 0.000273\alpha^4 & 5° \le \alpha \le 180° \end{cases}$ $K_P = 126172 - 2978.5\alpha + 20.53\alpha^2 + 0.02317\alpha^3 - 0.0003033\alpha^4$ $K_V = -156315.9 + 1123.8\alpha + 32.9756\alpha^2 - 0.322\alpha^3$ $\qquad + 0.0007267\alpha^4$ $\qquad\qquad\qquad\qquad\qquad\qquad\qquad\qquad 0 \le \alpha < 180°$

TABLE 16-5 (continued)   IN-PLANE RESPONSE OF UNIFORM CIRCULAR RINGS

Case	$\delta_x, \delta_z, \Delta R, M, P, V$

15. 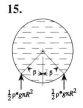
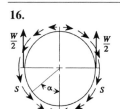
Unit length of pipe filled with liquid of mass density ρ^* and supported symmetrically at two locations
	$\delta_x = \left(K_x g\rho^* R^5/EI\right) \times 10^{-5}$ $\delta_z = -\left(K_z g\rho^* R^5/EI\right) \times 10^{-5}$
	 $\Delta R = \left(K_R g\rho^* R^5/(2EI)\right) \times 10^{-5}$
	 $M_1 = K_M g\rho^* R^3 \times 10^{-5}$
	 At $\alpha = \beta$
	 $P = K_P g\rho^* R^2 \times 10^{-5}$ $V = K_V g\rho^* R^2 \times 10^{-5}$
	 $K_x = 19949.5 + 466.516\beta - 35.2866\beta^2 + 0.54487\beta^3 - 0.003048\beta^4$
	    $+ 0.00000466\beta^5$
	 $K_z = 23085.7 + 133.53\beta - 25.868\beta^2 + 0.454537\beta^3 - 0.00299\beta^4$
	    $+ 0.000006647\beta^5$
	 $K_R = -19293.2 + 9.0767\beta + 10.897\beta^2 - 0.156025\beta^3$
	    $+ 0.00064066\beta^4$
	 $K_M = 77219.7 - 3285.01\beta + 48.839\beta^2 - 0.3098\beta^3 + 0.0006795\beta^4$
	 $K_P = 126224.9 - 537.37\beta + 66.636\beta^2 - 1.0577\beta^3 + 0.006257\beta^4$
	    $- 0.0000133084\beta^5$
	 $K_V = -3730.44 + 2317.25\beta - 48.5\beta^2 + 0.282\beta^3 - 0.0006166\beta^4$
	 $0 \le \beta < 180°$

16. 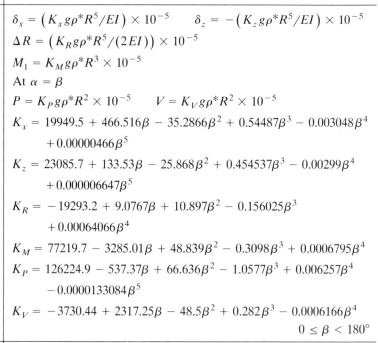 Bulkhead or supporting ring pipe, supported at sides and carrying a load W transferred by tangential shear S uniformly distributed as shown $S = \dfrac{W}{\pi R} \sin \alpha$	$\delta_x = 0$ $\delta_z = 0$ $M = K_M WR \times 10^{-5}$ $P = K_P W \times 10^{-5}$ $V = K_V W \times 10^{-5}$ $M_1 = -0.0113WR$ at bottom $\max(+M) = 0.0146WR$ at $\alpha = 66.8°$ $\max(-M) = -0.0146WR$ at $\alpha = 113.2°$ $P_1 = -0.0796W$ at $\alpha = 0$ $K_M = \begin{cases} -7945.12 - 7.4386\alpha + 6.66845\alpha^2 - 0.01605\alpha^3 \\ \quad - 0.00013\alpha^4 \qquad\qquad\qquad\qquad 0 \le \alpha < 90° \\ 34331 - 2513.46\alpha + 30.603\alpha^2 - 0.12524\alpha^3 \\ \quad - 0.000157083\alpha^4 \qquad\qquad\qquad 90 \le \alpha \le 180° \end{cases}$ $K_P = \begin{cases} -1121.98 - 4.2428\alpha + 1.5515\alpha^2 - 0.009163\alpha^3 \\ \quad - 0.0000669\alpha^4 \qquad\qquad\qquad\qquad 0 \le \alpha < 90° \\ 75826.2 - 1913.05\alpha + 16.577\alpha^2 - 0.058\alpha^3 \\ \quad + 0.00006837\alpha^4 \qquad\qquad\qquad 90 \le \alpha \le 180° \end{cases}$ $K_V = \begin{cases} 7.6305 + 134.445\alpha + 0.4064\alpha^2 - 0.04848\alpha^3 \\ \quad + 0.0001825\alpha^4 \qquad\qquad\qquad 90 \le \alpha < 90° \\ 54993.9 + 210.22\alpha + 9.3176\alpha^2 - 0.081\alpha^3 \\ \quad + 0.000179\alpha^4 \qquad\qquad\qquad 90 \le \alpha < 180° \end{cases}$

TABLE 16-5   **In-Plane Response of Uniform Circular Rings**   **822**

TABLE 16-5 (continued)   IN-PLANE RESPONSE OF UNIFORM CIRCULAR RINGS

Case	$\delta_x, \delta_z, \Delta R, M, P, V$
17. Same as case 16 except supported as shown $$S = \frac{W}{\pi R}\sin\alpha$$	$\delta_x = \left(K_x WR^3/EI\right) \times 10^{-5} \qquad \delta_z = \left(K_z WR^3/EI\right) \times 10^{-5}$ $\Delta R = \left(K_R WR^3/EI\right) \times 10^{-5}$ At $\alpha = \beta$ $M = K_M WR \times 10^{-5} \qquad P = K_P W \times 10^{-5}$ $V = K_V W \times 10^{-5}$ $K_x = 6797.16 + 18.722\beta - 4.423\beta^2 + 0.0502\beta^3$ $\qquad - 0.0001388\beta^4$ $K_z = -7423.29 - 10.27\beta + 5.947\beta^2 - 0.0873\beta^3$ $\qquad + 0.0003627\beta^4$ $K_R = -2977.03 - 5.925\beta + 1.579\beta^2 - 0.01651\beta^3$ $\qquad + 0.0000378\beta^4$ $K_M = 23961.7 - 927.9\beta + 16.0202\beta^2 - 0.11946\beta^3$ $\qquad + 0.0002517\beta^4$ $K_P = -23776.8 - 56.0366\beta + 18.56\beta^2 - 0.1682\beta^3$ $\qquad + 0.0003928\beta^4$ $K_V = -39.96 + 433.994\beta - 0.8944\beta^2 - 0.14\beta^3$ $\qquad + 0.000956\beta^4$ $0 \le \beta < 90°$
18. 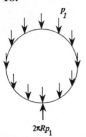	$\delta_x = 0.4292\dfrac{p_1 R^4}{EI} \qquad \delta_z = -0.4674\dfrac{p_1 R^4}{EI}$ $M_1 = 1.5p_1 R^2 \qquad M_2 = 0.5p_1 R^2$ $\max(+M) = 1.5p_1 R^2 \quad$ at bottom $V_2 = 0 \qquad P_2 = \tfrac{1}{2}p_1 R$

TABLE 16-5 (continued) IN-PLANE RESPONSE OF UNIFORM CIRCULAR RINGS

Case	$\delta_x, \delta_z, \Delta R, M, P, V$

19.

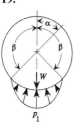

Radial pressure
varies with $(\alpha - \beta)^2$
from 0 at $\alpha = \beta$
to p_1 at $\alpha = \pi$

$$\begin{cases} 0 & \alpha < \beta \\ \dfrac{p_1}{(\pi - \beta)^2}(\alpha - \beta)^2 \\ \qquad \beta \le \alpha \le \pi \end{cases}$$

$W = 4p_1 R$

$\qquad \times \dfrac{(\pi - \beta - \sin \beta)}{(\pi - \beta)^2}$

$$\delta_x = \begin{cases} \dfrac{-p_1 R^4}{EI(\pi - \beta)^2}\Big\{ \beta^2(2 - \sin \beta) \\ \quad +\beta(\sin \beta - 3\cos \beta - 8.283) + 7\sin \beta + 2.115 \\ \quad + \cos \beta - 0.2122\big[(\pi - \beta)^3 - 6(\pi - \beta - \sin \beta)\big]\Big\} \\ \qquad\qquad \beta \le \tfrac{1}{2}\pi \\[2pt] \dfrac{-p_1 R^4}{EI(\pi - \beta)^2}\Big\{(2 - \cos \beta)(\pi - \beta) - 3\sin \beta \\ \quad -0.2122\big[(\pi - \beta)^3 - 6(\pi - \beta - \sin \beta)\big]\Big\} \\ \qquad\qquad \beta > \tfrac{1}{2}\pi \end{cases}$$

$$\delta_z = \dfrac{p_1 R^4}{EI(\pi - \beta)^2}\Big\{ 4 + 4\cos \beta - (\pi - \beta)\sin \beta - (\pi - \beta)^2$$
$$\quad + 0.2122\big[(\pi - \beta)^3 - 6(\pi - \beta - \sin \beta)\big]\Big\}$$

$$M_1 = \dfrac{-p_1 R^2}{\pi(\pi - \beta)^2}\Big[2\beta(2 - \cos \beta) + 6(\sin \beta - \pi)$$
$$\quad + \pi(\pi + \beta)^2 + 2(\pi - \beta - \sin \beta) - \tfrac{1}{3}(\pi - \beta)^3\Big]$$

$$M_z = \dfrac{-p_1 R^2}{\pi(\pi - \beta)^2}\Big[2(\pi - \beta)(2 - \cos \beta) - 6\sin \beta$$
$$\quad + 2(\pi - \beta - \sin \beta) - \tfrac{1}{3}(\pi - \beta)^3\Big]$$

$V_2 = 0$

max$(+M)$ occurs at angular position α_1 where $\alpha_1 > \beta$
and $\alpha_1 > 108.6°$ and satisfies $(\alpha_1 - \beta + \sin \beta \cos \alpha_1)$
$+ (3\sin \beta - 2\pi + 2\beta - \beta\cos \beta)\sin \alpha_1 = 0$
max$(-M) = M_1$

TABLE 16-5 In-Plane Response of Uniform Circular Rings   824

TABLE 16-5 (continued)   IN-PLANE RESPONSE OF UNIFORM CIRCULAR RINGS

Case	$\delta_x, \delta_z, \Delta R, M, P, V$

20.

$W = g\rho^*R^2(\pi - \beta$
$+ \sin\beta\cos\beta)$
Unit length of pipe
partially filled
with liquid
h-thickness of wall
of pipe

$$\delta_x = \begin{cases} \dfrac{3g\rho^*R^5(1 - \nu^2)}{2Eh^3\pi} \\ \times \left\{ \pi(\sin\beta\cos\beta + 2\pi - 3\beta + 2\beta\cos^2\beta) \right. \\ \quad + 8\pi(2\cos\beta - \sin\beta\cos\beta - \tfrac{1}{2}\pi + \beta) \\ \quad \left. + 8[(\pi - \beta)(1 - 2\cos\beta) + \sin\beta\cos\beta - 2\sin\beta] \right\} \\ \qquad \beta \le \tfrac{1}{2}\pi \\ \dfrac{3g\rho^*R^5(1 - \nu^2)}{2Eh^3\pi} \\ \times \left\{ \pi[(\pi - \beta)(1 + 2\cos^2\beta) + 3\cos\beta\sin\beta] \right. \\ \quad \left. + 8[(\pi - \beta)(1 - 2\cos\beta) + \sin\beta\cos\beta - 2\sin\beta] \right\} \\ \qquad \beta > \tfrac{1}{2}\pi \end{cases}$$

$$\delta_z = \dfrac{3g\rho^*R^5(1 - \nu^2)}{2Eh^3\pi} \left\{ \pi[\sin^2\beta + (\pi - \beta) \right.$$
$$\times (\pi - \beta - 2\sin\beta\cos\beta)] - 4\pi(1 + \cos\beta)^2$$
$$- 8[(\pi - \beta)(1 - 2\cos\beta) + \sin\beta\cos\beta - 2\sin\beta\}$$

$$M_2 = \dfrac{g\rho^*R^3}{4\pi}\{2\theta\sin^2\beta + 3\sin\beta\cos\beta - 3\beta + \pi + 2\pi\cos^2\beta$$
$$+ 2[\sin\beta\cos\beta - 2\sin\beta + (\pi - \beta)(1 - 2\cos\beta)]\}$$

$$P_2 = \dfrac{g\rho^*R^2}{4\pi}[3\sin\beta\cos\beta + (\pi - \beta)(1 - 2\cos^2\beta)]$$
$$V_2 = 0$$

21.

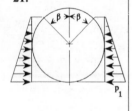

$\delta_x = (K_x p_1 R^4 / EI) \times 10^{-4}$ $\delta_z = (K_z p_1 R^4 / EI) \times 10^{-4}$
$M_1 = K_{M_1} p_1 R^2 \times 10^{-4}$ $M_2 = K_{M_2} p_1 R^2 \times 10^{-4}$
$P_2 = K_{P_2} p_1 R \times 10^{-4}$
$K_x = -831.109 - 2.65\beta + 0.162\beta^2 - 7.37 \times 10^{-4}\beta^3$
$\quad + 1.11 \times 10^{-7}\beta^4$
$K_z = 831.546 + 2.137\beta - 0.144\beta^2 + 5.83 \times 10^{-4}\beta^3$
$\quad + 2.53 \times 10^{-7}\beta^4$
$K_{M_1} = -1456.97 - 0.584\beta + 7.6986 \times 10^{-2}\beta^2$
$\quad + 7.58257 \times 10^{-4}\beta^3 - 5.478 \times 10^{-6}\beta^4$
$K_{M_2} = -1040.18 - 3.89612\beta + 0.288\beta^2$
$\quad - 2.036 \times 10^{-3}\beta^3 + 3.971 \times 10^{-6}\beta^4$
$K_{P_2} = -3122.7 - 7.929\beta + 0.968856\beta^2$
$\quad - 8.173 \times 10^{-3}\beta^3 + 1.99 \times 10^{-5}\beta^4$
$$30° \le \beta \le 120°$$

TABLE 16-5 (continued)   IN-PLANE RESPONSE OF UNIFORM CIRCULAR RINGS

Case	$\delta_x, \delta_z, \Delta R, M, P, V$
22.	$\delta_x = (K_x WR^3/EI) \times 10^{-4} \qquad \delta_z = (K_z WR^3/EI) \times 10^{-4}$ $M_1 = K_M WR \times 10^{-4} \qquad\qquad P_2 = K_P W \times 10^{-4}$ $\max(+M) = \begin{cases} \text{occurs at angular position } \alpha = \tan^{-1}(-\pi/\sin^2\beta) \\ \qquad \beta < 106.3° \\ \text{occurs at load } W \qquad \beta \geq 106.3° \end{cases}$ $\max(-M) = M_1$ $K_x = -4053.2 + 81.547\beta - 0.8616\beta^2 + 3.1913 \times 10^{-3}\beta^3$ $K_z = 4269.8 - 88.72\beta + 0.9634\beta^2 - 3.568 \times 10^{-3}\beta^3$ $K_M = -6052 - 29.9945\beta + 1.565\beta^2 - 1.795 \times 10^{-2}\beta^3$ $\qquad\quad + 6.35 \times 10^{-5}\beta^4$ $K_P = -801 + 66.6167\beta - 3.027\beta^2 + 2.952 \times 10^{-2}\beta^3$ $\qquad\quad - 8.2 \times 10^{-5}\beta^4$ $\hfill 30° \leq \beta \leq 150°$
23.	$\delta_x = (K_x WR^3/EI) \times 10^{-4} \qquad \delta_z = (K_z WR^3/EI) \times 10^{-4}$ $M_1 = K_{M_1} WR \times 10^{-4} \qquad\qquad P_2 = K_P W \times 10^{-4}$ $M_2 = K_{M_2} WR \times 10^{-4}$ $\max(+M) = \begin{cases} M_1 \qquad\qquad\qquad\quad \beta \leq 60° \\ K_M WR \times 10^{-4} \quad \text{at load if } \beta > 60° \end{cases}$ $K_x = 819 + 141.42\beta - 4.62\beta^2 + 0.040574\beta^3$ $\qquad\quad - 1.0957 \times 10^{-4}\beta^4$ $K_z = -2690 - 23.975\beta + 2.1817\beta^2 - 0.0205463\beta^3$ $\qquad\quad + 5.36523 \times 10^{-5}\beta^4$ $K_{M_1} = 5650 + 53.73\beta - 2.884\beta^2 + 0.0168\beta^3$ $\qquad\quad - 1.852 \times 10^{-5}\beta^4$ $K_P = 507.003 - 211.57\beta + 1.9037\beta^2 - 9.383 \times 10^{-4}\beta^3$ $\qquad\quad - 1.83 \times 10^{-5}\beta^4$ $K_{M_2} = 6753 - 202.622\beta + 1.184\beta^2 + 1.765 \times 10^{-3}\beta^3$ $\qquad\quad - 1.852 \times 10^{-5}\beta^4$ $K_M = 10578 - 496.41\beta + 9.175\beta^2 - 0.06416\beta^3$ $\qquad\quad + 1.484 \times 10^{-4}\beta^4$ $\hfill 30° \leq \beta \leq 150°$

TABLE 16-5   **In-Plane Response of Uniform Circular Rings**   **826**

TABLE 16-6, PART A   STATIC RESPONSE OF CIRCULAR CURVED BEAMS UNDER IN-PLANE LOADING: GENERAL RESPONSE EXPRESSIONS

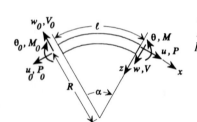

Notation

E = modulus of elasticity

A = area of the cross section

I^* = moment of inertia modified for curvature of bar, where z is measured from centroid of cross section,

$$= \int_A [z^2/(1 - z/R)] \, dA$$

Response

1. Extension:
$$u = u_0 \cos \alpha + w_0 \sin \alpha + \theta_0 R(\cos \alpha - 1)$$
$$+ V_0 \left[\frac{R}{2EA} \alpha \sin \alpha - \frac{R^3}{2EI^*} (2 - 2\cos \alpha - \alpha \sin \alpha) \right]$$
$$- M_0 \frac{R^2}{EI^*} (\alpha - \sin \alpha)$$
$$+ P_0 \left[\frac{R}{2EA} (\alpha \cos \alpha + \sin \alpha) + \frac{R^3}{2EI^*} (2\alpha + \alpha \cos \alpha - 3 \sin \alpha) \right]$$
$$+ F_u$$

2. Deflection:
$$w = - u_0 \sin \alpha + w_0 \cos \alpha - \theta_0 R \sin \alpha + V_0 \left(\frac{1}{EA} + \frac{R^2}{EI^*} \right)$$
$$\times \frac{R(\sin \alpha - \alpha \cos \alpha)}{2} + M_0 \frac{R^2}{EI^*} (\cos \alpha - 1)$$
$$+ P_0 \left[\frac{R}{2EA} \alpha \sin \alpha - \frac{R^3}{2EI^*} (2 - 2\cos \alpha - \alpha \sin \alpha) \right] + F_w$$

3. Slope:
$$\theta = \theta_0 + \frac{V_0 R^2 (1 - \cos \alpha)}{EI^*} + \frac{M_0 R \alpha}{EI^*} - \frac{P_0 R^2 (\alpha - \sin \alpha)}{EI^*} + F_\theta$$

4. Shear force:
$$V = V_0 \cos \alpha - P_0 \sin \alpha + F_V$$

5. Bending moment:
$$M = V_0 R \sin \alpha + M_0 + P_0 R(\cos \alpha - 1) + F_M$$

6. Axial force:
$$P = V_0 \sin \alpha + P_0 \cos \alpha + F_P$$

Loading functions F_u, F_w, F_θ, F_V, F_M, and F_P are defined in part B for a variety of applied loads.

To use these formulas, substitute the loading functions into the above formulas and calculate the initial parameters based on the boundary conditions in part C. Use the methodology of Appendix III.

TABLE 16-6, PART B   STATIC RESPONSE OF CIRCULAR CURVED BEAMS UNDER IN-PLANE LOADING: LOADING FUNCTIONS

1. For concentrated forces and moments: $\bar{\mathbf{z}}^i = \mathbf{f}$, where $\bar{\mathbf{z}}^i = [F_u \ \ F_w \ \ F_\theta \ \ F_V \ \ F_M \ \ F_P]^T$ and $\mathbf{f} = [f_u \ \ f_w \ \ f_\theta \ \ f_V \ \ f_M \ \ f_P]^T$.

2. For distributed forces and moments: $\bar{\mathbf{z}}^i = \mathbf{u}^i \mathbf{f}$, where $\mathbf{u}^i$ is the upper left 6×6 submatrix of the extended transfer matrix for massless circular bars of Table 16-13, part A, with $\psi = \alpha$. By definition:

$$\langle \alpha^n \sin \alpha \rangle = \begin{cases} 0 & \alpha < \alpha_1 \\ (\alpha - \alpha_1)^n \sin(\alpha - \alpha_1) & \alpha_1 \le \alpha < \alpha_2 \\ (\alpha_2 - \alpha_1)^n \sin(\alpha_2 - \alpha_1) & \alpha_2 \le \alpha \end{cases}$$

$$\langle \alpha^n \cos \alpha \rangle = \begin{cases} 0 & \alpha < \alpha_1 \\ (\alpha - \alpha_1)^n \cos(\alpha - \alpha_1) & \alpha_1 \le \alpha < \alpha_2 \\ (\alpha_2 - \alpha_1)^n \cos(\alpha_2 - \alpha_1) & \alpha_2 \le \alpha \end{cases}$$

$$\langle \alpha - \alpha_1 \rangle^0 = \begin{cases} 0 & \alpha < \alpha_1 \\ 1 & \alpha \ge \alpha_1 \end{cases}$$

$$\langle \alpha - \alpha_1 \rangle = \begin{cases} 0 & \alpha < \alpha_1 \\ (\alpha - \alpha_1) & \alpha \ge \alpha_1 \end{cases}$$

$$\langle \alpha^n \rangle = \begin{cases} 0 & \alpha < \alpha_1 \\ (\alpha - \alpha_1)^n & \alpha_1 \le \alpha < \alpha_2 \\ (\alpha_2 - \alpha_1)^n & \alpha_2 \le \alpha \end{cases}$$

$$n = 0, 1, 2$$

TABLE 16-6, Part B   Loading Functions for a Circular Beam   828

The elements of **f** are given below:

	W	p_1	C	c_1
f_u	$-\dfrac{WR}{2EA}\langle\alpha-\alpha_1\rangle\sin(\alpha-\alpha_1)$ $+\dfrac{WR^3}{EI^*}\{\langle\alpha-\alpha_1\rangle^0[1-\cos(\alpha-\alpha_1)]$ $-\tfrac{1}{2}[\langle\alpha-\alpha_1\rangle\sin(\alpha-\alpha_1)]\}$	$-\dfrac{p_1R^2}{2}\left[\left(\dfrac{1}{EA}+\dfrac{R^2}{EI^*}\right)\right.$ $\times(-\langle\alpha\cos\alpha\rangle+\langle\sin\alpha\rangle)-\dfrac{2R^2}{EI^*}\langle\alpha\rangle]$	$\dfrac{R^2C}{EI^*}(\langle\alpha-\alpha_1\rangle$ $-\sin\langle\alpha-\alpha_1\rangle)$	$\dfrac{c_1R^3}{EI^*}\left(\dfrac{1}{2}\langle\alpha^2\rangle+\langle\cos\alpha\rangle\right)$
f_w	$-\dfrac{WR}{2}\left(\dfrac{1}{EA}+\dfrac{R^2}{EI^*}\right)$ $\times[\sin\langle\alpha-\alpha_1\rangle-\langle\alpha-\alpha_1\rangle\cos(\alpha-\alpha_1)]$	$\dfrac{p_1R^2}{2}\left(\dfrac{1}{EA}+\dfrac{R^2}{EI^*}\right)$ $\times(2\langle\alpha-\alpha_1\rangle^0-\langle\alpha\sin\alpha\rangle-2\langle\cos\alpha\rangle)$	$-\dfrac{R^2C}{EI^*}\{\langle\alpha-\alpha_1\rangle^0$ $\times[\cos(\alpha-\alpha_1)-1]\}$	$\dfrac{c_1R^3}{EI^*}(\langle\alpha\rangle-\langle\sin\alpha\rangle)$
f_θ	$-\dfrac{WR^2}{EI^*}[\langle\alpha-\alpha_1\rangle^0-\langle\alpha-\alpha_1\rangle^0\cos(\alpha-\alpha_1)]$	$-\dfrac{p_1R^3}{EI^*}(\langle\alpha\rangle-\langle\sin\alpha\rangle)$	$-\dfrac{RC}{EI^*}\langle\alpha-\alpha_1\rangle$	$\dfrac{c_1R^2}{2EI^*}\langle\alpha^2\rangle$
f_V	$-W\langle\alpha-\alpha_1\rangle^0\cos(\alpha-\alpha_1)$	$p_1R\langle\sin\alpha\rangle$	0	0
f_M	$-WR\sin\langle\alpha-\alpha_1\rangle$	$p_1R^2\langle\cos\alpha\rangle$	$-C\langle\alpha-\alpha_1\rangle^0$	$Rc_1\langle\alpha\rangle$
f_P	$W\sin\langle\alpha-\alpha_1\rangle$	$-p_1R\langle\cos\alpha\rangle$	0	0

Case	Boundary Conditions
1. Clamped end	$u = w = \theta = 0$
2. Free end	$M = V = P = 0$
3. Pinned end	$u = w = M = 0$
4. Clamped–circumferentially guided end	$w = \theta = P = 0$
5. Clamped–radially guided end	$u = \theta = V = 0$

TABLE 16-6, Part C   In-Plane Boundary Conditions   830

Case	Boundary Conditions
6. Pinned–circumferentially guided end 	$w = M = P = 0$
7. Pinned–radially guided end 	$u = M = V = 0$
8. Clamped–free end 	$\theta = V = P = 0$

TABLE 16-7 BUCKLING LOADS FOR CIRCULAR ARCHES AND RINGS

Notation

All bars are uniform in cross section. Unless otherwise specified, the direction of uniform radial loading is not a function of the deformation of the bar.

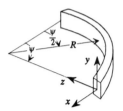

E = modulus of elasticity
G = shear modulus of elasticity
R = radius of curved beam
A = area of cross section
I, I_z = moments of inertia about y and z axes
I_{yz} = product of inertia
r_p = polar radius of gyration, I_p/A
I_p = polar moment of inertia
J = torsional constant
Γ = warping constant
z_S = z coordinate of shear center
α_s = shear correction factor
n = buckling mode number

$$D = \int_A Ez^2\, dA \qquad B = \int_A \alpha_s G\, dA \qquad C = \int_A E\, dA \qquad K_s = D/(BR^2)$$

Case	Buckling Loads
1. Hinged–Hinged In-plane loading 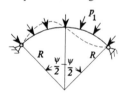	$$p_{1,cr} = \frac{EI}{R^3}\left(\frac{4\pi^2}{\psi^2} - 1\right)$$ Ref. 16.15
2. Fixed–Fixed In-plane loading 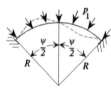	$$p_{1,cr} = \frac{EI}{R^3}(K^2 - 1)$$ Factor K is the solution of $$K \tan \frac{\psi}{2} \cot \frac{K\psi}{2} = 1$$ or use $K = 21.037 - 0.322439\psi + 0.0024401\psi^2$ $\quad - 9.69329 \times 10^{-6}\psi^3 + 1.94155 \times 10^{-8}\psi^4$ $\quad - 1.53892 \times 10^{-11}\psi^5 \quad 60° \le \psi \le 360°$ Ref. 16.15

TABLE 16-7 | Buckling Loads for Circular Arches and Rings | **832**

TABLE 16-7 (continued) **BUCKLING LOADS FOR CIRCULAR ARCHES AND RINGS**

Case	Buckling Loads
3. Simply supported arch Out-of-plane buckling 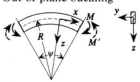 Boundary conditions: free to rotate about z and y axes but not free about arch axis (x axis)	$\left.\begin{matrix} M_{cr} \\ M'_{cr} \end{matrix}\right\} = \dfrac{EI_z + GJ}{2R} \pm \sqrt{\left(\dfrac{EI_z - GJ}{2R}\right)^2 + \dfrac{EI_z GJ}{R^2}\dfrac{\pi^2}{\psi^2}}$ 1. when $R \to \infty$ $\quad M_{cr} = \dfrac{\pi\sqrt{EI_z GJ}}{L} \qquad M'_{cr} = -\dfrac{\pi\sqrt{EI_z GJ}}{L}$ where L is the length of the beam. 2. When $\psi = \pi$ $\quad M'_{cr} = 0$ Ref. 16.15

4. In-plane buckling of rings subject to uniform external pressure: The pressure is initially directed to the center of the ring. After buckling, the direction of the pressure can be in one of the three directions. p_b and p_a indicate the loads before and after buckling. Before buckling $p_1 = p_b$ and after buckling, p_1 becomes p_a.

a. Pressure acts perpendicular to axis of ring section during buckling	If $I_{yz} = 0$ $p_{1,cr} = 3\dfrac{EI}{R^3}$ Otherwise $p_{1,cr} = 3\dfrac{EI_z}{R^3}\left(1 - \dfrac{I_{yz}^2}{I_z I}\right)$ Ref. 16.16
b. Direction of pressure does not change during buckling	$p_{1,cr} = 4\dfrac{EI}{R^3}$ Ref. 16.17
c. Pressure is directed toward initial center of curvature during buckling	$p_{1,cr} = 4.5\dfrac{EI}{R^3}$ Ref. 16.17

TABLE 16-7 (continued) BUCKLING LOADS FOR CIRCULAR ARCHES AND RINGS

Case	Buckling Loads
5. In-plane loading of rings: Circular rings subject to uniform pressure directed radially. The material can be nonuniform and shear deformation effects are considered. The load directions are the same as case 4. Ref. 16.18	
a. Force remains perpendicular to axis of ring section during buckling (see figure of case 4a)	$p_{1,cr} = \dfrac{3D}{R^3}\dfrac{1}{1 + 4K_s}$
b. Force remains parallel to its initial direction during buckling (see figure of case 4b)	$p_{1,cr} = \dfrac{4D}{R^3}\dfrac{1}{1 + 4K_s}$
c. Force remains directed toward initial center of curvature during buckling and $K_R^2 \ll 1$, (see figure of case 4c)	$p_{1,cr} = \dfrac{4.5D}{R^3}\dfrac{1}{1 + 4K_s}$ $K_R^2 = \dfrac{D}{CR^2}$
6. Out-of-plane buckling of rings: Circular rings subject to uniform external pressure which acts perpendicular to ring axis and remains parallel to plane of initial curvature during buckling	$p_{1,cr} = \dfrac{EI_z}{R^3}\dfrac{9}{4 + EI_z/GJ}$ When warping constant Γ should be considered $p_{1,cr} = \dfrac{EI_z}{R^3}\dfrac{n^2(n^2 - 1)}{n^4 + (EI_z/E\Gamma)R^2}$ $n \geq 2$ Ref. 16.16

TABLE 16-7 Buckling Loads for Circular Arches and Rings 834

TABLE 16-8   FUNDAMENTAL NATURAL FREQUENCY OF CIRCULAR ARCHES[a]

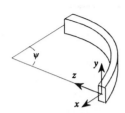

Notation

R = radius of arch
E = modulus of elasticity
G = shear modulus of elasticity
J = torsional constant
I, I_z = moments of inertia about y and z axes
ν = Poisson's ratio
ρ = mass per unit length

Case	Fundamental Natural Frequency
1. Clamped–clamped In-plane flexure	$$\dfrac{\lambda^2}{2\pi(R\psi)^2}\left[\dfrac{1 - 2\sigma^2(1 - 2/\sigma\lambda)(\psi/\lambda)^2 + (\psi/\lambda)^4}{1 + 5\sigma^2(1 - 2/\sigma\lambda)(\psi/\lambda)^2}\right]^{1/2}\left(\dfrac{EI}{\rho}\right)^{1/2}$$ $\lambda = 7.8532 \qquad \sigma = 1.00078$
2. Clamped–clamped Out-of-plane flexure	$$\dfrac{\pi}{2(R\psi)^2}\left[\dfrac{3.586(\psi/\pi)^2 + 1.246GJ\beta/EI_z}{(\psi/\pi)^2 + 1.246GJ/EI_z}\right]^{1/2}\left(\dfrac{EI_z}{\rho}\right)^{1/2}$$ $0 < EI_z/GJ < 2 \qquad \beta = (\psi/\pi)^4 - 2.492(\psi/\pi)^2 + 5.139$
3. Clamped–pinned Out-of-plane flexure	$$\dfrac{\pi}{2(R\psi)^2}\left[\dfrac{1.080(\psi/\pi)^2 + 1.166GJ\beta/EI_z}{(\psi/\pi)^2 + 1.166GJ/EI_z}\right]^{1/2}\left(\dfrac{EI_z}{\rho}\right)^{1/2}$$ $0 < EI_z/GJ < 2 \qquad \beta = (\psi/\pi)^4 - 2.332(\psi/\pi)^2 + 2.440$

[a]From Blevins [16.11], with permission.

TABLE 16-9 NATURAL FREQUENCIES AND MODE SHAPES OF CIRCULAR RINGS[a]

Notation

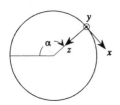

R = radius of ring.
E = modulus of elasticity
G = shear modulus of elasticity
ν = Poisson's ratio
J = torsional constant
I_p = polar moment of inertia
I, I_z = moments of inertia about y and z axes
ρ^* = mass per unit volume
ρ = mass per unit length
w, v, u, ϕ = mode shapes corresponding to deformations parallel to z, y, x axes and rotation about x axis

Case	Natural Frequency f_i (Hz)	Mode Shapes
1. Extension modes	$\dfrac{(1+i^2)^{1/2}}{2\pi R}\left(\dfrac{E}{\rho^*}\right)^{1/2}$ $i = 0, 1, 2, 3, \ldots$ Extension modes have natural frequencies above those of flexure modes of slender rings.	$\begin{bmatrix} w \\ v \\ u \\ \phi \end{bmatrix}_i = \begin{bmatrix} \cos i\alpha \\ 0 \\ -i \sin i\alpha \\ 0 \end{bmatrix}$
2. Torsion modes	$\dfrac{1}{2\pi R}\left(\dfrac{i^2 GJ + I_z E}{\rho I_p}\right)^{1/2}$ $i = 0, 1, 2, 3, \ldots$ For circular cross sections $\dfrac{(i^2 + \nu + 1)^{1/2}}{2\pi R}\left(\dfrac{G}{\rho^*}\right)^{1/2}$	$\begin{bmatrix} w \\ v \\ u \\ \phi \end{bmatrix}_i = \begin{bmatrix} 0 \\ -\varepsilon \\ 0 \\ \cos i\alpha \end{bmatrix}$ $\varepsilon \ll 1$ If $i = 0$, $\varepsilon = 0$
3. In-plane flexure modes	$\dfrac{i(i^2 - 1)}{2\pi R^2(i^2+1)^{1/2}}\left(\dfrac{EI}{\rho}\right)^{1/2}$ $i = 1, 2, 3, \ldots$	$\begin{bmatrix} w \\ v \\ u \\ \phi \end{bmatrix}_i = \begin{bmatrix} i \cos i\alpha \\ 0 \\ \sin i\alpha \\ 0 \end{bmatrix}$
4. Out-of-plane flexure modes	$\dfrac{i(i^2 - 1)}{2\pi R^2}\left[\dfrac{EI_z}{\rho(i^2 + EI_z/GJ)}\right]^{1/2}$ $i = 1, 2, 3, \ldots$	$\begin{bmatrix} w \\ v \\ u \\ \phi \end{bmatrix}_i = \begin{bmatrix} 0 \\ -\sin i\alpha \\ 0 \\ -\dfrac{i^2}{R}\left(\dfrac{1 + GJ/EI_z}{1 + i^2 GJ/EI_z}\right)\sin i\alpha \end{bmatrix}$
	For circular cross sections $\dfrac{i(i^2 - 1)}{2\pi R^2(i^2 + 1 + \nu)^{1/2}}\left(\dfrac{EI}{\rho}\right)^{1/2}$ $i = 1, 2, 3, \ldots$	$\begin{bmatrix} w \\ v \\ u \\ \phi \end{bmatrix}_i = \begin{bmatrix} 0 \\ -\sin i\alpha \\ 0 \\ -\dfrac{i^2}{R}\left(\dfrac{2 + \nu}{i^2 + 1 + \nu}\right)\sin i\alpha \end{bmatrix}$

[a]From Blevins [16.11], with permission.

TABLE 16-9 Frequencies and Mode Shapes of Circular Rings   **836**

TABLE 16-10 NATURAL FREQUENCIES OF SOME SIMPLE STRUCTURES CONTAINING CURVED BARS[a]

Notation

E = modulus of elasticity
G = shear modulus of elasticity
J = torsional constant
I, I_z = moments of inertia about y and z axes
I_p = polar moment of inertia
ρ = mass per unit length
$\gamma = \ell/R$
ℓ = length of straight members
R = radius of curved segment

Boundary conditions at points A and B:

P–P: pinned–pinned
C–P: clamped–pinned
C–C: clamped–clamped

Natural frequencies f_i (Hz):

In-plane motion: $f_i = \dfrac{\lambda_i}{2\pi R^2}\left(\dfrac{EI}{\rho}\right)^{1/2}$

Out-of-plane motion: $f_i = \dfrac{\lambda_i}{2\pi R^2}\left(\dfrac{EI_z}{\rho}\right)^{1/2}$

λ_i are defined below:

Case		Parameters λ_i
		In-plane motion
1.	P–P	$\lambda_1 = 23.0187 - 27.835\gamma + 77.106\gamma^2 - 127.155\gamma^3 + 92.67\gamma^4 - 30.475\gamma^5 + 3.702\gamma^6$
		$\lambda_2 = \begin{cases} 43.9286 - 37.202\gamma + 74.1072\gamma^2 - 83.33\gamma^3 & 0 \leq \gamma < 0.8 \\ 93.0596 - 175.348\gamma + 142.188\gamma^2 - 55.082\gamma^3 + 8.286\gamma^4 & 0.8 \leq \gamma \leq 2.0 \end{cases}$
	C–P	$\lambda_1 = 23.017 - 29.28\gamma + 86.0\gamma^2 - 131.486\gamma^3 + 84.51\gamma^4 - 22.89\gamma^5 + 1.98\gamma^6$
		$\lambda_2 = \begin{cases} 43.957 - 37.32\gamma + 85.71\gamma^2 - 93.75\gamma^3 & 0 \leq \gamma < 0.8 \\ 54.13 - 58.31\gamma + 23.8\gamma^2 - 3.47\gamma^3 & 0.8 \leq \gamma \leq 2.0 \end{cases}$
	C–C	$\lambda_1 = 23.013 - 21.315\gamma + 47.424\gamma^2 - 58.644\gamma^3 + 28.523\gamma^4 - 4.758\gamma^5$
		$\lambda_2 = \begin{cases} 44.0286 - 33.4524\gamma + 80.3572\gamma^2 - 83.3\gamma^3 & 0 \leq \gamma < 0.8 \\ 80.369 - 99.1865\gamma + 44.643\gamma^2 - 6.94\gamma^3 & 0.8 \leq \gamma \leq 2.0 \end{cases}$

1. Continued

Out-of-plane motion

P-P:
$\lambda_1 = 9.5367 - 8.327\gamma + 8.426\gamma^2 - 5.287\gamma^3 + 1.138\gamma^4$
$\lambda_2 = 27.0483 - 32.5\gamma + 109.85\gamma^2 - 212.817\gamma^3 + 180.284\gamma^4 - 69.854\gamma^5 + 10.244\gamma^6$

C-P:
$\lambda_1 = 9.5 - 8.736\gamma + 16.35\gamma^2 - 23.57\gamma^3 + 18.58\gamma^4 - 7.46\gamma^5 + 1.18\gamma^6$
$\lambda_2 = 27.032 - 29.7\gamma + 101.29\gamma^2 - 191.78\gamma^3 + 156.256\gamma^4 - 57.769\gamma^5 + 8.0423\gamma^6$

C-C:
$\lambda_1 = 9.50375 - 7.24\gamma + 15.447\gamma^2 - 29.65\gamma^3 + 29.4\gamma^4 - 13.67\gamma^5 + 2.3616\gamma^6$
$\lambda_2 = 27.05 - 21.782\gamma + 54.744\gamma^2 - 77.758\gamma^3 + 41.8283\gamma^4 - 7.64\gamma^5$

In-plane motion

P-P:
$\lambda_1 = 4.5 - 5.1125\gamma + 13.75\gamma^2 - 17.74\gamma^3 + 9.7656\gamma^4 - 1.912\gamma^5$
$\lambda_2 = \begin{cases} 17.8 - 6.67\gamma + 10.3125\gamma^2 - 9.1146\gamma^3 & 0 \le \gamma < 1.2 \\ 34.4 - 29.875\gamma + 7.1875\gamma^2 & 1.2 \le \gamma \le 2.0 \end{cases}$

C-P:
$\lambda_1 = 4.5 - 5.375\gamma + 15.078\gamma^2 - 19.954\gamma^3 + 11.23\gamma^4 - 2.238\gamma^5$
$\lambda_2 = \begin{cases} 17.8 - 7.417\gamma + 12.8125\gamma^2 - 10.677\gamma^3 & 0 \le \gamma < 1.2 \\ 28.85 - 22.625\gamma + 5.0\gamma^2 & 1.2 \le \gamma \le 2.0 \end{cases}$

C-C:
$\lambda_1 = 4.5 - 5.4875\gamma + 15.625\gamma^2 - 20.8\gamma^3 - 11.72\gamma^4 - 2.319\gamma^5$
$\lambda_2 = \begin{cases} 17.8 - 6.417\gamma + 10.31\gamma^2 - 7.552\gamma^3 & 0 \le \gamma < 1.2 \\ 36.65 - 27.825\gamma + 6.0\gamma^2 & 1.2 \le \gamma \le 2.0 \end{cases}$

2.

Out-of-plane motion

P-P:
$\lambda_1 = 5.81 - 1.605\gamma + 0.0217\gamma^2 + 0.5136\gamma^3 - 0.244\gamma^4$
$\lambda_2 = 9.33 - 2.936\gamma - 0.625\gamma^2 + 4.65\gamma^3 - 4.49\gamma^4 + 1.1556\gamma^5$

C-P:
$\lambda_1 = 5.784 - 1.727\gamma + 0.633\gamma^2 - 0.2662\gamma^3$
$\lambda_2 = 9.34 - 4.094\gamma + 6.625\gamma^2 - 5.9\gamma^3 + 1.49\gamma^4$

C-C:
$\lambda_1 = 5.8 - 1.3664\gamma - 0.2973\gamma^2 + 0.7849\gamma^3 - 0.2848\gamma^4$
$\lambda_2 = 9.3366 - 3.6\gamma + 4.8459\gamma^2 - 3.4469\gamma^3 + 0.675\gamma^4$

[Diagram: curved bar structure with radius R, segment ℓ, points A and B, with supports.]

"Based on the results of Lee [16.12].

TABLE 16-11 OUT-OF-PLANE RESPONSE OF UNIFORM CIRCULAR BARS

Notation

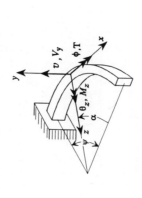

The boundary conditions are clamped-free.

R = radius of arch

E = modulus of elasticity

G = shear modulus of elasticity

J = torsional constant

I_z = moment of inertia about z axis

$\lambda = EI_z/GJ$

v, ϕ, θ_z = deflection, angle of rotation, and slope at free end of bar ($\alpha = \psi$)

M_z, T = internal bending and twisting moments

W_y, T_1, C_z = concentrated applied force, torque, and moment

TABLE 16-11 (continued)   OUT-OF-PLANE RESPONSE OF UNIFORM CIRCULAR BARS

Case	Internal Moments		Responses at Free End		
	M_z	T	v	ϕ	θ_z
1.	$-W_y R \sin\alpha$	$W_y R(1-\cos\alpha)$	$-\dfrac{W_y R^3}{EI_z}\left(\dfrac{1+3\lambda}{2}\psi\right)$ $+\dfrac{\lambda-1}{4}\sin 2\psi - 2\lambda\sin\psi$	$\dfrac{W_y R^2}{EI_z}\left[\dfrac{\lambda-1}{4}\sin 2\psi\right.$ $\left.+\dfrac{\lambda+1}{2}\psi - \lambda\sin\psi\right]$	$-\dfrac{W_y R^2}{EI_z}\left[\dfrac{\lambda-1}{2}\sin^2\psi\right.$ $\left.+\lambda(1-\cos\psi)\right]$
2.	$T_1 \sin\alpha$	$T_1 \cos\alpha$	$\dfrac{T_1 R^2}{EI_z}\left(\dfrac{\lambda-1}{4}\sin 2\psi\right.$ $\left.+\dfrac{\lambda+1}{2}\psi - \lambda\sin\psi\right)$	$\dfrac{T_1 R}{EI_z}\left(\dfrac{1+\lambda}{2}\psi\right.$ $\left.+\dfrac{\lambda-1}{4}\sin 2\psi\right)$	$-\dfrac{T_1 R}{EI_z}\dfrac{\lambda-1}{2}\sin^2\psi$
3.	$C_z \cos\alpha$	$-C_z \sin\alpha$	$\dfrac{C_z R^2}{EI_z}\left[\dfrac{\lambda-1}{2}\sin^2\psi\right.$ $\left.+\lambda(1-\cos\psi)\right]$	$-\dfrac{C_z R}{EI_z}\dfrac{\lambda-1}{2}\sin^2\psi$	$\dfrac{C_z}{EI_z}\left(\dfrac{\lambda+1}{2}\psi - \dfrac{\lambda-1}{4}\sin 2\psi\right)$
4.	$-p_y R^2$ $\times(1-\cos\alpha)$	$p_y R^2(\alpha - \sin\alpha)$	$-\dfrac{p_y R^4}{EI_z}\left[(1-\cos\psi)^2\right.$ $\left.+\lambda(\psi - \sin\psi)^2\right]$	$-\dfrac{p_y R^3}{EI_z}$ $\times[(\lambda+1)(1-\cos\psi)$ $-\dfrac{\lambda-1}{4}(1-\cos 2\psi)$ $-\lambda\psi\sin\psi]$	$-\dfrac{p_y R^3}{EI_z}\left[(\lambda+1)\left(\sin\psi - \dfrac{\psi}{2}\right)\right.$ $\left.+\dfrac{\lambda-1}{4}\sin 2\psi - \lambda\psi\cos\psi\right]$

TABLE 16-11   Out-of-Plane Response of Uniform Circular Bars   840

TABLE 16-12, PART A. STATIC RESPONSE OF A CIRCULAR BEAM UNDER OUT-OF-PLANE LOADING: GENERAL RESPONSE EXPRESSIONS

Notation

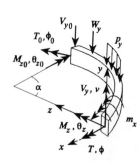

E = modulus of elasticity
G = shear modulus of elasticity
R = radius of arch
J = torsional constant
I_z = moment of inertia about z axis

$\lambda_1 = \dfrac{1}{GJ} + \dfrac{1}{EI_z}$

$\lambda_2 = \frac{1}{2}(\sin\alpha - \alpha\cos\alpha)$

$\lambda_3 = \frac{1}{2}(2 - 2\cos\alpha - \alpha\sin\alpha)$

$\lambda_4 = \frac{1}{2}(2\alpha + \alpha\cos\alpha - 3\sin\alpha)$

$\lambda_5 = \frac{1}{2}(\alpha\cos\alpha + \sin\alpha)$

$\lambda_6 = \frac{1}{2}(2 - 2\cos\alpha - \alpha\sin\alpha)$

$\lambda_7 = \frac{1}{2}(2\alpha + \alpha\cos\alpha - 3\sin\alpha)$

$\lambda_8 = 2\sin\alpha - \alpha\cos\alpha - \alpha$

Response

1. Angle of twist:
$$\phi = \phi_0\cos\alpha + \theta_{z_0}\sin\alpha - V_{y_0}\lambda_1 R^2\lambda_2 + M_{z_0}0.5\lambda_1 R\alpha\sin\alpha$$
$$+ T_0\left(\frac{R}{GJ}\lambda_5 - \frac{R}{EI_z}\lambda_2\right) + F_\phi$$

2. Deflection:
$$v = -\phi_0 R(\cos\alpha - 1) + v_0 + \theta_{z_0}R\sin\alpha$$
$$+ V_{y_0}\left(\frac{R^3}{GJ}\lambda_4 - \frac{R^3}{EI_z}\lambda_2\right)$$
$$+ M_{z_0}\left(\frac{R^2\alpha}{2EI_z}\sin\alpha - \frac{R^2}{GJ}\lambda_3\right)$$
$$- T_0\lambda_1 R^2\lambda_2 + F_v$$

3. Slope:
$$\theta_z = -\phi_0\sin\alpha + \theta_{z_0}\cos\alpha + V_{z_0}\left(\frac{R^2}{GJ}\lambda_3 - \frac{R^2\alpha}{2EI_z}\sin\alpha\right)$$
$$+ M_{z_0}\left(\frac{R}{EI_z}\lambda_5 - \frac{R}{GJ}\lambda_2\right) - T_0 0.5\lambda_1 R\alpha\sin\alpha + F_{\theta_z}$$

4. Shear force: $V_y = V_{y_0} + F_{V_y}$

5. Bending moment: $M_z = -V_{y_0}R\sin\alpha + M_{z_0}\cos\alpha - T_0\sin\alpha + F_{M_z}$

6. Torque: $T = V_{y_0}(\cos\alpha - 1) + M_{z_0}\sin\alpha + T_0\cos\alpha + F_T$

Loading functions F_ϕ, F_v, F_{θ_z}, F_{V_y}, F_{M_z}, and F_T are defined in part B for a variety of applied loads.

To use these formulas, substitute the loading functions into the above formulas and calculate the initial parameters based on the boundary conditions in part C. Use the methodology of Appendix III.

TABLE 16-12, PART B STATIC RESPONSE OF CIRCULAR BEAMS UNDER OUT-OF-PLANE LOADING: LOADING FUNCTIONS

1. For concentrated forces and moments: $\bar{\mathbf{z}}^i = \mathbf{f}$, where $\mathbf{z}^i = \begin{bmatrix} F_\phi & F_v & F_{\theta_z} & F_{V_y} & F_{M_z} & F_T \end{bmatrix}^T$ and $\mathbf{f} = \begin{bmatrix} f_\phi & f_v & f_{\theta_z} & f_{V_y} & f_{M_z} & f_T \end{bmatrix}^T$.

2. For distributed forces and moments: $\bar{\mathbf{z}}^i = \mathbf{u}^i\mathbf{f}$, where $\mathbf{u}^i$ is the upper left 6×6 submatrix of the extended transfer matrix for massless circular bars of Table 16-13, part B, with $\psi = \alpha$. By definition:

$$\langle \alpha^n \sin \alpha \rangle = \begin{cases} 0 & \alpha < \alpha_1 \\ (\alpha - \alpha_1)^n \sin(\alpha - \alpha_1) & \alpha_1 \le \alpha < \alpha_2 \\ (\alpha_2 - \alpha_1)^n \sin(\alpha_2 - \alpha_1) & \alpha_2 \le \alpha \end{cases}$$

$$\langle \alpha^n \cos \alpha \rangle = \begin{cases} 0 & \alpha < \alpha_1 \\ (\alpha - \alpha_1)^n \cos(\alpha - \alpha_1) & \alpha_1 \le \alpha < \alpha_2 \\ (\alpha_2 - \alpha_1)^n \cos(\alpha_2 - \alpha_1) & \alpha_2 \le \alpha \end{cases}$$

$$\langle \alpha - \alpha_1 \rangle^0 = \begin{cases} 0 & \alpha < \alpha_1 \\ 1 & \alpha \ge \alpha_1 \end{cases}$$

$$\langle \alpha - \alpha_1 \rangle = \begin{cases} 0 & \alpha < \alpha_1 \\ (\alpha - \alpha_1) & \alpha \ge \alpha_1 \end{cases}$$

$$\langle \alpha^n \rangle = \begin{cases} 0 & \alpha < \alpha_1 \\ (\alpha - \alpha_1)^n & \alpha_1 \le \alpha < \alpha_2 \\ (\alpha_2 - \alpha_1)^n & \alpha_2 \le \alpha \end{cases} \qquad n = 0, 1$$

TABLE 16-12, Part B Loading Functions for a Circular Beam 842

The elements of **f** are given below:

	Concentrated Vertical Load	Uniformly Distributed Transverse Force	Concentrated Bending Moment
f_ϕ	$\dfrac{W_y\lambda_1 R^2}{2}[\sin\langle\alpha-\alpha_1\rangle - \langle\alpha-\alpha_1\rangle\cos(\alpha-\alpha_1)]$	$-\dfrac{p_{y1}R^3}{2}\lambda_1(\langle\alpha-\alpha_1\rangle^0 - 2\langle\cos\alpha\rangle - \langle\alpha\sin\alpha\rangle)$	$-\dfrac{1}{2}\lambda_1 C_z R\langle\alpha-\alpha_1\rangle\sin(\alpha-\alpha_1)$
f_r	$-\dfrac{W_y R^3}{2}\left\{\dfrac{1}{GJ}[2\langle\alpha-\alpha_1\rangle + \langle\alpha-\alpha_1\rangle\cos(\alpha-\alpha_1) - 3\sin\langle\alpha-\alpha_1\rangle]\right.$ $\left. - \dfrac{1}{EI_z}[\sin\langle\alpha-\alpha_1\rangle - \langle\alpha-\alpha_1\rangle\cos(\alpha-\alpha_1)]\right\}$	$-\dfrac{p_{y1}R^4}{2}\left[\lambda_1(\langle\alpha\sin\alpha\rangle + \langle\cos\alpha\rangle - \langle\alpha-\alpha_1\rangle^0)\right.$ $\left. - \left(\dfrac{1}{EI_z}+\dfrac{3}{GJ}\right)\langle\cos\alpha\rangle - \dfrac{2}{GJ}\langle\alpha^2\rangle\right]$	$-C_z R^2\left\{\dfrac{1}{2EI_z}\sin\langle\alpha-\alpha_1\rangle\right.$ $+ \dfrac{1}{GJ}[\langle\alpha-\alpha_1\rangle^0 - \langle\alpha-\alpha_1\rangle^0\cos(\alpha-\alpha_1)]$ $\left. - \dfrac{1}{2}\langle\alpha-\alpha_1\rangle\sin(\alpha-\alpha_1)\right\}$
f_{θ_z}	$\dfrac{W_y R^2}{2}\left\{\dfrac{1}{EI_z}\langle\alpha-\alpha_1\rangle\sin(\alpha-\alpha_1)\right.$ $- \dfrac{1}{GJ}[2\langle\alpha-\alpha_1\rangle^0 - 2\langle\alpha-\alpha_1\rangle^0\cos(\alpha-\alpha_1)$ $\left. - \langle\alpha-\alpha_1\rangle\sin(\alpha-\alpha_1)]\right\}$	$-\dfrac{p_{y1}R^3}{2}\left[\lambda_1(\langle\sin\alpha\rangle - \langle\alpha\cos\alpha\rangle)\right.$ $\left. - \dfrac{2}{GJ}(\langle\alpha\rangle - \langle\sin\alpha\rangle)\right]$	$-C_z R\left\{\dfrac{1}{EI_z}[\langle\alpha-\alpha_1\rangle\cos(\alpha-\alpha_1) + \sin\langle\alpha-\alpha_1\rangle]\right.$ $\left. - \dfrac{1}{2GJ}[\sin\langle\alpha-\alpha_1\rangle - \langle\alpha-\alpha_1\rangle\cos(\alpha-\alpha_1)]\right\}$
f_{V_y}	$-W_y$	$-p_{y1}R\langle\alpha\rangle$	0
f_{M_z}	$W_y R\sin\langle\alpha-\alpha_1\rangle$	$p_{y1}R^2\langle\cos\alpha\rangle$	$-C_z\langle\alpha-\alpha_1\rangle^0\cos(\alpha-\alpha_1)$
f_T	$W_y R(\langle\alpha-\alpha_1\rangle^0\cos(\alpha-\alpha_1) - \langle\alpha-\alpha_1\rangle^0)$	$p_{1y}R^2(\langle\alpha\rangle - \langle\sin\alpha\rangle)$	$-C_z\sin\langle\alpha-\alpha_1\rangle$

	Uniformly Distributed Moment	Concentrated Torque	Uniformly Distributed Torque
f_ϕ	$-\dfrac{c_{z1}R^2}{2}\lambda_1(\langle\sin\alpha\rangle - \langle\alpha\cos\alpha\rangle)$	$-\dfrac{T_1 R}{2}\left\{\left(\dfrac{1}{GJ} - \dfrac{1}{EI_z}\right)\sin\langle\alpha-\alpha_1\rangle + \lambda_1[\langle\alpha-\alpha_1\rangle\cos(\alpha-\alpha_1)]\right\}$	$-\dfrac{m_{x1}R^2}{2}\left[\lambda_1(\langle\alpha-\alpha_1\rangle^0 - \langle\alpha\sin\alpha\rangle - \langle\cos\alpha\rangle) + \left(\dfrac{1}{GJ} - \dfrac{1}{EI_z}\right)\langle\cos\alpha\rangle\right]$
f_v	$-\dfrac{c_{z1}R^3}{2}\left[\lambda_1(\langle\sin\alpha\rangle - \langle\alpha\cos\alpha\rangle) - \dfrac{2}{GJ}(\langle\alpha\rangle - \langle\sin\alpha\rangle)\right]$	$-\dfrac{1}{2}T_1\lambda_1 R^2[\sin\langle\alpha-\alpha_1\rangle - \langle\alpha-\alpha_1\rangle\cos(\alpha-\alpha_1)]$	$\dfrac{1}{2}m_{x1}\lambda_1 R^3(\langle\alpha\sin\alpha\rangle + 2\langle\cos\alpha\rangle) - \langle\alpha-\alpha_1\rangle^0$
f_{θ_z}	$\dfrac{c_{z1}R^3}{2}\left[\lambda_1(\langle\alpha\sin\alpha\rangle + \langle\cos\alpha\rangle - \langle\alpha-\alpha_1\rangle^0) - \left(\dfrac{1}{EI_z} - \dfrac{1}{GJ}\right)\langle\cos\alpha\rangle\right]$	$\dfrac{1}{2}T_1\lambda_1 R\langle\alpha-\alpha_1\rangle\sin(\alpha-\alpha_1)$	$\dfrac{1}{2}m_{x1}\lambda_1 R^2(\langle\sin\alpha\rangle - \langle\alpha\cos\alpha\rangle)$
f_{v_y}	0	0	0
f_{M_z}	$-Rc_{z1}\langle\sin\alpha\rangle$	$T_1\sin\langle\alpha-\alpha_1\rangle$	$-m_{x1}R\langle\cos\alpha\rangle$
f_T	$-Rc_{z1}\langle\cos\alpha\rangle$	$-T_1\langle\alpha-\alpha_1\rangle^0\cos(\alpha-\alpha_1)$	$-m_{x1}R\langle\sin\alpha\rangle$

TABLE 16-12, Part B   Loading Functions for a Circular Beam   844

TABLE 16-12, PART C   STATIC RESPONSE OF CIRCULAR BEAMS UNDER OUT-OF-PLANE LOADING: OUT-OF-PLANE BOUNDARY CONDITIONS

Case	Boundary Conditions
1. Clamped end	$\phi = \upsilon = \theta_z = 0$
2. Free end	$M_z = V_y = T = 0$
3. Tangentially pinned–free end	$\theta_z = V_y = T = 0$
4. Tangentially pinned–fixed end	$\upsilon = \theta_z = T = 0$

Case	Boundary Conditions
5. Radially pinned–free end 	$M_z = \phi = V_y = 0$
6. Radially pinned–fixed end 	$M_z = \phi = v = 0$
7. Free–fixed end 	$T = M_z = v = 0$
8. Clamped–free end 	$\phi = \theta_z = V_y = 0$

TABLE 16-12, Part C   **Out-of-Plane Boundary Conditions**   **846**

TABLE 16-13, PART A   TRANSFER MATRICES FOR CIRCULAR SEGMENTS: IN-PLANE LOADING

Notation

R = radius of centroidal line of bar

I^* = moment of inertia modified for curvature of bar, $\int_A [z^2/(1 - z/R)]\,dA$

E = modulus of elasticity

ρ = mass per unit length

ω = natural frequency

p_x, p_z = distributed forces per unit length

m = distributed moment per unit length

u, w, θ, V, M, P = extension, deflection, slope, shear force, bending moment, and axial force

Bending and Extension

Definitions for λ_i:

$$\lambda_1 = \frac{1}{GJ} + \frac{1}{EI_z}$$

$$\lambda_2 = \tfrac{1}{2}(\sin\psi - \psi\cos\psi)$$

$$\lambda_3 = \tfrac{1}{2}(2 - 2\cos\psi - \psi\sin\psi)$$

$$\lambda_4 = \tfrac{1}{2}(2\psi + \psi\cos\psi - 3\sin\psi)$$

$$\lambda_5 = \tfrac{1}{2}(\psi\cos\psi + \sin\psi)$$

$$\lambda_6 = \tfrac{1}{2}(2 - 2\cos\psi - \psi\sin\psi)$$

$$\lambda_7 = \tfrac{1}{2}(2\psi + \psi\cos\psi - 3\sin\psi)$$

$$\lambda_8 = 2\sin\psi - \psi\cos\psi - \psi$$

State Variable:

$$\mathbf{z} = [u \quad w \quad \theta \quad V \quad M \quad P \quad 1]^T$$

Transfer Matrix:

$$\mathbf{U}^i = \begin{bmatrix}
U_{uu} & U_{uw} & U_{u\theta} & U_{uV} & U_{uM} & U_{uP} & F_u \\
U_{wu} & U_{ww} & U_{w\theta} & U_{wV} & U_{wM} & U_{wP} & F_w \\
U_{\theta u} & U_{\theta w} & U_{\theta\theta} & U_{\theta V} & U_{\theta M} & U_{\theta P} & F_\theta \\
U_{Vu} & U_{Vw} & U_{V\theta} & U_{VV} & U_{VM} & U_{VP} & F_V \\
U_{Mu} & U_{Mw} & U_{M\theta} & U_{MV} & U_{MM} & U_{MP} & F_M \\
U_{Pu} & U_{Pw} & U_{P\theta} & U_{PV} & U_{PM} & U_{PP} & F_P \\
0 & 0 & 0 & 0 & 0 & 0 & 1
\end{bmatrix}$$

Loading Vector:

$$\bar{\mathbf{z}}^i = \begin{bmatrix} F_u & F_w & F_\theta & F_V & F_M & F_P \end{bmatrix}^T \qquad \mathbf{f}^i = \begin{bmatrix} f_u & f_w & f_\theta & f_V & f_M & f_P \end{bmatrix}^T$$

Massless Circular Bars

Transfer Matrix:

$$
\mathbf{U}^i =
\begin{bmatrix}
\cos\psi & \sin\psi & -R(1-\cos\psi) & U_{uV} & -\dfrac{R^2}{EI^*}(\psi-\sin\psi) & \dfrac{R}{EA}\lambda_5 + \dfrac{R^3}{EI^*}\lambda_4 & F_u \\[2ex]
-\sin\psi & \cos\psi & -R\sin\psi & \dfrac{R^2}{EI^*}(1-\cos\psi) & -\dfrac{R^2}{EI^*}(1-\cos\psi) & -U_{wP} & F_w \\[2ex]
0 & 0 & 1 & \dfrac{R^2}{EI^*}(1-\cos\psi) & \dfrac{R}{EI^*}\psi & -\dfrac{R^2}{EI^*}(\psi-\sin\psi) & F_\theta \\[2ex]
0 & 0 & 0 & \cos\psi & 0 & -\sin\psi & F_V \\[1ex]
0 & 0 & 0 & R\sin\psi & 1 & -R(1-\cos\psi) & F_M \\[1ex]
0 & 0 & 0 & \sin\psi & 0 & \cos\psi & F_P \\[1ex]
0 & 0 & 0 & 0 & 0 & 0 & 1
\end{bmatrix}
$$

where the column element

$$U_{uV} = \left(\frac{1}{EA} + \frac{R^2}{EI^*}\right)R\lambda_2$$

$$U_{uV} = U_{wP} = \frac{R}{2AE}\,\psi\sin\psi - \frac{R^3}{EI^{*2}}\lambda_3$$

Loading vector: $\bar{\mathbf{z}}^i = \mathbf{u}^i \mathbf{f}$

$\mathbf{u}^i$ is the upper left 6×6 submatrix of $\mathbf{U}^i$.

TABLE 16-13, Part A   Transfer Matrices: In-Plane Loading   848

$$f_u = -R^2 \int_0^\psi \left[p_x\left(\frac{1}{EA}\lambda_5 + \frac{1}{EI^*}R^2\lambda_4\right) + m\frac{R}{EI^*}(\beta-\sin\beta) - p_z\left(\frac{1}{2}\frac{1}{EA}\beta\sin\beta - \frac{1}{EI^*}R^2\lambda_3\right)\right]d\beta$$

$$f_w = R^2 \int_0^\psi \left[p_x\left(\frac{1}{2}\frac{\beta}{EA}\sin\beta - \frac{1}{EI^*}R^2\lambda_3\right) - m\frac{1}{EI^*}R(1-\cos\beta) - p_z\left(\frac{\beta}{EA} + \frac{1}{EI^*}R^2\right)\lambda_2\right]d\beta$$

$$f_\theta = R^2 \int_0^\psi \left[p_x\frac{1}{EI^*}R(\beta-\sin\beta) + m\frac{1}{EI^*}\beta + p_z\frac{1}{EI^*}R(1-\cos\beta)\right]d\beta$$

$$f_V = R\int_0^\psi \left(p_x\sin\beta + p_z\cos\beta \right) d\beta$$

$$f_M = R\int_0^\psi \left[-p_xR(1-\cos\beta) - m - p_zR\sin\beta \right] d\beta$$

$$f_P = R\int_0^\psi \left(-p_x\cos\beta + p_z\sin\beta \right) d\beta$$

Circular Bars with Mass

$$\mathbf{U}^i =$$

$\cos\psi$	$\sin\psi$	0	$R(\cos\psi - 1)$	0	0	F_u
$-\sin\psi$	$\cos\psi$	0	$-R\sin\psi$	0	0	F_w
0	0	1	0	0	0	F_θ
$\rho\omega^2 R\psi\sin\psi$	$-\rho\omega^2 R\psi\cos\psi$	$\rho\omega^2 R^2(\psi\sin\psi + \cos\psi - 1)$	$\cos\psi$	0	$-\sin\psi$	F_V
$\rho\omega^2 R^2(\sin\psi - \psi\cos\psi)$	$-\rho\omega^2 R^2(\psi\sin\psi + \cos\psi - 1)$	$-\rho\omega^2 R\,r_y^2\psi + \rho\omega^2 R^3\lambda_8$	$R\sin\psi$	1	$R(\cos\psi - 1)$	F_M
$-\rho\omega^2 R\psi\cos\psi$	$-\rho\omega^2 R\psi\sin\psi$	$\rho\omega^2 R^2(\sin\psi - \psi\cos\psi)$	$\sin\psi$	0	$\cos\psi$	F_P
						1

The form of the loading vector is the same as that given above for a massless bar, i.e., $\bar{\mathbf{z}}^i = \mathbf{u}^i\mathbf{f}$, where $\mathbf{u}^i$ is the upper left 6×6 submatrix of this $\mathbf{U}^i$.

TABLE 16-13, PART B TRANSFER MATRICES FOR CIRCULAR SEGMENTS: OUT-OF-PLANE LOADING

Notation

I_z = moment of inertia about z axis
p_y = distributed force per unit length
m_x, m_z = distributed moments per unit length
G = shear modulus of elasticity
J = torsional constant
$\phi, v, \theta_z, V_y, M_z, T$ = angle of twist, deflection, slope, shear force, bending moment, and torque

See Table 16-13, Part A for further notation.

$$\mathbf{z} = [\phi \quad v \quad \theta_z \quad V_y \quad M_z \quad T \quad 1]^T$$

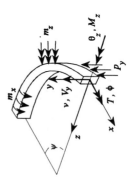

Bending and Torsion

State variable:

TABLE 16-13, Part B   Transfer Matrices: Out-of-Plane Loading   850

Transfer Matrix:

$$\mathbf{U}^i = \begin{bmatrix} U_{\phi\phi} & U_{\phi\upsilon} & U_{\phi\theta_z} & U_{\phi V_y} & U_{\phi M_z} & U_{\phi T} & F_\phi \\ U_{\upsilon\phi} & U_{\upsilon\upsilon} & U_{\upsilon\theta_z} & U_{\upsilon V_y} & U_{\upsilon M_z} & U_{\upsilon T} & F_\upsilon \\ U_{\theta_z\phi} & U_{\theta_z\upsilon} & U_{\theta_z\theta_z} & U_{\theta_z V_y} & U_{\theta_z M_z} & U_{\theta_z T} & F_{\theta_z} \\ U_{V_y\phi} & U_{V_y\upsilon} & U_{V_y\theta_z} & U_{V_y V_y} & U_{V_y M_z} & U_{V_y T} & F_{V_y} \\ U_{M_z\phi} & U_{M_z\upsilon} & U_{M_z\theta_z} & U_{M_z V_y} & U_{M_z M_z} & U_{M_z T} & F_{M_z} \\ U_{T\phi} & U_{T\upsilon} & U_{T\theta_z} & U_{TV_y} & U_{TM_z} & U_{TT} & F_T \\ 0 & 0 & 0 & 0 & 0 & 0 & 1 \end{bmatrix}$$

Loading Vector:

$$\bar{\mathbf{z}}^i = \begin{bmatrix} F_\phi & F_\upsilon & F_{\theta_z} & F_{V_y} & F_{M_z} & F_T \end{bmatrix}^T$$

$$\mathbf{f}^i = \begin{bmatrix} f_\phi & f_\upsilon & f_{\theta_z} & f_{V_y} & f_{M_z} & f_T \end{bmatrix}^T$$

Massless Circular Bars

Transfer Matrix:

$$U^i = \begin{bmatrix}
\cos\psi & 0 & \sin\psi & -\lambda_1 R^2\lambda_2 & \frac{1}{2}\lambda_1 R\psi\sin\psi & \frac{R}{GJ}\lambda_5 - \frac{R}{EI_z}\lambda_2 & F_\phi \\
R(\cos\psi-1) & 1 & R\sin\psi & -\frac{R^3}{EI_z}\lambda_2 + \frac{R^3}{GJ}\lambda_4 & \frac{R^2\psi}{2EI_z}\sin\psi - \frac{R^3}{GJ}\lambda_3 & -\lambda_1 R^2\lambda_2 & F_v \\
-\sin\psi & 0 & \cos\psi & \frac{R^2}{GJ}\lambda_3 - \frac{R^2\psi}{2EI_z}\sin\psi & \frac{R}{EI_z}\lambda_5 - \frac{R}{GJ}\lambda_2 & -\frac{1}{2}\lambda_1 R\psi\sin\psi & F_{\theta_z} \\
0 & 0 & 0 & 1 & 0 & 0 & F_{V_y} \\
0 & 0 & 0 & -R\sin\psi & \cos\psi & -\sin\psi & F_{M_z} \\
0 & 0 & 0 & R(\cos\psi-1) & \sin\psi & \cos\psi & F_T \\
0 & 0 & 0 & 0 & 0 & 0 & 1
\end{bmatrix}$$

Loading vector: $\bar{z}^i = \mathbf{u}^i\mathbf{f}$
$\mathbf{u}^i$ is the upper left 6×6 submatrix of U^i

$$f_\phi = R^2\int_0^\psi \left[-p_y\lambda_1 R\lambda_2 - \frac{\lambda_1}{2}m_z\beta\sin\beta + m_x\left(\frac{1}{GJ}\lambda_5 - \frac{1}{EI_z}\lambda_2\right)\right]d\beta$$

$$f_v = R^3\int_0^\psi \left[-p_y R\left(\frac{1}{EI_z}\lambda_2 - \frac{1}{GJ}\lambda_4\right) - m_z\left(\frac{1}{2EI_z}\beta\sin\beta - \frac{1}{GJ}\lambda_3\right) - m_x\lambda_1\lambda_2\right]d\beta$$

$$f_{\theta_z} = R^2\int_0^\psi \left[-p_y R\left(\frac{1}{GJ}\lambda_3 - \frac{1}{2EI_z}\beta\sin\beta\right) + m_z\left(\frac{1}{EI_z}\lambda_5 - \frac{1}{GJ}\lambda_2\right) + \frac{1}{2}m_x\lambda_1\beta\sin\beta\right]d\beta$$

$$f_{V_y} = R \int_0^\psi p_y \, d\beta$$

$$f_{M_z} = R \int_0^\psi \left(-p_y R \sin\beta - m_z \cos\beta - m_x \sin\beta \right) d\beta$$

$$f_T = -R \int_0^\psi \left[p_y R(\cos\beta - 1) - m_z \sin\beta + m_x \cos\beta \right] d\beta$$

Circular Bars with Mass

$\mathbf{U}^i =$

$\cos\psi$	0	$\sin\psi$	0	0	0	F_ϕ
$R(\cos\psi - 1)$	1	$R\cos\psi$	0	0	0	F_v
$-\sin\psi$	0	$\cos\psi$	0	0	0	F_{θ_z}
$\rho\omega^2 R^2(\psi - \sin\psi)$	$-\rho\omega^2 R\psi$	$-\rho\omega^2 R^2(1 - \cos\psi)$	1	0	0	F_{V_y}
$U_{M_z\phi}$	$\rho\omega^2 R^2(1 - \cos\psi)$	$U_{M_z\theta_z}$	$-R\sin\psi$	$\cos\psi$	$-\sin\psi$	F_{M_z}
$U_{T\phi}$	$\rho\omega^2 R^2(\psi - \sin\psi)$	$U_{T\theta_z}$	$R(\cos\psi - 1)$	$\sin\psi$	$\cos\psi$	F_T
0	0	0	0	0	0	1

$$U_{M_z\phi} = -\rho\omega^2 R^3 \lambda_6 + \rho\omega^2 R(r_p^2 + r_z^2)\tfrac{1}{2}\psi \sin\psi$$

$$U_{T\phi} = -\rho\omega^2 R r_p^2 \lambda_5 + \rho\omega^2 R r_z^2 \lambda_2 - \rho\omega^2 R^3 \lambda_7$$

$$U_{M_z\theta_z} = -\rho\omega^2 R r_z^2 \lambda_5 + \rho\omega^2 R(r_p^2 + R^2)\lambda_2$$

$$U_{T\theta_z} = -U_{M_z\phi}$$

The form of the loading vector is the same as that given above for the massless bar, i.e., $\bar{\mathbf{z}}^i = \mathbf{u}^i \mathbf{f}$, where $\mathbf{u}^i$ is the upper left 6×6 submatrix of this $\mathbf{U}^i$.

TABLE 16-14, PART A   STIFFNESS MATRICES FOR CIRCULAR SEGMENTS: IN-PLANE LOADING

Notation

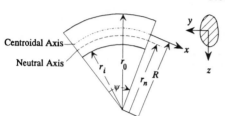

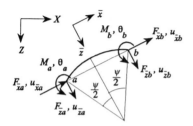

Bending and Extension

R = radius of centroidal line of bar
E = modulus of elasticity
G = shear modulus of elasticity
A = area of cross section
A_s = equivalent shear area
I = moment of inertia about y axis
$\beta = \frac{1}{2}\psi$
x, y, z = natural (local) curved element coordinates
$\bar{x}, \bar{y}, \bar{z}$ = generalized element coordinates
X, Y, Z = global coordinates
$F_{\bar{x}}, F_{\bar{z}}$ = forces on ends of element in $\bar{x}, \bar{z}$ directions
$u_{\bar{x}}, u_{\bar{z}}$ = displacements on ends of element in $\bar{x}, \bar{z}$ directions
θ, M = rotation and bending moment about $\bar{y}$ axis

Definitions for C_i:

$$C_1 = EA\frac{R - r_n}{r_n} \qquad C_2 = \frac{r_n}{GA_s}$$

$$C_3 = \frac{EA}{r_n} \qquad C_4 = \frac{r_n C_3 - C_1(1 - C_2 C_3)}{-r_n C_3 - C_1(1 + C_2 C_3)}$$

$$C_5 = 1 - C_4 - C_2 C_3(1 + C_4) \qquad C_6 = C_1 C_5$$

$$C_7 = C_3(1 + C_4) \qquad C_8 = \int_{r_i}^{r_0} r\, dA$$

$$C_9 = \int_{r_i}^{r_0}(r_0 - r)r\, dA \qquad C_{10} = \int_{r_i}^{r_0}(r_n - r)^2 r\, dA.$$

$$C_{11} = 2\frac{C_9}{C_{10}} - 2C_8 \qquad C_{12} = 2C_4 C_8 + \frac{C_5 C_9}{r_n}$$

$$C_{13} = 2C_4 C_8 - 2\frac{C_5 C_9}{r_n} \qquad C_{14} = C_4^2 C_8 + \frac{C_5^2 C_{10}}{r_n^2}$$

$$C_{15} = C_4^2 C_8 - \frac{C_5^2 C_{10}}{r_n^2}$$

For thin beams without shear deformation effects and rotary inertia
$$C_1 = EI/R^2 \qquad C_2 = C_9 = C_{10} = 0 \qquad r_n = R$$

TABLE 16-14, Part A | Stiffness Matrices: In-Plane Loading | 854

Nodal Variables:

$$\tilde{\mathbf{v}}^i = \begin{bmatrix} u_{\bar{x}a} & u_{\bar{z}a} & \theta_a & u_{\bar{x}b} & u_{\bar{z}b} & \theta_b \end{bmatrix}^T \qquad \tilde{\mathbf{p}}^i = \begin{bmatrix} F_{\bar{x}a} & F_{\bar{z}a} & M_a & F_{\bar{x}b} & F_{\bar{z}b} & M_b \end{bmatrix}^T$$

$$\tilde{\mathbf{p}}^i = \tilde{\mathbf{k}}^i \tilde{\mathbf{v}}^i$$

Element stiffness matrices $\mathbf{k}^i$ in the global coordinate system are obtained using $\mathbf{k}^i = \mathbf{T}^{iT}\tilde{\mathbf{k}}^i\mathbf{T}^i$, where $\mathbf{T}^i$ is the transformation matrix of Table 13-14 for in-plane loading.

Stiffness matrices:

$$\tilde{\mathbf{k}}^i = \mathbf{QS}$$

$$\mathbf{Q} = \begin{bmatrix} 0 & 0 & 0 & 0 & -C_7 & 0 \\ 0 & 0 & 0 & 0 & 0 & -C_7 \\ 0 & -C_1 & 0 & 0 & -C_6\cos\beta & C_6\sin\beta \\ 0 & 0 & 0 & 0 & C_7 & 0 \\ 0 & 0 & 0 & 0 & 0 & C_7 \\ 0 & C_1 & 0 & 0 & C_6\cos\beta & C_6\sin\beta \end{bmatrix}$$

$$\mathbf{S} = \begin{bmatrix}
0 & -(C_5\cos\beta)/\Delta_2 & -r_n(C_4\cos\beta\sin\beta - \beta)/\Delta_2 \\
-(C_5\sin\beta)/\Delta_1 & 0 & r_n(C_4\cos\beta\sin\beta + \beta)/\Delta_1 \\
\dfrac{1}{2} & \begin{array}{l}-(C_4\sin^2\beta \\ -C_5\cos^2\beta)/\Delta_2\end{array} & r_n(C_4\sin\beta - \beta\cos\beta)/\Delta_2 \\
\begin{array}{l}[C_5\sin\beta(\beta\sin\beta + \cos\beta) \\ -\beta C_4\cos^2\beta]/\Delta_1\end{array} & -\dfrac{1}{2} & \begin{array}{l}-r_n[\beta C_4\cos\beta + \beta(\beta\sin\beta \\ + \cos\beta)]/\Delta_1\end{array} \\
\dfrac{\beta}{\Delta_1} & 0 & -r_n(\sin\beta - \beta\cos\beta)/\Delta_1 \\
0 & 1/\Delta_2 & -(r_n\sin\beta)/\Delta_2 \\
0 & (C_5\cos\beta)/\Delta_2 & -r_n(C_4\sin\beta\cos\beta - \beta)/\Delta_2 \\
(C_5\sin\beta)/\Delta_1 & 0 & -r_n(C_4\cos\beta\sin\beta + \beta)/\Delta_1 \\
\dfrac{1}{2} & \begin{array}{l}-(C_5\cos^2\beta \\ -C_4\sin^2\beta)/\Delta_2\end{array} & r_n(C_4\sin\beta - \beta\cos\beta)/\Delta_2 \\
\begin{array}{l}-[C_5\sin\beta(\sin\beta + \cos\beta) \\ -\beta C_4\cos^2\beta]/\Delta_1\end{array} & -\dfrac{1}{2} & \begin{array}{l}r_n[\beta C_4\cos\beta + \beta(\beta\sin\beta \\ + \cos\beta)]/\Delta_1\end{array} \\
-\dfrac{\beta}{\Delta_1} & 0 & r_n(\sin\beta - \beta\cos\beta)/\Delta_1 \\
0 & -1/\Delta_2 & -(r_n\sin\beta)/\Delta_2
\end{bmatrix}$$

$$\Delta_1 = 2C_5\sin^2\beta - 2\beta(C_4 + C_5)\cos\beta\sin\beta - 2\beta^2$$
$$\Delta_2 = 2(C_4 + C_5)\cos\beta\sin\beta - 2\beta$$

Notation

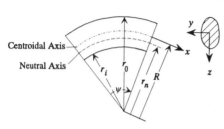

J = torsional constant
I_z = moment of inertia about z axis
I_p = polar moment of inertia about x axis
$\beta = \frac{1}{2}\psi$
x, y, z = natural (local) curved element coordinates
$\bar{x}, \bar{y}, \bar{z}$ = generalized element coordinates
X, Y, Z = global coordinates
$u_{\bar{y}}, F_{\bar{y}}$ = displacements, forces on ends of element in $\bar{y}$ direction
$\theta_{\bar{x}}, M_{\bar{x}}$ = rotation and moment about $\bar{x}$ axis
$\theta_{\bar{z}}, M_{\bar{z}}$ = rotation and moment about $\bar{z}$ axis

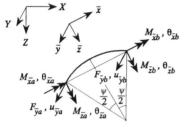

Bending and Torsion

See Table 16-14, Part A for further notation.

Definitions for C_i:

$$C_1 = 1/GA_s \qquad\qquad C_2 = GI_p/R^2$$

$$C_3 = EI_z/R^2 \qquad\qquad C_4 = 2RC_2C_3/(C_2 + C_3)$$

$$C_5 = 2C_3/(C_2 + C_3) \qquad C_6 = (C_2 - C_3)/(C_2 + C_3)$$

$$C_7 = I_z/A \qquad\qquad C_8 = I_p/A$$

$$C_9 = R^2 + C_7 + C_8 \qquad C_{10} = R^2 - C_7 + C_8$$

$$C_{11} = -C_{10} \qquad\qquad C_{12} = C_7 + C_5C_8$$

$$C_{13} = C_7 - C_5C_8 \qquad C_{14} = C_7 + C_5^2C_8$$

$$C_{15} = C_7 - C_5^2C_8$$

Nodal Variables:

$$\tilde{\mathbf{v}}^i = \begin{bmatrix} u_{\bar{y}a} & \theta_{\bar{x}a} & \theta_{\bar{z}a} & u_{\bar{y}b} & u_{\bar{x}b} & \theta_{\bar{z}b} \end{bmatrix}^T$$

$$\tilde{\mathbf{p}}^i = \begin{bmatrix} F_{\bar{y}a} & M_{\bar{x}a} & M_{\bar{z}a} & F_{\bar{y}b} & M_{\bar{x}b} & M_{\bar{z}b} \end{bmatrix}^T \qquad \tilde{\mathbf{p}}^i = \tilde{\mathbf{k}}^i\tilde{\mathbf{v}}^i$$

Element stiffness matrices $\mathbf{k}^i$ in the global coordinate system are obtained using $\mathbf{k}^i = \mathbf{T}^{iT}\tilde{\mathbf{k}}^i\mathbf{T}^i$, where $\mathbf{T}^i$ is the transformation matrix of Table 13-14 for out-of-plane loading.

Stiffness matrices:

$$\tilde{\mathbf{k}}^i = \mathbf{QS}$$

$$\mathbf{Q} = \begin{bmatrix} 0 & 0 & 0 & 0 & -C_2 & 0 \\ 0 & 0 & 0 & -C_4 & RC_2\cos\beta & 0 \\ 0 & 0 & 0 & 0 & -RC_2\sin\beta & -C_4 \\ 0 & 0 & 0 & 0 & C_2 & 0 \\ 0 & 0 & 0 & C_4 & -RC_2\cos\beta & 0 \\ 0 & 0 & 0 & 0 & -RC_2\sin\beta & C_4 \end{bmatrix}$$

$$\mathbf{S} =$$

$\dfrac{1}{2R}$	$-\dfrac{\cos\beta}{2}$	$(\beta\sin\beta - C_6\cos^3\beta + \cos\beta)/\Delta_1$
0	$\tfrac{1}{2}$	$(C_6\cos^2\beta - 1)/\Delta_1$
$-[\sin\beta(C_6 - 1)$ $+\beta\cos\beta]/(R\Delta_2)$	$-\beta[(1 + C_1C_2)(C_6\sin^2\beta - 1)$ $+\cos^2\beta]/\Delta_2$	$-\{[\beta(C_1C_2 + 1)C_6 - \beta]\cos\beta\sin\beta$ $-\beta^2(C_1C_2 + 1)\}/\Delta_2$
$-(\sin\beta)/R\Delta_2$	$[\cos\beta\sin\beta - (C_1C_2 + 1)\beta]/\Delta_2$	$-(\sin^2\beta)/\Delta_2$
$(C_6\cos\beta\sin\beta$ $-\beta)/(R\Delta_2)$	$[\sin\beta(1 + C_6\sin^2\beta)$ $+\beta\cos\beta]/\Delta_2$	$(C_6\cos\beta\sin^2\beta - \beta)\sin\beta/\Delta_2$
0	0	$-1/\Delta_1$

$\dfrac{1}{2R}$	$-\dfrac{\cos\beta}{2}$	$-(\beta\sin\beta - C_6\cos^3\beta + \cos\beta)/\Delta_1$
0	$\tfrac{1}{2}$	$-(C_6\cos^2\beta - 1)/\Delta_1$
$[\sin\beta(C_6 - 1)$ $-\beta\cos\beta]/(R\Delta_2)$	$\beta[(C_1C_2 + 1)C_6\sin^2\beta$ $+\cos^2\beta - C_1C_2 - 1]/\Delta_2$	$-[\beta(C_1C_2 + 1)C_6\sin^2\beta$ $-\beta\cos\beta\sin\beta$ $-\beta(C_1C_2 + 1)]/\Delta_2$
$\sin^2\beta/R\Delta_2$	$-[\cos\beta\sin\beta + \beta(1 + C_1C_2)]/\Delta_2$	$-\sin^2\beta/\Delta_2$
$-(C_6\cos\beta\sin\beta$ $-\beta)/(R\Delta_2)$	$-[\sin\beta(C_6\sin^2\beta - 1)$ $+\beta\cos\beta]/\Delta_2$	$[(C_6\cos\beta\sin\beta - \beta)\sin\beta]/\Delta_2$
0	0	$1/\Delta_1$

$\Delta_1 = 2[C_6\cos\beta\sin\beta + \beta]$

$\Delta_2 = 2[C_6 R\sin^4\beta + (C_6\cos^2\beta - 1)R\sin^2\beta - \beta(C_1C_2 - 1)C_6(\cos\beta) R\sin\beta + \beta^2(C_1C_2 + 1)]$

TABLE 16-15, PART A   IN-PLANE DEFORMATION: POINT MATRICES

Notation

ρ = mass per unit length
r_y = radius of gyration about y axis
a_i = location of point occurrence

$\mathbf{I}$ = unit diagonal matrix
ω = natural frequency
$\mathbf{z}$ = state vector = $[u \quad w \quad \theta \quad V \quad M \quad P \quad 1]^T$

See Part B for further notation

Case	Transfer Matrix	Stiffness and Mass Matrices
1. Concentrated applied forces	$$\mathbf{U}_i = \begin{bmatrix} \mathbf{I}_{6\times6} & \begin{matrix} \mathbf{0}_{3\times1} \\ -W \\ C \\ -P_i \end{matrix} \\ \hline \mathbf{0}_{1\times6} & 1 \end{bmatrix}$$	Traditionally these forces are taken as node forces.
2. Extension spring k_u, rotary spring k_1^*, and transverse elastic support k_1. See Part B of this table and Table 11-21.	$$\mathbf{U}_i = \begin{bmatrix} \mathbf{I}_{3\times3} & \begin{matrix} 0 & 0 & 1/k_u \\ 1/k_1 & 0 & 0 \\ 0 & 1/k_1^* & 0 \end{matrix} & \mathbf{0}_{6\times1} \\ \mathbf{0}_{3\times3} & \mathbf{I}_{3\times3} & \\ \hline \mathbf{0}_{1\times6} & & 1 \end{bmatrix}$$	$$\begin{bmatrix} P_a \\ V_a \\ M_a \end{bmatrix} = \mathbf{k}_{aa} \begin{bmatrix} u_a \\ w_a \\ \theta_a \end{bmatrix}$$ $$\mathbf{k}_{aa} = \begin{bmatrix} k_u & 0 & 0 \\ 0 & k_1 & 0 \\ 0 & 0 & k_1^* \end{bmatrix}$$

TABLE 16-15, Part A   **In-Plane Point Matrices**   858

3.		
Concentrated mass with different supports 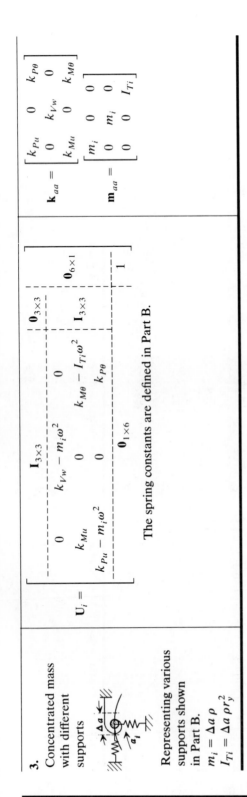 Representing various supports shown in Part B. $m_i = \Delta a\,\rho$ $I_{Ti} = \Delta a\,\rho\,r_y^2$	$$\mathbf{U}_i = \left[\begin{array}{ccc|ccc|c} & \mathbf{I}_{3\times3} & & & \mathbf{0}_{3\times3} & & \\ \hline k_{Mu} & k_{Vw} - m_i\omega^2 & 0 & & & & \\ k_{Pu} - m_i\omega^2 & 0 & k_{M\theta} - m_i\omega^2 & & \mathbf{I}_{3\times3} & & \mathbf{0}_{6\times1} \\ 0 & 0 & k_{P\theta} & & & & \\ \hline & \mathbf{0}_{1\times6} & & & & & 1 \end{array}\right]$$ The spring constants are defined in Part B.	$$\mathbf{k}_{aa} = \begin{bmatrix} k_{Pu} & 0 & k_{P\theta} \\ 0 & k_{Vw} & 0 \\ k_{Mu} & 0 & k_{M\theta} \end{bmatrix}$$ $$\mathbf{m}_{aa} = \begin{bmatrix} m_i & 0 & 0 \\ 0 & m_i & 0 \\ 0 & 0 & I_{Ti} \end{bmatrix}$$

TABLE 16-15, PART B IN-PLANE DEFORMATION: SPRING CONSTANTS [a]

	k_1	Rotary Spring k_1^*	k_u	$\overset{\leftarrow l_b \rightarrow}{E,I,A}$	l_b, E,I,A	l_b, E,I,A	l_b, E,I,A
k_{Mu}	0	0	0	$-\dfrac{6EI}{l_b^2}$	$-\dfrac{3EI}{l_b^2}$	0	0
k_{Pu}	0	0	k_u	$\dfrac{12EI}{l_b^3}$	$\dfrac{3EI}{l_b^3}$	$\dfrac{3EI}{l_b^3}$	0
k_{Vw}	k_1	0	0	$\dfrac{EA}{l_b}$	$\dfrac{EA}{l_b}$	$\dfrac{EA}{l_b}$	$\dfrac{EA}{l_b}$
$k_{M\theta}$	0	k_1^*	0	$\dfrac{4EI}{l_b}$	$\dfrac{3EI}{l_b}$	0	0
$k_{P\theta}$	0	0	0	$-\dfrac{6EI}{l_b^2}$	$-\dfrac{3EI}{l_b^2}$	0	0

[a] Units: k_1, k_u are in force/length; k_1^* is in force − length/rad. Circles in figures designate flexurally pinned ends.

TABLE 16-15, Part B | In-Plane Spring Constants | 860

TABLE 16-16, PART A   OUT-OF-PLANE DEFORMATION: POINT MATRICES

Notation

ρ = mass per unit length

r_p = polar radius of gyration about x axis

r_z = radius of gyration about z axis

a_i = location of occurrence

$\mathbf{I}$ = unit diagonal matrix

$m_i = \Delta a\, \rho$

$I_{pi} = \Delta a\, \rho r_p^2$

$I_{zi} = \Delta a\, \rho r_z^2$

$\mathbf{z}$ = state vector = $[\phi \quad v \quad \theta_z \quad V_y \quad M_z \quad T \quad 1]^T$

See Part B for further notation.

Case	Transfer Matrices	Stiffness and Mass Matrices
1. Concentrated applied forces	$$\mathbf{U}_i = \left[\begin{array}{c|c} \mathbf{I}_{6\times6} & \begin{array}{c} \mathbf{0}_{3\times1} \\ -W_y \\ C_z \\ -T_1 \end{array} \\ \hline \mathbf{0}_{1\times6} & 1 \end{array}\right]$$	Traditionally these forces are taken as node forces
2. Torsional spring k_ϕ, rotary spring k_1^* and transverse elastic support k_1. See Part B. 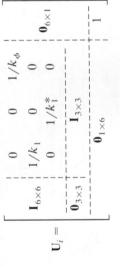	$$\mathbf{U}_i = \left[\begin{array}{c|c} \mathbf{I}_{6\times6} & \mathbf{0}_{6\times1} \\ \hline \begin{array}{ccc} 0 & 0 & 1/k_\phi \\ 1/k_1 & 0 & 0 \\ 0 & 1/k_1^* & 0 \\ \hline \mathbf{0}_{3\times3} & \mathbf{I}_{3\times3} \end{array} & \\ \hline \mathbf{0}_{1\times6} & 1 \end{array}\right]$$	$\begin{bmatrix} T_a \\ V_{ya} \\ M_{za} \end{bmatrix} = \mathbf{k}_{aa}\begin{bmatrix} \phi_a \\ v_a \\ \theta_{za} \end{bmatrix}$ $\mathbf{k}_{aa} = \begin{bmatrix} k_\phi & 0 & 0 \\ 0 & k_1 & 0 \\ 0 & 0 & k_1^* \end{bmatrix}$

3.

Concentrated mass
flexible supports

$m_i = \Delta a\,\rho$

$I_{pi} = \Delta a\,\rho r_p^2$

$I_{pi} = (I_{pi})_{\text{c.g. of } m_i} + m_i e^2$

c.g. = center of gravity

$$U_i = \left[\begin{array}{ccc|c}
k_{V_y\phi} - m_i\omega^2 e & k_{V_yv} - m_i\omega^2 & 0 & \\
0 & 0 & k_{M_z\theta_z} - I_{Ti}\omega^2 & \mathbf{0}_{3\times3} \\
k_{T\phi} - I_{pi}\omega^2 & k_{Tv} - m_i\omega^2 e & 0 & \\
\hline
& \mathbf{I}_{3\times3} & & \mathbf{0}_{6\times1} \\
\hline
& \mathbf{0}_{1\times6} & & 1
\end{array}\right]$$

The spring constants $k_{T\phi}$, k_{Tv}, $k_{V_y\phi}$, k_{V_yv}, and $k_{M_z\theta_z}$ are defined in Part B.

$$\mathbf{k}_{aa} = \begin{bmatrix}
k_{T\phi} & k_{Tv} & 0 \\
k_{V_y\phi} & k_{V_yv} & 0 \\
0 & 0 & k_{M_z\theta_z}
\end{bmatrix}$$

$$\mathbf{m}_{aa} = \begin{bmatrix}
I_{pi} & 0 & 0 \\
0 & m_i & 0 \\
0 & 0 & I_{Ti}
\end{bmatrix}$$

TABLE 16-16, Part A   **Out-of-Plane Point Matrices**   862

TABLE 16-16, PART B OUT-OF-PLANE DEFORMATION: SPRING CONSTANTS [a]

		Torsional Spring	Rotary Spring		
$k_{V_y \phi}$	0	0	0	0	0
$k_{T\phi}$	0	k_ϕ	0	$\dfrac{4EI}{l_b}$	$\dfrac{3EI}{l_b}$
$k_{V_y v}$	k_1	0	0	$\dfrac{EA}{l_b}$	$\dfrac{EA}{l_b}$
k_{Tv}	0	0	0	0	0
$k_{M_z \theta_z}$	0	0	k_1^*	$\dfrac{4EI}{l_b}$	$\dfrac{3EI}{l_b}$

TABLE 16-16, Part B | Out-of-Plane Spring Constants | 864

TABLE 16-16, PART B (continued) OUT-OF-PLANE DEFORMATION: SPRING CONSTANTS

	E, I	E, I	E, I, G, J	E, I, G, J	G, J
$k_{V_y\phi}$	0	0	$\dfrac{6EI}{\ell_b^2}$	$\dfrac{3EI}{\ell_b^2}$	0
$k_{T\phi}$	0	0	$\dfrac{4EI}{\ell_b}$	$\dfrac{3EI}{\ell_b}$	0
$k_{V_y v}$	$\dfrac{EA}{\ell_b}$	$\dfrac{EA}{\ell_b}$	$\dfrac{12EI}{\ell_b^3}$	$\dfrac{3EI}{\ell_b^3}$	0
k_{Tv}	0	0	$\dfrac{6EI}{\ell_b^2}$	$\dfrac{3EI}{\ell_b^2}$	0
$k_{M_z\theta_z}$	0	0	$\dfrac{GJ}{\ell_b}$	$\dfrac{GJ}{\ell_b}$	$\dfrac{GJ}{\ell_b}$

[a]Units: k_1 is in force/length; k_1^* and k_ϕ are in force − length/rad. Circles on figures designate the flexural pinned end.

TABLE 16-17   CONSISTENT MASS MATRICES FOR CIRCULAR SEGMENTS

Notation

$\rho^* $ = mass per unit volume ρ = mass per unit length

$\beta = \frac{1}{2}\psi$

Element variables and the sign convention are the same as those in Table 16-14. S and C_i are shown in Table 16-14.

H is symmetric and the elements not shown in this table are zero.

In-Plane Motion, $\mathbf{m}^i = \rho^* \mathbf{S}^T \mathbf{H} \mathbf{S}$

$$H_{11} = 2\beta\left(C_8 + \frac{C_{10}}{r_n^2} - \frac{2C_9}{r_n}\right)$$

$$H_{31} = -C_{11}\sin\beta$$

$$H_{61} = 2\sin\beta\left(C_8 + \frac{C_5 C_{10}}{r_n^2} - \frac{C_5 C_9}{r_n} - \frac{C_9}{r_n}\right) + C_{11}\beta\cos\beta$$

$$H_{22} = 2C_8\beta + \tfrac{2}{3}\beta^3\left(C_8 + \frac{C_{10}}{r_n^2} - \frac{2C_9}{r_n}\right)$$

$$H_{42} = 2\sin\beta\left(C_8 - \frac{C_9}{r_n}\right) + C_{11}\beta\cos\beta$$

$$H_{52} = \sin\beta\left(6C_8 + \frac{2C_5 C_{10}}{r_n^2} - \frac{4C_9}{r_n} + C_{13}\right) + C_{11}\beta^2\sin\beta$$

$$\qquad + 2\beta\cos\beta\left(\frac{C_5 C_9}{r_n} - \frac{C_5 C_{10}}{r_n^2} + \frac{2C_9}{r_n} - 3C_8\right)$$

$$H_{33} = 2C_8\beta$$

$$H_{63} = \frac{\beta C_{13}}{2} - \frac{C_{12}\sin\beta\cos\beta}{2}$$

$$H_{44} = 2C_8\beta$$

$$H_{54} = \frac{\beta C_{13}}{2} + \frac{C_{12}\sin\beta\cos\beta}{2}$$

$$H_{55} = C_8\frac{\beta^3}{3} + C_{14}\beta + \frac{C_{12}\sin 2\beta}{4} + C_{15}\sin\beta\cos\beta - \frac{C_{12}\beta\cos 2\beta}{2}$$

$$H_{66} = C_8\frac{\beta^3}{3} + C_{14}\beta - \frac{C_{12}\sin 2\beta}{4} - C_{15}\sin\beta\cos\beta + \frac{C_{12}\beta\cos 2\beta}{2}$$

TABLE 16-17 (continued) CONSISTENT MASS MATRICES FOR CIRCULAR SEGMENTS

$$\textit{Out-of-Plane Motion, } \mathbf{m}^i = \rho R S^T H S$$

$H_{11} = R^2\beta$

$H_{21} = R^2 \sin \beta$

$H_{22} = \frac{1}{2}(C_9\beta + C_{10} \sin \beta \cos \beta)$

$H_{33} = \frac{1}{2}(C_9\beta + C_{11} \sin \beta \cos \beta)$

$H_{43} = \frac{1}{8}(C_{10} \sin 2\beta + 2C_{11}\beta \cos 2\beta + 4C_{12}\beta + 4C_{13} \sin \beta \cos \beta)$

$H_{44} = \frac{1}{24}\big[4C_9\beta^3 + 6C_{10}\beta^2 \sin 2\beta + 12C_{14}\beta + 12C_{15} \sin \beta \cos \beta$

$\qquad + 6(C_{10} + 2C_{13})\beta \cos 2\beta + 3(C_{11} - 2C_{13})\sin 2\beta\big]$

$H_{53} = R^2(1 + C_1C_2)(\sin \beta - \beta \cos \beta) + C_7 \sin \beta$

$H_{54} = R^2(1 + C_1C_2)(\beta^2 \sin \beta + 2\beta \cos \beta - 2 \sin\beta) + C_7\beta \cos \beta$

$H_{55} = (1 + C_1C_2)^2\beta^3(\frac{1}{3}R^2) + C_7\beta$

$H_{61} = R^2(\sin \beta - \beta \cos \beta)$

$H_{62} = \frac{1}{8}C_{10} \sin 2\beta + \frac{1}{4}C_{11}\beta \cos 2\beta - C_{12}(\frac{1}{2}\beta)$

$\qquad + \frac{1}{2}C_{13}(\sin \beta \cos \beta)$

$H_{66} = \frac{1}{24}\big[4C_9\beta^3 - 6C_{10}\beta^2\sin 2\beta + 12C_{14}\beta - 12C_{15} \sin \beta \cos \beta$

$\qquad - 6(C_{10} + 2C_{13})\beta \cos 2\beta - 3(C_{11} - 2C_{13})\sin 2\beta\big]$

TABLE 16-17 | **Consistent Mass Matrices for Circular Segments**   866

Rotors

In this chapter the critical speeds of a rotor and the response of a rotor to unbalanced forces are treated. The transient response of the rotor to loadings on the shaft or through the bearing systems is also considered. The formulas presented are for shafts that are modeled primarily using the technical (Euler–Bernoulli) beam theory of shafts.

17.1 NOTATION

The notation in this chapter conforms to that normally employed in practice by engineers dealing with rotating-shaft systems. It differs somewhat from that used in the rest of this book.

A	Cross-sectional area (L^2)
A_s	Equivalent shear area, $= A/\alpha_s$ (L^2)
α_s	Shear correction factor (Table 2-4)
c	Coefficient of viscous damping (FT/L)
c_c	Critical damping coefficient (FT/L)
$c_{yy}, c_{yz}, c_{zy}, c_{zz}$	Damping coefficients for bearing or seal system (FT/L) (Fig. 17-1)

$c_{yy}^*, c_{yz}^*, c_{zy}^*, c_{zz}^*$	Rotary damping coefficients for bearing or seal system (FLT/rad)
$\bar{c}_{yy}, \bar{c}_{yz}, \bar{c}_{zy}, \bar{c}_{zz}$	Damping coefficients for pedestal of bearing or seal system (FT/L) (Fig. 17-1)
E	Modulus of elasticity of material (F/L^2)
e	Eccentricity arm for offset mass (L)
G	Shear modulus of elasticity (F/L^2)
$\mathbf{g}^i$	Gyroscopic matrix for ith element
I	Moment of inertia of cross-sectional area about transverse neutral axes (L^4)
I_p	Polar mass moment of inertia per unit length of shaft, $= \rho^* A r_p^2$ (ML); for hollow circular cross section $I_p = \frac{1}{2}\rho^* A(r_o^2 + r_i^2)$
I_{pi}	Polar mass moment of inertia of concentrated mass at station i (ML^2); can be calculated as $I_{pi} = \Delta a\, \rho^* A r_p^2$, where Δa is length of shaft lumped at station i; formulas for several configurations given in Table 12-5; for disk of concentrated mass M_i, $I_{pi} = \frac{1}{2}M_i(r_o^2 + r_i^2)$
I_T	Transverse or diametrical mass moment of inertia per unit length of shaft, $= \rho^* A r^2$ (ML); for hollow circular cross section $I_T = \frac{1}{4}\rho^* A(r_o^2 + r_i^2)$
I_{Ti}	Transverse or diametrical mass moment of inertia of concentrated mass at station i (ML^2); can be calculated as $I_{Ti} = \Delta a\, \rho^* A r^2$, where Δa is length of shaft lumped at station i; for hollow cylinder of length Δa and mass M_i, $I_{Ti} = \frac{1}{4}M_i(r_o^2 + r_i^2) + \frac{1}{12}M_i(\Delta a)^2$; for disk of concentrated mass M_i, $I_{Ti} = \frac{1}{4}M_i(r_o^2 + r_i^2)$
$\mathbf{k}^i$	Stiffness matrix for ith element
$k_{yy}, k_{yz}, k_{zy}, k_{zz}$	Stiffness coefficients for bearing or seal system (F/L) (Fig. 17-1)
$k_{yy}^*, k_{yz}^*, k_{zy}^*, k_{zz}^*$	Rotary stiffness coefficients for bearing system (FL/rad)
$\bar{k}_{yy}, \bar{k}_{yz}, \bar{k}_{zy}, \bar{k}_{zz}$	Stiffness coefficients for pedestal of bearing or seal system (F/L) (Fig. 17-1)
L	Length of shaft (L)
ℓ	Length of element; span of transfer matrix (L)
M_y, M_z	Bending moment components about y and z axes (FL)
m_b	Beam mass (shaft mass) (M)
M_i	Concentrated mass (M)
$\mathbf{m}^i$	Mass matrix for ith element
$\mathbf{m}_T^i$	Translation mass matrix for ith element
$\mathbf{m}_R^i$	Rotational mass matrix for ith element
$\mathbf{N}, N_n$	Shape functions
P	Axial force; plus for compression, minus for tension (F)
p_y, p_z	Applied loading intensities in y and z directions (F/L)

r	Radius of gyration of cross-sectional area about y or z axis (L)
r_p	Polar radius of gyration of cross-sectional area about x axis (L)
r_o	Outer radius of hollow circular cross section (L)
r_i	Inner radius of hollow circular cross section (L)
t	Time (T)
$\mathbf{U}^i$	Field transfer matrix of ith element
$\mathbf{U}_i$	Point transfer matrix at $x = a_i$
V_y, V_z	Shear force components in y and z directions (F)
w_x, w_y, w_z	Displacements in x, y, and z directions (L)
w_{yb}, w_{zb}	Bearing or seal displacements in y, z directions (L)
xyz	Fixed-reference coordinates
XYZ	Rotating-reference coordinates
$X\eta\zeta$	Rotating fluted coordinates with η and ζ as principal axes of inertia of sectional area of fluted shaft
$\xi\eta Z$	Rotating-reference coordinates with ξ and η as principal axes of inertia of sectional area of radial beam
Ω	Spin or rotational speed (rad/T)
ω	Whirl speed or natural frequency (rad/T)
ω_n	Natural frequency of nth mode (rad/T)
ω_c	Critical speed (rad/T)
ω_d	Damped critical speed (rad/T)
γ_n	Damping exponent of nth mode
ϕ_n	Mode shapes
ρ^*	Mass per unit volume (M/L^3)
ρ	Mass per unit length, $= \rho^* A$ (M/L)
θ_y, θ_z	Slope components of displacement curves about y and z axes (rad)
ζ	Damping ratio for single-degree-of-freedom system, $= c/c_c$
	A single overdot refers to the first derivative with respect to time t.
	A double overdot refers to the second derivative with respect to time t.
	A single prime refers to the first derivative with respect to space coordinate.
	A double prime refers to the second derivative with respect to space coordinate.

17.2 SIGN CONVENTION

Positive displacements, slopes, moments, shear forces, and applied loadings are indicated in Fig. 17-1. As with the notation, the sign convention for rotor systems conforms to that used in practice. See Appendix II, Fig. II-4.

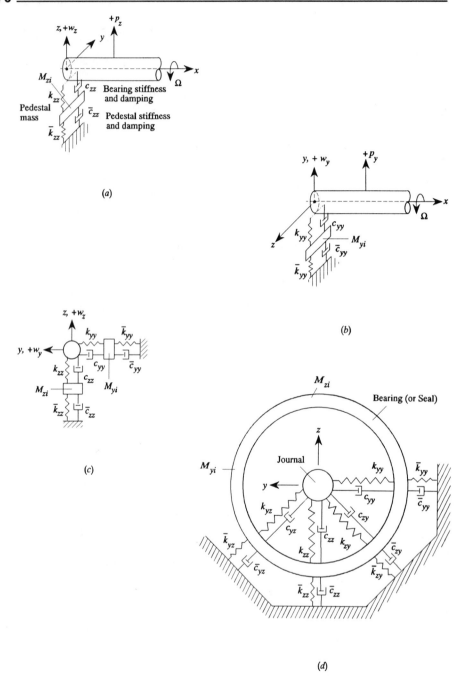

FIGURE 17-1 Notation and sign convention: (a) x, z plane; (b) x, y plane; (c, d) cross-sectional view; (e) positive forces, moments, and slopes for transfer matrices (sign convention 1), ($P > 0$ for compression, $P < 0$ for tension); (f) positive forces, moments, and slopes for stiffness, mass, and damping matrices (sign convention 2), ($P > 0$ for tension, $P < 0$ for compression).

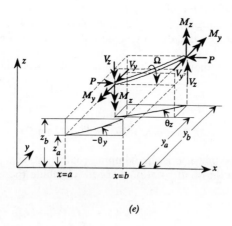

(e)

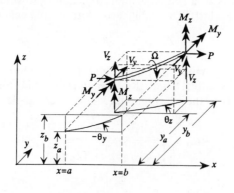

(f)

FIGURE 17-1 *Continued*

17.3 BENDING VIBRATION

Whirling of a Single-Mass Rotor

The fundamentals of rotor whirl due to residual rotor unbalance can be characterized using the simple rotor shown in Fig. 17-2. This rotor system, referred to as a *Jeffcott rotor*, consists of a massless elastic shaft on which a single disk is mounted at midspan, with both ends simply supported. The rotor mass is concentrated at the center of gravity G of the disk at a distance e from the geometric center (centroid) S of the disk. The center line $\overline{00}$ of the bearings intersects the plane of the disk at 0, and the shaft center is off center by a distance $0S = w$.

The equations of motion for the centroid S of the disk in the y and z directions are

$$M_i \ddot{w}_y + c\dot{w}_y + kw_y = M_i e\Omega^2 \cos \Omega t \qquad (17.1)$$

$$M_i \ddot{w}_z + c\dot{w}_z + kw_z = M_i e\Omega^2 \sin \Omega t \qquad (17.2)$$

where M_i is the disk mass, c is the damping coefficient, k is the shaft stiffness at midspan, and Ω is the shaft rotational speed. Combine these equations into the single equation

$$M_i \ddot{w} + c\dot{w} + kw = M_i e\Omega^2 e^{i\Omega t} \qquad (17.3)$$

where $w = w_y + iw_z$ is the whirl radius of the shaft geometric center, $i = \sqrt{-1}$. The solution consists of a complementary function (free whirl) and a particular solution (unbalance whirl). The unbalance whirl has the general form

$$w = w_0 e^{i(\Omega t - \phi)} \qquad (17.4)$$

where w_0 is the whirl radius amplitude and ϕ is the phase angle between the unbalance force $F = M_i e\Omega^2$ and the amplitude w_0. Substitution of Eq. (17.4) into Eq. (17.3) leads to the whirl amplitude and phase angle at the disk,

$$w_0 = \frac{e\eta^2}{\left[\left(1 - \eta^2\right)^2 + (2\zeta\eta)^2\right]^{1/2}} \qquad (17.5)$$

$$\phi = \tan^{-1} \frac{2\zeta\eta}{1 - \eta^2} \qquad (17.6)$$

where $\omega_c = \sqrt{k/M_i}$ is the undamped critical speed, $\eta = \Omega/\omega_c$ is the speed ratio, $\zeta = c/c_c$ is the damping ratio, and $c_c = 2M_i\omega_c = 2\sqrt{kM_i}$ is the critical damping.

The nondimensional whirl amplitude $\overline{w}_0$ is given by

$$\overline{w}_0 = \frac{w_0}{e} = \frac{\eta^2}{\left[\left(1 - \eta^2\right)^2 + (2\zeta\eta)^2\right]^{1/2}} \qquad (17.7a)$$

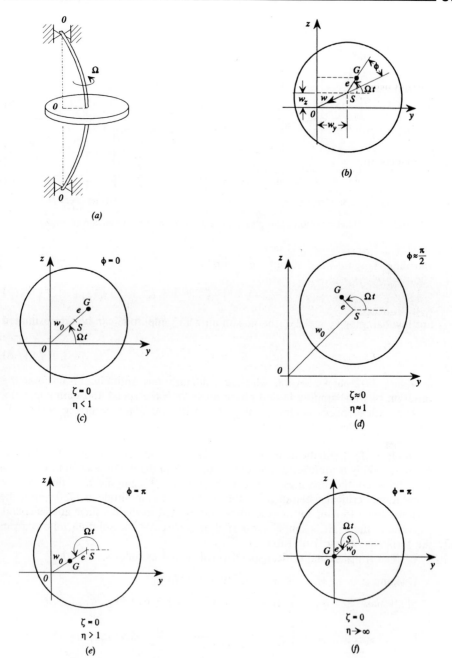

FIGURE 17-2 Whirling of a simple rotor in two radially rigid bearings.

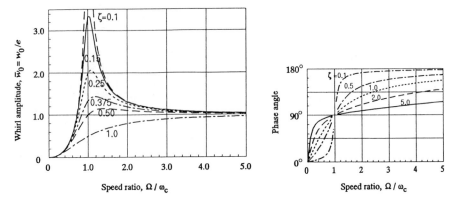

FIGURE 17-3 Whirl amplitude and phase for a whirling rotor as a function of speed.

and the maximum whirl amplitude occurs at

$$\eta = 1/\sqrt{1 - 2\zeta^2} \approx 1 + \zeta^2 \tag{17.7b}$$

For low damping ($\zeta < 0.25$), the maximum whirl amplitude can be approximated by

$$\overline{w}_0 \approx 1/2\zeta \tag{17.8}$$

Figure 17-2 shows several whirling configurations and Fig. 17-3 shows the variation of whirl amplitude and phase angle with the speed and damping ratio.

The critical speeds ω_c for simple rotors with commonly occurring end conditions are provided in Table 17-1.

Example 17.1 Centrifugal Pump The disk of a single-stage centrifugal pump weighing 300 N is attached to the center of a 0.1-m-diameter steel shaft of length 1 m between the bearings. Neglect the effects of damping and find (a) the translational critical speed of the rotor, (b) the whirl amplitude at 1750 rpm if the eccentricity is 15 μm, and (c) the force transmitted to the bearings at this speed.

Assume the shaft is simply supported at each end. The solution follows from the formulas of case 1 of Table 17-1.

(a) The translational critical speed is calculated as follows,

Disk mass: $M_i = W/g = 300/9.8 = 30.61$ kg

Shaft mass: $m_b = \rho^* AL = 7850\pi(0.1^2/4)1 = 61.65$ kg

Critical speed: $\omega_c = \left(\dfrac{48EI}{L^3(M_i + 0.49m_b)}\right)^{1/2} = 895.49$ rad/s $= 8551$ rpm

(b) From Eq. (17.7) the whirl amplitude at 1750 rpm is

$$w_0 = \frac{e\eta^2}{1 - \eta^2} = \frac{15 \times 10^{-6}(1750/8551)^2}{1 - (1750/8551)^2} = 0.656 \times 10^{-6} \text{ m} = 0.656 \ \mu\text{m}$$

(c) The force transmitted to the two bearings at 1750 rpm is equal to the centrifugal force $M_i(w_0 + e)\Omega^2$ acting in an outward direction through G (Fig. 17-2c):

$$F = M_i(w_0 + e)\Omega^2 = 30.61(0.656 \times 10^{-6} + 1.5 \times 10^{-5})(2\pi1750/60)^2$$
$$= 16.1 \text{ N}$$

The critical speeds of vertical shafts having an attached mass at an intermediate point and various end constraints with consideration of the axial force owing to the weight of the mass are shown in Table 17-2. Gyroscopic effects of the shaft and the attached mass are ignored.

Single-Mass Rotor on Elastic Supports

For a one-mass flexible rotor on two identical anisotropic bearings (Fig. 17-4), the equations of motion of the mass and bearings are

$$M_i\ddot{w}_y + k(w_y - w_{yb}) = M_ie\Omega^2 \cos \Omega t \qquad M_i\ddot{w}_z + k(w_z - w_{zb}) = M_ie\Omega^2 \sin \Omega t$$
$$(17.9)$$

$$k(w_{yb} - w_y) = -2F_y \qquad k(w_{zb} - w_z) = -2F_z \qquad (17.10)$$

where k is the shaft stiffness, F_y and F_z are the reacting forces of the bearings, and the shaft damping is ignored.

Coupled Systems   The bearing coefficients are denoted as

$$\begin{bmatrix} F_y \\ F_z \end{bmatrix} = \begin{bmatrix} k_{yy} & k_{yz} \\ k_{zy} & k_{zz} \end{bmatrix} \begin{bmatrix} w_{yb} \\ w_{zb} \end{bmatrix} + \begin{bmatrix} c_{yy} & c_{yz} \\ c_{zy} & c_{zz} \end{bmatrix} \begin{bmatrix} \dot{w}_{yb} \\ \dot{w}_{zb} \end{bmatrix} \qquad (17.11)$$

with $\dot{w}_{yb} = dw_{yb}/dt$ and $\dot{w}_{zb} = dw_{zb}/dt$.

Substitution of Eq. (17.11) into Eqs. (17.9) and (17.10) leads to the equation of motion

$$\mathbf{M\ddot{w}} + \mathbf{C\dot{w}} + \mathbf{Kw} = \mathbf{F} \qquad (17.12)$$

where

$$\mathbf{M} = \begin{bmatrix} M_i & 0 & 0 & 0 \\ 0 & M_i & 0 & 0 \\ 0 & 0 & 0 & 0 \\ 0 & 0 & 0 & 0 \end{bmatrix} \qquad \mathbf{C} = \begin{bmatrix} 0 & 0 & 0 & 0 \\ 0 & 0 & 0 & 0 \\ 0 & 0 & 2c_{yy} & 2c_{yz} \\ 0 & 0 & 2c_{zy} & 2c_{zz} \end{bmatrix}$$

$$\mathbf{K} = \begin{bmatrix} k & 0 & -k & 0 \\ 0 & k & 0 & -k \\ -k & 0 & k + 2k_{yy} & 2k_{yz} \\ 0 & -k & 2k_{zy} & k + 2k_{zz} \end{bmatrix} \qquad \mathbf{F} = M_ie\Omega^2 \begin{bmatrix} \cos \Omega t \\ \sin \Omega t \\ 0 \\ 0 \end{bmatrix}$$

$$\mathbf{w} = \begin{bmatrix} w_y & w_z & w_{yb} & w_{zb} \end{bmatrix}^T \qquad \dot{\mathbf{w}} = \frac{d\mathbf{w}}{dt} \qquad \ddot{\mathbf{w}} = \frac{d^2\mathbf{w}}{dt^2}$$

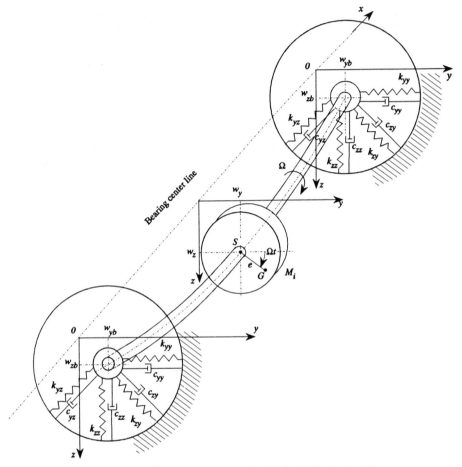

FIGURE 17-4 A single mass flexible rotor on two identical anisotropic bearings.

This equation can be solved for the critical speeds or for the response of the rotor to unbalance forces. For the latter case the solution for the motion of the mass can be of the form

$$
\begin{bmatrix} w_y(t) \\ w_z(t) \end{bmatrix} = \begin{bmatrix} w_y^c \\ w_z^c \end{bmatrix} \cos \Omega t + \begin{bmatrix} w_y^s \\ w_z^s \end{bmatrix} \sin \Omega t \tag{17.13}
$$

and then

$$
\begin{aligned}
w_0(t) &= w_y(t) + i w_z(t) \\
&= w_y^c \cos \Omega t + w_y^s \sin \Omega t + i(w_z^c \cos \Omega t + w_z^s \sin \Omega t) \\
&= w_0^+ e^{i\Omega t} + w_0^- e^{-i\Omega t}
\end{aligned} \tag{17.14}
$$

with

$$w_0^+ = \tfrac{1}{2}\left[\left(w_y^c + w_z^s\right) + i\left(w_z^c - w_y^s\right)\right] \qquad w_0^- = \tfrac{1}{2}\left[\left(w_y^c - w_z^s\right) + i\left(w_z^c + w_y^s\right)\right]$$

where w_0^+ is the whirl radius of the forward precession component, which is in the same direction as the rotation of the rotor, while w_0^- is that of the backward precession component.

The maximum whirl radius is defined by the major semiaxis of elliptic whirl orbit of the geometric shaft center,

$$\begin{aligned} w_{0,\,\mathrm{max}} &= |w_0^+| + |w_0^-| \\ &= \tfrac{1}{2}\left[\sqrt{\left(w_y^c + w_z^s\right)^2 + \left(w_z^c - w_y^s\right)^2} + \sqrt{\left(w_y^c - w_z^s\right)^2 + \left(w_z^c + w_y^s\right)^2}\right] \end{aligned}$$

$$(17.15)$$

Uncoupled Systems Consider the symmetric bearings without cross-coupling terms. A force balance at the bearings gives

$$k(w_y - w_{yb}) = 2(k_b w_{yb} + c_b \dot{w}_{yb}) \qquad k(w_z - w_{zb}) = 2(k_b w_{zb} + c_b \dot{w}_{zb})$$

$$(17.16)$$

where $k_b = k_{yy} = k_{zz}$, $c_b = c_{yy} = c_{zz}$. Insert Eq. (17.16) into Eqs. (17.9) and (17.10):

$$M_i \ddot{w}_y + k(w_y - w_{yb}) = M_i e\Omega^2 \cos \Omega t \qquad M_i \ddot{w}_z + k(w_z - w_{zb}) = M_i e\Omega^2 \sin \Omega t$$

$$k(w_y - w_{yb}) = 2(k_b w_{yb} + c_b \dot{w}_{yb}) \qquad k(w_z - w_{zb}) = 2(k_b w_{zb} + c_b \dot{w}_{zb})$$

$$(17.17a)$$

Combine these equations as

$$M_i \ddot{w} + k(w - w_b) = M_i e\Omega^2 e^{i\Omega t} \qquad k(w - w_b) = 2(k_b w_b + c_b \dot{w}_b) \quad (17.17b)$$

where $w = w_y + iw_z$ and $w_b = w_{yb} + iw_{zb}$ are the whirl radii of the shaft geometric center and the bearing journal center, respectively.

The solution to Eq. (17.17b) is of the form

$$w = \overline{w}e^{i\Omega t} \qquad w_b = \overline{w}_b e^{i\Omega t} \qquad (17.18)$$

The equations can be rewritten as

$$-M_i\Omega^2\overline{w} + k(\overline{w} - \overline{w}_b) = M_i e\Omega^2 \qquad (17.19a)$$

$$k(\overline{w} - \overline{w}_b) = 2(k_b + i\Omega c_b)\overline{w}_b \qquad (17.19b)$$

From Eq. (17.19b)

$$\bar{w}_b = \frac{k\bar{w}}{k + 2k_b + 2i\Omega c_b} \qquad (17.20)$$

Insert Eq. (17.20) into Eq. (17.19a) and rearrange terms,

$$\bar{w} = \frac{e\eta^2}{1 - \eta^2} \frac{(1 + \bar{K}) + i\Omega c_b/k_b}{\left[1 - \bar{K}\eta^2/(1 - \eta^2)\right] + i(\Omega c_b/k_b)} \qquad (17.21)$$

where $\bar{K} = k/2k_b$, $\eta^2 = (M_i/k)\Omega^2$

The whirl radius at the geometric shaft center is found to be

$$w_0 = |\bar{w}| = \left| \frac{e\eta^2}{1 - \eta^2} \sqrt{\frac{(1 + \bar{K})^2 + (\Omega c_b/k_b)^2}{\left[1 - \bar{K}\eta^2/(1 - \eta^2)\right]^2 + (\Omega c_b/k_b)^2}} \right| \qquad (17.22)$$

Substitution of Eq. (17.21) into Eq. (17.20) gives the shaft whirl radius at the bearings

$$w_{0b} = |\bar{w}_b| = \left| \frac{e\eta^2}{1 - \eta^2} \frac{\bar{K}}{\sqrt{\left[1 - \bar{K}\eta^2/(1 - \eta^2)\right]^2 + (\Omega c_b/k_b)^2}} \right| \qquad (17.23)$$

Example 17.2 Rotor with Flexible Supports Consider a midspan disk of weight $W = 300$ N and a steel shaft 0.1 m in diameter and 1 m in length between the bearings (Fig. 17-5). This rotor is assumed to operate with two identical isotropic end bearings. The coupled terms of bearing coefficients are ignored, i.e.,

$$k_{yz} = k_{zy} = 0 \qquad c_{yz} = c_{zy} = 0$$

Also,

$$k_{yy} = k_{zz} = k_b = 50 \text{ MN/m} \qquad c_{yy} = c_{zz} = c_b = 10 \text{ kN} \cdot \text{s/m}$$

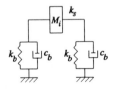

FIGURE 17-5 Example 17.2.

Find (a) the system undamped critical speed, (b) the damped critical speed, (c) the maximum shaft whirl radius at the disk for an eccentricity of 15 μm at the disk, and (d) the maximum journal whirl radius.

(a) *Undamped critical speed*: Shaft stiffness (Table 10-3, case 8) is calculated as

$$k_s = \frac{48EI}{L^3} = \frac{48(2.07 \times 10^{11})\pi 0.1^4}{1^3 \times 64} = 4.877 \times 10^7 \text{ N/m} \tag{1}$$

The bearing stiffness is k_b, so the combined shaft bearing stiffness is (Table 10-4, case 4)

$$\frac{1}{k} = \frac{1}{k_s} + \frac{1}{2k_b} = \frac{1}{4.877 \times 10^7} + \frac{1}{1 \times 10^8} \tag{2}$$

where

$$k = 3.278 \times 10^7 \text{ N/m}$$

The disk mass and shaft mass are calculated as

$$M_i = W/g = 30.61 \text{ kg} \quad \text{and} \quad m_b = \rho^* AL = 7850\pi(0.1^2/4) \times 1 = 61.65 \text{ kg}$$

The critical speed follows from the formulas of case 1 of Table 17-1 by replacing shaft stiffness $48EI/L^3$ with k,

$$\omega_c = \sqrt{\frac{k}{M_i + 0.49m_b}} = \sqrt{\frac{3.278 \times 10^7}{60.8185}} = 734.15 \text{ rad/s}$$

$$= \frac{60}{2\pi} \times 734.15 \text{ rpm}$$

$$= 7010.6 \text{ rpm} \tag{3}$$

(b) *Damped critical speed*: The total damping coefficient $c = 2c_b$. From the damping ratio $\zeta = c/c_c$ and $c_c = 2M_i\omega_c$ with M_i replaced by $M_i + 0.49m_b$:

$$\zeta = \frac{2c_b}{2(M_i + 0.49m_b)\omega_c} = \frac{2(10^4)}{2(60.8185)734.15} = 0.2240 \tag{4}$$

The damped critical speed is given by [Eq. (17.7b)]

$$\omega_d = \Omega = \omega_c\eta = \omega_c(1 + \zeta^2) = 7010.6(1 + 0.2240^2) = 7362.4 \text{ rpm}$$

$$= 770.98 \text{ rad/s} \tag{5}$$

(c) *Maximum shaft whirl radius at disk*: The maximum whirl motion produces the resonance ($\Omega = \omega_d$). From Eq. (17.22),

$$\bar{K} = \frac{k_s}{2k_b} = \frac{4.877 \times 10^7}{2(5 \times 10^7)} = 0.4877 \qquad \eta = \frac{\Omega}{\omega_c} = \frac{\omega_d}{\omega_c} = \frac{770.98}{734.15} = 1.05 \quad (6)$$

$$\frac{w_0}{e} = \left| \frac{\eta^2}{1 - \eta^2} \left(\frac{(1 + \bar{K})^2 + (\omega_d c_b / k_b)^2}{[1 - \bar{K}\eta^2/(1 - \eta^2)]^2 + (\omega_d c_b / k_b)^2} \right)^{1/2} \right|$$

$$= \left| \frac{1.05^2}{1 - 1.05^2} \left(\frac{(1 + 0.4877)^2 + (770.98 \times 10^4/(5 \times 10^7))^2}{\left[1 - \frac{(0.4877)1.05^2}{1 - 1.05^2} \right]^2 + \left(\frac{770.98 \times 10^4}{5 \times 10^7} \right)^2} \right)^{1/2} \right| = 2.58$$

$$(7)$$

and the maximum whirl radius is $w_0 = 2.58e = 3.87 \times 10^{-5}$ m $= 38.7 \ \mu$m.

(d) *Maximum journal whirl radius*: From Eq. (17.23), the dynamic magnification factor is given by

$$\frac{w_{0b}}{e} = \left| \frac{\eta^2}{1 - \eta^2} \left(\frac{\bar{K}^2}{[1 - \bar{K}\eta^2/(1 - \eta^2)]^2 + (\omega_d c_b / k_b)^2} \right)^{1/2} \right|$$

$$= \left| \frac{1.05^2}{1 - 1.05^2} \left(\frac{0.4877^2}{\left[1 - \frac{(0.4877)1.05^2}{1 - 1.05^2} \right]^2 + \left(\frac{770.98 \times 10^4}{5 \times 10^7} \right)^2} \right)^{1/2} \right| = 0.84$$

$$(8)$$

Then the maximum whirl radius is

$$w_{0b} = 0.84e = 1.26 \times 10^{-5} \text{ m} = 12.6 \ \mu\text{m} \qquad (9)$$

Uniform Rotating Shaft

In the Euler-Bernoulli shaft model the shear deformation effects are neglected, but the terms for the gyroscopic moment and the moment due to the inertia of rotation of the cross section are included [17.2],

$$EI\frac{\partial^4 w}{\partial x^4} + \rho^* A \frac{\partial^2 w}{\partial t^2} - \rho^* I\left(\frac{\partial^4 w}{\partial x^2 \partial t^2} - i2\Omega\frac{\partial^3 w}{\partial x^2 \partial t}\right) = p(x,t) \quad (17.24)$$

where $w = w_y + iw_z$ is the complex deflection and $p(x,t)$ is the external force. For a uniform shaft free from external forces, $p(x,t) = 0$, Eq. (17.24) has a solution of the form

$$w(x,t) = \sum_{n=1}^{\infty} \phi_n(x)\eta_n(t) = \sum_{n=1}^{\infty} W_n e^{(\sqrt{s_n}x + i\omega_n t)} \quad (17.25)$$

where W_n is the complex amplitude, s_n characterizes the mode shapes, and ω_n are the natural frequencies. This leads to the equation

$$EIs_n^2 + \rho^* I(\omega_n^2 - 2\Omega\omega_n)s_n - \rho^* A\omega_n^2 = 0 \quad (17.26)$$

From Eq. (17.26),

$$s_n = -\frac{\rho^* I(\omega_n^2 - 2\Omega\omega_n)}{2EI} \pm \left[\left(\frac{\rho^* I(\omega_n^2 - 2\Omega\omega_n)}{2EI}\right)^2 + \frac{\rho^* A\omega_n^2}{EI}\right]^{1/2} = \begin{cases} p_n^2 \\ -q_n^2 \end{cases}$$

$$\sqrt{s_n} = \begin{cases} p_n \\ iq_n \end{cases} \quad (17.27)$$

The eigenfunctions $\phi_n(x)$ will take the form

$$\phi_n(x) = C_{1n}\cosh p_n x + C_{2n}\sinh p_n x + C_{3n}\cos q_n x + C_{4n}\sin q_n x \quad (17.28)$$

where C_{1n}, C_{2n}, C_{3n}, and C_{4n} are integration constants to be determined from the boundary conditions, while p_n and q_n are the two values of $\sqrt{s_n}$ satisfying Eq. (17.27).

The corresponding natural frequencies are determined from Eq. (17.26):

$$\omega_{c1} = \frac{-\Omega\beta_n^2 + \sqrt{\Omega^2\beta_n^4 + (1 - \beta_n^2)\omega_0^2}}{1 - \beta_n^2}$$

$$\omega_{c2} = \frac{-\Omega\beta_n^2 - \sqrt{\Omega^2\beta_n^4 + (1 - \beta_n^2)\omega_0^2}}{1 - \beta_n^2}$$

$$(17.29)$$

where

$$\beta_n = \frac{\lambda_n r}{L} \qquad r^2 = \frac{I}{A} \qquad \omega_0 = \left(\frac{\lambda_n}{L}\right)^2 \left(\frac{EI}{\rho^* A}\right)^{1/2}$$

$$\lambda_n = p_n L \quad \text{or} \quad \lambda_n = q_n L$$

For a Rayleigh beam (gyroscopic effects are ignored)

$$\omega_{c1} = \frac{\omega_0}{\sqrt{|1 - \beta_n^2|}} \qquad \omega_{c2} = -\frac{\omega_0}{\sqrt{|1 - \beta_n^2|}} \tag{17.30}$$

There are two natural frequencies, one always positive and the other negative. The positive and negative natural frequencies are known to be associated with the forward and backward precessions, respectively. The critical speeds, mode shapes, and frequency equations for uniform rotors with various end conditions are provided in Table 17-3.

Example 17.3 Cylindrical Rotor A uniform cylindrical rotor is supported in undamped flexible end bearings of identical stiffness k_b in all radial directions (Fig. 17-6).
Deduce the frequency equation, and calculate the critical speeds using the frequency equation. Ignore the effect of the inertia of rotation of the cross section.
For this case

$$k_b = 1 \text{ GN/m} \qquad L = 1 \text{ m} \qquad d = 0.1 \text{ m}$$

$$E = 207 \text{ GN/m}^2 \qquad \rho^* = 7854 \text{ kg/m}^3$$

Refer to Fig. 17-1e in establishing the equations representing the boundary conditions. Use sign convention 1, Fig. 17-1e. The moments and shear forces on the boundaries are

$$\begin{aligned} x = 0: & \qquad M(0) = 0 \qquad V(0) = k_b w(0, t) \\ x = L: & \qquad M(L) = 0 \qquad V(L) = -k_b w(L, t) \end{aligned} \tag{1}$$

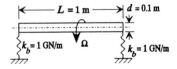

FIGURE 17-6 Example 17.3.

so

$$-EI\frac{\partial^2 w(0,t)}{\partial x^2} = 0 \qquad -EI\frac{\partial^3 w(0,t)}{\partial x^3} - k_b w(0,t) = 0$$

$$-EI\frac{\partial^2 w(L,t)}{\partial x^2} = 0 \qquad -EI\frac{\partial^3 w(L,t)}{\partial x^3} + k_b w(L,t) = 0 \tag{2}$$

If there is no gyroscopic moment and if the effect of the inertia of rotation of the cross section is ignored, Eq. (17.24) becomes

$$EI\frac{\partial^4 w}{\partial x^4} + \rho^* A\frac{\partial^2 w}{\partial t^2} = 0 \tag{3}$$

which has the solution

$$w(x,t) = \sum_{n=1}^{\infty} \phi_n e^{i\omega_n t} \tag{4}$$

with

$$\phi_n(x) = C_{1n}\cosh p_n x + C_{2n}\sinh p_n x + C_{3n}\cos p_n x + C_{4n}\sin p_n x \tag{5}$$

where $p_n^4 = \rho^* A\omega_n^2/EI$.

Use this general solution of (5) in conjunction with boundary conditions (2) to obtain the frequency determinant

$$\begin{vmatrix} \cos p_n L - \cosh p_n L + 2\bar{K}\sinh p_n L & \sin p_n L - \sinh p_n L \\ -\sin p_n L - \sinh p_n L + \bar{K}(\cos p_n L + \cosh p_n L) & \\ +2\bar{K}(\cosh p_n L - \bar{K}\sinh p_n L) & \cos p_n L - \cosh p_n L + \bar{K}(\sin p_n L + \sinh p_n L) \end{vmatrix} = 0 \tag{6}$$

where

$$\bar{K} = \frac{k_b}{EIp_n^3} = \frac{k_b L^3}{EI}\frac{1}{(p_n L)^3} = \frac{\bar{K}^*}{(p_n L)^3}$$

and $\bar{K}^* = k_b L^3/EI$ expresses the ratio of the bearing stiffness to the shaft stiffness. This leads to the frequency equation

$$(1 - \cos p_n L \cosh p_n L)(p_n L)^6 + 2\bar{K}^*(\cos p_n L \sinh p_n L$$

$$- \sin p_n L \cosh p_n L)(p_n L)^3 + 2\bar{K}^{*2}\sin p_n L \sinh p_n L = 0 \tag{7}$$

Alternatively, this relationship can be obtained from Table 17-3, case 11, by setting $\rho^*I = 0$ (gyroscopic effects are ignored) in the equations for p_n^2 and q_n^2.
For

$$\bar{K}^* = \frac{k_b L^3}{EI} = \frac{1 \times 10^9 \times 1^3}{(207 \times 10^9)(\pi 0.1^4/64)} = 984.1465$$

the eigenvalues, as determined by a computer solution of the frequency equation (7), are

$$p_1 L = 3.111, \, p_2 L = 6.033, \, p_3 L = 8.553, \, \ldots \tag{8}$$

From $p^4 = \rho^* A \omega_n^2 / EI$, the critical speeds are

$$\omega_{ci} = \frac{(p_i L)^2}{L^2} \left(\frac{EI}{\rho^* A} \right)^{1/2} = \frac{(p_i L)^2}{1^2} \left[\frac{(2.07 \times 10^{11})(\pi 0.1^4/64)}{(7854)(\pi 0.1^2/4)} \right]^{1/2}$$

$$= 128.349(p_i L)^2 \quad \text{rad/s} = 1225.6(p_i L)^2 \quad \text{rpm} \tag{9}$$

so that

$$\omega_{c1} = (1225.6)(p_1 L)^2 = 11{,}861.8 \text{ rpm} \quad \omega_{c2} = (1225.6)(p_2 L)^2 = 44{,}608.3 \text{ rpm}$$

$$\omega_{c3} = (1225.6)(p_3 L)^2 = 89{,}657.3 \text{ rpm}$$

If the gyroscopic moment and the effect of the inertia of rotation of the cross section are to be considered, the eigenvalues and eigenfunctions of the rotor can be determined using a computational solution of the frequency equation of Table 17-3, case 11.

========

Transfer Matrices

The transfer matrices for several commonly occurring rotor elements, for systems with constant rotating speeds, are provided in Tables 17-4 to 17-8. See Appendix III for the general theory of the transfer matrix method. In these tables, the displacements, slopes, shear forces, and moments are expressed as

$$\begin{aligned} w_z &= w_z^c \cos \Omega t + w_z^s \sin \Omega t & \theta_y &= \theta_y^c \cos \Omega t + \theta_y^s \sin \Omega t \\ V_z &= V_z^c \cos \Omega t + V_z^s \sin \Omega t & M_y &= M_y^c \cos \Omega t + M_y^s \sin \Omega t \\ w_y &= w_y^c \cos \Omega t + w_y^s \sin \Omega t & \theta_z &= \theta_z^c \cos \Omega t + \theta_z^s \sin \Omega t \\ V_y &= V_y^c \cos \Omega t + V_y^s \sin \Omega t & M_z &= M_z^c \cos \Omega t + M_z^s \sin \Omega t \end{aligned} \tag{17.31}$$

where, e.g., w_z^c, w_z^s are the cosine and sine terms, respectively, of w_z.

Rigid Disk From the equilibrium and compatibility conditions of a whirling disk,

$$w_z^R = w_z^L - \theta_y^L h$$

$$\theta_y^R = \theta_y^L$$

$$V_z^R = V_z^L + M_i\left(\ddot{w}_z^L - \tfrac{1}{2}\ddot{\theta}_y^L h\right) \qquad (17.32)$$

$$M_y^R = M_y^L + V_z^L h + M_i\left(\ddot{w}_z^L - \tfrac{1}{2}h\ddot{\theta}_y^L\right)\tfrac{1}{2}h + I_{Ti}\ddot{\theta}_y + I_{pi}\Omega\dot{\theta}_z$$

The quantities M_i, h, I_{Ti}, and I_{pi} are the mass, thickness, diametrical, and polar mass moments of inertia of the disk, respectively. The superscripts L and R indicate the left and right sides of the disk. When the whirling speed Ω is constant, the transfer matrices are given in Table 17-4 for a rigid disk and for a concentrated mass.

Uniform Shaft Element The equation of motion for a Timoshenko shaft element is

$$\underbrace{EI\frac{\partial^4 w}{\partial x^4} + \rho^* A\frac{\partial^2 w}{\partial t^2}}_{\text{Euler–Bernoulli theory}} - \underbrace{\rho^* I\left[\frac{\partial^4 w}{\partial x^2 \partial t^2}\right.}_{\substack{\text{principal rotary} \\ \text{inertia term}}} - \underbrace{i2\Omega\left.\frac{\partial^3 w}{\partial x^2 \partial t}\right]}_{\substack{\text{gyroscopic} \\ \text{moment term}}}$$

$$- \underbrace{\frac{\rho^* EI\alpha_s}{G}\frac{\partial^4 w}{\partial x^2 \partial t^2}}_{\substack{\text{principal shear} \\ \text{deformation term}}} + \underbrace{\frac{(\rho^*)^2 I\alpha_s}{G}\left(\frac{\partial^4 w}{\partial t^4} - i2\Omega\frac{\partial^3 w}{\partial t^3}\right)}_{\substack{\text{combined rotary inertia} \\ \text{and shear deformation}}} = 0 \quad (17.33)$$

where $w = w_y + iw_z$ is the complex deflection.

Assume that the shaft is whirling with a circular orbit, $w = w(x)e^{i\omega t}$. Then

$$w(x) = C_1 \sinh(px) + C_2 \cosh(px) + C_3 \sin(qx) + C_4 \cos(qx) \qquad \text{for } \gamma \geq 0$$
$$(17.34)$$

where

$$p^2 = \left|\sqrt{\beta^2 + \gamma} - \beta\right| \qquad q^2 = \left|\sqrt{\beta^2 + \gamma} + \beta\right|$$

$$\beta = \frac{\omega^2}{2}\left[\frac{\rho^* \alpha_s}{G} + \frac{\rho^*}{E}\left(\frac{2\Omega}{\omega} + 1\right)\right] \qquad \gamma = \omega^2\left[\frac{\rho^* A}{EI} - \frac{\rho^*}{E}\left(\frac{2\Omega}{\omega} + 1\right)\frac{\rho^* \omega^2 \alpha_s}{G}\right]$$

and constants C_i ($i = 1, 4$) are coefficients depending on the boundary conditions. If $\gamma < 0$, replace sinh and cosh with sin and cos, respectively. The transfer matrix for a uniform shaft with consideration of the effects of distributed mass and shear deformation is given in Table 17-5. Also, the transfer matrix for a uniform shaft section with static bow is given in this table. Table 17-6 shows the transfer matrix for a shaft element (shear and gyroscopic effects are ignored) with axial torque effects.

Bearing and Seal Elements The general bearings and seals are represented linearly by eight translational and eight rotary stiffness and damping coefficients. The bearing or seal forces and moments are

$$
\begin{bmatrix} V_y \\ V_z \end{bmatrix} = -\begin{bmatrix} k_{yy} & k_{yz} \\ k_{zy} & k_{zz} \end{bmatrix}\begin{bmatrix} w_y \\ w_z \end{bmatrix} - \begin{bmatrix} c_{yy} & c_{yz} \\ c_{zy} & c_{zz} \end{bmatrix}\begin{bmatrix} \dot{w}_y \\ \dot{w}_z \end{bmatrix}
$$

$$
\begin{bmatrix} M_z \\ M_y \end{bmatrix} = -\begin{bmatrix} k^*_{yy} & k^*_{yz} \\ k^*_{zy} & k^*_{zz} \end{bmatrix}\begin{bmatrix} \theta_z \\ \theta_y \end{bmatrix} - \begin{bmatrix} c^*_{yy} & c^*_{yz} \\ c^*_{zy} & c^*_{zz} \end{bmatrix}\begin{bmatrix} \dot{\theta}_z \\ \dot{\theta}_y \end{bmatrix}
$$

(17.35a)

From equilibrium

$$
V_z^R = V_z^L + k_{zz}w_z + k_{zy}w_y + c_{zz}\dot{w}_z + c_{zy}\dot{w}_y
$$

$$
M_y^R = M_y^L + k^*_{zz}\theta_y + k^*_{zy}\theta_z + c^*_{zz}\dot{\theta}_y + c^*_{zy}\dot{\theta}_z
$$

$$
V_y^R = V_y^L + k_{yz}w_z + k_{yy}w_y + c_{yz}\dot{w}_z + c_{yy}\dot{w}_y
$$

$$
M_z^R = M_z^L + k^*_{yz}\theta_y + k^*_{yy}\theta_z + c^*_{yz}\dot{\theta}_y + c^*_{yy}\dot{\theta}_z
$$

(17.35b)

where V_i^R, V_i^L and M_i^R, M_i^L ($i = y, z$) are the reaction forces and moments, respectively, to the right and the left of the bearing or seal. The responses w_i, θ_i and $\dot{w}_i, \dot{\theta}_i$ are the relative displacements and slopes and corresponding velocities between the journal and the bearing (Fig. 17-1d).

The transfer (point) matrix for a general bearing or seal is given in Table 17-7. Normally, in a rotor dynamic analysis, only reaction forces are considered and the reaction moments are ignored. In such cases the eight rotary stiffness and damping coefficients (Fig. 17-1d) are set equal to zero. Formulations based on the eight translational stiffness and damping coefficients are given in the next section ("Stiffness and Mass Matrices") about bearing and seal elements.

For an isotropic bearing ($k_{zz} = k_{yy}$, $c_{zz} = c_{yy}$), Table 17-7 is simplified as Table 17-8. This table includes the effect of a pedestal.

Example 17.4   Shaft on Isotropic Supports Consider the shaft of Fig. 17-7 that is rotating on isotropic supports.

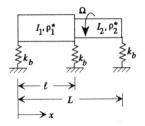

FIGURE 17-7 Example 17.4.

The state vector is $\mathbf{z} = [w \ \theta \ V \ M]^T$. In terms of transfer matrices, the state vectors along the shaft are given by

$$\mathbf{z} = \mathbf{U}^1 \mathbf{U}_1 \mathbf{z}_0 \qquad x < \ell \tag{1}$$

$$\mathbf{z} = \mathbf{U}^2 \mathbf{U}_2 \mathbf{U}^1 \mathbf{U}_1 \mathbf{z}_0 \qquad \ell < x < L$$

$$\mathbf{z}_{x=L} = \mathbf{U}_3 \mathbf{U}^2 \mathbf{U}_2 \mathbf{U}^1 \mathbf{U}_1 \mathbf{z}_0$$

$$= \mathbf{U}\mathbf{z}_0 \qquad x = L \tag{2}$$

From Table 17-8, set $k = k_b$, $c = \bar{c} = 0$, $M_i = 0$, and $\bar{k} \rightarrow \infty$ (no bearing pedestal motion), and

$$z_i = \frac{k_b \bar{k}}{k_b + \bar{k}} = \frac{k_b}{k_b/\bar{k} + 1} \rightarrow k_b$$

Then

$$\mathbf{U}_3 = \mathbf{U}_2 = \mathbf{U}_1 = \begin{bmatrix} 1 & 0 & 0 & 0 \\ 0 & 1 & 0 & 0 \\ -k_b & 0 & 1 & 0 \\ 0 & 0 & 0 & 1 \end{bmatrix} \tag{3}$$

and $\mathbf{U}^1, \mathbf{U}^2$ are the transfer matrices from Table 17-5 for the uniform shaft segments. Because of the isotropic properties, $\mathbf{U}^i = \mathbf{T}_s$. Since the left and right ends are considered to be free,

$$\begin{bmatrix} V \\ M \end{bmatrix}_{x=0} = \begin{bmatrix} V \\ M \end{bmatrix}_{x=L} = 0 \tag{4}$$

Equation (2), with boundary conditions taken into account, can be written as

Cancel rows because $w_{x=L}$ and $\theta_{x=L}$ are unknown

$$\begin{bmatrix} w \\ \theta \\ \hline V = 0 \\ M = 0 \end{bmatrix}_{x=L} = \left[\begin{array}{cc|cc} \bar{U}_{ww} & \bar{U}_{w\theta} & \bar{U}_{wV} & \bar{U}_{wM} \\ \bar{U}_{\theta w} & \bar{U}_{\theta\theta} & \bar{U}_{\theta V} & \bar{U}_{\theta M} \\ \hline \bar{U}_{Vw} & \bar{U}_{V\theta} & \bar{U}_{VV} & \bar{U}_{VM} \\ \bar{U}_{Mw} & \bar{U}_{M\theta} & \bar{U}_{MV} & \bar{U}_{MM} \end{array} \right] \begin{bmatrix} w \\ \theta \\ \hline V = 0 \\ M = 0 \end{bmatrix}_{x=0} \tag{5}$$

↑ Cancel columns because $M_0 = V_0 = 0$

where $\bar{U}_{ij} = (U_{ij})_{x=L}$ with $i, j = w, \theta, V, M$. Then $0 = \bar{U}_{Vw}w_0 + \bar{U}_{V\theta}\theta_0$ and $0 = \bar{U}_{Mw}w_0 + \bar{U}_{M\theta}\theta_0$. The determinant of these equations is

$$\nabla = \bar{U}_{M\theta}\bar{U}_{Vw} - \bar{U}_{Mw}\bar{U}_{V\theta} \tag{6}$$

The natural frequencies of the rotor are the roots of $\nabla = 0$. Since $\theta_0 = -w_0(\bar{U}_{Mw}/\bar{U}_{M\theta})$, from (5) the initial parameters are

$$
\mathbf{z}_0 = \begin{bmatrix} w_0 \\ -w_0(\bar{U}_{Mw}/\bar{U}_{M\theta}) \\ 0 \\ 0 \end{bmatrix} \tag{7}
$$

The mode shapes are found by inserting into (1) the natural frequencies and the initial parameters (7).

Stiffness and Mass Matrices

The rotor motion can be described with reference to the inertial frame xyz. The rotating reference XYZ is defined relative to the inertial reference system xyz by a single rotation Ωt about X with Ω denoting the whirl speed. Let the translations (deflections) w_y and w_z in the y and z directions locate the elastic centerline and the small-angle rotations θ_y and θ_z about the y and z axes, respectively, orient the plane of the cross section. The definitions of positive shear forces, moments, deflections, and slopes are shown in Fig. 17-1f.

Rotors are almost always modeled by circular shaft elements each having four degrees of freedom at each end, with rigid masses and rigid or flexible disks attached to model turbine disks, pump impellers, gears, seals, couplings, etc.

Rigid Disk The governing equations of motion for a rigid disk are [17.3]

$$
(\mathbf{m}_T + \mathbf{m}_R)\ddot{\mathbf{v}} - \Omega \mathbf{g}\dot{\mathbf{v}} = \mathbf{p} \tag{17.36}
$$

with $\mathbf{v} = [w_y \; w_z \; \theta_y \; \theta_z]^T$, where $\mathbf{m}_T$, $\mathbf{m}_R$, and $\mathbf{g}$ are provided in Table 17-4. The forcing term $\mathbf{p}$ contains the effects of the mass unbalance and other external effects on the disk.

For the unbalance force

$$
\mathbf{p} = M_i\Omega^2 \begin{bmatrix} \eta_a \\ \zeta_a \\ 0 \\ 0 \end{bmatrix} \cos \Omega t + M_i\Omega^2 \begin{bmatrix} -\zeta_a \\ \eta_a \\ 0 \\ 0 \end{bmatrix} \sin \Omega t
$$

$$
= \mathbf{p}_c \cos \Omega t + \mathbf{p}_s \sin \Omega t \tag{17.37}
$$

where M_i is the disk mass and (η_a, ζ_a) is the mass center eccentricity in the rotating coordinate system XYZ (see the figure in Table 17-4).

Shaft Elements For a uniform shaft element, the structural matrices can be obtained using the methodology for beam elements discussed in Appendix II. Use of the exact solution of the governing equation of a uniform shaft element as the shape function will lead to the "exact" dynamic force displacement relationship, the dynamic stiffness matrix. Alternatively, the deflection for an element can be taken to be the exact static response of a beam, leading to the consistent mass, stiffness, and gyroscopic matrices. For the latter case, with $\xi = x/\ell$ for an element of length ℓ, extending from $x = a$ to $x = b$,

$$\begin{bmatrix} w_y(\xi, t) \\ w_z(\xi, t) \end{bmatrix} = \mathbf{N}(\xi)\mathbf{v}(t) \tag{17.38}$$

$$\mathbf{v}(t) = \begin{bmatrix} w_{ya} & w_{za} & \theta_{ya} & \theta_{za} & w_{yb} & w_{zb} & \theta_{yb} & \theta_{zb} \end{bmatrix}^T$$

The static shape function matrix $\mathbf{N}(\xi)$ is calculated as

$$\mathbf{N}(\xi) = \begin{bmatrix} N_1 & 0 & 0 & N_2 & N_3 & 0 & 0 & N_4 \\ 0 & N_1 & -N_2 & 0 & 0 & N_3 & -N_4 & 0 \end{bmatrix} \tag{17.39}$$

where

$$N_1 = 1 - 3\xi^2 + 2\xi^3 \qquad N_2 = \ell(\xi - 2\xi^2 + \xi^3) \qquad N_3 = 3\xi^2 - 2\xi^3$$
$$N_4 = \ell(-\xi^2 + \xi^3) \tag{17.40}$$

The rotations (θ_y, θ_z) are related to the translations (w_y, w_z) by

$$\theta_y = -\frac{\partial w_z}{\partial x} \qquad \theta_z = \frac{\partial w_y}{\partial x} \tag{17.41}$$

Therefore, the rotations can be expressed as

$$\begin{bmatrix} \theta_y(\xi, t) \\ \theta_z(\xi, t) \end{bmatrix} = \overline{\mathbf{N}}(\xi)\mathbf{v}(t) \tag{17.42}$$

where

$$\overline{\mathbf{N}}(\xi) = \begin{bmatrix} \overline{\mathbf{N}}_y(\xi) \\ \overline{\mathbf{N}}_z(\xi) \end{bmatrix}$$

$$= \begin{bmatrix} 0 & -\dfrac{\partial N_1}{\partial x} & \dfrac{\partial N_2}{\partial x} & 0 & 0 & -\dfrac{\partial N_3}{\partial x} & \dfrac{\partial N_4}{\partial x} & 0 \\ \dfrac{\partial N_1}{\partial x} & 0 & 0 & \dfrac{\partial N_2}{\partial x} & \dfrac{\partial N_3}{\partial x} & 0 & 0 & \dfrac{\partial N_4}{\partial x} \end{bmatrix} \tag{17.43}$$

The governing equations of motion for a shaft element referred to the inertial reference frame are [17.3]

$$(\mathbf{m}_T + \mathbf{m}_R)\ddot{\mathbf{v}} - \Omega\mathbf{g}\dot{\mathbf{v}} + \mathbf{k}\mathbf{v} = \mathbf{p} \tag{17.44}$$

where

$$\mathbf{m}_T = \int_0^\ell \rho^* A \mathbf{N}^T \mathbf{N} \, dx \quad \text{(translational mass matrix)}$$

$$\mathbf{m}_R = \int_0^\ell I_T \overline{\mathbf{N}}^T \overline{\mathbf{N}} \, dx \quad \text{(rotary mass matrix)}$$

$$\mathbf{g} = \int_0^\ell I_p \overline{\mathbf{N}}_z^T \overline{\mathbf{N}}_y \, dx - \int_0^\ell I_p \overline{\mathbf{N}}_y^T \overline{\mathbf{N}}_z \, dx \quad \text{(gyroscopic matrix)}$$

$$\mathbf{k} = \int_0^\ell EI \mathbf{N}''^T \mathbf{N}'' \, dx \quad \text{(bending stiffness matrix)}$$

and $\mathbf{p}$ is the force vector including mass unbalance and other element external effects. For an element with distributed mass center eccentricity $(\eta(x), \zeta(x))$, the unbalance force is [17.3]

$$\mathbf{p} = \int_0^\ell \rho^* A \Omega^2 \mathbf{N}^T \left(\begin{bmatrix} \eta(\xi) \\ \zeta(\xi) \end{bmatrix} \cos \Omega t + \begin{bmatrix} -\zeta(\xi) \\ \eta(\xi) \end{bmatrix} \sin \Omega t \right) dx$$

$$= \mathbf{p}^c \cos \Omega t + \mathbf{p}^s \sin \Omega t \tag{17.45}$$

When the mass unbalance is distributed linearly, the mass center eccentricities are

$$\eta(\xi) = \eta_a(1 - \xi) + \eta_b \xi \qquad \zeta(\xi) = \zeta_a(1 - \xi) + \zeta_b \xi \tag{17.46}$$

where (η_a, ζ_a) and (η_b, ζ_b) express the mass center eccentricity at $x = a$ and $x = b$, respectively. The element matrices, including the dynamic stiffness matrix, are given in Table 17-5. Table 17-6 shows the element matrices for the shaft with axial torque effects. The conical shaft element matrices are given in Table 17-9.

Table 17-10 shows the element matrices for helically fluted shaft elements. The mass and stiffness matrices for an annular elastic thin-disk element are given in Table 17-11. See Chapters 18 ("Plates") and 19 ("Thick Shells and Disks") for the theory underlying thin disks.

Bearing Elements For aligned journal bearings, a general linearized mathematical model, now used in most rotor dynamic analyses, can be expressed by the eight spring and damping coefficients commonly employed to model the dynamic radial interaction force components between journal and bearings:

$$\begin{bmatrix} F_y \\ F_z \end{bmatrix} = -\begin{bmatrix} k_{yy} & k_{yz} \\ k_{zy} & k_{zz} \end{bmatrix} \begin{bmatrix} w_y^* \\ w_z^* \end{bmatrix} - \begin{bmatrix} c_{yy} & c_{yz} \\ c_{zy} & c_{zz} \end{bmatrix} \begin{bmatrix} \dot{w}_y^* \\ \dot{w}_z^* \end{bmatrix}$$

$$= -\mathbf{k} \mathbf{v}^b - \mathbf{c} \dot{\mathbf{v}}^b \tag{17.47}$$

where (F_y, F_z) are the dynamic radial force components, (w_y^*, w_z^*) are the relative radial displacements referenced to the static equilibrium state, and $(\dot{w}_y^*, \dot{w}_z^*)$ are the relative radial velocities between the journal and bearing.

The elements of matrices **k** and **c** are given in Table 17-12 for a short journal bearing and for a typical type of tilting pad bearing. See reference 17.8 for further details.

Seal Elements High-performance pumps, compressors, and turbines, i.e., those with high rotating speeds and high pressures, sometimes yield nonsynchronous vibrations induced by noncontacting annular and labyrinth sections. For the vibration analysis of these machines the dynamic characteristics of the seals should be included.

The linearized model for the seal is similar to that for a journal bearing,

$$\begin{bmatrix} F_y \\ F_z \end{bmatrix} = -\mathbf{k}\mathbf{v}^s - \mathbf{c}\dot{\mathbf{v}}^s \tag{17.48}$$

where $\mathbf{v}^s$ and $\dot{\mathbf{v}}^s$ are the relative displacement and velocity between journal and seal. The stiffness matrix **k** and damping matrix **c** are given in Table 17-12 for a short annular pressure seal including the effect of elastic deformation. See reference 17.9 for further discussion of seal elements.

Assembly of Global Matrices As explained in Appendix III, the global (system) mass, gyroscopic, stiffness and damping matrices or dynamic stiffness matrix of a rotor system can be assembled by summation of element matrices according to the positions of shaft elements, rigid disks, bearings and seals, etc. The global force vector can be obtained in the same manner.

The global mass and gyroscopic matrices are formed of contributions from the rigid disks and flexible shaft elements, while the global damping and stiffness matrices are formed from element matrices of shaft segments, bearings, seals and aerodynamic mechanisms which are commonly modeled with four spring and four damping coefficients.

Suppose a rotor system is discretized into M elements, with $M + 1$ nodes, along the axis of the rotor (Fig. 17-8a). Each node has 4 degrees of freedom w_y, w_z, θ_y and θ_z. In general, the global displacement vector would be defined as

$$\mathbf{V} = \begin{bmatrix} w_{y1} & w_{z1} & \theta_{y1} & \theta_{z1} & w_{y2} & w_{z2} & \theta_{y2} & \theta_{z2} & \cdots & w_{yi} & w_{zi} & \theta_{yi} & \theta_{zi} \\ & & & & \cdots & w_{y, M+1} & w_{z, M+1} & \theta_{y, M+1} & \theta_{z, M+1} \end{bmatrix} \tag{17.49}$$

The global matrices would be $4(M + 1) \times 4(M + 1)$ in dimension and the global force vector would be $4(M + 1) \times 1$.

First, consider the contributions of the shaft elements to the global matrices. The left and right ends of shaft element $i(i = 1, 2, \ldots, M)$ correspond to the nodes i and $i + 1$ of the rotor system, respectively. The left end of element i is the right end of element $i - 1$. The nodal displacements of element i, $w_{ya}, w_{za}, \theta_{ya}, \theta_{za}, w_{yb}, w_{zb}, \theta_{yb}$, and θ_{zb} in Table 17-5 correspond to $w_{yi}, w_{zi}, \theta_{yi}$, $\theta_{zi}, w_{y,i+1}, w_{z,i+1}, \theta_{y,i+1}$, and $\theta_{z,i+1}$ in Eq. (17.49), which are the contributions to **V** from $4(i - 1) + 1$ to $4i + 4$. The 8×8 element matrices of shaft element i, $\mathbf{m}_T + \mathbf{m}_R$, **g** and **k** in Table 17-5, should be added to the global mass, gyroscopic and stiffness matrices from row $4(i - 1) + 1$ to row $4i + 4$ and column $4(i - 1)$

+ 1 to column $4i + 4$, respectively. The 8×1 element force vector $\mathbf{p}^c \cos \Omega t +$ $\mathbf{p}^s \sin \Omega t$ of Table 17-5 should be added to the global force vector from row $4(i - 1) + 1$ to row $4i + 4$ (Fig. 17-8b).

Next consider the contribution of rigid disks to the global mass and gyroscopic matrices. Suppose there are M_d disks, with disk $j_d(j_d = 1, 2, \ldots, M_d)$ located at node i_d of the rotor system. Because the displacements and slopes w_y, w_z, θ_y, and θ_z in Table 17-4 for disk j_d corresponds to w_{yi_d}, w_{zi_d}, θ_{yi_d} and θ_{zi_d}, which are the entries from $4(i_d - 1) + 1$ to $4(i_d - 1) + 4$, of the global displacement vector $\mathbf{V}$ of Eq. (17.49), the 4×4 element matrices of the rigid disk, $\mathbf{m}_T + \mathbf{m}_R$ and $\mathbf{g}$ in Table 17-4 should be added to the global mass and gyroscopic matrices from row $4(i_d - 1) + 1$ to row $4(i_d - 1) + 4$ and from column $4(i_d - 1) + 1$ to column $4(i_d - 1) + 4$, respectively. The 4×1 force vector of the rigid disk, $\mathbf{p}^c \cos \Omega t +$ $\mathbf{p}^s \sin \Omega t$ in Table 17-4, should be added to the global force vector from row $4(i_d - 1) + 1$ to row $4(i_d - 1) + 4$ (Fig. 17-8).

Finally, consider the contribution of bearings, seals, or other mechanisms modeled as eight spring and damping coefficients to global stiffness and damping matrices. From Eq. (17.47)

$$\begin{bmatrix} F_y \\ F_z \end{bmatrix} = -\mathbf{k}\mathbf{v}^b - \mathbf{c}\dot{\mathbf{v}}^b = -\mathbf{k}(\mathbf{v}^j - \mathbf{v}^k) - \mathbf{c}(\dot{\mathbf{v}}^j - \dot{\mathbf{v}}^k) = -\mathbf{k}\mathbf{v}^j - \mathbf{c}\dot{\mathbf{v}}^j + \mathbf{k}\mathbf{v}^k + \mathbf{c}\dot{\mathbf{v}}^k$$

(17.50)

where $\mathbf{v}^j = [w_y^j \ \ w_z^j]^T$ is the displacement vector of the journal, which, generally, is the part of the rotor where the bearing or seal is located. The vector $\mathbf{v}^k = [w_y^k \ \ w_z^k]^T$ is the displacement vector of the bearing or seal.

Suppose there are M_b bearings or seals, one of which $j_b(j_b = 1, 2, \ldots, M_b)$ is at node i_b. Because the displacements w_y^j and w_z^j of the bearing j_b correspond to the displacement w_{yi_b} and w_{zi_b}, which are the $4(i_b - 1) + 1$ and $4(i_b - 1) + 2$ entries of the global displacement vector $\mathbf{V}$ (Eq. 17.49). The 2×2 stiffness and damping matrices $\mathbf{k}$ and $\mathbf{c}$ of the bearing in Eq. (17.50) or Table 17-12 should be added to the global stiffness and damping matrices, respectively, from row $4(i_b - 1) + 1$ to row $4(i_b - 1) + 2$ and from column $4(i_b - 1) + 1$ to $4(i_b - 1) +$ 2. The vector $\mathbf{k}\mathbf{v}^k + \mathbf{c}\dot{\mathbf{v}}^k$ in Eq. (17.50) can be added to the global force vector from row $4(i_b - 1) + 1$ to row $4(i_b - 1) + 2$ (Fig. 17-8).

Note that the global gyroscopic matrix multiplied by $-\Omega$ (Ω is the spin speed of the rotor) is a damping matrix (Eqs. (17.36) and (17.44)).

Example 17.5 Natural Frequencies: Complex Eigenvalue Analysis of a Pump System A pump rotor supported by two isotropic bearings is shown in Fig. 17-9. The elastic modulus is 210 GN/m². The mass and mass moment of inertia of the impeller are 308.268 kg and 51.378 kg · m², respectively. The bearing characteristics are modeled with stiffness coefficients of 1000 GN/m and damping coefficients of 10 kN · s/m.

(a) Compute the first two natural frequencies

(b) Compute the complex eigenvalues at a rotating speed of 1800 rpm.

(c) Compare the complex eigenvalues with the real eigenvalues (natural frequencies), which ignore the damping of the bearings.

(d) Graph the root loci of the first two modes as functions of rotating speed.

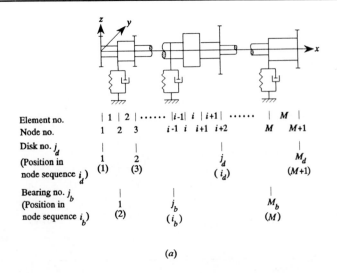

(a)

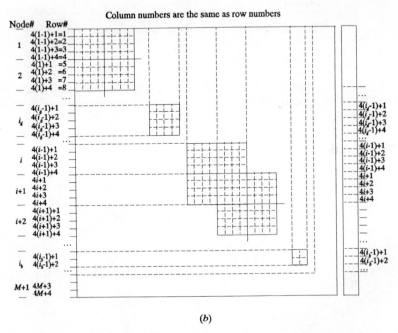

(b)

FIGURE 17-8 Rotor system and global matrices. (a) Rotor system discretization. (b) Assembly of global matrices.

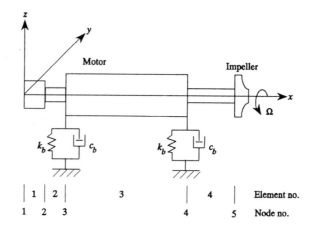

FIGURE 17-9 Rotor-bearing system of Example 17.5.

Element no.	1	2	3	4
Length (m)	0.2	0.2	1.0	0.4
Diameter (m)	0.2	0.16	0.3	0.16

The rotor is modeled with the four shaft elements and one disk shown in Fig. 17-9. The mass, gyroscopic, and stiffness matrices can be obtained from Table 17-4 and 17-5.

For the shaft elements, these matrices (expressed as 4×4 submatrices) have dimension 8×8 (Table 17-5):

$$\mathbf{m}^i = \mathbf{m}^i_T + \mathbf{m}^i_R = \begin{bmatrix} \mathbf{m}^i_{11} & \mathbf{m}^i_{12} \\ \mathbf{m}^i_{21} & \mathbf{m}^i_{22} \end{bmatrix}$$

$$\mathbf{g}^i = \begin{bmatrix} \mathbf{g}^i_{11} & \mathbf{g}^i_{12} \\ \mathbf{g}^i_{21} & \mathbf{g}^i_{22} \end{bmatrix} \tag{1}$$

$$\mathbf{k}^i = \begin{bmatrix} \mathbf{k}^i_{11} & \mathbf{k}^i_{12} \\ \mathbf{k}^i_{21} & \mathbf{k}^i_{22} \end{bmatrix}$$

where

$$i = 1, 2, 3, 4.$$

For the disk element, the 4×4 mass and gyroscopic matrices can be taken from Table 17-4,

$$
\mathbf{m}^d = \mathbf{m}_T^d + \mathbf{m}_R^d = \begin{bmatrix} M_i & & & \\ & M_i & & \\ & & I_{Ti} & \\ & & & I_{Ti} \end{bmatrix} \qquad \mathbf{g}^d = \begin{bmatrix} 0 & & & \\ & 0 & & \\ & & 0 & I_{pi} \\ & & -I_{pi} & 0 \end{bmatrix} \tag{2}
$$

where M_i, I_{Ti}, and I_{pi} are the mass, translational, and polar mass moments of the disk, respectively. Here $I_{Ti} = \frac{1}{2} I_{pi}$.

For the bearings, the stiffness and damping matrices have the forms [Eq. (17.47)]

$$
\mathbf{k}^b = \begin{bmatrix} k_b & 0 \\ 0 & k_b \end{bmatrix} \qquad \mathbf{c}^b = \begin{bmatrix} c_b & 0 \\ 0 & c_b \end{bmatrix} \tag{3}
$$

where $k_b = 1000$ GN/m and $c_b = 10$ kN $\cdot$ s/m. Define the global displacement vector

$$
\mathbf{V} = \begin{bmatrix} w_{y1} & w_{z1} & \theta_{y1} & \theta_{z1} & w_{y2} & w_{z2} & \theta_{y2} & \theta_{z2} & \cdots & w_{y5} & w_{z5} & \theta_{y5} & \theta_{z5} \end{bmatrix}^T
$$

According to the order of the shaft elements and the position of the disk, the global mass matrix is assembled as

From 5th to
12th column

$$
\mathbf{M} = \text{Mass Matrix} \quad \begin{bmatrix} \mathbf{m}_{11}^1 & \mathbf{m}_{12}^1 & & & \vdots & \\ \mathbf{m}_{21}^1 & \mathbf{m}_{22}^1 + \mathbf{m}_{11}^2 & \mathbf{m}_{12}^2 & & \\ & \mathbf{m}_{21}^2 & \mathbf{m}_{22}^2 + \mathbf{m}_{11}^3 & \mathbf{m}_{12}^3 & \\ & & \mathbf{m}_{21}^3 & \mathbf{m}_{22}^3 + \mathbf{m}_{11}^4 & \mathbf{m}_{12}^4 \\ & & & \mathbf{m}_{21}^4 & \mathbf{m}_{22}^4 + \mathbf{m}^d \end{bmatrix} \quad \begin{array}{l} \text{From 5th to} \\ \text{12th row} \end{array}
$$

$$\tag{4}$$

Here, for example, for the second shaft element with $i = 2$, $\mathbf{m}^i = \mathbf{m}^2$ is added from row and column $4(i - 1) + 1 = 5$ to row and column $4i + 4 = 12$, respectively. Because the disk is on the right side of shaft element 4, the position of the disk in the node sequence is $i_d = 5$ and $\mathbf{m}^d$ is added to $\mathbf{M}$ from row and column $4(i_d - 1) + 1 = 17$ to row and column $4(i_d - 1) + 4 = 20$. Thus the matrix $\mathbf{m}^d$ is added to the position of $\mathbf{m}_{22}^4$. Remember $\mathbf{m}_{pq}^i$ ($i = 1, 2, 3, 4$, $p, q = 1, 2$) are 4×4 matrices. If the disk is on the left side of element $4(i_d = 4)$, $\mathbf{m}^d$ should be added to the position of $\mathbf{m}_{11}^4$. The global gyroscopic matrix $\mathbf{G}$ has the same form as $\mathbf{M}$, with $\mathbf{m}_{pq}^i$ and $\mathbf{m}^d$ replaced by $\mathbf{g}_{pq}^i$ and $\mathbf{g}^d$ ($i = 1, 2, 3, 4$, $p, q = 1, 2$) respectively.

The global shaft stiffness matrix is assembled in the same fashion as **M**.

$$K = \begin{bmatrix} k_{11}^1 \cdots k_{12}^1 & & & & \\ k_{21}^1 & k_{22}^1 + k_{11}^2 \cdots k_{12}^2 & & & \\ & k_{21}^2 & k_{22}^2 + k_{11}^3 \cdots k_{12}^3 & & \\ & & k_{21}^3 \cdots\cdots\cdots k_{22}^3 + k_{11}^4 \cdots k_{12}^4 & \\ & & & k_{21}^4 \cdots\cdots\cdots k_{22}^4 \end{bmatrix} \tag{5}$$

According to the position of the bearings, the global bearing stiffness matrix is assembled as

$$K_b = \begin{bmatrix} k_b \cdots\cdots 0 \cdots\cdots\cdots\cdots\cdots\cdots\cdots & \text{row 9} \\ 0 \cdots\cdots k_b \cdots\cdots\cdots\cdots\cdots\cdots & \text{row 10} \\ & k_b \cdots\cdots 0 \cdots & \text{row 13} \\ & 0 \quad k_b \cdots & \text{row 14} \end{bmatrix} \tag{6}$$

Col. 9   Col. 10   Col. 13   Col. 14   (all other elements $= 0$)

and the global bearing damping matrix C_b is obtained by replacing k_b with c_b.

For the first bearing, the position in the node sequence is $i_b = 3$, so that k^b or c^b should be added to K_b or C_b from row and column $4(i_b - 1) + 1 = 9$ to row and column $4(i_b - 1) + 2 = 10$. For the second bearing, $i_b = 4$, so k^b and c^b are added to K_b and C_b from row and column 13 to row and column 14, respectively.

The system equation is

$$M\ddot{V} + (C_b - \Omega G)\dot{V} + (K + K_b)V = 0 \tag{7}$$

Transform this equation to first order state vector form

$$B\dot{w} + Aw = 0 \tag{8}$$

where

$$A = \begin{bmatrix} C_b - \Omega G & K + K_b \\ -(K + K_b) & 0 \end{bmatrix} \qquad B = \begin{bmatrix} M & 0 \\ 0 & K + K_b \end{bmatrix}$$

$$w = [\dot{V}^T \quad V^T]^T$$

For an assumed harmonic solution

$$w = w_0 e^{\lambda t} \tag{9}$$

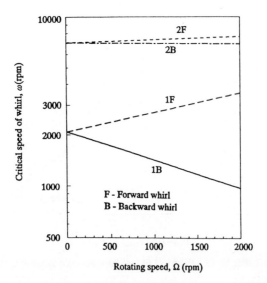

FIGURE 17-10 Critcal speeds of the pump system (damping is ignored).

where $\lambda = -\gamma_n + i\omega_d$ the associated eigenvalue problem is

$$(\mathbf{A} + \lambda\mathbf{B})\mathbf{w}_0 = \mathbf{0} \qquad (10)$$

By letting $\mathbf{C}_b = \mathbf{0}$, the critical speeds of the rotor system can be obtained. Otherwise, the complex eigenvalues can be computed. The results are shown in Fig. 17-10 to 17-12.

Mode	Damped Critical Speed ω_d (rad / s)		Damped Exponent γ_n (s^{-1})	
	Forward (F)	Backward (B)	Forward (F)	Backward (B)
1	410.74	96.02	−5.18	−0.66
2	784.14	724.84	−9.32	−12.73
3	2168.80	2148.51	−30.08	−29.92

(*a*)

Mode	Forward (F)		Backward (B)	
	Undamped ω_c	Damped ω_d	Undamped ω_c	Damped ω_d
1	310.124	310.156	130.470	130.471
2	758.713	758.541	730.550	730.408
3	2164.39	2164.17	2153.14	2152.93

(*b*)

FIGURE 17-11 Partial results for Example 17.5; (*a*) complex eigenvalues at 1800 rpm; (*b*) damped and undamped critical speeds (rad / s) at 1000 rpm.

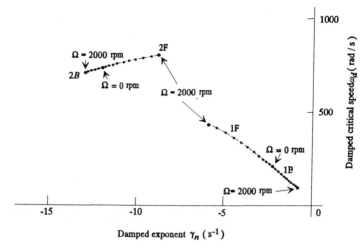

FIGURE 17-12 Root loci of the damped rotor-bearing system.

Further information on rotating-shaft systems can be found in references 17.10–17.13.

17.4 TORSIONAL VIBRATION

Since the speed of rotation has little effect on the torsional vibration of a shaft, the formulas of Chapter 12 can be employed to study the torsional behavior of rotor systems. A useful reference for torsional vibrations is reference 17.14.

17.5 VIBRATION OF RADIAL BEAM

The dynamic characteristics of rotating blades are important in the design of rotating structural elements such as turbine blades, aircraft propellers, cooling fans, helicopter rotors, and satellite booms. A radial rotating beam mounted in the rim of a disc is usually taken as the mathematical model for these kinds of structures.

The model is a uniform beam of length L and with a hub of radius R_0 rotating at a constant angular velocity of Ω (rad/s) about the x axis (Fig. 17-13). Here, xyz is a set of global fixed coordinate axes with the origin at the center of the hub, XYZ are global rotating coordinate axes with the origin at the center of the rotating hub, and $\xi\eta Z$ are rotating coordinates with the origin at the center of the rotating hub, with ξ, η as the principal axes of inertia of the sectional area of the beam. The neutral axis of the beam is inclined to the plane of rotation at an angle Ψ. For $\Psi = 0°$, the bending motion of the beam in the ξZ plane becomes purely out of plane (flapping), and for $\Psi = 90°$, the bending motion in the ξZ plane is purely in plane (lead–lag). The motion of the beam along the Z axis is called axial vibration.

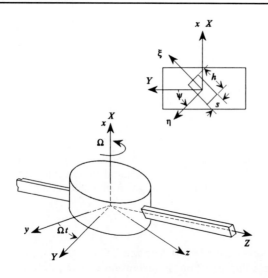

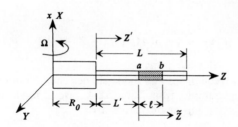

FIGURE 17-13 Coordinate system and geometry of a radial beam.

Bending Vibration

Natural Frequencies Table 17-13 lists the natural frequencies of a beam of circular and square cross section at angle $\Psi = 0°$ or $\Psi = 90°$ with various boundary conditions.

Stiffness and Mass Matrices For vibrations in the direction perpendicular to the neutral axis of the beam, i.e., for motion in the ξ direction (Fig. 17-13), the stiffness and mass matrices of the rotating beam element are defined as

$$\mathbf{k}_B = \ell \int_0^1 EI\mathbf{N}''^T\mathbf{N}'' \, d\zeta \quad \text{(bending stiffness matrix)}$$

$$\mathbf{k}_C = \ell \int_0^1 F_C\mathbf{N}'^T\mathbf{N}' \, d\zeta \quad \text{(centrifugal stiffness matrix)}$$

$$\mathbf{m}_T = \ell \int_0^1 \rho^*A\mathbf{N}^T\mathbf{N} \, d\zeta \quad \text{(translational mass matrix)}$$

$$\mathbf{m}_R = \ell \int_0^1 I_T\mathbf{N}'^T\mathbf{N}' \, d\zeta \quad \text{(rotary mass matrix)}$$

(17.51)

where $\mathbf{N}(\zeta) = [1 - 3\zeta^2 + 2\zeta^3 \quad \ell(\zeta - 2\zeta^2 + \zeta^3) \quad 3\zeta^2 - 2\zeta^3 \quad \ell(-\zeta^2 + \zeta^3)]$, with $\zeta = \tilde{Z}/\ell$, is the shape function matrix and ℓ is the length of the element. The centrifugal force F_c is given by

$$F_C = \rho^* A \Omega^2 \ell^2 \left[\frac{R_0}{\ell^2}(L - L') + \frac{1}{2\ell^2}(L^2 - L'^2) - \frac{1}{\ell}(R_0 + L')\zeta - \tfrac{1}{2}\zeta^2 \right]$$

(17.52)

with L equal to the length of the beam and L' equal to the total length of elements before and not including the element under consideration.

The mass and stiffness matrices are given in Table 17-14.

Axial Vibration

Natural Frequencies and Mode Shapes Since the formulas for axial vibrations are given in Chapter 12 on the extension of bars, only some problems that can occur for radial rotating bars are considered here.

The equation of motion is given by (Fig. 17-13)

$$-\rho^* A \frac{\partial^2 w_Z}{\partial t^2} + \rho^* A \Omega^2 (R_0 + Z' + w_Z) + EA \frac{\partial^2 w_Z}{\partial Z'^2} = 0 \qquad (17.53)$$

The natural frequencies and mode shapes can be obtained from the boundary conditions. For example, from Table 12-4, case 2, for a clamped–free beam, the natural frequencies and mode shapes are

$$\omega_n^2 = \frac{E}{\rho^*}\left(\frac{\lambda_n}{L}\right)^2 - \Omega^2 \qquad (17.54)$$

$$\phi_n = \sin(\lambda_n Z'/L) \qquad (17.55)$$

where

$$\lambda_n = \tfrac{1}{2}\pi, \tfrac{3}{2}\pi, \ldots = \tfrac{1}{2}\pi(2n - 1) \qquad n = 1, 2, 3, \ldots$$

Here, because of the rotating speed Ω, the formula for the natural frequency in Table 12-4 is changed to that shown in Eq. (17.54).

Transfer Matrices The fundamental equations of motion in first-order form for the axial vibration of a radial rotating bar (Fig. 17-13) are

$$\frac{\partial w_Z}{\partial Z'} = \frac{P}{AE}$$

$$\frac{\partial P}{\partial Z'} = \rho^* A \frac{\partial^2 w_Z}{\partial t^2} - \rho^* A \Omega^2 (R_0 + Z' + w_Z)$$

(17.56)

In this case, the centrifugal force represents a distributed axial force.

The transfer matrices, including exact-mass and lumped-mass modeling, for a bar segment of length ℓ are provided in Table 17-15.

Stiffness and Mass Matrices   The displacement of a typical point between the two ends of the element is

$$w_Z(\tilde{Z}, t) = \mathbf{N}(\zeta)\mathbf{v} \tag{17.57}$$

where $\mathbf{v} = [w_{Za} \ w_{Zb}]^T$, $\zeta = \tilde{Z}/\ell$, and $\mathbf{N}(\zeta)$ is the shape function matrix given by

$$
\begin{aligned}
\mathbf{N}(\zeta) &= \begin{bmatrix} N_1(\zeta) & N_2(\zeta) \end{bmatrix} \\
N_1(\zeta) &= \cos(\gamma\Omega\ell\zeta) - \cot(\gamma\Omega\ell)\sin(\gamma\Omega\ell\zeta) \\
N_2(\zeta) &= \csc(\gamma\Omega\ell)\sin(\gamma\Omega\ell\zeta)
\end{aligned}
\tag{17.58}
$$

where $\gamma = \sqrt{\rho^*/E}$.

The equivalent mass and stiffness matrices for a radial rotating-bar element are

$$
\mathbf{m} = \ell\int_0^1 \rho^* A \mathbf{N}^T \mathbf{N} \, d\zeta \quad \text{(translational mass matrix)}
$$

$$
\mathbf{k}_A = \ell\int_0^1 A E \mathbf{N}'^T \mathbf{N}' \, d\zeta \quad \text{(axial stiffness matrix)} \tag{17.59}
$$

$$
\mathbf{k}_C = \ell\int_0^1 F_C \mathbf{N}^T \mathbf{N} \, d\zeta \quad \text{(centrifugal stiffness matrix)}
$$

where F_C is the centrifugal force defined by Eq. (17.52). The mass and stiffness matrices are given in Table 17-15.

REFERENCES

17.1 Saito, H., and Chonan, S., "Vibrations of Vertical Rods with an Attached Mass," *J. Sound Vibrat.*, Vol. 84, No. 4, 1982, pp. 519–527.

17.2 Eshleman, R. L., and Eubanks, R. A., "On the Critical Speeds of a Continuous Rotor," *Trans. ASME J. Eng. Ind.*, Vol. 91, 1969, pp. 1180–1188.

17.3 Nelson, H. D., and McVaugh, J. M., "The Dynamics of Rotor-bearing Systems Using Finite Elements," *Trans. ASME J. Eng. Ind.*, Vol. 98, 1976, pp. 593–600.

17.4 Zorzi, E. S., and Nelson, H. D., "The Dynamics of Rotor-bearing Systems with Axial Torque—A Finite Element Approach," *Trans. ASME J. Mech. Des.*, Vol. 102, 1980, pp. 158–161.

17.5 Greenhill, L. M., Bickford, W. B., and Nelson, H. D., "A Conical Beam Finite Element for Rotor Dynamics Analysis," *Trans. ASME J. Vibrat. Acoust. Stress Reliabil. Des.*, Vol. 107, 1985, pp. 421–430.

17.6 Tekinalp, O., and Ulsoy, A. G., "Modelling and Finite Element Analysis of Drill Bit Vibrations," Proceedings of ASME Vibration Conference on Mechanical Vibration and Noise, Montreal, Canada, September 1989, pp. 61–68.

17.7 Kirkhope, J., and Wilson, G. J., "Vibration and Stress Analysis of Thin Rotating Discs Using Annular Finite Elements," *J. Sound Vibrat.*, Vol. 44, No. 4, 1976, pp. 461–474.

17.8 Nicholas, J. C., Gunter, E. J., Jr., and Allaire, P. E., "Stiffness and Damping Coefficients for the Five-Pad Tilting Pad Bearings," *ASLE J.*, Vol. 22, No. 2, 1979, pp. 113–124.

17.9 Iwatsubo, T., and Yang, B. S., "The Effects of Elastic Deformation on Seal Dynamics," *Trans. ASME J. Vibrat. Acoust. Stress Reliabil. Des.*, Vol. 110, 1988, pp. 59–64.

17.10 Pilkey, W. D., and Chang, P. Y. *Modern Formulas for Statics and Dynamics*, McGraw-Hill, New York, 1978.

17.11 Vance, J. M. *Rotordynamics of Turbomachinery*, Wiley, New York, 1988.

17.12 Ehrich, F. F., *Handbook of Rotordynamics*, McGraw-Hill, New York, 1992.

17.13 Smith, D. M. *Journal Bearings in Turbomachinery*, Chapman & Hall, London, 1969.

17.14 Doughty, S., and Vafaee, G. "Transfer Matrix Eigensolutions for Damped Torsional Systems," *Trans. ASME J. Vibrat. Acoust. Stress Reliabil. Des.*, Vol. 107, 1985, pp. 128–132.

17.15 Bauer, H. F., and Eidel, W. "Vibration of a Rotating Uniform Beam, Part II: Orientation Perpendicular to the Axis of Rotation," *J. Sound Vibrat.*, Vol. 122, No. 2, 1988, pp. 357–375.

17.16 Hoa, S. V. "Vibration of a Rotating Beam with Tip Mass," *J. Sound Vibrat.*, Vol. 67, No. 3, 1979, pp. 369–381.

Tables

TABLE 17-1 CRITICAL SPEEDS OF SIMPLE HORIZONTAL ROTORS

Notation

ω_{cj} = critical speed ($j = 1, 2, 3, 4$)
M_i = concentrated mass
m_b = beam mass (shaft mass)
I_{Ti} = transverse mass moment of inertia of concentrated mass
E = modulus of elasticity,

I_{pi} = polar mass moment of inertia of concentrated mass
Ω = rotational speed
I = moment of inertia of cross-sectional area

See Table 11-14 for a free–free rotor.

Case	Critical Speed (rad/s)
1. Center mass, pinned–pinned supports	$\omega_{c1} = \left[\dfrac{48EI}{L^3(M_i + 0.49m_b)}\right]^{1/2}$
2. Off-center mass, pinned–pinned supports	$\omega_{c1} = \left[\dfrac{3EIL}{L_1^2 L_2^2\{M_i + (\alpha + \beta)m_b\}}\right]^{1/2}$ $\alpha = \dfrac{L_1}{L}\left[\dfrac{(2L_2 + L_1)^2}{12L_2^2} + \dfrac{L_1^2}{28L_2^2} - \dfrac{L_1(2L_2 + L_1)}{10L_2^2}\right]$ $\beta = \dfrac{L_2}{L}\left[\dfrac{(2L_1 + L_2)^2}{12L_1^2} + \dfrac{L_2^2}{28L_1^2} - \dfrac{L_2(2L_1 + L_2)}{10L_1^2}\right]$
3. Overhung rotor	$\omega_{c1} = \left[\dfrac{3EI}{L^3(M_i + 0.24m_b)}\right]^{1/2}$
4. Overhung rotor with linear spring	$\omega_{c1} = \left(\dfrac{EI/(m_b L^3)}{1/(12.36236 + 3\xi \bar{k}) + \bar{m}/(3 + \bar{k})}\right)^{1/2}$ $\bar{m} = \dfrac{M_i}{m_b} \qquad \bar{k} = \dfrac{kL^3}{EI}$

TABLE 17-1 (continued) CRITICAL SPEEDS OF SIMPLE HORIZONTAL ROTORS

Case	Critical Speed (rad/s)
5. Center mass, clamped–clamped supports	$$\omega_{c1} = \left[\frac{192\,EI}{L^3(M_i + 0.37m_b)} \right]^{1/2}$$
6. Off-center mass, clamped–clamped supports	$$\omega_{c1} = \left(\frac{3EIL^3}{L_1^3 L_2^3[M_i + (\alpha + \beta)m_b]} \right)^{1/2}$$ $$\alpha = \frac{L_1}{L}\left[\frac{(3L_1 + L_2)^2}{28L_2^2} + \frac{9L^2}{20L_2^2} - \frac{L(3L_1 + L_2)}{4L_2^2} \right]$$ $$\beta = \frac{L_2}{L}\left[\frac{(3L_2 + L_1)^2}{28L_1^2} + \frac{9L^2}{20L_1^2} - \frac{L(3L_2 + L_1)}{4L_1^2} \right]$$
7. Long rigid rotor, elastic supports	$$\omega_{c1} = \omega_{c2} = \sqrt{\frac{2k}{M_i}}$$ $$\omega_{c3} = \frac{I_{pi}}{2I_{Ti}}\Omega + \sqrt{\frac{kL^2}{2I_{Ti}} + \left(\frac{I_{pi}}{2I_{Ti}}\Omega\right)^2}$$ $$\omega_{c4} = \frac{I_{pi}}{2I_{Ti}}\Omega - \sqrt{\frac{kL^2}{2I_{Ti}} + \left(\frac{I_{pi}}{2I_{Ti}}\Omega\right)^2}$$
8. Rigid rotor, anisotropic bearing $(k_y < k_z)$	$$\omega_{c1} = \left[\frac{2k_y}{M_i} + \frac{k_y}{\Delta k}\left(\frac{2c}{M_i}\right)^2 \right]^{1/2}$$ $$\omega_{c2} = \left[\frac{2k_z}{M_i} - \frac{k_z}{\Delta k}\left(\frac{2c}{M_i}\right)^2 \right]^{1/2}$$ where $\Delta k = k_z - k_y$

TABLE 17-2 CRITICAL SPEEDS OF SIMPLE VERTICAL ROTORS[a]

Notation

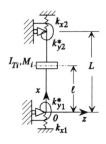

A = section area
A_s = equivalent shear area, $A_s = A/\alpha_s$
α_s = shear correction factor (Table 2-4)
E = modulus of elasticity
G = shear modulus of elasticity
g = gravitational acceleration
I = moment of inertia of cross-
     sectional area of shaft
I_{Ti} = transverse mass moment of
       inertia of concentrated mass
k_{x1}, k_{x2} = linear spring constants
k_{y1}^*, k_{y2}^* = rotary spring constants
M_i = mass of concentrated mass
ω_c = critical speed
ρ^* = mass per unit volume of shaft

Gyroscopic effects of the rotor are not taken into account.

The critical speeds are the roots of the following determinant set equal to zero (the frequency equation):

$$
\begin{vmatrix}
0 & 1 & 0 & 1 & 0 & 0 & 0 & 0 \\
a_{21} & a_{22} & a_{23} & a_{24} & 0 & 0 & 0 & 0 \\
a_{31} & a_{32} & a_{33} & a_{34} & a_{35} & a_{36} & a_{37} & a_{38} \\
a_{41} & a_{42} & a_{43} & a_{44} & a_{45} & a_{46} & a_{47} & a_{48} \\
a_{51} & a_{52} & a_{53} & a_{54} & a_{55} & a_{56} & a_{57} & a_{58} \\
a_{61} & a_{62} & a_{63} & a_{64} & a_{65} & a_{66} & a_{67} & a_{68} \\
0 & 0 & 0 & 0 & a_{75} & a_{76} & a_{77} & a_{78} \\
0 & 0 & 0 & 0 & a_{85} & a_{86} & a_{87} & a_{88}
\end{vmatrix} = 0
$$

$$a_{21} = \frac{k_{y1}^* L}{EI}\gamma_1 \qquad a_{22} = \alpha_1\gamma_1 \qquad a_{23} = \frac{k_{y1}^* L}{EI}\delta_1 \qquad a_{24} = -\beta_1\delta_1$$

$$a_{31} = \sin(\alpha_1 \ell/L) \qquad a_{32} = \cos(\alpha_1 \ell/L)$$
$$a_{33} = \sinh(\beta_1 \ell/L) \qquad a_{34} = \cosh(\beta_1 \ell/L)$$
$$a_{35} = -\sin(\alpha_2 \ell/L) \qquad a_{36} = -\cos(\alpha_2 \ell/L)$$
$$a_{37} = -\sinh(\beta_2 \ell/L) \qquad a_{38} = -\cosh(\beta_2 \ell/L)$$
$$a_{41} = \gamma_1 a_{32} \qquad a_{42} = -\gamma_1 a_{31}$$
$$a_{43} = \delta_1 a_{34} \qquad a_{44} = \delta_1 a_{33}$$
$$a_{45} = \gamma_2 a_{36} \qquad a_{46} = -\gamma_2 a_{35}$$
$$a_{47} = \delta_2 a_{38} \qquad a_{48} = \delta_2 a_{37}$$

TABLE 17-2 **Critical Speeds of Vertical Rotors** 906

TABLE 17-2 (continued)   CRITICAL SPEEDS OF SIMPLE VERTICAL ROTORS

$$a_{51} = \frac{M_i L}{GA_s} \omega_c^2 a_{31} + \left[\left(1 + \frac{P_1}{GA_s}\right)\gamma_1 - \alpha_1\right] a_{32}$$

$$a_{52} = \frac{M_i L}{GA_s} \omega_c^2 a_{32} - \left[\left(1 + \frac{P_1}{GA_s}\right)\gamma_1 + \alpha_1\right] a_{31}$$

$$a_{53} = \frac{M_i L}{GA_s} \omega_c^2 a_{33} + \left[\left(1 + \frac{P_1}{GA_s}\right)\gamma_1 - \beta_1\right] a_{34}$$

$$a_{54} = \frac{M_i L}{GA_s} \omega_c^2 a_{34} + \left[\left(1 + \frac{P_1}{GA_s}\right)\gamma_1 - \beta_1\right] a_{33}$$

$$a_{55} = \left(1 - \frac{P_2}{GA_s}\right) a_{45} - \alpha_2 a_{36} \qquad a_{56} = \left(1 - \frac{P_2}{GA_s}\right) a_{46} + \alpha_2 a_{35}$$

$$a_{57} = \left(1 - \frac{P_2}{GA_s}\right) a_{47} - \beta_2 a_{38} \qquad a_{58} = \left(1 - \frac{P_2}{GA_s}\right) a_{48} - \beta_2 a_{37}$$

$$a_{61} = \frac{I_{Ti} L}{EI} \omega_c^2 a_{41} - \alpha_1 a_{42} \qquad a_{62} = \frac{I_{Ti} L}{EI} \omega_c^2 a_{42} + \alpha_1 a_{41}$$

$$a_{63} = \frac{I_{Ti} L}{EI} \omega_c^2 a_{43} - \beta_1 a_{44} \qquad a_{64} = \frac{I_{Ti} L}{EI} \omega_c^2 a_{44} - \beta_1 a_{43}$$

$$a_{65} = -\alpha_2 a_{46} \qquad\qquad a_{66} = \alpha_2 a_{45}$$

$$a_{67} = -\beta_2 a_{48} \qquad\qquad a_{68} = -\beta_2 a_{47}$$

$$a_{75} = \sin \alpha_2 \qquad\qquad a_{76} = \cos \alpha_2$$

$$a_{77} = \sinh \beta_2 \qquad\qquad a_{78} = \cosh \beta_2$$

$$a_{85} = \frac{k_{y2}^* L}{EI} \gamma_2 a_{76} - \gamma_2 \alpha_2 a_{75} \qquad a_{86} = -\frac{k_{y2}^* L}{EI} \gamma_2 a_{75} - \gamma_2 \alpha_2 a_{76}$$

$$a_{87} = \frac{k_{y2}^* L}{EI} \delta_2 a_{78} + \delta_2 \beta_2 a_{77} \qquad a_{88} = \frac{k_{y2}^* L}{EI} \delta_2 a_{77} + \delta_2 \beta_2 a_{78}$$

where

$$\gamma_j = \frac{\alpha_j^2 - \rho^* A L^2 \omega_c^2 / GA_s}{(1 + P/GA_s)\alpha_j} \qquad (j = 1, 2)$$

$$\delta_j = \frac{\beta_j^2 + \rho^* A L^2 \omega_c^2 / GA_s}{(1 + P/GA_s)\beta_j} \qquad (j = 1, 2)$$

$$P_1 = M_i g\left[k_{x1} k_{x2}(L - \ell) + k_{x1} EA\right] / \left[k_{x1} k_{x2} L + (k_{x1} + k_{x2})EA\right]$$

$$P_2 = M_i g\left[k_{x1} k_{x2}\ell + k_{x2} EA\right] / \left[k_{x1} k_{x2} L + (k_{x1} + k_{x2})EA\right]$$

$$P = \begin{cases} P_1 & \text{for } j = 1 \\ -P_2 & \text{for } j = 2 \end{cases}$$

TABLE 17-2 (continued)   CRITICAL SPEEDS OF SIMPLE VERTICAL ROTORS

The constants α_j and β_j $(j = 1, 2)$ are determined by solving simultaneously the following two equations:

$$\alpha_j^2 - \beta_j^2 = \left(\frac{1}{EA} + \frac{1}{GA_s} \right) \rho^* A L^2 \omega_c^2 + \left(1 + \frac{P}{GA_s} \right) \frac{PL^2}{EI}$$

$$\alpha_j^2 \beta_j^2 = \frac{\rho^* A L^4}{EI} \omega_c^2 \left[\left(1 + \frac{P}{GA_s} \right) - \frac{\rho^* I}{GA_s} \omega_c^2 \right]$$

In the following, the stiffness coefficients $k_{x1}, \ldots, k_{y2}^*$ for several special support conditions are listed.

	Case 1	Case 2[b]	Case 3[b]	Case 4	Case 5	Case 6
	Hinged - Hinged	Hinged - Hinged	Hinged - Hinged	Clamped - Clamped	Clamped -Guided	Guided - Clamped
k_{x1}	∞	∞	0	∞	∞	0
k_{x2}	∞	0	∞	∞	0	∞
k_{y1}^*	0	0	0	∞	∞	∞
k_{y2}^*	0	0	0	∞	∞	∞

[a] Based on [17.1].
[b] Vertical (axial) motion permitted.

TABLE 17-2   **Critical Speeds of Vertical Rotors**   **908**

TABLE 17-3 CRITICAL SPEEDS AND MODE SHAPES FOR UNIFORM ROTORS[a] IN BENDING

Notation

A = cross-sectional area

E = modulus of elasticity

I = moment of inertia of cross-sectional area

n = mode number

r = radius of gyration of cross-sectional area, $r^2 = I/A$

ω_{c1} = forward whirling critical speed (positive)

ω_{c2} = backward whirling critical speed (negative)

ρ^* = mass per unit volume

Ω = rotating speed

$\xi = x/L$

Critical Speeds:

$$\omega_{c1,2} = \left[-\Omega\beta_n^2 \pm \sqrt{\Omega^2\beta_n^4 + (1 - \beta_n^2)\omega_0^2} \right] / (1 - \beta_n^2) \quad \text{(plus sign for } \omega_{c1}, \text{ minus sign for } \omega_{c2})$$

$$\beta_n = \lambda_n r/L \qquad \omega_0 = \left(\frac{\lambda_n}{L}\right)^2 \left[\frac{EI}{\rho^*A}\right]^{1/2}$$

Boundary Conditions	Frequency Equation	Constant λ_n	Mode Shapes ϕ_n
1. Free–free	$\cosh \lambda \cos \lambda = 1$	$\lambda_1 = 4.7300$ $\lambda_2 = 7.8532$ $\lambda_3 = 10.9956$ $\cdots$ For large n $\lambda_n \approx \frac{1}{2}(2n+1)\pi$	$(\cosh \lambda_n\xi + \cos \lambda_n\xi) - \dfrac{\cosh \lambda_n - \cos \lambda_n}{\sinh \lambda_n - \sin \lambda_n}(\sinh \lambda_n\xi + \sin \lambda_n\xi)$

TABLE 17-3 (continued)   CRITICAL SPEEDS AND MODE SHAPES FOR UNIFORM ROTORS IN BENDING

Boundary Conditions	Frequency Equation	Constant λ_n	Mode Shapes ϕ_n
2. Free–hinged	$\tan \lambda = \tanh \lambda$	$\lambda_1 = 3.9266$ $\lambda_2 = 7.0686$ $\lambda_3 = 10.2102$ $\cdots$ For large n $\lambda_n \approx \frac{1}{4}(4n+1)\pi$	$(\cosh \lambda_n \xi + \cos \lambda_n \xi) - \dfrac{\cosh \lambda_n + \cos \lambda_n}{\sinh \lambda_n + \sin \lambda_n}(\sinh \lambda_n \xi + \sin \lambda_n \xi)$
3. Free–guided	$\tan \lambda = -\tanh \lambda$	$\lambda_1 = 2.3650$ $\lambda_2 = 5.4978$ $\lambda_3 = 8.6394$ $\cdots$ For large n $\lambda \approx \frac{1}{4}(4n-1)\pi$	$(\cosh \lambda_n \xi + \cos \lambda_n \xi) - \dfrac{\sinh \lambda_n - \sin \lambda_n}{\cosh \lambda_n + \cos \lambda_n}(\sinh \lambda_n \xi + \sin \lambda_n \xi)$
4. Clamped–free	$\cosh \lambda \cos \lambda = -1$	$\lambda_1 = 1.8751$ $\lambda_2 = 4.6941$ $\lambda_3 = 7.8541$ $\cdots$ For large n $\lambda_n \approx \frac{1}{2}(2n-1)\pi$	$(\cosh \lambda_n \xi - \cos \lambda_n \xi) - \dfrac{\cosh \lambda_n + \cos \lambda_n}{\sinh \lambda_n + \sin \lambda_n}(\sinh \lambda_n \xi - \sin \lambda_n \xi)$

TABLE 17-3   Critical Speeds for Uniform Rotors   910

Critical Speeds for Uniform Rotors TABLE 17-3

5. Hinged–hinged	$\sin \lambda = 0$	$\lambda_n = n\pi$	$\sin n\pi\xi$
6. Hinged–guided	$\cos \lambda = 0$	$\lambda_n = \tfrac{1}{2}(2n - 1)\pi$	$\sin[\tfrac{1}{2}(2n - 1)\pi\xi]$
7. Guided–guided	$\sin \lambda = 0$	$\lambda_n = n\pi$	$\cos n\pi\xi$
8. Clamped–hinged	$\tan \lambda = \tanh \lambda$	$\lambda_1 = 3.9266$ $\lambda_2 = 7.0686$ $\lambda_3 = 10.2102$ $\cdots$ For large n $\lambda_n \approx \tfrac{1}{4}(4n + 1)\pi$	$(\cosh \lambda_n\xi - \cos \lambda_n\xi) - \dfrac{\cosh \lambda_n - \cos \lambda_n}{\sinh \lambda_n - \sin \lambda_n}(\sinh \lambda_n\xi - \sin \lambda_n\xi)$

Boundary Conditions	Frequency Equation	Constant λ_n	Mode Shapes ϕ_n
9. Clamped–guided	$\tan \lambda = -\tanh \lambda$	$\lambda_1 = 2.3650$ $\lambda_2 = 5.4978$ $\lambda_3 = 8.6394$ $\cdots$ For large n $\lambda_n \approx \frac{1}{4}(4n-1)\pi$	$(\cosh \lambda_n \xi - \cos \lambda_n \xi) - \dfrac{\sinh \lambda_n + \sin \lambda_n}{\cosh \lambda_n - \cos \lambda_n}(\sinh \lambda_n \xi - \sin \lambda_n \xi)$
10. Clamped–clamped	$\cosh \lambda \cos \lambda = 1$	$\lambda_1 = 4.7300$ $\lambda_2 = 7.8532$ $\lambda_3 = 10.9956$ $\cdots$ For large n $\lambda_n \approx \frac{1}{2}(2n+1)\pi$	$(\cosh \lambda_n \xi - \cos \lambda_n \xi) - \dfrac{\cosh \lambda_n - \cos \lambda_n}{\sinh \lambda_n - \sin \lambda_n}(\sinh \lambda_n \xi - \sin \lambda_n \xi)$

TABLE 17-3   **Critical Speeds for Uniform Rotors**   912

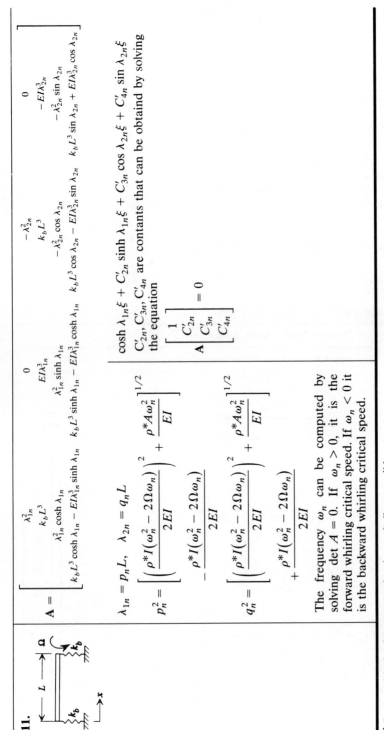

11.

$$\mathbf{A} = \begin{bmatrix} \lambda_{1n}^2 & 0 & -\lambda_{2n}^2 & 0 \\ k_b L^3 & EI\lambda_{1n}^3 & k_b L^3 & -EI\lambda_{2n}^3 \\ \lambda_{1n}^2 \cosh \lambda_{1n} & \lambda_{1n}^2 \sinh \lambda_{1n} & -\lambda_{2n}^2 \cos \lambda_{2n} & -\lambda_{2n}^2 \sin \lambda_{2n} \\ k_b L^3 \cosh \lambda_{1n} - EI\lambda_{1n}^3 \sinh \lambda_{1n} & k_b L^3 \sinh \lambda_{1n} - EI\lambda_{1n}^3 \cosh \lambda_{1n} & k_b L^3 \cos \lambda_{2n} - EI\lambda_{2n}^3 \sin \lambda_{2n} & k_b L^3 \sin \lambda_{2n} + EI\lambda_{2n}^3 \cos \lambda_{2n} \end{bmatrix}$$

$$\lambda_{1n} = p_n L, \quad \lambda_{2n} = q_n L$$

$$p_n^2 = \left[\left(\frac{\rho^* I(\omega_n^2 - 2\Omega\omega_n)}{2EI} \right)^2 + \frac{\rho^* A\omega_n^2}{EI} \right]^{1/2} - \frac{\rho^* I(\omega_n^2 - 2\Omega\omega_n)}{2EI}$$

$$q_n^2 = \left[\left(\frac{\rho^* I(\omega_n^2 - 2\Omega\omega_n)}{2EI} \right)^2 + \frac{\rho^* A\omega_n^2}{EI} \right]^{1/2} + \frac{\rho^* I(\omega_n^2 - 2\Omega\omega_n)}{2EI}$$

The frequency ω_n can be computed by solving det $A = 0$. If $\omega_n > 0$, it is the forward whirling critical speed. If $\omega_n < 0$ it is the backward whirling critical speed.

$$\cosh \lambda_{1n}\xi + C_{2n}' \sinh \lambda_{1n}\xi + C_{3n}' \cos \lambda_{2n}\xi + C_{4n}' \sin \lambda_{2n}\xi$$

$C_{2n}', C_{3n}', C_{4n}'$ are contants that can be obtaind by solving the equation

$$\mathbf{A} \begin{bmatrix} 1 \\ C_{2n}' \\ C_{3n}' \\ C_{4n}' \end{bmatrix} = 0$$

[a] Rotors with circular cross-sectional area, hollow or solid.

TABLE 17-4 TRANSFER, MASS, AND GYROSCOPIC MATRICES FOR RIGID DISK AND CONCENTRATED MASS

Notation

M_i = mass of rigid disk

I_{pi} = polar mass moment of inertia of disk,
  $I_{pi} = \frac{1}{2}M_i(r_o^2 + r_i^2)$,
  where r_i and r_o are inner and outer radii of disk

I_{Ti} = transverse mass moment of inertia of disk,
  $I_{Ti} = \frac{1}{2}I_{pi} + \frac{1}{12}M_i h^2$
  ($h = 0$ for concentrated mass)

e = eccentricity of mass center coordinates in the rotating of rigid disk

(η_a, ζ_a) = mass center coordinates in the rotating coordinate system XYZ

ω = whirl speed of rotor (for unbalanced response, $\omega = \Omega$)

Ω = spin speed

h = thickness of rigid disk

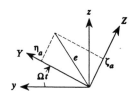

Superscripts: s, sine components; c, cosine components

Transfer Matrix (Sign Convention 1)

$\mathbf{z}_b = \mathbf{U}^i \mathbf{z}_a$

(All blanks indicate zeros)

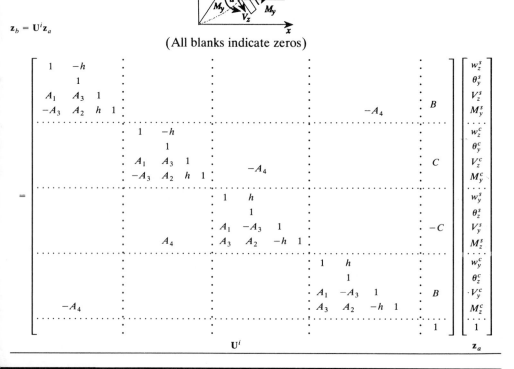

TABLE 17-4   Matrices for Rigid Disk and Concentrated Mass   914

$$A_1 = -M_i\omega^2$$
$$A_2 = -I_{Ti}\omega^2 + \tfrac{1}{4}M_i\omega^2 h^2$$
$$A_3 = \tfrac{1}{2}M_i\omega^2 h$$
$$A_4 = I_{pi}\Omega\omega$$
$$B = -M_i\Omega^2\eta_a$$
$$C = -M_i\Omega^2\zeta_a$$

For a concentrated mass, set h equal zero.

Mass and Gyroscopic Matrices (Sign Convention 2)

$$(\mathbf{m}_T + \mathbf{m}_R)\ddot{\mathbf{v}} - \Omega\mathbf{g}\dot{\mathbf{v}} = \mathbf{p}$$

with $\mathbf{v} = [w_y \quad w_z \quad \theta_y \quad \theta_z]^T$ $\mathbf{p} = [V_y \quad V_z \quad M_y \quad M_z]^T$

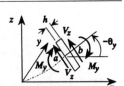

$$\mathbf{m}_T = \begin{bmatrix} M_i & & & \\ & M_i & & \\ & & 0 & \\ & & & 0 \end{bmatrix}$$

$$\mathbf{m}_R = \begin{bmatrix} 0 & & & \\ & 0 & & \\ & & I_{Ti} & \\ & & & I_{Ti} \end{bmatrix}$$

$$\mathbf{g} = \begin{bmatrix} 0 & & & \\ & 0 & & \\ & & 0 & I_{pi} \\ & & -I_{pi} & 0 \end{bmatrix}$$

For the unbalance force
$$\mathbf{p} = \mathbf{p}^c \cos\Omega t + \mathbf{p}^s \sin\Omega t$$

$$\mathbf{p}^c = M_i\Omega^2 \begin{bmatrix} \eta_a \\ \zeta_a \\ 0 \\ 0 \end{bmatrix} \qquad \mathbf{p}^s = M_i\Omega^2 \begin{bmatrix} -\zeta_a \\ \eta_a \\ 0 \\ 0 \end{bmatrix}$$

Notation

A = cross sectional area

E = modulus of elasticity

I = moment of inertia of section area

ℓ = length of shaft element

Ω = spin speed

r = Radius of gyration of cross sectional area about y or z axis

G = shear modulus of elasticity

α_s = shear correction factor (Table 2-4)

$A_s = A/\alpha_s$ = equivalent shear area

ρ^* = mass per unit volume

ω = whirl speed (For unbalanced response $\omega = \Omega$)

e = mass center eccentricity

$(\eta_a, \zeta_a), (\eta_b, \zeta_b)$ = mass center eccentricity at $x = a$, $x = b$ in the rotating coordinates XYZ.

The eccentricity is distributed linearly along the x-axis; $\eta(\xi) = \eta_a(1 - \xi) + \eta_b\xi$; $\zeta(\xi) = \zeta_a(1 - \xi) + \zeta_b\xi$ $(\xi = x/\ell)$.

$$\alpha = \sqrt{\beta^2 + \gamma} \qquad p^2 = |\alpha - \beta| \qquad q^2 = |\alpha + \beta| \qquad \varepsilon = \rho^*\omega^2/E \qquad \sigma = EI/GA_s$$

$$\beta = \tfrac{1}{2}\omega^2\left[\frac{\rho^*\alpha_s}{G} + \frac{\rho^*}{E}\left(\frac{2\Omega}{\omega} + 1\right)\right] \qquad \gamma = \omega^2\left[\frac{\rho^*A}{EI} - \frac{\rho^*}{E}\left(\frac{2\Omega}{\omega} + 1\right)\frac{\alpha_s\rho^*\omega^2}{G}\right]$$

$$\gamma \geq 0 \left(\omega \leq \sqrt{\Omega^2 + GA_s/(\rho^*I)} - \Omega\right)$$

$c_0 = [q^2\cosh(p\ell) + p^2\cos(q\ell)]/(p^2 + q^2)$

$c_1 = [q^2\sinh(p\ell)/(p\ell) + p^2\sin(q\ell)/(q\ell)]/(p^2 + q^2)$

$c_2 = [\cosh(p\ell) - \cos(q\ell)]/[\ell^2(p^2 + q^2)]$

$c_3 = [\sinh(p\ell)/(p\ell) - \sin(q\ell)/(q\ell)]/[\ell^2(p^2 + q^2)]$

$e_0 = [p^3\sinh(p\ell) - q^3\sin(q\ell)]/(p^2 + q^2)$

$e_1 = [p^2\cosh(p\ell) + q^2\cos(q\ell)]/(p^2 + q^2)$

$e_2 = [p\sinh(p\ell) + q\sin(q\ell)]/(p^2 + q^2)$

$e_3 = [\cosh(p\ell) - \cos(q\ell)]/(p^2 + q^2)$

$e_4 = [\sinh(p\ell)/p - \sin(q\ell)/q]/(p^2 + q^2)$

$e_5 = [\cosh(p\ell)/p^2 + \cos(q\ell)/q^2]/(p^2 + q^2) - \dfrac{1}{p^2q^2}$ (if $\gamma = 0$, set $e_5 = 0$)

$e_6 = [\sinh(p\ell)/p^3 + \sin(q\ell)/q^3]/(p^2 + q^2) - \dfrac{\ell}{p^2q^2}$ (if $\gamma = 0$, set $e_6 = 0$)

Set $\sinh(p\ell)/(p\ell) = 1$ if $p\ell = 0$

$$\gamma < 0 \left(\omega > \sqrt{\Omega^2 + GA_s/(\rho^*I)} - \Omega \right)$$

$$c_0 = [q^2 \cos(p\ell) - p^2 \cos(q\ell)]/(-p^2 + q^2)$$
$$c_1 = [q^2 \sin(p\ell)/(p\ell) - p^2 \sin(q\ell)/(q\ell)]/(-p^2 + q^2)$$
$$c_2 = [\cos(p\ell) - \cos(q\ell)]/[\ell^2(-p^2 + q^2)]$$
$$c_3 = [\sin(p\ell)/(p\ell) - \sin(q\ell)/(q\ell)]/[\ell^2(-p^2 + q^2)]$$

$$e_0 = [p^3 \sin(p\ell) - q^3 \sin(q\ell)]/(-p^2 + q^2)$$
$$e_1 = [-p^2 \cos(p\ell) + q^2 \cos(q\ell)]/(-p^2 + q^2)$$
$$e_2 = [-p \sin(p\ell) + q \sin(q\ell)]/(-p^2 + q^2)$$
$$e_3 = [\cos(p\ell) - \cos(q\ell)]/(-p^2 + q^2)$$
$$e_4 = [\sin(p\ell)/p - \sin(q\ell)/q]/(-p^2 + q^2)$$

$$e_5 = [-\cos(p\ell)/p^2 + \cos(q\ell)/q^2]/(-p^2 + q^2) + \frac{1}{p^2 q^2}$$

$$e_6 = [-\sin(p\ell)/p^3 + \sin(q\ell)/q^3]/(-p^2 + q^2) + \frac{\ell}{p^2 q^2}$$

Transfer Matrix: Shear Deformation Included (Sign Convention 1)

$$\mathbf{z}_b = \mathbf{U}^i \mathbf{z}_a$$
$$\mathbf{z} = \left[w_z^s \quad \theta_y^s \quad V_z^s \quad M_y^s \mid w_z^c \quad \theta_y^c \quad V_z^c \quad M_y^c \mid w_y^s \quad \theta_z^s \quad V_y^s \quad M_z^s \mid w_y^c \quad \theta_z^c \quad V_y^c \quad M_z^c \mid 1 \right]^T$$

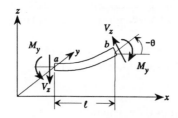

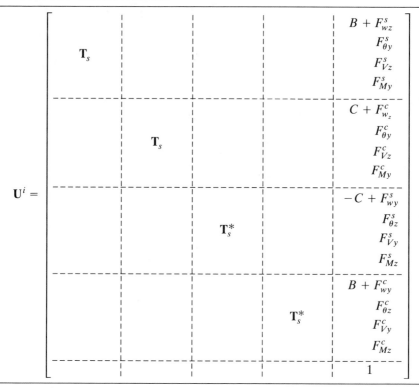

The gyroscopic coupling effects between x–y and x–z planes are ignored.

$\mathbf{T}_s = [T]_{4\times4}$

$\mathbf{T}_s^*$ is obtained by multiplying the 2nd and 4th rows of $\mathbf{T}_s$ by -1, then multiplying the 2nd and 4th columns of the resulting matrix by -1.

$B = (v_Y)_b - (v_Y)_a$

$C = (v_Z)_b - (v_Z)_a$

v_b = static bow at the right end of the segment

v_a = static bow at the left end of the segment

The static bow defines the initial permanent deformation of the geometric shaft center relative to the line of centers of the bearing system at rotating coordinates XYZ.

TABLE 17-5 **Matrices for a Uniform Shaft Element in Bending** **918**

$$T_{11} = c_0 - c_2 \frac{\rho^* \omega^2 \alpha_s \ell^2}{G}, \qquad T_{21} = -c_3 \frac{\rho^* \omega^2 A \ell^3}{EI}$$

$$T_{31} = -c_1 \rho^* \omega^2 A \ell + c_3 \frac{\rho^{*2} \omega^4 \alpha_s A \ell^3}{G}, \qquad T_{41} = -c_2 \rho^* \omega^2 A \ell^2$$

$$T_{12} = -c_1 \ell + c_3 \left[\frac{\rho^* \omega^2 \ell^3}{E} \left(\frac{2\Omega}{\omega} + 1 \right) + \frac{\rho^* \omega^2 \ell^3 \alpha_s}{G} \right], \qquad T_{22} = c_0 - c_2 \left[\frac{\rho^* \omega^2 \ell^2}{E} \left(\frac{2\Omega}{\omega} + 1 \right) \right]$$

$$T_{32} = c_2 \rho^* \omega^2 A \ell^2, \qquad T_{42} = -c_1 \rho^* \omega^2 I \ell \left(\frac{2\Omega}{\omega} + 1 \right) + c_3 \left[\frac{\rho^{*2} I \omega^4 \ell^3}{EI} \left(\frac{2\Omega}{\omega} + 1 \right)^2 + \rho^* \omega^2 A \ell^3 \right]$$

$$T_{13} = c_1 \frac{\alpha_s \ell}{GA} + c_3 \left(-\frac{\alpha_s^2 \rho^* \omega^2 \ell^3}{G^2 A} - \frac{\ell^3}{EI} \right), \qquad T_{23} = c_2 \frac{\ell^2}{EI}$$

$$T_{33} = c_0 - c_2 \frac{\rho^* \omega^2 \alpha_s \ell^2}{G} \qquad T_{43} = c_1 \ell - c_3 \left[\frac{\rho^* \omega^2 \ell^3}{E} \left(\frac{2\Omega}{\omega} + 1 \right) + \frac{\rho^* \omega^2 \alpha_s \ell^3}{G} \right]$$

$$T_{14} = -c_2 \frac{\ell^2}{EI}, \qquad T_{24} = c_1 \frac{\ell}{EI} - c_3 \frac{\rho^* \omega^2 \ell^3}{E^2 I} \left(\frac{2\Omega}{\omega} + 1 \right)$$

$$T_{34} = c_3 \frac{\rho^* \omega^2 A \ell^3}{EI}, \qquad T_{44} = c_0 - c_2 \frac{\rho^* \omega^2 \ell^2}{E} \left(\frac{2\Omega}{\omega} + 1 \right)$$

$$F_{wj}^k = \left[p_{aj}^k (e_5 - e_6/\ell) + p_{bj}^k e_6/\ell \right]/EI$$
$$+ \left[p_{aj}^k \{ e_3 + \varepsilon e_5 - (e_4 + \varepsilon e_6)/\ell \} + p_{bj}^k (e_4 + \varepsilon e_6)/\ell \right]/GA_s$$

$$F_{\theta j}^k = \left[p_{aj}^k (-e_4 + e_5/\ell) - p_{bj}^k e_5/\ell \right]/EI$$

$$F_{Vj}^k = p_{aj}^k \{ -(e_2 + \varepsilon e_4) + (e_3 + \varepsilon e_5)/\ell \} - p_{bj}^k (e_3 + \varepsilon e_5)/\ell$$

$$F_{Mj}^k = p_{aj}^k (-e_3 + e_4/\ell) - p_{bj}^k e_4/\ell$$

$$(k = s, c)$$
$$(j = z, y)$$

where

$$p_{az}^s = \rho^* A \eta_a \Omega^2; \qquad p_{bz}^s = \rho^* A \eta_b \Omega^2$$
$$p_{az}^c = \rho^* A \zeta_a \Omega^2; \qquad p_{bz}^c = \rho^* A \zeta_b \Omega^2$$
$$p_{ay}^s = -\rho^* A \zeta_a \Omega^2; \qquad p_{by}^s = -\rho^* A \zeta_b \Omega^2$$
$$p_{ay}^c = \rho^* A \eta_a \Omega^2; \qquad p_{by}^c = \rho^* A \eta_b \Omega^2$$

Stiffness, Mass, Gyroscopic, and Dynamic Stiffness Matrices (Sign Convention 2)

1. The matrices here, e.g. the consistent mass matrices, are based on the static deflection (shape function) of the beam (Eq. 17.40). The governing equation is

$$(\mathbf{m}_T + \mathbf{m}_R)\ddot{\mathbf{v}} - \Omega\mathbf{g}\dot{\mathbf{v}} + \mathbf{k}\mathbf{v} = \mathbf{p} = \mathbf{p}^c\cos\Omega t + \mathbf{p}^s\sin\Omega t$$

$$\mathbf{v} = \begin{bmatrix} w_{ya} & w_{za} & \theta_{ya} & \theta_{za} & w_{yb} & w_{zb} & \theta_{yb} & \theta_{zb} \end{bmatrix}^T$$

$$\mathbf{p} = \begin{bmatrix} V_{ya} & V_{za} & M_{ya} & M_{za} & V_{yb} & V_{zb} & M_{yb} & M_{zb} \end{bmatrix}^T$$

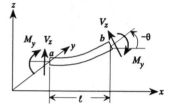

TRANSLATIONAL CONSISTENT MASS MATRIX

$$\mathbf{m}_T = \frac{\rho^* A\ell}{420} \begin{bmatrix} 156 & & & & & & & \\ 0 & 156 & & & & \text{Symmetric} & & \\ 0 & -22\ell & 4\ell^2 & & & & & \\ 22\ell & 0 & 0 & 4\ell^2 & & & & \\ 54 & 0 & 0 & 13\ell & 156 & & & \\ 0 & 54 & -13\ell & 0 & 0 & 156 & & \\ 0 & 13\ell & -3\ell^2 & 0 & 0 & 22\ell & 4\ell^2 & \\ -13\ell & 0 & 0 & -3\ell^2 & -22\ell & 0 & 0 & 4\ell^2 \end{bmatrix}$$

ROTARY CONSISTENT MASS MATRIX

$$\mathbf{m}_R = \frac{\rho^* A r^2}{30\ell} \begin{bmatrix} 36 & & & & & & & \\ 0 & 36 & & & & \text{Symmetric} & & \\ 0 & -3\ell & 4\ell^2 & & & & & \\ 3\ell & 0 & 0 & 4\ell^2 & & & & \\ -36 & 0 & 0 & -3\ell & 36 & & & \\ 0 & -36 & 3\ell & 0 & 0 & 36 & & \\ 0 & -3\ell & -\ell^2 & 0 & 0 & 3\ell & 4\ell^2 & \\ 3\ell & 0 & 0 & -\ell^2 & -3\ell & 0 & 0 & 4\ell^2 \end{bmatrix}$$

TABLE 17-5   **Matrices for a Uniform Shaft Element in Bending**   **920**

GYROSCOPIC MATRIX

$$
\mathbf{g} = \frac{\rho^* A r^2}{15\ell}
\begin{bmatrix}
0 & & & & & & & & \\
36 & 0 & & & & \text{Skew Symmetric} & & & \\
-3\ell & 0 & 0 & & & & & & \\
0 & -3\ell & 4\ell^2 & 0 & & & & & \\
0 & 36 & -3\ell & 0 & 0 & & & & \\
-36 & 0 & 0 & -3\ell & 36 & 0 & & & \\
-3\ell & 0 & 0 & \ell^2 & 3\ell & 0 & 0 & & \\
0 & -3\ell & -\ell^2 & 0 & 0 & 3\ell & 4\ell^2 & 0 &
\end{bmatrix}
$$

STIFFNESS MATRIX

$$
\mathbf{k} = \frac{EI}{(1+\phi)\ell^3}
\begin{bmatrix}
12 & & & & & & & \\
0 & 12 & & & & \text{Symmetric} & & \\
0 & -6\ell & (4+\phi)\ell^2 & & & & & \\
6\ell & 0 & 0 & (4+\phi)\ell^2 & & & & \\
-12 & 0 & 0 & -6\ell & 12 & & & \\
0 & -12 & 6\ell & 0 & 0 & 12 & & \\
0 & -6\ell & (2-\phi)\ell^2 & 0 & 0 & 6\ell & (4+\phi)\ell^2 & \\
6\ell & 0 & 0 & (2-\phi)\ell^2 & -6\ell & 0 & 0 & (4+\phi)\ell^2
\end{bmatrix}
$$

$$
\phi = \frac{12\,EI}{GA_s\ell^2}
$$

Set $\phi = 0$ if shear deformation effects are to be neglected.

EQUIVALENT UNBALANCE FORCE VECTORS

$$
\mathbf{p}^c = \frac{\rho^* A\ell}{60}\Omega^2
\begin{bmatrix}
21\eta_a + 9\eta_b \\
21\zeta_a + 9\zeta_b \\
-3\zeta_a\ell - 2\zeta_b\ell \\
3\eta_a\ell + 2\eta_b\ell \\
9\eta_a + 21\eta_b \\
9\zeta_a + 21\zeta_b \\
2\zeta_a\ell + 3\zeta_b\ell \\
-2\eta_a\ell - 3\eta_b\ell
\end{bmatrix},
\qquad
\mathbf{p}^s = \frac{\rho^* A\ell}{60}\Omega^2
\begin{bmatrix}
-21\zeta_a - 9\zeta_b \\
21\eta_a + 9\eta_b \\
-3\eta_a\ell - 2\eta_b\ell \\
-3\zeta_a\ell - 2\zeta_b\ell \\
-9\zeta_a - 21\zeta_b \\
9\eta_a + 21\eta_b \\
2\eta_a\ell + 3\eta_b\ell \\
2\zeta_a\ell + 3\zeta_b\ell
\end{bmatrix}
$$

2. The exact solution of the governing equation for the harmonic motion of a uniform shaft element is taken as the shape function. This gives the dynamic stiffness matrix k_{dyn}. The governing equation is

$$p = k_{dyn}v - \bar{p}$$

$$p = \begin{bmatrix} V_{ya} & V_{za} & M_{ya} & M_{za} & V_{yb} & V_{zb} & M_{yb} & M_{zb} \end{bmatrix}^T$$

$$v = \begin{bmatrix} w_{ya} & w_{za} & \theta_{ya} & \theta_{za} & w_{yb} & w_{zb} & \theta_{yb} & \theta_{zb} \end{bmatrix}^T$$

$$k_{dyn} = \begin{bmatrix}
k_{11} & & & & & & & \\
0 & k_{11} & & & & \text{Symmetric} & & \\
0 & k_{21} & k_{22} & & & & & \\
k_{21} & 0 & 0 & k_{22} & & & & \\
k_{31} & 0 & 0 & k_{32} & k_{33} & & & \\
0 & k_{31} & k_{32} & 0 & 0 & k_{33} & & \\
0 & k_{41} & k_{42} & 0 & 0 & k_{43} & k_{44} & \\
k_{41} & 0 & 0 & k_{42} & k_{43} & 0 & 0 & k_{44}
\end{bmatrix}$$

$$\bar{p} = \begin{bmatrix} \bar{p}_1 & \bar{p}_2 & \bar{p}_3 & \bar{p}_4 & \bar{p}_5 & \bar{p}_6 & \bar{p}_7 & \bar{p}_8 \end{bmatrix}^T$$

$$k_{11} = [(e_2 + 2\beta e_4)(e_1 + \varepsilon e_3) - \gamma e_3 e_4] EI/\Delta$$
$$k_{21} = [e_3(e_1 + 2\beta e_3) - e_2(e_2 + 2\beta e_4)] EI/\Delta$$
$$k_{31} = -(e_2 + 2\beta e_4) EI/\Delta$$
$$k_{41} = -e_3 EI/\Delta = -k_{32}$$
$$k_{22} = \{e_2 e_3 - (e_1 + 2\beta e_3)[e_4 - \sigma(e_2 + \varepsilon e_4)]\} EI/\Delta$$
$$k_{32} = e_3 EI/\Delta$$
$$k_{42} = [e_4 - \sigma(e_2 + \varepsilon e_4)] EI/\Delta$$
$$k_{33} = [(e_1 + \varepsilon e_3)(e_2 + 2\beta e_4) - \gamma e_3 e_4] EI/\Delta = k_{11}$$
$$k_{43} = \{(e_1 + \varepsilon e_3)e_3 - \gamma e_4[e_4 - \sigma(e_2 + \varepsilon e_4)]\} EI/\Delta$$
$$k_{44} = \{e_2 e_3 - (e_1 + 2\beta e_3)[e_4 - \sigma(e_2 + \varepsilon e_4)]\} EI/\Delta = k_{22}$$

TABLE 17-5   **Matrices for a Uniform Shaft Element in Bending**   **922**

$$\bar{p}_1 = V_{ay}^0 \qquad \bar{p}_5 = V_{by}^0$$
$$\bar{p}_2 = V_{az}^0 \qquad \bar{p}_6 = V_{bz}^0$$
$$\bar{p}_3 = M_{az}^0 \qquad \bar{p}_7 = M_{bz}^0$$
$$\bar{p}_4 = M_{ay}^0 \qquad \bar{p}_8 = M_{by}^0$$

$$V_{aj}^0 = -[(e_2 + 2\beta e_4)F_{wj} + e_3 F_{\theta j}]EI/\Delta$$
$$M_{aj}^0 = \{e_3 F_{wj} + [e_4 - \sigma(e_2 + \varepsilon e_4)]F_{\theta j}\}EI/\Delta$$
$$V_{bj}^0 = -F_{Vj} + \{[(e_1 + \varepsilon e_3)(e_2 + 2\beta e_4) - \gamma e_3 e_4]F_{wj}$$
$$\qquad + [(e_1 + \varepsilon e_3)e_3 - \gamma e_4[e_4 - \sigma(e_2 + \varepsilon e_4)]]F_{\theta j}\}EI/\Delta$$
$$M_{bj}^0 = -F_{Mj} + \{[(e_1 + \varepsilon e_3)e_3 + \sigma e_4[e_4 - \sigma(e_2 + \varepsilon e_4)]]F_{wj}$$
$$\qquad + [e_2 e_3 - (e_4 + 2\beta e_3)[e_4 - \sigma(e_2 + \varepsilon e_4)]]F_{\theta j}\}EI/\Delta$$
$$(j = y, z)$$
$$\Delta = e_3^2 - (e_2 + 2\beta e_4)[e_4 - \sigma(e_2 + \varepsilon e_4)]$$

$$F_{wj} = [p_{aj}(e_5 - e_6/\ell) + p_{bj}e_6/\ell]/EI$$
$$\qquad + \{p_{aj}[e_3 + \varepsilon e_5 - (e_4 + \varepsilon e_6)/\ell] + p_{bj}(e_4 + \varepsilon e_6)/\ell\}/GA_s$$
$$F_{\theta j} = [p_{aj}(-e_4 + e_5/\ell) - p_{bj}e_5/\ell]/EI$$
$$F_{Vj} = p_{aj}[-(e_2 + \varepsilon e_4) + (e_3 + \varepsilon e_5)]/\ell - p_{bj}(e_3 + \varepsilon e_5)/\ell$$
$$F_{Mj} = p_{aj}(-e_3 + e_4/\ell) - p_{bj}e_4/\ell$$
$$(j = y, z)$$

$$p_{ay} = \rho^* A\Omega^2(\eta_a \cos \Omega t - \zeta_a \sin \Omega t)$$

$$p_{by} = \rho^* A\Omega^2(\eta_b \cos \Omega t - \zeta_b \sin \Omega t)$$

$$p_{az} = \rho^* A\Omega^2(\zeta_a \cos \Omega t + \eta_a \sin \Omega t)$$

$$p_{bz} = \rho^* A\Omega^2(\zeta_b \cos \Omega t + \eta_b \sin \Omega t)$$

[a]Some of this table is based on [17.3].

TABLE 17-6 TRANSFER AND STIFFNESS MATRICES FOR A UNIFORM SHAFT ELEMENT IN BENDING WITH AXIAL TORQUE

Notation

Shear effects are ignored.

E = modulus of elasticity
ℓ = length of shaft section
Ω = spin speed
I = moment of inertia of cross-sectional area
T = axial torque

Superscripts: s, sine components; c, cosine components

Transfer Matrix (Sign Convention 1)

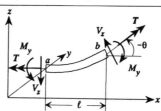

$\beta = EI/T^2 \qquad \gamma = T\ell/EI$

$\mathbf{z}_b = \mathbf{U}^i \mathbf{z}_a$

$\mathbf{z} = \begin{bmatrix} w_z^s & \theta_y^s & V_z^s & M_y^s & | & w_z^c & \theta_y^c & V_z^c & M_y^c & | & w_y^s & \theta_z^s & V_y^s & M_z^s & | & w_y^c & \theta_z^c & V_y^c & M_z^c & | & 1 \end{bmatrix}^T$

In this transfer matrix, the shaft is considered as a massless Euler–Bernoulli beam without shear and gyroscopic effects.

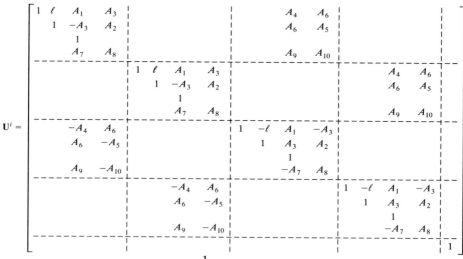

$$A_1 = \beta(\beta T \sin \gamma - \ell) \qquad A_2 = \frac{1}{T} \sin \gamma \qquad A_3 = \beta(1 - \cos \gamma)$$

$$A_4 = \frac{\ell^2}{2T} - \beta^2 T(1 - \cos \gamma) \qquad A_5 = \frac{1}{T}(1 - \cos \gamma) \qquad A_6 = \frac{\ell}{T} - \beta \sin \gamma$$

$$A_7 = -\beta T \sin \gamma \qquad A_8 = \cos \gamma \qquad A_9 = \beta T(1 - \cos \gamma) \qquad A_{10} = \sin \gamma$$

TABLE 17-6   **Matrices for an Element with Bending and Torsion**   **924**

Stiffness, Mass, and Gyroscopic Matrices (Sign Convention 2)

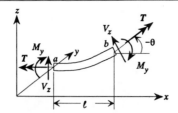

$$(\mathbf{m}_T + \mathbf{m}_R)\ddot{\mathbf{v}} - \Omega\mathbf{g}\dot{\mathbf{v}} + (\mathbf{k} - \mathbf{k}_T)\mathbf{v} = \mathbf{p} = \mathbf{p}^c \cos \Omega t + \mathbf{p}^s \sin \Omega t$$

$$\mathbf{v} = \begin{bmatrix} w_{ya} & w_{za} & \theta_{ya} & \theta_{za} & w_{yb} & w_{zb} & \theta_{yb} & \theta_{zb} \end{bmatrix}^T$$

$$\mathbf{p} = \begin{bmatrix} V_{ya} & V_{za} & M_{ya} & M_{za} & V_{yb} & V_{zb} & M_{yb} & M_{zb} \end{bmatrix}^T$$

The mass matrices $\mathbf{m}_T, \mathbf{m}_R$, gyroscopic matrix $\mathbf{g}$, stiffness matrix $\mathbf{k}$, and loading vectors $\mathbf{p}^c, \mathbf{p}^s$ can be taken from Table 17-5.

Axial torque incremental stiffness matrix:

$$\mathbf{k}_T = T \begin{bmatrix}
0 & 0 & 1/\ell & 0 & 0 & 0 & -1/\ell & 0 \\
0 & 0 & 0 & 1/\ell & 0 & 0 & 0 & -1/\ell \\
1/\ell & 0 & 0 & -\frac{1}{2} & -1/\ell & 0 & 0 & \frac{1}{2} \\
0 & 1/\ell & \frac{1}{2} & 0 & 0 & -1/\ell & -\frac{1}{2} & 0 \\
0 & 0 & -1/\ell & 0 & 0 & 0 & 1/\ell & 0 \\
0 & 0 & 0 & -1/\ell & 0 & 0 & 0 & 1/\ell \\
-1/\ell & 0 & 0 & -\frac{1}{2} & 1/\ell & 0 & 0 & \frac{1}{2} \\
0 & -1/\ell & \frac{1}{2} & 0 & 0 & 1/\ell & -\frac{1}{2} & 0
\end{bmatrix}$$

The total stiffness matrix of the shaft element is the difference $(\mathbf{k} - \mathbf{k}_T)$ between the stiffness matrix obtained in Table 17-5 and the incremental stiffness matrix $\mathbf{k}_T$.

[a]Some of the table is from [17.4].

TABLE 17-7

TABLE 17-7 TRANSFER MATRIX FOR BEARING OR SEAL SYSTEMS

Notation

c_{ij} = damping coefficients for bearing or seal system ($i, j = y, z$)

c_{ij}^* = rotary damping coefficients for bearing or seal system

k_{ij} = stiffness coefficients for bearing or seal system

k_{ij}^* = rotary stiffness coefficients for bearing or seal system

ω = whirl frequency of journal of bearing or seal system

Superscripts: s, sine components; c, cosine components

$$
\mathbf{z} =
\begin{Bmatrix}
w_z^s \\ \theta_y^s \\ V_z^s \\ M_y^s \\
w_z^c \\ \theta_y^c \\ V_z^c \\ M_y^c \\
w_y^s \\ \theta_z^s \\ V_y^s \\ M_z^s \\
w_y^c \\ \theta_z^c \\ V_y^c \\ M_z^c \\
1
\end{Bmatrix}
$$

$\mathbf{U}_i =$

1																
	1															
k'_{zz}	$k^{*\prime}_{zz}$	1		$-\omega c'_{zz}$	$-\omega c^{*\prime}_{zz}$			k'_{zy}	$k^{*\prime}_{zy}$			$-\omega c'_{zy}$	$-\omega c^{*\prime}_{zy}$			
			1													
				1												
					1											
$\omega c'_{zz}$	$\omega c^{*\prime}_{zz}$			k'_{zz}	$k^{*\prime}_{zz}$	1		$\omega c'_{zy}$	$\omega c^{*\prime}_{zy}$			k'_{zy}	$k^{*\prime}_{zy}$			
							1									
								1								
									1							
k'_{yz}	$k^{*\prime}_{yz}$			$-\omega c'_{yz}$	$-\omega c^{*\prime}_{yz}$			k'_{yy}	$k^{*\prime}_{yy}$	1		$-\omega c'_{yy}$	$-\omega c^{*\prime}_{yy}$			
											1					
												1				
													1			
$\omega c'_{yz}$	$\omega c^{*\prime}_{yz}$			k'_{yz}	$k^{*\prime}_{yz}$			$\omega c'_{yy}$	$\omega c^{*\prime}_{yy}$			k'_{yy}	$k^{*\prime}_{yy}$	1		
															1	
																1

All blanks indicate zeros.

TABLE 17-7 Transfer Matrix for Bearing or Seal Systems 926

	Col 1	Col 2	Col 3	Col 4	Col 5	Col 6 (Bearing/Seal)
k'_{yy}	k_{yy}	0	k_{yy}	λ_{0y}	$\lambda_{1y}/\lambda_{2y}$	k'_{yy}
k'_{yz}	0	0	0	0	0	k'_{yz}
k'_{zy}	0	0	0	0	0	k'_{zy}
k'_{zz}	k_{zz}	0	k_{zz}	λ_{0z}	$\lambda_{1z}\,\lambda_{2z}$	k'_{zz}
c'_{yy}	0	c_{yy}	c_{yy}	0	$\lambda_{3y}/\lambda_{2y}$	c'_{yy}
c'_{yz}	0	0	0	0	0	c'_{yz}
c'_{zy}	0	0	0	0	0	c'_{zy}
c'_{zz}	0	c_{zz}	c_{zz}	0	$\lambda_{3z}/\lambda_{2z}$	c'_{zz}

$$\lambda_{0j} = k_{jj}(\bar{k}_{jj} - M_{ji}\omega^2)/(k_{jj} + \bar{k}_{jj} - M_{ji}\omega^2)$$

$$\lambda_{1j} = \lambda_{0j} + \omega^2(k_{jj}\bar{c}_{jj}^2 + \bar{k}_{jj}c_{jj}^2) - \omega^4 c_{jj}^2 M_{ji}$$
$$j = y, z$$

$$k'_{yy} = \frac{1}{d_1^2 + \omega^2 d_2^2}(h_1 d_1 + \omega^2 h_2 d_2)$$

$$k'_{yz} = \frac{1}{d_1^2 + \omega^2 d_2^2}(h_3 d_1 + \omega^2 h_4 d_2)$$

$$\lambda_{2j} = \left(k_{jj} + \bar{k}_{jj} - M_{ji}\omega^2\right)^2 + \omega^2\left(c_{jj}^2\bar{c}_{jj} + c_{jj}\bar{c}_{jj}^2\right) + \bar{c}_{jj}k_{jj}^2$$

$$\lambda_{3j} = c_{jj}\left(\bar{k}_{jj} - M_{ji}\omega^2\right) + \omega^2\left(c_{jj}^2\bar{c}_{jj} + c_{jj}\bar{c}_{jj}^2\right) + \bar{c}_{jj}k_{jj}^2$$

$$c'_{yy} = \frac{1}{d_1^2 + \omega^2 d_2^2}(h_2 d_1 - h_1 d_2)$$

$$c'_{yz} = \frac{1}{d_1^2 + \omega^2 d_2^2}(h_4 d_1 - h_3 d_2)$$

TABLE 17-7 (continued)   TRANSFER MATRIX FOR BEARING OR SEAL SYSTEMS

$$k'_{zy} = \frac{1}{d_1^2 + \omega^2 d_2^2}(h_5 d_1 + \omega^2 h_6 d_2)$$

$$k'_{zz} = \frac{1}{d_1^2 + \omega^2 d_2^2}(h_7 d_1 + \omega^2 h_8 d_2)$$

$$d_1 = (e_1 e_4 - e_3 e_2) - \omega^2(f_1 f_4 - f_3 f_2)$$

$$h_1 = (\bar{k}_y g_1 + \bar{k}_{yz} g_5) - \omega^2(\bar{c}_{yy} g_2 + \bar{c}_{yz} g_6)$$

$$h_2 = (\bar{k}_y g_2 + \bar{k}_{yz} g_6) + (\bar{c}_{yy} g_1 + \bar{c}_{yz} g_5)$$

$$h_3 = (\bar{k}_y g_3 + \bar{k}_{yz} g_7) - \omega^2(\bar{c}_{yy} g_4 + \bar{c}_{yz} g_8)$$

$$h_4 = (\bar{k}_y g_4 + \bar{k}_{yz} g_8) + (\bar{c}_{yy} g_3 + \bar{c}_{yz} g_7)$$

$$\bar{k}_y = -\omega^2 M_{yi} + \bar{k}_{yy}$$

$$g_1 = (e_4 k_{yy} - e_2 k_{zy}) - \omega^2(f_4 c_{yy} - f_2 c_{zy})$$
$$g_2 = (e_4 c_{yy} - e_2 c_{zy}) + (f_4 k_{yy} - f_2 k_{zy})$$
$$g_3 = (e_4 k_{yz} - e_2 k_{zz}) - \omega^2(f_4 c_{yz} - f_2 c_{zz})$$
$$g_4 = (e_4 c_{yz} - e_2 c_{zz}) + (f_4 k_{yz} - f_2 k_{zz})$$

$$f_1 = \bar{c}_{yy} + c_{yy}$$
$$f_2 = \bar{c}_{yz} + c_{yz}$$
$$f_3 = \bar{c}_{zy} + c_{zy}$$
$$f_4 = \bar{c}_{zz} + c_{zz}$$

$$c'_{zy} = \frac{1}{d_1^2 + \omega^2 d_2^2}(h_6 d_1 - h_5 d_2)$$

$$c'_{zz} = \frac{1}{d_1^2 + \omega^2 d_2^2}(h_8 d_1 - h_7 d_2)$$

$$d_2 = (e_1 f_4 - e_3 f_2) + (e_4 f_1 - e_2 f_3)$$

$$h_5 = (\bar{k}_{zy} g_1 + \bar{k}_z g_5) - \omega^2(\bar{c}_{zy} g_2 + \bar{c}_{zz} g_6)$$

$$h_6 = (\bar{k}_{zy} g_2 + \bar{k}_z g_6) + (\bar{c}_{zy} g_1 + \bar{c}_{zz} g_5)$$

$$h_7 = (\bar{k}_{zy} g_3 + \bar{k}_z g_7) - \omega^2(\bar{c}_{zy} g_4 + \bar{c}_{zz} g_8)$$

$$h_8 = (\bar{k}_{zy} g_4 + \bar{k}_z g_8) + (\bar{c}_{zy} g_3 + \bar{c}_{zz} g_7)$$

$$\bar{k}_z = -\omega^2 M_{zi} + \bar{k}_{zz}$$

$$g_5 = (-e_3 k_{yy} + e_1 k_{zy}) - \omega^2(-f_3 c_{yy} + f_1 c_{zy})$$
$$g_6 = (-e_3 c_{yy} + e_1 c_{zy}) + (-f_3 k_{yy} + f_1 k_{zy})$$
$$g_7 = (-e_3 k_{yz} + e_1 k_{zz}) - \omega^2(-f_3 c_{yz} + f_1 c_{zz})$$
$$g_8 = (-e_3 c_{yz} + e_1 c_{zz}) + (-f_3 k_{yz} + f_1 k_{zz})$$

$$e_1 = -\omega^2 M_{yi} + \bar{k}_{yy} + k_{yy}$$
$$e_2 = \bar{k}_{yz} + k_{yz}$$
$$e_3 = \bar{k}_{zy} + k_{zy}$$
$$e_4 = -\omega^2 M_{zi} + \bar{k}_{zz} + k_{zz}$$

The values of $k^{*\prime}_{yy}$, $k^{*\prime}_{yz}$, $k^{*\prime}_{zy}$, $k^{*\prime}_{zz}$, $c^{*\prime}_{yy}$, $c^{*\prime}_{yz}$, $c^{*\prime}_{zy}$, $c^{*\prime}_{zz}$, and $c^{*\prime}_{zz}$ in U_i are taken from this table by replacing the coefficients $k_{yy}, k^*_{yz}, k^*_{zy}, \ldots$, with the corresponding rotary coefficients $k^*_{yy}, k^*_{yz}, c^*_{yy}, c^*_{yz}, \ldots$, respectively. $k_{yy}, k_{yz}, \ldots, c_{yy}, c_{yz}, \ldots$ in the above formulas with the corresponding rotary coefficients $k^*_{yy}, k^*_{yz}, \ldots, c^*_{yy}, c^*_{yz}, \ldots$, respectively.

TABLE 17-7   Transfer Matrix for Bearing or Seal Systems   928

TABLE 17-8   TRANSFER MATRIX FOR AN ISOTROPIC BEARING SYSTEM

Notation

Ω = spin speed of rotor
λ_n = damping ratio

$$s = \begin{cases} i\Omega & \text{for unbalanced response} \\ i\omega_n & \text{for undamped critical speed} \\ \lambda_n + i\omega_n & \text{for stability analysis or damped free vibration} \end{cases}$$

M_i = mass of bearing pedestal
ω_n = undamped critical speed

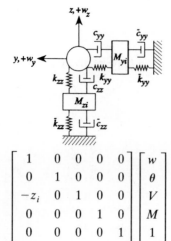

$$\mathbf{z} = \begin{bmatrix} w & \theta & V & M & 1 \end{bmatrix}^T$$
$$w = w_z + iw_y, \qquad \theta = \theta_y + i\theta_z$$
$$V = V_z + iV_y, \qquad M = M_y + iM_z$$
$$\mathbf{z}_b = \mathbf{U}_i \mathbf{z}_a$$

$$z_i = \frac{(k + sc)(\bar{k} + s\bar{c} + s^2 M_i)}{k + sc + \bar{k} + s\bar{c} + s^2 M_i}$$

$$k = k_{yy} = k_{zz}, \qquad c = c_{yy} = c_{zz}$$
$$\bar{k} = \bar{k}_{yy} = \bar{k}_{zz}, \qquad \bar{c} = \bar{c}_{yy} = \bar{c}_{zz}$$
$$M_i = M_{yi} = M_{zi}$$

$$\underbrace{\begin{bmatrix} 1 & 0 & 0 & 0 & 0 \\ 0 & 1 & 0 & 0 & 0 \\ -z_i & 0 & 1 & 0 & 0 \\ 0 & 0 & 0 & 1 & 0 \\ 0 & 0 & 0 & 0 & 1 \end{bmatrix}}_{\mathbf{U}_i} \underbrace{\begin{bmatrix} w \\ \theta \\ V \\ M \\ 1 \end{bmatrix}}_{\mathbf{z}}$$

929   **Transfer Matrix for an Isotropic Bearing System**   TABLE 17-8

TABLE 17-9 MASS, GYROSCOPIC, AND STIFFNESS MATRICES FOR A CONICAL SHAFT ELEMENT IN BENDING[a]

Notation

A_a = cross-sectional area at left end of element, $= \pi(r_{oa}^2 - r_{ia}^2)$

I_a = mass moment of gyration at left end of element,

$\qquad = \rho^* A_a(\pi/4)(r_{oa}^4 - r_{ia}^4)$

r_a^2 = radius of gyration of mass moment at left end of element, $= I_a/A_a$

r_i = inner radius of element at any x, $= r_{ia}[1 + (\varepsilon - 1)\xi]$

r_o = outer radius of element at any x, $= r_{oa}[1 + (\sigma - 1)\xi]$

ρ^* = mass per unit volume

Ω = spin speed

$\xi = x/\ell \qquad \varepsilon = r_{ib}/r_{ia} \qquad \sigma = r_{ob}/r_{oa}$

e = eccentricity of mass center

$(\eta_a, \zeta_a), (\eta_b, \zeta_b)$ = mass center eccentricity at $x = a$ and $x = b$, respectively

Superscripts: s, sine component; c, cosine component

$\alpha_1 = 2[r_{oa}^2(\sigma - 1) - r_{ia}^2(\varepsilon - 1)]/(r_{oa}^2 - r_{ia}^2)$

$\alpha_2 = [r_{oa}^2(\sigma - 1)^2 - r_{ia}^2(\varepsilon - 1)^2]/(r_{oa}^2 - r_{ia}^2)$

$\alpha = \rho^* A_a \ell /362,880$

$\beta = \rho^* I_a/(362,880\ell)$

$\gamma = EI_a/(5040\ell^3)$

$m = \rho^* A_a \Omega^2/5040$

$\delta_1 = 4[r_{oa}^4(\sigma - 1) - r_{ia}^4(\varepsilon - 1)]/(r_{oa}^4 - r_{ia}^4)$

$\delta_2 = 6[r_{oa}^4(\sigma - 1)^2 - r_{ia}^4(\varepsilon - 1)^2]/(r_{oa}^4 - r_{ia}^4)$

$\delta_3 = 4[r_{oa}^4(\sigma - 1)^3 - r_{ia}^4(\varepsilon - 1)^3]/(r_{oa}^4 - r_{ia}^4)$

$\delta_4 = [r_{oa}^4(\sigma - 1)^4 - r_{ia}^4(\varepsilon - 1)^4]/(r_{oa}^4 - r_{ia}^4)$

Coordinate System (Sign Convention 2)

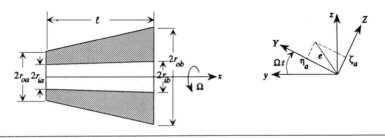

TABLE 17-9 | Matrices for a Conical Shaft Element in Bending | 930

Equations of Motion

$$(\mathbf{m}_T + \mathbf{m}_R)\ddot{\mathbf{v}} - \Omega \mathbf{g}\dot{\mathbf{v}} + \mathbf{k}\mathbf{v} = \mathbf{p}^c \cos \Omega t + \mathbf{p}^s \sin \Omega t$$

$$\mathbf{v} = [\,w_{ya} \quad w_{za} \quad \theta_{ya} \quad \theta_{za} \quad w_{yb} \quad w_{zb} \quad \theta_{yb} \quad \theta_{zb}\,]^T$$

All matrices are of size 8×8, and any term not explicitly defined is zero. Since all of the matrices are either symmetric or skew symmetric, only the lower triangle is given, with the type of symmetry indicated for each matrix.

Element Matrices

TRANSLATIONAL CONSISTENT MASS MATRIX $\mathbf{m}_T$ (SYMMETRIC)

$$m_T(1,1) = m_T(2,2) = \alpha(134,784 + 31,104\alpha_1 + 10,944\alpha_2)$$

$$m_T(4,1) = -m_T(3,2) = \alpha\ell(19,008 + 6048\alpha_1 + 2448\alpha_2)$$

$$m_T(5,1) = m_T(6,2) = \alpha(46,656 + 23,328\alpha_1 + 13,248\alpha_2)$$

$$m_T(7,2) = -m_T(8,1) = \alpha\ell(11,232 + 5184\alpha_1 + 2736\alpha_2)$$

$$m_T(3,3) = m_T(4,4) = \alpha\ell^2(3456 + 1296\alpha_1 + 576\alpha_2)$$

$$m_T(6,3) = -m_T(5,4) = -\alpha\ell(11,232 + 6048\alpha_1 + 3600\alpha_2)$$

$$m_T(7,3) = m_T(8,4) = -\alpha\ell^2(2592 + 1296\alpha_1 + 720\alpha_2)$$

$$m_T(5,5) = m_T(6,6) = \alpha(134,784 + 103,680\alpha_1 + 83,520\alpha_2)$$

$$m_T(8,5) = -m_T(7,6) = -\alpha\ell(19,008 + 12,960\alpha_1 + 9360\alpha_2)$$

$$m_T(7,7) = m_T(8,8) = \alpha\ell^2(3456 + 2160\alpha_1 + 1440\alpha_2)$$

ROTARY CONSISTENT MASS MATRIX $\mathbf{m}_R$ (SYMMETRIC)

$$m_R(1,1)$$
$$= m_R(2,2) = \beta(435,456 + 217,728\delta_1 + 124,416\delta_2 + 77,760\delta_3 + 51,840\delta_4)$$
$$m_R(4,1)$$
$$= -m_R(3,2) = \beta\ell(36,288 + 36,288\delta_1 + 25,920\delta_2 + 18,144\delta_3 + 12,960\delta_4)$$
$$m_R(8,1) = -m_R(7,2) = \beta\ell(36,288 - 10,368\delta_2 - 12,960\delta_3 - 12,960\delta_4)$$
$$m_R(3,3) = m_R(4,4) = \beta\ell^2(48,684 + 12,096\delta_1 + 6912\delta_2 + 4752\delta_3 + 3456\delta_4)$$
$$m_R(7,3) = m_R(8,4) = -\beta\ell^2(12,096 + 6048\delta_1 + 5184\delta_2 + 4752\delta_3 + 4320\delta_4)$$
$$m_R(7,7)$$
$$= m_R(8,8) = \beta\ell^2(48,384 + 36,288\delta_1 + 31,104\delta_2 + 28,080\delta_3 + 25,920\delta_4)$$
$$m_R(5,1) = m_R(6,2) = -m_R(5,5) = -m_R(6,6) = -m_R(1,1)$$
$$m_R(6,3) = -m_R(5,4) = m_R(4,1)$$
$$m_R(8,5) = -m_R(7,6) = -m_R(8,1)$$

GYROSCOPIC MATRIX **g** (SKEW SYMMETRIC)

$$g(2,1) = \beta(870,912 + 435,456\delta_1 + 248,832\delta_2 + 155,520\delta_3 + 103,680\delta_4)$$

$$g(3,1) = -\beta\ell(72,576 + 72,576\delta_1 + 51,840\delta_2 + 36,288\delta_3 + 25,920\delta_4)$$

$$g(7,1) = -\beta\ell(72,576 - 20,736\delta_2 - 25,920\delta_3 - 25,920\delta_4)$$

$$g(4,3) = \beta\ell^2(96,768 + 24,192\delta_1 + 13,824\delta_2 + 9504\delta_3 + 6912\delta_4)$$

$$g(8,3) = -\beta\ell^2(24,192 + 12,096\delta_1 + 10,368\delta_2 + 9504\delta_3 + 8640\delta_4)$$

$$g(8,7) = \beta\ell^2(96,768 + 72,576\delta_1 + 62,208\delta_2 + 56,160\delta_3 + 51,840\delta_4)$$

$$g(6,1) = -g(5,2) = -g(6,5) = -g(2,1)$$

$$g(4,2) = g(5,3) = g(6,4) = g(3,1)$$

$$g(8,2) = -g(7,5) = -g(8,6) = g(7,1)$$

$$g(7,4) = -g(8,3)$$

STIFFNESS MATRIX **k** (SYMMETRIC)

$$k(1,1) = \gamma(60,480 + 30,240\delta_1 + 24,192\delta_2 + 21,168\delta_3 + 19,008\delta_4)$$

$$k(4,1) = \gamma\ell(30,240 + 10,080\delta_1 + 7056\delta_2 + 6048\delta_3 + 5472\delta_4)$$

$$k(8,1) = \gamma\ell(30,240 + 20,160\delta_1 + 17,136\delta_2 + 15,120\delta_3 + 13,536\delta_4)$$

$$k(3,3) = k(4,4) = \gamma\ell^2(20,160 + 5040\delta_1 + 2688\delta_2 + 2016\delta_3 + 1728\delta_4)$$

$$k(7,3) = k(8,4) = \gamma\ell^2(10,080 + 5040\delta_1 + 4368\delta_2 + 4032\delta_3 + 3744\delta_4)$$

$$k(7,7) = k(8,8) = \gamma\ell^2(20,160 + 15,120\delta_1 + 12,768\delta_2 + 11,088\delta_3 + 9792\delta_4)$$

$$k(5,1) = k(6,2) = -k(2,2) = -k(5,5) = -k(6,6) = -k(1,1)$$

$$k(3,2) = k(5,4) = -k(6,3) = -k(4,1)$$

$$k(7,2) = k(8,5) = -k(7,6) = -k(8,1)$$

TABLE 17-9 | **Matrices for a Conical Shaft Element in Bending** | **932**

UNBALANCE FORCE VECTORS $\mathbf{p}^c, \mathbf{p}^s$

$$p^c(1) = m\eta_a(1764 + 420\alpha_1 + 156\alpha_2) + m\eta_b(756 + 336\alpha_1 + 156\alpha_2)$$

$$p^c(2) = m\zeta_a(1764 + 420\alpha_1 + 156\alpha_2) + m\zeta_b(756 + 336\alpha_1 + 156\alpha_2)$$

$$p^c(3) = -m\ell\zeta_a(252 + 84\alpha_1 + 36\alpha_2) - m\ell\zeta_b(168 + 84\alpha_1 + 48\alpha_2)$$

$$p^c(4) = m\ell\eta_a(252 + 84\alpha_1 + 36\alpha_2) + m\ell\eta_b(168 + 84\alpha_1 + 48\alpha_2)$$

$$p^c(5) = m\eta_a(756 + 420\alpha_1 + 264\alpha_2) + m\eta_b(1764 + 1344\alpha_1 + 1080\alpha_2)$$

$$p^c(6) = m\zeta_a(756 + 420\alpha_1 + 264\alpha_2) + m\zeta_b(1764 + 1344\alpha_1 + 1080\alpha_2)$$

$$p^c(7) = m\ell\zeta_a(168 + 84\alpha_1 + 48\alpha_2) + m\ell\zeta_b(252 + 168\alpha_1 + 120\alpha_2)$$

$$p^c(8) = -m\ell\eta_a(168 + 84\alpha_1 + 48\alpha_2) - m\ell\eta_b(252 + 168\alpha_1 + 120\alpha_2)$$

$$p^s(1) = -p^c(2) \qquad p^s(3) = -p^c(4)$$

$$p^s(2) = p^c(1) \qquad p^s(4) = p^c(3)$$

$$p^s(5) = -p^c(6) \qquad p^s(7) = -p^c(8)$$

$$p^s(6) = p^c(5) \qquad p^s(8) = p^c(7)$$

---

[a]Adopted from [17.5].

TABLE 17-10 MASS, GYROSCOPIC, AND STIFFNESS MATRICES FOR A HELICALLY FLUTED SHAFT ELEMENT IN BENDING[a]

Notation

A = cross sectional area

r_1 = radius of gyration of cross sectional area about ζ axis, = $\sqrt{I_1/A}$

r_2 = radius of gyration of cross sectional area about η axis, = $\sqrt{I_2/A}$

ρ^* = mass per unit volume

xyz = inertial frame

XYZ = rotating coordinates with constant rotational speed Ω

$X\eta\zeta$ = rotating fluted coordinates (η, ζ are principal axes of inertia of sectional area)

ℓ = length of element

Ω = spin speed

$P_x, P_\eta, P_\zeta, T = M_x$ = forces and torque acting on end of the shaft in $X\eta\zeta$ coordinates

I_1 = moment of inertia of section about ζ

I_2 = moment of inertia of section about η

β_0 = flute helix angle per unit length

Coordinate Systems (Sign Convention 2)

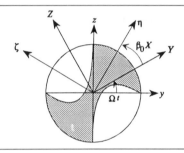
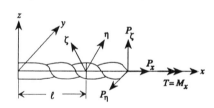

Equations of Motion

$$(\mathbf{m}_T + \mathbf{m}_R)\ddot{\mathbf{v}} + \mathbf{g}\Omega\dot{\mathbf{v}} + \mathbf{k}\mathbf{v} = 0$$

$$\mathbf{v} = [\, w_{\eta a} \quad w_{\zeta a} \quad \theta_{\eta a} \quad \theta_{\zeta a} \quad w_{\eta b} \quad w_{\zeta b} \quad \theta_{\eta b} \quad \theta_{\zeta b} \,]^T$$

Element Matrices

TRANSLATIONAL CONSISTENT MASS MATRIX $\mathbf{m}_T$:

$$\mathbf{m}_T = \frac{\rho^* A\ell}{420}
\begin{bmatrix}
156 & & & & & & & \\
0 & 156 & & & \text{Symmetric} & & & \\
0 & -22\ell & 4\ell^2 & & & & & \\
22\ell & 0 & 0 & 4\ell^2 & & & & \\
54 & 0 & 0 & 13\ell & 156 & & & \\
0 & 54 & -13\ell & 0 & 0 & 156 & & \\
0 & 13\ell & -3\ell^2 & 0 & 0 & 22\ell & 4\ell^2 & \\
-13\ell & 0 & 0 & -3\ell^2 & -22\ell & 0 & 0 & 4\ell^2
\end{bmatrix}$$

ROTARY CONSISTENT MASS MATRIX $\mathbf{m}_R$:

$$\mathbf{m}_R = \frac{\rho^*A}{30\ell}
\begin{bmatrix}
36r_1^2 \\
0 & 36r_2^2 & & & & & \text{Symmetric} \\
0 & -3\ell r_2^2 & 4\ell^2 r_2^2 \\
3\ell r_1^2 & 0 & 0 & 4\ell^2 r_1^2 \\
-36r_1^2 & 0 & 0 & -3\ell^2 r_1^2 & 36r_1^2 \\
0 & -36r_2^2 & 3\ell r_2^2 & 0 & 0 & 36r_2^2 \\
0 & -3\ell r_2^2 & -\ell^2 r_2 & 0 & 0 & 3\ell r_2^2 & 4\ell^2 r_2^2 \\
3\ell r_1^2 & 0 & 0 & -\ell^2 r_1^2 & -3\ell r_1^2 & 0 & 0 & 4\ell^2 r_1^2
\end{bmatrix}$$

GYROSCOPIC MATRIX $\mathbf{g}$:

$$\mathbf{g} = \frac{\rho^*A\ell}{210}
\begin{bmatrix}
0 \\
156 & 0 & & & & \text{Skew symmetric} \\
22\ell & 0 & 0 \\
0 & -22\ell & -4\ell^2 & 0 \\
0 & -54 & 13\ell & 0 & 0 \\
54 & 0 & 0 & 13\ell & 156 & 0 \\
-13\ell & 0 & 0 & -3\ell^2 & -22\ell & 0 & 0 \\
0 & 13\ell & 3\ell^2 & 0 & 0 & 22\ell & -4\ell^2 & 0
\end{bmatrix}$$

STIFFNESS MATRIX $\mathbf{k}$:

$$\mathbf{k} = \sum_{i=1}^{7} \mathbf{k}_i \quad \text{where } \mathbf{k}_1 = -\Omega^2 \mathbf{m}_T$$

$$\mathbf{k}_2 = \frac{E}{\ell^3}
\begin{bmatrix}
12I_1 \\
0 & 12I_2 & & & & & \text{Symmetric} \\
0 & -6\ell I_2 & 4\ell^2 I_2 \\
6\ell I_1 & 0 & 0 & 4\ell^2 I_1 \\
-12I_1 & 0 & 0 & -6\ell I_1 & 12I_1 \\
0 & -12I_2 & 6\ell I_2 & 0 & 0 & 12I_2 \\
0 & -6\ell I_2 & 2\ell^2 I_2 & 0 & 0 & 6\ell I_2 & 4\ell^2 I_2 \\
6\ell I_1 & 0 & 0 & 2\ell^2 I_1 & -6\ell I_1 & 0 & 0 & 4\ell^2 I_1
\end{bmatrix}$$

$$\mathbf{k}_3 = \frac{2E\beta_0}{\ell} \begin{bmatrix} 0 & & & & & & & \\ 0 & 0 & & & & \text{Symmetric} & & \\ p & 0 & 0 & & & & & \\ 0 & -p & -q & 0 & & & & \\ 0 & 0 & -p & 0 & 0 & & & \\ 0 & 0 & 0 & p & 0 & 0 & & \\ -p & 0 & 0 & \tfrac{1}{2}p\ell & p & 0 & 0 & \\ 0 & p & \tfrac{1}{2}p\ell & 0 & 0 & -p & q & 0 \end{bmatrix}$$

$$p = \frac{1}{\ell}(I_1 + I_2)$$

$$q = \tfrac{1}{2}(I_2 - I_1)$$

$$\mathbf{k}_4 = \begin{bmatrix} 36c & & & & & & & \\ 0 & 36d & & & & \text{Symmetric} & & \\ 0 & -3d\ell & 4d\ell^2 & & & & & \\ 3c\ell & 0 & 0 & 4c\ell^2 & & & & \\ -36c & 0 & 0 & -3c\ell & 36c & & & \\ 0 & -36d & 3d\ell & 0 & 0 & 36d & & \\ 0 & -3d\ell & -d\ell^2 & 0 & 0 & 3d\ell & 4d\ell^2 & \\ 3c\ell & 0 & 0 & -c\ell^2 & 3c\ell & 0 & 0 & 4c\ell^2 \end{bmatrix}$$

$$c = \frac{4EI_2\beta_0^2 + P_x}{30\ell}$$

$$d = \frac{4EI_1\beta_0^2 + P_x}{30\ell}$$

$$\mathbf{k}_5 = \frac{E\beta_0^2}{15\ell} \begin{bmatrix} 36I_1 & & & & & & & \\ 0 & 36I_2 & & & & \text{Symmetric} & & \\ 0 & -18\ell I_2 & 4\ell^2 I_2 & & & & & \\ 18\ell I_1 & 0 & 0 & 12\ell^2 I_1 & & & & \\ -36I_1 & 0 & 0 & -3\ell I_1 & 36I_1 & & & \\ 0 & -36I_2 & 3\ell I_2 & 0 & 0 & 36I_2 & & \\ 0 & -3\ell I_2 & -\ell^2 I_2 & 0 & 0 & 18\ell I_2 & 4\ell^2 I_2 & \\ 3\ell I_1 & 0 & 0 & -\ell^2 I_1 & -18\ell I_1 & 0 & 0 & 4\ell^2 I_1 \end{bmatrix}$$

TABLE 17-10   **Matrices for a Helically Fluted Shaft Element**   **936**

$$\mathbf{k}_6 = \frac{1}{60} \begin{bmatrix} 0 & & & & & & & \\ -30p' & 0 & & & & \text{Symmetric} & & \\ 3q'\ell & 0 & 0 & & & & & \\ 0 & -6q'\ell & 0 & 0 & & & & \\ 0 & -30q' & -6q'\ell & 0 & 0 & & & \\ 30q' & 0 & 0 & 6q'\ell & 30p' & 0 & & \\ -6q'\ell & 0 & 0 & -q'\ell^2 & 6q'\ell & 0 & 0 & \\ 0 & 6q'\ell & q'\ell^2 & 0 & 0 & -6q'\ell & 0 & 0 \end{bmatrix}$$

$$p' = 2E(I_1 + I_2)\beta_0^3 - P_x\beta_0$$
$$q' = 2E(I_1 - I_2)\beta_0^3$$

$$\mathbf{k}_7 = \frac{\ell}{420} \begin{bmatrix} 156c' & & & & & & & \\ 0 & 156d' & & & & \text{Symmetric} & & \\ 0 & -22\ell d' & 4\ell^2 d' & & & & & \\ 22\ell c' & 0 & 0 & 4\ell^2 c' & & & & \\ 54c' & 0 & 0 & 13\ell c' & 156c' & & & \\ 0 & 54d' & -13\ell d' & 0 & 0 & 156d' & & \\ 0 & 13\ell d' & -3\ell^2 d' & 0 & 0 & 22\ell d' & 4\ell^2 d' & \\ -13\ell c' & 0 & 0 & -3\ell^2 c' & -22\ell c' & 0 & 0 & 4\ell^2 c' \end{bmatrix}$$

$$c' = EI_1\beta_0^4 + P_x\beta_0^2$$
$$d' = EI_2\beta_0^4 + P_x\beta_0^2$$

[a]Adopted from [17.6].

TABLE 17-11   MASS AND STIFFNESS MATRICES FOR ANNULAR ELASTIC THIN DISC ELEMENTS[a]

Notation

C = constant expressing variation in radial stress; $C = \begin{cases} 2 & \text{for } m = 0 \\ 1 & \text{for } m \geq 1 \end{cases}$

E = modulus of elasticity

$h(r)$ = thickness

m = number of nodal diameters (see Chapter 18 for more complete definition of m)

r = radius

ν = Poisson's ratio

ρ^* = mass per unit volume

Ω = spin speed

If linear thickness variation within the element is assumed,

$$h(r) = \alpha + \beta r \qquad \alpha = \frac{h_a r_b - h_b r_a}{r_b - r_a} \qquad \beta = \frac{h_b - h_a}{r_b - r_a}$$

Coordinate System

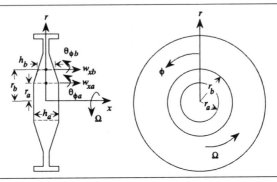

Equations of Motion

$$m\ddot{v} + (k_B + k_C)v = 0 \quad \text{with } v = [w_{xa} \quad w_{xb} \quad \theta_{\phi a} \quad \theta_{\phi b}]^T$$

Element Matrices

MASS MATRIX **m**:

$$\mathbf{m} = \mathbf{B}^T \mathbf{m}_d \mathbf{B}$$

where

$$\mathbf{m}_d = \begin{bmatrix} Q_1 & & \text{Symmetric} & \\ Q_2 & Q_3 & & \\ Q_3 & Q_4 & Q_5 & \\ Q_4 & Q_5 & Q_6 & Q_7 \end{bmatrix}$$

$$Q_k = C\pi\rho^* \int_{r_a}^{r_b} h(r) r^k \, dr$$

$$(k = 1, 2, \ldots, 7)$$

TABLE 17-11   Matrices for Annular Elastic Thin Disc Elements   938

BENDING STIFFNESS MATRIX $\mathbf{k}_B$:

$$\mathbf{k}_B = \mathbf{B}^T \mathbf{k}_b \mathbf{B}$$
where

$$k_b(1,1) = P_{-3}(m^4 + 2m^2 - 2\nu m^2) \qquad k_b(2,3) = P_0(m^4 - 3m^2 - 2\nu m^2 + 2\nu + 2)$$

$$k_b(1,2) = P_{-2}(m^4 - m^2) \qquad\qquad\qquad k_b(2,4) = P_1(m^4 - 4m^2 - 6\nu m^2 + 6\nu + 3)$$

$$k_b(1,3) = P_{-1}(m^4 - 4m^2) \qquad\qquad\; k_b(3,3) = P_1(m^4 - 2m^2 - 6\nu m^2 + 8\nu + 8)$$

$$k_b(1,4) = P_0(m^4 - 7m^2 - 2\nu m^2) \qquad k_b(3,4) = P_2(m^4 - m^2 - 12\nu m^2 + 18\nu + 18)$$

$$k_b(2,2) = P_{-1}(m^4 - 2m^2 + 1) \qquad\quad\; k_b(4,4) = P_3(m^4 + 2m^2 - 20\nu m^2 + 36\nu + 45)$$

$$k_b(i,j) = k_b(j,i) \qquad i \text{ and } j = 1,2,3,4$$

$$P_k = \frac{C\pi E}{12(1 - \nu^2)} \int_{r_a}^{r_b} h^3(r) r^k \, dr$$

$$(k = -3, -2, \ldots, 0, \ldots, 3)$$

CENTRIFUGAL STIFFNESS MATRIX $\mathbf{k}_C$:

$$\mathbf{k}_C = \mathbf{B}^T \mathbf{k}_G \mathbf{B}$$
where

$$\mathbf{k}_G = \begin{bmatrix} m^2 S_{-1} & & & \text{Symmetric} \\ m^2 S_0 & R_1 + m^2 S_1 & & \\ m^2 S_1 & 2R_2 + m^2 S_2 & 4R_3 + m^2 S_3 & \\ m^2 S_2 & 3R_3 + m^2 S_3 & 6R_4 + m^2 S_4 & 9R_5 + m^2 S_5 \end{bmatrix}$$

$$R_k = C\pi \int_{r_a}^{r_b} r^k h(r) \sigma_r(r) \, dr$$

$$(k = -1, 0, \ldots, 5)$$

$$S_k = C\pi \int_{r_a}^{r_b} r^k h(r) \sigma_\phi(r) \, dr$$

$$\sigma_r(r) = \frac{\sigma_{ra} r_b - \sigma_{rb} r_a}{r_b - r_a} + \frac{\sigma_{rb} - \sigma_{ra}}{r_b - r_a} r$$

$$\sigma_\phi(r) = \frac{\sigma_{\phi a} r_b - \sigma_{\phi b} r_a}{r_b - r_a} + \frac{\sigma_{\phi b} - \sigma_{\phi a}}{r_b - r_a} r$$

$\sigma_{ra}, \sigma_{\phi a}$ and $\sigma_{rb}, \sigma_{\phi b}$ are the radial and tangential stresses at the positions $r = r_a$ and $r = r_b$, respectively. These stresses can be calculated according to the dimensions and shape of the disc by using the formulas in Table 19-3.

$$B(1,1) = r_b^2(r_b - 3r_a)/(r_b - r_a)^3 \qquad B(3,1) = -3(r_a + r_b)/(r_b - r_a)^3$$

$$B(1,2) = r_a r_b^2/(r_b - r_a)^2 \qquad B(3,2) = (r_a + 2r_b)/(r_b - r_a)^2$$

$$B(1,3) = r_a^2(3r_b - r_a)/(r_b - r_a)^3 \qquad B(3,3) = 3(r_a + r_b)/(r_b - r_a)^3$$

$$B(1,4) = r_a^2 r_b/(r_b - r_a)^2 \qquad B(3,4) = (2r_a + r_b)/(r_b - r_a)^2$$

$$B(2,1) = 6r_a r_b/(r_b - r_a)^3 \qquad B(4,1) = 2/(r_b - r_a)^3$$

$$B(2,2) = -r_b(2r_a + r_b)/(r_b - r_a)^2 \qquad B(4,2) = -1/(r_b - r_a)^2$$

$$B(2,3) = -6r_a r_b/(r_b - r_a)^3 \qquad B(4,3) = -2/(r_b - r_a)^3$$

$$B(2,4) = -r_a(r_a + 2r_b)/(r_b - r_a)^2 \qquad B(4,4) = -1/(r_b - r_a)^2$$

The global matrices of the whole disk can be assembled using these element matrices along the radial direction as in beam calculation problems. See Appendix III for details.

[a]Adopted from [17.7].

TABLE 17-11   **Matrices for Annular Elastic Thin Disc Elements**   **940**

TABLE 17-12 STIFFNESS AND DAMPING MATRICES FOR SHORT JOURNAL BEARING, TILTING PAD BEARING, AND ANNULAR PLAIN SEAL ELEMENTS[a]

Notation

Ω = angular velocity of journal (spin speed)

F_y, F_z = bearing or seal forces in y and z directions, respectively

w_y^*, w_z^* = relative displacements between journal
and bearing pedestal or seal

Bearing or seal force:

$$\mathbf{F} = -\mathbf{kv} - \mathbf{c\dot{v}} \qquad \mathbf{F} = [F_y \quad F_z]^T \qquad \mathbf{v} = [w_y^* \quad w_z^*]^T$$

Stiffness matrix:

$$\mathbf{k} = \begin{bmatrix} k_{yy} & k_{yz} \\ k_{zy} & k_{zz} \end{bmatrix}$$

Damping matrix:

$$\mathbf{c} = \begin{bmatrix} c_{yy} & c_{yz} \\ c_{zy} & c_{zz} \end{bmatrix}$$

Bearings:

0_j = center of journal

0_b = center of bearing

μ = lubricant viscosity

R = bearing radius

c = nominal bearing clearance, $= C/R$

D = bearing diameter, $= 2R$

C = bearing radial clearance (L)

ℓ = bearing width or length

e_j = journal eccentricity

S = Sommerfeld number

ε = nominal eccentricity, $= e_j/C$

$\overline{W}$ = static load supported by bearing, $= \mu(30\Omega/\pi)\ell D(R/C)^2/S$

Seals:

c_0 = seal clearance before deformation (L)

p_e = seal outlet pressure of oil (F/L^2)

ℓ = seal width or length

p_i = seal inlet pressure of oil (F/L^2)

R = seal radius

λ = friction loss factor

ν_1, ν_2 = Poisson's ratio of seal and shaft

ρ_L^* = fluid (oil) density (M/L^3)

γ = inlet loss factor (≈ 0.5)

E_1, E_2 = Young's moduli of seal and shaft

c_1, c_2 = seal inlet and outlet clearance
after deformation (L)

Short Journal Bearing $(\pi\text{--film})$

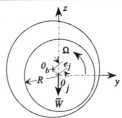

$$k_{yy} = \frac{\overline{W}}{C} \frac{4\left[\pi^2(2 - \varepsilon^2) + 16\varepsilon^2\right]}{\left[\pi^2(1 - \varepsilon^2) + 16\varepsilon^2\right]^{3/2}}$$

$$k_{yz} = -\frac{\overline{W}}{C} \frac{\pi\left[\pi^2(1 - \varepsilon^2)^2 - 16\varepsilon^4\right]}{\varepsilon(1 - \varepsilon^2)^{1/2}\left[\pi^2(1 - \varepsilon^2) + 16\varepsilon^2\right]^{3/2}}$$

$$k_{zy} = \frac{\overline{W}}{C} \frac{\pi\left[\pi^2(1 - \varepsilon^2)(1 + 2\varepsilon^2) + 32\varepsilon^2(1 + \varepsilon^2)\right]}{\varepsilon(1 - \varepsilon^2)^{1/2}\left[\pi^2(1 - \varepsilon^2) + 16\varepsilon^2\right]^{3/2}}$$

$$k_{zz} = \frac{\overline{W}}{C} \frac{4\left[\pi^2(1 - \varepsilon^2)(1 + 2\varepsilon^2) + 32\varepsilon^2(1 + \varepsilon^2)\right]}{(1 - \varepsilon^2)\left[\pi^2(1 - \varepsilon^2) + 16\varepsilon^2\right]^{3/2}}$$

$$c_{yy} = \frac{\overline{W}}{\Omega C} \frac{2\pi(1 - \varepsilon^2)^{1/2}\left[\pi^2(1 + 2\varepsilon^2) - 16\varepsilon^2\right]}{\varepsilon\left[\pi^2(1 - \varepsilon^2) + 16\varepsilon^2\right]^{3/2}}$$

$$c_{yz} = c_{zy} = -\frac{\overline{W}}{\Omega C} \frac{8\left[\pi^2(1 + 2\varepsilon^2) - 16\varepsilon^2\right]}{\left[\pi^2(1 - \varepsilon^2) + 16\varepsilon^2\right]^{3/2}}$$

$$c_{zz} = \frac{\overline{W}}{\Omega C} \frac{2\pi\left[\pi^2(1 - \varepsilon^2)^2 + 48\varepsilon^2\right]}{\varepsilon(1 - \varepsilon^2)^{1/2}\left[\pi^2(1 - \varepsilon^2) + 16\varepsilon^2\right]^{3/2}}$$

$$S = \frac{1}{(\ell/D)^2} \frac{1 - \varepsilon^2}{\pi\left[\pi^2 + (16 - \pi^2)\varepsilon^2\right]^{1/2}}$$

TABLE 17-12 | **Matrices for Bearing and Seal Elements** | **942**

Tilting pad bearing [b]

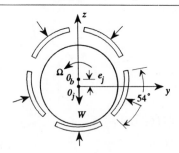

$$k_{yy} = \begin{cases} 4.01\dfrac{\overline{W}}{C}e^{-3.84\varepsilon} & (\varepsilon > 0.3) \\[2mm] \dfrac{\overline{W}}{C}(2.01 - 2.24\varepsilon) & (\varepsilon \le 0.3) \end{cases}$$

$$k_{zz} = 1.67\frac{\overline{W}}{C}e^{2.47\varepsilon}$$

$$k_{yz} = k_{zy} = 0$$

$$c_{yy} = \begin{cases} 13.5\dfrac{\overline{W}}{\Omega C}(1 - \varepsilon)^{2.11} & (\varepsilon > 0.3) \\[2mm] 35.0\dfrac{\overline{W}}{\Omega C}e^{-6.34\varepsilon} & (\varepsilon \le 0.3) \end{cases}$$

$$c_{zz} = \frac{\overline{W}}{\Omega C}(30 - 83.2\varepsilon + 76.4\varepsilon^2)$$

$$c_{yz} = c_{zy} = 0$$

$$S = \frac{6.125}{\pi(\ell/D)^2 e^{6.34\varepsilon}}$$

Annular Plain Seal

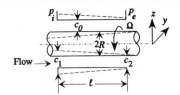

$$q_1 = \int_0^\ell \frac{1 + \gamma + c_1^2(\lambda + \delta)f(x)/\delta}{1 + \gamma + c_1^2(\lambda + \delta)f(c_2)/\delta} \, dx$$

$$\delta = \frac{R}{E_1\ell}(p_i - p_e)\left[(1 - \nu_1) + \left(\frac{E_1}{E_2}\right)(1 - \nu_2)\right]$$

$$f(x) = \frac{1}{x^2} - \frac{1}{c_1^2}$$

where

$$k_{yy} = -\frac{\pi}{\rho}Rm^2\left[(\delta + \lambda)d_1 - \left(\frac{\rho_L^*\Omega}{m}\right)^2 d_2\right]$$

$$k_{yz} = -\tfrac{1}{2}\pi R\Omega m\left[(\delta + \lambda)d_5 + d_3 - d_4\right]$$

$$k_{zy} = -k_{yz}$$

$$k_{zz} = k_{yy}$$

$$c_{yy} = -\pi Rm\left[(\delta + \lambda)d_5 + d_3 - d_4\right]$$

$$c_{yz} = -c_{zy} = -\rho_L^*\pi R\Omega d_2$$

$$c_{zz} = c_{yy}$$

TABLE 17-12   **Matrices for Bearing and Seal Elements**   **944**

$$m = c_1 \left(\frac{2\rho_L^*(p_i - p_e)}{1 + \gamma + (1 - \lambda/\delta)\left[(c_1/c_2)^2 - 1\right]} \right)^{1/2}$$

$$d_1 = \frac{\ell}{c_1^2 c_2^2} \left(\frac{\ell}{2} - q_1 \frac{c_1}{c_2} \right)$$

$$d_2 = \frac{\ell}{2\delta^2}(c_1 + c_2) + \frac{c_1 c_2}{\delta^3}\log\frac{c_2}{c_1} + \frac{q_1}{\delta^2}\left(\delta\ell + c_1\log\frac{c_2}{c_1}\right)$$

$$d_3 = \frac{\ell^2}{2c_1 c_2} - \frac{2\ell}{c_1\delta} - \frac{q_1\ell}{2c_1 c_2}\left(3 + \frac{c_1}{c_2}\right)$$

$$d_4 = \frac{\ell^2}{c_1\delta^2}(c_1 + c_2)\log\frac{c_2}{c_1} - \frac{q_1}{c_1\delta}\log\frac{c_2}{c_1}$$

$$d_5 = \frac{(c_1 + c_2)\ell}{2c_1 c_2\delta^2} + \frac{1}{\delta^3}\log\frac{c_2}{c_1} - \frac{q_1\ell^2}{2c_1 c_2^2}$$

When the effect of elastic deformation is ignored, set

$$c_1 = c_2 = c \qquad E_1 = E_2 = E \qquad \nu_1 = \nu_2 = \nu$$

[a]Some of this table is based on refs. 17.8 and 17.9.
[b]5 pads, zero preload, 54° pad arc length, 0.5 offset, $\ell/D < 0.5$, and static load on bottom pad.

TABLE 17-13 NATURAL FREQUENCIES OF VIBRATION OF A RADIAL BEAM OF CIRCULAR OR SQUARE CROSS SECTION[a]

Notation

Vibration occurs in the plane and perpendicular to the plane of rotation.

A = cross-sectional area

I = moment of inertia of cross-sectional area about transverse neutral axis (ξ or η axes); for circular cross section $I = \frac{1}{4}\pi r^4$, for square cross section $I = \frac{1}{12}h^4$

E = modulus of elasticity

R_0 = hub radius

Ω = spin speed

$\omega_{c\xi}$ = natural frequency for motion in ξ direction

$\omega_{c\eta}$ = natural frequency for motion in η direction

Coordinate Systems

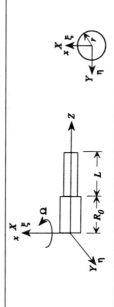

Critical Speeds

$$\omega_{c\xi}^2 = \alpha\,\frac{EI}{\rho^*AL^4} + \left[\frac{\beta I}{AL^2} + \gamma\right]\Omega^2 \qquad \omega_{c\eta}^2 = \alpha\,\frac{EI}{\rho^*AL^4} + \left[\frac{\beta I}{AL^2} + \gamma - 1\right]\Omega^2$$

where α, β, and γ are given below.

Boundary Conditions	α	β	γ
1. Free–hinged	244.08363	$-152.552 - 457.657\left(\frac{R_0}{L}\right) - 732.251\left(\frac{R_0}{L}\right)^2$	$-3.117 - 9.726\left(\frac{R_0}{L}\right) - 11.784\left(\frac{R_0}{L}\right)^2$

2. Hinged–free	244.08363	$305.105 + 457.657\left(\dfrac{R_0}{L}\right)$	$6.609 + 9.726\left(\dfrac{R_0}{L}\right)$
3. Free–guided	31.30813	$4.989 - 38.160\left(\dfrac{R_0}{L}\right) - 93.924\left(\dfrac{R_0}{L}\right)^2$	$-0.924 - 3.980\left(\dfrac{R_0}{L}\right) - 3.206\left(\dfrac{R_0}{L}\right)^2$
4. Guided–free	31.30813	$43.149 + 38.160\left(\dfrac{R_0}{L}\right)$	$3.056 + 3.980\left(\dfrac{R_0}{L}\right)$
5. Hinged–hinged	π^4	$\dfrac{3}{4}\pi^2 + \dfrac{\pi^4}{4}$	$-\dfrac{1}{4} + \dfrac{\pi^2}{12}$
6. Hinged–guided	$\dfrac{\pi^4}{16}$	$\dfrac{21}{16}\pi^2 + \dfrac{\pi^4}{64} + \dfrac{9}{8}\pi^2\left(\dfrac{R_0}{L}\right)$	$\dfrac{1}{4} + \dfrac{\pi^2}{48} + \dfrac{1}{2}\left(\dfrac{R_0}{L}\right)$

Boundary Conditions	α	β	γ
7. Guided–hinged	$\dfrac{\pi^4}{16}$	$\dfrac{3}{16}\pi^2 + \dfrac{\pi^4}{64} - \dfrac{9}{8}\pi^2\left(\dfrac{R_0}{L}\right)$	$-\dfrac{1}{4} + \dfrac{\pi^2}{48} - \dfrac{1}{2}\left(\dfrac{R_0}{L}\right)$
8. Guided–guided	π^4	$\dfrac{21}{4}\pi^2 + \dfrac{\pi^4}{4}$	$\dfrac{1}{4} + \dfrac{\pi^2}{12}$
9. Clamped–free	$\dfrac{162}{13}$	$\dfrac{1620}{91} + \dfrac{405}{13}\left(\dfrac{R_0}{L}\right)$	$\dfrac{61}{52} + \dfrac{81}{52}\left(\dfrac{R_0}{L}\right)$
10. Free–clamped	$\dfrac{162}{13}$	$-\dfrac{1215}{91} - \dfrac{405}{13}\left(\dfrac{R_0}{L}\right) - \dfrac{486}{13}\left(\dfrac{R_0}{L}\right)^2$	$-\dfrac{5}{13} - \dfrac{81}{52}\left(\dfrac{R_0}{L}\right) + \dfrac{135}{182}\left(\dfrac{R_0}{L}\right)^2$

TABLE 17-13 **Natural Frequencies of Vibration of Radial Beams** 948

11. Clamped–hinged	$\dfrac{4536}{19}$	$\dfrac{1620}{19} + \dfrac{1134}{19}\left(\dfrac{R_0}{L}\right)$	$\dfrac{1}{19} - \dfrac{27}{19}\left(\dfrac{R_0}{L}\right)$
12. Hinged–clamped	$\dfrac{4536}{19}$	$\dfrac{486}{19} - \dfrac{1134}{19}\left(\dfrac{R_0}{L}\right)$	$\dfrac{28}{19} + \dfrac{27}{19}\left(\dfrac{R_0}{L}\right)$
13. Clamped–guided	$\dfrac{63}{2}$	$\dfrac{1287}{32} + \dfrac{1323}{32}\left(\dfrac{R_0}{L}\right)$	$\dfrac{25}{64} + \dfrac{9}{64}\left(\dfrac{R_0}{L}\right)$
14. Guided–clamped	$\dfrac{63}{2}$	$-\dfrac{9}{8} - \dfrac{1323}{32}\left(\dfrac{R_0}{L}\right)$	$\dfrac{1}{4} - \dfrac{9}{64}\left(\dfrac{R_0}{L}\right)$
15. Clamped–clamped	504	90	1

[a]Based on [17.15].

TABLE 17-14   MASS AND STIFFNESS MATRICES FOR RADIAL BEAM ELEMENT[a]

Notation

Vibration is in the direction (ξ) perpendicular to the neutral axis of the beam.

A = cross-sectional area
ρ^* = mass per unit volume
I = moment of inertia of section
about η axis
E = modulus of elasticity
Ω = rotational speed (spin speed of rotor)
r = radius of gyration, $r^2 = I/A$

Coordinate System (Sign Convention 2)

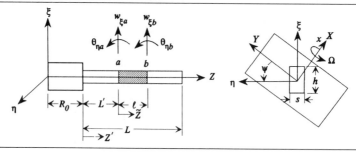

Equations of Motion

$$(\mathbf{m}_T + \mathbf{m}_R)\ddot{\mathbf{v}} + \left(\mathbf{k}_B + \mathbf{k}_C - \Omega^2 \sin \Psi \mathbf{m}_T\right)\mathbf{v} = 0$$

with $\mathbf{v} = [w_{\xi a} \quad \theta_{\eta a} \quad w_{\xi b} \quad \theta_{\eta b}]^T$, $w_{\xi a}$, $\theta_{\eta a}$ and $w_{\xi b}, \theta_{\eta b}$ are the displacements and slopes at nodes a and b, respectively.

Element Matrices

TRANSLATIONAL CONSISTENT MASS MATRIX $\mathbf{m}_T$:

$$\mathbf{m}_T = \frac{\rho^* A \ell}{420} \begin{bmatrix} 156 & & \text{Symmetric} & \\ 22\ell & 4\ell^2 & & \\ 54 & 13\ell & 156 & \\ -13\ell & -3\ell^2 & -22\ell & 4\ell^2 \end{bmatrix}$$

ROTARY CONSISTENT MASS MATRIX $\mathbf{m}_R$:

$$\mathbf{m}_R = \frac{\rho^* A r^2}{120\ell} \begin{bmatrix} 36 & & \text{Symmetric} & \\ 3\ell & 4\ell^2 & & \\ -36 & -3\ell & 36 & \\ 3\ell & -\ell^2 & -3\ell & 4\ell^2 \end{bmatrix}$$

TABLE 17-14   **Matrices for Radial Beam Element**   **950**

BENDING STIFFNESS MATRIX $\mathbf{k}_B$:

$$\mathbf{k}_B = \frac{EI}{\ell^3} \begin{bmatrix} 12 & & \text{Symmetric} & \\ 6\ell & 4\ell^2 & & \\ -12 & -6\ell & 12 & \\ 6\ell & 2\ell^2 & -6\ell & 4\ell^2 \end{bmatrix}$$

CENTRIFUGAL STIFFNESS MATRIX $\mathbf{k}_C$:

$$\mathbf{k}_C = \mathbf{k}_{C1} + \mathbf{k}_{C2} + \mathbf{k}_{C3}$$

where

$$\mathbf{k}_{C1} = \frac{\rho^* A\ell\Omega^2}{30} C_1 \begin{bmatrix} 36 & & \text{Symmetric} & \\ 3\ell & 4\ell^2 & & \\ -36 & -3\ell & 36 & \\ 3\ell & -\ell^2 & -3\ell & 4\ell^2 \end{bmatrix}$$

$$\mathbf{k}_{C2} = \frac{\rho^* A\ell\Omega^2}{60} C_2 \begin{bmatrix} 36 & & \text{Symmetric} & \\ 6\ell & 2\ell^2 & & \\ -36 & -6\ell & 36 & \\ 0 & -\ell^2 & 0 & 6\ell^2 \end{bmatrix}$$

$$\mathbf{k}_{C3} = \frac{\rho^* A\ell\Omega^2}{210} C_3 \begin{bmatrix} 72 & & \text{Symmetric} & \\ 15\ell & 4\ell^2 & & \\ -72 & -15\ell & 72 & \\ -6\ell & -3\ell^2 & 6\ell & 18\ell^2 \end{bmatrix}$$

$$C_1 = \frac{R_0}{\ell^2}(L - L') + \frac{1}{2\ell^2}(L^2 - L'^2), \qquad C_2 = \frac{1}{\ell}(R_0 + L'), \qquad C_3 = -\frac{1}{2}$$

[a]Based on [17.16].

TABLE 17-15 TRANSFER, STIFFNESS, AND MASS MATRICES FOR AXIAL VIBRATION OF RADIAL ROTATING BAR ELEMENT

Notation

A = cross-sectional area
Ω = rotational speed
R_0 = hub radius
E = modulus of elasticity
ρ^* = mass per unit volume
L = length of beam
L' = total length of elements before, not including the element under consideration

---

Transfer Matrix (Sign Convention 1)

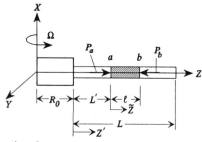

ω = vibration frequency
P = axial load
$\beta^2 = \rho^*(\omega^2 + \Omega^2)/E$

$\mathbf{z}_b = \mathbf{U}^i \mathbf{z}_a \qquad \mathbf{z} = [w_Z \quad P \quad 1]^T$

EXACT TRANSFER MATRIX:

$$\mathbf{U}^i = \begin{bmatrix} \cos \beta\ell & \sin \beta\ell/AE\beta & F_w \\ -AE\beta \sin \beta\ell & \cos \beta\ell & F_P \\ 0 & 0 & 1 \end{bmatrix}$$

$$F_w = -\frac{\rho^* \Omega^2}{E}\left[(R_0 + L')\frac{1 - \cos \beta\ell}{\beta^2} + \frac{\beta\ell - \sin \beta\ell}{\beta^3}\right]$$

$$F_P = -\rho^* A \Omega^2\left[(R_0 + L')\frac{\sin \beta\ell}{\beta} + \frac{1 - \cos \beta\ell}{\beta^2}\right]$$

LUMPED MASS TRANSFER MATRIX (mass lumped at middle point of element of length ℓ):

$$\mathbf{U}^i = \begin{bmatrix} 1 & \ell/EA & F_w \\ -\rho^* A(\omega^2 + \Omega^2)\ell & 1 & F_P \\ 0 & 0 & 1 \end{bmatrix}$$

$F_w = 0$

$F_P = -\rho^* A \Omega^2 [R_0 + L' + \tfrac{1}{2}\ell]\ell$

TABLE 17-15 | Matrices for Radial Rotating Bar Element   952

Stiffness and Mass Matrices (Sign Convention 2)

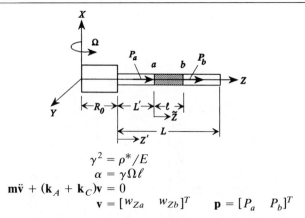

$$\gamma^2 = \rho^*/E$$
$$\alpha = \gamma\Omega\ell$$
$$\mathbf{m}\ddot{\mathbf{v}} + (\mathbf{k}_A + \mathbf{k}_C)\mathbf{v} = 0$$
$$\mathbf{v} = [w_{Za} \quad w_{Zb}]^T \qquad \mathbf{p} = [P_a \quad P_b]^T$$

TRANSLATIONAL CONSISTENT MASS MATRIX **m**:

$$\mathbf{m} = \rho^*A \begin{bmatrix} m_{11} & m_{12} \\ m_{21} & m_{22} \end{bmatrix}$$

where

$$m_{11} = \frac{(2\sin^2\alpha - 1)\sin(2\alpha) - 4\cos\alpha\sin^3\alpha + 2\alpha}{4\alpha\sin^2\alpha}$$

$$m_{12} = m_{21} = \frac{\cos\alpha\sin(2\alpha) + 2\sin^3\alpha - 2\alpha\cos\alpha}{4\alpha\sin^2\alpha}$$

$$m_{22} = \frac{2\alpha - \sin(2\alpha)}{4\alpha\sin^2\alpha}$$

AXIAL STIFFNESS MATRIX **k**$_A$:

$$\mathbf{k}_A = AE\frac{\alpha^2}{\ell} \begin{bmatrix} k_{a11} & k_{a12} \\ k_{a21} & k_{a22} \end{bmatrix}$$

where

$$k_{a11} = \frac{(1 - 2\sin^2\alpha)\sin(2\alpha) + 4\cos\alpha\sin^3\alpha + 2\alpha}{4\alpha\sin^2\alpha}$$

$$k_{a12} = k_{a21} = -\frac{\cos\alpha\sin(2\alpha) + 2\sin^3\alpha + 2\alpha\cos\alpha}{4\alpha\sin^2\alpha}$$

$$k_{a22} = \frac{2\alpha + \sin 2\alpha}{4\alpha\sin^2\alpha}$$

TABLE 17-15 Matrices for Radial Rotating Bar Element 954

TABLE 17-15 (continued) TRANSFER, STIFFNESS, AND MASS MATRICES FOR AXIAL VIBRATION OF RADIAL ROTATING BAR ELEMENT

Stiffness and Mass Matrices (Sign Convention 2)

$$\mathbf{k}_C = \rho^* A \Omega^2 \ell^3 \left(p\mathbf{m} - q\mathbf{d} - \tfrac{1}{2}\mathbf{e} \right)$$

CENTRIFUGAL STIFFNESS MATRIX $\mathbf{k}_C$:

where $p = \dfrac{R_0}{\ell^2}(L - L') + \dfrac{1}{2\ell^2}(L^2 - L'^2)$, $\quad q = \dfrac{1}{\ell}(R_0 + L')$

$\mathbf{m}$ is the translational consistent mass matrix

$$\mathbf{d} = \begin{bmatrix} d_{11} & d_{12} \\ d_{21} & d_{22} \end{bmatrix}$$

$$d_{11} = \frac{\left[4\alpha\sin^2\alpha - \sin(2\alpha) - 2\alpha\right]\sin(2\alpha) + \left[2\sin^2\alpha + 2\alpha\sin(2\alpha) - 1\right]\cos(2\alpha) - 2\sin^2\alpha + 2\alpha^2 + 1}{8\alpha^2\sin^2\alpha}$$

$$d_{12} = d_{21} = \frac{(\sin\alpha + 2\alpha\cos\alpha)\sin(2\alpha) + (\cos\alpha - 2\alpha\sin\alpha)\cos(2\alpha) - (1 + 2\alpha^2)\cos\alpha}{8\alpha^2\sin^2\alpha}$$

$$d_{22} = \frac{-2\alpha\sin(2\alpha) - \cos(2\alpha) + 2\alpha^2 + 1}{8\alpha^2\sin^2\alpha}$$

$$\mathbf{e} = \begin{bmatrix} e_{11} & e_{12} \\ e_{21} & e_{22} \end{bmatrix}$$

$$e_{11} = \frac{\left[6(2\alpha^2 - 1)\sin^2\alpha - 6\alpha\sin(2\alpha) - 6\alpha^2 + 3\right]\sin(2\alpha) + \left[12\alpha\sin^2\alpha + 3(2\alpha^2 - 1)\sin(2\alpha) - 6\alpha\right]\cos(2\alpha) + 3\sin(2\alpha) + 4\alpha^3}{24\alpha^3\sin^2\alpha}$$

$$e_{12} = e_{21} = \frac{\left[6\alpha\sin\alpha + 3(2\alpha^2 - 1)\cos\alpha\right]\sin(2\alpha) + \left[3(1 - 2\alpha^2)\sin\alpha + 6\alpha\cos\alpha\right]\cos(2\alpha) - 3\sin\alpha - 4\alpha^3\cos\alpha}{24\alpha^3\sin^2\alpha}$$

$$e_{22} = \frac{3(1 - 2\alpha^2)\sin(2\alpha) - 6\alpha\cos(2\alpha) + 4\alpha^3}{24\alpha^3\sin^2\alpha}$$

Plates

A plate is a flat structural member, the thickness of which is no larger than one-tenth of the length of the smallest lateral dimension. Most of the plate formulas presented here are based on the Kirchhoff–Love assumptions that the plate is thin; deflections and slopes are small; the plate material is linear, elastic, homogeneous, and isotropic; normal stresses transverse to the middle surface are negligible; and straight lines normal to the middle surface before deformation remain straight and normal to that surface after deformation. Some formulas for large deflections, nonuniform properties, and anisotropic and nonhomogeneous materials are provided. Other complications such as rotary inertia, shear deformation, and inelastic behavior are discussed in references 18.1 and 18.2. The sign convention for positive displacement, slopes, moments, and forces is shown in the figures and tables as appropriate.

18.1 NOTATION

All Plates

C	Applied concentrated moment per unit length (F)
D	Flexural rigidity of a plate, $= Eh^3/[12(1-\nu^2)]$ (FL)
E	Young's modulus (F/L^2)
h	Plate thickness (L)
p_1	Applied uniform transverse force per unit area (F/L^2)
p, p_s	Transverse applied load per unit area (F/L^2)
T	Temperature change (degrees), temperature change with respect to reference temperature
t	Time (T)
w	Transverse plate deflection (L)
W	Applied concentrated force per unit length (F/L)
W_T	Concentrated force applied at a point (F) or uniformly distributed force applied on a small area (F/L^2)
∇^4	Biharmonic differential operator
∇^2	Laplacian differential operator
ν	Poisson's ratio
ρ	Plate mass per unit area (FT^2/L^3)

Circular Plates

a_0	Inner radius of plate (L)
a_L	Outer radius of plate (L)
$D_r, D_\phi, D_{r\phi}$	Flexural rigidities (FL)
r_r, r_ϕ	Radii of gyration of mass about radial and tangential axes; for isotropic, homogeneous material, set $r_r^2 = r_\phi^2 = \frac{1}{12}h^2$.
M_r, M	Bending moment per unit length on planes normal to radial (r-axis) direction (FL/L)
M_ϕ	Bending moment per unit length on planes normal to azimuthal (tangential, ϕ-axis) direction (FL/L)
$M_{r\phi}$	Twisting moment per unit length on either an r- or ϕ-coordinate plane (FL/L)
P	Radial in-plane force per unit length (F/L)
Q_r	Transverse shear force per unit length on planes normal to the r-axis (F/L)
Q_ϕ	Transverse shear force per unit length on planes normal to the ϕ axis (F/L)

r, ϕ, z	Coordinates in a right-handed polar system
V_r, V	Equivalent shear force per unit length on planes normal to the r-axis (F/L)
V_ϕ	Equivalent shear force per unit length on planes normal to the ϕ-axis (F/L)
θ	Slope about ϕ axis (rad)

The material constants are defined in Table 18-1 for isotropic, orthotropic, composite, and layered circular plates.

Rectangular Plates

D, D_x, D_y	Plate flexural rigidities (FL)
D_{xy}	Torsional rigidity (FL)
L, L_y	Length of plate in x and y directions (L)
M_x	Bending moment per unit length parallel to y axis (FL/L)
M_y	Bending moment per unit length parallel to x axis (FL/L)
M_{xy}	Twisting moment per unit length (FL/L)
P_{cr}	Buckling load (F/L)
P_x	In-plane force per unit length acting in x direction (F/L)
P_y	In-plane force per unit length acting in y direction (F/L)
P_{xy}	In-plane shear force per unit length acting on planes normal to x or y axis (F/L)
Q_x	Shear force per unit length on surfaces normal to x axis (F/L)
Q_y	Shear force per unit length on surfaces normal to y axis (F/L)
V_x	Equivalent shear force acting on planes normal to x axis (F/L)
V_y	Equivalent shear force acting on planes normal to y axis (F/L)
w_{mn}	Deflection mode shape of plate in mode corresponding to ω_{mn} (L)
x, y, z	Right-handed system of coordinates
ω_{mn}	Natural frequency (rad/T)
σ_{cr}	Buckling stress (F/L^2)
σ_x	Normal stress on surfaces perpendicular to x axis (F/L^2)
σ_y	Normal stress on surfaces perpendicular to y axis (F/L^2)
θ	Slope of plate surface about line parallel to y axis (rad)
θ_y	Slope of plate surface about line parallel to x axis (rad)

The material constants are defined in Table 18-14 for isotropic, orthotropic, and stiffened rectangular plates.

18.2 CIRCULAR PLATES

Stresses

The tables of this chapter provide the deflection, slope, bending moment, and shear force. Once the internal moments and forces are known, the stresses can be determined with the formulas given in Table 18-2.

Simple Circular Plates

In polar coordinates, the governing equation of motion for an isotropic plate of uniform thickness and with no in-plane loading is

$$\nabla^4 w = \left(\frac{\partial^2}{\partial r^2} + \frac{1}{r^2} \frac{\partial^2}{\partial \phi^2} + \frac{1}{r} \frac{\partial}{\partial r} \right) \left(\frac{\partial^2}{\partial r^2} + \frac{1}{r^2} \frac{\partial^2}{\partial \phi^2} + \frac{1}{r} \frac{\partial}{\partial r} \right) w = \frac{p}{D} - \frac{\rho}{D} \frac{\partial^2 w}{\partial t^2}$$

(18.1)

where

$$\nabla^4 = \nabla^2 \nabla^2$$

$$\nabla^2 = \frac{\partial^2}{\partial r^2} + \frac{1}{r^2} \frac{\partial^2}{\partial \phi^2} + \frac{1}{r} \frac{\partial}{\partial r}$$

(18.2)

$$D = \frac{Eh^3}{12(1 - \nu^2)}$$

The expressions for the internal bending moments, twisting moment, and shear force per unit length are

$$M_r = -D \left[\frac{\partial^2 w}{\partial r^2} + \nu \left(\frac{1}{r^2} \frac{\partial^2 w}{\partial \phi^2} + \frac{1}{r} \frac{\partial w}{\partial r} \right) \right]$$

(18.3a)

$$M_\phi = -D \left(\frac{1}{r} \frac{\partial w}{\partial r} + \frac{1}{r^2} \frac{\partial^2 w}{\partial \phi^2} + \nu \frac{\partial^2 w}{\partial r^2} \right)$$

(18.3b)

$$M_{r\phi} = -(1 - \nu) D \left(\frac{1}{r} \frac{\partial^2 w}{\partial r \partial \phi} - \frac{1}{r^2} \frac{\partial w}{\partial \phi} \right)$$

(18.3c)

$$Q_r = -D \frac{\partial}{\partial r} (\nabla^2 w)$$

(18.3d)

$$Q_\phi = -D \frac{1}{r} \frac{\partial}{\partial \phi} (\nabla^2 w)$$

(18.3e)

$$V_r = Q_r + \frac{1}{r} \frac{\partial M_{r\phi}}{\partial \phi}$$

(18.3f)

$$V_\phi = Q_\phi + \frac{\partial M_{r\phi}}{\partial r}$$

(18.3g)

The equation of motion of Eq. (18.1) is solved readily for cases in which the loads and boundary conditions are rotationally symmetric (independent of ϕ). Then, with the inertia term ignored, the governing equation of motion reduces to

$$\frac{d^4 w}{dr^4} + \frac{2}{r} \frac{d^3 w}{dr^3} - \frac{1}{r^2} \frac{d^2 w}{dr^2} + \frac{1}{r^3} \frac{dw}{dr} = \frac{p}{D}$$

(18.4a)

or

$$\nabla^4 w = \left(\frac{d^2}{dr^2} + \frac{1}{r} \frac{d}{dr} \right) \left(\frac{d^2}{dr^2} + \frac{1}{r} \frac{d}{dr} \right) w = \frac{p}{D}$$

(18.4b)

The equations for the internal shear and moments are

$$M_r = -D\left(\frac{d^2w}{dr^2} + \frac{\nu}{r}\frac{\partial w}{dr}\right) \tag{18.5a}$$

$$M_\phi = -D\left(\nu\frac{d^2w}{dr^2} + \frac{1}{r}\frac{dw}{dr}\right) \tag{18.5b}$$

$$M_{r\phi} = 0 \tag{18.5c}$$

$$Q_r = V_r = -D\left(\frac{d^3w}{dr^3} + \frac{1}{r}\frac{d^2w}{dr^2} - \frac{1}{r^2}\frac{dw}{dr}\right) \tag{18.5d}$$

$$Q_\phi = 0 \tag{18.5e}$$

With derivatives with respect to r arranged to appear on the left-hand side, the above equations in first-order form can be written as

$$\frac{dw}{dr} = -\theta \tag{18.6a}$$

$$\frac{d\theta}{dr} = \frac{M}{D} - \nu\frac{\theta}{r} \tag{18.6b}$$

$$\frac{dV}{dr} = -\frac{V}{r} - p \tag{18.6c}$$

$$\frac{dM}{dr} = -(1 - \nu)\frac{M}{r} + V + \frac{D(1 - \nu^2)\theta}{r^2} \tag{18.6d}$$

with $M = M_r$ and $V = V_r$. The convention for positive displacement, slopes, moments, and shear forces is shown in Fig. 18-1.

Complex Circular Plates

The first-order governing differential equations for complex circular plates in polar coordinates are

$$\frac{\partial w}{\partial r} = -\theta$$

$$\frac{\partial \theta}{\partial r} = \frac{M}{D_r} + \nu_\phi\left(\frac{1}{r^2}\frac{\partial^2 w}{\partial\phi^2} - \frac{\theta}{r}\right)$$

$$\frac{\partial M}{\partial r} = -\left(1 - \frac{\nu_r D_\phi}{D_r}\right)\frac{M}{r} + V - \left[D_\phi(1 - \nu_r\nu_\phi) + 4D_{r\phi}\right]\frac{1}{r^3}\frac{\partial^2 w}{\partial\phi^2}$$

$$+ \frac{D_\phi(1 - \nu_r\nu_\phi)}{r^2}\theta - \frac{4D_{r\phi}}{r^2}\frac{\partial^2\theta}{\partial\phi^2} + \rho r_\phi^2\frac{\partial^2\theta}{\partial t^2} \tag{18.7}$$

$$\frac{\partial V}{\partial r} = -\frac{V}{r} - \frac{\nu_r D_\phi}{r^2 D_r}\frac{\partial^2 M}{\partial\phi^2} + \frac{D_\phi}{r^4}(1 - \nu_r\nu_\phi)\frac{\partial^4 w}{\partial\phi^4} - \frac{4D_{r\phi}}{r^4}\frac{\partial^2 w}{\partial\phi^2}$$

$$- \left[D_\phi(1 - \nu_r\nu_\phi) + 4D_{r\phi}\right]\frac{1}{r^3}\frac{\partial^2\theta}{\partial\phi^2} - \frac{\rho r_r^2}{r^2}\frac{\partial^4 w}{\partial t^2\partial\phi^2} + \rho\frac{\partial^2 w}{\partial t^2} - p(r, \phi, t)$$

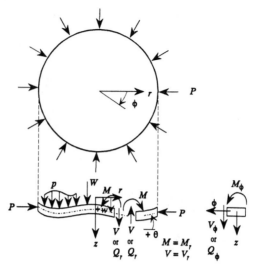

FIGURE 18-1 Positive displacement w, slope θ, moment M, shear force V, and applied loading.

where $M = M_r$, $V = V_r$. The internal forces are

$$M_\phi = -D_\phi\left(\frac{1}{r}\frac{\partial w}{\partial r} + \frac{1}{r^2}\frac{\partial^2 w}{\partial \phi^2} + \nu_r\frac{\partial^2 w}{\partial r^2}\right)$$

$$M_{r\phi} = M_{\phi r} = 2D_{r\phi}\left(\frac{1}{r}\frac{\partial^2 w}{\partial r\,\partial \phi} - \frac{1}{r^2}\frac{\partial w}{\partial \phi}\right)$$

$$(18.8)$$

Tabulated Formulas

Formulas for the deflections, moments, and shear forces for rather simple loadings are given in Table 18-3.

Example 18.1 Static Deflection of a Circular Plate Subjected to a Distributed Load A circular plate is subjected to a uniform load $p_1 = 26$ lb/in.^{2} At the center of the plate, compute the deflection, the radial and azimuthal bending moments per unit length, and the equivalent shear force per unit length acting on planes normal to the radius. Perform the computation for both pinned and fixed end conditions. Let $E = 3.0 \times 10^7$ lb/in.^{2}, $h = 1.0$ in., $\nu = 0.3$, and $a_L = 13.5$ in.

1. For pinned edges, case 1 of Table 18-3 applies:

$$D = \frac{Eh^3}{12(1-\nu^2)} = 2.75 \times 10^6 \text{ lb-in.} \tag{1}$$

For the center of the plate, $\alpha = r/a_L = 0$:

$$w = \frac{p_1 a_L^4(5 + \nu)}{64D(1 + \nu)}$$

$$= \frac{(26)(13.5)^4(5 + 0.3)}{64(2.75 \times 10^6)(1 + 0.3)} = 0.020 \text{ in.}$$

$$M_r = \frac{p_1 a_L^2}{16}(3 + \nu) = \frac{(26)(13.5)^2}{16}(3 + 0.3) = 977.3 \text{ in.-lb/in.} \tag{2}$$

$$M_\phi = \frac{p_1 a_L^2}{16}(3 + \nu) = 977.3 \text{ in.-lb/in.}$$

$$Q_r = -\frac{p_1 a_L}{2}\alpha = 0.0$$

2. For clamped edges, case 9 of Table 18-3 applies. At the center of the plate

$$\alpha = \frac{r}{a_L} = 0$$

$$w = \frac{(26)(13.5)^4(1)}{(64)(2.75 \times 10^6)} = 0.00491 \text{ in.}$$

$$M_r = \frac{(26)(13.5)^2}{16}(1 + 0.3) = 385.0 \text{ in.-lb/in.} \tag{3}$$

$$M_\phi = 385.0 \text{ in.-lb/in.}$$

$$Q_r = 0$$

====

Formulas for Plates with Arbitrary Loading

Table 18-3 gives the responses of circular plates for some simple loadings and boundary conditions. For more complicated uniform plates, the formulas in Table 18-4 can be used to calculate the deflections, slopes, bending moments, and responses.

Part A of 18-4 lists equations for the responses. The functions F_w, F_θ, F_V, and F_M are taken from Table 18-4, part B, by adding the appropriate terms for each load applied to the plate. The initial parameters w_0, θ_0, V_0, and M_0, which are values of w, θ, V and M at the inner edge ($r = a_0$) of the plate, are evaluated using the entry in Table 18-4, part C, for the appropriate edge conditions.

These general formulas are readily programmed for computer solution.

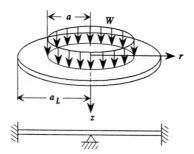

FIGURE 18-2   Circular plate with rigid center support.

Example 18.2   Plate with Concentrated Ring Load   Determine the deflection caused by a concentrated ring force in a plate fixed on the outer rim and rigidly supported at the center (Fig. 18-2).

From case 2 of Table 18-4, part A, the deflection is expressed by

$$
w = -\frac{R}{8\pi D}r^2(\ln r - 1) + C_1\frac{r^2}{4} + F_w \tag{1}
$$

Table 18-4, part B, gives F_w for the concentrated ring force W:

$$
F_w = \langle r - a\rangle^0\frac{Wa}{4D}\left[(r^2 + a^2)\ln\frac{r}{a} - (r^2 - a^2)\right] \tag{2}
$$

According Table 18-4, part C, the reaction R and constant C_1 are given by

$$
R = -\frac{16\pi D}{a_L^2}\bar{F}_w - \frac{8\pi D}{a_L}\bar{F}_\theta
$$

$$
C_1 = -\frac{8\left(\ln a_L - \frac{1}{2}\right)}{a_L^2}\bar{F}_w - \frac{4(\ln a_L - 1)}{a_L}\bar{F}_\theta \tag{3}
$$

Insertion of $\bar{F}_w = F_{w|r=a_L}$ and $\bar{F}_\theta = F_{\theta|r=a_L}$ from Table 18-4, part B, into (3) gives

$$
C_1 = \frac{Wa}{D}(1 - \beta^2 + 2\beta^2\ln\beta)\left[\ln a + (1 - \beta^2)\ln\beta\right]
$$

$$
R = 2\pi Wa(1 - \beta^2 + 2\beta^2\ln\beta) \tag{4}
$$

where $\beta = a/a_L$. Substitution of (2) and (4) into (1) provides the expression for the deflection at any radius r.

Buckling Loads

When a circular plate is subjected to a static in-plane radial force (per unit length) P, the plate equation is

$$D\,\nabla^4 w = P\,\nabla^2 w \tag{18.9}$$

in which the Laplacian operators are written in polar coordinates. For certain critical values of the in-plane load, the plate will buckle transversely even though transverse loads may not be present. A critical-load value is associated with each buckled mode shape.

Majumdar [18.3] studied the buckling of a circular plate clamped at the outer edge, free at the inner edge, and loaded with a uniform radial compressive force applied at the outside edge. It is shown that for small ratios of inner to outer radius the plate buckles in a radially symmetric mode. When the ratio of the inner to outer radius exceeds a certain value, the minimum buckling load corresponds to buckling modes with nodes, which are the loci of points for which the displacements in the buckling modes are zero, along a circumference. The number of nodes depends on the ratio of the inner and the outer radii.

Formulas for the critical load of several circular plate configurations are listed in Table 18-5. These critical loads are compressive in-plane forces per unit length applied at the outer edge. The stress corresponding to the critical load should be less than the yield strength of the material of the plate in order for the buckling load to be valid.

Techniques for obtaining solutions to circular plate stability problems are discussed in reference 18.1.

Example 18.3   Critical In-Plane Loads of a Circular Plate   Compute the critical in-plane compressive force for a circular plate with no center hole for both pinned and clamped outer edges if the buckled shape has neither nodal diameters nor circles. Let $E = 3.0 \times 10^7$ lb/in.^{2}, $h = 1.0$ in., $\nu = 0.3$, and $a_L = 36$ in., so that $D = Eh^3/[12(1 - \nu^2)] = 2.75 \times 10^6$ lb-in. Also, calculate the stress level corresponding to buckling to assure that the yield stress has not been reached.

1. For pinned edges, from case 1 of Table 18-5,

$$P_{cr} = 0.426\pi^2 \frac{D}{a_L^2} = 0.426\pi^2(2.75 \times 10^6)/(36)^2 \tag{1}$$

$$= 0.89 \times 10^4 \text{ lb/in.}$$

The stress corresponding to the stress resultant P_{cr} is given by

$$\sigma_{cr} = P_{cr}/h = 0.89 \times 10^4/1.0 = 8900 \text{ lb/in.}^2 \tag{2}$$

2. For clamped edges, from case 2 of Table 18-5,

$$P_{cr} = (1.49)\frac{\pi^2 D}{a_L^2} = 1.49\pi^2(2.75 \times 10^6)/(36)^2 = 3.12 \times 10^4 \text{ lb/in.} \tag{3}$$

$$\sigma_{cr} = P_{cr}/h = 3.12 \times 10^4/1.0 = 31{,}200 \text{ lb/in}^2 \tag{4}$$

Natural Frequencies

The formulas for natural frequencies in a number of cases of uniform thickness circular plates are listed in Table 18-6. The nodes, which are the loci of points along which the mode shape displacements are zero, occur along diameters, numbered n, of the plate or along concentric circles, numbered s, centered at the plate center. A particular mode is chosen by specifying the number of nodal diameters and nodal concentric circles. It can be observed that the fundamental frequency does not always correspond to the smallest s and n. Also, except for certain small values of s and n, the natural frequency increases as s increases for a fixed n or vice versa. Thus, to determine the fundamental, the second, the third natural frequency, etc., numerous combinations of s and n should be tested.

More complex cases can be treated using the transfer or stiffness matrices provided in other tables.

Example 18.4   Natural Frequencies of a Circular Plate   For the plate of Example 18.1 compute the natural frequencies of the first two rotationally symmetric mode shapes ($n = 0$). Perform the computation for free, pinned, and fixed outer edges. In all cases $D = 2.75 \times 10^6$ lb-in., $a_L = 13.5$ in., and $\rho = 7.273 \times 10^{-4}$ lb-s^2/in.^{3}

1. *Free Outer Boundary*. Use case 1 of Table 18-6. For the case of rotational symmetry $n = 0$ (no nodal diameters), and for the lowest frequency in case 1, $s = 1$ (one nodal circle). Thus,

$$\omega_{ns}|_{\substack{n=0 \\ s=1}} = \omega_{01} = \frac{\lambda_{01}}{a_L^2} \sqrt{\frac{D}{\rho}} \tag{1}$$

$$\lambda_{01} = 8.892$$

so that

$$\omega_{01} = 3000 \text{ rad/s} \quad \text{or} \quad f_{01} = 477.5 \text{ Hz} \tag{2}$$

For $s = 2$, $\lambda_{02} = 38.34$, so that the natural frequency corresponding to the second rotationally symmetric mode shape is

$$\omega_{ns}|_{\substack{n=0 \\ s=2}} = \omega_{02} = \frac{\lambda_{02}}{a_L^2} \sqrt{\frac{D}{\rho}} = 12{,}935.8 \text{ rad/s} \quad \text{or} \quad f_{02} = 2058.95 \text{ Hz} \tag{3}$$

Note that neither of these two frequencies is the fundamental frequency. For this problem, the fundamental frequency occurs at $s = 0$ and $n = 2$.

$$\omega_{ns}|_{\substack{n=2 \\ s=0}} = \omega_{20} = \frac{\lambda_{20}}{a_L^2} \sqrt{\frac{D}{\rho}} = 1857.32 \text{ rad/s} \quad \text{or} \quad f_{20} = 295.6 \text{ Hz} \tag{4}$$

2. *Pinned Outer Edge*. From case 2 of Table 18-6, it is evident that the lowest frequency occurs for $n = 0$ (no nodal diameters) and $s = 0$ (no nodal circles).

Then

$$\omega_{00} = \frac{\lambda_{00}}{a_L^2}\sqrt{\frac{D}{\rho}} \tag{5}$$

$$\lambda_{00} = 4.977$$

so that

$$\omega_{00} = 1679 \text{ rad/s} \quad \text{or} \quad f_{00} = 267.3 \text{ Hz} \tag{6}$$

Also,

$$\lambda_{01} = 29.76$$

and

$$\omega_{01} = 1.004 \times 10^4 \text{ rad/s} \quad \text{or} \quad f_{01} = 1598 \text{ Hz} \tag{7}$$

Here ω_{00} is the fundamental frequency but ω_{01} is the fourth frequency.

3. *Fixed Outer Boundary.* Use case 3 of Table 18-6. For the case that the mode shape contains no nodal diameter and no nodal circle,

$$\lambda_{00} = 10.216 \tag{8}$$

$$\omega_{00} = (\lambda_{00}/a_L^2)\sqrt{D/\rho} = 3448 \text{ rad/s} \quad \text{or} \quad f_{00} = 548.8 \text{ Hz}$$

and

$$\lambda_{01} = 39.771 \tag{9}$$

$$\omega_{01} = 13{,}419 \text{ rad/s} \quad \text{or} \quad f_{01} = 2135.6 \text{ Hz}$$

The position of ω_{00} and ω_{01} in the natural frequency sequence is the same as that in case 2 above.

General Circular Plates

The problems of determining the static deflection, critical in-plane load, natural frequencies, mode shapes, and steady sinusoidal response of complicated circular plates can be solved using the displacement method or the transfer matrix method, which are explained in Appendices II and III, as well as in reference 18.22. These approaches are well suited to computer implementation, and the techniques can be applied to circular plates without rotationally symmetric loads.

In the governing equations, the ϕ-dependence of state variables and of applied loads can be removed by expanding these quantities in a Fourier series. For the plate deflection this series is

$$w(r, \phi, t) = \sum_{m=0}^{\infty} \left[w_m^c(r, t)\cos m\phi + w_m^s(r, t)\sin m\phi \right] \tag{18.10}$$

An analogous series representation is used for the remaining state variables and for the applied loads. After these series have been introduced into the equations governing the plate motion, the ϕ-dependence of all quantities is eliminated, and the equations are integrated as functions of r and t.

Transfer Matrices The transfer matrices are obtained from the solution of the differential equations derived by substituting the Fourier series expansion of the state variables in the form of Eq. (18.10) into Eq. (18.1), (18.6), or (18.7). These differential equations are functions of r and t only. Table 18-7 provides the transfer matrix for a variety of plates under general loadings. In using these matrices, the loadings must be expanded in Fourier series of the form of Eq. (18.10) also. Together with the solutions for disks of Chapter 19, these matrices can be used to find the static response, buckling load, or natural frequencies for in-plane and transverse motion of circular plates. The methodology for using these matrices is detailed in Appendix III.

The notation for the transfer matrix for plate element i is

$$
\mathbf{U}^i = \begin{bmatrix}
U_{ww} & U_{w\theta} & U_{wV} & U_{wM} & F_w \\
U_{\theta w} & U_{\theta\theta} & U_{\theta V} & U_{\theta M} & F_\theta \\
U_{Vw} & U_{V\theta} & U_{VV} & U_{VM} & F_V \\
U_{Mw} & U_{M\theta} & U_{MV} & U_{MM} & F_M \\
0 & 0 & 0 & 0 & 1
\end{bmatrix}
$$

Table 18-9 provides some point transfer matrices for concentrated occurrences.

Stiffness Matrices Table 18-8 contains stiffness matrices and loading vectors. Use of these matrices in static, stability, and dynamic analyses is described in Appendix III.

For asymmetric bending of circular plates, displacement, force, and loading variables are the components of the Fourier expansion of the form of Eq. (18.10), i.e., the vector of nodal displacements for an element is

$$
\mathbf{v}^i = \begin{bmatrix} w_{ma}^j & \theta_{ma}^j & w_{mb}^j & \theta_{mb}^j \end{bmatrix}^T \tag{18.12}
$$

and the element force vector is

$$
\mathbf{p}^i = \begin{bmatrix} V_{ma}^j & M_{ma}^j & V_{mb}^j & M_{mb}^j \end{bmatrix}^T \qquad m = 0, 1, 2, \ldots \qquad j = s, c \quad [\text{Eq. (18.10)}] \tag{18.13}
$$

For simplicity, the subscript m and superscript j have been dropped in the tables. For example, w_{ma}^j becomes w_a and V_{ma}^j becomes V_a. The format for a stiffness matrix is

$$
\begin{bmatrix} V_a \\ M_a \\ V_b \\ M_b \end{bmatrix} = \begin{bmatrix} k_{11} & k_{12} & k_{13} & k_{14} \\ k_{21} & k_{22} & k_{23} & k_{24} \\ k_{31} & k_{32} & k_{33} & k_{34} \\ k_{41} & k_{42} & k_{43} & k_{44} \end{bmatrix} \begin{bmatrix} w_a \\ \theta_a \\ w_b \\ \theta_b \end{bmatrix} - \begin{bmatrix} V_a^0 \\ M_a^0 \\ V_b^0 \\ M_b^0 \end{bmatrix}
$$

$$
\mathbf{p}^i \quad = \quad\quad\quad \mathbf{k}^i \quad\quad\quad\quad \mathbf{v}^i \;-\; \overline{\mathbf{p}}^i
$$

(18.14)

Stiffness matrices for an infinite circular plate lying on an elastic foundation are presented in Table 18-10.

Table 18-9 provides some point stiffness matrices for concentrated occurrences.

Geometric Stiffness Matrices The geometric stiffness matrices used for the buckling analyses of circular plates are provided in Table 18-11.

The global geometric stiffness matrix $\mathbf{K}_G$ of a circular plate can be assembled from the element geometric stiffness matrices $\mathbf{k}_G^i$. Values for the in-plane force P and the in-plane displacement u_a at radius $r = a$ used in $\mathbf{k}_G^i$, the stiffness matrix for element i, are often calculated through an in-plane analysis, i.e., a disk analysis for the prescribed in-plane loading pattern as in Chapter 19. The in-plane forces for a buckling analysis are assumed to remain proportional to the distribution of in-plane forces found from the disk analysis for the initial pattern of applied in-plane forces. The *load factor* λ is the constant of proportionality. The critical (buckling) load for the circular plate can be obtained as the solution to the eigenvalue problem

$$
(\mathbf{K} - \lambda \mathbf{K}_G)\mathbf{V} = 0 \tag{18.15}
$$

where $\mathbf{K}$ and $\mathbf{V}$ are the global stiffness matrix and displacement vector and λ is an eigenvalue, the lowest value of which is the ratio of the buckling load to an initial applied in-plane force used in the disk analysis.

Mass Matrices Consistent and lumped mass matrices are given in Table 18-12. The nodal variables are the same as those in the stiffness matrices. See Appendix III for the use of mass matrices in dynamic analyses.

Large Deflections of Circular Plates

In general, the analytical solutions to the governing equations for large deflections of thin plates are difficult to obtain. However, the deflection and stresses at some special points of interest can be approximated by the formulas given in Table 18-13. These formulas for large deflections apply for linear elastic materials.

Example 18.5 Circular Plate with Large Deflections A solid circular steel plate 0.2 in. thick and 30 in. in diameter is fixed along the outer edge for transverse motion and remains restrained against radial movement. Also, it is uniformly loaded with $p_1 = 6.5$ lb/in.^{2}. Determine the maximum deflection and the maximum stress with $E = 3 \times 10^7$ lb/in.^{2}, $\nu = 0.3$.

From the given edge conditions, case 3 in Table 18-13 should be used to obtain the large-deflection solution.

$$\frac{p_1 a^4}{Eh^4} = \frac{6.5(30/2)^4}{3(10^7)(0.2)^4} = 6.86 = \frac{5.333}{1-\nu^2}\frac{w_0}{h} + 0.857\left(\frac{w_0}{h}\right)^3 \tag{1}$$

This cubic relationship between the load and deflection gives

$$\frac{w_0}{h} \approx 1.0$$

That is, the maximum deflection of the plate is

$$w_{max} = w_0 = h = 0.2 \text{ in.} \tag{2}$$

The stress at the center of the plate is

$$\sigma_{r0} = \frac{3(10^7)(0.2)^2}{(30/2)^2}\left[\left(\frac{2}{1-0.3^2}\right)(1.0) + 0.5(1.0)^2\right] \tag{3}$$

$$= 14.4 \times 10^3 \text{ lb/in.}^2$$

and the stress at the edge becomes

$$\sigma_{ra} = \frac{3(10^7)(0.2)^2}{(30/2)^2}\frac{4}{1-0.3^2}(1.0)^2 = 23.4 \times 10^3 \text{ lb/in.}^2 \tag{4}$$

Thus,

$$\sigma_{max} = \sigma_{ra} = 23.4 \times 10^3 \text{ lb/in.}^2 \tag{5}$$

18.3 RECTANGULAR PLATES

Stresses

This section contains several tables of formulas for the deflection, slope, shear force, and bending moment. Once the internal forces, i.e., bending moments, twisting moments, and transverse forces, are obtained from these formulas, the stresses can be calculated from the formulas in Table 18-15. The material properties are defined in Table 18-14.

Governing Differential Equations

The deflection of a simple plate, i.e., isotropic, uniform plate, is governed by a linear partial differential equation

$$\nabla^4 w = \frac{\partial^4 w}{\partial x^4} + 2\frac{\partial^4 w}{\partial x^2 \partial y^2} + \frac{\partial^4 w}{\partial y^4} = \frac{p}{D} - \frac{\rho}{D}\frac{\partial^2 w}{\partial t^2}$$

$$\nabla^4 = \frac{\partial^4}{\partial x^4} + 2\frac{\partial^4}{\partial x^2 \partial y^2} + \frac{\partial^4}{\partial y^4} = \left(\frac{\partial^2}{\partial x^2} + \frac{\partial^2}{\partial y^2}\right)^2 = \nabla^2\nabla^2 \quad (18.16)$$

$$D = \frac{Eh^3}{12(1 - \nu^2)}$$

This equation does not take into the account the effects of in-plane loading, shear deformation, or rotary inertia.

The governing differential equation for anisotropic plates (Table 18-14) is

$$D_x\frac{\partial^4 w}{\partial x^4} + 2B\frac{\partial^4 w}{\partial x^2 \partial y^2} + D_y\frac{\partial^4 w}{\partial y^4} = p - \rho\frac{\partial^2 w}{\partial t^2} \quad (18.17)$$

where

$$B = \tfrac{1}{2}(D_x\nu_y + D_y\nu_x + 4D_{xy})$$

The normal stresses are related to the deflections by

$$\sigma_x = \frac{-E_x z}{1 - \nu_x\nu_y}\left(\frac{\partial^2 w}{\partial x^2} + \nu_y\frac{\partial^2 w}{\partial y^2}\right) \quad (18.18)$$

$$\sigma_y = \frac{-E_y z}{1 - \nu_x\nu_y}\left(\frac{\partial^2 w}{\partial y^2} + \nu_x\frac{\partial^2 w}{\partial x^2}\right) \quad (18.19)$$

The bending and twisting moments per unit length are found from the deflection field through the equations

$$M_x = -D_x\left(\frac{\partial^2 w}{\partial x^2} + \nu_y\frac{\partial^2 w}{\partial y^2}\right) \quad (18.20)$$

$$M_y = -D_y\left(\frac{\partial^2 w}{\partial y^2} + \nu_x\frac{\partial^2 w}{\partial x^2}\right) \quad (18.21)$$

$$M_{xy} = 2D_{xy}\frac{\partial^2 w}{\partial x \partial y} = -M_{yx} \quad (18.22)$$

The transverse shear forces per unit length are expressed as

$$Q_x = \frac{\partial M_x}{\partial x} + \frac{\partial M_{xy}}{\partial y} \qquad (18.23a)$$

$$Q_y = \frac{\partial M_y}{\partial y} - \frac{\partial M_{yx}}{\partial x} \qquad (18.23b)$$

For simple plates

$$Q_x = -D\left(\frac{\partial^3 w}{\partial x^3} + \frac{\partial^3 w}{\partial x \, \partial y^2}\right) \qquad (18.24a)$$

$$Q_y = -D\left(\frac{\partial^3 w}{\partial y \, \partial x^2} + \frac{\partial^3 w}{\partial y^3}\right) \qquad (18.24b)$$

The *equivalent shearing forces* per unit length, which consist of the transverse shear forces plus the rate of change of the twisting moments, are

$$V_x = Q_x - \frac{\partial M_{xy}}{\partial y} \qquad (18.25a)$$

$$V_y = Q_y + \frac{\partial M_{yx}}{\partial x} \qquad (18.25b)$$

For simple plates

$$V_x = -D\left[\frac{\partial^3 w}{\partial x^3} + (2 - \nu)\frac{\partial^3 w}{\partial x \, \partial y^2}\right] \qquad V_y = -D\left[\frac{\partial^3 w}{\partial y^3} + (2 - \nu)\frac{\partial^3 w}{\partial x^2 \, \partial y}\right]$$
$$(18.26)$$

The signs of the quantities in Eqs. (18.16)–(18.26) correspond to sign convention 1 of Fig. 18-3.

Tabulated Formulas

Formulas for the deflection, internal moments, and forces for a variety of edge conditions and transverse loads are provided in Table 18-16.

Example 18.6   Uniform Pressure on a Steel Plate   A uniform pressure $p_1 = 26$ lb/in.^{2} acts on a square steel plate $24 \times 24 \times 1$ in. For this plate, $E = 3.0 \times 10^7$ lb/in.^{2}, $\nu = 0.3$, $L = L_y = 24$ in., and $h = 1$ in. (1) For four simply supported edges, compute the maximum deflection and bending moments per unit length. (2) For four clamped edges, compute the deflection and bending moments per unit length at the plate center. (3) For two opposite sides pinned and two opposite sides free, compute the deflection and the bending moments per unit length at the plate center.

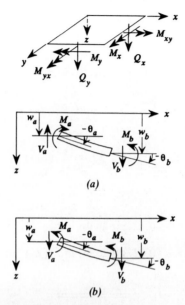

FIGURE 18-3 Sign conventions for rectangular plates; (a) sign convention 1 (transfer matrices); (b) sign convention 2 (stiffness matrices).

1. From case 1 of Table 18-16 for pinned edges, for $\beta = L/L_y = 1.0$. The formula

$$w_{max} = c_1 p_1 L^4 / Eh^3 \quad \text{with } c_1 = 0.0443$$

gives

$$w_{max} = 0.01274 \text{ in.}$$

Also,

$$(M_x)_{max} = (M_y)_{max} = c_2 p_1 L^2 \quad \text{with } c_2 = 0.0479$$

leads to

$$(M_x)_{max} = (M_y)_{max} = 717.35 \text{ lb-in./in.}$$

2. From case 10 of Table 18-16 for fixed edges, with $\alpha = L_y/L = 1$ and $c_1 = 0.0130$

$$w_{center} = c_1 p_1 L^4 / Eh^3 = 0.003738 \text{ in.}$$

and with $c_2 = c_3 = 0.0230$

$$(M_x)_{center} = (M_y)_{center} = c_2 p_1 L^2 = 344.45 \text{ lb-in./in.}$$

3. From case 11 of Table 18-16, with $\beta = L/L_y = 1.0$, for pinned–free edges $c_1 = 0.01309$, and

$$D = Eh^3/\left[12(1 - \nu^2)\right] = (3.0 \times 10^7 \text{ lb/in.}^2)(1.0 \text{ in.}^3)/\left[12\left(1 - (0.3)^2\right)\right]$$

$$= 2.747 \times 10^6 \text{ lb-in.}$$

Then

$$w_{\text{center}} = c_1 p_1 L^4/D = 0.0411 \text{ in.}$$

Also,

$$(M_x)_{\text{center}} = c_3 p_1 L^2 = (0.1225)(26 \text{ lb/in.}^2)(24)^2 \text{ in.}^2 = 1834.6 \text{ lb-in/in.}$$

$$(M_y)_{\text{center}} = c_4 p_1 L^2 = (0.0271)(26 \text{ lb/in.}^2)(24)^2 \text{ in.}^2 = 405.8 \text{ lb-in./in.}$$

Formulas for Plates with Arbitrary Loading

Responses of a plate with all four sides simply supported are given in Table 18-17. The parameters for various loadings are also provided in this table. These formulas are obtained by expanding the deflection w and loading p in the form

$$w = \sum_{n=1}^{\infty} \sum_{m=1}^{\infty} K_{mn} \sin \frac{n\pi x}{L} \sin \frac{m\pi y}{L_y} \tag{18.27}$$

$$p = \sum_{n=1}^{\infty} \sum_{m=1}^{\infty} a_{mn} \sin \frac{n\pi x}{L} \sin \frac{m\pi y}{L_y} \tag{18.28}$$

Substitute Eqs. (18.27) and (18.28) into the governing equation of Eq. (18.16) to obtain the expression for K_{mn}. Then the other responses are obtained from Eqs. (18.17)–(18.26). The convergence of the series in the formulas of Table 18-17 is usually fast for the case of distributed loads. The convergence, however, can be slow for concentrated and discontinuous loads.

Example 18.7 Response of a Simply Supported Plate Find the deflections of a simply supported rectangular plate subjected to a uniformly distributed load p_1. Determine the maximum moments and calculate the edge reactions. The lengths of the plate in the x and y directions are L and $L_y = 2L$, respectively.

First take the parameter a_{mn} for the distributed load from case 1 of Table 18-17:

$$a_{mn} = \frac{16p_1}{\pi^2 mn} \qquad m, n \text{ are odd integers} \tag{1}$$

The constant K_{mn} can then be determined from Table 18-17 as

$$K_{mn} = \frac{16p_1/(\pi^2 mn)}{D\pi^4\left(n^2/L^2 + m^2/L_y^2\right)^2} = \frac{16p_1 L^4}{D\pi^6 nm\left(n^2 + m^2/4\right)^2} \qquad (2)$$

Hence the deflection can be expressed as

$$w = \frac{16p_1 L^4}{D\pi^6} \sum_{n=1}^{\infty} \sum_{m=1}^{\infty} \frac{\sin(n\pi x/L)\sin(m\pi y/2L)}{nm\left(n^2 + m^2/4\right)^2} \qquad n, m = 1, 3, 5, \ldots \quad (3)$$

The maximum deflection occurs at the center, $x = \frac{1}{2}L$, $y = L$. From (3),

$$w_{max} \approx \frac{16p_1 L^4}{D\pi^6}(0.640 - 0.032 - 0.004 + 0.004) = 0.0101 p_1 L^4/D \qquad (4)$$

Note that this deflection series converges rapidly so that the summation of two terms provides accuracy sufficient for practical purposes.

The maximum moments are found in a similar fashion. They too occur at $x = \frac{1}{2}L$, $y = L$. Examination of the moment expressions shows that they converge more slowly than the deflection series. At the center, four terms provide sufficient accuracy. More terms are required as the moments are computed closer to the edges.

The shear forces are determined from response case 5 of Table 18-17. The reactions at the edge can be found from the resulting expression. For example, the reaction force along the $x = 0$ edge is

$$V_{|x=0} = \frac{16p_1 L}{\pi^3} \sum_{n=1}^{\infty} \sum_{m=1}^{\infty} \frac{n^2 + (2-\nu)(m^2/4)}{m\left(n^2 + m^2/4\right)^2}\sin\frac{m\pi y}{2L} \qquad m, n = 1, 3, 5, \ldots$$

$$(5)$$

Buckling Loads

The buckling of a plate subjected to in-plane forces is analogous to elastic buckling of axially loaded slender columns. The differential equation of a statically loaded rectangular plate with in-plane forces is

$$D\nabla^4 w = p + P_x\frac{\partial^2 w}{\partial x^2} + P_y\frac{\partial^2 w}{\partial y^2} + 2P_{xy}\frac{\partial^2 w}{\partial x\,\partial y} \qquad (18.29)$$

where ∇^4 is defined in Eq. (18.16) and P_x, P_y, and P_{xy} are the in-plane forces per unit length. Positive forces are shown in Fig. 18-4. Buckling may occur due to in-plane forces even if no transverse loads act. The expressions for the buckling loads for a variety of rectangular plates are shown in Table 18-18.

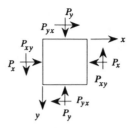

FIGURE 18-4   Positive in-plane forces per unit length P_x, P_y, and P_{xy}

Example 18.8   Critical In-Plane Load of a Steel Plate   Compute the critical in-plane load acting parallel to the x direction for a plate with $L = L_y = 30$ in., $h = 0.3$, $E = 3 \times 10^7$ lb/in.^{2}, and $\nu = 0.3$ when (1) all edges are simply supported and (2) two edges are simply supported and the edges at $x = 0$ and $x = L$ are fixed.

1. Use case 1 of Table 18-18:

$$D = \frac{Eh^3}{12(1 - \nu^2)} = 7.42 \times 10^4 \text{ lb-in.} \qquad \beta = L/L_y = 1 \qquad (1)$$

$$P_{cr} = kP' = (1 + 1)^2 \frac{\pi^2 D}{L_y^2}$$

$$= \frac{4(3.14)^2}{30^2} \times 7.42 \times 10^4 = 3.252 \times 10^3 \text{ lb/in.} \qquad (2)$$

2. From case 4 of Table 18-18, for $\beta = L/L_y = 1$, $k = 6.788$,

$$P_{cr} = kP' = \frac{k\pi^2}{L_y^2} D = 5.518 \times 10^3 \text{ lb/in.} \qquad (3)$$

Local Buckling   Instability is usually considered to be either *primary* or *local*. For example, a tube in compression can fail (1) through primary buckling when it acts like a column and there is an occurrence of an inordinate deflection or (2) by local buckling when the wall collapses at a stress level less than that needed to cause column failure. Other shapes, such as *I-beams*, can fail by the lateral deflection of a compression flange as a column, the local wrinkling of thin flanges, or torsional instability. Many shapes can be considered as being formed of flat plate elements. It is the behavior of these flat elements that is treated in this chapter.

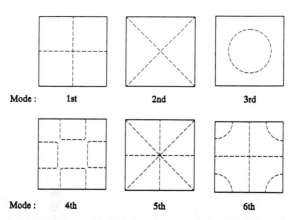

FIGURE 18-5 Mode shapes for a completely free square plate [18.2]. The dashed lines indicate nodal lines for which the displacements in the mode shapes are zero.

Natural Frequencies

The formulas for the natural frequencies of some rectangular plates are listed in Tables 18-19 and 18-20. Some mode shapes for a square plate are shown in Fig. 18-5. Additional data on frequencies and modes can be found in reference 18.2. For complex boundary conditions, approximate methods are used to estimate natural frequencies and mode shapes. Some solution techniques are discussed in Appendix III and others are considered in reference 18.1.

Example 18.9 Natural Frequencies of a Steel Plate Compute the fundamental frequency for a square plate for (1) pinned edges, (2) fixed edges, and (3) two opposite edges pinned and two free. Assume $L = L_y = 24$ in., the mass per unit area $\rho = 7.253 \times 10^{-4}$ lb-s.^{2}/in.^{3}, and $D = 2.747 \times 10^6$ lb-in.
   For this plate $\beta = L/L_y = 1$.

1. For pinned edges use case 1 of Table 18-19 with $m = n = 1$:

$$\lambda_{11} = \pi^2(1 + 1)$$
$$\omega_{11} = (\lambda_{11}/L^2)\sqrt{D/\rho} \qquad (1)$$
$$= 2109 \text{ rad/s} \quad \text{or} \quad f_{11} = \omega_{11}/2\pi = 335.6 \text{ Hz}$$

2. Take ω_i for all edges fixed from case 7 with $\beta = 1$. Then $\lambda_1 = 36.0$ and

$$\omega_1 = (\lambda_1/L^2)\sqrt{D/\rho} = 3846.4 \text{ rad/s} \quad \text{or} \quad f_1 = 612.2 \text{ Hz} \qquad (2)$$

3. Case 3, Table 18-19, provides the formula for two free edges and two pinned edges. For $\beta = 1$, $\lambda_{11} = 9.87$, and

$$\omega_{11} = (\lambda_{11}/L^2)\sqrt{D/\rho} = 1054.5 \text{ rad/s} \quad \text{or} \quad f_{11} = 167.8 \text{ Hz} \qquad (3)$$

General Rectangular Plates

The static and dynamic analysis of rectangular plates can also be performed with transfer matrices and stiffness matrices. For simply supported conditions at the $y = 0$ and $y = L_y$ edges of the plate, the y dependence of the variables w, θ, V, and M can be expressed in a sine series of the form

$$
\begin{bmatrix} w(x, y) \\ \theta(x, y) \\ V(x, y) \\ M(x, y) \end{bmatrix} = \sum_{m=1}^{\infty} \begin{bmatrix} w_m(x) \\ \theta_m(x) \\ V_m(x) \\ M_m(x) \end{bmatrix} \sin \frac{m\pi y}{L_y} \tag{18.30}
$$

The loadings are also expanded in a sine series. The transfer matrix and displacement methods are used to find the x dependence of the state variables, i.e., $w_m(x)$, $\theta_m(x)$, $V_m(x)$, and $M_m(x)$.

For the case of other boundary conditions, at $y = 0$ and $y = L_y$, a series expansion of the state variables w, θ, V, and M and the loadings such as M_{Tx}, M_{Ty}, and p in terms of the eigenfunctions of a vibrating beam with the same end conditions as the edges at $y = 0$ and $y = L_y$ can be assumed:

$$
\begin{bmatrix} w(x, y) \\ \theta(x, y) \\ V(x, y) \\ M(x, y) \end{bmatrix} = \sum_{m=1}^{\infty} \begin{bmatrix} w_m(x) \\ \theta_m(x) \\ V_m(x) \\ M_m(x) \end{bmatrix} \phi_m(y) \tag{18.31}
$$

The functions $\phi_m(y)$ satisfy the boundary conditions and the orthogonality conditions but may not necessarily be the actual deflected shape of the plate along the y direction. However, use of the eigenfunctions of a vibrating beam has been shown to produce accurate results.

The solution procedure using the transfer matrix and displacement method involves selecting a number of y positions at which the state variables $w(x, y)$, $\theta(x, y)$, $V(x, y)$, and $M(x, y)$ are to be computed and the number of terms to be employed in the series expansion of Eqs. (18.30) and (18.31). For each term in the expansion, a set of displacement and force variables $w_m(x)$, $\theta_m(x)$, $V_m(x)$, and $M_m(x)$ is computed according to the standard transfer matrix or displacement method procedure. These solutions are summed as indicated in Eq. (18.30) and (18.31) to give the resultant state variables $w(x, y)$, $\theta(x, y)$, $V(x, y)$, and $M(x, y)$ for a static response. The remaining state variables M_y, M_{xy}, Q_y, etc., can be found from the relationships given in Eqs. (18.17)–(18.21). Similar procedures apply for natural frequency, sinusoidal response, and stability calculations.

Transient dynamic responses can be included by using the techniques of Appendix III to compute $W_m(x, t)$, $\theta_m(x, t)$, $V_m(x, t)$, and $M_m(x, t)$. Then the left-hand sides of Eqs. (18.30) and (18.31) would contain $w(x, y, t)$, $\theta(x, y, t)$, $V(x, y, t)$, and $M(x, y, t)$.

Transfer Matrices Table 18-21 provides the transfer matrices for plates simply supported at $y = 0$ and $y = L_y$. The displacements and forces $w_m(x)$, $\theta_m(x)$, $V_m(x)$, and $M_m(x)$ are computed using the matrices of Table 18-21 and the state variables $w(x, y)$, $\theta(x, y)$, $V(x, y)$, and $M(x, y)$ are taken from the expansion of Eq. (18.30). Transfer matrices for various point occurrences are presented in Table 18-22.

A more general transfer matrix corresponding to the other boundary conditions is listed in Table 18-23.

Stiffness Matrices   Table 18-21 provides the stiffness matrices for rectangular plates simply supported both at $y = 0$ and $y = L_y$. Table 18-22 presents some stiffness matrices for several concentrated occurrences. Table 18-23 contains stiffness matrices for a more general plate with arbitrary boundary conditions at $y = 0$ and $y = L_y$. Note that mass is included in some of the matrices of Tables 18-21 and 18-23, and hence these matrices are dynamic stiffness matrices. Apart from the possible approximations for the y direction expansions, these matrices use exact shape functions and give exact results in the static, dynamic, buckling analyses. See Appendix II and III or reference 18.22.

Mass Matrices   The consistent mass matrix for a rectangular plate simply supported at $y = 0$ and $y = L_y$ is presented in Table 18-24. This matrix, together with the stiffness matrix of Table 18-21 can perform various kinds of dynamic analysis as described in Appendix III.

Large Deflections of Rectangular Plates

In some applications of thin plates, the maximum deflection may be larger than half of the plate thickness ($w \geq \frac{1}{2}h$). In such cases, the large-deflection theory of plates should be used. This theory assumes that the deflections are not small relative to the thickness h but are smaller than the remaining dimensions. The middle surface becomes strained and the corresponding stresses in it cannot be ignored. These stresses, termed diaphragm or membrane stresses, enable the plate to be stiffer than predicted by small-deflection theory. Furthermore, the square of the slope of the deflected surface is no longer negligible in comparison with unity.

Approximate Formulas   For aspect ratios $\beta = L/L_y \geq 0.75$, formulas for approximating deflections and corresponding moments and in-plane forces are given below. The accuracy of these formulas decreases as β becomes smaller. Given the moments and in-plane forces, the stresses can be calculated using

$$\sigma_x = P_x/h + 12 M_x z/h^3$$
$$\sigma_y = P_y/h + 12 M_y z/h^3 \qquad (18.32)$$
$$\tau_{xy} = P_{xy}/h + 12 M_{xy} z/h^3$$

1. *Simply supported with uniformly distributed loading p_1 with extension in x and y directions prevented.* An approximate value for the maximum deflection w_{max} can be obtained from solution of the cubic equation [18.11]

$$\frac{16 p_1 L^4}{\pi^6 D} = w_{max}(1 + \beta^2)^2 + \frac{3 w_{max}^3}{4h^2}\left[(3 - \nu^2)(1 + \beta^4) + 4\nu\beta^2\right] \quad (18.33)$$

where $\beta = L/L_y$.

Approximations to the bending moments and the tensile forces in the xy plane at the center of the plate are given by

$$M_{xc} = \left(\pi^2 D w_{max}/L^2\right)\left(1 + \nu\beta^2\right) \tag{18.34a}$$

$$M_{yc} = \left(\pi^2 D w_{max}/L^2\right)\left(\beta^2 + \nu\right) \tag{18.34b}$$

$$P_x = \left[\pi^2 E h w_{max}^2/[8(1 - \nu^2)L^2]\right]\left[(2 - \nu^2) + \nu\beta^2\right] \tag{18.34c}$$

$$P_y = \left[\pi^2 E h w_{max}^2/[8(1 - \nu^2)L^2]\right]\left[(2 - \nu^2)\beta^2 + \nu\right] \tag{18.34d}$$

2. *Simply supported plate with uniformly distributed loading p_1 with extension in x and y directions not prevented*:

$$\frac{16 p_1 L^4}{\pi^6 D} = w_{max}\left(1 + \beta^2\right)^2 + \frac{3.88\beta^2(1 - \nu^2)w_{max}^3}{\left(\beta^2 + 0.6 + 1/\beta^2\right)h^2} \tag{18.35}$$

$$M_{xc} = \left(\pi^2 D w_{max}/L^2\right)\left(1 + \nu\beta^2\right) \tag{18.36a}$$

$$M_{yc} = \left(\pi^2 D w_{max}/L^2\right)\left(\beta^2 + \nu\right) \tag{18.36b}$$

$$P_x = \frac{2.76 E h w_{max}^2}{\left(\beta^2 + 0.6 + 1/\beta^2\right)L^2} \tag{18.36c}$$

$$P_y = \frac{2.76 E h w_{max}^2}{\left(\beta^2 + 0.6 + 1/\beta^2\right)L_y^2} \tag{18.36d}$$

Example 18.10 Rectangular Plate with Large Deflection A rectangular steel plate with $L = L_y = 36$ in., $h = 0.3$ in., is subjected to a uniform load of $p_1 = 10$ lb/in.^{2}. Determine the maximum deflection and stress in the plate for the simply supported boundary conditions that allow rotation but no displacements at the edge. Also, $E = 3 \times 10^7$ lb/in.^{2}, $\nu = 0.316$.
    We find

$$D = \frac{E h^3}{12(1 - \nu^2)} = 7.5 \times 10^4 \text{ in.} \tag{1}$$

$$\beta = \frac{L}{L_y} = \frac{36}{36} = 1 \tag{}$$

Substitution of these results into Eq. (18.33) yields

$$4 w_{max} + 58.867 w_{max}^3 = 3.727 \tag{2}$$

The solution to this equation is

$$w_{max} \approx 0.342 \text{ in.} \tag{3}$$

The bending moments and tensile forces in the xy plane at the center of the plate are, by Eqs. (18.34a) and (18.34c).

$$M_{xc} = \frac{7.5(10^4)\pi^2(0.342)}{(36)^2}(1 + 0.316 \times 1) = 257.06 \text{ lb.in./in.}$$

$$= M_{yc} \tag{4}$$

$$P_x = P_y = \frac{\pi^2(3 \times 10^7)(0.30)(0.342^2)}{8(1 - 0.316^2)(36)^2}\left[(2 - 0.316^2) + 0.316 \times 1\right]$$

$$= 0.2467 \times 10^4 \text{ lb/in.} \tag{5}$$

Use of Eqs. (18.32), with $z = \frac{1}{2}h$, gives

$$\sigma_x = \sigma_y = \frac{0.2467 \times 10^4}{0.30} + \frac{12(257.06)}{(0.30)^2(2)} = 25{,}360.6 \text{ lb/in.}^2 \tag{6}$$

If we use the small-deflection theory to calculate the stresses (Table 18-16, case 1),

$$w_{max} = c_1\frac{p_1 L^4}{Eh^3} = 0.0443\frac{10 \times 36^4}{3 \times 10^7(0.3)^3} = 0.92 \text{ in.}$$

$$M_{max} = c_2 p_1 L^2 = 0.0479(10)(36)^2 = 620.784 \text{ lb-in./in.}$$

The stresses $\sigma_x = \sigma_y = \sigma_{max}$ would be

$$\sigma_{max} = \frac{12 M_{max} h}{2h^3} = \frac{6 M_{max}}{h^2}$$

$$= \frac{6(620.784)}{(0.3)^2} = 41{,}385.6 \text{ lb/in.}^2 \tag{7}$$

This value is about 1.6 times the stress obtained by large-deflection theory (6). Therefore, caution must be taken in utilizing the small- or large-deflection theory to calculate the stresses in a plate.

18.4 OTHER PLATES

Responses, buckling loads, and natural frequencies of plates of various shapes are given in Tables 18-25 to 18-27.

REFERENCES

18.1 Szilard, R., *Theory and Analysis of Plates*, Prentice-Hall, Englewood Cliffs, NJ, 1974.

18.2 Leissa, A. W., *Vibration of Plates*, NASA SP-160, Washington DC, 1969.

18.3 Majumdar, S., "Buckling of a Thin Annular Plate under Uniform Compression," *AIAA J.* Vol. 9, No. 9, 1971, pp. 1701–1707.

18.4   Column Research Committee of Japan (Ed.), *Handbook of Structural Stability*, Corona, Tokyo, 1971.

18.5   Nadai, A., "Über das Ausbeulen von Kreisformigen Platten," *ZVDI*, Bd. 59, 1915, pp. 169–175.

18.6   Sezawa, K., "The Stability of Thin Plates," *J. Soc. Naval Arch. Jpn*, Vol. 38, 1926, pp. 79–108.

18.7   Iwato, S., Ban, T., and Suzuki, M., "The Approximate Calculation of the Buckling Load of a Thin Circular Plate," the Memorial Number in Commemoration of the 50th Anniversary of Founding of JSME, Vol. 14, 1939, pp. 1–64.

18.8   Bryan, G. H., "On the Stability of a Plane Plate under Thrust in Its Own Plane with Application to the 'Buckling of the Side of A Ship,'" *Proc. Lond. Math. Soc.*, Vol. 22, 1891, pp. 54–67.

18.9   Kerr, A. D., "On the Instability of Circular Plates," *J. Aeronaut. Sci.*, Vol. 29, No. 4, 1962, pp. 486–487.

18.10 Han, S. and Pilkey, W. D., "Stiffness Matrices for the Static, Dynamic and Buckling Analysis of Circular Plates," *Finite Elem. Anal. Des.*, Vol. 7, No. 1, 1990, pp. 27–50.

18.11 Bares, R., *Tables for the Analysis of Plates, Slabs and Diaphragms Based on the Elastic Theory*, 2nd ed., Bauverlag, Wiesbaden, 1971 (German–English edition).

18.12 Timoshenko, S., and Woinowsky-Krieger, S., *Theory of Plates and Shells*, 2nd ed., McGraw-Hill, New York, 1959.

18.13 Lekhnitskii, S. G., *Anisotropic Plates*, Gordon & Breach, New York, 1968 (translated from the 2nd Russian edition).

18.14 Prescott, J., *Applied Elasticity*, Longmans, Green, 1924.

18.15 Blevins, R. D., *Formulas for Natural Frequency and Mode Shape*, Van Nostrand, New York, 1979.

18.16 Cook, I. T., and Parsons, H. W., "The Buckling of a Reinforced Circular Plate under Uniform Radial Thrust," *Aero. Quart.*, Vol. 12, No. 4, 1961, pp. 337–342.

18.17 Cox, H. L., "The Buckling of Thin Plates in Compression," Reports and Memoranda, Aero. Res. Coun., H. M. Stat. Off., No. 1554, 1933.

18.18 Timoshenko, S., "Einige Stabilitäts Probleme der Elastizitäts Theorie," *Z. f. Math. u. Phys.*, Bd. 58, 1910, p. 337.

18.19 Woinowsky-Krieger, S., "Berechnung der ringsum frei aufliegenden Dreiecksplatten", *Ing. Arohiv.*, Bd. 4, 1933, pp. 254–262.

18.20 Wakasugi, S., "Buckling of Right Angled Isosceles Triangular Plates Simply Supported at the Boundaries," *Trans. Japan Soc. Mech. Eng.* Vol. 19, 1953, pp. 59–65.

18.21 Burchard, W., "Beulspannungen der quadratichen Platte Mit Schragsteife unter Druck bzw. Schub," *Ing. Archiv*, Bd. 8, 1937, pp. 332–348.

18.22 Pilkey, W., and Wunderlich, W., *Mechanics of Structures, Variational and Computational Methods*, CRC, Boca Raton, FL, 1993.

Tables

TABLE 18-1 MATERIAL PROPERTIES FOR CIRCULAR PLATES

Notation

ν_r, ν_ϕ = Poisson's ratio in r and ϕ directions; for isotropic materials, $\nu_r = \nu_\phi = \nu$

E_r, E_ϕ = modulus of elasticity in r and ϕ directions; for isotropic materials, $E_r = E_\phi = E$

$D_r, D_\phi, D_{r\phi}$ = flexural rigidities

K_r, K_ϕ = extensional rigidities in r and ϕ directions

α_r, α_ϕ = thermal expansion coefficients in r and ϕ directions

h = thickness of plate

G = shear modulus of elasticity, same value for isotropic and orthotropic materials

Plate	Constants
1. Homogeneous isotropic	$\nu_r = \nu_\phi = \nu$ $D_r = D_\phi = D = Eh^3/\left[12(1 - \nu^2)\right]$ $G = E/[2(1 + \nu)]$ $D_{r\phi} = \frac{1}{2}D(1 - \nu) = \frac{1}{12}Gh^3$ $K_r = K_\phi = K = Eh/(1 - \nu^2)$ $\alpha_r = \alpha_\phi = \alpha$
2. Homogeneous orthotropic	$D_j = E_j h^3/\left[12(1 - \nu_r\nu_\phi)\right] \qquad j = r, \phi$ $K_j = E_j h/(1 - \nu_r\nu_\phi)$ $D_{r\phi} = \frac{1}{12}Gh^3$
3. Continuously composite, isotropic	$D_r = D_\phi = D = \displaystyle\int_{-h/2}^{h/2} \frac{Ez^2}{1 - \nu^2}\, dz$ $D_{r\phi} = \displaystyle\int_{-h/2}^{h/2} \frac{Ez^2}{2(1 + \nu)}\, dz$ $K_r = K_\phi = K = \displaystyle\int_{-h/2}^{h/2} \frac{E}{1 - \nu^2}\, dz$ $\alpha_r = \alpha_\phi = \alpha$

TABLE 18-1 (continued)   **MATERIAL PROPERTIES FOR CIRCULAR PLATES**

Plate	Constants
4. Continuously composite, orthotropic	$$D_j = \frac{1}{1 - \nu_r \nu_\phi} \int_{-h/2}^{h/2} \frac{E_j z^2}{1 - \nu_r \nu_\phi} \, dz \qquad j = r, \phi$$ $$K_j = \int_{-h/2}^{h/2} \frac{E_j}{1 - \nu_r \nu_\phi} \, dz$$ $$D_{r\phi} = \int_{-h/2}^{h/2} G z^2 \, dz$$
5. Layered	$$D_j = 2 \sum_i \left(\frac{E_j}{1 - \nu_r \nu_\phi} \right)_i \int_{\Delta h_i} z^2 \, dz \qquad j = r, \phi$$ $$K_j = 2 \sum_i \left(\frac{E_j}{1 - \nu_r \nu_\phi} \right)_i \Delta h_i$$ $$D_{r\phi} = 2 \sum_i (G)_i \int_{\Delta h_i} z^2 \, dz$$ Δh_i = thickness of ith layer The summation extends over half of the plate thickness.

TABLE 18-1 Material Properties for Circular Plates   **984**

TABLE 18-2 STRESSES OF CIRCULAR PLATES

Notation

M_r, M_ϕ = bending moments per unit length
$M_{r\phi}$ = twisting moment per unit length
P_r, P_ϕ = in-plane forces per unit length
T = temperature change
σ_r = radial normal stress (F/L^2)
σ_ϕ = circumferential normal stress (F/L^2)
$\tau_{r\phi}$ = shear stresses
ν_r, ν_ϕ = Poisson's ratio in r and ϕ directions. For isotropic materials, $\nu_r = \nu_\phi = \nu$
E_r, E_ϕ = modulus of elasticity in r and ϕ directions. For isotropic materials, $E_r = E_\phi = E$
$D_r, D_\phi, D_{r\phi}$ = flexural rigidities
K_r, K_ϕ = extensional rigidites in r and ϕ directions
α_r, α_ϕ = thermal expansion coefficients in r and ϕ directions
h = thickness of the plate
G = shear modulus of elasticity, $G_{r\phi} = \frac{1}{12}h^2 G$

Plate	Stresses
1. Homogeneous isotropic material	$\sigma_r = \dfrac{M_r z}{h^3/12}, \quad \sigma_\phi = \dfrac{M_\phi z}{h^3/12}, \quad \tau_{r\phi} = -\dfrac{z M_{r\phi}(1-\nu^2)}{h^3/12}$
2. Other materials; defined in cases 2 and 3 of Table 18-1	$\sigma_r = \bar{E}_r\left[-\dfrac{\bar{P}}{K_r} + \dfrac{\bar{M}_r z}{D_r} - (\alpha_r + \nu_\phi \alpha_\phi)T\right]$ $\sigma_\phi = \bar{E}_\phi\left[-\dfrac{\bar{P}_\phi}{K_\phi} + \dfrac{\bar{M}_\phi z}{D_\phi} - (\alpha_\phi + \nu_r \alpha_r)T\right]$ $\tau_{r\phi} = -G_{r\phi}\dfrac{M_{r\phi} z}{D_{r\phi}}$ where $\bar{P} = P_r + P_T \qquad \bar{P}_\phi = P_\phi + P_{T\phi}$ $\bar{M}_r = M_r + M_{Tr} \qquad \bar{M}_\phi = M_\phi + M_{T\phi}$ $P_T = \displaystyle\int_{-h/2}^{h/2}\dfrac{E_r(\alpha_r + \nu_\phi \alpha_\phi)}{1-\nu_r\nu_\phi}T\,dz, \quad M_{Tr} = \int_{-h/2}^{h/2}\dfrac{E_r(\alpha_r + \nu_\phi \alpha_\phi)}{1-\nu_r\nu_\phi}Tz\,dz$ $P_{T\phi} = \displaystyle\int_{-h/2}^{h/2}\dfrac{E_\phi(\alpha_\phi + \nu_r \alpha_r)}{1-\nu_r\nu_\phi}T\,dz, \quad M_{T\phi} = \int_{-h/2}^{h/2}\dfrac{E_\phi(\alpha_\phi + \nu_r \alpha_r)}{1-\nu_r\nu_\phi}Tz\,dz$ $\bar{E}_r = \dfrac{E_r}{1-\nu_r\nu_\phi} \qquad \bar{E}_\phi = \dfrac{E_\phi}{1-\nu_r\nu_\phi}$ and $\bar{P}$ and $\bar{P}_\phi$ are the in-plane compression forces. If the forces are tensile, replace $\bar{P}$ and $\bar{P}_\phi$ by $-\bar{P}$ and $-\bar{P}_\phi$.

TABLE 18-3 DEFLECTIONS AND INTERNAL FORCES FOR CIRCULAR PLATES WITH AXIALLY SYMMETRIC LOADS AND BOUNDARY CONDITIONS

Notation

w = deflection
M_r, M_ϕ = bending moments per unit length
r = radial coordinate
a_L = radius of outer boundary
$\alpha = r/a_L$
$D = Eh^3/[12(1 - \nu^2)]$

Q_r = transverse shear force per unit length
ν = Poisson's ratio
a_1 = radial location of loading
$\beta = a_1/a_L$
p, p_1 = distributed loading (F/L^2)

Structural System and Static Loading	Deflection and Internal Forces
1. 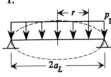	$$w = \frac{p_1 a_L^4}{64D}\left(\frac{5+\nu}{1+\nu} - \frac{6+2\nu}{1+\nu}\alpha^2 + \alpha^4\right)$$ $M_r = \frac{1}{16}p_1 a_L^2(3+\nu)(1-\alpha^2)$ $M_\phi = \frac{1}{16}p_1 a_L^2[3+\nu-(1+3\nu)\alpha^2]$ $Q_r = -\frac{1}{2}p_1 a_L \alpha$
2. 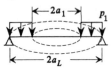	If $\alpha \leqq \beta$, $$w = \frac{p_1 a_L^4}{64D(1+\nu)}(C_1 - 2C_2\alpha^2)$$ $M_r = M_\phi = \frac{1}{16}p_1 a_L^2 C_2, \quad Q_r = 0$ where $C_1 = 5+\nu-4(3+\nu)\beta^2 + (7+3\nu)\beta^4$ $\qquad -4(1+\nu)\beta^4 \ln\beta$ $C_2 = 3+\nu-4\beta^2 + (1-\nu)\beta^4 + 4(1+\nu)\beta^2 \ln\beta$ If $\alpha \geqq \beta$, $$w = \frac{p_1 a_L^4}{32D(1+\nu)}\left[(3+\nu)(1-2\beta^2)-(1-\nu)\beta^4\right]$$ $\times(1-\alpha^2) - \frac{p_1 a_L^4}{64D}(1-\alpha^4 + 8\alpha^2\beta^2 \ln\alpha + 4\beta^4 \ln\alpha)$ $M_r = \frac{1}{16}p_1 a_L^2\left[\left(3+\nu-\frac{1-\nu}{\alpha^2}\cdot\beta^4\right)(1-\alpha^2)\right.$ $\qquad\qquad\left. + 4(1+\nu)\beta^2 \ln\alpha\right]$ $M_\phi = \frac{1}{16}p_1 a_L^2\left[\left(1+3\nu+\frac{1-\nu}{\alpha^2}\cdot\beta^4\right)(1-\alpha^2)\right.$ $\qquad\qquad\left. + 4(1+\nu)\beta^2 \ln\alpha + 2(1-\nu)(1-\beta^2)^2\right]$ $Q_r = -\frac{1}{2}p_1 a_1\left(\alpha - \frac{\beta^2}{\alpha}\right)$

TABLE 18-3 Deflections and Internal Forces for Circular Plates   **986**

Structural System and Static Loading	Deflection and Internal Forces
3. p = Maximum value of distributed load	$$w = \frac{7pa_L^4}{240(1+\nu)D}(1-\alpha^2)$$ $$+ \frac{pa_L^4}{14,400\,D}(129 - 290\alpha^2 + 225\alpha^4 - 64\alpha^5)$$ $$(M_r)_{\alpha=0} = (M_\phi)_{\alpha=0} = \tfrac{1}{720}pa_L^2(71 + 29\nu)$$ $$(Q_r)_{\alpha=1} = -\tfrac{1}{6}pa_L$$
4. 	$$w = \frac{pa_L^4}{24(1+\nu)D}(1-\alpha^2)$$ $$+ \frac{pa_L^4}{576D}(7 - 15\alpha^2 + 9\alpha^4 - \alpha^6)$$ $$M_r = \tfrac{1}{96}pa_L^2\left[13 + 5\nu - 6(3+\nu)\alpha^2 + (5+\nu)\alpha^4\right]$$ $$M_\phi = \tfrac{1}{96}pa_L^2\left[13 + 5\nu - 6(1+3\nu)\alpha^2 + (1+5\nu)\alpha^4\right]$$ $$Q_r = -\tfrac{1}{6}pa_L\alpha(2-\alpha^2)$$
5. Concentrated force: Units of total load W_T: force 	$$w = \frac{W_T a_L^2}{16\pi D}\frac{(3+\nu)}{(1+\nu)}(1-\alpha^2) + \frac{W_T a_L^2}{8\pi D}\alpha^2 \ln\alpha$$ $$M_r = -\frac{W_T}{4\pi}(1+\nu)\ln\alpha$$ $$M_\phi = M_r + \frac{W_T}{4\pi}(1-\nu)$$ $$Q_r = -\frac{W_T}{2\pi a_L\alpha}$$
6. Concentrated force applied on circle: Units of W: force/length 	If $\alpha \leq \beta$, $$w = \frac{Wa_L^2 a_1}{8D}(C_1 - C_2\alpha^2)$$ $$M_r = M_\phi = \tfrac{1}{4}Wa_1(1+\nu)C_2$$ where $$C_1 = \frac{3+\nu}{1+\nu}(1-\beta^2) + 2\beta^2\ln\beta$$ $$C_2 = \frac{1-\nu}{1+\nu}(1-\beta^2) - 2\ln\beta$$ $$Q_r = 0$$

Structural System and Static Loading	Deflection and Internal Forces
6. Continued	If $\alpha \geq \beta$, $$w = \frac{W a_L^2 a_1}{8D(1 + \nu)}\left(C_3 - C_3\alpha^2 + 2\beta^2 \ln \alpha + 2\alpha^2 \ln \alpha\right)$$ $$C_3 = \frac{3 + \nu}{1 + \nu} - \frac{1 - \nu}{1 + \nu}\beta^2$$ $$M_r = \tfrac{1}{4}\frac{W a_1}{\alpha^2}\left[(1 - \nu)(1 - \alpha^2)\beta^2 - 2(1 + \nu)\alpha^2 \ln \alpha\right]$$ $$M_\phi = \tfrac{1}{4}\frac{W a_1}{\alpha^2}$$ $$\times\left[2(1 - \nu)\alpha^2 - (1 - \nu)(1 + \alpha^2)\beta^2 - 2(1 + \nu)\alpha^2 \ln \alpha\right]$$ $$Q_r = -\frac{\beta}{\alpha}W$$
7. Concentrated moment applied on circle: Units of C: force $\times$ length/ length	$$w = \frac{C a_L^2}{2D(1 + \nu)}(1 - \alpha^2)$$ $$M_r = M_\phi = C$$ $$Q_r = 0$$
8.	If $\alpha \leq \beta$, $$w = \frac{C a_L^2}{4D}\left[2\left(\frac{1}{1 + \nu} - \ln \beta\right)\beta^2 - \left(1 + \frac{1 - \nu}{1 + \nu}\beta^2\right)\alpha^2\right]$$ $$M_r = M_\phi = \frac{C}{2}\left[1 + \nu + (1 - \nu)\beta^2\right]$$ $$Q_r = 0$$ If $\alpha \geq \beta$, $$w = \frac{C a_L^2}{D}\beta^2\left[\frac{1 - \nu}{1 + \nu}(1 - \alpha^2) - 2\ln \alpha\right]$$ $$M_r = -M_\phi = \frac{C}{2}(1 - \nu)(1 - \alpha^2)\frac{\beta^2}{\alpha^2}$$ $$Q_r = 0$$

TABLE 18-3 Deflections and Internal Forces for Circular Plates   **988**

Structural System and Static Loading	Deflection and Internal Forces
9. 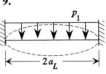	$$w = \frac{p_1 a_L^4}{64D}(1 - \alpha^2)^2$$ $$M_r = \tfrac{1}{16}p_1 a_L^2\left[1 + \nu - (3 + \nu)\alpha^2\right]$$ $$M_\phi = \tfrac{1}{16}p_1 a_L^2\left[1 + \nu - (1 + 3\nu)\alpha^2\right]$$ $$Q_r = -\tfrac{1}{2}p_1 a_L \alpha$$
10. 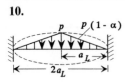	$$w = \frac{p a_L^4}{14{,}400D}(129 - 290\alpha^2 + 225\alpha^4 - 64\alpha^5)$$ $$(M_r)_{\alpha=0} = (M_\phi)_{\alpha=0} = \frac{29 p a_L^2}{720}(1 + \nu)$$ $$(Q_r)_{\alpha=1} = -\tfrac{1}{6}p a_L \quad (M_r)_{\alpha=1} = (M_\phi)_{\alpha=1} = -\frac{7 p a_L^2}{120}$$
11. 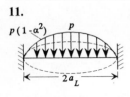	$$w = \frac{p a_L^4}{576D}(7 - 15\alpha^2 + 9\alpha^4 - \alpha^6)$$ $$M_r = \tfrac{1}{96}p a_L^2\left[5(1 + \nu) - 6(3 + \nu)\alpha^2 + (5 + \nu)\alpha^4\right]$$ $$M_\phi = \tfrac{1}{96}p a_L^2\left[5(1 + \nu) - 6(1 + 3\nu)\alpha^2 + (1 + 5\nu)\alpha^4\right]$$ $$Q_r = -\tfrac{1}{4}p a_L(2\alpha - \alpha^3)$$
12. 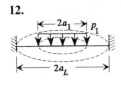	If $\alpha \leq \beta$, $$w = \frac{p_1 a_L^4}{64D}\left[C_1 + 2C_2(1 - \alpha^2) + \alpha^4\right]$$ where $$C_1 = 4\beta^2 - 5\beta^4 + 8\beta^2 \ln \beta + 4\beta^4 \ln \beta$$ $$C_2 = \beta^2(\beta^2 - 4 \ln \beta)$$ $$M_r = \frac{p_1 a_L^2}{16}\left[(1 + \nu)(\beta^4 - 4\beta^2 \ln \beta) - (3 + \nu)\alpha^2\right]$$ $$M_\phi = \frac{p_1 a_L^2}{16}\left[(1 + \nu)(\beta^4 - 4\beta^2 \ln \beta) - (1 + 3\nu)\alpha^2\right]$$ $$Q_r = -\frac{p_1 a_1}{2}\frac{\alpha}{\beta}$$

Structural System and Static Loading	Deflection and Internal Forces
12. Continued	If $\alpha \geq \beta$, $$w = \frac{p_1 a_L^2 a_1^2}{32D}\left[2 - 2\alpha^2 + \beta^2(1 - \alpha^2 + 2\ln\alpha) + 4\alpha^2\ln\alpha\right]$$ $$(M_r)_{\alpha=1} = -\tfrac{1}{8}p_1 a_1^2(2 - \beta^2)$$ $$(M_\phi)_{\alpha=1} = \nu(M_r)_{\alpha=1}$$ $$Q_r = -\frac{p_1 a_1}{2}\frac{\beta}{\alpha}$$
13.	If $\alpha \leq \beta$, $$w = \frac{Wa_L^2 a_1}{8D}(C_1 - C_2\alpha^2)$$ $$M_r = M_\phi = \tfrac{1}{4}Wa_1(1 + \nu)C_2$$ $$Q_r = 0$$ where $$C_1 = (1 - \beta^2) + 2\beta^2\ln\beta$$ $$C_2 = (\beta^2 - 1) - 2\ln\beta$$ If $\alpha \geq \beta$, $$w = \frac{Wa_L^2 a_1}{8D}\left[(1 + \beta^2)(1 - \alpha^2) + 2\beta\ln\alpha + 2\alpha^2\ln\alpha\right]$$ $$(M_r)_{\alpha=1} = -\tfrac{1}{2}W(1 - \beta^2) \qquad (M_\phi)_{\alpha=1} = \nu(M_r)_{\alpha=1}$$ $$Q_r = -\frac{W_T\beta}{\alpha}$$
14.	If $\alpha \leq \beta$, $$w = -\frac{Ca_L^2}{4D}\left[2\beta^2\ln\beta + (1 - \beta^2)\alpha^2\right]$$ $$M_r = M_\phi = \frac{C}{2}(1 + \nu)(1 - \beta^2)$$ If $\alpha \geq \beta$, $$w = -\frac{Ca_L^2}{4D}\beta^2(1 - \alpha^2 + 2\ln\alpha)$$ $$(M_r)_{\alpha=1} = -C\beta^2 \qquad (M_\phi)_{\alpha=1} = -\nu C\beta^2$$

TABLE 18-3   **Deflections and Internal Forces for Circular Plates**   **990**

TABLE 18-4, PART A   CIRCULAR PLATES WITH ARBITRARY LOADING: GENERAL RESPONSE EXPRESSIONS

Notation

w = deflection
θ = slope
V = shear force, $= V_r$
M = bending moment, $= M_r$
w_0, θ_0, V_0, M_0 = initial parameters for w, θ, V, and M
F_w, F_θ, F_V, F_M = loading functions defined in part B
p, p_1, p_2 = distributed applied loading in force/length^2
R = reaction force at center support defined in part C
C_1 = integration constant defined in part C
ν = Poisson's ratio
E = modulus of elasticity
h = thickness of plate
$D = Eh^3/[12(1 - \nu^2)]$

Plate[a]	Response Expressions
1. Plate without center hole	$w = w_0 - M_0 \dfrac{r^2}{2D(1 + \nu)} + F_w \qquad \theta = M_0 \dfrac{r}{D(1 + \nu)} + F_\theta$ $V = F_V \qquad M = M_0 + F_M$
2. Plate with rigid support at center	$w = -\dfrac{R}{8\pi D} r^2(\ln r - 1) + C_1 \dfrac{r^2}{4} + F_w$ $\theta = \dfrac{R}{4\pi D} r\left(\ln r - \dfrac{1}{2}\right) - C_1 \dfrac{r}{2} + F_\theta$ $V = \dfrac{R}{2\pi r} + F_V$ $M = \dfrac{R}{4\pi}\left[(1 + \nu)\ln r + \dfrac{1 - \nu}{2}\right] - \dfrac{C_1 D}{2}(1 + \nu) + F_M$

Plate	Response Expressions
3. Plate with center hole 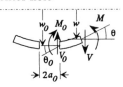	$w = w_0 + \theta_0 \left[-\dfrac{1}{2}(1+\nu)a_0 \ln \dfrac{r}{a_0} - (1-\nu)\dfrac{r^2 - a_0^2}{4a_0} \right]$ $\quad + M_0 \left(\dfrac{a_0^2}{2D} \ln \dfrac{r}{a_0} - \dfrac{r^2 - a_0^2}{4D} \right)$ $\quad + V_0 \dfrac{a_0}{4D} \left[-(a_0^2 + r^2)\ln \dfrac{r}{a_0} + (r^2 - a_0^2) \right] + F_w$ $\quad = w_0 U_{ww} + \theta_0 U_{w\theta} + M_0 U_{wM} + V_0 U_{wV} + F_w$ $\theta = \theta_0 \left[(1+\nu)\dfrac{a_0}{2r} + (1+\nu)\dfrac{r}{2a_0} \right]$ $\quad + M_0 \dfrac{1}{2Dr}(r^2 - a_0^2)$ $\quad + V_0 \left[\dfrac{a_0 r}{2D} \ln \dfrac{r}{a_0} - \dfrac{a_0}{4Dr}(r^2 - a_0^2) \right] + F_\theta$ $\quad = \theta_0 U_{\theta\theta} + M_0 U_{\theta M} + V_0 U_{\theta V} + F_\theta$ Note that $U_{\theta w} = 0$. $V = V_0 \dfrac{a_0}{r} + F_V = V_0 U_{VV} + F_V$ Note that $U_{Vw} = U_{V\theta} = U_{VM} = 0$. $M = \theta_0(1-\nu^2)\dfrac{Da_0}{2}\left(\dfrac{1}{a_0^2} - \dfrac{1}{r^2} \right) + M_0 \left(\dfrac{1-\nu}{2}\dfrac{a_0^2}{r^2} + \dfrac{1+\nu}{2} \right)$ $\quad + V_0 \dfrac{a_0}{2} \left[(1+\nu)\ln \dfrac{r}{a_0} + (1-\nu)\dfrac{r^2 - a_0^2}{2r^2} \right] + F_M$ $\quad = \theta_0 U_{M\theta} + V_0 U_{MV} + M_0 U_{MM} + F_M$ Note that $U_{Mw} = 0$. U_{ij} are transfer matrix elements

[a] Responses use the sign conventions in the figures. (Sign convention 1 of Appendix II).

TABLE 18-4, Part A   Circular Plates: Response Expressions   992

TABLE 18-4, PART B   CIRCULAR PLATES WITH ARBITRARY LOADING: LOADING FUNCTIONS

$$\text{Loading Functions } F_w(r), F_\theta(r), F_V(r), F_M(r) \qquad \langle r - a_i \rangle^0 = \begin{cases} 0 & \text{if } r < a_i \\ 1 & \text{if } r \geq a_i \end{cases}$$

	Concentrated Force (applies for plates with no center hole) W_T (force)	Concentrated Line Force W (force/length)	Uniform Loading	Ramp Loading $\Delta\ell = a_2 - a_1$	$\Delta\ell = a_2 - a_1$
$F_w(r)$	$\dfrac{W_T}{8\pi D} r^2(\ln r - 1)$	$\langle r - a \rangle^0 \dfrac{Wa}{4D}\left[(r^2+a^2)\ln\dfrac{r}{a} - (r^2 - a^2)\right]$	$\dfrac{p_1}{8D}\left[F_{wp_1}(r,a_1) - F_{wp_1}(r,a_2)\right]$	$\dfrac{p_2}{D\Delta\ell}\left[F_{wp_2}(r,a_1) - F_{wp_2}(r,a_2)\right] - p_2 F_{wp_1}(r,a_2)$	$-\dfrac{p_2}{D\Delta\ell}\left[F_{wp_2}(r,a_1) - F_{wp_2}(r,a_2)\right] + p_2 F_{wp_1}(r,a_1)$
$F_\theta(r)$	$-\dfrac{W_T}{4\pi D} r\left(\ln r - \tfrac{1}{2}\right)$	$-\langle r - a\rangle^0 \dfrac{Wa}{2D} \times \left[r\ln\dfrac{r}{a} - \dfrac{1}{2r}(r^2 - a^2)\right]$	$-\dfrac{p_1}{4D}\left[F_{\theta p_1}(r,a_1) - F_{\theta p_1}(r,a_2)\right]$	$-\dfrac{p_2}{D\Delta\ell}\left[F_{\theta p_2}(r,a_1) - F_{\theta p_2}(r,a_2)\right] - p_2 F_{\theta p_1}(r,a_2)$	$\dfrac{p_2}{D\Delta\ell}\left[F_{\theta p_2}(r,a_1) - F_{\theta p_2}(r,a_2)\right] + p_2 F_{\theta p_1}(r,a_1)$
$F_V(r)$	$-\dfrac{W_T}{2\pi r}$	$-\langle r-a\rangle^0 \dfrac{Wa}{r}$	$-\dfrac{p_1}{2}\left[F_{Vp_1}(r,a_1) - F_{Vp_1}(r,a_2)\right]$	$-\dfrac{p_2}{\Delta\ell}\left[F_{Vp_2}(r,a_1) - F_{Vp_2}(r,a_2)\right] - p_2 F_{Vp_1}(r,a_2)$	$\dfrac{p_2}{\Delta\ell}\left[F_{Vp_2}(r,a_1) - F_{Vp_2}(r,a_2)\right] + p_2 F_{Vp_1}(r,a_1)$
$F_M(r)$	$-\dfrac{W_T}{4\pi}\left[(1+v)\ln r + \dfrac{1-v}{2}\right]$	$-\langle r-a\rangle^0 \dfrac{Wa}{2}\left[(1+v)\ln\dfrac{r}{a} + \dfrac{1-v}{2}\left(1 - \dfrac{a^2}{r^2}\right)\right]$	$-\dfrac{p_1}{4}\left[F_{Mp_1}(r,a_1) - F_{Mp_1}(r,a_2)\right]$	$-\dfrac{p_2}{\Delta\ell}\left[F_{Mp_2}(r,a_1) - F_{Mp_2}(r,a_2)\right] - p_2 F_{Mp_1}(r,a_2)$	$\dfrac{p_2}{\Delta\ell}\left[F_{Mp_2}(r,a_1) - F_{Mp_2}(r,a_2)\right] + p_2 F_{Mp_1}(r,a_1)$

TABLE 18-4, PART B (continued)   CIRCULAR PLATES WITH ARBITRARY LOADING: LOADING FUNCTIONS

	Concentrated Ring Moment (force − length/length)	Arbitrary Loading	Loading Functions
$F_w(r)$	$-\langle r-a\rangle^0\,\dfrac{C}{2D}\left(a^2\ln\dfrac{r}{a}-\dfrac{r^2-a^2}{2}\right)$	$\displaystyle\int_{a_1}^r\frac{1}{r}\int r\int\frac{1}{r}\int\frac{pr}{D}\,dr$	$F_{wp_1}(r,a_i)=\langle r-a_i\rangle^0\left[\dfrac{r^4}{8}-\dfrac{5a_i^4}{8}+\dfrac{a_i^2 r^2}{2}-a_i^2\left(r^2+\dfrac{a_i^2}{2}\right)\ln\dfrac{r}{a_i}\right]$ $F_{\theta p_1}(r,a_i)=\langle r-a_i\rangle^0\left(\dfrac{r^3}{4}-\dfrac{a_i^4}{4r}-a_i^2 r\ln\dfrac{r}{a_i}\right)$ $F_{Mp_1}(r,a_i)=\langle r-a_i\rangle^0\left[\dfrac{3+\nu}{4}r^2-a_i^2+\dfrac{(1-\nu)a_i^4}{4r^2}-(1+\nu)a_i^2\ln\dfrac{r}{a_i}\right]$ $F_{Vp_1}(r,a_i)=\langle r-a_i\rangle^0\dfrac{r^2-a_i^2}{r}$ $F_{wp_2}(r,a_i)=\langle r-a_i\rangle^0\left[\dfrac{r^5}{225}-\dfrac{a_i r^4}{64}-\dfrac{a_i^3 r^2}{144}+\dfrac{29a_i^5}{1600}+\dfrac{a_i^3}{8}\left(\dfrac{r^2}{3}+\dfrac{a_i^2}{10}\right)\ln\dfrac{r}{a_i}\right]$ $F_{\theta p_2}(r,a_i)=\langle r-a_i\rangle^0\left(\dfrac{r^4}{45}-\dfrac{a_i r^3}{16}+\dfrac{a_i^3 r}{36}+\dfrac{a_i^5}{80r}+\dfrac{a_i^3}{12}r\ln\dfrac{r}{a_i}\right)$ $F_{Mp_2}(r,a_i)=\langle r-a_i\rangle^0\left(\dfrac{4+\nu}{45}r^3-\dfrac{3+\nu}{16}a_i r^2+\dfrac{4+\nu}{36}a_i^3-\dfrac{1-\nu}{80}\dfrac{a_i^5}{r^2}+\dfrac{1+\nu}{12}a_i^3\ln\dfrac{r}{a_i}\right)$ $F_{Vp_2}(r,a_i)=\langle r-a_i\rangle^0\left(\dfrac{r^2}{3}-\dfrac{a_i r}{2}+\dfrac{a_i^3}{6r}\right)$
$F_\theta(r)$	$-\langle r-a\rangle^0\,\dfrac{C}{2D}\dfrac{r^2-a^2}{r}$	$-\dfrac{1}{r}\displaystyle\int_{a_1}^r r\int\frac{1}{r}\int\frac{pr}{D}\,dr$	
$F_V(r)$	0	$-\dfrac{1}{r}\displaystyle\int_{a_1}^r pr\,dr$	
$F_M(r)$	$-\langle r-a\rangle^0\,\dfrac{C}{2}\left[(1-\nu)\left(\dfrac{a}{r}\right)^2+1+\nu\right]$	$-\displaystyle\int_{a_1}^r\frac{1}{r}\int pr\,dr+\dfrac{1-\nu}{r^2}\displaystyle\int_{a_1}^r r\int\frac{1}{r}\int pr\,dr$	

TABLE 18-4, Part B   **Circular Plates: Loading Functions**   994

TABLE 18-4, PART C   CIRCULAR PLATES WITH ARBITRARY LOADING: INITIAL PARAMETERS

Initial Parameters w_0, θ_0, V_0, M_0

Plates without Center Hole: $\bar{F}_w = F_w|_{r=a_L}$, $\bar{F}_\theta = F_\theta|_{r=a_L}$, $\bar{F}_V = F_V|_{r=a_L}$, $\bar{F}_M = F_M|_{r=a_L}$

Outer Edge / Center Condition	Simply Supported	Fixed	Free	Guided
1. No center support	$w_0 = -\dfrac{a_L^2}{2D(1+v)}\bar{F}_M - \bar{F}_w$ $M_0 = -\bar{F}_M$	$w_0 = -\dfrac{1}{2}a_L\bar{F}_\theta - \bar{F}_w$ $M_0 = -\dfrac{D(1+v)}{a_L}\bar{F}_\theta$	Kinematically unstable	Kinematically unstable
2. Center support	$R = -\dfrac{16(1+v)\pi D}{(3+v)a_L^2}\bar{F}_w - \dfrac{8\pi}{3+v}\bar{F}_M$ $C_1 = \dfrac{8(1+v)\ln a_L + 4(1-v)a_L^2}{(3+v)a_L^2}\bar{F}_w$ $\quad + \dfrac{4(\ln a_L - 1)}{D(3+v)}\bar{F}_M$	$R = -\dfrac{16\pi D}{a_L^2}\bar{F}_w - \dfrac{8\pi D}{a_L}\bar{F}_\theta$ $C_1 = -\dfrac{8(\ln a_L - \frac{1}{2})}{a_L^2}\bar{F}_w$ $\quad - \dfrac{4(\ln a_L - 1)}{a_L}\bar{F}_\theta$	$R = -2\pi a_L\bar{F}_V$ $C_1 = \dfrac{2}{D(1+v)}\bar{F}_M$ $\quad - \dfrac{a_L}{D(1+v)}\left[(1+v)\ln a_L \right.$ $\quad \left. + \dfrac{1-v}{2}\right]\bar{F}_V$	$R = -2\pi a_L\bar{F}_V$ $C_1 = -\dfrac{a_L}{D}\left(\ln a_L - \dfrac{1}{2}\right)\bar{F}_V$ $\quad + \dfrac{2}{a_L}\bar{F}_\theta$

Plates with Center Hole: $\bar{F}_w = F_w|_{r=a_L}$, $\bar{F}_\theta = F_\theta|_{r=a_L}$, $\bar{F}_V = F_V|_{r=a_L}$, $\bar{F}_M = F_M|_{r=a_L}$, $\bar{U}_{ij} = U_{ij}|_{r=a_L}$

Inner Edge \ Outer Edge	Simply Supported	Fixed	Free	Guided
1. Simply supported $w_0 = 0,\ M_0 = 0$	$\theta_0 = (\bar{F}_M \bar{U}_{wV} - \bar{F}_w \bar{U}_{MV})/\nabla$ $V_0 = (\bar{F}_w \bar{U}_{M\theta} - \bar{F}_M \bar{U}_{w\theta})/\nabla$ $\nabla = \bar{U}_{w\theta} \bar{U}_{MV} - \bar{U}_{M\theta} \bar{U}_{wV}$	$\theta_0 = (\bar{F}_\theta \bar{U}_{wV} - \bar{F}_w \bar{U}_{\theta V})/\nabla$ $V_0 = (\bar{F}_w \bar{U}_{\theta\theta} - \bar{F}_\theta \bar{U}_{w\theta})/\nabla$ $\nabla = \bar{U}_{w\theta} \bar{U}_{\theta V} - \bar{U}_{\theta\theta} \bar{U}_{wV}$	$\theta_0 = (\bar{F}_V \bar{U}_{MV} - \bar{F}_M \bar{U}_{VV})/\nabla$ $V_0 = (\bar{F}_M \bar{U}_{V\theta} - \bar{F}_V \bar{U}_{M\theta})/\nabla$ $\nabla = \bar{U}_{M\theta} \bar{U}_{VV} - \bar{U}_{V\theta} \bar{U}_{MV}$	$\theta_0 = (\bar{F}_V \bar{U}_{\theta V} - \bar{F}_\theta \bar{U}_{VV})/\nabla$ $V_0 = (\bar{F}_\theta \bar{U}_{V\theta} - \bar{F}_V \bar{U}_{\theta\theta})/\nabla$ $\nabla = \bar{U}_{\theta\theta} \bar{U}_{VV} - \bar{U}_{V\theta} \bar{U}_{\theta V}$
2. Fixed $w_0 = 0,\ \theta_0 = 0$	$M_0 = (\bar{F}_M \bar{U}_{wV} - \bar{F}_w \bar{U}_{MV})/\nabla$ $V_0 = (\bar{F}_w \bar{U}_{MM} - \bar{F}_M \bar{U}_{wM})/\nabla$ $\nabla = \bar{U}_{wM} \bar{U}_{MV} - \bar{U}_{MM} \bar{U}_{wV}$	$M_0 = (\bar{F}_\theta \bar{U}_{wV} - \bar{F}_w \bar{U}_{\theta V})/\nabla$ $V_0 = (\bar{F}_w \bar{U}_{\theta M} - \bar{F}_\theta \bar{U}_{wM})/\nabla$ $\nabla = \bar{U}_{wM} \bar{U}_{\theta V} - \bar{U}_{\theta M} \bar{U}_{wV}$	$M_0 = (\bar{F}_V \bar{U}_{MV} - \bar{F}_M \bar{U}_{VV})/\nabla$ $V_0 = (\bar{F}_M \bar{U}_{VM} - \bar{F}_V \bar{U}_{MM})/\nabla$ $\nabla = \bar{U}_{MM} \bar{U}_{VV} - \bar{U}_{VM} \bar{U}_{MV}$	$M_0 = (\bar{F}_V \bar{U}_{\theta V} - \bar{F}_\theta \bar{U}_{VV})/\nabla$ $V_0 = (\bar{F}_\theta \bar{U}_{VM} - \bar{F}_V \bar{U}_{\theta M})/\nabla$ $\nabla = \bar{U}_{\theta M} \bar{U}_{VV} - \bar{U}_{\theta V} \bar{U}_{VM}$
3. Free $M_0 = 0,\ V_0 = 0$	$w_0 = (\bar{F}_M \bar{U}_{w\theta} - \bar{F}_w \bar{U}_{M\theta})/\nabla$ $\theta_0 = (\bar{F}_w \bar{U}_{Mw} - \bar{F}_M \bar{U}_{ww})/\nabla$ $\nabla = \bar{U}_{ww} \bar{U}_{M\theta} - \bar{U}_{Mw} \bar{U}_{w\theta}$	$w_0 = (\bar{F}_\theta \bar{U}_{w\theta} - \bar{F}_w \bar{U}_{\theta\theta})/\nabla$ $\theta_0 = (\bar{F}_w \bar{U}_{\theta w} - \bar{F}_\theta \bar{U}_{ww})/\nabla$ $\nabla = \bar{U}_{ww} \bar{U}_{\theta\theta} - \bar{U}_{\theta w} \bar{U}_{w\theta}$	$w_0 = (\bar{F}_V \bar{U}_{M\theta} - \bar{F}_M \bar{U}_{V\theta})/\nabla$ $\theta_0 = (\bar{F}_M \bar{U}_{Vw} - \bar{F}_V \bar{U}_{Mw})/\nabla$ $\nabla = \bar{U}_{Mw} \bar{U}_{V\theta} - \bar{U}_{M\theta} \bar{U}_{Vw}$	$w_0 = (\bar{F}_V \bar{U}_{\theta\theta} - \bar{F}_\theta \bar{U}_{V\theta})/\nabla$ $\theta_0 = (\bar{F}_\theta \bar{U}_{Vw} - \bar{F}_V \bar{U}_{\theta w})/\nabla$ $\nabla = \bar{U}_{\theta w} \bar{U}_{V\theta} - \bar{U}_{\theta\theta} \bar{U}_{Vw}$

TABLE 18-4, Part C   Circular Plates: Initial Parameters   996

Case				
4. Guided $\theta_0=0,\ V_0=0$	$w_0 = (\bar{F}_M \bar{U}_{wM} - \bar{F}_w \bar{U}_{MM})/\nabla$ $M_0 = (\bar{F}_w \bar{U}_{Mw} - \bar{F}_M \bar{U}_{ww})/\nabla$ $\nabla = \bar{U}_{ww} \bar{U}_{MM} - \bar{U}_{Mw} \bar{U}_{wM}$	$w_0 = (\bar{F}_\theta \bar{U}_{wM} - \bar{F}_w \bar{U}_{\theta M})/\nabla$ $M_0 = (\bar{F}_w \bar{U}_{\theta w} - \bar{F}_\theta \bar{U}_{ww})/\nabla$ $\nabla = \bar{U}_{ww} \bar{U}_{\theta M} - \bar{U}_{wM} \bar{U}_{\theta w}$	$w_0 = (\bar{F}_V \bar{U}_{MM} - \bar{F}_M \bar{U}_{VM})/\nabla$ $M_0 = (\bar{F}_M \bar{U}_{Vw} - \bar{F}_V \bar{U}_{Mw})/\nabla$ $\nabla = \bar{U}_{Mw} \bar{U}_{VM} - \bar{U}_{MM} \bar{U}_{Vw}$	$w_0 = (\bar{F}_V \bar{U}_{\theta M} - \bar{F}_\theta \bar{U}_{VM})/\nabla$ $M_0 = (\bar{F}_\theta \bar{U}_{Vw} - \bar{F}_V \bar{U}_{\theta w})/\nabla$ $\nabla = \bar{U}_{\theta w} \bar{U}_{VM} - \bar{U}_{\theta M} \bar{U}_{Vw}$
5. Rigid insert with total load W_T 	$w_0 = \left[\dfrac{W_T}{2\pi a_0}\right.$ $\left(\bar{U}_{MM}\bar{U}_{wV} - \bar{U}_{wM}\bar{U}_{MV}\right)$ $\left. + \bar{U}_{wM}\bar{F}_M - \bar{U}_{MM}\bar{F}_w\right]/\nabla$ $M_0 = \left[\dfrac{W_T}{2\pi a_0}\right.$ $\left(\bar{U}_{ww}\bar{U}_{MV} - \bar{U}_{Mw}\bar{U}_{wV}\right)$ $\left. + \bar{U}_{Mw}\bar{F}_V - \bar{U}_{ww}\bar{F}_M\right]/\nabla$ $V_0 = -W_T/(2\pi a_0) \quad \theta_0 = 0$ $\nabla = \bar{U}_{ww}\bar{U}_{MM} - \bar{U}_{Mw}\bar{U}_{wM}$	$w_0 = \left[\dfrac{W_T}{2\pi a_0}\right.$ $\left(\bar{U}_{\theta M}\bar{U}_{wV} - \bar{U}_{wM}\bar{U}_{\theta V}\right)$ $\left. + \bar{U}_{wM}\bar{F}_\theta - \bar{U}_{\theta M}\bar{F}_w\right]/\nabla$ $M_0 = \left[\dfrac{W_T}{2\pi a_0}\right.$ $\left(\bar{U}_{ww}\bar{U}_{\theta V} - \bar{U}_{wV}\bar{U}_{\theta w}\right)$ $\left. + \bar{U}_{\theta w}\bar{F}_w - \bar{U}_{ww}\bar{F}_\theta\right]/\nabla$ $V_0 = -W_T/(2\pi a_0) \quad \theta_0 = 0$ $\nabla = \bar{U}_{ww}\bar{U}_{\theta M} - \bar{U}_{\theta w}\bar{U}_{wM}$	$w_0 = \left[\dfrac{W_T}{2\pi a_0}\right.$ $\left(\bar{U}_{MV}\bar{U}_{VM} - \bar{U}_{MM}\bar{U}_{VV}\right)$ $\left. + \bar{U}_{MM}\bar{F}_V - \bar{U}_{VM}\bar{F}_M\right]/\nabla$ $M_0 = \left[\dfrac{W_T}{2\pi a_0}\right.$ $\left(\bar{U}_{Mw}\bar{U}_{VV} - \bar{U}_{MV}\bar{U}_{Vw}\right)$ $\left. + \bar{U}_{Vw}\bar{F}_M - \bar{U}_{MM}\bar{F}_V\right]/\nabla$ $V_0 = -W_T/(2\pi a_0) \quad \theta_0 = 0$ $\nabla = \bar{U}_{Mw}\bar{U}_{VM} - \bar{U}_{MM}\bar{U}_{Vw}$	$w_0 = \left[\dfrac{W_T}{2\pi a_0}\right.$ $\left(\bar{U}_{\theta V}\bar{U}_{VM} - \bar{U}_{\theta M}\bar{U}_{VV}\right)$ $\left. + \bar{U}_{\theta M}\bar{F}_V - \bar{U}_{VM}\bar{F}_\theta\right]/\nabla$ $M_0 = \left[\dfrac{W_T}{2\pi a_0}\right.$ $\left(\bar{U}_{\theta w}\bar{U}_{VV} - \bar{U}_{\theta V}\bar{U}_{Vw}\right)$ $\left. + \bar{U}_{Vw}\bar{F}_\theta - \bar{U}_{\theta w}\bar{F}_w\right]/\nabla$ $V_0 = -W_T/(2\pi a_0) \quad \theta_0 = 0$ $\nabla = \bar{U}_{\theta w}\bar{U}_{VM} - \bar{U}_{\theta M}\bar{U}_{Vw}$

TABLE 18-5 CRITICAL IN-PLANE FORCES FOR CIRCULAR PLATES

Notation

E = modulus of elasticity

ν = Poisson's ratio

c = buckling coefficient

P_{cr} = buckling load (force per unit length)

$$\beta = a_0/a_L \qquad P' = \frac{\pi^2 D}{a_L^2} \qquad D = \frac{Eh^3}{12(1 - \nu^2)}$$

Nodal circle or nodal diameter refer to the circle or diameter in the plane of the plate for which the displacement is zero in a buckling mode shape.

Conditions	Buckling Loads
1. Simply supported outer boundary 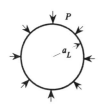	Applies for no nodal circles or nodal diameters present in the buckling mode shapes, $P_{cr} = 0.426P'$ Ref. 18.5
2. Clamped outer boundary 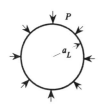	**1.** Applies for no nodal circles or nodal diameters present in the buckling mode shapes, $P_{cr} = 1.49P'$ **2.** Applies when nodal circles and/or nodal diameters are present in the buckling mode shapes, $P_{cr} = \eta \times 1.49P'$ 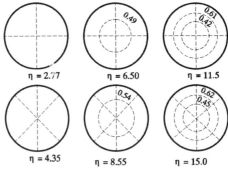 Nodal circles and diameters are shown. The numbers on the dashed nodal circles refer to the percentage of a_L, e.g., $0.49a_L$. Refs. 18.5–18.8

TABLE 18-5 Buckling Loads for Circular Plates   998

TABLE 18-5 (continued)   CRITICAL IN-PLANE FORCES FOR CIRCULAR PLATES

Conditions	Results

3.
Elastically restrained boundary
 	
c = buckling
    coefficient
k = spring constant
(FL/L)
$\gamma = \left(\dfrac{ka_L}{D}\right)^{1/2}$
$P_{cr} = c^2\dfrac{D}{a_L^2}$
	1.
	First mode

	Elastically restrained edge:
	$c = 1.99421 + 0.4488381\gamma + 0.156167\gamma^2 - 0.04576\gamma^3,$
	   $0 \le \gamma \le 12$
	Clamped edge: $c = 3.832$
	Simply supported edge: $c = 2.049$
	
	2.
	Second mode, where a nodal line exists as a diameter
	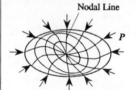
	Elastically restrained edge:
	$c = 3.618543 + 0.1795\gamma + 0.189377\gamma^2 - 0.0433434\gamma^3,$
	   $0 \le \gamma \le 12$
	Clamped edge: $c = 5.136$
	Simply supported edge: $c = 3.625$
	
	Ref. 18.9

4. Simply supported outer boundary, inner boundary free 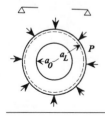	$P_{cr} = cP'$ Symmetrical buckling $c = \begin{cases} 0.436 - 0.4387\beta + 1.35\beta^2 - 5.72\beta^3 + 6.67\beta^4 \\ \quad 0 \le \beta < 0.5 \\ -35.86 + 209.79\beta - 448.35\beta^2 + 418.75\beta^3 - 144.58\beta^4 \\ \quad 0.5 \le \beta \le 0.9 \end{cases}$

Conditions	Results
5. Clamped outer boundary, inner boundary free 	$P_{cr} = cP'$ Symmetrical buckling $c = 1.482 - 0.452\beta - 4.484\beta^2 + 19.1\beta^3$ $0 \le \beta < 0.5$
6. Simply supported outer boundary of plate of variable thickness 	$$P_{cr} = \frac{cEh^3}{12(1 - \nu^2)a_L^2}$$ Average thickness $h = h_1\left(1 - \beta^2 + \dfrac{h_0}{h_1}\beta^2\right)$ For values of c, see table footnote 1. Ref. 18.4, 18.16
7. Clamped outer boundary of plate of variable thickness 	$$P_{cr} = \frac{cEh^3}{12(1 - \nu^2)a_L^2}$$ $h = h_1\left(1 - \beta^2 + \dfrac{h_0}{h_1}\beta^2\right)$ For values of c, see table footnote 2. Ref. 18.4, 18.16

TABLE 18-5 (continued)   FOOTNOTES

1. For Case 6, values of c (for a uniform plate, $c = 4.28$) are

$\dfrac{h_0}{h_1}$	$\beta = 0.03$	0.2	0.4	0.6	0.7	0.75	0.8	0.85	0.9	0.97
0.3	4.27	3.71	3.05	2.72			3.13			
0.5	4.27	3.78	3.14	2.96			3.27			4.08
0.7	4.28	3.99	3.53	3.34			3.55		3.84	4.13
0.8									3.98	
0.9							4.03			
1.1	4.29	4.34	4.47	4.58	4.58		4.53		4.42	
1.3				5.01	5.07		4.96		4.67	4.40
1.5	4.29	4.39	4.70	5.17	5.35	5.36	5.29		4.88	4.47
1.6				5.15	5.41		5.41			
1.7				5.07	5.40	5.50	5.49	5.33		4.52
1.8				4.95	5.34	5.49	5.53	5.40		
1.9				4.79			5.54			
2.0	4.28	4.26	4.25	4.61	5.07	5.33	5.52	5.49	5.21	4.57
2.2								5.49		
2.5				3.60	4.01		4.89	5.33	5.33	
3.0	4.27	3.86	3.07	2.74	2.98		3.85	4.64	5.22	4.69

2. For Case 7, values of c (for a uniform plate, $c = 14.68$) are

$\dfrac{h_0}{h_1}$	$\beta = 0.03$	0.1	0.2	0.3	0.4	0.45	0.5	0.55	0.6	0.7	0.8	0.9	0.97
0.3	14.58	13.71	12.04		5.47				3.82		4.65	6.88	11.05
0.4					9.14								
0.5	14.59	13.84	12.42	11.69	11.27		10.36		9.48		9.61	11.12	13.28
0.6					12.07							12.60	13.89
0.7	14.64	14.21	13.24		12.63				13.21		13.25	13.66	
0.9	14.66	14.56	14.25		13.94				14.35	14.57	14.66	14.66	
1.1	14.68	14.77	15.00		15.39	15.37		15.16	15.00		14.43	14.45	
1.3					16.49	16.55	16.45	16.16	15.71			13.53	
1.4						16.94	16.88	16.58	16.06			12.96	
1.5	14.70	14.91	15.49		16.94	17.15	17.16	16.91	16.38		12.99	12.39	
1.6					16.93	17.19	17.28	17.12	16.60				
1.7					16.79	17.07	17.24	17.16					
1.8					16.54	16.83	17.04	17.05	16.72				
1.9					16.23		16.70	16.79					
2.0	14.69	14.83	15.13		15.85	16.07	16.27	16.40	16.32	14.85	12.03	10.09	11.32
3.0	14.66	14.45	13.73		11.58		10.80		10.62	11.22	11.11	8.26	7.87

TABLE 18-6 NATURAL FREQUENCIES OF SOME CIRCULAR PLATES AND MEMBRANES [a]

Notation

ν = Poisson's ratio
n = number of nodal diameters
h = thickness of the plate
$\beta = a_0/a_L$

E = modulus of elasticity
ρ = mass per unit area of the plate
s = number of nodal circles, not including the boundary circle when the boundary is constrained
D = flexural rigidity of plate, $= Eh^3/[12(1 - \nu^2)]$

Nodal circles and diameters refer to loci of points along which the mode shape displacements are zero.

$$\omega_{ns}(\text{rad/T}) = \frac{\lambda_{ns}}{a_L^2}\sqrt{\frac{D}{\rho}} \qquad f_{ns}(\text{Hz}) = \frac{\lambda_{ns}}{2\pi a_L^2}\sqrt{\frac{D}{\rho}}$$

The values of λ_{ns} are independent of ν except where indicated otherwise.
All results are for the transverse vibration of plates unless otherwise indicated. For membranes, the natural frequencies for transverse vibrations are given by $f_{ns}(\text{Hz}) = \dfrac{\lambda_{ns}}{2}\left(\dfrac{P}{\rho A}\right)^{1/2}$, where A is the area of the membrane and P is the tension per unit length.

Configuration and Boundary Conditions	Constants

1. Free

2a_L

λ_{ns} for $\nu = 0.25$

n \ s	0	1	2	3
0	—	—	5.513	12.75
1	8.892	20.41	35.28	53.16
2	38.34	59.74	84.38	112.36
3	87.65	118.88	153.29	191.02
4	156.73	196.67	241.99	289.51
5	245.52	296.46	350.48	408.16

TABLE 18-6 Natural Frequencies of Circular Plates   1002

2. Simply supported

λ_{ns} for $\nu = 0.3$

s \ n	0	1	2
0	4.977	13.94	25.65
1	29.76	48.51	70.14
2	74.20	102.80	134.33
3	138.34	176.84	218.24

3. Fixed

λ_{ns}

s \ n	0	1	2	3
0	10.216	21.26	34.88	51.04
1	39.771	60.82	84.58	111.01
2	89.104	120.08	153.81	190.30
3	158.18	199.06	242.71	289.17
4	247.00	297.77	351.38	407.72
5	355.57	416.20	479.65	545.97

MEMBRANE λ_{ns}

s \ n	0	1	2	3
1	1.357	2.162	2.897	3.600
2	3.114	3.958	4.749	5.507
3	4.882	5.740	6.556	7.343
4	6.653	7.517	8.348	9.153

TABLE 18-6 (continued)   NATURAL FREQUENCIES OF SOME CIRCULAR PLATES

Configuration and Boundary Conditions	Constants
4. Fixed both on outer and inner edge	Number of nodal circles $s = 0$ $\lambda_{00} = 28.944 - 49.5495\beta + 338.977\beta^2$ $\lambda_{10} = 26.537 + 2.40\beta + 222.672\beta^2$ $\lambda_{20} = 41.8525 - 80.131\beta + 360.317\beta^2$ $0.1 \leq \beta \leq 0.6$ MEMBRANE For $a_0/a_L > 1/2$: $\lambda_{ns} = \dfrac{1}{\pi^{1/2}}\left[4s^2\left(\dfrac{a_L - a_0}{a_L + a_0}\right) + n^2\pi^2\left(\dfrac{a_L + a_0}{a_L - a_o}\right)\right]^{1/2}$ $s = 0, 1, 2, \ldots$ $n = 1, 2, 3, \ldots$
5. Free outer edge, simply supported inner edge	$\lambda_{10} = 4.4725 - 39.829\beta + 219.9576\beta^2 - 417.08215\beta^3 + 279.1661\beta^4$ $\lambda_{00} = 4.8375 - 22.9475\beta + 114.5183\beta^2 - 220.6284\beta^3 + 154.0753\beta^4$ $\lambda_{20} = 9.7589 - 74.3830\beta + 375.4257\beta^2 - 701.6646\beta^3 + 468.2281\beta^4$ $\lambda_{30} = 18.4765 - 102.7286\beta + 510.7268\beta^2 - 977.0797\beta^3 + 661.4566\beta^4$ $\lambda_{01} = 2.5633 - 1.4277\dfrac{1}{1-\beta} + 15.4673\dfrac{1}{(1-\beta)^2}$ $\lambda_{11} = 13.8536 - 10.7728\dfrac{1}{1-\beta} + 18.05616\dfrac{1}{(1-\beta)^2} - 0.1757\dfrac{1}{(1-\beta)^3}$ $\lambda_{21} = 38.9306 - 27.1254\dfrac{1}{1-\beta} + 21.9928\dfrac{1}{(1-\beta)^2} - 0.4260\dfrac{1}{(1-\beta)^3}$

TABLE 18-6   **Natural Frequencies of Circular Plates**   1004

$$\lambda_{31} = 71.7981 - 45.8024\frac{1}{1-\beta} + 26.2399\frac{1}{(1-\beta)^2} - 0.6898\frac{1}{(1-\beta)^3}$$

$$0.1 \le \beta \le 0.9 \qquad \nu = 0.33$$

6.
Fixed inner edge, free outer edge

$\lambda_{10} = -3.2345 + 2.7374\gamma + 2.6634\gamma^2 + 0.05452\gamma^3$

$\lambda_{00} = 6.08616 - 9.3994\gamma + 7.7058\gamma^2 - 0.6312\gamma^3$

$\lambda_{20} = 1.4048 - 0.2623\gamma + 3.4822\gamma^2$

$\lambda_{30} = 18.2517 - 12.5105\gamma + 6.7362\gamma^2 - 0.2148\gamma^3$

$\lambda_{01} = 17.3288 - 31.446\gamma + 37.7419\gamma^2 - 2.5071\gamma^3$

$\lambda_{11} = 2.1547 - 1.8353\gamma + 22.05198\gamma^2$

$\lambda_{21} = 24.9692 - 18.5025\gamma + 26.4093\gamma^2 - 0.2869\gamma^3$

$\lambda_{31} = 68.4516 - 52.6774\gamma + 36.0325\gamma^2 - 0.9449\gamma^3$

$$0.1 \le \beta \le 0.9 \qquad \nu = 0.33 \qquad \gamma = \frac{1}{1-\beta}$$

7.
Free outer edge with concentrated mass at center

$$\omega = \frac{\lambda}{a_L^2}\sqrt{\frac{D}{\rho}}$$

$$\lambda = \begin{cases} 8.8626 - 4.4412\mu - 0.9465\mu^2 + 1.4061\mu^3 & s = 1 \\ 36.005 - 32.127\mu + 24.2798\mu^2 - 6.2115\mu^3 & s = 2 \\ 78.0257 - 49.8526\mu + 48.4639\mu^2 - 14.7959\mu^3 & s = 3 \\ 137.3952 - 61.4408\mu + 73.1068\mu^2 - 29.2415\mu^3 & s = 4 \end{cases}$$

$$\mu = \left(\frac{m}{\pi\rho a_L^2}\right)^{1/2} \qquad \nu = 0.3 \qquad m = \text{magnitude of concentrated mass}$$

[a]Adopted from Leissa Ref. 18.2.

TABLE 18-7 **Transfer Matrices for Circular Plates** 1006

TABLE 18-7   TRANSFER MATRICES FOR CIRCULAR PLATE ELEMENTS[a]

Notation

The plate response is given by the Fourier series expansion of Eq. (18.10).

w, θ, V, M = deflection, slope, effective shear force per unit length, and bending moment per unit length

E = modulus of elasticity

h = thickness of plate

$$D = \frac{Eh^3}{12(1 - \nu^2)}$$

ν = Poisson's ratio

ε_m^j = provided in Part C of this table

Circular Plate Element with Center Hole

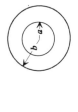

Notation

$$\frac{\Delta p}{\Delta \ell} = \frac{p_b - p_a}{b - a} \qquad \frac{\Delta c}{\Delta \ell} = \frac{c_b - c_a}{b - a} \qquad q_1 = p_a - a\frac{\Delta p}{\Delta \ell}$$

$$\gamma = b/a$$

p_a, p_b = magnitude of applied distributed forces at $r = a$ and $r = b$ (F/L^2)
c_a, c_b = magnitude of applied distributed moments at $r = a$ and $r = b$ (FL/L^2)

Transfer Matrices (Sign Convention 1)

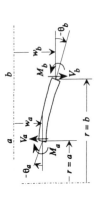

The transfer matrix relations $\mathbf{z}_b = \mathbf{U}^i\mathbf{z}_a + \bar{\mathbf{z}}^i$ are extended to include the loading vector $\bar{\mathbf{z}}^i$ in the basic matrices (Appendix II), so that

$$\mathbf{z}_b = \mathbf{U}^i\mathbf{z}_a$$

$\mathbf{z}_b = [w_b \quad \theta_b \quad V_b \quad M_b \quad 1]^T \qquad \mathbf{z}_a = [w_a \quad \theta_a \quad V_a \quad M_a \quad 1]^T$

$m = 0$ (SYMMETRIC DEFORMATION)

$$
\mathbf{z}_a =
\begin{bmatrix}
1 & -\dfrac{1+\nu}{2}a\ln\dfrac{b}{a}-\dfrac{1-\nu}{4a}(b^2-a^2) & -\dfrac{a(b^2+a^2)}{4D}\ln\dfrac{b}{a}+\dfrac{a}{4D}(b^2-a^2) & -\dfrac{a^2}{2D}\ln\dfrac{b}{a}-\dfrac{1}{4D}(b^2-a^2) & F_w \\[4mm]
0 & \dfrac{1+\nu}{2}\dfrac{a}{b}+\dfrac{1-\nu}{2}\dfrac{b}{a} & \dfrac{ab}{2D}\ln\dfrac{b}{a}-\dfrac{a}{4Db}(b^2-a^2) & \dfrac{b^2-a^2}{2Db} & F_\theta \\[4mm]
0 & 0 & \dfrac{a}{b} & 0 & F_V \\[4mm]
0 & \dfrac{D(1-\nu^2)}{2}\dfrac{b^2-a^2}{ab^2} & \dfrac{1+\nu}{2}a\ln\dfrac{b}{a}+\dfrac{(1-\nu)a}{4b^2}(b^2-a^2) & \dfrac{1-\nu}{2}\left(\dfrac{a}{b}\right)^2+\dfrac{1+\nu}{2} & F_M \\[4mm]
0 & 0 & 0 & 0 & 1
\end{bmatrix}
\begin{bmatrix} w_a \\[2mm] \theta_a \\[2mm] V_a \\[2mm] M_a \\[2mm] 1 \end{bmatrix}
$$

$$\mathbf{U}^i$$

$$F_w = \varepsilon_0^c\left(\frac{p_a}{8D}e_{11} + \frac{\Delta p}{\Delta\ell}\frac{1}{D}e_{12} + \frac{c_a}{4D}e_{13} + \frac{\Delta c}{\Delta\ell}\frac{1}{4D}e_{14}\right)$$

$$F_\theta = \varepsilon_0^c\left(-\frac{p_a}{4D}e_{21} - \frac{\Delta p}{\Delta\ell}\frac{1}{D}e_{22} + \frac{c_a}{4D}e_{23} + \frac{\Delta c}{\Delta\ell}\frac{1}{4D}e_{24}\right)$$

$$F_V = \varepsilon_0^c\left(-\frac{p_a}{2}e_{31} - \frac{\Delta p}{\Delta\ell}e_{32}\right)$$

$$F_M = -\varepsilon_0^c\left(-\frac{p_a}{4}e_{41} - \frac{\Delta p}{\Delta\ell}e_{42} + c_a e_{43} + \frac{\Delta c}{\Delta\ell}e_{44}\right)$$

$$e_{11} = \frac{b^4}{8} - \frac{5a^4}{8} + \frac{a^2b^2}{2} - a^2\left(b^2 + \frac{a^2}{2}\right)\ln\frac{b}{a}$$

$$e_{12} = \frac{b^5}{225} - \frac{ab^4}{64} + \frac{a^3b^2}{144} - \frac{29a^5}{1600} + \frac{a^3}{8}\left(\frac{b^2}{3} + \frac{a^2}{10}\right)\ln\frac{b}{a}$$

$$e_{13} = -\frac{2a^3}{3}\ln\frac{a}{b} + \frac{1}{9}(5a^3 + 4b^3 - 9b^2a)$$

$$e_{14} = \frac{b^4}{8} - \frac{4}{9}ab^3 + \frac{a^2b^2}{2} - \frac{13}{72}a^4 - \frac{1}{6}a^4\ln\frac{b}{a}$$

$$e_{21} = \frac{a^4}{4} - a^2b^2\ln\frac{b}{a}$$

$$e_{22} = \frac{b^4}{45} - \frac{ab^3}{16} + \frac{a^3b}{36} + \frac{a^5}{80b} + \frac{a^3b}{12}\ln\frac{b}{a}$$

$$e_{23} = -\left(\frac{4b^2}{3} - 2ab + \frac{2a^3}{3b}\right)$$

$$e_{24} = -\left(\frac{b^3}{2} - \frac{4}{3}ab^2 + a^2b - \frac{a^4}{6b}\right)$$

$$e_{31} = \frac{1}{b}(b^2-a^2)$$

$$e_{32} = \frac{b^2}{3} - \frac{ab}{2} + \frac{a^3}{6b}$$

$$e_{41} = \frac{3+\nu}{4}b^2 - a^2 + \frac{1}{4b^2}(1-\nu)a^4 - (1+\nu)a^2\ln\frac{b}{a}$$

$$e_{42} = \frac{4+\nu}{45}b^3 - \frac{3+\nu}{16}ab^2 + \frac{4+\nu}{36}a^3 - \frac{1-\nu}{80}\frac{a^5}{b^2} + \frac{1+\nu}{12}a^3\ln\frac{b}{a}$$

$$e_{43} = -\left(\frac{2+\nu}{3}b - \frac{1+\nu}{2}a - \frac{1-\nu}{6}\frac{a^3}{b^2}\right)$$

$$e_{44} = -\left(\frac{3+\nu}{8}b^2 - \frac{2+\nu}{3}ab + \frac{1+\nu}{4}a^2 + \frac{1-\nu}{24}\frac{a^4}{b^2}\right)$$

TABLE 18-7 (continued) TRANSFER MATRICES FOR CIRCULAR PLATE ELEMENTS

m = 1 (ASYMMETRIC DEFORMATION)

$$
\mathbf{U}^i=
\begin{bmatrix}
\dfrac{(3-\nu)b}{4a}-\dfrac{(1-\nu)b^3}{8a^3}+\dfrac{(3+\nu)a}{8b} &
\dfrac{(1+\nu)b}{4}-\dfrac{(1-\nu)b^3}{8a^2}+\dfrac{(3+\nu)a^2}{8b} &
\dfrac{a^2b}{4D}\ln\dfrac{b}{a}-\dfrac{1}{16Db}(b^4-a^4) &
-\dfrac{ab}{4D}\ln\dfrac{b}{a}-\dfrac{(b^2-a^2)(b^2-3a^2)}{16Dba} & F_w\\[12pt]
-\dfrac{3-\nu}{4a}+\dfrac{3(1-\nu)b^2}{8a^3}+\dfrac{(3+\nu)a}{8b^2} &
\dfrac{1+\nu}{4}+\dfrac{3(1-\nu)b^2}{8a^2}+\dfrac{(3+\nu)a^2}{8b^2} &
-\dfrac{a^2}{4D}\ln\dfrac{b}{a}+\dfrac{(b^2-a^2)(3b^2-a^2)}{16Db^2} &
-\dfrac{a}{4D}\ln\dfrac{b}{a}+\dfrac{3(b^4-a^4)}{16Dab^2} & F_\theta\\[12pt]
\dfrac{D(3+\nu)(1-\nu)}{4}\dfrac{b^4-a^4}{a^3b^4} &
\dfrac{D(3+\nu)(1-\nu)}{4}\dfrac{b^4-a^4}{a^2b^4} &
\dfrac{3+\nu}{8}+\dfrac{(1-\nu)a^4}{8b^4}-\dfrac{(3-\nu)a^2}{4b^2} &
\dfrac{3+\nu}{8a}+\dfrac{3(1-\nu)a^3}{8b^4}-\dfrac{(3-\nu)a}{4b^2} & F_V\\[12pt]
\dfrac{D(3+\nu)(1-\nu)}{4}\dfrac{b^4-a^4}{a^3b^3} &
\dfrac{D(3+\nu)(1-\nu)}{4}\dfrac{b^4-a^4}{a^2b^3} &
\dfrac{(3+\nu)b}{8}-\dfrac{(1-\nu)a^4}{8b^3}-\dfrac{(1+\nu)a^2}{4b} &
\dfrac{(3+\nu)b}{8a}+\dfrac{3(1-\nu)a^3}{8b^3}+\dfrac{(1+\nu)a}{4b} & F_M\\[12pt]
0 & 0 & 0 & 0 & 1
\end{bmatrix}
\begin{bmatrix}w_a\\[10pt]\theta_a\\[10pt]V_a\\[10pt]M_a\\[10pt]1\end{bmatrix}=\mathbf{z}_a
$$

$$
F_w=\varepsilon_1^j\left[\frac{a^4 p_a}{4D}\left(\frac{4\gamma^4}{45}-\frac{\gamma^3}{4}+\frac{\gamma}{9}+\frac{1}{20\gamma}+\frac{\gamma}{3}\ln\gamma\right)+\frac{a^5}{4D}\frac{\Delta p}{\Delta\ell}\left(\frac{\gamma^5}{48}-\frac{4\gamma^4}{45}+\frac{\gamma^3}{8}-\frac{7\gamma}{144}-\frac{1}{120\gamma}-\frac{\gamma}{12}\ln\gamma\right)\right]
$$

$$
F_\theta=\varepsilon_1^j\left[\frac{a^3 p_a}{4D}\left(\frac{16\gamma^3}{45}-\frac{3\gamma^2}{4}+\frac{4}{9}-\frac{1}{20\gamma^2}+\frac{1}{3}\ln\gamma\right)-\frac{a^4}{4D}\frac{\Delta p}{\Delta\ell}\left(\frac{5\gamma^4}{48}-\frac{16\gamma^3}{45}+\frac{3\gamma^2}{8}-\frac{19}{144}+\frac{1}{120\gamma^2}-\frac{1}{12}\ln\gamma\right)\right]
$$

$$
F_V=\varepsilon_1^j\left[\frac{a p_a}{4}\left(\frac{4(9+\nu)\gamma}{15}-\frac{3+\nu}{2}+\frac{3+\nu}{3\gamma}+\frac{1-\nu}{3\gamma}-\frac{1-\nu}{10\gamma^3}\right)-\frac{a^2}{4}\frac{\Delta p}{\Delta\ell}\left(\frac{17+\nu}{12}-\frac{4(9+\nu)}{15}\gamma+\frac{3+\nu}{4}-\frac{3-\nu}{12\gamma^2}-\frac{1-\nu}{60\gamma^4}\right)\right]
$$

$$
F_M=\varepsilon_1^j\left[\frac{a^2 p_a}{4}\left(\frac{4(4+\nu)\gamma^2}{15}-\frac{3+\nu}{2}\gamma+\frac{1+\nu}{2}+\frac{1+\nu}{3\gamma}+\frac{1-\nu}{10\gamma^3}\right)-\frac{a^3}{4}\frac{\Delta p}{\Delta\ell}\left(\frac{5+\nu}{12}\gamma^3-\frac{4(4+\nu)}{15}\gamma^2+\frac{3+\nu}{4}\gamma^2+\frac{1+\nu}{4}\gamma-\frac{1-\nu}{12\gamma}-\frac{1-\nu}{60\gamma^3}\right)\right]
$$

$\gamma = b/a$
$j = c, s$

TABLE 18-7 Transfer Matrices for Circular Plates 1008

$m \geq 2$

$$\mathbf{U}^i = \begin{bmatrix} \mathbf{H}(b)\,\mathbf{H}^{-1}(a) & \bar{\mathbf{z}}^i \\ 0 & 1 \end{bmatrix} = \begin{bmatrix} \mathbf{H}(b)\,\mathbf{H}^{-1}(a) & -\varepsilon_m^j[\mathbf{G}(b) - \mathbf{H}(b)\,\mathbf{H}^{-1}(a)\,\mathbf{G}(a)] \\ 0 & 1 \end{bmatrix} = \begin{bmatrix} \mathbf{H}(b)\,\mathbf{H}^{-1}(a) & -\varepsilon_m^j\mathbf{H}(b)[\mathbf{R}(b) - \mathbf{R}(a)] \\ 0 & 1 \end{bmatrix}$$

WHERE $\mathbf{G}(r) = \mathbf{H}(r)\mathbf{R}(r)$

$$\mathbf{H}(b) = \begin{bmatrix} b^m & b^{-m} & b^{m+2} & b^{-m+2} \\ -mb^{m-1} & mb^{-m-1} & -(m+2)b^{m+1} & (m-2)b^{-m+1} \\ Dm^2(m-1)(1-v)b^{m-3} & -Dm^2(m+1)(1-v)b^{-m-3} & Dm(m+1)(m-vm-4)b^{m-1} & -Dm(m-1)(m-vm+4)b^{-m-1} \\ -Dm(m-1)(1-v)b^{m-2} & -Dm(m-1)(1-v)b^{-m-2} & -D(m+1)(m+2-vm+2v)b^m & -D(m-1)(m-2-vm-2v)b^{-m} \end{bmatrix}$$

$$\mathbf{H}^{-1}(a) = \begin{bmatrix} \dfrac{m(1-v)+4}{8}a^{-m} & \dfrac{(m+2)(1-v)-4}{8m}a^{-m+1} & \dfrac{a^{-m+3}}{8m(m-1)D} & \dfrac{m-2}{8m(m-1)D}a^{-m+2} \\[3mm] -\dfrac{m(1-v)-4}{8}a^m & \dfrac{(m-2)(1-v)+4}{8m}a^{m+1} & \dfrac{a^{m+3}}{8m(m+1)D} & \dfrac{m+2}{8m(m+1)D}a^{m+2} \\[3mm] -\dfrac{m(1-v)}{8}a^{-m-2} & -\dfrac{1-v}{8}a^{-m-1} & -\dfrac{a^{-m+1}}{8m(m+1)D} & -\dfrac{a^{-m}}{8(m+1)D} \\[3mm] \dfrac{m(1-v)}{8}a^{m-2} & -\dfrac{1-v}{8}a^{m-1} & -\dfrac{a^{m+1}}{8m(m-1)D} & \dfrac{a^m}{8(m-1)D} \end{bmatrix}$$

$$\bar{\mathbf{z}}^i = \begin{bmatrix} \bar{F}_w \\ \bar{F}_\theta \\ \bar{F}_V \\ \bar{F}_M \end{bmatrix} = -\varepsilon_m^j\left[\mathbf{G}(b) - \mathbf{H}(b)\,\mathbf{H}^{-1}(a)\,\mathbf{G}(a)\right] = -\varepsilon_m^j\mathbf{H}(b)[\mathbf{R}(b) - \mathbf{R}(a)]$$

$j = c, s$

Transfer Matrices for Circular Plates | TABLE 18-7

TABLE 18-7 (continued)   TRANSFER MATRICES FOR CIRCULAR PLATE ELEMENTS

$$q_1 = p_a - a\frac{\Delta p}{\Delta \ell}$$

m = 2

$$\mathbf{R}(r) = \frac{1}{48D}\begin{bmatrix} r^2\left(\dfrac{3}{2}q_1 + r\dfrac{\Delta p}{\Delta \ell}\right) \\[2ex] r^6\left(\dfrac{1}{6}q_1 + r\dfrac{\Delta p}{7\,\Delta \ell}\right) \\[2ex] -\left(q_1 \ln r + r\dfrac{\Delta p}{\Delta \ell}\right) \\[2ex] -3r^4\left(\dfrac{1}{4}q_1 + r\dfrac{\Delta p}{5\,\Delta \ell}\right) \end{bmatrix}$$

m = 3

$$\mathbf{R}(r) = \frac{1}{96D}\begin{bmatrix} r\left(2q_1 + r\dfrac{\Delta p}{\Delta \ell}\right) \\[2ex] r^7\left(\dfrac{1}{7}q_1 + r\dfrac{\Delta p}{8\,\Delta \ell}\right) \\[2ex] \dfrac{1}{r}q_1 - \dfrac{\Delta p}{\Delta \ell}\ln r \\[2ex] -r^5\left(\dfrac{2}{5}q_1 + r\dfrac{\Delta p}{3\,\Delta \ell}\right) \end{bmatrix}$$

m = 4

$$\mathbf{R}(r) = \frac{1}{32D}\begin{bmatrix} \dfrac{1}{3}\left(q_1 \ln r + \dfrac{\Delta p}{\Delta \ell}r\right) \\[2ex] \dfrac{r^8}{5}\left(\dfrac{1}{8}q_1 + \dfrac{r}{9}\dfrac{\Delta p}{\Delta \ell}\right) \\[2ex] \dfrac{1}{5r^2}\left(\dfrac{1}{2}q_1 + \dfrac{\Delta p}{\Delta \ell}r\right) \\[2ex] -\dfrac{r^6}{3}\left(\dfrac{1}{6}q_1 + \dfrac{r}{7}\dfrac{\Delta p}{\Delta \ell}\right) \end{bmatrix}$$

m = 5

$$\mathbf{R}(r) = \frac{1}{80D}\begin{bmatrix} \dfrac{1}{2}\left(-\dfrac{1}{r}q_1 + \dfrac{\Delta p}{\Delta \ell}\ln r\right) \\[2ex] \dfrac{1}{3}r^9\left(\dfrac{1}{9}q_1 + \dfrac{r}{10}\dfrac{\Delta p}{\Delta \ell}\right) \\[2ex] \dfrac{1}{3r^3}\left(\dfrac{1}{3}q_1 + \dfrac{r}{2}\dfrac{\Delta p}{\Delta \ell}\right) \\[2ex] -\dfrac{1}{2}r^7\left(\dfrac{1}{7}q_1 + \dfrac{r}{8}\dfrac{\Delta p}{\Delta \ell}\right) \end{bmatrix}$$

m ≥ 6

$$\mathbf{R}(r) = \begin{bmatrix} \alpha_1 r^{4-m}\left(\dfrac{q_1}{4-m} + \dfrac{r}{5-m}\dfrac{\Delta p}{\Delta \ell}\right) \\[2ex] \alpha_2 r^{4+m}\left(\dfrac{q_1}{4+m} + \dfrac{r}{5+m}\dfrac{\Delta p}{\Delta \ell}\right) \\[2ex] \alpha_3 r^{2-m}\left(\dfrac{q_1}{2-m} + \dfrac{r}{3-m}\dfrac{\Delta p}{\Delta \ell}\right) \\[2ex] \alpha_4 r^{2+m}\left(\dfrac{q_1}{2+m} + \dfrac{r}{3+m}\dfrac{\Delta p}{\Delta \ell}\right) \end{bmatrix}$$

$$\alpha_1 = \frac{1}{8m(m-1)D} \qquad \alpha_2 = \frac{1}{8m(m+1)D} \qquad \alpha_3 = -\alpha_2 \qquad \alpha_4 = -\alpha_1$$

TABLE 18-7 | Transfer Matrices for Circular Plates | 1010

TABLE 18-7 (continued)   TRANSFER MATRICES FOR CIRCULAR PLATE ELEMENTS

Circular Plate Element without Center Hole

Notation

p = magnitude of applied distributed forces at r

c = magnitude of applied distributed moments at r

W_T = concentrated force (F)

$$\frac{\Delta p}{\Delta \ell} = \frac{p_b - p_0}{b} \quad \frac{\Delta c}{\Delta \ell} = \frac{c_b - c_0}{b}$$

Transfer Matrices (Sign Convention 1)

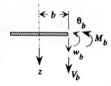

The $\mathbf{U}^i$ in this table for $m \geq 1$ are not truly transfer matrices as the vector at radius $r = 0$ contains arbitrary constants rather than state variables. However, the vector at other radii is the state vector. All operations remain the same as those for the transfer matrices.

$m = 0$ (SYMMETRIC DEFORMATION)

$$\begin{bmatrix} 1 & 0 & 0 & -\dfrac{b^2}{2D(1+\nu)} & F_w \\[2ex] 0 & 0 & 0 & \dfrac{b}{D(1+\nu)} & F_\theta \\[2ex] 0 & 0 & 0 & 0 & F_V \\[1ex] 0 & 0 & 0 & 1 & F_M \\[1ex] 0 & 0 & 0 & 0 & 1 \end{bmatrix}$$

$$\mathbf{U}^i$$

TABLE 18-7 (continued)   TRANSFER MATRICES FOR CIRCULAR PLATE ELEMENTS

$$F_w = \varepsilon_0^c \left[p_0 \frac{b^4}{64D} + \frac{\Delta p}{\Delta \ell} \frac{b^5}{225D} + c_0 \frac{b^3}{9D} + \frac{\Delta c}{\Delta \ell} \frac{b^4}{32D} \right] + \frac{W_T}{8\pi D} b^2 (\ln b - 1)$$

$$F_\theta = -\varepsilon_0^c \left[-p_0 \frac{b^3}{16D} - \frac{\Delta p}{\Delta \ell} \frac{b^4}{45D} - c_0 \frac{b^2}{3D} - \frac{\Delta c}{\Delta \ell} \frac{b^3}{8D} \right] - \frac{W_T}{4\pi D} b \left(\ln b - \tfrac{1}{2} \right)$$

$$F_V = \varepsilon_0^c \left[-p_0 \frac{b}{2} - \frac{\Delta p}{\Delta \ell} \frac{b^2}{3} \right] - \frac{W_T}{2\pi b}$$

$$F_M = -\varepsilon_0^c \left[-p_0 \frac{b^2}{16}(3 + \nu) - \frac{\Delta p}{\Delta \ell} \frac{b^3}{45}(4 + \nu) - c_0 b(2 + \nu) - \frac{\Delta c}{\Delta \ell} \frac{b^2}{8}(3 + \nu) \right]$$
$$- \frac{W_T}{4\pi} \left[(1 + \nu)\ln b + \frac{1 - \nu}{2} \right]$$

$m = 1$ (ASYMMETRIC DEFORMATION)

$$\begin{bmatrix} 0 & -b & -\dfrac{b^3}{2D(3 + \nu)} & 0 & F_w \\[2mm] 0 & 1 & \dfrac{3b^2}{2D(3 + \nu)} & 0 & F_\theta \\[2mm] 0 & 0 & 1 & 0 & F_V \\[1mm] 0 & 0 & r & 0 & F_M \\[1mm] 0 & 0 & 0 & 0 & 1 \end{bmatrix}$$
$$\mathbf{U}^i$$

F_w, F_θ, F_V, and F_M are the loading functions ($m = 1$) of page 1008 with $a = 0$.

$m \geq 2$

$$\begin{bmatrix} b^m & b^{m+2} & 0 & 0 & F_w \\ -mb^{m-1} & -(m + 2)b^{m+1} & 0 & 0 & F_\theta \\ Dm^2(m - 1)(1 - \nu)b^{m-3} & Dm(1 + m)[m(1 - \nu) - 4]b^{m-1} & 0 & 0 & F_V \\ -Dm(m - 1)(1 - \nu)b^{m-2} & -D(1 + m)(m + 2 - \nu m + 2\nu)b^m & 0 & 0 & F_M \\ 0 & 0 & 0 & 0 & 1 \end{bmatrix}$$

F_w, F_θ, F_V, and F_M are the corresponding loading functions ($m \geq 2$) of page 1009 with $a = 0$.

TABLE 18-7   **Transfer Matrices for Circular Plates**   **1012**

TABLE 18-7 (continued)   TRANSFER MATRICES FOR CIRCULAR PLATE ELEMENTS

C. *Parameter* ε_m^j *in Loading Functions* $(j = c \text{ or } s)$

Case	Parameter
1. Distributed load constant in ϕ direction	$\varepsilon_0^c = 1 \qquad \varepsilon_m^j = 0 \qquad m > 0$ $c = j, s$
2. Distributed load constant in ϕ direction, covering $\phi = \phi_1$ to $\phi = \phi_2$	$\varepsilon_0^c = (\phi_2 - \phi_1)/2\pi$ $\varepsilon_m^c = \dfrac{1}{m\pi}(\sin m\phi_2 - \sin m\phi_1) \qquad m > 0$ $\varepsilon_m^s = -\dfrac{1}{m\pi}(\cos m\phi_2 - \cos m\phi_1) \qquad m > 0$
3. Distributed load ramp in ϕ direction	$\varepsilon_0^c = \dfrac{1}{4\pi}(\phi_2 - \phi_1)^2$ $\varepsilon_m^c = \dfrac{1}{m\pi}\left[(\phi_2 - \phi_1)\sin m\phi_2 + \dfrac{1}{m}(\cos m\phi_2 - \cos m\phi_1)\right]$ $m > 0$ $\varepsilon_m^s = \dfrac{1}{m\pi}\left[(\phi_1 - \phi_2)\cos m\phi_2 + \dfrac{1}{m}(\sin m\phi_2 - \sin m\phi_1)\right]$ $m > 0$
4. Harmonic load, $\cos \phi$	$\varepsilon_0^c = 0 \qquad \varepsilon_1^c = 1$ $\varepsilon_m^c = 0 \qquad m > 1 \qquad \varepsilon_m^s = 0 \qquad m > 0$

[a]From Han and Pilkey Ref. 18.10.

TABLE 18-8 | Stiffness Matrices for Circular Plates | 1014

TABLE 18-8 STIFFNESS MATRICES FOR CIRCULAR PLATE ELEMENTS

Circular Plate Element with Center Hole

Notation

The plate response is given by the Fourier series expansion of Eq. (18.10).

w, θ, V, M = deflection, slope, effective shear force per unit length, and bending moment per unit length

E = modulus of elasticity

h = thickness of plate

ν = Poisson's ratio

$\gamma = b/a$

$$D = \frac{Eh^3}{12(1 - \nu^2)}$$

ε_m^j = provided in Part C of Table 18-7

F_w, F_θ, F_V, F_M = loading functions defined in Table 18-7

Stiffness Matrices (Sign Convention 2)

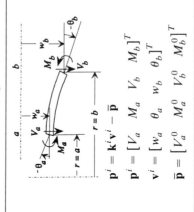

$\mathbf{p}^i = \mathbf{k}^i \mathbf{v}^i - \bar{\mathbf{p}}$

$\mathbf{p}^i = [V_a \quad M_a \quad V_b \quad M_b]^T$

$\mathbf{v}^i = [w_a \quad \theta_a \quad w_b \quad \theta_b]^T$

$\bar{\mathbf{p}} = [V_a^0 \quad M_a^0 \quad V_b^0 \quad M_b^0]^T$

$m = 0$ (SYMMETRIC DEFORMATION)

$$k_{11} = \frac{16\pi D}{a^2 H_0}(\gamma^2 - 1) \qquad k_{12} = -\frac{8\pi D}{aH_0}(2\gamma^2 \ln\gamma - \gamma^2 + 1)$$

$$k_{22} = 2\pi D\left[4\gamma^2(1+\nu)\ln^2\gamma + (1-\nu)\gamma^4 + 2\gamma^2(1+\nu-4\ln\gamma) - 3 - \nu\right]/H_0$$

$$k_{13} = -k_{11} \qquad k_{14} = -\frac{8\pi D\gamma}{aH_0}(\gamma^2 - 1 - 2\ln\gamma) \qquad k_{23} = -k_{12}$$

$$k_{24} = 8\pi D\gamma(\gamma^2\ln\gamma + \ln\gamma + 1 - \gamma^2)/H_0 \qquad k_{33} = k_{11} \qquad k_{34} = -k_{14}$$

$$k_{44} = 2\pi D\left[-4\gamma^2(1+\nu)\ln^2\gamma + (3+\nu)\gamma^4 - 2\gamma^2(1+\nu+4\ln\gamma) - 1 + \nu\right]/H_0$$

$$\gamma = b/a \qquad H_0 = (\gamma^2 - 1)^2 - 4\gamma^2\ln^2\gamma \qquad k_{ij} = k_{ji}$$

$$V_a^0 = k_{13}F_w + k_{14}F_\theta \qquad M_a^0 = k_{23}F_w + k_{24}F_\theta$$

$$V_b^0 = -2\pi a\gamma F_V + k_{33}F_w + k_{34}F_\theta \qquad M_b^0 = -2\pi a\gamma F_M + k_{43}F_w + k_{44}F_\theta$$

F_w, F_θ, F_V, and F_M are the loading functions ($m = 0$) of page 1007.

$m = 1$ (ASYMMETRIC DEFORMATION)

$$k_{11} = \frac{4\pi D\gamma}{a^2 H_1}\left[(\gamma^4 + 3)\ln\gamma + 6(\gamma^2 - 1) - \nu(\gamma^4 - 1)\ln\gamma + \nu(\gamma^2 - 1)^2\right]$$

$$k_{12} = \frac{4\pi D\gamma}{aH_1}\left[(\gamma^4 + 3)\ln\gamma - 2\gamma^2(\gamma^2 - 1) - \nu(\gamma^4 - 1)\ln\gamma + \nu(\gamma^2 - 1)^2\right]$$

$$k_{22} = \frac{4\pi D\gamma}{H_1}\left[(\gamma^4 + 3)\ln\gamma - 2(\gamma^2 - 1) - \nu(\gamma^4 - 1)\ln\gamma + \nu(\gamma^2 - 1)^2\right]$$

$$k_{33} = \frac{4\pi D\gamma^{-1}}{a^2 H_1}\left[6\gamma^2(\gamma^2 - 1) + (3\gamma^4 + 1)\ln\gamma + \nu(\gamma^4 - 1)\ln\gamma - \nu(\gamma^2 - 1)^2\right]$$

$$k_{34} = -\frac{4\pi D}{aH_1}\left[2(\gamma^2 - 1) - (3\gamma^4 + 1)\ln\gamma - \nu(\gamma^4 - 1)\ln\gamma + \nu(\gamma^2 - 1)^2\right]$$

TABLE 18-8 (continued)   STIFFNESS MATRICES FOR CIRCULAR PLATE ELEMENTS

$$k_{44} = \frac{4\pi D\gamma}{H_1}\left[(3\gamma^4 + 1)\ln\gamma - 2\gamma^2(\gamma^2 - 1) + \nu(\gamma^4 - 1)\ln\gamma - \nu(\gamma^2 - 1)^2\right]$$

$$k_{13} = -\frac{4\pi D}{a^2 H_1}\left[3(\gamma^4 - 1) + 4\gamma^2 \ln\gamma\right] \qquad k_{14} = -\frac{4\pi D\gamma}{aH_1}\left[4\gamma^2(\ln\gamma - 1) + \gamma^4 + 3\right]$$

$$k_{23} = -\frac{4\pi D}{aH_1}\left[4\gamma^2(1 + \ln\gamma) - 3\gamma^4 - 1\right] \qquad k_{24} = \frac{4\pi D\gamma}{H_1}\left[\gamma^4 - 1 - 4\gamma^2 \ln\gamma\right]$$

$$H_1 = 2\gamma\left[(\gamma^4 - 1)\ln\gamma - (\gamma^2 - 1)^2\right] \qquad \gamma = b/a \qquad k_{ij} = k_{ji}$$

$$V_a^0 = \varepsilon_1^j(k_{13}F_w + k_{14}F_\theta) \qquad M_a^0 = \varepsilon_1^j(k_{23}F_w + k_{24}F_\theta) \qquad j = c, s$$

$$V_b^0 = \varepsilon_1^j[-2\pi bF_V + k_{33}F_w + k_{34}F_\theta]$$

$$M_b^0 = \varepsilon_1^j[-2\pi bF_M + k_{34}F_w + k_{44}F_\theta]$$

F_w, F_θ, F_V, and F_M are the loading functions ($m = 1$) of page 1008.

$m \geq 2$

$$k_{11} = \frac{2\pi Dm^2}{a^2 H_m}\left[4(\gamma^2 - 1)(m^2 - 2) + 2m\gamma^2(\gamma^{2m} - \gamma^{-2m}) - (1 + \nu)\gamma^2(\gamma^m - \gamma^{-m})^2 - (1 - \nu)m^2(\gamma^2 - 1)^2\right]$$

$$k_{12} = -\frac{2\pi D}{aH_m}\left[m^2\gamma^2(\gamma^m + \gamma^{-m})^2 + m^4(1 - \nu)(\gamma^2 - 1)^2 - 4m^2 - 2m\gamma^2(\gamma^{2m} - \gamma^{-2m}) + \nu m^2\gamma^2(\gamma^m - \gamma^{-m})^2\right]$$

$$k_{22} = \frac{2\pi D}{H_m}\left[2m\gamma^2(\gamma^{2m} - \gamma^{-2m}) - m^2(\gamma^2 - 1)(\gamma^2 + 3) - (1 + \nu)\gamma^2(\gamma^m - \gamma^{-m})^2 + m^2\nu(\gamma^2 - 1)^2\right]$$

$$k_{13} = -\frac{4\pi Dm^2}{a^2 H_m}\left[m(\gamma^m - \gamma^{-m})(\gamma^2 + 1) + (\gamma^m + \gamma^{-m})(\gamma^2 - 1)(m^2 - 2)\right]$$

TABLE 18-8   Stiffness Matrices for Circular Plates   1016

$$k_{23} = \frac{4\pi Dm}{aH_m}\left[(\gamma^m - \gamma^{-m})\left[m^2(\gamma^2 - 1) - 2\gamma^2\right] + m(\gamma^m + \gamma^{-m})(\gamma^2 - 1)\right]$$

$$k_{33} = \frac{2\pi Dm^2}{a^2 H_m}\left[(1 + \nu)(\gamma^m - \gamma^{-m})^2 + 2m(\gamma^{2m} - \gamma^{-2m}) + (\gamma^2 - 1)(5m^2 - 8 - \nu m^2) - m^2\gamma^{-2}(\gamma^2 - 1)(1 - \nu)\right]$$

$$k_{14} = -\frac{4\pi Dm\gamma}{aH_m}\left[(\gamma^m - \gamma^{-m})\left[m^2(\gamma^2 - 1) + 2\right] - m(\gamma^m + \gamma^{-m})(\gamma^2 - 1)\right]$$

$$k_{24} = -\frac{4\pi Dm\gamma}{H_m}\left[(\gamma^m - \gamma^{-m})(\gamma^2 + 1) - m(\gamma^m + \gamma^{-m})(\gamma^2 - 1)\right]$$

$$k_{34} = \frac{2\pi D\gamma}{aH_m}\left[m^4(\gamma^2 - 1)^2\gamma^{-2}(1 - \nu) + m^2(1 + \nu)(\gamma^m - \gamma^{-m})^2 + 2m(\gamma^{2m} - \gamma^{-2m}) - 4m^2(\gamma^2 - 1)\right]$$

$$k_{44} = \frac{2\pi D}{H_m}\left[2m\gamma^2(\gamma^{2m} - \gamma^{-2m}) + \gamma^2(1 + \nu)(\gamma^m - \gamma^{-m})^2 - m^2(\gamma^2 - 1)(3\gamma^2 + 1) - \nu m^2(\gamma^2 - 1)^2\right]$$

$$H_m = \left[\gamma^2(\gamma^m - \gamma^{-m})^2 - m^2(\gamma^2 - 1)^2\right] \qquad \gamma = b/a \qquad k_{ij} = k_{ji}$$

$$V_a^0 = \varepsilon_m^j(k_{13}F_w + k_{14}F_\theta) \qquad M_a^0 = \varepsilon_m^j(k_{23}F_w + k_{24}F_\theta)$$

$$V_b^0 = \varepsilon_m^j[-2\pi bF_V + k_{33}F_w + k_{34}F_\theta] \qquad M_b^0 = \varepsilon_m^j[-2\pi bF_M + k_{43}F_w + k_{44}F_\theta]$$

$$j = c, s$$

F_w, F_θ, F_V, and F_M are the loading functions ($m \geq 2$) of page 1009.

Stiffness Matrices for Circular Plates   TABLE 18-8

TABLE 18-8 (continued)   STIFFNESS MATRICES FOR CIRCULAR PLATE ELEMENTS

Circular Plate Element without Center Hole

Stiffness Matrices (Sign Convention 2)

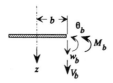

$m = 0$ (SYMMETRIC DEFORMATION)

$$\begin{bmatrix} V_b \\ M_b \end{bmatrix} = \underbrace{\begin{bmatrix} 0 & 0 \\ 0 & 2\pi D(1+\nu) \end{bmatrix}}_{\mathbf{k}^i} \begin{bmatrix} w_b \\ \theta_b \end{bmatrix} - \underbrace{\begin{bmatrix} V_b^0 \\ M_b^0 \end{bmatrix}}_{\overline{\mathbf{p}}}$$

$$\overline{\mathbf{p}} = \begin{bmatrix} -2\pi b F_V \\ 2\pi b \left(-F_M + \dfrac{D(1+\nu)}{b} F_\theta \right) \end{bmatrix}$$

$m = 1$ (ASYMMETRIC DEFORMATION)

$$\begin{bmatrix} V_b \\ M_b \end{bmatrix} = \underbrace{\begin{bmatrix} \dfrac{2\pi D(3+\nu)}{b^2} & \dfrac{2\pi D(3+\nu)}{b} \\ \dfrac{2\pi D(3+\nu)}{b} & 2\pi D(3+\nu) \end{bmatrix}}_{\mathbf{k}^i} \begin{bmatrix} w_b \\ \theta_b \end{bmatrix} - \underbrace{\begin{bmatrix} V_b^0 \\ M_b^0 \end{bmatrix}}_{\overline{\mathbf{p}}}$$

$$\overline{\mathbf{p}} = -2\pi b \begin{bmatrix} F_V \\ F_M \end{bmatrix} + \mathbf{k}^i \begin{bmatrix} F_w \\ F_\theta \end{bmatrix}$$

$m \geq 2$

$$\begin{bmatrix} V_b \\ M_b \end{bmatrix} = \underbrace{\begin{bmatrix} k_{11} & k_{12} \\ k_{21} & k_{22} \end{bmatrix}}_{\mathbf{k}^i} \begin{bmatrix} w_b \\ \theta_b \end{bmatrix} - \underbrace{\begin{bmatrix} V_b^0 \\ M_b^0 \end{bmatrix}}_{\overline{\mathbf{p}}}$$

$k_{11} = 2\pi D m^2 (1 + 2m + \nu)/b^2$
$k_{12} = k_{21} = 2\pi D m(2 + m + m\nu)/b$
$k_{22} = 2\pi D(1 + 2m + \nu)$

$$\overline{\mathbf{p}} = -2\pi b \begin{bmatrix} F_V \\ F_M \end{bmatrix} + \mathbf{k}^i \begin{bmatrix} F_w \\ F_\theta \end{bmatrix}$$

[a]From Han and Pilkey Ref. 18.10.

TABLE 18-8   Stiffness Matrices for Circular Plates   1018

Notation

The plate response is given by the Fourier series expansion of Eq. (18.10). See Table 11-21 for values of $k_1, k_1^*, \ldots$ for more complex spring systems, including those with masses.

$$w, \theta, V, M = \text{deflection, slope, effective shear force per unit length,}$$
and bending moment per unit length

W_m^j, C_m^j = loading functions in transfer matrix

j = c or s

$m = 0, 1, 2, \ldots$ = identifies term in Fourier series expansion (Eq. 18.10)

Ω = speed of rotation of rotating circular plate

m_i = lumped mass or mass of ring per unit length

r_s = radius of gyration of ring about circumferential axis

r_r, r_ϕ = radii of gyration of mass per unit length
about radial and tangential axes,
for homogeneous material set $r_r^2 = r_\phi^2 = \frac{1}{12}h^2$

h = thickness of plate

ω = natural frequency

I_s = area moment of inertia of ring
about middle plane of plate

E_s = modulus of elasticity of ring

Transfer Matrix:

$$\begin{bmatrix} 1 & 0 & 0 & 0 & 0 \\ 0 & 1 & 0 & 0 & 0 \\ k_1 - m_i\omega^2 - m_i\omega^2 m^2 r_r^2 & -m_i\Omega^2 a_i & 1 & 0 & -W_m^j \\ 0 & -m_i\omega^2 r_\phi^2 + k_1^* & 0 & 1 & C_m^j \\ 0 & 0 & 0 & 0 & 1 \end{bmatrix}$$
$$\mathbf{U}_i$$

Stiffness and Mass Matrices:

$$\begin{bmatrix} V_a \\ M_a \end{bmatrix} = \begin{bmatrix} k_{11} & k_{12} \\ k_{21} & k_{22} \end{bmatrix} \begin{bmatrix} w_a \\ \theta_a \end{bmatrix} \qquad \begin{bmatrix} V_a \\ M_a \end{bmatrix} = -\omega^2 \begin{bmatrix} m_{11} & m_{12} \\ m_{21} & m_{32} \end{bmatrix} \begin{bmatrix} w_a \\ \theta_a \end{bmatrix}$$

Case	Point Matrices
1. Concentrated force W_T (force) 	TRANSFER MATRIX: $W_0^c = W_T/2\pi a_i$ and $W_0^s = 0$ $W_m^c = (W_T \cos m\phi_1)/\pi a_i \qquad m > 0$ $W_m^s = (W_T \sin m\phi_1)/\pi a_i \qquad m > 0$ STIFFNESS AND MASS MATRICES: Traditionally, this applied load is implemented as nodal conditions.

Case	Point Matrices
2. Uniform line ring load W (force/length) $\rightarrow a_i \leftarrow$	TRANSFER MATRIX: $W_0^c = W \qquad W_0^s = 0 \qquad W_m^c = W_m^s = 0 \qquad m > 0$ STIFFNESS MATRIX: Traditionally, this applied load is implemented as nodal conditions.
3. Uniform line load W (force/length in ϕ direction) $\phi_2 \quad \phi_1$	TRANSFER MATRIX: $W_0^c = W(\phi_2 - \phi_1)/2\pi \qquad W_0^s = 0$ $W_m^c = W(\sin m\phi_2 - \sin m\phi_1)/m\pi \qquad m > 0$ $W_m^s = -W(\cos m\phi_2 - \cos m\phi_1)/m\pi \qquad m > 0$ STIFFNESS MATRIX: Traditionally, this applied load is implemented as nodal conditions.
4. Concentrated moment C_T (force − length) $a_i \quad C_T$ ϕ_1	TRANSFER MATRIX: $C_0^c = C_T/2\pi a_i \qquad C_0^s = 0$ $C_m^c = (C_T \cos m\phi_1)/\pi a_i \qquad m > 0$ $C_m^s = (C_T \sin m\phi_1)/\pi a_i \qquad m > 0$ STIFFNESS MATRIX: Traditionally, this applied load is implemented as nodal conditions.
5. Line ring spring k_1 (force/length squared)[*] $\rightarrow a_i \leftarrow$	TRANSFER MATRIX: See transfer matrix under notation. STIFFNESS MATRIX: $a = a_i$ $\begin{bmatrix} V_a \\ M_a \end{bmatrix} = \begin{bmatrix} k_1 & 0 \\ 0 & 0 \end{bmatrix} \begin{bmatrix} w_a \\ \theta_a \end{bmatrix}$

[*]These matrices apply for $m = 0$ (symmetric deformation).

TABLE 18-9   **Point Matrices for Circular Plates**   **1020**

Case	Point Matrices
6. Rotary line ring spring $(\text{force} - \text{length}/\text{length})^*$ 	TRANSFER MATRIX: See transfer matrix under notation. STIFFNESS MATRIX: $a = a_i$ $$\begin{bmatrix} V_a \\ M_a \end{bmatrix} = \begin{bmatrix} 0 & 0 \\ 0 & k_1^* \end{bmatrix} \begin{bmatrix} w_a \\ \theta_a \end{bmatrix}$$
7. Ring lumped mass m_i $(\text{mass}/\text{length})^*$ $m_i = \Delta a \rho$ ρ = mass per unit area Rotary inertia terms contain r_r, r_ϕ	TRANSFER MATRIX: See transfer matrix under notation. MASS MATRIX: $a = a_i$ $$\begin{bmatrix} V_a \\ M_a \end{bmatrix} = -\omega^2 \begin{bmatrix} m_i & 0 \\ 0 & m_i r_\phi^2 \end{bmatrix} \begin{bmatrix} w_a \\ \theta_a \end{bmatrix}$$
8. Circular reinforcing ring at $r = a_i$ Reinforcing ring must be symmetric in z direction about middle plane of plate*	TRANSFER MATRIX: $$\mathbf{U}_i = \begin{bmatrix} 1 & 0 & 0 & 0 & 0 \\ 0 & 1 & 0 & 0 & 0 \\ -m_i\omega^2 & 0 & 1 & 0 & 0 \\ 0 & E_s I_s/a_i^2 - m_i r_s^2 \omega^2 & 0 & 1 & 0 \\ 0 & 0 & 0 & 0 & 1 \end{bmatrix}$$ STIFFNESS AND MASS MATRICES: $a = a_i$ $$\begin{bmatrix} V_a \\ M_a \end{bmatrix} = \begin{bmatrix} 0 & 0 \\ 0 & E_s I_s/a_i^2 \end{bmatrix} \begin{bmatrix} w_a \\ \theta_a \end{bmatrix}$$ $$\begin{bmatrix} V_a \\ M_a \end{bmatrix} = -\omega^2 \begin{bmatrix} m_i & 0 \\ 0 & m_i r_s^2 \end{bmatrix} \begin{bmatrix} w_a \\ \theta_a \end{bmatrix}$$

*These matrices apply for $m = 0$ (symmetric deformation).

TABLE 18-10   STIFFNESS MATRICES FOR INFINITE CIRCULAR PLATE ELEMENT WITH ELASTIC FOUNDATION[a]

Notation

Use the stiffness matrix of this table for the outer infinite plate element. Use stiffness matrices of Table 18-8 for inner elements. Loadings should be placed on inner elements, represented by the loading vectors of Table 18-8. The plate response is given by the Fourier series expansion of Eq. (18.10).

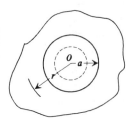

$m = 0, 1, 2, \ldots,$ identifies term in Fourier series expansion

E = modulus of elasticity

ν = Poisson's ratio

D = plate rigidity, $= Eh^3/[12(1 - \nu^2)]$

$\ker_m, \mathrm{kei}_m$ = Kelvin functions of mth order

V_a^j, M_a^j = shear force and bending moment per unit length along circumference at $r = a$

w_a^j, θ_a^j = deflection and slope at $r = a$.

$j = c, s$

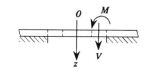

Stiffness Matrices

Infinite Circular Element with Center Hole on Elastic Foundation

$m = 0$ (SYMMETRIC BENDING)

$$\begin{bmatrix} V_a \\ M_a \end{bmatrix} = \begin{bmatrix} k_{11} & k_{12} \\ k_{21} & k_{22} \end{bmatrix} \begin{bmatrix} w_a \\ \theta_a \end{bmatrix} \qquad w_a = w_a^c, \theta_a = \theta_a^c, M_a = M_a^c, V_a = V_a^c$$

$$\mathbf{k}_{aa}^i$$

$$k_{11} = 2\pi a[b_{33}(a\lambda)e_{23}(a\lambda) - b_{34}(a\lambda)e_{23}(a\lambda)]/B_0$$

$$k_{12} = k_{21} = 2\pi a[b_{33}(a\lambda)e_{14}(a\lambda) - b_{34}(a\lambda)e_{13}(a\lambda)]/B_0$$

$$k_{22} = 2\pi a[b_{43}(a\lambda)e_{14}(a\lambda) - b_{44}(a\lambda)e_{13}(a\lambda)]/B_0$$

$$B_0 = e_{14}(a\lambda)e_{23}(a\lambda) - e_{13}(a\lambda)e_{24}(a\lambda)$$

$$e_{13} = \ker x \quad e_{14} = \mathrm{kei}\, x \qquad x = \lambda r \qquad \lambda = (k/D)^{1/4}$$

where $\ker x, \mathrm{kei}\, x$ are Kelvin functions of order 0.

$$e_{23} = \frac{\lambda}{\sqrt{2}}(\ker_1 x + \mathrm{kei}_1 x) \qquad e_{24} = \frac{\lambda}{\sqrt{2}}(-\ker_1 x + \mathrm{kei}_1 x)$$

$$b_{33} = D\lambda^2 \left[\frac{\lambda}{\sqrt{2}}(\mathrm{kei}_1 x - \ker_1 x) - \frac{2}{r}\mathrm{kei}\, x \right]$$

$$b_{34} = \frac{-D\lambda^3}{\sqrt{2}}(\ker_1 x + \mathrm{kei}_1 x)$$

$$b_{43} = D\lambda \left[\frac{1 - \nu}{\sqrt{2}\, r}(\ker_1 x + \mathrm{kei}_1 x) - \lambda\, \mathrm{kei}\, x \right]$$

$$b_{44} = -D\lambda \left[\frac{1 - \nu}{\sqrt{2}\, r}(\ker_1 x - \mathrm{kei}_1 x) + \lambda\, \ker x \right]$$

TABLE 18-10   Stiffness Matrix for Infinite Circular Plate Element   1022

Stiffness Matrices

$m \geq 1$ (ASYMMETRIC BENDING)

$$\begin{bmatrix} V_a^j \\ M_a^j \end{bmatrix} = \begin{bmatrix} k_{11} & k_{12} \\ k_{21} & k_{22} \end{bmatrix}^i \begin{bmatrix} w_a^j \\ \theta_a^j \end{bmatrix} \qquad j = s, c$$

$$\mathbf{k}_{aa}^i$$

$k_{11} = 2\pi a[b_{33}(a\lambda)e_{24}(a\lambda) - b_{34}(a\lambda)e_{23}(a\lambda)]/B_m$

$k_{12} = k_{21} = 2\pi a[b_{33}(a\lambda)e_{14}(a\lambda) - b_{34}(a\lambda)e_{13}(a\lambda)]/B_m$

$k_{22} = 2\pi a[b_{43}(a\lambda)e_{14}(a\lambda) - b_{44}(a\lambda)e_{13}(a\lambda)]/B_m$

$B_m = e_{14}(a\lambda)e_{23}(a\lambda) - e_{13}(a\lambda)e_{24}(a\lambda)$

$e_{13} = \ker_m x \qquad e_{14} = \kei_m x \qquad x = \lambda r \qquad \lambda = (k/D)^{1/4}$

$e_{23} = -\lambda \left[\dfrac{m}{x} \ker_m x + \dfrac{1}{\sqrt{2}} (\ker_{m-1} x + \kei_{m-1} x) \right]$

$e_{24} = -\lambda \left[\dfrac{m}{x} \kei_m x + \dfrac{1}{\sqrt{2}} (\kei_{m-1} x - \ker_{m-1} x) \right]$

$e_{33} = \lambda^2 \dfrac{d}{dx} e_{23} \qquad e_{34} = \lambda^2 \dfrac{d}{dx} e_{24} \qquad e_{43} = \lambda \dfrac{d}{dx} e_{33} \qquad e_{44} = \lambda \dfrac{d}{dx} e_{34}$

$b_{33} = D\left[\dfrac{m^2(\nu - 3)}{r^3} e_{13} + \dfrac{1 + m^2(2 - \nu)}{r^2} e_{23} - \dfrac{e_{33}}{r} - e_{43} \right]$

$b_{34} = D\left[\dfrac{m^2(\nu - 3)}{r^3} e_{14} + \dfrac{1 + m^2(2 - \nu)}{r^2} e_{24} - \dfrac{e_{34}}{r} - e_{44} \right]$

$b_{43} = D\left[\dfrac{\nu m^2}{r^2} e_{13} - \dfrac{\nu}{r} e_{23} - e_{33} \right] \qquad b_{44} = D\left[\dfrac{\nu m^2}{r^2} e_{14} - \dfrac{\nu}{r} e_{24} - e_{34} \right]$

[a]From Han and Pilkey Ref. 18.10.

TABLE 18-11 GEOMETRIC STIFFNESS MATRICES FOR CIRCULAR PLATES: SYMMETRIC CASE[a]

Notation

$$w, \theta = \text{deflection and slope} \qquad \gamma = b/a$$
$$h = \text{thickness of plate}$$
$$u_a = \text{radial displacement at } r = a;$$
$$\text{obtained from a disk analysis using the formulas of Chapter 19.}$$

Sign convention is the same as that for stiffness matrices in Table 18-8.

Case	Geometric Stiffness Matrices for Symmetric Bending ($m = 0$)
1. Circular plate element with center hole 	This matrix corresponds to a radial compressive force (per unit length) P in plane of plate; this in-plane force in an element is obtained from a disk analysis (Chapter 19) of the whole disk subject to a system of prescribed applied in-plane loadings.

$$\mathbf{k}_G^i = \begin{bmatrix} g_{11} & g_{12} & g_{13} & g_{14} \\ g_{21} & g_{22} & g_{23} & g_{24} \\ g_{31} & g_{32} & g_{33} & g_{34} \\ g_{41} & g_{42} & g_{43} & g_{44} \end{bmatrix} \qquad \mathbf{v}^i = \begin{bmatrix} w_a \\ \theta_a \\ w_b \\ \theta_b \end{bmatrix}$$

$$g_{ij} = g_{ji} = (-1)^{i+j} 2\pi \sum_{k=1}^{4} \sum_{l=1}^{4} \gamma_{ik}\gamma_{jl}\zeta_{kl} \qquad i, j = 1, 2, 3, 4$$

where γ_{ik}, γ_{jl} are given in Table 18-12 for $m = 0$.

$$\zeta_{kl} = \int_a^b \left[1 - \left(1 - \frac{a^2}{r^2} \right) C_1 \right] \beta_k'(r)\beta_l'(r) r\, dr \qquad k, l = 1, 2, 3, 4$$

$$\beta_1'(r) = 0 \qquad \beta_2'(r) = \frac{2r}{a^2} \qquad \beta_3'(r) = \frac{1}{r}$$

$$\beta_4'(r) = \frac{r}{a^2}\left(1 + 2\ln\frac{r}{a} \right)$$

$$C_1 = \frac{(1 - \nu)}{2} + u_a\frac{Eh}{2aP} \qquad \zeta_{kl} = \zeta_{lk}$$

TABLE 18-11 Geometric Stiffness Matrices for Circular Plates **1024**

Case	Geometric Stiffness Matrices for Symmetric Bending ($m = 0$)
1. Continued	In explicit form: $\zeta_{11} = \zeta_{12} = \zeta_{13} = \zeta_{14} = 0.$ $\zeta_{22} = (\gamma^2 - 1)[1(\gamma^2 + 1) - C_1(\gamma^2 - 1)]$ $\gamma = b/a$ $\zeta_{23} = (1 - C_1)(\gamma^2 - 1) + 2C_1 \ln \gamma$ $\zeta_{24} = (1 - C_1)[\gamma^4(\frac{1}{4} + \ln \gamma) - \frac{1}{4}] + 2C_1\gamma^2 \ln \gamma$ $\zeta_{33} = (1 - C_1)\ln \gamma + \frac{1}{2}C_1(1 - \gamma^{-2})$ $\zeta_{34} = (1 - C_1)\gamma^2 \ln \gamma + C_1(1 + \ln \gamma)\ln \gamma$ $\zeta_{44} = (1 - C_1)[\gamma^4(\ln^2 \gamma + \frac{1}{2}\ln \gamma + \frac{1}{8}) - \frac{1}{8}]$ $+ C_1[\gamma^2(2\ln^2 \gamma + \frac{1}{2}) - \frac{1}{2}]$
2. Circular plate without center hole	This matrix corresponds to an in-plane compressive force P_a (force/length) applied at $r = a$. $\mathbf{k}_G^i = \begin{bmatrix} 0 & 0 \\ 0 & \frac{1}{2}\pi a^2 \end{bmatrix}$ $\mathbf{v}^i = \begin{bmatrix} w_a \\ \theta_a \end{bmatrix}$

[a]From Han and Pilkey Ref. 18.10.

TABLE 18-12 MASS MATRICES FOR CIRCULAR PLATE ELEMENTS IN BENDING[a]

Notation

w, θ = deflection and slope

ρ = mass per unit area

$\gamma = b/a$

M = total mass of circular plate element with center hole, $= \rho\pi(b^2 - a^2)$

The sign convention is the same as that for stiffness matrices in Table 18-8.

Mass Matrix:

$$\mathbf{m}^i = \begin{bmatrix} m_{11} & m_{12} & m_{13} & m_{14} \\ m_{21} & m_{22} & m_{23} & m_{24} \\ m_{31} & m_{32} & m_{33} & m_{34} \\ m_{41} & m_{42} & m_{43} & m_{44} \end{bmatrix} \qquad \mathbf{v}^i = \begin{bmatrix} w_a \\ \theta_a \\ w_b \\ \theta_b \end{bmatrix}$$

Mass Matrices

Circular Plate with Center Hole, Consistent Mass

$m = 0$ (SYMMETRIC DEFORMATION) $\gamma = b/a$

$$m_{kj} = (-1)^{k+j} 2\pi \sum_{n=1}^{4} \sum_{l=1}^{4} \alpha_{kn} \alpha_{jl} \lambda_{nl}$$

$$\lambda_{nl} = \int_{a}^{b} \rho \beta_n(r) \beta_l(r) r\, dr$$

$\alpha_{11} = \gamma^2[\gamma^2 - 1 + 2\ln\gamma - 4\ln^2\gamma]/H_0$

$\alpha_{12} = (1 - 2\gamma^2\ln\gamma - \gamma^2)/H_0$

$\alpha_{13} = (4\gamma^2\ln\gamma)/H_0$

$\alpha_{14} = 2(\gamma^2 - 1)/H_0$

$\alpha_{21} = (2\gamma a\ln\gamma)/H_0$

$\alpha_{22} = -\alpha_{21} \qquad \alpha_{23} = a\gamma^2(\gamma^2 - 1 - 2\ln\gamma)/H_0$

$\alpha_{24} = a(2\gamma^2\ln\gamma - \gamma^2 + 1)/H_0$

$\alpha_{31} = \alpha_{12} \qquad \alpha_{32} = -\alpha_{12} \qquad \alpha_{33} = -\alpha_{13} \qquad \alpha_{34} = -\alpha_{14}$

$\alpha_{41} = [a\gamma(\gamma^2 - 1)\ln\gamma]/H_0$

$\alpha_{42} = -\alpha_{41} \qquad \alpha_{43} = a\gamma(2\gamma^2\ln\gamma - \gamma^2 + 1)/H_0$

$\alpha_{44} = a\gamma(\gamma^2 - 1 - 2\ln\gamma)/H_0$

$H_0 = (\gamma^2 - 1)^2 - 4\gamma^2\ln^2\gamma \qquad \beta_1(r) = 1, \qquad \beta_2(r) = (r/a)^2$

$\beta_3(r) = \ln(r/a) \qquad \beta_4(r) = (r/a)^2\ln(r/a)$

TABLE 18-12 | **Mass Matrices for Circular Plate Elements** | **1026**

Mass Matrices

For uniform mass density ρ,

$\lambda_{11} = \rho a^2(\gamma^2 - 1)/2 \qquad \lambda_{21} = \rho a^2(\gamma^4 - 1)/4 \qquad \lambda_{22} = \rho a^2(\gamma^6 - 1)/6$

$\lambda_{31} = \rho a^2[2\gamma^2 \ln \gamma - (\gamma^2 - 1)]/4 \qquad \lambda_{32} = \rho a^2[4\gamma^2 \ln \gamma - (\gamma^4 - 1)]/16$

$\lambda_{33} = \rho a^2[2\gamma^2 \ln \gamma(\ln \gamma - 1) + (\gamma^2 - 1)]/4 \qquad \lambda_{41} = \lambda_{32}$

$\lambda_{42} = \rho a^2[6\gamma^6 \ln \gamma - (\gamma^6 - 1)]/36$

$\lambda_{43} = \rho a^2[4\gamma^4 \ln \gamma(2 \ln \gamma - 1) + (\gamma^4 - 1)]/32$

$\lambda_{44} = \rho a^2[6\gamma^6 \ln \gamma(3 \ln \gamma - 1) + (\gamma^6 - 1)]/108 \qquad \lambda_{nl} = \lambda_{ln} \qquad (n, l = 1, 2, 3, 4)$

$m \geq 1$ (ASYMMETRIC DEFORMATION)

$$m_{kj} = (-1)^{k+j}\pi \sum_{n=1}^{4} \sum_{l=1}^{4} \alpha_{kn}\alpha_{jl}\lambda_{nl}$$

$$\lambda_{nl} = \int_a^b \rho r \beta_n(r)\beta_l(r)\, dr$$

$m = 1$, $\rho = $ CONST., $\gamma = b/a$

$\alpha_{11} = \gamma[2(\gamma^4 - 3)\ln \gamma - (\gamma^2 - 1)(\gamma^2 - 3)]/H_1$

$\alpha_{12} = \gamma^3[\gamma^2(2 \ln \gamma - 3) + 3]/H_1$

$\alpha_{13} = \gamma(\gamma^2 + 2 \ln \gamma - 1)/H_1$

$\alpha_{14} = 2\gamma(3 + \gamma^2)(1 - \gamma^2)/H_1$

$\alpha_{21} = a\gamma[2 + (\gamma^4 + 1)(2 \ln \gamma - 1)]/H_1$

$\alpha_{22} = a\gamma^3[\gamma^2(1 - 2 \ln \gamma) - 1]/H_1$

$\alpha_{23} = a\gamma(\gamma^2 - 2 \ln \gamma - 1)/H_1 \qquad \alpha_{24} = -2a\gamma(\gamma^2 - 1)^2/H_1$

$\alpha_{31} = [4\gamma^2(1 + \ln \gamma) - 3\gamma^4 - 1]/H_1 \qquad \alpha_{32} = \gamma^2(3\gamma^2 - 3 - 2 \ln \gamma)/H_1$

$\alpha_{33} = [1 - \gamma^2(1 + 2 \ln \gamma)]/H_1 \qquad \alpha_{34} = 2(\gamma^2 - 1)(3\gamma^2 + 1)/H_1$

$\alpha_{41} = a\gamma(\gamma^4 - 4\gamma^2 \ln \gamma - 1)/H_1 \qquad \alpha_{42} = a\gamma^3(2 \ln \gamma + 1 - \gamma^2)/H_1$

$\alpha_{43} = a\gamma[\gamma^2(2 \ln \gamma - 1) + 1]/H_1 \qquad \alpha_{44} = -2a\gamma(\gamma^2 - 1)^2/H_1 = \alpha_{24}$

$H_1 = 4\gamma[(\gamma^4 - 1)\ln \gamma - (\gamma^2 - 1)^2]$

$\beta_1(r) = r/a \qquad \beta_2(r) = (r/a)^{-1} \qquad \beta_3(r) = (r/a)^3$

$\beta_4(r) = (r/a)\ln(r/a)$

$\lambda_{11} = \rho a^2(\gamma^4 - 1)/4 \qquad \lambda_{22} = \rho a^2 \ln \gamma$

$\lambda_{21} = \rho a^2(\gamma^2 - 1)/2 \qquad \lambda_{32} = \rho a^2(\gamma^4 - 1)/4$

$\lambda_{31} = \rho a^2(\gamma^6 - 1)/6 \qquad \lambda_{42} = \rho a^2(2\gamma^2 \ln \gamma - \gamma^2 + 1)/4$

$\lambda_{41} = \rho a^2(4\gamma^4 \ln \gamma - \gamma^4 + 1)/16 \qquad \lambda_{33} = \rho a^2(\gamma^8 - 1)/8$

$\lambda_{43} = \rho a^2(6\gamma^6 \ln \gamma - \gamma^6 + 1)/36$

$\lambda_{44} = \rho a^2(8\gamma^4 \ln^2 \gamma - 4\gamma^4 \ln \gamma + \gamma^4 - 1)/32$

$\lambda_{nl} = \lambda_{ln} \qquad (n, l = 1, 2, 3, 4)$

Mass Matrices

$m > 1;\ \rho = \text{CONST}$

$\alpha_{11} = \gamma^2[m^2(1 - \gamma^2) - (m + 2)(1 - \gamma^{-2m})]/H_m$

$\alpha_{12} = \gamma^2[m^2(1 - \gamma^2) + (m - 2)(1 - \gamma^{2m})]/H_m$

$\alpha_{13} = m[(m - 1)(\gamma^2 - 1) - (\gamma^{-2m+2} - 1)]/H_m$

$\alpha_{14} = m[(m + 1)(\gamma^2 - 1) + (\gamma^{2m+2} - 1)]/H_m$

$\alpha_{21} = a\gamma^2[m(1 - \gamma^2) + (1 - \gamma^{-2m})]/H_m$

$\alpha_{22} = a\gamma^2[m(\gamma^2 - 1) + (1 - \gamma^{2m})]/H_m$

$\alpha_{23} = a[m(\gamma^2 - 1) - \gamma^2(1 - \gamma^{-2m})]/H_m$

$\alpha_{24} = a[m(1 - \gamma^2) + \gamma^2(\gamma^{2m} - 1)]/H_m$

$\alpha_{31} = \gamma^2[m^2\gamma^{-m-2}(\gamma^2 - 1) + (m + 2)(\gamma^m - \gamma^{-m})]/H_m$

$\alpha_{32} = \gamma^2[m^2\gamma^{m-2}(\gamma^2 - 1) + (m - 2)(\gamma^m - \gamma^{-m})]/H_m$

$\alpha_{33} = m[(m - 2)\gamma^{-m}(1 - \gamma^2) - (\gamma^m - \gamma^{-m})]/H_m$

$\alpha_{34} = m[(m + 2)\gamma^m(1 - \gamma^2) - (\gamma^m - \gamma^{-m})]/H_m$

$\alpha_{41} = a\gamma^2[m\gamma^{-m-1}(\gamma^2 - 1) - \gamma(\gamma^m - \gamma^{-m})]/H_m$

$\alpha_{42} = a\gamma^2[m\gamma^{m-1}(1 - \gamma^2) + \gamma(\gamma^m - \gamma^{-m})]/H_m$

$\alpha_{43} = a[m\gamma^{-m+1}(1 - \gamma^2) + \gamma(\gamma^m - \gamma^{-m})]/H_m$

$\alpha_{44} = a[m\gamma^{m+1}(\gamma^2 - 1) - \gamma(\gamma^m - \gamma^{-m})]/H_m$

$H_m = 2[\gamma^2(\gamma^m - \gamma^{-m})^2 - m^2(\gamma^2 - 1)^2]$

$\beta_1(r) = \left(\dfrac{r}{a}\right)^m$

$\beta_2(r) = \left(\dfrac{r}{a}\right)^{-m}$

$\beta_3(r) = \left(\dfrac{r}{a}\right)^{2+m}$

$\beta_4(r) = \left(\dfrac{r}{a}\right)^{2-m}$

$\lambda_{11} = \rho a^2(\gamma^{2m+2} - 1)/(2m + 2)$

$\lambda_{21} = \rho a^2(\gamma^2 - 1)/2$

$\lambda_{22} = \rho a^2(1 - \gamma^{2-2m})/(2m - 2)$

$\lambda_{31} = \rho a^2(\gamma^{2m+4} - 1)/(2m + 4)$

$\lambda_{32} = \rho a^2(\gamma^4 - 1)/4$

$\lambda_{33} = \rho a^2(\gamma^{2m+6} - 1)/(2m + 6)$

$\lambda_{41} = \rho a^2(\gamma^4 - 1)/4$

$\lambda_{42} = \rho a^2(1 - \gamma^{4-2m})/(2m - 4)\quad (m \neq 2)$

$\lambda_{42} = \rho a^2 \ln \gamma \quad (m = 2)$

$\lambda_{43} = \rho a^2(\gamma^6 - 1)/6$

$\lambda_{44} = \rho a^2(1 - \gamma^{6-2m})/(2m - 6)\quad (m \neq 3)$

$\lambda_{44} = \rho a^2 \ln \gamma \quad (m = 3)$

$\gamma = b/a \qquad \lambda_{nl} = \lambda_{ln} \qquad (n, l = 1, 2, 3, 4)$

TABLE 18-12   **Mass Matrices for Circular Plate Elements**   **1028**

Mass Matrices

Circular Plate without Center Hole, Consistent Mass

$m = 0$ (SYMMETRIC DEFORMATION)

$$\mathbf{m}^i = \begin{bmatrix} \pi \rho b^2 & \pi \rho b^3/4 \\ \pi \rho b^3/4 & \pi \rho b^4/12 \end{bmatrix}$$

$$\mathbf{v}^i = \begin{bmatrix} w_b \\ \theta_b \end{bmatrix}$$

$m \geq 1$ (ASYMMETRIC DEFORMATION)

$$\mathbf{m}^i = \begin{bmatrix} m_{11} & m_{12} \\ m_{21} & m_{22} \end{bmatrix}$$

$$\mathbf{v}^i = \begin{bmatrix} w_b \\ \theta_b \end{bmatrix}$$

$m_{11} = \pi \rho b^2 [(m + 2)^2/(2m + 2) - m + m^2/(2m + 6)]/4$

$m_{12} = m_{21} = -\pi \rho b^3 [(m + 1)/(m + 2) - (m + 2)/(2m + 2) - m/(2m + 6)]/4$

$m_{22} = \pi \rho b^4 [1/(2m + 2) - 2/(2m + 4) + 1/(2m + 6)]/4$

Circular Plate with Center Hole, Lumped Mass

MASS LUMPED ALONG CIRCUMFERENCE AT $r = a$:

$$\mathbf{m}^i = \begin{bmatrix} M & 0 & 0 & 0 \\ 0 & 0 & 0 & 0 \\ 0 & 0 & 0 & 0 \\ 0 & 0 & 0 & 0 \end{bmatrix}$$

$$\mathbf{v}^i = \begin{bmatrix} w_a \\ \theta_a \\ w_b \\ \theta_b \end{bmatrix}$$

Mass Matrices

MASS LUMPED ALONG CIRCUMFERENCE AT $r = b$:

$$\mathbf{m}^i = \begin{bmatrix} 0 & 0 & 0 & 0 \\ 0 & 0 & 0 & 0 \\ 0 & 0 & M & 0 \\ 0 & 0 & 0 & 0 \end{bmatrix}$$

$$\mathbf{v}^i = \begin{bmatrix} w_a \\ \theta_a \\ w_b \\ \theta_b \end{bmatrix}$$

MASS LUMPED AT BOTH $r = a$ AND $r = b$:

$$\mathbf{m}^i = \begin{bmatrix} \frac{1}{2}M & 0 & 0 & 0 \\ 0 & 0 & 0 & 0 \\ 0 & 0 & \frac{1}{2}M & 0 \\ 0 & 0 & 0 & 0 \end{bmatrix}$$

$$\mathbf{v}^i = \begin{bmatrix} w_a \\ \theta_a \\ w_b \\ \theta_b \end{bmatrix}$$

Circular Plate without Center Hole, Lumped Mass

MASS LUMPED AT $r = b$:

$$\mathbf{m}^i = \begin{bmatrix} M_0 & 0 \\ 0 & 0 \end{bmatrix}$$

$$\mathbf{v}^i = \begin{bmatrix} w_b \\ \theta_b \end{bmatrix}$$

$$M_0 = \pi b^2 \rho$$

---

[a]From Han and Pilkey Ref. 18.10.

TABLE 18-12   **Mass Matrices for Circular Plate Elements**   **1030**

TABLE 18-13   DISPLACEMENTS AND STRESSES FOR LARGE DEFLECTION OF UNIFORMLY LOADED CIRCULAR PLATES

Notation

E = modulus of elasticity
ν = Poisson's ratio
h = thickness of plate
a = radius of plate
w_0 = maximum deflection at center
p_1 = uniformly distributed load on plate

The stresses in this table are at the lower surface. To use this table, solve the relationship between load and deflection for w_0 and then calculate the other responses.

Plates	Deflection and Stresses
1. Simply supported outer edge with in-plane force [18.11] 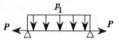 The radial displacement at the outer edge is zero.	Relationship between load and deflection: $$\frac{p_1 a^4}{Eh^4} = \frac{5.333}{1-\nu^2}\frac{w_0}{h} + \frac{2(23-9\nu)}{21(1-\nu)}\left(\frac{w_0}{h}\right)^3$$ Extensional radial stress at center: $$(\sigma_{re})_0 = \frac{Eh^2}{a^2}\frac{(5-3\nu)}{6(1-\nu)}\left(\frac{w_0}{h}\right)^2$$ Maximum stress in bending at center: $$(\sigma_{rb})_0 = \frac{Eh^2}{a^2}\frac{2}{1-\nu}\left(\frac{w_0}{h}\right)$$ Maximum total radial stress at center: $$\sigma_{r0} = (\sigma_{re})_0 + (\sigma_{rb})_0$$ Maximum total radial stress at outer boundary: $$\sigma_{ra} = \frac{Eh^2}{a^2}\left[\frac{1}{3(1-\nu)}\left(\frac{w_0}{h}\right)^2 + \frac{4}{1-\nu^2}\frac{w_0}{h}\right]$$ Tensile in-plane radial outer boundary reaction per unit length: $$P = \frac{1}{3(1-\nu)}\frac{Eh^3}{a^2}\left(\frac{w_0}{h}\right)^2$$

Plates	Deflection and Stresses
2. Simply supported outer edge without in-plane force [18.14] The radial displacement at the outer edge is not restrained. Hence the radial force P is zero.	Relationship between load and deflection: $$\frac{p_1 a^4}{Eh^4} = \frac{1.016}{1 - \nu}\frac{w_0}{h} + 0.376\left(\frac{w_0}{h}\right)^3$$ Maximum radial stress at center: $$\sigma_{r0} = \frac{Eh^2}{a^2}\left[\frac{1.238}{1 - \nu^2}\frac{w_0}{h} + 0.294\left(\frac{w_0}{h}\right)^2\right]$$
3. Fixed at outer edge with in-plane force [18.14] The radial displacement at the outer edge is zero.	Relationship between load and deflection: $$\frac{p_1 a^4}{Eh^4} = \frac{5.333}{1 - \nu^2}\frac{w_0}{h} + 0.857\left(\frac{w_0}{h}\right)^3$$ Maximum radial stress at center: $$\sigma_{r0} = \frac{Eh^2}{a^2}\left[\frac{2}{1 - \nu^2}\frac{w_0}{h} + 0.5\left(\frac{w_0}{h}\right)^2\right]$$ Maximum radial stress at outer boundary: $$\sigma_{ra} = \frac{Eh^2}{a^2}\left[\frac{4}{1 - \nu^2}\frac{w_0}{h}\right]$$

TABLE 18-13   **Large Deflection of Circular Plates**   **1032**

TABLE 18-14   MATERIAL PROPERTIES FOR RECTANGULAR PLATES

Notation

E_x, E_y = moduli of elasticity in x and y directions, for isotropic
materials $E_x = E_y = E$
h = thickness of plate
D_x, D_y = flexural rigidities in x and y directions
ν_x, ν_y = Poisson's ratios in x and y directions
$B = \frac{1}{2}(D_x\nu_y + D_y\nu_x + 4D_{xy})$
D_{xy} = torsional rigidity, $= \frac{1}{4}(2B - D_x\nu_y - D_y\nu_x)$

Plate	Constants
1. Homogeneous isotropic	$\nu_x = \nu_y = \nu \qquad D_x = D_y = D \qquad D_{xy} = \frac{1}{2}(1 - \nu)D$ $D = Eh^3/[12(1 - \nu^2)] \qquad B = D$
2. Homogeneous orthotropic	$D_x = \dfrac{E_x h^3}{12(1 - \nu_x\nu_y)} \qquad D_y = \dfrac{E_y h^3}{12(1 - \nu_x\nu_y)} \qquad D_{xy} = \dfrac{Gh^3}{12}$
3. Isotropic plate with equidistant stiffeners in one direction [18.12] 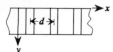	Stiffeners on two sides: $D_x = \dfrac{Eh^3}{12(1 - \nu^2)} \qquad D_y = D_x + \dfrac{E_s I_s}{d}$ where I_s = moment of inertia of stiffener taken about middle axis of cross section of plate E_s = modulus of elasticity of stiffeners Stiffeners on one side: T section $D_x = \dfrac{Eh^3 d}{12\left[d - c + c(h/H)^3\right]} \qquad D_y = \dfrac{EI_s}{d}$ $D_{xy} = D'_{xy} + \dfrac{GJ}{2d}$ where GJ = torsional rigidity of a rib D'_{xy} = torsional rigidity of slab without ribs I_s = moment of inertia of T section of width d about its neutral axis

Plate	Constants
4. Isotropic plate with equidistant stiffeners in two directions; axes of ribs are parallel to principal directions [18.12]	$$\nu_x = \nu_y = \nu \qquad D_x = \frac{Eh^3}{12(1-\nu^2)} + \frac{E_1 I_1}{d_1}$$ $$D_y = \frac{Eh^3}{12(1-\nu^2)} + \frac{E_2 I_2}{d_2} \qquad D_{xy} = \frac{Eh^3}{12(1-\nu^2)}$$ where I_1 = moment of inertia about plate's middle surface of stiffener in x direction E_1 = modulus of elasticity of this stiffener d_1 = spacing of these stiffeners I_2, E_2, d_2 = corresponding constants for stiffeners lying in y direction
5. Corrugated plate [18.12] $z = H\sin(\pi x/\ell)$ h = thickness of sheet $s = \ell\,[1 + \pi^2 H^2/(4\ell^2)]$	$$\nu_x = \nu_y = \nu \qquad D_x = \frac{\ell}{s}\frac{Eh^3}{12(1-\nu^2)}$$ $$D_y = EI \qquad B = \frac{s}{\ell}\frac{Eh^3}{12(1+\nu)}$$ where I = mean moment of inertia in x, y plane per unit length. $I = 0.5hH^2(1 - 0.81/C_1) \qquad C_1 = 1 + 2.5(H/2\ell)^2$
6. Open gridworks [18.1]	$$\nu_x = \nu_y = \nu \qquad D_x = \frac{EI_1}{d_1} \qquad D_y = \frac{EI_2}{d_2}$$ $$B = \frac{GJ_1}{2d_1} + \frac{GJ_2}{2d_2} \qquad D_{xy} = \sqrt{D_x D_y}$$ where GJ_1, GJ_2 = torsional rigidities of beams parallel to x and y axes EI_1, EI_2 = bending rigidities of beams parallel to x and y axes

TABLE 18-14 (continued)   MATERIAL PROPERTIES FOR RECTANGULAR PLATES

Plate	Constants
7. Concrete slab with steel reinforcement bars in both x and y directions [18.1] 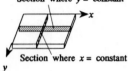	$v_x = v_y = v \qquad D_x = \dfrac{E_c}{1 - v_c^2}\left[I_{cx} + \left(\dfrac{E_s}{E_c} - 1\right)I_{sx}\right]$ $D_y = \dfrac{E_c}{1 - v_c^2}\left[I_{cy} + \left(\dfrac{E_s}{E_c} - 1\right)I_{sy}\right]$ $D_{xy} = \tfrac{1}{2}(1 - v_c)\sqrt{D_x D_y}$ where v_c = Poisson's ratio for concrete E_c, E_s = moduli of elasticity for concrete and steel, respectively I_{cx}, I_{cy} = moments of inertia of slab material about neutral axis in section where x, y = const I_{sx}, I_{sy} = moments of inertia of reinforcement bars about neutral axis in section where x, y = const
8. Concrete slab stiffened by concrete ribs [18.1] 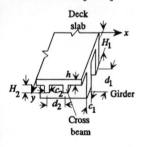	$D_x = \dfrac{EI_x}{d_1} \qquad D_y = \dfrac{EI_y}{d_2}$ $B = \dfrac{Eh^3}{12(1 - v_x v_y)} + \dfrac{G}{2}\dfrac{H_1 c_1^3 \alpha_1 k_1}{d_1} + \dfrac{H_2 c_2^3 \alpha_2 k_2}{d_2}$ where I_x, I_y = moments of inertia of slab section where x, y = const (see case 7), with respect to neutral axis k_i = reduction factors, depending on c_i/H_i, which are inserted to reduce the torsional rigidities of the concrete ribs after cracks have developed Values of α_i are provided for some values of c_i/H_i:

c_i/H_i	1.0	1.2	1.5	2.0	2.5	3.0
α_i	0.140	0.166	0.196	0.229	0.249	0.263

c_i/H_i	4.0	6.0	8.0	10.0	∞
α_i	0.281	0.299	0.307	0.313	0.333

TABLE 18-14 (continued) MATERIAL PROPERTIES FOR RECTANGULAR PLATES

Plate	Constants
9. Steel deck plate [18.1] with multiple stiffeners 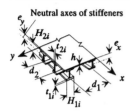 Neutral axes of stiffeners	$$D_x = \frac{Eh^3}{12(1 - \nu_x\nu_y)} + \frac{Ehe_x^2}{(1 - \nu_x\nu_y)} + \frac{EI_x}{d_1}$$ $$D_y = \frac{Eh^3}{12(1 - \nu_x\nu_y)} + \frac{Ehe_y^2}{(1 - \nu_x\nu_y)} + \frac{EI_y}{d_2}$$ $$B = \frac{Eh^3}{12(1 - \nu_x\nu_y)} + \frac{G}{6}\left(\frac{\sum_{i=1}^{M_1} H_{1i}t_{1i}^3}{d_1} + \frac{\sum_{i=1}^{M_2} H_{2i}t_{2i}^3}{d_2} \right)$$ where I_x, I_y = moments of inertia of stiffeners with respect to neutral axes in x, y directions e_x, e_y = distances from middle plane of plate to neutral axes of stiffeners $M_1(M_2)$ = number of segments, i.e., flanges and webs, of thickness t_{1i} (t_{2i}) and length $H_{1i}(H_{2i})$ composing a single stiffener section in the $x(y)$ direction.
10. Composite (concrete-steel) slab 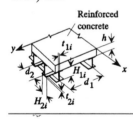 Reinforced concrete	Use the formulas for a steel deck plate. First transform the concrete part of the slab into an equivalent steel plate.
11. Continuously composite, isotropic	$\nu_x = \nu_y = \nu \qquad D_x = D_y = D \qquad D_{xy} = \frac{1}{2}(1 - \nu)D$ $$D = \frac{1}{1 - \nu^2} \int_{-h/2}^{h/2} Ez^2 \, dz$$
12. Continuously composite, orthotropic	$$D_x = \frac{1}{1 - \nu_x\nu_y} \int_{-h/2}^{h/2} E_x z^2 \, dz$$ $$D_y = \frac{1}{1 - \nu_x\nu_y} \int_{-h/2}^{h/2} E_y z^2 \, dz$$ $$D_{xy} = \int_{-h/2}^{h/2} Gz^2 \, dz$$

TABLE 18-14 **Material Properties for Rectangular Plates** **1036**

TABLE 18-14 (continued)   MATERIAL PROPERTIES FOR RECTANGULAR PLATES

Plate	Constants
13. Multiple isotropic layers [18.1] 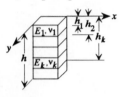	$\nu_x = \nu_y = \nu \qquad D_x = D_y = D$ $D = (Q_1 C - Q_2^2)/Q_1$ $Q_1 = \sum\limits_{k=1} \dfrac{E_k}{1 - \nu_k^2}(h_k - h_{k-1})$ $Q_2 = \dfrac{1}{2}\sum\limits_{k=1} \dfrac{E_k}{1 - \nu_k^2}(h_k^2 - h_{k-1}^2)$ $C = \dfrac{1}{3}\sum\limits_{k=1} \dfrac{E_k}{1 - \nu_k^2}(h_k^3 - h_{k-1}^3)$
14. Symmetrically constructed with isotropic layers [18.13] 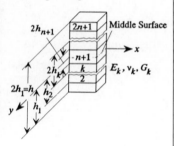	$\nu_x = \nu_y = \nu \qquad D_x = D_y = D$ $D_{xy} = \frac{1}{2}(1 - \nu)D$ $D = \dfrac{2}{3}\sum\limits_{m=1}^{n+1} c_m \qquad \nu = \dfrac{2}{3D}\sum\limits_{m=1}^{n+1}\nu_m c_m$ $c_m = \dfrac{E_m(h_m^3 - h_{m+1}^3)}{1 - \nu_m^2}$ with h_{n+2} set equal to zero where E_m, ν_m are Young's modulus and Poisson's ratio for mth layer This plate is constructed of an odd number of homogeneous layers symmetrically located about middle layer.
15. Symmetrically constructed with orthotropic layers [18.13] 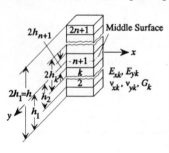	$\nu_x = \dfrac{2}{3D_y}\sum\limits_{m=1}^{n+1} c_{xm}\nu_{ym} \qquad \nu_y = \dfrac{\nu_x D_y}{D_x}$ $D_x = \dfrac{2}{3}\sum\limits_{m=1}^{n+1} c_{xm} \qquad D_y = \dfrac{2}{3}\sum\limits_{m=1}^{n+1} c_{ym}$ $D_{xy} = \dfrac{2}{3}\left[\sum\limits_{m=1}^{n+1} G_m(h_m^3 - h_{m+1}^3)\right]$ $c_{xm} = \dfrac{E_{xm}(h_m^3 - h_{m+1}^3)}{1 - \nu_{xm}\nu_{ym}} \qquad c_{ym} = \dfrac{E_{ym}(h_m^3 - h_{m+1}^3)}{1 - \nu_{xm}\nu_{ym}}$ with $h_{n+2} = 0$ where $E_{xm}, E_{ym}, \nu_{xm}, \nu_{ym}$ = Young's moduli and Poisson's ratios of mth layer

TABLE 18-14 (continued)   MATERIAL PROPERTIES FOR RECTANGULAR PLATES

Plate	Constants

16.
Symmetrically constructed with identical orthotropic layers.

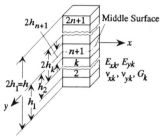

$$\nu_x = \frac{2(2n+1)^3\nu_2}{Q_2} \qquad \nu_2 = \frac{2(2n+1)^3\nu_2}{Q_1}$$

$$D_x = \frac{E_x h^3}{12(1-\nu_x\nu_y)} \qquad D_y = \frac{E_y h^3}{12(1-\nu_x\nu_y)}$$

$$D_{xy} = \frac{Gh^3}{12}$$

$$E_x = \frac{E_1}{2(2n+1)^3}\frac{1-\nu_x\nu_y}{1-\nu_1\nu_2}Q_1$$

The principal directions of adjacent layers are mutually perpendicular. That is, if the material properties in perpendicular directions are E_1, ν_1 and E_2, ν_2, then for odd numbered layers ($k = 1, 3, 5, \ldots, 2n+1$) $E_{xk} = E_1, \nu_{xk} = \nu_1, E_{yk} = E_2, \nu_{yk} = \nu_2$ and for even numbered layers ($k = 2, 4, 6, \ldots, 2n$) $E_{xk} = E_2, \nu_{xk} = \nu_2, E_{yk} = E_1, \nu_{yk} = \nu_1$.

$$E_y = \frac{E_2}{2(2n+1)^3}\frac{1-\nu_x\nu_y}{(1-\nu_1\nu_2)E_2/E_1}Q_2$$

$$Q_1 = (2n+1)^3(1+E_2/E_1)$$
$$+\left[3(2n+1)^2-2\right](1-E_2/E_1)$$

$$Q_2 = (2n+1)^3(1+E_2/E_1)$$
$$-\left[3(2n+1)^2-2\right](1-E_2/E_1)$$

17.
Plywood [18.12]; x axis parallel to face grain.

$$\nu_x = E''/E_y' \qquad \nu_y = E''/E_x' \qquad D_x = \tfrac{1}{12}E_x'h^3$$
$$D_y = \tfrac{1}{12}E_y'h^3 \qquad D_{xy} = \tfrac{1}{12}Gh^3$$

	E_x'	E_y'	E''	G
Maple, 5-ply	1.87	0.60	0.073	0.159
Afara, 3-ply	1.96	0.165	0.043	0.110
Gaboon (Okoume), 3-ply	1.28	0.11	0.014	0.085
Birch, 3- and 5-ply	2.00	0.167	0.077	0.17
Birch with bakelite membranes	1.70	0.85	0.061	0.10

TABLE 18-14   **Material Properties for Rectangular Plates**   **1038**

TABLE 18-15   STRESSES OF RECTANGULAR PLATES

Notation

w = deflection

σ_x, σ_y = normal stresses in x and y directions

τ_{xy} = shear stress

E_x, E_y = modulus of elasticity in x and y directions, for isotropic materials $E_x = E_y = E$

ν_x, ν_y = Poisson's ratio in x and y directions

α_x, α_y = thermal expansion coefficients in x and y directions

$P = P_x, P_y, P_{xy}$ = in-plane forces (Fig. 18-4), P_x and P_y are in compression

G = shear modulus of elasticity, has the same value in x and y directions

h = thickness of plate

$M_x = M, M_y$ = bending moments per unit length on surface normal to x and y axes

M_{xy} = twisting moment per unit length

Q_x, Q_y = shear force per unit length on surfaces normal to x and y axes

Plate	Stresses
1. Isotropic or orthotropic with material properties that do not vary through depth of plate	$\sigma_x = -\dfrac{P}{h} + \dfrac{Mz}{h^3/12}$ $\qquad$ $\sigma_y = -\dfrac{P_y}{h} + \dfrac{M_y z}{h^3/12}$ $\tau_{xy} = \dfrac{P_{xy}}{h} - \dfrac{M_{xy} z}{h^3/12}$ $\qquad$ P_{xy} is an in-plane shear force $\tau_{xz} = \dfrac{3Q_x}{2h}\left[1 - \left(\dfrac{z}{h/2}\right)^2\right]$ $\tau_{yz} = \dfrac{3Q_y}{2h}\left[1 - \left(\dfrac{z}{h/2}\right)^2\right]$

TABLE 18-15 (continued)   STRESSES OF RECTANGULAR PLATES

Plate	Stresses
2. Orthotropic with material properties that vary through depth of plate; variation must be symmetric about middle surface; in-plane force P_{xy} ignored	$$\sigma_x = -\frac{E_x P}{\int_{-h/2}^{h/2} E_x\, dz} + \frac{E_x Mz}{\int_{-h/2}^{h/2} E_x z^2\, dz}$$ $$\sigma_y = -\frac{E_y P_y}{\int_{-h/2}^{h/2} E_y\, dz} + \frac{E_y M_y z}{\int_{-h/2}^{h/2} E_y z^2\, dz}$$ $$\tau_{xy} = \frac{-GM_{xy}z}{\int_{-h/2}^{h/2} Gz^2\, dz}$$ $$\tau_{xz} = \frac{\partial}{\partial x}\left[\frac{\partial^2 w}{\partial x^2}\int_{-h/2}^{z}\frac{E_x z}{1-\nu_x\nu_y}\,dz\right.$$ $$+\frac{\partial^2 w}{\partial y^2}\int_{-h/2}^{z}\left(\frac{E_x\nu_y}{1-\nu_x\nu_y}+2G\right)z\,dz\Bigg]$$ $$\tau_{yz} = \frac{\partial}{\partial y}\left[\frac{\partial^2 w}{\partial x^2}\int_{-h/2}^{z}\left(\frac{E_x\nu_y}{1-\nu_x\nu_y}+2G\right)z\,dz\right.$$ $$+\frac{\partial^2 w}{\partial y^2}\int_{-h/2}^{z}\frac{E_y z}{1-\nu_x\nu_y}\,dz\Bigg]$$
3. Layered plate; stresses in mth layer given [18.13] In-plane forces are ignored	$$\sigma_{xm} = -zC_{xm}\left(\frac{\partial^2 w}{\partial x^2}+\nu_{xm}\frac{\partial^2 w}{\partial y^2}\right)$$ $$\sigma_{ym} = -zC_{ym}\left(\frac{\partial^2 w}{\partial y^2}+\nu_{ym}\frac{\partial^2 w}{\partial x^2}\right)$$ $$\tau_{xym} = -2zG_m\frac{\partial^2 w}{\partial x\,\partial y}$$ For a symmetrically constructed plate the other shear stresses for $m=1,2,3,\ldots,n+1$ (stresses are symmetrically distributed) are $$\tau_{xzm} = \frac{z^2}{2}\frac{\partial}{\partial x}\left[C_{xm}\frac{\partial^2 w}{\partial x^2}+(C_{xm}\nu_{ym}+2G_m)\frac{\partial^2 w}{\partial y^2}\right]$$ $$-\frac{h^2}{2}\frac{\partial}{\partial x}\left[C_{11m}\frac{\partial^2 w}{\partial x^2}+(C_{12m}+2C_{66m})\frac{\partial^2 w}{\partial y^2}\right]$$ $$\tau_{yzm} = \frac{z^2}{2}\frac{\partial}{\partial y}\left[(C_{xm}\nu_{ym}+2G_m)\frac{\partial^2 w}{\partial x^2}+C_{ym}\frac{\partial^2 w}{\partial y^2}\right]$$ $$-\frac{h^2}{2}\frac{\partial}{\partial y}\left[(C_{12m}+2C_{66m})\frac{\partial^2 w}{\partial x^2}+C_{22m}\frac{\partial^2 w}{\partial y^2}\right]$$

TABLE 18-15   **Stresses of Rectangular Plates**   **1040**

TABLE 18-15 (continued)   STRESSES OF RECTANGULAR PLATES

Plate	Stresses
3. Continued	$$C_{xm} = \frac{E_{xm}}{1 - \nu_{xm}\nu_{ym}} \qquad C_{ym} = \frac{E_{ym}}{1 - \nu_{xm}\nu_{ym}}$$
	$$C_{11m} = \begin{cases} C_{x1} & m = 1 \\ \dfrac{1}{h^2}\left[\displaystyle\sum_{k=1}^{m-1} C_{xk}\left(h_k^2 - h_{k+1}^2\right) + C_{xm}h_m^2 \right] & m \geq 2 \end{cases}$$
	$$C_{12m} = \begin{cases} C_{x1}\nu_{y1} & m = 1 \\ \dfrac{1}{h^2}\left[\displaystyle\sum_{k=1}^{m-1} C_{xk}\nu_{ym}\left(h_k^2 - h_{k+1}^2\right) + C_{xm}\nu_{ym}h_m^2 \right] \\ \hspace{6cm} m \geq 2 \end{cases}$$
	$$C_{22m} = \begin{cases} C_{y1} & m = 1 \\ \dfrac{1}{h^2}\left[\displaystyle\sum_{k=1}^{m-1} C_{yk}\left(h_k^2 - h_{k+1}^2\right) + C_{ym}h_m^2 \right] & m \geq 2 \end{cases}$$
	$$C_{66m} = \begin{cases} G_1 & m = 1 \\ \dfrac{1}{h^2}\left[\displaystyle\sum_{k=1}^{m-1} G_k\left(h_k^2 - h_{k+1}^2\right) + G_m h_m^2 \right] & m \geq 2 \end{cases}$$
4. Thermal loading	$$\sigma_x = \frac{Mz}{h^3/12} - \frac{E_x(\alpha_x + \nu_y\alpha_y)}{1 - \nu_x\nu_y}T$$ $$\sigma_y = \frac{M_y z}{h^3/12} - \frac{E_y(\alpha_y + \nu_x\alpha_x)}{1 - \nu_x\nu_y}T$$

TABLE 18-16   DEFLECTIONS AND INTERNAL FORCES
OF RECTANGULAR PLATES

Notation

w = deflection
E = modulus of elasticity
ν = Poisson's ratio
h = thickness of plate
$D = Eh^3/[12(1 - \nu^2)]$
W_T = concentrated loading (Force)
p_1 = uniformly distributed loading (Force/length^2)
$\alpha = L_y/L$
$\beta = L/L_y$
M_x, M_y = bending moment per unit length
   on surface normal to x and y axes
Q_x, Q_y = shear force per unit length on
   surfaces normal to x and y axes
V_x, V_y = equivalent shear force per unit length
   acting on planes normal to x and y axes

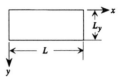

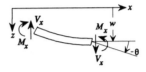

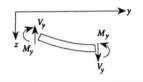

Structural System and Static Loading	Deflection and Internal Forces
1. Simply supported on all edges, uniform loading 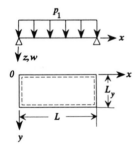	$$w = \frac{16p_1L^4}{\pi^6 D}\sum_m\sum_n\frac{\sin(n\pi x/L)\sin(m\pi y/L_y)}{mn\left(n^2/L^2 + m^2/L_y^2\right)^2}$$ $$M_x = \frac{16p_1L^2}{\pi^4}\sum_m\sum_n\gamma_1\frac{\sin(n\pi x/L)\sin(m\pi y/L_y)}{mn\left(n^2 + m^2/\alpha^2\right)^2}$$ $$M_y = \frac{16p_1L^2}{\pi^4}\sum_m\sum_n\gamma_2\frac{\sin(n\pi x/L)\sin(m\pi y/L_y)}{mn\left(n^2 + m^2/\alpha^2\right)^2}$$ $$w_{max} = c_1\frac{p_1L^4}{Eh^3} = w_{center}$$ $\gamma_1 = (n^2 + \nu m^2/\alpha^2), \quad \gamma_2 = (m^2/\alpha^2 + \nu n^2)$ $(M_x)_{max} = c_2 p_1 L^2 \quad \text{(center)}$ $(M_y)_{max} = c_3 p_2 L^2 \quad \text{(center)}$ $(Q_x)_{max} = c_4 p_1 L \quad (x = 0, L \text{ edges})$ $(Q_y)_{max} = c_5 p_1 L_y \quad (y = 0, L_y \text{ edges})$ $(V_x)_{max} = c_6 p_1 L \quad \text{(lateral edge forces)}$ $(V_y)_{max} = c_7 p_1 L_y \quad \text{(lateral edge forces)}$ $R_0 = c_8 p_1 L L_y \quad \text{(corner forces)}$

TABLE 18-16   **Deflections and Forces of Rectangular Plates**   **1042**

Structural System and Static Loading	Deflection and Internal Forces

1. Continued

CONSTANTS:

$m = 1, 3, 5, \ldots, \infty \qquad n = 1, 3, 5, \ldots, \infty$

$\nu = 0.3$

$c_1 = 0.1421 + 0.08204\beta - 0.3985\beta^2 + 0.219\beta^3$

$c_2 = 0.1247 + 0.05524\beta - 0.2762\beta^2 + 0.1441\beta^3$

$c_3 = 0.03726 - 0.01504\beta + 0.1028\beta^2 - 0.07756\beta^3$

$c_4 = 0.4967 + 0.1686\beta - 0.6849\beta^2 + 0.3976\beta^3$

$c_5 = -0.0007805 + 0.3679\beta + 0.04574\beta^2 - 0.07488\beta^3$

$c_6 = 0.4983 + 0.02879\beta + 0.004425\beta^2 - 0.1137\beta^3$

$c_7 = -0.003591 + 0.5091\beta + 0.05874\beta^2 - 0.1445\beta^3$

$c_8 = -0.003545 + 0.1179\beta - 0.02755\beta^2 - 0.02202\beta^3$

2.
Simply supported on all edges, concentrated load

$$w = \frac{4W_T}{D\pi^4 LL_y} \sum_m \sum_n$$

$$\times \frac{\sin(n\pi a/L)\sin(m\pi b/L_y)\sin(n\pi x/L)\sin(m\pi y/L_y)}{\left(n^2/L^2 + m^2/L_y^2\right)^2}$$

$m = 1, 2, 3, \ldots; n = 1, 2, 3, \ldots$

or

$$w = \frac{W_T L^2}{D\pi^3} \sum_{n=1}^{\infty} \left(1 + \beta_n \coth \beta_n - \frac{\beta_n y_1}{L_y} \coth \frac{\beta_n y_1}{L_y}\right.$$

$$\left. - \frac{\beta_n b_1}{L_y} \coth \frac{\beta_n b_1}{L_y}\right)$$

$$\times \frac{\sinh\dfrac{\beta_n b_1}{L_y}\sinh\dfrac{\beta_n y_1}{L_y}\sin\dfrac{n\pi a}{L}\sin\dfrac{m\pi x}{L}}{n^3 \sinh \beta_n}$$

If $y \geq b$, use $y_1 = L_y - y$ and $b_1 = b$

If $y < b$, use $y_1 = y$ and $b_1 = L_y - b$

$$\beta_n = \frac{n\pi L_y}{L}$$

$n = 1, 2, 3, \ldots$

Structural System and Static Loading	Deflection and Internal Forces
3. Simply supported on all edges, uniform load on a small circle of radius r_0 at center of plate 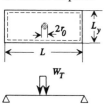	At center: $$(w)_{max} = k_1 \frac{W_T L_y^2}{Eh^2}$$ $$(\sigma)_{max} = \frac{3W_T}{2\pi h^2}\left[(1 + \nu)\ln \frac{2L_y}{\pi r_e} + k_2\right]$$ where $$r_e = \begin{cases} \sqrt{1.6r_0^2 + h^2} - 0.675h & r_0 < 0.5h \\ r_0 & r_0 > 0.5h \end{cases}$$ W_T = total load on plate $$k_1 = 0.1851 + 0.06342\alpha - 0.1643\alpha^2 + 0.04232\alpha^3$$ $$k_2 = 0.9998 + 0.5195\alpha - 1.29\alpha^2 + 0.2042\alpha^3$$
4. Three edges simply supported, one edge free, uniform load 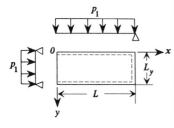	$$(M_x)_{\substack{x=L/2 \\ y=L_y/2}} = c_1 p_1 L_y^2$$ $$(M_y)_{\substack{x=L/2 \\ y=L_y/2}} = c_2 p_1 L_y^2$$ $$(M_y)_{\substack{x=0 \\ y=L_y/2}} = c_3 p_1 L_y^2$$ $$\nu = 0.15$$ CONSTANTS: $$c_1 = \begin{cases} -2.7975 + 16.537\beta^{1/2} - 36.423\beta \\ \quad + 35.523\beta^{3/2} - 12.913\beta^2 & 0.3 \le \beta < 0.8 \\ -0.2714 + 0.6371\beta^{1/2} - 0.4533\beta \\ \quad + 0.1021\beta^{3/2} & 0.8 \le \beta \le 2.0 \end{cases}$$ $$c_2 = 0.1488 - 0.7829\beta^{1/2} + 1.415\beta - 0.9113\beta^{3/2}$$ $$\quad + 0.2039\beta^2 \qquad 0.3 \le \beta \le 2.0$$ $$c_3 = 0.2258 - 1.3186\beta^{1/2} + 2.6532\beta - 1.9313\beta^{3/2}$$ $$\quad + 0.4817\beta^2 \qquad 0.3 \le \beta \le 2.0$$

TABLE 18-16   **Deflections and Forces of Rectangular Plates**   1044

Structural System and Static Loading	Deflection and Internal Forces

5.

Three edges clamped, one edge free, uniform load

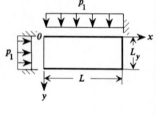

$$(M_x)_{\substack{x=L/2 \\ y=L_y/2}} = c_1 p_1 L_y^2$$

$$(M_y)_{\substack{x=L/2 \\ y=L_y/2}} = c_2 p_1 L_y^2$$

$$(M_y)_{\substack{x=0 \\ y=L_y/2}} = c_3 p_1 L_y^2$$

$$(M_x)_{\substack{x=L \\ y=L_y/2}} = c_4 p_1 L^2$$

$$(M_y)_{\substack{x=L/2 \\ y=0, L_y}} = c_5 p_1 L_y^2$$

$$(M_y)_{\substack{x=0 \\ y=0, L_y}} = c_6 p_1 L_y^2$$

$$\nu = 0.15$$

CONSTANTS:

$$c_1 = \begin{cases} -0.01247 + 0.05532\alpha - 0.04778\alpha^2 \\ \quad + 0.01519\alpha^3 - 0.001687\alpha^4 & 1.25 < \alpha \le 3.3 \\ -0.006967 + 0.01615\alpha + 0.01147\alpha^2 \\ \quad - 0.01205\alpha^3 & 0.5 \le \alpha \le 1.25 \end{cases}$$

$$c_2 = 0.05474 - 0.03243\alpha + 0.006359\alpha^2$$
$$\qquad - 0.0003985\alpha^3 \qquad\qquad\qquad 0.5 \le \alpha \le 3.3$$

$$c_3 = 0.03724 + 0.02528\alpha - 0.02263\alpha^2$$
$$\qquad + 0.003719\alpha^3 \qquad\qquad\qquad 0.5 \le \alpha \le 3.3$$

$$c_4 = -0.006349 + 0.02708\alpha - 0.09165\alpha^2$$
$$\qquad + 0.01488\alpha^3 \qquad\qquad\qquad 0.5 \le \alpha \le 3.3$$

$$c_5 = -0.1062 + 0.04391\alpha - 0.00235\alpha^2$$
$$\qquad - 0.0007418\alpha^3 \qquad\qquad\qquad 0.5 \le \alpha \le 3.3$$

$$c_6 = -0.06827 - 0.04205\alpha + 0.02756\alpha^2$$
$$\qquad - 0.003528\alpha^3 \qquad\qquad\qquad 0.5 \le \alpha \le 3.3$$

Structural System and Static Loading	Deflection and Internal Forces

6.

Simply supported
on all edges,
linearly varying load

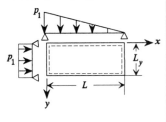

If $L/L_y < 1.00$,

$$(M_x)_{\substack{x=L/2 \\ y=L_y/2}} = c_1 p_1 L^2$$

$$(M_x)_{max} = c_2 p_1 L^2$$

$$(M_y)_{\substack{x=L/2 \\ y=L_y/2}} = c_3 p_1 L^2$$

$$(M_y)_{max} = c_4 p_1 L^2$$

If $(L/L_y) \geq 1.00$, replace $p_1 L^2$ by $p_1 L_y^2$.

$\nu = 0.15$

CONSTANTS:

$$c_1 = \begin{cases} 0.08438 - 0.06242\beta - 0.03592\beta^2 \\ \quad + 0.03239\beta^3 \qquad\qquad\qquad 0.5 \leq \beta \leq 1.0 \\ -0.003636 + 0.05639\beta - 0.04352\beta^2 \\ \quad + 0.009206\beta^3 \qquad\qquad\quad 1.0 < \beta \leq 2.0 \end{cases}$$

$$c_2 = \begin{cases} 0.085096 - 0.0635\beta - 0.02495\beta^2 \\ \quad + 0.02498\beta^3 \qquad\qquad\qquad 0.5 \leq \beta \leq 1.0 \\ -0.01872 + 0.07831\beta - 0.04657\beta^2 \\ \quad + 0.008649\beta^3 \qquad\qquad\quad 1.0 < \beta \leq 2.0 \end{cases}$$

$$c_3 = \begin{cases} -0.01873 + 0.07069\beta - 0.02985\beta^2 \\ \quad - 0.003716\beta^3 \qquad\qquad\quad 0.5 < \beta \leq 1.0 \\ -0.04394 + 0.08744\beta - 0.03008\beta^2 \\ \quad + 0.004946\beta^3 \qquad\qquad\quad 1.0 < \beta \leq 2.0 \end{cases}$$

$$c_4 = \begin{cases} \quad 0.01756 - 0.05004\beta + 0.1028\beta^2 \\ \quad - 0.05186\beta^3 \qquad\qquad\quad 0.5 \leq \beta \leq 1.0 \\ -0.01646 + 0.02599\beta + 0.01385\beta^2 \\ \quad - 0.004938\beta^3 \qquad\qquad\quad 1.0 < \beta \leq 2.0 \end{cases}$$

TABLE 18-16   **Deflections and Forces of Rectangular Plates**   **1046**

TABLE 18-16 (continued)   DEFLECTIONS AND INTERNAL FORCES OF RECTANGULAR PLATES

Structural System and Static Loading	Deflection and Internal Forces
7. Two edges simply supported, two edges clamped, linearly varying load 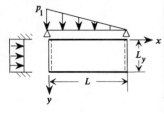	If $\beta = L/L_y < 1.00$, $-\left(M_y\right)_{max} = c_1 p_1 L^2$ $\left(M_y\right)_{\substack{x=L/2 \\ y=0, L_y}} = c_2 p_1 L^2$ $\left(M_x\right)_{max} = c_3 p_1 L^2$ $\left(M_x\right)_{\substack{x=L/2 \\ y=L_y/2}} = c_4 p_1 L^2$ $\left(M_y\right)_{max} = c_5 p_1 L^2$ $\left(M_y\right)_{\substack{x=L/2 \\ y=L_y/2}} = c_6 p_1 L^2$ If $\beta \geq 1.00$, replace $p_1 L^2$ by $p_1 L_y^2$. $\nu = 0.15$ CONSTANTS: $0.5 \leq \beta \leq 1.0$ $c_1 = -0.0947 + 0.09607\beta - 0.0758\beta^2 + 0.03702\beta^3$ $c_2 = -0.07833 + 0.01816\beta + 0.04384\beta^2 - 0.01853\beta^3$ $c_3 = 0.147 - 0.3292\beta + 0.286\beta^2 - 0.09076\beta^3$ $c_4 = 0.120 - 0.218\beta + 0.1254\beta^2 - 0.01945\beta^3$ $c_5 = -0.04249 + 0.196\beta - 0.2114\beta^2 + 0.07222\beta^3$ $c_6 = -0.04249 + 0.196\beta - 0.2114\beta^2 + 0.07222\beta^3$ $1.0 < \beta \leq 2.0$ $c_1 = -0.001352 - 0.04739\beta + 0.01262\beta^2 - 0.001445\beta^3$ $c_2 = 0.01942 - 0.09533\beta + 0.04973\beta^2 - 0.008762\beta^3$ $c_3 = -0.005903 + 0.04346\beta - 0.03065\beta^2 + 0.00615\beta^3$ $c_4 = 0.02255 - 0.01713\beta + 0.001814\beta^2 + 0.0006693\beta^3$ $c_5 = -0.0156 + 0.04088\beta - 0.01164\beta^2 + 0.000638\beta^3$ $c_6 = -0.02249 + 0.06271\beta - 0.03123\beta^2 + 0.005338\beta^3$

Structural System and Static Loading	Deflection and Internal Forces
8. Clamped on all edges, linearly varying load 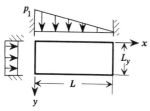	If $\beta = L/L_y < 1.00$,

$$(M_x)_{\substack{x=0 \\ y=L_y/2}} = c_1 p_1 L^2$$

$$(M_x)_{\substack{x=L \\ y=L_y/2}} = c_2 p_1 L^2$$

$$-(M_y)_{max} = c_3 p_1 L^2$$

$$(M_y)_{\substack{x=L/2 \\ y=0, L_y}} = c_4 p_1 L^2$$

$$(M_x)_{max} = c_5 p_1 L^2$$

$$(M_x)_{\substack{x=L/2 \\ y=L_y/2}} = c_6 p_1 L^2$$

$$(M_y)_{max} = c_7 p_1 L^2$$

$$(M_y)_{\substack{x=L/2 \\ y=L_y/2}} = c_8 p_1 L^2$$

If $\beta \geq 1.00$, replace

$p_1 L$ by $p_1 L_y^2$.

$\nu = 0.15$

CONSTANTS:

$0.5 \leq \beta \leq 1.0$
$c_1 = -0.03596 - 0.08083\beta + 0.1319\beta^2 - 0.04816\beta^3$
$c_2 = -0.01030 - 0.116\beta + 0.1791\beta^2 - 0.07038\beta^3$
$c_3 = -0.02636 - 0.004588\beta - 0.009978\beta^2 + 0.01388\beta^3$
$c_4 = -0.02993 + 0.01295\beta - 0.02866\beta^2 + 0.02036\beta^3$
$c_5 = 0.0261 - 0.00459\beta - 0.01879\beta^2 + 0.007399\beta^3$
$c_6 = 0.01218 + 0.05503\beta - 0.09913\beta^2 + 0.04073\beta^3$
$c_7 = 0.03321 - 0.1267\beta + 0.1792\beta^2 - 0.07687\beta^3$
$c_8 = -0.005465 + 0.003079\beta + 0.03526\beta^2 - 0.02409\beta^3$

$1.0 < \beta \leq 2.0$
$c_1 = 0.02841 - 0.1064\beta + 0.0549\beta^2 - 0.01001\beta^3$
$c_2 = 0.005924 - 0.05511\beta + 0.03986\beta^2 - 0.00827\beta^3$
$c_3 = 0.03386 - 0.08143\beta + 0.02145\beta^2 - 0.0008475\beta^3$
$c_4 = 0.04529 - 0.1128\beta + 0.04949\beta^2 - 0.007329\beta^3$
$c_5 = -0.00482 + 0.03546\beta - 0.02663\beta^2 + 0.006204\beta^3$
$c_6 = -0.00233 + 0.03325\beta - 0.02862\beta^2 + 0.006528\beta^3$
$c_7 = -0.01682 + 0.02965\beta - 0.002959\beta^2 - 0.001043\beta^3$
$c_8 = -0.03277 + 0.06381\beta - 0.02579\beta^2 + 0.003542\beta^3$

TABLE 18-16   Deflections and Forces of Rectangular Plates   1048

Structural System and Static Loading	Deflection and Internal Forces

9.

Two edges simply supported, two edges flexibly supported, uniform loading

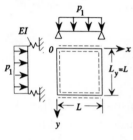

$$w_{max} = c_1 \frac{p_1 L^4}{D} = w_{center}$$

$$(M_x)_{max} = c_2 p_1 L^2 \quad \text{(center)}$$

$$(M_y)_{max} = c_3 p_1 L^2 \quad \text{(center)}$$

$$\nu = 0.3$$

EI = stiffness of edge beam supports

CONSTANTS:

$$c_1 = 0.004037 + 0.003975 \frac{1}{1+\eta} - 0.003862 \frac{1}{(1+\eta)^2}$$
$$+ 0.008931 \frac{1}{(1+\eta)^3}$$

$$c_2 = 0.04756 + 0.0356 \frac{1}{1+\eta} - 0.05788 \frac{1}{(1+\eta)^2}$$
$$+ 0.1001 \frac{1}{(1+\eta)^3}$$

$$c_3 = 0.04759 - 0.00504 \frac{1}{1+\eta} - 0.05026 \frac{1}{(1+\eta)^2}$$
$$+ 0.03471 \frac{1}{(1+\eta)^3}$$

$$\eta = \frac{EI}{LD} \geq 0$$

Structural System and Static Loading	Deflection and Internal Forces

10.

Clamped on all edges, uniform loading

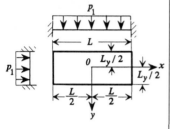

$$w_{max} = c_1 \frac{p_1 L^4}{Eh^3} = w_{center}$$

$$(M_x)_{\substack{x=0 \\ y=0}} = c_2 p_1 L^2 \qquad (M_y)_{\substack{x=0 \\ y=0}} = c_3 p_1 L^2$$

$$(M_x)_{\substack{x=L/2 \\ y=0}} = -c_4 p_1 L^2 \qquad (M_y)_{\substack{x=0 \\ y=L_y/2}} = -c_5 p_1 L^2$$

$$\nu = 0.3$$

CONSTANTS:

$$1.0 \le \alpha \le 2.0$$
$$c_1 = -0.06479 + 0.1327\alpha - 0.0665\alpha^2 + 0.01162\alpha^3$$
$$c_2 = -0.07859 + 0.1748\alpha - 0.09038\alpha^2 + 0.01652\alpha^3$$
$$c_3 = 0.0009449 + 0.04083\alpha - 0.0212\alpha^2 + 0.001949\alpha^3$$
$$c_4 = -0.03425 + 0.1083\alpha - 0.02085\alpha^2 - 0.002018\alpha^3$$
$$c_5 = 0.04247 + 0.002192\alpha + 0.01065\alpha^2 - 0.003921\alpha^3$$

11.

Two edges simply supported, two edges free, uniform loading

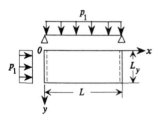

$$(w)_{\substack{x=L/2 \\ y=L_y/2}} = c_1 p_1 L^4 / D$$

$$(w)_{\substack{x=L/2 \\ y=0, L_y}} = c_2 p_1 L^4 / D$$

$$(M_x)_{\substack{x=L/2 \\ y=L_y/2}} = c_3 p_1 L^2$$

$$(M_y)_{\substack{x=L/2 \\ y=L_y/2}} = c_4 p_1 L^2$$

$$(M_x)_{\substack{x=L/2 \\ y=0, L_y}} = c_5 p_1 L^2$$

$$\nu = 0.3$$

CONSTANTS:

$$0 \le \beta \le 2.0$$
$$c_1 = 0.01302 - 0.0007094\beta + 0.001017\beta^2 - 0.0002372\beta^3$$
$$c_2 = 0.01522 + 0.0000742\beta - 0.0001711\beta^2$$
$$\qquad - 0.000033\beta^3$$
$$c_3 = 0.125 - 0.003249\beta + 0.0002489\beta^2 + 0.0005003\beta^3$$
$$c_4 = 0.0375 + 0.01038\beta - 0.02955\beta^2 + 0.008767\beta^3$$
$$c_5 = 0.133 + 0.0006847\beta - 0.001652\beta^2 - 0.0002326\beta^3$$

TABLE 18-16   **Deflections and Forces of Rectangular Plates**   **1050**

TABLE 18-17 RESPONSE OF RECTANGULAR PLATES WITH FOUR SIDES SIMPLY SUPPORTED

Notation

w = deflection

θ, θ_y = slopes about lines parallel to y and x directions

M, M_y, M_{xy} = bending moments per unit length on planes normal to x and y directions and twisting moment per unit length

V, V_y = equivalent shear forces per unit length acting on planes normal to x and y axes

L, L_y = length of plate in x and y directions

h = thickness of plate

E = modulus of elasticity

ν = Poisson's ratio

$$D = \frac{Eh^3}{12(1-\nu^2)} \qquad K_{mn} = \frac{a_{mn}}{D\pi^4\left[(n^2/L^2)+(m^2/L_y)\right]^2} \qquad n, m = 1, 2, 3, \ldots$$

The parameters a_{mn} are given on the following two pages for various loadings.

General Response Expressions

1. Deflection:

$$w = \sum_{n=1}^{\infty}\sum_{m=1}^{\infty} K_{mn} \sin\frac{n\pi x}{L} \sin\frac{m\pi y}{L_y}$$

2. Slopes:

$$\theta = -\sum_{n=1}^{\infty}\sum_{m=1}^{\infty} K_{mn}\frac{n\pi}{L}\cos\frac{n\pi x}{L}\sin\frac{m\pi y}{L_y}$$

$$\theta_y = \sum_{n=1}^{\infty}\sum_{m=1}^{\infty} K_{mn}\frac{m\pi}{L_y}\sin\frac{n\pi x}{L}\cos\frac{m\pi y}{L_y}$$

3. Bending moments:

$$M = \pi^2 D\sum_{n=1}^{\infty}\sum_{m=1}^{\infty} K_{mn}\left[\left(\frac{n}{L}\right)^2 + \nu\left(\frac{m}{L_y}\right)^2\right]\sin\frac{n\pi x}{L}\sin\frac{m\pi y}{L_y}$$

$$M_y = \pi^2 D\sum_{n=1}^{\infty}\sum_{m=1}^{\infty} K_{mn}\left[\left(\frac{m}{L_y}\right)^2 + \nu\left(\frac{n}{L}\right)^2\right]\sin\frac{n\pi x}{L}\sin\frac{m\pi y}{L_y}$$

4. Twisting moment:

$$M_{xy} = \pi^2 D(1-\nu)\sum_{n=1}^{\infty}\sum_{m=1}^{\infty} K_{mn}\frac{mn}{L_y L}\cos\frac{n\pi x}{L}\cos\frac{m\pi y}{L_y}$$

5. Shear forces:

$$V = \pi^3 D\sum_{n=1}^{\infty}\sum_{m=1}^{\infty} K_{mn}$$
$$\times\left[\left(\frac{n}{L}\right)^3 + (2-\nu)\frac{n}{L}\left(\frac{m}{L_y}\right)^2\right]\cos\frac{n\pi x}{L}\sin\frac{m\pi y}{L_y}$$

$$V_y = \pi^3 D\sum_{n=1}^{\infty}\sum_{m=1}^{\infty} K_{mn}$$
$$\times\left[\left(\frac{m}{L_y}\right)^3 + (2-\nu)\frac{m}{L_y}\left(\frac{n}{L}\right)^2\right]\sin\frac{n\pi x}{L}\cos\frac{m\pi y}{L_y}$$

	Parameters a_{mn} for Various Loadings
Loading	Parameter
1. Uniform load p_1 over whole plate	$a_{nm} = \dfrac{16p_1}{\pi^2 nm}$ $n, m = 1, 3, 5, \ldots$
2. Linearly varying load	$a_{nm} = (-1)^n \dfrac{8L}{mn\pi} \dfrac{\Delta p}{\Delta \ell}$ $\begin{array}{l} m = 1, 3, 5, \ldots \\ n = 1, 2, 3, 4, 5, \ldots \end{array}$
3. Uniform rectangular load	$a_{nm} = \dfrac{4p_1}{mn\pi^2}\left(\cos\dfrac{n\pi a_1}{L} - \cos\dfrac{n\pi a_2}{L}\right)$ $\times \left(\cos\dfrac{m\pi b_1}{L_y} - \cos\dfrac{m\pi b_2}{L_y}\right)$
4.	$a_{nm} = \dfrac{4}{nm\pi}\dfrac{\Delta p}{\Delta \ell}\left(\cos\dfrac{m\pi b_1}{L_y} - \cos\dfrac{m\pi b_2}{L_y}\right)$ $\times \left[(a_1 - a_2)\cos\dfrac{n\pi a_2}{L} + \dfrac{L}{m\pi}\left(\sin\dfrac{n\pi a_2}{L} - \sin\dfrac{n\pi a_1}{L}\right)\right]$
5. Line force W (force/length)	$a_{nm} = \dfrac{8W}{\pi Lm}\sin\dfrac{n\pi a}{L}$ $\begin{array}{l} m = 1, 3, 5, \ldots \\ n = 1, 2, 3, \ldots \end{array}$ If this line load begins at $y = b_1$ and ends at $y = b_2$, then for a_{nm} use $\dfrac{4W}{m\pi L}\sin\dfrac{n\pi a}{L}\left(\cos\dfrac{m\pi b_1}{L_y} - \cos\dfrac{m\pi b_2}{L_y}\right)$
6. Line force W	$a_{nm} = \dfrac{4W}{LL_y}\left[\dfrac{L_y}{m\pi}\left(\sin\dfrac{n\pi a_1}{L} + \sin\dfrac{n\pi a_2}{L}\right)\right.$ $\times \left(\cos\dfrac{m\pi b_1}{L_y} - \cos\dfrac{m\pi b_2}{L_y}\right) + \dfrac{L}{n\pi}\left(\cos\dfrac{n\pi a_1}{L} - \cos\dfrac{n\pi a_2}{L}\right)$ $\left. \times \left(\sin\dfrac{m\pi b_1}{L_y} + \sin\dfrac{m\pi b_2}{L_y}\right)\right]$

TABLE 18-17   **Simply Supported Rectangular Plates**   **1052**

Loading	Parameter
7. Line force W 	$$a_{nm} = \frac{2W}{LL_y c_3}\left\{\left[\sin\left(c_3 b_2 - \frac{n\pi c_1}{L}\right) - \sin\left(c_3 b_1 - \frac{n\pi c_1}{L}\right)\right]\right.$$ $$\left. - \left[\sin\left(c_4 b_2 + \frac{n\pi c_1}{L}\right) - \sin\left(c_4 b_1 + \frac{n\pi c_1}{L}\right)\right]\right\}$$ $$c_3 = \frac{m\pi L - n\pi c_2 L_y}{LL_y} \qquad c_4 = \frac{m\pi L + n\pi c_2 L_y}{LL_y}$$
8. 	$$a_{nm} = \frac{4W_T}{LL_y}\sin\frac{n\pi a}{L}\sin\frac{m\pi b}{L_y}$$
9. 	$$a_{nm} = \frac{4W_T}{LL_y}\left(\sin\frac{n\pi a_1}{L} + \sin\frac{n\pi a_2}{L}\right)\left(\sin\frac{m\pi b_1}{L_y} + \sin\frac{m\pi b_2}{L_y}\right)$$
10. Line moment (force-length/length) 	$$a_{nm} = -\frac{4nC}{mL^2}\cos\frac{n\pi a}{L}\left(\cos\frac{m\pi b_1}{L_y} - \cos\frac{m\pi b_2}{L_y}\right)$$
11. Concentrated moment (force-length) 	$$a_{nm} = -\frac{4nC_T}{L^2 L_y}\cos\frac{n\pi a}{L}\sin\frac{m\pi b}{L_y}$$
12. 	$$a_{nm} = \frac{4\pi C_T}{LL_y}\frac{\left(\dfrac{c_2 m}{L_y}\cos\dfrac{m\pi b}{L_y}\sin\dfrac{n\pi a}{L} + \dfrac{n}{L}\sin\dfrac{m\pi b}{L_y}\cos\dfrac{n\pi a}{L}\right)}{\left(1 + c_2^2\right)^{1/2}}$$
13. 	$$a_{nm} = \frac{4}{LL_y}\int_0^L\int_0^{L_y} p_z(x,y)\sin\frac{n\pi x}{L}\sin\frac{m\pi y}{L_y}\,dx\,dy$$

TABLE 18-18   CRITICAL IN-PLANE LOADS OF RECTANGULAR PLATES

Notation

E = modulus of elasticity
h = thickness of plate
k = buckling coefficient
ν = Poisson's ratio
L, L_y = length of plate in x and y directions
$(W_T)_{cr}$ = concentrated buckling loads (F)
σ_{cr} = normal stress at buckling (F/L^2). For σ_{cr} to be applicable,
$\quad \sigma_{cr} < \sigma_{ys}$ (yield strength).
P_{cr} = buckling load, $= h\sigma_{cr}$ (F/L)
$(P_{xy})_{cr}$ = in-plane shear buckling load (F/L)

Half wave refers to half of a complete cycle of a sinusoidal curve. For example, $\sin(n\pi x/L)\sin(m\pi y/L_y)$ $(0 \le x \le L,\ 0 \le y \le L_y)$ defines $n = 1, 2, \ldots$ and $m = 1, 2, \ldots$ half waves in x and y directions.

$$P' = \frac{\pi^2 D}{L_y^2}$$

$$D = \frac{Eh^3}{12(1 - \nu^2)} \qquad \beta = \frac{L}{L_y} \qquad \alpha = \frac{L_y}{L}$$

D_x, D_y, and B are given in Table 18-14.

Conditions	Buckling Loads
1. All edges simply supported	$P_{cr} = kP'$ $k = \left(\dfrac{\beta}{m} + \dfrac{m}{\beta} \right)^2$ where $m = 1$ $\left(\text{for } \beta \le \sqrt{2} \right)$ $= 2$ $\left(\text{for } \sqrt{2} \le \beta \le \sqrt{6} \right)$ $= 3$ $\left(\text{for } \sqrt{6} \le \beta \le \sqrt{12} \right)$ $= 4$ $\left(\text{for } \sqrt{12} \le \beta \le \sqrt{20} \right)$ For $\beta > 4$, $k \cong 4.00$. Ref. 18.8
2. All edges clamped	$P_{cr} = kP'$ $k = \dfrac{4}{3} \left[\dfrac{4\beta^2}{n^2 + 1} + 2 + \dfrac{3n^2\beta^2}{4} \dfrac{1 + 6/n^2 + 1/n^4}{1 + 1/n^2} \right]$ where n is the number of half waves. Ref. 18.17

TABLE 18-18 | **Critical In-Plane Loads of Rectangular Plates** | **1054**

TABLE 18-18 (continued)   CRITICAL IN-PLANE LOADS OF RECTANGULAR PLATES

Conditions	Buckling Loads
3. Two edges clamped, two edges simply supported 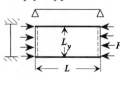	FOR ISOTROPIC PLATE: $P_{cr} = kP'$ $$k = \frac{4}{3}\left(\frac{4L^2}{n^2L_y^2} + 2 + \frac{3n^2L_y^2}{4L^2}\right)$$ FOR ORTHOTROPIC PLATE: $$P_{cr} = k\frac{\pi^2\sqrt{D_xD_y}}{L^2}$$ where $\quad k = \sqrt{\dfrac{D_y}{D_x}}\left(\dfrac{n}{\beta}\right)^2 + \dfrac{8Bm^2}{3\sqrt{D_xD_y}} + \dfrac{16}{3}\sqrt{\dfrac{D_x}{D_y}}\left(\dfrac{\beta m^2}{n}\right)^2$ $n, m = 1, 2, 3, \ldots$
4. Two edges simply supported, two edges clamped 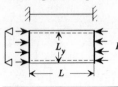	$P_{cr} = kP'$ $k = 2.964 + 6.774\alpha - 9.380\alpha^2 + 7.908\alpha^3 - 1.478\alpha^4$ $\quad 0.33 \le \alpha \le 2.5$
5. One edge clamped, other three edges simply supported 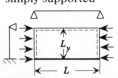	$P_{cr} = kP'$ $k = 13.2387 - 19.6159\beta + 12.2847\beta^2$ $\quad 0.728 \le \beta \le 0.889$ $k_{min} = 5.41 \quad$ at $\beta = 0.79$
6. One edge free, other three edges simply supported 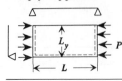	$P_{cr} = kP'$ $k = 0.4376 + 0.06992\alpha + 0.8983\alpha^2 + 0.02869\alpha^3 \quad\quad \alpha \le 2$ Approximate formula: $\quad k = 0.42 + \alpha^2$

TABLE 18-18 (continued) **CRITICAL IN-PLANE LOADS OF RECTANGULAR PLATES**

Conditions	Buckling Loads
7. One edge free, one clamped, and two edges simply supported	$P_{cr} = kP'$ $k = 4.561 - 4.851\beta + 2.315\beta^2 - 0.3423\beta^3$ $\quad 1.0 \le \beta \le 2.5$ Approximate formulas: $L/L_y = \beta \le 1.64 \quad k = 0.559 + \dfrac{1}{\beta^2} + 0.13\beta^2$ $\qquad\quad \beta > 1.64 \quad k = 1.28$ $\qquad\quad k_{min} = 1.28 \quad$ at $\beta = 1.635$
8. All edges simply supported	$(W_T)_{cr} = \dfrac{\pi^2 D}{2L_y}\left(\beta^3 \displaystyle\sum_{m=1,3,5,\ldots} \dfrac{1}{(\beta + m^2)^2}\right)^{-1}$ For $\beta > 2$: $\quad (W_T)_{cr} \approx \dfrac{\pi E h^3}{3(1 - \nu^2)L_y}$ Ref. 18.18
9. Two clamped, two edges simply supported	$(W_T)_{cr} = \dfrac{\pi^2 D}{2L_y}\left((2\beta)^3 \displaystyle\sum_{m=1,3,5,\ldots} \dfrac{1}{\{(2\beta)^2 + m^2\}^2}\right)^{-1}$ For $\beta \ge 2$: $\quad (W_T)_{cr} \approx \dfrac{2\pi E h^3}{3(1 - \nu^2)L_y}$ Ref. 18.18
10. All edges simply supported $P_{xy} = P_{yx}$	$(P_{xy})_{cr} = kP'$ $k = 6.393 - 3.249\alpha + 6.67\alpha^2 - 0.09172\alpha^3$ $\quad 0.33 \le \alpha \le 3$ Exact solution: $k = 5.348 + 2.299\alpha - 1.8406\alpha^2 + 3.544\alpha^3 \qquad \alpha \le 1$

TABLE 18-18 **Critical In-Plane Loads of Rectangular Plates** **1056**

TABLE 18-18 (continued)   CRITICAL IN-PLANE LOADS OF RECTANGULAR PLATES

Conditions	Buckling Loads
11. All edges clamped 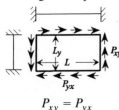 $P_{xy} = P_{yx}$	$(P_{xy})_{cr} = kP'$ $k = 8.942 + 30.89\alpha^{1/2} - 75.36\alpha + 50.20\alpha^{3/2}$ $\quad 0.4 \le \alpha \le 2.5$ Exact solution: For $\beta \to \infty \quad k = 8.98$
12. Two edges clamped: two edges simply supported 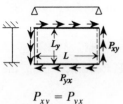 $P_{xy} = P_{yx}$	$(P_{xy})_{cr} = kP'$ $k = 8.905 + 3.674\alpha - 4.499\alpha^2 + 5.090\alpha^3 - 0.7569\alpha^4$ $\quad \alpha \le 3$
13. All edges simply supported 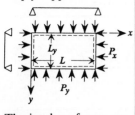 The in-plane forces are assumed to remain proportional to each other. Hence, $P_x/P_y = \lambda$ is a known (prescribed) constant ratio of the in-plane forces. $(P_y)_{cr}$ is treated as the critical load to be calculated.	$(P_y)_{cr} = kP'$ ISOTROPIC PLATE: $k = \dfrac{[m + n^2/(m\beta^2)]^2}{1 + (P_x/P_y)[n/(\beta m)]^2}$ ORTHOTROPIC PLATE: $k = \dfrac{\sqrt{D_y/D_x}\,m^2 + 2Bn^2/(\beta^2\sqrt{D_xD_y}) + \sqrt{D_x/D_y}\,[n^2/(\beta^2 m)]^2}{1 + \lambda[n/(\beta m)]^2}$ $P' = \dfrac{\pi^2\sqrt{D_xD_y}}{L_y^2}$ $\lambda = P_x/P_y$ where m, n are the number of half waves in x and y direction

TABLE 18-19 NATURAL FREQUENCIES OF ISOTROPIC RECTANGULAR PLATES AND MEMBRANES[a]

Notation

E = modulus of elasticity
h = thickness of plate
ρ = mass per unit area
$\beta = L/L_y$
ν = Poisson's ratio
L, L_y = length of plate in x and y directions

$$D = \frac{Eh^3}{12(1 - \nu^2)}$$

The natural frequencies in this table are defined in two ways:

1. $\omega_{nm} = \dfrac{\lambda_{nm}}{L^2}\sqrt{\dfrac{D}{\rho}}$ rad/s

where n and m $(n, m = 1, 2, 3, \ldots)$ are the numbers of half waves in the mode shapes in the x and y directions; ω_{11} is the fundamental frequency.

2. $\omega_i = \dfrac{\lambda_i}{L^2}\sqrt{\dfrac{D}{\rho}}$ rad/s

where i $(i = 1, 2, 3, \ldots)$ is the number of the natural frequency. For the fundamental frequency, $i = 1$.

$$f_{nm} \text{ (Hz)} = \frac{\omega_{nm}}{2\pi} \quad \text{and} \quad f_i \text{ (Hz)} = \frac{\omega_i}{2\pi}$$

The values of λ_{nm} or λ_i are independent of ν except where specifically indicated.

Configuration and Boundary Conditions	Natural Frequencies
1. All edges simply supported	$\lambda_{nm} = \pi^2(n^2 + \beta^2 m^2)$ $m, n = 1, 2, 3, \ldots$ First mode is ω_{11}; second is ω_{12}, etc.
2. Three edges simply supported, one edge free	$\lambda_{11} = 6.238 + 16.07\beta - 21.2\beta^2 + 15.51\beta^3 - 5.3936\beta^4$ $\qquad + 0.8995\beta^5 - 0.0576\beta^6 \qquad 0.5 \leq \beta \leq 5.0$ $\nu = 0.25$

TABLE 18-19   **Frequencies of Isotropic Rectangular Plates**   **1058**

Configuration and Boundary Conditions	Natural Frequencies
3. Two edges simply supported, two edges free 	$\lambda_{11} = 9.87 \ (\beta = 1)$ $\lambda_{21} = 8.54 + 4.31\beta + 3.952\beta^2 - 0.6751\beta^3$ $\quad 0.5 \le \beta \le 2.0$ $\lambda_{12} = 39.48 \ (\beta = 1)$ $\lambda_{22} = 39.2 - 0.273\beta + 9.123\beta^2 - 1.3256\beta^3$ $\quad 0.5 \le \beta \le 2.0$ $\nu = 0.3$
4. One edge built in, other edges simply supported 	$\lambda_{11} = \beta^2(50.4 - 41.85\beta + 17.65\beta^2 - 2.5\beta^3)$ $\quad 1 \le \beta \le 3$ $\lambda_{12} = 51.7 \ \ (\beta = 1)$ $\lambda_{21} = 58.7 \ \ (\beta = 1)$ $\lambda_{22} = 86.12 \ \ (\beta = 1)$
5. Two opposite edges clamped, other edges simply supported 	$\lambda_{11} = \beta^2(122.0 - 188.567\beta + 121.2\beta^2 - 25.73\beta^3)$ $\lambda_{21} = \beta^2(159.4 - 181.2\beta + 115.4\beta^2 - 24.4\beta^3)$ $\lambda_{31} = \beta^2(219.6 - 180.7\beta + 114.2\beta^2 - 24.0\beta^3)$ $\lambda_{12} = \beta^2(463.9 - 820.167\beta + 520.8\beta^2 - 109.73\beta^3)$ $\quad 0.5 \le \beta \le 2.0$
6. Three edges clamped, one edge simply supported 	APPROXIMATE: $\lambda_1 = 22.4 + 0.85\beta + 4.85\beta^2 + 3.7\beta^3$ $\beta \le 2$

Configuration and Boundary Conditions	Natural Frequencies
7. All edges clamped	$\lambda_1 = \beta^2(89.3 - 84.73\beta + 36.7\beta^2 - 5.27\beta^3)$ $\lambda_2 = \beta^2(107.2 - 51.9\beta + 21.5\beta^2 - 3.0\beta^3)$ $\lambda_3 = \beta^2(262.7 - 241.3\beta + 102.1\beta^2 - 14.47\beta^3)$ $\quad 1.0 \leq \beta \leq 3.0$ MEMBRANES: $f_{nm}(\text{Hz}) = \dfrac{\lambda_{nm}}{2}\left(\dfrac{P}{\rho A}\right)$ are the natural frequencies for transverse vibrations, where A is the cross-sectional area and P is the tension force per unit length. $\lambda_{nm} = (n^2\alpha + m^2\beta)^{1/2}$
8. One edge clamped all other edges free	FOR SYMMETRIC MODES: $\lambda_1 = 3.52 - 0.0967\beta + 0.03\beta^2 - 4.76 \times 10^{-3}\beta^3$ $\lambda_2 = -12.6 + 58.82\beta - 29.26\beta^2 + 4.166\beta^3$ $\lambda_3 = 31.086 - 37.1\beta + 40.35\beta^2 - 7.32\beta^3$ $\beta = 0.5, 1, 2,$ or 4 $\nu = 0.3$

[a]Adopted from Leissa Ref. 18.2.

TABLE 18-19 **Frequencies of Isotropic Rectangular Plates** **1060**

TABLE 18-20   NATURAL FREQUENCIES OF ORTHOTROPIC RECTANGULAR PLATES[a]

Notation

D_x, D_y, D_{xy} = flexural rigidities defined in Table 18-14

L, L_y = length of plate in x and y directions

ρ = mass per unit area

$\alpha = L_y/L$

$\gamma_0 = n\pi$ $\gamma_1 = (n + 1/4)\pi$ $\gamma_2 = (n + 1/2)\pi$

$\gamma_3 = m\pi$ $\gamma_4 = (m + 1/4)\pi$ $\gamma_5 = (m + 1/2)\pi$

n, m = number of half waves in the mode shapes in the x, y directions $(n, m = 1, 2, 3, \ldots)$

Natural frequencies:

$$\omega_{nm} = \frac{1}{L_y^2}\sqrt{\frac{1}{\rho}\left[D_x(\alpha k_1)^4 + 2D_{xy}\alpha^2 k_3 + D_y(k_2)^4\right]}$$

Except for the case of all four sides simply supported, the values of k_1, k_2, and k_3 are approximate.

Configuration and Boundary Conditions	k_1	k_2	k_3	Conditions
1. All edges clamped	4.730	4.730	151.3	$n = 1, m = 1$
	4.730	γ_5	$12.30\gamma_5(\gamma_5 - 2)$	$n = 1, m = 2, 3, 4, \ldots$
	γ_2	4.730	$12.30\gamma_2(\gamma_2 - 2)$	$n = 2, 3, 4, \ldots, m = 1$
	γ_2	γ_5	$\gamma_2\gamma_5(\gamma_2 - 2)(\gamma_5 - 2)$	$n = 2, 3, 4, \ldots, m = 2, 3, 4, \ldots$

Configuration and Boundary Conditions	k_1	k_2	k_3	Conditions
2. Three edges clamped, one edge simply supported	4.730 γ_2	γ_4 γ_4	$12.30\gamma_4(\gamma_4 - 1)$ $\gamma_2\gamma_4(\gamma_2 - 2)(\gamma_4 - 1)$	$n = 1, 2, 3, \ldots, m = 1$ $n = 1, 2, 3, \ldots, m = 2, 3, 4, \ldots$
3. Two opposite edges clamped, other edges simply supported	4.730 γ_2	γ_3 γ_3	$12.30\gamma_3^2$ $\gamma_2\gamma_3^2(\gamma_2 - 2)$	$n = 1, m = 1, 2, 3, \ldots$ $n = 2, 3, 4, \ldots, m = 1, 2, 3, \ldots$

TABLE 18-20   **Frequencies of Orthotropic Rectangular Plates**   1062

4. Two consecutive edges clamped, other edges simply supported	γ_1	γ_4	$\gamma_1\gamma_4(\gamma_1 - 1)(\gamma_4 - 1)$	$n = 1, 2, 3, \ldots$ $m = 1, 2, 3, \ldots$
5. One edge clamped, other edges simply supported	γ_1	γ_3	$\gamma_1\gamma_3^2(\gamma_1 - 1)$	$n = 1, 2, 3, \ldots$ $m = 1, 2, 3, \ldots$
6. All edges simply supported	γ_0	γ_3	$\gamma_0^2\gamma_3^2$	$n = 1, 2, 3, \ldots$ $m = 1, 2, 3, \ldots$

[a]Adopted from Leissa Ref. 18.2.

TABLE 18-21 TRANSFER AND STIFFNESS MATRICES FOR RECTANGULAR PLATE SEGMENT SIMPLY SUPPORTED ON TWO OPPOSITE EDGES

Notation

Simple supports are at $y = 0$ and $y = L_y$.
Element extends from $x = a$ to $x = b$.

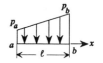

w = deflection
θ = slope about y axis
E = modulus of elasticity
ℓ = length of element in x direction, span of matrices
L_y = length of element in y direction
ε_m = constant that accounts for variation of distributed loads in y direction (Table 18-22)

ν = Poisson's ratio
h = thickness of plate
ρ = mass per unit area
P_y = compressive force per unit length in y direction
ω = natural frequency (rad/s)
M = bending moment per unit length, about y axis
V = equivalent shear force per unit length, in z direction

$$M_T(x) = \int_{-h/2}^{h/2} \frac{E\alpha}{1-\nu} T(x, z)z\,dz$$

$$M_{T_a} = \int_{-h/2}^{h/2} \frac{E\alpha}{1-\nu} T(a, z)z\,dz$$

$$M_{T_b} = \int_{-h/2}^{h/2} \frac{E\alpha}{1-\nu} T(b, z)z\,dz$$

T = change in temperature
$D = Eh^3/[12(1 - \nu^2)]$
$\beta = m\pi/L_y$

$$c_h = \begin{cases} \cosh \beta\ell & \text{for massless isotropic segment} \\ \cosh d\ell & \text{for isotropic plate segment} \\ & \quad \text{with mass or in-plane force} \end{cases}$$

$c = \cos q\ell$

$$s_h = \begin{cases} \sinh \beta\ell & \text{for massless isotropic segment} \\ \sinh d\ell & \text{for isotropic plate segment} \\ & \quad \text{with mass or in-plane force} \end{cases}$$

$s = \sin ql$

TABLE 18-21 Matrices for Rectangular Plate Segment   1064

ε_m = constant that accounts for variation of distributed loads in y direction. Values of ε_m are given below.

Loading (Force or Moment) Distribution in y Direction	Constant ε_m
1. Distributed load constant in y direction 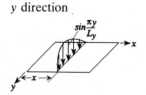	$\varepsilon_m = \dfrac{2}{m\pi}\left(\cos\dfrac{m\pi b_1}{L_y} - \cos\dfrac{m\pi b_2}{L_y}\right)$ If $b_1 = 0$, $b_2 = L_y$: $\varepsilon_m = \dfrac{2}{m\pi}(1 - \cos m\pi)$ $= \begin{cases} 4/(m\pi) & \text{if } m = 1, 3, 5, 7, \ldots \\ 0 & \text{if } m = 2, 4, 6, 8, \ldots \end{cases}$
2. Distributed load ramp in y direction	$\varepsilon_m = \dfrac{2}{m^2\pi^2}\left[-\ell_y m\pi \cos\dfrac{m\pi b_2}{L_y} \right.$ $\left. + L_y\left(\sin\dfrac{m\pi b_2}{L_y} - \sin\dfrac{m\pi b_1}{L_y}\right)\right]$
3. Sinusoidal load in y direction	$\varepsilon_1 = 1$ $\varepsilon_m = 0 \qquad m > 1$

For static response of plate with in-plane compressive force P_y

$$d^2 = \frac{m\pi}{L_y}\sqrt{\frac{P_y}{D}} + \left(\frac{m\pi}{L_y}\right)^2 \qquad q^2 = \frac{m\pi}{L_y}\sqrt{\frac{P_y}{D}} - \left(\frac{m\pi}{L_y}\right)^2$$

$$\lambda = \frac{m\pi}{L_y}\sqrt{\frac{P_y}{D}} \qquad \eta_1 = \frac{m\pi}{L_y}\sqrt{\frac{P_y}{D}} - (1-\nu)\left(\frac{m\pi}{L_y}\right)^2$$

$$\eta_2 = \frac{m\pi}{L_y}\sqrt{\frac{P_y}{D}} + (1-\nu)\left(\frac{m\pi}{L_y}\right)^2 \qquad \zeta = \nu\left(\frac{m\pi}{L_y}\right)^2$$

For vibrating plate:

$$d^2 = \sqrt{\frac{\rho}{D}}\,\omega + \left(\frac{m\pi}{L_y}\right)^2 \qquad q^2 = \sqrt{\frac{\rho}{D}}\,\omega - \left(\frac{m\pi}{L_y}\right)^2$$

$$\lambda = \sqrt{\frac{\rho}{D}}\,\omega \qquad \eta_1 = \sqrt{\frac{\rho}{D}}\,\omega - (1-\nu)\left(\frac{m\pi}{L_y}\right)^2$$

$$\eta_2 = \sqrt{\frac{\rho}{D}}\,\omega + (1-\nu)\left(\frac{m\pi}{L_y}\right)^2 \qquad \zeta = \nu\left(\frac{m\pi}{L_y}\right)^2$$

The variables at x, y are given by

$$\begin{bmatrix} w(x,y) \\ \theta(x,y) \\ V(x,y) \\ M(x,y) \end{bmatrix} = \sum_{m=1}^{\infty} \begin{bmatrix} w_m(x) \\ \theta_m(x) \\ V_m(x) \\ M_m(x) \end{bmatrix} \sin\frac{m\pi y}{L_y}$$

where $w_m(x), \theta_m(x), V_m(x), M_m(x)$ are computed using the matrices of this table and the methodology of Appendix III.

TABLE 18-21   **Matrices for Rectangular Plate Segment**   **1066**

Transfer Matrix (Sign Convention 1):

The transfer matrix method can be used to compute $w_m(x), \theta_m(x), V_m(x), M_m(x)$ for each m. To simplify the notation, the subscript m will be dropped.

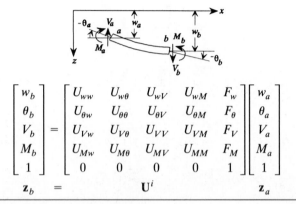

$$
\begin{bmatrix} w_b \\ \theta_b \\ V_b \\ M_b \\ 1 \end{bmatrix} = \begin{bmatrix} U_{ww} & U_{w\theta} & U_{wV} & U_{wM} & F_w \\ U_{\theta w} & U_{\theta\theta} & U_{\theta V} & U_{\theta M} & F_\theta \\ U_{Vw} & U_{V\theta} & U_{VV} & U_{VM} & F_V \\ U_{Mw} & U_{M\theta} & U_{MV} & U_{MM} & F_M \\ 0 & 0 & 0 & 0 & 1 \end{bmatrix} \begin{bmatrix} w_a \\ \theta_a \\ V_a \\ M_a \\ 1 \end{bmatrix}
$$

$$\mathbf{z}_b \quad = \quad \mathbf{U}^i \quad \mathbf{z}_a$$

Stiffness Matrix (Sign Convention 3):

The responses $w_m(x), \theta_m(x), V_m(x), M_m(x)$ can be computed for each m using the displacement method. To simplify the notation, the subscript m will be dropped from $w_m, \theta_m, V_m,$ and M_m.

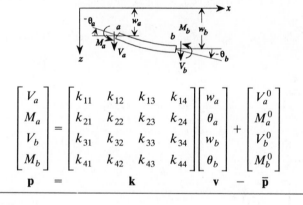

$$
\begin{bmatrix} V_a \\ M_a \\ V_b \\ M_b \end{bmatrix} = \begin{bmatrix} k_{11} & k_{12} & k_{13} & k_{14} \\ k_{21} & k_{22} & k_{23} & k_{24} \\ k_{31} & k_{32} & k_{33} & k_{34} \\ k_{41} & k_{42} & k_{43} & k_{44} \end{bmatrix} \begin{bmatrix} w_a \\ \theta_a \\ w_b \\ \theta_b \end{bmatrix} + \begin{bmatrix} V_a^0 \\ M_a^0 \\ V_b^0 \\ M_b^0 \end{bmatrix}
$$

$$\mathbf{p} \quad = \quad \mathbf{k} \quad \mathbf{v} \quad - \quad \overline{\mathbf{p}}$$

Matrices

Massless, Isotropic Segment

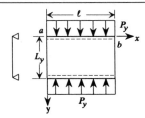

TRANSFER MATRIX:

$$U_{ww} = -\frac{\beta}{2}(1-\nu)\ell s_h + c_h$$

$$U_{Vw} = \frac{D\beta^4}{2}\left[(3 - 2\nu - \nu^2)\frac{s_h}{\beta} - (1-\nu)^2\ell c_h\right]$$

$$U_{w\theta} = -\frac{1}{2}\left[(1+\nu)\frac{s_h}{\beta} + (1-\nu)\ell c_h\right]$$

$$U_{V\theta} = -\frac{D\beta^3}{2}(1-\nu)^2\ell s_h$$

$$U_{wV} = \frac{1}{2D\beta^2}\left(\frac{s_h}{\beta} - \ell c_h\right)$$

$$U_{VV} = -\frac{\beta}{2}(1-\nu)\ell s_h + c_h$$

$$U_{wM} = -\frac{\ell s_h}{2D\beta}$$

$$U_{VM} = \frac{\beta^2}{2}\left[(1+\nu)\frac{s_h}{\beta} - (1-\nu)\ell c_h\right]$$

$$U_{\theta w} = -\frac{\beta^2}{2}\left[(1+\nu)\frac{s_h}{\beta} - (1-\nu)\ell c_h\right]$$

$$U_{Mw} = \frac{D}{2}\beta^3(1-\nu)^2\ell s_h$$

$$U_{\theta\theta} = \frac{\beta}{2}(1-\nu)\ell s_h + c_h$$

$$U_{M\theta} = \frac{D\beta^2}{2}\left[(3 - 2\nu - \nu^2)\frac{s_h}{\beta} + (1-\nu)^2\ell c_h\right]$$

$$U_{\theta V} = \frac{\ell}{2D\beta}s_h$$

$$U_{MV} = \tfrac{1}{2}\left[(1+\nu)\frac{s_h}{\beta} + (1-\nu)\ell c_h\right]$$

$$U_{\theta M} = \frac{1}{2D}\left(\frac{s_h}{\beta} + \ell c_h\right)$$

$$U_{MM} = \tfrac{1}{2}\beta(1-\nu)\ell s_h + c_h$$

LOADING FUNCTIONS:

$$F_w = \frac{\varepsilon_m}{2D\beta^4}\left[p_a(\beta\ell s_h - 2c_h + 2) + \frac{p_b - p_a}{\ell}\left(-\frac{3}{\beta}s_h + \ell c_h + 2\ell\right)\right.$$

$$-c_a\beta^2\left(\frac{s_h}{\beta} - \ell c_h\right) + \frac{c_b - c_a}{\ell}(\beta\ell s_h - 2c_h + 2)$$

$$\left.-2M_{T_a}\beta^2\left(-\frac{\beta\ell}{2}s_h - 2 + 2c_h\right) - \frac{1}{\ell}(M_{T_b} - M_{T_a})2\beta(2s_h - 2\beta\ell - \beta\ell c_h)\right]$$

$$F_\theta = \frac{\varepsilon_m}{2D\beta^4}\left[p_a\beta^2\left(\frac{s_h}{\beta} - \ell c_h\right) - \frac{p_b - p_a}{\ell}(\beta\ell s_h - 2c_h + 2) - c_a\beta^3\ell s_h\right.$$

$$+ \frac{c_b - c_a}{\ell}\beta^2\left(\frac{s_h}{\beta} - \ell c_h\right) - M_{T_a}2\beta^3\left(\frac{3}{2}s_h - \frac{\beta\ell}{2}c_h\right)$$

$$\left.-\frac{M_{T_b} - M_{T_a}}{\ell}2\beta^4\left(2c_h - 2 - \frac{\beta\ell}{2}s_h\right)\right]$$

| TABLE 18-21 | **Matrices for Rectangular Plate Segment** | 1068 |

Matrices

$$F_V = \varepsilon_m \left[p_a \left(\frac{1-\nu}{2} \ell c_h - \frac{3-\nu}{2\beta} s_h \right) + \frac{p_b - p_a}{\ell} \left(\frac{1-\nu}{2\beta} \ell s_h - \frac{2-\nu}{\beta^2} c_h + \frac{2-\nu}{\beta^2} \right) \right.$$

$$+ c_a \left(\frac{1-\nu}{2} \beta \ell s_h - c_h + 1 \right) - \frac{c_b - c_a}{\ell} \left(\frac{3-\nu}{2} \frac{s_h}{\beta} - \frac{1-\nu}{2} \ell c_h - \ell \right)$$

$$\left. - M_{T_a} 2(1-\nu)\beta s_h - \frac{1}{\ell} (M_{T_b} - M_{T_a}) 2(1-\nu)(c_h - 1) \right]$$

$$F_M = \frac{\varepsilon_m}{\beta^2} \left[-p_a \left(\frac{1-\nu}{2} \beta \ell s_h + \nu c_h - \nu \right) + \frac{p_b - p_a}{\ell} \left(\frac{1-3\nu}{2} \frac{s_h}{\beta} - \frac{1-\nu}{2} \ell c_h + \nu \ell \right) \right.$$

$$- c_a \beta^2 \left(\frac{1+\nu}{2} \frac{s_h}{\beta} + \frac{1-\nu}{2} \ell c_h \right) - \frac{c_b - c_a}{\ell} \left(\frac{1-\nu}{2} \beta \ell s_h + \nu c_h - \nu \right)$$

$$\left. - M_{T_a} \beta^2 (1-\nu) \left(\frac{\beta \ell s_h}{2} + 2 - 2c_h \right) - \frac{1}{\ell}(M_{T_b} - M_{T_a}) \beta^2 (1-\nu) \left(2\ell - \frac{2}{\beta} s_h \right) \right]$$

STIFFNESS MATRIX:

$k_{11} = -2D\beta^3 (\ell\beta + c_h s_h)/H_s$

$k_{21} = D\beta^2 \left[(1+\nu)\ell^2\beta^2 + (1 - \nu + 2\ell^2\beta^2\nu) s_h^2 \right]/H_s$

$k_{22} = 2D\beta(\ell\beta - c_h s_h)/H_s$

$k_{31} = 2D\beta^3 (s_h + \ell\beta c_h)/H_s$

$k_{32} = -2D\ell\beta^3 s_h/H_s$

$k_{33} = k_{11}$

$k_{41} = -k_{32}$

$k_{42} = -2D\beta(\ell\beta c_h - s_h)/H_s$

$k_{43} = -D\beta^2 \left[(1-\nu)\ell^2\beta^2 + (1+\nu) s_h^2 \right]/H_s$

$k_{44} = k_{22}$ This matrix is symmetric.

$H_s = \ell^2\beta^2 - sh^2$

$k_{ij} = k_{ji}$

$V_a^0 = (U_{\theta M} F_w - U_{wM} F_\theta)/\Delta_0$

$M_a^0 = (-U_{\theta V} F_w + U_{wV} F_\theta)/\Delta_0$

$V_b^0 = F_V - \{ (U_{VV} U_{\theta M} + U_{VM} U_{\theta V}) F_w - (U_{VM} U_{wV} - U_{VV} U_{wM}) F_\theta \}/\Delta_0$

$M_b^0 = F_M - \{ (U_{MV} U_{\theta M} + U_{MM} U_{\theta V}) F_w - (U_{MM} U_{wV} - U_{MV} U_{wM}) F_\theta \}/\Delta_0$

$\Delta_0 = U_{wV} U_{\theta M} - U_{\theta V} U_{wM}$

Matrices

Isotropic Plate Segment with Mass or In-Plane Force

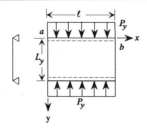

TRANSFER MATRIX:

$$
\begin{bmatrix}
\dfrac{1}{2\lambda}(\eta_1 c_h + \eta_2 c) & -\dfrac{1}{2\lambda}\left(\dfrac{\eta_2}{d}s_h + \dfrac{\eta_1}{q}s\right) & -\dfrac{1}{2\lambda D}\left(\dfrac{s_h}{d} - \dfrac{s}{q}\right) & -\dfrac{1}{2\lambda D}(c_h - c) & F_w \\[2ex]
-\dfrac{1}{2\lambda}(\eta_1 s_h d - \eta_2 q s) & \dfrac{1}{2\lambda}(\eta_2 c_h + \eta_1 c) & \dfrac{1}{2\lambda D}(c_h - c) & \dfrac{1}{2\lambda D}(d s_h + q s) & F_\theta \\[2ex]
-\dfrac{D}{2\lambda}(\eta_1^2 d s_h + \eta_2^2 q s) & \dfrac{D}{2\lambda}\eta_1\eta_2(c_h - c) & \dfrac{1}{2\lambda}(\eta_1 c_h + \eta_2 c) & \dfrac{1}{2\lambda}(\eta_1 d s_h - \eta_2 q s) & F_V \\[2ex]
-\dfrac{D}{2\lambda}\eta_1\eta_2(c_h - c) & \dfrac{D}{2\lambda}\left(\dfrac{\eta_2^2}{d}s_h - \dfrac{\eta_1^2}{q}s\right) & \dfrac{1}{2\lambda}\left(\dfrac{\eta_2}{d}s_h + \dfrac{\eta_1}{d}s\right) & \dfrac{1}{2\lambda}(\eta_2 c_h + \eta_1 c) & F_M \\[2ex]
0 & 0 & 0 & 0 & 1
\end{bmatrix}
$$

$$\mathbf{U}^i$$

$$
F_w = \frac{\varepsilon_m}{2\lambda D}\left\{ p_a\left(\frac{c_h}{d^2} + \frac{c}{q^2} - \frac{2}{d^2}\frac{\lambda}{q^2}\right) + \frac{p_b - p_a}{\ell}\left(\frac{s_h}{d^3} + \frac{s}{q^3} - \frac{2}{d^2}\frac{\lambda\ell}{q^2}\right) + c_a\left(\frac{s_h}{d} - \frac{s}{q}\right) \right.
$$

$$
+ \frac{c_b - c_a}{\ell}\left(\frac{c_h}{d^2} + \frac{c}{q^2} - \frac{2}{d^2}\frac{\lambda}{q^2}\right) - M_{T_a}\left[\eta_2\frac{c_h - 1}{d^2} + \eta_1\frac{1 - c}{q^2}\right.
$$

$$
\left. + (1 - \nu)\left(\frac{m\pi}{L_y}\right)^2\left(\frac{c_h - 1}{d^2} - \frac{1 - c}{q^2}\right)\right]
$$

$$
- \frac{M_{T_b} - M_{T_a}}{\ell}\left[\eta_2\frac{s_h - d\ell}{d^3} + \eta_1\frac{q\ell + s_h}{q^3}\right.
$$

$$
\left.\left. + (1 - \nu)\left(\frac{m\pi}{L_y}\right)^2\left(\frac{s_h - d\ell}{d^3} - \frac{q\ell + s_h}{q^3}\right)\right]\right\}
$$

TABLE 18-21   **Matrices for Rectangular Plate Segment**   **1070**

Matrices

$$F_\theta = \frac{-\varepsilon_m}{2\lambda D}\left\{ p_a\left(\frac{s_h}{d} - \frac{s}{q}\right) + \frac{p_b - p_a}{\ell}\left(\frac{c_h}{d^2} + \frac{c}{q^2} - \frac{2\lambda}{d^2 q^2}\right)\right.$$

$$+ c_a(c_h - c) + \frac{c_b - c_a}{\ell}\left(\frac{s_h}{d} - \frac{s}{q}\right)$$

$$- M_{T_a}\left[\eta_2\frac{s_h}{d} - \eta_1\frac{s}{q} + (1 - \nu)\left(\frac{m\pi}{L_y^2}\right)^2\left(\frac{s_h}{d} + \frac{s}{q}\right)\right]$$

$$- \frac{M_{T_b} - M_{T_a}}{\ell}\left[\eta_2\frac{c_h - 1}{d^2} - \eta_1\frac{c - 1}{q^2}\right.$$

$$\left.\left. + (1 - \nu)\left(\frac{m\pi}{L_y}\right)^2\left(\frac{c_h - 1}{d^2} - \frac{c - 1}{q^2}\right)\right]\right\}$$

$$F_V = \frac{-\varepsilon_m}{2\lambda}\left\{ p_a\left(\frac{\eta_2}{d}s_h + \frac{\eta_1}{q}s\right) + \frac{p_b - p_a}{\ell}\left(\frac{\eta_1}{d^2}c_h - \frac{\eta_2}{q^2}c + \frac{d^4 - q^4 - 2\lambda s}{d^2 q^2}\right)\right.$$

$$+ c_a(\eta_1 c_h + \eta_2 c - 2\lambda) + \frac{c_b - c_a}{\ell}\left(\frac{\eta_1}{d}s_h + \frac{\eta_2}{q}s - 2\lambda\ell\right)$$

$$- M_{T_a}\left[\eta_1\eta_2\left(\frac{s_h}{d} + \frac{s}{q}\right) + (1 - \nu)\left(\frac{m\pi}{L_y^2}\right)\left(\eta_1\frac{s_h}{d} - \eta_s\frac{s}{q}\right)\right]$$

$$- \frac{M_{T_b} - M_{T_a}}{\ell}\left[\eta_1\eta_2\left(\frac{c_h - 1}{d^2} + \frac{1 - c}{q^2}\right)\right.$$

$$\left.\left. + (1 - \nu)\left(\frac{m\pi}{L_y}\right)^2\left(\eta_1\frac{c_h - 1}{d^2} + \eta_2\frac{1 - c}{q^2}\right)\right]\right\}$$

$$F_M = \frac{-\varepsilon_m}{2\lambda}\left\{ p_a\left(\frac{\eta_2}{d^2}c_h - \frac{\eta_1}{q^2}c + \frac{2\lambda\zeta}{d^2 q^2}\right) + \frac{p_b - p_a}{\ell}\left(\frac{\eta_2}{d^3}s_h - \frac{\eta_1}{q^3}s + \frac{2\lambda\zeta\ell}{d^2 q^2}\right)\right.$$

$$+ c_a\left(\frac{\eta_2}{d}s_h + \frac{\eta_1}{q}s\right) + \frac{c_b - c_a}{\ell}\left(\frac{\eta_2}{d^2}c_h - \frac{\eta_1}{q^2}c + \frac{2\lambda\zeta}{d^2 q^2}\right)$$

$$- M_{T_a}\left[\eta_2^2\frac{c_h - 1}{d^2} + \eta_1^2\frac{c - 1}{q^2} + (1 - \nu)\left(\frac{m\pi}{L_y^2}\right)^2\left(\eta_2\frac{c_h - 1}{d^2} - \eta_1\frac{c - 1}{q^2}\right)\right]$$

$$- \frac{M_{T_b} - M_{T_a}}{\ell}\left[\eta_2^2\frac{s_h - d\ell}{d^3} + \eta_1^2\frac{s - q\ell}{q^3} + (1 - \nu)\left(\frac{m\pi}{L_y}\right)^2\left(\eta_2\frac{s_h - d\ell}{d^2} - \eta_1\frac{s_h - q\ell}{q^3}\right)\right]\right\}$$

Matrices

STIFFNESS MATRIX:

$$H_0 = 2(1 - c_h c)$$

$$k_{11} = (\eta_1 + \eta_2)(ds_h c + qc_h s)D/H_0$$

$$k_{12} = \left[(\eta_2 - \eta_1)(1 - c_h c) - (d^2\eta_1 + q^2\eta_2)\frac{s_h s}{dq}\right]D/H_0$$

$$k_{13} = -2\lambda(ds_h + qs)D/H_0$$

$$k_{14} = 2\lambda(c - c_h)D/H_0$$

$$k_{21} = k_{12}$$

$$k_{22} = (\eta_1 + \eta_2)\left(\frac{c_h s}{q} - \frac{cs_h}{d}\right)D/H_0$$

$$k_{23} = 2\lambda(c_h - c)D/H_0$$

$$k_{24} = 2\lambda\left(\frac{s_h}{d} - \frac{s}{q}\right)D/H_0$$

$$k_{31} = k_{13} \qquad k_{32} = k_{23}$$

$$k_{33} = (\eta_1 + \eta_2)(ds_h c + qsc_h c)D/H_0$$

$$k_{34} = \left[(\eta_2 - \eta_1)(-1 + c_h c) + \frac{s_h s}{dq}(d^2\eta_1 + q^2\eta_2)\right]D/H_0$$

$$k_{41} = k_{14} \qquad k_{42} = k_{24} \qquad k_{43} = k_{34}$$

$$k_{44} = (\eta_1 + \eta_2)\left(\frac{s}{q}c_h - \frac{c}{d}s_h\right)D/H_0$$

$$V_a^0 = 2\lambda D[(ds_h + qs)F_w + (c_h - c)F_\theta]/H_0$$

$$M_a^0 = -2\lambda D\left[(c_h - c)F_w + \left(\frac{s_h}{d} - \frac{s}{q}\right)F_\theta\right]/H_0$$

$$V_b^0 = F_V - D\left\{(\eta_1 + \eta_2)(dcs_h + qsc_h)F_w\right.$$

$$\left. + \left[(\eta_2 - \eta_1)(cc_h - 1) + \frac{ss_h}{aq}(d^2\eta_1 + q^2\eta_2)\right]F_\theta\right\}/H_0$$

$$M_b^0 = F_M - D\left\{\left[(\eta_2 - \eta_1)(cc_h - 1) + \frac{ss_h}{dq}(d^2\eta_1 + q^2\eta_2)\right]F_w\right.$$

$$\left. + (\eta_1 + \eta_2)\left(\frac{sc_h}{q} - \frac{cs_h}{d}\right)F_\theta\right\}/H_0$$

TABLE 18-21   **Matrices for Rectangular Plate Segment**   **1072**

TABLE 18-22   TRANSFER AND STIFFNESS MATRICES FOR POINT OCCURRENCES FOR RECTANGULAR PLATE SEGMENT WHERE TWO OPPOSITE EDGES ARE SIMPLY-SUPPORTED

Case	Matrices

Case

1.
Concentrated force W_T

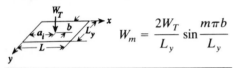

$$W_m = \frac{2W_T}{L_y} \sin \frac{m\pi b}{L_y}$$

2.
Line force W
(force/length in y direction)

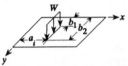

$$W_m = \frac{2W}{m\pi}\left(\cos \frac{m\pi b_1}{L_y} - \cos \frac{m\pi b_2}{L_y}\right)$$

3.
Linearly varying line force

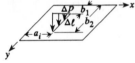

$$W_m = \frac{\Delta p}{\Delta \ell}\frac{2}{m^2\pi^2}$$
$$\times \left[(b_1 - b_2)m\pi \cos \frac{m\pi b_2}{L_y}\right.$$
$$\left. + L_y\left(\sin \frac{m\pi b_2}{L_y} - \sin \frac{m\pi b_1}{L_y}\right)\right]$$

4.
Line moment C
(force-length/length in y direction)

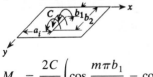

$$M_m = \frac{2C}{m\pi}\left(\cos \frac{m\pi b_1}{L_y} - \cos \frac{m\pi b_2}{L_y}\right)$$

Matrices

TRANSFER MATRIX FOR CASES 1–5:

$$\mathbf{U}_i = \begin{bmatrix} 1 & 0 & 0 & 0 & w_m \\ 0 & 1 & 0 & 0 & -\alpha_m \\ 0 & 0 & 1 & 0 & -W_m \\ 0 & 0 & 0 & 1 & -M_m \\ 0 & 0 & 0 & 0 & 1 \end{bmatrix}$$

STIFFNESS MATRIX FOR CASES 1–6:

Traditionally, these applied loads are implemented as nodal conditions.

Case	Matrices

5.

Jump in level w_1 (length)
and change in slope
α (radians)

$$w_m = \frac{2w_1}{m\pi}(1 - \cos m\pi)$$

$$\alpha_m = \frac{2\alpha}{m\pi}(1 - \cos m\pi)$$

6.
Linear and rotary hinges k_2, k_2^*

TRANSFER MATRIX FOR CASES 1–8:

$$\mathbf{U}_i = \begin{bmatrix} 1 & 0 & 1/k_2 & 0 & w_m \\ 0 & 1 & 0 & 1/k_2^* & -\alpha_m \\ k_1 - m_i\omega^2 & 0 & 1 & 0 & -W_m \\ 0 & k_1^* & 0 & 1 & -M_m \\ 0 & 0 & 0 & 0 & 1 \end{bmatrix}$$

7.
Springs: k_1 (force/length squared)
and k_1^* (force-length/length
squared). Values of k_1, k_1^* can be
taken from Table 11-21 for various spring, flexible support combinations; for example, Line spring
k_1.

STIFFNESS MATRIX FOR CASE 7:

$$\begin{bmatrix} V_a \\ M_a \end{bmatrix} = \begin{bmatrix} k_1 & 0 \\ 0 & k_1^* \end{bmatrix} \begin{bmatrix} w_a \\ \theta_a \end{bmatrix}$$

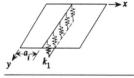

STIFFNESS MATRIX FOR CASE 8:

$$\begin{bmatrix} V_a \\ M_a \end{bmatrix} = -\omega^2 \begin{bmatrix} M_i & 0 \\ 0 & 0 \end{bmatrix} \begin{bmatrix} w_a \\ \theta_a \end{bmatrix}$$

8.
Line lumped mass M_i
(mass/length in y direction)

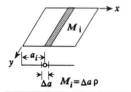

$M_i = \Delta a\, \rho$

TABLE 18-22   **Matrices for Point Occurrences**   1074

Notation

w = deflection

θ = slope about y axis

ℓ = length of element in the x direction

P, P_y = compressive forces per unit length in x and y directions

M = bending moment per unit length, about y axis

V = equivalent shear force per unit length, in z direction

D_x, D_y, D_{xy} = defined in Table 18-14

ρ = mass per unit area

ω = natural frequency

k = modulus of elastic foundation (F/L^3)

$$\eta_m = \begin{cases} \text{simply–simply supported} & m\pi \\ \text{fixed–simply supported} & (4m+1)\pi/4 \\ \text{fixed–fixed} & (2m+1)\pi/2 \end{cases}$$

These supports are at $y = 0$ and $y = L_y$:

L_y = length of plate in y direction

ν_x, ν_y = Poisson's ratio in x and y directions

$B = \frac{1}{2}(D_x\nu_y + D_y\nu_x + 4D_{xy})$

$\beta_m = \eta_m/L_y$

$\lambda_m = \dfrac{1}{D_x}\left\{D_y\beta_m^4\left[1 + \nu_x\nu_y(\varphi_m^2 - 1)\right] + \beta_m^2\varphi_m P_y - \rho\omega^2 + k_y\right\}$

$\alpha_{1_m} = -\nu_y\beta_m^2\varphi_m$

$\alpha_{3_m} = -\dfrac{P}{D_x} - \left(1 - \dfrac{\nu_x D_y}{D_x}\right)\beta_m^2\varphi_m$

$\zeta_m = \dfrac{1}{D_x}\left(P - 2B\beta_m^2\varphi_m\right)$

$\alpha_{2_m} = -\left(\dfrac{4D_{xy}}{D_x} + \nu_y\right)\beta_m^2\varphi_m$

φ_m = modal constant defined below

	Boundary Conditions		Values for $m = 1, 2, 3, \ldots, \infty$			
Case	$y = 0$	$y = L_y$	φ_1	φ_2	φ_3	$\varphi_m \ (m \geq 4)$
1	Simply supported	Simply supported	1	1	1	1
2	Simply supported	Fixed	$\dfrac{2.9317}{\eta_1}$	$\dfrac{6.0686}{\eta_2}$	$\dfrac{9.2095}{\eta_3}$	$\dfrac{\eta_m - 1}{\eta_m}$
3	Fixed	Fixed	$\dfrac{2.6009}{\eta_1}$	$\dfrac{5.8634}{\eta_2}$	$\dfrac{8.9984}{\eta_3}$	$\dfrac{\eta_m - 2}{\eta_m}$

To use this table to obtain a matrix for a particular element, follow the steps:

1. Evaluate the desired number of terms of β_m and φ_m with the boundary conditions at $y = 0$ and $y = L_y$ considered.
2. Calculate the parameters $\lambda_m, \zeta_m, \alpha_{1_m}, \alpha_{2_m}, \alpha_{3_m}$.
3. Look up the appropriate e_i functions in Part B of this table.
4. Substitute these e_i expressions into the matrices below.
5. According to the distribution of the loading in the y direction, insert ε_m of Part D of this table into the loading functions of Part C of this table.
6. Use the matrices of this table and the techniques of Appendix III to compute $w_m(x)$, $\theta_m(x)$, $V_m(x)$, and $M_m(x)$ and calculate $w(x, y)$, $\theta(x, y)$, $V(x, y)$, and $M(x, y)$ from

$$
\begin{bmatrix}
w(x, y) \\
\theta(x, y) \\
V(x, y) \\
M(x, y)
\end{bmatrix}
=
\sum_{m=1}^{\infty}
\begin{bmatrix}
w_m(x) \\
\theta_m(x) \\
V_m(x) \\
M_m(x)
\end{bmatrix}
\varphi_m(y)
$$

To simplify the notation in this table, the subscript m will be dropped from $w_m, \theta_m, V_m,$ and M_m. The quantity $\varphi_m(y)$ is given by

Case	Boundary Conditions		$\varphi_m(y)$
	$y = 0$	$y = L_y$	
1	Simply Supported	Simply Supported	$\sin \beta_m y$
2	Simply Supported	Fixed	$\cosh \beta_m y - \cos \beta_m y + E_m[\sinh \beta_m y - \sin \beta_m y]$
3	Fixed	Fixed	$\cosh \beta_m y - \cos \beta_m y - E_m[\sinh \beta_m y - \sin \beta_m y]$

$E_m = [\cosh \eta_m - \cos \eta_m]/[\sinh \eta_m - \sin \eta_m]$

Matrices

Transfer Matrix (Sign Convention 1)

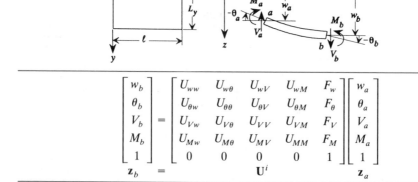

$$
\begin{bmatrix}
w_b \\
\theta_b \\
V_b \\
M_b \\
1
\end{bmatrix}
=
\begin{bmatrix}
U_{ww} & U_{w\theta} & U_{wV} & U_{wM} & F_w \\
U_{\theta w} & U_{\theta\theta} & U_{\theta V} & U_{\theta M} & F_\theta \\
U_{Vw} & U_{V\theta} & U_{VV} & U_{VM} & F_V \\
U_{Mw} & U_{M\theta} & U_{MV} & U_{MM} & F_M \\
0 & 0 & 0 & 0 & 1
\end{bmatrix}
\begin{bmatrix}
w_a \\
\theta_a \\
V_a \\
M_a \\
1
\end{bmatrix}
$$

$$
\mathbf{z}_b \quad = \quad \mathbf{U}^i \quad\quad \mathbf{z}_a
$$

TABLE 18-23   **Matrices for General Rectangular Plate Segment**   **1076**

$$U_{ww} = e_1 + (\zeta_m + \alpha_{1m})e_3$$
$$U_{w\theta} = -e_2 - (\zeta_m + \alpha_{2m})e_4$$
$$U_{wV} \doteq -e_4/D_x$$
$$U_{wM} = -e_3/D_x$$
$$U_{Vw} = D_x[(\lambda_m + \alpha_{1m}\alpha_{2m} + \alpha_{1m}\zeta_m)e_2 - \lambda_m(\alpha_{2m} - \alpha_{1m})e_4]$$
$$U_{V\theta} = -D_x[\lambda_m + \alpha_{2m}(\zeta_m + \alpha_{2m})]e_3$$
$$U_{VV} = e_1 - \alpha_{2m}e_3$$
$$U_{VM} = e_0 - \alpha_{2m}e_2$$
$$U_{Mw} = D_x[\lambda_m + \alpha_{1m}(\zeta_m + \alpha_{1m})]e_3$$
$$U_{M\theta} = D_x[e_0 - (\alpha_{1m} - \zeta_m - \alpha_{2m})e_2 - \alpha_{1m}(\zeta_m + \alpha_{2m})e_4]$$
$$U_{MV} = e_2 - \alpha_{1m}e_4$$
$$U_{MM} = e_1 - \alpha_{1m}e_3$$

$$U_{\theta w} = -e_0 - (\zeta_m + \alpha_{1m})e_2$$
$$U_{\theta\theta} = e_1 + (\zeta_m + \alpha_{2m})e_3$$
$$U_{\theta V} = e_3/D_x$$
$$U_{\theta M} = e_2/D_x$$

Stiffness Matrix (Sign Convention 2)

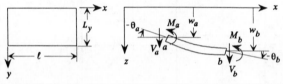

$$
\begin{bmatrix} V_a \\ M_a \\ V_b \\ M_b \end{bmatrix}
=
\begin{bmatrix}
H_0 H_1 & & \text{Symmetric} & \\
H_0[H_3 + H_4 + H_5] & H_0 H_2 & & \\
-H_0 H_6 & H_0 H_7 & H_0 H_1 & \\
-H_0 H_7 & H_0 H_3 & H_0[H_1 - \alpha_{2m}\nabla] & H_0 H_2
\end{bmatrix}
\begin{bmatrix} w_a \\ \theta_a \\ w_b \\ \theta_b \end{bmatrix}
+
\begin{bmatrix} V_a^0 \\ M_a^0 \\ V_b^0 \\ M_b^0 \end{bmatrix}
$$

$$\mathbf{p}^i \quad = \qquad\qquad \mathbf{k}^i \qquad\qquad\qquad\qquad\qquad\qquad \mathbf{v}^i \quad - \quad \bar{\mathbf{p}}^i$$

$$\nabla = e_3^2 - e_2 e_4$$
$$H_0 = D_x/\nabla$$
$$H_1 = e_1 e_2 - e_0 e_3$$
$$H_2 = e_2 e_3 - e_1 e_4$$
$$H_3 = e_1 e_3 - e_2^2$$
$$H_4 = [e_3^2(\zeta_m + \alpha_{2m})]$$
$$H_5 = e_2 e_4(\zeta_m + \alpha_{2m})$$
$$H_6 = e_2$$
$$H_7 = e_3$$
$$H_8 = e_4$$
$$H_9 = e_1 e_3 - e_0 e_4$$
$$V_a^0 = H_0[H_6 F_w + H_7 F_\theta]$$
$$M_a^0 = -H_0[H_7 F_w + H_8 F_\theta]$$
$$V_b^0 = F_V - [H_1 F_w + (H_9 - \alpha_{2m}\nabla)F_\theta]H_0$$
$$M_b^0 = F_M - [(\alpha_{1m}\nabla - H_5)F_w + H_2 F_\theta]H_0$$

TABLE 18-23 (continued)   TRANSFER AND STIFFNESS MATRICES FOR A GENERAL RECTANGULAR PLATE SEGMENT

Constants for the Matrices

Column	1	2	3	4	5
	$\lambda_m < 0$	$\lambda_m = 0$	$\lambda_m = \frac{1}{4}\zeta_m^2$	$\lambda_m > 0$	
				$\lambda_m < \frac{1}{4}\zeta_m^2$	$\lambda_m > \frac{1}{4}\zeta_m^2$
e_0	$\dfrac{1}{g}(d^3 s_h - q^3 s)$	$-\zeta_m \delta$	$-\dfrac{\zeta_m}{4}(3\delta_2 + c\ell)$	$-\dfrac{1}{g}(q^3\delta_2 - d^3 s_1)$	$-\lambda_m e_4 - \zeta_m e_2$
e_1	$\dfrac{1}{g}(d^2 c_h + q^2 c)$	c	$\dfrac{1}{2}(2c - \delta_1 \ell)$	$\dfrac{p}{g}(q^2\delta - d^2 c)$	$c_h c - \dfrac{q^2 - d^2}{2dq} s_h s$
e_2	$\dfrac{1}{g}(ds_h + qs)$	δ	$\dfrac{1}{2}(\delta_2 + c\ell)$	$\dfrac{p}{g}(qs_2 - ds_1)$	$\dfrac{1}{2dq}(dc_h s + qcs_h)$
e_3	$\dfrac{1}{g}(c_h - c)$	$\dfrac{1}{\zeta_m}(1 - c)$	$\dfrac{\delta_2 \ell}{2}$	$+\dfrac{1}{g}(c - s)$	$\dfrac{1}{2dq} s_h s$
e_4	$\dfrac{1}{g}\left(\dfrac{s_h}{d} - \dfrac{s}{q}\right)$	$\dfrac{1}{\zeta_m}(\ell - \delta)$	$\dfrac{1}{\zeta_m}(\delta_2 - c\ell)$	$+\dfrac{1}{g}\left(\dfrac{s_1}{d} - \dfrac{s_2}{q}\right)$	$\dfrac{1}{2(d^2+q^2)} \times \left(\dfrac{c_h s}{q} - \dfrac{cs_h}{d}\right)$
e_5	$\dfrac{1}{g}\left(\dfrac{c_h}{d^2} + \dfrac{c}{q^2}\right) - \dfrac{1}{d^2 q^2}$	$\dfrac{1}{\zeta_m}\left(\dfrac{\ell^2}{2} - e_3\right)$	$\dfrac{2}{\zeta_m^2}(-2c - \delta_1\ell + 2)$	$\dfrac{p}{g}\left(\dfrac{\delta}{q^2} - \dfrac{c}{d^2}\right) + \dfrac{1}{d^2 q^2}$	$\dfrac{1 - e_1}{\lambda_m} - \dfrac{\zeta_m}{\lambda_m} e_3$
e_6	$\dfrac{1}{g}\left(\dfrac{s_h}{d^3} + \dfrac{s}{q^3}\right) - \dfrac{\ell}{d^2 q^2}$	$\dfrac{1}{\zeta_m}\left(\dfrac{\ell^2}{6} - e_4\right)$	$\dfrac{2}{\zeta_m^2}(-3\delta_2 + c\ell + 2\ell)$	$\dfrac{p}{g}\left(\dfrac{s_2}{q^3} - \dfrac{s_1}{d^3}\right) + \dfrac{1}{d^2 q^2}$	$\dfrac{\ell - e_2}{\lambda_m} - \dfrac{\zeta_m}{\lambda_m} e_1$

TABLE 18-23   Matrices for General Rectangular Plate Segment   1078

Columns 1, 2, … below are to be used with columns 1, 2, …, respectively, above.

Column	1	2a $\zeta_m > 0$:	3a $\zeta_m > 0$:	4a $\zeta_m > 0$:	5
	$c_h = \cosh d\ell$	$\alpha^2 = \zeta_m$	$\beta_m^2 = -\tfrac{1}{2}\zeta_m$	$g = q^2 - d^2 \quad p = 1$	$c_h = \cosh d\ell$
	$c = \cos q\ell$	$c = \cos \alpha\ell$	$c = \cos \beta_m\ell$	$c = \cos d\ell$	$c = \cos q\ell$
	$s_h = \sinh d\ell$	$\delta = \dfrac{\sin \alpha\ell}{\alpha}$	$\delta_1 = \beta_m \sin \beta_m\ell$	$\delta = \cos q\ell$	$s_h = \sinh d\ell$
	$s = \sin q\ell$		$\delta_2 = \dfrac{\sin \beta_m\ell}{\beta_m}$	$s_1 = \sin d\ell$	$s = \sin q\ell$
	$g = d^2 + q^2$			$s_2 = \sin q\ell$	$d^2 = \tfrac{1}{2}\sqrt{\lambda_m} - \tfrac{1}{4}\zeta_m$
	$d^2 = \sqrt{\beta_m^4 + \tfrac{1}{4}\zeta_m^2} - \tfrac{1}{2}\zeta_m$			$d^2 = \tfrac{1}{2}\zeta_m - \sqrt{\tfrac{1}{4}\zeta_m^2 - \lambda_m}$	$q^2 = \tfrac{1}{2}\sqrt{\lambda_m} + \tfrac{1}{4}\zeta_m$
	$q^2 = \sqrt{\beta_m^4 + \tfrac{1}{4}\zeta_m^2} + \tfrac{1}{2}\zeta_m$			$q^2 = \tfrac{1}{2}\zeta_m + \sqrt{\tfrac{1}{4}\zeta_m^2 - \lambda_m}$	
		2b $\zeta_m < 0$:	3b $\zeta_m < 0$:	4b $\zeta_m < 0$:	
	$\beta_m^4 = -\lambda_m$	$\alpha^2 = -\zeta_m$	$\beta_m^2 = -\tfrac{1}{2}\zeta_m$	$p = -1$	
		$c = \cosh \alpha\ell$	$c = \cosh \beta_m\ell$	$g = q^2 - d^2$	
		$\delta = \dfrac{\sinh \alpha\ell}{\alpha}$	$\delta_1 = -\beta_m \sinh \beta_m\ell$	$c = \cosh d\ell, \ \delta = \cosh q\ell$	
			$\delta_2 = \dfrac{\sinh \beta_m\ell}{\beta_m}$	$s_1 = \sinh d\ell, \ s_2 = \sinh q\ell$	
				$d^2 = -\tfrac{1}{2}\zeta_m + \sqrt{\tfrac{1}{4}\zeta_m^2 - \lambda_m}$	
				$q^2 = -\tfrac{1}{2}\zeta_m - \sqrt{\tfrac{1}{4}\zeta_m^2 - \lambda_m}$	

TABLE 18-23 (continued) TRANSFER AND STIFFNESS MATRICES FOR A GENERAL RECTANGULAR PLATE SEGMENT

Notation

$$\bar{e}_j = e_j\big|_{\ell = \ell - a_1} \qquad \hat{e}_j = e_j\big|_{\ell = \ell - a_2} \qquad \Delta e_j = \bar{e}_j - \hat{e}_j$$

$$\frac{\Delta p}{\Delta \ell} = \frac{p_{a_2} - p_{a_1}}{a_2 - a_1} \qquad \frac{\Delta c}{\Delta \ell} = \frac{c_{a_2} - c_{a_1}}{a_2 - a_1}$$

Loading Functions for the Matrices

Loading	F_w	F_θ	F_V	F_M
1.	$\dfrac{\varepsilon_m}{D_x} W_T \bar{e}_4$	$\dfrac{\varepsilon_m}{D_x} W_T \bar{e}_3$	$\varepsilon_m W_T[\alpha_{2m}\bar{e}_3 - \bar{e}_1]$	$\varepsilon_m W_T[\alpha_{1m}\bar{e}_4 - \bar{e}_2]$
2.	$\dfrac{\varepsilon_m}{D_x} C \bar{e}_3$	$\dfrac{\varepsilon_m}{D_x} C \bar{e}_2$	$\varepsilon_m C[\alpha_{2m}\bar{e}_2 - \bar{e}_0]$	$\varepsilon_m C[\alpha_{1m}\bar{e}_3 - \bar{e}_1]$
3.	$\dfrac{\varepsilon_m}{D_x}\left[p_{a_1}\bar{e}_5 - p_{a_2}\hat{e}_5 + \dfrac{\Delta p}{\Delta \ell}\Delta e_6 \right]$ $+ \dfrac{\varepsilon_m}{D_x}\left[c_a \bar{e}_4 - c_b \hat{e}_4 \right]$ $+ \dfrac{\Delta c}{\Delta \ell}\Delta e_5$	$\dfrac{\varepsilon_m}{D_x}\left[p_{a_1}\hat{e}_4 - p_{a_1}\bar{e}_4 - \dfrac{\Delta p}{\Delta \ell}\Delta e_5 \right]$ $+ \dfrac{\varepsilon_m}{D_x}\left[c_{a_2}\hat{e}_3 - c_{a_1}\bar{e}_3 \right]$ $- \dfrac{\Delta c}{\Delta \ell}\Delta e_4$	$\varepsilon_m\left\{ p_{a_1}(\alpha_{2m}\hat{e}_4 - \hat{e}_2) \right.$ $- p_{a_2}(\alpha_{2m}\hat{e}_4 - \hat{e}_2)$ $+ \dfrac{\Delta p}{\Delta \ell}(\alpha_{2m}\Delta e_5 - \Delta e_3) \Big\}$ $+ \varepsilon_m\Big\{ c_{a_1}(\alpha_{2m}\bar{e}_3 - \bar{e}_1 + 1)$ $+ \dfrac{\Delta c}{\Delta \ell}(\alpha_{2m}\Delta e_4 - \Delta e_2$ $+ \alpha_{2m} - \alpha_{1m})$ $\left. - c_{a_2}(\alpha_{2m}\hat{e}_3 - \hat{e}_1) \right\}$	$\varepsilon_m\left\{ p_{a_1}(\alpha_{1m}\hat{e}_5 - \hat{e}_3) \right.$ $- p_{a_2}(\alpha_{1m}\hat{e}_5 - \hat{e}_3)$ $+ \dfrac{\Delta p}{\Delta \ell}(\alpha_{1m}\Delta e_6 - \Delta e_4)$ $+ \varepsilon_m\big\{ c_{a_1}(\alpha_{1m}\bar{e}_4 - \bar{e}_2)$ $- c_{a_2}(\alpha_{1m}\hat{e}_4 - \hat{e}_2)$ $\left. + \dfrac{\Delta c}{\Delta \ell}(\alpha_{1m}\Delta e_5 - \Delta e_3) \right\}$

TABLE 18-23 Matrices for General Rectangular Plate Segment 1080

Constants ε_m for the Loading Functions

Notation

$$y_k = \tfrac{1}{2}(b_1 + b_2) \qquad \Delta y = \tfrac{1}{2}(b_2 - b_1)$$

$$c_1 = c_2 = c_3 = 0, \qquad c_4 = 1 \quad \text{for simply supported at } y = 0 \text{ and } y = L_y$$

$$c_1 = c_3 = 1, \qquad c_2 = c_4 = E_m \qquad \text{for fixed at } y = 0 \text{ and simply supported or fixed at } y = L_y$$

$$E_m = [\cosh \eta_m - \cos \eta_m]/[\sinh \eta_m - \sin \eta_m]$$

Loading	Constant, ε_m
1. 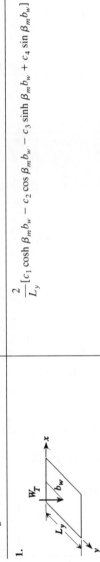	$\dfrac{2}{L_y}[c_1 \cosh \beta_m b_w - c_2 \cos \beta_m b_w - c_3 \sinh \beta_m b_w + c_4 \sin \beta_m b_w]$
2. 	$\dfrac{2}{L_y}[c_1 \cosh \beta_m b_c - c_2 \cos \beta_m b_c - c_3 \sinh \beta_m b_c + c_4 \sin \beta_m b_c]$
3. Line force W and moment C 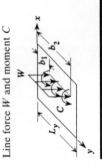	$\dfrac{4}{\eta_m}\{[c_1 \cosh \beta_m y_k - c_2 \sinh \beta_m y_k]\sinh \beta_m \Delta y - [c_3 \cos \beta_m y_k - c_4 \sin \beta_m y_k]\sin \beta_m \Delta y\}$

TABLE 18-24 CONSISTENT MASS MATRIX FOR RECTANGULAR PLATE ELEMENT SIMPLY SUPPORTED ON TWO OPPOSITE EDGES

Notation

ρ = mass per unit area

ℓ, L_y = length of element in x and y directions

$\alpha = \ell\, m\pi$

$s_h = \sinh \beta\ell$

$H_m = (\alpha^2 - L_y^2 s_h^2)$

$\beta = m\pi/L_y$

$c_h = \cosh \beta\ell$

Element variables:

$$\mathbf{v}^i = [w_a \quad \theta_a \quad w_b \quad \theta_b]^T$$

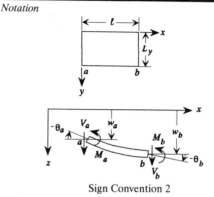

Sign Convention 2

Mass Matrix

$$\mathbf{m}^i = \begin{bmatrix} m_{11} & & \text{Symmetric} & \\ m_{21} & m_{22} & & \\ m_{31} & m_{32} & m_{33} & \\ m_{41} & m_{42} & m_{43} & m_{44} \end{bmatrix}$$

$$m_{11} = \frac{-\ell\rho}{12\alpha}\left\{2\alpha^5 - 5\alpha^3 L_y^3\left(1 + 2c_h^2\right) + 15\alpha L_y^4 s_h^2 - \alpha^2 c_h s_h L_y\left(15L_y^2 + 2\alpha^2\right)\right.$$
$$\left. + 15L_y^5 c_h s_h^3\right\}/H_m$$

$$m_{21} = -\frac{L_y \ell^2 \rho}{6\alpha^2}\left\{\alpha^4 L_y\left(2 + c_h^2\right)\left(1 + \nu s_h^2\right) - 3\alpha^2 L_y^3 s_h^2 - 3L_y^3 s_h^4\left(2\alpha^2\nu + L_y^2\right)\right.$$
$$\left. + 3\alpha^3\left(\alpha^2\nu + L_y^2\right)c_h s_h\right\}/H_m$$

$$m_{22} = \frac{L_y^2 \ell^3 \rho}{12\alpha^2}\left\{2\alpha^5 + L_y\left[\alpha^2\left(2\alpha^2 - 9L_y^2\right) - 3L_y^4 s_h^2\right]c_h s_h + \alpha^3 L_y^2\left(1 + 2c_h^2\right)\right.$$
$$\left. + 9\alpha L_y^4 s_h^2\right\}/H_m$$

$$m_{31} = -\frac{\ell\rho}{12\alpha}\left\{L_y s_h\left[2\alpha^4 + 3\alpha^2 L_y^2\left(3 + 2c_h^2\right) - 15L_y^4 s_h^2\right]\right.$$
$$\left. + \alpha c_h\left[\alpha^4 + \alpha^2 L_y^2\left(14 + c_h^2\right) - 15L_y^4 c_h^2\right]\right\}/H_m$$

$$m_{32} = -m_{41}$$

$$m_{33} = m_{11}$$

$$m_{41} = -\frac{\ell^2\rho}{12\alpha}\left\{\alpha^2 L_y s_h\left[\alpha^2 + L_y^2\left(8 + c_h^2\right)\right] - 15L_y^5 s_h^3 + 3\alpha L_y^2 c_h\left(\alpha^2 + L_y^2 s_h^2\right)\right\}/H_m$$

$$m_{42} = \frac{L_y^2 \ell^3 \rho}{12\alpha^3}\left\{3L_y^5 s_h^3 + \alpha^2 L_y\left(4\alpha^2 + 9L_y^2\right)s_h + \alpha\left[\alpha^2 L_y^2\left(c_h^2 - 4\right)\right.\right.$$
$$\left.\left. + \alpha^4 - 9L_y^4 s_h^2\right]c_h\right\}/H_m$$

$$m_{43} = \frac{L_y^2 \ell^2 \rho}{6\alpha^2}\left\{-3\alpha^2 L_y^2 s_h^2 + \alpha^4\left(2 + c_h^2\right) - 3L_y^4 s_h^4 + 3\alpha^3 L_y c_h s_h\right\}/H_m$$

$$m_{44} = m_{22}$$

TABLE 18-24 | Mass Matrix for Rectangular Plate Element | 1082

Notation

$$w = \text{deflection}$$
$$M_r, M_\phi, M_x, M_y = \text{moments per unit length}$$
$$Q_r = \text{shear force per unit length}$$
$$E = \text{modulus of elasticity}$$
$$\nu = \text{Poisson's ratio}$$
$$h = \text{thickness of plate}$$
$$r = \text{radial coordinate}$$
$$p_1 = \text{distributed applied force per area}$$
$$D = \frac{Eh^3}{12(1 - \nu^2)}$$

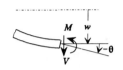

Case	Deflection and Internal Forces
1. Simply supported semicircular plate	$$w = \frac{p_1 a_L^4}{D} \sum_{m=1,3,5,\ldots} \left\{ \frac{4r^4}{a_L^4} \frac{1}{m\pi(16 - m^2)(4 - m^2)} \right.$$ $$+ \frac{r^m}{a_L^m} \frac{m + 5 + \nu}{m\pi(16 - m^2)(2 + m)\left[m + \frac{1}{2}(1 + \nu)\right]}$$ $$\left. - \frac{r^{m+2}}{a_L^{m+2}} \frac{m + 3 + \nu}{m\pi(4 + m)(4 - m^2)\left[m + \frac{1}{2}(1 + \nu)\right]} \right\} \sin m\varphi$$ $$M_r = c_1 p_1 a_L^2 \qquad M_\phi = c_2 p_1 a_L^2$$ CONSTANTS: $\nu = 0.3$ $$c_1 = -0.045 + 0.2122\left(\frac{r}{a_L}\right) - 0.0392\left(\frac{r}{a_L}\right)^2 - 0.128\left(\frac{r}{a_L}\right)^3$$ $$c_2 = 0.022 + 0.0484\left(\frac{r}{a_L}\right) - 0.0264\left(\frac{r}{a_L}\right)^2 - 0.0352\left(\frac{r}{a_L}\right)^3$$ $$0.25 \le \frac{r}{a_L} \le 1$$
2. Fixed semicircular plate 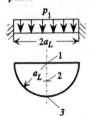	$$w_2 = 0.002021 \frac{p_1 a_L^4}{D} \quad \text{at point 2}$$ $$(M_r)_{\max} = 0.069 p_1 a_L^2 \quad \text{at point 1}$$ $$(M_\phi)_2 = -0.019 p_1 a_L^2$$ $$(M_r)_3 = 0.06 p_1 a_L^2$$ $$(Q_r)_1 = -0.497 p_1 a_L$$ $$(Q_r)_3 = -0.380 p_1 a_L$$ $$\nu = 0.3$$

Case	Deflection and Internal Forces
3. Wedge plate 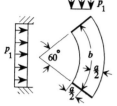 Only the boundary conditions relevant to Case 1 are shown.	$$M_{max} = c_i \frac{p_1 a^2}{6}$$ Case (i) | Boundary conditions 1 | Straight edges clamped, curved edges free 2 | All edges fixed 3 | Straight edges simply supported, curved edges fixed CONSTANTS: $\beta = b/a$ $c_1 = 0.8939 - 0.2013\beta + 0.01312\beta^2 + 0.006661\beta^3$ $\qquad 0.8 \le \beta \le 2.5$ $c_2 = \begin{cases} 9.0967 - 44.5356\beta + 85.997\beta^2 - 79.4533\beta^3 \\ + 35.4362\beta^4 - 6.1197\beta^5 \quad 0.6 \le \beta \le 1.75 \\ c_1 \qquad 1.75 < \beta \le 2.5 \end{cases}$ $c_3 = 0.04652 + 0.8334\beta - 0.3994\beta^2 + 0.06026\beta^3$ $\qquad 0.8 \le \beta \le 2.5$
4. Wedge plate 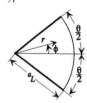 Clamped on straight edges, free on curved boundary.	$$w = \frac{p_1 r^4}{64D}\left(1 + \frac{\cos 4\varphi - 4\cos\theta\cos 2\varphi}{2\cos^2\theta + 1}\right)$$ $$M_r = \frac{-p_1}{16}r^2 \nu\left(1 - \frac{3\cos 4\phi}{2\cos^2\theta + 1}\right)$$ $$\quad - \frac{3p_1}{16}r^2\left(1 + \frac{\cos 4\phi - 4\cos\theta\cos 2\phi}{2\cos^2\theta + 1}\right)$$ $$M_\phi = \frac{-p_1 r^2}{16}\left(1 - \frac{3\cos 4\phi}{2\cos^2\theta + 1}\right)$$ $$\quad - \frac{3p_1 r^2}{16}\nu\left(1 + \frac{\cos 4\phi - 4\cos\theta\cos 2\phi}{2\cos^2\theta + 1}\right)$$ $$Q_r = \frac{-p_1 r}{2}\left(1 + \frac{3\cos\theta\cos 2\phi}{2\cos^2\theta + 1}\right)$$

TABLE 18-25   **Deflections and Internal Forces of Plates**   **1084**

Case	Deflection and Internal Forces
5. Simply supported ellipse with distributed load Equivalent Rectangular Plate Ref. 18.12	$$(w_{max})_{center} \simeq \frac{(0.146\alpha - 0.1)p_1 b^4}{\alpha E h^3}$$ $$(\sigma_{max})_{center} = \pm \frac{0.3125(2\alpha - 1)p_1 b^2}{\alpha h^2}$$ $$M_x = c_1 \frac{p_1 b^2}{4} \qquad M_y = c_2 \frac{p_1 b^2}{4} \qquad \text{(center)}$$ $$\nu = 0.3$$ CONSTANTS: $\alpha = a/b \quad 1.0 \le \alpha \le 2.0$ $c_1 = -0.1565 + 0.7066\alpha - 0.4263\alpha^2 + 0.0823\alpha^3$ $c_2 = -0.2882 + 0.720\alpha - 0.258\alpha^2 + 0.03241\alpha^3$ Alternatively, use of an equivalent rectangular plate may give acceptable approximation.
6. Simply supported ellipse with concentrated force Equivalent Rectangular Plate 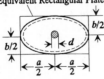	For small d only: $$w_{center} = \frac{W_T b^2}{E h^3}(0.6957 + 0.2818\beta - 0.422\beta^2)$$ where $$\nu = 0.3$$ $$\beta = \frac{b}{a}$$ Alternatively, use of an equivalent rectangular plate may give acceptable approximation.

Case	Deflection and Internal Forces
7. Fixed elliptical plate with uniformly distributed load 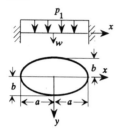 Ref. 18.12	$$W_0 = \frac{P_1}{D\left(\dfrac{24}{a^4} + \dfrac{24}{b^4} + \dfrac{16}{a^2b^2}\right)}$$ $$w = W_0\left(1 - \frac{x^2}{a^2} - \frac{y^2}{b^2}\right)^2$$ $$M_x = -4W_0D\left[\left(\frac{\nu}{a^2b^2} + \frac{3}{a^4}\right)x^2 \right.$$ $$\left. + \left(\frac{1}{a^2b^2} + \frac{3\nu}{b^4}\right)y^2 - \left(\frac{\nu}{b^2} + \frac{1}{a^2}\right)\right]$$ $$M_y = -4W_0D\left[\left(\frac{\nu}{a^2b^2} + \frac{3}{b^4}\right)y^2 \right.$$ $$\left. + \left(\frac{1}{a^2b^2} + \frac{3\nu}{a^4}\right)x^2 - \left(\frac{\nu}{a^2} + \frac{1}{b^2}\right)\right]$$
8. Fixed elliptical plate with linearly varying distributed load 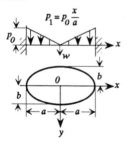 Ref. 18.12	$$W_0 = \frac{P_0}{24D\left(\dfrac{5}{a^4} + \dfrac{1}{b^4} + \dfrac{2}{a^2b^2}\right)a}$$ $$w = W_0x\left(1 - \frac{x^2}{a^2} - \frac{y^2}{b^2}\right)^2$$ $$M_x = 4W_0Dx\left[\left(\frac{3}{a^2} - \frac{5}{a^4}x^2 - \frac{3}{a^2b^2}y^2\right) \right.$$ $$\left. + \nu\left(\frac{1}{b^2} - \frac{x^2}{a^2b^2} - \frac{3y^2}{b^4}\right)\right]$$ $$M_y = 4W_0Dx\left[\left(\frac{1}{b^2} - \frac{x^2}{a^2b^2} - \frac{3}{b^4}y^2\right) \right.$$ $$\left. + \nu\left(\frac{3}{a^2} - \frac{5}{a^4}x^2 - \frac{3}{a^2b^2}y^2\right)\right]$$

Case	Deflection and Internal Forces

9.
Parallelogram plate,
simply supported

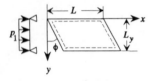

$(w)_{center} = c_1 p_1 L_y^4 / Eh^3$

$(M_x)_{center} = c_2 p_1 L_y^2$

$(M_y)_{center} = c_3 p_1 L_y^2$

$\nu = 0.3$

CONSTANTS:

$L = 2L_y$

ϕ	c_1	c_2	c_3
$0°$	0.1096	0.0461	0.1020
$30°$	0.1059	0.0485	0.0990
$45°$	0.0989	0.0489	0.0940
$60°$	0.0718	0.0531	0.0764
$75°$	0.0097	0.0370	0.0113

10.
Triangular plate

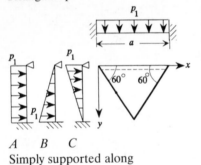

A B C
Simply supported along
$y = 0$ edge, other two
edges clamped.
Ref. 18.12

$w_{max} = c_1 \dfrac{p_1 a^4}{Eh^3}$

$(M_x)_{max} = c_2 p_1 a^2$

$(M_y)_{max} = c_3 p_1 a^2$

$\nu = 0.3$

CONSTANTS:

Load	A	B	C
c_1	0.00268	0.00082	0.00193
c_2	0.0106	0.00395	0.00700
c_3	0.00992	0.00390	0.00790

Case	Deflection and Internal Forces

11.
Triangular plate

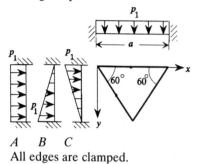

A   B   C
All edges are clamped.

$$w_{max} = c_1 \frac{p_1 a^4}{Eh^3}$$

$$(M_x)_{max} = c_2 p_1 a^2$$

$$(M_y)_{max} = c_3 p_1 a^2$$

$$\nu = 0.3$$

CONSTANTS:

Load	A	B	C
c_1	0.00195	0.00063	0.00131
c_2	0.00885	0.00356	0.00533
c_3	0.00806	0.00379	0.00583

TABLE 18-25   Deflections and Internal Forces of Plates   1088

TABLE 18-26 BUCKLING LOADS FOR PLATES OF VARIOUS SHAPES

Notation

E = modulus of elasticity
h = thickness of plate
ν = Poisson's ratio
$D = Eh^3/[12(1 - \nu)^2]$
P_{cr} = in-plane normal buckling load per unit length
$(P_{xy})_{cr}$ = buckling shear force per unit length

$$P' = \frac{E\pi^2 h}{12(1 - \nu^2)}\left(\frac{h}{a}\right)^2$$

P = uniformly distributed compressive in-plane load per unit length
P_{xy} = uniformly distributed shear force per unit length

Case	Buckling Load
1. Simply supported equilateral triangular plate with uniform pressure	$P_{cr} = 4.00P'$ Ref. 18.4, 18.19
2. Simply supported right-angled isosceles triangular plate with uniform pressure	$P_{cr} = 5.00P'$ Ref. 18.4, 18.20
3. Simply supported right-angled isosceles triangular plate	$P_{cr} = 9.11P'$ Ref. 18.21

TABLE 18-26 (continued)   BUCKLING LOADS FOR PLATES OF VARIOUS SHAPES

Case	Buckling Load
4. Simply supported parallelogram 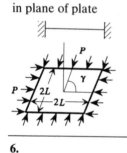	$P_{cr} = k\dfrac{\pi^2 D}{L^2}$ $k = \beta^2(\phi_k/\sin^3 \gamma)$ When $\beta = L/L_y = 1$, $\phi_k = 2.15 + 0.004556\gamma + 6.667 \times 10^{-4}\gamma^2$ $\qquad - 5.432 \times 10^{-6}\gamma^3$ $45° \le \gamma \le 90°$
5. Clamped rhomboidal plate, under uniform pressure perpendicular to sides in plane of plate 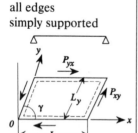	$P_{cr} = k\dfrac{\pi^2 D}{L^2}$ $k = 9.85 - 0.281\gamma + 3.1267 \times 10^{-3}\gamma^2 - 1.175 \times 10^{-5}\gamma^3$ $45° \le \gamma \le 90°$
6. Parallelogram plate, all edges simply supported $P_{xy} = P_{yx}$	$(P_{xy})_{cr} = k\dfrac{\pi^2 D}{L_y^2}$ $k = \dfrac{\psi}{4\beta \sin^3 \gamma}$ $\beta = L/L_y$ When $\beta = 1$, $\psi = 122.51 + 0.223\gamma - 0.03055\gamma^2 + 2.472 \times 10^{-4}\gamma^3$ $\qquad - 5.75 \times 10^{-7}\gamma^4$ When $\beta = 2$, $\psi = 126.27 + 1.44\gamma - 0.0548\gamma^2 + 4.257 \times 10^{-4}\gamma^3$ $\qquad - 1.046 \times 10^{-6}\gamma^4$ $\qquad\qquad 45° \le \gamma \le 135°$

TABLE 18-26   **Buckling Loads for Plates**

TABLE 18-27   NATURAL FREQUENCIES OF PLATES AND MEMBRANES OF VARIOUS SHAPES[a]

Notation

E = modulus of elasticity
h = thickness of plate
ν = Poisson's ratio
ρ = mass per unit area
$\alpha = L_y/L$
$\beta = L/L_y$

Natural frequencies (Hz):

$$f_i = \frac{\lambda_i}{2\pi L^2}\left[\frac{Eh^3}{12\rho(1-\nu^2)}\right]^{1/2}$$

The values of λ_i (i = mode number) are independent of ν except where specially indicated.

All results are for the transverse vibration of plates unless otherwise indicated. For membranes the natural frequencies are for transverse vibrations.

A = area of membrane
P = tension per unit length

Case	Parameter λ_i
1. Simply supported rhombus	$\lambda_1 = 132.24 - 3.725\gamma + 0.0418\gamma^2 - 1.59 \times 10^{-4}\gamma^3$ $\lambda_2 = 206.15 - 5.314\gamma + 0.0584\gamma^2 - 2.08 \times 10^{-4}\gamma^3$ $\lambda_3 = -1937.6 + 87.3\gamma - 1.241\gamma^2 + 5.737 \times 10^{-3}\gamma^3$ $\lambda_4 = 448.9 - 14.204\gamma + 0.182\gamma^2 - 7.73 \times 10^{-4}\gamma^3$ $\lambda_5 = -2.1 + 8.93\gamma - 0.1688\gamma^2 + 9.1 \times 10^{-4}\gamma^3$ $\lambda_6 = 1005.06 - 36.6\gamma + 0.5056\gamma^2 - 2.34 \times 10^{-3}\gamma^3$ $45° \le \gamma \le 90°$
2. Simply supported parallelogram	$\beta = 1/2, \quad 45° \le \gamma \le 70°$ $\lambda_1 = 71.062 - 1.525\gamma + 0.01009\gamma^2$ $\lambda_2 = 87.352 - 1.746\gamma + 0.01149\gamma^2$ $\beta = 1/3, \quad 45° \le \gamma \le 70°$ $\lambda_1 = 67.47 - 1.469\gamma + 0.009733\gamma^2$ $\lambda_2 = 73.736 - 1.544\gamma + 0.01021\gamma^2$ $\beta = 2/3, \quad 45° \le \gamma \le 90°$ $\lambda_1 = 101.038 - 2.8904\gamma + 0.03264\gamma^2 - 1.24889 \times 10^{-4}\gamma^3$ $\lambda_2 = 137.212 - 3.6264\gamma + 0.040565\gamma^2 - 1.5363 \times 10^{-4}\gamma^3$

TABLE 18-27 (continued)   **NATURAL FREQUENCIES OF PLATES OF VARIOUS SHAPES**

Case	Parameter λ_i
3. Clamped–simply supported, simply supported, simply supported rhombus	$\lambda_1 = 345.31 - 15.5\gamma + 0.2965\gamma^2 - 2.63 \times 10^{-3}\gamma^3 + 8.98 \times 10^{-6}\gamma^4$ $\lambda_2 = 461.18 - 19.63\gamma + 0.3713\gamma^2 - 3.256 \times 10^{-3}\gamma^3$ $\quad + 1.1 \times 10^{-5}\gamma^4$ $\lambda_3 = 139.43 + 5.884\gamma - 0.2564\gamma^2 + 3.106 \times 10^{-3}\gamma^3$ $\quad - 1.2164 \times 10^{-5}\gamma^4$ $\lambda_4 = 3443.2622 - 1163.3586\gamma^{1/2} + 134.4698\gamma - 5.1802\gamma^{3/2}$ $\lambda_5 = 845.5009 - 177.752\gamma^{1/2} + 13.6238\gamma - 0.3341\gamma^{3/2}$ $\lambda_6 = 740.78 - 14.71\gamma - 0.0845\gamma^2 + 4.3 \times 10^{-3}\gamma^3$ $\quad - 2.67 \times 10^{-5}\gamma^4$ $40° \le \gamma \le 90°$
4. Clamped–simply supported, clamped–simply supported rhombus 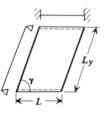	$\lambda_1 = 410.53 - 18.48\gamma + 0.3554\gamma^2 - 3.17 \times 10^{-3}\gamma^3$ $\quad + 1.089 \times 10^{-5}\gamma^4$ $\lambda_2 = 538.72 - 23.53\gamma + 0.454\gamma^2 - 4.056 \times 10^{-3}\gamma^3$ $\quad + 1.39 \times 10^{-5}\gamma^4$ $\lambda_3 = 486.22 - 17.147\gamma + 0.3196\gamma^2 - 3.117 \times 10^{-3}\gamma^3$ $\quad + 1.234 \times 10^{-5}\gamma^4$ $\lambda_4 = 1251.43 - 57.33\gamma + 1.0705\gamma^2 - 8.917 \times 10^{-3}\gamma^3$ $\quad + 2.793 \times 10^{-5}\gamma^4$ $\lambda_5 = 786.41 - 33.80\gamma + 0.7296\gamma^2 - 7.56 \times 10^{-3}\gamma^3$ $\quad + 2.987 \times 10^{-5}\gamma^4$ $\lambda_6 = 1134.6 - 40.13\gamma + 0.5387\gamma^2 - 2.41 \times 10^{-3}\gamma^3$ $40° \le \gamma \le 90°$
5. Parallelogram plate, clamped–simply supported, clamped–simply supported	$\beta = 2/3 \quad 45° \le \gamma \le 90°$ $\lambda_1 = 327.06 - 13.84\gamma + 0.2514\gamma^2 - 2.126 \times 10^{-3}\gamma^3$ $\quad + 6.965 \times 10^{-6}\gamma^4$ $\lambda_2 = 349.6 - 14.22\gamma + 0.2554\gamma^2 - 2.1414 \times 10^{-3}\gamma^3$ $\quad + 6.98 \times 10^{-6}\gamma^4$ $\lambda_3 = 405.6 - 15.97\gamma + 0.288\gamma^2 - 2.42 \times 10^{-3}\gamma^3$ $\quad + 7.9 \times 10^{-6}\gamma^4$ $\beta = 1/2$ $\lambda_1 = 129.6 - 2.52763\gamma + 0.01506\gamma^2$ $\lambda_2 = 207.8099 - 5.9092\gamma + 0.06589\gamma^2 - 0.0002479\gamma^3$ $\lambda_3 = 227.31 - 6.2092\gamma + 0.06913\gamma^2 - 0.0002598\gamma^3$

TABLE 18-27   **Natural Frequencies of Plates**   **1092**

TABLE 18-27 (continued)   NATURAL FREQUENCIES OF PLATES OF VARIOUS SHAPES

Case	Parameter λ_i
6. Clamped–simply supported, simply supported–clamped rhombus 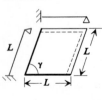	$\lambda_1 = 457.28 - 24.94\gamma + 0.6339\gamma^2 - 9.12 \times 10^{-3}\gamma^3$ $\qquad + 7.4729 \times 10^{-5}\gamma^4 - 3.3 \times 10^{-7}\gamma^5 + 6.136 \times 10^{-10}\gamma^6$ $\lambda_2 = 435.3 - 16.2\gamma + 0.265\gamma^2 - 1.947 \times 10^{-3}\gamma^3$ $\qquad + 5.44 \times 10^{-6}\gamma^4 - 1.11 \times 10^{-10}\gamma^5$ $\lambda_3 = 369.53 - 8.0135\gamma + 0.0696\gamma^2 - 2.745 \times 10^{-4}\gamma^3$ $\qquad + 7.447 \times 10^{-7}\gamma^4$ $\lambda_4 = 2206.26 - 140.757\gamma + 3.93\gamma^2 - 0.0586\gamma^3 + 4.896 \times 10^{-4}\gamma^4$ $\qquad - 2.173 \times 10^{-6}\gamma^5 + 4.0 \times 10^{-9}\gamma^6$ $\lambda_5 = 1367.43 - 85.0579\gamma + 2.5565\gamma^2 - 0.0412287\gamma^3$ $\qquad + 3.65634 \times 10^{-4}\gamma^4 - 1.67497 \times 10^{-6}\gamma^5 + 3.1051 \times 10^{-9}\gamma^6$ $\lambda_6 = 2856.2 - 184.832\gamma + 5.35065\gamma^2 - 0.0829448\gamma^3$ $\qquad + 7.15537 \times 10^{-4}\gamma^4 - 3.23696 \times 10^{-6}\gamma^5$ $\qquad + 6.00311 \times 10^{-9}\gamma^6$ $40° \leq \gamma \leq 140°$
7. Clamped–free, free–free rhombus	$\lambda_1 = 12.95 - 0.3066\gamma + 3.358 \times 10^{-3}\gamma^2 - 1.24 \times 10^{-5}\gamma^3$ $\lambda_2 = 50.59 - 1.4242\gamma + 0.01628\gamma^2 - 6.2765 \times 10^{-5}\gamma^3$ $45° \leq \gamma \leq 90°$ $\nu = 0.3$
8. Clamped–clamped, simply supported–clamped rhombus 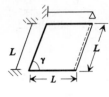	$\lambda_1 = 425.87 - 18.906\gamma + 0.360033\gamma^2 - 3.2003 \times 10^{-3}\gamma^3$ $\qquad + 1.0946 \times 10^{-5}\gamma^4$ $\lambda_2 = 558.58 - 23.58\gamma + 0.444\gamma^2 - 3.881 \times 10^{-3}\gamma^3$ $\qquad + 1.31 \times 10^{-5}\gamma^4$ $\lambda_3 = 352.69 - 6.228\gamma + 0.0314\gamma^2 + 3.39 \times 10^{-5}\gamma^3$ $\lambda_4 = 1622.79 - 82.28\gamma + 1.693\gamma^2 - 0.0156\gamma^3$ $\qquad + 5.413 \times 10^{-5}\gamma^4$ $\lambda_5 = 1321.28 - 69.31\gamma + 1.608\gamma^2 - 0.0169\gamma^3$ $\qquad + 6.65 \times 10^{-5}\gamma^4$ $\lambda_6 = 771.246 - 12.375\gamma - 0.192\gamma^2 + 5.79 \times 10^{-3}\gamma^3$ $\qquad - 3.35 \times 10^{-5}\gamma^4$ $40° \leq \gamma \leq 90°$

TABLE 18-27 (continued)   NATURAL FREQUENCIES OF PLATES OF VARIOUS SHAPES

Case	Parameter λ_i
9. Clamped rhombus 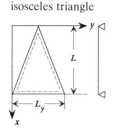	$\lambda_1 = 264.064 - 7.6\gamma + 0.086\gamma^2 - 3.295 \times 10^{-4}\gamma^3$ $\lambda_2 = 386.36 - 10.92\gamma + 0.1262\gamma^2 - 4.84 \times 10^{-4}\gamma^3$ $\lambda_3 = 469.58 - 11.184\gamma + 0.105\gamma^2 - 3.3 \times 10^{-4}\gamma^3$ $\lambda_4 = 696.52 - 22.1\gamma + 0.278\gamma^2 - 1.17 \times 10^{-3}\gamma^3$ $\lambda_5 = 167.84 + 5.1263\gamma - 0.1352\gamma^2 + 8.2 \times 10^{-4}\gamma^3$ $\lambda_6 = 1381.06 - 49.76\gamma + 0.6778\gamma^2 - 3.1 \times 10^{-3}\gamma^3$ $45° \leq \gamma \leq 90°$

Case	Parameter λ_i	
10. Simply supported isosceles triangle 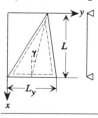	$\lambda_1 = 10.18 + 22.34\beta + 13.32\beta^2$ $\lambda_2 = 8.51 + 54.53\beta + 69.36\beta^2$ $\quad - 29.5\beta^3$ $\lambda_3 = 96.85 - 201.1\beta + 318.786\beta^2$ $\quad - 93.53\beta^3$ $\lambda_4 = 14.436 + 78.54\beta + 159.24\beta^2$ $\quad - 74.6\beta^3$ $\lambda_5 = 18.81 + 140.97\beta + 40.3\beta^2$ $0.5 \leq \beta \leq 1.5$	MEMBRANE All edges supported. P = tension on all edges. $f_1(\text{Hz}) = \dfrac{\lambda_1}{2}\left(\dfrac{P}{\rho A}\right)^{1/2}$ $\lambda_1 = 4.0609 - 7.3317\sqrt{\alpha}$ $\quad + 7.9241\alpha$ $\quad - 3.8893\alpha^{3/2}$ $\quad + 0.7603\alpha^2$ $0.35 \leq \alpha \leq 2.0$ $A = \frac{1}{2}LL_y$

11.
Simply supported
asymmetric triangle

$\lambda_i\ (= \lambda_1)$ for fundamental mode:

β	$\gamma = 0°$	$\gamma = 10°$	$\gamma = 20°$	$\gamma = 30°$	$\gamma = 45°$
0.5	24.69	24.78	25.06	25.64	27.78
1.0	45.85	46.28	47.71	50.57	60.22
1.5	73.66	74.64	77.85	84.21	105.1

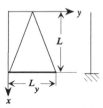

Case	Parameter λ_i
12. Clamped–free–free isosceles triangle	$\lambda_1 = 7.178 - 0.0297\beta + 3.665 \times 10^{-4}\beta^2 + 2.3335 \times 10^{-4}\beta^3$ $\lambda_2 = 30.92 - 0.145\beta + 0.025\beta^2 - 1.44 \times 10^{-3}\beta^3$ $\lambda_3 = 36.88 + 21.5\beta + 2.94\beta^2 - 0.19\beta^3$ $\lambda_4 = 44.49 + 100.653\beta + 3.9\beta^2 - 0.253\beta^3$ $\nu = 0.30 \quad 1 \leq \beta \leq 7$

TABLE 18-27   **Natural Frequencies of Plates**   **1094**

TABLE 18-27 (continued) **NATURAL FREQUENCIES OF PLATES OF VARIOUS SHAPES**

Case	Parameter λ_i
13. Clamped–free–free right triangle 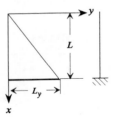	MEMBRANE All edges supported. P = tension on all edges $\lambda_1 = 4.7223 + 0.691\beta$ $\quad - 0.05433\beta^2$ $\lambda_2 = 20.07 + 3.148\beta$ $\quad - 0.2413\beta^2$ $\nu = 0.30 \quad 2 \le \beta \le 7$ $f_{nm}(\text{Hz}) = \dfrac{\lambda_{nm}}{2}\left(\dfrac{P}{\rho A}\right)^{1/2}$ $\lambda_{nm} = \left(\dfrac{m^2 + n^2}{2}\right)^{1/2}$ $n, m = 1, 2, 3, \ldots$ $L = L_y, \ A = \tfrac{1}{2}L^2$
14. Simply supported symmetric trapezoid 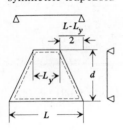	λ_1 for fundamental mode: $\quad 0 \le \alpha \le 1.0$ $d/L = 0.5$ $\lambda_1 = 98.7375 - 136.5055\alpha + 140.0249\alpha^2 - 52.9631\alpha^3$ $d/L = 2/3$ $\lambda_1 = 69.7416 - 87.4303\alpha + 71.751\alpha^2 - 21.9561\alpha^3$ $d/L = 1.0$ $\lambda_1 = 45.8944 - 45.3731\alpha + 19.9653\alpha^2 - 0.706\alpha^3$ $d/L = 1.5$ $\lambda_1 = 32.7435 - 24.014\alpha + 1.4861\alpha^2 + 4.0509\alpha^3$
15. Simply supported unsymmetric trapezoid 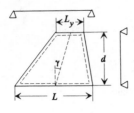	λ_1 for fundamental mode: $\quad 10° \le \gamma \le 45°$ $d/L = 0.5, \quad \alpha = 0.4$ $\lambda_1 = 62.3913 + 0.1493\gamma - 0.006487\gamma^2 + 0.0001848\gamma^3$ $d/L = 0.5, \quad \alpha = 0.8$ $\lambda_1 = 52.5236 + 0.06484\gamma - 0.002564\gamma^2 + 0.0000844\gamma^3$ $d/L = 1.0, \quad \alpha = 0.4$ $\lambda_1 = 30.1211 + 0.1341\gamma - 0.004489\gamma^2 + 0.0001865\gamma^3$ $d/L = 1.0, \quad \alpha = 0.8$ $\lambda_1 = 21.5645 + 0.1095\gamma - 0.004355\gamma^2 + 0.0001459\gamma^3$

TABLE 18-27 (continued)   **NATURAL FREQUENCIES OF PLATES OF VARIOUS SHAPES**

Case	Parameter λ_i
15. Continued	$d/L = 1.5, \quad \alpha = 0.4$ $\lambda_1 = 22.9914 + 0.1254\gamma - 0.003636\gamma^2 + 0.0001681\gamma^3$ $d/L = 1.5, \quad \alpha = 0.8$ $\lambda_1 = 15.9585 + 0.1209\gamma - 0.004465\gamma^2 + 0.0001586\gamma^3$

16.
Simply supported
regular polygon
with n sides

Number of sides =	4	5	6	7	8
$\lambda_1 =$	19.74	11.01	7.152	5.068	3.794

$R_1 = L/(2\sin\gamma) \qquad R_2 = L/(2\tan\gamma)$

MEMBRANE   All edges supported.

P = tension on all edges

No. of sides =	4	5	6	7	8
$\lambda_1 =$	1.414	1.385	1.372	1.366	1.362

$$f_1(\text{Hz}) = \frac{\lambda_1}{2}\left(\frac{P}{\rho A}\right)^{1/2} \qquad A = \frac{n}{4}L^2 \cot\gamma$$

17.
Clamped regular
polygon with n sides

Number of sides: =	4	5	6	7	8
$\lambda_1 =$	35.08	19.71	12.81	9.081	6.787

$R_1 = L/(2\sin\gamma) \qquad R_2 = L/(2\tan\gamma)$

18. Clamped ellipse

λ_1 for fundamental mode:

$\lambda_1 = 132.88 - 301.31\sqrt{\alpha} + 269.89\alpha - 107.02\alpha^{1.5}$
$\qquad + 15.771\alpha^2 \quad 1.0 \le \alpha = a/b \le 5.0$

MEMBRANE   All edges supported.

P = tension on all edges

$$f_1(\text{Hz}) = \frac{\lambda_1}{2}\left(\frac{P}{\rho A}\right)^{1/2} \qquad \begin{array}{l} A = \pi ab \\ \alpha = a/b \end{array}$$

$$\lambda_1 = 2.405\left[\left(\alpha + \frac{1}{\alpha}\right)/2\pi\right]^{1/2}$$

[a]Adopted from Blevins Ref. 18.15.

TABLE 18-27   **Natural Frequencies of Plates**   **1096**

Thick Shells and Disks

Formulas for the displacements, stresses, and free vibration characteristics of thick-walled cylinders, thick spheres, and disks are provided in this chapter. These formulas are based on the linear theory of elasticity. The loading and responses are axially symmetric for cylinders and disks and spherically symmetric for spheres. Computer programs for the determination of the static and dynamic responses of thick shells and disks have been prepared to accompany this book.

19.1   DEFINITIONS AND NOTATION

If the wall thickness of a shell of revolution is more than about one-tenth the radius, the shell is usually called a thick shell, e.g., a thick-walled cylinder (or thick cylinder).

Thick Cylinders

The formulas presented in this chapter for thick cylinders are applicable for sections some distance from the ends of the cylinder. In other words, the effects of end constraints are negligible. The applied loading as well as the resulting displacements and stresses are axially symmetric. The solution is based on generalized plane strain models for which the strain ε_z in the axial direction is a constant.

a_0, a_L	Radii of inner and outer surfaces (L)
E	Modulus of elasticity (F/L^2)
$G = E/[2(1 + \nu)]$	Shear modulus of elasticity, Lamé coefficient (F/L^2)
M_i	Concentrated cylindrical mass, i.e., mass is lumped as thin cylindrical shell (M/L^2)
N	Axial force (F)
p	Applied radial pressure (F/L^2); applied pressure in positive radial (increasing r) direction taken to be positive
p_1	Applied internal pressure (F/L^2), positive as shown in Fig. 19-1.
p_2	Applied external pressure (F/L^2), positive as shown in Fig. 19-1
p_r	Radial loading intensity (F/L^3); positive if its direction is along positive r direction
p_a^0, p_b^0	Loading components for stiffness equation at $r = a$ and $r = b$ (F/L)
p_0, p_a, p_b	Radial forces at $r = 0$, $r = a$, and $r = b$ (F/L)
r, ϕ	Radial, circumferential coordinates
T	Change in temperature (degrees), i.e., temperature rise or loss with respect to reference temperature
T_1	Magnitude of temperature change, uniform in r direction
t	Time (T)
u	Radial displacement (L)
u_0, u_a, u_b	Radial displacement at $r = 0$, $r = a$, $r = b$
α	Coefficient of thermal expansion $(L/L \cdot \text{degree})$
ρ^*	Mass per unit volume (M/L^3)

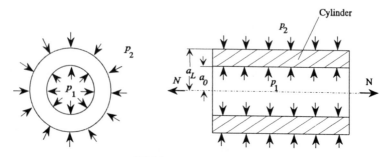

FIGURE 19-1 Cylinder.

$\lambda = E\nu/[(1 + \nu)(1 - 2\nu)]$ Lamé coefficient (F/L^2)
ν Poisson's ratio
ω Natural frequency (rad/T)
Ω Angular velocity of rotation (rad/T)
$\sigma = \sigma_r$ Radial stress (F/L^2)
σ_ϕ Circumferential (tangential) stress (F/L^2)
σ_z Axial stress, positive in tension (F/L^2)

Thick Spherical Shells

The notation that differs from that for cylinders is defined here.

M_i Concentrated spherical mass (M/L^2); lumped as thin spherical shell
p_a^0, p_b^0 Loading components for stiffness equation at $r = a$ and $r = b$ (F)
p_0, p_a, p_b Radial forces at $r = 0$, $r = a$, and $r = b$ (F)

Disks

The formulas for disks are based on plane stress models for which the axial normal stress is zero, as are axially oriented shear stresses. The notation that differs from that for cylinders and spheres is defined here.

h Thickness of disk (L)
M_i Concentrated ring mass (M/L)
p_a^0, p_b^0 Loading components for stiffness equation at $r = a$ and $r = b$ (F)
p_0, p_a, p_b Radial forces for stiffness equation at $r = 0$, $r = a$, and $r = b$ (F)
P_0, P_a, P_b Radial forces per unit circumferential length at $r = 0$, $r = a$, and $r = b$ (F/L)
P Internal radial force per unit circumferential length (normal force per unit length on r face), $= \sigma_r h$ (F/L)
P_ϕ Circumferential force per unit radial length, $= \sigma_\phi h$ (F/L)
p_r Radial loading density; positive if its direction is along positive r direction (F/L^2)

19.2 STRESSES

The stress formulas for thick cylinders and spheres under uniform pressure on the inner and outer circumferential surfaces are given in this section. Also, the stresses in a rotating disk are listed. The positive stresses σ_r and σ_ϕ are as in Fig. 19-2.

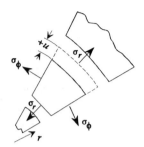

FIGURE 19-2 Positive radial displacement u and stresses σ_r, σ_ϕ.

Thick Cylinders

For a cylinder with both inner and outer pressures

$$\sigma_r = \frac{p_1 a_0^2}{a_L^2 - a_0^2}\left(1 - \frac{a_L^2}{r^2}\right) - \frac{p_2 a_L^2}{a_L^2 - a_0^2}\left(1 - \frac{a_0^2}{r^2}\right) \tag{19.1a}$$

$$\sigma_\phi = \frac{p_1 a_0^2}{a_L^2 - a_0^2}\left(1 + \frac{a_L^2}{r^2}\right) - \frac{p_2 a_L^2}{a_L^2 - a_0^2}\left(1 + \frac{a_0^2}{r^2}\right) \tag{19.1b}$$

Figure 19-3 shows the relative magnitude of σ_ϕ and σ_r under inner and outer pressure. If only the loading p_1 is applied on the inner circumferential surface of the cylinder, the maximum σ_ϕ and σ_r both occur at the inner surface, where σ_ϕ is $[1 + (a_L/a_0)^2]/[1 - (a_L/a_0)^2]$ times σ_r. Also if only p_2 is applied on the outer surface, the maximum σ_ϕ occurs at the inner surface while the maximum σ_r occurs at the outer surface. The ratio of σ_ϕ to σ_r at $r = a_L$ is $[(a_L/a_0)^2 + 1]/[(a_L/a_0)^2 - 1]$. For thick cylinders with an applied axial force N, positive in tension, axial stress σ_z may develop. This stress can be expressed as

$$\sigma_z = N/\left[\pi\left(a_L^2 - a_0^2\right)\right] = \text{const} \tag{19.1c}$$

Thick Spherical Shells

The stress formulas for thick spheres are

$$\sigma_r = \frac{p_1 a_0^3}{a_L^3 - a_0^3}\left(1 - \frac{a_L^3}{r^3}\right) - \frac{p_2 a_L^3}{a_L^3 - a_0^3}\left(1 - \frac{a_0^3}{r^3}\right) \tag{19.2a}$$

$$\sigma_\phi = \frac{p_1 a_0^3}{a_L^3 - a_0^3}\left(1 + \frac{a_L^3}{2r^3}\right) - \frac{p_2 a_L^3}{a_L^3 - a_0^3}\left(1 + \frac{a_0^3}{2r^3}\right) \tag{19.2b}$$

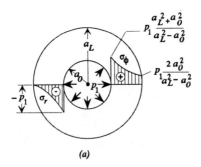

(a)

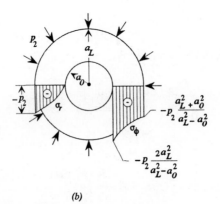

(b)

FIGURE 19-3 Stresses σ_r and σ_ϕ in thick-walled cylinders under internal and external pressure: (a) internal pressure; (b) external pressure.

Nonpressurized Rotating Disk of Constant Thickness

Solid Disk (No Center Hole)

$$\sigma_r = \frac{3 + \nu}{8} \rho^* \Omega^2 a_L^2 \left(1 - \frac{r^2}{a_L^2} \right) \tag{19.3a}$$

$$\sigma_\phi = \frac{3 + \nu}{8} \rho^* \Omega^2 a_L^2 \left(1 - \frac{1 + 3\nu}{3 + \nu} \frac{r^2}{a_L^2} \right) \tag{19.3b}$$

Disk with Central Hole

$$\sigma_r = \frac{3 + \nu}{8} \rho^* \Omega^2 \left(a_L^2 + a_0^2 - \frac{a_0^2 a_L^2}{r^2} - r^2 \right) \tag{19.4a}$$

$$\sigma_\phi = \frac{3 + \nu}{8} \rho^* \Omega^2 \left(a_L^2 + a_0^2 + \frac{a_L^2 a_0^2}{r^2} - \frac{1 + 3\nu}{3 + \nu} r^2 \right) \tag{19.4b}$$

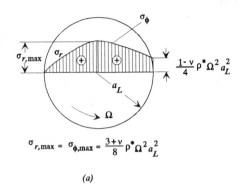

(a)

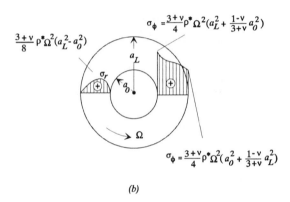

(b)

FIGURE 19-4 Stresses σ_r and σ_ϕ in rotating disks of constant thickness: (a) solid disk; (b) disk with a center hole.

The maximum value of σ_r occurs at $r = \sqrt{a_0 a_L}$ and is

$$\sigma_{r,\max} = \tfrac{1}{8}(3 + \nu)\rho^*\Omega^2(a_L - a_0)^2 \tag{19.5}$$

The maximum σ_r is always less than the maximum value of σ_ϕ regardless of the ratio a_L/a_0.

Equations (19.3) and (19.4) are illustrated in Fig. 19-4. Comparison of Eqs. (19.3) and (19.4) shows that the peak stresses in a disk with a hole are always greater than those in a disk without a hole. A small central hole, even a pinhole, doubles the peak normal stress σ_ϕ over the case of no hole. This can be seen by setting $r = a_0$ in Eqs. (19.4) and letting a_0 approach zero. The resulting σ_ϕ is twice that given by Eq. (19.3).

Note that these two formulas apply for thin disks only, where the plane stress assumption holds.

Example 19.1 Cylinder Design A long cylinder without an applied axial force is to have an inner radius of 40 mm and carry an internal pressure of 82 MPa with a safety factor of 2. Determine the outside radius according to the Tresca theory of failure (yield). Also, find the principal stresses at $r = a_0$. For this material, $\sigma_{ys} = 500$ MPa.

From Eqs. (19.1a) and (19.1b), the largest circumferential and radial stresses occur on the inner surface of this cylinder. Also, at this point, the maximum principal stress is σ_ϕ and the minimum is σ_r. From Eq. (3.22b)

$$\tau_{max} = \frac{1}{2}(\sigma_{max} - \sigma_{min}) = \frac{1}{2}(\sigma_\phi - \sigma_r) = \frac{1}{2}\frac{2p_1a_L^2}{a_L^2 - a_0^2} = \frac{p_1a_L^2}{a_L^2 - a_0^2} \tag{1}$$

The maximum allowable shear stress in tension is

$$\tau_{max} = \tfrac{1}{2}\sigma_{ys} = 250 \text{ MPa} \tag{2}$$

The Tresca theory of failure leads to

$$p_1a_L^2/(a_L^2 - a_0^2) = 250 \tag{3}$$

With the factor of safety of 2, $p_1 = 2 \times 82 = 164$ MPa. Then, with $a_0 = 40$ mm, (3) yields

$$a_L = \left(\frac{250}{250 - 164}\right)^{1/2}(40) = 68.20 \text{ mm} \tag{4}$$

The principal stresses at $r = a_0$ under the working pressure $p_1 = 82$ MPa are, from Eqs. (19.1a)–(19.1c),

$$\sigma_{\phi|_{r=a_0}} = \sigma_1 = \frac{(82)(40)^2}{68.20^2 - 40^2}\left(1 + \frac{68.20^2}{40^2}\right) = 168 \text{ MPa}$$

$$\sigma_{z|_{r=a_0}} = \sigma_2 = 0 \tag{5}$$

$$\sigma_{r|_{r=a_0}} = \sigma_3 = -82 \text{ MPa}$$

═══════════

19.3 DESIGN OF CYLINDERS WITH INTERNAL PRESSURE

A frequently occurring design problem occurs for cylinders subject to internal pressure (p_1) only. For this kind of loading, the maximum values of σ_r and σ_ϕ occur at the inner circumferential surface of the cylinder. From Eqs. (19.1) and Fig. 19-3a, $\sigma_r = p_1$ and σ_ϕ varies with the thickness of the cylinder (Fig. 19-5). Controlling the maximum value of σ_ϕ is the major concern of the design. Let $t = a_L - a_0$ and $p_2 = 0$. Then the expression for σ_ϕ at $r = a_0$ is

$$\sigma_{\phi|_{r=a_0}} = \frac{1 + (1 + t/a_0)^2}{(t/a_0)(2 + t/a_0)}p_1 \tag{19.6}$$

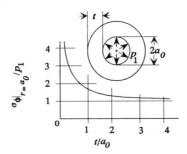

FIGURE 19-5 Stress σ_ϕ at $r = a_0$ of cylinder due to internal pressure only.

which can be shown to be equal to the value given in Fig. 19-3a. It can be seen from this expression that this stress approaches p_1 as the ratio t/a_0 approaches infinity, as shown in Fig. 19-5. Therefore, if the allowable stress of the cylinder is σ_{ys}, the internal pressure p_1 must never exceed σ_{ys} no matter how thick (t) the wall is made. To overcome this limitation, the cylinder can be prestressed to generate a state of initial compression (i.e., residual stress) at and near the inner surface. There are two common methods to produce residual stresses in cylinders. One is to press the cylinder from the inner surface until it deforms plastically to some distance in the radial direction. This procedure is called *autofrettage* or *self-hooping*. Another is to make a composite cylinder by shrink fitting one or more jackets over a cylinder.

For two shrink-fit cylinders of the same material (Fig. 19-6a) subjected to internal pressure p_1, a logical residual stress distribution would be one that results in the composite cylinder failing simultaneously at the inner radii of the inner and the outer cylinders. To achieve this, take two cylinders with outer and inner radii c and $c - \Delta$ (or $c + \Delta$ and c). Preheat the outer cylinder (or cool the inner one) and make them fit. Then at room temperature, stresses will develop at the inner and outer surfaces of the cylinders. After p_1 is applied, an interface pressure p_c is developed between the inner and outer cylinders (Fig. 19-6b). The pressure p_c and the radius c should be determined such that the maximum shear stresses from a theory of failure (e.g., Tresca) are minimized. Reference 19.1 presents a solution for this problem. Let the two maximum shear stresses have the same magnitude at the inner radii of the inner and the outer cylinders. This maximum shear stress is expressed as

$$\tau_{max} = p_1 a_L / [2(a_L - a_0)] \tag{19.7}$$

With a_0 and τ_{max} known, a_L can be determined from this relationship. The expressions for p_c and c from the Tresca theory are then found as

$$p_c = p_1(a_L - a_0) / [2(a_L + a_0)] \tag{19.8}$$

$$c = \sqrt{a_L a_0} \tag{19.9}$$

After p_c and c are computed, the outer radius r_{out} of the inner cylinder and the

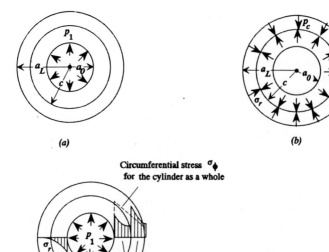

FIGURE 19-6 Stress distribution in composite cylinders: (a) composite configuration; (b) initial stress distribution due to shrink-fit contact pressure at interface $r = c$; (c) combined stress distribution due to shrink-fit and internal pressure (dashed lines represent stresses due to p_1 alone).

inner radius r_{in} of the outer cylinder can then be determined as

$$r_{out} = c \qquad r_{in} = c - \Delta \quad \text{(outer cylinder should be heated)}$$

or

$$r_{out} = c + \Delta \qquad r_{in} = c \quad \text{(inner cylinder should be cooled)}$$

with

$$\Delta = \frac{2c^3 p_c}{E} \frac{a_L^2 - a_0^2}{\left(a_L^2 - c^2\right)\left(c^2 - a_0^2\right)} \tag{19.10a}$$

or

$$\Delta = p_1 \sqrt{a_L a_0} / E \tag{19.10b}$$

The temperature change T needed to heat the outer cylinder (or to cool the inner cylinder) in order to fit the cylinders together is

$$T = \frac{\Delta}{c\alpha} = \frac{p_1}{\alpha E} \tag{19.11}$$

Example 19.2 Composite Cylinder If the cylinder in Example 19.1 is composite with the inner and outer cylinders as shown in Fig. 19-6, determine the outer
radius of the composite cylinder under the design pressure 164 MPa, the
required contact pressure p_c, the radius c of the interface, and the circumferential stresses at $r = a_0 = 40$ mm and at $r = c$ under the working pressure 82 MPa.
The inner and outer cylinders are made from steel with $E = 200$ GPa, $\alpha = 12 \times 10^{-6}/°C$, and the allowable τ_{max} at 250 MPa.

With $a_0 = 40$ mm and a design pressure of 164 MPa, the outer radius of the
composite cylinder is obtained from [Eq. (19.7)]

$$\tau_{max} = 250 = 164 a_L / 2(a_L - 40) \tag{1}$$

as $a_L = 59.52$ mm.

The required contact pressure is, from Eq. (19.8),

$$p_c = \frac{164(59.52 - 40)}{2(59.52 + 40)} = 16.08 \text{ MPa} \tag{2}$$

From Eq. (19.9), the radius of the interface is

$$c = \sqrt{40(59.52)} = 48.79 \text{ mm} \tag{3}$$

The circumferential stress at $r = a_0$ is due to the working pressure $p_1 = 82$
MPa and the contact pressure p_c.

The state of stress under combined internal working pressure and shrink-fit
loading is shown in Fig. 19-6c. From Eq. (19.1b), for the inner radius of the inner
cylinder,

$$\sigma_{\phi|_{r=a_0}} = -p_c \frac{2c^2}{c^2 - a_0^2} + p_1 \frac{a_0^2 + a_L^2}{a_L^2 - a_0^2} = -p_c \frac{2}{1 - (a_0/c)^2} + p_1 \frac{1 + (a_L/a_0)^2}{(a_L/a_0)^2 - 1} \tag{4}$$

$$= -\frac{2(16.08)}{1 - (40/48.79)^2} + 82 \frac{1 + (59.52/40)^2}{(59.52/40) - 1}$$

$$= -98.09 + 217.07 = 118.98 \text{ MPa}$$

At the inner radius of the outer cylinder

$$\sigma_{\phi|_{r=c}} = p_c \frac{a_L^2 + c^2}{a_L^2 - c^2} + \frac{p_1 a_0^2}{a_L^2 - a_0^2} \left(1 + \frac{a_L^2}{c^2}\right) = 250.0 \text{ MPa} \tag{5}$$

The required interference Δ in radius at room temperature is, from Eq.
(19.10),

$$\Delta = p_1 \sqrt{a_0 a_L} / E \tag{6}$$

For the design pressure $p_1 = 164$ MPa,

$$\Delta = 164\sqrt{40(59.52)}/200(10^3) = 0.04 \text{ mm} \qquad (7)$$

For steel with $\alpha = 12(10)^{-6}/°C$, the temperature change T needed is, by Eq. (19.11),

$$T = 164/\left[12(10)^{-6}(200)(10)^3\right] = 68.33°C \qquad (8)$$

It is interesting to note that the weight of the cylinder in this example is less than that in Example 19.1, since the area of the cross sections are, respectively,

$$A_2 = \pi(59.52^2 - 40^2) = 6099.86 \text{ mm}^2 \qquad (9)$$

$$A_1 = \pi(68.20^2 - 40^2) = 9580.89 \text{ mm}^2 \qquad (10)$$

This is an $(A_1 - A_2)/A_1 = 36.3\%$ reduction in weight per unit axial length of the cylinder. In fact, the reduction will be greater if the design pressure is higher.

19.4 SIMPLE SHELLS AND DISKS

Thick Cylinders

The governing equations for the radial displacement and stresses in a thick cylinder are given by

$$\frac{d^2u}{dr^2} + \frac{1}{r}\frac{du}{dr} - \frac{u}{r^2} = -\frac{1}{\lambda + 2G}p_r + \frac{3\lambda + 2G}{\lambda + 2G}\alpha\frac{dT}{dr} \qquad (19.12\text{a})$$

$$(\lambda + 2G)\frac{du}{dr} + \frac{\lambda}{r}u - (3\lambda + 2G)\alpha T = \sigma_r \qquad (19.12\text{b})$$

The radial displacement and stress for a cylinder are provided in Table 19-1, along with the tangential and axial stresses σ_ϕ and σ_z. Part A of the table lists formulas for the radial displacements and stresses. The loading functions are taken from part B of table 19-1 by adding the appropriate terms for each applied load. The initial parameters are provided in part C of the table for particular inner and outer surface conditions.

Example 19.3 Thick Cylinder with Internal Pressure Calculate the displacement and stresses of a thick cylinder of inner and outer radii $a_0 = 10$ in. and $a_L = 20$ in. The cylinder is subjected to internal uniform pressure $p_1 = 100$ lb/in.^{2} The material constants are $E = 3 \times 10^7$ lb/in.^{2}, $\nu = 0.3$, $G = E/[2(1 + \nu)] = 1.1538 \times 10^7$ lb/in.^{2}, and $\lambda = E\nu/[(1 + \nu)(1 - 2\nu)] = 1.730 \times 10^7$ lb/in.^{2}
The response of the cylinder can be calculated from the formulas in Table 19-1. First find the initial parameters u_0 and σ_0. Both the inner and outer

circumferential boundaries of the cylinder cross section are free, so that, from Table 19-1, part C,

$$\sigma_0 = 0 \tag{1}$$

$$u_0 = -\frac{\bar{F}_\sigma}{[G/(1-\nu)][(a_L^2 - a_0^2)/(a_0 a_L^2)]} \tag{2}$$

From Table 19-1, part B, if the radial pressure is applied at $a_1 = a_0$, the loading functions are

$$F_u(r) = -\frac{p_1 \nu}{2(1-\nu)\lambda} \frac{r^2 - a_0^2}{r} \tag{3}$$

$$F_\sigma(r) = -\frac{p_1}{1-\nu}\left[\frac{1}{2} + \frac{G\nu}{\lambda}\left(\frac{a_0}{r}\right)^2\right] \tag{4}$$

Then

$$\bar{F}_\sigma = F_\sigma|_{r=a_L} = -\frac{p_1}{1-\nu}\left[\frac{1}{2} + \frac{G\nu}{\lambda}\left(\frac{a_0}{a_L}\right)^2\right] = -78.57 \tag{5}$$

Substitute (5) into (2) to find $u_0 = 6.36 \times 10^{-5}$ in.

The displacement u and stresses σ_r, σ_ϕ, and σ_z are then found from part A of Table 19-1 to be

$$u = \frac{u_0[(G\nu/\lambda)(r/a_0) + a_0/2r]}{1-\nu} - \frac{p_1 \nu}{2(1-\nu)\lambda}\frac{r^2 - a_0^2}{r} \tag{6a}$$

$$\sigma_r = \frac{u_0 G(r^2 - a_0^2)}{(1-\nu)a_0 r^2} - \frac{p_1}{1-\nu}\left[\frac{1}{2} + \frac{G\nu}{\lambda}\left(\frac{a_0}{r}\right)^2\right] \tag{6b}$$

$$\sigma_\phi = \frac{4G(\lambda + G)}{\lambda + 2G}\frac{u}{r} + \frac{\lambda}{\lambda + 2G}\sigma_r \tag{6c}$$

$$\sigma_z = \frac{2\lambda G}{\lambda + 2G}\frac{u}{r} + \frac{\lambda}{\lambda + 2G}\sigma_r \tag{6d}$$

Substitution of the numerical values of λ, G, ν, and u_0 into (6) leads to the following responses:

r	u	σ_r	σ_ϕ	σ_z
10	6.36×10^{-5}	-100.0	166.667	20
12	5.508×10^{-5}	-59.26	125.926	20
14	4.936×10^{-5}	-34.69	101.36	20
16	4.536×10^{-5}	-18.75	85.42	20
18	4.25×10^{-5}	-7.82	74.486	20
20	4.04×10^{-5}	0	66.667	20

The values of σ_r and σ_ϕ are the same as the stress values obtained from Eqs. (19.1a) and (19.1b). Note that no axial force is involved in this example but the axial stress σ_z does not vanish. This is because the formulas in Table 19-1 are based on the plane strain assumption of $\varepsilon_z = 0$; i.e., either the cylinder is very long or the ends of the cylinder are constrained. This contrasts to the situation for Eq. (19.1c), which is based on the generalized plane strain assumption of $\varepsilon_z = $ const. Equation (19.1c) describes the axial stress in the cylinder when the ends of the cylinder are free.

======

Example 19.4 Thermally Loaded Cylinder Find the stress and displacements in a thermally loaded cylinder with zero tractions on the inner and outer boundaries. The temperature change along the radial direction for a cylinder with temperature changes T_{a_0}, T_{a_L} on the inner (a_0) and outer (a_L) surfaces is [19.2]

$$T(r) = \frac{T_{a_0} \ln(a_L/r) - T_{a_L} \ln(a_0/r)}{\ln(a_L/a_0)} \tag{1}$$

For isotropic material, the radial stress and displacement are given by cases 1 and 2 of Table 19-1, part A, as

$$u = \frac{u_0[(Gv/\lambda)(r/a_0 + a_0/2r)]}{1 - v} + \frac{\sigma_0 v[(r^2 - a_0^2)/r]}{2(1 - v)\lambda} + F_u \tag{2}$$

$$\sigma_r = \frac{u_0 G(r^2 - a_0^2)}{(1 - v)a_0 r^2} + \frac{\sigma_0[\frac{1}{2} + (Gv/\lambda)(a_0/r)^2]}{1 - v} + F_\sigma \tag{3}$$

From Table 19-1, part B, the loading functions for an arbitrary temperature change T are

$$F_u(r) = \frac{1 + v}{1 - v} \frac{\alpha}{r} \int_{a_0}^{r} T(\xi)\xi \, d\xi \tag{4}$$

$$F_\sigma(r) = -\frac{E}{1 - v} \frac{\alpha}{r^2} \int_{a_0}^{r} T(\xi)\xi \, d\xi \tag{5}$$

The inner and outer boundaries are free, so that from Table 19-1, part C, the initial parameters are $\sigma_0 = 0$ and

$$u_0 = -\frac{F_\sigma|_{r=a_L}}{[G/(1 - v)][(a_L^2 - a_0^2)/(a_0 a_L^2)]} = \frac{[E/(1 - v)](\alpha/a_L^2)\int_{a_0}^{a_L} T(\xi)\xi \, d\xi}{[G/(1 - v)][(a_L^2 - a_0^2)/(a_0 a_L^2)]}$$

$$= \frac{a_0 E\alpha \int_{a_0}^{a_L} T(\xi)\xi \, d\xi}{G(a_L^2 - a_0^2)} \tag{6}$$

To complete the solution, calculate the integrals containing the temperature change,

$$\int_{a_0}^r T(\xi)\xi \, d\xi = \int_{a_0}^r \frac{\left[T_{a_0} \ln(a_L/\xi) - T_{a_L} \ln(a_0/\xi) \right]\xi}{\ln(a_L/a_0)} \, d\xi$$

$$= \tfrac{1}{4}r^2 \frac{T_{a_0} - T_{a_L}}{\ln(a_L/a_0)} + \tfrac{1}{2}r^2 T(r) + \tfrac{1}{4}\left(T_{a_L} - T_{a_0} \right)a_0^2 - \tfrac{1}{2}T_{a_0}a_0^2 \ln \frac{a_L}{a_0}$$

(7)

Then

$$\int_{a_0}^{a_L} T(\xi)\xi \, d\xi = \tfrac{1}{4}a_L^2 \frac{T_{a_0} - T_{a_L}}{\ln(a_L/a_0)} + \tfrac{1}{2}a_L^2 \frac{T_{a_L}}{\ln(a_L/a_0)}$$

$$+ \tfrac{1}{4}\left(T_{a_L} - T_{a_0} \right)a_0^2 - \tfrac{1}{2}T_{a_0}a_0^2 \ln \frac{a_L}{a_0}$$

(8)

Equations (7) and (8) placed in (4), (5), and (6) complete the response represented by the displacement of (2) and the stress of (3).

===

Thick Spherical Shells

The governing equations for the radial displacement and stresses are given by

$$\frac{d^2u}{dr^2} + \frac{2}{r}\frac{du}{dr} - 2\frac{u}{r^2} = -\frac{1}{\lambda + 2G}P_r + \frac{3\lambda + 2G}{\lambda + 2G}\alpha\frac{dT}{dr} \qquad (19.13a)$$

$$(\lambda + 2G)\frac{du}{dr} + \frac{2\lambda}{r}u - (3\lambda + 2G)\alpha T = \sigma_r \qquad (19.13b)$$

Formulas for the radial displacement and radial stress as well as the tangential stress are given in Table 19-2, part A. The loading functions and initial parameters are provided in parts B and C of the table.

Disks

The governing equations for the radial displacement and stresses in terms of the applied loadings are given by

$$\frac{d^2u}{dr^2} + \frac{1}{r}\frac{du}{dr} - \frac{1}{r^2}u = -\frac{1 - \nu^2}{Eh}P_r + (1 + \nu)\alpha\frac{dT}{dr} \qquad (19.14a)$$

$$\frac{Eh}{1 - \nu^2}\frac{du}{dr} + \frac{\nu Eh}{1 - \nu^2}\frac{u}{r} - \frac{Eh\alpha}{1 - \nu}T = P \qquad (19.14b)$$

where $P = h\sigma_r$. Table 19-3 provides formulas for the radial displacement, radial force per circumferential length, and tangential force per circumferential length of disks of constant thickness.

Example 19.5 Rotating Disk with Internal Pressure Find the displacement and internal forces in a disk of constant thickness rotating at angular velocity Ω and subject to an internal pressure p_1 (force per length) on the inner periphery at $r = a_0$.

The radial displacement and force are given by cases 1 and 2 of Table 19-3, part A, as

$$u = u_0 \frac{r}{a_0}\left[1 - \frac{1 + \nu}{2}\frac{r^2 - a_0^2}{r^2}\right] + P_0 \frac{1 - \nu^2}{2Eh}\frac{r^2 - a_0^2}{r} + F_u \tag{1}$$

$$P = u_0 \frac{Eh}{2a_0 r^2}(r^2 - a_0^2) + P_0\left[1 - \frac{1 - \nu}{2}\frac{r^2 - a_0^2}{r^2}\right] + F_P \tag{2}$$

From part B of Table 19-3, the loading functions for the centrifugal loading and internal pressure are given by

$$F_u = -\frac{p_1}{Eh}\frac{1 - \nu}{2}\frac{r^2 - a_0^2}{r} - \frac{(r^2 - a_0^2)^2}{r}\frac{1 - \nu^2}{8}\frac{\rho^*\Omega^2}{E} \tag{3}$$

$$F_P = -p_1\left(1 - \frac{1 - \nu}{2}\frac{r^2 - a_0^2}{r^2}\right)$$
$$- (r^2 - a_0^2)\frac{\rho^*\Omega^2 h}{4}\left[(1 + \nu) + \frac{1 - \nu}{2}\frac{(r^2 + a_0^2)}{r^2}\right] \tag{4}$$

The initial parameters of (1) and (2) are provided in Table 19-3, part C. The boundaries are taken to be free–free. The inner boundary is treated as being free since the pressure is accounted for as a loading and not as a boundary condition. From Table 19-3, part C,

$$P_0 = 0 \tag{5}$$

$$u_0 = -\frac{2F_{P|r=a_L}a_0 a_L^2}{Eh(a_L^2 - a_0^2)} = \frac{2p_1 a_0 a_L^2\left[1 - (1 - \nu)(a_L^2 - a_0^2)/2a_L^2\right]}{Eh(a_L^2 - a_0^2)}$$
$$+ \frac{a_0 a_L^2}{E}\frac{\rho^*\Omega^2}{2}\left[(1 + \nu) + \frac{1 - \nu}{2}\frac{a_L^2 + a_0^2}{a_L^2}\right] \tag{6}$$

With (3), (4), (5), and (6), Eqs. (1) and (2) now provide the radial displacement and force throughout the disk. The tangential force is given by case 3 of Table 19-3, part A, using the u and P found above.

19.5 NATURAL FREQUENCIES

The natural frequencies for simple thick cylinders, spheres, and disks are presented in Table 19-4. For more complicated cases, use the procedures given in Appendix III and the transfer, stiffness, and mass matrices in Tables 19-5 to 19-9 to compute the natural frequencies.

19.6 GENERAL SHELLS AND DISKS

The formulas of Tables 19-1 to 19-3 apply to rather simple shells and disks. For more general members, e.g., those formed of several members, it is advisable to use the displacement method or the transfer matrix procedure, which are explained technically at the end of this book (Appendices II and III).

Several transfer and stiffness matrices are tabulated in Tables 19-4 to 19-7. Mass matrices for use in a displacement method analysis are given in Table 19-8.

These responses are based on Eqs. (19.12)–(19.14), as appropriate. For dynamic problems, substitute $p_r - \rho^* \, \partial^2 u / \partial t^2$ for p_r in Eqs. (19.12) and (19.13) for cylinders and shells and $p_r - h\rho^* \, \partial^2 u / \partial t^2$ for p_r in Eq. (19.14) for disks.

Example 19.6 Disk of Hyperbolic Profile A steel disk of hyperbolic profile is loosely attached to a rigid post as shown in Fig. 19-7. The configuration rotates at frequency Ω. Find the radial stress distribution in the disk.

The transfer matrix of Table 19-7, case 3, applies to this problem. For this disk $n = 1$, $h_k = h_a a = 15$, $E = 3 \times 10^7$ lb/in.^{2}, $\nu = 0.3$, $n_1 = 1.75$, and $n_2 = -0.75$. The stress is given by

$$\sigma = \frac{P}{h} = \frac{u_0 U_{Pu}}{h} + \frac{P_0 U_{PP}}{h} + \frac{F_P}{h} \tag{1}$$

where, from Table 19-7 with $a = 3$ in.,

$$U_{Pu} = \frac{Eh_k(n_1 + \nu)(n_2 + \nu)}{(1 - \nu^2)(b^{n+1})(n_2 - n_1)} \left[\left(\frac{b}{a}\right)^{n_1} - \left(\frac{b}{a}\right)^{n_2} \right]$$

$$= 182.5 \times 10^6 \times \frac{1}{b^2} \left[\left(\frac{b}{3}\right)^{1.75} - \left(\frac{b}{3}\right)^{-0.75} \right] \tag{2}$$

$$F_P = -\frac{\rho^* \Omega^2 h_k b^{2-n}}{8 - (3 + \nu)n} \left\{ (3 + \nu) \right.$$

$$\left. - \frac{1}{n_2 - n_1} \left[(n_1 + \nu)(n_2 - 3)\left(\frac{b}{a}\right)^{n_1 - 3} - (n_2 + \nu)(n_1 - 3)\left(\frac{b}{a}\right)^{n_2 - 3} \right] \right\}$$

$$= -3.19 \rho^* \Omega^2 b \left[3.3 - 3.075 \left(\frac{b}{3}\right)^{-1.25} - 0.225 \left(\frac{b}{3}\right)^{-3.75} \right]$$

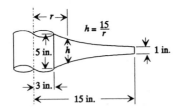

FIGURE 19-7 Example 19.6.

It is not necessary to look up U_{PP} since P_0 of Eq. (1) is zero as a result of the loose fit at the inner boundary. Also, at the outer circumference, where $r = b = 15$ in., the force P is zero. From Table 19-3 for free–free conditions,

$$P_0 = 0 \qquad u_0 = -\bar{F}_P/\nabla_{|b=15} \tag{3}$$

where $\bar{F}_P = (F_P)_{b=15}$ and $\nabla = U_{Pu|b=15}$. Substitution of $b = 15$ in. into (2) gives

$$U_{Pu|b=15} = 13.32 \times 10^6 \qquad \bar{F}_P = -138.2 \, \rho^* \Omega^2 \tag{4}$$

Equations (3) and (4) lead to $u_0 = 1.04 \times 10^{-5} \rho^* \Omega^2$. If this value of u_0 is placed in (1), we find

$$\sigma = (-10.527b + 316.27b^{-0.25} - 4282.33b^{-2.75})(\rho^* \Omega^2/h) \tag{5}$$

This expression applies for any solution if b is replaced by r so that the radial stress in the disk is completely defined by (5).

=====

Example 19.7 Shrink-Fit Disk–Shaft System A disk–shaft system (Fig. 19-8a) can be used to illustrate both the solution to a complicated disk problem and the treatment of shrink fit.

Suppose the shaft can be modeled as a solid (holeless) disk of thickness equal to that (h_1) of the outer disk where it is connected to the shaft (Fig. 19-8b). The outer disk, which possesses a variable thickness, will be handled as a succession of disks, each with a constant thickness (Fig. 19-8c). Assume the disk to be shrink fit to the shaft.

Expressions for the displacements and forces in the shaft can be treated separately and then the results are joined with the disk response to account for shrink-fit results. Let c be the inner radius of the disk and the outer radius of the shaft after shrinking. In the case of the shaft, the responses are taken from case 4 of Table 19-7 as

$$u_{|r=c} = P_0 U_{uP|r=c} + F_{u|r=c} = P_0 \frac{c}{Eh_1}(1-\nu) - (1-\nu^2)\frac{\rho^* \Omega^2 c^3}{8E} \tag{1a}$$

$$P_{|r=c} = P_0 U_{PP|r=c} + F_{P|r=c} = P_0 - (3+\nu)\frac{\rho^* h_1 \Omega^2 c^2}{8} \tag{1b}$$

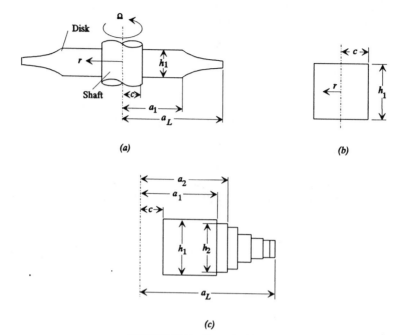

(a)

(b)

(c)

FIGURE 19-8 Example 19.7.

For the disk of variable cross section which is treated as a succession of disks of constant thickness, the overall transfer matrix $\mathbf{U}$ can be developed as

$$\mathbf{U} = \mathbf{U}^1 \mathbf{U}^2 \mathbf{U}^3 \cdots \mathbf{U}^n \tag{2}$$

where $\mathbf{U}^i$ is the transfer matrix for the ith disk of constant thickness. The expression for $\mathbf{U}^i$ is taken from case 1, Table 19-7, with $a = a_{i-1}$, $b = a_i$ as

$$\mathbf{U}^i =
\begin{bmatrix}
\dfrac{a_i}{a_{i-1}}\left[1 - \dfrac{1+\nu}{2}\dfrac{a_i^2 - a_{i-1}^2}{a_i^2}\right] & \dfrac{1-\nu^2}{2Eh_i}\dfrac{a_i^2 - a_{i-1}^2}{a_i} & -\dfrac{(a_i^2 - a_{i-1}^2)^2}{a_i}\dfrac{1-\nu^2}{8}\dfrac{\rho^*\Omega^2}{E} \\
\dfrac{Eh_i}{2a_{i-1}a_i^2}(a_i^2 - a_{i-1}^2) & 1 - \dfrac{1-\nu}{2}\dfrac{a_i^2 - a_{i-1}^2}{a_i^2} & -\dfrac{(a_i^2 - a_{i-1}^2)\rho^*\Omega^2 h_i}{4}\left[(1+\nu) + \dfrac{1-\nu}{2}\dfrac{a_i^2 + a_{i-1}^2}{a_i^2}\right] \\
0 & 0 & 1
\end{bmatrix} \tag{3}$$

where $i = 1, 2, \ldots, n$.

In nonmatrix form, the displacement and force at $r = a_L$ are

$$u_{|r=a_L} = u_c U_{uu}(a_L, c) + P_c U_{uP}(a_L, c) + F_u(a_L, c) \tag{4a}$$

$$P_{|r=a_L} = u_c U_{Pu}(a_L, c) + P_c U_{PP}(a_L, c) + F_P(a_L, c) \tag{4b}$$

where $U_{jk}(a_L, c)$ are components of the overall transfer matrix from $r = c$ to $r = a_L$.

Generation of a solution requires knowledge of initial parameters P_0 of (1) and u_c, P_c of (4). For a member without shrink fit, these can be obtained as the initial parameters of Table 19-3, part C. In the case of a shrink-fit problem, the interaction between the shaft and the disk provides the condition necessary to evaluate the initial parameters. Suppose that r_s is the outer radius of the shaft before shrinking and r_D is the inner radius of the disk before shrinking. Also let $u_{D|r=c}$ be the displacement of the inner radius of the disk upon shrink fitting and $u_{s|r=c}$ the displacement of the outer radius of the shaft after shrinking. The shrink-fit deformation, or shrinkage, is

$$\Delta_{sf} = r_s - r_D = u_{D|r=c} - u_{s|r=c} \tag{5}$$

The shaft displacement $u_{s|r=c}$ is found in terms of $P_{|r=c} = P_c$ by eliminating P_0 from (1a) and (1b):

$$u_{s|r=c} = \frac{P_c - F_{P|r=c}}{U_{PP|r=c}} U_{uP|r=c} + F_{u|r=c} \tag{6}$$

The disk displacement $u_{D|r=c}$ is given by (4b) as a function of P_c,

$$u_{D|r=c} = \frac{P_{|r=a_L} - F_P(a_L, c)}{U_{Pu}(a_L, c)} - P_c \frac{U_{PP}(a_L, c)}{U_{Pu}(a_L, c)} \tag{7}$$

Substitution of (6) and (7) into (5) provides a relationship sufficient to solve a variety of shrink-fit problems. For example, for a prescribed shrinkage Δ_{sf} the pressure P_c between the shaft and disk can be computed from (5). Or the shrinkage Δ_{sf} necessary to achieve an interaction pressure P_c can be calculated. In each case, a specified external pressure on the disk can be included. In problems involving cylinders, it is common to include the effect of an internal pressure on an inner cylinder. With either Δ_{sf} or P_c given, it is possible to calculate all displacements and stresses in the shaft and disk.

REFERENCES

19.1 Cook, R. D., and Young, W. C., *Advanced Mechanics of Materials*, Macmillan, New York, 1985.

19.2 Zudans, Z., Yen, T. C., and Steigelmann, W. H., *Thermal Stress Techniques*, American Elsevier, New York, 1965.

19

Tables

TABLE 19-1, PART A   THICK CYLINDERS: GENERAL RESPONSE EXPRESSIONS

Notation

a_0 = radius of inner boundary
G = shear modulus of elasticity
ν = Poisson's ratio
λ = Lamé coefficient
α = coefficient of thermal expansion
T = change of temperature along length
of wall thickness

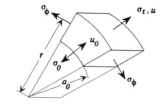

$$\sigma_0 = \sigma_{r|r=a_0} \qquad u_0 = u_{|r=a_0}$$

The formulas in this table are based on the plane strain assumption of $\varepsilon_z = 0$.

Response

1.
Displacement:

$$u = \frac{u_0\left[(G\nu/\lambda)(r/a_0) + a_0/2r\right]}{1 - \nu} + \frac{\sigma_0\nu\left(r^2 - a_0^2\right)}{2(1 - \nu)\lambda r} + F_u$$

2.
Radial stress:

$$\sigma_r = \frac{u_0 G\left(r^2 - a_0^2\right)}{(1 - \nu)a_0 r^2} + \frac{\sigma_0\left[\frac{1}{2} + (G\nu/\lambda)(a_0/r)^2\right]}{1 - \nu} + F_\sigma$$

3.
Tangential stress:

$$\sigma_\phi = \frac{4G(\lambda + G)}{\lambda + 2G}\frac{u}{r} + \frac{\lambda}{(\lambda + 2G)}\sigma_r - \frac{E\alpha T}{1 - \nu}$$

4.
Axial stress:

$$\sigma_z = \frac{2\lambda G}{\lambda + 2G}\frac{u}{r} + \frac{\lambda}{\lambda + 2G}\sigma_r - \frac{E\alpha T}{1 - \nu}$$

TABLE 19-1, Part A   Thick Cylinders: Response Expressions   1118

TABLE 19-1, PART B   THICK CYLINDERS: LOADING FUNCTIONS

Notation

E = modulus of elasticity
ν = Poisson's ratio
ρ^* = mass per unit volume
Ω = angular velocity of rotation
α = coefficient of thermal expansion

$$\langle r - a_1 \rangle^0 = \begin{cases} 0 & \text{if } r < a_1 \\ 1 & \text{if } r \geq a_1 \end{cases}$$

	$F_u(r)$	$F_\sigma(r)$
1. Radial pressure applied at $r = a_1$ (force/length^2)	$-\dfrac{p\nu}{2(1-\nu)\lambda}$ $\times \dfrac{r^2 - a_1^2}{r} \langle r - a_1 \rangle^0$	$-\dfrac{p}{(1-\nu)}$ $\times \left[\dfrac{1}{2} + \dfrac{G\nu}{\lambda} \left(\dfrac{a_1}{r} \right)^2 \right] \langle r - a_1 \rangle^0$
2. Constant temperature change T_1 (independent of r)	$\dfrac{r^2 - a_0^2}{2r} \dfrac{1+\nu}{1-\nu} \alpha T_1$	$-\dfrac{r^2 - a_0^2}{2r^2} \dfrac{\alpha E}{1-\nu} T_1$
3. Centrifugal loading due to rotation of cylinder at angular velocity Ω	$-\dfrac{\rho^* \Omega^2}{8} \dfrac{\left(r^2 - a_0^2 \right)^2}{r}$ $\times \dfrac{(1+\nu)(1-2\nu)}{E(1-\nu)}$	$-\left(r^2 - a_0^2 \right) \dfrac{\rho^* \Omega^2}{4}$ $\times \left(2 - \dfrac{1-2\nu}{1-\nu} \dfrac{r^2 - a_0^2}{2r^2} \right)$
4. Arbitrary temperature change $T(r)$	$\dfrac{1+\nu}{1-\nu} \dfrac{\alpha}{r} \int_{a_0}^{r} T(\xi)\xi \, d\xi$	$-\dfrac{E}{1-\nu} \dfrac{\alpha}{r^2} \int_{a_0}^{r} T(\xi)\xi \, d\xi$

TABLE 19-1, PART C   THICK CYLINDERS: INITIAL PARAMETERS

Notation

ν = Poisson's ratio
λ = Lamé coefficient
G = shear modulus of elasticity
a_L = radius of outer surface of cylinder
$\bar{F}_u = F_{u|r=a_L}$ $\bar{F}_\sigma = F_{\sigma|r=a_L}$

Inner Edge / Outer Edge	Fixed	Free
1. Fixed $u_0 = 0$	$\sigma_0 = -\dfrac{\bar{F}_u}{\nabla}$ $\nabla = \dfrac{\nu}{2(1-\nu)\lambda}\dfrac{a_L^2 - a_0^2}{a_L}$	$\sigma_0 = -\dfrac{\bar{F}_\sigma}{\nabla}$ $\nabla = \dfrac{1}{1-\nu}\left[\dfrac{1}{2} + \dfrac{G\nu}{\lambda}\left(\dfrac{a_0}{a_L}\right)^2\right]$
2. Free $\sigma_0 = 0$	$u_0 = -\dfrac{\bar{F}_u}{\nabla}$ $\nabla = \dfrac{1}{1-\nu}\left(\dfrac{G\nu}{\lambda}\dfrac{a_L}{a_0} + \dfrac{a_0}{2a_L}\right)$	$u_0 = -\dfrac{\bar{F}_\sigma}{\nabla}$ $\nabla = \dfrac{G}{1-\nu}\dfrac{a_L^2 - a_0^2}{a_0 a_L^2}$

TABLE 19-1. Part C   **Thick Cylinders: Initial Parameters**   **1120**

TABLE 19-2, PART A   SPHERES: GENERAL RESPONSE EXPRESSIONS

Notation

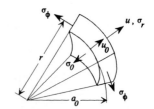

a_0 = radius of inner boundary
G = shear modulus of elasticity
λ = Lamé coefficient
α = coefficient of thermal expansion
T = change of temperature
   in radial direction
$\sigma_0 = \sigma_{r|r=a_0}$ $u_0 = u_{|r=a_0}$

Response

1.
Displacement:
$$u = u_0 \frac{4Gr^3 + (3\lambda + 2G)a_0^3}{3(\lambda + 2G)a_0 r^2} + \sigma_0 \frac{a_0}{3(\lambda + 2G)}\left(\frac{r}{a_0} - \frac{a_0^2}{r^2}\right) + F_u$$

2.
Radial stress:
$$\sigma_r = u_0 \frac{4G(3\lambda + 2G)}{3(\lambda + 2G)a_0 r^3}(r^3 - a_0^3) + \sigma_0 \frac{3\lambda + 2G + 4Ga_0^3/r^3}{3(\lambda + 2G)} + F_\sigma$$

3.
Tangential stress:
$$\sigma_\phi = 2G\frac{2 + 3\lambda}{2G + \lambda}\frac{u}{r} + \frac{\lambda}{2G + \lambda}\sigma_r - \frac{2G(3\lambda + 2G)}{2G + \lambda}\alpha T$$

TABLE 19-2, PART B   SPHERES: LOADING FUNCTIONS

Notation

λ = Lamé coefficient
G = shear modulus of elasticity
α = coefficient of thermal expansion

$$\langle r - a_1 \rangle^0 = \begin{cases} 0 & \text{if } r < a_1 \\ 1 & \text{if } r \geq a_1 \end{cases}$$

	$F_u(r)$	$F_\sigma(r)$
1. Radial pressure applied at $r = a_1$ (force/length^2)	$-\dfrac{pa_1}{3(\lambda + 2G)}\left(\dfrac{r}{a_1} - \dfrac{a_1^2}{r^2}\right)$ $\times \langle r - a_1 \rangle^0$	$-p\dfrac{3\lambda + 2G + 4Ga_1^3/r^3}{3(\lambda + 2G)}\langle r - a_1 \rangle^0$
2. Constant temperature change T_1 (independent of r)	$\dfrac{3\lambda + 2G}{3(\lambda + 2G)}\dfrac{(r^3 - a_0^3)}{r^2}\alpha T_1$	$\dfrac{-4G(3\lambda + 2G)}{3(\lambda + 2G)}\dfrac{(r^3 - a_0^3)}{r^3}\alpha T_1$
3. Arbitrary temperature change $T(r)$	$\dfrac{3\lambda + 2G}{\lambda + 2G}\dfrac{\alpha}{r^2}\displaystyle\int_{a_0}^r T(\xi)\xi^2\, d\xi$	$\dfrac{-4G\alpha}{r^3}\dfrac{3\lambda + 2G}{\lambda + 2G}\displaystyle\int_{a_0}^r T(\xi)\xi^2\, d\xi$

TABLE 19-2, Part B | **Spheres: Loading Functions** | **1122**

TABLE 19-2, PART C   SPHERES: INITIAL PARAMETERS

Notation

a_L = radius of outer boundary
λ = Lamé coefficient
G = shear modulus of elasticity

$$\bar{F}_u = F_{u|r=a_L} \qquad \bar{F}_\sigma = F_{\sigma|r=a_L}$$

Inner Edge \ Outer Edge	Fixed	Free
1. Fixed $u_0 = 0$	$\sigma_0 = -\dfrac{\bar{F}_u}{\nabla}$ $\nabla = \dfrac{a_0}{3(\lambda + 2G)}\left(\dfrac{a_L}{a_0} - \dfrac{a_0^2}{a_L^2}\right)$	$\sigma_0 = -\dfrac{\bar{F}_\sigma}{\nabla}$ $\nabla = \dfrac{3\lambda + 2G + 4Ga_0^3/a_L^3}{3(\lambda + 2G)}$
2. Free $\sigma_0 = 0$	$u_0 = -\dfrac{\bar{F}_u}{\nabla}$ $\nabla = \dfrac{4Ga_L^3 + (3\lambda + 2G)a_0^3}{3(\lambda + 2G)a_0 a_L^2}$	$u_0 = -\dfrac{\bar{F}_\sigma}{\nabla}$ $\nabla = \dfrac{4G(3\lambda + 2G)\left(a_L^3 - a_0^3\right)}{3(\lambda + 2G)a_L^3 a_0}$

TABLE 19-3, PART A   DISKS: GENERAL RESPONSE EXPRESSIONS

Notation

a_0 = radius of inner boundary
E = modulus of elasticity
ν = Poisson's ratio
α = coefficient of thermal expansion
T = change of temperature
 in radial direction
h = thickness of disk
σ_r, σ_ϕ = radial and circumferential stresses,
 respectively

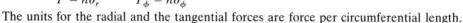

$$P_0 = P_{|r=a_0} \qquad u_0 = u_{|r=a_0}$$
$$P = h\sigma_r \qquad P_\phi = h\sigma_\phi$$

The units for the radial and the tangential forces are force per circumferential length.

Response

1.
Displacement:
$$u = u_0 \frac{r}{a_0}\left(1 - \frac{1+\nu}{2}\frac{r^2 - a_0^2}{r^2}\right) + P_0\frac{1-\nu^2}{2Eh}\frac{r^2 - a_0^2}{r} + F_u$$

2.
Radial force:
$$P = u_0\frac{Eh}{2a_0 r^2}(r^2 - a_0^2) + P_0\left(1 - \frac{1-\nu}{2}\frac{r^2 - a_0^2}{r^2}\right) + F_P$$

3.
Tangential force:
$$P_\phi = \frac{Eh}{r}u + \nu P - Eh\alpha T$$

TABLE 19-3, Part A   **Disks: General Response Expressions**   **1124**

TABLE 19-3, PART B   DISKS: LOADING FUNCTIONS

Notation

E = modulus of elasticity
ν = Poisson's ratio
G = shear modulus of elasticity
h = thickness of disk
ρ^* = mass per unit volume
Ω = angular velocity of rotation

$$\langle r - a_1 \rangle^0 = \begin{cases} 0 & \text{if } r < a_1 \\ 1 & \text{if } r \geq a_1 \end{cases}$$

	$F_u(r)$	$F_P(r)$
1. Radial force per unit circumferential length applied at $r = a_1$	$-\dfrac{P^*}{Eh}\dfrac{1-\nu^2}{2}\dfrac{r^2-a_1^2}{r}\langle r-a_1\rangle^0$	$-P^*\left(1-\dfrac{1-\nu}{2}\dfrac{r^2-a_1^2}{r^2}\right)\langle r-a_1\rangle^0$
2. Constant temperature change T_1 (independent of r)	$\dfrac{r^2-a_0^2}{2r}(1+\nu)\alpha T_1$	$-h\dfrac{r^2-a_0^2}{2r^2}E\alpha T_1$
3. Centrifugal loading due to rotation of disk at angular velocity Ω	$-\dfrac{\left(r^2-a_0^2\right)^2}{r}\dfrac{1-\nu^2}{8}\dfrac{\rho^*\Omega^2}{E}$	$-\left(r^2-a_0^2\right)\dfrac{\rho^*\Omega^2 h}{4}$ $\times\left[(1+\nu)+\dfrac{1-\nu}{2}\dfrac{r^2+a_0^2}{r^2}\right]$
4. Arbitrary temperature change $T(r)$	$\dfrac{1+\nu}{r}\alpha\displaystyle\int_{a_0}^{r}\xi T(\xi)\,d\xi$	$-\dfrac{hE\alpha}{r^2}\displaystyle\int_{a_0}^{r}\xi T(\xi)\,d\xi$

TABLE 19-3, PART C   DISKS: INITIAL PARAMETERS

Notation

E = modulus of elasticity
ν = Poisson's ratio
h = thickness of disk
a_L = radius of outer surface
$\overline{F}_u = F_{u|r=a_L}$ $\overline{F}_P = F_{P|r=a_L}$

Outer Edge / Inner Edge	Fixed	Free
1. Fixed $u_0 = 0$	$P_0 = -\dfrac{\overline{F}_u}{\nabla}$ $\nabla = \dfrac{1-\nu^2}{2Eh}\dfrac{a_L^2 - a_0^2}{a_L}$	$P_0 = -\dfrac{\overline{F}_P}{\nabla}$ $\nabla = 1 - \dfrac{1-\nu}{2}\dfrac{a_L^2 - a_0^2}{a_L^2}$
2. Free $P_0 = 0$	$u_0 = -\dfrac{\overline{F}_u}{\nabla}$ $\nabla = \dfrac{a_L}{a_0}\left(1 - \dfrac{1+\nu}{2}\dfrac{a_L^2 - a_0^2}{a_L^2}\right)$	$u_0 = -\dfrac{\overline{F}_P}{\nabla}$ $\nabla = \dfrac{Eh}{2a_0 a_L^2}\left(a_L^2 - a_0^2\right)$

TABLE 19-3, Part C   Disks: Initial Parameters   1126

TABLE 19-4 NATURAL FREQUENCIES FOR THICK SHELLS AND DISKS

Notation

E = modulus of elasticity
G = shear modulus of elasticity
ρ^* = mass per unit volume
ω_n = nth natural frequency
ν = Poisson's ratio
$\lambda = E\nu/[(1 + \nu)(1 - 2\nu)]$ = Lamé coefficient
a_0, a_L = inner and outer radii
λ_n = frequency parameter

$$\omega_n^2 = \begin{cases} \lambda_n^2 \dfrac{\lambda + 2G}{\rho^* a_L^2} & \text{(isotropic thick cylinders and spheres)} \\[3mm] \lambda_n^2 \dfrac{c_{11}}{\rho^* a_L^2} & \text{(anisotropic cylinder of case 3)} \\[3mm] \lambda_n^2 \dfrac{E}{\rho^*(1 - \nu^2) a_L^2} & \text{(disks)} \end{cases}$$

$$\alpha = \frac{a_L}{a_0}$$

$c_{11}, c_{12}, c_{13}, c_{22}, c_{23}$ = Elastic constants for anisotropic material (Eq. 4.13)
$\nu = 0.3$ for all formulas in this table.

Configuration	Frequency Parameters
1. Cylinder segment, free inner and outer surfaces.	$\lambda_1 = 0.337629 + 0.703004\alpha - 0.15434\alpha^2 + 0.017533\alpha^3$ $\lambda_2 = 3.383514 + \dfrac{3.021092}{(\alpha - 1)} + \dfrac{0.021578}{(\alpha - 1)^2} - \dfrac{0.0011806}{(\alpha - 1)^3}$ $\lambda_3 = 6.398913 + \dfrac{6.22643}{(\alpha - 1)} + \dfrac{0.010095}{(\alpha - 1)^2} - \dfrac{0.000550699}{(\alpha - 1)^3}$ $1.1 \leq \alpha \leq 2.0$
2. Cylinder without center hole, free outer surface	$\lambda_1 = 2.125748$ $\lambda_2 = 5.41389$ $\lambda_3 = 8.5870$

TABLE 19-4 (continued) **NATURAL FREQUENCIES FOR THICK SHELLS AND DISKS**

Configuration	Frequency Parameters
3. Anisotropic cylinder without center hole free outer surface 	Fundamental Frequency Parameter: $\lambda_1 = 0.4423 + 0.2049\gamma - 0.0237\gamma^2 + 0.0017\gamma^3$ $\qquad \eta = 0.1$ $\lambda_1 = 0.9421 + 0.4587\gamma - 0.05199\gamma^2 + 0.00377\gamma^3$ $\qquad \eta = 0.5$ $\lambda_1 = 1.2575 + 0.644\gamma - 0.07038\gamma^2 + 0.0049926\gamma^3$ $\qquad \eta = 1.0$ $\lambda_1 = 1.6013 + 0.8791\gamma - 0.08693\gamma^2 + 0.005831\gamma^3$ $\qquad \eta = 2.0$ $\lambda_1 = 1.9914 + 1.1814\gamma - 0.08385\gamma^2 + 0.0049185\gamma^3$ $\qquad \eta = 5.0$ $\gamma = c_{22}/c_{11} \qquad \eta = (c_{22}/c_{11})^{1/2} + c_{12}/c_{11}$
4. Spherical segment, free inner and outer surfaces 	$\lambda_1 = 0.626 + 0.9725\alpha - 0.1375\alpha^2 - 1.32453\alpha^3$ $\lambda_2 = 4.01637 + \dfrac{2.45702}{(\alpha - 1)} + \dfrac{0.210243}{(\alpha - 1)^2} - \dfrac{0.021077}{(\alpha - 1)^3}$ $\lambda_3 = 6.69786 + \dfrac{5.964807}{(\alpha - 1)} + \dfrac{0.096849}{(\alpha - 1)^2} - \dfrac{0.0096571}{(\alpha - 1)^3}$ $1.2 \le \alpha \le 2.0$
5. Spherical segment without center hole, free outer surface 	$\lambda_1 = 2.67021$ $\lambda_2 = 6.091989$ $\lambda_3 = 9.300894$
6. Disk Ring, free inner and outer boundries 	$\lambda_1 = 0.34598 + 0.7691\alpha - 0.1804\alpha^2 + 0.020096\alpha^3$ $\lambda_2 = 3.5347 + \dfrac{2.8678}{(\alpha - 1)} + \dfrac{0.07512}{(\alpha - 1)^2} - \dfrac{0.0069235}{(\alpha - 1)^3}$ $\lambda_3 = 6.4717 + \dfrac{6.15398}{(\alpha - 1)} + \dfrac{0.03511}{(\alpha - 1)^2} - \dfrac{0.003218}{(\alpha - 1)^3}$ $1.2 \le \alpha \le 2.0$

TABLE 19-4   **Frequencies for Thick Shells and Disks**   **1128**

TABLE 19-4 (continued)   NATURAL FREQUENCIES FOR THICK SHELLS AND DISKS

Configuration	Frequency Parameters
7. Disk without center hole, free outer boundary	$\lambda_1 = 2.04885$ $\lambda_2 = 5.38936$ $\lambda_3 = 8.57816$
8. Cylinder segment, fixed inner surface and free outer surface	$\lambda_1 = 1.79138 + \dfrac{1.45157}{(\alpha - 1)} + \dfrac{0.020484}{(\alpha - 1)^2} - \dfrac{0.0010994}{(\alpha - 1)^3}$ $\lambda_2 = 4.82918 + \dfrac{4.64203}{(\alpha - 1)} + \dfrac{0.013076}{(\alpha - 1)^2} - \dfrac{0.00072803}{(\alpha - 1)^3}$ $\lambda_3 = 7.92344 + \dfrac{7.81376}{(\alpha - 1)} + \dfrac{0.0073076}{(\alpha - 1)^2} - \dfrac{0.0040230}{(\alpha - 1)^3}$
9. Cylinder segment, free inner surface and fixed outer surface	$\lambda_1 = 2.16852 + \dfrac{1.28071}{(\alpha - 1)} + \dfrac{0.052985}{(\alpha - 1)^2} - \dfrac{0.0029207}{(\alpha - 1)^3}$ $\lambda_2 = 4.89249 + \dfrac{4.62673}{(\alpha - 1)} + \dfrac{0.015628}{(\alpha - 1)^2} - \dfrac{0.00086171}{(\alpha - 1)^3}$ $\lambda_3 = 7.95983 + \dfrac{7.80404}{(\alpha - 1)} + \dfrac{0.0090911}{(\alpha - 1)^2} - \dfrac{0.00050073}{(\alpha - 1)^3}$
10. Spherical segment, fixed inner surface and free outer surface	$\lambda_1 = 2.15852 + \dfrac{1.27428}{(\alpha - 1)} + \dfrac{0.050321}{(\alpha - 1)^2} - \dfrac{0.0026847}{(\alpha - 1)^3}$ $\lambda_2 = 5.015518 + \dfrac{4.54511}{(\alpha - 1)} + \dfrac{0.030062}{(\alpha - 1)^2} - \dfrac{0.0016459}{(\alpha - 1)^3}$ $\lambda_3 = 8.044102 + \dfrac{7.74788}{(\alpha - 1)} + \dfrac{0.019202}{(\alpha - 1)^2} - \dfrac{0.0010549}{(\alpha - 1)^3}$

TABLE 19-4 (continued) NATURAL FREQUENCIES FOR THICK SHELLS AND DISKS

Configuration	Frequency Parameters
11. Spherical segment, free inner surface and fixed outer surface	$\lambda_1 = 7.26960 - \dfrac{6.84825}{(\alpha - 1)}$ $+ \dfrac{3.991179}{(\alpha - 1)^2} - \dfrac{0.68994}{(\alpha - 1)^3} + \dfrac{0.036929}{(\alpha - 1)^4}$ $\lambda_2 = 5.0054713 + \dfrac{4.47794}{(\alpha - 1)} + \dfrac{0.056356}{(\alpha - 1)^2} - \dfrac{0.0035934}{(\alpha - 1)^3}$ $\lambda_3 = 8.070949 + \dfrac{7.60964}{(\alpha - 1)} + \dfrac{0.062880}{(\alpha - 1)^2} - \dfrac{0.0040676}{(\alpha - 1)^3}$
12. Disk ring, fixed inner boundary and free outer boundary	$\lambda_1 = 1.70971 + \dfrac{1.45025}{(\alpha - 1)} + \dfrac{0.020784}{(\alpha - 1)^2} - \dfrac{0.0011170}{(\alpha - 1)^3}$ $\lambda_2 = 4.79584 + \dfrac{4.64898}{(\alpha - 1)} + \dfrac{0.011415}{(\alpha - 1)^2} - \dfrac{0.00062540}{(\alpha - 1)^3}$ $\lambda_3 = 7.90695 + \dfrac{7.81390}{(\alpha - 1)} + \dfrac{0.0072626}{(\alpha - 1)^2} - \dfrac{0.00039914}{(\alpha - 1)^3}$
13. Disk ring, free inner boundary and fixed outer boundary	$\lambda_1 = 2.33085 + \dfrac{1.24567}{(\alpha - 1)} + \dfrac{0.058895}{(\alpha - 1)^2} - \dfrac{0.0032348}{(\alpha - 1)^3}$ $\lambda_2 = 4.96403 + \dfrac{4.60411}{(\alpha - 1)} + \dfrac{0.019722}{(\alpha - 1)^2} - \dfrac{0.0010864}{(\alpha - 1)^3}$ $\lambda_3 = 8.0032775 + \dfrac{7.79008}{(\alpha - 1)} + \dfrac{0.011628}{(\alpha - 1)^2} - \dfrac{0.00064035}{(\alpha - 1)^3}$
14. Disk without center hole, fixed outer boundary	$\lambda_1 = 3.8317$ $\lambda_2 = 7.0156$ $\lambda_3 = 10.1735$

TABLE 19-4 **Frequencies for Thick Shells and Disks** **1130**

TABLE 19-5   TRANSFER AND STIFFNESS MATRICES FOR CYLINDERS

Notation

E = modulus of elasticity

ν = Poisson's ratio

ρ^* = mass per unit volume

$T(r)$ = arbitrary temperature change; in expressions for $\overline{F}_u$ and $\overline{F}_\sigma$ set $T(\xi) = 0$ if only constant temperature change is present

$p_r(r)$ = arbitrary loading intensity in r direction (F/L^3)

G = shear modulus of elasticity

λ = Lamé constant

T_1 = constant temperature change

u_0, u_a, u_b = radial displacements at $r = 0, a, b$

$\sigma_0, \sigma_a, \sigma_b$ = radial stresses at $r = 0, a, b$ (F/L^2)

$p_0, p_a = 2\pi a \sigma_a, \ p_b = 2\pi b \sigma_b$ = radial forces (stress resultants) at $r = 0, a, b$ (F/L)

p_a^0, p_b^0 = loading components for stiffness equation at $r = a, b$ (F/L)

$\overline{F}_u = F_{u|r=b} \qquad \overline{F}_\sigma = F_{\sigma|r=b}$

Ω = angular velocity of rotation that leads to centrifugal loading force

$$\beta^2 = \begin{cases} \rho^*\Omega^2/(\lambda + 2G) & \text{isotropic material} \\ \rho^*\Omega^2/c_{11} & \text{anisotropic} \\ & \text{material (Eq. 4.13)} \end{cases}$$

Replace Ω by ω if the natural frequencies ω_i are of interest.

$$e_1 = e_3(a)J_\gamma(\beta a) - e_2(a)Y_\gamma(\beta a)$$

$$e_2(r) = \begin{cases} \dfrac{\lambda + 2G}{r}\left[\dfrac{2(\lambda + G)}{\lambda + 2G} J_1(\beta r) - \beta r J_2(\beta r) \right] & \text{isotropic material} \\[3mm] \dfrac{c_{11}}{r}\left[\left(\gamma + \dfrac{c_{12}}{c_{11}}\right) J_\gamma(\beta r) - \beta r J_{\gamma+1}(\beta r) \right] & \text{anisotropic material} \end{cases}$$

$$e_3(r) = \begin{cases} \dfrac{\lambda + 2G}{r}\left[\dfrac{2(\lambda + G)}{\lambda + 2G} Y_1(\beta r) - \beta r Y_2(\beta r) \right] & \text{isotropic material} \\[3mm] \dfrac{c_{11}}{r}\left[\left(\gamma + \dfrac{c_{12}}{c_{11}}\right) Y_\gamma(\beta r) - \beta r Y_{\gamma+1}(\beta r) \right] & \text{anisotropic material} \end{cases}$$

$J_\gamma(\beta r)$ and $Y_\gamma(\beta r)$ are Bessel functions of order γ of the first and second kind, respectively

$$\gamma = \begin{cases} 1 & \text{isotropic material} \\ (c_{22}/c_{11})^{1/2} & \text{anisotropic material (Eq. 4.13)} \end{cases}$$

TABLE 19-5 (continued)   TRANSFER AND STIFFNESS MATRICES FOR CYLINDERS

Matrices

Case	Transfer Matrices	Stiffness Matrices

Transfer Matrices

$z_b = U^i z_a$

$z_b = \begin{bmatrix} u_b & \sigma_b & 1 \end{bmatrix}^T$

$z_a = \begin{bmatrix} u_a & \sigma_a & 1 \end{bmatrix}^T$

$U^i = \begin{bmatrix} U_{uu} & U_{u\sigma} & F_u \\ U_{\sigma u} & U_{\sigma\sigma} & F_\sigma \\ 0 & 0 & 1 \end{bmatrix}$

Stiffness Matrices

$p^i = k^i v^i - \bar{p}^i$

$p^i = \begin{bmatrix} p_a & p_b \end{bmatrix}^T$

$v^i = \begin{bmatrix} u_a & u_b \end{bmatrix}^T$

$\bar{p}^i = \begin{bmatrix} p_a^0 & p_b^0 \end{bmatrix}^T$

$k^i = \begin{bmatrix} k_{11} & k_{12} \\ k_{21} & k_{22} \end{bmatrix}$

Element i

Case 1. Massless cylinder, with center hole

$$U^i = \begin{bmatrix} \dfrac{1}{1-\nu}\left(\dfrac{G\nu b}{\lambda a} + \dfrac{a}{2b}\right) & \dfrac{\nu}{2(1-\nu)\lambda}\,\dfrac{b^2-a^2}{b} & F_u \\[2ex] \dfrac{G}{1-\nu}\,\dfrac{b^2-a^2}{ab^2} & \dfrac{1}{1-\nu}\left[\dfrac{1}{2} + \dfrac{G\nu}{\lambda}\left(\dfrac{a}{b}\right)^2\right] & F_\sigma \\[2ex] 0 & 0 & 1 \end{bmatrix}\begin{bmatrix} u_a \\ \sigma_a \\ 1 \end{bmatrix}$$

$k_{11} = 2\pi(2G\nu\beta_0^2 + \lambda)/H_0$

$k_{12} = k_{21} = -4\pi\lambda(1-\nu)\beta_0/H_0$

$k_{22} = 2\pi(2G\nu + \lambda\beta_0^2)/H_0$

$H_0 = \nu(\beta_0^2 - 1)$

$\beta_0 = b/a$

$p_a^0 = k_{12}\bar{F}_u$

$p_b^0 = -2\pi b\bar{F}_\sigma + k_{22}\bar{F}_u$

TABLE 19-5   Matrices for Cylinders   1132

Matrices for Cylinders   TABLE 19-5

2. Cylinder with center hole and mass	$U_{uu} = \dfrac{1}{e_1}\left[e_3(a)J_\gamma(\beta b) - e_2(a)Y_\gamma(\beta b)\right]$ $U_{u\sigma} = \dfrac{1}{e_1}\left[J_\gamma(\beta a)Y_\gamma(\beta b) - Y_\gamma(\beta a)J_\gamma(\beta b)\right]$ $U_{\sigma u} = \dfrac{1}{e_1}\left[e_3(a)e_2(b) - e_2(a)e_3(b)\right]$ $U_{\sigma\sigma} = \dfrac{1}{e_1}\left[J_\gamma(\beta a)e_3(b) - Y_\gamma(\beta a)e_2(b)\right]$	$k_{11} = 2\pi a\left[e_3(a)J_\gamma(\beta b) - e_2(a)Y_\gamma(\beta b)\right]/H_2$ $k_{12} = k_{21} = -2\pi ae_1/H_2$ $k_{22} = 2\pi b\left[e_3(b)J_\gamma(\beta a) - e_2(b)Y_\gamma(\beta a)\right]/H_2$ $H_2 = J_\gamma(\beta a)Y_\gamma(\beta b) - Y_\gamma(\beta a)J_\gamma(\beta b)$ $p_a^0 = k_{12}F_u \qquad p_b^0 = -2\pi bF_\sigma + k_{22}F_u$
3. Massless cylinder without center hole	$z_b = U^i z_0$ $z_0 = \begin{bmatrix} u_0 & \sigma_0 & 1 \end{bmatrix}^T$ $\begin{bmatrix} 0 & \dfrac{\nu b}{\lambda} & F_u \\ 0 & 1 & F_\sigma \\ 0 & 0 & 1 \end{bmatrix}\begin{bmatrix} u_0 \\ \sigma_0 \\ 1 \end{bmatrix}$ $\qquad\quad U^i \qquad\quad z_0$	$p^i = k^i v^i - \bar{p}^i \quad p^i = \begin{bmatrix} p_0 & p_b \end{bmatrix}^T \quad \bar{p}^i = \begin{bmatrix} 0 & p_b^0 \end{bmatrix}^T$ $v^i = \begin{bmatrix} u_0 & u_b \end{bmatrix}^T$ $k^i = \begin{bmatrix} 0 & 0 \\ 0 & 2\pi\lambda/\nu \end{bmatrix}$ $p_b^0 = -2\pi bF_\sigma + (2\pi\lambda/\nu)F_u$
4. Cylinder without center hole including mass	$z_b = U^i z_0$ $z_0 = \begin{bmatrix} u_0 & \sigma_0 & 1 \end{bmatrix}^T$ $\begin{bmatrix} \dfrac{\lambda+2G}{\beta(\lambda+G)b}\left[\dfrac{2(\lambda+G)}{\lambda+2G}J_1(\beta b) - \beta bI_2(\beta b)\right] & \dfrac{J_1(\beta b)}{(\lambda+G)\beta} & F_u \\ 0 & 1 & F_\sigma \\ 0 & 0 & 1 \end{bmatrix}\begin{bmatrix} u_0 \\ \sigma_0 \\ 1 \end{bmatrix}$ $\qquad\qquad\qquad U^i \qquad\qquad z_0$	$p^i = k^i v^i - \bar{p}^i \quad p^i = \begin{bmatrix} p_0 & p_b \end{bmatrix}^T \quad \bar{p}^i = \begin{bmatrix} 0 & p_b^0 \end{bmatrix}^T$ $v^i = \begin{bmatrix} u_0 & u_b \end{bmatrix}^T$ $k^i = \begin{bmatrix} 0 & 0 \\ 0 & k_{22} \end{bmatrix}$ $k_{22} = \dfrac{2\pi(\lambda+2G)}{J_1(\beta b)}$ $\qquad \times\left[\dfrac{2(\lambda+G)}{\lambda+2G}J_1(\beta b) - \beta bI_2(\beta b)\right]$ $p_b^0 = -2\pi bF_\sigma + k_{22}F_u$

Loading Vectors

Case	Loading Vectors
1. Massless cylinder with center hole	$F_u = \dfrac{b^2 - a^2}{2b}\dfrac{1+\nu}{1-\nu}\alpha T_1 + \dfrac{1+\nu}{1-\nu}\dfrac{\alpha}{b}\displaystyle\int_a^b \xi T(\xi)\,d\xi$ $\qquad -\dfrac{\rho^*\Omega^2}{8}\dfrac{(b^2-a^2)^2}{b}\dfrac{(1+\nu)(1-2\nu)}{E(1-\nu)}$ $\qquad -\dfrac{(1+\nu)(1-2\nu)}{E(1-\nu)}\dfrac{1}{b}\displaystyle\int_a^b \eta\left[\int_a^\eta p_r(\xi)\,d\xi\,d\eta\right]$ $F_\sigma = -\dfrac{b^2-a^2}{2b^2}\dfrac{\alpha E}{1-\nu}T_1 - \dfrac{E}{1-\nu}\dfrac{\alpha}{b^2}\displaystyle\int_a^b \xi T(\xi)\,d\xi$ $\qquad -(b^2-a^2)\dfrac{\rho^*\Omega^2}{4}\left(2-\dfrac{1-2\nu}{1-\nu}\dfrac{b^2-a^2}{2b^2}\right)$ $\qquad +\dfrac{1-2\nu}{1-\nu}\dfrac{1}{b^2}\displaystyle\int_a^b \eta\left[\int_a^\eta p_r(\xi)\,d\xi\right]d\eta - \int_a^b p_r(\xi)\,d\xi$
2. Cylinder with center hole and mass	$F_u = -\displaystyle\int_a^b p_r(\xi)U_{u\sigma}(\xi,a)\,d\xi$ $F_\sigma = -\displaystyle\int_a^b p_r(\xi)U_{\sigma\sigma}(\xi,a)\,d\xi$ $U_{u\sigma}(\xi,a)$ and $U_{\sigma\sigma}(\xi,a)$ are obtained by replacing b with ξ in $U_{u\sigma}$ and $U_{\sigma\sigma}$ of case 2, page 1133.

TABLE 19-5 (continued)   TRANSFER AND STIFFNESS MATRICES FOR CYLINDERS

Case	Loading Vectors
3. Massless cylinder without center hole	$F_u = (1 + v)b\alpha T_1 + \dfrac{1 + v}{1 - v}\dfrac{\alpha}{b}\displaystyle\int_0^b \xi T(\xi)\, d\xi$ $+ \dfrac{\alpha b E v T_{\|r=0}}{2\lambda(1 - v)} - \dfrac{\rho^*\Omega^2 b^3}{8}\dfrac{(1 + v)(1 + 2v)}{E(1 - v)}$ $- \dfrac{(1 + v)(1 - 2v)}{E(1 - v)b}\displaystyle\int_0^b \eta\left[\int_0^\eta p_r(\xi)\, d\xi\right] d\eta$ $F_\sigma = -\dfrac{E}{1 - v}\dfrac{\alpha}{b^2}\displaystyle\int_0^b \xi T(\xi)\, d\xi$ $+ \dfrac{\alpha E T_{\|r=0}}{2(1 - v)} - \dfrac{\rho^*\Omega^2 b^2}{8}\dfrac{(3 - 2v)}{1 - v}$ $+ \dfrac{1 - 2v}{1 - v}\dfrac{1}{b^2}\displaystyle\int_0^b \eta\left[\int_0^\eta p_r(\xi)\, d\xi\right] d\eta - \int_0^b p_r(\xi)\, d\xi$
4. Cylinder without center hole including mass	$F_u = -\displaystyle\int_0^b p_r(\xi) U_{u\sigma}(\xi)\, d\xi$ $F_\sigma = -\displaystyle\int_0^b p_r(\xi) U_{\sigma\sigma}(\xi)\, d\xi$ $U_{u\sigma}(\xi)$ and $U_{\sigma\sigma}(\xi)$ are obtained by replacing b with ξ in $U_{u\sigma}$ and $U_{\sigma\sigma}$ of case 4, page 1133.

TABLE 19-6   TRANSFER AND STIFFNESS MATRICES FOR SPHERES

Notation

E = modulus of elasticity

ν = Poisson's ratio

ρ^* = mass per unit volume

$T(r)$ = arbitrary temperature change; in expressions for $\bar{F}_u$ and $\bar{F}_\sigma$ set $T(\xi) = 0$ if only a constant temperature change is present.

$p_r(r)$ = arbitrary loading intensity in r direction (F/L^3)

G = shear modulus of elasticity

λ = Lamé constant

T_1 = constant temperature change

Ω = angular velocity of rotation that leads to centrifugal loading force

u_0, u_a, u_b = radial displacements at $r = 0, a, b$

$\sigma_0, \sigma_a, \sigma_b$ = radial stresses at $r = 0, a, b$

$p_0, p_a = 4\pi a^2 \sigma_a$, $p_b = 4\pi b^2 \sigma_b$ = total radial forces at $r = 0, a, b$

p_a^0, p_b^0 = loading components for stiffness equation at $r = a, b$

$$
\beta^2 = \begin{cases} \rho^*\Omega^2/(\lambda + 2G) & \text{isotropic material} \\ \rho^*\Omega^2/c_{11} & \begin{array}{c}\text{anisotropic}\\\text{material (Eq. 4.13)}\end{array} \end{cases} \quad \begin{array}{l}\text{Replace } \Omega \text{ by } \omega \text{ if the}\\ \text{natural frequencies } \omega_i\\ \text{are of interest.}\end{array}
$$

$$
e_1 = e_3(a)\frac{J_\gamma(\beta a)}{\sqrt{\beta a}} - e_2(a)\frac{Y_\gamma(\beta a)}{\sqrt{\beta a}}
$$

$$
e_2(r) = \begin{cases} \dfrac{\beta(\lambda + 2G)}{(\beta r)^{3/2}}\left[\dfrac{2(\lambda + G)}{\lambda + 2G}J_{3/2}(\beta r) - \beta r J_{5/2}(\beta r)\right] & \text{isotropic material} \\[4mm] \dfrac{\beta c_{11}}{(\beta r)^{3/2}}\left[\left(\gamma - \dfrac{1}{2} + \dfrac{2c_{12}}{c_{11}}\right)J_\gamma(\beta r) - \beta r J_{\gamma+1}(\beta r)\right] & \begin{array}{c}\text{anisotropic}\\\text{material}\end{array} \end{cases}
$$

$$
e_3(r) = \begin{cases} \dfrac{\beta(\lambda + 2G)}{(\beta r)^{3/2}}\left[\dfrac{2(\lambda + G)}{\lambda + 2G}Y_{3/2}(\beta r) - \beta r Y_{5/2}(\beta r)\right] & \text{isotropic material} \\[4mm] \dfrac{\beta c_{11}}{(\beta r)^{3/2}}\left[\left(\gamma - \dfrac{1}{2} + \dfrac{2c_{12}}{c_{11}}\right)Y_\gamma(\beta r) - \beta r Y_{\gamma+1}(\beta r)\right] & \begin{array}{c}\text{anisotropic}\\\text{material}\end{array} \end{cases}
$$

$J_\gamma(\beta r)$ and $Y_\gamma(\beta r)$ are Bessel functions of order γ of the first and second kind, respectively.

$$
\gamma = \begin{cases} \dfrac{3}{2} & \text{isotropic material} \\[4mm] \dfrac{1}{2}\left[\dfrac{8(c_{22} + c_{23} - c_{12})}{c_{11}} + 1\right]^{1/2} & \text{anisotropic material (Eq. 4.13)} \end{cases}
$$

TABLE 19-6 | **Matrices for Spheres** | **1136**

Matrices

Case	Transfer Matrices	Stiffness Matrices

Element i

Top of Transfer Matrices column:

$$z_b = U^i z_a$$
$$z_b = [u_b \quad \sigma_b \quad 1]^T$$
$$z_a = [u_a \quad \sigma_a \quad 1]^T$$
$$U^i = \begin{bmatrix} U_{uu} & U_{u\sigma} & F_u \\ U_{\sigma u} & U_{\sigma\sigma} & F_\sigma \\ 0 & 0 & 1 \end{bmatrix}$$

Top of Stiffness Matrices column:

$$p^i = k^i v^i - \bar{p}^i$$
$$p^i = [p_a \quad p_b]^T$$
$$v^i = [u_a \quad u_b]^T$$
$$\bar{p}^i = [p_a^0 \quad p_b^0]^T$$
$$k^i = \begin{bmatrix} k_{11} & k_{12} \\ k_{21} & k_{22} \end{bmatrix}$$

1. Massless spherical segment

Transfer Matrices:

$$\begin{bmatrix} \dfrac{4Gb^3+(3\lambda+2G)a^3}{3(\lambda+2G)ab^2} & \dfrac{a}{3(\lambda+2G)}\left(\dfrac{b}{a}-\dfrac{a^2}{b^2}\right) & F_u \\[2mm] \dfrac{4G(3\lambda+2G)(b^3-a^3)}{3ab^3(\lambda+2G)} & \dfrac{3\lambda+2G+4Ga^3/b^3}{3(\lambda+2G)} & F_\sigma \\[2mm] 0 & 0 & 1 \end{bmatrix} \begin{bmatrix} u_a \\ \sigma_a \\ 1 \end{bmatrix}$$

$$U^i \qquad z_a$$

Stiffness Matrices:

$$k_{11} = 4\pi a[3\lambda + 2(2\beta_0^3+1)G]/H_0$$
$$k_{21}=k_{12} = -12\pi a\beta_0^2(\lambda+2G)/H_0$$
$$k_{22} = 4\pi b[4G+\beta_0^3(2G+3\lambda)]/H_0$$
$$p_a^0 = k_{12}F_u$$
$$p_b^0 = -4\pi b^2 F_\sigma + k_{22}F_u$$
$$\beta_0 = b/a$$
$$H_0 = \beta_0^3 - 1$$

2. Spherical segment with mass (with center hole)

Transfer Matrices:

$$U_{uu} = \frac{1}{e_1\sqrt{\beta b}}\left[e_3(a)J_\gamma(\beta b) - e_2(a)Y_\gamma(\beta b)\right]$$
$$U_{u\sigma} = \frac{1}{e_1\beta\sqrt{ab}}\left[J_\gamma(\beta a)Y_\gamma(\beta b) - Y_\gamma(\beta a)J_\gamma(\beta b)\right]$$
$$U_{\sigma u} = \frac{1}{e_1}\left[e_3(a)e_2(b) - e_2(a)e_3(b)\right]$$
$$U_{\sigma\sigma} = \frac{1}{e_1\sqrt{\beta a}}\left[e_3(b)J_\gamma(\beta a) - e_2(b)Y_\gamma(\beta a)\right]$$

Stiffness Matrices:

$$k_{11} = 4\pi a^2\sqrt{\beta a}\left[e_3(a)J_\gamma(\beta b) - e_2(a)Y_\gamma(\beta b)\right]/H_3$$
$$k_{12} = k_{21} = -4\pi a^2 e_1\beta\sqrt{ba}/H_3$$
$$k_{22} = 4\pi b^2\sqrt{\beta b}\left[e_3(b)J_\gamma(\beta a) - e_2(b)Y_\gamma(\beta a)\right]/H_3$$
$$H_3 = J_\gamma(\beta a)Y_\gamma(\beta b) - Y_\gamma(\beta a)J_\gamma(\beta b)$$
$$p_a^0 = k_{12}F_u \qquad p_b^0 = -4\pi b^2 F_\sigma + k_{22}F_u$$

Matrices for Spheres TABLE 19-6

TABLE 19-6 (continued) TRANSFER AND STIFFNESS MATRICES FOR SPHERES

3. Massless spherical segment without center hole including mass 	$z_b = U^i z_0$ $z_0 = [u_0 \quad \sigma_0 \quad 1]^T$ $$\underbrace{\begin{bmatrix} 0 & \dfrac{b}{3\lambda + 2G} & F_u \\[2mm] 0 & 1 & F_\sigma \\[1mm] 0 & 0 & 1 \end{bmatrix}}_{U^i} \begin{bmatrix} u_0 \\ \sigma_0 \\ 1 \end{bmatrix} \; z_0$$	$p^i = k^i v^i - \bar{p}^i$ $p^i = [p_0 \quad p_b]^T \qquad v^i = [u_0 \quad u_b]^T$ $\bar{p}^i = [0 \quad p_b^0]^T$ $k^i = \begin{bmatrix} 0 & 0 \\ 0 & k_{22} \end{bmatrix}$ $k_{22} = 4\pi b(3\lambda + 2G)$ $p_b^0 = -4\pi b^2 F_\sigma + k_{22} F_u$
4. Spherical segment without center hole including mass 	$z_b = U^i z_0$ $z_0 = [u_0 \quad \sigma_0 \quad 1]^T$ $U_{u\sigma} = \dfrac{3\sqrt{\pi}}{\sqrt{2}\,(3\lambda + 2G)} \dfrac{J_{3/2}(\beta b)}{\beta^{3/2} b^{1/2}}$ $U_{\sigma\sigma} = \dfrac{3(\lambda + 2G)\sqrt{\pi}}{\sqrt{2}\,(3\lambda + 2G)(\beta b)^{3/2}}$ $\qquad \times \left[\dfrac{3\lambda + 2G}{\lambda + 2G} J_{3/2}(\beta b) - \beta b J_{5/2}(\beta b) \right]$ $U_{uu} = U_{\sigma u} = 0$	$p^i = k^i v^i - \bar{p}^i$ $p^i = [p_0 \quad p_b]^T \qquad v^i = [u_0 \quad u_b]^T$ $\bar{p}^i = [0 \quad p_b^0]^T$ $k^i = \begin{bmatrix} 0 & 0 \\ 0 & k_{22} \end{bmatrix}$ $k_{22} = 4\pi b \left[\dfrac{3\lambda + 2G}{b} - \dfrac{(\lambda + 2G)\beta b}{J_{3/2}(\beta b)} J_{5/2}(\beta b) \right]$ $p_b^0 = -4\pi b^2 F_\sigma + k_{22} F_u$

TABLE 19-6 Matrices for Spheres 1138

TABLE 19-6 (continued)   TRANSFER AND STIFFNESS MATRICES FOR SPHERES

Loading Vectors

Case	Loading Vectors
1. Massless spherical segment	$F_u = \dfrac{3\lambda + 2G}{3(\lambda + 2G)} \dfrac{(b^3 - a^3)}{b^2} \alpha T_1$ $+ \dfrac{3\lambda + 2G}{(\lambda + 2G)} \dfrac{\alpha}{b^2} \displaystyle\int_a^r \xi^2 T(\xi)\, d\xi$ $- \dfrac{1}{b^2(\lambda + 2G)} \displaystyle\int_a^b \eta^2 \left[\int_a^\eta p_r(\xi)\, d\xi \right] d\eta$ $F_\sigma = -\dfrac{4G(3\lambda + 2G)}{3(\lambda + 2G)} \dfrac{b^3 - a^3}{b^3} \alpha T_1$ $- \dfrac{4G(3\lambda + 2G)}{\lambda + 2G} \dfrac{\alpha}{b^3} \displaystyle\int_a^b \xi^2 T(\xi)\, d\xi$ $+ \dfrac{4G}{b^3(\lambda + 2G)} \displaystyle\int_a^b \eta^2 \left[\int_a^\eta p_r(\xi)\, d\xi \right] d\eta$ $- \displaystyle\int_a^b p_r(\xi)\, d\xi$
2. Spherical segment with mass (with center hole)	$F_u = -\displaystyle\int_a^b p_r(\xi) U_{u\sigma}(\xi, a)\, d\xi$ $F_\sigma = -\displaystyle\int_a^b p_r(\xi) U_{\sigma\sigma}(\xi, a)\, d\xi$ $U_{u\sigma}(\xi, a)$ and $U_{\sigma\sigma}(\xi, a)$ are obtained by replacing b with ξ in $U_{u\sigma}$ and $U_{\sigma\sigma}$ of case 2, page 1137.

TABLE 19-6 (continued)   TRANSFER AND STIFFNESS MATRICES FOR SPHERES

Case	Loading Vectors
3. Massless spherical segment without center hole including mass	$F_u = b\alpha T_1 + \dfrac{3\lambda + 2G}{\lambda + 2G} \dfrac{\alpha}{b^2} \displaystyle\int_0^b \xi^2 T(\xi)\, d\xi$ $+ \dfrac{4G\alpha b}{3(\lambda + 2G)} T_{r=0}$ $+ \dfrac{1}{3(\lambda + 2G)} \left(\dfrac{1}{b^2} \displaystyle\int_0^b p_r \xi^3\, d\xi - b \displaystyle\int_0^b p_r\, d\xi \right)$ $F_\sigma = -\dfrac{4G(3\lambda + 2G)}{\lambda + 2G} \dfrac{\alpha}{b^3} \displaystyle\int_0^b \xi^2 T(\xi)\, d\xi$ $+ \dfrac{4G(3\lambda + 2G)\alpha}{3(\lambda + 2G)} T_{r=0}$ $- \dfrac{1}{3(\lambda + 2G)} \left[\dfrac{4G}{b^3} \displaystyle\int_0^b p_r \xi^3\, d\xi \right.$ $\left. + (3\lambda + 2G) \displaystyle\int_0^b p_r\, d\xi \right]$
4. Spherical segment without center hole including mass	$F_u = -\displaystyle\int_0^b p_r(\xi) U_{u\sigma}(\xi)\, d\xi$ $F_\sigma = -\displaystyle\int_0^b p_r(\xi) U_{\sigma\sigma}(\xi)\, d\xi$ $U_{u\sigma}(\xi)$ and $U_{\sigma\sigma}(\xi)$ are obtained by replacing b with ξ in $U_{u\sigma}$ and $U_{\sigma\sigma}$ of case 4, page 1138.

TABLE 19-6 **Matrices for Spheres**   **1140**

TABLE 19-7 **TRANSFER AND STIFFNESS MATRICES FOR DISKS**

Notation

E = modulus of elasticity

ν = Poisson's ratio

ρ^* = mass per unit volume

$T(r)$ = arbitrary temperature change; in expressions for $\bar{F}_u$ and $\bar{F}_P$ set $T(\xi) = 0$ if only a constant temperature change is present.

$p_r(r)$ = arbitrary loading intensity in r direction (F/L^2)

G = shear modulus of elasticity

λ = Lamé constant

T_1 = constant temperature change

Ω = angular velocity of rotation that leads to centrifugal loading force

h = thickness

u_0, u_a, u_b = radial displacements at $r = 0, a, b$

P_0, P_a, P_b = radial forces per unit length at $r = 0, a, b$

$P_0, P_a = 2\pi a P_a, \ P_b = 2\pi b P_b$ = total radial forces at $r = 0, a, b$

P_a^0, P_b^0 = loading components for stiffness equation at $r = a, b$

$\beta^2 = \rho^* \Omega^2 (1 - \nu^2)/E$

$e_1 = e_3(a) J_1(\beta a) - e_2(a) Y_1(\beta a)$

$e_2(r) = \dfrac{1}{r} \dfrac{Eh}{1 - \nu^2} [(1 + \nu) J_1(\beta r) - \beta r J_2(\beta r)]$

$e_3(r) = \dfrac{1}{r} \dfrac{Eh}{1 - \nu^2} [(1 + \nu) Y_1(\beta r) - \beta r Y_2(\beta r)]$

$J_1(\beta r)$ and $J_2(\beta r)$ are Bessel functions of the first kind of order 1 and 2, respectively.

$Y_1(\beta r)$ and $Y_2(\beta r)$ are Bessel functions of the second kind of order 1 and 2, respectively.

TABLE 19-7 (continued) **TRANSFER AND STIFFNESS MATRICES FOR DISKS**

Matrices

Case	Tranfer matrices	Stiffness Matrices

Tranfer matrices

$$\mathbf{z}_b = \mathbf{U}^i \mathbf{z}_b$$

$$\mathbf{z}_b = [u_b \quad P_b \quad 1]^T$$

$$\mathbf{z}_a = [u_a \quad P_a \quad 1]^T$$

$$\mathbf{U}^i = \begin{bmatrix} U_{uu} & U_{uP} & F_u \\ U_{Pu} & U_{PP} & F_P \\ 0 & 0 & 1 \end{bmatrix}$$

Stiffness Matrices

$$\mathbf{p}^i = \mathbf{k}^i \mathbf{v}^i - \overline{\mathbf{p}}^i$$

$$\mathbf{p}^i = [p_a \quad p_b]^T$$

$$\mathbf{v}^i = [u_a \quad u_b]^T$$

$$\overline{\mathbf{p}}^i = [p_a^0 \quad p_b^0]^T$$

$$\mathbf{k}^i = \begin{bmatrix} k_{11} & k_{12} \\ k_{21} & k_{22} \end{bmatrix}$$

Element i

$k_{11} = 2\pi Eh[\beta_0^2(1-\nu) + (1+\nu)]/H$

$k_{12} = k_{21} = -4\pi Eh\beta_0/H$

$k_{22} = 2\pi[\beta_0^2(1+\nu) + (1-\nu)]/H$

$H = (1-\nu^2)(\beta_0^2 - 1)$

$p_a^0 = k_{12}F_u$

$p_b^0 = -2\pi b F_P + k_{22}F_u$

$\beta_0 = b/a$

Case 1.
Annular element without mass

$$\begin{bmatrix} \dfrac{b}{a}\left[1 - \dfrac{1+\nu}{2}\dfrac{(b^2-a^2)}{b^2}\right] & \dfrac{1}{Eh}\dfrac{1-\nu^2}{2}\dfrac{(b^2-a^2)}{b} & \overline{F}_u \\ \dfrac{Eh}{2ab^2}(b^2-a^2) & 1 - \dfrac{1-\nu}{2}\dfrac{(b^2-a^2)}{b^2} & \overline{F}_P \\ 0 & 0 & 1 \end{bmatrix}$$

TABLE 19-7 **Matrices for Disks** 1142

2.

Annular element with mass

$$U_{uu} = \frac{1}{e_1}\left[e_3(a)J_1(\beta b) - e_2(a)Y_1(\beta b)\right]$$

$$U_{uP} = \frac{1}{e_1}\left[J_1(\beta a)Y_1(\beta b) - Y_1(\beta a)J_1(\beta b)\right]$$

$$U_{Pu} = \frac{1}{e_1}\left[e_3(a)e_2(b) - e_2(a)e_3(b)\right]$$

$$U_{PP} = \frac{1}{e_1}\left[J_1(\beta a)e_3(b) - Y_1(\beta a)e_2(b)\right]$$

$$k_{11} = 2\pi a\left[e_3(a)J_1(\beta b) - e_2(a)Y_1(\beta b)\right]/H_4$$
$$k_{12} = k_{21} = -2\pi ae_1/H_4$$
$$k_{22} = 2\pi b\left[J_1(\beta a)e_3(b) - Y_1(\beta a)e_2(b)\right]/H_4$$
$$H_4 = J_1(\beta a)Y_1(\beta b) - Y_1(\beta a)J_1(\beta b)$$
$$p_a^0 = k_{12}F_u \qquad p_b^0 = -2\pi bF_P + k_{22}F_u$$

3.

Disk of variable thickness

$h = h_k/r^n$
h_k is a reference thickness
$h_k = h_a a^n$

Hyperbolic profile

$$n_1 = \frac{n}{2} + \sqrt{\frac{n^2}{4} + vn + 1}$$

$$n_2 = \frac{n}{2} - \sqrt{\frac{n^2}{4} + vn + 1}$$

$$U_{uu} = \frac{1}{n_2 - n_1}\left[(n_2 + v)\left(\frac{b}{a}\right)^{n_1} - (n_1 + v)\left(\frac{b}{a}\right)^{n_2}\right]$$

$$U_{uP} = \frac{b^n(1-v^2)a}{h_kE(n_2-n_1)}\left[\left(\frac{a}{b}\right)^{n_1} - \left(\frac{a}{b}\right)^{n_2}\right]$$

$$U_{Pu} = \frac{Eh_k(n_1+v)(n_2+v)}{(1-v^2)(b^{n+1})(n_2-n_1)}\left[\left(\frac{b}{a}\right)^{n_1} - \left(\frac{b}{a}\right)^{n_2}\right]$$

$$U_{PP} = \frac{a}{b}\frac{1}{n_2-n_1}\left[(n_2+v)\left(\frac{a}{b}\right)^{n_1} - (n_1+v)\left(\frac{a}{b}\right)^{n_2}\right]$$

$$k_{11} = 2\pi ah_kE\left[(n_2+v)\beta_0^{n_1} + (n_1+v)\beta_0^{n_2}\right]/H_4$$
$$k_{21} = k_{12} = -2\pi ah_kE(n_2 - n_1)/H_4$$
$$k_{22} = 2\pi bh_kE\beta_0^{-1}\left[(n_2+v)\beta_0^{-n_2}\right.$$
$$\left. - (n_1+v)\beta_0^{-n_1}\right]/H_4$$
$$H_4 = h^n(1-v^2)a(\beta_0^{-n_1} - \beta_0^{-n_2})$$
$$p_a^0 = k_{12}F_u$$
$$p_b^0 = -2\pi bF_P + k_{22}F_u$$
$$\beta_0 = b/a$$

TABLE 19-7 (continued)   TRANSFER AND STIFFNESS MATRICES FOR DISKS

Case	Transfer Matrices	Stiffness Matrices
4. Massless disk element without center hole (sketch of disk with radius b)	$z_b = U^i z_0$ $z_0 = [u_0 \quad P_0 \quad 1]^T$ $$\begin{bmatrix} 1 & \dfrac{b}{Eh}(1-\nu) & F_u \\[2mm] 0 & 1 & F_P \\[1mm] 0 & 0 & 1 \end{bmatrix}\begin{bmatrix} u_0 \\ P_0 \\ 1 \end{bmatrix}$$ $\qquad\qquad U^i$	$p^i = k^i v^i - \bar{p}^i$ $p^i = [p_0 \quad p_b]^T \qquad v^i = [u_0 \quad u_b]^T$ $\bar{p}^i = [0 \quad p_b^0]^T$ $k^i = \begin{bmatrix} 0 & 0 \\ 0 & k_{22} \end{bmatrix}$ $k_{22} = 2\pi Eh/(1-\nu)$ $p_b^0 = -2\pi b F_P + k_{22} F_u$
5. Disk element without center hole including mass (sketch of disk with radius b)	$z_b = U^i z_0$ $z_0 = [u_0 \quad P_0 \quad 1]^T$ $$\begin{bmatrix} 0 & \dfrac{2J_1(\beta b)}{\beta Eh}(1-\nu) & F_u \\[3mm] \dfrac{2}{\beta(1+\nu)b}\big[(1+\nu)J_1(\beta b) - \beta b J_2(\beta b)\big] & 0 & F_P \\[2mm] 0 & 0 & 1 \end{bmatrix}\begin{bmatrix} u_0 \\ P_0 \\ 1 \end{bmatrix}$$ $\qquad\qquad\qquad U^i$	$p^i = k^i v^i - \bar{p}^i$ $p^i = [p_0 \quad p_b]^T \qquad v^i = [u_0 \quad u_b]^T$ $\bar{p}^i = [0 \quad p_b^0]^T \qquad k^i = \begin{bmatrix} 0 & 0 \\ 0 & k_{22} \end{bmatrix}$ $k_{22} = 2\pi Eh\{(1+\nu)J_1(\beta b)$ $\qquad\quad - \beta b J_2(\beta a)/[(1-\nu^2)J_1(\beta b)]\}$ $p_b^0 = -2\pi b F_P + k_{22} F_u$

TABLE 19-7   **Matrices for Disks**   1144

TABLE 19-7 (continued) **TRANSFER AND STIFFNESS MATRICES FOR DISKS**

Loading Vectors

Case	Loading Vectors
1. Annular element without mass	$$F_u = \frac{b^2 - a^2}{2b}(1 + v)\alpha T_1 + \frac{1 + v}{b}\alpha \int_a^b \xi T(\xi)\, d\xi$$ $$- \frac{(b^2 - a^2)^2}{b}\frac{1 - v^2}{8}\frac{\rho^* \Omega^2}{E}$$ $$- \frac{1 - v^2}{Eh}\int_a^b \left[\eta \int_a^\eta p_r(\xi)\, d\xi\right] d\eta$$ $$F_P = -\frac{h(b^2 - a^2)}{2b^2}E\alpha T_1 - \frac{hE\alpha}{b^2}\int_a^b \xi T(\xi)\, d\xi$$ $$- (b^2 - a^2)\frac{\rho^* \Omega^2 h}{4}\left[(1 + v) + \frac{1 - v}{2}\frac{(b^2 + a^2)}{b^2}\right]$$ $$+ \frac{1 - v}{b^2}\int_a^b \eta\left[\int_a^\eta p_r(\xi)\, d\xi\right] d\eta - \int_a^b p_r(\xi)\, d\xi$$
2. Annular element with mass	$$F_u = -\int_a^b p_r(\xi)U_{uP}(\xi, a)\, d\xi$$ $$F_P = -\int_a^b p_r(\xi)U_{PP}(\xi, a)\, d\xi$$ $U_{uP}(\xi, a)$ and $U_{PP}(\xi, a)$ are obtained by replacing b with ξ in U_{uP} and U_{PP} of case 2, page 1143.
3. Disk of variable thickness $h = h_k/r^n$ h_k is a reference thickness $h_k = h_a a^n$ Hyperbolic profile	$$F_u = \frac{b^3 \Omega^2 \rho^*}{E}\left[\frac{v^2 - 1}{8 - (3 + v)n}\right]$$ $$\times \left\{1 - \frac{1}{n_2 - n_1}\left[(n_2 - 3)\left(\frac{b}{a}\right)^{n_1 - 3} - (n_1 - 3)\left(\frac{b}{a}\right)^{n_2 - 3}\right]\right\}$$ $$F_P = -\frac{\rho^* \Omega^2 h_k b^{2-n}}{8 - (3 + v)n}\left\{(3 + v) - \frac{1}{n_2 - n_1}\right.$$ $$\left.\times \left[(n_1 + v)(n_2 - 3)\left(\frac{b}{a}\right)^{n_1 - 3} - (n_2 + v)(n_1 - 3)\left(\frac{b}{a}\right)^{n_2 - 3}\right]\right\}$$

TABLE 19-7 (continued)   TRANSFER AND STIFFNESS MATRICES FOR DISKS

Vectors

Case	Loading Vectors
4. Massless disk element without center hole	$F_u = b\alpha T_1 + (1 + \nu)\dfrac{\alpha}{b}\displaystyle\int_0^b \xi T(\xi)\, d\xi$ $\qquad + b\alpha \dfrac{1-\nu}{2} T\big|_{r=0} - (1 - \nu^2)\dfrac{\rho^*\Omega^2 b^3}{8E}$ $\qquad - \dfrac{1-\nu^2}{Ehb}\displaystyle\int_0^b\left[\eta\int_0^\eta p_r(\xi)\,d\xi\right]d\eta$ $F_P = -\dfrac{hE\alpha}{b^2}\displaystyle\int_0^b \xi T(\xi)\,d\xi + \dfrac{h\alpha}{2}ET\Big|_{r=0}$ $\qquad - (3 + \nu)\dfrac{\rho^* h\Omega^2 b^2}{8} + \dfrac{1-\nu}{b^2}\displaystyle\int_0^b r\left[\int_0^b p_r(r)\,dr\right]dr$ $\qquad - \displaystyle\int_0^b p_r(r)\,dr$
5. Disk element without center hole including mass	$F_u = -\displaystyle\int_0^b p_r(\xi) U_{uP}(\xi)\, d\xi$ $F_P = -\displaystyle\int_0^b p_r(\xi) U_{PP}(\xi)\, d\xi$ $U_{uP}(\xi)$ and $U_{PP}(\xi)$ are obtained by replacing b with ξ in U_{uP} and U_{PP} of case 5, page 1144.

TABLE 19-7   **Matrices for Disks**   **1146**

TABLE 19-8   POINT MATRICES OF CYLINDERS, SPHERES, AND DISKS

Notation

$\rho^* = $ mass per unit volume
$\omega = $ natural frequency
$a = $ radial coordinate of point
$h = $ thickness of disk
$M_i = $ mass per unit circumferential length or area

For cylinders: $M_i = \dfrac{(a^+)^2 - (a^-)^2}{2a}\rho^* = \Delta a\,\rho^*$ and $p_a = 2\pi a\sigma_a$

For spheres: $M_i = \dfrac{(a^+)^3 - (a^-)^3}{3a^2}\rho^* = \Delta a\,\rho^*$ and $p_a = 4\pi a^2\sigma_a$

For disks: $M_i = \dfrac{(a^+)^2 - (a^-)^2}{2a}h\rho^* = \Delta a\,h\rho^*$ and $p_a = 2\pi a P_a$

Case	Transfer Matrices $z_a^+ = U_i z_a^-$	Stiffness Matrix and Loading Vector[a] $p_a = k_a u_a - p_a^0$
1. Pressure (radial force per unit circumferential length or area) applied at $r = a$	CYLINDERS: $$\begin{bmatrix} 1 & 0 & 0 \\ 0 & 1 & -p \\ \hline 0 & 0 & 1 \end{bmatrix}\begin{bmatrix} u_a \\ \sigma_a \\ 1 \end{bmatrix}$$ $\qquad U_i \qquad z_a$	$p_a^0 = -p$ Often these pressures are incorporated in the displacement method solution as nodal loading.
	SPHERES: $$\begin{bmatrix} 1 & 0 & 0 \\ 0 & 1 & -p \\ \hline 0 & 0 & 1 \end{bmatrix}\begin{bmatrix} u_a \\ \sigma_a \\ 1 \end{bmatrix}$$ $\qquad U_i \qquad z_a$	
	DISKS: $$\begin{bmatrix} 1 & 0 & 0 \\ 0 & 1 & -p \\ \hline 0 & 0 & 1 \end{bmatrix}\begin{bmatrix} u_a \\ P_a \\ 1 \end{bmatrix}$$ $\qquad U_i \qquad z_a$	

TABLE 19-8 (continued)   **POINT MATRICES OF CYLINDERS, SPHERES, AND DISKS**

Case	$\mathbf{z}_a^+ = \mathbf{U}_i \mathbf{z}_a^-$	Stiffness and Mass Matrices $p_a = (k_a - \omega^2 m_a) u_a$
2. Concentrated mass and elastic support	CYLINDERS: $$\underbrace{\begin{bmatrix} 1 & 0 & 0 \\ -M_i\omega^2 & 1 & 0 \\ \hline 0 & 0 & 1 \end{bmatrix}}_{\mathbf{U}_i} \underbrace{\begin{bmatrix} u_a \\ \sigma_a \\ 1 \end{bmatrix}}_{\mathbf{z}_a}$$	$m_a = M_i$ The lumped mass can be incorporated as a nodal condition.
$\Delta a = a^+ - a^-$	SPHERES: $$\underbrace{\begin{bmatrix} 1 & 0 & 0 \\ -M_i\omega^2 & 1 & 0 \\ \hline 0 & 0 & 1 \end{bmatrix}}_{\mathbf{U}_i} \underbrace{\begin{bmatrix} u_a \\ \sigma_a \\ 1 \end{bmatrix}}_{\mathbf{z}_a}$$	$m_a = M_i$
For disks $\leftarrow r \rightarrow$ Elastic support k	DISKS: $$\underbrace{\begin{bmatrix} 1 & 0 & 0 \\ k - M_i\omega^2 & 1 & 0 \\ \hline 0 & 0 & 1 \end{bmatrix}}_{\mathbf{U}_i} \underbrace{\begin{bmatrix} u_a \\ P_a \\ 1 \end{bmatrix}}_{\mathbf{z}_a}$$	$m_a = M_i,\ k_a = k$

[a]Traditionally stiffness and mass matrices for case 1 are implemented as nodal conditions.

TABLE 19-8   **Point Matrices of Cylinders, Spheres, and Disks**   **1148**

TABLE 19-9 MASS MATRICES FOR CYLINDERS, SPHERES, AND DISKS

Notation

$$\begin{bmatrix} m_{11} & m_{12} \\ m_{21} & m_{22} \end{bmatrix} \begin{bmatrix} u_a \\ u_b \end{bmatrix}$$

$\mathbf{m}^i \qquad\qquad \mathbf{v}^i$

ρ^* = mass per unit volume $\qquad \beta_0 = b/a \qquad h$ = thickness of disk

Case	Mass Matrices
1. Thick cylinder with center hole (consistent mass) 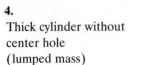	$m_{11} = \pi a^2 \rho^* \big[\beta_0^2 (4\beta_0^2 \ln \beta_0 - 3\beta_0^2 + 4)$ $\qquad -1 \big]/2(\beta_0^2 - 1)^2$ $m_{12} = m_{21} = \pi a^2 \rho^* \beta_0 \big[\beta_0^4 - 4\beta_0^2 \ln \beta_0$ $\qquad -1 \big]/2(\beta_0^2 - 1)^2$ $m_{22} = \pi a^2 \rho^* \beta_0^2 \big[\beta_0^4 - 4\beta_0^2 + 4\ln \beta_0 + 3 \big]/2(\beta_0^2 - 1)^2$
2. Thick sphere element (consistent mass) 	$m_{11} = 4\pi \rho^* a^3 \big[5\beta_0^4 + \beta_0^3 - 3\beta_0^2 - 2\beta_0 - 1 \big]/5 \big[\beta_0^4$ $\qquad +2\beta_0^3 + 3\beta_0^2 + 2\beta_0 + 1 \big]$ $m_{12} = m_{21} = 6\pi \rho^* a^3 \beta_0^2 \big[\beta_0^3 + 2\beta_0^2 - 2\beta_0 - 3 \big]/5B$ $m_{22} = 4\pi \rho^* a^3 \beta_0^3 \big[\beta_0^4 + 2\beta_0^3 + 3\beta_0^2 - \beta_0 - 5 \big]/5B$ $B = \beta_0^4 + 2\beta_0^3 + 3\beta_0^2 + 2\beta_0 + 1$
3. Disk element with center hole (consistent mass) 	$m_{11} = \pi h \rho^* a^2 \big[\beta_0^2 (4\beta_0^2 \ln \beta_0 - 3\beta_0^2 + 4)$ $\qquad -1 \big]/2(\beta_0^2 - 1)^2$ $m_{12} = m_{21} = \pi h \rho^* a^2 \beta_0 \big[\beta_0^4 - 4\beta_0^2 \ln \beta_0$ $\qquad -1 \big]/2(\beta_0^2 - 1)^2$ $m_{22} = \pi h \rho^* a^2 \beta_0^2 \big[\beta_0^4 - 4\beta_0^2 + 4\ln \beta_0$ $\qquad +3 \big]/2(\beta_0^2 - 1)^2$
4. Thick cylinder without center hole (lumped mass) 	$\begin{bmatrix} 0 & 0 \\ 0 & \frac{1}{2}\pi b^2 \rho^* \end{bmatrix} \begin{bmatrix} u_0 \\ u_b \end{bmatrix}$ $\mathbf{m}^i \qquad\qquad \mathbf{v}^i$

TABLE 19-9 (continued)   MASS MATRICES FOR CYLINDERS, SPHERES, AND DISKS

Case	Mass Matrices
5. Sphere (lumped mass)	$$\begin{bmatrix} 0 & 0 \\ 0 & \frac{4}{5}\pi b^3 \rho^* \end{bmatrix} \begin{bmatrix} u_0 \\ u_b \end{bmatrix}$$ $$\qquad \mathbf{m}^i \qquad\qquad \mathbf{v}^i$$
6. Disk (lumped mass)	$$\begin{bmatrix} 0 & 0 \\ 0 & \frac{1}{2}\pi b^2 h \rho^* \end{bmatrix} \begin{bmatrix} u_0 \\ u_b \end{bmatrix}$$ $$\qquad \mathbf{m}^i \qquad\qquad \mathbf{v}^i$$

TABLE 19-9 Mass Matrices for Cylinders, Spheres, and Disks   **1150**

Thin Shells

A shell is a three-dimensional body bounded by two curved surfaces. Most of the formulas here apply to *shells of revolution*, a special but commonly occurring shell. To generate a thin shell of revolution, a plane that passes longitudinally through a polar axis is rotated about that axis; two lines that lie close to each other and are called generators of the shell form the inner and outer surfaces of the shell as the generating plane is rotated.

Formulas for membrane shells and shells with bending are provided in this chapter. In the case of a membrane shell, the shell's middle surface is free of bending and twisting moments as well as transverse shear forces. In many shell problems the presence of moments and shear forces is necessary to accept the type of loading and to satisfy the shell boundary conditions. This need has led to bending shell theory, which is a more comprehensive theory. When the term *bending* is used, it often applies to a shell with both membrane and bending deformations.

20.1   DEFINITIONS

The *middle surface* of a shell is the surface that is everywhere equidistant between the inner and outer surfaces. The mechanical state of the shell is specified by the values of certain stress resultants that act on the middle surface.

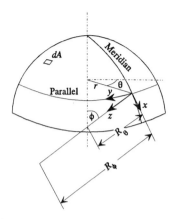

FIGURE 20-1 Coordinates specifying points on a shell of revolution.

A *meridian* of the shell is formed by the intersection of the generating plane and the middle surface. A *parallel* (azimuth) of the shell is the intersection of a plane perpendicular to the polar axis with the middle surface. The principal radii of curvature of the middle surface are the radii of curvature of the meridians and parallels R_ϕ and R_θ, respectively. The coordinates that specify a point on the surface of the shell are shown in Fig. 20-1. The radius R_θ is the distance to the polar axis along the shell normal at the surface point. The angle θ is the angle between any arbitrary reference line and the radius of the parallel that passes through the point. The angle ϕ is the angle between the polar axis and R_ϕ. An additional coordinate system is defined in the shell surface; x is tangent to the meridian, y is tangent to the parallel, and z coincides with the surface normal. The positive directions of the x, y, z coordinates are shown in Fig. 20-1.

Notation

B	Extensional rigidity of shell surface (F/L)
D	Flexural rigidity of shell surface (FL)
E	Young's modulus (F/L^2)
f	Natural frequency of a shell, $= \omega/2\pi$ (Hz)
h	Thickness of shell wall (L)
H	Horizontal force per unit length on a shell edge (F/L)
L, ℓ	Shell length (L)
m	Number of longitudinal nodes in a given mode of vibration
M	Bending moment per unit length on a shell edge (FL/L)
M_x	Bending moment per unit length on shell edge, in plane perpendicular to x axis (FL/L)
M_ϕ	Bending moment per unit length on parallel planes (FL/L)
M_θ	Bending moment per unit length on meridional planes (FL/L)
n	Number of circumferential nodes in given mode of vibration
N_ϕ	Normal force per unit length in meridional direction (F/L)

N_θ	Normal force per unit length in azimuthal (parallel) direction, which is normal to the meridional direction (F/L)
N_x, N_y	Normal forces in x and y directions (F/L)
N_{xy}	Inplane shear force per unit length (F/L)
$N_{\phi\theta}, N_{\theta\phi}$	Inplane shear forces per unit length (F/L)
P	Applied axial force (F)
p_1	Applied uniform load on an area (F/L^2)
p_2	Uniform vertical load on a projected area (F/L^2)
p_x, p_y, p_z	Components of loads applied to shell surface (F/L^2)
q	Dead weight of shell (F/L^2)
Q_θ	Transverse shear force per unit length on meridional planes (F/L)
Q_ϕ	Transverse shear force per unit length on parallel planes (F/L)
Q	Transverse shear force per unit length on a parallel plane shell edge supplied by shear diagram
r	Radius of a parallel circle, $= R_\theta \sin\phi$ (L)
R	Radius of a circular cylinder or sphere (L)
R_θ, θ, ϕ	Coordinates that locate a point on middle surface of a shell
R_ϕ	Radius of curvature of meridian at a point (L)
s, ξ	Nondimensional length coordinate of a cylindrical or conical shell, $s = x/L$
t	Time (T)
u, v, w	Displacements of a point in middle surface in x, y, z directions, respectively (L)
W	Lantern-loading, load per unit length on shell edge (F/L)
x, y, z	Coordinate system in surface of a shell
β	Rotation of tangent to meridian during deflection (degrees)
ΔR	Displacement in direction of radius of the parallel (L)
$\varepsilon_\phi, \varepsilon_\theta$	Strains in meridional and parallel directions in middle surface (L/L)
κ	$= h^2/12R^2$
ν	Poisson's ratio
ω	Natural (circular) frequency of a shell (rad/T)
Ω^2	Frequency parameter, $= \rho^*(1 - \nu^2)R^2\omega^2/E$
Ω_p	Frequency parameter for cylindrical shells, $p = \begin{cases} a & \text{for axial modes} \\ t & \text{for torsional modes} \\ at & \text{for coupled axial–radial modes} \\ rt & \text{for coupled radial–torsional modes} \end{cases}$
ρ^*	Mass per unit volume of a shell (FT^2/L^4)
ρ_w	Specific weight of liquid (F/L^3)
σ_i	Component of normal stress in i direction (F/L^2)
$\sigma_\phi, \sigma_\theta$	Normal stresses in meridional and azimuthal (parallel) directions (F/L^2)

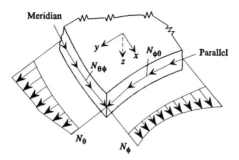

FIGURE 20-2   Membrane forces acting on a shell element.

Subscripts

C	Cylinder
D	Dome

20.2   MEMBRANE SHELLS OF REVOLUTION

The membrane hypothesis produces the simplest and most readily solvable system of shell equations. If the wall of the shell is thin and there are no abrupt changes in thickness, slope, or curvature and if the loading is uniformly distributed or smoothly varying and symmetric, the bending responses can be very small and can be neglected. Hence it can be assumed that the shell's middle surface is free of bending moments, twisting moments, and transverse shear forces. The stress resultants that are assumed to be present are the in-plane normal and shear forces per unit length of shell surface. These membrane forces are depicted in Fig. 20-2. Because there are four stress resultants to be found and four equilibrium conditions to satisfy, the membrane shell problem is statically determinate. Since the membrane forces cannot produce moments about the x and y axes, the conditions $\Sigma M_x = 0$ and $\Sigma M_y = 0$ are automatically satisfied. The remaining equilibrium equations are

$$\Sigma M_x = 0, \; \Sigma F_x = \Sigma F_y = \Sigma F_z = 0$$

Applying the equilibrium conditions to a shell element produces the four equations for the static deformation of a membrane shell [20.1]:

$$N_{\phi\theta} = N_{\theta\phi} \tag{20.1a}$$

$$\frac{\partial}{\partial \phi}(N_\phi R_\theta \sin \phi) + \frac{\partial N_{\phi\theta}}{\partial \theta}R_\theta - N_\theta R_\phi \cos \phi + p_x R_\phi R_\theta \sin \phi = 0 \tag{20.1b}$$

$$\frac{\partial N_\theta}{\partial \theta}R_\phi + \frac{\partial}{\partial \phi}(N_{\phi\theta} R_\theta \sin \phi) + N_{\phi\theta} R_\phi \cos \phi + p_y R_\phi R_\theta \sin \phi = 0 \tag{20.1c}$$

$$N_\phi R_\theta + N_\theta R_\phi + p_z R_\phi R_\theta = 0 \tag{20.1d}$$

FIGURE 20-3 Boundary condition which membrane theory satisfies.

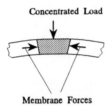

Concentrated Load

Membrane Forces

FIGURE 20-4 Loading condition incompatible with membrane theory.

Because the bending stresses are zero, the normal stresses are simply

$$\sigma_\phi = N_\phi/h \tag{20.2}$$

$$\sigma_\theta = N_\theta/h \tag{20.3}$$

From considerations of the nature of the deformation, Hooke's law, and the definition of the stress resultant, the equations for the strains in terms of stress resultants are found to be

$$\varepsilon_\phi = (N_\phi - \nu N_\theta)/Eh \tag{20.4a}$$

$$\varepsilon_\theta = (N_\theta - \nu N_\phi)/Eh \tag{20.4b}$$

The membrane theory of shells is not strictly applicable to cases in which boundary conditions and loading conditions cannot be countered by in-plane forces. See Figs. 20-3 and 20-4. Figure 20-4 shows a situation in which the membrane theory is inapplicable; however, approximate methods are available for treating such cases without invoking the full bending theory of shells. Also, in most cases in which significant bending stresses occur, the stresses are confined to small regions of the shell near the boundaries. Tables 20-1 to 20-5 present formulas for the stresses and deformation for membrane shells with several types of loads.

Example 20.1 Membrane Analysis of a Hemispherical Shell Subjected to a Uniformly Distributed Load Suppose a uniformly distributed vertical load (case 2, Table 20-1, part A) $p_2 = 40$ psi acts on a hemispherical dome. Compute the maximum values of the membrane forces, the corresponding shell stresses, and the displacement at the base. The shell is constructed of steel with $h = 2$ in. and $R = 30$ ft.

Take a differential area dA on the middle surface of the shell. The projection of dA on the horizontal plane is $dA \cos \phi$. The total load on this differential area is then

$$P = p_2 \, dA \cos \phi$$

The load on the unit area of the shell surface is

$$P/dA = p_2 \cos \phi$$

The components of this load in the x, y, and z directions are

$$p_x = p_2 \cos \phi \sin \phi, \qquad p_y = 0, \qquad p_z = p_2 \cos^2 \phi$$

This confirms case 2 of Table 20-1, part A. From case 1 of Table 20-1, part B,

$$N_\phi = -\tfrac{1}{2} p_2 R, \qquad N_\theta = -\left(\tfrac{1}{2} p_2 R\right) \cos 2\phi$$

At $\phi = 90°$

$$N_\phi = -N_{\theta, \max} = \tfrac{1}{2}(-p_2 R) = \tfrac{1}{2}[(-40)(30) \times 12] = -7200 \text{ lb/in.}$$

$$\sigma_\phi = -\sigma_{\theta, \max} = \frac{N_{\theta, \max}}{h} = -\frac{7200 \text{ lb/in.}}{2 \text{ in.}} = -3600 \text{ lb/in.}^2$$

At the base of the shell

$$\Delta R(\phi = 90°) = \frac{R^2 p_2}{Eh} \frac{1+\nu}{2} = \frac{(30 \times 12)^2 (40)(1 + 0.3)}{(3 \times 10^7)(2)(2)} = 0.05616 \text{ in.}$$

$$\beta(\phi = 90°) = 0$$

Example 20.2 Membrane Analysis of a Cylinder Filled with Water A cylinder 20 ft high with a radius of 3 ft is filled with water. The specific weight ρ_w of water is 62.4 lb/ft³. The other constants are $E = 3 \times 10^7$ psi and $h = 2$ in. Compute the stress resultants and displacement at the base.

The load of this problem corresponds to case 1 of Table 20-3 with $\lambda_p = 0$. The horizontal water pressure is distributed linearly along the x direction with $p_z = -p_0(1 - \xi)$, $\xi = x/L$, and

$$p_0 = L\rho_w = (20 \text{ ft})(62.4 \text{ lb/ft}^3) = 1248 \text{ lb/ft}^2 = 8.6667 \text{ lb/in}^2$$

where L is the height of the cylinder. Then at the base ($\xi = 0$)

$$N_\theta = p_0 R = (1248 \text{ lb/ft}^2)(3 \text{ ft}) = 3744 \text{ lb/ft} = 312 \text{ lb/in.}$$

$$w = -\frac{p_0 R^2}{Eh} = -\frac{1}{(3 \times 10^7)(2)}\left[(8.6667)(3 \times 12)^2\right] = -1.872 \times 10^{-4} \text{ in.}$$

The displacement u is zero at the base where $\xi = 0$. The azimuthal normal stress at $\xi = 0$ is

$$\sigma_\theta = N_\theta/h = 312/2 = 156 \text{ lb/in.}^2$$

20.3 SHELLS OF REVOLUTION WITH BENDING

There are a variety of formulations for shells of revolution with bending. Usually 10 stress resultants are considered to act on a shell element. Membrane stress resultants remain as shown in Fig. 20-2, while non–membrane stress resultants are depicted in Fig. 20-5. The same system of coordinates is used for the bending theory as was used for the membrane shells. The forces shown in Fig. 20-5 are positive. Because there are 10 stress resultants and only 6 equilibrium equations, in contrast to the membrane shell, the problem of the bending of a shell is statically indeterminate.

If the loads are axisymmetric, the response quantities do not vary with the θ coordinate. In addition, axial symmetry dictates that the twisting moments, inplane shear forces, and the transverse shear forces on the meridional planes are zero:

$$M_{\theta\phi} = M_{\phi\theta} = 0, \qquad N_{\phi\theta} = N_{\theta\phi} = 0, \quad \text{and} \quad Q_\theta = 0 \qquad (20.5)$$

Application of the laws of equilibrium to a differential shell element yields the following three equations in five unknowns [20.1]:

$$\frac{d}{d\phi}(N_\phi R_\theta \sin \phi) - N_\theta R_\phi \cos \phi - Q_\phi R_\theta \sin \phi + p_x R_\phi R_\theta \sin \phi = 0 \quad (20.6a)$$

$$N_\phi R_\theta \sin \phi + N_\theta R_\phi \sin \phi + \frac{d}{d\phi}(Q_\phi R_\theta \sin \phi) + p_z R_\phi R_\theta \sin \phi = 0 \quad (20.6b)$$

$$\frac{d}{d\phi}(M_\phi R_\theta \sin \phi) - Q_\phi R_\phi R_\theta \sin \phi + M_\theta R_\phi \cos \phi = 0 \quad (20.6c)$$

By considering the deformation of the shell, three additional variables ($\varepsilon_\theta, \varepsilon_\phi, \beta$) are introduced and five additional equations are obtained [20.1].

$$\beta = (\varepsilon_\phi - \varepsilon_\theta)\cot \phi - \frac{R_\theta}{R_\phi}\frac{d\varepsilon_\theta}{d\phi} \qquad (20.7a)$$

$$N_\phi = B(\varepsilon_\phi + \nu\varepsilon_\theta) \qquad (20.7b)$$

$$N_\theta = B(\varepsilon_\theta + \nu\varepsilon_\phi) \qquad (20.7c)$$

$$M_\phi = -D(\chi_\phi + \nu\chi_\theta) \qquad (20.7d)$$

$$M_\theta = D(\chi_\theta + \nu\chi_\phi) \qquad (20.7e)$$

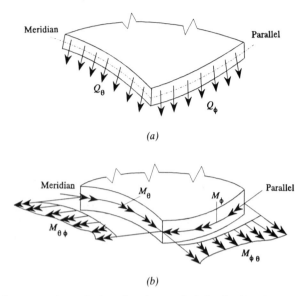

FIGURE 20-5 Stress resultants for bending theory of shells. Positive forces and moments are shown. (a) Transverse shear forces per unit length. (b) Bending and twisting moments per unit length.

in which

$$\chi_\phi = \frac{1}{R_\phi} \frac{d\beta}{d\phi} \tag{20.7f}$$

$$\chi_\theta = \frac{\beta \cot \phi}{R_\theta} \tag{20.7g}$$

$$B = \frac{Eh}{1 - \nu^2} \quad \text{(extensional rigidity)} \tag{20.7h}$$

$$D = \frac{Eh^3}{12(1 - \nu^2)} \quad \text{(flexural rigidity)} \tag{20.7i}$$

This final set of eight equations in eight unknowns forms the equations of motion for the static axisymmetric deformation of a bending shell. The normal stresses are given in terms of the stress resultants as

$$\sigma_\phi = \frac{N_\phi}{h} + \frac{M_\phi}{h^3/12} z \tag{20.8a}$$

$$\sigma_\theta = \frac{N_\theta}{h} - \frac{M_\theta}{h^3/12} z \tag{20.8b}$$

where z is measured from the middle surface.

Although the eight relations of Eqs. (20.6)–(20.7) are sufficient to solve the problem of a shell with bending, the process of solution is so complicated that

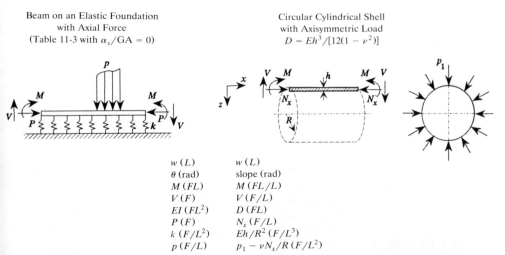

Beam on an Elastic Foundation
with Axial Force
(Table 11-3 with $\alpha_s/GA = 0$)

Circular Cylindrical Shell
with Axisymmetric Load
$D = Eh^3/[12(1 - \nu^2)]$

$w(L)$	$w(L)$
θ (rad)	slope (rad)
M (FL)	M (FL/L)
V (F)	V (F/L)
EI (FL²)	D (FL)
P (F)	N_x (F/L)
k (F/L²)	Eh/R^2 (F/L³)
p (F/L)	$p_1 - \nu N_x/R$ (F/L²)

FIGURE 20-6 Equivalence of a circular cylindrical shell with axisymmetric load and a beam on elastic foundation.

approximate methods are commonly employed. One approximate technique is the force method [20.1, 20.2]. In this method, the edge forces on a shell are treated separately from the applied loadings (e.g., dead weight and normal pressure). Begin with a membrane analysis of the shell with the applied loads (without the boundary forces) and the membrane boundary conditions. This is followed by a bending analysis with the edge forces as the loads. The boundary conditions for this analysis are consistent with the assumptions of the membrane theory; i.e., the boundaries are free to displace and rotate in the manner necessary to satisfy the membrane hypothesis. The edge displacements of these two analyses are calculated, with the displacements of the bending analysis expressed in terms of the unknown edge forces. Neither of the displacements from these two analyses is compatible with the actual boundary conditions of the shell, so they are superimposed to satisfy the actual boundary conditions. The resulting relationships are conditions that can be solved for the heretofore unknown edge forces. Once the edge forces are determined, the complete solution of the shell can be obtained by the superposition of the membrane and bending analysis results.

For a circular cylindrical shell with axially symmetric loading, the governing differential equations are the same as those for a beam on an elastic foundation of modulus k and with axial force P [Eq. (11.7)]. Hence simple cylinder problems can be solved using the beam formulas. To do this, substitute the shell parameters

$$\frac{Eh}{R^2} \quad \text{and} \quad D = \frac{Eh^3}{12(1 - \nu^2)} \tag{20.9}$$

in Table 11-3 for k and EI, respectively, and set $\alpha_s/GA = 0$. Here R is the radius of the cylinder and h is the wall thickness. The deflection, slope, shear

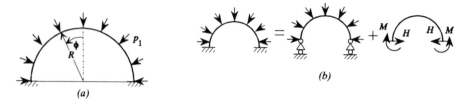

FIGURE 20-7 Hemispherical shell with external pressure. (a) Hemisphere under uniform radial pressure. (b) Superposition of solutions.

force, and moments of the cylinder along the x axis are given in Table 11-3, part A, with loadings of Table 11-3, part C, and the boundary conditions of Table 11-3, part D. This equivalence is illustrated in Fig. 20-6.

Responses for bending shells of revolution, which are subject to edge loads, are listed in Tables 20-6 to 20-8. Definitions of the $\bar{F}_i$, $F_i(\xi)$ factors are contained in Table 20-9.

Example 20.3 Hemispherical Shell under Uniform External Pressure The hemispherical shell shown in Fig. 20-7a is subjected to a uniform radial pressure of 100 psi. For this shell, the edges are built-in, $R = 10$ ft, $h = 3.0$ in., $E = 3 \times 10^7$ psi, and $\nu = 0.3$. Compute the responses at $\phi = 15°$.

The shell can be treated (Fig. 20-7b) as the superposition of a simply supported membrane shell under uniform normal pressure of case 1 of Table 20-1, part B, and a shell with bending deformation with forces and moments at the lower edge of cases 1 and 2 of Table 20-6. The first step of the computation is to determine the reaction forces H and M. These can be determined from the geometric conditions that the displacement and rotation at the lower edges are zero, i.e.,

$$\Delta R = \Delta R^m + \Delta R^b = 0 \tag{1}$$

$$\beta = \beta^m + \beta^b = 0 \tag{2}$$

where the superscripts m and b stand for membrane and bending, respectively. From case 1 of Table 20-1, part B, and cases 1 and 2 of Table 20-6, at $\phi = 90°$

$$\Delta R^m = -\frac{p_1}{2}\frac{R^2}{Eh}(1 - \nu) \tag{3}$$

$$\Delta R^b = -\frac{2Rk}{Eh}H + \frac{2k^2}{Eh}M \tag{4}$$

$$\beta^m = 0 \tag{5}$$

$$\beta^b = \frac{2k^2}{Eh}H - \frac{4k^3}{EhR}M \tag{6}$$

Substitute Eqs. (3)–(6) into (1) and (2) to form

$$\Delta R = -\frac{p_1}{2}\frac{R^2}{Eh}(1 - \nu) - \frac{2Rk}{Eh}H + \frac{2k^2}{Eh}M = 0 \tag{7}$$

$$\beta = \frac{2k^2}{Eh}H - \frac{4k^3}{EhR}M = 0 \tag{8}$$

Solve these equations to find

$$M = -p_1 R^2 (1 - \nu)/4k^2 \tag{9}$$

$$H = 2kM/R \tag{10}$$

With M and H known, the responses can be found from Tables 20-1 and 20-6 with $\phi = 15°$ and $\alpha = \frac{1}{2}\pi - \phi = 75°$ as

$$\Delta R|_{\phi=15°} = \Delta R^m|_{\phi=15°} + \Delta R^b|_{\phi=15°}$$

$$= -\frac{p_1 R^2}{2 Eh}(1-\nu)\sin 15°$$

$$-\frac{H}{Eh}\left\{ \mathrm{Re}^{-k\alpha}\left[2k\sin 15° \cos k\alpha - \sqrt{2}\,\nu \cos 15° \cos\left(k\alpha + \tfrac{1}{4}\pi\right)\right]\right\}$$

$$+\frac{2Mk}{Eh}e^{-k\alpha}\left[\sqrt{2}\,k\sin 15° \cos\left(k\alpha + \tfrac{1}{4}\pi\right) + \nu\cos 15° \sin k\alpha\right]$$

$$= -1.449 \times 10^{-3} \text{ in.} \tag{11}$$

$$\beta|_{\phi=15°} = \beta^m + \beta^b$$

$$= 0 + \frac{H}{Eh}\left[2\sqrt{2}\,k^2 e^{-k\alpha}\sin\left(k\alpha + \tfrac{1}{4}\pi\right)\right] - \frac{M}{Eh}\left(\frac{4k^3}{R}e^{-k\alpha}\cos k\alpha\right)$$

$$= 1.701 \times 10^{-8} \text{ rad} \tag{12}$$

$$Q_\phi|_{\phi=15°} = \sqrt{2}\,He^{-k\alpha}\cos\left(k\alpha + \tfrac{1}{4}\pi\right) + \frac{2kM}{R}e^{-k\alpha}\sin k\alpha$$

$$= 4.2791 \times 10^{-3} \text{ lb/in.} \tag{13}$$

$$N_\phi^m|_{\phi=15°} = N_\theta^m|_{\phi=15°} = -\tfrac{1}{2}Rp_1 = -6000 \text{ lb/in.}$$

$$N_\phi^b|_{\phi=15°} = -Q_\phi \cot 15° = -1.59 \times 10^{-2} \text{ lb/in.}$$

$$N_\theta^b|_{\phi=15°} = -2Hke^{-k\alpha}\cos k\alpha + 2\sqrt{2}\,M\frac{k^2}{R}e^{-k\alpha}\cos\left(k\alpha + \tfrac{1}{4}\pi\right)$$

$$= -0.1289 \text{ lb/in.}$$

$$N_\phi|_{\phi=15°} = N_\phi^m|_{\phi=15°} + N_\phi^b|_{\phi=15°} = -6000.0159 \text{ lb/in.} \tag{14}$$

$$N_\theta|_{\phi=15°} = N_\theta^m|_{\phi=15°} + N_\theta^b|_{\phi=15°} = -6000.1289 \text{ lb/in.} \tag{15}$$

$$M_\phi = -H\frac{R}{k}e^{-k\alpha}\sin k\alpha + \sqrt{2}\,Me^{-k\alpha}\sin\left(k\alpha + \tfrac{1}{4}\pi\right)$$

$$= -5.39 \times 10^{-2} \text{ lb-in./in.} \tag{16}$$

$$M_\theta|_{\phi=15°} = H\frac{R}{k^2\sqrt{2}}e^{-k\alpha}\cot 15° \sin\left(k\alpha + \tfrac{1}{4}\pi\right) + \nu M_\phi$$

$$+ \frac{M}{k}e^{-k\alpha}\cot 15° \cos k\alpha$$

$$= 5.21 \times 10^{-2} \text{ lb-in/in.} \tag{17}$$

The stresses at the outer surface $(z = -0.5h)$ are [Eq. (20.8)]

$$\sigma_\phi|_{\phi=15°} = \frac{N_\phi}{h} + \frac{M_\phi}{h^3/12} z = -1999.96 \text{ psi} \tag{18}$$

$$\sigma_\theta|_{\phi=15°} = \frac{N_\theta}{h} - \frac{M_\theta}{h^3/12} z = -2000.0078 \text{ psi} \tag{19}$$

It can be seen that the internal forces due to bending are relatively small so that the effect of bending for this load on this shell can be neglected.

20.4 MULTIPLE-SEGMENT SHELLS OF REVOLUTION

Shells can often be modeled as a succession of simple shell elements. A cylindrical shell with spherical bulkheads is an example. Multiple-segment shells of revolution are to be considered here.

The force method, which was discussed in the previous section, can be applied to multiple-segment shells. They are analyzed by combining the relations that must apply at the junctions of the shell with the knowledge of the influence coefficients that relate the deformations at the junctions with the forces and moments that act on the shell edges. This method involves the following steps:

1. Divide the shell into separate segments, and solve the membrane problem for the applied loading for each segment to obtain the displacements and rotations at the connecting lines for each pair of segments. The calculated displacements and rotations are not compatible for adjacent segments at the common connection lines. Hence correction forces (usually H and M) at these edges are needed to hold them together.

2. Calculate the displacements and rotations at the common connection lines due to unit correction forces $(H = 1$ and $M = 1)$ only.

3. For each segment establish relationships between the actual displacements and rotations at the edges and the correction forces and membrane responses. For example, for the segment in Fig. 20-8, the displacements and rotations are

$$\begin{bmatrix} \beta_1 \\ \Delta R_1 \\ \beta_2 \\ \Delta R_2 \end{bmatrix} = \mathbf{f} \begin{bmatrix} M_1 \\ H_1 \\ M_2 \\ H_2 \end{bmatrix} + \begin{bmatrix} \beta_1^0 \\ \Delta R_1^0 \\ \beta_2^0 \\ \Delta R_2^0 \end{bmatrix} \tag{20.10}$$

where $\beta_i, \Delta R_i, i = 1, 2$, are actual rotations and displacements at edges 1 and 2; $H_i, M_i, i = 1, 2$, are correction forces and moments at edges 1 and 2; $\beta_i^0, \Delta R_i^0$, $i = 1, 2$, are rotations and displacements at edges 1 and 2 from the membrane

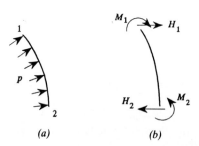

FIGURE 20-8   Single segment of a multiple-segment shell. (*a*) Shell segment for membrane analysis to find $\beta_i^0, \Delta R_i^0$, $i = 1, 2$, in Eq. (20.10). (*b*) Shell segment for bending analysis to find the elements in Eq. (20.11).

analysis of the shell segment with applied loading; and

$$
\mathbf{f} =
\begin{bmatrix}
\beta_1^{M_1} & \beta_1^{H_1} & \beta_1^{M_2} & \beta_1^{H_2} \\
\Delta R_1^{M_1} & \Delta R_1^{H_1} & \Delta R_1^{M_2} & \Delta R_1^{H_2} \\
\beta_2^{M_1} & \beta_2^{H_1} & \beta_2^{M_2} & \beta_2^{H_2} \\
\Delta R_2^{M_1} & \Delta R_2^{H_1} & \Delta R_2^{M_2} & \Delta R_2^{H_2}
\end{bmatrix}
\tag{20.11}
$$

is the *flexibility matrix*. The elements in the flexibility matrix are the deformations at edge 1 or 2 (as indicated by the subscripts) due to the unit loadings $M = 1$ and $H = 1$ at these edges. For example, $\beta_1^{M_2}$ is the rotation at edge 1 due to the unit moment at edge 2. These elements are also called *influence coefficients*.

4. Establish the equilibrium and compatibility equations at each connection line.

5. Combine the equations of step 3 [Eq. (20.10)] using the conditions in step 4 to form a set of algebraic equations with as many unknowns (deformations and correction forces) as there are equations.

6. Solve the equations to find the unknowns and find the responses at the points of interest.

The formulas in Tables 20-10 and the equations for membrane deformations and influence coefficients in Tables 20-11 to 20-13 provide the information necessary to apply this method. Table 20-10, part A, gives relations [Eq. (20.10)] for the deformations of the edges of the shell segments, and Table 20-10, part B, lists the equilibrium and compatibility equations at the edges. The influence coefficients are found in Tables 20-11 to 20-13. For the sign convention of the correction forces as shown in the figures of Table 20-10, part A, positive moments cause tension in the inner shell surface. Positive horizontal forces cause tension in the inner shell surface at the upper edge and compression in the inner shell surface at the lower edge.

For a shell segment greater than a certain length, the deformation at one edge of the segment is not significantly affected by edge loads that act on the opposite edge. For a cylinder, when $kL > 4$, where $k = [3(1 - \nu^2)]^{1/4}/\sqrt{Rh}$, or $L \geq 3.1\sqrt{Rh}$, the influence of the correction force at one edge on the other edge is usually negligible. These cases are much simpler than those in which edge forces and moments exert a measurable effect on the deformations of the other edge. This will be illustrated in the following example.

Example 20.4 Cylindrical Shell with a Spherical Dome The steel shell shown in Fig. 20-9a is subjected to an internal pressure of $p_1 = 300$ psi. The constants of the shell are R_C (cylinder) = 1.5 ft, R_D (dome) = 3.0 ft, $\phi_1 = 30°$, $L = 5$ ft, $h = 1$ in., $\nu = 0.3$, and $E = 3 \times 10^7$ psi.

The following notation is used for the actual and membrane deformations:

β_{Di} = actual rotation of dome at junction i due to applied load and edge correction forces

β_{Di}^0 = rotation of dome at junction i due to applied load from membrane analysis

Similar notation applies for other responses: β_{Ci}, ΔR_{Di}, ΔR_{Ci}, β_{Ci}^0, ΔR_{Di}^0, ΔR_{Ci}^0, etc. For the influence coefficients

$\beta_{Di}^{M_j}$ = rotation of dome at junction i caused by unit moment at junction j

$\Delta R_{Di}^{M_j}$ = horizontal displacement of dome at junction i caused by unit moment at junction j

This pattern applies for the remainder of the influence coefficients $\beta_{Di}^{H_j}$, $\Delta R_{Di}^{H_j}$, $\beta_{Ci}^{M_j}$, $\Delta R_{Ci}^{M_j}$, $\beta_{Ci}^{H_j}$, $\Delta R_{Ci}^{H_j}$, in which H denotes a unit horizontal load and C refers to the cylinder.

Proceed as follows to solve this shell problem:

Step 1: Divide the whole shell into two segments, i.e., a dome and a cylinder. The segments and junctions as well as the edge loads that act on each segment are shown in Fig. 20-9b. Calculate the deformations at the edges of the dome and cylinder due to the applied loads through a membrane analysis.

The formulas for the membrane deformations of the dome are read from Table 20-11, part A, case 4:

$$\beta_{D1}^0 = 0$$

$$\Delta R_{D1}^0 = -\frac{R_D^2(-p_1)}{2Eh}(1 - \nu)\sin\phi_1 = 2.268 \times 10^{-3} \text{ in.} \tag{1}$$

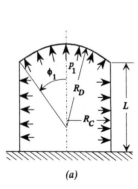

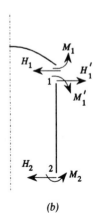

(a)   (b)

FIGURE 20-9 Cylinder with spherical dome subject to internal pressure. (a) The configuration of Example 20.4. (b) Edge loads acting on segments.

For the cylinder, membrane deformations are from Table 20-13, part A, case 1:

$$\beta_{C1}^0 = 0$$

$$\Delta R_{C1}^0 = \frac{R_C^2 p_1}{Eh} = 3.24 \times 10^{-3} \text{ in.} \tag{2}$$

It is seen that at the connection edge between the dome and cylinder, the displacements are not compatible. There is a gap

$$\Delta = \Delta R_{D1}^0 - \Delta R_{C1}^0 = -9.72 \times 10^{-4} \text{ in.} \tag{3}$$

between the dome and the cylinder.

The membrane deformations at the lower edge of the cylinder can also be found from case 1 of Table 20-13, part A, to be

$$\beta_{C2}^0 = 0$$

$$\Delta R_{C2} = \frac{p_1 R_C^2}{Eh} = 3.24 \times 10^{-3} \tag{4}$$

It is evident that ΔR_{C2} is not compatible with the clamped boundary condition.

In order for the deformations to be compatible at the connection lines, correction forces of M and H (Fig. 20-9b) at the edges are needed.

Step 2: Calculate the deformations at the edges of the dome and cylinder due to unit correction forces, i.e., calculate the influence coefficients.

The rotation of the dome lower edge due to the unit edge moment is found in case 6 of Table 20-11, part C:

$$\beta_{D1}^{M_1} = -\frac{4k_D^3}{EhR_D} \tag{5}$$

From Table 20-11, for the sphere, $k = k_D = [3(1 - \nu^2)(R_D/h)^2]^{1/4} = 7.712$, so that

$$\beta_{D1}^{M_1} = -1.699 \times 10^{-6}$$

Also, from the same table,

$$\beta_{D1}^{H_1} = \frac{2k_D^2}{Eh} \sin \phi_1 = 1.9825 \times 10^{-6}$$

$$\Delta R_{D1}^{M_1} = \frac{2k_D^2}{Eh} \sin \phi_1 = 1.9825 \times 10^{-6} \text{ in.} \tag{6}$$

$$\Delta R_{D1}^{H_1} = -\frac{R_D}{Eh} \sin \phi_1 (2k_D \sin \phi_1 - \nu \cos \phi_1) = -4.4713 \times 10^{-6} \text{ in.}$$

For the cylinder, if $kL > 4$ the influence of the edge forces of one edge on the deformations on other edge of the cylinder being treated is usually negligible. From Table 20-13,

$$k = k_C = \left[3(1 - \nu^2)\right]^{1/4}/\sqrt{R_C h} = 0.303 \text{ in.}^{-1} \tag{7}$$

and $kL = 18.18$. To verify the weakness of the effect of the opposing edge, several of the influence coefficients will be computed using Table 20-13, part B. The influence coefficients that connect opposite cylinder edges are $\Delta R_{C1}^{H_2}$ and $\beta_{C1}^{H_2}$. The influence coefficient $\Delta R_{C1}^{H_2}$ is the horizontal displacement of the cylinder at junction 1 caused by a unit horizontal force at junction 2. It is obtained from page 1248, case 1 (corresponding to H_2) and column 3 (corresponding to ΔR_{C1}). This gives

$$\Delta R_{C1}^{H_2} = \frac{2 R_C^2 k}{Eh} \frac{\overline{F}_9}{\overline{F}_1} \tag{8a}$$

Similarly, from case 1, column 4

$$\beta_{C1}^{H_2} = \frac{2 R_C^2 k^2}{Eh} \frac{2\overline{F}_8}{\overline{F}_1} \tag{8b}$$

where $\overline{F}_1$, $\overline{F}_8$, and $\overline{F}_9$ are read from Table 20-9:

$$\overline{F}_1 = \sinh^2 kL - \sin^2 kL = 1.545 \times 10^{15}$$
$$\overline{F}_8 = \sinh kL \sin kL = -2.439 \times 10^7$$
$$\overline{F}_9 = \cos kL \sin kL - \sinh kL \cos kL = -5.521 \times 10^7$$

The ratios $\overline{F}_9/\overline{F}_1$ and $\overline{F}_8/\overline{F}_1$ are of the order 10^{-8}, which makes the influence coefficients in (8) very small. Thus the influence of loads at one edge on the deformations at the other is negligible. Hence, only the coefficients for the influence of the unit forces on their own edges need to be calculated.

Table 20-13, part B, lists the influence coefficients

$$\beta_{C1}^{H_1'} = \beta_{C2}^{H_2} = \frac{2 R_C^2 k_C^2}{Eh} \frac{\overline{F}_2}{\overline{F}_1} = 1.983 \times 10^{-6}$$

$$\Delta R_{C1}^{H_1'} = -\Delta R_{C2}^{H_2} = \frac{2 R_C^2 k_C}{Eh} \frac{\overline{F}_4}{\overline{F}_1} = 6.5448 \times 10^{-6} \text{ in.}$$

$$\beta_{C1}^{M_1'} = -\beta_{C2}^{M_2} = \frac{2 R_C^2 k_C^3}{Eh} 2\frac{\overline{F}_3}{\overline{F}_1} = 1.202 \times 10^{-6} \tag{9}$$

$$\Delta R_{C1}^{M_1'} = \Delta R_{C2}^{M_2} = \frac{2 R_C^2 k_C^2}{Eh} \frac{\overline{F}_2}{\overline{F}_1} = 1.983 \times 10^{-6} \text{ in.}$$

where, from Table 20-9,

$$\bar{F}_2 = \sinh^2 k_C L + \sin^2 k_C L = 1.545 \times 10^{15}, \qquad \bar{F}_2/\bar{F}_1 = 1$$

$$\bar{F}_3 = \sinh k_C L \cosh k_C L + \sin k_C L \cos k_C L = 1.545 \times 10^{15}, \qquad \bar{F}_3/\bar{F}_1 = 1$$

$$\bar{F}_4 = \sinh k_C L \cosh k_C L - \sin k_C L \cos k_C L = 1.545 \times 10^{15}, \qquad \bar{F}_4/\bar{F}_1 = 1$$

Step 3: Find the expressions of the actual deformations at the edges for each segment.

The actual deformations of the dome at junction 1 are (case 2 of Table 20-10, part A)

$$\begin{bmatrix} \beta_{D1} \\ \Delta R_{D1} \end{bmatrix} = \mathbf{f}_D \begin{bmatrix} M_1 \\ H_1 \end{bmatrix} + \begin{bmatrix} \beta_{D1}^0 \\ \Delta R_{D1}^0 \end{bmatrix} = \begin{bmatrix} \beta_{D1}^{M_1} & \beta_{D1}^{H_1} \\ \Delta R_{D1}^{M_1} & \Delta R_{D1}^{H_1} \end{bmatrix} \begin{bmatrix} M_1 \\ H_1 \end{bmatrix} + \begin{bmatrix} \beta_{D1}^0 \\ \Delta R_{D1}^0 \end{bmatrix} \quad (10)$$

$$\mathbf{v}_{D1} = \mathbf{f}_D \mathbf{p}_{D1} + \mathbf{v}_{D1}^0$$

where

$$\mathbf{v}_{D1} = [\beta_{D1} \quad \Delta R_{D1}]^\mathrm{T}, \qquad \mathbf{p}_{D1} = [M_1 \quad H_1]^\mathrm{T}, \qquad \mathbf{v}_{D1}^0 = [\beta_{D1}^0 \quad \Delta R_{D1}^0]^\mathrm{T}$$

The deformations of the cylinder at its two edges are (case 1 of table 20-10, part A)

$$\begin{bmatrix} \beta_{C1} \\ \Delta R_{C1} \\ \beta_{C2} \\ \Delta R_{C2} \end{bmatrix} = \mathbf{f}_C \begin{bmatrix} M_1' \\ H_1' \\ M_2 \\ H_2 \end{bmatrix} + \begin{bmatrix} \beta_{C1}^0 \\ \Delta R_{C1}^0 \\ \beta_{C2}^0 \\ \Delta R_{C2}^0 \end{bmatrix} \quad (11)$$

or

$$\begin{bmatrix} \mathbf{v}_{C1} \\ \mathbf{v}_{C2} \end{bmatrix} = \begin{bmatrix} \mathbf{f}_{C11} & \mathbf{f}_{C12} \\ \mathbf{f}_{C21} & \mathbf{f}_{C22} \end{bmatrix} \begin{bmatrix} \mathbf{p}_{C1} \\ \mathbf{p}_{C2} \end{bmatrix} + \begin{bmatrix} \mathbf{v}_{C1}^0 \\ \mathbf{v}_{C2}^0 \end{bmatrix}$$

where

$$\mathbf{v}_{C1} = [\beta_{C1} \quad \Delta R_{C1}]^\mathrm{T}, \qquad \mathbf{v}_{C2} = [\beta_{C2} \quad \Delta R_{C2}]^\mathrm{T}$$

$$\mathbf{p}_{C1} = [M_1' \quad H_1']^\mathrm{T}, \qquad \mathbf{p}_{C2} = [M_2 \quad H_2]^\mathrm{T}$$

$$\mathbf{f}_{C11} = \begin{bmatrix} \beta_{C1}^{M_1'} & \beta_{C1}^{H_1'} \\ \Delta R_{C1}^{M_1'} & \Delta R_{C1}^{H_1'} \end{bmatrix}, \qquad \mathbf{f}_{C12} = \begin{bmatrix} \beta_{C1}^{M_2} & \beta_{C1}^{H_2} \\ \Delta R_{C1}^{M_2} & \Delta R_{C1}^{H_2} \end{bmatrix}$$

$$\mathbf{f}_{C21} = \begin{bmatrix} \beta_{C2}^{M_1'} & \beta_{C2}^{H_1'} \\ \Delta R_{C2}^{M_1'} & \Delta R_{C2}^{H_1'} \end{bmatrix}, \qquad \mathbf{f}_{C22} = \begin{bmatrix} \beta_{C2}^{M_2} & \beta_{C2}^{H_2} \\ \Delta R_{C2}^{M_2} & \Delta R_{C2}^{H_2} \end{bmatrix}$$

Step 4: Establish the equilibrium and compatibility equations at the junctions.

At junction 1, the equilibrium and compatibility equations are (case 1 of Table 20-10, part B)

$$\mathbf{p}_{D1} = -\mathbf{p}_{C1}, \qquad \mathbf{v}_{D1} = \mathbf{v}_{C1} \quad (12)$$

At junction 2 (case 3 of Table 20-10, part B)

$$\mathbf{v}_{C2} = \mathbf{0} \tag{13}$$

Step 5: Form the algebraic equations for the unknown correction forces. Rearrange (10) and (11) using the conditions of (12) and (13) to obtain

$$\mathbf{v}_{C1} + \mathbf{f}_D \mathbf{p}_{C1} = \mathbf{v}_{D1}^0 \tag{14a}$$

$$\mathbf{v}_{C1} - \mathbf{f}_{C11}\mathbf{p}_{C1} - \mathbf{f}_{C12}\mathbf{p}_{C2} = \mathbf{v}_{C1}^0 \tag{14b}$$

$$- \mathbf{f}_{C21}\mathbf{p}_{C1} - \mathbf{f}_{C22}\mathbf{p}_{C2} = \mathbf{v}_{C2}^0 \tag{14c}$$

This is a set of linear algebraic equations with unknowns $\mathbf{v}_{C1}$, $\mathbf{p}_{C1}$, and $\mathbf{p}_{C2}$. Note that there are six equations for six unknowns.

As is mentioned earlier, the influence of the forces on one end of the cylinder on the deformations of the other end can be ignored. This condition leads to $\mathbf{f}_{C12} = \mathbf{f}_{C21} = \mathbf{0}$. The omission of the coupling between edges separates the problem into one set of two equations for M_1 and H_1 and another set of two equations for M_2 and H_2. Subtract (14b) from (14a) to find

$$(\mathbf{f}_D + \mathbf{f}_{C11})\mathbf{p}_{C1} = \mathbf{v}_{D1}^0 - \mathbf{v}_{C1}^0 \tag{15}$$

i.e.,

$$\left(\beta_{D1}^{M_1} + \beta_{C1}^{M_1'}\right)M_1 + \left(\beta_{D1}^{H_1} + \beta_{C1}^{H_1'}\right)H_1 = \beta_{D1}^0 - \beta_{C1}^0$$

$$\left(\Delta R_{D1}^{M_1} + \Delta R_{C1}^{M_1'}\right)M_1 + \left(\Delta R_{D1}^{H_1} + \Delta R_{C1}^{H_1'}\right)H_1 = \Delta R_{D1}^0 - \Delta R_{C1}^0$$

or

$$(-4.97 \times 10^{-7})M_1 + (3.966 \times 10^{-6})H_1 = 0$$

$$(3.966 \times 10^{-6})M_1 + (2.0735 \times 10^{-6})H_1 = -9.72 \times 10^{-4} \tag{16}$$

and (14c) becomes

$$\mathbf{f}_{C22}\mathbf{p}_{C2} = -\mathbf{v}_{C2}^0$$

or

$$\beta_{C2}^{M_2}M_2 + \beta_{C2}^{H_2}H_2 = -\beta_{C2}^0, \qquad \Delta R_{C2}^{M_2}M_2 + \Delta R_{C2}^{H_2}H_2 = -\Delta R_{C2}^0 \tag{17}$$

or

$$(-1.202 \times 10^{-6})M_2 + (1.9831 \times 10^{-6})H_2 = 0$$

$$(1.9831 \times 10^{-6})M_2 + (-6.5448 \times 10^{-6})H_2 = -3.24 \times 10^{-3}$$

Step 6: Solve the linear equations to find the correction forces. Equations (16) and (17) yield

$$M_1 = -230.0 \text{ lb-in./in.}, \qquad H_1 = -28.82 \text{ lb/in.} \tag{18}$$

and

$$M_2 = 990.1 \text{ lb-in./in.}, \qquad H_2 = 1633.9 \text{ lb/in.} \tag{19}$$

20.5 OTHER SHELLS

In the previous sections, responses of shells of revolution are presented. In this section, membrane responses of other types of shells will be given. Table 20-14 gives the stress resultants for some membrane shells.

Note that in the figure of case 2 of Table 20-14, the two dashed lines are parabolas, one of which opens downward and another upward. The shell is formed by sliding the parabola in the $y0z$ plane along the parabola in the $x0z$ plane, or vice versa. The boundaries of the shell are all parabolas.

Such cases as the cylindrical shells in cases 3–7 of Table 20-14 are said to have *shear diaphragms* at the ends $(x = \pm \frac{1}{2}L)$. For these boundaries, the radial and circumferential displacements, the force in the axial direction, and the moment about the tangent of the circumferential wall contour are all zero at the boundary. These conditions can be closely approximated in physical applications by rigidly attaching a thin, flat, cover plate at each end. The plate would have considerable stiffness in its own plane such that the displacements v and w are restrained. However, the plate, by virtue of its thinness, would have very little stiffness in the x direction transverse to its plane. Consequently, a plate generates a negligible longitudinal membrane force N_x in the shell as the shell deforms (Fig. 20-10). Also, corresponding to $M_x = 0$ is the condition that there is no restraint against rotation about the circumferential boundary. The name shear diaphragm reflects the capability of the plate to supply shearing forces $N_{x\theta}$ to the shell. The shear diaphragm boundary condition is often called *simply supported*. The term simply supported is borrowed from linear beam and plate theory. However, the shear diaphragm explanation is usually considered to be more appropriate for shell theory [20.6].

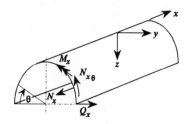

FIGURE 20-10 Boundary forces for a cylindrical shell.

20.6 STABILITY

When thin shells are subject to axial compression, torsion, bending, lateral pressure, or a combination of these loads, elastic buckling of the shell wall can occur for certain critical values of the applied loads. Unlike columns and thin plates for which the buckling loads from classical small-deflection theory are considered to be reasonably realistic, the buckling loads for some types of shells may be much less than the load predicted by the theory. Sources of this deviation of buckling loads may be from the dependence of the buckling loads on the deviations from the nominal shape of the structure or on the local edge conditions. In many cases empirical formulas are used to predict the buckling loads.

Table 20-15 presents formulas for the recommended design-allowable buckling loads for spheres, truncated cones, and cylindrical shells.

The simply supported boundary condition is defined in Section 20.5. If the shell is under bending deformation, in addition to N_x, the shear diaphragm generates negligible bending moment M_x. Also, the shear diaphragm supplies shear force Q_x together with $N_{x\theta}$ (Fig. 20-10).

Many additional cases of buckling loads are provided in reference 20.5.

Example 20.5 Axial Compression of a Simply Supported Circular Cylindrical Shell Compute the theoretical and empirical critical axial loads for a simply supported cylindrical shell with the properties $L = 0.6096$ m, $R = 0.198$ m, $h = 39\,624$ mm, $E = 207$ GPa, and $\nu = 0.3$. These give $R/h = 500$ and $D = Eh^3/12(1 - \nu^2) = 1.179$ Nm. From Table 20-15 for cylindrical shells

$$Z = L^2(1 - \nu^2)^{1/2}/Rh$$

$$= (0.6096)^2(1 - 0.3^2)^{1/2}/[(0.198)(3.9624 \times 10^{-4})] = 4515.7 \quad (1)$$

Then, case 9 gives K theoretical of 3.503 and K empirical of 3.059 for $R/h = 500$. Then,

$$\begin{aligned} K_c &= 3187 \qquad \text{for the theoretical formula} \\ K_c &= 1145 \qquad \text{for the empirical formula} \end{aligned} \quad (2)$$

Thus the critical loads are found to be as follows:

Theoretical formula:

$$\sigma_{\text{cr}} = K_c \frac{\pi^2 D}{L^2 h} = 3187 \times \frac{\pi^2 \times 1.179}{0.6096^2 \times 3.9624 \times 10^{-4}}$$

$$= 2.517 \times 10^8 \text{ N/m}^2 = 251.7 \text{ MPa} \quad (3)$$

$$N_{x,\text{cr}} = \sigma_{\text{cr}} h = 2.517 \times 10^8 \times 3.9624 \times 10^{-4} = 99.734 \text{ N/m}$$

Empirical formula:

$$\sigma_{\text{cr}} = K_c \frac{\pi^2 D}{L^2 h} = 1331 \times \frac{\pi^2 \times 1.179}{0.6096^2 \times 3.9624 \times 10^{-4}}$$

$$= 1.05 \times 10^8 \text{ N/m}^2 = 105.1 \text{ MPa} \quad (4)$$

$$N_{x,\text{cr}} = \sigma_{\text{cr}} h = 1.051 \times 10^8 \times 3.9624 \times 10^{-4} = 41\,645 \text{ N/m}$$

In this case the theoretical buckling load is 2.38 times as large as the value of the empirical load. More credence is usually given to results based on the empirical formula.

═══════

Example 20.6 Simply Supported Circular Cylindrical Shell Subjected to Axial Compressive Load and Internal Pressure Suppose a simply supported cylindrical shell of Example 20.5 is subjected to an internal pressure of 6.895 MPa in addition to an axial load. Compute the critical value of the axial load.

Case 10 of Table 20-15 provides the critical axial load.

$$\frac{R}{h} = \frac{0.198}{3.9624 \times 10^{-4}} = 500, \; \bar{p} = \frac{p_1}{E(R/h)^2} = \frac{6.895 \times 10^6}{(2.07 \times 10^{11})500^2}$$

$$= 1.333 \times 10^{-10},$$

$K_c = 0.2786$

$$\sigma_{cr} = K_c(Eh/R) = (0.2786)(2.07 \times 10^{11})/(500) = 115.4 \; \text{MPa} \qquad (1)$$

$$P_{cr} = \sigma_{cr}(2\pi Rh) + p_1\pi R^2 = (115.4 \times 10^6)(2\pi)(0.198)(3.9624 \times 10^{-4})$$

$$+ (6.895 \times 10^6)(0.198^2)(\pi)$$

$$= 919 \; \text{MN} \qquad (2)$$

$N_{x,cr} = P_{cr}/2\pi R$

$$= 9.19 \times 10^5/(2\pi)(0.198) = 738\,704 \; \text{N/m} \qquad (3)$$

The effect of the internal pressure is to increase the critical axial load by a factor of $738\,704/41\,645 = 17.7$ over the empirical load with no internal pressure.

═══════

Example 20.7 Simply Supported Truncated Conical Shell under Axial Compression Compute the compressive axial load for axisymmetric and asymmetric buckling for a truncated cone that has the properties (see figure in case 3, Table 20-15)

$$h = 0.0156 \; \text{in.}, \qquad \alpha = 30°, \qquad E = 3 \times 10^7 \; \text{psi}, \qquad \nu = 0.3$$

$$x_1 = 1 \; \text{ft}, \qquad x_2 = 3 \; \text{ft}, \qquad r_1 = x_1 \sin \alpha = 0.5 \; \text{ft}$$

From case 3 in Table 20-15 the axial load for axisymmetric buckling is found to be

$$P_{cr} = 2Eh^2\pi \cos^2 \alpha/\sqrt{3(1 - \nu^2)}$$

$$= 2(3 \times 10^7)(0.0156)^2\pi \cos^2(30°)/\sqrt{3(1 - 0.3^2)} = 20{,}822 \; \text{lb} \qquad (1)$$

Also, from case 3 of Table 20-15 the minimum axial load for asymmetric buckling is

$$\sigma_{cr} = \sigma_{cr,A}\left(\frac{1}{2}\frac{1 + x_1/x_2}{1 - x_1/x_2} \log \frac{x_2}{x_1}\right)^{1/2} \qquad (2)$$

where

$$\sigma_{cr, A} = E \frac{h}{r_1} \frac{1}{\sqrt{3(1 - \nu^2)}} \cos \alpha = 23{,}603.9 \text{ lb/in.}^2 \tag{3}$$

Then

$$\sigma_{cr} = \sigma_{cr, A} \left(\frac{1}{2} \frac{1 + x_1/x_2}{1 - x_1/x_2} \log \frac{x_2}{x_1} \right)^{1/2} = 16{,}304.15 \text{ lb/in.}^2 \tag{4}$$

Finally

$$(P_{cr})_{\text{asymmetric}} = \sigma_{cr}(\pi x_1 h \sin 2\alpha) = 8303.95 \text{ lb} \tag{5}$$

Example 20.8 Complete Spherical Shell under External Pressure Compute the empirical critical values of the external pressure for a sphere for which

$$R = 3 \text{ ft}, \qquad h = 0.0156 \text{ in.}, \qquad E = 3 \times 10^7 \text{ psi}, \qquad \nu = 0.3$$

From case 1 of Table 20-15 the empirical buckling pressure is

$$(p_1)_{cr} = \frac{(0.8)E}{\sqrt{1 - \nu^2}} \left(\frac{h}{R} \right)^2 = 4.725 \text{ lb/in.}^2 \tag{1}$$

20.7 NATURAL FREQUENCIES

For the dynamics of shells, the governing equations of motion are a set of partial differential equations, including derivatives with respect to time. Since many simplifications can be invoked, numerous forms of governing equations of motion for shells have been derived.

Leissa [20.6] discusses a variety of forms of the equations of motion of thin shells. All of these small displacement theories utilize the Love–Kirchhoff hypothesis that:

1. The shell is thin, i.e., $h/R \ll 1$.
2. The problem is linear, which allows all calculations to be referred to the original configuration of the shell.
3. Transverse stresses normal to the middle surface are negligible.
4. Straight lines initially normal to the middle surface remain normal to that surface after deformation, and they undergo no extension.

Residual stresses, anisotropy, variable thickness, shear deformation, rotary inertia, nonlinearities, and the influence of the external environment are ignored.

References 20.7 and 20.8 are representative of the literature providing approximate frequencies for shells.

Circular Cylindrical Shells

The simplest set of equations for the bending of a cylindrical shell is probably that of Donnell-Mushtari [20.6, 20.9]. With respect to the coordinates shown in Fig. 20-11, these equations of motion are

$$\frac{\partial^2 u}{\partial s^2} - \frac{1-\nu}{2}\frac{\partial^2 u}{\partial \theta^2} + \frac{1+\nu}{2}\frac{\partial^2 v}{\partial s\, \partial \theta} - \nu\frac{\partial w}{\partial s} = \rho^* \frac{(1-\nu^2)R^2}{E}\frac{\partial^2 u}{\partial t^2}$$

$$-\frac{1+\nu}{2}\frac{\partial^2 u}{\partial s\, \partial \theta} - \frac{1-\nu}{2}\frac{\partial^2 v}{\partial s^2} + \frac{\partial^2 v}{\partial \theta^2} + \frac{\partial w}{\partial \theta} = -\rho^* \frac{(1-\nu^2)R^2}{E}\frac{\partial^2 v}{\partial t^2} \quad (20.12)$$

$$\nu\frac{\partial u}{\partial s} + \frac{\partial v}{\partial \theta} - w - \kappa\, \nabla^4 w = \rho^*(1-\nu^2)\frac{R^2}{E}\frac{\partial^2 w}{\partial t^2}$$

$$s = x/R$$

where $\kappa = h^2/(12R^2)$. When the term containing κ is ignored, these equations reduce to those for membrane shells. Modifications to Eqs. (20.12) for other theories are presented in reference 20.6. The discussion here will be limited to the theory of Donnell-Mushtari and that of Flügge [20.9], which involves a modified version of Eqs. (20.12) [20.6].

Tables 20-16 and 20-17 give the natural frequencies of cylindrical shells under various end conditions. The basic assumption for the mode shapes u, v, and w in the x, y, and z directions are that they are formed of n and m half waves in the circumferential and longitudinal directions, i.e.,

$$u = A \cos \lambda s \cos n\theta \cos \omega t, \qquad v = B \sin \lambda s \sin n\theta \cos \omega t,$$
$$w = C \sin \lambda s \cos n\theta \cos \omega t \qquad\qquad\qquad (20.13)$$

where A, B, and C are constants and $\lambda = m\pi R/L$. Substitution of the assumed

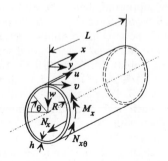

FIGURE 20-11 Coordinates for a cylindrical shell.

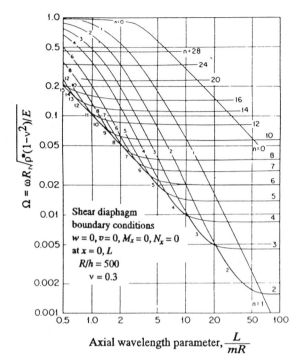

FIGURE 20-12 Variation of the frequency parameter Ω for a cylinder [20.6].

shapes into the governing equations of motion of Eq. (20.12) yields the equations for the natural frequencies and mode shapes.

The cylindrical shell of infinite length is rather simple to analyze. It is assumed that the wavelength in the x direction is infinitely long, i.e., the mode shapes are independent of x, and hence the terms containing λ in Eqs. (20.13) are removed. For this shell, the axial motion is independent of the radial and torsional displacements. The shell only vibrates radially and torsionally. Three natural frequencies exist for each n. For $n = 0$, the three modes are independent, but otherwise the radial and torsional modes are coupled. For very large n, the radial and torsional modes tend to become independent again.

For the case of simply supported shells, the end conditions are such that the v,w displacements as well as the force N_x in the longitudinal direction and the moment M_x (Fig. 20-11) in the circumferential direction are zero. These boundary conditions correspond to the shear diaphragm discussed in Section 20.5.

Note that in case 2 of Tables 20-16 and 20-17, the cubic equation for the frequency parameter Ω^2 ($= \rho^*(1 - v^2)R^2\omega^2/E$) will have three roots for fixed values of n and λ ($= m\pi R/L$). Thus a shell of a given length may vibrate in any of three distinct modes, each with the same number of circumferential and longitudinal waves, and each with its own distinct frequency. The modes associated with each frequency can be classified as primarily radial (or flexural), longitudinal (or axial), or circumferential (or torsional). The lowest frequency is usually associated with a motion that is primarily radial.

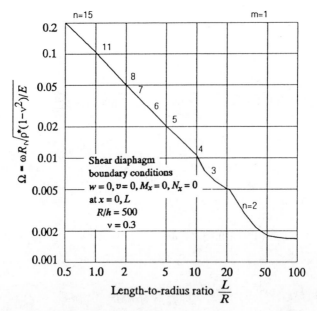

FIGURE 20-13 Fundamental frequency parameter Ω for various L / R ratios of a cylinder with $R / h = 500$ and $m = 1$ [20.6].

Unlike most of the other structural members, such as beams and bars, the fundamental frequency of circular cylindrical shells is not always associated with the smallest numbers for n and m. Also, the frequency does not necessarily increase monotonically with increasing values of the number of half waves m and n in Eq. (20.13). Figure 20-12 shows the relationship between the frequency parameter Ω and the indices n and m for a fixed ratio R/h. It is seen that for a particular value of the number of circumferential waves n, the smaller the m, the smaller the Ω, so the fundamental frequency always occurs for $m = 1$. With $m = 1$, the index number n that is associated with the fundamental frequency strongly depends on the value L/R. In Fig. 20-13, which shows the fundamental frequency in terms of length-to-radius ratio L/R and the index number n, for $L/R = 5$, the fundamental frequency corresponds to $m = 1$ and $n = 5$, while for $L/R = 1$, the fundamental frequency corresponds to $n = 11$. Thus for a specific thin cylinder, numerous n should be scrutinized to determine which one is associated with the fundamental frequency.

Example 20.9 Natural Frequencies of an Infinite Membrane Cylinder Compute the natural frequencies at $n = 0, 1, 2, 3, 4$ for an infinitely long circular cylindrical membrane. For this shell $R = 0.762$ m, $L \rightarrow \infty$, $\nu = 0.3$, $\rho^* = 7747.6$ kg/m^3, and $E = 207$ GPa. The frequency parameter Ω^2 is taken from Table 20-16, part A, case 1, and then the frequency is obtained using

$$\omega^2 = E\Omega^2 / \left[\rho^* (1 - \nu^2) R^2 \right] \qquad (1)$$

For $n = 0$, case 1 of Table 20-16 gives

$$\Omega_a^2 = 0, \qquad \Omega_{rt}^2 = 1$$

$$\omega_{rt}^2 = \frac{E\Omega_{rt}^2}{\rho^*(1 - \nu^2)R^2} = \frac{(2.07 \times 10^{11})\Omega_{rt}^2}{(7747.6)(1 - 0.3^2)(0.762)^2} = 5.056 \times 10^7$$

Then

$$\omega_{rt} = 7110.6 \text{ rad/s} \quad \text{and} \quad f_{rt} = \omega_{rt}/2\pi = 1131.68 \text{ Hz} \tag{2}$$

Similarly, for $n = 1$,

$$\Omega_a^2 = \tfrac{1}{2}(1 - 0.3)(1)^2 = 0.35 \qquad \omega_a^2 = 1.7696 \times 10^7 \quad \text{or} \quad \omega_a = 4206.66 \text{ rad/s}$$

and

$$f_a = 669.51 \tag{3}$$

Note that the lowest frequency is not associated with $n = 0$. Also, for $n = 1$,

$$\Omega_{rt}^2 = (1 + 1^2) = 2$$

$$\omega_{rt}^2 = 1.0112 \times 10^8 \quad \text{or} \quad \omega_{rt} = 10\,055.8 \text{ rad/s}$$

and

$$f_{rt} = 1\,600.43 \text{ Hz} \tag{4}$$

Continue the computation for $n = 2, 3, 4$:

n	f_{rt} (Hz)	f_a (Hz)	
2	2529.54	1338.51	
3	3577.31	2007.76	(5)
4	4664.24	2677.02	

Example 20.10 Radial (Bending)–Torsional Frequencies of an Infinite Cylinder with Bending Compute the radial–torsional natural frequencies of an infinitely long circular cylindrical shell of $R = 30$ in., $\rho^* = 72.5 \times 10^{-5}$ lb-s^2/in.^{4}, $h = 3.16$ in., $E = 3 \times 10^7$ psi, and $\nu = 0.3$ using the bending formulas for $n = 0, 1, 2, 3, 4$. From Table 20-17

$$\kappa = h^2/12R^2 = 9.25 \times 10^{-4}$$

The natural frequencies ω_{rt} are obtained from

$$\omega_{rt}^2 = \frac{E\Omega_{rt}^2}{\rho^*(1 - \nu^2)R^2}$$

where the formulas for Ω_{rt} are taken from case 1 of Table 20-17. With $f_{rt} = \omega_{rt}/2\pi$:

n	Donnell-Mushtari f_{rt} (Hz)	Flügge f_{rt} (Hz)
0	0.0	0.0
	1131.3	1131.3
1	24.3	29.8
	1600.1	1599.96
2	123.18	130.67
	2530.37	2529.99
3	294.12	302.22
	3578.74	3578.07
4	535.14	543.52
	4666.21	4665.24

For each n there are two natural frequencies that correspond to the minus and plus signs in the equations for Ω_{rt} in case 1 of Table 20-17, part A. The higher values of these two frequencies for $n = 0, \ldots, 4$ do not differ significantly from those values computed using the membrane theory. This is because k is very small for this example, and the terms kn^4 and kn^6 in case 1 of Table 20-17, part A, are negligible for small n. For large n, the difference between the two theories increases.

=====

Example 20.11 Simply Supported (by Shear Diaphragms) Circular Cylindrical Membrane Suppose the shell treated in Example 20.9 has a length of 9.144 m. Compute the frequencies associated with the axisymmetric mode ($n = 0$) if $m = 1$.

The parameter λ of Table 20-16, part A, is

$$\lambda = m\pi R/L = \pi \times 0.762/9.144 = 0.2618$$

From case 2 of Table 20-16, part A, for torsional motion with $n = 0$ ___ (1)

$$\Omega_t^2 = \tfrac{1}{2}(1 - \nu)\lambda^2 = 0.024$$

and with $\omega_t^2 = E\Omega_t^2/[\rho^*(1 - \nu^2)R^2]$,

$$\omega_t^2 = 1.212 \times 10^6 \quad \text{or} \quad f_t = 175.2 \text{ Hz} \qquad (2)$$

For axial–radial modes with $n = 0$

$$\Omega^2_{ar} = \tfrac{1}{2}\left(\left[1 + (0.2618)^2\right] \pm \left\{\left[1 - (0.2618)^2\right]^2 + 4(0.3)^2(0.2618)^2\right\}^{1/2}\right)$$

so that $\Omega^2_{ar} = 0.0620$ and 1.0066. Then, from $\omega^2_{ar} = E\Omega^2_{ar}/[\rho^*(1 - \nu^2)R^2]$,

$$\omega^2_{ar} = (5.056 \times 10^7)(0.0620) = 3.135 \times 10^6$$

and

$$\omega^2_{ar} = (5.056 \times 10^7)(1.0066) = 5.085 \times 10^7$$

Thus,

$$f_{ar} = 281.8 \text{ Hz} \quad \text{and} \quad 1134.9 \text{ Hz} \tag{3}$$

The two frequencies given here correspond to the minus and plus signs of the equation in case 2 of Table 20-16, part A, for the frequency parameter Ω^2_{ar} at $n = 0$. In this case two distinct frequencies have the same coupled axial–radial mode shapes.

Example 20.12 Simply Supported Cylindrical Shell with Bending Repeat the frequency computation of Example 20.11 using the bending formulas for $n = 0$ and $m = 1$.

From case 2 of Table 20-17, part A, it is seen that the torsional mode with $f_t = 175.2$ Hz is the same as it was under the membrane hypothesis in Example 20.11.

For coupled axial–radial modes, with $\kappa = h^2/12R^2 = 92.5 \times 10^{-5}$, the relations of case 2 of Table 20-17, part A, lead to identical results for the Flügge and Donnell-Mushtari theories. These are

$$f_{ar} = 281.5 \text{ Hz} \quad \text{corresponds to minus sign}$$
$$f_{ar} = 1134.7 \text{ Hz} \quad \text{corresponds to plus sign}$$

Thus, for $n = 0$, the bending theory yields results essentially the same as those obtained in Example 20.11 for the membrane case of coupled axial–radial motion.

Conical Shells

The coordinates used to describe the conical shell are shown in Fig. 20-14. The term *conical* refers here to a right circular cone that may be truncated.

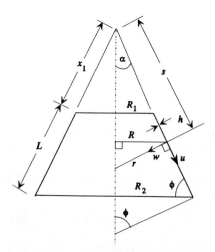

FIGURE 20-14 Coordinates for a conical shell.

Information for computing the natural frequencies of cones and conical frustums is presented in Table 20-18. It should be noted that the lowest frequency of the shell does not necessarily occur for $n = 0$.

Example 20.13 Frequencies of Axisymmetric Modes for a Complete Conical shell with a Clamped Base Compute the frequencies of the first three axisymmetric modes of a conical Shell with a clamped base and $h = 0.125$ in. $= 3.175 \times 10^{-3}$ m, $R = 4.0$ in. $= 0.1016$ m, $\alpha = 30°$, $\nu = 0.3$, $E = 3 \times 10^7$ psi $= 2.07 \times 10^{11}$ Pa, and $\rho^* = 725.4 \times 10^{-6}$ lb-s^2/in.$^4 = 7752$ kg/m^3.

From case 2 of Table 20-18,

$$\eta = \frac{12(1 - \nu^2)}{\tan^4 \alpha} \left(\frac{R}{h} \right)^2 = \frac{12(1 - 0.3^2)}{\tan^4(30°)} \left(\frac{0.1016}{3.175 \times 10^{-3}} \right)^2 = 100{,}638.7 \quad (1)$$

For $j = 1$, it is seen from case 2 that $\Omega_1 \approx 1.796$ ($\eta = 100{,}638.7 \approx 100{,}000$). Finally,

$$f_1 = \frac{\Omega_1}{2\pi R} \left(\frac{E}{\rho^*} \right)^{1/2} = \frac{1.796}{(2)(\pi)(0.1016)} \left(\frac{2.07 \times 10^{11}}{7752} \right)^{1/2} = 14{,}538.2 \text{ Hz} \quad (2)$$

Similarly, if $j = 2$, $\Omega_2 \approx 2.429$ and

$$f_2 = \frac{\Omega_2}{2\pi R} \left(\frac{E}{\rho^*} \right)^{1/2} = \frac{\Omega_2}{\Omega_1} f_1 = \frac{2.429}{1.796}(14{,}538.2) = 19{,}662.2 \text{ Hz} \quad (3)$$

Also, for $j = 3$, $\Omega_3 \approx 3.447$ and

$$f_3 = \frac{\Omega_3}{\Omega_1} f_1 = \frac{3.447}{1.796}(14{,}538.2) = 27{,}902.7 \text{ Hz} \tag{4}$$

Example 20.14   Conical Shell with Free Boundary   Assume that the shell treated in the previous example has a free base and compute the frequencies for the first three axisymmetric modes.

From case 1 of Table 20-18 the estimated frequency parameters are $\Omega_1 \approx$ 1.251, $\Omega_2 \approx 1.981$, and $\Omega_3 \approx 2.906$. Use

$$f_i = \frac{\Omega_i}{2\pi R_2}\left(\frac{E}{\rho^*}\right)^{1/2} = \Omega_i \frac{1}{2\pi(0.1016)}\left(\frac{2.07 \times 10^{11}}{7752}\right)^{1/2} = 8094.8\Omega_i \tag{1}$$

to find

$$f_1 = (1.251)(8094.8) = 10{,}126.59 \text{ Hz} \tag{2}$$
$$f_2 = (1.997)(8094.8) = 16{,}165.32 \text{ Hz} \tag{3}$$
$$f_3 = (2.906)(8094.8) = 23{,}523.49 \text{ Hz} \tag{4}$$

Example 20.15   Lowest Frequency of the Axisymmetric Mode of a Clamped–Free Conical Frustum   Compute the lowest frequency of the axisymmetric mode of a clamped–free conical frustum for which $\alpha = 75.49°$, $h = 0.03125$ in., $R_1 = 0.4$ in., $R_2 = 1.0$ in., $E = 1.04 \times 10^7$ psi, $\rho^* = 259.2 \times 10^{-6}$ lb-s²/in.⁴, and $\nu = 0.33$.

First calculate γ of case 6 of Table 20-18 as

$$\gamma = \frac{12(1 - \nu^2)(R_2/h)^2}{\tan^4 \alpha} = \frac{12(1 - 0.33^2)(1/0.03125)^2}{\tan^4(75.49°)} = 49.124 \tag{1}$$

Since $R_1/R_2 = 0.4$, case 6 gives $\Omega^2 = 2.473$, so that

$$f_1 = \frac{\sqrt{2.473}}{2\pi(1)}\left(\frac{1.04 \times 10^7}{259.2 \times 10^{-6}}\right)^{1/2} = 50.134 \times 10^3 \text{ Hz} \tag{2}$$

Spherical Shells

The frequencies of the axisymmetric modes of a complete spherical membrane are found using the formulas in Table 20-19.

Example 20.16 Spherical Membrane Shell Compute the fundamental radial frequency for a spherical shell with $R = 30$ ft, $h = 2$ in., $E = 3 \times 10^7$ psi, $\nu = 0.3$, and $\rho^* = 75.5 \times 10^{-5}$ lb-s^2/in.[4] From case 1 of Table 20-19, the frequency parameter for membrane theory is

$$\Omega_r^2 = \frac{2(1 + \nu)}{1 + h^2/(12R^2)}$$

so that the natural frequency is given as

$$f_1 = \frac{1}{2\pi R} \left[\frac{\Omega_r^2 E}{\rho^*(1 - \nu^2)} \right]^{1/2}$$

$$= \left[2(1 + \nu)/(1 + h^2/12R^2) \right]^{1/2} \frac{1}{2\pi R} \left[\frac{E}{\rho^*(1 - \nu^2)} \right]^{1/2} = 149.0 \text{ Hz} \quad (1)$$

REFERENCES

20.1   Baker, E. H., Kovalevsky, L., and Rish, F. L., *Structural Analysis of Shells*, Krieger, Malabar, FL, 1986.

20.2   Baker, E. H., Capelli, A. P., Kovalevsky, L., Rish, F. L., and Verette, R. M., *Shell Analysis Manual*, CR-912 NASA, Washington, DC, 1968.

20.3   Kelkar, V. S., and Sewell, R. T., *Fundamentals of the Analysis and Design of Shell Structures*, Prentice-Hall, Englewood Cliffs, NJ, 1987.

20.4   Ramaswamy, G. S., *Design and Construction of Concrete Shell Roofs*, McGraw-Hill, New York, 1968.

20.5   Column Research Committee of Japan, *Handbook of Structural Stability*, Corona, Tokyo, 1971.

20.6   Leissa, A. W., *Vibration of Shells*, NASA SP-288, Washington, DC, 1973.

20.7   Bräutigam, C., and Schnell, W. "Free Vibrations of Hyperboloidal Shells," in *Transactions of Ninth International Conference on Structural Mechanics in Reactor Technology*, A. A. Balkema, Rotterdam, 1987, pp. 445–450.

20.8   Schnell, W., and Schwarte, J. "Eigenfrequenzen von Kreiszylinderschales nach der Halbbiegetheorie," *Z. Flugwiss. Weltraumforsch.*, Vol. 13, 1989, pp. 320–325.

20.9   Flügge, W., *Stresses in Shells*, Springer-Verlag, Berlin, 1960.

20.10 Hu, W. C. L., Gormley, J. F., and Lindholm, U. S., "An Experimental Study and Inextensional Analysis of Free–Free Conical Shells," *Int. J. Mech. Sci.*, Vol. 9, 1967, pp. 123–135.

20.11 Blevins, R. D., *Formulas for Natural Frequency and Mode Shape*, Van Nostrand Reinhold, New York, 1979.

20.12 Reynolds, T. E. "A Survey of Research on the Stability of Hydrostatically-loaded Shell Structures Conducted at the David Taylor Model Basin," *NASA*, TND-1510, 1962, pp. 551–560.

20.13 Seide, P., "Axisymmetrical Buckling of Circular Cones Under Axial Compression," *J. Appl. Mech.*, 1956, pp. 625–628.

20

Tables

TABLE 20-1   MEMBRANE SPHERICAL SHELLS[a]

Notation

E = modulus of elasticity
ρ_w = specific weight of liquid
h = thickness of shell
ϕ_1, ϕ_2 = angle ϕ corresponding to lower and upper edges
q = dead weight
∇ = surface of liquid
ν = Poisson's ratio
R = radius of shell
ϕ = meridional angle of point of interest
p_1 = applied uniform pressure
p_2 = uniformly distributed loading on projected area
d = distance from surface of liquid to top or bottom of shell
N_ϕ, N_θ = normal forces per unit length (stress resultants) in meridional and circumferential (parallel) directions
p_x, p_y, p_z = components of applied load in meridional, circumferential (parallel), and normal directions
β = rotation of tangent to meridian during deformation
ΔR = displacement in direction of radius of curvature of a parallel

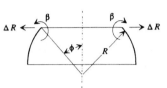

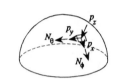

The stresses are defined as

$$\sigma_\phi = N_\phi/h \qquad \sigma_\theta = N_\theta/h$$

Membrane forces are positive if causing tension in the shell.

Part A. Loadings

Case	Components
1. Dead weight	$p_x = q \sin \phi$ $p_y = 0$ $p_z = q \cos \phi$

TABLE 20-1 (continued) MEMBRANE SPHERICAL SHELLS

Part A. Loadings

Case	Components
2. Uniformly distributed loading on projected area (snow load)	$p_x = p_2 \cos \phi \sin \phi$ $p_y = 0$ $p_z = p_2 \cos^2 \phi$
3. Hydrostatic pressure loading	$p_x = p_y = 0$ $p_z = \rho_w[d + R(1 - \cos \phi)]$ For reversed spherical shell, d is the distance from the surface of the liquid to the apex of the reversed shell, and $p_z = \rho_w[d - R(1 - \cos \phi)]$.
4. Uniform loading in z direction (pressurization)	$p_x = p_y = 0 \qquad p_z = p_1$
5. Lantern loading W, load per unit length on upper shell edge	$p_x = p_y = p_z = 0$

TABLE 20-1 **Membrane Spherical Shells** **1186**

TABLE 20-1 (continued)   **MEMBRANE SPHERICAL SHELLS**

Part B. Stress Resultants, Deformations, and Rotations

1. *Simply supported spherical cap*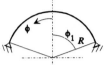

Internal Forces	Deformations
DEAD-WEIGHT LOADING: $$N_\phi = -\frac{Rq}{1 + \cos\phi}$$ $$N_\theta = -Rq\left(\cos\phi - \frac{1}{1 + \cos\phi}\right)$$	$$\Delta R = \frac{R^2 q}{Eh}\sin\phi\left[-\cos\phi + \frac{1+\nu}{\sin^2\phi}(1 - \cos\phi)\right]$$ $$\beta = -\frac{Rq}{Eh}(2 + \nu)\sin\phi$$
UNIFORM LOADING ON PROJECTED AREA: $$N_\phi = -\tfrac{1}{2}p_2 R$$ $$N_\theta = -\tfrac{1}{2}p_2 R\cos 2\phi$$	$$\Delta R = \frac{R^2 p_2}{Eh}\sin\phi\left[-\cos^2\phi + \frac{1+\nu}{2}\right]$$ $$\beta = -\frac{Rp_2}{Eh}(3 + \nu)\sin\phi\cos\phi$$
HYDROSTATIC PRESSURE LOADING: $$N_\phi = -\frac{\rho_w R^2}{6}\left(-1 + 3\frac{d}{R} - \frac{2\cos^2\phi}{1 + \cos\phi}\right)$$ $$N_\theta = -\frac{\rho_w R^2}{6}\left(-1 + 3\frac{d}{R} - \frac{4\cos^2\phi - 6}{1 + \cos\phi}\right)$$	$$\Delta R = -\frac{\rho_w R^3}{6Eh}\sin\phi\left[3(1 + \frac{d}{R})(1 - \nu)\right.$$ $$\left. - 6\cos\phi - \frac{2(1+\nu)}{\sin^2\phi}(\cos^3\phi - 1)\right]$$ $$\beta = \frac{\rho_w R^2}{Eh}\sin\phi$$
NORMAL UNIFORM LOADING: $$N_\phi = -\tfrac{1}{2}Rp_1$$ $$N_\theta = -\tfrac{1}{2}Rp_1$$	$$\Delta R = -\frac{R^2 p_1}{2Eh}(1 - \nu)\sin\phi$$ $$\beta = 0$$

TABLE 20-1 (continued)   MEMBRANE SPHERICAL SHELLS

Part B. Stress Resultants, Deformations, and Rotations

2. Reversed, simply supported

Internal Forces	Deformations
DEAD-WEIGHT LOADING:	
$$N_\phi = \dfrac{Rq}{1 + \cos\phi}$$ $$N_\theta = Rq\left(\cos\phi - \dfrac{1}{1 + \cos\phi}\right)$$	$$\Delta R = -\dfrac{R^2 q}{Eh}\sin\phi\left[-\cos\phi + \dfrac{1 + \nu}{\sin^2\phi}(1 - \cos\phi)\right]$$ $$\beta = -\dfrac{Rq}{Eh}(2 + \nu)\sin\phi$$
UNIFORM LOADING ON PROJECTED AREA:	
$$N_\phi = \tfrac{1}{2}p_2 R$$ $$N_\theta = \tfrac{1}{2}p_2 R \cos 2\phi$$	$$\Delta R = -\dfrac{R^2 p_2}{Eh}\sin\phi\left(\dfrac{1 + \nu}{2} - \cos^2\phi\right)$$ $$\beta = -\dfrac{Rp_2}{Eh}(3 + \nu)\sin\phi\cos\phi$$
HYDROSTATIC PRESSURE LOADING:	
$$N_\phi = -\dfrac{\rho_w R^2}{6}\left(-1 - 3\dfrac{d}{R} - \dfrac{2\cos^2\phi}{1 + \cos\phi}\right)$$ $$N_\theta = \dfrac{\rho_w R^2}{6}\left(1 + 3\dfrac{d}{R} + \dfrac{4\cos^2\phi - 6}{1 + \cos\phi}\right)$$	$$\Delta R = -\dfrac{\rho_w R^3}{6Eh}\sin\phi\left[3\left(1 - \dfrac{d}{R}\right)(1 - \nu)\right.$$ $$\left. -6\cos\phi - \dfrac{2(1 + \nu)}{\sin^2\phi}(\cos^3\phi - 1)\right]$$ $$\beta = -\dfrac{\rho_w R^2}{Eh}\sin\phi$$
NORMAL UNIFORM LOADING:	
$$N_\phi = \tfrac{1}{2}Rp_1$$ $$N_\theta = \tfrac{1}{2}Rp_1$$	$$\Delta R = \dfrac{R^2 p_1}{2Eh}(1 - \nu)\sin\phi$$ $$\beta = 0$$

TABLE 20-1   **Membrane Spherical Shells**   1188

TABLE 20-1 (continued)   MEMBRANE SPHERICAL SHELLS

Part B. Stress Resultants, Deformations, and Rotations

3. *Simply supported, open spherical shell*

Internal Forces	Deformations
DEAD-WEIGHT LOADING:	

$$N_\phi = -\frac{Rq}{\sin^2\phi}(\cos\phi_2 - \cos\phi)$$

$$N_\theta = -Rq\left[\cos\phi - \frac{1}{\sin^2\phi}(\cos\phi_2 - \cos\phi)\right]$$

$$\Delta R = \frac{R^2 q}{Eh}\sin\phi\left[-\cos\phi + \frac{1+\nu}{\sin^2\phi}(\cos\phi_2 - \cos\phi)\right]$$

$$\beta = -\frac{Rq}{Eh}(2+\nu)\sin\phi$$

UNIFORM LOAD ON PROJECTED AREAS:

$$N_\phi = -\frac{p_2 R}{2}\left(1 - \frac{\sin^2\phi_2}{\sin^2\phi}\right)$$

$$N_\theta = -\frac{p_2 R}{2}\left(2\cos^2\phi - 1 + \frac{\sin^2\phi_2}{\sin^2\phi}\right)$$

$$\Delta R = \frac{R^2 p_2}{Eh}\sin\phi\left[-\cos^2\phi + \frac{1+\nu}{2}\left(1 - \frac{\sin^2\phi_2}{\sin^2\phi}\right)\right]$$

$$\beta = -\frac{Rp_2}{Eh}(3+\nu)\sin\phi\cos\phi$$

HYDROSTATIC PRESSURE LOADING:

$$N_\phi = -\frac{\rho_w R^2}{6}\left[3\left(1 + \frac{d}{R}\right)\left(1 - \frac{\sin^2\phi_2}{\sin^2\phi}\right) - 2\frac{\cos^3\phi_2 - \cos^3\phi}{\sin^2\phi}\right]$$

$$N_\theta = -\frac{\rho_w R^2}{6}\left[3\left(1 + \frac{d}{R}\right)\left(1 + \frac{\sin^2\phi_2}{\sin^2\phi}\right) + \frac{2(2\cos^3\phi + \cos^3\phi_2) - 6\cos\phi}{\sin^2\phi}\right]$$

$$\Delta R = -\frac{\rho_w R^3}{6Eh}\sin\phi\left\{3\left(1 + \frac{d}{R}\right)\left[1 - \nu + (1+\nu)\frac{\sin^2\phi_2}{\sin^2\phi}\right] - 6\cos\phi + 2(1+\nu)\frac{\cos^3\phi_2 - \cos^3\phi}{\sin^2\phi}\right\}$$

$$\beta = \frac{\rho_w R^2}{Eh}\sin\phi$$

TABLE 20-1 (continued)   MEMBRANE SPHERICAL SHELLS

Part B. Stress Resultants, Deformations, and Rotations

Internal Forces	Deformations
UNIFORM NORMAL LOADING:	

$$N_\phi = -\frac{Rp_1}{2}\left(1 - \frac{\sin^2\phi_2}{\sin^2\phi}\right)$$

$$N_\theta = -\frac{Rp_1}{2}\left(1 + \frac{\sin^2\phi_2}{\sin^2\phi}\right)$$

$$\Delta R = -\frac{R^2 p_1}{Eh}\sin\phi\left[1 \right.$$

$$\left. -\frac{1+\nu}{2}\left(1 - \frac{\sin^2\phi_2}{\sin^2\phi}\right)\right]$$

$$\beta = 0$$

LANTERN LOADING:

$$N_\phi = -\frac{W\sin\phi_2}{\sin^2\phi}$$

$$N_\theta = \frac{W\sin\phi_2}{\sin^2\phi}$$

$$\Delta R = \frac{WR}{Eh}(1+\nu)\frac{\sin\phi_2}{\sin\phi}$$

$$\beta = 0$$

4. *Reversed, simply supported, open spherical shell*

DEAD-WEIGHT LOADING:

$$N_\phi = \frac{Rq}{\sin^2\phi}(\cos\phi_2 - \cos\phi)$$

$$N_\theta = Rq\left[\cos\phi - \frac{1}{\sin^2\phi}(\cos\phi_2\right.$$

$$\left. - \cos\phi)\right]$$

$$\Delta R = -\frac{R^2 q}{Eh}\sin\phi\left[-\cos\phi\right.$$

$$\left. + \frac{1+\nu}{\sin^2\phi}(\cos\phi_2 - \cos\phi)\right]$$

$$\beta = -\frac{Rq}{Eh}(2+\nu)\sin\phi$$

UNIFORM LOAD ON PROJECTED AREAS:

$$N_\phi = \frac{p_2 R}{2}\left(1 - \frac{\sin^2\phi_2}{\sin^2\phi}\right)$$

$$N_\theta = \frac{p_2 R}{2}\left(2\cos^2\phi - 1 + \frac{\sin^2\phi_2}{\sin^2\phi}\right)$$

$$\Delta R = -\frac{R^2 p_2}{Eh}\sin\phi\left[-\cos^2\phi\right.$$

$$\left. + \frac{1+\nu}{2}\left(1 - \frac{\sin^2\phi_2}{\sin^2\phi}\right)\right]$$

$$\beta = -\frac{Rp_2}{Eh}(3+\nu)\sin\phi\cos\phi$$

TABLE 20-1   **Membrane Spherical Shells**   **1190**

TABLE 20-1 (continued) MEMBRANE SPHERICAL SHELLS

Part B. Stress Resultants, Deformations, and Rotations

Internal Forces	Deformations

HYDROSTATIC PRESSURE LOADING:

$$N_\phi = -\frac{\rho_w R^2}{6}\left[3\left(1 - \frac{d}{R}\right)\left(1 - \frac{\sin^2 \phi_2}{\sin^2 \phi}\right) - \frac{2(\cos^3 \phi_2 - \cos^3 \phi)}{\sin^2 \phi}\right]$$

$$N_\theta = -\frac{\rho_w R^2}{6}\left[3\left(1 - \frac{d}{R}\right)\left(1 + \frac{\sin^2 \phi_2}{\sin^2 \phi}\right) + \frac{2(2\cos^3 \phi + \cos^3 \phi_2) - 6\cos \phi}{\sin^2 \phi}\right]$$

$$\Delta R = -\frac{\rho_w R^3}{6Eh}\sin \phi \left\{3\left(1 - \frac{d}{R}\right)\left[1 - \nu + (1 + \nu)\frac{\sin^2 \phi_2}{\sin^2 \phi}\right] - 6\cos \phi + 2(1 + \nu)\frac{\cos^3 \phi_2 - \cos^3 \phi}{\sin^2 \phi}\right\}$$

$$\beta = -\frac{\rho_w R^2}{Eh}\sin \phi$$

UNIFORM NORMAL LOADING:

$$N_\phi = \frac{Rp_1}{2}\left(1 - \frac{\sin^2 \phi_2}{\sin^2 \phi}\right)$$

$$N_\theta = \frac{Rp_1}{2}\left(1 + \frac{\sin^2 \phi_2}{\sin^2 \phi}\right)$$

$$\Delta R = \frac{R^2 p_1}{Eh}\sin \phi \left[1 - \frac{1 + \nu}{2}\left(1 - \frac{\sin^2 \phi_2}{\sin^2 \phi}\right)\right]$$

$$\beta = 0$$

LANTERN LOADING:

$$N_\phi = \frac{W \sin \phi_2}{\sin^2 \phi}$$

$$N_\theta = -\frac{W \sin \phi_2}{\sin^2 \phi}$$

$$\Delta R = -\frac{WR(1 + \nu)}{Eh}\frac{\sin \phi_2}{\sin \phi}$$

$$\beta = 0$$

TABLE 20-1 (continued)   MEMBRANE SPHERICAL SHELLS

Part B. Stress Resultants, Deformations, and Rotations

5. *External liquid over portion of simply supported shell*

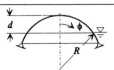

Internal Forces	Deformations
LOADING: $p_z = \rho_w(R - R\cos\phi - d)$. FOR POINTS ABOVE THE LIQUID LEVEL: $N_\phi = N_\theta = 0$ FOR POINTS BELOW THE LIQUID LEVEL: $N_\phi = -\rho_w \dfrac{R^2}{6}\left\{\dfrac{d}{R}\left[\dfrac{1}{\sin^2\phi}\dfrac{d}{R}\left(3 - \dfrac{d}{R}\right) - 3\right]\right.$ $\left. + 1 - \dfrac{2\cos^2\phi}{1 + \cos\phi}\right\}$ $N_\theta = -\rho_w R^2\left(1 - \cos\phi - \dfrac{d}{R}\right) - N_\phi$	

6. *Internal liquid filling portion of simply supported shell*
$$\phi' = 180° - \phi$$

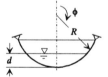

LOADING: $p_z = -\rho_w(R - R\cos\phi' - d)$. FOR POINTS ABOVE THE LIQUID LEVEL: $N_\phi = \rho_w \dfrac{d^2}{6}\left(3 - \dfrac{d}{R}\right)\dfrac{1}{\sin^2\phi'}$ $N_\theta = -\rho_w \dfrac{d^2}{6}\left(3 - \dfrac{d}{R}\right)\dfrac{1}{\sin^2\phi'}$ FOR POINTS BELOW THE LIQUID LEVEL: $N_\phi = \rho_w \dfrac{R^2}{6}\left[3\dfrac{d}{R} - 1 + \dfrac{2\cos^2\phi'}{1 + \cos\phi'}\right]$ $N_\theta = \rho_w R^2\left[\dfrac{d}{R} - 1 + \cos\phi'\right] - N_\phi$	

[a]Adopted from Baker et al. [20.2].

TABLE 20-1   Membrane Spherical Shells   **1192**

TABLE 20-2 MEMBRANE CONICAL SHELL[a]

Notation

E = modulus of elasticity
ν = Poisson's ratio
h = thickness of shell
ρ_w = specific weight of liquid
∇ = surface of liquid
d = distance from surface of liquid to vertex of shell
q = weight of shell per unit area
p_1 = applied uniform pressure
p_2 = uniformly distributed loading on projected area
p_x, p_y, p_z = components of applied force in shell
ΔR = horizontal displacement of shell
β = rotation of tangent of meridian
N_ϕ, N_θ = normal forces per unit length (stress resultants) in meridional and circumferential (parallel) directions

Loading and internal forces

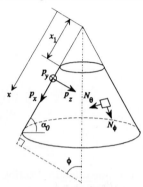

Deformations

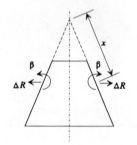

TABLE 20-2 (continued)   MEMBRANE CONICAL SHELL

Part A. Loadings

Case	Components
1. Uniform normal pressure p_1 	$p_x = 0$ $p_y = 0$ $p_z = p_1$
2. Hydrostatic pressure loading (a) External liquid (b) Liquid inside	$p_x = 0$ $p_y = 0$ $p_z = \rho_w(d + x \sin \alpha_0)$ for (a) $p_z = \rho_w(d - x \sin \alpha_0)$ for (b)
3. Uniformly distributed loading over the base 	$p_x = p_2 \cos \alpha_0 \sin \alpha_0$ $p_y = 0$ $p_z = p_2 \cos^2 \alpha_0$

TABLE 20-2 **Membrane Conical Shell**   **1194**

TABLE 20-2 (continued)   MEMBRANE CONICAL SHELL

Part A. Loadings

Case	Components
4. Dead-weight loading 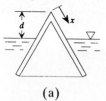	$p_x = q \sin \alpha_0$ $p_y = 0$ $p_z = q \cos \alpha_0$ q = weight of shell per unit area
5. Hydrostatic pressure over portion of shell (a) (b)	$p_x = 0$ $p_y = 0$ $p_z = \rho_w(x \sin \alpha_0 - d)$ for (a) $p_z = \rho_w(d - x \sin \alpha_0)$ for (b)
6. Equally distributed loading along opening edge (lantern load) 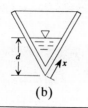	W is in force per length. $p_x = p_y = p_z = 0$

TABLE 20-2 (continued)   MEMBRANE CONICAL SHELL

Part B. Stress Resultants, Deformations, and Rotations

1. *Closed conical shell (supported)*

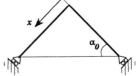

Internal Forces	Deformations
DEAD-WEIGHT LOADING:	
$$N_\theta = -\frac{qx\cos^2\alpha_0}{\sin\alpha_0}$$ $$N_x = -\frac{1}{x}\left(\frac{qx^2}{2\sin\alpha_0}\right)$$	$$\Delta R = -\frac{x^2}{Eh}q\cot\alpha_0\left(\cos^2\alpha_0 - \frac{\nu}{2}\right)$$ $$\beta = \frac{qx\cos\alpha_0}{Eh\sin^2\alpha_0}\left[(2+\nu)\cos^2\alpha_0 - \frac{1}{2} - \nu\right]$$
UNIFORM LOADING ON PROJECTED AREA:	
$$N_\theta = -p_2\frac{\cos^3\alpha_0}{\sin\alpha_0}x$$ $$N_x = -p_2\frac{x}{2}\cot\alpha_0$$	$$\Delta R = -p_2\frac{x^2}{Eh}\cos\alpha_0\cot\alpha_0\left(\cos^2\alpha_0 - \frac{\nu}{2}\right)$$ $$\beta = \frac{p_2 x}{Eh}\cot^2\alpha_0\left[(2+\nu)\cos^2\alpha_0 - \nu - \frac{1}{2}\right]$$
UNIFORM NORMAL LOADING:	
$$N_\theta = -p_1 x\cot\alpha_0$$ $$N_x = -p_1\frac{x}{2}\cot\alpha_0$$	$$\Delta R = -p_1\frac{x^2}{Eh}\cos\alpha_0\cot\alpha_0\left(1 - \frac{\nu}{2}\right)$$ $$\beta = \frac{3}{2}\frac{p_1 x}{Eh}\cot^2\alpha_0$$
HYDROSTATIC PRESSURE LOADING:	
$$N_\theta = -\rho_w x\cos\alpha_0\left(\frac{d}{\sin\alpha_0} + x\right)$$ $$N_x = -\rho_w x\cos\alpha_0\left(\frac{d}{2\sin\alpha_0} + \frac{x}{3}\right)$$	$$\Delta R = \frac{\rho_w x^2}{Eh}\cos^2\alpha_0\left[\frac{d}{\sin\alpha_0}\left(\frac{\nu}{2} - 1\right) + x\left(\frac{\nu}{3} - 1\right)\right]$$ $$\beta = \frac{\rho_w x^2}{Eh}\frac{\cos^2\alpha_0}{\sin\alpha_0}\left(\frac{3}{2}\frac{d}{\sin\alpha_0} + \frac{8}{3}x\right)$$

HYDROSTATIC PRESSURE OVER PORTION OF SHELL:

$$N_\theta = \begin{cases} 0 & \text{for points above } \nabla \\ -\rho_w x(x\cos\alpha_0 - d\cot\alpha_0) & \text{for points below } \nabla \end{cases}$$

$$N_x = \begin{cases} 0 & \text{for points above } \nabla \\ -\dfrac{\rho_w}{6x}\left[\dfrac{\cos\alpha_0}{\sin^3\alpha_0}d^3 + x^2(2x\cos\alpha_0 - 3d\cot\alpha_0)\right] & \text{for points below } \nabla \end{cases}$$

TABLE 20-2 **Membrane Conical Shell**   **1196**

TABLE 20-2 (continued)   MEMBRANE CONICAL SHELL

Part B. Stress Resultants, Deformations, and Rotations

2. *Closed conical shell (hanging)*

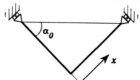

Internal Forces	Deformations
DEAD-WEIGHT LOADING:	
$N_\theta = q \dfrac{\cos^2 \alpha_0}{\sin \alpha_0} x$ $N_x = \dfrac{q}{2 \sin \alpha_0} x$	$\Delta R = \dfrac{qx^2}{Eh} \cot \alpha_0 \left(\cos^2 \alpha_0 - \dfrac{\nu}{2} \right)$ $\beta = \dfrac{qx \cos \alpha_0}{Eh \sin^2 \alpha_0} \left[(2 + \nu)\cos^2\alpha_0 - \dfrac{1}{2} - \nu \right]$
UNIFORM LOADING ON PROJECTED AREA:	
$N_\theta = p_2 x \dfrac{\cos^3 \alpha_0}{\sin \alpha_0}$ $N_x = p_2 \dfrac{x}{2} \cot \alpha_0$	$\Delta R = p_2 \dfrac{x^2}{Eh} \cos \alpha_0 \cot \alpha_0 \left(\cos^2 \alpha_0 - \dfrac{\nu}{2} \right)$ $\beta = \dfrac{p_2 x}{Eh} \cot^2 \alpha_0 \left[(2 + \nu)\cos^2 \alpha_0 - \nu - \dfrac{1}{2} \right]$
UNIFORM NORMAL PRESSURE:	
$N_\theta = -px \cot \alpha_0$ $N_x = -p_1 \dfrac{x}{2} \cot \alpha_0$	$\Delta R = -p_1 \dfrac{x^2}{Eh} \cos \alpha_0 \cot \alpha_0 \left(1 - \dfrac{\nu}{2} \right)$ $\beta = -\dfrac{3}{2} \dfrac{p_1 x}{Eh} \cot^2 \alpha_0$
HYDROSTATIC PRESSURE LOADING:	
$N_\theta = -\rho_w x \cos \alpha_0 \left(x - \dfrac{d}{\sin \alpha_0} \right)$ $N_x = -\rho_w x \cos \alpha_0 \left(\dfrac{x}{3} - \dfrac{d}{2 \sin \alpha_0} \right)$	$\Delta R = \dfrac{\rho_w x^2}{Eh} \cos^2 \alpha_0 \left[x\left(\dfrac{\nu}{3} - 1 \right) - \dfrac{d}{\sin \alpha_0}\left(\dfrac{\nu}{2} - 1 \right) \right]$ $\beta = \dfrac{\rho_w x \cos^2 \alpha_0}{Eh \sin \alpha_0} \left(-\dfrac{8}{3}x + \dfrac{3}{2}\dfrac{d}{\sin \alpha_0} \right)$
HYDROSTATIC PRESSURE OVER PORTION OF SHELL:	

$$N_\theta = \begin{cases} 0 & \text{for points above } \nabla \\ \rho_w x (d \cot \alpha_0 - x \cos \alpha_0) & \text{for points below } \nabla \end{cases}$$

$$N_x = \begin{cases} \dfrac{\rho_w d^3}{6x} \dfrac{\cos \alpha_0}{\sin^3 \alpha_0} & \text{for points above } \nabla \\ \dfrac{\rho_w x}{2}(3d \cot \alpha_0 - 2x \cos \alpha_0) & \text{for points below } \nabla \end{cases}$$

TABLE 20-2 (continued)   MEMBRANE CONICAL SHELL

Part B. Stress Resultants, Deformations, and Rotations

3. *Open conical shell (supported)*

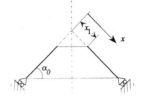

Internal Forces	Deformations
DEAD-WEIGHT LOADING:	
$$N_\theta = -q\frac{\cos^2 \alpha_0}{\sin \alpha_0}x$$ $$N_x = -\frac{qx}{2\sin \alpha_0}\left[1 - \left(\frac{x_1}{x}\right)^2\right]$$	$$\Delta R = -\frac{qx^2}{2Eh}\cot \alpha_0\left\{2\cos^2 \alpha_0 - \nu\left[1 - \left(\frac{x_1}{x}\right)^2\right]\right\}$$ $$\beta = \frac{qx\cos \alpha_0}{2Eh\sin^2 \alpha_0}\left[2(2 + \nu)\cos^2 \alpha_0 - 1 + \left(\frac{x_1}{x}\right)^2 - 2\nu\right]$$
UNIFORM LOADING ON PROJECTED AREA:	
$$N_\theta = -\frac{p_2 x\cos^3 \alpha_0}{\sin \alpha_0}$$ $$N_x = -\frac{1}{2}p_2 x\left[1 - \left(\frac{x_1}{x}\right)^2\right]\cot \alpha_0$$	$$\Delta R = -\frac{p_2 x^2}{2Eh}\frac{\cos^2 \alpha_0}{\sin \alpha_0}\left\{2\cos^2 \alpha_0 - \nu\left[1 - \left(\frac{x_1}{x_2}\right)^2\right]\right\}$$ $$\beta = \frac{p_2 x}{2Eh}\cot^2 \alpha_0\left[2(2 + \nu)\cos^2 \alpha_0 - 2\nu + \left(\frac{x_1}{x}\right)^2 - 1\right]$$
UNIFORM NORMAL PRESSURE:	
$$N_\theta = -p_1 x\cot \alpha_0$$ $$N_x = -\frac{p_1}{2}x\cot \alpha_0\left[1 - \left(\frac{x_1}{x}\right)^2\right]$$	$$\Delta R = -\frac{p_1}{Eh}x^2\cos \alpha_0\cot \alpha_0\left\{1 - \frac{\nu}{2}\left[1 - \left(\frac{x_1}{x}\right)^2\right]\right\}$$ $$\beta = \frac{p_1 x}{2Eh}\cot^2 \alpha_0\left[3 + \left(\frac{x_1}{x}\right)^2\right]$$
EQUALLY DISTRIBUTED LOADING ALONG OPENING EDGE (LANTERN LOAD):	
$$N_\theta = 0$$ $$N_x = -\frac{W}{\sin \alpha_0}\frac{x_1}{x}$$	$$\Delta R = -\frac{\nu W x_1\cot \alpha_0}{Eh}$$ $$\beta = -\frac{W}{Eh}\frac{x_1}{x}\frac{\cot \alpha_0}{\sin \alpha_0}$$
HYDROSTATIC PRESSURE LOADING:	
$$N_\theta = -\rho_w x\cos \alpha_0\left(\frac{d}{\sin \alpha_0} + x\right)$$ $$N_x = -\rho_w x\cos \alpha_0\left\{\frac{d}{2\sin \alpha_0}\left[1 - \left(\frac{x_1}{x}\right)^2\right]\right.$$ $$\left. + \frac{x}{3}\left[1 - \left(\frac{x_1}{x}\right)^3\right]\right\}$$	$$\Delta R = \frac{\rho_w x^2}{Eh}\cos^2 \alpha_0\left\{\nu\left[\frac{d}{2\sin \alpha_0}\left[1 - \left(\frac{x_1}{x}\right)^2\right]\right.\right.$$ $$\left.\left. + \frac{x}{3}\left[1 - \left(\frac{x_1}{x}\right)^3\right]\right] - \frac{d}{\sin \alpha_0} - x\right\}$$ $$\beta = \frac{\rho_w x}{Eh}\frac{\cos^2 \alpha_0}{\sin \alpha_0}\left\{\frac{d}{2\sin \alpha_0}\left[3 + \left(\frac{x_1}{x}\right)^2\right]\right.$$ $$\left. + \frac{x}{3}\left[8 + \left(\frac{x_1}{x}\right)^3\right]\right\}$$

TABLE 20-2   **Membrane Conical Shell**   **1198**

TABLE 20-2 (continued)   MEMBRANE CONICAL SHELL

Part B. Stress Resultants, Deformations, and Rotations

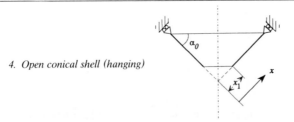

4. *Open conical shell (hanging)*

Internal Forces	Deformations
DEAD-WEIGHT LOADING: $$N_\theta = q\frac{\cos^2\alpha_0}{\sin\alpha_0}x$$ $$N_x = \frac{qx}{2\sin\alpha_0}\left[1 - \left(\frac{x_1}{x}\right)^2\right]$$	$$\Delta R = \frac{qx^2\cot\alpha_0}{2Eh}\left\{2\cos^2\alpha_0 - \nu\left[1 - \left(\frac{x_1}{x}\right)^2\right]\right\}$$ $$\beta = \frac{qx\cos\alpha_0}{2Eh\sin^2\alpha_0}\left[2(2+\nu)\cos^2\alpha_0 - 1 + \left(\frac{x_1}{x}\right)^2 - 2\nu\right]$$
UNIFORM LOADING ON PROJECTED AREA: $$N_\theta = \frac{p_2 x\cos^3\alpha_0}{\sin\alpha_0}$$ $$N_x = \frac{1}{2}p_2 x\left[1 - \left(\frac{x_1}{x}\right)^2\right]\cot\alpha_0$$	$$\Delta R = \frac{p_2 x^2}{2Eh}\frac{\cos^2\alpha_0}{\sin\alpha_0}\left\{2\cos^2\alpha_0 - \nu\left[1 - \left(\frac{x_1}{x}\right)^2\right]\right\}$$ $$\beta = \frac{p_2 x^2}{2Eh}\cot^2\alpha_0\left[2(2+\nu)\cos^2\alpha_0 - 2\nu + \left(\frac{x_1}{x}\right)^2 - 1\right]$$
UNIFORM NORMAL PRESSURE: $$N_\theta = -p_1 x\cot\alpha_0$$ $$N_x = -\frac{p_1}{2}x\cot\alpha_0\left[1 - \left(\frac{x_1}{x}\right)^2\right]$$	$$\Delta R = -\frac{p_1 x^2}{Eh}\cos\alpha_0\cot\alpha_0\left\{1 - \frac{\nu}{2}\left[1 - \left(\frac{x_1}{x}\right)^2\right]\right\}$$ $$\beta = -\frac{p_1 x}{2Eh}\cot^2\alpha_0\left[3 + \left(\frac{x_1}{x}\right)^2\right]$$
EQUALLY DISTRIBUTED LOADING ALONG OPENING EDGE (LANTERN LOAD): $$N_\theta = 0 \qquad N_x = \frac{W}{\sin\alpha_0}\frac{x_1}{x}$$	$$\Delta R = -\nu\frac{Wx_1}{Eh}\cot\alpha_0 \qquad \beta = -\frac{W}{Eh}\frac{x_1}{x}\frac{\cot\alpha_0}{\sin\alpha_0}$$
HYDROSTATIC PRESSURE LOADING: $$N_\theta = -\rho_w x\cos\alpha_0\left(x - \frac{d}{\sin\alpha_0}\right)$$ $$N_x = -\rho_w x\cos\alpha_0\left\{\frac{x}{3}\left[1 - \left(\frac{x_1}{x}\right)^3\right]\right.$$ $$\left. -\frac{d}{2\sin\alpha_0}\cdot\left[1 - \left(\frac{x_1}{x}\right)^2\right]\right\}$$	$$\Delta R = \frac{\rho_w x^2}{Eh}\cos^2\alpha_0\left\{\nu\left[\frac{x}{3}\left[1 - \left(\frac{x_1}{x}\right)^3\right]\right.\right.$$ $$\left. -\frac{d}{2\sin\alpha_0}\left[1 - \left(\frac{x_1}{x}\right)^2\right]\right] + \frac{d}{\sin\alpha_0} - x\Big\}$$ $$\beta = \frac{\rho_w x\cos^2\alpha_0}{Eh\sin\alpha_0}\left\{\frac{d}{2\sin\alpha_0}\left[3 + \left(\frac{x_1}{x}\right)^2\right]\right.$$ $$\left. -\frac{x}{3}\left[8 + \left(\frac{x_1}{x}\right)^3\right]\right\}$$

^{a}Adopted from Baker et al. [20.2].

TABLE 20-3   MEMBRANE CYLINDRICAL SHELLS[a]

Notation

$$E = \text{modulus of elasticity}$$
$$R = \text{radius of cylinder}$$
$$\lambda_p, \alpha, \beta, \lambda_{pi}, \alpha_i = \text{constants}$$
$$\xi = x/L$$
$$\nu = \text{Poisson's ratio}$$
$$h = \text{thickness of shell}$$
$$L = \text{length of cylinder}$$
$$u, v, w = \text{displacement in } x, y, \text{ and } z \text{ directions}$$
$$N_x, N_\theta = \text{normal forces per unit length (stress resultants)}$$
in x and circumferential (parallel) directions
$$N_{x\theta} = \text{shear forces per unit length (stress resultants)}$$
with respect to x and circumferential directions

The boundaries of the shell are free to deflect normal to the shell middle surface and to rotate.

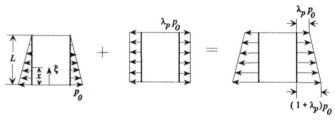

Linear loading $p_z = -p_0(1 + \lambda_p - \xi)$

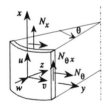

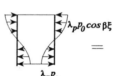

Trigonometric loading $p_z = -p_0(\sin \alpha\xi + \lambda_p \cos \beta\xi)$

Exponential loading: $p_z = -p_0 \sum_i \lambda_{pi} e^{-\alpha_i \xi}$

Deadweight loading: $p_x = -p_0(1 - \xi)$

Periodical loading: $p_z = -p_0 \sum_i \lambda_{pi} \cos \alpha_i \theta$

TABLE 20-3   Membrane Cylindrical Shells   **1200**

TABLE 20-3 (continued)   MEMBRANE CYLINDRICAL SHELLS

Case	Internal Forces	Deformations
1. Cylindrical shell under linear loading	$N_\theta = p_0(1 + \lambda_p - \xi)R$	$u = \dfrac{1}{Eh}\left[-\nu p_0 RL\xi\left(1 + \lambda_p - \tfrac{1}{2}\xi\right)\right]$ $w = -\dfrac{1}{Eh}\left[p_0 R^2(1 + \lambda_p - \xi)\right]$
2. Cylindrical shell under trigonometric loading	$N_\theta = p_0 R(\sin\alpha\xi + \lambda_p\cos\beta\xi)$	$u = \dfrac{1}{Eh}\nu p_0 RL\left[\dfrac{\cos\alpha\xi}{\alpha} - \lambda_p\dfrac{\sin\beta\xi}{\beta}\right]$ $w = -\dfrac{1}{Eh}p_0 R^2(\sin\alpha\xi + \lambda_p\cos\beta\xi)$
3. Cylindrical shell under exponential loading	$N_\theta = p_0 R\sum\limits_i \lambda_{px}\,e^{-\alpha_i\xi}$	$u = \dfrac{1}{Eh}\nu p_0 RL\sum\limits_i \dfrac{\lambda_{pi}}{\alpha_i}\,e^{-\alpha_i\xi}$ $w = -\dfrac{1}{Eh}p_0 R^2\sum\limits_i \lambda_{pi}\,e^{-\alpha_i\xi}$

TABLE 20-3 (continued)   MEMBRANE CYLINDRICAL SHELLS

Case	Internal Forces	Deformations
4. Linearly varying loading in x direction $P_x = -p_0(1-\xi)$	$N_x = -p_0 L\left(\frac{1}{2} - \xi + \frac{1}{2}\xi^2\right)$	$u = -\frac{1}{Eh}\left[p_0 L^2 \xi\left(\frac{1}{2} - \frac{1}{2}\xi + \frac{1}{6}\xi^2\right)\right]$ $w = \frac{1}{Eh}\nu p_0 RL\left(-\frac{1}{2} + \xi - \frac{1}{2}\xi^2\right)$
5. Cylindrical shell under circumferential loading Loads reacted at lower base.	$N_{x\theta} = p_y L(1 - \xi)$	$v = \frac{1}{Eh}\left[2(1+\nu)p_y L^2\left(\xi - \frac{1}{2}\xi^2\right)\right]$

[a]Adopted from Baker et al. [20.2].

TABLE 20-3 | **Membrane Cylindrical Shells** | **1202**

TABLE 20-4 MEMBRANE TOROIDAL SHELLS[a]

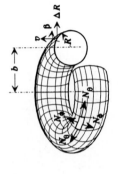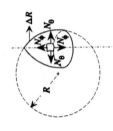

Notation

E = modulus of elasticity
ρ_w = specific weight of liquid
h = thickness of shell
ϕ_1, ϕ_2 = meridional angle of lower and upper edge
q = dead weight
p_1 = applied uniform pressure
p_2 = uniformly distributed loading on projected area
ν = Poisson's ratio
R = radius of curvature of meridian at point
ϕ = meridional angle of point of interest
$\Delta R, v$ = horizontal and vertical displacements
β = rotation of tangent of meridian
∇ = surface of liquid
W = lantern loading
d = distance from center of curvature of meridian to surface of liquid
N_ϕ, N_θ = normal force per unit length (stress resultants) in meridional and circumferential (parallel) directions

The shells in this table are formed by rotating a circle (or a segment of a circle) around an axis; b is the distance from the axis to the center of the circle.
See Table 20-1, part A, for definitions of the loadings.

TABLE 20-4 (continued)   MEMBRANE TOROIDAL SHELLS

Part A. Stress Resultants, Deformations, and Rotations for Toroidal Shells

Case	Internal Forces	Deformations
1. Toroidal shell with internal pressure	$N_\phi = \dfrac{p_1 R}{2}\left[\dfrac{2b + R\sin\phi}{b + R\sin\phi}\right]$ $N_\theta = \dfrac{p_1 R}{2}$	$\Delta R = \dfrac{p_1 R^2}{2Eh}\left[\dfrac{b}{R}(1-2\nu) + (1-\nu)\sin\phi\right]$
2. Toroidal segment Approximate useful range $35° \le \phi \le 90°$	$N_\phi = \dfrac{p_1 R}{2}\left(\dfrac{b}{R\sin\phi} + 1\right)$ $N_\theta = \dfrac{p_1 R}{2}\left(1 - \dfrac{b^2}{R^2\sin^2\phi}\right)$	$\Delta R = \dfrac{p_1 R^2}{2Eh}\left[\dfrac{b}{R}(1-2\nu) + (1-\nu)\sin\phi + \dfrac{b^2(1+\nu)}{R^2\sin\phi} - \dfrac{b^3}{R^3\sin^2\phi}\right]$ $v = \dfrac{p_1 R^2}{2Eh}\left[(1-\nu)\cos\phi + \dfrac{b}{R}\left(1 + \dfrac{2b^2}{3R^2}\right)\cot\phi\right.$ $\left. + \dfrac{b^2}{R^2}\left(\dfrac{3}{2} + 2\nu\right)\dfrac{\cot\phi}{\sin\phi} - \dfrac{b^2}{R^2}\left(1 - \dfrac{b}{3R}\right)\dfrac{\cot\phi}{\sin^2\phi}\right.$ $\left. - \dfrac{b^2}{2R^2}(1+2\nu)\ln\left(\tan\dfrac{\phi}{2}\right)\right]$ $\beta = -\dfrac{p_1 R\cot\phi}{2Eh\sin\phi}\left[\dfrac{b}{R}\left(\dfrac{2b}{R\sin\phi} - 1\right)\left(\dfrac{b}{R\sin\phi} + 1\right)\right]$

TABLE 20-4   **Membrane Toroidal Shells**   1204

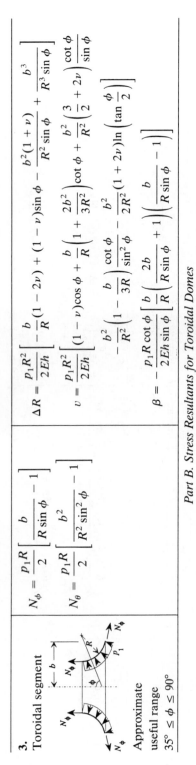

3. Toroidal segment

$$N_\phi = \frac{p_1 R}{2}\left[\frac{b}{R\sin\phi} - 1\right]$$

$$N_\theta = \frac{p_1 R}{2}\left[\frac{b^2}{R^2\sin^2\phi} - 1\right]$$

Approximate useful range $35° \leq \phi \leq 90°$

$$\Delta R = \frac{p_1 R^2}{2Eh}\left[-\frac{b}{R}(1-2\nu) + (1-\nu)\sin\phi - \frac{b^2(1+\nu)}{R^2\sin\phi} + \frac{b^3}{R^3\sin\phi}\right]$$

$$v = \frac{p_1 R^2}{2Eh}\left[(1-\nu)\cos\phi + \frac{b}{R}\left(1+\frac{2b^2}{3R^2}\right)\cot\phi + \frac{b^2}{R^2}\left(\frac{3}{2}+2\nu\right)\frac{\cot\phi}{\sin\phi}\right.$$
$$\left. -\frac{b^2}{R^2}\left(1-\frac{b}{3R}\right)\frac{\cot\phi}{\sin^2\phi} - \frac{b^2}{2R^2}(1+2\nu)\ln\left(\tan\frac{\phi}{2}\right)\right]$$

$$\beta = -\frac{p_1 R\cot\phi}{2Eh\sin\phi}\left[\frac{b}{R}\left(\frac{2b}{R\sin\phi}+1\right)\left(\frac{b}{R\sin\phi}-1\right)\right]$$

Part B. Stress Resultants for Toroidal Domes

1. Pointed dome

Loading Condition	N_ϕ	N_θ
DEAD-WEIGHT LOADING: $p_x = q\sin\phi$ $p_z = q\cos\phi$	$-qR\dfrac{c_0 - c - (\phi - \phi_0)s_0}{(s-s_0)s}$ $s = \sin\phi,\ s_0 = \sin\phi_0,$ $c = \cos\phi,\ c_0 = \cos\phi_0$	$-q\dfrac{R}{\sin^2\phi}\big[(\phi-\phi_0)\sin\phi_0 - (\cos\phi_0 - \cos\phi)$ $+ (\sin\phi - \sin\phi_0)\sin\phi\cos\phi\big]$
UNIFORMLY DISTRIBUTED VERTICAL LOADING ON PROJECTED AREA: $p_x = p_2\sin\phi\cos\phi$ $p_z = p_2\cos^2\phi$	$-p_2\dfrac{R}{2}\left(1 - \dfrac{\sin^2\phi_0}{\sin^2\phi}\right)$	$-p_2\dfrac{R}{2}\left(\cos 2\phi + 2\sin\phi\sin\phi_0 - \dfrac{\sin^2\phi_0}{\sin^2\phi}\right)$

TABLE 20-4 (continued) MEMBRANE TOROIDAL SHELLS

Part B. Stress Resultants for Toroidal Domes

2. Toroid surface

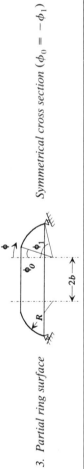

Loading Condition	N_ϕ	N_θ
DEAD-WEIGHT LOADING: $p_x = q \sin \phi$ $p_z = q \cos \phi$	$-qR \dfrac{1 - \cos \phi + \phi \sin \phi_0}{\sin \phi (\sin \phi + \sin \phi_0)}$	$-qR\left[\cos \phi - \dfrac{1 - \cos \phi}{\sin^2 \phi} + \sin \phi_0 \left(\cot \phi - \dfrac{\phi}{\sin^2 \phi} \right) \right]$
UNIFORMLY DISTRIBUTED VERTICAL LOADING ON PROJECTED AREA: $p_x = p_2 \sin \phi \cos \phi$ $p_z = p_2 \cos^2 \phi$	$-p_2 R \dfrac{\sin \phi + 2 \sin \phi_0}{2 \sin \phi + \sin \phi_0}$	$-p_2 \dfrac{R}{2} (\cos 2\phi - 2 \sin \phi \sin \phi_0)$

3. Partial ring surface

Symmetrical cross section $(\phi_0 = -\phi_1)$

| DEAD-WEIGHT LOADING: $p_x = q \sin \phi$ $p_z = q \cos \phi$ | $-q \dfrac{bR(\phi - \phi_0) + R^2(\cos \phi_0 - \cos \phi)}{(b + R \sin \phi) \sin \phi}$ For $\phi_0 = -\phi_1,$ $-q \dfrac{bR\phi + R^2(1 - \cos \phi)}{(b + R \sin \phi) \sin \phi}$ | $-\dfrac{q}{\sin \phi} \left[(b + R \sin \phi) \cos \phi - b(\phi - \phi_0) - R(\cos \phi_0 - \cos \phi) \right]$ For $\phi_0 = -\phi_1,$ $-\dfrac{q}{\sin \phi} \left[(b + R \sin \phi) \cos \phi - b\phi - R(1 - \cos \phi) \right]$ |

TABLE 20-4 Membrane Toroidal Shells 1206

UNIFORMLY DISTRIBUTED NORMAL LOADING: $p_z = p_1$	$-\dfrac{p_1}{2(b+R\sin\phi)\sin\phi}\left[(b+R\sin\phi)^2 - (b+R\sin\phi_0)^2\right]$ For $\phi_0 = -\phi_1$,[b] $-\dfrac{p_1 R}{2}\dfrac{2b+R\sin\phi}{b+R\sin\phi}$	$-\dfrac{p_1}{2\sin^2\phi}\left[2b\sin\phi_0 + R(\sin^2\phi_0 + \sin^2\phi)\right]$ For $\phi_0 = -\phi_1$,[b] $-\tfrac{1}{2}p_1 R$
LANTERN LOADING: W (F/L) 	$-W\dfrac{b+R\sin\phi_0}{(b+R\sin\phi)\sin\phi}$	$\dfrac{W_1}{R}\dfrac{b+R\sin\phi_0}{\sin^2\phi}$

TABLE 20-4 (continued) MEMBRANE TOROIDAL SHELLS

Part B. Stress Resultants for Toroidal Domes

Loading Condition	N_ϕ	N_θ
HYDROSTATIC PRESSURE LOADING: $p_z = \rho_w(d - R\cos\phi)$	$-\dfrac{\rho_w R}{(b + R\sin\phi)\sin\phi}\left[-bd(\sin\phi_0 - \sin\phi) \right.$ $+\dfrac{Rd}{2}(\cos^2\phi_0 - \cos^2\phi)$ $+\dfrac{bR}{2}(\sin\phi_0\cos\phi_0 - \sin\phi\cos\phi - \phi + \phi_0)$ $\left. -\dfrac{R^2}{3}(\cos^3\phi_0 - \cos^3\phi) \right]$ For $\phi_0 = -\phi_1$,^{c} $-\dfrac{\rho_w R}{(b + R\sin\phi)\sin\phi}\left[bd\sin\phi + \dfrac{Rd}{2}\sin^2\phi \right.$ $-\dfrac{bR}{2}(\sin\phi\cos\phi + \phi)$ $\left. -\dfrac{R^2}{3}(1 - \cos^3\phi) \right]$	$-\dfrac{\rho_w}{\sin^2\phi}\left[(d - R\cos\phi)(b + R\sin\phi)\sin\phi \right.$ $+bd(\sin\phi_0 - \sin\phi) - \dfrac{Rd}{2}(\cos^2\phi_0 - \cos^2\phi)$ $-\dfrac{bR}{2}(\sin\phi_0\cos\phi_0 - \sin\phi\cos\phi - \phi + \phi_0)$ $\left. +\dfrac{R^2}{3}(\cos^3\phi_0 - \cos^3\phi) \right]$ For $\phi_0 = -\phi_1$,^{c} $-\dfrac{\rho_w}{\sin^2\phi}\left[\dfrac{Rd}{2}\sin^2\phi - \dfrac{bR}{2}(\sin\phi\cos\phi - \phi) \right.$ $\left. -R^2\left(\cos\phi\sin^2\phi - \dfrac{1 - \cos^3\phi}{3}\right) \right]$

^{a}Adopted from Baker et al. [20.2].
^{b}Symmetrical cross section.

TABLE 20-4 **Membrane Toroidal Shells** **1208**

TABLE 20-5   VARIOUS MEMBRANE SHELLS OF REVOLUTION[a]

Notation

R = radius of curvature of meridian
    at a point
ϕ = meridional angle of point of interest
ρ_w = specific weight of liquid
p_1 = applied uniformly distributed load
p_2 = uniformly distributed loading
    on projected area
q = dead weight
r_0 = radius of curvature at vertex
ϕ_0 = meridional angle of apex
x_0 = radius of top opening
∇ = surface of liquid
d = distance from surface
    of liquid to vertex of shell
p_x, p_y, p_z = components of applied load in x (meridional),
    y (parallel), and z (normal) directions
N_ϕ, N_θ = membrane forces per unit length
    (positive if carrying tension in wall)

See Table 20-1, part A, for definitions of loadings.

1. Parabola

Loading Condition	N_ϕ	N_θ
DEAD-WEIGHT LOADING: $p_x = q \sin \phi$ $p_z = q \cos \phi$	$-q \dfrac{r_0(1 - \cos^3 \phi)}{3 \sin^2 \phi \cos^2 \phi}$	$-q \dfrac{r_0(2 - 3\cos^2 \phi + \cos^3 \phi)}{3 \sin^2 \phi}$
UNIFORMLY DISTRIBUTED VERTICAL LOADING ON THE PROJECTED AREA: $p_x = p_2 \sin \phi \cos \phi$ $p_z = p_2 \cos^2 \phi$	$-p_2 \dfrac{r_0}{2} \dfrac{1}{\cos \phi}$	$-p_2 \dfrac{r_0}{2} \cos \phi$
UNIFORM NORMAL LOADING: $p_z = p_1$	$-p_1 \dfrac{r_0}{2} \dfrac{1}{\cos \phi}$	$-p_1 \dfrac{r_0}{2} \dfrac{1 + \sin^2 \phi}{\cos \phi}$

TABLE 20-5 (continued)   VARIOUS MEMBRANE SHELLS OF REVOLUTION

Loading Condition	N_ϕ	N_θ
HYDROSTATIC PRESSURE LOADING: $$P_z = \rho_w\left(d + \frac{r_0}{2}\tan^2\phi\right)$$ 	$$-\rho_w\frac{r_0}{2}\left(d + \frac{r_0}{4}\tan^2\phi\right)\frac{1}{\cos\phi}$$	$$-\rho_w\frac{r_0}{2}\left[d(2\tan^2\phi + 1) + r_0\tan^2\phi\left(\tan^2\phi + \frac{3}{4}\right)\right]\cos\phi$$

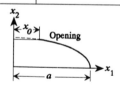

2. Cycloid

DEAD-WEIGHT LOADING: $$P_x = q\sin\phi$$ $$P_z = q\cos\phi$$	$$-2qr_0\frac{\phi s + c - \frac{1}{3}c^3 - \frac{2}{3}}{(2\phi + \sin 2\phi)\sin\phi}$$ $$s = \sin\phi,\ c = \cos\phi$$	$$-qr_0\left[\frac{1}{3}\frac{1 - \cos^3\phi}{\sin^2\phi\cos\phi} - \frac{\phi}{2}\tan\phi - \frac{1}{2}\sin^2\phi\right]$$
UNIFORMLY DISTRIBUTED VERTICAL LOADING ON, PROJECTED AREA: $$P_x = p_2\sin\phi\cos\phi$$ $$P_z = p_2\cos^2\phi$$	$$-p_2\frac{r_0}{8}\frac{2\phi + \sin 2\phi}{\sin\phi}$$	$$-p_2\frac{r_0}{16}\frac{2\phi + \sin 2\phi}{\sin\phi}\times\left(4\cos^2\phi - \frac{2\phi}{\sin 2\phi} - 1\right)$$

3. Modified elliptical shell

Equation of meridian: $$x_2 = -\int\frac{x_1^3\,dx_1}{\sqrt{(1 - x_1^2)(x_1^2 - a_1)(x_1^2 - a_2)}}$$

where

$$a_1 = \frac{1}{2}\left(\sqrt{1 - \frac{4x_0^2}{1 - x_0^2}} - 1\right) \qquad a_2 = -\frac{1}{2}\left(\sqrt{1 + \frac{4x_0^2}{1 - x_0^2}} + 1\right)$$

$$m = (1 - x_1^2)(x_1^2 - a_1)(x_1^2 - a_2) \qquad n = 1 + \frac{x_1^6}{m}$$

TABLE 20-5 | Various Membrane Shells of Revolution   **1210**

TABLE 20-5 (continued)   **VARIOUS MEMBRANE SHELLS OF REVOLUTION**

Loading Condition	N_ϕ	N_θ
UNIFORM NORMAL LOADING: $p_z = p_1$	$\dfrac{ap_1 n^{1/2} m^{1/2}}{2x_1^2}$ For the bottom edge $\frac{1}{2}p_1 a$	$N_\phi\left(2 - \dfrac{x}{n}\right)$ $N_\phi\left(2 - \dfrac{x}{n}\right)$

4. Pointed shell

Solution is not valid at apex.

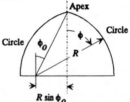

$$m = 1 - \frac{\sin \phi_0}{\sin \phi}$$

UNIFORM NORMAL LOADING: $p_z = p_1$	$\frac{1}{2}p_1 R m$	$p_1 m\left(1 - \frac{1}{2}m\right)R$

[a]Adopted from Baker et al. [20.2].

TABLE 20-6 SPHERICAL SHELLS WITH AXIALLY SYMMETRIC LOADING: BENDING THEORY[a]

Notation

ν = Poisson's ratio
h = thickness of shell
β = angle of rotation
M = applied moment per unit length
ϕ_1, ϕ_2 = meridional angles of lower and upper edges
$F_i(\alpha), \overline{F}_i$ = factors defined in Table 20-9
Q_ϕ = shear force per unit length
R = radius of shell
ΔR = horizontal displacement
H = applied horizontal force per unit length
ϕ = meridional angle of point of interest
M_ϕ, M_θ = moments per unit length
N_ϕ, N_θ = membrane forces per unit length

$$\alpha = \phi_1 - \phi \qquad \alpha_0 = \phi_1 - \phi_2 \qquad k = [3(1 - \nu^2)(R/h)^2]^{1/4}$$

Case	Internal Forces	Deformations
1. Spherical cap subject to edge horizontal load	$Q_\phi = H\left[\sqrt{2}\,e^{-k\alpha}\sin\phi_1\cos\left(k\alpha + \tfrac{1}{4}\pi\right)\right]$	$Eh\beta = H\left[2\sqrt{2}\,k^2 e^{-k\alpha}\sin\phi_1\sin\left(k\alpha + \tfrac{1}{4}\pi\right)\right]$
	$N_\phi = -Q_\phi\cot\phi$	$Eh(\Delta R) = -H\{\mathrm{Re}^{-k\alpha}\sin\phi_1[2k\sin\phi\cos k\alpha$
	$N_\theta = -H\left(2ke^{-k\alpha}\sin\phi_1\cos k\alpha\right)$	$\qquad - \sqrt{2}\,\nu\cos\phi\cos\left(k\alpha + \tfrac{1}{4}\pi\right)]\}$
	$M_\phi = -H\left(\dfrac{R}{k}e^{-k\alpha}\sin\phi_1\sin k\alpha\right)$	For $\alpha = 0$ and $\phi = \phi_1$:
	$M_\theta = H\left[\dfrac{R}{k^2\sqrt{2}}e^{-k\alpha}\sin\phi_1\cot\phi\sin\left(k\alpha + \tfrac{1}{4}\pi\right)\right] + \nu M_\phi$	$Eh\beta = H(2k^2\sin\phi_1)$
		$Eh(\Delta R) = -H[R\sin\phi_1(2k\sin\phi_1 - \nu\cos\phi_1)]$
		For $\phi_1 = 90°$:
		$Eh\beta = 2k^2H$
		$Eh(\Delta R) = -2RkH$

TABLE 20-6 **Bending Theory of Spherical Shells** 1212

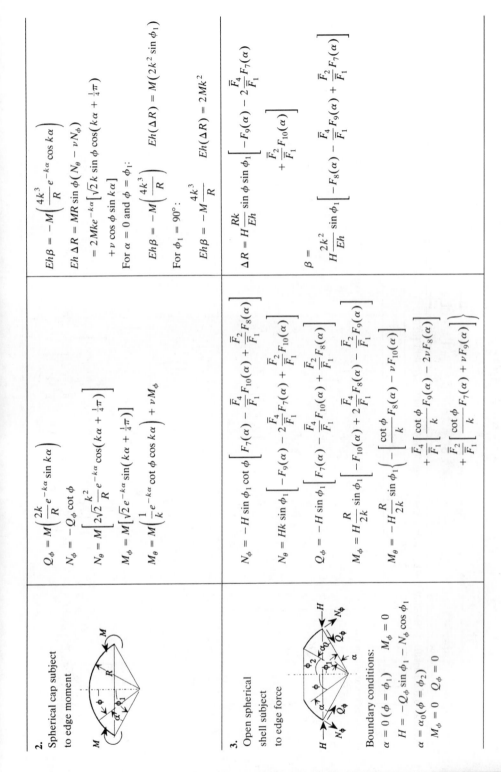

2.
Spherical cap subject to edge moment

$$Q_\phi = M\left(\frac{2k}{R}e^{-k\alpha}\sin k\alpha\right)$$

$$N_\phi = -Q_\phi \cot\phi$$

$$N_\theta = M\left[2\sqrt{2}\,\frac{k^2}{R}e^{-k\alpha}\cos\left(k\alpha + \tfrac{1}{4}\pi\right)\right]$$

$$M_\phi = M\left[\sqrt{2}\,e^{-k\alpha}\sin\left(k\alpha + \tfrac{1}{4}\pi\right)\right]$$

$$M_\theta = M\left(\frac{1}{k}e^{-k\alpha}\cot\phi\cos k\alpha\right) + \nu M_\phi$$

$$Eh\beta = -M\left(\frac{4k^3}{R}e^{-k\alpha}\cos k\alpha\right)$$

$$Eh\,\Delta R = MR\sin\phi\,(N_\theta - \nu N_\phi)$$

$$= 2Mke^{-k\alpha}\left[\sqrt{2}\,k\sin\phi\cos\left(k\alpha + \tfrac{1}{4}\pi\right) + \nu\cos\phi\sin k\alpha\right]$$

For $\alpha = 0$ and $\phi = \phi_1$:

$$Eh\beta = -M\left(\frac{4k^3}{R}\right) \qquad Eh(\Delta R) = M\left(2k^2\sin\phi_1\right)$$

For $\phi_1 = 90°$:

$$Eh\beta = -M\frac{4k^3}{R} \qquad Eh(\Delta R) = 2Mk^2$$

3.
Open spherical shell subject to edge force

$$N_\phi = -H\sin\phi_1\cot\phi\left[F_7(\alpha) - \frac{\bar{F}_4}{\bar{F}_1}F_{10}(\alpha) + \frac{\bar{F}_2}{\bar{F}_1}F_8(\alpha)\right]$$

$$N_\theta = Hk\sin\phi_1\left[-F_9(\alpha) - 2\frac{\bar{F}_4}{\bar{F}_1}F_7(\alpha) + \frac{\bar{F}_2}{\bar{F}_1}F_{10}(\alpha)\right]$$

$$Q_\phi = -H\sin\phi_1\left[F_7(\alpha) - \frac{\bar{F}_4}{\bar{F}_1}F_{10}(\alpha) + \frac{\bar{F}_2}{\bar{F}_1}F_8(\alpha)\right]$$

$$M_\phi = H\frac{R}{2k}\sin\phi_1\left[-F_{10}(\alpha) + 2\frac{\bar{F}_4}{\bar{F}_1}F_8(\alpha) - \frac{\bar{F}_2}{\bar{F}_1}F_9(\alpha)\right]$$

$$M_\theta = -H\frac{R}{2k}\sin\phi_1\left\{-\left[\frac{\cot\phi}{k}F_8(\alpha) - \nu F_{10}(\alpha)\right] + \frac{\bar{F}_4}{\bar{F}_1}\left[\frac{\cot\phi}{k}F_9(\alpha) - 2\nu F_8(\alpha)\right] + \frac{\bar{F}_2}{\bar{F}_1}\left[\frac{\cot\phi}{k}F_7(\alpha) + \nu F_9(\alpha)\right]\right\}$$

Boundary conditions:

$\alpha = 0\,(\phi = \phi_1)$ $M_\phi = 0$

$H = -Q_\phi\sin\phi_1 - N_\phi\cos\phi_1$

$\alpha = \alpha_0(\phi = \phi_2)$

$M_\phi = 0$ $Q_\phi = 0$

$$\Delta R = H\frac{Rk}{Eh}\sin\phi\sin\phi_1\left[-F_9(\alpha) - 2\frac{\bar{F}_4}{\bar{F}_1}F_7(\alpha) + \frac{\bar{F}_2}{\bar{F}_1}F_{10}(\alpha)\right]$$

$$\beta = H\frac{2k^2}{Eh}\sin\phi_1\left[-F_8(\alpha) - \frac{\bar{F}_4}{\bar{F}_1}F_9(\alpha) + \frac{\bar{F}_2}{\bar{F}_1}F_7(\alpha)\right]$$

TABLE 20-6 (continued) SPHERICAL SHELLS WITH AXIALLY SYMMETRIC LOADING: BENDING THEORY

Case	Internal Forces	Deformations
4. Open spherical shell with force at top edge Boundary conditions: $\alpha = 0 \; (\phi = \phi_1)$ $M_\phi = 0 \quad Q_\phi = 0$ $\alpha = \alpha_0 \; (\phi = \phi_2)$ $M_\phi = 0$ $H = Q\sin\phi_2 - N\cos\phi_2$	$N_\phi = H\cot\phi\,\sin\phi_2 \left[\dfrac{\bar{F}_9}{\bar{F}_1} F_{10}(\alpha) - 2\dfrac{\bar{F}_8}{\bar{F}_1} F_8(\alpha) \right]$ $N_\theta = H2k\sin\phi_2 \left[-\dfrac{\bar{F}_9}{\bar{F}_1} F_7(\alpha) + \dfrac{\bar{F}_8}{\bar{F}_1} F_{10}(\alpha) \right]$ $Q_\phi = H\sin\phi_2 \left[\dfrac{\bar{F}_9}{\bar{F}_1} F_{10}(\alpha) - \dfrac{2\bar{F}_8}{\bar{F}_1} F_8(\alpha) \right]$ $M_\phi = H\dfrac{R}{k}\sin\phi_2 \left[\dfrac{\bar{F}_9}{\bar{F}_1} F_8(\alpha) - \dfrac{\bar{F}_8}{\bar{F}_1} F_9(\alpha) \right]$ $M_\theta = H\dfrac{R}{2k}\sin\phi_2 \left\{ \dfrac{\bar{F}_9}{\bar{F}_1} \left[-\dfrac{\cot\phi}{k} F_9(\alpha) + \nu 2F_8(\alpha) \right] \right.$ $\left. - \dfrac{2\bar{F}_8}{\bar{F}_1} \left[\dfrac{\cot\phi}{k} F_7(\alpha) + \nu F_9(\alpha) \right] \right\}$	$\Delta R = -H\sin\phi_2 \dfrac{Rk}{Eh} 2\sin\phi \left[\dfrac{\bar{F}_9}{\bar{F}_1} F_7(\alpha) - \dfrac{\bar{F}_8}{\bar{F}_1} F_{10}(\alpha) \right]$ $\beta = H\sin\phi_2 \dfrac{2k^2}{Eh} \left[\dfrac{\bar{F}_9}{\bar{F}_1} F_9(\alpha) + \dfrac{2\bar{F}_8}{\bar{F}_1} F_7(\alpha) \right]$
5. Open spherical shell with moment at bottom edge	$N_\phi = M\cot\phi\,\dfrac{2k}{R} \left[\dfrac{\bar{F}_6}{\bar{F}_1} F_{15}(\alpha) + \dfrac{\bar{F}_5}{\bar{F}_1} F_{16}(\alpha) - \dfrac{\bar{F}_3}{\bar{F}_1} F_8(\alpha) \right]$ $N_\theta = -M\dfrac{2k^2}{R} \left[\dfrac{\bar{F}_6}{\bar{F}_1} F_{14}(\alpha) + \dfrac{\bar{F}_5}{\bar{F}_1} F_{13}(\alpha) - \dfrac{\bar{F}_3}{\bar{F}_1} F_{10}(\alpha) \right]$ $Q_\phi = M\dfrac{2k}{R} \left[\dfrac{\bar{F}_6}{\bar{F}_1} F_{15}(\alpha) + \dfrac{\bar{F}_5}{\bar{F}_1} F_{16}(\alpha) - \dfrac{\bar{F}_3}{\bar{F}_1} F_8(\alpha) \right]$	$\Delta R = -M\dfrac{2k^2}{Eh}\sin\phi \left[\dfrac{\bar{F}_6}{\bar{F}_1} F_{14}(\alpha) + \dfrac{\bar{F}_5}{\bar{F}_1} F_{13}(\alpha) - \dfrac{\bar{F}_3}{\bar{F}_1} F_{10}(\alpha) \right]$ $\beta = -M\dfrac{4k^3}{EhR} \left[\dfrac{\bar{F}_6}{\bar{F}_1} F_{16}(\alpha) - \dfrac{\bar{F}_5}{\bar{F}_1} F_{15}(\alpha) - \dfrac{\bar{F}_3}{\bar{F}_1} F_7(\alpha) \right]$

TABLE 20-6 Bending Theory of Spherical Shells 1214

Boundary conditions:

$\alpha = 0 \quad M_\phi = -M$

$Q_\phi = 0$

$\alpha = \alpha_0$

$M_\phi = 0 \qquad Q_\phi = 0$

$$M_\phi = -M\left[\frac{\bar{F}_6}{\bar{F}_1}F_{13}(\alpha) - \frac{\bar{F}_5}{\bar{F}_1}F_{14}(\alpha) + \frac{\bar{F}_3}{\bar{F}_1}F_9(\alpha)\right]$$

$$M_\theta = M\left\{\frac{\bar{F}_6}{\bar{F}_1}\left[\frac{\cot\phi}{k}F_{16}(\alpha) - \nu F_{13}(\alpha)\right]\right.$$
$$\left. - \frac{\bar{F}_5}{\bar{F}_1}\left[\frac{\cot\phi}{k}F_{15}(\alpha) - \nu F_{14}(\alpha)\right]\right.$$
$$\left. - \frac{\bar{F}_3}{\bar{F}_1}\left[\frac{\cot\phi}{k}F_7(\alpha) + \nu F_9(\alpha)\right]\right\}$$

$$\Delta R = M\frac{2k^2}{Eh}\sin\phi\left[-2\frac{\bar{F}_8}{\bar{F}_1}F_7(\alpha) + \frac{\bar{F}_{10}}{\bar{F}_1}F_{10}(\alpha)\right]$$

$$\beta = M\frac{4k^3}{EhR}\left[\frac{\bar{F}_8}{\bar{F}_1}F_9(\alpha) + \frac{\bar{F}_{10}}{\bar{F}_1}F_7(\alpha)\right]$$

6. Open spherical shell with top moment

$$N_\phi = M\frac{2k}{R}\cot\phi\left[\frac{\bar{F}_8}{\bar{F}_1}F_{10}(\alpha) - \frac{\bar{F}_{10}}{\bar{F}_1}F_8(\alpha)\right]$$

$$N_\theta = M\frac{2k^2}{R}\left[-2\frac{\bar{F}_8}{\bar{F}_1}F_7(\alpha) + \frac{\bar{F}_{10}}{\bar{F}_1}F_{10}(\alpha)\right]$$

$$Q_\phi = -M\frac{2k}{R}\left[-\frac{\bar{F}_8}{\bar{F}_1}F_{10}(\alpha) + \frac{\bar{F}_{10}}{\bar{F}_1}F_8(\alpha)\right]$$

$$M_\phi = M\left[2\frac{\bar{F}_8}{\bar{F}_1}F_8(\alpha) - \frac{\bar{F}_{10}}{\bar{F}_1}F_9(\alpha)\right]$$

$$M_\theta = -M\left\{\frac{\bar{F}_8}{\bar{F}_1}\left[\frac{\cot\phi}{k}F_9(\alpha) - \nu 2F_8(\alpha)\right]\right.$$
$$\left. + \frac{\bar{F}_{10}}{\bar{F}_1}\left[\frac{\cot\phi}{k}F_7(\alpha) + \nu F_9(\alpha)\right]\right\}$$

Boundary conditions:

$\alpha = 0 \; (\phi = \phi_1)$

$M_\phi = 0 \qquad Q_\phi = 0$

$\alpha = \alpha_0 \; (\phi = \phi_2)$

$M_\phi = M \qquad Q_\phi = 0$

^{a}Adopted from Baker et al. [20.2].

TABLE 20-7 Bending Theory of Conical Shells

1216

TABLE 20-7 CONICAL SHELLS WITH AXIALLY SYMMETRIC LOADING: BENDING THEORY[a]

Notation

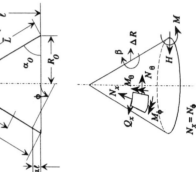

E = modulus of elasticity

h = thickness of shell

R = distance from point of interest to shell axis

ΔR = horizontal displacement at ϕ

H = applied horizontal force per unit length

$F_i(\xi), \overline{F}_i$ = factors defined in Table 20-9

ν = Poisson's ratio

ϕ = angle between normal of shell and axis

R_0 = radius of base circle

β = rotation of meridian

M = applied moment per unit length

α = coefficient that locates section of interest

$$
k = \begin{cases} \ell \left[3(1 - \nu^2)\right]^{1/4} / \left(\sqrt{R_0 h \sin \phi}\right) & \text{for conical cap} \\ \left[3(1 - \nu^2)\right]^{1/4} / \left(h x_m \cot \alpha_0\right)^{1/2} & \text{for open conical shell} \end{cases}
$$

$$
x_m = L - \frac{\overline{L}}{2} \qquad \xi = \frac{\overline{x}}{L}
$$

$$
D = \frac{Eh^3}{12(1 - \nu^2)}
$$

Case	Internal Forces	Deformations
1. Conical shell with horizontal load	$N_\phi = -H\left[\sqrt{2}\cos\phi\, e^{-k\alpha}\cos\left(k\alpha + \tfrac{1}{4}\pi\right)\right]$ $N_\theta = -H\left(\dfrac{2Rk\sin^2\phi}{\ell}e^{-k\alpha}\cos k\alpha\right)$ $M_\phi = -H\dfrac{\ell}{k}e^{-k\alpha}\sin k\alpha$ $M_\theta = \dfrac{H\ell^2}{\sqrt{2}Rk^2}\dfrac{\cot\phi}{\sin\phi}e^{-k\alpha}\sin\left(k\alpha + \tfrac{1}{4}\pi\right) - \nu M_\phi$ $Q_x = H\sqrt{2}\sin\phi\, e^{-k\alpha}\cos\left(k\alpha + \tfrac{1}{4}\pi\right)$	$\Delta R = \dfrac{-H\ell^3 e^{-k\alpha}}{2Dk^3\sin\phi}\left[\cos k\alpha - \nu\,\dfrac{\ell}{\sqrt{2}Rk}\dfrac{\cot\phi}{\sin\phi}\cos\left(k\alpha + \tfrac{1}{4}\pi\right)\right]$ $\beta = \dfrac{H\ell^2 e^{-k\alpha}\sin\left(k\alpha + \tfrac{1}{4}\pi\right)}{\sqrt{2}Dk^2\sin\phi}$ For $\alpha = 0$: $\Delta R = \dfrac{-H\ell^3}{2Dk^3\sin\phi}\left(1 - \dfrac{\nu\ell\cot\phi}{2Rk\sin\phi}\right)$ $\beta = \dfrac{H\ell^2}{2Dk^2\sin\phi}$
2. Conical shell with moment at lower edge	$N_\phi = M\left(\dfrac{2k\cos\phi}{\ell}e^{-k\alpha}\sin k\alpha\right)$ $N_\theta = -M\left[\dfrac{2\sqrt{2}Rk^2\sin^2\phi}{\ell^2}e^{-k\alpha}\cos\left(k\alpha + \tfrac{1}{4}\pi\right)\right]$ $M_\phi = M\left[\sqrt{2}e^{-k\alpha}\sin\left(k\alpha + \tfrac{1}{4}\pi\right)\right]$ $M_\theta = -M\dfrac{\ell\cot\phi}{Rk\sin\phi}e^{-k\alpha}\cos k\alpha - \nu M_\phi$ $Q_x = M\left(\dfrac{2k\sin\phi}{\ell}e^{-k\alpha}\sin k\alpha\right)$	$\Delta R = M\dfrac{\ell^3 e^{-k\alpha}}{2Dk^2\sin\phi}\left[\sqrt{2}\cos\left(k\alpha + \tfrac{1}{4}\pi\right) + \nu\dfrac{\ell}{R}\dfrac{\cos\phi\sin k\alpha}{k\sin^2\phi}\right]$ $\beta = -M\dfrac{\ell e^{-k\alpha}\cos k\alpha}{Dk\sin\phi}$ For $\alpha = 0$: $\Delta R = M\dfrac{\ell^2}{2Dk^2\sin\phi}$ $\beta = \dfrac{-M\ell}{Dk\sin\phi}$

Case	Internal Forces	Deformations
3. Open conical shell with upper edge load $Q = H \sin \alpha_0$ $N_x = -H \cos \alpha_0$	$N_x = -H \cos \alpha_0 \left[F_7(\xi) - \dfrac{\overline{F}_4}{\overline{F}_1} F_{10}(\xi) + \dfrac{\overline{F}_2}{\overline{F}_1} F_8(\xi) \right]$ $N_\theta = Hx_m k \cos \alpha_0 \left[F_9(\xi) + \dfrac{2\overline{F}_4}{\overline{F}_1} F_7(\xi) - \dfrac{\overline{F}_2}{\overline{F}_1} F_{10}(\xi) \right]$ $M_x = H\dfrac{\sin \alpha_0}{2k} \left[F_{10}(\xi) - \dfrac{2\overline{F}_4}{\overline{F}_1} F_8(\xi) + \dfrac{\overline{F}_2}{\overline{F}_1} F_9(\xi) \right]$ $Q_x = H \sin \alpha_0 \left[F_7(\xi) - \dfrac{\overline{F}_4}{\overline{F}_1} F_{10}(\xi) + \dfrac{\overline{F}_2}{\overline{F}_1} F_8(\xi) \right]$	$\Delta R = H\dfrac{\sin^2 \alpha_0}{4Dk^3} \left[F_9(\xi) + \dfrac{2\overline{F}_4}{\overline{F}_1} F_7(\xi) - \dfrac{\overline{F}_2}{\overline{F}_1} F_{10}(\xi) \right]$ $\beta = H\dfrac{\sin \alpha_0}{2Dk^2} \left[-F_8(\xi) + \dfrac{\overline{F}_4}{\overline{F}_1} F_9(\xi) + \dfrac{\overline{F}_2}{\overline{F}_1} F_7(\xi) \right]$
4. Open conical shell with lower edge load $Q_{x_2} = H \sin \alpha_0$ $N_{x_2} = -H \cos \alpha_0$	$N_x = -H \cos \alpha_0 \left[-\dfrac{\overline{F}_9}{\overline{F}_1} F_{10}(\xi) + \dfrac{2\overline{F}_8}{\overline{F}_1} F_8(\xi) \right]$ $N_\theta = -2Hx_m k \cos \alpha_0 \left[-\dfrac{\overline{F}_9}{\overline{F}_1} F_7(\xi) + \dfrac{\overline{F}_8}{\overline{F}_1} F_{10}(\xi) \right]$ $M_x = -H\dfrac{\sin \alpha_0}{k} \left[\dfrac{\overline{F}_9}{\overline{F}_1} F_8(\xi) - \dfrac{\overline{F}_8}{\overline{F}_1} F_9(\xi) \right]$ $Q_x = H \sin \alpha_0 \left[-\dfrac{\overline{F}_9}{\overline{F}_1} F_{10}(\xi) + \dfrac{2\overline{F}_8}{\overline{F}_1} F_8(\xi) \right]$	$\Delta R = -H\dfrac{\sin^2 \alpha_0}{2Dk^3} \left[\dfrac{\overline{F}_9}{\overline{F}_1} F_7(\xi) + \dfrac{\overline{F}_8}{\overline{F}_1} F_{10}(\xi) \right]$ $\beta = H\dfrac{\sin \alpha_0}{2Dk^2} \left[\dfrac{\overline{F}_9}{\overline{F}_1} F_9(\xi) + \dfrac{2\overline{F}_8}{\overline{F}_1} F_7(\xi) \right]$

TABLE 20-7   Bending Theory of Conical Shells   1218

5. Open conical shell with upper edge moment 	$N_x = M2k \cot \alpha_0 \left[\dfrac{\overline{F}_6}{\overline{F}_1} F_{15}(\xi) + \dfrac{\overline{F}_5}{\overline{F}_1} F_{16}(\xi) - \dfrac{\overline{F}_3}{\overline{F}_1} F_8(\xi) \right]$ $N_\theta = M2k^2 x_m \cot \alpha_0 \left[\dfrac{\overline{F}_6}{\overline{F}_1} F_{14}(\xi) + \dfrac{\overline{F}_5}{\overline{F}_1} F_{13}(\xi) - \dfrac{\overline{F}_3}{\overline{F}_1} F_{10}(\xi) \right]$ $M_x = M \left[\dfrac{\overline{F}_6}{\overline{F}_1} F_{13}(\xi) - \dfrac{\overline{F}_5}{\overline{F}_1} F_{14}(\xi) + \dfrac{\overline{F}_3}{\overline{F}_1} F_9(\xi) \right]$ $Q_x = M2k \left[\dfrac{\overline{F}_6}{\overline{F}_1} F_{15}(\xi) + \dfrac{\overline{F}_5}{\overline{F}_1} F_{16}(\xi) - \dfrac{\overline{F}_3}{\overline{F}_1} F_8(\xi) \right]$	$\Delta R = M \dfrac{\sin \alpha_0}{2Dk^2} \left[\dfrac{\overline{F}_6}{\overline{F}_1} F_{14}(\xi) + \dfrac{\overline{F}_5}{\overline{F}_1} F_{13}(\xi) - \dfrac{\overline{F}_3}{\overline{F}_1} F_{10}(\xi) \right]$ $\beta = -\dfrac{M}{Dk} \left[\dfrac{\overline{F}_6}{\overline{F}_1} F_{16}(\xi) - \dfrac{\overline{F}_5}{\overline{F}_1} F_{15}(\xi) - \dfrac{\overline{F}_3}{\overline{F}_1} F_7(\xi) \right]$
6. Open conical shell with lower edge moment 	$N_x = -M2k \cot \alpha_0 \left[\dfrac{\overline{F}_8}{\overline{F}_1} F_{10}(\xi) - \dfrac{\overline{F}_{10}}{\overline{F}_1} F_8(\xi) \right]$ $N_\theta = -M2k^2 x_m \cot \alpha_0 \left[\dfrac{2\overline{F}_8}{\overline{F}_1} F_7(\xi) - \dfrac{\overline{F}_{10}}{\overline{F}_1} F_{10}(\xi) \right]$ $M_x = M \left[\dfrac{2\overline{F}_8}{\overline{F}_1} F_8(\xi) - \dfrac{\overline{F}_{10}}{\overline{F}_1} F_9(\xi) \right]$ $Q_x = -M2k \left[\dfrac{\overline{F}_8}{\overline{F}_1} F_{10}(\xi) - \dfrac{\overline{F}_{10}}{\overline{F}_1} F_8(\xi) \right]$	$\Delta R = -M \dfrac{\sin \alpha_0}{2Dk^2} \left[\dfrac{2\overline{F}_8}{\overline{F}_1} F_7(\xi) - \dfrac{\overline{F}_{10}}{\overline{F}_1} F_{10}(\xi) \right]$ $\beta = -M \dfrac{1}{Dk} \left[\dfrac{\overline{F}_8}{\overline{F}_1} F_9(\xi) + \dfrac{\overline{F}_{10}}{\overline{F}_1} F_7(\xi) \right]$

[a]Adopted from Baker et al. [20.2].

TABLE 20-8 | Bending Theory of Cylindrical Shells | 1220

TABLE 20-8 CYLINDRICAL SHELLS WITH AXIALLY SYMMETRIC LOADING: BENDING THEORY[a]

Notation

E = modulus of elasticity

h = thickness of shell

L = length of shell

β = angle of rotation along shell

M = applied moment per unit length

ν = Poisson's ratio

R = radius of shell

ΔR = horizontal displacement along shell

H = applied horizontal force per unit length

$F_i(\xi), \overline{F}_i$ = factors defined in Table 20-9

N_x, N_θ = normal forces per unit length (stress resultants) in axial and circumferential (parallel) directions

M_ϕ, M_θ = moments per unit length about circumferential and longitudinal (x) axes

Q_x = transverse shear force per unit length, in plane perpendicular to x axis

$$k = L[3(1-\nu^2)]^{1/4}/\sqrt{Rh} \qquad D = \frac{Eh^3}{12(1-\nu^2)} \qquad \xi = \frac{x}{L}$$

Case	Internal Forces	Deformations
1. Long cylindrical shell with upper edge load	$N_x = 0$ $N_\theta = -H\dfrac{2Rk}{L}e^{-k\lambda}\cos k\lambda$ $M_x = H\dfrac{L}{k}e^{-k\lambda}\sin k\lambda$ $M_\theta = \nu M_x$ $Q_x = H\sqrt{2}\,e^{-k\lambda}\cos\left(k\lambda + \tfrac{1}{4}\pi\right)$	$\beta = H\dfrac{L^2}{\sqrt{2}\,k^2 D}e^{-k\lambda}\sin\left(k\lambda + \tfrac{1}{4}\pi\right)$ $\Delta R = -\dfrac{R}{Eh}\left(N_\theta - \nu N_x\right)$ $\quad = H\dfrac{L^3}{2Dk^3}e^{-k\lambda}\cos k\lambda$ For the case $\lambda = 0$: $\beta = HL^2/2k^2D$ $\Delta R = HL^3/2k^3D$
2. Long cylindrical shell with upper edge moment	$N_x = 0$ $N_\theta = -M\dfrac{2\sqrt{2}\,Rk^2}{L^2}e^{-k\lambda}\cos\left(k\lambda + \tfrac{1}{4}\pi\right)$ $M_x = M\left[\sqrt{2}\,e^{-k\lambda}\sin\left(k\lambda + \tfrac{1}{4}\pi\right)\right]$ $M_\theta = \nu M_x$ $Q_x = -M\dfrac{2k}{L}e^{-k\lambda}\sin k\lambda$	$\beta = M\left(\dfrac{L}{Dk}e^{-k\lambda}\cos k\lambda\right)$ $\Delta R = -\dfrac{R}{Eh}\left(N_\theta - \nu N_x\right)$ $\quad = M\left[\dfrac{L^2}{\sqrt{2}\,Dk^2}e^{-k\lambda}\cos\left(k\lambda + \tfrac{1}{4}\pi\right)\right]$ For the case $\lambda = 0$: $\beta = M(L/Dk)$ $\Delta R = M(L^2/2Dk^2)$

Case	Internal Forces	Deformations
3. Cylindrical shell with upper edge load 	$N_\theta = H2kR\left[-\dfrac{\bar{F}_9}{\bar{F}_1}F_7(\xi) + \dfrac{\bar{F}_8}{\bar{F}_1}F_{10}(\xi)\right]$ $M_x = \dfrac{H}{k}\left[\dfrac{\bar{F}_9}{\bar{F}_1}F_8(\xi) - \dfrac{\bar{F}_8}{\bar{F}_1}F_9(\xi)\right]$ $Q_x = -H\left[\dfrac{\bar{F}_9}{\bar{F}_1}F_{10}(\xi) - \dfrac{2\bar{F}_8}{\bar{F}_1}F_8(\xi)\right]$	$\Delta R = \dfrac{H}{2Dk^3}\left[-\dfrac{\bar{F}_9}{\bar{F}_1}F_7(\xi) + \dfrac{\bar{F}_8}{\bar{F}_1}F_{10}(\xi)\right]$ $\beta = \dfrac{H}{2Dk^2}\left[\dfrac{\bar{F}_9}{\bar{F}_1}F_9(\xi) + \dfrac{2\bar{F}_8}{\bar{F}_1}F_7(\xi)\right]$
4. Cylindrical shell with lower edge load 	$N_\theta = -H2kR\left[\dfrac{\bar{F}_4}{\bar{F}_1}F_7(\xi) - \dfrac{\bar{F}_5}{\bar{F}_1}F_{15}(\xi)\right.$ $\left. - \dfrac{\bar{F}_6}{\bar{F}_1}F_{16}(\xi)\right]$ $M_x = \dfrac{-H}{k}\left[-\dfrac{\bar{F}_4}{\bar{F}_1}F_8(\xi) - \dfrac{\bar{F}_5}{\bar{F}_1}F_{16}(\xi)\right.$ $\left. + \dfrac{\bar{F}_6}{\bar{F}_1}F_{15}(\xi)\right]$ $Q_x = -H\left[\dfrac{\bar{F}_4}{\bar{F}_1}F_{10}(\xi) + \dfrac{\bar{F}_5}{\bar{F}_1}F_{13}(\xi)\right.$ $\left. - \dfrac{\bar{F}_6}{\bar{F}_1}F_{14}(\xi)\right]$	$\Delta R = \dfrac{-H}{2Dk^3}\left[\dfrac{\bar{F}_4}{\bar{F}_1}F_7(\xi) - \dfrac{\bar{F}_5}{\bar{F}_1}F_{15}(\xi)\right.$ $\left. - \dfrac{\bar{F}_6}{\bar{F}_1}F_{16}(\xi)\right]$ $\beta = \dfrac{H}{2Dk^2}\left[\dfrac{\bar{F}_4}{\bar{F}_1}F_9(\xi) + \dfrac{\bar{F}_5}{\bar{F}_1}F_{14}(\xi)\right.$ $\left. + \dfrac{\bar{F}_6}{\bar{F}_1}F_{13}(\xi)\right]$

TABLE 20-8   **Bending Theory of Cylindrical Shells**   **1222**

5. Cylindrical shell with upper edge moment	$N_\theta = M2k^2R\left[-\dfrac{2\bar{F}_8}{\bar{F}_1}F_7(\xi) + \dfrac{\bar{F}_{10}}{\bar{F}_1}F_{10}(\xi)\right]$ $M_x = M\left[\dfrac{2\bar{F}_8}{\bar{F}_1}F_8(\xi) - \dfrac{\bar{F}_{10}}{\bar{F}_1}F_9(\xi)\right]$ $Q_x = -Mk\left[\dfrac{2\bar{F}_8}{\bar{F}_1}F_{10}(\xi) - \dfrac{2\bar{F}_{10}}{\bar{F}_1}F_8(\xi)\right]$	$\Delta R = \dfrac{M}{2Dk^2}\left[-\dfrac{2\bar{F}_8}{\bar{F}_1}F_7(\xi) + \dfrac{\bar{F}_{10}}{\bar{F}_1}F_{10}(\xi)\right]$ $\beta = \dfrac{M}{2Dk}\left[\dfrac{2\bar{F}_8}{\bar{F}_1}F_9(\xi) + \dfrac{2\bar{F}_{10}}{\bar{F}_1}F_7(\xi)\right]$
6. Cylindrical shell with lower edge moment	$N_\theta = -M2k^2R\left[\dfrac{\bar{F}_2}{\bar{F}_1}F_7(\xi) + \dfrac{\bar{F}_3}{\bar{F}_1}F_{10}(\xi) - F_8(\xi)\right]$ $M_x = -M\left[\dfrac{\bar{F}_2}{\bar{F}_1}F_8(\xi) - \dfrac{\bar{F}_3}{\bar{F}_1}F_9(\xi) - F_7(\xi)\right]$ $Q_x = kM\left[\dfrac{\bar{F}_2}{\bar{F}_1}F_{10}(\xi) - \dfrac{2\bar{F}_3}{\bar{F}_1}F_8(\xi) + F_8(\xi)\right]$	$\Delta R = \dfrac{-M}{2Dk^2}\left[\dfrac{\bar{F}_2}{\bar{F}_1}F_7(\xi) + \dfrac{\bar{F}_3}{\bar{F}_1}F_{10}(\xi) - F_8(\xi)\right]$ $\beta = \dfrac{-M}{2Dk}\left[\dfrac{\bar{F}_2}{\bar{F}_1}F_9(\xi) + \dfrac{2\bar{F}_3}{\bar{F}_1}F_7(\xi) - F_{10}(\xi)\right]$

[a]Adopted from Baker et al. [20.2].

TABLE 20-9   FACTORS FOR USE IN TABLES 20-6 TO 20-8[a]

Notation

h = thickness of shell
ν = Poisson's ratio

For cylindrical shells:

$$k = [3(1 - \nu^2)]^{1/4}/\sqrt{Rh}$$
$$\xi = x/L$$

For conical shells:

$$k = \begin{cases} \ell\,[3(1 - \nu^2)]^{1/4}/\left(\sqrt{R_0 h}\,\sin\phi\right) & \text{for conical cap} \\ [3(1 - \nu^2)]^{1/4}/(hx_m \cot\alpha_0)^{1/2} & \\ \qquad\qquad \text{for open conical shell} \end{cases}$$

$$\xi = \begin{cases} x/L & \text{for conical cap} \\ \bar{x}/\bar{L} & \text{for open conical shell} \end{cases}$$

$$x_m = L - \tfrac{1}{2}\bar{L}$$

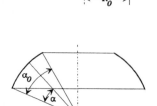

For spherical shells:

$$k = [3(1 - \nu^2)(R/h)^2]^{1/4}$$

Replace L by α_0

$$\xi = \alpha/\alpha_0$$

TABLE 20-9   Factors for Use in Tables 20-6 to 20-8   1224

TABLE 20-9 (continued) FACTORS FOR USE IN TABLES 20-6 TO 20-8

i	$F_i(\xi)$	$\bar{F}_i$
1^b	$\sinh^2 kL\xi - \sin^2 kL\xi$	$\sinh^2 kL - \sin^2 kL$
2	$\sinh^2 kL\xi + \sin^2 kL\xi$	$\sinh^2 kL + \sin^2 kL$
3	$\sinh kL\xi \cosh kL\xi + \sin kL\xi \cos kL\xi$	$\sinh kL \cosh kL + \sin kL \cos kL$
4	$\sinh kL\xi \cosh kL\xi - \sin kL\xi \cos kL\xi$	$\sinh kL \cosh kL - \sin kL \cos kL$
5	$\sin^2 kL\xi$	$\sin^2 kL$
6	$\sinh^2 kL\xi$	$\sinh^2 kL$
7	$\cosh kL\xi \cos kL\xi$	$\cosh kL \cos kL$
8	$\sinh kL\xi \sin kL\xi$	$\sinh kL \sin kL$
9	$\cosh kL\xi \sin kL\xi - \sinh kL\xi \cos kL\xi$	$\cosh kL \sin kL - \sinh kL \cos kL$
10	$\cosh kL\xi \sin kL\xi + \sinh kL\xi \cos kL\xi$	$\cosh kL \sin kL + \sinh kL \cos kL$
11	$\sin kL\xi \cos kL\xi$	$\sin kL \cos kL$
12	$\sinh kL\xi \cosh kL\xi$	$\sinh kL \cosh kL$
13	$\cosh kL\xi \cos kL\xi - \sinh kL\xi \sin kL\xi$	$\cosh kL \cos kL - \sinh kL \sin kL$
14	$\cosh kL\xi \cos kL\xi + \sinh kL\xi \sin kL\xi$	$\cosh kL \cos kL + \sinh kL \sin kL$
15	$\cosh kL\xi \sin kL\xi$	$\cosh kL \sin kL$
16	$\sinh kL\xi \cos kL\xi$	$\sinh kL \cos kL$
17^c	$\exp(-kL\xi \cos kL\xi)$	$\exp(-kL \cos kL)$
18	$\exp(-kL\xi \sin kL\xi)$	$\exp(-kL \sin kL)$
19	$\exp[-kL\xi(\cos kL\xi + \sin kL\xi)]$	$\exp[-kL(\cos kL + \sin kL)]$
20	$\exp[-kL\xi(\cos kL\xi - \sin kL\xi)]$	$\exp[-kL(\cos kL - \sin kL)]$

[a] From Baker et al. [20.2].
[b] For sphere: $F_1(\alpha) = \sinh^2 k\alpha_0(\alpha/\alpha_0) - \sin^2 k\alpha_0(\alpha/\alpha_0) = \sinh^2 k\alpha - \sin^2 k\alpha$.
[c] $\exp(\beta) = e^\beta$.

Factors for Use in Tables 20-6 to 20-8 TABLE 20-9

TABLE 20-10 RELATIONS FOR MULTIPLE SEGMENT SHELLS OF REVOLUTION[a]

Notation

$\mathbf{f}$ = flexibility matrix

$\beta_i^f, \Delta R_i^f$ = influence coefficients; rotation and displacement at edge i due to force f ($i = 1, 2$, $f = M_1, H_1, M_2, H_2$)

$\beta_i^0, \Delta R_i^0$ = rotation and displacement at edge i obtained from membrane analysis of segment with actual applied load ($i = 1, 2$)

$\beta_i, \Delta R_i$ = rotation and displacement at edge i from membrane solution and correction forces

Part A. Equations for Deformation of Shell Segments

Geometry	Equation	Flexibility Matrix $\mathbf{f}$
1. Two-edged segment	$$\begin{bmatrix} \beta_1 \\ \Delta R_1 \\ \beta_2 \\ \Delta R_2 \end{bmatrix} = \mathbf{f} \begin{bmatrix} M_1 \\ H_1 \\ M_2 \\ H_2 \end{bmatrix} + \begin{bmatrix} \beta_1^0 \\ \Delta R_1^0 \\ \beta_2^0 \\ \Delta R_2^0 \end{bmatrix}$$	$$\begin{bmatrix} \beta_1^{M_1} & \beta_1^{H_1} & \beta_1^{M_2} & \beta_1^{H_2} \\ \Delta R_1^{M_1} & \Delta R_1^{H_1} & \Delta R_1^{M_2} & \Delta R_1^{H_2} \\ \beta_2^{M_1} & \beta_2^{H_1} & \beta_2^{M_2} & \beta_2^{H_2} \\ \Delta R_2^{M_1} & \Delta R_2^{H_1} & \Delta R_2^{M_2} & \Delta R_2^{H_2} \end{bmatrix}$$
2. Segments with one edge	FOR UPPER SEGMENT: $$\begin{bmatrix} \beta_1 \\ \Delta R_1 \end{bmatrix} = \mathbf{f} \begin{bmatrix} M_1 \\ H_1 \end{bmatrix} + \begin{bmatrix} \beta_1^0 \\ \Delta R_1^0 \end{bmatrix}$$ FOR LOWER SEGMENT: $$\begin{bmatrix} \beta_2 \\ \Delta R_2 \end{bmatrix} = \mathbf{f} \begin{bmatrix} M_2 \\ H_2 \end{bmatrix} + \begin{bmatrix} \beta_2^0 \\ \Delta R_2^0 \end{bmatrix}$$	$$\begin{bmatrix} \beta_1^{M_1} & \beta_1^{H_1} \\ \Delta R_1^{M_1} & \Delta R_1^{H_1} \end{bmatrix}$$ $$\begin{bmatrix} \beta_2^{M_2} & \beta_2^{H_2} \\ \Delta R_2^{M_2} & \Delta R_2^{H_2} \end{bmatrix}$$

TABLE 20-10 Multiple Segment Shells 1226

Part B. Compatibility and Equilibrium Equations at Junctions

Junction	Equations
1. Junction of two shell elements	$$\begin{bmatrix} M_2 \\ H_2 \\ \beta_2 \\ \Delta R_2 \end{bmatrix} + \begin{bmatrix} M_2' \\ H_2' \\ 0 \\ 0 \end{bmatrix} = \begin{bmatrix} 0 \\ 0 \\ \beta_2' \\ \Delta R_2' \end{bmatrix}$$
2. Junction of three shell elements	$$\begin{bmatrix} M_2 \\ H_2 \\ \beta_2 \\ \beta_2 \\ \Delta R_2 \\ \Delta R_2 \end{bmatrix} + \begin{bmatrix} M_2' \\ H_2' \\ 0 \\ 0 \\ 0 \\ 0 \end{bmatrix} + \begin{bmatrix} M_2'' \\ H_2'' \\ 0 \\ 0 \\ 0 \\ 0 \end{bmatrix} = \begin{bmatrix} 0 \\ 0 \\ \beta_2' \\ \beta_2'' \\ \Delta R_2' \\ \Delta R_2'' \end{bmatrix}$$
3. Fixed edge	$\beta_1 = 0 \qquad \Delta R_1 = 0$
4. Pinned edge	$M_1 = 0 \qquad \Delta R_1 = 0$
5. Sliding edge	$M_1 = 0 \qquad H_1 = 0$

[a]Adopted from Baker et al. [20.2].

TABLE 20-11 MEMBRANE EDGE DEFORMATIONS AND INFLUENCE COEFFICIENTS FOR SPHERICAL ELEMENTS[a]

Notation

E = modulus of elasticity
ν = Poisson's ratio
R = radius of shell
h = thickness of shell
$\Delta R_i, \Delta R_j$ = horizontal displacements at i, j edges (rad), positive in same direction as in figures.
β_i, β_j = angles of rotation of meridional tangents at i, j edges
H_i, H_j = horizontal forces per unit length on i, j edges
M_i, M_j = moments per unit length on i, j edges
ϕ_1, ϕ_2 = meridional angles of lower and upper edges
$\overline{F}_i$ = factors defined in Table 20-9
ρ_w = specific weight of liquid (F/L^3)
W = lantern loading (see Table 20-1, part A, for definition)
$\Delta R_i^0, \beta_i^0$ = membrane deformations at edge i due to applied loads
$\Delta R_i^f, \beta_i^f$ = influence coefficients, i.e., deformations at edge i due to unit force f $(f = M_i, H_i, M_j, H_j)$
$k = [3(1 - \nu^2)(R/h)^2]^{1/4}$

Positive moments cause tension in the inner shell surface.

Positive horizontal forces cause tension in the inner shell surface at the upper edge and compression in the inner shell surface at the lower edge.

The deformations in each case (column) in the table are due to the applied loading shown on the left.

Part A. Membrane Deformations of Spherical Caps

Loading Condition	Edge Deformation	
	$\Delta R_i^0 \quad\quad \Delta R_i^0$	$\beta_i^0 \quad\quad \beta_i^0$
1. Dead-weight loading	$\dfrac{R^2 q}{Eh} \sin \phi_1 \left[-\cos \phi_1 + \dfrac{1 + \nu}{1 + \cos \phi_1} \right]$	$-\dfrac{Rq}{Eh}(2 + \nu)\sin \phi_1$

TABLE 20-11   Influence Coefficients for Spherical Elements   1228

Part A. Membrane Deformations of Spherical Caps

Loading Condition	Edge Deformation	
	$\Delta R_i^0 \overset{\frown}{} \Delta R_i^0$	$\beta_i^0 \overset{\frown}{} \beta_i^0$
2. Uniform vertical pressure on projected area p_2	$\dfrac{R^2 p_2}{Eh} \sin\phi_1 \left(-\cos^2\phi_1 + \dfrac{1+\nu}{2} \right)$	$-\dfrac{R p_2}{Eh}(3+\nu)\sin\phi_1 \cos\phi_1$
3. Hydrostatic loading d	$-\dfrac{\rho_w R^3}{6Eh}\sin\phi_1\left[3\left(1+\dfrac{d}{R}\right)(1-\nu)\right.$ $\left. -6\cos\phi_1 - 2\dfrac{1+\nu}{\sin^2\phi_1}\left(\cos^3\phi_1 - 1\right)\right]$	$\dfrac{\rho_w R^2}{Eh}\sin\phi_1$
4. Uniform normal pressure p_1	$-\dfrac{R^2 p_1}{2Eh}(1-\nu)\sin\phi_1$	0

Part A. Membrane Deformations of Spherical Caps

Loading Condition	ΔR_i^0	β_i^0
5. Dead-weight loading	$\dfrac{R^2 q}{Eh} \sin \phi_1 \left[\cos \phi_1 - \dfrac{1 + \nu}{1 + \cos \phi_1} \right]$	$-\dfrac{Rq}{Eh}(2 + \nu)\sin \phi_1$
6. Uniform vertical pressure on projected area	$-\dfrac{R^2 p_2}{Eh} \sin \phi_1 \left(-\cos^2 \phi_1 + \dfrac{1 + \nu}{2} \right)$	$-\dfrac{Rp_2}{Eh}(3 + \nu)\sin \phi_1 \cos \phi_1$
7. Hydrostatic loading	$-\dfrac{\rho_w R^3}{6Eh} \sin \phi_1 \left[3 \left(1 - \dfrac{d}{R} \right)(1 - \nu) \right.$ $\left. - 6 \cos \phi_1 - 2 \dfrac{1 + \nu}{\sin^2 \phi_1} (\cos^3 \phi_1 - 1) \right]$	$-\dfrac{\rho_w R^2}{Eh} \sin \phi_1$
8. Uniform normal pressure	$\dfrac{R^2 p_1}{2Eh}(1 - \nu)\sin \phi_1$	0

Part B. Membrane Deformations of Spherical Segments

Loading Condition	Edge Deformation			
1. Dead-weight loading	$\dfrac{R^2 q}{Eh}\sin\phi_1\left[-\cos\phi_1 + \dfrac{1+\nu}{\sin^2\phi_1}\left(-\cos\phi_1+\cos\phi_2\right)\right]$	$-\dfrac{Rq}{Eh}(2+\nu)\sin\phi_1$	$-\dfrac{R^2 q}{Eh}\sin\phi_2\cos\phi_2$	$-\dfrac{Rq}{Eh}(2+\nu)\sin\phi_2$
2. Uniform vertical pressure on projected area p_2	$\dfrac{R^2 p_2}{Eh}\sin\phi_1\left[-\cos^2\phi_1 + \dfrac{1+\nu}{2}\left(1-\dfrac{\sin^2\phi_2}{\sin^2\phi_1}\right)\right]$	$-\dfrac{Rp_2}{Eh}(3+\nu)\sin\phi_1\cos\phi_1$	$-\dfrac{R^2 p_2}{Eh}\sin\phi_2\cos^2\phi_2$	$-\dfrac{Rp_2}{Eh}(3+\nu)\sin\phi_2\cos\phi_2$

TABLE 20-11 (continued) MEMBRANE EDGE DEFORMATIONS AND INFLUENCE COEFFICIENTS FOR SPHERICAL ELEMENTS

Part B. Membrane Deformations of Spherical Segments

Loading Condition	Edge Deformation			
	ΔR_i^0	β_i^0	ΔR_j^0	β_j^0
3. Hydrostatic loading	$-\dfrac{\rho_w R^3}{6Eh}\sin\phi_1\left\{3\left(1+\dfrac{d}{R}\right)\left[1-\nu+(1+\nu)\dfrac{\sin^2\phi_2}{\sin^2\phi_1}\right]-6\cos\phi_1+2(1+\nu)\dfrac{\cos^3\phi_2-\cos^3\phi_1}{\sin^2\phi_1}\right\}$	$\dfrac{\rho_w R^2}{Eh}\sin\phi_1$	$-\dfrac{\rho_w R^3}{Eh}\sin\phi_2\left(1+\dfrac{d}{R}-\cos\phi_2\right)$	$\dfrac{\rho_w R^2}{Eh}\sin\phi_2$
4. Uniform normal load	$-\dfrac{R^2 p_1}{Eh}\sin\phi_1\left[1-\dfrac{1+\nu}{2}\left(1-\dfrac{\sin^2\phi_2}{\sin^2\phi_1}\right)\right]$	0	$-\dfrac{R^2 p_1}{Eh}\sin\phi_2$	0

TABLE 20-11 Influence Coefficients for Spherical Elements   1232

Loading Condition	Edge Deformation			
	ΔR_j^0	β_j^0	ΔR_i^0	β_i^0
5. Lantern load	$\dfrac{WR}{Eh}(1+\nu)\dfrac{\sin\phi_2}{\sin\phi_1}$	0	$\dfrac{WR}{Eh}(1+\nu)$	0
6. Dead-weight loading	$-\dfrac{R^2 q}{Eh}\sin\phi_1\left[-\cos\phi_1 + \dfrac{1+\nu}{\sin^2\phi_1}\left(-\cos\phi_1+\cos\phi_2\right)\right]$	$-\dfrac{Rq}{Eh}(2+\nu)\sin\phi_1$	$\dfrac{R^2 q}{Eh}\sin\phi_2\cos\phi_2$	$-\dfrac{Rq}{Eh}(2+\nu)\sin\phi_2$
7. Uniform vertical load on projected area	$\dfrac{R^2 p_2}{Eh}\sin\phi_1\left[-\cos^2\phi_1 + \dfrac{1+\nu}{2}\left(1-\dfrac{\sin^2\phi_2}{\sin^2\phi_1}\right)\right]$	$-\dfrac{Rp_2}{Eh}(3+\nu)\sin\phi_1\cos\phi_1$	$\dfrac{R^2 p_2}{Eh}\sin\phi_2\cos^2\phi_2$	$-\dfrac{Rp_2}{Eh}(3+\nu)\sin\phi_2\cos\phi_2$

TABLE 20-11 (continued) MEMBRANE EDGE DEFORMATIONS AND INFLUENCE COEFFICIENTS FOR SPHERICAL ELEMENTS

Part B. Membrane Deformations of Spherical Segments

Loading Condition	Edge Deformation			
	ΔR_j^0	β_j^0	ΔR_i^0	β_i^0
8. Hydrostatic load	$-\dfrac{\rho_w R^3}{6Eh}\sin\phi_1\left\{3\left(1-\dfrac{d}{R}\right)\right.$ $\times\left[1-\nu+(1+\nu)\dfrac{\sin^2\phi_2}{\sin^2\phi_1}\right]$ $-6\cos\phi_1+2(1+\nu)$ $\left.\times\dfrac{\cos^3\phi_2-\cos^3\phi_1}{\sin^2\phi_1}\right\}$	$-\dfrac{\rho_w R^2}{Eh}\sin\phi_1$	$-\dfrac{\rho_w R^3}{Eh}\sin\phi_2\left(1-\dfrac{d}{R}\right.$ $\left.-\cos\phi_2\right)$	$-\dfrac{\rho_w R^2}{Eh}\sin\phi_2$
9. Uniform normal pressure	$\dfrac{R^2 p_1}{Eh}\sin\phi_1\left[1\right.$ $\left.-\dfrac{1+\nu}{2}\left(1-\dfrac{\sin^2\phi_2}{\sin\phi_1}\right)\right]$	0	$\dfrac{R^2 p_1}{Eh}\sin\phi_2$	0
10. Lantern load	$-\dfrac{WR}{Eh}(1+\nu)\dfrac{\sin\phi_2}{\sin\phi_1}$	0	$-\dfrac{WR}{Eh}(1+\nu)$	0

TABLE 20-11   Influence Coefficients for Spherical Elements   1234

Part C. Influence Coefficients (Bending Deformations at Edges) of Spherical Shells

Load Condition	Edge Deformation			
	ΔR_i^f	β_i^f	ΔR_j^f	β_j^f
1. $f = H_i$	$-\dfrac{2Rk}{Eh}\sin^2\phi_1 \dfrac{\bar{F}_4}{\bar{F}_1}$	$\dfrac{2k^2}{Eh}\sin\phi_1 \dfrac{\bar{F}_2}{\bar{F}_1}$	$\dfrac{2Rk}{Eh}\sin\phi_1\sin\phi_2 \dfrac{\bar{F}_9}{\bar{F}_1}$	$\dfrac{2k^2}{Eh}\sin\phi_1 \dfrac{2\bar{F}_8}{\bar{F}_1}$
2. $f = M_i$	$\dfrac{2k^2}{Eh}\sin\phi_1 \dfrac{\bar{F}_2}{\bar{F}_1}$	$-\dfrac{2k^3}{EhR}\dfrac{2\bar{F}_3}{\bar{F}_1}$	$-\dfrac{2k^2}{Eh}\sin\phi_2 \dfrac{2\bar{F}_8}{\bar{F}_1}$	$-\dfrac{2k^3}{EhR}\dfrac{2\bar{F}_{10}}{\bar{F}_1}$
3. $f = H_j$	$-\dfrac{2Rk}{Eh}\sin\phi_1\sin\phi_2 \dfrac{\bar{F}_9}{\bar{F}_1}$	$\dfrac{2k^2}{Eh}\sin\phi_2 \dfrac{2\bar{F}_8}{\bar{F}_1}$	$\dfrac{2Rk}{Eh}\sin^2\phi_2 \dfrac{\bar{F}_4}{\bar{F}_1}$	$\dfrac{2k^2}{Eh}\sin\phi_2 \dfrac{\bar{F}_2}{\bar{F}_1}$
4. $f = M_j$	$\dfrac{2k^2}{Eh}\sin\phi_1 \dfrac{2\bar{F}_8}{\bar{F}_1}$	$-\dfrac{2k^3}{EhR}\dfrac{2\bar{F}_{10}}{\bar{F}_1}$	$\dfrac{2k^2}{Eh}\sin\phi_2 \dfrac{\bar{F}_2}{\bar{F}_1}$	$\dfrac{2k^3}{EhR}\dfrac{2\bar{F}_3}{\bar{F}_1}$

Part C. Influence Coefficients (Bending Deformations at Edges) of Spherical Shells

Shell Geometry and Loading Condition	Edge Deformation	
	ΔR_i^f	β_i^f
5.	$-\dfrac{R}{Eh}\sin\phi_1(2k\sin\phi_1 - \nu\cos\phi_1)$	$\dfrac{2k^2}{Eh}\sin\phi_1$
6.	$\dfrac{2k^2}{Eh}\sin\phi_1$	$-\dfrac{4k^3}{EhR}$

[a] Adopted from Baker et al. [20.2].

TABLE 20-11   Influence Coefficients for Spherical Elements   1236

TABLE 20-12 MEMBRANE EDGE DEFORMATIONS AND INFLUENCE COEFFICIENTS FOR CONICAL SHELL ELEMENTS[a]

Notation

E = modulus of elasticity

ν = Poisson's ratio

h = thickness of shell

ρ_w = specific weight of liquid

$\Delta R_i, \Delta R_j$ = horizontal displacements at i, j edges

β_i, β_j = angles of rotation of meridional tangents at i, j edges (rad), positive in same direction as in figures

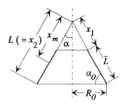

$\Delta R_i^0, \beta_i^0$ = membrane deformations at edge i due to applied loads

$\Delta R_i^f, \beta_i^f$ = influence coefficients, i.e., deformations at edge i due to force f ($f = M_i, H_i, M_j, H_j$)

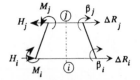

H_i, H_j = horizontal forces per unit length on i, j edges

M_i, M_j = moments per unit length on i, j edges

$\overline{F}_i$ = factors defined in Table 20-9

p_1 = applied uniform pressure

p_2 = uniformly distributed loading on projected area

q = dead weight

W = lantern load; see Table 20-2, part A, for definition

$k = [3(1 - \nu^2)]^{1/4}/(hx_m \cot \alpha)^{1/2}$ For open conical shell

$$D = \frac{Eh^3}{12(1 - \nu^2)}$$

$x_m = L - \tfrac{1}{2}\overline{L}$

Positive edge moments cause tension in the inner shell surface.

Positive horizontal edge forces cause tension in the inner shell surface at the upper edge and compression in the inner shell surface at the lower edge.

The deformations in each case (columns) in the table are due to the applied loading shown on the left.

Part A. Membrane Deformations of Conical Shells

Loading Condition	Edge Deformation	
	ΔR_i^0 (i) ΔR_i^0	β_i^0 (i) β_i^0
1. Dead-weight loading	$-\dfrac{qx_2^2}{Eh}\cot\alpha_0\left(\cos^2\alpha_0 - \dfrac{\nu}{2}\right)$	$\dfrac{qx_2}{Eh}\dfrac{\cos\alpha_0}{\sin^2\alpha_0}[(2+\nu)\cos^2\alpha_0 - \tfrac{1}{2} - \nu]$
2. Uniform vertical load on projected area P_2	$-\dfrac{p_2 x_2^2}{Eh}\cos\alpha_0\cot\alpha_0\left(\cos^2\alpha_0 - \dfrac{\nu}{2}\right)$	$\dfrac{p_2 x_2}{Eh}\cot^2\alpha_0[(2+\nu)\cos^2\alpha_0 - \tfrac{1}{2} - \nu]$
3. Hydrostatic load	$\dfrac{\rho_w x_2^2}{Eh}\cos^2\alpha_0\left[\dfrac{d}{\sin\alpha_0}\left(\dfrac{\nu}{2}-1\right)\right.$ $\left. + x_2\left(\dfrac{\nu}{3}-1\right)\right]$	$\dfrac{\rho_w x_2}{Eh}\dfrac{\cos^2\alpha_0}{\sin\alpha_0}\left(\dfrac{3}{2}\dfrac{d}{\sin\alpha_0} + \dfrac{8}{3}x_2\right)$
4. Uniform normal pressure	$-\dfrac{p_1 x_2^2}{Eh}\left(1 - \dfrac{\nu}{2}\right)\cos\alpha_0\cot\alpha_0$	$\dfrac{p_1 x_2}{Eh}\dfrac{3}{2}\cot^2\alpha_0$

TABLE 20-12   Influence Coefficients for Conical Shell Elements   **1238**

Part A. Membrane Deformations of Conical Shells

Loading Condition	ΔR_i^0 ⟵ (i) ⟶ ΔR_i^0	β_i^0 (i) β_i^0
5. Dead-weight loading	$\dfrac{qx_2^2}{Eh}\cot\alpha_0\left(\cos^2\alpha_0 - \dfrac{\nu}{2}\right)$	$\dfrac{qx_2}{Eh}\dfrac{\cos\alpha_0}{\sin^2\alpha_0}[(2+\nu)\cos^2\alpha_0 - \tfrac12 - \nu]$
6. Uniform vertical load on projected area P_2	$\dfrac{p_2 x_2^2}{Eh}\cos\alpha_0\cot\alpha_0\left(\cos^2\alpha_0 - \dfrac{\nu}{2}\right)$	$\dfrac{p_2 x_2}{Eh}\cot^2\alpha_0[(2+\nu)\cos^2\alpha_0 - \tfrac12 - \nu]$
7. Hydrostatic load	$-\dfrac{\rho_w x_2^2}{Eh}\cos^2\alpha_0\left[\dfrac{d}{\sin\alpha_0}\left(\dfrac{\nu}{2}-1\right) - x_2\left(\dfrac{\nu}{3}-1\right)\right]$	$\dfrac{\rho_w x_2}{Eh}\dfrac{\cos^2\alpha_0}{\sin\alpha_0}\left(\dfrac{3}{2}\dfrac{d}{\sin\alpha_0} - \dfrac{8}{3}x_2\right)$
8. Uniform normal pressure	$\dfrac{p_1 x_2^2}{Eh}\cos\alpha_0\cot\alpha_0\left(1-\dfrac{\nu}{2}\right)$	$\dfrac{p_1 x_2}{Eh}\dfrac{3}{2}\cot^2\alpha_0$

TABLE 20-12 (continued) MEMBRANE EDGE DEFORMATIONS AND INFLUENCE COEFFICIENTS FOR CONICAL SHELL ELEMENTS

Part B. Membrane Deformations of Truncated Cones

Loading Condition	Edge Deformation			
	ΔR_i^0	β_i^0	ΔR_j^0	β_j^0
1. Dead-weight loading	$-\dfrac{qx_2^2}{Eh}\cot\alpha_0\left[2\cos^2\alpha_0 - \nu\left(1 - \dfrac{x_1^2}{x_2^2}\right)\right]$	$\dfrac{qx_2}{Eh}\dfrac{\cos\alpha_0}{2\sin^2\alpha_0}\times\left[2(2+\nu)\cos^2\alpha_0 - 1 - 2\nu - \left(\dfrac{x_1}{x_2}\right)^2\right]$	$-\dfrac{qx_1^2}{Eh}\dfrac{\cos^3\alpha_0}{2\sin\alpha_0}$	$\dfrac{qx_1}{Eh}\dfrac{\cos\alpha_0}{\sin^2\alpha_0}[(2+\nu)\cos^2\alpha_0 - \nu]$
2. Uniform vertical load on projected area	$-\dfrac{p_2 x_2^2}{Eh}\dfrac{\cos^2\alpha_0}{2\sin\alpha_0}\left[2\cos^2\alpha_0 - \nu\left(1 - \dfrac{x_1^2}{x_2^2}\right)\right]$	$\dfrac{p_2 x_2}{Eh}\dfrac{\cot^2\alpha_0}{2}\times\left[2(2+\nu)\cos^2\alpha_0 - 1 - 2\nu + \left(\dfrac{x_1}{x_2}\right)^2\right]$	$-\dfrac{p_2 x_1^2}{Eh}\dfrac{\cos^4\alpha_0}{\sin\alpha_0}$	$\dfrac{p_2 x_1}{Eh}\cot^2\alpha_0[(2+\nu)\cos^2\alpha_0 - \nu]$

TABLE 20-12 Influence Coefficients for Conical Shell Elements 1240

Load				
3. Hydrostatic load	$\dfrac{\rho_w x_2^2}{Eh}\cos^2\alpha_0 \times \left\{ \nu\left[\dfrac{d}{2\sin\alpha_0}\left(1 - \dfrac{x_1^2}{x_2^2}\right) + \dfrac{x_2}{3}\left(1 - \dfrac{x_1^3}{x_2^3}\right)\right] - \dfrac{d}{\sin\alpha_0} - x_2 \right\}$	$\dfrac{\rho_w x_2 \cos^2\alpha_0}{Eh \sin\alpha_0} \times \left[\dfrac{d}{2\sin\alpha_0}\left(3 + \dfrac{x_1^2}{x_2^2}\right) + \dfrac{x_2}{3}\left(8 + \dfrac{x_1^3}{x_2^3}\right)\right]$	$-\dfrac{\rho_w x_1^2}{Eh}\cos^2\alpha_0\left(\dfrac{d}{\sin\alpha_0} + x_1\right)$	$\dfrac{\rho_w x_1 \cos^2\alpha_0}{Eh \sin\alpha_0}\left(\dfrac{2d}{\sin\alpha_0} + 3x_1\right)$
4. Uniform vertical load	$-\dfrac{p_1 x_2^2}{Eh}\cos\alpha_0\cot\alpha_0 \times \left[1 - \dfrac{\nu}{2}\left(1 - \dfrac{x_1^2}{x_2^2}\right)\right]$	$\dfrac{p_1 x_2}{Eh}\dfrac{\cos^2\alpha_0}{2}\left[3 + \left(\dfrac{x_1}{x_2}\right)^2\right]$	$-\dfrac{p_1 x_1^2}{Eh}\cos\alpha_0\cot\alpha_0$	$2\dfrac{p_1 x_1}{Eh}\cot^2\alpha_0$
5. Lantern load	$\dfrac{W}{Eh}x_1\nu\cot\alpha_0$	$\dfrac{W}{Eh}\dfrac{x_1\cot\alpha_0}{x_2\sin\alpha_0}$	$-\dfrac{W}{Eh}x_1\nu\cot\alpha_0$	$-\dfrac{W}{Eh}\dfrac{\cot\alpha_0}{\sin\alpha_0}$

Part B. Membrane Deformations of Truncated Cones

Loading Condition	Edge Deformation			
6. Dead-weight load	$\dfrac{qx_2^2 \cot \alpha_0}{Eh} \left[\dfrac{2\cos^2 \alpha_0}{2} - \nu\left(1 - \dfrac{x_1^2}{x_2^2}\right) \right]$	$\dfrac{qx_2 \cos \alpha_0}{Eh\, 2\sin^2 \alpha_0} \times \left[2(2+\nu)\cos^2 \alpha_0 - 1 - 2\nu + \left(\dfrac{x_1}{x_2}\right)^2 \right]$	$\dfrac{qx_1^2 \cos^3 \alpha_0}{Eh\, \sin \alpha_0}$	$\dfrac{qx_1 \cos \alpha_0}{Eh\, \sin^2 \alpha_0}[(2+\nu)\cos^2 \alpha - \nu]$
7. Uniform vertical load on projected area	$\dfrac{p_2 x_2^2 \cos^2 \alpha_0}{Eh\, 2\sin \alpha_0} \left[2\cos^2 \alpha_0 - \nu\left(1 - \dfrac{x_1^2}{x_2^2}\right) \right]$	$\dfrac{p_2 x_2 \cot^2 \alpha_0}{Eh\, 2} \times \left[2(2+\nu)\cos^2 \alpha_0 - 1 - 2\nu + \left(\dfrac{x_1}{x_2}\right)^2 \right]$	$\dfrac{p_2 x_1^2 \cos^4 \alpha_0}{Eh\, \sin \alpha_0}$	$\dfrac{p_2 x_1}{Eh} \cot^2 \alpha_0[(2+\nu)\cos^2 \alpha_0 - \nu]$

TABLE 20-12   Influence Coefficients for Conical Shell Elements   **1242**

Loading	(1)	(2)	(3)	(4)
8. Hydrostatic load 	$\dfrac{\rho_w x_2^2}{Eh}\cos^2\alpha_0\left\{\nu\left[\dfrac{x_2}{3}\left(1-\dfrac{x_1^3}{x_2^3}\right)\right]\right.$ $-\dfrac{d}{2\sin\alpha_0}\left(1-\dfrac{x_1^2}{x_2^2}\right)$ $\left.+\dfrac{d}{\sin\alpha_0}-x_2\right\}$	$\dfrac{\rho_w x_2}{Eh}\dfrac{\cos^2\alpha_0}{\sin\alpha_0}$ $\times\left[\dfrac{d}{2\sin\alpha_0}\left(3+\dfrac{x_1^2}{x_2^2}\right)\right.$ $\left.-\dfrac{x_2}{3}\left(8+\dfrac{x_1^3}{x_2^3}\right)\right]$	$\dfrac{\rho_w x_1^2}{Eh}\cos^2\alpha_0\left(\dfrac{d}{\sin\alpha_0}-x_1\right)$	$\dfrac{\rho_w x_1}{Eh}\dfrac{\cos^2\alpha_0}{\sin\alpha_0}\left(\dfrac{2d}{\sin\alpha_0}-3x_1\right)$
9. Uniform normal pressure 	$\dfrac{p_1 x_2^2}{Eh}\cos\alpha_0\cot\alpha_0$ $\times\left[1-\dfrac{\nu}{2}\left(1-\dfrac{x_1^2}{x_2^2}\right)\right]$	$\dfrac{p_1 x_2}{Eh}\dfrac{\cot^2\alpha_0}{2}\left[3+\left(\dfrac{x_1}{x_2}\right)^2\right]$	$\dfrac{p_1 x_1^2}{Eh}\cos\alpha_0\cot\alpha_0$	$\dfrac{p_1 x_1}{Eh}2\cot^2\alpha_0$
10. Lantern load 	$-\dfrac{W}{Eh}x_1\nu\cot\alpha_0$	$-\dfrac{W}{Eh}\dfrac{x_1\cot\alpha_0}{x_2\sin\alpha_0}$	$-\dfrac{W}{Eh}x_1\nu\cot\alpha_0$	$-\dfrac{W}{Eh}\dfrac{\cot\alpha_0}{\sin\alpha_0}$

Part C. Influence Coefficients (Bending Deformation) of Truncated Cones

Loading Condition	Edge Deformation			
	ΔR_i^f	β_i^f	ΔR_j^f	β_j^f
1. $f = H_i$	$-\dfrac{\sin^2\alpha_0}{2Dk^3}\dfrac{\bar{F}_4}{\bar{F}_1}$	$\dfrac{\sin\alpha_0}{2Dk^2}\dfrac{\bar{F}_2}{\bar{F}_1}$	$\dfrac{\sin^2\alpha_0}{2Dk^3}\dfrac{\bar{F}_9}{\bar{F}_1}$	$\dfrac{\sin\alpha_0}{2Dk^2}\dfrac{2\bar{F}_8}{\bar{F}_1}$
2. $f = M_i$	$\dfrac{\sin\alpha_0}{2Dk^2}\dfrac{\bar{F}_2}{\bar{F}_1}$	$-\dfrac{1}{2Dk}\dfrac{2\bar{F}_3}{\bar{F}_1}$	$-\dfrac{\sin\alpha_0}{2Dk^2}\dfrac{2\bar{F}_8}{\bar{F}_1}$	$-\dfrac{1}{2Dk}\dfrac{2\bar{F}_{10}}{\bar{F}_1}$
3. $f = H_j$	$-\dfrac{\sin^2\alpha_0}{2Dk^3}\dfrac{\bar{F}_9}{\bar{F}_1}$	$\dfrac{\sin\alpha_0}{2Dk^2}\dfrac{2\bar{F}_8}{\bar{F}_1}$	$\dfrac{\sin^2\alpha_0}{2Dk^3}\dfrac{\bar{F}_4}{\bar{F}_1}$	$\dfrac{\sin\alpha_0}{2Dk^2}\dfrac{\bar{F}_2}{\bar{F}_1}$
4. $f = M_j$	$-\dfrac{\sin\alpha_0}{2Dk^2}\dfrac{2\bar{F}_8}{\bar{F}_1}$	$\dfrac{1}{2Dk}\dfrac{2\bar{F}_{10}}{\bar{F}_1}$	$\dfrac{\sin\alpha_0}{2Dk^2}\dfrac{\bar{F}_2}{\bar{F}_1}$	$\dfrac{1}{2Dk}\dfrac{2\bar{F}_3}{\bar{F}_1}$

[a]Adopted from Baker et al. [20.2].

TABLE 20-12 | Influence Coefficients for Conical Shell Elements | 1244

TABLE 20-13 MEMBRANE EDGE DEFORMATIONS AND INFLUENCE COEFFICIENTS FOR CYLINDRICAL SHELL ELEMENTS[a]

Notation

E = modulus of elasticity

ν = Poisson's ratio

h = thickness of shell

R = radius of shell

L = length of shell

$\Delta R_i, \Delta R_j$ = horizontal displacements at i, j edges (rad), positive in same direction as in figures

β_i, β_j = angles of rotation of meridional tangents at i, j edges

$\Delta R_i^0, \beta_i^0$ = membrane deformations at edge i due to applied loads

$\Delta R_i^f, \beta_i^f$ = influence coefficients, i.e., deformations at edge i due to force f ($f = M_i, H_i, M_j, H_j$)

H_i, H_j = horizontal forces per unit length on i, j edges

M_i, M_j = moments per unit length on i, j edges

$\overline{F}_i$ = factors defined in Table 20-9

$k = [3(1 - \nu^2)]^{1/4}/\sqrt{Rh}$

Deformations in each case (column) of the table are due to the applied loading on the left.

Positive edge moments cause tension in the inner shell surface.

Positive edge horizontal forces cause tension in the inner shell surface at the upper edge and compression in the inner shell surface at the lower edge.

TABLE 20-13 (continued)   **MEMBRANE EDGE DEFORMATIONS AND INFLUENCE COEFFICIENTS FOR CYLINDRICAL SHELL ELEMENTS**

A. Membrane Deformations

Loading Conditions	Edge Deformation			
	ΔR_j^0	β_j^0	ΔR_i^0	β_i^0
1. $p(\xi) = p_0 = \text{constant}$	$\dfrac{p_0 R^2}{Eh}$	0	$\dfrac{p_0 R^2}{Eh}$	0
2. $p(\xi) = p_0(1-\xi)$	0	$-\dfrac{p_0 R^2}{EhL}$	$\dfrac{p_0 R^2}{Eh}$	$-\dfrac{p_0 R^2}{EhL}$

Case / Load				
3. $p(\xi)=p_0\sin\alpha\xi$	$\dfrac{p_0R^2}{Eh}\dfrac{4(kL)^4}{\alpha^4+4(kL)^4}\left[\sin\alpha-\dfrac{\alpha^2}{2(kL)^2}\left(\dfrac{\alpha}{kL}\dfrac{\bar F_9}{\bar F_1}-\dfrac{\bar F_2}{\bar F_1}\sin\alpha\right)+\dfrac{\alpha}{kL}\dfrac{\bar F_4}{\bar F_1}\cos\alpha\right]$	$\dfrac{p_0R^2}{Eh}\dfrac{4(kL)^4}{\alpha^4+4(kL)^4}k\left[\dfrac{\alpha}{kL}\cos\alpha+\dfrac{\alpha^2}{2(kL)^2}\left(\dfrac{2\bar F_3}{\bar F_1}\right)\sin\alpha-\dfrac{\alpha}{kL}\dfrac{\bar F_2}{\bar F_1}\cos\alpha\right]$	$\dfrac{p_0R^2}{Eh}\dfrac{\alpha^2}{2(kL)^2}\dfrac{4(kL)^4}{\alpha^4+4(kL)^4}\dfrac{2\bar F_8}{\bar F_1}\sin\alpha\times\left(\dfrac{\alpha}{kL}\dfrac{\bar F_4}{\bar F_1}-\dfrac{2\bar F_8}{\bar F_1}\cos\alpha\right)+\dfrac{\alpha}{kL}\dfrac{\bar F_9}{\bar F_1}\cos\alpha$	$\dfrac{p_0R^2}{Eh}\dfrac{4(kL)^4}{\alpha^4+4(kL)^4}k\left[\dfrac{\alpha}{kL}-\dfrac{2\bar F_{10}}{\bar F_1}\sin\alpha-\dfrac{\alpha^2}{2(kL)^2}\left(\dfrac{\alpha}{kL}\dfrac{\bar F_2}{\bar F_1}-\dfrac{2\bar F_{10}}{\bar F_1}\sin\alpha\right)+\dfrac{\alpha}{kL}\dfrac{\bar F_8}{\bar F_1}\cos\alpha\right]$
4. $p(\xi)=p_0\cos\alpha\xi$	$\dfrac{p_0R^2}{Eh}\dfrac{4(kL)^4}{\alpha^4+4(kL)^4}\left[\cos\alpha-\dfrac{\alpha^2}{2(kL)^2}\left(\dfrac{2\bar F_8}{\bar F_1}-\dfrac{\bar F_3}{\bar F_1}\cos\alpha\right)-\dfrac{\alpha}{kL}\dfrac{\bar F_4}{\bar F_1}\sin\alpha\right]$	$\dfrac{p_0R^2}{Eh}\dfrac{4(kL)^4}{\alpha^4+4(kL)^4}k\left[\dfrac{\alpha}{kL}\sin\alpha+\dfrac{\alpha^2}{2(kL)^2}\left(\dfrac{2\bar F_{10}}{\bar F_1}-\dfrac{2\bar F_3}{\bar F_1}\cos\alpha\right)-\dfrac{\alpha}{kL}\dfrac{\bar F_2}{\bar F_1}\sin\alpha\right]$	$\dfrac{p_0R^2}{Eh}\dfrac{4(kL)^4}{\alpha^4+4(kL)^4}\left[1+\dfrac{\alpha^2}{2(kL)^2}\left(\dfrac{\bar F_2}{\bar F_1}-\dfrac{2\bar F_8}{\bar F_1}\cos\alpha\right)-\dfrac{\alpha}{kL}\dfrac{\bar F_9}{\bar F_1}\sin\alpha\right]$	$-\dfrac{p_0R^2}{Eh}\dfrac{4(kL)^4}{\alpha^4+4(kL)^4}k\left(\dfrac{2\bar F_3}{\bar F_1}-\dfrac{2\bar F_{10}}{\bar F_1}\cos\alpha\right)\times\left[\dfrac{\alpha^2}{2(kL)^2}\dfrac{\alpha}{kL}\right]$
5. $p(\xi)=p_0e^{-\alpha\xi}$	$\dfrac{p_0R^2}{Eh}\dfrac{4(kL)^4}{\alpha^4+4(kL)^4}\left[e^{-\alpha}-\dfrac{\alpha^2}{2(kL)^2}\left(\dfrac{\bar F_2}{\bar F_1}\right)+\dfrac{\alpha}{kL}\dfrac{\bar F_4}{\bar F_1}e^{-\alpha}\right]$	$\dfrac{p_0R^2}{Eh}\dfrac{4(kL)^4}{\alpha^4+4(kL)^4}k\left[\dfrac{\alpha}{kL}e^{-\alpha}+\dfrac{\alpha^2}{2(kL)^2}\left(\dfrac{2\bar F_8}{\bar F_1}-\dfrac{\alpha}{kL}\dfrac{\bar F_8}{\bar F_1}\right)+\dfrac{\alpha}{kL}\dfrac{2\bar F_3}{\bar F_1}\right]$	$\dfrac{p_0R^2}{Eh}\dfrac{4(kL)^4}{\alpha^4+4(kL)^4}\left[1-\dfrac{\alpha^2}{2(kL)^2}\left(\dfrac{\bar F_2}{\bar F_1}-\dfrac{\alpha}{kL}\dfrac{\bar F_4}{\bar F_1}\right)\dfrac{2\bar F_8}{\bar F_1}+\dfrac{\alpha}{kL}\dfrac{\bar F_9}{\bar F_1}e^{-\alpha}\right]$	$-\dfrac{p_0R^2}{Eh}\dfrac{4(kL)^4}{\alpha^4+(kL)^4}k\left[\dfrac{\alpha}{kL}\dfrac{\bar F_2}{\bar F_1}-\dfrac{\alpha^2}{2(kL)^2}\left(\dfrac{2\bar F_3}{\bar F_1}-\dfrac{\alpha}{kL}\dfrac{\bar F_2}{\bar F_1}\right)\dfrac{2\bar F_8}{\bar F_1}-\dfrac{\alpha}{kL}\dfrac{2\bar F_8}{\bar F_1}e^{-\alpha}\right]$

B. Influence Coefficients (Bending Deformations)

Loading Condition	Edge Deformation			
	ΔR_i^f	β_i^f	ΔR_j^f	β_j^f
1. $f = H_i$ $H_i = 1$	$-\dfrac{2R^2 k}{Eh}\,\dfrac{\bar{F}_4}{\bar{F}_1}$	$\dfrac{2R^2 k^2}{Eh}\,\dfrac{\bar{F}_2}{\bar{F}_1}$	$\dfrac{2R^2 k}{Eh}\,\dfrac{\bar{F}_9}{\bar{F}_1}$	$\dfrac{2R^2 k^2}{Eh}\,\dfrac{2\bar{F}_8}{\bar{F}_1}$
2. $f = M_i$ $M_i = 1$	$\dfrac{2R^2 k^2}{Eh}\,\dfrac{\bar{F}_2}{\bar{F}_1}$	$-\dfrac{2R^2 k^3}{Eh}\,\dfrac{2\bar{F}_3}{\bar{F}_1}$	$-\dfrac{2R^2 k^2}{Eh}\,\dfrac{2\bar{F}_8}{\bar{F}_1}$	$-\dfrac{2R^2 k^3}{Eh}\,\dfrac{2\bar{F}_{10}}{\bar{F}_1}$

Loading Condition			
3. $f = H_j$	$-\dfrac{2R^2k}{Eh}\dfrac{\bar{F}_9}{\bar{F}_1}$	$\dfrac{2R^2k^2}{Eh}\dfrac{2\bar{F}_8}{\bar{F}_1}$	$\dfrac{2R^2k}{Eh}\dfrac{\bar{F}_4}{\bar{F}_1}$	$\dfrac{2R^2k^2}{Eh}\dfrac{\bar{F}_2}{\bar{F}_1}$
4. $f = M_j$	$-\dfrac{2R^2k^2}{Eh}\dfrac{2\bar{F}_8}{\bar{F}_1}$	$\dfrac{2R^2k^3}{Eh}\dfrac{2\bar{F}_{10}}{\bar{F}_1}$	$\dfrac{2R^2k^2}{Eh}\dfrac{\bar{F}_2}{\bar{F}_1}$	$\dfrac{2R^2k^3}{Eh}\dfrac{2\bar{F}_3}{\bar{F}_1}$

Loading Condition	ΔR_j^f	β_j^f
5. $f = M_j$	$\dfrac{R^2k^2}{Eh}\dfrac{2\bar{F}_2}{\bar{F}_1+2}$	$\dfrac{R^2k^3}{Eh}\dfrac{4\bar{F}_3}{\bar{F}_1+2}$
6. $f = H_j$	$\dfrac{R^2k}{Eh}\dfrac{2\bar{F}_4}{\bar{F}_1+2}$	$\dfrac{R^2k^2}{Eh}\dfrac{2\bar{F}_2}{\bar{F}_1+2}$

B. Influence Coefficients (Bending Deformations)

Loading Condition	Edge Deformation	
	ΔR_j^f	β_j^f
7. $f = M_j$	$\dfrac{R^2 k^2}{Eh}\dfrac{2\bar{F}_3}{\bar{F}_4}$	$\dfrac{R^2 k^3}{Eh}\dfrac{4(\bar{F}_1+1)}{\bar{F}_4}$
8. $f = H_j$	$\dfrac{R^2 k}{Eh}\dfrac{2\bar{F}_2}{\bar{F}_4}$	$\dfrac{R^2 k^2}{Eh}\dfrac{2\bar{F}_3}{\bar{F}_4}$

[a]From Baker et al. [20.2].

TABLE 20-14 STRESS RESULTANTS OF MEMBRANE SHELLS OF VARIOUS SHAPES

Notation

h_1, h_2 = constants shown in equations of shells

α = angle between tangent of curve and horizontal line

p_2 = uniform vertical load (snow load)

q = dead-weight load

$\lambda_n = (n\pi/b)\sqrt{h_2/h_1}$

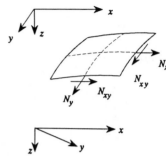

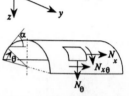

Boundary conditions for the cylindrical membrane shells (cases 3–7) are equivalent to being simply supported (shear diaphragms) at $x = \pm \frac{1}{2}L$. Normally, the development of these membrane solutions for stress resultants requires that only the force boundary conditions be invoked. The boundary conditions must be compatible with the conditons of equilibrium.

See Table 20-1, part A, for the definitions of applied loads.

Description	Stress Resultants
1. Elliptic paraboloid shell under uniform vertical load p_2  Shell geometry: $z = x^2/(2h_1) + y^2/(2h_2)$ $h_1 = a^2/8h_x$ $h_2 = b^2/8h_y$ Boundary conditions: $N_x = 0$ at $x = \pm\frac{1}{2}a$ $N_y = 0$ at $y = \pm\frac{1}{2}b$	$N_x = \dfrac{h_2}{h_1}\sqrt{\dfrac{h_1^2 + x^2}{h_2^2 + y^2}}\dfrac{4p_2 h_1}{\pi}\displaystyle\sum_{n=1,3,\ldots}^{\infty}(-1)^{(n+1)/2}\dfrac{1}{n}\Bigg[1$ $\qquad - \dfrac{\cosh \lambda_n x}{\cosh(\lambda_n a/2)}\Bigg]\cos\dfrac{n\pi y}{b}$ $N_y =$ $h_3\dfrac{4p_2 h_2}{\pi}\displaystyle\sum_{n=1,3,\ldots}^{\infty}(-1)^{(n+1)/2}\dfrac{1}{n}\dfrac{\cosh \lambda_n x}{\cosh(\lambda_n a/2)}\cos\dfrac{n\pi y}{b}$ $N_{xy} =$ $\dfrac{4p_2}{\pi}\sqrt{h_1 h_2}\displaystyle\sum_{n=1,3,\ldots}^{\infty}(-1)^{(n+1)/2}\dfrac{1}{n}\dfrac{\sinh \lambda_n x}{\sinh(\lambda_n a/2)}\sin\dfrac{n\pi y}{b}$ $h_3 = \dfrac{h_1}{h_2}\sqrt{\dfrac{h_2^2 + y^2}{h_1^2 + x^2}}$ Ref. 20.3

Description	Stress Resultants

2.
Hyperbolic paraboloid shell with generating parabolas as boundaries and under uniform vertical load p_2

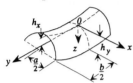

Shell geometry:
$z = x^2/(2h_1) - y^2/(2h_2)$
$h_1 = a^2/8h_y$,
$h_2 = b^2/8h_y$
Boundary conditions:
free at $x = \pm \frac{1}{2}a$ and
$y = \pm \frac{1}{2}b$

$$N_x = \frac{h_2}{h_1} \sqrt{\frac{h_1^2 + x^2}{h_2^2 + y^2}} \frac{4p_2 h_1}{\pi} \sum_{n=1,3,\dots}^{\infty} (-1)^{(n+1)/2} \frac{1}{n} \left[1 - \frac{\cosh \lambda_n x}{\cosh(\lambda_n a/2)} \right] \cos \frac{n\pi y}{b}$$

$$N_y = h_3 \frac{4p_2 h_2}{\pi} \sum_{n=1,3,\dots}^{\infty} (-1)^{(n-1)/2} \frac{1}{n} \frac{\cosh \lambda_n x}{\cosh(\lambda_n a/2)} \cos \frac{n\pi y}{b}$$

$$N_{xy} = \frac{4p_2}{\pi} \sqrt{h_1 h_2} \sum_{n=1,3,\dots}^{\infty} (-1)^{(n-1)/2} \frac{1}{n} \frac{\sinh \lambda_n x}{\sinh(\lambda_n a/2)} \sin \frac{n\pi y}{b}$$

with

$$a = \sqrt{h_1/h_2}\, 2b \qquad h_3 = \frac{h_1}{h_2} \sqrt{\frac{h_2^2 + y^2}{h_1^2 + x^2}}$$

Ref. 20.3

3.
Cylindrical shell with semielliptic cross section

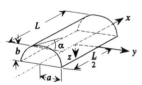

Boundary condition:
simply supported at
$x = \pm \frac{1}{2}L$

DEAD WEIGHT:

$$N_\theta = -q \left[\frac{a^2 b^2 \cos \alpha}{(a^2 \sin^2 \alpha + b^2 \cos^2 \alpha)^{3/2}} \right]$$

$$N_{x\theta} = -qx \left[2 + \frac{3(a^2 - b^2)\cos^2 \alpha}{a^2 \sin^2 \alpha + b^2 \cos^2 \alpha} \right] \sin \alpha$$

$$N_x = -\frac{q}{2} \left(\frac{L^2}{4} \right.$$

$$\left. - x^2 \right) \left[\frac{2ab}{k^3} + \frac{3(a^2 - b^2)}{abk} \left(\cos^2 \alpha - \frac{2a^2}{b^2} \sin^2 \alpha \right) \right] \cos \alpha$$

where

$$k = \frac{ab}{(a^2 \sin^2 \alpha + b^2 \cos^2 \alpha)^{1/2}}$$

TABLE 20-14   **Stress Resultants of Membrane Shells**   **1252**

Description	Stress Resultants
3. Continued	UNIFORMLY DISTRIBUTED LOADING ON PROJECTED AREA (SNOW LOAD): $$N_\theta = -p_2\left[\frac{a^2 b^2 \cos^2 \alpha}{(a^2 \sin^2 \alpha + b^2 \cos^2 \alpha)^{3/2}}\right]$$ $$N_{x\theta} = -3p_2 x\left(\frac{a^2 \sin \alpha \cos \alpha}{a^2 \sin^2 \alpha + b^2 \cos^2 \alpha}\right)$$ $$N_x = -\frac{3}{2}p_2\left(\frac{L^2}{4} - x^2\right)\left[\frac{-a^2 \sin^2 \alpha + b^2 \cos^2 \alpha}{b^2(a^2 \sin^2 \alpha + b^2 \cos^2 \alpha)^{1/2}}\right]$$ Ref. 20.4
4. Cylindrical shell with circular cross section 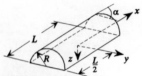 Boundary condition: simply supported at $x = \pm\frac{1}{2}L$	DEAD WEIGHT: $N_\theta = -qR \cos \alpha$ $N_{x\theta} = -2qx \sin \alpha$ $$N_x = -\frac{q}{R}\left(\frac{L^2}{4} - x^2\right)\cos \alpha$$ SNOW LOAD: $N_\theta = -p_2 R \cos^2 \alpha$ $N_{x\theta} = -1.5 p_2 x \sin 2\alpha$ $$N_x = -1.5\frac{p_2}{R}\left(\frac{L^2}{4} - x^2\right)\cos 2\alpha$$ Ref. 20.4
5. Cylindrical shell with catenary cross section 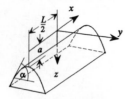 Geometry of shell: $z = a \cosh(y/a)$ where a is a constant that is the distance from the vertex of the catenary to the x axis. Boundary condition: simply supported at $x = \pm\frac{1}{2}L$	DEAD WEIGHT: $$N_\theta = -\frac{qa}{\cos \alpha}$$ $N_{x\theta} = N_x = 0$ SNOW LOAD: $N_\theta = -p_2 a$ $N_{x\theta} = -0.5 p_2 x \sin 2\alpha$ $$N_x = -0.5\frac{p_2}{a}\left(\frac{L^2}{4} - x^2\right)\cos 2\alpha \cos^2 \alpha$$ Ref. 20.4

Description	Stress Resultants
6. Cylindrical shell with cycloidal cross section 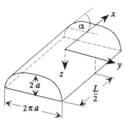 Boundary conditions: simply supported at $x = \pm \frac{1}{2}L$ 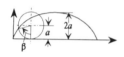	$y = a(\beta - \pi - \sin \beta)$ $R_0 = 4a$ $z = a(1 + \cos \beta)$ $0 \le \beta \le 2\pi$ DEAD WEIGHT: $N_\theta = -qR_0 \cos^2 \alpha$ $N_{x\theta} = -3qx \sin \alpha$ $N_x = -\dfrac{3}{2}\dfrac{q}{R_0}\left(\dfrac{L^2}{4} - x^2\right)$ SNOW LOAD: $N_\theta = -p_2 R_0 \cos^2 \alpha$ $N_{x\theta} = -2p_2 x \sin \alpha \cos \alpha$ $N_x = -2\dfrac{p_2}{R_0}\left(\dfrac{L^2}{4} - x^2\right)\dfrac{\cos^2 \alpha - \sin^2 \alpha}{\cos \alpha}$ Ref. 20.4
7. Cylindrical shells with parabolic cross section 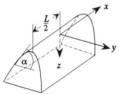 Geometry of shell: $y^2 = 4az$ $a = $ const Boundary conditions: simply supported at $x = \pm \frac{1}{2}L$	DEAD WEIGHT: $N_\theta = -\dfrac{qR_0}{\cos^2 \alpha}$ $N_{x\theta} = qx \sin \alpha$ $N_x = 0.5\dfrac{q}{R_0}\left(\dfrac{L^2}{4} - x^2\right)\cos^4 \alpha$ SNOW LOAD: $N_\theta = -\dfrac{p_2 R_0}{\cos \alpha}$ $N_{x\theta} = N_x = 0$ where $R_0 = 2a$ Ref. 20.4

TABLE 20-14   **Stress Resultants of Membrane Shells**   **1254**

TABLE 20-15   CRITICAL LOADS FOR VARIOUS SHELLS OF REVOLUTION

Notation

E = modulus of elasticity
ν = Poisson's ratio
h = thickness of shell
L = length of shell along generator
b = width of panel in circumferential direction
R = radii of cylinders and panels
R_e = constant given in various cases for cones
p_1 = uniform pressure (F/L^2)
p_{cr} = critical pressure at buckling (F/L^2)
P_{cr} = critical concentrated force at buckling (F/L)
σ_{cr} = stress at buckling; for truncated cones, stress at small end
τ_{cr} = shear stress at buckling; for truncated cones, shear stress at small end
$D = Eh^3/12(1 - \nu^2)$

Unless otherwise specified,

$$Z = \begin{cases} \dfrac{L^2}{R_e h}(1 - \nu^2)^{1/2} & \text{for conical shells (for case 5, use } L_e \text{ instead of } L) \\[2ex] \dfrac{L^2}{Rh}(1 - \nu^2)^{1/2} & \text{for cylindrical shells} \\[2ex] \dfrac{b^2}{Rh}(1 - \nu^2)^{1/2} & \text{for panels} \end{cases}$$

For a simply supported (shear diaphragm) boundary condition, the radial and circumferential displacements are zero, the force in the axial direction and the moment about the tangent of the circumferential wall contour are zero, and there is no restraint against translation in the axial direction and rotation about the circumferential boundary.

Unless otherwise specified, the boundary conditions of the shells are simply supported.

Description	Critical Load
1. Spherical shell with external pressure p_1 	Empirical buckling formula: $$(p_1)_{cr} = \frac{0.80E}{\sqrt{1 - \nu^2}}\left(\frac{h}{R}\right)^2$$ Ref. 20.12

TABLE 20-15 (continued)   **CRITICAL LOADS FOR VARIOUS SHELLS OF REVOLUTION**

Description	Critical Load
2. Clamped spherical cap with external pressure p_1 	$$(p_1)_{cr} = \left(0.14 + \frac{3.2}{\lambda^2}\right) K_c$$ where $$K_c = \frac{2}{[3(1 - \nu^2)]^{1/2}} E\left(\frac{h}{R}\right)^2$$ $$\lambda = [12(1 - \nu^2)]^{1/4}\left(\frac{R}{h}\right)^{1/2} 2\sin\frac{\phi}{2}$$ Ref. 20.2
3. Truncated conical shell subjected to concentrated axial compression force 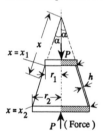 Boundary condition: simply supported at upper and lower edges	For axisymmetrical buckling: $$P_{cr} = K_c \cos^2 \alpha$$ where $$K_c = \frac{2Eh^2\pi}{\sqrt{3(1 - \nu^2)}}$$ For asymmetric buckling: $$P_{cr} = \pi x_1 h \sin 2\alpha \cdot \sigma_{cr}$$ where $$\sigma_{cr} = \frac{Eh\cos\alpha}{r_1\sqrt{3(1 - \nu^2)}}\sqrt{\frac{1}{2}\frac{1 + x_1/x_2}{1 - x_1/x_2}\log\frac{x_2}{x_1}}$$ Ref. 20.13
4. Truncated conical shell with concentrated force and internal pressure p_1 	$$P_{cr} = 2\pi R_e \sigma_{cr} h \cos^2 \alpha + \pi R_e^2 p_1 \cos^2 \alpha \quad (\alpha < 75°)$$ where $$R_e = \frac{R_1}{\cos\alpha} \qquad \sigma_{cr} = (\gamma K_c + K_b)\frac{Eh}{R_e}$$ $$\gamma = \frac{1}{[3(1 - \nu^2)]^{1/2}}$$ $$K_c = 6.1424 + 5.9264\log\eta_1 - 4.3154\log^2\eta_1$$ $$+ 0.6357\log^3\eta_1$$

TABLE 20-15   **Critical Loads for Various Shells**   **1256**

TABLE 20-15 (continued) **CRITICAL LOADS FOR VARIOUS SHELLS OF REVOLUTION**

Description	Critical Load
4. Continued	$K_b = 10^K$
	$K = -0.6869 + 0.1846 \log \eta_2 - 0.1452 \log^2 \eta_2$
	$\qquad + 0.030019 \log^3 \eta_2$
	where
	$\eta_1 = \dfrac{R_e}{h} \qquad \eta_2 = \dfrac{p_1}{E}\left(\dfrac{R_e}{h}\right)^2$
	This formula is valid for
	$R_e/h > 700$
	$Z > 25$ for simply supported edges
	$Z > 80$ for clamped edges
	Ref. 20.2
5. Truncated conical shell with torsional moment (Force • Length)	$T_{cr} = 2\pi R_1^2 h \tau_{cr} \qquad (\alpha < 60°)$
	where
	$\tau_{cr} = \dfrac{R_e^2}{R_1^2} K_c \dfrac{Eh}{R_e Z^{1/4}}$
	$Z = \dfrac{L_e^2}{R_e h}(1 - \nu^2)^{1/2}$
	$R_e = \left\{ 1 + \left(\dfrac{1 + R_2/R_1}{2}\right)^{1/2} \right.$
	$\qquad \left. - \left(\dfrac{1 + R_2/R_1}{2}\right)^{-1/2} \right\} R_1 \cos \alpha$
	$K_c = 0.4218 + 83.1595\eta^{-1} - 13{,}710.7197\eta^{-2}$
	$\qquad + 810{,}673.75\eta^{-3}$
	$\eta = \dfrac{R_e}{h}$
	This formula is valid for
	$Z > 100$ for simply supported edges and clamped edges
	Ref. 20.2

TABLE 20-15 (continued)   CRITICAL LOADS FOR VARIOUS SHELLS OF REVOLUTION

Description	Critical Load
6. Truncated conical shell with bending moment 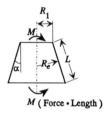 M (Force • Length)	$M_{cr} = \pi R_1^2 \sigma_{cr} h \cos \alpha \qquad (\alpha < 60°)$ where $$\sigma_{cr} = \gamma K_c \frac{Eh}{R_e} \qquad R_e = \frac{R_1}{\cos \alpha} \qquad \gamma = \frac{1}{[3(1 - \nu^2)]^{1/2}}$$ $$K_c = 15.9069 - 14.7607 \log \eta + 11.1455 \log^2 \eta$$ $$- 3.9906 \log^3 \eta + 0.4778 \log^4 \eta$$ $\eta = R_e/h$ This formula is valid for $Z > 20$ for simply supported edges $Z > 80$ for clamped edges Ref. 20.2
7. Truncated conical shell with bending moment and internal pressure  M (Force • Length)	$M_{cr} = \pi R_1^2 h \sigma_{cr} \cos \alpha \qquad (\alpha < 60°)$ where $$\sigma_{cr} = (\gamma K_c + K_b) \frac{Eh}{R_e} \qquad \left(\frac{R_e}{h} > 500\right)$$ $$R_e = \frac{R_1}{\cos \alpha} \qquad \gamma = \frac{1}{[3(1 - \nu^2)]^{1/2}}$$ K_c is taken from case 6. $$K_b = \begin{cases} 10^{K_1} & 0.01 \leq \eta < 0.2 \\ 10^{K_2} & 0.2 \leq \eta \leq 10 \end{cases}$$ $$K_1 = -0.4166 + 0.1882 \log \eta - 0.03687 \log^2 \eta$$ $$- 0.01516 \log^3 \eta$$ $$K_2 = -0.1376 + 0.8051 \log \eta + 0.1946 \log^2 \eta$$ $$- 0.1374 \log^3 \eta$$ $$\eta = \frac{p_1}{E} \left(\frac{R_e}{h}\right)^2$$ This formula is valid for $Z > 20$ for simply supported edges $Z > 80$ for clamped edges Ref. 20.2

TABLE 20-15   **Critical Loads for Various Shells**   1258

TABLE 20-15 (continued) CRITICAL LOADS FOR VARIOUS SHELLS OF REVOLUTION

Description	Critical Load
8. Truncated conical shell with external pressure 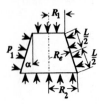	$$(p_1)_{cr} = \frac{\sigma_{cr} h \cos \alpha}{R_2} \qquad (\alpha < 75°)$$ where $$R_e = \frac{R_1 + R_2}{2 \cos \alpha}$$ $$\sigma_{cr} = K_c \frac{\pi^2 E}{12(1 - \nu^2)} \left(\frac{h}{L} \right)^2 \frac{R_2}{R_e \cos \alpha}$$ $$K_c = 10^{-0.186 + 0.535 \log Z}$$ This formula is for simply supported edges and is conservative for clamped edges. Ref. 20.2
9. Cylindrical shell axial compression 	$$\sigma_{cr} = K_c \frac{\pi^2 D}{L^2 h} \quad \text{where } K_c = 10^K, \quad Z = \frac{L^2}{Rh} \sqrt{1 - \nu^2}$$ For simply supported edges Empirical design curves: $0.03967 + 0.3788 \log Z + 0.1301 \log^2 Z$ $\quad - 0.01108 \log^3 Z \qquad (R/h = 3000)$ $0.03839 + 0.4002 \log Z + 0.142 \log^2 Z$ $\quad - 0.01422 \log^3 Z \qquad (R/h = 2000)$ $K = \Big\{$ $0.03345 + 0.4701 \log Z + 0.1298 \log^2 Z$ $\quad - 0.01229 \log^3 Z \qquad (R/h = 1000)$ $0.03909 + 0.5159 \log Z + 0.1366 \log^2 Z$ $\quad - 0.01414 \log^3 Z \qquad (R/h = 500)$ Theoretical: $0.01920 + 0.7378 \log Z + 0.1206 \log^2 Z$ $\quad - 0.01686 \log^3 Z$

TABLE 20-15 (continued)   CRITICAL LOADS FOR VARIOUS SHELLS OF REVOLUTION

Description	Critical Load
9. Continued	For clamped edges

$$K = \begin{cases} \text{Empirical design curves:} \\ 0.6046 + 0.02859 \log Z + 0.1935 \log^2 Z \\ \qquad -0.01388 \log^3 Z \qquad\qquad\qquad (R/h = 3000) \\ 0.6078 - 0.002595 \log Z + 0.2260 \log^2 Z \\ \qquad -0.01884 \log^3 Z \qquad\qquad\qquad (R/h = 2000) \\ 0.6128 - 0.05537 \log Z + 0.2816 \log^2 Z \\ \qquad -0.02637 \log^3 Z \qquad\qquad\qquad (R/h = 1000) \\ 0.6148 - 0.1096 \log Z + 0.3535 \log^2 Z \\ \qquad -0.03833 \log^3 Z \qquad\qquad\qquad (R/h = 500) \\ \text{Theoretical:} \\ 0.5900 - 0.03724 \log Z + 0.4361 \log^2 Z \\ \qquad -0.05646 \log^3 Z \end{cases}$$

10.
Clamped
cylindrical
shell axial
compression and
internal pressure

$$P_{cr} = \pi R (2h\sigma_{cr} + p_1 R)$$
where
$$\sigma_{cr} = (K_c Eh)/R$$

$$K_c = \begin{cases} 0.09988 + 0.1235\sqrt{\bar{p}} + 0.004858\,\bar{p} \\ \qquad -0.001976\left(\sqrt{\bar{p}}\,\right)^3 \qquad\qquad (R/h \to \infty) \\ 0.1630 + 0.0947\sqrt{\bar{p}} + 0.0118\,\bar{p} \\ \qquad -0.00267\left(\sqrt{\bar{p}}\,\right)^3 \qquad\qquad (R/h = 2000) \\ 0.1874 + 0.1339\sqrt{\bar{p}} - 0.006912\,\bar{p} \\ \qquad -0.0007018\left(\sqrt{\bar{p}}\,\right)^3 \qquad\quad (R/h = 1333) \\ 0.2462 + 0.1316\sqrt{\bar{p}} - 0.01203\,\bar{p} \\ \qquad -0.00005\left(\sqrt{\bar{p}}\,\right)^3 \qquad\qquad (R/h = 800) \\ 0.2786 + 0.1277\sqrt{\bar{p}} - 0.01156\,\bar{p} \\ \qquad -0.0001997\left(\sqrt{\bar{p}}\,\right)^3 \qquad\quad (R/h = 500) \\ 0.3295 + 0.1165\sqrt{\bar{p}} - 0.01187\,\bar{p} \\ \qquad -0.0000728\left(\sqrt{\bar{p}}\,\right)^3 \qquad\quad (R/h = 400) \end{cases}$$

$$\bar{p} = \frac{p_1}{E(R/h)^2}$$

TABLE 20-15   **Critical Loads for Various Shells**   **1260**

TABLE 20-15 (continued)   CRITICAL LOADS FOR VARIOUS SHELLS OF REVOLUTION

Description	Critical Load
11. Cylindrical shell with torsional moment at ends (Force · Length)	$$T_{cr} = 2\pi R^2 K_c \frac{Eh^2}{RZ^{1/4}}$$ where $$K_c = 0.4233 + 79.9779\eta^{-1} - 12{,}759.6621\eta^{-2}$$ $$+ 755{,}633.25\eta^{-3}$$ $\eta = R/h$ This formula is valid for $Z < 78(R/h)^2(1 - \nu^2)$ $Z > 100$ for simply supported edges $Z > 100$ for clamped edges Ref. 20.2
12. Cylindrical shell with torsional moments at ends and internal pressure	$$T_{cr} = 2\pi R^2 h^2 \left(K_c \frac{E}{RZ^{1/4}} + K_b \frac{E}{R} \right)$$ where K_c is defined in case 11 $K_b = 10^K$ $$K = -0.1318 + 0.8758 \log \eta + 0.07645 \log^2 \eta$$ $$+ 0.01314 \log^4 \eta$$ $$\eta = \frac{p_1}{E}\left(\frac{R}{h}\right)^2$$ The formula in this case is valid under the same conditions as for case 11. Ref. 20.2
13. Cylindrical shell with bending moment at ends M (Force · Length)	$M_{cr} = \pi R^2 \sigma_{cr} h$ where $$\sigma_{cr} = \gamma K_c \frac{Eh}{R}$$ $$\gamma = \frac{1}{\left[3(1 - \nu^2)\right]^{1/2}}$$ $$K_c = 15.3914 - 13.8791 \log \eta + 10.6439 \log^2 \eta$$ $$- 3.8693 \log^3 \eta + 0.4669 \log^4 \eta$$ $\eta = R/h$ Ref. 20.2

TABLE 20-15 (continued)   CRITICAL LOADS FOR VARIOUS SHELLS OF REVOLUTION

Description	Critical Load
14. Cylindrical shell with bending moment and internal pressure *M* (Force • Length)	$M_{cr} = \pi R^2 \sigma_{cr} h \quad (R/h > 500)$ where $\sigma_{cr} = (\gamma K_c + K_b) \dfrac{Eh}{R}$ γ and K_c are defined in case 13 $K_b = 10^K$ $K = \begin{cases} \left. \begin{array}{l} -0.866 - 0.926 \log \eta \\ -0.869 \log^2 \eta - 0.2073 \log^3 \eta \end{array} \right\} & 0.01 \le \eta < 0.2 \\[1em] \left. \begin{array}{l} -0.1377 + 0.8275 \log \eta \\ +0.1908 \log^2 \eta - 0.1383 \log^3 \eta \end{array} \right\} & 0.2 \le \eta \le 10 \end{cases}$ $\eta = \dfrac{p_1}{E}\left(\dfrac{R}{h}\right)^2$ Ref. 20.2
15. Cylindrical shell with external pressure	$(p_1)_{cr} = \sigma_{cr} h / R$ where $\sigma_{cr} = K_c \dfrac{\pi^2 E}{12(1-\nu^2)}\left(\dfrac{h}{L}\right)^2$ $K_c = 10^K$ $K = 0.6337 - 0.1455 \log Z + 0.1977 \log^2 Z$ $\qquad - 0.01915 \log^3 Z$ Ref. 20.2
16. Curved panel with axial compression	$(p_1)_{cr} = K_c \dfrac{Eh}{R} \quad (a/b > 0.5)$ where $K_c = 0.22195 + 29.7611\eta^{-1} - 2322.08667\eta^{-2}$ $\qquad + 65{,}832.1484\,\eta^{-3}$ $\eta = R/h$ This formula is valid for $Z > 30$ for simply supported edges $Z > 50$ for clamped edges Ref. 20.2

TABLE 20-15   **Critical Loads for Various Shells**   **1262**

Description	Critical Load
17. Cylindrical panel with shear forces $\tau(F/L^2)$	$\tau_{cr} = K_c \dfrac{\pi^2 E}{12(1-\nu^2)}\left(\dfrac{h}{b}\right)^2 \qquad (a > b)$ where $K_c = 10^K$ $K = \begin{cases} 0.7171 - 0.2427\log Z + 0.2613\log^2 Z \\ \quad -0.02906\log^3 Z & (a/b \to \infty) \\ 0.7545 - 0.3151\log Z + 0.3261\log^2 Z \\ \quad -0.03461\log^3 Z & (a/b = 3.0) \\ 0.8056 - 0.3378\log Z + 0.35\log^2 Z \\ \quad -0.03825\log^3 Z & (a/b = 2.0) \\ 0.8653 - 0.3246\log Z + 0.3336\log^2 Z \\ \quad -0.03492\log^3 Z & (a/b = 1.5) \\ 0.9643 - 0.2683\log Z + 0.2949\log^2 Z \\ \quad -0.02898\log^3 Z & (a/b = 1.0) \end{cases}$ Ref. 20.2
18. Curved panel subject to bending at ends	$\sigma_{cr} = K_c \dfrac{\pi^2 E}{12(1-\nu^2)}\left(\dfrac{h}{b}\right)^2$ where $K_c = 10^K$ For simply supported edges $K = \begin{cases} 1.3838 + 0.0672\log Z - 0.04973\log^2 Z \\ \quad +0.04021\log^3 Z & (R/h = 2000) \\ 1.387 + 0.01058\log Z - 0.01525\log^2 Z \\ \quad +0.04497\log^3 Z & (R/h = 1000) \\ 1.395 - 0.2141\log Z + 0.1874\log^2 Z \\ \quad +0.01432\log^3 Z & (R/h = 500) \end{cases}$ For clamped edges $K = \begin{cases} 1.667 + 0.1819\log Z - 0.2028\log^2 Z \\ \quad +0.06786\log^3 Z & (R/h = 2000) \\ 1.6705 + 0.1575\log Z - 0.2131\log^2 Z \\ \quad +0.084\log^3 Z & (R/h = 1000) \\ 1.6779 - 0.008678\log Z - 0.0779\log^2 Z \\ \quad +0.06963\log^3 Z & (R/h = 500) \end{cases}$ Ref. 20.2

TABLE 20-16 NATURAL FREQUENCIES OF MEMBRANE CIRCULAR CYLINDRICAL SHELLS[a]

Notation

E = modulus of elasticity
ρ^* = mass per unit volume
h = thickness of shell
ν = Poisson's ratio
R = radius of cylinder
L = length of cylinder
ω = natural frequency
Ω = frequency parameter
C_1 = constant given in Part B
n = number of waves in mode shapes in circumferential direction
m = number of half waves in mode shapes in longitudinal direction
m_1 = number of circumferential nodal circles in mode shapes along longitudinal direction; nodal circles are circles that have zero displacements in the mode shapes
Ω_a, Ω_{rt} = frequency parameters for axial and coupled radial–torsional modes, respectively
Ω_t, Ω_{ar} = frequency parameters for torsional and coupled axial–radial modes, respectively

Examples of longitudinal mode patterns

Nodal circles

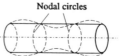

$m=3, m_1=2$, for clamped - clamped, simply supported - simply supported, clamped - simply supported cases

Example of circuferential mode pattern

$n=3$

Nodal circles

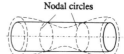

$m_1=2$ for free - free case

Nodal circles

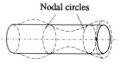

$m_1=2$ for clamped - free case

The simply supported boundary condition is defined in Section 20.5 (shear diaphragm).

$$\omega^2 = \frac{E\Omega^2}{\rho^*(1-\nu^2)R^2} \quad \text{where } \Omega^2 = \Omega^2 \text{ or } \Omega_a^2 \text{ or } \Omega_{rt}^2 \text{ or } \Omega_{ar}^2, \text{ as appropriate}$$

$$\eta = \frac{R}{nL} \qquad \lambda = \frac{m\pi R}{L}$$

Unless specified otherwise, the vibration modes are general responses, i.e., they do not pertain to axial, radial, or torsional modes in particular.

Cases 3–6 give lower bounds for the frequency parameters.

Part A. Frequencies	
Case	Frequency Parameter Ω
1. Infinite 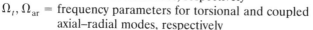	Axial modes: $\Omega_a^2 = \frac{1}{2}(1-\nu)n^2$ Coupled radial–torsional modes: $\Omega_{rt}^2 = 1 + n^2$ For the coupled radial–torsional modes, for each n there is a rigid body mode such that $\Omega_{rt}^2 = 0$.

TABLE 20-16 | Frequencies of Membrane Cylinders | 1264

Case	Frequency Parameter Ω
2. Simply supported (shear diaphragms)	In general, Ω^2 is taken from the polynomial $$\Omega^6 - K_2\Omega^4 + K_1\Omega^2 - K_0 = 0$$ with $$K_0 = \tfrac{1}{2}(1 - \nu)\left[(1 - \nu^2)\lambda^4\right]$$ $$K_1 = \tfrac{1}{2}(1 - \nu)\left[(3 + 2\nu)\lambda^2 + n^2 + (n^2 + \lambda^2)^2\right]$$ $$K_2 = 1 + \tfrac{1}{2}(3 - \nu)(n^2 + \lambda^2)$$ For $n = 0$ $$\Omega_t^2 = \tfrac{1}{2}(1 - \nu)\lambda^2$$ for the torsional modes and $$\Omega_{ar}^2 = \tfrac{1}{2}\left\{(1 + \lambda^2) \pm \left[(1 - \lambda^2)^2 + 4\nu^2\lambda^2\right]^{1/2}\right\}$$ for coupled axial–radial modes
3. Clamped–clamped	$\Omega^2 = (1 - \nu^2)C_1$ C_1 is given in part B
4. Clamped–simply supported	$\Omega^2 = (1 - \nu^2)C_1$ C_1 is given in part B
5. Clamped–free	$\Omega^2 = (1 - \nu^2)C_1$ C_1 is given in part B
6. Free–free	$\Omega^2 = (1 - \nu^2)C_1$ C_1 is given in part B

Part B. Values of C_1 in Part A

Clamped–Clamped

m	$\eta = R/nL$	C_1
1	$0.05 \leq \eta \leq 0.5$	$0.007635 - 0.3997\eta + 05.6899\eta^2 + 3.6782\eta^3 - 38.02963\eta^4 + 39.3571\eta^5$
	$0.02 \leq \eta < 0.05$	$-0.0000928 + 0.01661\eta - 1.1437\eta^2 + 36.8874\eta^3$
2	$0.05 \leq \eta \leq 0.5$	$0.03535 - 2.1611\eta + 41.7726\eta^2 - 150.9171\eta^3 + 229.5575\eta^4 - 129.7816\eta^5$
	$0.02 \leq \eta < 0.05$	$0.002662 - 0.2738\eta + 7.4659\eta^2 + 52.0798\eta^3$
3	$0.03 \leq \eta \leq 0.5$	$0.04294 - 3.5672\eta + 94.2065\eta^2 - 482.5035\eta^3 + 1016.7775\eta^4 - 772.1803\eta^5$
	$0.02 \leq \eta < 0.03$	$0.00300 - 0.3341\eta + 7.7305\eta^2 + 327.2686\eta^3$
4	$0.11 \leq \eta \leq 0.5$	$-0.2947 + 9.0206\eta - 23.0492\eta^2 + 19.9852\eta^3$
	$0.05 \leq \eta < 0.11$	$-0.06306 - 0.1960\eta + 83.05571\eta^2 - 361.1688\eta^3$
	$0.02 \leq \eta < 0.05$	$0.04067 - 4.5017\eta + 155.2445\eta^2 - 902.207\eta^3$
5	$0.11 \leq \eta \leq 0.5$	$-0.0815 + 8.1183\eta - 21.7869\eta^2 + 19.5029\eta^3$
	$0.05 \leq \eta < 0.11$	$-0.1334 + 2.9094\eta + 79.5741\eta^2 - 439.1654\eta^3$
	$0.02 \leq \eta < 0.05$	$0.03745 - 4.6983\eta + 185.8469\eta^2 - 869.5823\eta^3$

Clamped–Simply Supported

m_1	$\eta = R/nL$	C_1
0	$0.07 \leq \eta \leq 0.5$	$0.01077 - 0.3552\eta + 2.9869\eta^2 + 15.041\eta^3 - 55.2075\eta^4 + 48.2361\eta^5$
	$0.02 \leq \eta < 0.07$	$-0.000144 + 0.01955\eta - 1.002849\eta^2 + 23.8347\eta^3$
1	$0.08 \leq \eta \leq 0.5$	$-0.01327 - 1.032\eta + 28.657\eta^2 - 82.15\eta^3 + 71.34\eta^4$
	$0.02 \leq \eta < 0.08$	$-0.001386 + 0.2027\eta - 11.49\eta^2 + 311.98\eta^3 - 1222.9\eta^4$
2	$0.08 \leq \eta \leq 0.5$	$-0.4030 + 7.962\eta - 17.1523\eta^2 + 13.0029\eta^3$
	$0.02 \leq \eta < 0.08$	$0.00146 - 0.003047\eta - 13.304\eta^2 + 777.5\eta^3 - 4198.7\eta^4$
3	$0.08 \leq \eta \leq 0.5$	$-0.3694 + 9.8843\eta - 26.04476\eta^2 + 23.1253\eta^3$
	$0.02 \leq \eta < 0.08$	$0.01973 - 2.0216\eta + 55.382\eta^2 + 485.28\eta^3 - 4895.0\eta^4$
4	$0.08 \leq \eta \leq 0.5$	$-0.2154 + 9.7156\eta - 27.3906\eta^2 + 25.3478\eta^3$
	$0.02 \leq \eta < 0.08$	$0.05062 - 5.8587\eta + 212.6672\eta^2 - 1100.2689\eta^3$

TABLE 20-16 **Frequencies of Membrane Cylinders** **1266**

TABLE 20-16 (continued)   **VALUES OF C_1 IN PART A**

		Clamped–Free
m_1	$\eta = R/nL$	C_1
0	$0.1 \leq \eta \leq 0.5$	$0.00092 + 0.002226\eta - 0.4103\eta^2 + 4.5762\eta^3 - 4.6157\eta^4$
	$0.035 \leq \eta < 0.1$	$-0.0001156 + 0.008336\eta - 0.2215\eta^2 + 2.6565\eta^3$
	$0.02 \leq \eta < 0.035$	$-0.0000021 + 0.0005052\eta - 0.03834\eta^2 + 1.15745\eta^3$
1	$0.08 \leq \eta \leq 0.5$	$0.060995 - 1.855\eta + 18.704\eta^2 - 36.66\eta^3 + 23.329\eta^4$
	$0.027 \leq \eta < 0.08$	$0.0004134 - 0.013664\eta - 0.70137\eta^2 + 37.08687\eta^3$
	$0.02 \leq \eta < 0.027$	$0.0004057 - 0.046776\eta + 1.51436\eta^2$
2	$0.08 \leq \eta \leq 0.5$	$-0.075256 + 0.10096\eta + 28.599\eta^2 - 93.341\eta^3 + 86.742\eta^4$
	$0.02 \leq \eta < 0.08$	$0.0018112 - 0.15584\eta + 1.7872\eta^2 + 146.712\eta^3$
3	$0.08 \leq \eta \leq 0.5$	$-0.43498 + 9.2468443\eta - 22.2211\eta^2 + 18.4197\eta^3$
	$0.027 \leq \eta < 0.08$	$0.01085 - 0.80057\eta + 8.04\eta^2 + 692.4\eta^3 - 4262.9\eta^4$
	$0.02 \leq \eta < 0.027$	$-0.0001266 + 0.071225\eta - 10.3095\eta^2 + 615.065\eta^3$
4	$0.07 \leq \eta \leq 0.5$	$-0.3339 + 9.929\eta - 25.77\eta^2 + 22.202\eta^3$
	$0.02 \leq \eta < 0.07$	$0.007042 - 0.5893\eta - 5.747\eta^2 + 1838.4\eta^3 - 14{,}025\eta^4$

		Free–Free
m_1	$\eta = R/nL$	C_1
2	$0.08 \leq \eta \leq 0.5$	$0.18343 - 5.233\eta + 48.646\eta^2 - 118.47\eta^3 + 94.264\eta^4$
	$0.02 \leq \eta < 0.08$	$-0.0004718 + 0.057847\eta - 2.609\eta^2 + 52.783\eta^3 + 71.018\eta^4$
3	$0.08 \leq \eta \leq 0.5$	$-0.24301 + 3.2871\eta + 14.394\eta^2 - 68.137\eta^3 + 70.793\eta^4$
	$0.02 \leq \eta < 0.08$	$-0.0042972 + 0.565\eta - 28.073\eta^2 + 649.04\eta^3 - 2551.1\eta^4$
4	$0.08 \leq \eta \leq 0.5$	$-0.4655 + 10.45823\eta - 27.1505\eta^2 + 23.7978\eta^3$
	$0.02 \leq \eta < 0.08$	$0.000248 + 0.27803\eta - 34.487\eta^2 + 1439.1\eta^3 - 8039\eta^4$

[a]Adopted from Leissa [20.6].

Notation

E = modulus of elasticity
ρ^* = mass per unit volume
h = thickness of shell
D_1 = constant given in this table, part B
C_1 = constant given in Table 20-16, part B
ω = natural frequency
Ω = frequency parameter
$\Omega_a, \Omega_{rt}, \Omega_t, \Omega_{at}$ = see definitions of Table 20-16
ν = Poisson's ratio
R = radius of cylinder
L = length of cylinder
n = number of waves in mode shapes in circumferential direction
m = number of half waves in mode shapes in longitudinal direction
m_1 = number of circumferential nodal circles in mode shapes
along longitudinal direction; nodal circles are circles that have
zero displacements in mode shapes. See figure of page 1264.

$$\omega^2 = \frac{E\Omega^2}{\rho^*(1-\nu^2)R^2} \quad \text{where } \Omega^2 = \Omega^2 \text{ or } \Omega_a^2 \text{ or } \Omega_{rt}^2 \text{ or } \Omega_t^2$$

or Ω_{at}^2 as appropriate.

$$\eta = \frac{R}{nL} \qquad \lambda = \frac{m\pi R}{L} \qquad \kappa = \frac{h^2}{12R^2}$$

Unless specified otherwise, the vibration modes are general responses; i.e., they do not pertain to axial, radial, or torsional modes in particular.
Cases 4–7 give lower bounds for the frequency parameters.

Part A. Frequencies	
Case	Frequency Parameter Ω
1. Infinite 	FROM DONNELL–MUSHTARI THEORY: Axial modes: $\Omega_a^2 = \frac{1}{2}(1-\nu)n^2$ Coupled radial–torsional modes: $\Omega_{rt}^2 = \frac{1}{2}\left\{(1+n^2+\kappa n^4) \mp \left[(1+n^2)^2 + 2\kappa n^4(1-n^2)\right]^{1/2}\right\}$ FROM FLÜGGE THEORY: Axial modes: $\Omega_a^2 = \frac{1}{2}(1-\nu)n^2$ Coupled radial–torsional modes $\Omega_{rt}^2 = \frac{1}{2}\left\{(1+n^2+\kappa n^4) \mp \left[(1+n^2)^2 - 2\kappa n^6\right]^{1/2}\right\}$

TABLE 20-17   Frequencies of Cylinders with Bending   1268

Case	Frequency Parameter Ω
2. Simply supported (shear diaphragms)	In general Ω^2 is taken from the polynomial $$\Omega^6 - K_2\Omega^4 + K_1\Omega^2 - K_0 = 0$$ The coefficients K_0, K_1, and K_2 are: FROM DONNELL-MUSHTARI THEORY: $$K_0 = \tfrac{1}{2}(1 - \nu)\left[(1 - \nu^2)\lambda^4 + \kappa(n^2 + \lambda^2)^4\right]$$ $$K_1 = \tfrac{1}{2}(1 - \nu)\left[(3 + 2\nu)\lambda^2 + n^2 + (n^2 + \lambda^2)^2\right.$$ $$\left. + \frac{3 - \nu}{1 - \nu}\kappa(n^2 + \lambda^2)^3\right]$$ $$K_2 = 1 + \tfrac{1}{2}(3 - \nu)(n^2 + \lambda^2) + \kappa(n^2 + \lambda^2)^2$$ FROM FLÜGGE THEORY: $$K_0 = \tfrac{1}{2}(1 - \nu)\left[(1 - \nu^2)\lambda^4 + \kappa(n^2 + \lambda^2)^4\right] + \kappa\,\Delta K_0$$ $$\Delta K_0 = \tfrac{1}{2}(1 - \nu)[2(2 - \nu)\lambda^2 n^2 + n^4 - 2\nu\lambda^6 - 6\lambda^4 n^2$$ $$-2(4 - \nu)\lambda^2 n^4 - 2n^6]$$ K_1 and K_2 are the same as in Donnell-Mushtari theory. For $n = 0$: FROM DONNELL-MUSHTARI THEORY: Torsional modes: $$\Omega_t^2 = \tfrac{1}{2}(1 - \nu)\lambda^2$$ Coupled axial–radial modes: $$\Omega_{ar}^2 = \left\{(1 + \lambda^2 + \kappa\lambda^4)\right.$$ $$\left. \mp\left[(1 - \lambda^2)^2 + 2\lambda^2(2\nu^2 + \kappa\lambda^2 - \kappa\lambda^4)\right]^{1/2}\right\}$$ FROM FLÜGGE THEORY: $$\Omega_t^2 = \tfrac{1}{2}(1 - \nu)\lambda^2$$ $$\Omega_{ar}^2 = \tfrac{1}{2}\left\{[1 + \lambda^2 + \kappa\lambda^4] \mp \left[(1 + \lambda^2)^2 + 4\nu^2\lambda^2 - 2\kappa\lambda^6\right]^{1/2}\right\}$$
3. Approximations for simply supported cylindrical shell	1. Neglect the tangential inertia (not accurate for $n = 1$) $$\Omega^2 = \frac{K_0 + \kappa\,\Delta K_0}{[(1 - \nu)/2](\lambda^2 + n^2)^2}$$ in which K_0 and ΔK_0 are defined in the previous case.

Case	Frequency Parameter Ω
3. Continued	2. Assume $\lambda^2 \ll n^2$. (The circumferential wave length is small relative to the axial wave length) The constants in case 2 now become $$K_0 = \tfrac{1}{2}(1 - \nu)[(1 - \nu^2)\lambda^4 + \kappa n^8]$$ $$K_1 = \tfrac{1}{2}[(1 - \nu)n^2(n^2 + 1) + (3 - \nu)\kappa n^6]$$ $$K_2 = 1 + \tfrac{1}{2}(3 - \nu)n^2 + \kappa n^4$$ The modification for the Flügge theory is $$\Delta K_0 = \tfrac{1}{2}(1 - \nu)n^4(1 - 2n^2)$$ 3. Combination of approximations 1 and 2 leads for given (n, λ) to $$\Omega^2 = \frac{(1 - \nu^2)\lambda^4}{\left(n^2 - \lambda^2\right)^2} + \kappa(n^2 - \lambda^2)^2$$ 4. Neglect Ω^6 and Ω^4 in the equation of case 2: $$\Omega^2 = \frac{K_0 + \kappa\,\Delta K_0}{K_1}$$ in which K_0, K_1, and ΔK_0 are given in case 2.
4. Clamped–clamped	$\Omega^2 = (1 - \nu^2)C_1 + \kappa n^4 D_1^2$ C_1 is given in Table 20-16, part B, and D_1 is given in this table, part B
5. Clamped–simply supported	$\Omega^2 = (1 - \nu^2)C_1 + \kappa n^4 D_1^2$ C_1 is given in Table 20-16, part B, and D_1 is given in this table, part B
6. Clamped–free	$\Omega^2 = (1 - \nu^2)C_1 + \kappa n^4 D_1^2$ C_1 is given in Table 20-16, part B, and D_1 is given in this table, part B
7. Free–free	$\Omega^2 = (1 - \nu^2)C_1 + \kappa n^4 D_1^2$ C_1 is given in Table 20-16, part B, and D_1 is given in this table, part B

TABLE 20-17 **Frequencies of Cylinders with Bending**

Part B. Values of D_1 in Part A

Clamped–Clamped

m	$\eta = R/nL$	D_1
1	$0.08 \le \eta \le 0.5$	$1.1026 - 2.22216\eta + 25.3243\eta^2 - 0.9368\eta^3$
	$0.02 \le \eta < 0.08$	$0.9967 + 0.2264\eta + 4.7564\eta^2 + 64.967\eta^3$
2	$0.08 \le \eta \le 0.5$	$1.1176 - 3.0116\eta + 70.4188\eta^2 - 8.2918\eta^3$
	$0.02 \le \eta < 0.08$	$0.9988 + 0.078665\eta + 38.21754\eta^2 + 119.6485\eta^3$
3	$0.08 \le \eta \le 0.5$	$1.0147 - 1.3757\eta + 123.966\eta^2 - 2.146\eta^3$
	$0.02 \le \eta < 0.08$	$1.0029 - 0.2573\eta + 97.6849\eta^2 + 152.3918\eta^3$
4	$0.08 \le \eta \le 0.5$	$0.9364 - 0.1667\eta + 198.895\eta^2 + 1.7888\eta^3$
	$0.02 \le \eta < 0.08$	$1.1055 - 8.3424\eta + 355.4656\eta^2 - 979.03424\eta^3$
5	$0.08 \le \eta \le 0.5$	$9.085 - 238.8638\eta + 2817.9036\eta^2 - 11{,}918.832\eta^3$ $+ 25{,}521.979\eta^4 - 19{,}963.1504\eta^5$
	$0.02 \le \eta < 0.08$	$1.0121 - 1.3277\eta + 294.845\eta^2 + 79.63435\eta^3$

Clamped–Simply Supported

m_1	$\eta = R/nL$	D_1
0	$0.08 \le \eta \le 0.5$	$1.0953 - 1.83523\eta + 20.696\eta^2 - 5.6876\eta^3$
	$0.02 \le \eta < 0.08$	$0.98544 + 0.7798\eta - 4.61999\eta^2 + 101.65366\eta^3$
1	$0.08 \le \eta \le 0.5$	$1.06957 - 1.8352\eta + 55.9259\eta^2 - 6.4323\eta^3$
	$0.02 \le \eta < 0.08$	$0.9648 + 3.09396\eta - 42.9602\eta^2 + 663.6377\eta^3$
2	$0.08 \le \eta \le 0.5$	$0.9798 - 0.3857\eta + 104.6534\eta^2$
	$0.02 \le \eta < 0.08$	$0.9950 + 0.31014\eta + 83.92687\eta^2 + 129.6919\eta^3$
3	$0.08 \le \eta \le 0.5$	$2.0005 - 17.6278\eta + 260.8745\eta^2 - 103.5345\eta^3$
	$0.02 \le \eta < 0.08$	$0.9995 - 0.2958\eta + 172.6369\eta^2 + 30.4299\eta^3$
4	$0.08 \le \eta \le 0.5$	$1.3992 - 6.9458\eta + 299.01544\eta^2 - 30.2670\eta^3$
	$0.02 \le \eta < 0.08$	$2.32243 - 175.344\eta + 8953.7051\eta^2 - 201{,}717.40625\eta^3$ $+ 2{,}190{,}977.5\eta^4 - 8{,}950{,}106\eta^5$

TABLE 20-17 (continued)   NATURAL FREQUENCIES OF CIRCULAR CYLINDRICAL SHELLS WITH BENDING

		Clamped–Free
m_1	$\eta = R/nL$	D_1
0	$0.08 \leq \eta \leq 0.5$	$0.993996 + 0.0221\eta + 1.8443\eta^2 + 0.054795\eta^3$
	$0.02 \leq \eta < 0.08$	$1.003234 - 0.22555\eta + 3.7473\eta^2 - 2.1798\eta^3$
1	$0.08 \leq \eta \leq 0.5$	$0.9567 + 0.8161\eta + 16.36077\eta^2 + 0.9839\eta^3$
	$0.02 \leq \eta < 0.08$	$0.99833 + 0.04249\eta + 19.4784\eta^2 + 6.6541\eta^3$
2	$0.08 \leq \eta \leq 0.5$	$1.0087 + 0.26014\eta + 54.2387\eta^2$
	$0.02 \leq \eta < 0.08$	$0.9072 + 7.1194\eta - 100.8743\eta^2 + 1019.1997\eta^3$
3	$0.08 \leq \eta \leq 0.5$	$1.0206 + 0.10173\eta + 110.7527\eta^2$
	$0.02 \leq \eta < 0.08$	$0.999521 - 0.03371\eta + 119.5925\eta^2 - 55.66833\eta^3$
4	$0.08 \leq \eta \leq 0.5$	$1.01137 + 0.080428\eta + 186.8723\eta^2$
	$0.02 \leq \eta < 0.08$	$0.9932 + 0.4689\eta + 185.9586\eta^2$

		Free–Free
m_1	$\eta = R/nL$	D_1
0	$0.08 \leq \eta \leq 0.5$	$0.9754 + 0.2547\eta + 7.5513\eta^2 - 5.8221\eta^3$
	$0.02 \leq \eta < 0.08$	$0.99117 + 0.44896\eta - 4.610324\eta^2 + 87.0146\eta^3$
1	$0.08 \leq \eta \leq 0.5$	$0.8894 + 2.4549\eta + 20.2022\eta^2 + 0.018356\eta^3$
	$0.02 \leq \eta < 0.08$	$0.9828 + 1.0068\eta + 15.86037\eta^2 + 102.00745\eta^3$
2	$0.08 \leq \eta \leq 0.05$	$0.8703 + 3.5628\eta + 53.1368\eta^2 + 7.2333\eta^3$
	$0.02 \leq \eta < 0.08$	$0.9849 + 0.8886\eta + 67.5611\eta^2 + 23.1573\eta^3$
3	$0.08 \leq \eta \leq 0.05$	$0.6701 + 7.8683\eta + 85.67425\eta^2 + 52.2036\eta^3$
	$0.02 \leq \eta < 0.08$	$0.9950 + 0.0677115\eta + 156.3289\eta^2 - 205.7957\eta^3$
4	$0.08 \leq \eta \leq 0.5$	$1.08086 + 0.9024\eta + 198.6602\eta^2$
	$0.02 \leq \eta < 0.08$	$0.98223 + 1.1859\eta + 216.1208\eta^2 - 106.05356\eta^3$

[a]Adopted from Leissa [20.6].

TABLE 20-17   **Frequencies of Cylinders with Bending**   **1272**

TABLE 20-18   NATURAL FREQUENCIES OF CONICAL SHELLS WITH BENDING[a]

Notation

E = modulus of elasticity

h = thickness of shell

s = distance from apex

$\Omega, \Omega_j, \Omega_n$ = frequency parameters

R = radius of cone base

J_1 = Bessel function of first kind

Y_1 = Bessel function of second kind

ν = Poisson's ratio

n = number of circumferential waves in mode shapes. See figure of page 1264.

ρ^* = mass per unit volume

α = half angle of cone

j = number of root of characteristic equation for specific mode, i.e., for a specific n; this means that there are several distinct natural frequencies corresponding to a fixed n

ω = natural frequency (rad/s)

1. *Complete cone with free base*

Natural Frequency (ω)	Frequency Parameter
$$\frac{\Omega_j}{R}\left(\frac{E}{\rho^*}\right)^{1/2}$$	Axisymmetric Mode ($n = 0$)

$\eta = 12(1 - \nu^2)(R/h)^2/\tan^4 \alpha$ with $\nu = 0.3$

$$\Omega_1 = \begin{cases} 3.2003 + 81.0593\eta^{-1} & 0.1 \le \eta \le 6 \\[2mm] 1.4362 - 3.7517(\log \eta)^{-1} + 17.861(\log \eta)^{-2} \\ \quad -4.3278(\log \eta)^{-3} & 8 \le \eta \le 800 \\[2mm] 1.0598 - 1.9098(\log \eta)^{-1} + 14.6899(\log \eta)^{-2} \\ \quad -1.8035(\log \eta)^{-3} & 800 < \eta \le 100,000 \end{cases}$$

$$\Omega_2 = \begin{cases} 8.26823 + 1477.901\eta^{-1} & 0.1 \le \eta \le 80 \\[2mm] 1.0811 + 1.7538(\log \eta)^{-1} - 31.967(\log \eta)^{-2} \\ \quad +228.46(\log \eta)^{-3} & 100 \le \eta \le 100,000 \end{cases}$$

$$\Omega_3 = \begin{cases} 13.7244 + 7700.08936\eta^{-1} & 0.1 \le \eta \le 80 \\[2mm] -8.608 \times 10^4 + 0.844 \times 10^5(\log \eta)^{-1} - 3.0847 \times 10^5(\log \eta)^{-2} \\ \quad +4.9798 \times 10^5(\log \eta)^{-3} \\ \quad -2.9809 \times 10^5(\log \eta)^{-4} & 100 \le \eta < 1000 \\[2mm] -453.16 + 7472.7(\log \eta)^{-1} + 45,745(\log \eta)^{-2} \\ \quad +123,420(\log \eta)^{-3} - 122,520(\log \eta)^{-4} & 1000 < \eta \le 100,000 \end{cases}$$

$$\Omega_4 = \begin{cases} 18.57213 + 24,591.9844\eta^{-1} & 0.1 \le \eta \le 6 \\[2mm] 15,066.074 - 121,113.0469(\log \eta)^{-1} + 380,252.875(\log \eta)^{-2} \\ \quad -584,762(\log \eta)^{-3} + 448,058.344(\log \eta)^{-4} \\ \quad -135,024.52(\log \eta)^{-5} & 8 \le \eta \le 200 \\[2mm] -64.661 + 906.32715(\log \eta)^{-1} \\ \quad -4358.1699(\log \eta)^{-2} \\ \quad +7734.356(\log \eta)^{-3} & 200 < \eta \le 100,000 \end{cases}$$

TABLE 20-18   Natural Frequencies of Conical Shells   1274

2. *Complete cone with clamped base*

Natural Frequency (ω)	Frequency Parameter
$\dfrac{\Omega_j}{R}\left(\dfrac{E}{\rho^*}\right)^{1/2}$	Axisymmetric Mode ($n = 0$)

$$\eta = 12(1 - \nu^2)(R/h)^2/\tan^4 \alpha \quad \text{with } \nu = 0.3$$

$$\Omega_1 = \begin{cases} 5.90356 + 104.38289\eta^{-1} & 0.1 \leq \eta \leq 40 \\ -0.29463 + 8.39616(\log \eta)^{-1} + 11.2915(\log \eta)^{-2} \\ \quad -5.0101(\log \eta)^{-3} & 60 \leq \eta \leq 100{,}000 \end{cases}$$

$$\Omega_2 = \begin{cases} 19{,}665.967 - 43{,}630.3125\eta^{-1} + 36{,}566.6133\eta^{-2} \\ \quad -12{,}819.2305\eta^{-3} + 1903.352\eta^{-4} \\ \quad -94.385\eta^{-5} & 0.1 \leq \eta \leq 1 \\ 66.963 - 398.13(\log \eta)^{-1} + 775.51(\log \eta)^{-2} \\ \quad -320.38(\log \eta)^{-3} + 43.05(\log \eta)^{-4} & 2 \leq \eta \leq 600 \\ -1.577 + 31.5281(\log \eta)^{-1} - 126.1601(\log \eta)^{-2} \\ \quad +343.7357(\log \eta)^{-3} & 600 < \eta \leq 100{,}000 \end{cases}$$

$$\Omega_3 = \begin{cases} 14.0355 + 7939.58398\eta^{-1} & 0.1 \leq \eta < 400 \\ -9.51032 + 161.3258(\log \eta)^{-1} - 859.5827(\log \eta)^{-2} \\ \quad +1885.0045(\log \eta)^{-3} & 400 \leq \eta < 100{,}000 \end{cases}$$

$$\Omega_4 = \begin{cases} 18.5415 + 25022.31445\eta^{-1} & 0.1 \leq \eta \leq 800 \\ -43.1926 + 664.6763(\log \eta)^{-1} - 3464.3162(\log \eta)^{-2} \\ \quad +6677.5742(\log \eta)^{-3} & 1000 \leq \eta \leq 100{,}000 \end{cases}$$

3. *Frustum of cone (open conical shell) with simply supported edges*

Natural Frequency (ω)	Frequency Parameter
$\dfrac{\Omega}{L}\left[\dfrac{E}{\rho^*(1-\nu^2)}\right]^{1/2}$	Lowest Mode for $\nu = 0.3$

$h/R_2 = 0.03 \qquad \beta = 90° - \alpha$

$\Omega = 0.16498 + 0.01997\beta - 0.000256\beta^2$

$\qquad + 0.000001525\beta^3 \hspace{3cm} 15° \le \beta \le 85°$

$h/R_2 = 0.01$

$\Omega = 0.04757 + 0.021399\beta - 0.0008059\beta^2$

$\qquad + 0.000020133\beta^3 - 0.0000002478\beta^4$

$\qquad + 0.0000000011634\beta^5 \hspace{2.3cm} 5° \le \beta \le 87°$

$h/R_2 = 0.005$

$\Omega = 0.03094 + 0.016305\beta - 0.0006938\beta^2$

$\qquad + 0.00001935\beta^3 - 0.0000002575\beta^4$

$\qquad + 0.0000000012755\beta^5 \hspace{2.2cm} 3° \le \beta \le 87°$

$h/R_2 = 0.001$

$\Omega = 0.0089055 + 0.009064\beta - 0.00042673\beta^2$

$\qquad + 0.00001244\beta^3 - 0.0000001694\beta^4$

$\qquad - 0.0000000008554\beta^5 \hspace{2.2cm} 3° \le \beta \le 87°$

TABLE 20-18   **Natural Frequencies of Conical Shells**   **1276**

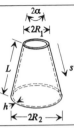

4. Frustum of cone (open conical shell) with clamped edges

Natural Frequency (ω)	Frequency Parameter
$\dfrac{\Omega_j}{S_1}\left[\dfrac{E}{2(1+\nu)\rho^*}\right]^{1/2}$ $S_1 = R_1/\sin\alpha$ Solution also applicable to annular disks: $\alpha = 90°$	**Axisymmetric Torsional Mode** $\eta = R_2/R_1 \quad 1.0 \le \eta \le 50$ $\overline{\Omega}_1 = 3.0909 + 0.06098\eta - 0.001923\eta^2 + 0.000019652\eta^3$ $\overline{\Omega}_2 = 6.23083 + 0.046874\eta - 0.00125\eta^2 + 0.000011594\eta^3$ $\overline{\Omega}_3 = 9.38016 + 0.036212\eta - 0.0008112\eta^2 + 0.000006657\eta^3$ $\overline{\Omega}_4 = 12.52926 + 0.028825\eta - 0.0005425\eta^2 + 0.000003902\eta^3$ $\overline{\Omega}_5 = 15.6766 + 0.02366\eta - 0.000374482\eta^2$ $\qquad + 0.000002283\eta^3$ $\Omega_j = \dfrac{\overline{\Omega}_j}{(\eta - 1)}$ Ω_j is independent of α and is the solution of $J_1(\Omega)Y_1(\eta\Omega) = J_1(\eta\Omega)Y_1(\Omega)$ j indicates the jth root of the equation

5. Frustum of cone (open conical shell) with free edges

$\dfrac{\Omega_n}{R_2}\left[\dfrac{E}{\rho^*(1-\nu^2)}\right]^{1/2}$	For one half wave in the mode shape along L $\left(\dfrac{h^2}{12R_2^2}\right)\dfrac{n(n^2-1)}{(n^2+\cos^2\alpha)^{1/2}}\left[1 + \dfrac{6(1-R_1/R_2)}{n-2}\sin\dfrac{3\alpha}{2}\right]$ $n = 2, 3, 4, \ldots \qquad \alpha < 60°$ See ref. 20.10 for other modes

6. *Frustum of cone (open conical shell) with clamped–free edges (upper edge clamped)*

Natural Frequency (ω)	Frequency Parameter
$\dfrac{\Omega}{R_2}\left(\dfrac{E}{\rho^*}\right)^{1/2}$	For Lowest Axisymmetric Mode for $\nu = 0.3$

$\gamma = 12(1 - \nu^2)(R_2/h)^2/\tan^4 \alpha \qquad \eta = \log \gamma \qquad \Omega^2 = 10^K$

$R_1/R_2 = 0.1$

$$K = \begin{cases} 1.2568 - 1.0406\eta - 0.0442826\eta^2 + 0.07858\eta^3 \\ \qquad + 0.06301\eta^4 & -1 \le \eta < 1.25 \\ 1.38211 - 1.5298\eta + 0.57211\eta^2 - 0.09466\eta^3 \\ \qquad + 0.0054478\eta^4 & 1.25 \le \eta \le 2.75 \\ -0.1731 & 2.75 < \eta \le 4 \end{cases}$$

$R_1/R_2 = 0.2$

$$K = \begin{cases} 1.434614 - 1.0208\eta + 0.02603\eta^2 \\ \qquad + 0.070555\eta^3 & -1 \le \eta < 1.25 \\ 2.4628 - 2.8499\eta + 1.1124\eta^2 - 0.1499\eta^3 & 1.25 < \eta \le 2.5 \\ -0.7692 & 2.75 < \eta \le 4 \end{cases}$$

$R_1/R_2 = 0.3$

$$K = \begin{cases} 1.66798 - 1.03017\eta + 0.004725\eta^2 \\ \qquad + 0.064098\eta^3 & -1 \le \eta < 1.75 \\ 15.554 - 26.4455\eta + 18.0405\eta^2 - 6.1084\eta^3 \\ \qquad + 1.01534\eta^4 - 0.0657\eta^5 & 1.75 \le \eta \le 3.25 \\ -0.7692 & 3.25 < \eta \le 4 \end{cases}$$

$R_1/R_2 = 0.4$

$$K = \begin{cases} 1.9051 - 1.02472\eta + 0.017246\eta^2 + 0.02796\eta^3 \\ \qquad + 0.004479\eta^4 & -1 \le \eta < 2.5 \\ 3.4879 - 3.01358\eta + 0.86266\eta^2 - 0.08218\eta^3 & 2.5 \le \eta \le 3.25 \\ -0.01538 & 3.25 < \eta \le 4 \end{cases}$$

Natural Frequency (ω)	Frequency Parameter

$R_1/R_2 = 0.5$

$$K = \begin{cases} 2.25325 - 1.00933\eta - 0.015067\eta^2 \\ \quad + 0.02883\eta^3 & -1 \leq \eta < 2 \\ -100.85114 + 189.11171\eta - 138.6594\eta^2 \\ \quad + 49.9954\eta^3 - 8.8953\eta^4 + 0.6261\eta^5 & 2 \leq \eta \leq 3.5 \end{cases}$$

$R_1/R_2 = 0.6$

$$K = \begin{cases} 2.6292 - 1.03156\eta - 0.02928\eta^2 + 0.01584\eta^3 \\ \quad + 0.007319\eta^4 & -1 \leq \eta < 2.25 \\ -277.2174 + 506.6892\eta - 365.6982\eta^2 \\ \quad + 130.7203\eta^3 - 23.17983\eta^4 \\ \quad + 1.632534\eta^5 & 2.25 \leq \eta \leq 3.5 \end{cases}$$

$R_1/R_2 = 0.7$

$$K = \begin{cases} 3.1251 - 1.00322\eta - 0.002791\eta^2 \\ \quad + 0.001848\eta^3 & -1 \leq \eta < 2.5 \\ -11.5472 + 12.7924\eta - 4.3464\eta^2 \\ \quad + 0.4693\eta^3 & 2.5 \leq \eta \leq 3.25 \end{cases}$$

$R_1/R_2 = 0.8$

$K = 3.8484 - 1.0395\eta - 0.01532\eta^2 + 0.01456\eta^3$ $-1 \leq \eta \leq 3$

[a]Adopted from Leissa [20.6].

TABLE 20-19   NATURAL FREQUENCIES OF SPHERICAL SHELLS[a]

Notation

E = modulus of elasticity

ρ^* = mass per unit volume

h = thickness of shell

ω = natural frequency

i, j = integer indices, $i, j = 0, 1, 2, \ldots$

ν = Poisson's ratio

R = radius of sphere or spherical segment

Ω^2 = frequency parameter

$\Omega_r, \Omega_t, \Omega_{rt}$ = frequency parameters for radial, torsional, and coupled radial-torsional modes, respectively.

$(\omega_{ij})_p$ = natural frequency of plate in bending corresponding to projection of shell and with same boundary conditions as shell; $(\omega_{ij})_p$ can be obtained from Chapter 18

Unless specified otherwise, the vibration modes are general responses; i.e., they do not pertain to torsional, radial, or tangential modes in particular.

$$\omega^2 = \frac{E\Omega^2}{\rho^*(1 - \nu^2)R^2} \quad \text{where } \Omega^2 = \Omega^2 \text{ or } \Omega_r^2 \text{ or } \Omega_t^2 \text{ or } \Omega_{rt}^2 \text{ as appropriate.}$$

$$\kappa = \frac{h^2}{12R^2} \qquad \xi = 1/\kappa$$

$$k_1 = 1 + \kappa \qquad k_r = 1 + 1.8\kappa \qquad r = i(i + 1) \qquad i = 0, 1, 2, \ldots$$

Case	Frequency Parameter
1. Complete spherical shells	FUNDAMENTAL RADIAL MODE Membrane analysis $\Omega_r^2 = \dfrac{2(1 + \nu)}{1 + h^2/(12R^2)}$ Bending analysis (solve for Ω_r^2): $2.4\Omega_r^6 k_1(k_r k_1 - 4\kappa)/(1 - \nu)$ $\quad - \Omega_r^4\{(k_r k_1 - 4\kappa)[r + 4.8(1 + \nu)/(1 - \nu)]$ $\qquad + k_1[\xi(1 + 3\kappa) + 3 + 1.8\kappa$ $\qquad + 4.8(1 + 1.4\kappa)(r/(1 - \nu) - 1]\}$ $\quad + \Omega_r^2\{4(1 + \nu)(2 - r)$ $\qquad + k_r[r(r - 3 - \nu) + 4.4(1 + \nu)(r - 2)]$ $\qquad + k_1[2.4r(r + 4\nu)/(1 - \nu) + r(r + \xi + \nu)$ $\qquad + (1 + 3\nu)(\xi - 2.4) - (1 - \nu)]\}$ $\quad - (r - 2)[r(r - 2) + 2.4(1 + \nu)(r - 1 + \nu)$ $\qquad + (1 - \nu^2)(\xi + 1)] = 0$

TABLE 20-19 **Natural Frequencies of Spherical Shells** 1280

TABLE 20-19 (continued) NATURAL FREQUENCIES OF SPHERICAL SHELLS

Case	Frequency Parameter
1. Continued	TORSIONAL MODES

$$\Omega_t^2 = \frac{(1 - \nu)(i^2 + i - 2)}{2 + 5h^2/(6R^2)}$$

RADIAL–TANGENTIAL MODES

Membrane analysis:
$$\Omega_{rt}^2 = \tfrac{1}{2}\{(i^2 + i + 1 + 3\nu) \mp [(i^2 + i + 1 + 3\nu)^2 - 4(1 - \nu^2)(i^2 + i - 2)]^{1/2}\}$$

Bending analysis (solve for Ω_{rt}^2):
$$4.8\Omega_{rt}^4(k_r k_1 - 4\kappa)/(1 - \nu) - 2\Omega_{rt}^2[\xi(1 + 3\kappa)$$
$$+ 3 + 1.8\kappa + 2.4(1 + 1.4\kappa)(r - 2)]$$
$$+ (1 - \nu)(r - 2)[\xi + 1 + 1.2(r - 2)] = 0$$

2.
Deep spherical
shell segments

$$\Omega^2 = \frac{(i^2 - 1)^2 i^2 (1 - \nu^2)}{3(1 + \nu)} \left(\frac{h}{R}\right)^2 \frac{s_1(i, \phi_0)}{s_2(i, \phi_0)}$$

$$s_1(i, \phi_0) = \frac{1}{8}\left[\frac{[\tan(\phi_0/2)]^{2i-2}}{n - 1} + \frac{2[\tan(\phi_0/2)]^{2i}}{n} \right.$$

$$\left. + \frac{[\tan(\phi_0/2)]^{2i+2}}{n + 1} \right]$$

$$s_2(i, \phi_0) = \int_0^{\phi_0}\left(\tan\frac{\phi_0}{2}\right)^{2i}\left[(i + \cos\phi)^2 \right.$$

$$\left. + 2(\sin\phi)^2\right]\sin\phi \, d\phi$$

3.
Shallow spherical
shell segments

$$\Omega^2 = \left[(\omega_{ij})_p^2 + \frac{E}{\rho^* R^2}\right]^{1/2}\left[\frac{\rho^*(1 - \nu^2)R^2}{E}\right]$$

For a segment to be shallow, the rise of the shell d must be less than about $\frac{1}{8}$ of the diameter D, which is the diameter of the smallest circle that contains the projection. The projection can have various shapes. If the segment is not shallow, use the formulas of case 2.

[a]Adopted from Blevins [20.11], with permission.

Natural Frequencies of Spherical Shells | **TABLE 20-19**

Fundamental Mathematics

Frequently used mathematics formulas are provided in this appendix along with brief outlines of some useful solution procedures.

I.1 ALGEBRAIC OPERATIONS

Algebraic Laws

$$\text{Commutative law:} \quad x + y = y + x, \qquad xy = yx$$
$$\text{Associative law:} \quad x + (y + z) = (x + y) + z,$$
$$x(yz) = (xy)z \tag{I.1}$$
$$\text{Distributive law:} \quad z(x + y) = zx + zy$$

Exponents

Here a = basis and n = exponent:

$$a^n = a \cdot a \cdots a \ (n \text{ times})$$
$$a^0 = 1 \quad \text{if } a \neq 0, \quad a^n \cdot b^n = (ab)^n$$
$$a^m \cdot a^n = a^{m+n}, \qquad a^m / a^n = a^{m-n}$$
$$(a^m)^n = a^{m \cdot n}, \quad (a + b)^2 = a^2 + 2ab + b^2$$
$$a^2 - b^2 = (a + b) \cdot (a - b), \qquad (a + b)^3 = a^3 + 3a^2b + 3ab^2 + b^3$$

The Binomial Theorem

$$(a \pm b)^n = \binom{n}{0}a^n \pm \binom{n}{1}a^{n-1}b^1 + \binom{n}{2}a^{n-2}b^2 \pm \binom{n}{3}a^{n-3}b^3 + \cdots \tag{I.2a}$$
$$+ (\pm 1)^{n-1}\binom{n}{n-1}a^1 b^{n-1} + (\pm 1)^n b^n, \qquad n = 1, 2, 3, \ldots$$

$$\binom{n}{0} = 1, \quad \binom{n}{1} = n, \quad \binom{n}{2} = \frac{n \cdot (n-1)}{2 \cdot 1}, \quad \binom{n}{3} = \frac{n(n-1)(n-2)}{3 \cdot 2 \cdot 1}, \cdots$$

$$\binom{n}{n} = 1, \quad \binom{n}{k} = \frac{n!}{k!(n-k)!}, \quad n! = n \cdot (n-1) \cdots 2 \cdot 1$$

$$0! = 1, \quad \binom{n}{n-k} = \binom{n}{k}, \quad \binom{n}{n-1} = n$$

If x, m are real numbers, k is an integer:

$$(1 \pm x)^m = 1 \pm \binom{m}{1}x + \binom{m}{2}x^2 \pm \binom{m}{3}x^3 + \cdots \qquad (I.2b)$$

where

$$\binom{m}{k} = \begin{cases} \dfrac{m(m-1)\cdots(m-k+1)}{k!} & \text{for } k > 0 \\ 1 & \text{for } k = 0 \\ 0 & \text{for } k < 0 \end{cases}$$

This series, for

$$\begin{cases} m = 0, 1, 2 \ldots \text{ (i.e., } m \text{ is an integer) and } x \text{ of any value} & \text{is a finite series} \\ m \neq 0, 1, 2 \ldots \text{ (i.e., } m \text{ is not an integer) and } |x| < 1 & \text{is an infinite} \\ & \quad \text{convergent series} \\ m \neq 0, 1, 2 \ldots \text{ (i.e., } m \text{ is not an interger) and } |x| > 1 & \text{is an infinite} \\ & \quad \text{divergent series} \end{cases}$$

Roots

Here n = exponent and c = root:

$$\sqrt[n]{a} = c$$

$$\sqrt[n]{a} = a^{1/n}, \qquad \sqrt[n]{a^n} = a, \qquad \sqrt[n]{a^{m \cdot n}} = a^m, \qquad \sqrt[n]{a \cdot b} = \sqrt[n]{a} \cdot \sqrt[n]{b}$$

$$\sqrt[n]{a/b} = \frac{\sqrt[n]{a}}{\sqrt[n]{b}}, \qquad \sqrt[n]{1/a} = \frac{1}{\sqrt[n]{a}} = a^{-1/n}, \qquad \sqrt[n]{a^m} = a^{m/n} \qquad (I.3)$$

$$\sqrt[n]{\sqrt[m]{a}} = {}^{(n \cdot m)}\sqrt{a} = \sqrt[m]{\sqrt[n]{a}}$$

$$\sqrt[m]{a} \cdot \sqrt[n]{a} = a^{1/m} \cdot a^{1/n} = a^{(1/m + 1/n)} = a^{(m+n)/(m \cdot n)} = {}^{m \cdot n}\sqrt{a^{n+m}}$$

Logarithms

Basic Relationships Here b = base, a = number, and c = logarithm:

$$\log_b a = c$$

$$\log_b a = c \leftrightarrow b^c = a$$

$$\log_b b = 1 \text{ (because } b^1 = b), \qquad \log_b 1 = 0 \text{ (because } b^0 = 1) \qquad (I.4)$$

$$\log_b 0 = -\infty \left(\text{because } \lim_{c \to -\infty} b^c = 0 \right)$$

Rules for Calculation

$$\log_b(c \cdot d) = \log_b c + \log_b d, \qquad \log_b a^n = n \cdot \log_b a$$

$$\log_b\left(\frac{c}{d}\right) = \log_b c - \log_b d, \qquad \log_b\sqrt[n]{a} = (1/n)\log_b a \tag{I.5}$$

Natural Logarithm

$$\log_e a = \ln a \quad \text{where } e = \lim_{n \to \infty}(1 + 1/n)^n = 2.718281828\ldots$$

Transformation from One Logarithm System to Another

$$\log_b a = \frac{1}{\log_c b} \cdot \log_c a = \log_b c \cdot \log_c a \tag{I.6}$$

Series

Arithmetic series: $a + (a + d) + (a + 2d) + \cdots + [a + (n - 1)d]$
$$= \tfrac{1}{2}n[2a + (n - 1) \cdot d]$$

Geometric series: $a + a \cdot q + a \cdot q^2 + \cdots + a \cdot q^{n-1} = a \cdot \dfrac{q^n - 1}{q - 1}, \quad q \neq 1$

Infinite geometric series: $a + a \cdot q + a \cdot q^2 + \cdots = \dfrac{a}{1 - q}$ for $|q| < 1$

$$\sum_{i=1}^{n} i = \tfrac{1}{2}n(n + 1), \qquad \sum_{i=1}^{n} i^2 = \tfrac{1}{3}n(n + 1)\left(n + \tfrac{1}{2}\right), \qquad \sum_{i=1}^{n} i^3 = \tfrac{1}{4}n^2(n + 1)^2$$

$$\tag{I.7}$$

I.2 COMPLEX NUMBERS

Here $z = x + iy$, where $i = \sqrt{-1}$, x = real number (real part), and y = real number (iy is the imaginary part). The complex number is designated z, and $z = x + iy$, $\bar{z} = x - iy$ are conjugate complex numbers.

A complex number may be represented as a vector in the x, y plane (complex plane z) as shown in Fig. I-1.

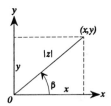

$|z|$ = Absolute Value (modulus) of the
complex number
β = Orientation of z

FIGURE I-1   Complex plane z.

Rules for Calculations

$$z_1 + z_2 = (x_1 + x_2) + i(y_1 + y_2)$$
$$z = x + iy = |z|(\cos \beta + i \sin \beta) = |z|e^{i\beta} \qquad (I.8)$$
$$\bar{z} = x - iy = |z|(\cos \beta - i \sin \beta) = |z|e^{-i\beta}$$
$$z_1 z_2 = (x_1 + iy_1)(x_2 + iy_2) = x_1 x_2 - y_1 y_2 + i(x_1 y_2 + x_2 y_1)$$
$$z^n = (x + iy)^n = [|z|(\cos \beta + i \sin \beta)]^n = |z|^n e^{in\beta}$$
$$\bar{z}^n = (x - iy)^n = [|z|(\cos \beta - i \sin \beta)]^n = |z|^n e^{-in\beta}$$
$$\sqrt[n]{z} = \sqrt[n]{x + iy} = \sqrt[n]{|z|}\left(\cos \frac{\beta + 2k\pi}{n} + i \sin \frac{\beta + 2k\pi}{n}\right) = \sqrt[n]{|z|}\, e^{i(\beta + 2k\pi)/n}$$

for $k = 0, 1, 2, \ldots, n - 1$

$$(I.9)$$

I.3   PLANE TRIGONOMETRY

Definitions

One degree equals $\frac{1}{360}$ of one complete rotation.

One hundred eighty degrees equals π radians.

One radian equals the angle at the center of a circle corresponding to an arc
of length equal to the radius of the circle. See Fig. I-2.

An acute angle is an angle between 0 and 90°.

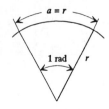

FIGURE I-2   A radian.

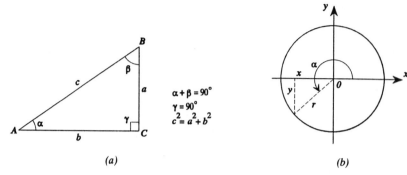

$$\alpha + \beta = 90°$$
$$\gamma = 90°$$
$$c^2 = a^2 + b^2$$

(a)   (b)

FIGURE I-3   (a) Acute angle α, right triangle. (b) Arbitrary angle α.

An obtuse angle is an angle between 90° and 180°.
Acute angle α (Fig. I-3a):

$$\sin \alpha = a/c, \quad \cos \alpha = b/c, \quad \tan \alpha = a/b$$
$$\cot \alpha = b/a, \quad \sec \alpha = c/b, \quad \csc \alpha = c/a \tag{I.10a}$$

Arbitrary angle α (Fig. 1-3b):

$$\sin \alpha = y/r, \quad \cos \alpha = x/r, \quad \tan \alpha = y/x$$
$$\cot \alpha = x/y, \quad \sec \alpha = r/x, \quad \csc \alpha = r/y \tag{I.10b}$$

The graphs of the basic trigonometric functions are shown in Fig. I-4.

Laws of Sines and Cosines

Refer to Fig. I-5:

Law of sines: $a/\sin \alpha = b/\sin \beta = c/\sin \gamma$
Law of cosines: $a^2 = b^2 + c^2 - 2bc \cos \alpha$
$$b^2 = a^2 + c^2 - 2ac \cos \beta$$
$$c^2 = a^2 + b^2 - 2ab \cos \gamma \tag{I.11}$$

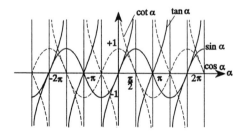

FIGURE I-4   Basic trigonometric functions.

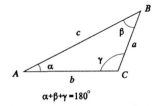

$\alpha+\beta+\gamma=180°$

FIGURE I-5   Arbitrary triangle.

Identities

$$\sin(\alpha \pm \beta) = \sin \alpha \cos \beta \pm \cos \alpha \sin \beta,$$

$$\cos(\alpha \pm \beta) = \cos \alpha \cos \beta \mp \sin \alpha \sin \beta$$

$$\tan(a \pm \beta) = \frac{\tan \alpha \pm \tan \beta}{1 \mp \tan \alpha \tan \beta},$$

$$\cot(\alpha \pm \beta) = \frac{\cot \alpha \cot \beta \mp 1}{\cot \beta \pm \cot \alpha} \tag{I.12}$$

$$\sin(\alpha + \beta)\sin(\alpha - \beta) = \cos^2\beta - \cos^2\alpha,$$

$$\cos(\alpha + \beta)\cos(\alpha - \beta) = \cos^2\alpha - \sin^2\beta$$

$$\sin^2\alpha + \cos^2\alpha = 1$$

$$\sin 2\alpha = 2 \sin \alpha \cos \alpha$$

$$\cos 2\alpha = \cos^2\alpha - \sin^2\alpha = 1 - 2 \sin^2\alpha = 2 \cos^2\alpha - 1$$

$$\tan 2\alpha = \frac{2 \tan \alpha}{1 - \tan^2\alpha} = \frac{2}{\cot \alpha - \tan \alpha} \tag{I.13a}$$

$$\cot 2\alpha = \frac{\cot^2\alpha - 1}{2 \cot \alpha} = \frac{\cot \alpha - \tan \alpha}{2}$$

$$\sin \alpha = 2 \sin \frac{\alpha}{2} \cos \frac{\alpha}{2}$$

$$\cos \alpha = \cos^2\frac{\alpha}{2} - \sin^2\frac{\alpha}{2}$$

$$= 1 - 2 \sin^2\frac{\alpha}{2} = 2 \cos^2\frac{\alpha}{2} - 1 \tag{I.13b}$$

$$\tan \alpha = \frac{2 \tan(\alpha/2)}{1 - \tan^2(\alpha/2)} = \frac{2}{\cot(\alpha/2) - \tan(\alpha/2)}$$

$$\cot \alpha = \frac{\cot^2(\alpha/2) - 1}{2 \cot(\alpha/2)} = \frac{\cot(\alpha/2) - \tan(\alpha/2)}{2}$$

$$\sin \alpha = \pm \sqrt{\frac{1 - \cos 2\alpha}{2}}$$

$$\cos \alpha = \pm \sqrt{\frac{1 + \cos 2\alpha}{2}} \tag{I.13c}$$

$$\tan \alpha = \pm \sqrt{\frac{1 - \cos 2\alpha}{1 + \cos 2\alpha}} = \frac{\sin 2\alpha}{1 + \cos 2\alpha} = \frac{1 - \cos 2\alpha}{\sin 2\alpha}$$

$$\sin \frac{a}{2} = \pm \sqrt{\frac{1 - \cos \alpha}{2}}$$

$$\cos \frac{\alpha}{2} = \pm \sqrt{\frac{1 + \cos \alpha}{2}} \tag{I.13d}$$

$$\tan \frac{\alpha}{2} = \pm \sqrt{\frac{1 - \cos \alpha}{1 + \cos \alpha}} = \frac{1 - \cos \alpha}{\sin \alpha} = \frac{\sin \alpha}{1 + \cos \alpha}$$

$$\sin \alpha + \sin \beta = 2 \sin \frac{\alpha + \beta}{2} \cos \frac{\alpha - \beta}{2}$$

$$\sin \alpha - \sin \beta = 2 \cos \frac{\alpha + \beta}{2} \sin \frac{\alpha - \beta}{2}$$

$$\cos \alpha + \cos \beta = 2 \cos \frac{\alpha + \beta}{2} \cos \frac{\alpha - \beta}{2} \tag{I.14}$$

$$\cos \alpha - \cos \beta = -2 \sin \frac{\alpha + \beta}{2} \sin \frac{\alpha - \beta}{2}$$

$$\sin^2 \alpha = \tfrac{1}{2}(1 - \cos 2\alpha), \qquad \cos^2 \alpha = \tfrac{1}{2}(1 + \cos 2\alpha)$$

Inverse Trigonometric Functions

Definitions

Trigonometric Functions	Corresponding Inverse Trigonometric Functions	Principal Values	
$y = \sin \alpha$	$\alpha = \sin^{-1} y = \arcsin y$	$-\tfrac{1}{2}\pi \leq \alpha \leq +\tfrac{1}{2}\pi$	
$y = \cos \alpha$	$\alpha = \cos^{-1} y = \arccos y$	$0 \leq \alpha \leq \pi$	
$y = \tan \alpha$	$\alpha = \tan^{-1} y = \arctan y$	$-\tfrac{1}{2}\pi < \alpha < +\tfrac{1}{2}\pi$	(I.15)
$y = \cot \alpha$	$\alpha = \cot^{-1} y = \text{arccot } y$	$0 < \alpha < \pi$	

Thus, α is the arc of an angle for which the trigonometric function is y.

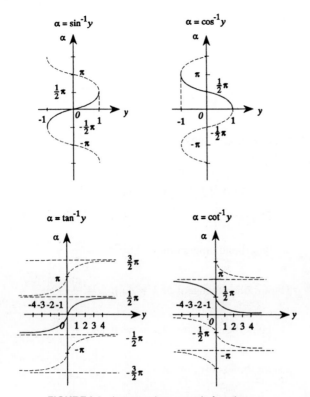

FIGURE I-6   Inverse trigonometric functions.

Identities

$$\sin^{-1}y + \cos^{-1}y = \tfrac{1}{2}\pi \qquad \tan^{-1}y + \cot^{-1}y = \tfrac{1}{2}\pi$$
$$\sin^{-1}(-y) = -\sin^{-1}y \qquad \tan^{-1}(-y) = -\tan^{-1}y \qquad (\text{I.16})$$
$$\cos^{-1}(-y) = \pi - \cos^{-1}y \qquad \cot^{-1}(-y) = \pi - \cot^{-1}y$$

Graphs of the inverse trigonometric functions are shown in Fig. I-6. The solid lines in Fig. I-6 correspond to the principal values of the argument α. The dashed lines, which correspond to other values of the argument α, are based on the relationships

$$\sin(\alpha + 2k\pi) = \sin\alpha \qquad \cos(\alpha + 2k\pi) = \cos\alpha$$
$$\tan(\alpha + k\pi) = \tan\alpha \qquad \cot(\alpha + k\pi) = \cot\alpha \qquad (\text{I.17})$$

where $k = \pm 0, 1, 2, 3, \ldots$.

Exponential Relations: Euler's Equation

$$e^{i\alpha} = \cos \alpha + i \sin \alpha \qquad i = \sqrt{-1}$$

$$\sin \alpha = \frac{e^{i\alpha} - e^{-i\alpha}}{2i} \qquad \cos \alpha = \frac{e^{i\alpha} + e^{-i\alpha}}{2}$$

I.4 HYPERBOLIC FUNCTIONS

Definitions

$$\text{Hyperbolic sine of } x = \sinh x = \tfrac{1}{2}(e^x - e^{-x})$$

$$\text{Hyperbolic cosine of } x = \cosh x = \tfrac{1}{2}(e^x + e^{-x})$$

$$\text{Hyperbolic tangent of } x = \tanh x = \frac{e^x - e^{-x}}{e^x + e^{-x}}$$

$$\text{Hyperbolic cotangent of } x = \coth x = \frac{e^x + e^{-x}}{e^x - e^{-x}} \qquad (I.18)$$

$$\text{Hyperbolic secant of } x = \operatorname{sech} x = \frac{2}{e^x + e^{-x}}$$

$$\text{Hyperbolic cosecant of } x = \operatorname{csch} x = \frac{2}{e^x - e^{-x}}$$

The graphs of the hyperbolic functions are shown in Fig. I-7.

Identities

$$\cosh^2 x - \sinh^2 x = 1 \qquad \tanh x = \frac{\sinh x}{\cosh x}$$

$$\operatorname{sech} x \cosh x = 1 \qquad \tanh^2 x + \operatorname{sech}^2 x = 1$$

$$\coth x = \frac{\cosh x}{\sinh x}$$

$$\operatorname{csch} x \sinh x = 1 \qquad \coth^2 x - \operatorname{csch}^2 x = 1$$

$$\tanh x \coth x = 1 \qquad \sinh(-x) = -\sinh x$$

$$\tanh(-x) = -\tanh x,$$

$$\cosh(-x) = \cosh x \qquad \operatorname{sech}(-x) = \operatorname{sech} x \qquad (I.19)$$

$$\coth(-x) = -\coth x \qquad \operatorname{csch}(-x) = -\operatorname{csch} x$$

$$\sinh(x \pm y) = \sinh x \cosh y \pm \cosh x \sinh y$$

$$\cosh(x \pm y) = \cosh x \cosh y \pm \sinh x \sinh y$$

$$e^{\pm x} = \cosh x \pm \sinh x$$

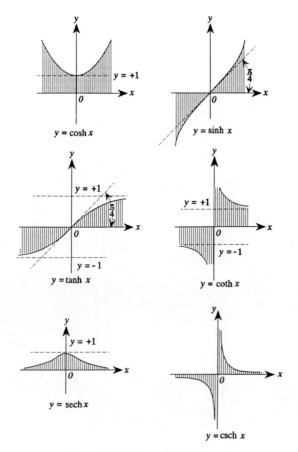

FIGURE I-7 Hyperbolic functions.

I.5 COORDINATE SYSTEMS

Rectangular Coordinate System

The rectangular coordinates of a point P are x, y, z, the distances of P from the yz, xz, and xy planes, respectively (Fig. I-8). If there are three points $P_1, P_2, P_3 = P_1(x_1, y_1, z_1), P_2(x_2, y_2, z_2), P_3(x_3, y_3, z_3)$, the following definitions are useful:

Distance between P_1 and P_2 is calculated as

$$\sqrt{(x_2 - x_1)^2 + (y_2 - y_1)^2 + (z_2 - z_1)^2} \tag{I.20}$$

P_1, P_2, P_3 are colinear if and only if

$$\frac{x_2 - x_1}{x_3 - x_1} = \frac{y_2 - y_1}{y_3 - y_1} = \frac{z_2 - z_1}{z_3 - z_1} \tag{I.21a}$$

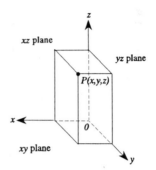

FIGURE I-8 Coordinates.

P_1, P_2, P_3, P_4 are coplanar if and only if a determinant vanishes, that is,

$$
\begin{bmatrix}
x_1 & y_1 & z_1 & 1 \\
x_2 & y_2 & z_2 & 1 \\
x_3 & y_3 & z_3 & 1 \\
x_4 & y_4 & z_4 & 1
\end{bmatrix} = 0
\tag{1.21b}
$$

If point P_3 divides P_1, P_2 as shown in Fig. I-9, P_3 would have the coordinates

$$
\frac{ax_2 + bx_1}{a + b}, \quad \frac{ay_2 + by_1}{a + b}, \quad \frac{az_2 + bz_1}{a + b}
\tag{1.22}
$$

In particular, when $a = b$, P_3 is the midpoint of $P_1 P_2$ given by $\frac{1}{2}(x_1 + x_2), \frac{1}{2}(y_1 + y_2), \frac{1}{2}(z_1 + z_2)$.

The relationships between rectangular and other common coordinate systems are listed in Table I-1.

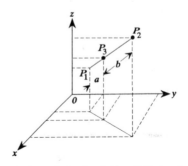

FIGURE I-9 Straight lines. Point P_3 divides $P_1 P_2$ into lengths a and b.

Direction Cosines

The three angles between a line P_1P_2 and the coordinate axes x, y, z are called the *direction angles* of the line, denoted α, β, and γ.

The *direction cosines* of line P_1P_2 are designated

$$n_x = \cos \alpha = \frac{x_2 - x_1}{d}, \qquad n_y = \cos \beta = \frac{y_2 - y_1}{d}, \qquad n_z = \cos \gamma = \frac{z_2 - z_1}{d}$$

$$(I.23)$$

where d is the length of the line (distance between P_1 and P_2). Here x_1, y_1, z_1 are the coordinates of P_1 and x_2, y_2, z_2 are those of P_2.

Identity:

$$\cos^2\alpha + \cos^2\beta + \cos^2\gamma = 1 \qquad (I.24)$$

or

$$n_x^2 + n_y^2 + n_z^2 = 1$$

Basic Formulas in Plane Analytic Geometry

Area of triangle made of three points $P_1(x_1, y_1)$, $P_2(x_2, y_2)$, and $P_3(x_3, y_3)$ in (Fig. I-10):

$$A = \frac{1}{2}\begin{vmatrix} x_1 & y_1 & 1 \\ x_2 & y_2 & 1 \\ x_3 & y_3 & 1 \end{vmatrix}$$

Distance between two points $P_1(x_1, y_1)$ and $P_2(x_2, y_2)$ (Fig. I-11):

$$d = \sqrt{(x_2 - x_1)^2 + (y_2 - y_1)^2}$$

$$\tan \alpha = \frac{y_2 - y_1}{x_2 - x_1}$$

$$\cos \alpha = \frac{x_2 - x_1}{d} \qquad (I.25)$$

$$\cos \beta = \frac{y_2 - y_1}{d}$$

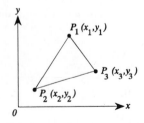

FIGURE I-10   Triangle made of three points P_1, P_2, and P_3.

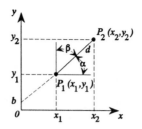

FIGURE I-11   Notation.

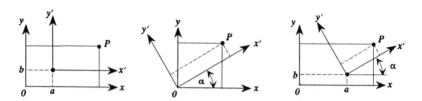

FIGURE I-12   (a) Translation and (b) rotation; (c) translation and rotation.

Equation of a line:

$$y = mx + b, \qquad \text{Slope } m = \tan \alpha \tag{I.26}$$

Translation of coordinates (Fig. I-12a):

$$\begin{aligned} x &= x' + a \quad \text{or} \quad x' = x - a \\ y &= y' + b \quad \text{or} \quad y' = y - b \end{aligned} \tag{I.27}$$

Rotation (Fig. I-12b)

$$\begin{aligned} x &= x' \cos \alpha - y' \sin \alpha \qquad x' = x \cos \alpha + y \sin \alpha \\ &\qquad\qquad\qquad \text{or} \\ y &= x' \sin \alpha + y' \cos \alpha \qquad y' = -x \sin \alpha + y \cos \alpha \end{aligned} \tag{I.28}$$

Translation and rotation (Fig. I-12c):

$$\begin{aligned} x &= x' \cos \alpha - y'\sin \alpha + a \qquad x' = (x - a)\cos \alpha + (y - b)\sin \alpha \\ &\qquad\qquad\qquad \text{or} \\ y &= x' \sin \alpha + y'\cos \alpha + b \qquad y' = -(x - a)\sin \alpha + (y - b)\cos\alpha \end{aligned} \tag{I.29}$$

I.6   QUADRATIC EQUATIONS

$$\begin{aligned} Ax^2 + Bx + C = 0, \qquad x_{1,2} &= \frac{1}{2A}\left(-B \pm \sqrt{B^2 - 4AC}\right) \\ x^2 + px + q = 0, \qquad x_{1,2} &= \tfrac{1}{2}\left(-p \pm \sqrt{p^2 - 4q}\right) \end{aligned} \tag{I.30}$$

I.7   SYSTEM OF LINEAR EQUATIONS

Determinants

$$D = |a_{ik}| = \begin{vmatrix} a_{11} & a_{12} & \cdots & a_{1n} \\ a_{21} & a_{22} & \cdots & a_{2n} \\ \vdots & \vdots & & \vdots \\ a_{n1} & a_{n2} & \cdots & a_{nn} \end{vmatrix} \tag{I.31}$$

where i = row number and k = column number. Here a_{ik} indicates the element in the ith row and the kth column. An exchange of the columns and rows does not affect the value of the determinant:

$$|a_{ik}| = |a_{ki}|$$

An exchange of two rows or two columns changes the sign of the determinant. If all the elements of one row (column) are k times the corresponding elements of another row (column), then $D = 0$. The addition of the elements of one row (column) to the elements of another row (column) does not change the value of the determinant.

The minor D_{ik} of the element a_{ik} is the determinant obtained from D by removing the ith row and kth column from Eq. (I.31). The *cofactor* A_{ik} of the element a_{ik} is its minor multiplied by $(-1)^{i+k}$, or

$$A_{ik} = (-1)^{i+k} \begin{vmatrix} a_{1,1} & \cdots & a_{1,k-1} & a_{1,k+1} & \cdots & a_{1,n} \\ \vdots & \cdots & \vdots & \vdots & \cdots & \vdots \\ a_{i-1,1} & \cdots & a_{i-1,k-1} & a_{i-1,k+1} & \cdots & a_{i-1,n} \\ a_{i+1,1} & \cdots & a_{i+1,k-1} & a_{i+1,k+1} & \cdots & a_{i+1,n} \\ \vdots & & \vdots & \vdots & & \vdots \\ a_{n,1} & \cdots & a_{n,k-1} & a_{n,k+1} & \cdots & a_{n,n} \end{vmatrix} \tag{I.32}$$

where $a_{1,1} = a_{11}$, $a_{1,n} = a_{1n}$, $a_{n,n} = a_{nn}$, etc.

Expansion of Terms of Cofactors   A determinant can be represented in terms of the elements and cofactors of any row j or column j as follows:

$$D = \det(a_{ik}) = |a_{ik}| = \sum_{i=1}^{n} a_{ij}A_{ij} = \sum_{k=1}^{n} a_{jk}A_{jk} \qquad (j = 1, 2, 3, \ldots, n) \tag{I.33}$$

For example,

$$D = \begin{vmatrix} a_{11} & a_{12} \\ a_{21} & a_{22} \end{vmatrix} = a_{11}a_{22} - a_{21}a_{12}$$

$$D = \begin{vmatrix} a_{11} & a_{12} & a_{13} \\ a_{21} & a_{22} & a_{23} \\ a_{31} & a_{32} & a_{33} \end{vmatrix} = a_{11}\begin{vmatrix} a_{22} & a_{23} \\ a_{32} & a_{33} \end{vmatrix} - a_{12}\begin{vmatrix} a_{21} & a_{23} \\ a_{31} & a_{33} \end{vmatrix} + a_{13}\begin{vmatrix} a_{21} & a_{22} \\ a_{31} & a_{32} \end{vmatrix}$$

$$= a_{11}(a_{22}a_{33} - a_{32}a_{23}) - a_{12}(a_{21}a_{33} - a_{31}a_{23}) + a_{13}(a_{21}a_{32} - a_{31}a_{22})$$

$$= a_{11}a_{22}a_{33} + a_{12}a_{23}a_{31} + a_{13}a_{21}a_{32} - a_{12}a_{21}a_{33} - a_{11}a_{23}a_{32} - a_{13}a_{22}a_{31} \tag{I.34}$$

Sarrus Scheme for 3 × 3 Determinant The first two columns are written to the right of the determinant. The three-term products of the main diagonals are summed and the three-term products of the opposing diagonals are subtracted:

$$
|a_{ik}| = a_{11}a_{22}a_{33} + a_{12}a_{23}a_{31} \\
+ a_{13}a_{21}a_{32} - a_{12}a_{21}a_{33} \quad (\text{I.35}) \\
- a_{11}a_{23}a_{32} - a_{13}a_{22}a_{31}
$$

Cramer's Rule

Consider the system of linear equations

$$
\begin{aligned}
a_{11}\cdot x_1 + a_{12}\cdot x_2 + \cdots + a_{1n}\cdot x_n &= b_1 \\
a_{21}\cdot x_1 + a_{22}\cdot x_2 + \cdots + a_{2n}\cdot x_n &= b_2 \\
\vdots \qquad \vdots \qquad \qquad \vdots \quad &\ \ \vdots \\
a_{n1}\cdot x_1 + a_{n2}\cdot x_2 + \cdots + a_{nn}\cdot x_n &= b_n
\end{aligned} \quad (\text{I.36a})
$$

The system of equations has a unique solution if $D = |a_{ik}| \neq 0$.

Cramer's rule provides a solution in the form

$$
x_1 = \frac{D_1}{D}, x_2 = \frac{D_2}{D}, \ldots, x_n = \frac{D_n}{D} \quad (\text{I.36b})
$$

with the determinants

$$
D_1 = \begin{vmatrix} b_1 & a_{12} & \cdots & a_{1n} \\ b_2 & a_{22} & \cdots & a_{2n} \\ \vdots & \vdots & & \vdots \\ b_n & a_{n2} & \cdots & a_{nn} \end{vmatrix}
$$

(first column of D is replaced by right side of system of equations)

$$
D_2 = \begin{vmatrix} a_{11} & b_1 & \cdots & a_{1n} \\ a_{21} & b_2 & \cdots & a_{2n} \\ \vdots & \vdots & & \vdots \\ a_{n1} & b_n & \cdots & a_{nn} \end{vmatrix}
$$

(second column of D is replaced by right side of system of equations)

etc.

Example I.1   Cramer's Rule Find the solution of the following equations:

$$
2.5x_1 - 3.1x_2 = 7.2 \\
1.5x_1 + 4.2x_2 = 5.0
$$

From Eqs. (I.36b),

$$
D = \begin{vmatrix} 2.5 & -3.1 \\ 1. & 4.2 \end{vmatrix} = 2.5 \times 4.2 - (-3.1) \times 1.5 = 15.15
$$

$$
D_1 = \begin{vmatrix} 7.2 & -3.1 \\ 5.0 & 4.2 \end{vmatrix} = 7.2 \times 4.2 - (-3.1) \times 5.0 = 45.74 \quad (1)
$$

$$
D_2 = \begin{vmatrix} 2.5 & 7.2 \\ 1.5 & 5.0 \end{vmatrix} = 2.5 \times 5.0 - 7.2 \times 1.5 = 1.7
$$

we obtain

$$x_1 = \frac{D_1}{D} = 3.01914, \qquad x_2 = \frac{D_2}{D} = 0.1122 \qquad (2)$$

I.8   DIFFERENTIAL AND INTEGRAL CALCULUS

Basic Operations

The derivative of the sum of two functions equals the sum of derivatives:

$$(f + g)' = f' + g'$$

where the superscript prime indicates a derivative. A constant multiplier a is factored out of the derivative,

$$(af)' = a(f')$$

For given constants a and b and functions f and g there exist the following operations:

Addition of functions: $(af + bg)' = af' + bg'$

Product of functions: $(fg)' = f'g + fg'$

Division of functions: $\left(\dfrac{f}{g}\right)' = \dfrac{f'g - fg'}{g^2}$

Function of a function: $\dfrac{d}{dx} f[g(x)] = \dfrac{df(g)}{dg} \dfrac{dg(x)}{dx}$
(composite function)

Inverse function: If $y'(x)$ exists, then $x'(y) = 1/y'(x)$

Refer to Table I-2 for differentiation formulas of common functions.

Differentiation of Function with Multiple Variables

$$df(x_1, x_2, \ldots, x_n) = \left(\frac{\partial f}{\partial x_1}\right) dx_1 + \left(\frac{\partial f}{\partial x_2}\right) dx_2 + \cdots + \left(\frac{\partial f}{\partial x_n}\right) dx_n \quad (I.37a)$$

If $x_2, \ldots, x_n$ are functions of x_1

$$\frac{df(x_1, x_2, \ldots, x_n)}{dx_1} = \frac{\partial f}{\partial x_1} + \frac{\partial f}{\partial x_2} \frac{dx_2}{dx_1} + \cdots + \frac{\partial f}{\partial x_n} \frac{dx_n}{dx_1} \quad (I.37b)$$

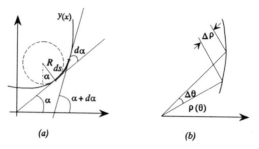

FIGURE I-13 Curvatures.

If the derivative $df(x)/dx$ has the value zero at x_0, i.e.,

$$\left.\frac{df}{dx}\right|_{x=x_0} = 0$$

then f has a *stationary value* at x_0. There are three possible types of such points: a minimum, a maximum, and a point of inflection. To find out which occurs at $x = x_0$:

$$\left.\begin{array}{l}\text{Minimum if } d^2f/dx^2 > 0 \\ \text{Maximum if } d^2f/dx^2 < 0 \\ \text{Point inflection if } d^2f/dx^2 = 0\end{array}\right\} \text{ evaluated at } x = x_0$$

Curvature Formulas

With respect to rectangular coordinates (Fig. I-13a) curvature κ is calculated as

$$\kappa = \lim_{\Delta s \to 0} \frac{\Delta \alpha}{\Delta s} = \frac{d\alpha}{ds} = \frac{y''}{\left(1 + y'^2\right)^{3/2}} \quad \text{where } y = y(x) \qquad (\text{I.38})$$

With respect to polar coordinates (Fig. I-13b)

$$\kappa = \frac{\rho^2 + 2\rho'^2 - \rho\rho''}{\left(\rho^2 + \rho'^2\right)^{3/2}} \quad \text{where } \rho = \rho(\theta)$$

Radius of curvature R is calculated as

$$R = \frac{1}{|\kappa|} = \left|\frac{ds}{d\alpha}\right|$$

With respect to rectangular coordinates

$$R = \left| \frac{\left(1 + y'^2\right)^{3/2}}{y''} \right| \tag{I.39a}$$

With respect to polar coordinates

$$R = \left| \frac{\left(\rho^2 + \rho'^2\right)^{3/2}}{\rho^2 + 2\rho'^2 - \rho\rho''} \right| \tag{I.39b}$$

Integral Theorems

See Table I-3 for common integral formulas.

The integration-by-parts formula

$$\int_a^b u(x)v'(x)\, dx = u(x)v(x)\big|_a^b - \int_a^b v(x)u'(x)\, dx \tag{I.40a}$$

or

$$\int_a^b u\, dv = uv\big|_a^b - \int_a^b v\, du \tag{I.40b}$$

where $u'(x) = du/dx$, $v'(x) = dv/dx$, and $u(x)$ and $v(x)$ must be differentiable for $a \le x \le b$.

Green's Formula

$$\oint_C P(x, y)\, dx + \oint_C Q(x, y)\, dy = \iint_A \left(\frac{\partial Q}{\partial x} - \frac{\partial P}{\partial y} \right) dx\, dy \tag{I.41}$$

where C is the boundary of region A, $\oint_C$ is the line integration along C, and P, Q, $\partial P/\partial y$, and $\partial Q/\partial x$ are continuous in the region A.
Based on Green's formula the integration-by-parts formula for two-dimensional problems can be written as

$$\iint_A R \frac{\partial Q}{\partial x}\, dA = \oint_C QRn_x\, ds - \iint_A Q \frac{\partial R}{\partial x}\, dA \tag{I.42}$$

where $Q = Q(x, y)$, $R = R(x, y)$, and $dA = dx\, dy$, a surface element. The quantity s is a coordinate along the contour of the cross section. Also, n_x is the direction cosine between the outward normal and the x axis.

Gauss Integral Theorem (Divergence Theorem)

$$\int_V \operatorname{div} \mathbf{v}\, dV = \int_S \mathbf{v} \cdot \mathbf{n}\, dS \tag{I.43a}$$

where V is the volume enclosed by surface S, $\mathbf{v} = [v_x\ v_y\ v_z]^T$ is an arbitrary vector differentiable in V with continuous partial derivatives, $\mathbf{n} = [n_x\ n_y\ n_z]^T = [\cos\alpha\ \cos\beta\ \cos\gamma]^T$ is a unit vector at a point on S and is the outward normal to S, and

$$\operatorname{div} \mathbf{v} = \frac{\partial v_x}{\partial x} + \frac{\partial v_y}{\partial y} + \frac{\partial v_z}{\partial z}$$

is the divergence of $\mathbf{v}$. Also, α, β, and γ are the angles between the outward normal and the coordinates x, y, and z, and $\mathbf{v} \cdot \mathbf{n}$ represents the dot product $\mathbf{v} \cdot \mathbf{n} = v_x n_x + v_y n_y + v_z n_z$. On S it is sufficient that the integral exists.

In expanded notation, Eq. (I.43a) becomes

$$\int_V \left(\frac{\partial v_x}{\partial x} + \frac{\partial v_y}{\partial y} + \frac{\partial v_z}{\partial z} \right) dV = \int_S (v_x n_x + v_y n_y + v_z n_z)\, dS \tag{I.43b}$$

where $dV = dx\, dy\, dz$ in rectangular coordinates and dS is a surface element area on S.

I.9 REPRESENTATION OF FUNCTIONS BY SERIES

Taylor's Series for Single Variable

If a function $f(x)$ is continuous and single-valued and has all derivatives on an interval including $x = x_0 + h$, then

$$f(x_0 + h) = f(x_0) + \frac{f'(x_0)}{1!}h + \frac{f''(x_0)}{2!}h^2 + \cdots + \frac{f^{(n)}(x_0)}{n!}h^n + R_n \tag{I.44}$$

where $f' = df/dx$,

$$R_n = \frac{f^{(n+1)}(x_0 + \theta h)}{(n+1)!}h^{n+1}, \qquad 0 < \theta < 1$$

Maclaurin's Series

The Taylor expansion for the special case $x_0 = 0$ and $h = x$ gives Maclaurin's series expansion:

$$f(x) = f(0) + \frac{x}{1!}f'(0) + \frac{x^2}{2!}f''(0) + \cdots + \frac{x^n}{n!}f^{(n)}(0) + R_n \tag{I.45}$$

where

$$R_n = \frac{x^{n+1}}{(n+1)!} f^{(n+1)}(\theta x), \qquad 0 < \theta < 1$$

Taylor's Series for Two Variables

Taylor's series for a function of two variables is

$$f(x+a, y+b) = f(x,y) + \frac{1}{1!} D_1[f(x,y)]$$

$$+ \frac{1}{2!} D_2[f(x,y)] + \cdots + \frac{1}{n!} D_n[f(x,y)] + R_n \quad (\mathrm{I}.46)$$

where

$$D_n[f(x,y)] = \left(a\frac{\partial}{\partial x} + b\frac{\partial}{\partial y}\right)^n [f(x,y)] = \sum_{i=1}^{n} \binom{n}{i} a^{n-i} b^i \frac{\partial^n f(x,y)}{\partial x^{n-i}\,\partial y^i}$$

and

$$R_n = \frac{1}{(n+1)!} D_{n+1}[f(x+\theta_1 a, y + \theta_2 b)]$$

$$0 < \theta_1 < 1, \qquad 0 < \theta_2 < 1$$

Let $x = y = 0$ in the above equation and $a = x$, $b = y$. Then

$$f(x,y) = f(0,0) + \frac{1}{1!} D_1[f(0,0)] + \frac{1}{2!}[f(0,0)] + \cdots + \frac{1}{n!} D_n[f(0,0)] + R_n$$

where

$$D_n = \left(x\frac{\partial}{\partial x} + y\frac{\partial}{\partial y}\right)^n \quad \text{and} \quad R_n = \frac{1}{(n+1)!} D_{n+1}[f(\theta_1 x, \theta_2 y)]$$

$$0 < \theta_1 < 1, \qquad 0 < \theta_2 < 1$$

Fourier Series

Let $f(x)$ be a periodic function in the interval $[-\ell, \ell]$ and let $\int_{-\ell}^{\ell} |f(x)|\,dx$ exist. Then the Fourier series of the function $f(x)$ is

$$f(x) = \frac{a_0}{2} + \sum_{n=1}^{\infty}\left(a_n \cos\frac{n\pi}{\ell}x + b_n \sin\frac{n\pi}{\ell}x\right) \qquad (\mathrm{I}.47)$$

where

$$a_n = \frac{1}{\ell} \int_{-\ell}^{\ell} f(x)\cos\frac{n\pi}{\ell}x\,dx \qquad (n = 0,1,2,\dots)$$

$$b_n = \frac{1}{\ell} \int_{-\ell}^{\ell} f(x)\sin\frac{n\pi}{\ell}x\,dx \qquad (n = 1,2,3,\dots)$$

Series Expansions of Some Common Functions

The series expansion of functions can be obtained by using either the Taylor or Maclaurin theorem. Series expansions of some common functions are given here:

$$e^x = 1 + \frac{x}{1!} + \frac{x^2}{2!} + \cdots + \frac{x^n}{n!} + \cdots \qquad (-\infty < x < +\infty)$$

$$\sin x = x - \frac{x^3}{3!} + \frac{x^5}{5!} - \frac{x^7}{7!} + \cdots \qquad (-\infty < x < +\infty)$$

$$\cos x = 1 - \frac{x^2}{2!} + \frac{x^4}{4!} - \frac{x^6}{6!} + \cdots \qquad (-\infty < x < +\infty)$$

$$\tan x = x + \frac{x^3}{3} + \frac{2x^5}{15} + \frac{17x^7}{315} + \frac{62x^9}{2835} + \cdots \qquad \left(|x| < \tfrac{1}{2}\pi\right)$$

$$\sinh x = x + \frac{x^3}{3!} + \frac{x^5}{5!} + \frac{x^7}{7!} + \cdots \qquad (-\infty < x < +\infty)$$

$$\cosh x = 1 + \frac{x^2}{2!} + \frac{x^4}{4!} + \frac{x^6}{6!} + \cdots \qquad (-\infty < x < \infty) \tag{I.48}$$

$$\ln x = 2\left[\frac{x-1}{x+1} + \frac{1}{3}\left(\frac{x-1}{x+1}\right)^3 + \frac{1}{5}\left(\frac{x-1}{x+1}\right)^5 + \cdots\right] \qquad (0 < x < +\infty)$$

$$\ln x = \frac{x-1}{x} + \frac{1}{2}\left(\frac{x-1}{x}\right)^2 + \frac{1}{3}\left(\frac{x-1}{x}\right)^3 + \cdots \qquad (x > \tfrac{1}{2})$$

$$\ln x = (x-1) - \tfrac{1}{2}(x-1)^2 + \tfrac{1}{3}(x-1)^3 - \cdots \qquad (0 < x \le 2)$$

$$\ln(1+x) = x - \tfrac{1}{2}x^2 + \tfrac{1}{3}x^3 - \tfrac{1}{4}x^4 + \cdots \qquad (-1 < x < +1)$$

$$(1 \pm x)^n = 1 \pm nx + \frac{n(n-1)x^2}{2!} \pm \frac{n(n-1)(n-2)x^3}{3!} + \cdots$$

$$(|x| < 1, n > 0)$$

$$(1 \pm x)^{-n} = 1 \mp nx + \frac{n(n+1)x^2}{2!} \mp \frac{n(n+1)(n+2)x^3}{3!} + \cdots$$

$$(|x| < 1, n > 0)$$

where n is a real number.

I.10   MATRIX ALGEBRA

Definitions

A matrix is an array of elements consisting of m rows and n columns:

$$\mathbf{A} = \begin{bmatrix} a_{11} & a_{12} & \cdots & a_{1n} \\ a_{21} & a_{22} & \cdots & a_{2n} \\ \vdots & \vdots & & \vdots \\ a_{m1} & a_{m2} & \cdots & a_{mn} \end{bmatrix} \tag{I.49}$$

If $m = n$, the matrix is called a square matrix; otherwise it is a rectangular matrix. The principal diagonal is the diagonal spanning between the upper left corner and the lower right corner.

The transpose of a matrix $\mathbf{A}$, denoted by $\mathbf{A}^T$, is the matrix with elements defined as

$$a_{ij}^T = a_{ji}$$

That is, the rows have become corresponding columns and the columns have become corresponding rows.

> The *symmetric* square matrix is the matrix with the property $\mathbf{A} = \mathbf{A}^T$, or $a_{ij} = a_{ji}$.
> The *skew-symmetric* (or antisymmetric) square matrix is the matrix with the property $\mathbf{A} = -\mathbf{A}^T$, or $a_{ij} = -a_{ji}$.
> A *lower triangular matrix* has all elements equal to zero above the principal diagonal.
> An *upper triangular matrix* has all zeros as elements below the principal diagonal.
> A *diagonal matrix* has all elements equal to zero except those along the principal diagonal.
> A *null matrix* is a matrix with all its elements equal to zero.
> An *identity matrix* is a diagonal matrix with all its diagonal elements equal to 1.

Singularity, Inverse, and Rank

A matrix $\mathbf{A}$ is *singular* if its determinant is zero, $|\mathbf{A}| = 0$. Otherwise it is *nonsingular*. $\tag{I.50}$

If $\mathbf{A}$ is a square matrix and its determinant $|\mathbf{A}| \neq 0$ (nonsingular), the matrix $\mathbf{A}^{-1}$ satisfying the relation $\mathbf{A}\mathbf{A}^{-1} = \mathbf{A}^{-1}\mathbf{A} = \mathbf{I}$ is called the *inverse* matrix. The inverse matrix is unique.

The order of the largest nonzero determinant that can be obtained from the elements of a matrix is called the *rank* of the matrix. The *trace* of a square matrix $\mathbf{A}$ is the sum of the diagonal elements:

$$\operatorname{tr}\mathbf{A} = \sum_{i=1}^{n} a_{ii}$$

Laws

$$\mathbf{A} + \mathbf{B} = \mathbf{B} + \mathbf{A}, \qquad \mathbf{AB} \neq \mathbf{BA},$$
$$\mathbf{A}[\mathbf{B} + \mathbf{C}] = \mathbf{AB} + \mathbf{AC}, \qquad [\mathbf{A} + \mathbf{B}]\mathbf{C} = \mathbf{AC} + \mathbf{BC}$$
$$\mathbf{A} + (\mathbf{B} + \mathbf{C}) = (\mathbf{A} + \mathbf{B}) + \mathbf{C}, \qquad \mathbf{A}(\mathbf{BC}) = (\mathbf{AB})\mathbf{C},$$
$$(\mathbf{A} + \mathbf{B})^{\mathsf{T}} = \mathbf{A}^{\mathsf{T}} + \mathbf{B}^{\mathsf{T}}, \qquad (\mathbf{A} + \mathbf{B})^{-1} = \mathbf{A}^{-1} + \mathbf{B}^{-1}$$
$$(\mathbf{AB})^{\mathsf{T}} = \mathbf{B}^{\mathsf{T}}\mathbf{A}^{\mathsf{T}}, \qquad (\mathbf{AB})^{-1} = \mathbf{B}^{-1}\mathbf{A}^{-1}, \qquad (\mathbf{B}^{\mathsf{T}})^{\mathsf{T}} = \mathbf{B}, \qquad (\mathbf{B}^{-1})^{-1} = \mathbf{B}$$
$$\operatorname{tr}(\mathbf{AB}) = \operatorname{tr}(\mathbf{BA}), \qquad \operatorname{tr}(\mathbf{ABC}) = \operatorname{tr}(\mathbf{BCA}) = \operatorname{tr}(\mathbf{CAB}) \tag{I.51}$$

$$\mathbf{AI} = \mathbf{IA} = \mathbf{A} \quad \text{where } \mathbf{I} = \begin{bmatrix} 1 & 0 & \cdots & 0 \\ 0 & 1 & \cdots & 0 \\ \vdots & \vdots & & \vdots \\ 0 & 0 & \cdots & 1 \end{bmatrix}, \qquad \mathbf{I}^{-1} = \mathbf{I}$$

$$a(\mathbf{AB}) = (a\mathbf{A})\mathbf{B} = \mathbf{A}(a\mathbf{B}), \qquad a(\mathbf{A} + \mathbf{B}) = a\mathbf{A} + a\mathbf{B}, \qquad [0] = \begin{bmatrix} 0 & \cdots & 0 \\ \vdots & & \vdots \\ 0 & \cdots & 0 \end{bmatrix}$$

$$(a + b)\mathbf{A} = a\mathbf{A} + b\mathbf{A}, \qquad a(b\mathbf{A}) = (ab)\mathbf{A}$$

where a, b are real numbers

$$a\mathbf{A} = \mathbf{A}a \qquad (a \cdot b \neq 0)$$

If $\mathbf{AB} = 0$, then $\mathbf{A}$ and/or $\mathbf{B}$ may or may not be zero.

Two square matrices $\mathbf{A}$ and $\mathbf{B}$ related by a transformation $\mathbf{A} = \mathbf{T}^{\mathsf{T}}\mathbf{BT}$, where $\mathbf{T}$ is nonsingular, are *congruent*. The congruent transformation has the property that if $\mathbf{B}$ is symmetric, $\mathbf{A}$ will be symmetric.

Basic Operations

Addition / Subtraction

$$\mathbf{A}_{mn} \pm \mathbf{B}_{mn} = [a_{ij}] \pm [b_{ij}] = \begin{bmatrix} a_{11} \pm b_{11} & a_{12} \pm b_{12} & \cdots & a_{in} \pm b_{in} \\ \vdots & \vdots & & \vdots \\ a_{m1} \pm b_{m1} & a_{m2} \pm b_{m2} & \cdots & a_{mn} \pm b_{mn} \end{bmatrix}$$
$$\tag{I.52a}$$

where $\mathbf{A}_{mn}$ and $\mathbf{B}_{mn}$ are matrices $\mathbf{A}$ and $\mathbf{B}$, each consisting of m rows and n columns.

$$\mathbf{A}_{mn} = [a_{ij}], \qquad \mathbf{B}_{mn} = [b_{ij}]$$

Multiplication of Two Matrices

$$\mathbf{A}_{mn} \cdot \mathbf{B}_{nl} = \mathbf{C}_{ml} = \left[C_{ij} \right]$$

where

$$C_{ij} = \sum_{k=1}^{n} a_{ik} b_{kj} \quad (i = 1, 2, \ldots, m, \; j = 1, 2, \ldots, l) \quad \text{(I.52b)}$$

Inverse Matrix

$$\mathbf{A}^{-1} = \begin{bmatrix} a_{11} & a_{12} & \cdots & a_{1n} \\ a_{21} & a_{22} & \cdots & a_{2n} \\ \vdots & \vdots & & \vdots \\ a_{n1} & a_{n2} & \cdots & a_{nn} \end{bmatrix}^{-1} = \frac{1}{|\mathbf{A}|} \begin{bmatrix} A_{11} & A_{21} & \cdots & A_{n1} \\ A_{12} & A_{22} & \cdots & A_{n2} \\ \vdots & \vdots & & \vdots \\ A_{1n} & A_{2n} & \cdots & A_{nn} \end{bmatrix} = \frac{\text{adj } \mathbf{A}}{|\mathbf{A}|}$$

$$\text{(I.53)}$$

where

$$|\mathbf{A}| = \det \mathbf{A}, \quad A_{jk} = \text{cofactor of element } a_{jk} \quad [\text{Eq. (I.32)}]$$

and the transpose of the matrix whose elements are A_{jk} is called the *adjoint matrix* adj $\mathbf{A}$.

Example I.2 Inverse of a Matrix Determine the inverse matrix of

$$\mathbf{A} = \begin{bmatrix} 2.5 & -3.1 \\ 1.5 & 4.2 \end{bmatrix}$$

First we find

$$A_{11} = (-1)^{1+1} |4.2| = 4.2, \quad A_{21} = (-1)^{2+1} |-3.1| = -(-3.1) = 3.1$$

$$A_{12} = (-1)^{2+1} |1.5| = -1.5, \quad A_{22} = (-1)^{2+2} |2.5| = 2.5$$

$$|\mathbf{A}| = \begin{vmatrix} 2.5 & -3.1 \\ 1.5 & 4.2 \end{vmatrix} = 15.15$$

From Eq. (I.53),

$$\mathbf{A}^{-1} = \frac{\text{adj } \mathbf{A}}{|\mathbf{A}|} = \frac{1}{|\mathbf{A}|} \begin{bmatrix} A_{11} & A_{21} \\ A_{12} & A_{22} \end{bmatrix}$$

$$= \frac{1}{15.15} \begin{bmatrix} 4.2 & 3.1 \\ -1.5 & 2.5 \end{bmatrix} = \begin{bmatrix} 0.2772 & 0.2046 \\ -0.0990 & 0.1650 \end{bmatrix} \quad \text{(1)}$$

The inverse matrix is often used to solve linear equations of the form

$$\mathbf{Ax} = \mathbf{b} \tag{I.54}$$

$$\mathbf{A} = \begin{bmatrix} a_{11} & \cdots & a_{1n} \\ \vdots & & \vdots \\ a_{n1} & \cdots & a_{nn} \end{bmatrix}, \quad |\mathbf{A}| \neq 0, \quad \mathbf{x} = \begin{bmatrix} x_1 \\ x_2 \\ \vdots \\ x_n \end{bmatrix}, \quad \mathbf{b} = \begin{bmatrix} b_1 \\ b_2 \\ \vdots \\ b_n \end{bmatrix} \tag{I.55}$$

$$\mathbf{x} = \mathbf{A}^{-1}\mathbf{b}$$

Example I.3   Solution of a System of Equations   Solve the problem of Example I.1 by computing the inverse matrix i.e., use Eq. (I.55) and Eq. (1) of Example I.2. From Example I.1

$$\mathbf{A} = \begin{bmatrix} 2.5 & -3.1 \\ 1.5 & 4.2 \end{bmatrix}$$

and from Example I.2

$$\mathbf{A}^{-1} = \begin{bmatrix} 0.2772 & 0.2046 \\ -0.0990 & 0.1650 \end{bmatrix}$$

Then

$$\mathbf{x} = \begin{bmatrix} x_1 \\ x_2 \end{bmatrix} = \mathbf{A}^{-1}\mathbf{b} = \begin{bmatrix} 0.2772 & 0.2046 \\ -0.0990 & 0.1650 \end{bmatrix}\begin{bmatrix} b_1 \\ b_2 \end{bmatrix}$$

$$= \begin{bmatrix} 0.2772 & 0.2046 \\ -0.0990 & 0.1650 \end{bmatrix}\begin{bmatrix} 7.2 \\ 5.0 \end{bmatrix} = \begin{bmatrix} 3.01884 \\ 0.1122 \end{bmatrix}$$

Determinants
$\mathbf{A} = \mathbf{A}_{nn}$ matrix:

$$|\mathbf{A}^T| = |\mathbf{A}|, \quad |\mathbf{A}^{-1}| = \frac{1}{|\mathbf{A}|}, \quad |\mathbf{I}| = 1$$

$$|\mathbf{AB}| = |\mathbf{A}||\mathbf{B}|, \quad |\mathbf{AB}^{-1}| = |\mathbf{A}|/|\mathbf{B}| \tag{I.56}$$

$$|\mathbf{A}^{-1}\mathbf{B}^{-1}| = 1/(|\mathbf{A}||\mathbf{B}|)$$

Eigenvalues and Eigenvectors

If $\mathbf{A}$ is a matrix of order n, then the equation

$$|\mathbf{A} - \lambda\mathbf{I}| = 0$$

is called the characteristic equation of the matrix $\mathbf{A}$, and it is a polynomial of degree n in λ. The roots of this equation are called *eigenvalues* of $\mathbf{A}$ or the *characteristic values* of $\mathbf{A}$.

Any vector $\mathbf{x}$ satisfying $\mathbf{Ax} = \lambda\mathbf{x}$ corresponding to the characteristic values λ is called a characteristic vector or eigenvector of $\mathbf{A}$. Often the characteristic vector is normalized to have a unit length, $\mathbf{x}^T\mathbf{x} = 1$.

Example I.4 Eigenvalue Problem Determine the eigenvalues and their corresponding eigenvectors of the square matrix

$$\mathbf{A} = \begin{bmatrix} -2 & 4 \\ -6 & 8 \end{bmatrix}$$

From $|\mathbf{A} - \lambda\mathbf{I}| = 0$, the characteristic equation is

$$\begin{vmatrix} -2 - \lambda & 4 \\ -6 & 8 - \lambda \end{vmatrix} = (-2 - \lambda)(8 - \lambda) - (4)(-6) = (\lambda - 2)(\lambda - 4) = 0$$

The eigenvalues are the solutions of the equation

$$\lambda_1 = 2, \qquad \lambda_2 = 4$$

Let the eigenvector be

$$\mathbf{x} = \begin{bmatrix} x_1 \\ x_2 \end{bmatrix}$$

Substitute λ_1, λ_2 into the equation $\mathbf{Ax} = \lambda\mathbf{x}$:

$$\begin{bmatrix} -2 & 4 \\ -6 & 8 \end{bmatrix}\begin{bmatrix} x_1 \\ x_2 \end{bmatrix} = \lambda\begin{bmatrix} x_1 \\ x_2 \end{bmatrix}$$

i.e.,

$$(-2 - \lambda)x_1 + 4x_2 = 0 \tag{2}$$
$$-6x_1 + (8 - \lambda)x_2 = 0$$

For $\lambda = 2$

$$-4x_1 + 4x_2 = 0$$
$$-6x_1 + 6x_2 = 0$$

It is clear that $x_1 = x_2$ satisfies these two equations. There are an infinite number of solutions and an additional condition is necessary for the solutions to be unique. For example, if the eigenvectors are chosen such that $\mathbf{x}^T\mathbf{x} = 1$,

$$[x_1 \ x_2]\begin{bmatrix} x_1 \\ x_2 \end{bmatrix} = x_1^2 + x_2^2 = 1$$

Hence $x_1 = x_2 = \sqrt{2}/2$, or

$$\mathbf{x} = \begin{bmatrix} x_1 \\ x_2 \end{bmatrix} = \begin{bmatrix} \sqrt{2}/2 \\ \sqrt{2}/2 \end{bmatrix} \tag{3}$$

Similarly, for $\lambda = 4$, the eigenvector is

$$\mathbf{x} = \begin{bmatrix} x_1 \\ x_2 \end{bmatrix} = \begin{bmatrix} 2\sqrt{13}/13 \\ 3\sqrt{13}/13 \end{bmatrix} \tag{4}$$

I.11 NUMERICAL METHODS

Linear Interpolation Method to Solve $f(x) = 0$

The roots of an algebraic equation are desired. If $f(x)$ is a continuous function and two values x_{10}, x_{20} ($x_{10} < x_{20}$) are chosen such that $f(x_{10})$ and $f(x_{20})$ are of opposite sign (Fig. I-14), then the line segment between $[x_{10}, f(x_{10})]$ and $[x_{20}, f(x_{20})]$ intersects the x axis at

$$x_1 = \frac{x_{10}f(x_{20}) - x_{20}f(x_{10})}{f(x_{20}) - f(x_{10})} \tag{I.57}$$

Next x_1 is taken as x_{20}, and the same iteration is followed for x_{10} and x_1 to find x_2, then x_3, etc., until x_n is obtained close to the x_{n-1} value within the acceptable tolerance, which means $x_n \approx x^*$, $f(x^*) = 0$ (Fig. I-14).

Newton's Method

Assume x_0 is an approximate root of $f(x) = 0$. Calculate a better approximation by means of

$$x_1 = x_0 - \frac{f(x_0)}{f'(x_0)} \tag{I.58}$$

where $f' = df/dx$, and then use x_1 as x_0 and repeat this process until a desired accuracy is achieved. The general form of Eq. (I.58) can be written as

$$x_{n+1} = x_n - \frac{f(x_n)}{f'(x_n)}, \qquad n = 0, 1, 2, \cdots \tag{I.59}$$

and $x_n \to x^*$. Note that Newton's method is used to find only the real roots of $f(x) = 0$, not the complex roots.

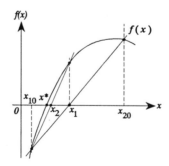

FIGURE I-14 Linear interpolation.

Zeros of a Polynomial

A polynomial of degree n can be expressed as

$$f(x) = a_n x^n + a_{n-1} x^{n-1} + \cdots + a_1 x + a_0 \qquad (I.60)$$

where $a_n \neq 0$, $a_{n-1}, \ldots, a_0$ are real or complex numbers, and n is a positive integer.

Based on Newton's method, an iterative approximation for the location of a zero, i.e., the roots of the polynomial equation $f = 0$, beginning with the approximate value x_0, can be obtained using

$$x_{i+1} = x_i - \frac{a_n x_i^n + a_{n-1} x_i^{n-1} + \cdots + a_1 x_i + a_0}{n a_n x_i^{n-1} + (n-1) a_{n-1} x_i^{n-2} + \cdots + a_1}, \qquad i = 0, 1, 2, \cdots$$

$$(I.61)$$

According to a fundamental theorem of algebra, Eq. (I.60) has exactly n zero locations (roots), which may be real, complex, and not necessarily distinct. Equation (I.61) is used for approximating real roots.

Example I.5   Zeros of a polynomial   Find the location of a zero of $y = 2x^3 - 3x^2 + x - 1$. From Eq. (I.61)

$$x_{i+1} = x_i - \frac{2x_i^3 - 3x_i^2 + x_i - 1}{6x_i^2 - 6x_i + 1}$$

Choose $x_0 = 2.0$. Then

$$x_1 = 1.615385, \qquad x_2 = 1.440559, \qquad x_3 = 1.400239$$
$$x_4 = 1.398166, \qquad x_5 = 1.398161, \qquad x_6 = 1.398161$$

Further iteration will not change the location significantly. Hence, it is concluded that the location of the zero is about 1.398161. In this example, there is only one real root, but in other cases there may be more than one real root. These roots can be found using the same method by beginning with other approximate values. Complex roots cannot be found by using Eq. (I.61) or Newton's method.

Gauss Algorithm

Gauss elimination solves a system of linear equations [Eqs. (I.36a)] using the following method: Multiply the first equation of the system by the coefficient $-a_{21}/a_{11}$ and then add this equation to the second equation. The first coefficient of the second equation, denoted by a'_{21}, is now zero. This process continues similarly with the other equations until $a'_{31}, \ldots, a'_{n1}$ are zero. That is, the second equation and each following equation does not contain x_1 since its coefficient is zero. Now begin with the second equation to eliminate the coefficient of x_2 in the remaining equations. In short, for the nth equation in such a system, its first $n - 1$ coefficients are zero after Gauss elimination. The end result is a system which takes the form

$$
\begin{aligned}
a_{11}x_1 + a_{12}x_2 + a_{13}x_3 + \cdots + a_{1n}x_n &= b_1 \\
a'_{22}x_2 + a'_{23}x_3 + \cdots + a'_{2n}x_n &= b'_2 \\
a''_{33}x_3 + \cdots + a''_{3n}x_n &= b''_3 \\
\ddots \quad \vdots \quad &\vdots \\
a_{nn}^{(n-1)}x_n &= b_n^{(n-1)}
\end{aligned}
\tag{I.62}
$$

Now solve the final equation for x_n, then the second to the last equation for x_{n-1}, etc.

It can be seen that Gauss elimination contains two phases, forward elimination and backward substitution.

Example I.6   Gauss Elimination   Solve the following system of linear equations by Gauss elimination:

$$
\begin{aligned}
5x_1 + 2x_2 + 3x_3 + 2x_4 &= -1 \tag{1} \\
2x_1 + 4x_2 + x_3 - 2x_4 &= 5 \tag{2} \\
x_1 - 3x_2 + 4x_3 + 3x_4 &= 4 \tag{3} \\
3x_1 + 2x_2 + 2x_3 + 8x_4 &= -6 \tag{4}
\end{aligned}
$$

First, use forward elimination to transform (1) to (4) into the form of Eq. (I.62). The initial step is to eliminate the first unknown x_1 from (2)–(4):

$$(2) - (1) \times \tfrac{2}{5}: \qquad \tfrac{16}{5}x_2 - \tfrac{1}{5}x_3 - \tfrac{14}{5}x_4 = \tfrac{27}{5}$$

or

$$16x_2 - x_3 - 14x_4 = 27 \tag{5}$$

where $(2) - (1) \times \tfrac{2}{5}$ means Eq. (2) minus Eq. (1) multiplied by $\tfrac{2}{5}$. Similarly,

$$(3) - (1) \times \tfrac{1}{5}: \qquad -\tfrac{17}{5}x_2 + \tfrac{17}{5}x_3 + \tfrac{13}{5}x_4 = \tfrac{21}{5}$$

or

$$-17x_2 + 17x_3 + 13x_4 = 21 \tag{6}$$

$$(4) - (1) \times \tfrac{3}{5}: \qquad \tfrac{4}{5}x_2 + \tfrac{1}{5}x_3 + \tfrac{34}{5}x_4 = -\tfrac{27}{5}$$

or

$$4x_2 + x_3 + 34x_4 = -27 \tag{7}$$

It is seen that there are three unknowns in (5), (6) and (7). Keep (5) and eliminate unknown x_2 from (6) and (7),

$$(6) - (5) \times \left(-\tfrac{17}{16}\right): \qquad \tfrac{255}{16}x_3 - \tfrac{30}{16}x_4 = \tfrac{795}{16}$$

or

$$17x_3 - 2x_4 = 53 \qquad (8)$$

$$(7) - (5) \times \tfrac{4}{16}: \qquad \tfrac{5}{16}x_3 + \tfrac{150}{16}x_4 = -135$$

or

$$x_3 + 30x_4 = -27 \qquad (9)$$

Finally, from (8) and (9), eliminate the unknown x_3 in (9),

$$(9) - (8) \times \tfrac{1}{17}: \qquad 512x_4 = -512 \qquad (10)$$

Consequently, (1), (5), (8), and (10) together lead to the form of Eq. (I.62):

$$\begin{aligned}
5x_1 + 2x_2 + 3x_3 + 2x_4 &= -1 \\
16x_2 - x_3 - 14x_4 &= 27 \\
17x_3 - 2x_4 &= 53 \\
512x_4 &= -512
\end{aligned} \qquad (11)$$

Use back substitution to solve for the unknowns from (11). This gives $x_4 = -1$, $x_3 = 3$, $x_2 = 1$, $x_1 = -2$.

===

Numerical Integration

Trapezoidal rule (n is even or odd) (Fig. I-15): Define $h = \dfrac{b-a}{n}$, $y_i = f(x_i)$

$$\int_a^b y(x)\,dx = \int_a^b f(x)\,dx \approx \tfrac{1}{2}h(y_0 + 2y_1 + 2y_2 + \cdots + 2y_{n-1} + y_n) + \varepsilon_T$$

$$(I.63)$$

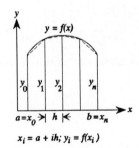

$$x_i = a + ih;\ y_i = f(x_i)$$

FIGURE I-15   Trapezoidal rule.

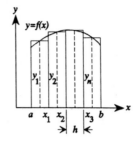

FIGURE I-16 Rectangular formula.

Truncation error:

$$\varepsilon_T \approx -\tfrac{1}{12}\left[nh^3 f''(\xi)\right], \qquad a \le \xi \le b$$

Rectangular formula (Fig. I-16): Define

$$h = \frac{b-a}{n}, \qquad y_i = f(x_i), \qquad x_i = a + \tfrac{1}{2}(2i-1)h$$

$$\int_a^b y\,dx \approx h(y_1 + y_2 + \cdots + y_n) \tag{1.64}$$

Simpson's rule [n is even] (Fig. I-17): Define $h = \dfrac{b-a}{n}$

$$\int_a^b y(x)\,dx = \int_a^b f(x)\,dx \approx \tfrac{1}{3}h(y_0 + 4y_1 + 2y_2 + 4y_3 + \cdots + 4y_{n-3}$$
$$+ 2y_{n-2} + 4y_{n-1} + y_n) \tag{1.65}$$

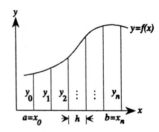

FIGURE I-17 Simpson's rule.

Tables

TABLE I-1   RELATIONSHIPS BETWEEN COMMON COORDINATE SYSTEMS

Translation: If the rectangular coordinate system (x, y, z) translates to a point $0'(a, b, c)$ and forms a new rectangular coordinae system (x', y', z'), then

$$x = x' + a, \qquad y = y' + b, \qquad z = z' + c$$

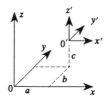

Rotation: If the new system x', y', z' rotates about the origin and has the following relationship with the old one

New Axis	Direction Cosine with Old Axis		
	x	y	z
x'	l_1	m_1	n_1
y'	l_2	m_2	n_2
z'	l_3	m_3	n_3

then

$$x = l_1 x' + l_2 y' + l_3 z'$$
$$y = m_1 x' + m_2 y' + m_3 z'$$
$$z = n_1 x' + n_2 y' + n_3 z'$$

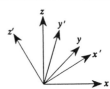

A *cylindrical coordinate system* (r, θ, z) has the following relationship with the Cartesian (rectangular) coordinates (x, y, z):

$$x = r \cos \theta, \qquad r = \sqrt{x^2 + y^2}$$
$$y = r \sin \theta, \qquad \theta = \arctan(y/x)$$
$$z = z \qquad\qquad z = z$$

A *spherical coordinate system* (ρ, θ, ϕ) has the following relationship with the Cartesian (rectangular) coordinates (x, y, z):

$$x = \rho \cos \theta \sin \phi$$
$$y = \rho \sin \theta \sin \phi$$
$$z = \rho \cos \phi$$
$$\phi = \arccos \frac{z}{\sqrt{x^2 + y^2 + z^2}}$$
$$\theta = \arctan \frac{y}{x}$$
$$\rho^2 = x^2 + y^2 + z^2$$

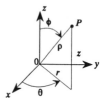

TABLE I-1 | Relationships between Coordinate Systems   1316

TABLE I-2   DIFFERENTIATION FORMULAS

$$a = \text{constant} \neq 0$$

$$\frac{da}{dx} = 0, \qquad da = 0, \qquad \frac{d(x)}{dx} = 1, \qquad d(x) = dx$$

$$\frac{d}{dx}(u + v - w) = \frac{du}{dx} + \frac{dv}{dx} - \frac{dw}{dx}, \qquad d(u + v - w) = du + dv - dw$$

$$\frac{d}{dx}(av) = a\frac{dv}{dx}, \qquad d(av) = a\,dv$$

$$\frac{d}{dx}(uv) = u\frac{dv}{dx} + v\frac{du}{dx}, \qquad d(uv) = u\,dv + v\,du$$

$$\frac{d}{dx}(v^n) = nv^{n-1}\frac{dv}{dx}, \qquad \frac{d}{dx}(x^n) = nx^{n-1}, \qquad d(v^n) = nv^{n-1}\,dv$$

$$\frac{d}{dx}\left(\frac{u}{v}\right) = \frac{v(du/dx) - u(dv/dx)}{v^2}, \qquad d\left(\frac{u}{v}\right) = \frac{v\,du - u\,dv}{v^2}$$

$$\frac{d}{dx}\left(\frac{a}{v}\right) = -\frac{a(dv/dx)}{v^2}, \qquad d\left(\frac{a}{v}\right) = -\frac{a\,dv}{v^2}$$

$$d\ln(v_1 v_2 \cdots v_n) = \frac{dv_1}{v_1} + \frac{dv_2}{v_2} + \cdots + \frac{dv_n}{v_n}$$

$$\frac{dy}{dx} = \frac{dy}{dv}\frac{dv}{dx}, \qquad \frac{dy}{dx} = \frac{1}{dx/dy}, \qquad \frac{dy}{dx} = \frac{dy/dt}{dx/dt}$$

$$\frac{d^n}{dx^n}(uv) = \frac{d^n u}{dx^n}v + n\frac{d^{n-1}u}{dx^{n-1}}\frac{dv}{dx} + \frac{n(n-1)}{2!}\frac{d^{n-2}u}{dx^{n-2}}\frac{d^2 v}{dx^2}$$

$$+ \frac{n(n-1)(n-2)}{3!}\frac{d^{n-3}u}{dx^{n-3}}\frac{d^3 v}{dx^3} + \cdots + u\frac{d^n v}{dx^n}$$

$$\frac{d}{dx}(\sin u) = \cos u\frac{du}{dx}, \qquad \frac{d}{dx}(\cos u) = -\sin u\frac{du}{dx}$$

$$\frac{d}{dx}(\tan u) = \sec^2 u\frac{du}{dx}, \qquad \frac{d}{dx}(\cot u) = -\csc^2 u\frac{du}{dx}$$

$$\frac{d}{dx}(\sec u) = \sec u \tan u\frac{du}{dx}, \qquad \frac{d}{dx}(\csc u) = -\csc u \cot u\frac{du}{dx}$$

$$\frac{d}{dx}(\sinh u) = \cosh u\frac{du}{dx}, \qquad \frac{d}{dx}(\cosh u) = \sinh u\frac{du}{dx}$$

$$\frac{d}{dx}(\tanh u) = \text{sech}^2 u\frac{du}{dx}, \qquad \frac{d}{dx}(\coth u) = -\text{csch}^2 u\frac{du}{dx}$$

$$\frac{d}{dx}(\text{sech } u) = -\tanh u \,\text{sech } u\frac{du}{dx}, \qquad \frac{d}{dx}(\text{csch } u) = -\coth u \,\text{csch } u\frac{du}{dx}$$

TABLE I-2 (continued)   DIFFERENTIATION FORMULAS

$$\frac{d}{dx}\left(\arcsin\frac{u}{a}\right) = \frac{1}{\sqrt{a^2 - u^2}}\frac{du}{dx}, \qquad \frac{d}{dx}\left(\arccos\frac{u}{a}\right) = -\frac{1}{\sqrt{a^2 - u^2}}\frac{du}{dx}$$

$$\frac{d}{dx}\left(\arctan\frac{u}{a}\right) = \frac{a}{a^2 + u^2}\frac{du}{dx}, \qquad \frac{d}{dx}\left(\text{arccot}\frac{u}{a}\right) = -\frac{a}{a^2 + u^2}\frac{du}{dx}$$

$$\frac{d}{dx}\left(\text{arcsec}\frac{u}{a}\right) = \frac{a}{|u|\sqrt{u^2 - a^2}}\frac{du}{dx}, \qquad \frac{d}{dx}\left(\text{arccsc}\frac{u}{a}\right) = -\frac{a}{|u|\sqrt{u^2 - a^2}}\frac{du}{dx}$$

$$\frac{d}{dx}(u^v) = u^v(\ln u)\frac{dv}{dx} + vu^{v-1}\frac{du}{dx}$$

$$\frac{d}{dx}(\text{arcsinh } u) = \frac{1}{\sqrt{1 + u^2}}\frac{du}{dx},$$

$$\frac{d}{dx}(\text{arccosh } u) = \pm\frac{1}{\sqrt{u^2 - 1}}\frac{du}{dx}, \qquad u > 1$$

$$\frac{d}{dx}(\text{arctanh } u) = \frac{1}{1 - u^2}\frac{du}{dx}, \qquad |u| < 1$$

$$\frac{d}{dx}(\text{arccoth } u) = \frac{1}{1 - u^2}\frac{du}{dx}, \qquad |u| > 1$$

$$\frac{d}{dx}(\text{arcsech } u) = \pm\frac{1}{u\sqrt{1 - u^2}}\frac{du}{dx}, \qquad 0 < u < 1$$

$$\frac{d}{dx}(\text{arccsch } u) = -\frac{1}{|u|\sqrt{1 + u^2}}\frac{du}{dx}$$

$$\frac{d}{dx}\log_a u = \frac{1}{\ln a}\frac{1}{u}\frac{du}{dx}, \qquad \frac{d}{dx}\ln u = \frac{1}{u}\frac{du}{dx}$$

$$\frac{d}{dx}(a^u) = (\ln a)a^u\frac{du}{dx}, \qquad \frac{d}{dx}(e^u) = e^u\frac{du}{dx}$$

TABLE I-2   **Differentiation Formulas**   1318

TABLE I-3   INTEGRAL FORMULAS

a, b = constants c = constant of integration

$$\int a \, dx = ax \qquad \int af(x) \, dx = a \int f(x) \, dx$$

$$\int \phi(y) \, dx = \int \frac{\phi(y)}{y'} \, dy \quad \text{where } y' = \frac{dy}{dx}$$

$$\int (u + v) \, dx = \int u \, dx + \int v \, dx \quad \text{where } u \text{ and } v \text{ are any functions of } x$$

$$\int u \, dv = u \int dv - \int v \, du = uv - \int v \, du$$

$$\int u \frac{dv}{dx} \, dx = uv - \int v \frac{du}{dx} \, dx$$

$$\int x^n \, dx = \frac{x^{n+1}}{n+1} + c \quad \text{except } n = -1$$

$$\int \frac{f'(x) \, dx}{f(x)} = \ln f(x) + c, \qquad df(x) = f'(x) \, dx$$

$$\int \frac{dx}{x} = \ln x + c$$

$$\int \frac{f'(x) \, dx}{2\sqrt{f(x)}} = \sqrt{f(x)} + c, \qquad df(x) = f'(x) \, dx$$

$$\int e^x \, dx = e^x + c \qquad \int e^{ax} \, dx = e^{ax}/a + c$$

$$\int b^{ax} \, dx = \frac{b^{ax}}{a \ln b} + c \qquad (b > 0)$$

$$\int \ln x \, dx = x \ln x - x + c \qquad \int a^x \ln a \, dx = a^x + c \qquad (a > 0)$$

$$\int \frac{dx}{a^2 + x^2} = \frac{1}{a} \tan^{-1} \frac{x}{a} + c$$

$$\int \frac{dx}{a^2 - x^2} = \begin{cases} \dfrac{1}{a} \tanh^{-1} \dfrac{x}{a} + c \\ \text{or} \\ \dfrac{1}{2a} \ln \dfrac{a + x}{a - x} + c \quad (x^2 < a^2) \end{cases}$$

$$\int \frac{dx}{x^2 - a^2} = \begin{cases} -\dfrac{1}{a} \coth^{-1} \dfrac{x}{a} + c \\ \text{or} \\ \dfrac{1}{2a} \ln \dfrac{x - a}{x + a} + c \quad (x^2 > a^2) \end{cases}$$

TABLE I-3 (continued)   INTEGRAL FORMULAS

$$\int \frac{dx}{\sqrt{a^2 - x^2}} = \sin^{-1} \frac{x}{|a|} + c \quad (a > 0)$$

$$\int \frac{dx}{\sqrt{x^2 \pm a^2}} = \ln\left(x + \sqrt{x^2 \pm a^2}\right)$$

$$\int \frac{dx}{x\sqrt{x^2 - a^2}} = \frac{1}{a}\sec^{-1} \frac{x}{a} + c \quad (a > 0)$$

$$\int \frac{dx}{x\sqrt{a^2 \pm x^2}} = -\frac{1}{a}\ln\left(\frac{a + \sqrt{a^2 \pm x^2}}{x}\right) + c$$

$$\int \sinh u\, du = \cosh u + c, \qquad \int \cosh u = \sinh u + c$$

$$\int \tanh u\, du = \ln \cosh u + c, \qquad \int \coth u\, du = \ln \sinh u + c$$

$$\int \operatorname{sech} u\, du = \tan^{-1}(\sinh u) + c = 2\tan^{-1} e^u + c$$

$$\int \operatorname{csch} u\, du = \ln \tanh \tfrac{1}{2}u + c$$

$$\int \operatorname{sech}^2 u\, du = \tanh u + c, \qquad \int \operatorname{csch}^2 u\, du = -\coth u + c$$

$$\int \operatorname{sech} u \tanh u\, du = -\operatorname{sech} u + c$$

$$\int u \sinh u\, du = u \cosh u - \sinh u + c, \qquad \int u \cosh u\, du = u \sinh u - \cosh u + c$$

$$\int \sinh^{-1} u\, du = u \sinh^{-1} u - \sqrt{1 + u^2} + c,$$

$$\int \cosh^{-1} u\, du = u \cosh^{-1} u - \sqrt{u^2 - 1} + c$$

$$\int \tanh^{-1} u\, du = u \tanh^{-1} u + \tfrac{1}{2}\ln(1 - u^2) + c$$

$$\int \coth^{-1} u\, du = u \coth^{-1} u + \tfrac{1}{2}\ln(u^2 - 1) + c$$

$$\int \operatorname{sech}^{-1} u\, du = u \operatorname{sech}^{-1} u + \sin^{-1} u + c$$

$$\int \operatorname{csch}^{-1} u\, du = u \operatorname{csch}^{-1} u + \sinh^{-1} u + c$$

Note: $\sin^{-1} u = \arcsin u, \qquad \cos^{-1} u = \arccos u, \qquad \tan^{-1} u = \arctan u$

TABLE I-3   Integral Formulas   1320

Structural Members

In this and the following appendix we will outline modern structural analysis methodology. Many references, such as II.1, are available to provide a more thorough background in structural mechanics. We begin with the study of structural members, with primary emphasis given to the beam element. In the next appendix, the transfer matrix and displacement methods of joining the elements into structural systems will be outlined.

The fundamental equations describing the deformation of a solid can be placed in three distinct categories: the equations of equilibrium, the material law, and the strain–displacement relations (conditions of geometric fit, kinematic relations). For structural members these relationships can be expressed as differential equations.

Ideally, the complete set of these three basic relationships for an elastic body would be solved for an "exact" solution. For arbitrary configurations general closed-form solutions do not exist. However, for special structures or structural

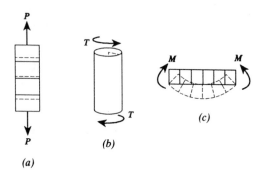

FIGURE II-1 Some basic structural members: (*a*) extension bar; (*b*) torsion bar; (*c*) beam.

members, such as beams and plates, approximate theories have been developed. Usually these approximate theories are based on assumptions made with respect to the distribution of strains or displacements within the structure, and sometimes they are supplemented by assumptions on the influence of certain stress components. For example, for the extension, torsion, and bending of a bar, it is assumed that transverse cross sections simply translate, rotate, and remain plane, respectively, while deforming (Fig. II-1). As is to be expected, these approximate theories may contain some inconsistencies. For instance, in the case of the bending of a bar, shear stresses are introduced and the corresponding strains lead to deformations that violate the basic deformation assumptions. Solutions based on these approximations can be no better than the deformation model permits.

II.1 ENGINEERING BEAM THEORY: DIFFERENTIAL FORM OF GOVERNING EQUATIONS

We will derive briefly the governing differential equations for a beam.

Geometric Relationships

For beams, the geometric relationships will be strain–displacement equations. The bending strain is chosen to be the curvature $\kappa = 1/\rho$, where ρ is the radius of curvature of the beam axis through the centroids of the cross section. From analytical geometry, the curvature of the deflection curve is given by

$$\kappa = \frac{\partial\theta/\partial x}{\left(1 + \theta^2\right)^{3/2}}$$

or for θ small relative to unity

$$\kappa = \frac{\partial\theta}{\partial x} \tag{II.1}$$

where θ is the slope of the deflection curve.

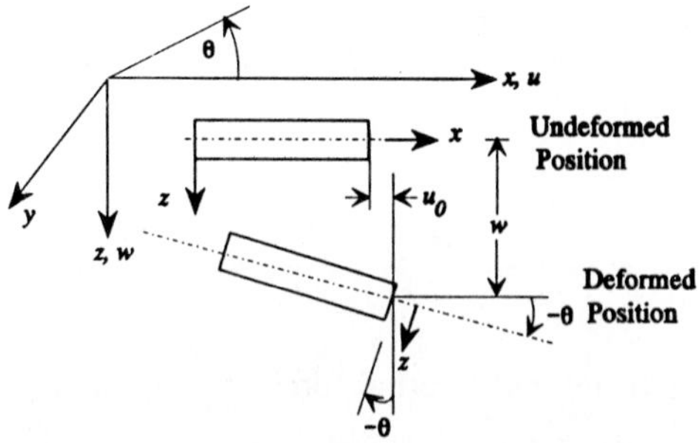

FIGURE II-2 Element of a beam in bending.

It is assumed that in all cases the beam deforms as though it were undergoing pure bending (a constant moment along the beam). This contention (Bernoulli's or Navier's hypothesis) implies that cross sections of the beam remain plane under bending. This means that for a beam with variable cross-sectional properties or applied loading it is assumed that a "flat" cross section remains flat as it deforms (Fig. II-1c), as it would for a uniform beam subjected only to a constant moment along the beam. See an elementary strength-of-materials text, e.g., reference II.2, for a more detailed discussion of beam theory. For this deformation the axial displacement $u(x, z)$ of a point on a cross-sectional plane is (Fig. II-2)

$$u(x, z) = u_0(x) + z\theta(x) \tag{II.2}$$

where u_0 is the axial displacement of the centroidal x axis. Displacements in the positive coordinate directions are considered to be positive. Rotations (slopes) are positive if their vectors, according to the right-hand rule, lie in the positive coordinate direction. The shear strain γ_{xz} is given as

$$\gamma_{xz} = \frac{\partial u}{\partial z} + \frac{\partial w}{\partial x} = \theta + \frac{\partial w}{\partial x} = \gamma \tag{II.3}$$

For the cross sections to remain plane, it is necessary that this shear strain be zero, i.e., shear deformation effects are neglected. Then

$$\theta = -\frac{\partial w}{\partial x} \quad \text{or} \quad \kappa = -\frac{\partial^2 w}{\partial x^2} \tag{II.4}$$

This is the desired geometric relation for bending. The component w is the displacement of the beam axis, i.e., the deflection of the centerline of the beam.

Material Laws

The material law of the beam should reflect the assumption that the elongation and contraction of longitudinal fibers are the dominant deformations. It follows that the material should be assumed to be rigid in the z direction. This implies that there will be no contribution to the longitudinal strain ε_x by stresses in the z direction. Also, assume that $\sigma_y = 0$ as the loading is in the x, z plane. Then the material law would simply be $\varepsilon_x = \sigma_x/E$ or $\sigma_x = E\varepsilon_x$. From Eq. (II.2) $u = u_0 + z\theta$, so that

$$\varepsilon_x = \frac{\partial u}{\partial x} = \frac{\partial u_0}{\partial x} + z\frac{\partial \theta}{\partial x} = \frac{\partial u_0}{\partial x} + z\kappa \tag{II.5}$$

The stress resultants or the net internal forces are the axial force P,

$$P = \int_A \sigma_x \, dA = \int_A E\varepsilon_x \, dA = \int_A E\left(\frac{\partial u_0}{\partial x} + z\kappa\right) dA = EA\frac{\partial u_0}{\partial x} \tag{II.6a}$$

and the bending moment M,

$$M = \int_A \sigma_x z \, dA = \int_A E\varepsilon_x z \, dA = \int_A E\left(\frac{\partial u_0}{\partial x} + z\kappa\right) z \, dA = \kappa E \int_A z^2 \, dA = \kappa EI \tag{II.6b}$$

where I is the moment of inertia about the y axis. The integral $\int z \, dA$ is zero if z is measured from a centroidal axis of the beam.

If the shear deformation effects are to be taken into account, the material equation relating the shear strain and the net internal shear force should supplement Eqs. (II.6). Hooke's law for shear takes the form $\tau_{xz} = \tau = G\gamma$. Let $V = \tau_{av} A$ be the stress force relationship, where V is the shear force and A is the cross-sectional area. Select $\tau_{av} = k_s\tau$, where k_s is a dimensionless *shear form factor* that depends on the cross-sectional shape and τ is the shear stress at the centroid of the cross section. Normally, the structural equations are expressed in terms of the *shear correction factor* $\alpha_s = 1/k_s$, where values of α_s are given in Table 2-4. Also, it is convenient to define a shear corrected area $A_s = k_s A = A/\alpha_s$. The material law relationship becomes

$$V = GA_s\gamma \tag{II.7}$$

Now θ is not the slope of the deflection curve but is an angle of rotation of the beam cross section which is not perpendicular to the beam axis.

Equations of Equilibrium

Figure II-3 illustrates a beam element with internal forces. The forces and moments shown in Fig. II-3 are positive, including the applied loads which are positive if their corresponding vectors lie in positive coordinate directions.

The summation of forces in the vertical direction provides

$$-V + p_z \, dx + V + \frac{\partial V}{\partial x} \, dx = 0 \quad \text{or} \quad \frac{\partial V}{\partial x} + p_z = 0 \tag{II.8a}$$

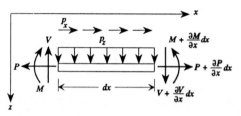

FIGURE II-3   Undeformed beam element on which the conditions of equilibrium are based. Both net internal forces and applied loading shown here are positive.

Similarly,

$$\frac{\partial P}{\partial x} + p_x = 0 \tag{II.8b}$$

is the equilibrium relation for the axial (x) direction. Sum moments about the left end of the element,

$$-M + M + \frac{\partial M}{\partial x}\, dx - dx\left(V + \frac{\partial V}{\partial x}\, dx\right) - \tfrac{1}{2}\, dx\, p_z\, dx = 0$$

or

$$\frac{\partial M}{\partial x} - V - \frac{\partial V}{\partial x}\, dx - \tfrac{1}{2}p_z\, dx = 0$$

In the limit as $dx \to 0$, $(\partial V/\partial x)\, dx$ and $\tfrac{1}{2}p_z\, dx$ approach zero. Then

$$\frac{\partial M}{\partial x} - V = 0 \tag{II.9a}$$

The equilibrium condition of Eq. (II.9a) can also be expressed in stress components as

$$\frac{\partial \sigma_x}{\partial x} + \frac{\partial \tau_{xz}}{\partial z} = 0 \tag{II.9b}$$

Eliminate the shear force V from Eqs. (II.8a) and (II.9),

$$\frac{\partial^2 M}{\partial x^2} + p_z = 0 \tag{II.10}$$

Displacement Form of Governing Differential Equations

The usual beam equations are the Euler–Bernoulli beam in which the shear deformation has been neglected. Ignore the axial extension relationships and find, from Eqs. (II.1), (II.4), and (II.6b),

$$M = EI\kappa = EI\frac{\partial\theta}{\partial x} = -EI\frac{\partial^2 w}{\partial x^2} \tag{II.11}$$

Place this relationship in the equilibrium conditions, Eqs. (II.8a) and (II.9a),

$$V = \frac{\partial M}{\partial x} = -\frac{\partial}{\partial x}EI\frac{\partial^2 w}{\partial x^2} \tag{II.12}$$

and

$$-\frac{\partial V}{\partial x} = \frac{\partial^2}{\partial x^2}EI\frac{\partial^2 w}{\partial x^2} = p_z \tag{II.13}$$

The complete set of governing differential equations is then

$$\frac{\partial^2}{\partial x^2}EI\frac{\partial^2 w}{\partial x^2} = p_z \tag{II.14a}$$

$$V = -\frac{\partial}{\partial x}EI\frac{\partial^2 w}{\partial x^2} \tag{II.14b}$$

$$M = -EI\frac{\partial^2 w}{\partial x^2} \tag{II.14c}$$

$$\theta = -\frac{\partial w}{\partial x} \tag{II.14d}$$

Other effects are readily included. To take into account mass and thereby to include dynamic effects, d'Alembert's principle is useful. In spite of being objectionable to some, this principle, which is studied in elementary dynamics courses, permits the mass to be introduced and to be physically interpreted. If the transverse displacement of a beam is given by w, the velocity and acceleration will be

$$\frac{\partial w}{\partial t} = \dot{w} \quad \text{and} \quad \frac{\partial^2 w}{\partial t^2} = \ddot{w}$$

The acceleration $\ddot{w}$ produces the d'Alembert force

$$p_z = -\rho\ddot{w}$$

where ρ is the mass per unit length along the beam. The minus sign indicates that the inertia force is in the direction opposite to the motion.

The governing equations with dynamics included appear as

$$\frac{\partial^2}{\partial x^2} EI \frac{\partial^2 w}{\partial x^2} + \frac{\partial}{\partial x} P \frac{\partial w}{\partial x} + kw + \rho \frac{\partial^2 w}{\partial t^2} = p_z(x, t) \qquad \text{(II.15a)}$$

$$V = -\frac{\partial}{\partial x} EI \frac{\partial^2 w}{\partial x^2} - P \frac{\partial w}{\partial x} \qquad \text{(II.15b)}$$

$$M = -EI \frac{\partial^2 w}{\partial x^2} \qquad \text{(II.15c)}$$

$$\theta = -\frac{\partial w}{\partial x} \qquad \text{(II.15d)}$$

Also included here is the effect of a compressive axial force P (force) and a Winkler elastic foundation of modulus k (force per length squared).

Retention of shear deformation terms leads to a more general form of the displacement formulation equations. Substitute the strain–displacement relations of Eq. (II.1) and (II.3) into the material law relations $V = GA_s \gamma$ and $M = EI\kappa$. Place the resulting force–displacement equations in the conditions of equilibrium to eliminate the force terms

$$p_z = -\frac{\partial}{\partial x}\left[GA_s \left(\frac{\partial w}{\partial x} + \theta \right)\right] \qquad \text{(II.16a)}$$

$$0 = -\frac{\partial}{\partial x}\left(EI \frac{\partial \theta}{\partial x} \right) + GA_s \frac{\partial w}{\partial x} + GA_s \theta \qquad \text{(II.16b)}$$

$$V = GA_s \left(\frac{\partial w}{\partial x} + \theta \right) \qquad \text{(II.16c)}$$

$$M = EI \frac{\partial \theta}{\partial x} \qquad \text{(II.16d)}$$

The inertia term for dynamics is readily included in this relationship.
Even more effects can be included. Define the following:

$\bar{c}$ Moment intensity, applied moment per unit length along beam (FL/L)

I_T Rotary inertia, transverse or diametrical mass moment of inertia per unit length, $= \rho r_y^2 (ML)$

k Winkler (elastic) foundation modulus (F/L^2)

k^* Rotary foundation modulus (FL/L)

M_T Thermal moment, $\int_A E\alpha\, Tz\, dA$ (FL)

α Coefficient of thermal expansion ($L/(L \cdot \text{degree})$)

T Temperature change (degrees), i.e., the temperature rise with respect to a reference temperature

c External or viscous damping coefficient [FT/L^2, $M/(TL)$]

r_y Radius of gyration, $r_y^2 = I/A$

The more complete governing equations of motion become

$$p_z(x,t) = -\frac{\partial}{\partial x}\left[GA_s\left(\frac{\partial w}{\partial x} + \theta\right)\right] + kw + c\frac{\partial w}{\partial t} + \rho\frac{\partial^2 w}{\partial t^2} \qquad \text{(II.17a)}$$

$$\bar{c}(x,t) = -\frac{\partial}{\partial x}\left(EI\frac{\partial\theta}{\partial x}\right) + GA_s\frac{\partial w}{\partial x} + (GA_s + k^* - P)\theta + \frac{\partial}{\partial x}M_T + \rho r_y^2\frac{\partial^2\theta}{\partial t^2} \qquad \text{(II.17b)}$$

$$V = GA_s\left(\frac{\partial w}{\partial x} + \theta\right) \qquad \text{(II.17c)}$$

$$M = EI\frac{\partial\theta}{\partial x} - M_T \qquad \text{(II.17d)}$$

In addition to bending, this beam, the so-called *Timoshenko beam*, includes the effects of shear deformation and rotary inertia. The expressions are reduced to those for a Rayleigh beam (bending, rotary inertia) by setting $1/GA_s = 0$, for a shear beam (bending, shear deformation) by setting $\rho r_y^2\partial^2\theta/\partial t^2 = 0$, and for an Euler–Bernoulli beam (bending) by setting $1/GA_s = 0$ and $\rho r_y^2\partial^2\theta/\partial t^2 = 0$. Equations (II.17) are appropriate for a beam with a tensile axial force if P is replaced by $-P$. In their present form they apply to beams with a compressive axial force P.

Mixed Form of Governing Differential Equations

A frequently used form of the governing equations is a mixed form involving both forces and displacements. These relations are found in a straightforward fashion by eliminating the strain between the strain–displacement and the constitutive relationships. For an Euler–Bernoulli beam without inertia effects but including shear deformation, this leads to

$$\frac{\partial w}{\partial x} = -\theta + \frac{V}{GA_s} \qquad \text{(II.18a)}$$

$$\frac{\partial\theta}{\partial x} = \frac{M}{EI} \qquad \text{(II.18b)}$$

Supplement these relationships with the equilibrium conditions in the form

$$\frac{\partial V}{\partial x} = -p_z \qquad \text{(II.18c)}$$

$$\frac{\partial M}{\partial x} = V \qquad \text{(II.18d)}$$

where the inertia terms are not included explicitly, or

$$\frac{\partial \mathbf{z}}{\partial x} = \mathbf{A}\mathbf{z} + \mathbf{P} \tag{II.19a}$$

where

$$\mathbf{z} = \begin{bmatrix} w \\ \theta \\ V \\ M \end{bmatrix}, \quad \mathbf{A} = \begin{bmatrix} 0 & -1 & \dfrac{1}{GA_s} & 0 \\ 0 & 0 & 0 & \dfrac{1}{EI} \\ 0 & 0 & 0 & 0 \\ 0 & 0 & 1 & 0 \end{bmatrix}, \quad \mathbf{P} = \begin{bmatrix} 0 \\ 0 \\ -p_z \\ 0 \end{bmatrix} \tag{II.19b}$$

Note that these mixed methods governing equations for a beam do not involve derivatives of geometric or material parameters and all derivatives are of first order. Both of these characteristics contrast to the displacement governing equations (II.14) and (II.15). These characteristics can be advantageous when solving the equations. For example, numerical integration schemes often operate with first-order derivatives only and equations with higher order derivatives must first be transformed to this form.

If the axial terms are included, Eq. (II.18) would be defined as

$$\mathbf{z} = \begin{bmatrix} u_0 \\ w \\ \theta \\ P \\ V \\ M \end{bmatrix}, \quad \mathbf{A} = \begin{bmatrix} 0 & 0 & 0 & \dfrac{1}{EA} & 0 & 0 \\ 0 & 0 & -1 & 0 & \dfrac{1}{GA_s} & 0 \\ 0 & 0 & 0 & 0 & 0 & \dfrac{1}{EI} \\ 0 & 0 & 0 & 0 & 0 & 0 \\ 0 & 0 & 0 & 0 & 0 & 0 \\ 0 & 0 & 0 & 0 & 1 & 0 \end{bmatrix}, \quad \mathbf{P} = \begin{bmatrix} 0 \\ 0 \\ 0 \\ -p_x \\ -p_z \\ 0 \end{bmatrix} \tag{II.20}$$

The Timoshenko beam equations in first-order form are

$$\frac{\partial w}{\partial x} = -\theta + \frac{V}{GA_s} \tag{II.21a}$$

$$\frac{\partial \theta}{\partial x} = \frac{M}{EI} + \frac{M_T}{EI} \tag{II.21b}$$

$$\frac{\partial V}{\partial x} = kw + c\frac{\partial w}{\partial t} + \rho\frac{\partial^2 w}{\partial t^2} - p_z(x,t) \tag{II.21c}$$

$$\frac{\partial M}{\partial x} = V + (k^* - P)\theta + \rho r_y^2 \frac{\partial^2 \theta}{\partial t^2} - \bar{c}(x,t) \tag{II.21d}$$

For harmonic motion, e.g., assume all forces and displacements vary in time as $\sin \omega t$,

$$\frac{dw}{dx} = -\theta + \frac{V}{GA_s} \tag{II.22a}$$

$$\frac{d\theta}{dx} = \frac{M}{EI} + \frac{M_T}{EI} \tag{II.22b}$$

$$\frac{dV}{dx} = kw - \rho\omega^2 w - p_z(x) \tag{II.22c}$$

$$\frac{dM}{dx} = V + (k^* - P)\theta - \rho r_y^2 \omega^2 \theta - \bar{c}(x) \tag{II.22d}$$

where damping has been ignored. The variables, w, θ, V, and M are functions of x only, i.e., $w = w(x)$, $\theta = \theta(x)$, $V = V(x)$, and $M = M(x)$, since it was assumed that the responses are harmonic in time, e.g., $w(x, t) = w(x)\sin \omega t$.

Stress Formulas

The governing equations are solved to provide the state variables w, θ, V, and M along the beam. With the moment M and shear force V known, the normal and shear stresses in the beam can be computed. Substitution of Eqs. (II.6a) and (II.6b) into $\sigma_x = E\varepsilon_x = E(\partial u_0/\partial x + z \, \partial\theta/\partial x)$ gives

$$\sigma_x = \frac{P}{A} + \frac{Mz}{I} \tag{II.23}$$

To find the shear stress τ_{xz}, substitute $\sigma_x = Mz/I$ into the conditions of equilibrium. This leads to $-\partial\tau_{xz}/\partial z = Vz/I$. For a rectangular cross section of height h, suppose the shear stresses are distributed uniformly across the width. Integrate $-\partial\tau_{xz}/\partial z = Vz/I$ with respect to z from the level $z = z_1$ to $z = \frac{1}{2}h$, where z_1 is the position where τ_{xz} is to be evaluated and $\frac{1}{2}h$ defines the top (or bottom) surface of the beam. Then

$$-\tau_{xz}\Big|_{z_1}^{h/2} = \frac{V}{I} \int_{z_1}^{h/2} z \, dz$$

For the x direction on the upper or lower surfaces, τ_{xz} is zero at $z = \frac{1}{2}h$, and hence, for the shear stress τ_{xz} at z_1

$$\tau_{xz} = \frac{V}{I} \int_{z_1}^{h/2} z \, dz = \frac{V}{2I}\left(\frac{h^2}{4} - z_1^2\right) \tag{II.24}$$

II.2 SIGN CONVENTION FOR BEAMS

Traditionally, two distinct sign conventions are employed: one for analytical formulas and the other for computational solutions. The sign convention of the

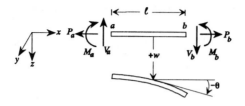

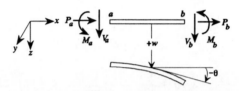

FIGURE II-4   Sign convention for a beam element. (*a*) Sign convention 1: Positive forces, moments, slopes, and displacements are shown. Used for analytical formulas and structural members. (*b*) Sign convention 2: Forces and moments on both ends of the beam element are positive if they (their vectors) lie in the positive coordinate directions. Positive forces, moments, slopes, and displacements are shown. Positive deflection and slope are the same as for sign convention 1. Sign convention 2 is convenient to use in the study of network structural systems using matrix methods.

previous section, which is frequently employed for formulas and structural members, where the distribution of the *internal* bending moment and shear force are of concern, is illustrated in Fig. II-4*a*, sign convention 1.

The other sign convention is better suited for use in many matrix analyses of structures. This sign convention, which will be referred to as sign convention 2, is shown in Fig. II-4*b*. For this second convention, for both ends ($x = a$ and $x = b$) of the beam element, the forces and moments along the positive coordinate directions are considered to be positive. Comparing the ends of the beam elements in Figs. I-4*a, b*, the forces of the two sign conventions are related as follows:

Sign Convention 1	Sign Convention 2	
V_b	V_b	
M_b	M_b	
P_b	P_b	(II.25)
V_a	$-V_a$	
M_a	$-M_a$	
P_a	$-P_a$	

Since deflections and slopes remain the same according to both sign conventions, no special displacement transformation is required.

II.3   SOLUTION OF GOVERNING EQUATIONS FOR A BEAM ELEMENT

We begin the study of the solution of the governing beam equations by employing simple integration for a simple Euler–Bernoulli beam. Integration of Eqs. (II.14) leads to (sign convention 1)

$$V = -EI\frac{d^3w}{dx^3} = -C_1 - \int_0^x p_z(\tau)\, d\tau$$

$$M = -EI\frac{d^2w}{dx^2} = -C_2 - C_1 x - \iint_0^x p_z(\tau)\, d\tau$$

$$\theta = -\frac{dw}{dx} = -\frac{C_3}{EI} - \frac{C_2 x}{EI} - \frac{C_1}{EI}\frac{x^2}{2} - \iiint_0^x \frac{p_z(\tau)}{EI}\, d\tau$$

$$w = \frac{C_4}{EI} + \frac{C_3}{EI}x + \frac{C_2}{EI}\frac{x^2}{2} + \frac{C_1}{EI}\frac{x^3}{3!} + \iiiint_0^x \frac{p_z(\tau)}{EI}\, d\tau$$

(II.26)

A more useful form of the solution is obtained by expressing the arbitrary constants of integration C_1, C_2, C_3, and C_4 in terms of physically meaning constants. We choose to replace C_1, C_2, C_3, and C_4 by values of the displacements and forces at the left end of the beam element. That is, we wish to reorganize the constants of integration C_1, C_2, C_3, and C_4 in terms of the variables at a, i.e., w_a, θ_a, V_a, M_a. Suppose $x = 0$ corresponds to the left end a of the beam element. Let there be no loading at $x = 0$ so that the integrals of p_z vanish at $x = 0$. From Eq. (II.26) for $x = 0$.

$$w_a = w_{x=0} = \frac{C_4}{EI}, \qquad \theta_a = \theta_{x=0} = -\frac{C_3}{EI}$$

$$M_a = M_{x=0} = -C_2, \qquad V_a = V_{x=0} = -C_1$$

(II.27)

Use Eq. (II.27) to replace the constants C_1, C_2, C_3, and C_4 in Eq. (II.26) by the state variables and set $x = \ell$:

$$w_b = w_a - \theta_a\ell - V_a\frac{\ell^3}{3!EI} - M_a\frac{\ell^2}{2EI} + \iiiint_0^\ell \frac{p_z(\tau)}{EI}\, d\tau$$

$$\theta_b = \theta_a + V_a\frac{\ell^2}{2EI} + M_a\frac{\ell}{EI} - \iiint_0^\ell \frac{p_z(\tau)}{EI}\, d\tau$$

$$V_b = V_a \qquad\qquad\qquad - \int_0^\ell p_z(\tau)\, d\tau$$

$$M_b = V_a\ell + M_a \qquad\quad - \iint_0^\ell p_z(\tau)\, d\tau$$

(II.28)

In matrix notation this appears as

$$\mathbf{z}_b = \mathbf{U}^i \mathbf{z}_a + \bar{\mathbf{z}}^i \tag{II.29}$$

where $\mathbf{z} = [w \quad \theta \quad V \quad M]^T$

$$\mathbf{U}^i = \mathbf{U}^i(\ell) = \begin{bmatrix} 1 & -\ell & -\dfrac{\ell^3}{6EI} & -\dfrac{\ell^2}{2EI} \\ 0 & 1 & \dfrac{\ell^2}{2EI} & \dfrac{\ell}{EI} \\ 0 & 0 & 1 & 0 \\ 0 & 0 & \ell & 1 \end{bmatrix} \tag{II.30}$$

$$\bar{\mathbf{z}}^i = \begin{bmatrix} \displaystyle\iiint\int_0^\ell \dfrac{p_z(\tau)}{EI}\,d\tau \\[6pt] -\displaystyle\iiint_0^\ell \dfrac{p_z(\tau)}{EI}\,d\tau \\[6pt] -\displaystyle\int_0^\ell p_z(\tau)\,d\tau \\[6pt] -\displaystyle\iint_0^\ell p_z(\tau)\,d\tau \end{bmatrix} = \begin{bmatrix} F_w \\[6pt] F_\theta \\[6pt] F_V \\[6pt] F_M \end{bmatrix} \tag{II.31}$$

The matrix $\mathbf{U}^i$, which is sometimes denoted by $\mathbf{U}^i(\ell) = \mathbf{U}^i(b - a)$, is referred to as a *transfer matrix* since it "transfers" the variables w, θ, V, and M from $x = a$ to $x = b$. The vector $\mathbf{z}$ of displacements and forces is called the *state vector* as these variables fully describe the response, or "state," of the beam.

If the loading is ignored, the transfer matrix appears as

$$\begin{bmatrix} w \\ \theta \\ V \\ M \end{bmatrix}_b = \left[\begin{array}{cc:cc} 1 & -\ell & -\ell^3/6EI & -\ell^2/2EI \\ 0 & 1 & \ell^2/2EI & \ell/EI \\ \hdashline 0 & 0 & 1 & 0 \\ 0 & 0 & \ell & 1 \end{array}\right] \begin{bmatrix} w \\ \theta \\ V \\ M \end{bmatrix}_a \tag{II.32}$$

$$\mathbf{z}_b \quad\quad = \quad\quad\quad\quad\quad\quad \mathbf{U}^i \quad\quad\quad\quad\quad\quad\quad \mathbf{z}_a$$

It can be shown that the partitions of this relationship can be identified with the basic relations for a beam:

$$\mathbf{U}^i = \left[\begin{array}{c:c} \text{Geometry} & \text{Material} \\ \text{(rigid-body} & \text{law} \\ \text{displacements)} & \\ \hdashline \text{(Influence of} & \\ \text{springs, foundations, etc.)} & \text{Equilibrium} \end{array}\right] \tag{II.33}$$

First-Order Form of Governing Equations A typical method of developing transfer matrices, which applies to both simple and difficult problems, is that of integration of first-order equations in the state variables. Integration of Eqs. (II.19) gives

$$\mathbf{z}_b = \mathbf{U}^i\left[\mathbf{z}_a + \int_a^b (\mathbf{U}^i)^{-1}\mathbf{P}\,d\tau\right] = \mathbf{U}^i\mathbf{z}_a + \bar{\mathbf{z}}^i \tag{II.34}$$

where

$$\bar{\mathbf{z}}^i = \bar{\mathbf{z}}_b^i = \mathbf{U}^i\int_0^\ell [\mathbf{U}^i(\tau)]^{-1}\mathbf{P}(\tau)\,d\tau \tag{II.35}$$

and for a constant coefficient matrix $\mathbf{A}$,

$$\mathbf{U}^i = \mathbf{U}^i(b - a) = e^{\mathbf{A}(b-a)} \tag{II.36}$$

with $b - a = \ell$. Substitution of Eq. (II.34) into Eq. (II.19) will verify Eq. (II.34). The exponential representation of the transfer matrix of Eq. (II.36) can be expanded in the series

$$\mathbf{U}^i = e^{\mathbf{A}\ell} = \mathbf{I} + \frac{\mathbf{A}\ell}{1!} + \frac{\mathbf{A}^2\ell^2}{2!} + \cdots = \sum_{s=0}^\infty \frac{\mathbf{A}^s\ell^s}{s!} \tag{II.37}$$

where $\mathbf{I}$ is the identity matrix, a square matrix with diagonal values of 1 as the only nonzero elements. This expansion lends itself well for numerical calculations for complicated members, as it is often possible to control the error. Analogous to the solution of a first-order scalar differential equation, the loading term would be of the form

$$\bar{\mathbf{z}}^i = e^{\mathbf{A}(b-a)}\int_a^b e^{-\mathbf{A}(x-a)}\mathbf{P}\,dx \tag{II.38}$$

Since $[\mathbf{U}^i(x)]^{-1} = e^{-\mathbf{A}x}$, it follows that for constant $\mathbf{A}$

$$[\mathbf{U}^i(x)]^{-1} = \mathbf{U}^i(-x) \tag{II.39}$$

This relationship can be useful when finding the loading vector $\mathbf{Z}^i$.

Example II.1 Transfer Matrix for an Euler–Bernoulli Beam For the Euler–Bernoulli beam (no shear deformation), with the governing equations of Eq. (II.19), the transfer matrix is obtained from Eq. (II.37) using

$$\mathbf{A} = \begin{bmatrix} 0 & -1 & 0 & 0 \\ 0 & 0 & 0 & 1/EI \\ 0 & 0 & 0 & 0 \\ 0 & 0 & 1 & 0 \end{bmatrix}$$

$$\mathbf{A}^2 = \begin{bmatrix} 0 & 0 & 0 & -1/EI \\ 0 & 0 & 1/EI & 0 \\ 0 & 0 & 0 & 0 \\ 0 & 0 & 0 & 0 \end{bmatrix} \tag{1}$$

$$\mathbf{A}^3 = \begin{bmatrix} 0 & 0 & -1/EI & 0 \\ 0 & 0 & 0 & 0 \\ 0 & 0 & 0 & 0 \\ 0 & 0 & 0 & 0 \end{bmatrix}$$

$$\mathbf{A}^4 = \mathbf{0}$$

Several methods for computing transfer matrices are treated in references II.1, II.3 and II.4. For example, the solution $e^{A\ell}$ can be represented as a matrix polynomial using the Cayley–Hamilton theorem, i.e., the minimal polynomial, which requires knowledge of the eigenvalues of $\mathbf{A}$. The number of terms needed in an expansion of $e^{A\ell}$ can be reduced by using Padé approximations.

For a nonconstant $\mathbf{A}$, Picard iteration [II.4] and other methods are available. Numerical integration techniques, such as Runge–Kutta, are available to solve differential equations. Since state space control methods often involve the solution of a system of first-order differential equations, the relevant control theory literature is a fruitful source of information on the calculation of transfer matrices.

Two General Analytical Techniques

Two procedures suitable for finding the transfer matrices for general forms of the governing equations of motion will be presented here, one based on the Cayley–Hamilton theorem mentioned above and the other on the Laplace transform.

We begin with first-order partial differential equations for the static and dynamic responses of a beam with axial load P, displacement foundation k, and rotary foundation k^* [Eqs. (II.22), with $\bar{c} = M_T = 0$]:

$$\frac{d\mathbf{z}}{dx} = \mathbf{z}' = \mathbf{A}\mathbf{z} + \mathbf{P} \tag{II.40a}$$

with

$$\mathbf{A} = \begin{bmatrix} 0 & -1 & 1/GA_s & 0 \\ 0 & 0 & 0 & 1/EI \\ k - \rho\omega^2 & 0 & 0 & 0 \\ 0 & k^* - P - \rho r_y^2\omega^2 & 1 & 0 \end{bmatrix}, \quad \mathbf{P} = \begin{bmatrix} 0 \\ 0 \\ -p_z \\ 0 \end{bmatrix} \tag{II.40b}$$

It follows from Eq. (II.35) that the loading elements $\bar{\mathbf{z}}^i$ can be computed if the transfer matrix $\mathbf{U}^i$ is available. Thus, it is necessary only to find $\mathbf{U}^i$ in order to complete a solution and p_z can be set equal to zero in Eq. (II.40).

Solution to First-Order Form of Equations

We wish to solve the homogeneous differential equations

$$\frac{d\mathbf{z}}{dx} = \mathbf{A}\mathbf{z}$$

The solution can be in the form [Eq. (II.37)]

$$\mathbf{z}_b = \mathbf{U}^i \mathbf{z}_a = e^{A\ell}\mathbf{z}_a$$

Since a function of a square matrix $\mathbf{A}$ of order n is equal to a polynomial in $\mathbf{A}$ of order $n - 1$, the 4×4 matrix $\mathbf{A}$ can be expanded as

$$e^{A\ell} = c_0\mathbf{I} + c_1\mathbf{A}\ell + c_2(\mathbf{A}\ell)^2 + c_3(\mathbf{A}\ell)^3 \tag{II.41a}$$

Because a matrix satisfies its own characteristic equation (the Cayley–Hamilton theorem), $\mathbf{A}$ of Eq. (II.41a) can be replaced by its characteristic values λ_i. Then

$$e^{\lambda_i \ell} = c_0 + c_1 \lambda_i \ell + c_2 (\lambda_i \ell)^2 + c_3 (\lambda_i \ell)^3, \qquad i = 1, 2, 3, 4 \qquad \text{(II.41b)}$$

are four equations that can be solved for the functions c_0, c_1, c_2, and c_3. Place these values into Eq. (II.41a) to obtain the desired transfer matrix.

The characteristic values λ_i are found by solving the characteristic equation

$$|\mathbf{I}\lambda_i - \mathbf{A}| = 0$$

For the $\mathbf{A}$ of Eq. (II.40b)

$$|\mathbf{I}\lambda_i - \mathbf{A}| = \begin{vmatrix} \lambda_i & 1 & -1/GA_s & 0 \\ 0 & \lambda_i & 0 & -1/EI \\ -k + \rho\omega^2 & 0 & \lambda_i & 0 \\ 0 & -k^* + P + \rho r_y^2 \omega^2 & 1 & \lambda_i \end{vmatrix} = 0$$

This determinant has the roots $\lambda_1, \lambda_2, \lambda_3, \lambda_4 = (\pm n_1, \pm i n_2)$,

$$n_{1,2} = \left\{ \left[\tfrac{1}{4}(\zeta + \eta)^2 - \lambda \right]^{1/2} \mp \tfrac{1}{2}(\zeta - \eta) \right\}^{1/2}$$

where

$$\lambda = (k - \rho\omega^2)/EI, \qquad \eta = (k - \rho\omega^2)/GA_s,$$
$$\zeta = (P - k^* + \rho r_y^2 \omega^2)/EI$$

Substitution of these characteristic values into Eq. (II.41b) gives

$$e^{n_1 \ell} = c_0 + c_1 n_1 \ell + c_2 (n_1 \ell)^2 + c_3 (n_1 \ell)^3$$
$$e^{-n_1 \ell} = c_0 - c_1 n_1 \ell + c_2 (n_1 \ell)^2 - c_3 (n_1 \ell)^3$$
$$e^{i n_2 \ell} = c_0 + i c_1 n_2 \ell - c_2 (n_2 \ell)^2 - i c_3 (n_2 \ell)^3$$
$$e^{-i n_2 \ell} = c_0 - i c_1 n_2 \ell - c_2 (n_2 \ell)^2 + i c_3 (n_2 \ell)^3$$

The constants c_0, c_1, c_2, c_3 from these four equations are

$$c_0 = \left(n_2^2 \cosh n_1 \ell + n_1^2 \cos n_2 \ell \right) / \left(n_1^2 + n_2^2 \right)$$
$$c_1 = \left[(n_2^2/n_1) \sinh n_1 \ell + (n_1^2/n_2) \sin n_2 \ell \right] / \left[\ell \left(n_1^2 + n_2^2 \right) \right]$$
$$c_2 = \left(\cosh n_1 \ell - \cos n_2 \ell \right) / \left[\ell^2 \left(n_1^2 + n_2^2 \right) \right]$$
$$c_3 = \left[(1/n_1) \sinh n_1 \ell - (1/n_2) \sin n_2 \ell \right] / \left[\ell^3 \left(n_1^2 + n_2^2 \right) \right]$$

From Eq. (II.41a) the transfer matrix becomes

$$\mathbf{U}^i = c_0\mathbf{I} + c_1(\mathbf{A}\ell) + c_2(\mathbf{A}\ell)^2 + c_3(\mathbf{A}\ell)^3$$

$$= \begin{bmatrix} c_0 + \ell^2 c_2\eta & -\ell c_1 - \ell^3 c_3(\eta - \zeta) & (c_1\ell + c_3\ell^3\eta)/k_sGA - c_3\ell^3/EI & -\ell^2 c_2/EI \\ \lambda c_3\ell^3 & c_0 - \ell^2 c_2\zeta & \ell^2 c_2/EI & (\ell c_1 - c_3\ell^3\zeta)/EI \\ \lambda EI(\ell c_1 + \eta\ell^3 c_3) & -\lambda EI c_2\ell^2 & c_0 + c_2\ell^2\eta & -\ell^3 c_3\lambda \\ \lambda EI c_2\ell^2 & EI[-c_1\ell\zeta + c_3\ell^3(\zeta^2 - \lambda)] & \ell c_1 + c_3\ell^3(\eta - \zeta) & c_0 - c_2\ell^2\zeta \end{bmatrix}$$

$$(\text{II.42})$$

This transfer matrix is presented in Table 11-22, where the influence of inertia has been included.

Laplace Transform

Another viable technique for deriving transfer matrices is to use the Laplace transform. Although this transform can be applied to the first-order equations of Eqs. (II.40), we choose to utilize a single fourth-order equation obtained from the homogeneous form of Eqs. (II.40):

$$\frac{d^4 w}{dx^4} + (\zeta - \eta)\frac{d^2 w}{dx^2} + (\lambda - \zeta\eta)w = 0 \qquad (\text{II.43})$$

where

$$\zeta = (P - k^* + \rho r_y^2\omega^2)/(EI), \qquad \eta = (k - \rho\omega^2)/(GA_s),$$
$$\lambda = (k - \rho\omega^2)/(EI)$$

The Laplace transform of Eq. (II.43) gives

$$w(s)\left[s^4 + (\zeta - \eta)s^2 + (\lambda - \zeta\eta)\right] = s^3 w(0) + s^2 w'(0) + sw''(0) + w'''(0)$$
$$+ (\zeta - \eta)w'(0) + (\zeta - \eta)sw(0)$$

where s is the transform variable. The inverse transform is

$$w(x) = \left[e_1(x) + (\zeta - \eta)e_3(x)\right]w(0) + \left[e_2(x) + (\zeta - \eta)e_4(x)\right]w'(0)$$
$$+ e_3(x)w''(0) + e_4(x)w'''(0)$$

where

$$e_i(x) = L^{-1}\frac{s^{4-i}}{s^4 + (\zeta - \eta)s^2 + \lambda - \zeta\eta} \qquad (\text{II.44a})$$

L^{-1} indicating the inverse Laplace transform. Equation (II.44a) leads to several useful identities:

$$e_i(x) = \frac{d}{dx}e_{i+1}(x), \qquad i = -2, -1, 0, 1, 2, 3$$
$$\qquad\qquad\qquad\qquad\qquad\qquad\qquad\qquad (\text{II.44b})$$
$$e_{i+1}(x) = \int_0^x e_i(u)\, du, \qquad i = 4, 5, 6$$

Arrange $w(x)$ and its three derivatives $w' = dw/dx$, $w'' = d^2w/dx^2$, and $w''' = d^3w/dx^3$ as

$$
\begin{bmatrix} w(x) \\ w'(x) \\ w''(x) \\ w'''(x) \end{bmatrix} = \begin{bmatrix} e_1 + (\zeta - \eta)e_3 & e_2 + (\zeta - \eta)e_4 & e_3 & e_4 \\ e_0 + (\zeta - \eta)e_2 & e_1 + (\zeta - \eta)e_3 & e_2 & e_3 \\ e_{-1} + (\zeta - \eta)e_1 & e_0 + (\zeta - \eta)e_2 & e_1 & e_2 \\ e_{-2} + (\zeta - \eta)e_0 & e_{-1} + (\zeta - \eta)e_1 & e_0 & e_1 \end{bmatrix} \begin{bmatrix} w(0) \\ w'(0) \\ w''(0) \\ w'''(0) \end{bmatrix}
$$

$$
\mathbf{w}(x) = \mathbf{Q}(x) \qquad \mathbf{w}(0)
$$

$$
(11.45a)
$$

By taking the derivatives d^2w/dx^2 and d^3w/dx^3 of $dw/dx = -\theta + V/GA_s$ of Eq. (II.40) form $\mathbf{w}(x) = \mathbf{Rz}(x)$, which relates the deflection $w(x)$ and its derivatives to the state variables $\mathbf{z}(x)$. In this equation

$$
\mathbf{R} = \begin{bmatrix} 1 & 0 & 0 & 0 \\ 0 & -1 & 1/GA_s & 0 \\ \eta & 0 & 0 & -1/EI \\ 0 & \zeta - \eta & -1/EI + \eta/GA_s & 0 \end{bmatrix} \qquad (II.45b)
$$

The transfer matrix is obtained from $\mathbf{z}(x) = \mathbf{R}^{-1}\mathbf{w}(x)$ and Eq. (II.45a):

$$
\mathbf{z}(x) = \mathbf{R}^{-1}\mathbf{Q}(x)\mathbf{w}(0) = \mathbf{R}^{-1}\mathbf{Q}(x)\mathbf{R}\,\mathbf{z}(0)
$$

or

$$
\mathbf{z}_b = \mathbf{R}^{-1}\mathbf{Q}(\ell)\mathbf{R}\,\mathbf{z}_a = \mathbf{U}^i\mathbf{z}_a \qquad (II.46)
$$

This procedure readily leads to the general transfer matrix of Table 11-22.

Effect of Applied Loading

The influence of a prescribed loading $\mathbf{P}$ can be incorporated in the response expressions using Eq. (II.35). It is apparent that this effect can be calculated if the transfer matrix for the element is known either analytically or numerically.

Example II.2 The Effect of a Linearly Varying Distributed Load Demonstrate the use of Eq. (II.35) to compute loading functions F_w, F_θ, F_V, and F_M for an Euler–Bernoulli beam segment of constant cross section and length ℓ loaded with a linearly increasing force described by $p_z = p_0 x/\ell$.

Since [Eq. (II.39)] for a beam of constant cross section $[\mathbf{U}^i(x)]^{-1} = \mathbf{U}^i(-x)$,

$$\int_0^\ell (\mathbf{U}^i)^{-1}\mathbf{P}\,dx = \int_0^\ell \mathbf{U}^i(-x)\mathbf{P}\,dx = \int_0^\ell \begin{bmatrix} 1 & x & \dfrac{x^3}{6EI} & -\dfrac{x^2}{2EI} \\[2mm] 0 & 1 & \dfrac{x^2}{2EI} & \dfrac{-x}{EI} \\[2mm] 0 & 0 & 1 & 0 \\[2mm] 0 & 0 & -x & 1 \end{bmatrix}\begin{bmatrix} 0 \\ 0 \\ -p_0 x/\ell \\ 0 \end{bmatrix}dx$$

$$= \int_0^\ell \frac{1}{\ell}\begin{bmatrix} \dfrac{-p_0 x^4}{6EI} \\[2mm] \dfrac{-p_0 x^3}{2EI} \\[2mm] -p_0 x \\[2mm] p_0 x^2 \end{bmatrix}dx \tag{1}$$

From Eq. (II.35) the vector $\bar{\mathbf{z}}^i = [F_w \quad F_\theta \quad F_V \quad F_M]^T$ is given by

$$\bar{\mathbf{z}}^i = \mathbf{U}^i(\ell)\int_0^\ell [\mathbf{U}^i(x)]^{-1}\mathbf{P}\,dx = p_0\left[\frac{\ell^4}{120EI} \quad -\frac{\ell^3}{24EI} \quad -\frac{\ell}{2} \quad -\frac{\ell^2}{6}\right]^T \tag{2}$$

which applies for $x = \ell$. For values of x less ℓ,

$$\bar{\mathbf{z}}^i = \mathbf{U}^i(x)\int_0^x [\mathbf{U}^i(\tau)]^{-1}\mathbf{P}(\tau)\,d\tau = \frac{p_0}{\ell}\left[\frac{x^5}{120EI} \quad \frac{-x^4}{24EI} \quad -\frac{x^2}{2} \quad -\frac{x^3}{6}\right]^T \tag{3}$$

With $\mathbf{U}^i = e^{\mathbf{A}x}$ the loading vector $\bar{\mathbf{z}}^i$ can be written in the series form

$$\bar{\mathbf{z}}^i = \sum_{j=0}^{\infty} \frac{\mathbf{A}^j x^{(j+1)}}{(j+k+1)!}(k!)\mathbf{P} \tag{4}$$

where $k = 0$ for a uniform load, $k = 1$ for a linearly varying load, etc.
For the case of our linearly varying load

$$\bar{\mathbf{z}}^i = \left(\frac{\mathbf{I}x}{2} + \frac{\mathbf{A}x^2}{3!} + \frac{\mathbf{A}^2 x^3}{4!} + \frac{\mathbf{A}^3 x^4}{5!}\right)\mathbf{P} \tag{5}$$

since $\mathbf{A}^j = 0$ for $j \geq 4$ and $\mathbf{A}^0 = \mathbf{I}$, the unit diagonal matrix. At $x = \ell$, this expression leads to (2).

A technique for finding the effect of applied loading, that is particularly useful if the transfer matrix elements are known analytically, will be presented here.

It is useful to define a general notation for a transfer matrix:

$$
\begin{bmatrix} w \\ \theta \\ V \\ M \end{bmatrix}_b =
\begin{bmatrix}
U_{ww} & U_{w\theta} & U_{wV} & U_{wM} \\
U_{\theta w} & U_{\theta\theta} & U_{\theta V} & U_{\theta M} \\
U_{Vw} & U_{V\theta} & U_{VV} & U_{VM} \\
U_{Mw} & U_{M\theta} & U_{MV} & U_{MM}
\end{bmatrix}
\begin{bmatrix} w \\ \theta \\ V \\ M \end{bmatrix}_a +
\begin{bmatrix} F_w \\ F_\theta \\ F_V \\ F_M \end{bmatrix}_{\ell = b-a}
\tag{II.47}
$$

$$
\mathbf{z}_b = \mathbf{U}^i(\ell) \qquad \mathbf{z}_a + \bar{\mathbf{z}}^i
$$

where U_{ij} represents a transfer matrix element and F_w, F_θ, F_V, and F_M are loading functions.

A transfer matrix is such that often the effect of various types of loading can be identified by observation. For example, it is apparent from the first row of Eq. (II.47) that the contribution of a shear force V at $x = a$ to the deflection w at $x = b$ is $VU_{wV}(\ell)$. Quite similar to a shear force at a point is an applied concentrated load. That is, the effect on the deflection at $x = b$ of a downward concentrated force W at $x = a$ would be expressed as $-WU_{wV}(\ell)$ in the case of sign convention 1. It follows that the contribution of a concentrated force at $x = a$ to the other responses θ, V, and M is similar, so that the loading function vector becomes

$$
\begin{bmatrix} F_w \\ F_\theta \\ F_V \\ F_M \end{bmatrix} =
\begin{bmatrix}
-WU_{wV}(\ell) \\
-WU_{\theta V}(\ell) \\
-WU_{VV}(\ell) \\
-WU_{MV}(\ell)
\end{bmatrix}
\tag{II.48}
$$

Use of a Duhamel or convolution integral permits distributed loads to be treated [II.2, II.3]:

$$
F_j = -\int_0^\ell p_z(x)U_{jV}(\ell - x)\,dx = -\int_0^\ell p_z(\ell - x)U_{jV}(x)\,dx
\tag{II.49}
$$

with $j = w, \theta, V, M$.

Example II.3 Loading Functions for a Linearly Varying Load Calculate the loading function component F_w for the linearly varying distributed applied load shown in Fig. II-5.

For this distributed load

$$
p_z(x) = \frac{p_0}{L}(L - x)
\tag{1}
$$

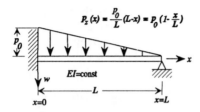

FIGURE II-5   Beam with ramp loading.

so that Eq. (II.49), with $\ell = L$, gives

$$F_w = -\int_0^L p_z(x) U_{wV}(L - x)\, dx = -\int_0^L \frac{p_0}{L}(L - x)\left[-\frac{(L - x)^3}{3!EI} \right] dx = \frac{p_0 L^4}{30 EI}$$

$$(2)$$

II.4   PRINCIPLE OF VIRTUAL WORK: INTEGRAL FORM OF GOVERNING EQUATIONS

Essential to the development of structural mechanics theory are the variational theorems, the principle of virtual work, and the principle of complementary virtual work. Here we will summarize briefly the fundamentals of the principle of virtual work.

Virtual Work

It is useful to define the work done by the loads on a body during a small, admissible change in the displacements. An admissible or possible change is a displacement that varies continuously as a function of the coordinates and does not violate displacement boundary conditions. Although the actual displacements may be large, the *change* in the displacements must be small. Traditionally, these infinitesimal, admissible changes in displacements have been named *virtual displacements*. Virtual displacements are designated by δu_i, indicating that they correspond to a variation of a function as defined in the calculus of variations.

The definition of virtual work follows directly from the definition of ordinary work, which is the product of a force and the displacement of its point of application in the direction of the force. In the case of beams, the curvature κ is taken as the measure of bending strain and the bending moment M is the corresponding force. Then the internal virtual work (δW_i) due to bending of a beam would be

$$\delta W_i = -\int_x \delta \kappa\, M\, dx \qquad (II.50)$$

where $\delta \kappa$ represents the virtual strain and the negative sign is chosen to reflect that the work of the internal moment is the negative of that due to the bending stress. For a beam with no shear deformation considered, the curvature is given by [Eq. (II.4)] $\kappa = -\partial^2 w/\partial x^2$ and the bending moment $M = \kappa EI$ [Eq. (II.6b)]. The internal virtual work would then be

$$\delta W_i = -\int_x \delta\kappa \, M \, dx = -\int_x \delta\kappa \, EI\kappa \, dx = -\int_x \left(\delta\frac{\partial^2 w}{\partial x^2}\right) EI\frac{\partial^2 w}{\partial x^2} \, dx \quad \text{(II.51)}$$

For a beam segment from $x = a$ to $x = b$, the external virtual work (δW_e) would be

$$\delta W_e = \int_x \delta w \, p_z \, dx + [M \, \delta\theta + V \, \delta w]_a^b \quad \text{(II.52a)}$$

where δw is the virtual deflection, p_z is the applied loading intensity along the beam, and M, V are concentrated moments, shear forces on the ends a, b of the element,

$$[M \, \delta\theta + V \, \delta w]_a^b = (M \, \delta\theta)_b + (V \, \delta w)_b - (M \, \delta\theta)_a - (V \, \delta w)_a \quad \text{(II.52b)}$$

Statement of Principle of Virtual Work

The principle of virtual work for a solid can be derived from the equations of equilibrium and vice versa. They are, in a sense, equivalent in that the principle of virtual work is a global (integral) form of the conditions of equilibrium. As is shown in textbooks on structural mechanics, an integral form of the equations of equilibrium, with the help of integration by parts (or the divergence theorem if more than one dimension is involved), leads to the relationship

$$\delta W = \delta W_i + \delta W_e = 0 \quad \text{(II.53)}$$

which embodies the principle of virtual work.

This principle can be stated as follows: *A deformable system is in equilibrium if the sum of the total external virtual work and the internal virtual work is zero for virtual displacements that satisfy the strain–displacement equations and displacement boundary conditions.*

The fundamental unknowns for the principle of virtual work are displacements. Although stresses or forces often appear in the equations representing the principle, these variables should be considered as being expressed as functions of the displacements. Also, the variations are always taken on the displacements in the principle of virtual work. In fact, this principle is also known as the *principle of virtual displacements.*

In the case of a beam, with δW_i and δW_e given by Eqs. (II.51) and (II.52), respectively, the principle of virtual work [Eq. (II.53)] takes the form

$$\delta W = \delta W_i + \delta W_e$$

$$= -\int_a^b \left(\delta\frac{\partial^2 w}{\partial x^2}\right) EI\frac{\partial^2 w}{\partial x^2} \, dx + \int_a^b \delta w \, p_z \, dx + [M \, \delta\theta + V \, \delta w]_a^b = 0 \quad \text{(II.54)}$$

If dynamic effects are included, then this relationship, supplemented with $p_z = -\rho \ddot{w}$, becomes

$$\delta W = -\int_a^b \delta w'' \, EIw'' \, dx + \int_a^b \delta w \, p_z \, dx - \int_a^b \delta w \, \rho \ddot{w} \, dx + [M \, \delta\theta + V \, \delta w]_a^b = 0$$

(II.55)

where $\partial^2 w / \partial x^2 = w''$ has been employed.

II.5   STIFFNESS MATRIX

Definition of Stiffness Matrices

For a beam element from $x = a$ to $x = b$, a stiffness matrix provides a relationship between the displacements at a and b $(w_a, \theta_a, w_b, \theta_b)$ to all the forces (V_a, M_a, V_b, M_b). For the ith element, the stiffness matrix $\mathbf{k}^i$ is defined as

$$\mathbf{p}^i = \mathbf{k}^i \mathbf{v}^i$$

(II.56)

where

$$\mathbf{p}^i = \begin{bmatrix} \mathbf{p}_a \\ \mathbf{p}_b \end{bmatrix} = \begin{bmatrix} V_a \\ M_a \\ V_b \\ M_b \end{bmatrix}, \qquad \mathbf{v}^i = \begin{bmatrix} \mathbf{v}_a \\ \mathbf{v}_b \end{bmatrix} = \begin{bmatrix} w_a \\ \theta_a \\ w_b \\ \theta_b \end{bmatrix}$$

$$\mathbf{k}^i = \begin{bmatrix} \mathbf{k}_{aa} & \mathbf{k}_{ab} \\ \mathbf{k}_{ba} & \mathbf{k}_{bb} \end{bmatrix} = \begin{bmatrix} k_{11} & k_{12} & k_{13} & k_{14} \\ k_{21} & k_{22} & k_{23} & k_{24} \\ k_{31} & k_{32} & k_{33} & k_{34} \\ k_{41} & k_{42} & k_{43} & k_{44} \end{bmatrix}$$

The stiffness matrix is the most essential ingredient in the analysis of structural systems.

An important technique for finding a stiffness matrix for a structural member is to simply reorganize a transfer matrix to form the stiffness matrix. This is to be expected since both the transfer and stiffness matrices are relationships between the same eight variables $w_a, \theta_a, w_b, \theta_b, V_a, M_a, V_b, M_b$. Of course, there are numerous other methods for finding the stiffness matrix, some of which will be considered later in this section.

Sometimes, as was done previously for the transfer matrix, it is useful to include with the stiffness matrix a vector to account for applied loading. Normally this vector would account for only the loading applied between the ends since end loadings are inserted using the vector $\mathbf{p}^i$. To derive a stiffness matrix with a loading vector appended, begin by writing a transfer matrix in the notation of sign convention 2.

Consider first the rearrangement of the transfer matrix into a stiffness matrix. Begin by writing the transfer matrix in the partitioned form

$$\begin{bmatrix} \mathbf{v}_b \\ \mathbf{p}_b \end{bmatrix} = \begin{bmatrix} \mathbf{U}_{vv} & \mathbf{U}_{vp} \\ \mathbf{U}_{pv} & \mathbf{U}_{pp} \end{bmatrix} \begin{bmatrix} \mathbf{v}_a \\ \mathbf{p}_a \end{bmatrix} + \begin{bmatrix} \mathbf{F}_v \\ \mathbf{F}_p \end{bmatrix}$$

$$\mathbf{z}_b \quad = \quad \mathbf{U}^i \quad \mathbf{z}_a \quad + \quad \bar{\mathbf{z}}^i$$

(II.57)

where

$$\mathbf{F}_v = \begin{bmatrix} F_w \\ F_\theta \end{bmatrix}, \qquad \mathbf{F}_p = \begin{bmatrix} F_V \\ F_M \end{bmatrix}$$

It is assumed here that sign convention 2 (Fig. II.4b) applies for the forces. From Eq. (II.57)

$$\mathbf{p}_b = \mathbf{U}_{pv}\mathbf{v}_a + \mathbf{U}_{pp}\mathbf{p}_a + \mathbf{F}_p, \qquad \mathbf{v}_b = \mathbf{U}_{vv}\mathbf{v}_a + \mathbf{U}_{vp}\mathbf{p}_a + \mathbf{F}_v$$

It follows that

$$\mathbf{p}_a = \mathbf{U}_{vp}^{-1}\mathbf{v}_b - \mathbf{U}_{vp}^{-1}\mathbf{U}_{vv}\mathbf{v}_a - \mathbf{U}_{vp}^{-1}\mathbf{F}_v$$

and

$$\mathbf{p}_b = \mathbf{U}_{pv}\mathbf{v}_a + \mathbf{U}_{pp}\mathbf{p}_a + \mathbf{F}_p$$

$$= \left(\mathbf{U}_{pv} - \mathbf{U}_{pp}\mathbf{U}_{vp}^{-1}\mathbf{U}_{vv}\right)\mathbf{v}_a + \mathbf{U}_{pp}\mathbf{U}_{vp}^{-1}\mathbf{v}_b + \mathbf{F}_p - \mathbf{U}_{pp}\mathbf{U}_{vp}^{-1}\mathbf{F}_v \qquad (II.58)$$

In matrix form

$$\begin{bmatrix} \mathbf{p}_a \\ \mathbf{p}_b \end{bmatrix} = \left[\begin{array}{c|c} -\mathbf{U}_{vp}^{-1}\mathbf{U}_{vv} & \mathbf{U}_{vp}^{-1} \\ \hline \mathbf{U}_{pv} - \mathbf{U}_{pp}\mathbf{U}_{vp}^{-1}\mathbf{U}_{vv} & \mathbf{U}_{pp}\mathbf{U}_{vp}^{-1} \end{array} \right] \begin{bmatrix} \mathbf{v}_a \\ \mathbf{v}_b \end{bmatrix} + \begin{bmatrix} -\mathbf{U}_{vp}^{-1}\mathbf{F}_v \\ \mathbf{F}_p - \mathbf{U}_{pp}\mathbf{U}_{vp}^{-1}\mathbf{F}_v \end{bmatrix} \qquad (II.59)$$

$$\begin{bmatrix} V_a \\ M_a \\ V_b \\ M_b \end{bmatrix} = \begin{bmatrix} k_{11} & k_{12} & k_{13} & k_{14} \\ k_{21} & k_{22} & k_{23} & k_{24} \\ k_{31} & k_{32} & k_{33} & k_{34} \\ k_{41} & k_{42} & k_{43} & k_{44} \end{bmatrix} \begin{bmatrix} w_a \\ \theta_a \\ w_b \\ \theta_b \end{bmatrix} - \begin{bmatrix} V_a^0 \\ M_a^0 \\ V_b^0 \\ M_b^0 \end{bmatrix}$$

$$\mathbf{p}^i \quad = \quad \mathbf{k}^i \quad\quad \mathbf{v}^i \quad - \quad \bar{\mathbf{p}}^i$$

(II.60)

For a beam, Eq. (II.58) leads to the stiffness matrix

$$
\begin{bmatrix} V_a \\ M_a \\ V_b \\ M_b \end{bmatrix}
=
\begin{bmatrix}
\dfrac{12EI}{\ell^3} & -\dfrac{6EI}{\ell^2} & -\dfrac{12EI}{\ell^3} & -\dfrac{6EI}{\ell^2} \\[2mm]
-\dfrac{6EI}{\ell^2} & \dfrac{4EI}{\ell} & \dfrac{6EI}{\ell^2} & \dfrac{2EI}{\ell} \\[2mm]
-\dfrac{12EI}{\ell^3} & \dfrac{6EI}{\ell^2} & \dfrac{12EI}{\ell^3} & \dfrac{6EI}{\ell^2} \\[2mm]
-\dfrac{6EI}{\ell^2} & \dfrac{2EI}{\ell} & \dfrac{6EI}{\ell^2} & \dfrac{4EI}{\ell}
\end{bmatrix}
\begin{bmatrix} w_a \\ \theta_a \\ w_b \\ \theta_b \end{bmatrix}
\tag{II.61}
$$

$$
= \begin{bmatrix} \mathbf{k}_{aa} & \mathbf{k}_{ab} \\ \mathbf{k}_{ba} & \mathbf{k}_{bb} \end{bmatrix} \mathbf{v}^i
$$

$$
\mathbf{p}^i = \mathbf{k}^i \qquad \mathbf{v}^i
$$

It is evident from this relationship that a stiffness element k_{ij}, e.g., $k_{11} = 12EI/\ell^3$, is the force developed at coordinate i due to a unit displacement at coordinate j, with all other displacements equal to zero. These "coordinates" are usually called the *degrees of freedom* (DOF), which are normally defined as the independent coordinates (displacement components) necessary to fully describe the spatial position of a structure.

It is often helpful to scale the stiffness matrix of Eq. (II.61) as

$$
\begin{bmatrix} V_a \\ M_a/\ell \\ V_b \\ M_b/\ell \end{bmatrix}
= \frac{EI}{\ell^3}
\begin{bmatrix}
12 & -6 & -12 & -6 \\
-6 & 4 & 6 & 2 \\
-12 & 6 & 12 & 6 \\
-6 & 2 & 6 & 4
\end{bmatrix}
\begin{bmatrix} w_a \\ \ell\theta_a \\ w_b \\ \ell\theta_b \end{bmatrix}
\tag{II.62}
$$

$$
\mathbf{p}^i = \mathbf{k}^i \qquad \mathbf{v}^i
$$

A very general stiffness matrix, including the effect of elastic foundations, inertia, and shear deformation, can now be obtained by inserting the general transfer matrix components of Table 11-22 in Eq. (II.60). This leads to the generalized dynamic stiffness matrix of Table 11-22.

Determination of Stiffness Matrices

In addition to the conversion of a transfer matrix into a stiffness matrix described above, other analysis techniques, such as the use of the unit load method or Castigliano's theorem, will also lead to the stiffness matrix. Many such methods are described in standard textbooks on structural mechanics, such as reference II.1. The use of trial functions to derive a stiffness matrix is of special interest as this approach provides insight into other structural matrices such as mass matrices.

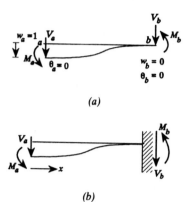

FIGURE II-6 Beam element for computing the first column of the stiffness matrix: (a) configuration for computing k_{i1}; (b) configuration equivalent to a.

Before turning to the trial function approach, we will illustrate with a beam the direct evaluation of a stiffness matrix using the governing differential equations. This entails the application of unit displacements (deflection or rotation) at the ends of a beam element.

To compute k_{i1}, $i = 1, 2, 3, 4$, corresponding to the first column of the stiffness matrix, use the configuration of Fig. II-6a. the forces in this model with displacements $w_a = 1$, $\theta_a = 0$, $w_b = 0$, and $\theta_b = 0$ correspond to the stiffness coefficients

$$k_{11} = V_a, \qquad k_{21} = M_a, \qquad k_{31} = V_b, \qquad k_{41} = M_b$$

The cantilevered beam of Fig. II-6b, which models the prescribed displacements, can be used to find V_a, M_a, V_b, and M_b based on the displacement conditions $w_a = 1$ and $\theta_a = 0$. For this beam, with sign convention 2, $M = -M_a - V_a x$. Integrate $d^2w/dx^2 = -M/EI$ to find

$$\frac{dw}{dx} = \frac{1}{EI}\left(M_a x + V_a \frac{x^2}{2}\right) + C_1 = -\theta$$

$$w = \frac{1}{EI}\left(M_a \frac{x^2}{2} + V_a \frac{x^3}{6}\right) + C_1 x + C_2$$

Use $\theta_a = 0$ and $w_b = 0$ to find $C_1 = 0$ and $C_2 = -M_a \ell^2/2EI - V_a \ell^3/6EI$. Then impose $\theta_b = 0$ on the first equation and $w_a = 1$ on the second, giving

$$V_a = 12EI/\ell^3 = k_{11}, \qquad M_a = -6EI/\ell^2 = k_{21}$$

From the conditions of equilibrium for the beam of Fig. II-6b the forces M_b and V_b can be evaluated:

$$k_{31} = V_b = -V_a = -12EI/\ell^3, \qquad k_{41} = M_b = -V_a\ell - M_a = -6EI/\ell^2$$

The second, third, and fourth columns of the stiffness matrix are computed in a similar fashion.

Approximation-by-Trial Function   The stiffness matrices considered thus far are "exact" in the sense that the exact solution of the engineering beam theory has been placed in the form of a stiffness matrix. For structural elements other than beams it is often not possible to establish an exact solution. In these cases a technique involving an *assumed* or *trial series solution* is employed to obtain an approximate solution. This is the approach for finding stiffness matrices for the finite-element method.

Assume the deflection of a beam element can be approximated by the polynomial

$$w = C_1 + C_2x + C_3x^2 + C_4x^3 + \cdots = \hat{w}_1 + \hat{w}_2x + \hat{w}_3x^2 + \hat{w}_4x^3 + \cdots$$

where $C_j = \hat{w}_j$, $j = 1, 2\ldots$ are unknown constants. This is a trial function, often referred to as a *basis* function. Express the first four terms as

$$w = \begin{bmatrix} 1 & x & x^2 & x^3 \end{bmatrix} \begin{bmatrix} \hat{w}_1 \\ \hat{w}_2 \\ \hat{w}_3 \\ \hat{w}_4 \end{bmatrix} = \mathbf{N}_u\hat{\mathbf{w}} = \hat{\mathbf{w}}^T\mathbf{N}_u^T$$

where

$$\mathbf{N}_u = \begin{bmatrix} 1 & x & x^2 & x^3 \end{bmatrix}$$

Rewrite the series as an *interpolation function* by expressing it in terms of the unknown displacements at the end of the beam element rather than in terms of the constants $\hat{w}_j$. To accomplish this, transform the vector of unknowns $\hat{w}_j$ into the unknown nodal displacement vector for the ith element

$$\mathbf{v}^i = \begin{bmatrix} w_a \\ \theta_a = -w'_a \\ w_b \\ \theta_b = -w'_b \end{bmatrix}$$

The derivative of w is given by

$$w' = \mathbf{N}'_u\hat{\mathbf{w}} = \hat{\mathbf{w}}^T(\mathbf{N}'_u)^T$$

where

$$\mathbf{N}'_u = \begin{bmatrix} 0 & 1 & 2x & 3x^2 \end{bmatrix}$$

Evaluate w and $\theta = -w'$ at $x = a$ and $x = b$:

$$\begin{bmatrix} w_a \\ \theta_a \\ w_b \\ \theta_b \end{bmatrix} = \begin{bmatrix} w(0) \\ -w'(0) \\ w(\ell) \\ -w'(\ell) \end{bmatrix} = \begin{bmatrix} 1 & 0 & 0 & 0 \\ 0 & -1 & 0 & 0 \\ 1 & \ell & \ell^2 & \ell^3 \\ 0 & -1 & -2\ell & -3\ell^2 \end{bmatrix} \begin{bmatrix} \hat{w}_1 \\ \hat{w}_2 \\ \hat{w}_3 \\ \hat{w}_4 \end{bmatrix}$$

$$\mathbf{v}^i \qquad\qquad = \qquad\qquad \hat{\mathbf{N}}_u \qquad\qquad \hat{\mathbf{w}}$$

It follows that

$$\hat{\mathbf{w}} = \hat{\mathbf{N}}_u^{-1}\mathbf{v}^i = \mathbf{G}\mathbf{v}^i$$

where

$$\mathbf{G} = \hat{\mathbf{N}}_u^{-1} = \begin{bmatrix} 1 & 0 & 0 & 0 \\ 0 & -1 & 0 & 0 \\ -3/\ell^2 & 2/\ell & 3/\ell^2 & 1/\ell \\ 2/\ell^3 & -1/\ell^2 & -2/\ell^3 & -1/\ell^2 \end{bmatrix}$$

Finally, the desired relationship between w and $\mathbf{v}^i$ is

$$w = \mathbf{N}_u\hat{\mathbf{w}} = \mathbf{N}_u\mathbf{G}\mathbf{v}^i = \mathbf{N}\mathbf{v}^i \tag{II.64}$$

where $\mathbf{N} = \mathbf{N}_u\mathbf{G}$. This expression is the interpolation form of the assumed series and is usually referred to as a *shape* function. If the normalized coordinate $\xi = x/\ell$ is introduced, $\mathbf{v}^i$ can be redefined as in Eq. (II.62), and

$$w = \begin{bmatrix} 1 & \xi & \xi^2 & \xi^3 \end{bmatrix} \begin{bmatrix} 1 & 0 & 0 & 0 \\ 0 & -1 & 0 & 0 \\ -3 & 2 & 3 & 1 \\ 2 & -1 & -2 & -1 \end{bmatrix} \begin{bmatrix} w_a \\ \ell\theta_a \\ w_b \\ \ell\theta_b \end{bmatrix} = \mathbf{N}\mathbf{v}^i \tag{II.65a}$$

$$= \qquad\qquad \mathbf{N}_u \qquad\qquad\qquad \mathbf{G} \qquad\qquad \mathbf{v}^i$$

or

$$w = \left(1 - 3\xi^2 + 2\xi^3\right)w_a + \left(-\xi + 2\xi^2 - \xi^3\right)\theta_a\ell + \left(3\xi^2 - 2\xi^3\right)w_b$$
$$+ \left(\xi^2 - \xi^3\right)\theta_b\ell \tag{II.65b}$$

The quantities in brackets are Hermitian polynomials, which are well-known tabulated functions. The polynomials of Eq. (II.64) or (II.65) can be used with the principle of virtual work to generate stiffness matrices.

Evaluation of Stiffness Matrix Using Principle of Virtual Work   The following procedure is quite general in that it can be used to derive stiffness matrices for any element. If axial, dynamic, and shear deformation effects are not taken into account, the principle of virtual work ($\delta W_i = -\delta W_e$) for a beam appears as Eq. (II.54)

$$\int_a^b \delta w'' \, EIw'' \, dx = \int_a^b \delta w \, p_z \, dx + [M \, \delta\theta + V \, \delta w]_a^b \tag{II.66}$$

where $w' = dw/dx$.

To find the element stiffness matrix, substitute the assumed polynomial for w in Eq. (II.66). First find the variational quantities δw and $\delta w''$ expressed in terms of the trial series. In Eq. (II.64) **G** contains constant elements and $\mathbf{N}_u$ is a function of the axial coordinate x. Thus

$$\delta w = \delta(\mathbf{N}\mathbf{v}^i) = \mathbf{N} \, \delta\mathbf{v}^i = \delta\mathbf{v}^{iT} \, \mathbf{N}^T$$

where $\delta\mathbf{v}^{iT}$ are the virtual end displacements, and

$$w'' = \mathbf{N}_u''\mathbf{G}\mathbf{v}^i = \mathbf{B}_u\mathbf{G}\mathbf{v}^i = \mathbf{N}''\mathbf{v}^i = \mathbf{B}\mathbf{v}^i$$

$$\delta w'' = \mathbf{B} \, \delta\mathbf{v}^i = \delta\mathbf{v}^{iT} \, \mathbf{B}^T \tag{II.67}$$

with $\mathbf{N}_u'' = \mathbf{B}_u = [0 \quad 0 \quad 2 \quad 6\xi]/\ell^2$, $\mathbf{B} = \mathbf{B}_u\mathbf{G}$, and $\mathbf{B} = \mathbf{N}''$.

Substitute these expressions into Eq. (II.66):

$$\int_a^b \overbrace{\delta\mathbf{v}^{iT} \, \mathbf{B}^T}^{\delta w''} EI \overbrace{\mathbf{B} \quad \mathbf{v}^i}^{w''} \, dx = \int_a^b \overbrace{\delta\mathbf{v}^{iT} \, \mathbf{N}^T}^{\delta w} p_z \, dx + \delta\mathbf{v}^{iT} \, \mathbf{p}^i \tag{II.68}$$

where $\mathbf{p}^i$ contains the applied loading M and V at the ends, i.e.,

$$\mathbf{p}^i = \begin{bmatrix} -V_a \\ -M_a \\ V_b \\ M_b \end{bmatrix} = \begin{bmatrix} V_a \\ M_a \\ V_b \\ M_b \end{bmatrix}$$

$$\begin{array}{cc} \text{Sign} & \text{Sign} \\ \text{convention} & \text{convention} \\ 1 & 2 \end{array}$$

Then

$$\delta\mathbf{v}^{iT}\left(\int_a^b \mathbf{B}^T EI\mathbf{B} \, dx \, \mathbf{v} - \int_a^b \mathbf{N}^T p_z \, dx - \mathbf{p}^i\right) = 0 \tag{II.69}$$

or since $\mathbf{G}$ is not a function of x,

$$\delta \mathbf{v}^{iT} \left\{ \underbrace{\mathbf{G}^T \int_a^b \mathbf{B}_u^T EI \mathbf{B}_u \, dx \, \mathbf{G}}_{\mathbf{k}^i} \underbrace{\mathbf{v}^i}_{\mathbf{v}^i} - \left(\underbrace{\mathbf{G}^T \int_a^b \mathbf{N}_u^T p_z \, dx + \mathbf{p}^i}_{\bar{\mathbf{p}}^i} \right) \right\} = 0 \qquad \text{(II.70)}$$

or $\delta \mathbf{v}^{iT} (\mathbf{k}^i \mathbf{v}^i - \bar{\mathbf{p}}^i - \mathbf{p}^i) = 0$, with

$$\bar{\mathbf{p}}^i = \int_a^b \mathbf{N}^T p_z \, dx = \mathbf{G}^T \int_a^b \mathbf{N}_u^T p_z \, dx = \ell \, \mathbf{G}^T \int_0^1 \mathbf{N}_u^T p_z(\xi) \, d\xi \qquad \text{(II.71)}$$

Note that the stiffness matrix is given by

$$\mathbf{k}^i = \mathbf{G}^T \int_a^b \mathbf{B}_u^T EI \mathbf{B}_u \, dx \, \mathbf{G} = \int_a^b \mathbf{B}^T EI \mathbf{B} \, dx \qquad \text{(II.72)}$$

and for constant EI

$$\mathbf{k}^i = EI \int_a^b \mathbf{B}^T \mathbf{B} \, dx \qquad \text{(II.73)}$$

Suppose there is no applied distributed loading, i.e., $p_z = 0$. Then $\bar{\mathbf{p}}^i = 0$, and

$$\delta \mathbf{v}^{iT} \left\{ \mathbf{G}^T \int_a^b \mathbf{B}_u^T EI \mathbf{B}_u \, dx \, \mathbf{G} \mathbf{v}^i - \mathbf{p}^i \right\} = 0 \qquad \text{(II.74)}$$

Thus, for element i

$$\delta \mathbf{v}^{iT} (\mathbf{k}^i \mathbf{v}^i - \mathbf{p}^i) = 0 \quad \text{or} \quad \mathbf{k}^i \mathbf{v}^i = \mathbf{p}^i \qquad \text{(II.75)}$$

As expected, the principle of virtual work expresses the conditions of equilibrium ($\mathbf{k}^i \mathbf{v}^i = \mathbf{p}^i$) between the forces $\mathbf{k}^i \mathbf{v}^i$ representing the element properties and the load vector $\mathbf{p}^i$ representing the applied loads at the ends.

If element i is a portion of a structural system, then this relationship represents the contribution of the ith element to the equilibrium of the whole system, expressed as the virtual work of the ith element that is a part of the virtual work of the whole structural system.

The evaluation of the stiffness matrix of Eq. (II.72) is readily carried out. Remember that $\mathbf{v}^i = [w_a \;\; \ell\theta_a \;\; w_b \;\; \ell\theta_b]^T$, $\mathbf{p}^i = [V_a \;\; M_a/\ell \;\; V_b \;\; M_b/\ell]^T$ and use $\boldsymbol{\Phi}'' = \mathbf{H}$ and $dx = \ell \, d\xi$. The integral in $\mathbf{k}^i$ is integrated over 0 to 1 rather than a to b:

$$\int_0^1 \mathbf{B}_u^T EI \mathbf{B}_u \ell \, d\xi = \begin{bmatrix} 0 & 0 & 0 & 0 \\ 0 & 0 & 0 & 0 \\ 0 & 0 & 4 & 6 \\ 0 & 0 & 6 & 12 \end{bmatrix} \frac{EI}{\ell^3} \tag{II.76}$$

Then $\mathbf{k}^i$ of Eq. (II.72) becomes

$$\mathbf{G}^T \int_0^1 \mathbf{B}_u^T EI \mathbf{B}_u \ell \, d\xi \, \mathbf{G} = \underbrace{\begin{bmatrix} 12 & -6 & -12 & -6 \\ -6 & 4 & 6 & 2 \\ -12 & 6 & 12 & 6 \\ -6 & 2 & 6 & 4 \end{bmatrix}}_{\mathbf{k}^i} \frac{EI}{\ell^3} \tag{II.77}$$

Observe that the use of the polynomial of Eq. (II.64) to represent w results in the exact [Eq. (II.62)], rather than an approximate, stiffness matrix. Use of a different polynomial can lead to a different stiffness matrix. Stiffness matrices for many elements are not exact.

The loading vector $\bar{\mathbf{p}}^i$ is evaluated using Eq. (II.71):

$$\bar{\mathbf{p}}^i = \ell \int_0^1 \mathbf{N}^T p_z(\xi) \, d\xi$$

$$= \ell \, \mathbf{G}^T \int_0^1 \mathbf{N}_u^T p_z(\xi) \, d\xi = \ell \int_0^1 \begin{bmatrix} 1 - 3\xi^2 + 2\xi^3 \\ (-\xi + 2\xi^2 - \xi^3)\ell \\ 3\xi^2 - 2\xi^3 \\ (\xi^2 - \xi^3)\ell \end{bmatrix} p_z(\xi) \, d\xi \tag{II.78}$$

If p_z is a constant of magnitude p_0,

$$\bar{\mathbf{p}}^i = -p_0\ell \begin{bmatrix} -\frac{1}{2} \\ \frac{1}{12}\ell \\ -\frac{1}{2} \\ -\frac{1}{12}\ell \end{bmatrix} \tag{II.79}$$

For hydrostatic loading with p_z varying linearly from $\xi = 0$ to $\xi = 1$ where its magnitude is p_0, $p_z = p_0\xi$ and

$$\bar{\mathbf{p}}^i = -\frac{p_0\ell}{60} \begin{bmatrix} 9 \\ -2\ell \\ 21 \\ 3\ell \end{bmatrix} \tag{II.80}$$

Properties of Stiffness Matrices

It is not difficult to show that all stiffness matrices have several characteristics in common. Stiffness matrices are symmetric,

$$k_{ij} = k_{ji} \quad \text{and} \quad \mathbf{k}_{ab} = \mathbf{k}_{ba}^{T} \tag{II.81}$$

Also, the diagonal elements of a stiffness matrix are positive. Furthermore, after elimination of rigid-body motion, a stiffness matrix is positive definite.

A particularly interesting property can be observed by studying the stiffness matrix of Eq. (II.61). The sum of rows 1 and 3 is [0 0 0 0]. Thus, $\mathbf{k}^{i}$ with a value of zero for its determinant is *singular*. We say that rows 1 and 3 are *linearly dependent*, as are several other rows and columns such as columns 1 and 3. The application of the boundary conditions has the effect of rendering the stiffness matrix nonsingular.

II.6 MASS MATRICES

The incorporation of mass in an analysis raises several interesting questions become some mass models lead to inefficient and even ineffective numerical solutions. In the case of a transfer matrix, the mass per length ρ can be retained in its distributed form. This leads to the transfer matrices of Chapter 11 (Table 11-22) in which ρ appears nonlinearly in transcendental expressions. If such transfer matrices are converted [Eq. (II.59)] to stiffness matrices, the results are called *dynamic stiffness matrices*. In such a case, as can be seen in Table 11-22, the mass continues to appear nonlinearly in transcendental terms. As will be shown in Appendix III, this nonlinear representation of mass, although it constitutes exact modeling, tends to be difficult to handle efficiently in a dynamic analysis.

One technique to avoid having ρ appear in transcendental functions is to employ lumped-mass modeling, in which the mass is considered to act at distinct points only. The mass distributed to each side of a point is considered to be concentrated at the point. The transfer matrix to take into account a lumped mass can be derived from a transfer matrix containing distributed mass ρ by going to the limit as $x \to 0$ and $\rho x \to m$, where m is the magnitude of the mass (units of mass). In addition, this *point matrix* can be found from the conditions of continuity and equilibrium of the mass. Thus, taking only translational motion into account (Fig. II-7), $w_{+} = w_{-}$, $\theta_{+} = \theta_{-}$, $M_{+} = M_{-}$, and $V_{+} = V_{-} - m\omega^{2}w_{j}$,

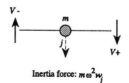

Inertia force: $m\omega^{2}w_{j}$

FIGURE II-7 A concentrated mass.

where ω is the frequently of the mass motion. This leads to the transfer point matrix

$$\mathbf{U}_j = \begin{bmatrix} 1 & 0 & 0 & 0 \\ 0 & 1 & 0 & 0 \\ -m\omega^2 & 0 & 1 & 0 \\ 0 & 0 & 0 & 1 \end{bmatrix} \tag{II.82}$$

The subscript j indicates that the lumped mass occurs at point j.

In stiffness matrix form (with ω^2 factored out) the lumped mass would be the diagonal matrix

$$\mathbf{m}^i = \begin{bmatrix} \frac{1}{2}m & 0 & 0 & 0 \\ 0 & 0 & 0 & 0 \\ 0 & 0 & \frac{1}{2}m & 0 \\ 0 & 0 & 0 & 0 \end{bmatrix} \tag{II.83}$$

Another approach for deriving a mass matrix, which is quite common in practice, is based on the principle of virtual work as expressed by Eq. (II.55):

$$\int_a^b \delta w'' \, EIw'' \, dx - \int_a^b \delta w \, p_z \, dx + \int_a^b \delta w \, \rho \ddot{w} \, dx - [M \, \delta\theta + V \, \delta w]_a^b = 0 \tag{II.84}$$

The third term on the left-hand side, which includes ρ, is of interest here. Recall from Eq. (II.64) that $w = \mathbf{N}\mathbf{v}^i$. Then $\ddot{w} = \mathbf{N}\ddot{\mathbf{v}}^i$. Since $\delta w = \delta \mathbf{v}^{iT}\mathbf{N}^T$, the third integral in Eq. (II.84) becomes

$$\delta \mathbf{v}^{iT} \int_a^b \rho \mathbf{N}^T \mathbf{N} \, dx \, \ddot{\mathbf{v}}^i \tag{II.85}$$

The integral

$$\mathbf{m}^i = \int_a^b \rho \mathbf{N}^T \mathbf{N} \, dx \tag{II.86}$$

defines a mass matrix that can be employed in the dynamic analysis of large systems.

If the same $\mathbf{N}$ is chosen for Eq. (II.86) as is employed to compute the stiffness matrix $\mathbf{k}^i$ of Eq. (II.72), then the mass matrix $\mathbf{m}^i$ is said to be *consistent*. Substitution of $\mathbf{N}$ in Eq. (II.64) in the integral of Eq. (II.86) gives

$$\mathbf{m}^i = \frac{\rho\ell}{420} \begin{bmatrix} 156 & -22\ell & 54 & 13\ell \\ -22\ell & 4\ell^2 & -13\ell & -3\ell^2 \\ 54 & -13\ell & 156 & 22\ell \\ 13\ell & -3\ell^2 & 22\ell & 4\ell^2 \end{bmatrix} \tag{II.87}$$

where it has been assumed that ρ is constant. The corresponding $\mathbf{v}^i$ is defined as in Eq. (II.61), i.e., $\mathbf{v}^i = [w_a \; \theta_a \; w_b \; \theta_b]^T$. Note that this mass matrix contains only simple linear or quadratic expressions in ℓ and that ρ, the mass per length, has been extracted from the matrix. Although this mass matrix is not diagonal, it is symmetric and often leads to a computationally efficient dynamic solution.

A more exact mass matrix can be obtained by using a more exact N in Eq. (II.86). Such an N can be taken from Table 11-22. Often this results in ρ appearing inside the mass matrix in transcendental form and results in a less efficient, but more accurate, dynamic analysis.

II.7 DYNAMIC STIFFNESS MATRICES

As mentioned above, transfer matrices such as those of Eq. (II.42) can contain the mass ρ without any approximations. This is referred to as "exact" mass modeling, which contrasts with the approximate lumped and consistent mass modeling. If such a transfer matrix is converted to a stiffness matrix, the resulting dynamic stiffness matrix k^i_{dyn} provides *exact* modeling of the mass.

II.8 GEOMETRIC STIFFNESS MATRICES

The treatment of the axial force P in a beam analysis is very similar to that of ρ, the mass per length. First, the axial force can be considered as being continuous, leading to the exact transfer and stiffness matrices of Table 11-22. These can be difficult to utilize numerically. Second, the axial force can be lumped at particular locations, providing a computationally attractive diagonal stiffness matrix.

Finally, the principle of virtual work can give a matrix for axial forces that is similar to the matrix of the previous section for mass. If axial force is taken into account explicitly, then the principle of virtual work of Eq. (II.54) would appear as

$$\int_a^b \delta w'' \, EIw'' \, dx - \int_a^b \delta w \, p_z \, dx - \int_a^b \delta w' \, Pw' \, dx - [M \, \delta\theta + V \, \delta w]_a^b = 0 \quad (II.88)$$

where P is in compression. The third integral on the left-hand side leads to the stiffness matrix

$$k^i_G = \int_a^b N'^T N' \, dx = G^T \int_a^b N'^T_u N'_u \, dx \, G \quad (II.89)$$

called the *geometric, differential,* or *stress stiffness matrix*. As explained in Appendix III, this is a very useful matrix for studies of structural stability.

If the same displacement trial function is employed in forming k^i_G that is used in deriving k^i, the geometric stiffness matrix is said to be *consistent*. If N of Eq. (II.64) is used,

$$k^i_G = \frac{1}{30\ell} \begin{bmatrix} 36 & -3\ell & -36 & -3\ell \\ -3\ell & 4\ell^2 & 3\ell & -\ell^2 \\ -36 & 3\ell & 36 & 3\ell \\ -3\ell & -\ell^2 & 3\ell & 4\ell^2 \end{bmatrix} \quad (II.90)$$

This symmetric matrix is the most commonly used geometric stiffness matrix.

More accurate, but computationally less favorable, geometric stiffness matrices can be obtained by utilizing a more accurate $\mathbf{N}$ such as that which can be taken from Table 11-22.

Also, $\mathbf{k}^i$ and $\mathbf{k}_G^i$ need not be based on the same displacement functions. Since only first-order derivatives of w appear in $\mathbf{k}_G^i$ of Eq. (II.89), whereas second-order derivatives of w appear in the $\mathbf{k}^i$ term [the first integral of Eq. (II.88)], a "simpler" displacement function is sometimes employed in forming $\mathbf{k}_G^i$.

REFERENCES

II.1   Pilkey, W. D., and Wunderlich, W., *Mechanics of Structures, Variational and Computational Methods*, CRC., Florida, 1993.

II.2   Pilkey, W. D., and Pilkey, O. H., *Mechanics of Solids*, Krieger, Melbourne, FL, 1986.

II.3   Pilkey, W. D., and Chang, P. Y., *Modern Formulas for Statics and Dynamics*, McGraw-Hill, New York, 1978.

II.4   Pestel, E., and Leckie, F., *Matrix Methods in Elastomechanics*, McGraw-Hill, New York, 1963.

II.5   Gallagher, R., *Finite Element Analysis Fundamentals*, Prentice-Hall, Englewood Cliffs, NJ, 1975.

Structural Systems

A structure will be considered to be composed of structural elements connected at *nodes* (joints). This structure is analyzed by assembling the characteristics of each element.

The first global analysis procedure to be considered is the *transfer matrix method*. This method is characterized by progressive matrix multiplications along

1357

a line system, resulting in a final matrix of size that does not depend on the number of elements in the structure.

A network structure such as a framework is normally treated using the *force* or *displacement method*. Unlike the transfer matrix method, the force and displacement methods lead to final system matrices that increase in size as the number of elements composing the system increases.

The force method (the *flexibility*, *influence coefficient*, or *compatibility* method) is based on the principle of complementary virtual work, which leads to global compatibility conditions and a system flexibility matrix relating redundant forces to applied loadings.

Computer-oriented structural analysis is dominated by the displacement method (the *stiffness method* or *equilibrium method*). Most of this appendix is concerned with the displacement method, which is based on the principle of virtual work and leads to global equilibrium equations. These relations are solved for the nodal displacements as functions of applied forces. For structural systems the displacement method is normally considered to be simpler to automate than the force method.

Structural mechanics textbooks such as reference III.1 contain detailed developments of the methods of analysis. This appendix provides a description of the most important techniques of structural analysis.

III.1   TRANSFER MATRIX METHOD

The transfer matrix giving the state variables $\mathbf{z}$ at point b in terms of the state variables at point a of a structure appears as

$$\mathbf{z}_b = \mathbf{U}^i \mathbf{z}_a + \bar{\mathbf{z}}^i \tag{III.1}$$

in which the transfer matrix $\mathbf{U}^i$ for a beam element for sign convention 1 is given by [Eqs. (II.32)]

$$
\begin{bmatrix} w \\ \theta \\ V \\ M \end{bmatrix}_b =
\begin{bmatrix}
1 & -\ell & -\ell^3/6EI & -\ell^2/2EI \\
0 & 1 & \ell^2/2EI & \ell/EI \\
0 & 0 & 1 & 0 \\
0 & 0 & \ell & 1
\end{bmatrix}
\begin{bmatrix} w \\ \theta \\ V \\ M \end{bmatrix}_a +
\begin{bmatrix} F_w \\ F_\theta \\ F_V \\ F_M \end{bmatrix}^i \tag{III.2}
$$

$$\mathbf{z}_b \qquad = \qquad\qquad \mathbf{U}^i \qquad\qquad\qquad \mathbf{z}_a \quad + \quad \bar{\mathbf{z}}^i$$

Frequently it is helpful to incorporate the loading terms in the transfer matrix by defining an *extended state vector* $\mathbf{z}$ and an *extended transfer matrix* $\mathbf{U}^i$.

$$
\begin{bmatrix} w \\ \theta \\ V \\ M \\ \hline 1 \end{bmatrix}_b =
\left[
\begin{array}{cccc|c}
1 & -\ell & -\ell^3/6EI & -\ell^2/2EI & F_w \\
0 & 1 & \ell^2/2EI & \ell/EI & F_\theta \\
0 & 0 & 1 & 0 & F_V \\
0 & 0 & \ell & 1 & F_M \\
\hline
0 & 0 & 0 & 0 & 1
\end{array}
\right]
\begin{bmatrix} w \\ \theta \\ V \\ M \\ \hline 1 \end{bmatrix}_a \tag{III.3}
$$

$$\mathbf{z}_b \qquad = \qquad\qquad\qquad \mathbf{U}^i \qquad\qquad\qquad\qquad \mathbf{z}_a$$

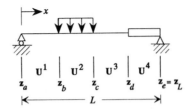

FIGURE III-1   System transfer matrix.

The system of Fig. III-1 is formed of several beam elements with transfer matrices

$$z_b = U^1 z_a \tag{III.4a}$$

$$z_c = U^2 z_b \tag{III.4b}$$

$$z_d = U^3 z_c \tag{III.4c}$$

$$z_e = U^4 z_d \tag{III.4d}$$

where z_i is the state vector at location i. Each transfer matrix is given by Eq. (III.3) using the appropriate EI, ℓ, and loading functions F_j. In Eqs. (III.4a)–(III.4d) the state vectors z_b, z_c, z_d, and z_e can be written in terms of the initial state vector z_a by replacing z_b in Eq. (III.4b) by z_b of Eq. (III.4a), z_c of Eq. (III.4c) by z_c of Eq. (III.4b), etc. Thus

$$
\begin{aligned}
z_b &= U^1 z_a \\
z_c &= U^2 z_b = U^2 U^1 z_a \\
z_d &= U^3 z_c = U^3 U^2 U^1 z_a \\
z_e &= U^4 z_d = U^4 U^3 U^2 U^1 z_a
\end{aligned}
\tag{III.5}
$$

It has been shown here that the state vector at any point along the beam is obtained by progressive multiplication of the transfer matrices for the elements from left to right up to that point. That is, at any point j

$$z_j = U^j U^{j-1} \cdots U^2 U^1 z_a \tag{III.6}$$

For a system with a total of M elements

$$z_{x=L} = z_L = U^M U^{M-1} \cdots U^2 U^1 z_a = U z_a \tag{III.7}$$

where z_L is the state vector at the right end and U is the *global*, or *overall*, *transfer matrix* extending from the left to the right end of the system.

Two "sweeps" along the structure are needed to complete a transfer matrix solution. First the overall, or global, transfer matrix U of Eq. (III.7) is established, normally using a computer program that calls up stored transfer matrices and performs the matrix multiplications of Eq. (III.7). For a beam the four boundary conditions are applied to Eq. (III.7), leading to equations for the four unknown state variables w_a, θ_a, V_a, and M_a. Thus, z_a is known. Now, a second sweep along the system using Eq. (III.6) permits the responses w, θ, V, and M to be

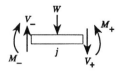

FIGURE III-2   Beam segment showing M and V to each side of j for sign convention 1.

calculated and printed out along the beam. Between stations the responses are calculated by adjusting the x coordinate (ℓ) in the transfer matrix for that element. A later section introduces some techniques that provide greater computational economy in developing a complete transfer matrix solution.

The transfer matrix procedure is characterized by simplicity and systemization. It involves system matrices of equal dimension to the element matrices. It is a *mixed method* in that both force and displacement responses are computed simultaneously as the calculations proceed. The primary disadvantage of the transfer matrix method is that it is numerically sensitive, particularly when the boundaries are far enough apart to have little influence on each other. It is apparent that the transfer matrix method applies only to structural systems possessing a chainlike topology.

Loading and In-Span Conditions

The incorporation in the transfer matrix solution of the effects of such occurrences as springs, lumped masses, and supports requires special consideration. Formulas for the calculation of the loading functions for distributed applied loading were developed in the previous appendix. The introduction of concentrated applied loadings will be treated here. One case, lumped masses, was considered in Appendix II.

Suppose a concentrated transverse force W is applied at point (node) j. Consider the short segment spanning j shown in Fig. III-2. The deflection w and slope θ will be continuous across j so that, $w_+ = w_-$, $\theta_+ = \theta_-$. For an infinitesimally short element a summation of moments about j shows that the bending moment is also continuous, giving $M_+ = M_-$. However, the condition of equilibrium for the vertical forces gives $V_- - W - V_+ = 0$ or $V_+ = V_- - W$, which shows that the shear force changes by a magnitude W in moving across the load. In summary,

$$
\begin{bmatrix} w \\ \theta \\ V \\ M \end{bmatrix}_j^+ = \begin{bmatrix} w \\ \theta \\ V \\ M \end{bmatrix}_j^- + \begin{bmatrix} 0 \\ 0 \\ -W \\ 0 \end{bmatrix}_j \tag{III.8a}
$$

$$
\mathbf{z}_j^+ = \mathbf{z}_j^- + \bar{\mathbf{z}}_j
$$

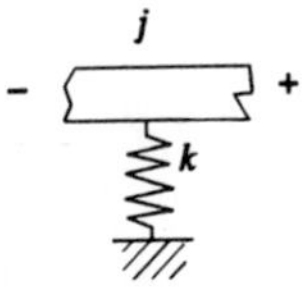

FIGURE III-3 Beam and spring.

or in transfer matrix form

$$
\begin{bmatrix} w \\ \theta \\ V \\ M \\ 1 \end{bmatrix}_j^{+} = \begin{bmatrix} 1 & 0 & 0 & 0 & 0 \\ 0 & 1 & 0 & 0 & 0 \\ 0 & 0 & 1 & 0 & -W \\ 0 & 0 & 0 & 1 & 0 \\ 0 & 0 & 0 & 0 & 1 \end{bmatrix} \begin{bmatrix} w \\ \theta \\ V \\ M \\ 1 \end{bmatrix}_j^{-}
$$

$$
\mathbf{z}_j^{+} \quad = \quad\quad\quad \mathbf{U}_j \quad\quad\quad \mathbf{z}_j^{-}
$$

(IIII.8b)

Such a transfer matrix is referred to as a *point matrix* while the transfer matrix for an element with distributed properties is called a *field matrix*.

Example III.1 Spring The point matrix for an extension spring is rather simple to derive. The force in the spring of Fig. III-3 is proportional to the beam deflection w at j, i.e., the force is kw_j. Utilize the point matrix of Eq. (III.8) with $W = -kw_j$, where the minus sign indicates that the force due to the spring is upward while W of Fig. III-2 is in the downward direction. The point matrix for the spring is given by

$$
\begin{bmatrix} w \\ \theta \\ V \\ M \\ 1 \end{bmatrix}_j^{+} = \begin{bmatrix} 1 & 0 & 0 & 0 & 0 \\ 0 & 1 & 0 & 0 & 0 \\ 0 & 0 & 1 & 0 & kw_j \\ 0 & 0 & 0 & 1 & 0 \\ 0 & 0 & 0 & 0 & 1 \end{bmatrix} \begin{bmatrix} w \\ \theta \\ V \\ M \\ 1 \end{bmatrix}_j^{-}
$$

$$
= \begin{bmatrix} 1 & 0 & 0 & 0 & 0 \\ 0 & 1 & 0 & 0 & 0 \\ k & 0 & 1 & 0 & 0 \\ 0 & 0 & 0 & 1 & 0 \\ 0 & 0 & 0 & 0 & 1 \end{bmatrix} \begin{bmatrix} w \\ \theta \\ V \\ M \\ 1 \end{bmatrix}_j^{-}
$$

(1)

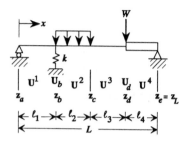

FIGURE III-4   Point occurrences.

Point and field matrices are incorporated in the same manner in the progressive matrix multiplications of a transfer matrix solution. For the beam of Fig. III-4, for example, the state variable z_e is given by

$$z_{x=L} = z_e = U^4 U_d U^3 U^2 U_b U^1 z_a$$

Introduction of Boundary Conditions

Formulas for transfer matrices are provided throughout this book. Transfer matrix notation is summarized in Table III-1. A solution for a static problem begins with the modeling of the structural system in terms of elements that connect locations of point occurrences such as applied concentrated forces or jumps in cross-sectional area. Determine the section properties such as the element moment of inertia and calculate the field matrix for each element as well as the point matrices for the concentrated occurrences. Then form the global transfer matrix by multiplying progressively the transfer matrices from the left end to the right end of the system. Thus, for a system with M elements, calculate U of

$$z_{x=L} = z_L = U^M U^{M-1} \cdots U_k \cdots U^2 U^1 z_a = U z_0 \qquad (III.9a)$$

From this relationship evaluate the initial variables w_a, θ_a, V_a, and M_a of

$$z_a = \begin{bmatrix} w \\ \theta \\ V \\ M \\ 1 \end{bmatrix}_a$$

by applying the boundary conditions to Eq. (III.9a). In implementing this, eliminate the unnecessary rows and columns of Eq. (III.9a) and solve the remaining equations. The solution is completed by calculating the deflection,

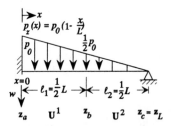

FIGURE III-5 Beam with linearly varying loading.

slope, shear force, and internal moment at all points of interest using

$$\mathbf{z}_j = \mathbf{U}^j \mathbf{U}^{j-1} \cdots \mathbf{U}_k \cdots \mathbf{U}^2 \mathbf{U}^1 \mathbf{z}_a \tag{III.9b}$$

The responses can be computed between the ends of the elements by adjusting the x coordinate (ℓ) in the transfer matrix for that element.

The transfer matrix method is described in many texts such as references III.1–III.4. Techniques for improving the computational efficiency and the numerical stability of transfer matrix solutions are considered in the following sections.

Example III.2 Beam with Linearly Varying Loading Since beam solutions are exact, the uniform beam of Fig. III-5 can be modeled with one element. We choose, however, to consider the beam as being modeled with two elements.

From Table 11-19, with shear deformation ignored, the transfer matrices for each element are

$$\mathbf{U}^1 = \begin{bmatrix} 1 & -\ell & -\ell^3/6EI & -\ell^2/2EI & (4p_0 + \frac{1}{2}p_0)\ell^4/(120EI) \\ 0 & 1 & \ell^2/2EI & \ell/EI & -(3p_0 + \frac{1}{2}p_0)\ell^3/(24EI) \\ 0 & 0 & 1 & 0 & -\frac{1}{2}(p_0 + \frac{1}{2}p_0)\ell \\ 0 & 0 & \ell & 1 & -\frac{1}{6}(2p_0 + \frac{1}{2}p_0)\ell^2 \\ 0 & 0 & 0 & 0 & 1 \end{bmatrix}_{\ell=\ell_1} \tag{1}$$

$$\mathbf{U}^2 = \begin{bmatrix} 1 & -\ell & -\ell^3/6EI & -\ell^2/2EI & p_0\ell^4/60EI \\ 0 & 1 & \ell^2/2EI & \ell/EI & -p_0\ell^3/16EI \\ 0 & 0 & 1 & 0 & -\frac{1}{4}p_0\ell \\ 0 & 0 & \ell & 1 & -\frac{1}{6}p_0\ell^2 \\ 0 & 0 & 0 & 0 & 1 \end{bmatrix}_{\ell=\ell_2} \tag{2}$$

The overall transfer matrix **U** is calculated as

$$\mathbf{z}_{x=L} = \mathbf{z}_c = \mathbf{U}^2 \mathbf{U}^1 \mathbf{z}_a = \mathbf{U}\mathbf{z}_a \tag{3}$$

The vector $\mathbf{z}_a$ of initial values is evaluated by applying the boundary conditions to (3). For beams, two of the four initial responses w_a, θ_a, V_a, and M_a are determined by observation. Thus, for the beam of Fig. III-5, it is evident that $w_a = 0$ and $\theta_a = 0$ since the left end is fixed. At the other end, which is simply supported, the conditions are $w_{x=L} = 0$ and $M_{x=L} = 0$. These conditions are applied to (3):

$$
\begin{bmatrix} w = 0 \\ \theta \\ V \\ M = 0 \\ 1 \end{bmatrix}_{x=L}
=
\left[\begin{array}{cc|cc} & & U_{wV} & U_{wM} & F_w \\ \hline & & & & \\ & & U_{MV} & U_{MM} & F_M \\ & & 0 & 0 & 1 \end{array}\right]_{x=L}
\begin{bmatrix} w = 0 \\ \theta = 0 \\ V \\ M \\ 1 \end{bmatrix}_{x=0}
\tag{4}
$$

$$
\mathbf{z}_{x=L} = \mathbf{U}^i \quad \mathbf{z}_a
$$

| Cancel columns 1 and 2 since $w_0 = \theta_0 = 0$ | Ignore rows 2 and 3 since $\theta_{x=L}$ and $V_{x=L}$ are unknown |

where U_{kj} and F_k are the elements of $\mathbf{U}$ of (3). The equations $w_{x=L} = 0$ and $M_{x=L} = 0$ in (4) are used to compute M_a and V_a:

$$
\begin{bmatrix} w \\ M \end{bmatrix}_{x=L} = \begin{bmatrix} 0 \\ 0 \end{bmatrix} = \begin{bmatrix} U_{wV} & U_{wM} \\ U_{MV} & U_{MM} \end{bmatrix}_{x=L} \begin{bmatrix} V_a \\ M_a \end{bmatrix} + \begin{bmatrix} F_w \\ F_M \end{bmatrix}_{x=L}
\tag{5}
$$

or

$$
0 = V_a \bar{U}_{wV} + M_a \bar{U}_{wM} + \bar{F}_w \qquad 0 = V_a \bar{U}_{MV} + M_a \bar{U}_{MM} + \bar{F}_M
\tag{6}
$$

where

$$
\bar{U}_{ij} = U_{ij}|_{x=L} \qquad \bar{F}_i = F_i|_{x=L}
$$

That is, an overbar is used to indicate that the transfer matrix or loading component are global components in that they are evaluated at the right end of the member. Equation (6) can be solved for M_a, V_a

$$
V_a = (F_M U_{wM} - F_w U_{MM})|_{x=L}/\nabla = \tfrac{2}{5} p_0 L
$$
$$
M_a = (F_w U_{MV} - F_M U_{wV})|_{x=L}/\nabla = -\tfrac{1}{15} p_0 L^2
\tag{7}
$$

where

$$
\nabla = (U_{wV} U_{MM} - U_{wM} U_{MV})|_{x=L} = \bar{U}_{wV} \bar{U}_{MM} - \bar{U}_{wM} \bar{U}_{MV}
\tag{8}
$$

The vector $\mathbf{z}_a$ is fully determined now, since $w_a = 0$, $\theta_a = 0$, and M_a, V_a are given by (7). The variables w, θ, V, and M can be computed using Eq. (III.6) and can be printed out at selected locations. For example, the responses at nodes b and c are

$$\mathbf{z}_b = \mathbf{z}_{x=\ell_1} = \mathbf{U}^1\mathbf{z}_a \qquad \mathbf{z}_c = \mathbf{z}_{x=L} = \mathbf{U}^2\mathbf{U}^1\mathbf{z}_a = \mathbf{U}\mathbf{z}_a \tag{9}$$

Between the ends of elements, responses are computed by adjusting the coordinate in the appropriate transfer matrix. For example, if the responses are desired at the midpoint of the second element,

$$\mathbf{z}_{x=\ell_1+\frac{1}{2}\ell_2} = \mathbf{U}^2(\tfrac{1}{2}\ell_2)\mathbf{U}^1(\ell_1)\mathbf{z}_a \tag{10}$$

Example III.3 Beam on Flexible Supports The beam of Fig. III-6b is a free–free beam model with 10 extension springs and 1 concentrated force of the beam of Fig. III-6a.

In transfer matrix form the response at the right end is given by

$$\mathbf{z}_{x=L} = \mathbf{U}_k\mathbf{U}^{10}\mathbf{U}_j\mathbf{U}^9 \cdots \mathbf{U}^6\mathbf{U}_f\mathbf{U}^5 \cdots \mathbf{U}^2\mathbf{U}_b\mathbf{U}^1\mathbf{U}_a\mathbf{z}_0 = \mathbf{U}\mathbf{z}_0 \tag{1}$$

The transfer matrices are given by

$$\mathbf{U}^i = \begin{bmatrix} 1 & -\ell & -\ell^3/6EI & -\ell^2/2EI & 0 \\ 0 & 1 & \ell^2/2EI & \ell/EI & 0 \\ 0 & 0 & 1 & 0 & 0 \\ 0 & 0 & \ell & 1 & 0 \\ 0 & 0 & 0 & 0 & 1 \end{bmatrix} \qquad \begin{array}{l} i = 1, 2, \ldots, 9, 10 \\ \text{(Table 11-19)} \end{array} \tag{2}$$

$$\mathbf{U}_i = \begin{bmatrix} 1 & 0 & 0 & 0 & 0 \\ 0 & 1 & 0 & 0 & 0 \\ k_i & 0 & 1 & 0 & 0 \\ 0 & 0 & 0 & 1 & 0 \\ 0 & 0 & 0 & 0 & 1 \end{bmatrix} \qquad i = a, b, c, d, e, g, \ldots, k \qquad \text{(Table 11-21)} \tag{3}$$

$$\mathbf{U}_f = \begin{bmatrix} 1 & 0 & 0 & 0 & 0 \\ 0 & 1 & 0 & 0 & 0 \\ 0 & 0 & 1 & 0 & -W \\ 0 & 0 & 0 & 1 & 0 \\ 0 & 0 & 0 & 0 & 1 \end{bmatrix} \qquad \text{(Table 11-21)} \tag{4}$$

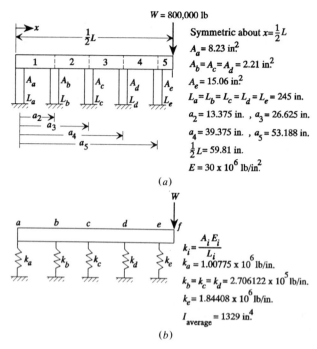

FIGURE III-6   Beam on flexible supports: (a) half of a beam; (b) model of beam in (a).

The ends of the beam are free. Thus, the boundary conditions are $M_{x=0} = V_{x=0} = M_{x=L} = V_{x=L} = 0$. These are applied to (1), $\mathbf{z}_{x=L} = \mathbf{U}\mathbf{z}_{x=0}$, where matrices with an overbar are evaluated at $x = L$,

$$
\begin{bmatrix}
w \\
\theta \\
V = 0 \\
M = 0 \\
1
\end{bmatrix}_{x=L}
=
\begin{bmatrix}
\cdots & \cdots & \vdots & \bar{F}_w \\
\cdots & \cdots & \vdots & \bar{F}_\theta \\
\bar{U}_{Vw} & \bar{U}_{V\theta} & \vdots & \bar{F}_V \\
\bar{U}_{Mw} & \bar{U}_{M\theta} & \vdots & \bar{F}_M \\
0 & 0 & \vdots & 1
\end{bmatrix}
\begin{bmatrix}
w \\
\theta \\
V = 0 \\
M = 0 \\
1
\end{bmatrix}_{x=0}
\tag{5}
$$

Cancel columns 3 and 4 because $V_0 = M_0 = 0$

Ignore rows 1 and 2 because $w_{x=L}, \theta_{x=L}$ are unknown

where U_{kj} and F_k are the elements of $\mathbf{U}$ of (1), and $\bar{U}_{kj} = U_{kj}|_{x=L}$ and $\bar{F}_k = F_k|_{x=L}$. The equations $M_{x=L} = 0$ and $V_{x=L} = 0$ are used to compute the unknown initial parameters w_0, θ_0. Thus, from (5),

$$V_{x=L} = 0 = w_0 \overline{U}_{Vw} + \theta_0 \overline{U}_{V\theta} + \overline{F}_V \qquad M_{x=L} = 0 = w_0 \overline{U}_{Mw} + \theta_0 \overline{U}_{M\theta} + \overline{F}_M$$

so that

$$w_0 = \left(-\overline{F}_V \overline{U}_{M\theta} + \overline{F}_M \overline{U}_{V\theta}\right)/\nabla \qquad \theta_0 = \left(-\overline{F}_M \overline{U}_{Vw} + \overline{F}_V \overline{U}_{Mw}\right)/\nabla \tag{6}$$

$$\nabla = \overline{U}_{Vw} \overline{U}_{M\theta} - \overline{U}_{V\theta} \overline{U}_{Mw}$$

Now that z_0 (i.e., w_0, θ_0, V_0, and M_0) is known, the state vector can be printed out along the beam, e.g.,

$$z_{x=a_2} = \mathbf{U}^1 \mathbf{U}_a z_0 \qquad z_{x=a_3} = \mathbf{U}^2 \mathbf{U}_b \mathbf{U}^1 \mathbf{U}_a z_0 \tag{7}$$

The reactions at the springs are found by monitoring the change in shear force across the springs or by forming the product of the beam deflections at the springs and the spring constants. The reactions are computed to be

$$R_a = 34{,}230 \text{ lb} \qquad R_b = 19{,}180 \text{ lb} \qquad R_c = 27{,}560 \text{ lb} \qquad R_d = 36{,}330 \text{ lb}$$
$$R_e = 281{,}800 \text{ lb} \tag{8}$$

The above approach permits the inclusion of a variable moment of inertia along the beam axis.

Example III.4 Thermal Stress Analysis Suppose a beam of rectangular cross section with the left end fixed and right end simply supported (Fig. III-7) is subjected to a change in temperature,

$$T = T(z) = T_0 z/h \tag{1}$$

where T_0 is a known constant. The reaction R_L of Fig. III-7 is produced by the temperature change imposed on the beam. Find the displacements and bending stresses due to this thermal loading. The thermal moment is given by

$$M_T = E\alpha \int_A Tz \, dA = \frac{E\alpha T_0}{h} \int_A z^2 \, dA = \frac{EI\alpha T_0}{h} \tag{2}$$

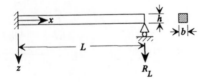

FIGURE III-7 Beam subjected to transverse temperature change.

Since the origin of the y, z axes is the centroid of the cross section, the thermal axial force is

$$P_T = E\alpha \int_A T \, dA = E\alpha T_0 \frac{1}{h} \int_A z \, dA = 0 \tag{3}$$

Suppose $E = 200$ GN/m², $L = 2$ m, $h = 0.15$ m, $b = 0.07$ m, $T_0 = 20°C$, and $\alpha = 11 \times 10^{-6}$ (1/°C). Insertion of these values into (2) gives

$$M_T = 5.775 \text{ kN} \cdot \text{m} \tag{4}$$

where $I = \frac{1}{12}bh^3$.

In transfer matrix notation the response takes the form

$$\mathbf{z}_x = \mathbf{U}^1 \mathbf{z}_0 \tag{5}$$

with $\mathbf{z}$ given by $[w \ \theta \ V \ M \ 1]^T$. The transfer matrix $\mathbf{U}^1$ can be taken directly from Table 11-19. Alternatively, $\mathbf{U}^1$ can be extracted from Table 11-22. In doing so, note that there is no elastic foundation ($k = 0$), no axial force ($P = 0$), no rotary foundation ($k^* = 0$), and mass and shear deformation are not to be considered ($\rho = 0$ and $1/GA_S = 0$). Thus

$$\lambda = \eta = \zeta = 0 \tag{6}$$

For this case, the e_i functions with $x = L$ are given by case 2 of Table 11-22, page 586,

$$e_0 = 0 \quad e_1 = 1 \quad e_2 = L \quad e_3 = \tfrac{1}{2}L^2$$
$$e_4 = \tfrac{1}{6}L^3 \quad e_5 = \tfrac{1}{24}L^4 \quad e_6 = \tfrac{1}{120}L^5 \tag{7}$$

The loading column is formed using Table 11-22, page 585, with

$$M_{Ta} = M_{Tb} = M_T \quad P_a = P_b = c_a = c_b = 0 \tag{8}$$

Then

$$\bar{F}_w = -M_T L^2/2EI \quad \bar{F}_\theta = M_T L/EI \quad \bar{F}_V = 0 \quad \bar{F}_M = 0 \tag{9}$$

Substitution of e_i from (7) and the loading functions of (9) into the transfer matrix gives, at $x = L$,

$$\begin{bmatrix} 1 & -L & -L^3/6EI & -L^2/2EI & -M_T L^2/2EI \\ 0 & 1 & L^2/2EI & L/EI & M_T L/EI \\ 0 & 0 & 1 & 0 & 0 \\ 0 & 0 & L & 1 & 0 \\ 0 & 0 & 0 & 0 & 1 \end{bmatrix} \tag{10}$$

LOCATION (m)	DEFLECTION (m)	SLOPE	MOMENT (N · m)	SHEAR (N)
.00000	.00000	.00000	− 8662.5	4331.3
.20000	1.32000E-05	− 1.24667E-04	− 7796.3	4331.3
.40000	4.69333E-05	− 2.05333E-04	− 6930.0	4331.3
.60000	9.24000E-05	− 2.42000E-04	− 6063.8	4331.3
.80000	1.40800E-04	− 2.34667E-04	− 5197.5	4331.3
1.0000	1.83333E-04	− 1.83333E-04	− 4331.3	4331.3
1.2000	2.11200E-04	− 8.80000E-05	− 3465.0	4331.3
1.4000	2.15600E-04	5.13333E-05	− 2598.8	4331.3
1.6000	1.87733E-04	2.34667E-04	− 1732.5	4331.3
1.8000	1.18800E-04	4.62000E-04	− 866.25	4331.3
2.0000	.00000	7.33333E-04	.00000	4331.3

FIGURE III-8 Response of beam with thermal loading.

Applied to (10), the boundary conditions $w_{x=0} = w_{x=L} = 0$, $\theta_{x=0} = 0$, and $M_{x=L} = 0$ appear as

$$\begin{bmatrix} w = 0 \\ \theta \\ V \\ M = 0 \\ 1 \end{bmatrix}_{x=L} = \bar{U}^1 \begin{bmatrix} w = 0 \\ \theta = 0 \\ V \\ M \\ 1 \end{bmatrix}_{x=0} \tag{11}$$

Solution of the equations $w_{x=L} = 0$ and $M_{x=L} = 0$ of (11) gives the initial conditions

$$V_0 = 3M_T/2L = 4331.3 \text{ N} \qquad M_0 = -\tfrac{3}{2}M_T = -8662.5 \text{ N} \cdot \text{m} \tag{12}$$

Equation (5) can be used to print out the responses along the beam by using the initial conditions of (12) and the transfer matrix of (10) with L replaced by x. The results are shown in Fig. III-8.

If an analytical expression is desired, Eq. (5) is readily shown to reduce to the deflection

$$w = \frac{M_T}{4EI}\left(x^2 - \frac{x^3}{L}\right) = \frac{\alpha T_0}{4h}\left(x^2 - \frac{x^3}{L}\right) \tag{13}$$

and the internal moment is given by

$$M = \frac{3M_T}{2}\left(\frac{x}{L} - 1\right) = \frac{3EI\alpha T_0}{2h}\left(\frac{x}{L} - 1\right) \tag{14}$$

The normal stress due to this thermal loading is (Table 15-1, Case 1)

$$\sigma = -\alpha ET + \frac{Mz}{I} = \frac{\alpha ET_0 z}{2h}\left(3\frac{x}{L} - 5\right) \tag{15}$$

Stability

Structural members can reach a critical state that is quite different from the usual critical strength or stiffness levels set as criteria for structural failure. This state is referred to as *instability* or *buckling* and is the result of the ordinary equilibrium mode of deformation becoming unstable. The state of buckling is usually caused by an axial or in-plane force being of such a value (the critical load) that the response, e.g., displacement, begins to increase inordinately as the load is increased slightly. The governing equations of motion for a simple beam with a compressive axial force P are [Eq. (11.7a)], with $k = 0$,

$$EI\frac{d^4w}{dx^4} + P\frac{d^2w}{dx^2} = p_z$$

$$V = -EI\frac{d^3w}{dx^3} - P\frac{dw}{dx} \qquad M = -EI\frac{d^2w}{dx^2} \qquad \theta = -\frac{dw}{dx}$$

These relations, with $p_z = 0$, can be solved in the transfer matrix form, giving

$$
\begin{bmatrix} w \\ \theta \\ V \\ M \end{bmatrix}_b
=
\begin{bmatrix}
1 & -\dfrac{s}{\alpha} & \dfrac{\alpha\ell - s}{\alpha dEI} & \dfrac{1-c}{dEI} \\
0 & c & -\dfrac{1-c}{dEI} & \dfrac{s}{\alpha EI} \\
0 & 0 & 1 & 0 \\
0 & \dfrac{EIsd}{\alpha} & \dfrac{s}{\alpha} & c
\end{bmatrix}
\begin{bmatrix} w \\ \theta \\ V \\ M \end{bmatrix}_a
\qquad (\text{III.10})
$$

$$\mathbf{z}_b \qquad\qquad = \qquad\qquad\qquad \mathbf{U}^i \qquad\qquad\qquad\qquad \mathbf{z}_a$$

For a compressive axial force

$$\alpha^2 = P/EI \qquad s = \sin\alpha\ell \qquad c = \cos\alpha\ell \qquad d = -\alpha^2 \qquad (\text{III.11})$$

and for a tensile axial force

$$\alpha^2 = -P/EI \qquad s = \sinh\alpha\ell \qquad c = \cosh\alpha\ell \qquad d = \alpha^2 \qquad (\text{III.12})$$

This same transfer matrix is given in Table 11-22.

Buckling can be identified by determining the axial load for which expressions w, θ, V, M experience unrestrained growth. These expressions become large if the values of the initial parameters w_0, θ_0, V_0, M_0 (or w_a, θ_a, V_a, M_a) become large. Thus, the critical level of axial force is reached if the denominators of the w_a, θ_a, V_a, M_a expressions, obtained by application of the boundary conditions to Eq. (III.7), approach zero. In solving for w_a, θ_a, V_a, and M_a, it is found that these initial values increase inordinately in magnitude for specific values of the axial force. The lowest value is the *critical* or *buckling load*.

This sort of problem involving particular values of a parameter (here the axial force) is called an *eigenvalue* problem. These special discrete values are called *characteristic values* or *eigenvalues*, and the corresponding responses are the *characteristic functions*, *eigenfunctions*, or *mode shapes*. The expression that leads to the critical values is called the *characteristic equation*. Eigenvalue problems also arise in the study of the dynamics of structural members, with the *natural frequencies* being the eigenvalues.

It should be understood that this classical approach to instability, which involves "unrestrained growth" of the response, is based on fundamental equations of motion that were derived for linearly elastic material and small deflections. Strictly speaking, it is improper to think of truly large deflections. More accurate theories are required to describe large deflections.

Example III.5   Buckling Load for a Fixed–Fixed Column   Find the critical axial force for a uniform column of length L with fixed ends.

The transfer matrix $\mathbf{U}^i = \mathbf{U}$ for a uniform beam element with axial force P is given by Eq. (III.10). Since the column is fixed on both ends, the boundary conditions are

$$w_{x=a} = \theta_{x=a} = w_{x=L} = \theta_{x=L} = 0 \tag{1}$$

Applied to $\mathbf{z}_L = \mathbf{U}\mathbf{z}_a$, give these conditions

$$
\begin{bmatrix} w = 0 \\ \theta = 0 \\ V \\ M \end{bmatrix}_b
=
\left[
\begin{array}{cc|cc}
 & & \bar{U}_{wV} & \bar{U}_{wM} \\
 & & \bar{U}_{\theta V} & \bar{U}_{\theta M} \\
\hline
 & & & \\
 & & &
\end{array}
\right]
\begin{bmatrix} w = 0 \\ \theta = 0 \\ V \\ M \end{bmatrix}_a
\tag{2}
$$

or

$$0 = V_a \bar{U}_{wV} + M_a \bar{U}_{wM} \qquad 0 = V_a \bar{U}_{\theta V} + M_a \bar{U}_{\theta M} \tag{3}$$

where $\bar{U}_{ij} = U_{ij}|_{x=L}$.

The determinant of the equations on the right-hand side is

$$\nabla = \left(\bar{U}_{wV} \bar{U}_{\theta M} - \bar{U}_{wM} \bar{U}_{\theta V} \right)_{x=L} \tag{4}$$

If these were not homogeneous relations and Cramer's rule were employed to solve these equations, then the denominator of the responses would be a determinant, say ∇. See, for example, Eqs. (7) of Example III.2 or Eqs. (6) of Example III.3. Inordinately large responses correspond to $\nabla = 0$.

The buckling criterion (characteristic equation) of $\nabla = 0$ can also be reasoned to be the condition for finding nontrivial solutions of a system of homogeneous equations.

Substitution of the appropriate transfer matrix elements of Eq. (III.10) in $\nabla = 0$ gives

$$-\frac{(1 - \cos \alpha L)^2}{P^2} - \frac{\sin^2 \alpha L}{P^2} + \frac{\alpha L}{P^2} \sin \alpha L = 0 \tag{5}$$

which for the meaningful case of $P \neq 0$ reduces to

$$2 - 2\cos \alpha L - \alpha L \sin \alpha L = 0$$

This expression is satisfied by

$$\alpha L = 2n\pi \qquad n = 0, 1, 2, \ldots \qquad \alpha^2 = P/EI \qquad (6)$$

The desired critical or buckling load is given by the lowest meaningful value of P in (6), i.e., for $n = 1$,

$$P|_{\nabla = 0} = \frac{4\pi^2 EI}{L^2} = P_{\text{cr}} \qquad (7)$$

The value $n = 0$ is ruled out as it implies that $P = 0$, which is a trivial or meaningless case. Although ∇ assumes a value of zero for other values of n ($n = 2, 3, \ldots$), the corresponding unstable loads P are of limited engineering interest since the column has essentially already failed.

An interesting characteristic of this theory of instability is that the critical axial load of a beam is the same regardless of the transverse loading on the bar because the loading does not affect the values of the determinant of the equations. For example, in Eqs. (7) of Example III.2 or Eqs. (6) of Example III.3, the loading functions F_w, F_θ, F_V, and F_M appear only in the numerators of the initial-value expressions and not in ∇.

The stability of a member with abrupt in-span changes is treated in a fashion similar to that of the member of constant cross section. The characteristic equation is formed of the global, or overall, transfer matrix elements developed by the usual progressive multiplication of the transfer matrices of the various elements of the member.

It should be noted that the axial force in various elements of a member may be calculated by applying an equilibrium of forces in the axial direction. Thus for the member of Fig. III-9

$$(P)_{\text{element } 1} = P_0$$
$$(P)_{\text{element } 2} = P_0 + P_1$$
$$\vdots$$
$$(P)_{\text{element } j} = P_0 + P_1 + \cdots + P_{j-1}$$

As often as not, it is difficult, if not impossible, to find an analytical expression for P that satisfies the characteristic equation, e.g., $\nabla = 0$ in Eq. (5) of Example III.5. Thus P_{cr} is usually found by computationally searching for the lowest value of P, i.e., the root, for which $\nabla = 0$. A computer program is usually written that evaluates ∇ numerically for trial values of P in the search routine. The magnitude of P is increased until ∇ changes sign. Then employ the logic of one of the many root-finding techniques to close in on P_{cr}.

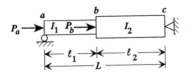

FIGURE III-9 Column of stepped cross section.

FIGURE III-10 Stepped column.

In the case of a member with a variable axial load (e.g., the beam of Fig. III-9) the ratio of the load applied at each element to a nominal value, e.g., P_0, is usually known. Then the nominal value leading to instability is sought by setting $\nabla = 0$.

Example III.6 Column of Variable Cross Section Find the buckling load for the stepped column of Fig. III-10.

The global transfer matrix is defined by

$$\mathbf{z}_L = \mathbf{U}\mathbf{z}_a \tag{1}$$

with

$$\mathbf{U} = \mathbf{U}^2\mathbf{U}^1 \tag{2}$$

where $\mathbf{U}^1$ is given by Eq. (III.10) with ℓ, EI, and P replaced by ℓ_1, EI_1, and P_a. The transfer matrix $\mathbf{U}^2$ is given by Eq. (III.10) using ℓ_2, EI_2, and $P_a + P_b$. Note that the axial force in element 2 is $P_a + P_b$, not just P_b.

The column is simply supported on both ends so that the boundary conditions are

$$w_{x=a} = M_{x=a} = w_{x=L} = M_{x=L} = 0 \tag{3}$$

These conditions applied to (1) appear as

$$\underbrace{\begin{bmatrix} w = 0 \\ \theta \\ V \\ M = 0 \end{bmatrix}_L}_{\mathbf{z}_L} = \underbrace{\begin{bmatrix} \overline{U}_{w\theta} & \overline{U}_{wV} \\ \hline \overline{U}_{M\theta} & \overline{U}_{MV} \end{bmatrix}}_{\mathbf{U}} \underbrace{\begin{bmatrix} w = 0 \\ \theta \\ V \\ M = 0 \end{bmatrix}_a}_{\mathbf{z}_a} \tag{4}$$

or

$$0 = \theta_a \overline{U}_{w\theta} + V_a \overline{U}_{wV} \qquad 0 = \theta_a \overline{U}_{M\theta} + V_a \overline{U}_{MV} \tag{5}$$

where $\overline{U}_{ij} = U_{ij}|_{x=L}$.

The determinant (set equal to zero) of these homogeneous equations constitutes the characteristic equation. That is,

$$\nabla = \overline{U}_{w\theta}\overline{U}_{MV} - \overline{U}_{M\theta}\overline{U}_{wV} = 0 \tag{6}$$

leads to the buckling load.

Relation (6) becomes

$$\frac{1}{\alpha_1^2} - \frac{\alpha_1^2 LEI_1 P_b + \ell_1}{\alpha_1 \tan \alpha_1 \ell_1} - \frac{I_1/I_2}{\alpha_2^2} - \frac{\alpha_2^2 LEI_1 P_b - \ell_2 I_1/I_2}{\alpha_2 \tan \alpha_2 \ell_2} = 0 \tag{7}$$

where

$$\alpha_1^2 = P_a/EI_1 \qquad \alpha_2^2 = (P_a + P_b)/EI_2$$

Combinations of P_a and P_b that satisfy $\nabla = 0$ define the conditions of instability. Normally P_a and P_b are not independent. Typically, P_b is known to be proportional to P_a, i.e., $P_b = cP_a$, where c is a known constant. Then (6) is the characteristic equation for a single unknown, the lowest value of which is the buckling load.

As mentioned above, the buckling loads for complicated stepped columns must be found by a numerical search for the lowest root of $\nabla = 0$. Typically, this process begins by evaluating ∇ for an estimated buckling load that is believed to be below the actual buckling load. Increase the estimate and repeat the evaluation of ∇. Continue the process until ∇ changes sign. The desired buckling load, which lies between the two previous estimates, is then found to a prescribed accuracy by utilizing a root-finding scheme such as Newton–Raphson.

More information about eigenvalue problems is given in the following section on dynamics in which the natural frequencies rather than the critical loads are the eigenvalues.

Free Vibrations

Special consideration must be given to structural problems when acceleration effects cannot be neglected. More specifically, problems in dynamics arise when the inertia of the acceleration of the structural mass must be taken into account. The response or solution that is sought will now be given by time-dependent state variables. Some terminology for the fundamentals of dynamics is provided in Chapter 10.

If a structural member possesses only a small amount of damping or if the dynamic response is desired for only a short period of time, an assumption of no damping may be imposed upon the equations governing the motion. As an example, for an undamped Euler–Bernoulli beam the equations of motion become

$$\frac{\partial^2}{\partial x^2}\left(EI\frac{\partial^2 w}{\partial x^2}\right) = -\rho\frac{\partial^2 w}{\partial t^2} + p_z(x,t)$$

$$V(x,t) = -\frac{\partial}{\partial x}\left(EI\frac{\partial^2 w}{\partial x^2}\right) \quad \text{(higher order form)} \quad \text{(III.13a)}$$

$$M(x,t) = -EI\frac{\partial^2 w}{\partial x^2}$$

$$\theta(x,t) = -\frac{\partial w}{\partial x}$$

and

$$\frac{\partial w}{\partial x} = -\theta$$

$$\frac{\partial\theta}{\partial x} = \frac{M}{EI}$$

$$\frac{\partial V}{\partial x} = \rho\frac{\partial^2 w}{\partial t^2} - p_z(x,t) \quad \text{(first-order form)} \quad \text{(III.13b)}$$

$$\frac{\partial M}{\partial x} = V$$

where ρ is the mass per unit length and $w = w(x,t)$.

Even without the consideration of damping, these partial differential equations are difficult to solve. Indeed, no simple solution exists except for a few elementary structural dynamics problems. Numerical integration of the equations of motion is hazardous at best. Even if they are successfully integrated, no information is provided concerning the natural frequencies of the structure. *Normal-mode theory*, however, does involve the computation of the natural frequencies and appears to be the most logical approach for coupling with transfer matrix methodology. It will be examined in some detail here.

It is possible for a structure to respond in one of many so-called natural (normal, principal, characteristic, free) modes, that is, deformation configurations, which are characteristic of the member. Natural vibration occurs under the action of innate forces of the member and is not due to external impressed forces. Motion in a natural mode can be generated by imposing appropriate initial conditions of displacement and velocity. Normal-mode theory for structural response uses these natural mode shapes to construct a solution for a structure subject to any time-dependent loading, initial, boundary, and in-span conditions.

Elastic structures exhibit many natural modes; fortunately, usually information on only a limited number of them is necessary to represent most dynamic responses. Apart from the use of natural modes and frequencies in the construction of the transient dynamic solution, the modal characteristics are valuable information in their own right. For example, if a member is excited by a harmonic loading function, the member will respond with a motion at the frequency of the loading function. If this loading frequency coincides with one of the natural frequencies of the system, large dangerous amplitudes may occur. This is the so-called *resonance* condition.

Consider now the free motion of a beam. "Free" refers to the absence of externally applied loads. Set the loading function $p_z(x, t)$ equal to zero in the first of Eqs. (III.13a). Then

$$\frac{\partial^2}{\partial x^2}\left(EI\frac{\partial^2 w}{\partial x^2}\right) = -\rho\frac{\partial^2 w}{\partial t^2} \tag{III.14}$$

Assume the variables can be separated in space (x) and time (t) giving

$$w(x, t) = w(x)q(t) \tag{III.15}$$

where q is a function of time. Substitution of this expression into Eq. (III.14) yields

$$\frac{d^2}{dx^2}\left(EI\frac{d^2 w(x)}{dx^2}\right)q(t) = -\rho w(x)\frac{d^2}{dt^2}q(t)$$

or

$$\frac{(d^2/dx^2)EI\left[d^2 w(x)/dx^2\right]}{\rho w(x)} = -\frac{d^2 q(t)/dt^2}{q(t)}$$

In order for the left-hand side, which is a function only of x, to be equal to the right-hand side, which is dependent only on t, both sides must be equal to the same constant, say ω^2. Then

$$\frac{d^2}{dx^2}EI\frac{d^2 w}{dx^2} = \rho\omega^2 w \tag{III.16}$$

and

$$\frac{d^2 q(t)}{dt^2} + \omega^2 q(t) = 0$$

The solution to this latter relation is given by

$$q(t) = A\sin\omega t + B\cos\omega t \tag{III.17}$$

where A, B are constants. It is seen that ω is the frequency, in radians per unit of time, of the free motion. This is the undamped natural frequency of the beam.

The assumed value of $w(x, t)$ of Eq. (III.15) placed in Eqs. (III.13) leads to

$$\theta(x, t) = \theta(x)q(t) \qquad V(x, t) = V(x)q(t)$$
$$M(x, t) = M(x)q(t) \tag{III.18}$$

with

$$\frac{d^2}{dx^2}\left(EI\frac{d^2w}{dx^2}\right) = \rho\omega^2 w$$

$$V(x) = -\frac{d}{dx}\left(EI\frac{d^2w}{dx^2}\right)$$ (higher order form)   (III.19a)

$$M(x) = -EI\frac{d^2w}{dx^2}$$

$$\theta(x) = -\frac{dw}{dx^2}$$

and

$$\frac{dw}{dx} = -\theta$$

$$\frac{d\theta}{dx} = \frac{M}{EI}$$ (first-order form)   (III.19b)

$$\frac{dV}{dx} = -\rho\omega^2 w$$

$$\frac{dM}{dx} = V$$

where $w = w(x)$.

These equations can be solved in the same fashion as the equations of motion for the static response of a beam. For a uniform beam the transfer matrix solution to Eqs. (III.19) appears as

$$
\begin{bmatrix} w \\ \theta \\ V \\ M \end{bmatrix}_b =
\begin{bmatrix}
\dfrac{\cosh\beta\ell + \cos\beta\ell}{2} & -\dfrac{\sinh\beta\ell + \sin\beta\ell}{2\beta} \\[2mm]
-\dfrac{\beta(\sinh\beta\ell - \sin\beta\ell)}{2} & \dfrac{\cosh\beta\ell + \cos\beta\ell}{2} \\[2mm]
-\dfrac{EI\beta^3(\sinh\beta\ell + \sin\beta\ell)}{2} & \dfrac{EI\beta^2}{2}(\cosh\beta\ell - \cos\beta\ell) \\[2mm]
-\dfrac{EI\beta^2}{2}(\cosh\beta\ell - \cos\beta\ell) & \dfrac{EI\beta}{2}(\sinh\beta\ell - \sin\beta\ell)
\end{bmatrix}
$$

$$
\begin{array}{cc}
-\dfrac{\sinh\beta\ell - \sin\beta\ell}{2EI\beta^3} & -\dfrac{\cosh\beta\ell - \cos\beta\ell}{2EI\beta^2} \\[2mm]
\dfrac{\cosh\beta\ell - \cos\beta\ell}{2EI\beta^2} & \dfrac{\sinh\beta\ell + \sin\beta\ell}{2EI\beta} \\[2mm]
\dfrac{\cosh\beta\ell + \cos\beta\ell}{2} & \dfrac{\beta}{2}(\sinh\beta\ell - \sin\beta\ell) \\[2mm]
\dfrac{\sinh\beta\ell + \sin\beta\ell}{2\beta} & \dfrac{\cosh\beta\ell + \cos\beta\ell}{2}
\end{array}
\begin{bmatrix} w \\ \theta \\ V \\ M \end{bmatrix}_a
$$

(III.20)

where $\beta^4 = \rho\omega^2/EI$.

Equations (III.20) contain five unknowns, the four constants of integration w_0, θ_0, V_0, and M_0 and the undamped natural frequency ω. Recall that there are only four known conditions—the boundary conditions—available to apply to these equations to find the unknowns. Apparently, one quantity remains unknown in the study of free motion; however, it will be shown that motion generated by prescribed initial conditions or applied loading is fully determined. The study of free vibrations bears a close resemblance to the study of instability wherein the boundary conditions are applied to find the critical loading rather than the initial parameters, as in the case with static equilibrium problems. Both problems belong to the class of so-called eigenvalue problems. Frequencies and mode shapes are referred to as eigenvalues and eigenfunctions, respectively.

Consider the case of a beam simply supported at both ends. The boundary conditions are $w_{x=0} = M_{x=0} = w_{x=L} = M_{x=L} = 0$. These conditions applied to Eq. (II.47) with $a = 0$, $b = L$, and with the loading terms set equal to zero, give

$$w_0 = 0 \qquad M_0 = 0 \tag{III.21a}$$

$$w_{x=L} = \theta_0 \overline{U}_{w\theta} + V_0 \overline{U}_{wV} = 0 \tag{III.21b}$$

$$M_{x=L} = \theta_0 \overline{U}_{M\theta} + V_0 \overline{U}_{MV} = 0 \tag{III.21c}$$

where $\overline{U}_{w\theta}$, $\overline{U}_{wV}$, $\overline{U}_{M\theta}$, and $\overline{U}_{MV}$ are $U_{w\theta}$, U_{wV}, $U_{M\theta}$, and U_{MV}, respectively, evaluated at $x = L$. Equations such as these are said to be homogeneous. Nontrivial solutions to these homogeneous equations exist when the determinant of the initial parameters goes to zero. That is, when

$$\nabla = \begin{vmatrix} U_{w\theta} & U_{wV} \\ U_{M\theta} & U_{MV} \end{vmatrix}_{x=L} = (U_{w\theta} U_{MV} - U_{M\theta} U_{wV})_{x=L}$$

$$= \overline{U}_{w\theta} \overline{U}_{MV} - \overline{U}_{M\theta} \overline{U}_{wV} = 0 \tag{III.22}$$

Equation (III.22) is a function of the unknown frequency ω. The determinant ∇ can be zero for many different values of ω; these values are the desired natural frequencies of the system and are designated by ω_n, $n = 1, 2, \ldots$.

Reference here to transfer matrix relations for static responses is informative in that a rough (nonrigorous) concept of the philosophy behind setting $\nabla = 0$ is obtained. In the case of static equilibrium each initial parameter appears in the form [see e.g., Eqs. (7) of Example III.2 or Eqs. (6) of Example III.3]

$$\frac{\begin{bmatrix} \text{Loading} \\ \text{function} \end{bmatrix} \begin{bmatrix} \text{transfer} \\ \text{matrix element} \end{bmatrix} - \begin{bmatrix} \text{loading} \\ \text{function} \end{bmatrix} \begin{bmatrix} \text{transfer} \\ \text{matrix element} \end{bmatrix}}{\nabla}$$

Since the loading functions are zero by definition of free vibration, the numerator of this expression is zero. Then the only possibility for a nontrivial solution is for the denominator to be zero also. Thus, ∇ is set equal to zero.

It is noteworthy that it is not possible to determine both of the unknowns θ_0, V_0 of Eq. (III.21). However, there is sufficient information to find the ratio θ_0/V_0. From Eq. (III.21b)

$$\theta_0/V_0 = -(U_{wV}/U_{w\theta})_{x=L} \tag{III.23}$$

and from Eq. (III.21c)

$$\theta_0/V_0 = -(U_{MV}/U_{M\theta})_{x=L} \tag{III.24}$$

Recall that the frequencies have been selected such that $\nabla = 0$, or from Eq. (III.22)

$$(U_{wV}/U_{w\theta})_{x=L} = (U_{MV}/U_{M\theta})_{x=L} \tag{III.25}$$

Thus the values of θ_0/V_0 expressed by Eqs. (III.23) and (III.24) are equivalent.

Alternatively, the problem can be viewed from the standpoint of the four boundary conditions leading to the four relations: $w_0 = 0$, $M_0 = 0$, and θ_0/V_0 as given in Eq. (III.23) and θ_0/V_0 of Eq. (III.24). Since the two expressions for θ_0/V_0 must be equal, Eq. (III.25) [or equivalently, Eq. (III.22)] is again obtained as a condition that must be satisfied. Thus, application of the boundary conditions again leads to the condition $\nabla = 0$.

As just observed, the ratio of the initial parameters can be determined but not the values of all of the initial parameters themselves. Thus, one of the initial parameters can be found in terms of the other. This latter initial parameter is chosen to remain as the only unknown of the problem. If θ_0 is assumed to be the arbitrary constant of unknown magnitude, then the initial parameters for this hinged–hinged beam appear as

$$w_0 = 0 \qquad M_0 = 0 \qquad \theta_0 = \theta_0$$
$$V_0 = -\theta_0(U_{M\theta}/U_{MV})_{x=L} \tag{III.26}$$

Formulas of this sort can be derived and tabulated for any set of boundary conditions.

In summary it is seen that the four conditions $w_{x=0} = M_{x=0} = w_{x=L} = M_{x=L} = 0$ applied to the transfer matrix equations with five unknowns [Eqs. (II.47) with $a = 0$, $b = L$, and $\bar{z}^i = 0$] result in a situation with one unknown remaining. That is, only the ratio of two of the parameters can be determined and not the values of the initial parameters themselves. Fortunately this situation occurs only in the study of free motion because the specification of loading and initial conditions suffices to eliminate the unknown.

To obtain the undamped mode shapes, place the frequencies ω_n, $n = 1, 2, \ldots$, and the expressions for the initial parameters [Eqs. (III.26)] in the response expressions [Eqs. (II.47)]:

$$w_n(x) = \theta_0\big[U_{w\theta}|_x - U_{wV}|_x(U_{M\theta}/U_{MV})_{x=L}\big]_{\omega=\omega_n} \tag{III.27a}$$

$$\theta_n(x) = \theta_0\big[U_{\theta\theta}|_x - U_{\theta V}|_x(U_{M\theta}/U_{MV})_{x=L}\big]_{\omega=\omega_n} \tag{III.27b}$$

$$V_n(x) = \theta_0\big[U_{V\theta}|_x - U_{VV}|_x(U_{M\theta}/U_{MV})_{x=L}\big]_{\omega=\omega_n} \tag{III.27c}$$

$$M_n(x) = \theta_0\big[U_{M\theta}|_x - U_{MV}|_x(U_{M\theta}/U_{MV})_{x=L}\big]_{\omega=\omega_n} \tag{III.27d}$$

where $\omega_n = \omega_1, \omega_2, \ldots$. These are really "shape" functions as they contain the unknown quantity or amplitude θ_0. Often this amplitude is assigned a unit value.

Consider the simply supported beam of length L in more detail. Place the transfer matrix elements of Eqs. (III.20) in ∇ of Eq. (III.22):

$$\nabla = -\frac{\sinh \beta L + \sin \beta L}{2\beta} \frac{\sinh \beta L + \sin \beta L}{2\beta}$$

$$+ EI\beta \frac{\sinh \beta L - \sin \beta L}{2} \frac{\sinh \beta L - \sin \beta L}{2EI\beta^3}$$

$$= -\frac{\sinh \beta L \sin \beta L}{\beta^2} = 0 \tag{III.28}$$

The values of β that make this expression equal to zero are desired. Other than $\beta = 0$, it is clear that the zeros of the function $\sinh \beta L \sin \beta L$ are the zeros of $\sin \beta L$. This follows because $\sinh \beta L$ is not zero except at $\beta L = 0$. Thus, $\sin \beta L = 0$ or $\beta L = n\pi$, $n = 1, 2, \ldots$. The notation is improved somewhat if ω and β are replaced by ω_n and β_n. Then $\beta_n L = n\pi$. Since $\beta_n^4 = \rho \omega_n^2 / EI$,

$$\omega_n = (n\pi)^2 \frac{\sqrt{EI/\rho}}{L^2} \qquad n = 1, 2, \ldots \tag{III.29}$$

That is,

$$\omega_1 = \frac{\pi^2}{L^2} \sqrt{\frac{EI}{\rho}} \qquad \omega_2 = \frac{4\pi^2}{L^2} \sqrt{\frac{EI}{\rho}}, \quad \text{etc.}$$

These are the undamped natural frequencies for the uniform, simply supported beam. The deflection mode shape [Eq. (III.27a)] becomes

$$w_n(x) = \theta_0 \left(-\frac{\sinh \beta_n x + \sin \beta_n x}{2\beta_n} + \frac{\sinh \beta_n x - \sin \beta_n x}{2\beta_n} \frac{\sinh \beta_n L - \sin \beta_n L}{\sinh \beta_n L + \sin \beta_n L} \right)$$

$$= \left(-\frac{\theta_0}{\beta_n} \right) \sin \beta_n x \tag{III.30}$$

with $\omega_n^2 = (n\pi/L)^4 EI/\rho$, $n = 1, 2, \ldots$. This completes the determination of the undamped frequencies and the mode shapes for the uniform, simply supported beam. In the case of more complicated configurations, e.g., a beam of variable cross section, it is not always possible to obtain explicit expressions for the frequencies. Then the roots (frequencies or eigenvalues) of ∇, which are formed from the global transfer matrix elements, can be found computationally.

Example III.7   Transfer Matrix Method for a Beam with Continuous Mass
The natural frequencies of a uniform beam simply supported at the ends will be found using several methods in this appendix. For this beam let $L = 80$ in., $E = 3 \times 10^7$ lb/in.^{2}, $I = 1.3333$ in.^{4}, and $\rho = 2.912 \times 10^{-3}$ lb-s^2/in. Begin by considering the mass to be continuously distributed.

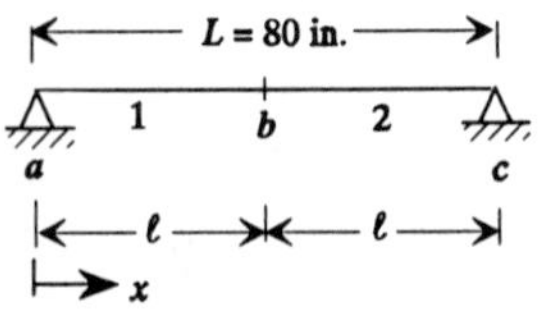

FIGURE III-11 Two-element model of uniform simply supported beam.

Suppose the beam is modeled as two segments, as shown in Fig. III-11. Apart from the desire to illustrate the transfer matrix procedure, there is no reason that this uniform beam should be modeled with two elements. Of course, a single-element model is substantially simpler to utilize.

For the model of Fig. III-11

$$\mathbf{z}_c = \mathbf{U}^2 \mathbf{U}^1 \mathbf{z}_a \tag{1}$$

where $\mathbf{U}^2 = \mathbf{U}^1$, and from Eq. (III.20)

$$\mathbf{U}^1 = \mathbf{U}^2 = \begin{bmatrix} \frac{1}{2}(C_4 + C_2) & -\frac{1}{2\beta}(C_3 + C_1) & -\frac{1}{2EI\beta^3}(C_3 - C_1) & -\frac{1}{2EI\beta^2}(C_4 - C_2) \\ -\frac{1}{2}\beta(C_3 - C_1) & \frac{1}{2}(C_4 + C_2) & \frac{1}{2EI\beta^2}(C_4 - C_2) & \frac{1}{2EI\beta}(C_3 + C_1) \\ -\frac{1}{2}EI\beta^3(C_3 + C_1) & \frac{1}{2}EI\beta^2(C_4 - C_2) & \frac{1}{2}(C_4 + C_2) & \frac{1}{2}\beta(C_3 - C_1) \\ -\frac{1}{2}EI\beta^2(C_4 - C_2) & \frac{1}{2}EI\beta(C_3 - C_1) & \frac{1}{2\beta}(C_3 + C_1) & \frac{1}{2}(C_4 + C_2) \end{bmatrix} \tag{2}$$

where

$$C_1 = \sin \beta\ell \qquad C_2 = \cos \beta\ell \qquad C_3 = \sinh \beta\ell \qquad C_4 = \cosh \beta\ell$$

$$\beta^4 = \rho\omega^2/EI$$

Then

$$
\mathbf{U} = \mathbf{U}^2\mathbf{U}^1 =
\begin{bmatrix}
\overline{U}_{ww} & \overline{U}_{w\theta} & \overline{U}_{wV} & \overline{U}_{wM} \\
\overline{U}_{\theta w} & \overline{U}_{\theta\theta} & \overline{U}_{\theta V} & \overline{U}_{\theta M} \\
\overline{U}_{Vw} & \overline{U}_{V\theta} & \overline{U}_{VV} & \overline{U}_{VM} \\
\overline{U}_{Mw} & \overline{U}_{M\theta} & \overline{U}_{MV} & \overline{U}_{MM}
\end{bmatrix}
\tag{3}
$$

where $\overline{U}_{ij}$ are the transfer matrix components evaluated at $x = L$.

Insert (3) and the simply supported boundary conditions $w_a = M_a = w_c = $, $M_c = 0$ into (1),

$$
\begin{bmatrix} w = 0 \\ \theta \\ V \\ M = 0 \end{bmatrix}_c =
\begin{bmatrix}
\overline{U}_{ww} & \overline{U}_{w\theta} & \overline{U}_{wV} & \overline{U}_{wM} \\
\overline{U}_{\theta w} & \overline{U}_{\theta\theta} & \overline{U}_{\theta V} & \overline{U}_{\theta M} \\
\overline{U}_{Vw} & \overline{U}_{V\theta} & \overline{U}_{VV} & \overline{U}_{VM} \\
\overline{U}_{Mw} & \overline{U}_{M\theta} & \overline{U}_{MV} & \overline{U}_{MM}
\end{bmatrix}
\begin{bmatrix} w = 0 \\ \theta \\ V \\ M = 0 \end{bmatrix}_a
\tag{4}
$$

The relations $w_c = 0$ and $M_c = 0$ appear as

$$
\overline{U}_{w\theta}\theta_a + \overline{U}_{wV}V_a = 0 \qquad \overline{U}_{M\theta}\theta_a + \overline{U}_{MV}V_a = 0
\tag{5}
$$

For a nontrivial solution for θ_a and V_a, the condition

$$
\nabla =
\begin{vmatrix}
\overline{U}_{w\theta} & \overline{U}_{wV} \\
\overline{U}_{M\theta} & \overline{U}_{MV}
\end{vmatrix}
= \overline{U}_{w\theta}\overline{U}_{MV} - \overline{U}_{wV}\overline{U}_{M\theta} = 0
\tag{6}
$$

must hold. From (2) and (3)

$$
\nabla = \frac{1}{\beta^2}(2C_3C_4)(-2C_1C_2) = 0
$$

where

$$
C_3C_4 = \sinh \beta\ell \cosh \beta\ell \qquad C_1C_2 = \sin \beta\ell \cos \beta\ell
\tag{7}
$$

For a nontrivial solution of β, only $\sin \beta\ell \cos \beta\ell = 0$ is possible. Because of the double-angle identities, Eqs. (I.13a), this leads to

$$
\sin 2\beta\ell = 0
$$

which implies that

$$
\beta = n\pi/2\ell = n\pi/L \qquad n = 1, 2, 3, \ldots
\tag{9}
$$

Finally,

$$\rho\omega^2/EI = (n\pi/2\ell)^4$$

or

$$\omega_n = \sqrt{\frac{EI}{\rho}} \left(\frac{n\pi}{2\ell}\right)^2 = \sqrt{\frac{EI}{\rho}} \left(\frac{n\pi}{L}\right)^2 \tag{10}$$

which is the same expression in Eq. (III.29) for a single-element model.

For this beam

$$\omega_1 = \sqrt{\frac{1.3333(3 \times 10^7)}{2.912 \times 10^{-3}}} \left(\frac{\pi}{2 \times 40}\right)^2 = 180.74 \text{ rad/s} \quad \text{or} \quad f_1 = \frac{\omega_1}{2\pi} = 28.767 \text{ Hz}$$

$$\omega_2 = \sqrt{\frac{1.3333(3 \times 10^7)}{2.912 \times 10^{-3}}} \left(\frac{2\pi}{2 \times 40}\right)^2 = 722.96 \text{ rad/s} = 115.066 \text{ Hz}$$

$$\omega_3 = 1626.66 \text{ rad/s} = 258.899 \text{ Hz}$$

$$\omega_4 = 2891.84 \text{ rad/s} = 460.264 \text{ Hz} \tag{11}$$

$$\vdots$$

Often the difficulties in finding a solution, including the computational burden of identifying the frequencies, can be reduced if certain approximations are made in modeling the structure. One of the most popular models involves the discretization, or "lumping," of physical parameters along the member. These lumped parameter models, which lead to solutions of widely varying degrees of accuracy, usually result in $\nabla = 0$ being a polynomial expression for the frequency. Computationally the polynomial is a highly tractable form when roots are desired, especially when compared to the often unwieldy transcendental functions incurred in the distributed parameter representation. Discretization of parameters limits the degrees of freedom of motion. There will usually be only as many frequencies as there are "lumps." Naturally, care must be taken in forming the lumped parameter model of the structural member. Fortunately, the lower frequencies, which are normally those of greatest engineering concern, are usually found with adequate accuracy using a reasonable number of lumps in the lumped parameter model. A point matrix representing the most common of lumped parameter models—the lumped mass model—was derived in Section II.6, Eq. (II.82). Lumped mass matrices are given in Table 11-21.

In a lumped mass beam model a beam with distributed mass is replaced by point masses joined by massless beams (Fig. III-12). The global transfer matrix $\mathbf{U}$ for a lumped mass model is pieced together in the familiar fashion

$$\mathbf{z}_{x=L} = \mathbf{U}^{M+1}\mathbf{U}_M\mathbf{U}^M \cdots \mathbf{U}_k\mathbf{U}^k \cdots \mathbf{U}_1\mathbf{U}^1\mathbf{z}_a = \mathbf{U}\mathbf{z}_a \tag{III.31}$$

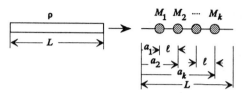

FIGURE III-12 Lumped mass modeling.

where $\mathbf{U}_k$ are the point matrices [Eq. (II.82)] and the $\mathbf{U}^k$ are the field matrices for the massless beam segments, i.e., Eqs. (II.32).

Example III.8   Transfer Matrix Method for a Lumped Mass Beam Model
Suppose the mass of the beam of Example III.7 is lumped as shown in Fig. III-13. Find the natural frequencies using the transfer matrix method.
   In transfer matrix form

$$\mathbf{z}_L = \mathbf{z}_d = \mathbf{U}^3\mathbf{U}_c\mathbf{U}^2\mathbf{U}_b\mathbf{U}^1\mathbf{z}_a = \mathbf{U}\mathbf{z}_a \tag{1}$$

where

$$\mathbf{U}^1 = \begin{bmatrix} 1 & -\ell_1 & -\ell_1^3/6EI & -\ell_1^2/2EI \\ 0 & 1 & \ell_1^2/2EI & \ell_1/EI \\ 0 & 0 & 1 & 0 \\ 0 & 0 & \ell_1 & 1 \end{bmatrix} \tag{2}$$

$$\mathbf{U}^2 = \begin{bmatrix} 1 & -2\ell_1 & -8\ell_1^3/6EI & -4\ell_1^2/2EI \\ 0 & 1 & 4\ell_1^2/2EI & 2\ell_1/EI \\ 0 & 0 & 1 & 0 \\ 0 & 0 & 2\ell_1 & 1 \end{bmatrix} \tag{3}$$

$$\mathbf{U}^3 = \mathbf{U}^1 \tag{4}$$

$$\mathbf{U}_b = \begin{bmatrix} 1 & 0 & 0 & 0 \\ 0 & 1 & 0 & 0 \\ -m_1\omega^2 & 0 & 1 & 0 \\ 0 & 0 & 0 & 1 \end{bmatrix} \tag{5}$$

$$\mathbf{U}_c = \mathbf{U}_b \quad \text{since } m_1 = m_2 \tag{6}$$

Substitution of the above results into (1) gives

$$\mathbf{U} = \mathbf{U}^3\mathbf{U}_c\mathbf{U}^2\mathbf{U}_b\mathbf{U}^1 = \begin{bmatrix} \bar{U}_{ww} & \bar{U}_{w\theta} & \bar{U}_{wV} & \bar{U}_{wM} \\ \bar{U}_{\theta w} & \bar{U}_{\theta\theta} & \bar{U}_{\theta V} & \bar{U}_{\theta M} \\ \bar{U}_{Vw} & \bar{U}_{V\theta} & \bar{U}_{VV} & \bar{U}_{VM} \\ \bar{U}_{Mw} & \bar{U}_{M\theta} & \bar{U}_{MV} & \bar{U}_{MM} \end{bmatrix} \tag{7}$$

where $\bar{U}_{ij} = U_{ij}|_{x=L}$ are components of $\mathbf{U}$ evaluated at $x = L$, i.e., at d.

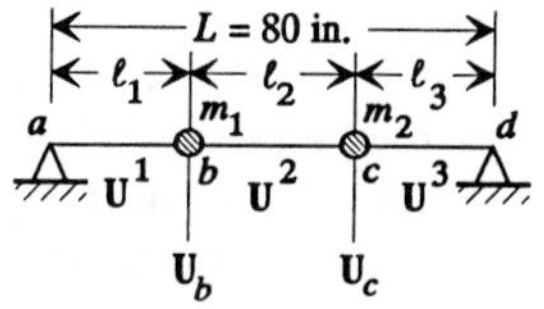

$\ell_1 = \ell_3 = \frac{1}{2}\ell_2 = 20$ in.

$m_1 = m_2 = (\ell_1 + \frac{1}{2}\ell_2)\rho = (20 + 20) \times 2.912 \times 10^{-3} = 0.1165$ lb-s^2/in.

FIGURE III-13 Lumped mass model of simply supported beam of Fig. III-11.

Application of the boundary conditions $w_a = M_a = w_d = M_d = 0$ to $z_d = Uz_a$ leads to the homogeneous relations

$$\bar{U}_{w\theta}\theta_a + \bar{U}_{wV}V_a = 0$$
$$\bar{U}_{M\theta}\theta_a + \bar{U}_{MV}V_a = 0 \tag{8}$$

For nontrivial solutions, the determinant of the coefficients θ_a and V_a must be zero, i.e.,

$$\nabla = \bar{U}_{w\theta}\bar{U}_{MV} - \bar{U}_{wV}\bar{U}_{M\theta} = 0 \tag{9}$$

From (7) it is found that

$$\bar{U}_{w\theta} = -4\ell_1 - \frac{\ell_1^4}{2EI}(9m_1 + m_2)\omega^2 - \frac{2m_1 m_2 \ell_1^7}{9E^2 I^2}\omega^4$$

$$\bar{U}_{MV} = 4\ell_1 + \frac{\ell_1^4}{2EI}(9m_2 + m_1)\omega^2 + \frac{2m_1 m_2 \ell_1^7}{9E^2 I^2}\omega^4$$

$$\bar{U}_{wV} = -\frac{32\ell_1^3}{3EI} - \frac{3(m_1 + m_2)\ell_1^6}{4E^2 I^2}\omega^2 - \frac{1}{27}\frac{m_1 m_2 \ell_1^9}{E^3 I^3}\omega^4 \tag{10}$$

$$\bar{U}_{M\theta} = 3\ell_1^2(m_1 + m_2)\omega^2 + \frac{4m_1 m_2 \ell_1^5}{3EI}\omega^4$$

Substitute the numerical values of E, I, $m_1 = m_2$, and ℓ_1 into (10). Then (9) becomes the frequency equation

$$\omega^4 - (2.896 \times 10^5)\omega^2 + 0.82855 \times 10^{10} = 0 \tag{11}$$

The roots of this polynomial are

$$\omega_1 = 179.415 \text{ rad/s} = 28.555 \text{ Hz} \qquad \omega_2 = 507.356 \text{ rad/s} = 80.748 \text{ Hz} \quad (12)$$

Note that the first natural frequency is quite close to the more precise results of Example III.7, although slightly lower. The "rule of thumb" is that one-half of the frequencies obtained using a lumped mass model should be considered as being sufficiently accurate to be useful. Also, normally the lumped mass model provides values somewhat lower than the correct natural frequencies.

Note that the frequency equation of Eq. (11), Example III.8 is a polynomial in ω^2. It is of interest that the global transfer matrix $\mathbf{U}$ and hence ∇ can always be expressed in terms of a polynomial in ω^2 for lumped mass systems. This leads to a global transfer matrix expressed as a polynomial in ω^2. Since the frequency equation is also a polynomial, the cumbersome determinant search technique for finding natural frequencies can be replaced by efficient, standard polynomial root-solving routines [III.2]. However, for large systems and higher frequencies numerical instabilities may occur.

Responses due to arbitrary time-dependent applied loading or prescribed initial conditions are considered in Section III.7. In the following section we will examine the response of a member to cyclic loading. After some time the initial displacement and velocity are of no concern and the member responds in a cyclic fashion. This is referred to as *steady-state motion*.

Steady-State Motion

Assume an undamped beam is excited by a cyclic loading, in particular, by the harmonic loading $p_z(x, t) = p_z(x)\sin \Omega t$. Permit the motion to continue until the effect of any irregular starting (initial) conditions dies out. Then, all of the state variables of this elastic member will respond with the same harmonic motion. Thus

$$w(x, t) = w(x)\sin \Omega t \qquad \theta(x, t) = \theta(x)\sin \Omega t$$
$$V(x, t) = V(x)\sin \Omega t \qquad M(x, t) = M(x)\sin \Omega t \tag{III.32}$$

This is one form of steady-state motion. Unlike the response frequencies for free motion that are found from the condition $\nabla = 0$, the frequency Ω of the loading and response is specified input information for the problem and as such is a known variable.

Substitution of Eqs. (III.32) in Eqs. (III.13) leads to

$$\frac{d^2}{dx^2}\left(EI\frac{d^2w}{dx^2}\right) - \rho\Omega^2 w = p_z(x)$$

$$V(x) = -\frac{d}{dx}\left(EI\frac{d^2w}{dx^2}\right) \quad \text{(higher order form)} \quad \text{(III.33a)}$$

$$M(x) = -EI\frac{d^2w}{dx^2}$$

$$\theta(x) = -\frac{dw}{dx}$$

and

$$\frac{dw}{dx} = -\theta$$

$$\frac{d\theta}{dx} = \frac{M}{EI}$$

$$\frac{dV}{dx} + \rho\Omega^2 w = -p_z(x) \qquad \text{(first-order form)} \quad \text{(III.33b)}$$

$$\frac{dM}{dx} = V$$

The problem has been reduced to one of statics. Since the governing equations (III.33) with $p_z(x) = 0$ are identical to those for the free motion of a beam [Eqs. (III.19)] if $\omega = \Omega$, the solution to Eqs. (III.33) is given by the transfer matrices tabulated in this book in the dynamic response sections. These transfer matrices (with $\omega = \Omega$) are then used in precisely the fashion employed in the static response problems. That is, the loading functions are developed, field changes taken into account, and initial parameters evaluated using the static response tables. Upon completion of the static solution to find $w(x)$, $\theta(x)$, $V(x)$, and $M(x)$, these variables are inserted in Eqs. (III.32) to give the full spatial and temporal solution [i.e., $w(x, t), \theta(x, t), V(x, t), M(x, t)$].

It is noteworthy that this steady-state response can encompass any loading, in-span condition, change in cross section or material, and nonhomogeneous boundary condition that can be accounted for by a static solution.

If the forcing frequency Ω coincides with one of the natural frequencies, then, as in the case of free motion, the denominator (∇) in the initial parameter equations will be zero. This means that the initial parameters are indeterminate, as well they should be since physically this situation corresponds to a state of resonance response.

Indeterminate In-Span Conditions

Such in-span conditions as rigid supports and releases require special attention because, unlike the in-span concentrated force, spring, and mass considered earlier in this appendix, there is insufficient information available at the location of the condition to take it into account in the response expressions as the progressive matrix multiplications proceed across the condition. This becomes evident if a rigid support is considered to be an infinitely stiff spring. The reaction force R_b in the spring at x_b is proportional to the compression of the spring, i.e., $R_b = kw_b$, where k is the spring rate. This reaction becomes an unknown force at a rigid in-span support since then $R_b = kw_b = (k = \infty)(w_b = 0)$. Unfortunately, the condition $w_b = 0$ at x_b cannot be used to evaluate the unknown R_b as the transfer matrix multiplication moves across x_b. A condition such as this is sometimes referred to as being an in-span "indeterminate." Other in-span indeterminates for beams are moment releases, shear releases, and angle guides, as shown in Fig. III-14. In each case, for each new unknown created there is a new condition to be satisfied.

In-Span Indeterminate Condition	Fixed State Variable[a]	Discontinuous State Variable[b]
1. Rigid Support	w	V
2. Moment Release, (Hinge)	M	θ
3. Shear Release	V	w
4. Angle Guide	θ	M

[a]Often equal to zero
[b]Of unknown magnitude at the in-span condition

FIGURE III-14   In-span indeterminate conditions.

There are several methods available for incorporating these in-span conditions into the solution.

Increase in Number of Unknowns   One method for including these in-span conditions is to simply expand the initial state vector to introduce each new unknown as it occurs. For example, consider the portion of the beam shown in Fig. III-15a where the left end of the beam is hinged and a rigid in-span support occurs at $x = x_b$. In this section, superscripts indicate the element involved and the subscripts denote locations. For the left-end conditions of $w_a = M_a = 0$, the

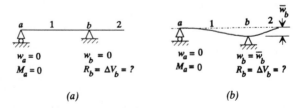

(a)   *(b)*

FIGURE III-15   Portion of beam with in-span rigid support: (*a*) zero deflection at support; (*b*) imposed deflection of magnitude $\overline{w}_b$.

initial state vector can be written as

$$
\mathbf{z}_a =
\begin{bmatrix} 0 \\ \theta_a \\ V_a \\ 0 \end{bmatrix}
=
\underbrace{\begin{bmatrix} 0 & 0 \\ 1 & 0 \\ 0 & 1 \\ 0 & 0 \end{bmatrix}}_{\mathbf{V}_a}
\begin{bmatrix} \theta_a \\ V_a \end{bmatrix}
= \mathbf{V}_a \hat{\mathbf{z}}_a
\tag{III.34}
$$

The state vector just to the left of node b would be

$$
\mathbf{z}_b^- = \mathbf{U}^1 \mathbf{V}_a \hat{\mathbf{z}}_a + \bar{\mathbf{z}}_b^1 = \mathbf{V}_b^1 \hat{\mathbf{z}}_a + \bar{\mathbf{z}}_b^1
$$

$$
=
\begin{bmatrix} V_{11} & V_{12} \\ V_{21} & V_{22} \\ V_{31} & V_{32} \\ V_{41} & V_{42} \end{bmatrix}
\hat{\mathbf{z}}_a +
\begin{bmatrix} z_1 \\ z_2 \\ z_3 \\ z_4 \end{bmatrix}_b
=
\begin{bmatrix} 0 \\ \theta \\ V \\ M \end{bmatrix}_b
\tag{III.35}
$$

where $\mathbf{V}_b^1 = \mathbf{U}^1 \mathbf{V}_a$. Just to the right of node b, upon inclusion of the effect of the unknown reaction of magnitude $R_b = \Delta V_b$, the state vector becomes

$$
\mathbf{z}_b^+ =
\begin{bmatrix} 0 \\ \theta \\ V \\ M \end{bmatrix}_b
+
\begin{bmatrix} 0 \\ 0 \\ \Delta V_b \\ 0 \end{bmatrix}
=
\underbrace{\begin{bmatrix} V_{11} & V_{12} & 0 \\ V_{21} & V_{22} & 0 \\ V_{31} & V_{32} & 1 \\ V_{41} & V_{42} & 0 \end{bmatrix}}_{\mathbf{V}_b^2}
\begin{bmatrix} \theta_a \\ V_a \\ \Delta V_b \end{bmatrix}
+
\begin{bmatrix} z_1 \\ z_2 \\ z_3 \\ z_4 \end{bmatrix}_b
\tag{III.36}
$$

This procedure can continue, adding a new unknown at each new in-span indeterminate. The unknown in-span variables and the two unknown initial state variables are obtained from the equations arising from the prescribed conditions at the in-span indeterminates, e.g., $w_b = 0$ in Eq. (III.36), as well as the two right-end boundary conditions.

Progressive Elimination of Unknowns   Another viable approach for the inclusion of in-span indeterminates is to proceed as just outlined but to eliminate an unknown each time an indeterminate condition occurs. As is described in reference III.2, this procedure is readily generalized and automated for use in a structural member analysis computer program.

To illustrate the progressive elimination of unknowns, consider the beam of Fig. III-15b, which is the same beam portion just treated with the support of $x = x_b$ lowered a height $\bar{w}_b$. This seemingly complicated constraint is included to illustrate that quite general conditions are readily taken into account. For this case, the condition of a prescribed displacement inserted in Eq. (III.35) can be employed to eliminate one of the initial state variables, either θ_a or V_a. That is,

Eq. (III.35) would appear as

$$
\mathbf{z}_b^- = \mathbf{V}_b^1 \hat{\mathbf{z}}_a + \bar{\mathbf{z}}_b^1 =
\begin{bmatrix}
V_{11} & V_{12} \\
V_{21} & V_{22} \\
V_{31} & V_{32} \\
V_{41} & V_{42}
\end{bmatrix}
\begin{bmatrix}
\theta_a \\
V_a
\end{bmatrix}
+
\begin{bmatrix}
z_1 \\
z_2 \\
z_3 \\
z_4
\end{bmatrix}_b
=
\begin{bmatrix}
\bar{w}_b \\
\theta_b \\
V_b \\
M_b
\end{bmatrix}
\qquad (III.37)
$$

and if we choose to eliminate V_a, the first equation in Eq. (III.37) gives

$$
V_a = -(1/V_{12})(V_{11}\theta_a - \bar{w}_b + z_1) \qquad (III.38)
$$

Substitute this into Eq. (III.37) to eliminate V_a and proceed beyond the support, introducing the new unknown reaction ΔV_b:

$$
\mathbf{z}_b^+ =
\begin{bmatrix}
0 \\
0 \\
\Delta V_b \\
0
\end{bmatrix}
+ \mathbf{z}_b^- =
\begin{bmatrix}
V_{11} - V_{12}V_{11}/V_{12} & 0 \\
V_{21} - V_{22}V_{11}/V_{12} & 0 \\
V_{31} - V_{32}V_{11}/V_{12} & 1 \\
V_{41} - V_{42}V_{11}/V_{12} & 0
\end{bmatrix}
\begin{bmatrix}
\theta_a \\
\Delta V_b
\end{bmatrix}
$$

$$
+
\begin{bmatrix}
\bar{w}_b \\
(\bar{w}_b - z_1)V_{22}/V_{12} + z_2 \\
(\bar{w}_b - z_1)V_{32}/V_{12} + z_3 \\
(\bar{w}_b - z_1)V_{42}/V_{12} + z_4
\end{bmatrix}
\mathbf{V}_b^2 \hat{\mathbf{z}}_b + \bar{\mathbf{z}}_b^2 \qquad (III.39)
$$

Now proceed to the next in-span indeterminate condition (or the right boundary, whichever occurs first) and utilize the new known condition to eliminate the unknown ΔV_b. This process continues until the right boundary is reached and the two current unknowns can be evaluated.

Numerical Difficulties

The transfer matrix method with its assembly of the overall transfer matrix by progressive multiplication of element matrices tends to encounter numerical difficulties. This is hardly surprising as an applied loading at one end of a chainlike structure may have little effect on the response at a distant end. The sources of the numerical problems and techniques for overcoming them are treated in depth in such references as III.2–III.6.

Sources of Numerical Difficulties The accumulation of roundoff and truncation errors by the progressive multiplication of the form

$$
\mathbf{z}_j = \mathbf{U}^j \cdots \mathbf{U}^2 \mathbf{U}^1 \mathbf{z}_a = \mathbf{U}\mathbf{z}_a \qquad (III.40)
$$

is one source of numerical difficulties of the transfer matrix method. Rather than converging to $\mathbf{z}_j$, this expression can converge to the eigenvector (say $\hat{\mathbf{z}}$) of the first eigenvalue of $\mathbf{U}$. This can occur if the vectors in $\mathbf{U}$ become linearly dependent and, as a result, the determinant of the system of equations will approach the value zero. Even if $\mathbf{z}_a$ is known precisely, the solution can converge to $\tilde{\mathbf{z}}$.

Another source of numerical difficulties occurs when the transfer matrix procedure involves the difference of large numbers. For computer calculations in which a limited number of digits are retained, this can lead to inaccuracies. This problem can arise, for example, if a very stiff spring is included in the model or if high natural frequencies are being calculated. Also, this difficulty can be anticipated if the effect of occurrences on one boundary is small on the other boundary so that the calculation for the determination of the initial conditions $(\mathbf{z}_a)$ can involve the difference between large numbers.

Corrective Measures The procedures to overcome the numerical difficulties inherent in the transfer matrix process are characterized by a division of the structural system into intervals and the application of a solution technique that avoids the numerical difficulties. This would appear to be counterproductive since it violates the fundamental "spirit" of the transfer matrix method and its sequence of matrix manipulations. However, this new subdivision solution can be more efficient as measured by operation counts than the usual transfer matrix procedure as well as leading to a sufficiently accurate solution. Several corrective methods will be outlined here.

In-span State Variables as New Unknowns Let the structural system be divided into subintervals as shown in Fig. III-16. These new intervals need not correspond to the elements originally employed in setting up the transfer matrix formulation. However, each interval should be chosen to be sufficiently small so as to assure that no numerical error is introduced in expressing a state vector at one station, say c, in terms of the state vector at the preceding station, say b, using transfer matrices. The in-span state variables of $\mathbf{z}_b, \mathbf{z}_c, \mathbf{z}_d, \ldots$ are then treated as unknowns.

Suppose the overall transfer matrices between the new in-span stations $b, c, d, \ldots$, each of which may involve several progressive transfer matrix multiplications corresponding to the elements of the original model, are given by

$$\mathbf{z}_b = \mathbf{U}^1 \mathbf{z}_a \qquad \mathbf{z}_c = \mathbf{U}^2 \mathbf{z}_b \qquad \mathbf{z}_d = \mathbf{U}^3 \mathbf{z}_c \qquad \cdots \qquad (\text{III.41})$$

In order to establish a system of equations with $\mathbf{z}_a, \mathbf{z}_b, \ldots$ as the unknowns,

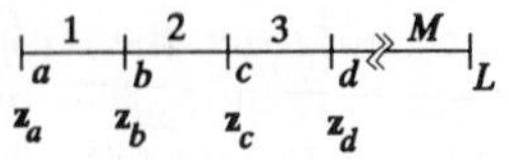

FIGURE III-16 Division of system into intervals.

rewrite Eq. (III.41) as

$$
\begin{aligned}
-\mathbf{U}^1 \mathbf{z}_a + \mathbf{I}\mathbf{z}_b \qquad\qquad &= 0 \\
-\mathbf{U}^2 \mathbf{z}_b + \mathbf{I}\mathbf{z}_c \qquad &= 0 \\
-\mathbf{U}^3 \mathbf{z}_c + \mathbf{I}\mathbf{z}_d &= 0 \\
\vdots
\end{aligned}
\tag{III.42}
$$

or in matrix form

$$
\begin{bmatrix}
-\mathbf{U}^1 & \mathbf{I} & & \\
& -\mathbf{U}^2 & \mathbf{I} & \\
& & -\mathbf{U}^3 & \mathbf{I} \\
& & &
\end{bmatrix}
\begin{bmatrix}
\mathbf{z}_a \\ \mathbf{z}_b \\ \mathbf{z}_c \\ \mathbf{z}_d \\ \vdots
\end{bmatrix} = \mathbf{0}
\tag{III.43}
$$

This relationship is readily modified to show the applied loadings. This set of equations can be solved for $\mathbf{z}_a, \mathbf{z}_b, \ldots$ using any reliable linear equation solver. Boundary and in-span conditions of any sort are readily incorporated into Eq. (III.43).

Use of a Displacement or Force Method   One of the simplest procedures to overcome the numerical difficulties of the transfer matrix method is to convert the transfer matrices into stiffness or flexibility matrices, as explained in Appendix II, and then utilize the displacement or force method, as appropriate, to solve for the unknowns. The transfer matrices chosen can correspond in length to the original element models or to the other intervals, e.g., those of Fig. III-16. Boundary and in-span conditions are often easier to incorporate during the displacement or force method stage of the solution than during the development of the transfer matrices. The displacement and force methods are to be discussed in the following sections. Since, in practice, the displacement method is employed more than the force method for large systems and hence is the more familiar of the two, *the displacement method is usually the best choice.* After the displacements at the nodes are computed and the forces at the nodes are determined using the stiffness matrices, the transfer matrices can be used to print out the displacements and forces along the member. An operations count shows that the combined transfer Matrix–displacement method is more efficient than the use of pure transfer matrices.

A Frontal Solution Procedure   One of the better solution procedures for solving the system of equations for a displacement method is the *frontal* approach, which is an element-by-element technique proceeding like a "wave front" spreading over the system. This type of procedure can be formulated explicitly in terms of transfer matrix equations.

With a frontal transfer matrix approach the initial unknowns $\mathbf{z}_a$ at the left end a are replaced by new unknowns at point b. This process is continued, from b to c, from c to d, etc. This is a condensation procedure that can be considered as the progressive replacement of the structure by equivalent springs (Fig. III-17). See reference III.1 for details.

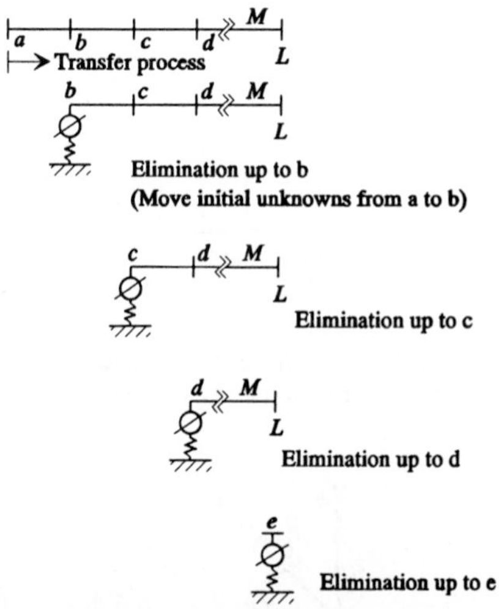

FIGURE III-17 Frontal procedure for solving transfer matrix problems.

III.2 GENERAL STRUCTURAL SYSTEMS

The displacement method is the dominant technique currently in use for analyzing general structural systems. An alternative solution technique is the force
method. Although the transfer matrix method applies only for structures with a
linelike geometry, the force and displacement methods are appropriate for any
geometry. For the transfer matrix method the system matrix remains small
regardless of the system complexity, while the force and displacement methods
lead to large system matrices whose size depends on the complexity of the
structure. We begin the study of the force and displacement methods by defining
rather cumbersome notation. See Table III-2.

Coordinate Systems, Definitions, and Degrees of Freedom

Network structures are usually modeled as a finite number of elements connected
at nodes. Only nodal variables such as forces and displacements will occur in the
governing equations. This is said to be a spatially discretized model (Fig. III-18).
The model may contain one-, two-, or three-dimensional elements, as required by
the system. Such models are often called *finite-element* models, and the solution
technology is called the finite-element method.

Nodal Variables   The location of the nodes is described in a global coordinate
system (X, Y, Z). At each node, nodal forces and displacements are defined (Fig.
III-19). These system forces and displacements are the unknowns. After the
nodal forces and displacements are calculated, the internal forces and displacements between the nodes of an element are computed.

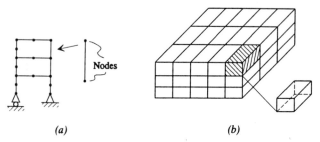

FIGURE III-18 Spatially discretized model with elements: (a) framework model with rod elements connected at nodes; (b) three-dimensional model with solid elements.

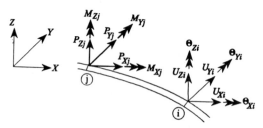

FIGURE III-19 Global coordinates X, Y, Z and nodal displacements and forces at nodes i and j.

The *degrees of freedom* (DOF) of a node are the independent coordinates (displacements) essential for completely describing the motion of a node. For a general solid, each node can have six DOF: three translations and three rotations. Three forces and three moments correspond to these DOF. Thus, at each node of a solid the following displacements and forces occur:

$$
\begin{array}{llll}
U_X & U_Y & U_Z & \text{3 translations} \\
\Theta_X & \Theta_Y & \Theta_Z & \text{3 rotations} \\
P_X & P_Y & P_Z & \text{3 forces} \\
M_X & M_Y & M_Z & \text{3 moments}
\end{array}
\tag{III.44}
$$

Systems forces and displacements are designated by capital letters. In general, in solid mechanics terminology, the terms *forces* and *displacements* include moments and rotations, respectively.

The forces and displacements at each node are written in vector form as

$$
\mathbf{P}_j = \begin{bmatrix} P_X \\ P_Y \\ P_Z \\ M_X \\ M_Y \\ M_Z \end{bmatrix}_j
\qquad
\mathbf{V}_j = \begin{bmatrix} U_X \\ U_Y \\ U_Z \\ \Theta_X \\ \Theta_Y \\ \Theta_Z \end{bmatrix}_j
\tag{III.45}
$$

The subscript j designates the jth node. For the whole structure, the nodal forces $\mathbf{P}$ and nodal displacements $\mathbf{V}$ are assembled as

$$\mathbf{P} = \begin{bmatrix} \mathbf{P}_1 \\ \mathbf{P}_2 \\ \vdots \\ \mathbf{P}_j \\ \vdots \\ \mathbf{P}_N \end{bmatrix} \qquad \mathbf{V} = \begin{bmatrix} \mathbf{V}_1 \\ \mathbf{V}_2 \\ \vdots \\ \mathbf{V}_j \\ \vdots \\ \mathbf{V}_N \end{bmatrix} \qquad (III.46)$$

where N is the number of nodes.

Element Variables Consider element forces and displacements aligned with the system coordinates X, Y, Z. At nodes a and b of element i, element forces $\mathbf{p}^i$ and corresponding displacements $\mathbf{v}^i$ act (Fig. III-20):

$$\mathbf{p}^i = \begin{bmatrix} \mathbf{p}_a^i \\ \mathbf{p}_b^i \end{bmatrix} = \begin{bmatrix} F_{Xa}^i \\ F_{Ya}^i \\ F_{Za}^i \\ F_{Xb}^i \\ F_{Yb}^i \\ F_{Zb}^i \end{bmatrix} \qquad \mathbf{v}^i = \begin{bmatrix} \mathbf{v}_a^i \\ \mathbf{v}_b^i \end{bmatrix} = \begin{bmatrix} u_{Xa}^i \\ u_{Ya}^i \\ u_{Za}^i \\ u_{Xb}^i \\ u_{Yb}^i \\ u_{Zb}^i \end{bmatrix} \qquad (III.47)$$

Moments and rotations can be included in $\mathbf{p}^i$ and $\mathbf{v}^i$, respectively.

The elements stiffness matrix $\mathbf{k}^i$ relates the element forces $\mathbf{p}^i$ and element displacements $\mathbf{v}^i$:

$$\begin{bmatrix} \mathbf{p}_a \\ \mathbf{p}_b \end{bmatrix}^i = \begin{bmatrix} \mathbf{k}_{aa} & \mathbf{k}_{ab} \\ \mathbf{k}_{ba} & \mathbf{k}_{bb} \end{bmatrix}^i \begin{bmatrix} \mathbf{v}_a \\ \bar{\mathbf{v}}_b \end{bmatrix}^i \qquad (III.48)$$
$$\mathbf{p}^i \quad = \quad \mathbf{k}^i \qquad \mathbf{v}^i$$

In Eq. (III.48), the first subscript of the submatrices designates the node or location for which the equation is established, while the second subscript identifies the DOF "causing" or corresponding to the force.

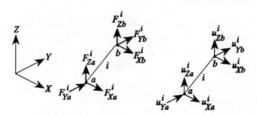

FIGURE III-20 Some forces and displacements aligned with global coordinates X, Y, Z at ends a and b of element i are shown. Moments and rotations could also have been shown.

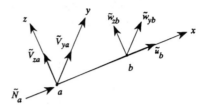

FIGURE III-21 Local coordinate system, force, and displacements of element i. Moments and rotations could also have been shown.

In addition to the global reference frame X, Y, Z, a new coordinate system along with corresponding forces and displacements is defined. A *local* reference frame x, y, z is aligned in a natural direction along the element. The element forces, displacements, and stiffness matrix in the local element coordinate system appear as (Fig. III-21)

$$\begin{bmatrix} \tilde{\mathbf{p}}_a \\ \tilde{\mathbf{p}}_b \end{bmatrix}^i = \begin{bmatrix} \tilde{\mathbf{k}}_{aa} & \tilde{\mathbf{k}}_{ab} \\ \tilde{\mathbf{k}}_{ba} & \tilde{\mathbf{k}}_{bb} \end{bmatrix}^i \begin{bmatrix} \tilde{\mathbf{v}}_a \\ \tilde{\mathbf{v}}_b \end{bmatrix}^i$$

$$\tilde{\mathbf{p}}^i = \tilde{\mathbf{k}}^i \quad \tilde{\mathbf{v}}^i \qquad \text{(III.49)}$$

The local coordinate quantities are indicated with a tilde.

For a bar (Fig. III-21)

$$\tilde{\mathbf{p}}^i = \begin{bmatrix} \tilde{\mathbf{p}}_a^i \\ \tilde{\mathbf{p}}_b^i \end{bmatrix} = \begin{bmatrix} \tilde{N}_a \\ \tilde{V}_{ya} \\ \tilde{V}_{za} \\ \tilde{N}_b \\ \tilde{V}_{yb} \\ \tilde{V}_{zb} \end{bmatrix} \qquad \tilde{\mathbf{v}}^i = \begin{bmatrix} \tilde{\mathbf{v}}_a \\ \tilde{\mathbf{v}}_b \end{bmatrix} = \begin{bmatrix} \tilde{u}_a \\ \tilde{w}_{ya} \\ \tilde{w}_{za} \\ \tilde{u}_b \\ \tilde{w}_{yb} \\ \tilde{w}_{zb} \end{bmatrix}$$

which could also include moments and rotations. In the notation of Table 13-15, $\tilde{w}_y = \tilde{v}$ and $\tilde{w}_z = \tilde{w}$.

Coordinate Transformations

All forces and displacements for the elements are referred to a common reference frame by transforming to the global coordinates the nodal forces and displacements expressed in the local coordinates.

To transform global to local coordinates in two dimensions, use (Fig. III-22)

$$\begin{bmatrix} x \\ y \\ z \end{bmatrix} = \begin{bmatrix} \cos xX & \cos xY & \cos xZ \\ \cos yX & \cos yY & \cos yZ \\ \cos zX & \cos zY & \cos zZ \end{bmatrix} \begin{bmatrix} X \\ Y \\ Z \end{bmatrix} = \begin{bmatrix} \cos \alpha & 0 & -\sin \alpha \\ 0 & 1 & 0 \\ \sin \alpha & 0 & \cos \alpha \end{bmatrix} \begin{bmatrix} X \\ Y \\ Z \end{bmatrix} \quad \text{(III.50)}$$

where, for example, xX is the angle between the x axis and the X axis. Forces

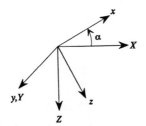

FIGURE III-22 Right-handed global (X, Y, Z) and local (x, y, z) coordinate systems. Positive angle α is shown. The vector corresponding to α is positive along the positive y direction.

and displacements transform in a similar fashion. Thus,

$$
\begin{bmatrix} \tilde{\mathbf{p}}_a \\ \tilde{\mathbf{p}}_b \end{bmatrix}^i = \begin{bmatrix} \mathbf{T}^i_{aa} & \mathbf{0} \\ \mathbf{0} & \mathbf{T}^i_{bb} \end{bmatrix} \begin{bmatrix} \mathbf{p}_a \\ \mathbf{p}_b \end{bmatrix}^i \quad \text{and} \quad \tilde{\mathbf{v}}^i = \mathbf{T}^i \mathbf{v}^i \qquad \text{(III.51)}
$$

$$
\tilde{\mathbf{p}}^i \quad = \quad \mathbf{T}^i \quad \quad \mathbf{p}^i
$$

where the transformation matrix $\mathbf{T}^i$ is defined by

$$
\mathbf{T}^i = \begin{bmatrix} \mathbf{T}_{aa} & \mathbf{T}_{ab} \\ \mathbf{T}_{ba} & \mathbf{T}_{bb} \end{bmatrix}^i
$$

$$
\text{with} \quad \mathbf{T}^i_{aa} = \mathbf{T}^i_{bb} = \begin{bmatrix} \cos xX & 0 & -\cos xZ \\ 0 & 1 & 0 \\ \cos zX & 0 & \cos zZ \end{bmatrix}^i
$$

$$
= \begin{bmatrix} \cos \alpha & 0 & -\sin \alpha \\ 0 & 1 & 0 \\ \sin \alpha & 0 & \cos \alpha \end{bmatrix}^i \quad \mathbf{T}^i_{ba} = \mathbf{T}^i_{ab} = 0 \qquad \text{(III.52)}
$$

with α (or xX), the angle between the (global) X coordinate and the (local) x coordinate. It is evident that $\mathbf{T}^i_{jj}$, $j = a$ or b of Eq. (III.52) satisfies

$$
\mathbf{T}^{iT}_{jj} \mathbf{T}^i_{jj} = \mathbf{I}, \quad \text{also} \quad \mathbf{T}^{iT} \mathbf{T}^i = \mathbf{I}
$$

where the superscript T indicates a transpose and $\mathbf{I}$ is the unit diagonal matrix. Since $(\mathbf{T}^i)^{-1}\mathbf{T}^i = \mathbf{I}$, it is observed that

$$
(\mathbf{T}^i)^{-1} = \mathbf{T}^{iT} \qquad \text{(III.53)}
$$

This relationship permits the transformation of forces and displacements from local to global coordinates to be written as

$$
\begin{bmatrix} \mathbf{p}_a \\ \mathbf{p}_b \end{bmatrix}^i = \begin{bmatrix} \mathbf{T}^{iT}_{aa} & \mathbf{0} \\ \mathbf{0} & \mathbf{T}^{iT}_{bb} \end{bmatrix} \begin{bmatrix} \tilde{\mathbf{p}}_a \\ \tilde{\mathbf{p}}_b \end{bmatrix}^i \quad \text{and} \quad \mathbf{v}^i = \mathbf{T}^{iT}\tilde{\mathbf{v}}^i \qquad \text{(III.54)}
$$

$$
\mathbf{p}^i \quad = \quad \mathbf{T}^{iT} \quad \quad \tilde{\mathbf{p}}^i
$$

It is essential that we are able to transform from one coordinate system to another. Observe that

$$\mathbf{p}^i = \mathbf{T}^{iT}\tilde{\mathbf{p}}^i = \mathbf{T}^{iT}\tilde{\mathbf{k}}^i\tilde{\mathbf{v}}^i = \mathbf{T}^{iT}\tilde{\mathbf{k}}^i\mathbf{T}^i\mathbf{v}^i$$

Since $\mathbf{p}^i = \mathbf{k}^i\mathbf{v}^i$, it is seen that the stiffness matrix transforms according to

$$\mathbf{k}^i = \mathbf{T}^{iT}\tilde{\mathbf{k}}^i\mathbf{T}^i \tag{III.55}$$

which is referred to as a *congruent transformation*. Under this transformation $\mathbf{k}^i$ will be a symmetric matrix since $\tilde{\mathbf{k}}^i$ is symmetric.

III.3 DISPLACEMENT METHOD

In practice, the displacement method is considered to be the "standard" method for the analysis of large structural systems. Although the displacement method can be formulated directly (*direct stiffness method*), it is normally considered to be a variationally based approach. It follows from the principle of virtual work. Since this principle is equivalent to the equations of equilibrium, the displacement method is also referred to as the *equilibrium method*.

Displacement Method Based on Principle of Virtual Work

Suppose the structure is modeled in terms of elements for which the responses are represented by forces and displacements at their ends. If there are M elements with internal end forces $\mathbf{p}^i$ and displacements $\mathbf{v}^i$ and applied end forces $\bar{\mathbf{p}}^i$ (including the effect of loads distributed along the element) and displacements $\bar{\mathbf{v}}^i$, the principle of virtual work (Appendix II), $\delta W_i + \delta W_e = 0$, appears as

$$
\begin{aligned}
-\sum_{i=1}^{M} (\delta W_i + \delta W_e)^i &= -(\delta W_i + \delta W_e) \\
&= \sum_{i=1}^{M} \delta\mathbf{v}^{iT}\mathbf{k}^i\mathbf{v}^i - \sum_{i=1}^{M} \delta\mathbf{v}^{iT}\mathbf{p}^i \\
&= \sum_{i=1}^{M} \delta\mathbf{v}^{iT}(\mathbf{k}^i\mathbf{v}^i - \mathbf{p}^i) = 0
\end{aligned}
\tag{III.56}
$$

where $\delta\mathbf{v}^i$ is a vector of "virtual displacements." As explained in Appendix II, virtual displacements are small variations in the displacements that behave according to the principles of *variational calculus*. For our purposes, variations can be considered to behave like ordinary derivatives, although a virtual displacement does not represent a rate of change along a direction as does an ordinary derivative of a displacement.

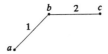

FIGURE III-23 Three-node (a, b, c) two-element $(1, 2)$ structure.

Equation (III.56) represents the summation of internal and external virtual work done by the element forces at the nodes. Alternatively, this total virtual work can be expressed in terms of system responses **P** and **V**.

In order to ensure *compatibility* at the nodes, the end displacements of the various elements joined at each node must match the values of the displacements of the node. Assume that the local (element) end displacements $\bar{\mathbf{v}}^i$ have been transformed to the directions of the global coordinate system $(\mathbf{v}^i)$.

For the two-element, three-node structure of Fig. III-23, the nodal compatibility conditions take the form

$$\mathbf{v}_a^1 = \mathbf{V}_a \qquad \mathbf{v}_b^1 = \mathbf{v}_b^2 = \mathbf{V}_b \qquad \mathbf{v}_c^2 = \mathbf{V}_c \qquad\qquad \text{(III.57)}$$

which, in matrix notation, appear as

$$
\begin{bmatrix} \mathbf{v}^1 \\ \mathbf{v}^2 \end{bmatrix} =
\begin{bmatrix} \mathbf{v}_a^1 \\ \mathbf{v}_b^1 \\ \mathbf{v}_b^2 \\ \mathbf{v}_c^2 \end{bmatrix} =
\begin{bmatrix} \mathbf{I} & & \\ \hline & \mathbf{I} & \\ \hline & \mathbf{I} & \\ \hline & & \mathbf{I} \end{bmatrix}
\begin{bmatrix} \mathbf{V}_a \\ \mathbf{V}_b \\ \mathbf{V}_c \end{bmatrix} \qquad\qquad \text{(III.58)}
$$

$$\mathbf{v} \quad = \quad \mathbf{a} \qquad \mathbf{V}$$

where **I** is the unit diagonal matrix. The Boolean matrix (containing null or unit values) **a**, known as the *global kinematic, connectivity, locator,* or *incidence matrix*, indicates which element is connected to which node. These compatibility conditions in essence transform a displacement vector **v** containing some duplicate displacements, i.e., $\mathbf{v}_b^1$ and $\mathbf{v}_b^2$, into a displacement vector **V** with no redundant variables.

Write Eq. (III.56) in matrix rather than index summation form,

$$\sum_{i=1}^{M} \delta\mathbf{v}^{iT}(\mathbf{k}^i\mathbf{v}^i - \mathbf{p}^i) = \delta\mathbf{v}^T(\mathbf{k}\mathbf{v} - \mathbf{p}) = 0 \qquad\qquad \text{(III.59)}$$

where

$$\mathbf{v} = \begin{bmatrix} \mathbf{v}^1 \\ \mathbf{v}^2 \\ \vdots \\ \mathbf{v}^M \end{bmatrix} \quad \text{is an unassembled displacement vector}$$

(III.60)

$$\mathbf{p} = \begin{bmatrix} \mathbf{p}^1 \\ \mathbf{p}^2 \\ \vdots \\ \mathbf{p}^M \end{bmatrix} \quad \text{is an unassembled load vector}$$

(III.61)

$$\mathbf{k} = \begin{bmatrix} \mathbf{k}^1 & & & \\ & \mathbf{k}^2 & & \\ & & \ddots & \\ & & & \mathbf{k}^M \end{bmatrix} = \text{diag}[\mathbf{k}^i] \quad \text{is an unassembled global stiffness matrix}$$

Substitute $\mathbf{v} = \mathbf{aV}$ in Eq. (III.59) in order to write the principle of virtual work expression in terms of the system nodal displacements:

$$\delta \mathbf{v}^T (\mathbf{kv} - \mathbf{p}) = \delta \mathbf{V}^T \mathbf{a}^T (\mathbf{kaV} - \mathbf{p}) = \delta \mathbf{V}^T (\mathbf{a}^T \mathbf{kaV} - \mathbf{a}^T \mathbf{p}) = 0 \quad (\text{III.62})$$

The quantity

$$\mathbf{K} = \mathbf{a}^T \mathbf{ka} \tag{III.63}$$

is defined as the "assembled" system stiffness matrix, and

$$\overline{\mathbf{P}} = \mathbf{a}^T \mathbf{p} \tag{III.64}$$

is the assembled applied load vector. Thus Eq. (III.62) becomes

$$\delta \mathbf{V}^T (\mathbf{KV} - \overline{\mathbf{P}}) = 0$$

which implies

$$\mathbf{KV} = \overline{\mathbf{P}} \tag{III.65}$$

This is a set of algebraic equations for the unknown nodal displacements that represent the global statement of equilibrium. The matrices involved here are assembled in the sense that the duplications that occurred in $\mathbf{v}$ [e.g., in Eq. (III.58), where $\mathbf{v}_b^1 = \mathbf{v}_b^2$] were removed by introducing the compatibility conditions such that $\mathbf{v}$ is replaced by $\mathbf{aV}$. The system nodal displacements obtained from Eq. (III.65) can be used in computing forces, stresses, and other displacements.

The connectivity matrix $\mathbf{a}$ governs the assembly of the global stiffness matrix $\mathbf{K}$. However, the assembled matrix relationships $\mathbf{K} = \mathbf{a}^T \mathbf{ka}$ and $\overline{\mathbf{P}} = \overline{\mathbf{a}}^T \mathbf{p}$ are not of great practical value, as this assembly is normally implemented as a superposition, i.e., an addition process.

The rationale underlying the assembly by summation is readily visualized using the two-element, three-node system of Fig. III-23. The stiffness matrix for element 1, which spans from a to b, can be written as [Eq. (III.48)]

$$\mathbf{k}^1 = \begin{bmatrix} \mathbf{k}_{aa} & \mathbf{k}_{ab} \\ \mathbf{k}_{ba} & \mathbf{k}_{bb} \end{bmatrix} \tag{III.66}$$

where it is assumed that the necessary coordinate transformations have been implemented so that all variables and matrices are referred to the global coordinates. For element 2, which begins at node b and ends at node c,

$$\mathbf{k}^2 = \begin{bmatrix} \mathbf{k}_{bb} & \mathbf{k}_{bc} \\ \mathbf{k}_{cb} & \mathbf{k}_{cc} \end{bmatrix} \tag{III.67}$$

To obtain the global stiffness matrix $\mathbf{K}$, form $\mathbf{K} = \mathbf{a}^T \mathbf{k} \mathbf{a}$:

$$
\overset{\mathbf{a}^T}{\begin{bmatrix} \mathbf{I} & & \\ & \mathbf{I} & \mathbf{I} \\ & & \mathbf{I} \end{bmatrix}}
\overset{\mathbf{k}}{\begin{bmatrix} \mathbf{k}^1 & \mathbf{0} \\ \mathbf{0} & \mathbf{k}^2 \end{bmatrix}}
\overset{\mathbf{a}}{\begin{bmatrix} \mathbf{I} & & \\ & \mathbf{I} & \\ & \mathbf{I} & \\ & & \mathbf{I} \end{bmatrix}}
$$

$$
= \begin{bmatrix} \mathbf{k}^1_{aa} & \mathbf{k}^1_{ab} & \\ \mathbf{k}^1_{ba} & \mathbf{k}^1_{bb} + \mathbf{k}^2_{bb} & \mathbf{k}^2_{bc} \\ & \mathbf{k}^2_{cb} & \mathbf{k}^2_{cc} \end{bmatrix} = \mathbf{K} \tag{III.68}
$$

It is evident from this matrix that the global stiffness matrix is assembled by summing element stiffness matrices of like subscripts. The unassembled stiffness matrix appears as

$$
\underset{\mathbf{p}}{\begin{bmatrix} \mathbf{p}^1 \\ \mathbf{p}^2 \end{bmatrix}} = \begin{bmatrix} \mathbf{p}^1_a \\ \mathbf{p}^1_b \\ \mathbf{p}^2_b \\ \mathbf{p}^2_c \end{bmatrix} = \underset{\mathbf{k}}{\begin{bmatrix} \mathbf{k}^1 & \\ & \mathbf{k}^2 \end{bmatrix}} \underset{\mathbf{v}}{\begin{bmatrix} \mathbf{v}^1 \\ \mathbf{v}^2 \end{bmatrix}} = [\mathbf{k}] \begin{bmatrix} \mathbf{v}^1_a \\ \mathbf{v}^1_b \\ \mathbf{v}^2_b \\ \mathbf{v}^2_c \end{bmatrix} \tag{III.69}
$$

whereas the assembled global stiffness matrix of Eq. (III.68) is

$$
\begin{bmatrix} \bar{\mathbf{P}}_a \\ \bar{\mathbf{P}}_b \\ \bar{\mathbf{P}}_c \end{bmatrix} = \begin{bmatrix} \boxed{\mathbf{k}^1} & \\ & \boxed{\mathbf{k}^2} \end{bmatrix} \begin{bmatrix} \mathbf{V}_a \\ \mathbf{V}_b \\ \mathbf{V}_c \end{bmatrix} = \begin{bmatrix} \mathbf{k}^1_{aa} & \mathbf{k}^1_{ab} & \\ \mathbf{k}^1_{ba} & \mathbf{k}^1_{bb} + \mathbf{k}^2_{bb} & \mathbf{k}^2_{bc} \\ & \mathbf{k}^2_{cb} & \mathbf{k}^2_{cc} \end{bmatrix} \begin{bmatrix} \mathbf{V}_a \\ \mathbf{V}_b \\ \mathbf{V}_c \end{bmatrix}
$$

$$
\bar{\mathbf{P}} \quad = \quad \mathbf{K} \qquad \mathbf{V} \tag{III.70}
$$

Apparently all coefficients of $\mathbf{K}$ either are taken directly from $\mathbf{k}^1$ or $\mathbf{k}^2$ or, as seen by the overlapping boxes, are the sum of certain $\mathbf{k}^1$ and $\mathbf{k}^2$ coefficients. The process of assembling the stiffness matrix $\mathbf{K}$ by summation of those element stiffness matrix coefficients with identical subscripts, is written $\mathbf{K}_{ij} = \mathbf{k}^1_{ij} + \mathbf{k}^2_{ij}$. This summation process is made possible by carefully fitting the element stiffness matrix into the global nodal numbering system. This will be discussed in a later section.

Direct Derivation of Global Displacement Equations

The principle of virtual work leads to equations of equilibrium for displacements that satisfy compatibility requirements. The displacement relations of Eqs. (III.65) can be derived directly from the conditions of equilibrium. Return to the two-element, three-node system of Fig. III-23. At each node, sum all forces contributed by the elements joined at this node. These nodal equilibrium relations appear as

$$
\mathbf{p}^1_a = \bar{\mathbf{P}}_a
$$
$$
\mathbf{p}^1_b + \mathbf{p}^2_b = \bar{\mathbf{P}}_b
$$
$$
\mathbf{p}^2_c = \bar{\mathbf{P}}_c
$$

or

$$
\begin{bmatrix} \mathbf{I} & & & \\ \hline & \mathbf{I} & \mathbf{I} & \\ \hline & & & \mathbf{I} \end{bmatrix} \begin{bmatrix} \mathbf{p}^1_a \\ \mathbf{p}^1_b \\ \mathbf{p}^2_b \\ \mathbf{p}^2_c \end{bmatrix} = \begin{bmatrix} \bar{\mathbf{P}}_a \\ \bar{\mathbf{P}}_b \\ \bar{\mathbf{P}}_c \end{bmatrix} \tag{III.71}
$$

$$
\mathbf{b}^* \qquad\quad \mathbf{p} \qquad\quad \bar{\mathbf{P}}
$$

where $\mathbf{b}^*$ is the *global statics* or *equilibrium matrix*. These relationships constitute the conditions of equilibrium between $\mathbf{p}$ and $\bar{\mathbf{P}}$. The reciprocal relation is

$$
\mathbf{p} = \mathbf{b}\bar{\mathbf{P}} \tag{III.72}
$$

Since $\mathbf{b}^*$ is not necessarily a square matrix, in general, $\mathbf{b}^* \neq \mathbf{b}^{-1}$. However,

$$
\mathbf{b}^*\mathbf{b} = \mathbf{I} \tag{III.73}
$$

whereas $\mathbf{bb}^* \neq \mathbf{I}$. A comparison of Eqs. (III.59) and (III.71) indicates that[†]

$$\mathbf{b}^* = \mathbf{a}^T \qquad\qquad \text{(III.74)}$$

From Eqs. (III.71) and (III.74)

$$\mathbf{b}^*\mathbf{p} = \bar{\mathbf{P}} = \mathbf{a}^T\mathbf{p} \qquad\qquad \text{(III.75)}$$

Introduce into this relationship the set of unassembled stiffness equations,

$$\mathbf{p} = \mathbf{kv} \qquad\qquad \text{(III.76)}$$

where $\mathbf{p}, \mathbf{k}, \mathbf{v}$ are defined in Eq. (III.61),

$$\mathbf{a}^T\mathbf{kv} = \bar{\mathbf{P}}$$

Finally, from the nodal connectivity equations of Eq. (III.59)

$$\mathbf{a}^T\mathbf{kaV} = \bar{\mathbf{P}} \quad \text{or} \quad \mathbf{KV} = \bar{\mathbf{P}} \qquad\qquad \text{(III.77)}$$

which are the desired displacement equilibrium relations of Eqs. (III.65).

System Stiffness Matrix Assembled by Summation

Fundamental to the process of assembling a global stiffness matrix by summation is the proper identification of where an element fits into the system with its node-numbering system. This can be accomplished with the aid of an *incidence table* that identifies each end of each element with a global node. For the two-bar-element, three-node system of Fig. III-23, the incidence table appears as follows:

	Global Node Numbers Corresponding to Element End Numbers	
Element	System Node Number where Element Begins	System Node Number where Element Ends
1	a	b
2	b	c

After the element ends are assigned to system nodes, the element stiffness coefficients are summed to provide the corresponding system stiffness coefficient. The formation of $\mathbf{K}$ is equivalent to a loop summation calculation over all elements i,

$$\mathbf{K}_{jk} \leftarrow \mathbf{K}_{jk} + \mathbf{k}^i_{jk} \qquad\qquad \text{(III.78)}$$

where j and k are taken from the incidence table for each element i.

[†]It is possible to show that $\mathbf{a}^* = \mathbf{b}^T$, with $\mathbf{a}^*$ given by $\mathbf{V} = \mathbf{a}^*\mathbf{v}$. Also, $\mathbf{aa}^* \neq \mathbf{I}$, $\mathbf{a}^*\mathbf{a} = \mathbf{I}$.

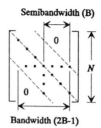

FIGURE III-24 Bandwidth. Typically, coefficients in the band are nonzero.

As is to be expected, a host of numerical schemes have been proposed for efficiently treating the matrices during the assembly process. Most of the procedures are concerned with how to avoid the need to fully develop all of the matrices.

Characteristics of Stiffness Matrices

Element and global stiffness matrices are symmetric. They are also positive definite. The symmetry property is useful since only terms on and to one side of the main diagonal need to be utilized in a computer program. Several solution techniques for systems of linear equations can take advantage of the sparseness that tends to occur in stiffness matrices, especially if the nonzero terms occur close to the diagonal. This is a *banded* matrix. Proper choice of a numbering system for the degrees of freedom can lead to the nonzero terms being clusted close to the diagonal. Basically, in order to minimize the bandwidth (Fig. III-24), the degrees of freedom should be numbered such that the distance (difference between nodal numbers) from the main diagonal to the most remote nonzero term in a particular row of the stiffness matrix is minimized. Normally, consecutive numbering of nodes across the shorter dimension of a structure results in a small bandwidth.

There is a large literature [e.g., III.7, III.8] on efficient methods of retaining only the essential information in an analysis involving stiffness matrices. Methods have been developed [e.g., III.8, III.9] for the automatic renumbering of nodes so that the bandwidth is reduced.

Incorporation of Boundary Conditions

The boundary conditions can be introduced rather simply by ignoring those columns in **K** of $\overline{\mathbf{P}} = \mathbf{KV}$ that correspond to zero (prescribed) displacements and ignoring those rows in $\overline{\mathbf{P}} = \mathbf{KV}$ for the corresponding unknown reactions. Proceed, then, to solve the remaining equations (a square matrix) for the unknown nodal displacements.

Several techniques are available whereby the boundary conditions are applied to the element stiffness matrices before they are assembled into global matrices [e.g., III.7].

Reactions and Internal Forces, Stress Resultants, and Stresses

The equations (rows) for reactions that are usually ignored as the system equations are solved can be reconstructed (after the rest of the equations are solved) and then utilized to calculate the reactions.

Element internal forces, stress resultants, or stress distributions are sometimes a bit difficult to calculate since only nodal displacements $\mathbf{V}$ are determined directly in a displacement method analysis. The element joint forces can be computed from

$$\mathbf{p}^i = \mathbf{k}^i \mathbf{v}^i \tag{III.79}$$

if the displacement vector $\mathbf{v}^i$ is available, perhaps from $\mathbf{v} = \mathbf{aV}$ [Eq. (III.58)]. Element forces and displacements in the local coordinate system are obtained using Eq. (III.51).

The distribution of response variables along a member, e.g., a beam or bar, can be computed using the transfer matrix method. Since the locally oriented displacements and forces are found by postprocessing the results of a global displacement analysis, the state vector is known at the ends of a member (or element). It is then a straightforward process to use transfer matrices to determine the distribution of these variables along a member.

Frames

To demonstrate the application of the displacement method, consider a *frame* or rigid frame, which is composed of beam elements in which bending and axial (extension and/or torsion) effects occur. The term *frame* sometimes includes pin-jointed trusses as well as rigid-jointed frames. Frame formulas are provided in Chapter 13.

Element Coordinate Transformations Consider two-dimensional frames with in-plane loading that lie in the x, z plane. Components of local and global forces and displacements in the local xz and global XZ coordinate systems are given in Fig. III-25. Transformation relations are also shown. As referred to the global coordinates, the forces and displacements at end a of the ith element of a plane frame can be represented as

$$\mathbf{p}_a^i = \begin{bmatrix} F_X \\ F_Z \\ M \end{bmatrix}_a^i \qquad \mathbf{v}_a^i = \begin{bmatrix} u_X \\ u_Z \\ \theta \end{bmatrix}_a^i \tag{III.80}$$

where $M_a^i = M_{Ya}^i$ and $\theta_a^i = \theta_{Ya}^i$. In terms of the local coordinates, the corresponding forces are designated as $\tilde{\mathbf{p}}_a^i$. The transformations from global to local coordinates follows from the geometry of Fig. III-25:

$$\tilde{N}_a = F_{Xa} \cos \alpha - F_{Za} \sin \alpha = F_{Xa} \cos xX + F_{Za} \cos xZ$$

$$\tilde{V}_a = F_{Xa} \sin \alpha + F_{Za} \cos \alpha = F_{Xa} \cos zX + F_{Za} \cos zZ$$

$$\tilde{M}_a = M_a$$

where, for example xX is the angle between the x axis and the X axis. In matrix notation, the local and global forces for end a of the ith element are related by

$$\tilde{\mathbf{p}}_a^i = \mathbf{T}_{aa}^i \mathbf{p}_a^i$$

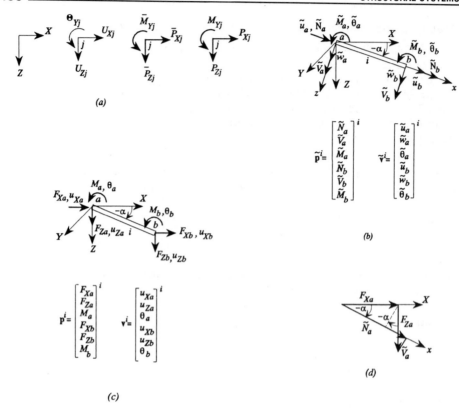

FIGURE III-25 Forces and displacements for a frame element: (a) Global coordinate system, nodal DOF, and system applied nodal loading. (b) Local coordinates, forces, and displacements on the ends of an element (Sign convention 2). For right-handed global (X, Y, Z) and local (x, y, z) coordinate systems, the vector corresponding to a positive α is along the y axis. The angle α is measured from a global coordinate axis to the corresponding local coordinate axis. (c) Forces and displacements aligned along global coordinates on the ends of an element (Sign convetion 2). (d) Components of forces.

where

$$
\mathbf{T}_{aa}^{i} = \begin{bmatrix} \cos\alpha & -\sin\alpha & 0 \\ \sin\alpha & \cos\alpha & 0 \\ 0 & 0 & 1 \end{bmatrix}^{i} = \begin{bmatrix} \cos xX & \cos xZ & 0 \\ \cos zX & \cos zZ & 0 \\ 0 & 0 & 1 \end{bmatrix}^{i} \tag{III.81}
$$

Displacements transform in the same fashion. These can be generalized as in Eq. (III.51).

The stiffness matrix in global coordinates [Eq. (III.55)] is represented as

$$
\mathbf{k}^{i} = \mathbf{T}^{iT}\tilde{\mathbf{k}}^{i}\mathbf{T}^{i} = \begin{bmatrix} \mathbf{k}_{jj}^{i} & \mathbf{k}_{jk}^{i} \\ \hline \mathbf{k}_{kj}^{i} & \mathbf{k}_{kk}^{i} \end{bmatrix} \tag{III.82}
$$

where

$$\mathbf{T}^i = \begin{bmatrix} \mathbf{T}^i_{aa} & 0 \\ 0 & \mathbf{T}^i_{bb} \end{bmatrix} \quad \text{with } \mathbf{T}^i_{aa} = \mathbf{T}^i_{bb} \text{ given by Eq. (III.81)}$$

$\mathbf{k}^i_{kj}$ are 3×3 matrices, and $\tilde{\mathbf{k}}^i$ is the element stiffness matrix of Eq. (II.61) with the addition of axial extension terms,

$$\tilde{\mathbf{k}}^i = \begin{bmatrix} EA/\ell & 0 & 0 & -EA/\ell & 0 & 0 \\ 0 & 12EI/\ell^3 & -6EI/\ell^2 & 0 & -12EI/\ell^3 & -6EI/\ell^2 \\ 0 & -6EI/\ell^2 & 4EI/\ell & 0 & 6EI/\ell^2 & 2EI/\ell \\ -EA/\ell & 0 & 0 & EA/\ell & 0 & 0 \\ 0 & -12EI/\ell^3 & 6EI/\ell^2 & 0 & 12EI/\ell^3 & 6EI/\ell^2 \\ 0 & -6EI/\ell^2 & 2EI/\ell & 0 & 6EI/\ell^2 & 4EI/\ell \end{bmatrix}^i$$

(III.83)

This stiffness matrix, which is referred to local coordinates, applies for frame elements with in-plane loading. Now that this is available, a displacement method frame analysis can proceed.

Example III.9   Displacement Method for a Frame   The simple two-dimensional frame of Fig. III-26a is idealized as shown in Fig. III-26b. Both legs are fixed and concentrated loadings are applied, as shown in Fig. III-27. All bars are made of steel with $E = 200$ GN/m^2. For bars 1 and 2, $I = 2.056 \times 10^{-5}$ m^4 and $A = 4.25 \times 10^{-3}$ m^2. For bar 3, $I = 4.412 \times 10^{-5}$ m^4 and $A = 7.16 \times 10^{-3}$ m^2.

Begin by correlating the element stiffness matrices with global node numbers. To assist in this process, prepare an incidence table:

Beam	System Node Number where Beam Begins	System Node Number where Beam Ends
1	a	b
2	b	c
3	c	d

The entries here are used to assign subscripts to the stiffness matrix elements corresponding to the global node numbers.

(a)

(b)

FIGURE III-26   A simple plane frame: (a) the framework; (b) the model with elements (bars) 1, 2, 3 and nodes (joints) a, b, c, d.

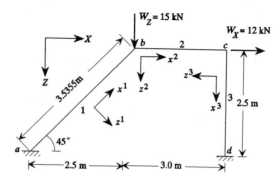

FIGURE III-27 A plane frame. Positive Y and y axes are directed out of the page.

Transform the element stiffness matrix to global coordinates.

Element 1

Use $\ell = 3.536$ m, $EI = 4.112$ MN $\cdot$ m^2, and $EA = 850$ MN. From Eq. (III.83)

$$\bar{\mathbf{k}}^1 = \begin{bmatrix} 240,416,300 & 0 & 0 & -240,416,300 & 0 & 0 \\ 0 & 1,116,527 & -1,973,760 & 0 & -1,116,527 & -1,973,760 \\ 0 & -1,973,760 & 4,652,197 & 0 & 1,973,760 & 2,326,099 \\ -240,416,300 & 0 & 0 & 240,416,300 & 0 & 0 \\ 0 & -1,116,527 & 1,973,760 & 0 & 1,116,527 & 1,973,760 \\ 0 & -1,973,760 & 2,326,099 & 0 & 1,973,760 & 4,652,197 \end{bmatrix} \tag{1}$$

For a global coordinate system XZ placed at a in Fig. III-27, the angles between the local and global coordinates are $x^1X = 45°$, $x^1Z = 135°$, $z^1X = 45°$, and $z^1Z = 45°$. Alternatively, $\alpha = 45°$. From $\mathbf{T}^i$ of Eq. (III.82)

$$\mathbf{T}^1 = \begin{bmatrix} 0.70711 & -0.70711 & 0 & 0 & 0 & 0 \\ 0.70711 & 0.70711 & 0 & 0 & 0 & 0 \\ 0 & 0 & 1.00000 & 0 & 0 & 0 \\ 0 & 0 & 0 & 0.70711 & -0.70711 & 0 \\ 0 & 0 & 0 & 0.70711 & 0.70711 & 0 \\ 0 & 0 & 0 & 0 & 0 & 1.00000 \end{bmatrix} \tag{2}$$

From Eq. (III.82), the element stiffness matrix in global coordinates is

$\mathbf{k}^1 = \mathbf{T}^{1T}\bar{\mathbf{k}}^1\mathbf{T}^1$

$$= \begin{bmatrix} 120,766,400 & -119,649,900 & -1,395,659 & -120,766,400 & 119,649,900 & -1,395,659 \\ -119,649,900 & 120,766,400 & -1,395,659 & 119,649,900 & -120,766,400 & -1,395,659 \\ -1,395,659 & -1,395,659 & 4,652,197 & 1,395,659 & 1,395,659 & 2,326,099 \\ -120,766,400 & 119,649,900 & 1,395,659 & 120,766,400 & -119,649,900 & 1,395,659 \\ 119,649,900 & -120,766,400 & 1,395,659 & -119,649,900 & 120,766,400 & 1,395,659 \\ -1,395,659 & -1,395,659 & 2,326,099 & 1,395,659 & 1,395,659 & 4,652,197 \end{bmatrix}$$

$$= \begin{bmatrix} \mathbf{k}^1_{aa} & \mathbf{k}^1_{ab} \\ \hline \mathbf{k}^1_{ba} & \mathbf{k}^1_{bb} \end{bmatrix} \tag{3}$$

where the subscripts have been taken from the incidence table.

Element 2

Since the local and global coordinates coincide for this case, $\mathbf{k}^2 = \tilde{\mathbf{k}}^2$. Substitute $\ell = 3.0$ m, $EI = 4.112$ MN $\cdot$ m^2, and $EA = 850$ MN in Eq. (III.83):

$$\mathbf{k}^2 = \begin{bmatrix} 283{,}333{,}300 & 0 & 0 & -283{,}333{,}300 & 0 & 0 \\ 0 & 1{,}827{,}556 & -2{,}741{,}333 & 0 & -1{,}827{,}556 & -2{,}741{,}333 \\ 0 & -2{,}741{,}333 & 5{,}482{,}667 & 0 & 2{,}741{,}333 & 2{,}741{,}333 \\ -283{,}333{,}300 & 0 & 0 & 283{,}333{,}300 & 0 & 0 \\ 0 & -1{,}827{,}556 & 2{,}741{,}333 & 0 & 1{,}827{,}556 & 2{,}741{,}333 \\ 0 & -2{,}741{,}333 & 2{,}741{,}333 & 0 & 2{,}741{,}333 & 5{,}482{,}667 \end{bmatrix}$$

$$= \begin{bmatrix} \mathbf{k}^2_{bb} & \mathbf{k}^2_{bc} \\ \mathbf{k}^2_{cb} & \mathbf{k}^2_{cc} \end{bmatrix} \tag{4}$$

Element 3

For this case, use $\ell = 2.5$ m, $EI = 8.824$ MN $\cdot$ m^2, and $EA = 1432$ MN in Eq. (III.83). For a global coordinate system XZ placed at point c, the angles between the local and global coordinates are $x^3X = 90°$, $x^3Z = 0°$, $z^3X = 180°$, $z^3Z = 90°$. Alternatively, $\alpha = -90°$. From Eq. (III.82)

$$\mathbf{k}^3 = \mathbf{T}^{3T}\tilde{\mathbf{k}}^3\mathbf{T}^3 = \begin{bmatrix} 6{,}776{,}832.0 & 42.7 & 8{,}471{,}040.0 \\ 42.7 & 572{,}800{,}000.0 & -0.6 \\ 8{,}471{,}040.0 & -0.6 & 14{,}118{,}400.0 \\ -6{,}776{,}832.0 & -42.7 & -8{,}471{,}040.0 \\ -42.7 & -572{,}800{,}000.0 & 0.6 \\ 8{,}471{,}040.0 & -0.6 & 7{,}059{,}200.0 \end{bmatrix}$$

$$\begin{bmatrix} -6{,}776{,}832.0 & -42.7 & 8{,}471{,}040.0 \\ -42.7 & -572{,}800{,}000.0 & -0.6 \\ -8{,}471{,}040.0 & 0.6 & 7{,}059{,}200.0 \\ 6{,}776{,}832.0 & 42.7 & -8{,}471{,}040.0 \\ 42.7 & 572{,}800{,}000.0 & 0.6 \\ -8{,}471{,}040.0 & 0.6 & 14{,}118{,}400.0 \end{bmatrix}$$

$$= \begin{bmatrix} \mathbf{k}^3_{cc} & \mathbf{k}^3_{cd} \\ \mathbf{k}^3_{dc} & \mathbf{k}^3_{dd} \end{bmatrix} \tag{5}$$

Assemble the global stiffness matrix using

$$\overline{\mathbf{K}}_{jk} = \sum_{i=1}^{M} \mathbf{k}^i_{jk} \tag{6}$$

where the summation is taken over all beam elements (M). It is clear that

$$\mathbf{K}_{jk} = \mathbf{k}^i_{jk} \qquad i = 1, 2, 3 \tag{7}$$

except for $\mathbf{K}_{bb}$ and $\mathbf{K}_{cc}$, which are given by

$$\mathbf{K}_{bb} = \mathbf{k}^1_{bb} + \mathbf{k}^2_{bb} \qquad \mathbf{K}_{cc} = \mathbf{k}^2_{cc} + \mathbf{k}^3_{cc} \tag{8}$$

The assembled global stiffness matrix will appear as

$$\mathbf{K} = \begin{bmatrix} \boxed{\mathbf{k}^1} & & \\ & \boxed{\mathbf{k}^2} & \\ & & \boxed{\mathbf{k}^3} \end{bmatrix} \tag{9}$$

Thus, $\mathbf{KV} = \overline{\mathbf{P}}$ has now been established, where

$$\mathbf{V} = \begin{bmatrix} U_{Xa} \\ U_{Za} \\ \Theta_{Ya} \\ U_{Xb} \\ U_{Zb} \\ \Theta_{Yb} \\ U_{Xc} \\ U_{Zc} \\ \Theta_{Yc} \\ U_{Xd} \\ U_{Zd} \\ \Theta_{Yd} \end{bmatrix} = \begin{bmatrix} 0 \\ 0 \\ 0 \\ U_{Xb} \\ U_{Zb} \\ \Theta_{Yb} \\ U_{Xc} \\ U_{Zc} \\ \Theta_{Yc} \\ 0 \\ 0 \\ 0 \end{bmatrix} \qquad \overline{\mathbf{P}} = \begin{bmatrix} P_{Xa} \\ P_{Za} \\ M_{Ya} \\ P_{Xb} \\ P_{Zb} \\ M_{Yb} \\ P_{Xc} \\ P_{Zc} \\ M_{Yc} \\ P_{Xd} \\ P_{Zd} \\ M_{Yd} \end{bmatrix} = \begin{bmatrix} P_{Xa} = ? \\ P_{Za} = ? \\ M_{Ya} = ? \\ 0 \\ W_Z \\ 0 \\ W_X \\ 0 \\ 0 \\ P_{Xd} = ? \\ P_{Zd} = ? \\ M_{Yd} = ? \end{bmatrix} \tag{10}$$

The displacement boundary conditions and the applied forces are shown above. Question marks indicate reactions at the supports at a and d. Delete the columns corresponding to the zero displacements at the supports and ignore the rows corresponding to the reactions. This results in

$$\begin{bmatrix} .4041 \times 10^9 & -.1196 \times 10^9 & .1396 \times 10^7 \\ -.1196 \times 10^9 & .1226 \times 10^9 & -.1346 \times 10^7 \\ .1396 \times 10^7 & -.1346 \times 10^7 & .1013 \times 10^8 \\ -.2833 \times 10^9 & .0000 & .0000 \\ .0000 & -.1828 \times 10^7 & .2741 \times 10^7 \\ .0000 & -.2741 \times 10^7 & .2741 \times 10^7 \end{bmatrix}$$

$$\begin{bmatrix} -.2833 \times 10^9 & .0000 & .0000 \\ .0000 & -.1828 \times 10^7 & -.2741 \times 10^7 \\ .0000 & .2741 \times 10^7 & .2741 \times 10^7 \\ .2901 \times 10^9 & .4273 \times 10^2 & .8471 \times 10^7 \\ .4273 \times 10^2 & .5746 \times 10^9 & .2741 \times 10^7 \\ .8471 \times 10^7 & .2741 \times 10^7 & .1960 \times 10^8 \end{bmatrix} \begin{bmatrix} U_{Xb} \\ U_{Zb} \\ \Theta_{Yb} \\ U_{Xc} \\ U_{Zc} \\ \Theta_{Yc} \end{bmatrix}$$

$$= \begin{bmatrix} 0 \\ 0.015 \times 10^6 \\ 0 \\ 0.012 \times 10^6 \\ 0 \\ 0 \end{bmatrix} \tag{11}$$

which yields the displacements

$$
\begin{bmatrix}
U_{Xb} \\
U_{Zb} \\
\Theta_{Yb} \\
U_{Xc} \\
U_{Zc} \\
\Theta_{Yc}
\end{bmatrix}
=
\begin{bmatrix}
.29559 \times 10^{-2} \text{ m} \\
.29900 \times 10^{-2} \text{ m} \\
.22786 \times 10^{-3} \text{ rad} \\
.29543 \times 10^{-2} \text{ m} \\
.12679 \times 10^{-4} \text{ m} \\
- .89223 \times 10^{-3} \text{ rad}
\end{bmatrix}
\tag{12}
$$

The displacements at the nodes are given by (12). To find the reactions, return to $\mathbf{KV} = \overline{\mathbf{P}}$ using $\mathbf{K}$ of (9), $\mathbf{V}$ of (10) with the numerical values of (12), and $\overline{\mathbf{P}}$ of (10). This leads to the reactions (in global coordinates)

$$
\begin{aligned}
P_{Xa} &= .46260 \times 10^3 \text{ N} & P_{Xd} &= - .12463 \times 10^5 \text{ N} \\
P_{Za} &= - .77374 \times 10^4 \text{ N} & P_{Zd} &= - .72626 \times 10^4 \text{ N} \\
M_{Ya} &= .88286 \times 10^4 \text{ N} & M_{Yd} &= .18728 \times 10^5 \text{ N}
\end{aligned}
\tag{13}
$$

The responses in local coordinates on the ends of the elements require some extra effort. For example, to find the forces on the ends of an element, calculate the forces in global coordinates first and then transform these forces into the local coordinate system. The forces referred to global coordinates for element 1

$$
\mathbf{p}^1 =
\begin{bmatrix}
F_{Xa} \\
F_{Za} \\
M_a \\
F_{Xb} \\
F_{Zb} \\
M_b
\end{bmatrix}
= \mathbf{k}^1 \mathbf{v}^1 = \mathbf{k}^1
\begin{bmatrix}
u_{Xa} \\
u_{Za} \\
\theta_a \\
u_{Xb} \\
u_{Zb} \\
\theta_b
\end{bmatrix}
$$

$$
= \mathbf{k}^1
\begin{bmatrix}
U_{Xa} \\
U_{Za} \\
\Theta_{Ya} \\
U_{Xb} \\
U_{Zb} \\
\Theta_{Yb}
\end{bmatrix}
= [\text{matrix of Eq. (3)}]
\begin{bmatrix}
0 \\
0 \\
0 \\
.29559 \times 10^{-2} \text{ m} \\
.29900 \times 10^{-2} \text{ m} \\
.22786 \times 10^{-3} \text{ rad}
\end{bmatrix}
$$

$$
=
\begin{bmatrix}
.46260 \times 10^3 \text{ N} \\
- .77374 \times 10^4 \text{ N} \\
.88286 \times 10^4 \text{ N} \cdot \text{m} \\
- .46260 \times 10^3 \text{ N} \\
.77374 \times 10^4 \text{ N} \\
.93586 \times 10^4 \text{ N} \cdot \text{m}
\end{bmatrix}
\tag{14}
$$

Referred to local coordinates, element 1 forces are

$$
\tilde{\mathbf{p}}^1 = \mathbf{T}^1 \mathbf{p}^1 =
\begin{bmatrix}
\tilde{N}_a \\
\tilde{V}_a \\
\tilde{M}_a \\
\tilde{N}_b \\
\tilde{V}_b \\
\tilde{M}_b
\end{bmatrix}
=
\begin{bmatrix}
.57983 \times 10^4 \text{ N} \\
-.51441 \times 10^4 \text{ N} \\
.88286 \times 10^4 \text{ N} \cdot \text{m} \\
-.57983 \times 10^4 \text{ N} \\
.51441 \times 10^4 \text{ N} \\
.93586 \times 10^4 \text{ N} \cdot \text{m}
\end{bmatrix}
\tag{15}
$$

Structures with Distributed Loads

Concentrated loads applied between the ends of an element can be accommodated by adding new nodes, a practice that increases the size of the system of equations to be solved. An alternative is to include another loading vector in the stiffness equations. This addition of a new loading vector is particularly appropriate for distributed loads applied between the nodes. For a beam element this loading vector was considered in Section II.5. The additional loading vector for the global stiffness matrices will be treated here.

The element stiffness matrix including the extra loading vector appears as (Appendix II)

$$
\mathbf{p}^i = \mathbf{k}^i \mathbf{v}^i - \overline{\mathbf{p}}^i
\tag{III.84}
$$

Expressions for the components of the loading vector $\mathbf{p}^i$ for a variety of types of loading are provided in the tables of Chapter 11. For example, for an Euler–Bernoulli beam element with a uniformly distributed load of magnitude p_0 (Table 11-19)

$$
\overline{\mathbf{p}}^i =
\begin{bmatrix}
V_a^0 \\
M_a^0 \\
V_b^0 \\
M_b^0
\end{bmatrix}
=
\frac{p_0 \ell}{2}
\begin{bmatrix}
1 \\
-\tfrac{1}{6} \\
1 \\
\tfrac{1}{6}
\end{bmatrix}
\tag{III.85}
$$

For displacement and forces referred to a local coordinate system

$$
\tilde{\mathbf{p}}^i = \tilde{\mathbf{k}}^i \tilde{\mathbf{v}}^i - \tilde{\overline{\mathbf{p}}}^i
\tag{III.86}
$$

Transformation from local to global coordinates systems is implemented using

$$
\mathbf{p}^i = \mathbf{T}^{iT} \tilde{\mathbf{p}}^i \qquad \mathbf{k}^i = \mathbf{T}^{iT} \tilde{\mathbf{k}}^i \mathbf{T}^i \qquad \overline{\mathbf{p}}^i = \mathbf{T}^{iT} \tilde{\overline{\mathbf{p}}}^i
\tag{III.87}
$$

The global stiffness matrix is still assembled using

$$\mathbf{K}_{jk} = \sum_{i=1}^{M} \mathbf{k}_{jk}^{i} \tag{III.88a}$$

where M is the number of elements. The new global loading vector is assembled as

$$\bar{\mathbf{P}}_j = \sum_{i=1}^{M} \bar{\mathbf{p}}_j^i \tag{III.88b}$$

The $\mathbf{P}_j$ for node j can be formed into $\bar{\mathbf{P}}$ for the whole system. This can be incorporated into $\bar{\mathbf{P}}$ of Eq. (III.77) or, alternatively, one can distinguish between nodal loading terms due to distributed loading and direct nodal loading. The system equilibrium equations now appear as

$$\mathbf{P} = \mathbf{KV} - \bar{\mathbf{P}} \tag{III.89}$$

where $\mathbf{P}$ is a nodal vector containing direct nodal loads and $\bar{\mathbf{P}}$ is a nodal vector due to applied distributed loading.

Example III.10 Beam with Linearly Varying Loading Consider the fixed–simply supported beam with linearly varying applied load of Fig. III-5. Although we choose to model the beam with two elements, the most logical model would contain a single element as the stiffness matrix for beams is exact and hence can span any length. The local and global coordinate systems coincide for this horizontal beam; consequently, there is no need to transform variables from local to global coordinates.

Table 11-19 can supply the element loading vectors $\bar{\mathbf{p}}^i$. In using the formulas of this table, note that for the first element the distributed load begins with a magnitude of p_0 and ends with $\frac{1}{2}p_0$. For the second element the load begins with $\frac{1}{2}p_0$ and ends with a magnitude of zero.

An alternative approach for determining loading vectors is to utilize Eq. (II.71). To illustrate this, consider the linearly varying distributed load $p_z(\xi)$ of Fig. III-28:

$$p_z(\xi) = p_a + (p_b - p_a)\xi \tag{1}$$

with $\xi = x/\ell$. Rewrite this as

$$p_z(\xi) = \mathbf{N}_p \mathbf{G}_p \bar{\mathbf{p}}_p \tag{2}$$

where

$$\mathbf{N}_p = \begin{bmatrix} 1 & \xi \end{bmatrix} \qquad \mathbf{G}_p = \begin{bmatrix} 1 & 0 \\ -1 & 1 \end{bmatrix} \qquad \bar{\mathbf{p}}_p = \begin{bmatrix} p_a \\ p_b \end{bmatrix} \tag{3}$$

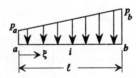

FIGURE III-28 Linearly distributed load.

From Eq. (II.71)

$$\bar{\mathbf{p}}^i = \mathbf{G}^T \int_a^b \mathbf{N}_u^T p_z \, dx = \mathbf{G}^T \int_0^1 \mathbf{N}_u^T \mathbf{N}_p \, d\xi \, \ell \, \mathbf{G}_p \bar{\mathbf{p}}_p \tag{4}$$

Take $\mathbf{G}$ and $\mathbf{N}_u$ from Eq. (II.65a). Then

$$\int_0^1 \mathbf{N}_u^T \mathbf{N}_p \, d\xi \, \ell = \int_0^1 \begin{bmatrix} 1 \\ \xi \\ \xi^2 \\ \xi^3 \end{bmatrix} [1 \quad \xi] \, d\xi \, \ell = \begin{bmatrix} 1 & \frac{1}{2} \\ \frac{1}{2} & \frac{1}{3} \\ \frac{1}{3} & \frac{1}{4} \\ \frac{1}{4} & \frac{1}{5} \end{bmatrix} \ell \tag{5}$$

and from (4)

$$\mathbf{G}^T \int_0^1 \mathbf{N}_u^T \mathbf{N}_p \, d\xi \, \ell \, \mathbf{G}_p \bar{\mathbf{p}}_p = \overset{\mathbf{G}^T}{\begin{bmatrix} 1 & 0 & -3 & 2 \\ 0 & -1 & 2 & -1 \\ 0 & 0 & 3 & -2 \\ 0 & 0 & 1 & -1 \end{bmatrix}} \ell \begin{bmatrix} 1 & \frac{1}{2} \\ \frac{1}{2} & \frac{1}{3} \\ \frac{1}{3} & \frac{1}{4} \\ \frac{1}{4} & \frac{1}{5} \end{bmatrix} \overset{\mathbf{G}_p}{\begin{bmatrix} 1 & 0 \\ -1 & 1 \end{bmatrix}}$$

$$= \ell \begin{bmatrix} \frac{7}{20} & \frac{3}{20} \\ -\frac{1}{20} & -\frac{1}{30} \\ \frac{3}{20} & \frac{7}{20} \\ \frac{1}{30} & \frac{1}{20} \end{bmatrix} \tag{6}$$

Finally,

$$\bar{\mathbf{p}}^i = \ell \begin{bmatrix} \frac{7}{20} & \frac{3}{20} \\ -\frac{1}{20} & -\frac{1}{30} \\ \frac{3}{20} & \frac{7}{20} \\ \frac{1}{30} & \frac{1}{20} \end{bmatrix} \begin{bmatrix} p_a \\ p_b \end{bmatrix} \tag{7}$$

For our beam of Fig. III-5 with $\ell = \ell_1 = \ell_2 = \frac{1}{2}L$, for element 1, with $p_a = p_0$ and $p_b = \frac{1}{2}p_0$,

$$\bar{\mathbf{p}}^1 = \frac{p_0 \ell}{120} \begin{bmatrix} 51 \\ -8 \\ 39 \\ -7 \end{bmatrix} = \begin{bmatrix} \bar{p}_a^1 \\ \bar{p}_b^1 \end{bmatrix} \tag{8}$$

For element 2, with $p_a = \frac{1}{2}p_0$ and $p_b = 0$,

$$\bar{\mathbf{p}}^2 = \frac{p_0 \ell}{120} \begin{bmatrix} 21 \\ -3 \\ 9 \\ 2 \end{bmatrix} = \begin{bmatrix} \bar{p}_b^2 \\ \bar{p}_c^2 \end{bmatrix} \tag{9}$$

The global matrix can now be formed as

$$\mathbf{KV} - \bar{\mathbf{P}} = \mathbf{P} \tag{10}$$

where

$$\mathbf{V} = \begin{bmatrix} w_a \\ \theta_a \ell \\ w_b \\ \theta_b \ell \\ w_c \\ \theta_c \ell \end{bmatrix} \tag{11}$$

and from Eq. (II.62)

$$
\mathbf{K} = \begin{bmatrix} \mathbf{k}_{aa}^1 & \mathbf{k}_{ab}^1 & 0 \\ \mathbf{k}_{ba}^1 & \mathbf{k}_{bb}^1 + \mathbf{k}_{bb}^2 & \mathbf{k}_{bc}^2 \\ 0 & \mathbf{k}_{cb}^2 & \mathbf{k}_{cc}^2 \end{bmatrix}
$$

$$
= \frac{EI}{\ell^3} \begin{bmatrix} 12 & -6 & -12 & -6 & 0 & 0 \\ -6 & 4 & 6 & 2 & 0 & 0 \\ -12 & 6 & 24 & 0 & -12 & -6 \\ -6 & 2 & 0 & 8 & 6 & 2 \\ 0 & 0 & -12 & 6 & 12 & 6 \\ 0 & 0 & -6 & 2 & 6 & 4 \end{bmatrix}
$$

Assemble the global loading vector $\bar{\mathbf{P}}$ in the same manner as for the global stiffness matrix. It follows from Eq. (III.88b),

$$\bar{\mathbf{P}}_j = \sum_i \bar{\mathbf{p}}_j^i$$

that

$$
\bar{\mathbf{P}} = \begin{bmatrix} \bar{\mathbf{P}}_a \\ \bar{\mathbf{P}}_b \\ \bar{\mathbf{P}}_c \end{bmatrix} = \begin{bmatrix} \bar{\mathbf{p}}_a^1 \\ \bar{\mathbf{p}}_b^1 + \bar{\mathbf{p}}_b^2 \\ \bar{\mathbf{p}}_c^2 \end{bmatrix} = \frac{p_0 \ell}{120} \begin{bmatrix} 51 \\ -8 \\ 39 + 21 \\ 7 - 3 \\ 9 \\ 2 \end{bmatrix} = \frac{p_0 \ell}{120} \begin{bmatrix} 51 \\ -8 \\ 60 \\ 4 \\ 9 \\ 2 \end{bmatrix} \tag{12}
$$

The vector $\mathbf{P}$ contains concentrated loads applied at the nodes as well as the unknown reactions. Since for this beam there are no loads applied at the nodes and the rows corresponding to the unknown reactions are not used in calculating the displacements $\mathbf{V}, \mathbf{P}$ will be ignored and the system equations will appear as $\mathbf{KV} - \bar{\mathbf{P}} = \mathbf{0}$.

Introduction of the boundary conditions $w_a = \theta_a = w_c = 0$ reduces $\mathbf{KV} = \bar{\mathbf{P}}$ to the nonsingular system of equations

$$\frac{EI}{\ell^3}\begin{bmatrix} 24 & 0 & -6 \\ 0 & 8 & 2 \\ -6 & 2 & 4 \end{bmatrix}\begin{bmatrix} w_b \\ \theta_b\ell \\ \theta_c\ell \end{bmatrix} = \frac{p_0\ell}{120}\begin{bmatrix} 60 \\ 4 \\ 2 \end{bmatrix} \tag{13}$$

which has the solution

$$\begin{bmatrix} w_b \\ \theta_b \\ \theta_c \end{bmatrix} = \frac{p_0\ell^3}{120EI}\begin{bmatrix} 4.5\ell \\ -1.5 \\ 8.0 \end{bmatrix} \tag{14}$$

Since the vector of displacements $\mathbf{V}$ has now been determined, the beam reactions can be computed using $\mathbf{KV} - \bar{\mathbf{P}} = \mathbf{P}$.

To calculate the forces at the ends of the elements, use

$$\mathbf{k}^i\mathbf{v}^i - \bar{\mathbf{p}}^i = \mathbf{p}^i \tag{15}$$

For this horizontally oriented beam, the local displacements at the ends of the elements are equal to the node (global) displacements so that a coordinate transformation will not be necessary. Then, (15) leads to

$$\mathbf{p}^1 = \begin{bmatrix} V_a \\ M_a \\ V_b \\ M_b \end{bmatrix} = \mathbf{k}^1\mathbf{v}^1 - \bar{\mathbf{p}}^1 = \frac{p_0\ell^3}{120EI}\frac{EI}{\ell^3}\begin{bmatrix} -12 & -6\ell \\ 6\ell & 2\ell^2 \\ 12 & 6\ell \\ 6\ell & 4\ell^2 \end{bmatrix}\begin{bmatrix} 4.5\ell \\ -1.5 \end{bmatrix}$$

$$- \frac{p_0\ell}{120}\begin{bmatrix} 51 \\ -8\ell \\ 39 \\ 7\ell \end{bmatrix} = \frac{p_0\ell}{120}\begin{bmatrix} -96 \\ 32\ell \\ 6 \\ 14\ell \end{bmatrix} \tag{16}$$

$$\mathbf{p}^2 = \begin{bmatrix} V_b \\ M_b \\ V_c \\ M_c \end{bmatrix} = \mathbf{k}^2\mathbf{v}^2 - \bar{\mathbf{p}}^2 = \frac{p_0\ell^3}{120EI}\frac{EI}{\ell^3}\begin{bmatrix} 12 & -6\ell & -6\ell \\ -6\ell & 4\ell^2 & 2\ell^2 \\ -12 & 6\ell & 6\ell \\ -6\ell & 2\ell^2 & 4\ell^2 \end{bmatrix}\begin{bmatrix} 4.5\ell \\ -1.5 \\ 8.0 \end{bmatrix}$$

$$- \frac{p_0\ell}{120}\begin{bmatrix} 21 \\ -3\ell \\ 9 \\ 2\ell \end{bmatrix} = \frac{p_0\ell}{120}\begin{bmatrix} -6 \\ -14\ell \\ -24 \\ 0 \end{bmatrix} \tag{17}$$

Although stiffness equations can be employed to find the variation of the responses, e.g., the deflection, along the beam, it is often simpler to use transfer

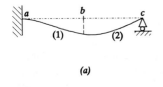

(a)

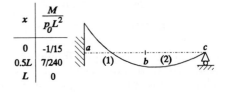

(b)

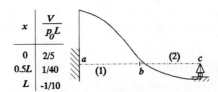

(c)

FIGURE III-29 Response of beam of Fig. III-5 with linearly varying loading: (a) deflection; (b) bending moment, (c) shear force.

matrices. Hence,

$$\mathbf{z}_j = \mathbf{U}^1 \mathbf{z}_a \quad \text{for element 1} \tag{18}$$

$$\mathbf{z}_j = \mathbf{U}^2 \mathbf{z}_b \quad \text{for element 2} \tag{19}$$

The vectors $\mathbf{z}_a$ and $\mathbf{z}_b$ contain known displacements and forces at $x = a$ and $x = b$. With the transfer matrices $\mathbf{U}^1$ and $\mathbf{U}^2$ expressed in terms of the variable x, (18) and (19) provide the desired responses. In the case of the deflection, with $\xi = x/L$,

$$w(\xi) = \begin{cases} \dfrac{p_0 \ell^4}{120EI}\left(16\xi_1^2 - 16\xi_1^3 + 5\xi_1^4 - 0.5\xi_1^5\right)\xi_1 = 2\xi & 0 \le x \le L/2 \\[2mm] \dfrac{p_0 \ell^4}{120EI}\Big[4.5 + 1.5(2\xi - 1) - 7(2\xi - 1)^2 - (2\xi - 1)^3 \\[2mm] \qquad + 2.5(2\xi - 1)^4 - 0.5(2\xi - 1)^5\Big] & L/2 \le x \le L \end{cases} \tag{20}$$

or, for any ξ, $w(\xi) = [p_0 L^4/(120EI)](4\xi^2 - 8\xi^3 + 5\xi^4 - \xi^5)$. These exact deflections, along with the bending moment and shear force, are plotted in Fig. III-29.

Special Intermediate Conditions

In-span conditions, e.g., hinges or supports, such as those illustrated in Fig. III-14 require special attention as one response variable is constrained (usually the value is zero) while a discontinuity (a reaction) in the complementary variable is generated. In general, this type of occurrence is more readily incorporated in the solution when using the displacement method than it is with the transfer matrix method. Normally, a global degree of freedom is constrained by these conditions. For example, a rigid support at a node completely restrains the displacement in the direction of a component of **V**, and then the condition that must be imposed is a prescribed global displacement (usually zero). This is implemented by simply setting one displacement in **V** equal to zero. See reference III.1 for details.

III.4 FORCE METHOD

Virtually all of the currently available general-purpose computer programs are based on the displacement method. The force method, although it is popular for hand calculations, is not the method of choice for solving large-scale problems

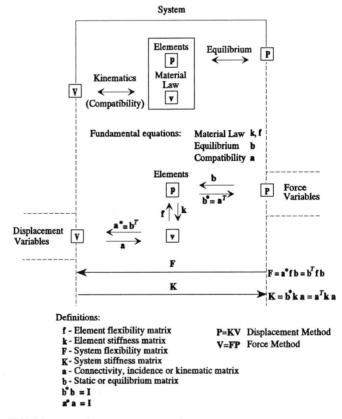

FIGURE III-30 System matrices for force and displacement methods.

[III.10]. The cause of the domination by the displacement method is that at the time when most of the general-purpose analysis computer programs were being developed, it was felt that the force method could not be easily automated for large-scale problems. The principle of complementary virtual work provides the basis of the force method. This principle is equivalent to the global form of the kinematic admissibility conditions (compatibility). As a consequence, the force method is sometimes referred to as the *compatibility method* as well as the *flexibility* or *influence coefficient method*.

The force method equations will not be derived here, but the derivation in a structural mechanics text such as reference [III.1] will usually follow closely the derivation of the displacement method equations. The similarity of the formulations of the displacement and force methods is symbolic of the dual nature of the two techniques. This same duality exists between the principle of virtual work and the principle of complementary virtual work. Whereas the displacement method and the principle of virtual work require kinematically admissible displacements (i.e., **a** must be formed) and provide equilibrium equations, the force method and the principle of complementary virtual work begin with equilibrium conditions (i.e., **b** must be formed) and lead to kinematic equations. This process is illustrated in Fig. III-30.

The force, displacement, and transfer matrix approaches are compared in Table III-3. In contrast to the displacement and force methods, as mentioned earlier the transfer matrix method does not involve the assembly of a system matrix for which the size increases with the degrees of freedom of the system. Moreover, the system matrix for the transfer matrix method is formed by progressive element matrix multiplications rather than superposition.

III.5   STABILITY BASED ON DISPLACEMENT METHOD

In the same fashion that the principle of virtual work of Eq. (II.66) leads to the equilibrium equations

$$\mathbf{KV} = \bar{\mathbf{P}}$$

the principle of virtual work including an explicit in-plane force, e.g., Eq. (II.88), provides the set of equations

$$(\mathbf{K} - \lambda \mathbf{K}_G)\mathbf{V} = \bar{\mathbf{P}} \qquad\qquad (\text{III.90})$$

where $\mathbf{K}_G$ is the system (global) geometric stiffness matrix that can be assembled from element matrices in the same manner as the system stiffness matrix $\mathbf{K}$. In an instability study of a structural system, the axial forces throughout the system must be defined relative to each other. For example, the axial forces can remain fixed in magnitude relative to one another. Thus, if $\mathbf{K}_G$ is the global geometric stiffness matrix for a reference level of axial forces, then $\lambda \mathbf{K}_G$ corresponds to another level of axial forces, where λ is a scalar multiplier called the *load factor*. For a single member, e.g. a uniform column, λ is simply the axial force P.

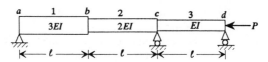

FIGURE III-31 Stepped column.

For classical instability theory of linearly elastic structures, the applied loadings $\overline{\mathbf{P}}$ of Eq. (III.90) do not affect the buckling load of a structure. The buckling load is found by solving the homogeneous problem

$$(\mathbf{K} - \lambda\mathbf{K}_G)\mathbf{V} = 0 \tag{III.91}$$

The critical force problem represented by this relationship is in the form of a classical eigenvalue problem and can be solved efficiently with considerable reliability by using standard, readily available eigenvalue problem software. One of the least efficient methods for solving for the buckling load (yet necessary for certain types of $\mathbf{K}_G$) is to perform a critical value search of $\det(\mathbf{K} - \lambda\mathbf{K}_G) = 0$.

Example III.11   A Stepped Column   The stepped column of Fig. III-31 will be used to illustrate several of the techniques for computing buckling loads for a structural system. The boundary and in-span conditions are $w_a = w_c = w_d = 0$ and $M_a = M_d = 0$.

Transfer Matrix Method
Use the transfer matrix of Eq. (III.10) along with the methodology presented in Section III.1 for incorporating the in-span support to develop a global transfer matrix. The boundary conditions applied to the global transfer matrix equations lead to the characteristic equation and the critical axial load of

$$P_{cr} = 7.064\frac{EI}{\ell^2} \tag{1}$$

Displacement Method
The goal here is to set up the eigenvalue problem in the form of Eq. (III.91). First establish the ordinary stiffness matrix $\mathbf{K}$. To do so, assemble element stiffness matrices $\mathbf{k}^i$, $i = 1, 2, 3$, of Eq. (II.62):

$$
\begin{bmatrix} V_a \\ M_{a/\ell} \\ V_b \\ M_{b/\ell} \\ V_c \\ M_{c/\ell} \\ V_d \\ M_{d/\ell} \end{bmatrix}
= \frac{EI}{\ell^3}
\begin{bmatrix}
36 & -18 & -36 & -18 & 0 & 0 & 0 & 0 \\
-18 & 12 & 18 & 6 & 0 & 0 & 0 & 0 \\
-36 & 18 & 60 & 6 & -24 & -12 & 0 & 0 \\
-18 & 6 & 6 & 20 & 12 & 4 & 0 & 0 \\
0 & 0 & -24 & 12 & 36 & 6 & -12 & -6 \\
0 & 0 & -12 & 4 & 6 & 12 & 6 & 2 \\
0 & 0 & 0 & 0 & -12 & 6 & 12 & 6 \\
0 & 0 & 0 & 0 & -6 & 2 & 6 & 4
\end{bmatrix}
\begin{bmatrix} w_a \\ \ell\theta_a \\ w_b \\ \ell\theta_b \\ w_c \\ \ell\theta_c \\ w_d \\ \ell\theta_d \end{bmatrix}
\tag{2}
$$

Next the global geometric stiffness matrix should be assembled using the element geometric stiffness matrices $\mathbf{k}_G^i$. If the consistent geometric stiffness matrix of Eq. (II.90) is utilized, the global geometric stiffness matrix will be

$$
\mathbf{K}_G = \frac{1}{\ell}
\begin{bmatrix}
\frac{18}{15} & -\frac{1}{10} & -\frac{18}{15} & -\frac{1}{10} & 0 & 0 & 0 & 0 \\
-\frac{1}{10} & \frac{2}{15} & \frac{1}{10} & -\frac{1}{30} & 0 & 0 & 0 & 0 \\
-\frac{18}{15} & \frac{1}{10} & \frac{36}{15} & 0 & -\frac{18}{15} & -\frac{1}{10} & 0 & 0 \\
-\frac{1}{10} & -\frac{1}{30} & 0 & \frac{4}{15} & \frac{1}{10} & -\frac{1}{30} & 0 & 0 \\
0 & 0 & -\frac{18}{15} & \frac{1}{10} & \frac{36}{15} & 0 & -\frac{18}{15} & -\frac{1}{10} \\
0 & 0 & -\frac{1}{10} & -\frac{1}{30} & 0 & \frac{4}{15} & \frac{1}{10} & -\frac{1}{30} \\
0 & 0 & 0 & 0 & -\frac{18}{15} & \frac{1}{10} & \frac{18}{15} & \frac{1}{10} \\
0 & 0 & 0 & 0 & -\frac{1}{10} & -\frac{1}{30} & \frac{1}{10} & \frac{2}{15}
\end{bmatrix}
\tag{3}
$$

Apply the displacement boundary conditions and ignore the rows corresponding to the unknown reactions. Then the eigenvalue problem appears as

$$
\left(
\begin{bmatrix}
12 & 18 & 6 & 0 & 0 \\
18 & 60 & 6 & -12 & 0 \\
6 & 6 & 20 & 4 & 0 \\
0 & -12 & 4 & 12 & 2 \\
0 & 0 & 0 & 2 & 4
\end{bmatrix}
- \lambda
\begin{bmatrix}
\frac{2}{15} & \frac{1}{10} & -\frac{1}{30} & 0 & 0 \\
\frac{1}{10} & \frac{12}{5} & 0 & -\frac{1}{10} & 0 \\
-\frac{1}{30} & 0 & \frac{4}{15} & -\frac{1}{30} & 0 \\
0 & -\frac{1}{10} & -\frac{1}{30} & \frac{4}{15} & -\frac{1}{30} \\
0 & 0 & 0 & -\frac{1}{30} & \frac{2}{15}
\end{bmatrix}
\right)\mathbf{V} = 0
$$

$$\tag{4}$$

where

$$
\lambda = \frac{P\ell^2}{EI} \quad \text{and} \quad \mathbf{V} =
\begin{bmatrix}
\ell\,\theta_a \\
w_b \\
\ell\,\theta_b \\
\ell\,\theta_c \\
\ell\,\theta_d
\end{bmatrix}
$$

The solution of these equations, which constitute a generalized (linear) eigenvalue problem, provides a critical solution of magnitude

$$
P_{cr} = 7.298\left(EI/\ell^2\right)
\tag{5}
$$

This consistent matrix model leads to a result that is above the "exact" value of (1). Use of this type of geometric stiffness matrix always leads to a value above the correct eigenvalue.

III.6 FREE VIBRATIONS BASED ON DISPLACEMENT METHOD

The governing equations $\mathbf{KV} = \bar{\mathbf{P}}$ for the nodal displacements of a system under static loading are now familiar. If dynamic effects are to be taken into account, then according to D'Alembert's principle, we simply assure that equilibrium of forces at the nodes includes the effect of inertia. Then the governing equations become

$$\mathbf{M}\ddot{\mathbf{V}} + \mathbf{KV} = \bar{\mathbf{P}} \tag{III.92}$$

where $\mathbf{M}$, the global mass matrix, is assembled from element mass matrices $\mathbf{m}^i$ in the same fashion that $\mathbf{K}$ is obtained from $\mathbf{k}^i$. For example, $\mathbf{M}$ can be assembled using the consistent mass element matrices $\mathbf{m}^i$ of Eq. (II.87).

To study the free vibrations, set the applied loading $\bar{\mathbf{P}}$ to zero and assume that each displacement performs harmonic motion in phase with all other displacements. Thus, as explained in Section III.1 in the discussion of free vibrations, assume that

$$\mathbf{V}(x, y, z, t) = \mathbf{V}(x, y, z)\sin \omega t \tag{III.93}$$

so that Eq. (III.92) reduces to

$$(\mathbf{K} - \omega^2\mathbf{M})\mathbf{V} = \mathbf{0} \tag{III.94}$$

or

$$(\mathbf{K} - \lambda\mathbf{M})\mathbf{V} = \mathbf{0} \tag{III.95}$$

where $\lambda = \omega^2$. Equation (III.95) represents a generalized eigenvalue problem.

The trivial solution of Eq. (III.95) would be $\mathbf{V} = \mathbf{0}$. If $\mathbf{V} \neq \mathbf{0}$, only particular values λ_n satisfy Eq. (III.95). These λ_n are the *characteristics* or *eigenvalues* of the problem, which correspond to the natural frequencies of the structure. The lowest value of λ_n is the fundamental natural frequency. To each λ_n corresponds an *eigenvector* $\mathbf{V}_n$, which defines the mode shape of the nth mode of motion for the structure. It is important to understand that $\mathbf{V}_n$ defines the shape and not the magnitude of the motion. The eigenvalue problem is to extract from Eq. (III.95) the solution pairs λ_n and $\mathbf{V}_n$.

Mass Matrix

The most common element mass matrices are the consistent mass matrices $\mathbf{m}^i$ of Eq. (II.87) and the lumped mass matrix of Eq. (II.83).

The lumped mass element matrix is diagonal, as is the corresponding assembled global matrix $\mathbf{M}$. The lumped mass matrix is positive semidefinite when zeros occur on the diagonal, whereas the consistent mass (element and global) matrices are positive definite. The zeros on the diagonal can complicate certain numerical algorithms. It is clear that a lumped mass matrix would require less storage space than a consistent mass matrix. It is also more economical to form and to manipulate. Consistent mass matrices lead to eigenvalues that are higher than the exact value, whereas lumped mass matrices will usually approach the exact eigenvalues from below. See reference III.7 for a succinct comparison of lumped and consistent mass matrices along with some interesting variations on these two types of mass modeling.

Eigenvalue Problem

The characteristic equation for the eigenvalue problem of Eq. (III.95) is

$$\det(\mathbf{K} - \lambda\mathbf{M}) = 0 \tag{III.96}$$

in which the displacement boundary conditions have reduced the size of $\mathbf{K}$ and $\mathbf{M}$. Equation (III.96) takes the form of a polynomial in λ, whose roots are the desired eigenvalues λ_n. However, irrespective of whether a numerical determinant search of Eq. (III.96) is implemented or the characteristic polynomial is formed and solved, use of $\det(\mathbf{K} - \lambda\mathbf{M}) = 0$ for problems of many degrees of freedom is normally found to be a cumbersome process that should be avoided. Normally, as indicated later in this subsection, a number of eigenvalue solution routines provide preferred solution methodology.

Example III.12 Displacement Method for a Beam Using a Consistent Mass Matrix Use the displacement method to find the natural frequencies of the beam of Fig. III-11 and Example III.7. The beam is modeled with two elements of equal length.

Begin by assembling the global stiffness and mass matrices. The notation is shown in Fig. III-11. The element stiffness matrix $\mathbf{k}^i$ is given by Eq. (II.61), while the consistent mass matrix $\mathbf{m}^i$ is given by Eq. (II.87):

$$\mathbf{k}^i = \frac{EI}{\ell^3}\begin{bmatrix} 12 & -6\ell & -12 & -6\ell \\ -6\ell & 4\ell^2 & 6\ell & 2\ell^2 \\ -12 & 6\ell & 12 & 6\ell \\ -6\ell & 2\ell^2 & 6\ell & 4\ell^2 \end{bmatrix} \quad i = 1, 2$$

$$= \begin{bmatrix} \mathbf{k}^1_{aa} & \mathbf{k}^1_{ab} \\ \mathbf{k}^1_{ba} & \mathbf{k}^1_{bb} \end{bmatrix} = \begin{bmatrix} \mathbf{k}^2_{bb} & \mathbf{k}^2_{bc} \\ \mathbf{k}^2_{cb} & \mathbf{k}^2_{cc} \end{bmatrix} \tag{1}$$

$$\mathbf{m}^i = \frac{\rho\ell}{420}\begin{bmatrix} 156 & -22\ell & 54 & 13\ell \\ -22\ell & 4\ell^2 & -13\ell & -3\ell^2 \\ 54 & -13\ell & 156 & 22\ell \\ 13\ell & -3\ell^2 & 22\ell & 4\ell^2 \end{bmatrix} \quad i = 1, 2$$

$$= \begin{bmatrix} \mathbf{m}^1_{aa} & \mathbf{m}^1_{ab} \\ \mathbf{m}^1_{ba} & \mathbf{m}^1_{bb} \end{bmatrix} = \begin{bmatrix} \mathbf{m}^2_{bb} & \mathbf{m}^2_{bc} \\ \mathbf{m}^2_{cb} & \mathbf{m}^2_{cc} \end{bmatrix} \tag{2}$$

The global stiffness matrix is assembled as

$$
\mathbf{K} = \begin{bmatrix} \mathbf{k}_{aa}^1 & \mathbf{k}_{ab}^1 & \mathbf{0} \\ \mathbf{k}_{ba}^1 & \mathbf{k}_{bb}^1 + \mathbf{k}_{bb}^2 & \mathbf{k}_{bc}^2 \\ \mathbf{0} & \mathbf{k}_{cb}^2 & \mathbf{k}_{cc}^2 \end{bmatrix}
$$

$$
= \frac{EI}{\ell^3} \begin{bmatrix} 12 & -6\ell & -12 & -6\ell & 0 & 0 \\ -6\ell & 4\ell^2 & 6\ell & 2\ell^2 & 0 & 0 \\ -12 & 6\ell & 24 & 0 & -12 & -6\ell \\ -6\ell & 2\ell^2 & 0 & 8\ell^2 & 6\ell & 2\ell^2 \\ 0 & 0 & -12 & 6\ell & 12 & 6\ell \\ 0 & 0 & -6\ell & 2\ell^2 & 6\ell & 4\ell^2 \end{bmatrix} \tag{3}
$$

with the corresponding global displacement vector

$$
\mathbf{V} = \begin{bmatrix} w_a & \theta_a & w_b & \theta_b & w_c & \theta_c \end{bmatrix}^T
$$

Similarly, the global mass matrix is

$$
\mathbf{M} = \begin{bmatrix} \mathbf{m}_{aa}^1 & \mathbf{m}_{ab}^1 & \mathbf{0} \\ \mathbf{m}_{ba}^1 & \mathbf{m}_{bb}^1 + \mathbf{m}_{bb}^2 & \mathbf{m}_{bc}^2 \\ \mathbf{0} & \mathbf{m}_{cb}^2 & \mathbf{m}_{cc}^2 \end{bmatrix}
$$

$$
= \frac{\rho\ell}{420} \begin{bmatrix} 156 & -22\ell & 54 & 13\ell & 0 & 0 \\ -22\ell & 4\ell^2 & -13\ell & -3\ell^2 & 0 & 0 \\ 54 & -13\ell & 312 & 0 & 54 & 13\ell \\ 13\ell & -3\ell^2 & 0 & 8\ell^2 & -13\ell & -3\ell^2 \\ 0 & 0 & 54 & -13\ell & 156 & 22\ell \\ 0 & 0 & 13\ell & -3\ell^2 & 22\ell & 4\ell^2 \end{bmatrix} \tag{4}
$$

The frequencies can be determined by solving the generalized eigenvalue problem of Eq. (III.95). Equation (III.95) is first modified by applying the displacement boundary conditions to $\mathbf{V}$ and ignoring the rows in $(\mathbf{K} - \lambda\mathbf{M})\mathbf{V} = 0$ corresponding to the unknown reactions.

Thus, with $w_a = w_c = 0$

$$
\mathbf{V} = \begin{bmatrix} w_a & \theta_a & w_b & \theta_b & w_c & \theta_c \end{bmatrix}^T \tag{5}
$$

reduces to

$$
\begin{bmatrix} \theta_a & w_b & \theta_b & \theta_c \end{bmatrix}^T \tag{6}
$$

so that the columns in $\mathbf{K}$ and $\mathbf{M}$ corresponding to w_a and w_c can be canceled. Furthermore, ignore the rows for the reactions V_a and V_c, which are unknown.

The eigenvalue problem becomes

$$
\left\{
\frac{EI}{\ell^3}
\begin{bmatrix}
4\ell^2 & 6\ell & 2\ell^2 & 0 \\
6\ell & 24 & 0 & -6\ell \\
2\ell^2 & 0 & 8\ell^2 & 2\ell^2 \\
0 & -6\ell & 2\ell^2 & 4\ell^2
\end{bmatrix}
\right.
$$

$$
\left.
-\lambda
\begin{bmatrix}
4\ell^2 & -13\ell & -3\ell^2 & 0 \\
-13\ell & 312 & 0 & 13\ell \\
-3\ell^2 & 0 & 8\ell^2 & -3\ell^2 \\
0 & 13\ell & -3\ell^2 & 4\ell^2
\end{bmatrix}
\frac{\rho\ell}{420}
\right\}
\begin{bmatrix}
\theta_a \\
w_b \\
\theta_b \\
\theta_c
\end{bmatrix}
= 0
\qquad (7)
$$

Use of a classical eigenvalue solution procedure will lead to the desired frequencies. Computer software for eigenvalue problems is readily available.

An alternative technique for finding the frequencies is to establish the characteristic equation from the determinant of the coefficients of $(\mathbf{K} - \lambda\mathbf{M})\mathbf{V} = 0$. This can be implemented for problems of limited degrees of freedom. Let

$$
\mathbf{K} - \omega^2\mathbf{M} = [D_{ij}] \qquad i, j = 1, 2, \ldots, 6 \qquad (8)
$$

and apply the displacement boundary conditions ($w_a = w_c = 0$). This leads to

$$
\begin{bmatrix}
D_{11} & D_{12} & D_{13} & D_{14} & D_{15} & D_{16} \\
D_{21} & D_{22} & D_{23} & D_{24} & D_{25} & D_{26} \\
D_{31} & D_{32} & D_{33} & D_{34} & D_{35} & D_{36} \\
D_{41} & D_{42} & D_{43} & D_{44} & D_{45} & D_{46} \\
D_{51} & D_{52} & D_{53} & D_{54} & D_{55} & D_{56} \\
D_{61} & D_{62} & D_{63} & D_{64} & D_{65} & D_{66}
\end{bmatrix}
\begin{bmatrix}
w_a = 0 \\
\theta_a \\
w_b \\
\theta_b \\
w_c = 0 \\
\theta_c
\end{bmatrix}
= 0
\qquad (9)
$$

The first and fifth equations (rows) correspond to unknown reactions. The remaining equations appear as

$$
\begin{bmatrix}
D_{22} & D_{23} & D_{24} & D_{26} \\
D_{32} & D_{33} & D_{34} & D_{36} \\
D_{42} & D_{43} & D_{44} & D_{46} \\
D_{62} & D_{63} & D_{64} & D_{66}
\end{bmatrix}
\begin{bmatrix}
\theta_a \\
w_b \\
\theta_b \\
\theta_c
\end{bmatrix}
= 0
\qquad (10)
$$

The characteristic equation is obtained from the determinant of the coefficients of (10), i.e.,

$$
\nabla =
\begin{vmatrix}
D_{22} & D_{23} & D_{24} & D_{26} \\
D_{32} & D_{33} & D_{34} & D_{36} \\
D_{42} & D_{43} & D_{44} & D_{46} \\
D_{62} & D_{63} & D_{64} & D_{66}
\end{vmatrix}
= 0
\qquad (11)
$$

This relationship can be obtained directly from (7).

Insert the numerical values

$$D_{22} = 3.9999 \times 10^6 - 1.7752\omega^2$$

$$D_{32} = D_{23} = 1.49996 \times 10^5 + 0.1442\omega^2$$

$$D_{33} = 1.4999 \times 10^4 - 0.0865\omega^2$$

$$D_{42} = D_{24} = 1.9999 \times 10^6 + 1.3314\omega^2$$

$$D_{43} = D_{34} = 0$$

$$D_{44} = 7.9998 \times 10^6 + 3.5505\omega^2 \qquad (12)$$

$$D_{62} = D_{26} = 0$$

$$D_{36} = D_{63} = -1.49996 \times 10^5 - 0.1442\omega^2$$

$$D_{64} = D_{46} = D_{42}$$

$$D_{66} = D_{22}$$

Substitution of (12) into (11) and use of factorization leads to two equations:

$$\omega^4 - 41.0093 \times 10^5\omega^2 + 13.3967 \times 10^{10} = 0$$

and $\qquad\qquad\qquad\qquad\qquad\qquad\qquad\qquad\qquad\qquad (13)$

$$\omega^4 - 14.1631 \times 10^6\omega^2 + 8.7036 \times 10^{12} = 0$$

The roots of these equations are

$$\omega_1 = 181.48 \quad \text{or} \quad f_1 = \omega_1/2\pi = 28.88 \text{ Hz}$$

$$\omega_2 = 802.37 \quad \text{or} \quad f_2 = 127.70 \text{ Hz}$$

$$\omega_3 = 2016.92 \quad \text{or} \quad f_3 = 321.00 \text{ Hz} \qquad (14)$$

$$\omega_4 = 3676.86 \quad \text{or} \quad f_4 = 585.19 \text{ Hz}$$

Note that the consistent mass matrix leads to higher frequencies than the "exact" values of Example III.7. The approximation introduced in this example is the use of the consistent mass matrix, which is approximate. This contrasts with Example III.7, where the exact mass was employed. In order to make the higher frequencies more accurate when consistent mass matrices are employed, more elements should be included in the model.

===============

Example III.13 Displacement Method for a Lumped Mass Model of a Beam
Use the displacement method to find the natural frequencies of the beam of Fig. III-11 along with the lumped parameter idealization of Fig. III-32. Begin by assembling the global stiffness and mass matrices. From Eq. (II.83), for a lumped

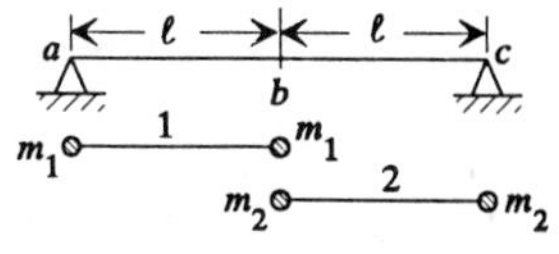

$$m_1 = m_2 = \tfrac{1}{2}\rho\ell$$
$$\ell = 40 \text{ in.}$$
$$\rho = 2.912 \times 10^{-3} \text{ lb-s}^2/\text{in.}^2$$
$$E = 3 \times 10^7 \text{ lb/in.}^2$$
$$I = 1.3333 \text{ in.}^4$$

FIGURE III-32 Lumped parameter model. Note that this model differs from that of Fig. III-13.

mass model of a beam element

$$\mathbf{m}^i = \frac{\rho\ell}{2}\begin{bmatrix} 1 & 0 & 0 & 0 \\ 0 & 0 & 0 & 0 \\ 0 & 0 & 1 & 0 \\ 0 & 0 & 0 & 0 \end{bmatrix} \qquad i = 1, 2 \tag{1}$$

so that the assembled mass matrix will be

$$\mathbf{M} = \frac{\rho\ell}{2}\begin{bmatrix} 1 & 0 & 0 & 0 & 0 & 0 \\ 0 & 0 & 0 & 0 & 0 & 0 \\ 0 & 0 & 1+1 & 0+0 & 0 & 0 \\ 0 & 0 & 0+0 & 0+0 & 0 & 0 \\ 0 & 0 & 0 & 0 & 1 & 0 \\ 0 & 0 & 0 & 0 & 0 & 0 \end{bmatrix} \tag{2}$$

The global stiffness matrix is still given by Eq. (3) of Example III.12.
Apply the displacement boundary conditions to $(\mathbf{K} - \lambda\mathbf{M})\mathbf{V} = \mathbf{0}$ and ignore the rows corresponding to the unknown reactions. This leads to the linear eigenvalue problem

$$\left\{ \frac{EI}{\ell^3}\begin{bmatrix} 4\ell^2 & 6\ell & 2\ell^2 & 0 \\ 6\ell & 24 & 0 & -6\ell \\ 2\ell^2 & 0 & 8\ell^2 & 2\ell^2 \\ 0 & -6\ell & 2\ell^2 & 4\ell^2 \end{bmatrix} - \lambda\begin{bmatrix} 0 & 0 & 0 & 0 \\ 0 & 2 & 0 & 0 \\ 0 & 0 & 0 & 0 \\ 0 & 0 & 0 & 0 \end{bmatrix}\frac{\rho\ell}{2} \right\}\begin{bmatrix} \theta_a \\ w_b \\ \theta_b \\ \theta_c \end{bmatrix} = \mathbf{0} \tag{3}$$

where $\lambda = \omega^2$. The frequencies can be found from this relationship using a standard eigenvalue solution procedure.
Alternatively, the characteristic equation can be established from the determinant of the system of equations. Let

$$(\mathbf{K} - \omega^2\mathbf{M}) = \begin{bmatrix} H_{ij} \end{bmatrix} \qquad i, j = 1, 2, \ldots, 6 \tag{4}$$

Follow the procedure of the previous example. This leads to

$$\nabla = \begin{vmatrix} H_{22} & H_{23} & H_{24} & H_{26} \\ H_{32} & H_{33} & H_{34} & H_{36} \\ H_{42} & H_{43} & H_{44} & H_{46} \\ H_{62} & H_{63} & H_{64} & H_{66} \end{vmatrix} = 0 \tag{5}$$

where

$$H_{22} = 3.9999 \times 10^6$$
$$H_{32} = H_{23} = 1.49996 \times 10^5$$
$$H_{33} = 1.49996 \times 10^4 - 0.1165\omega^2$$
$$H_{42} = H_{24} = 1.9999 \times 10^6$$
$$H_{43} = H_{34} = 0 \tag{6}$$
$$H_{44} = 7.9998 \times 10^6$$
$$H_{62} = H_{26} = 0$$
$$H_{63} = H_{36} = -1.49996 \times 10^5$$
$$H_{64} = H_{46} = H_{42}$$
$$H_{66} = H_{22}$$

With these numbers, (5) gives the fundamental natural frequency

$$\omega = 179.408 \text{ rad/s} \quad \text{or} \quad f = \omega/2\pi = 28.55 \text{ Hz} \tag{7}$$

As is generally the case, the fundamental frequency derived using the lumped mass model is lower than the exact value of Example III.7.

There are a variety of effective and efficient algorithms, along with full-developed software, that attack the eigenvalue problem of Eq. (III.95) directly without resorting to solving Eq. (III.96). Often eigenvalue problem computer programs require that Eq. (III.95) be converted to the so-called *standard* eigenvalue problem

$$(\mathbf{A} - \lambda\mathbf{I})\mathbf{Y} = \mathbf{0} \tag{III.97}$$

where $\mathbf{I}$ is the unit diagonal matrix, $\mathbf{A}$ is symmetric, and $\mathbf{Y}$ is the eigenvector. Eigenvalues λ that satisfy this relationship are said to be the eigenvalues of $\mathbf{A}$. To obtain the standard form, premultiply Eq. (III.95) by $\mathbf{M}^{-1}$:

$$(\mathbf{M}^{-1}\mathbf{K} - \lambda\mathbf{I})\mathbf{V} = \mathbf{0} \tag{III.98}$$

However, $\mathbf{M}^{-1}\mathbf{K}$ is in general not symmetric, so further manipulations are required to achieve a standard form. In particular, Choleski decomposition of $\mathbf{M}$

or **K** is performed, resulting in the standard form. Furthermore, since practical models often involve large mass matrices **M**, it is desirable to avoid implementation of the inverse $\mathbf{M}^{-1}$. In fact, in some cases, for lumped mass models $\mathbf{M}^{-1}$ does not exist. For in depth discussions of these issues, see reference III.7.

Consult a structural dynamics textbook for such popular eigenvalue solution routines as Jacobi, Householder, inverse iteration or power, subspace iteration, or the Lanczo's method. Frequently, it is helpful to use condensation techniques, e.g., *Guyan reduction*, to reduce the number of degrees of freedom of a dynamics problem.

A zero eigenvalue λ_i should be obtained for each possible rigid-body motion of a structure that is not restrained. Since the mass can hold the structure together, a singular stiffness matrix **K** is more palatable for a dynamic problem than for a static solution, although some operations may not be suitable. For a real, symmetric, and nonsingular **K**, the rank of **M** is equal to the number of nonzero independent eigenvalues of Eq. (III.95). It follows that for an **M** formed using consistent mass element matrices, the number of frequencies available from an analysis is equal to the number of unrestrained nodal displacements.

Frequency-dependent Stiffness and Mass Matrices

The mass matrices considered thus far are clearly approximate. In the case of the lumped mass matrix, a user-orchestrated physical discretization is imposed on the system. For the consistent mass matrix $\mathbf{m}^i$, the shape function **N**, almost always a polynomial, that was employed to form the stiffness matrix for a static analysis is inserted in

$$\mathbf{m}^i = \int_a^b \rho \mathbf{N}^T \mathbf{N}\, dx \qquad (\text{III.99})$$

The methodology and formulas of this book permit a much more general and exact approach, albeit a less efficient one. Some of the transfer matrices, e.g., those of Table 11-22, contain distributed mass without approximation. If such a transfer matrix is converted to a stiffness matrix format using the transformation of Eq. (II.59), a so-called *dynamic stiffness matrix* $\mathbf{k}^i_{\text{dyn}}$ results. Table 11-22 contains such a matrix. The distributed mass in the dynamic stiffness matrix is modeled exactly and the resulting stiffness matrix is a function of the frequency. Suppose the assembled global stiffness matrix is $\mathbf{K}_{\text{dyn}}$. The characteristic equation from which the natural frequencies can be computed can be written as

$$\det \mathbf{K}_{\text{dyn}}(\omega) = 0 \qquad (\text{III.100})$$

where the boundary conditions have been applied and have resulted in a reduced $\mathbf{K}_{\text{dyn}}$. A numerical determinant search in which the determinant is evaluated for trial frequencies can be utilized to find the frequencies.

Example III.14   Use of a Dynamic Stiffness Matrix for the Natural Frequencies of a Beam   Use the dynamic stiffness matrix to find the natural frequencies of the beam of Fig. III-11. Since an exact mass model is to be employed, this beam can be treated as a single-element beam. We choose, however, to continue with the two-element model shown in Fig. III-11. Assuming the use of a long element does not lead to numerical instabilities, both models provide the same exact solution (frequencies and mode shapes).

The element stiffness relationship is

$$\mathbf{k}^i \mathbf{v}^i = \mathbf{p}^i \quad \text{where } i = 1, 2 \tag{1}$$

$$\mathbf{k}^i_{\text{dyn}} = \mathbf{k}^i = \begin{bmatrix} k^i_{11} & k^i_{12} & k^i_{13} & k^i_{14} \\ k^i_{21} & k^i_{22} & k^i_{23} & k^i_{24} \\ k^i_{31} & k^i_{32} & k^i_{33} & k^i_{34} \\ k^i_{41} & k^i_{42} & k^i_{43} & k^i_{44} \end{bmatrix} \quad \mathbf{v}^i = \begin{bmatrix} w_a \\ \theta_a \\ w_b \\ \theta_b \end{bmatrix} \quad \mathbf{p}^i = \begin{bmatrix} V_a \\ M_a \\ V_b \\ M_b \end{bmatrix}$$

The stiffness elements for a dynamic stiffness matrix for a beam, which includes the effects of the mass, are given in Table 11-22:

$$k^i_{11} = (EI/\Delta)[(e_2 - \eta e_4)(e_1 + \zeta e_3) + \lambda e_3 e_4]$$

$$k^i_{12} = (EI/\Delta)[e_3(e_1 - \eta e_3) - e_2(e_2 - \eta e_4)]$$

$$k^i_{13} = -(EI/\Delta)(e_2 - \eta e_4)$$

$$k^i_{14} = -(EI/\Delta)(e_3)$$

$$k^i_{21} = k^i_{12}$$

$$k^i_{22} = (EI/\Delta)\{e_3 e_2 - (e_1 - \eta e_3)[e_4 - \xi(e_2 + \zeta e_4)]\}$$

$$k^i_{23} = (EI/\Delta)(e_3) = -k^i_{14}$$

$$k^i_{24} = (EI/\Delta)[e_4 - \xi(e_2 + \zeta e_4)]$$

$$k^i_{31} = k^i_{13} \tag{2}$$

$$k^i_{32} = k^i_{23}$$

$$k^i_{33} = (EI/\Delta)[(e_1 + \zeta e_3)(e_2 - \eta e_4) + \lambda e_3 e_4] = k^i_{11}$$

$$k^i_{34} = (EI/\Delta)\{(e_1 + \zeta e_3)e_3 + \lambda e_4[e_4 - \xi(e_2 + \zeta e_4)]\}$$

$$k^i_{41} = k^i_{14}$$

$$k^i_{42} = k^i_{24}$$

$$k^i_{43} = k^i_{34}$$

$$k^i_{44} = (EI/\Delta)\{e_2 e_3 - (e_1 - \eta e_3)[e_4 - \xi(e_2 + \zeta e_4)]\} = k^i_{22}$$

$$\Delta = e_3^2 - (e_2 - \eta e_4)[e_4 - \xi(e_2 + \zeta e_4)]$$

Also from Table 11-22,

$$\lambda = (k - \rho\omega^2)/EI = -\rho\omega^2/EI < 0$$
$$\eta = (k - \rho\omega^2)/GA_s = 0 \quad \text{if shear deformation}$$
$$\text{is not included} \quad (1/GA_s = 0) \quad (3)$$
$$\xi = EI/GA_s = 0$$
$$\zeta = (P - k^* + \rho r_y^2\omega^2)/EI = 0$$

since $P = 0$, $k^* = 0$, and rotary effects are to be ignored. For $\lambda < 0$, Table 11-22, p. 586, case 1, provides

$$e_1 = (d^2A + q^2B)/g \qquad e_2 = (dC + qD)/g$$
$$e_3 = (A - B)/g \qquad\qquad e_4 = (C/d - D/q)/g \qquad (4)$$

in which the parameters are obtained from Table 11-22, p. 587:

$$d^2 = \sqrt{(\zeta + \eta)^2/4 - \lambda} - (\zeta - \eta)/2 = \sqrt{\rho/EI}\,\omega$$
$$q^2 = \sqrt{(\zeta + \eta)^2/4 - \lambda} + (\zeta + \eta)/2 = \sqrt{\rho/EI}\,\omega = d^2$$
$$g = d^2 + q^2 = 2\sqrt{\rho/EI}\,\omega \qquad (5)$$
$$A = \cosh(d\ell) \qquad B = \cos(q\ell) = \cos(d\ell)$$
$$C = \sinh(d\ell) \qquad D = \sin(q\ell) = \sin(d\ell)$$

Insertion of (5) into (4) gives

$$e_1 = \tfrac{1}{2}[\cosh(d\ell) + \cos(d\ell)]$$
$$e_2 = \frac{1}{2\sqrt{\omega}}\left(\frac{EI}{\rho}\right)^{1/4}[\sinh(d\ell) + \sin(d\ell)]$$
$$e_3 = \frac{1}{2\omega}\sqrt{\frac{EI}{\rho}}\,[\cosh(d\ell) - \cos(d\ell)] \qquad (6)$$
$$e_4 = \frac{1}{2\omega\sqrt{\omega}}\left(\frac{EI}{\rho}\right)^{3/4}[\sinh(d\ell) - \sin(d\ell)]$$

and

$$\Delta = e_3^2 - (e_2 - 0)(e_4 - 0)$$
$$= e_3^2 - e_2 e_4$$
$$= \frac{1}{2\omega^2}\left(\frac{EI}{\rho}\right)[1 - \cosh(d\ell)\cos(d\ell)]$$

Substitution of (3) into (2) provides

$$k_{11}^i = (EI/\Delta)(e_1e_2 + \lambda e_3e_4)$$
$$k_{12}^i = (EI/\Delta)(e_1e_3 - e_2^2)$$
$$k_{13}^i = -(EI/\Delta)e_2$$
$$k_{14}^i = -(EI/\Delta)e_3$$
$$k_{21}^i = k_{12}^i$$
$$k_{22}^i = (EI/\Delta)(e_3e_2 - e_1e_4)$$
$$k_{23}^i = (EI/\Delta)e_3 = -k_{14}^i$$
$$k_{24}^i = (EI/\Delta)e_4$$
$$k_{31}^i = k_{13}^i \tag{7}$$
$$k_{32}^i = k_{23}^i$$
$$k_{33}^i = (EI/\Delta)(e_1e_2 + \lambda e_3e_4) = k_{11}^i$$
$$k_{34}^i = (EI/\Delta)(e_1e_3 + \lambda e_4^2)$$
$$k_{41}^i = k_{14}^i$$
$$k_{42}^i = k_{24}^i$$
$$k_{43}^i = k_{34}^i$$
$$k_{44}^i = (EI/\Delta)(e_3e_2 - e_1e_4) = k_{22}^i$$

The global dynamic stiffness matrix assembled for the two-element beam model appears as

$$\mathbf{K}_{dyn} = \begin{bmatrix} k_{11}^1 & k_{12}^1 & k_{13}^1 & k_{14}^1 & 0 & 0 \\ k_{21}^1 & k_{22}^1 & k_{23}^1 & k_{24}^1 & 0 & 0 \\ k_{31}^1 & k_{32}^1 & k_{33}^1 + k_{11}^2 & k_{34}^1 + k_{12}^2 & k_{13}^2 & k_{14}^2 \\ k_{41}^1 & k_{42}^1 & k_{43}^1 + k_{21}^2 & k_{44}^1 + k_{22}^2 & k_{23}^2 & k_{24}^2 \\ 0 & 0 & k_{31}^2 & k_{32}^2 & k_{33}^2 & k_{34}^2 \\ 0 & 0 & k_{41}^2 & k_{42}^2 & k_{43}^2 & k_{44}^2 \end{bmatrix} \tag{8}$$

with the corresponding global displacement vector

$$\mathbf{V} = \begin{bmatrix} w_a & \theta_a & w_b & \theta_b & w_c & \theta_c \end{bmatrix}^T$$

Introduce the displacement boundary conditions $w_a = w_c = 0$ to eliminate columns 1 and 5 of (8) and delete the rows (1 and 5) for the reactions complementary to these prescribed displacements. The characteristic equation

[Eq. (III.100)] now appears as

$$
\begin{vmatrix}
k_{22}^1 & k_{23}^1 & k_{24}^1 & 0 \\
k_{32}^1 & k_{33}^1 + k_{11}^2 & k_{34}^1 + k_{12}^2 & k_{14}^2 \\
k_{42}^1 & k_{43}^1 + k_{21}^2 & k_{44}^1 + k_{22}^2 & k_{24}^2 \\
0 & k_{41}^2 & k_{42}^2 & k_{44}^2
\end{vmatrix} = 0
$$

A frequency search applied to this determinant relationship leads to the natural frequencies.

As expected, this dynamic stiffness matrix yields the same (exact) natural frequencies obtained in Example III.7. This is because no approximation such as the use of consistent or lumped mass modeling was made. The advantage of exact mass modeling permits a coarser mesh, i.e., fewer elements, to be employed in the model. The disadvantage is that the determinant search for frequencies can be a difficult, inefficient procedure, especially for complex systems.

———————

As mentioned in Example III.14, a determinant search, such as would be required by Eq. (III.100), is often a numerically cumbersome, inefficient process that perhaps should be avoided. The inefficiency results from the need to calculate the value of the determinant for each trial value of the frequency. A detailed review of the problems associated with the use of the dynamic stiffness matrix is provided in reference III.11. The determinant search can be circumvented by creating an eigenvalue problem in which the structural matrices are not functions of the frequency parameter ω.

Typically one begins by generating a frequency-dependent mass matrix in a manner similar to the formation of the dynamic stiffness matrix $\mathbf{k}_{dyn}^i$. A frequency dependent mass matrix can be obtained by placing the exact (frequency-dependent) shape function $\mathbf{N}$ in Eq. (III.99). The same exact mass matrix is available by differentiating the element dynamic stiffness matrix by the frequency parameter ω [III.12]:

$$
\mathbf{m}^i = -\partial \mathbf{k}_{dyn}^i / \partial \omega^2 \tag{III.101}
$$

An economical alternative is to calculate a frequency-dependent "quasi-static" mass matrix $\mathbf{m}^i$ defined as

$$
\tilde{\mathbf{m}}^i = \int_V \mathbf{N}_0^T \rho \mathbf{N} \, dV \tag{III.102}
$$

where $\mathbf{N}_0$ is an element static shape function such as the $\mathbf{N}$ given in Eq. (II.65).

Define an associated frequency-dependent stiffness matrix $\mathbf{k}^i(\omega)$ in terms of the dynamic stiffness matrix $\mathbf{k}_{dyn}^i(\omega)$ and a frequency-dependent mass matrix $\mathbf{m}^i(\omega)$,

$$
\mathbf{k}_{dyn}^i(\omega) = \mathbf{k}^i(\omega) - \omega^2 \mathbf{m}^i(\omega) \tag{III.103}
$$

The operations involved in forming $\mathbf{k}_{dyn}^i$, $\mathbf{m}^i$, and $\mathbf{k}^i$ can be performed using symbolic manipulation software.

Assemble the global matrices $\mathbf{M}(\omega)$ and $\mathbf{K}(\omega)$ using the element matrices $\mathbf{m}^i(\omega)$ and $\mathbf{k}^i(\omega)$, giving the eigenvalue problem

$$\left[\mathbf{K}(\omega) - \omega^2 \mathbf{M}(\omega)\right]\mathbf{V} = \mathbf{0} \tag{III.104}$$

A simple iterative solution of this problem usually converges rapidly to the exact eigenvalue solution:

$$\left[\mathbf{K}(0) - \omega^2 \mathbf{M}(0)\right]\mathbf{V} = 0 \Rightarrow \omega = \omega_1^0, \omega_2^0, \omega_3^0, \ldots, \qquad \mathbf{v} = \mathbf{v}^0$$

$$\left[\mathbf{K}(\omega_1^0) - \omega^2 \mathbf{M}(\omega_1^0)\right]\mathbf{V} = 0 \Rightarrow \omega = \omega_1^1, \omega_2^1, \omega_3^1, \ldots \qquad \mathbf{v} = \mathbf{v}^1$$

$$\vdots$$

$$\left[\mathbf{K}(\omega_1^{j-1}) - \omega^2 \mathbf{M}(\omega_1^{j-1})\right]\mathbf{V} = 0 \Rightarrow \omega = \omega_1^j, \omega_2^j, \omega_3^j, \ldots \qquad \mathbf{v} = \mathbf{v}^j \tag{III.105}$$

where superscript j designates the eigensolution of the jth iteration. The frequencies $\omega_1^0, \omega_2^0, \omega_3^0, \ldots$ are those that are obtained using a consistent mass matrix and the usual static stiffness matrix.

This approach, although less efficient than solving the generalized eigenvalue problem of Eqs. (III.95), is normally more efficient than the determinant search required to solve Eq. (III.100). This procedure, represented by Eqs. (III.105), has the advantage of being able to utilize readily available, reliable, standard eigenvalue solvers, which should result in an accurate set of frequencies and mode shapes. Furthermore, remember that the frequency-dependent mass and stiffness matrices permit a model to be employed with fewer (larger) elements than is possible with consistent or lumped mass matrices. Even more precise higher frequencies are obtained from Eqs. (III.105) if $\mathbf{K}$ and $\mathbf{M}$ are evaluated at ω_n^{j-1}, $n > 1$, rather than at the lowest natural frequency ω_1^{j-1}.

More common than the use of the iterative procedure of Eq. (III.105) is the establishment of a quadratic, generalized eigenvalue problem using matrices expanded in series [III.13]. Expand the mass, stiffness, and dynamic stiffness matrices in Taylor series:

$$\mathbf{m}^i = \sum_{n=0}^{\infty} \mathbf{m}_n \omega^{2n} \quad \text{or} \quad \tilde{\mathbf{m}}^i = \sum_{n=0}^{\infty} \tilde{\mathbf{m}}_n \omega^{2n} \tag{III.106a}$$

$$\mathbf{k}^i = \sum_{n=0}^{\infty} \mathbf{k}_n \omega^{2n} \tag{III.106b}$$

$$\mathbf{k}_{\text{dyn}}^i = \sum_{n=0}^{\infty} \left(\mathbf{k}_{\text{dyn}}\right)_n \omega^{2n} \tag{III.106c}$$

From Eq. (III.103) define

$$\left(\mathbf{k}_{\text{dyn}}\right)_n = \mathbf{k}_n - \mathbf{m}_{n-1} \qquad n \geq 1 \tag{III.107}$$

A simple relationship between the series terms $\mathbf{m}_0, \mathbf{m}_1, \ldots$ and $\mathbf{k}_1, \mathbf{k}_2, \ldots$ is quite useful [III.13]:

$$(n + 1)\mathbf{k}_{n+1} = n\mathbf{m}_n = n(n + 1)\tilde{\mathbf{m}}_n = -n(n + 1)(\mathbf{k}_{dyn})_{n+1} \qquad n \geq 1 \tag{III.108}$$

Also,

$$(\mathbf{k}_{dyn})_0 = \mathbf{k}_0 \quad \text{(traditional stiffness matrix)}$$
$$(\mathbf{k}_{dyn})_1 = -\mathbf{m}_0 \quad \text{(traditional consistent mass matrix)} \tag{III.109}$$

and $\mathbf{k}_1 = 0$. Usually enough terms in the series expansions are retained to create a quadratic eigenvalue problem. As detailed in reference III.14, there is a sizable literature on the solution of higher order eigenvalue problems, especially on the efficient solution of quadratic eigenvalue problems.

III.7 TRANSIENT RESPONSES

The response of a structural system to prescribed time-dependent loading involves the solution of

$$\mathbf{M}\ddot{\mathbf{V}} + \mathbf{K}\mathbf{V} = \overline{\mathbf{P}} \tag{III.110}$$

This is an ordinary differential equation in time that can be integrated directly. Textbooks on vibrations or structural dynamics, e.g., reference III.1, describe a variety of time integration techniques for solving Eq. (III.110). However, by far the most frequently used technique in practice is the modal superposition method, which employs the free vibration responses. Since the natural frequencies and mode shapes are often calculated at the offset of a dynamic analysis of a structural system, it is a relatively simple procedure to proceed to compute the transient response using modal superposition. This method will be described in brief here for use with structural systems. See reference III.1 for the details on using modal superposition to find the response of structural members. Also, further information on this method is available in structural dynamics textbooks.

With modal superposition the solution to Eqs. (III.110) are expressed as a summation involving the eigenvectors $\mathbf{V}_n$. Actually, the solution is written in terms of normalized eigenvectors. One common method of normalizing the eigenvectors is to choose the unknown constants such that

$$\mathbf{V}_j^T \mathbf{M} \mathbf{V}_j = \mathbf{I} \tag{III.111}$$

Arrange the newly normalized eigenvectors as columns in a matrix $\mathbf{\Phi}$. This is called the *modal matrix* and has the orthogonality properties

$$\mathbf{\Phi}^T \mathbf{M} \mathbf{\Phi} = \mathbf{I} \qquad \mathbf{\Phi}^T \mathbf{K} \mathbf{\Phi} = \boldsymbol{\omega}^2 \tag{III.112}$$

where $\mathbf{I}$ is the unit diagonal matrix and $\boldsymbol{\omega}^2$ is a diagonal matrix of the squared natural frequencies, called the *spectral matrix*.

To find the solution to Eq. (III.110), express the displacements $\mathbf{V}$ as the sum of the normal modes,

$$\mathbf{V} = \mathbf{\Phi}\mathbf{q} \tag{III.113}$$

where $\mathbf{q}$ is a vector of time-dependent modal amplitudes. The values of $\mathbf{q}(t)$ are found by substituting $\mathbf{V}$ of Eq. (III.113) in Eq. (III.110) and premultiplying by $\mathbf{\Phi}^T$. It follows from Eqs. (III.112) that

$$\ddot{q}_n + \omega_n^2 q_n = p_n \tag{III.114a}$$

where

$$p_n = \mathbf{\phi}_n^T \overline{\mathbf{P}} \tag{III.114b}$$

The vector $\mathbf{\phi}_n$ is the nth column of $\mathbf{\Phi}$. Equation (III.114a) represents a set of uncoupled ordinary differential equations in time, with the solution

$$q_n(t) = q_n(0)\cos \omega_n t + \dot{q}_n(0)\frac{\sin \omega_n t}{\omega_n} + \int_0^t p_n(\tau)\frac{\sin \omega_n(t-\tau)}{\omega_n}\,d\tau \tag{III.115}$$

where $p_n(t)$ is a known function of time. There are as many q_n equations as there are degrees of freedom. To find initial values of q_n and $\dot{q}_n$ of Eq. (III.115) evaluate $\mathbf{V} = \mathbf{\Phi}\mathbf{q}$ and $\dot{\mathbf{V}} = \mathbf{\Phi}\dot{\mathbf{q}}$ at $t = 0$ and premultiply both sides by $\mathbf{\Phi}^T\mathbf{M}$. This gives

$$\mathbf{q}(0) = \mathbf{\Phi}^T\mathbf{M}\mathbf{V}(0) \qquad \dot{\mathbf{q}}(0) = \mathbf{\Phi}^T\mathbf{M}\dot{\mathbf{V}}(0) \tag{III.116}$$

$$q_n(0) = \mathbf{\phi}_n^T\mathbf{M}\mathbf{V}(0) \qquad \dot{q}_n(0) = \mathbf{\phi}_n^T\mathbf{M}\dot{\mathbf{V}}(0) \tag{III.117}$$

This modal solution may appear to be computationally inefficient. However, $\mathbf{\Phi}$ is usually not difficult to form as only a few modes may be needed for a sufficiently accurate solution. If m modes are employed, then there are m columns in $\mathbf{\Phi}$ and only m solutions to Eq. (III.114) required. For a problem with multiple loading histories, the modal summation approach is particularly advantageous as the second, third, etc., loadings are handled very economically.

Treatises on structural dynamics contain considerable insight into making the modal superposition approach more efficient and accurate.

REFERENCES

III.1 Pilkey, W. D., and Wunderlich, W., *Structural Mechanics, Variational and Computational Methods*, CRC, Florida, 1993.

III.2 Pilkey, W. D., and Okada, Y., *Matrix Methods in Mechanical Vibrations*, Corona, Tokyo, 1989.

III.3 Pilkey, W. D., and Pilkey, O. H., *Mechanics of Solids*, Krieger, Melbourne, FL, 1986.

III.4 Pestel, E., and Leckie, F., *Matrix Methods in Elastomechanics*, McGraw-Hill, New York, 1963.

III.5 Marguerre, V. K., and Uhrig, R., "Berechnung vielgliedriger Gelenkketten I. Ubertragungsverfahren und seine Grenzen, *Zeitschrift fur Angewandte Mathematik und Mechanik*," Bd. 44, 1964, pp. 1–21.

III.6 Horner, G. C., and Pilkey, W. D., "The Riccati Transfer Matrix Method," *J. Mechan. Des.*, Vol. 1, No. 2, 1978, pp. 297–302.

III.7 Cook, R. D., *Concepts and Applications of Finite Element Analysis*, 2nd ed., Wiley, New York, 1981.

III.8 Everstine, G. C., "A Comparison of Three Resequencing Algorithms for the Reduction of Matrix Profile and Wavefront," *IJNME*, Vol. 14, No. 6, 1979, pp. 837–858.

III.9 Gibbs, N. E., Poole, W. G., Jr., and Stockmeyer, P. K., "An Algorithm for Reducing the Bandwidth and Profile of a Sparse Matrix," *SIAM J. Numer. Anal.*, Vol. 13, No. 2, 1976, pp. 236–250.

III.10 Gallegher, R. H., "Finite Element Structural Analysis and Complementary Energy," *Finite Elements in Analysis and Design*, Vol. 13, 1993, pp. 115–126.

III.11 Fergusson, N., and Pilkey, W. D., "The Dynamic Stiffness Method, A Survey of the Published Literature, Part I: DSM Elements," *Shock Vibrat. Technol. Rev.*, Vol. 1, No. 7, 1991, pp. 4–13.

III.12 Richards, T. H., and Leung, Y. T., "An Accurate Method in Structural Vibration Analysis," *J. Sound Vibrat.*, Vol. 55, No. 3, 1977, pp. 363–376.

III.13 Fergusson, N., and Pilkey, W. D., "Frequency-dependent Element Mass Matrices," *J. Appl. Mechan.*, Vol. 59, 1992, pp. 136–139.

III.14 Fergusson, N., and Pilkey, W. D., "The Dynamic Stiffness Method, A Survey of the Published Literature, Part II: Frequency Extraction Methods and Modal Analysis," *Shock Vibrat Technol Rev.*, Vol. 1, No. 8, 1991, pp. 14–20.

Tables

TABLE III-1   NOTATION FOR TRANSFER MATRIX METHOD

Symbol	Definition
$\mathbf{U}^i$	Transfer matrix for ith element (field) $\mathbf{z}_b = \mathbf{U}^i\mathbf{z}_a$
$\mathbf{U}$	Global or overall transfer matrix that spans several (M) elements, $\mathbf{U} = \mathbf{U}^M\mathbf{U}^{M-1} \cdots \mathbf{U}^2\mathbf{U}^1$
$\mathbf{U}_k$	Point matrix to account for concentrated occurrence, e.g., a point force or discrete spring, at location k
$\mathbf{z}_k$	State vector at location k; contains all displacement and force state variables
$\bar{\mathbf{z}}^i$	Applied loading function vector for ith element

TABLE III-1   Notation for Transfer Matrix Method   **1440**

TABLE III-2 NOTATION FOR DISPLACEMENT AND FORCE METHODS

Notation

Indices: Superscript for distributed quantity, e.g., an element index; subscript for point quantity, e.g., a node index. As an example, $\mathbf{v}_k^i$ is the displacement $\mathbf{v}$ of the ith element at the kth node.

Local coordinate system: Variables referred to a local coordinate system are indicated by a tilde, e.g., $\tilde{\mathbf{k}}^i$ is the local stiffness matrix of the ith element.

Prescribed (applied) variables will be indicated with a line over the letter, e.g., $\overline{\mathbf{P}}_k$ is the applied force at node k.

Symbol	Definition
$\mathbf{V}$	Vector of global nodal displacements
$\mathbf{P}$	Vector of global nodal forces
$\overline{\mathbf{V}}$	Applied global displacements
$\overline{\mathbf{P}}$	Applied global forces
$\mathbf{v}^i$	Vector of nodal displacements of ith element
$\mathbf{p}^i$	Vector of nodal forces of ith element
$\overline{\mathbf{v}}^i, \overline{\mathbf{p}}^i$	Element loading vectors
$\mathbf{k}^i$	Stiffness matrix of ith element
$\mathbf{K}$	System (global) stiffness matrix
$\mathbf{f}^i$	Flexibility matrix of ith element
$\mathbf{F}$	System flexibility matrix
$\mathbf{a}$	Kinematic transformation (incident) matrix, $\mathbf{v} = \mathbf{a}\mathbf{V}$
$\mathbf{b}$	Static transformation matrix, $\mathbf{p} = \mathbf{b}\mathbf{P}$
$\mathbf{T}^i$	Transformation matrix; e.g., $\tilde{\mathbf{v}}^i = \mathbf{T}^i \mathbf{v}^i$
$\mathbf{T}_{jj}^i, j = a, b$	Transformation matrix, element i, node j; $$\mathbf{T}^i = \begin{bmatrix} \mathbf{T}_{aa} & 0 \\ \hline 0 & \mathbf{T}_{bb} \end{bmatrix}^i$$ For nodes a, b of element i

TABLE III-2 (continued)   NOTATION FOR DISPLACEMENT AND FORCE METHODS

$\tilde{N}$	Internal axial force in local coordinates
$\tilde{V}_y, \tilde{V}_z$ or $\tilde{V}$	Internal shear forces in local coordinates
$\tilde{T} = \tilde{M}_x,\ \tilde{M} = \tilde{M}_y,\ \tilde{M}_z$	Internal moments in local coordinates
$\tilde{u} = \tilde{u}_x$	Axial displacement in local coordinates
$\tilde{v} = \tilde{u}_y,\ \tilde{w} = \tilde{u}_z$	Transverse displacements in local coordinates
$\tilde{\phi} = \tilde{\theta}_x,\ \tilde{\theta} = \tilde{\theta}_y,\ \tilde{\theta}_z$	Rotations about x, z, and y axes in local coordinates
F_X, F_Y, F_Z	Forces on the ends of element in global X, Y, and Z directions, respectively.
u_X, u_Y, u_Z	Displacements on the ends of element in global X, Y, and Z directions, respectively.
$\theta_X, \theta_Y = \theta, \theta_Z$	Rotations on the ends of element in global X, Y and Z directions, respectively
P_X, P_Y, P_Z	Nodal forces in global coordinates
M_X, M_Y, M_Z or M	Nodal moments in global coordinates
U_X, U_Y, U_Z	Nodal displacements in global coordinates
$\Theta_X, \Theta_Y, \Theta_Z$	Nodal rotation in global coordinates X, Y and Z directions, respectively
$\bar{P}_X, \bar{P}_Y, \bar{P}_Z$	Applied nodal loading forces in global coordinates
$\bar{M}_X, \bar{M}_Y, \bar{M}_Z$	Applied nodal moments in global coordinates X, Y and Z directions, respectively

For plane structure with in-plane loading:

$$\tilde{\mathbf{p}} = \begin{bmatrix} \tilde{N} \\ \tilde{V} \\ \tilde{M} \end{bmatrix}, \quad \tilde{\mathbf{v}} = \begin{bmatrix} \tilde{u} \\ \tilde{w} \\ \tilde{\theta} \end{bmatrix}, \quad \mathbf{p} = \begin{bmatrix} F_X \\ F_Z \\ M \end{bmatrix}, \quad \mathbf{v} = \begin{bmatrix} u_X \\ u_Z \\ \theta \end{bmatrix}$$

TABLE III-2   **Notation for Displacement and Force Methods**   1442

TABLE III-3   COMPARISON OF METHODS FOR STRUCTURAL ANALYSIS

Analysis	Method		
	Transfer Matrix	Displacement	Force
Unknowns in analysis	Displacement and force variables	Displacement variables	Force variables
Characterization of ith element	Transfer matrix $\mathbf{U}^i$	Stiffness matrix $\mathbf{k}^i$	Flexibility matrix $\mathbf{f}^i$
Matrix characterizing system	$\mathbf{U} = \prod_i \mathbf{U}^i$	$\mathbf{K} = \sum_i \mathbf{k}^i = \mathbf{a}^T \mathbf{k} \mathbf{a}$	$\mathbf{F} = \mathbf{b}^T \mathbf{f} \mathbf{b}$
Conditions fulfilled at outset of formulation	—	Compatibility	Equilibrium
System equations satisfy	Equilibrium and compatibility	Equilibrium	Compatibility

Index